Borrelia

Molecular Biology, Host Interaction and Pathogenesis

Edited by

D. Scott Samuels

Division of Biological Sciences
The University of Montana
Missoula, MT
USA

and

Justin D. Radolf

Department of Medicine and
Department of Genetics and Developmental Biology
University of Connecticut Health Center
Farmington, CT
USA

Caister Academic Press

Caister Academic Press
Norfolk, UK

www.caister.com

British Library Cataloguing-in-Publication Data
A catalogue record for this book is available from the British Library

ISBN: 978-1-904455-58-5

Cover designed by Scott Samuels and Stuart Goldstein. The background image, provided by Ursula Norman, Tara Moriarty, Paul Kubes and George Chaconas, is a three-dimensional reconstruction of GFP-expressing *B. burgdorferi* adhering to and escaping from the postcapillary venule in the skin of a living mouse (see also Moriarty *et al.*, 2008, PLoS Pathog. *4*, e1000169). The insert images (clockwise from top) are the crystal structure of Gac (see also Corbett and Berger, 2004, Proc. Natl. Acad. Sci. U.S.A. *101*, 7293–7298), provided by Scott Samuels; *B. duttonii* in mouse blood, provided by Sven Bergström and Marie Andersson; three-dimensional view of a *B. burgdorferi* cell illustrating the periplasmic flagellar ribbon (see also Charon *et al.*, 2009, J. Bacteriol. *191*, 600–607), provided by Nyles Charon and Stuart Goldstein; and *B. burgdorferi* in human serum stained with SYTO® 9 and propidium iodide, provided by Rich Marconi and Beth Rogers.

The index was compiled by Sara Lynn Eastler.

Printed and bound in Great Britain

Contents

Contributors

Maria E. Aguero-Rosenfeld
Department of Pathology
New York University Langone Medical Center
New York, NY
USA

maria.aguero-rosenfeld@nyumc.org

Darrin R. Akins
Department of Microbiology and Immunology
Oklahoma University Health Sciences Center
Oklahoma City, OK
USA

darrin-akins@ouhsc.edu

Alan G. Barbour
Department of Medicine and
Department of Microbiology & Molecular Genetics
University of California Irvine
Irvine, CA
USA

abarbour@uci.edu

Stephen W. Barthold
Center for Comparative Medicine
Schools of Medicine and Veterinary Medicine
University of California, Davis
One Shields Avenue
Davis, CA
USA

swbarthold@ucdavis.edu

Jorge L. Benach
Department of Molecular Genetics and Microbiology
Stony Brook University
Stony Brook, NY
USA

jbenach@notes.cc.sunysb.edu

Sven Bergström
Department of Molecular Biology
Umeå University
Umeå
Sweden

sven.bergstrom@molbiol.umu.se

Linda K. Bockenstedt
Section of Rheumatology, Department of Internal Medicine
Yale University School of Medicine
New Haven, CT
USA

linda.bockenstedt@yale.edu

Julie Boylan
Laboratory of Zoonotic Pathogens
Rocky Mountain Laboratories
National Institute of Allergy and Infectious Diseases
National Institutes of Health
Hamilton, MT
USA

jboylan@niaid.nih.gov

Dustin Brisson
Department of Biology
University of Pennsylvania
Leidy Laboratories
Philadelphia, PA
USA

dbrisson@sas.upenn.edu

Felipe Cabello
Department of Microbiology and Immunology
New York Medical College
Valhalla, NY
USA

felipe_cabello@nymc.edu

Diego Cadavid
Center for Immunology and Inflammatory Diseases
Massachusetts General Hospital
Boston, MA
USA

dcadavid@partners.org

James A. Carroll
Laboratory of Persistent Viral Diseases
Rocky Mountain Laboratories
National Institute of Allergy and Infectious Diseases
National Institutes of Health
Hamilton, MT
USA

carollja2@niaid.nih.gov

Sherwood R. Casjens
Department of Pathology
University of Utah
Salt Lake City, UT
USA

sherwood.casjens@path.utah.edu

George Chaconas
Department of Biochemistry & Molecular Biology and
Department of Microbiology & Infectious Diseases
The University of Calgary
Calgary, Alberta
Canada

chaconas@ucalgary.ca

Nyles W. Charon
Department of Microbiology, Immunology and Cell Biology
Robert C. Byrd Health Sciences Center
West Virginia University
Morgantown, WV
USA

ncharon@hsc.wvu.edu

Jenifer Coburn
Department of Medicine, Division of Infectious Diseases
Center For Biopreparedness and Infectious Diseases
Medical College of Wisconsin
Milwaukee, WI
USA

jcoburn@mcw.edu

Raymond J. Dattwyler
Department of Medicine, Division of Allergy, Immunology and Rheumatology and
Department of Microbiology and Immunology
New York Medical College
Valhalla, NY
USA

raymond_dattwyler@nymc.edu

Daniel Dykhuizen
Department of Ecology and Evolution
Stony Brook University
Stony Brook, NY
USA

dandyk@life.bio.sunysb.edu

Christopher G. Earnhart
Department of Microbiology and Immunology
Medical College of Virginia at Virginia Commonwealth University
Richmond, VA
USA

cgearnhart@vcu.edu

Christian H. Eggers
Department of Biomedical Sciences
Quinnipiac University
Hamden, CT
USA

christian.eggers@quinnipiac.edu

Erol Fikrig
Section of Infectious Diseases, Department of Internal Medicine
Yale University School of Medicine
New Haven, CT
USA

erol.fikrig@yale.edu

Juan Carlos García-Moncó
Department of Neurology
Hospital de Galdacano
Galdacano
Vizcaya
Spain

hospit05@sarenet.es

Frank Gherardini
Laboratory of Zoonotic Pathogens
Rocky Mountain Laboratories
National Institute of Allergy and Infectious Diseases
National Institutes of Health
Hamilton, MT
USA

fgherardini@niaid.nih.gov

Stuart F. Goldstein
Department of Genetics, Cell Biology, and Development
University of Minnesota
Minneapolis, MN
USA

golds004@umn.edu

Betty P. Guo
Department of Molecular Biology
Umeå University
Umeå
Sweden

bp_guo@yahoo.com

David A. Haake
Division of Infectious Diseases
Veterans Affairs Greater Los Angeles Healthcare System and
Departments of Medicine, Urology, and
Microbiology, Immunology & Molecular Genetics
The David Geffen School of Medicine
University of California
Los Angeles, CA
USA

dhaake@ucla.edu

Magnus Höök
Center for Extracellular Matrix Biology
Institute of Biosciences and Technology
Texas A & M University
Houston, TX
USA

mhook@ibt.tamhsc.edu

Linden T. Hu
Department of Medicine
Tufts Medical Center
Boston, MA
USA

lhu@tuftsmedicalcenter.org

Kevin Lawrence
Laboratory of Zoonotic Pathogens
Rocky Mountain Laboratories
National Institute of Allergy and Infectious Diseases
National Institutes of Health
Hamilton, MT
USA

klawrence@niaid.nih.gov

John M. Leong
Department of Molecular Genetics & Microbiology
University of Massachusetts Medical Center
North Worcester, MA
USA

john.leong@umassmed.edu

Chunhao Li
Department of Oral Biology
State University of New York at Buffalo
Buffalo, NY
USA

cli9@buffalo.edu

Jun Liu
Department of Pathology and Laboratory Medicine
University of Texas Medical School at Houston
Houston, TX
USA

jun.liu.1@uth.tmc.edu

Richard T. Marconi
Department of Microbiology and Immunology and Center for the Study of Biological Complexity
Medical College of Virginia at Virginia Commonwealth University
Richmond, VA
USA

rmarconi@vcu.edu

Michael Miller
Department of Biochemistry
Robert C. Byrd Health Sciences Center
West Virginia University
Morgantown, WV
USA

mmiller@hsc.wvu.edu

Md. Abdul Motaleb
Department of Microbiology and Immunology
Brody School of Medicine
East Carolina University
Greenville, NC
USA

motalebm@ecu.edu

Steven J. Norris
Department of Pathology and Laboratory Medicine
University of Texas Medical School at Houston
Houston, TX
USA

Steven.J.Norris@uth.tmc.edu

Utpal Pal
Department of Veterinary Medicine
University of Maryland, College Park
College Park, MD
USA

upal@umd.edu

Mario T. Philipp
Department of Bacteriology and Parasitology
Tulane National Primate Research Center
Tulane University Health Sciences Center
Covington, LA
USA

philipp@tulane.edu

Joseph Piesman
Bacterial Diseases Branch
Division of Vector-Borne Infectious Diseases
Centers for Disease Control and Prevention
Fort Collins, CO
USA

jfp2@cdc.gov

Justin D. Radolf
Department of Medicine and
Department of Genetics and Developmental Biology
University of Connecticut Health Center
Farmington, CT
USA

JRadolf@up.uchc.edu

Patricia A. Rosa
Laboratory of Zoonotic Pathogens
Rocky Mountain Laboratories
National Institute of Allergy and Infectious Diseases
National Institutes of Health
Hamilton, MT
USA

prosa@niaid.nih.gov

Juan C. Salazar
Department of Pediatrics
Division of Pediatric Infectious Diseases
University of Connecticut School of Medicine
Connecticut Children's Medical Center
Hartford, CT
USA

jsalazar@nso1.uchc.edu

D. Scott Samuels
Division of Biological Sciences and
Center for Biomolecular Structure and Dynamics
The University of Montana
Missoula, MT
USA

scott.samuels@umontana.edu

Tom G. Schwan
Laboratory of Zoonotic Pathogens
Rocky Mountain Laboratories
National Institute of Allergy and Infectious Diseases
National Institutes of Health
Hamilton, MT
USA

tschwan@niaid.nih.gov

Ira Schwartz
Department of Microbiology and Immunology
New York Medical College
Valhalla, NY
USA

schwartz@nymc.edu

Ruth E. Silversmith
Department of Microbiology and Immunology
University of North Carolina
Chapel Hill, NC
USA

silversr@med.unc.edu

Jon T. Skare
Department of Microbial and Molecular Pathogenesis
Texas A&M Health Science Center
College Station, TX
USA

jskare@medicine.tamhsc.edu

Guiqing Wang
Department of Pathology and
Department of Microbiology & Immunology
New York Medical College
Valhalla, NY
USA

gwang@mednet.ucla.edu

Janis J. Weis
Department of Pathology
University of Utah
Salt Lake City, UT
USA

janis.weis@path.utah.edu

Charles W. Wolgemuth
Department of Cell Biology
University of Connecticut Health Center
Farmington, CT
USA

cwolgemuth@uchc.edu

Gary P. Wormser
Department of Medicine, Division of Infectious Diseases
New York Medical College
Valhalla, NY
USA

gary_wormser@nymc.edu

X. Frank Yang
Department of Microbiology and Immunology
Indiana University School of Medicine
Indianapolis, IN
USA

xfyang@iupui.edu

Wolfram R. Zückert
Department of Microbiology, Molecular Genetics & Immunology
University of Kansas Medical Center
Kansas City, KS
USA

wzueckert@kumc.edu

This book is dedicated to Nyles Charon, Phil Baker, Russ Johnson, and Herb Stoenner for their lifetimes of accomplishments in the field of *Borrelia* research and for inspiring successive generations of colleagues and trainees.

Preface

Since the discovery of *Borrelia burgdorferi*, the Lyme disease spirochaete, borreliology has evolved from a microbiological curiosity to a major branch of bacterial pathogenesis research. The rapid expansion of our field has been fuelled by the continual emergence ('metamorphosis' might be a better word) of Lyme disease as a global public health problem; indeed, few infectious diseases have garnered more sustained attention from the scientific and, notably, the lay media. The Federal government and numerous private foundations have marshalled considerable resources to combat this illness on both the basic scientific and clinical fronts. These monies have not been wasted: our knowledge of the disease and its causative agent has grown exponentially, a trend that continues despite constrained NIH budgets and dwindling paylines. This same period has also witnessed a resurgence of interest and, consequently, scientific advancement in relapsing fever, the other major human disease caused by borrelia species, the agents of which have long been known for their uncanny ability to alter their surface antigen profile.

We, the editors, believe that the field of borreliology was not just ripe for, but also in great need of, the comprehensive treatment that this book represents. Although other books have been written about Lyme disease, their focus generally has been clinical or epidemiological; this is the first volume, to our knowledge, with a nearly encyclopaedic coverage of the microbiological, immunological, genetic, ecological, and epizootic facets of pathogenic borrelia. *Borrelia: Molecular Biology, Host Interaction and Pathogenesis* coalesces and illuminates more than 25 years of tenacious study of the spirochaetes that cause Lyme disease and relapsing fever. Our goal was to integrate the diverse aspects of our field towards a unified view of how these vector-borne, enzootic diseases have developed into such serious threats to humans. We are gratified that this project was embraced by numerous experts, some of whom participated in the dramatic events surrounding the discovery of the Lyme disease spirochaete, one of the all-time great stories of medical sleuthing.

The field of Lyme disease research has been under assault from self-proclaimed 'Lyme literati' that have used misinterpretations, misrepresentations and bold fabrications to promote their ill-conceived and self-serving agenda regarding diagnosis and treatment. This book is intended to dispel accusations that scientists have conspired with their clinical colleagues to conceal from public view the 'real truths' about Lyme disease and its aetiological agent. In so doing, we endeavour to emphasize that well-controlled, carefully executed, and, above all, stringently peer-reviewed research is the only path to unravelling the mysteries of *B. burgdorferi* and its strategies for establishing persistence and inflicting damage on its incidental human host. In essence, this volume is, in part, a response to the repeated attacks on the purposes and integrity of scientists who have dedicated themselves to the study of this recalcitrant bacterium. Our objective is for this compendium to provide a foundation not only for further research progress, but also for sound policy decisions.

Unquestionably, there remains a substantial amount to learn about *Borrelia* and their host interactions. Our field will maintain its present trajectory only if it continues to attract bright and passionate researchers. In addition to providing a much needed reference work for borreliologists and infectious disease specialists, we hope this book will explain the enigma of how these cunning bacteria have managed to hold so many of us in thrall for so long! We also hope, and expect, that the tangled web these spirochaetes weave, as expounded in this book, will ensnare our successors.

D. Scott Samuels and Justin D. Radolf

Introduction

David A. Haake

The discovery of *Borrelia burgdorferi* and the subsequent elucidation of its unique molecular and cellular biology ranks among the greatest stories in the annals of microbial pathogenesis research. In 1976, a geographic clustering of children was identified in the area of Lyme, Connecticut, with a mysterious syndrome initially thought to be juvenile rheumatoid arthritis. It became apparent that Lyme arthritis was actually a late manifestation of a tick-transmitted multisystem disease, frequently preceded by an advancing erythematous rash at the site of a tick bite that responded to antibiotic therapy. In 1981, a medical entomologist named Willy Burgdorfer, along with Alan Barbour and Jorge Benach, both authors of chapters in this book, discovered a spirochaete in the midgut of hard-body *Ixodes* ticks from Shelter Island, New York, where Lyme disease was known to be endemic. They cultured the tick samples in a medium developed for the relapsing fever spirochaetes and found it to be a species of *Borrelia.* The new spirochaete, subsequently designated *Borrelia burgdorferi,* was then cultured from patients with early Lyme disease, and patients' immune responses were linked conclusively with that organism. Since those early days of proving the spirochaetal aetiology of the infection, there has been an explosion of new insights into the genetics, physiology, pathogenesis, and ecology of this bacterium and its tick/mammal life cycle. We now know that Lyme borreliosis-like illnesses can be caused by a large number of species related to *B. burgdorferi,* which are referred to broadly as *B. burgdorferi* sensu lato. Most of these organisms, including *B. garinii* and *B. afzelii* are widely distributed across Europe and Asia. The specific species, *B. burgdorferi* sensu stricto, arrived in North America quite a long time ago. *Borrelia burgdorferi* DNA has been found in museum specimens of ticks and mice from Long Island dating from the nineteenth century. The disease was rare after the European colonization of North America because woodland was cleared for farming and deer were hunted virtually to extinction. However, conditions improved for the ecology of Lyme borreliosis during the twentieth century. As farmland reverted to woodland, deer proliferated and white-footed mice became plentiful. Soil moisture and land cover were favourable for the *Ixodes* ticks to thrive, with *B. burgdorferi* colonizing tick midguts at high prevalence rates in certain areas. Suburbanization led to increased exposure of humans to infected ticks, resulting in resurgence of disease. Today, more than 30 years later, the incidence of Lyme borreliosis continues its steady upwards climb, with over 27,000 cases reported in the USA in 2007. And back where it all began 30 years ago, Connecticut continues to have the highest prevalence of Lyme borreliosis in the USA, with a whopping 87 cases per 100,000 population, 10 times higher than the national incidence.

Many aspects of *B. burgdorferi* research have benefited from previous studies on the relapsing fever borreliae. With the exception of *B. recurrentis,* which is transmitted by the human body louse, the relapsing fever borreliae are transmitted by 'soft body' *Ornithodoros* ticks. These 'soft body' ticks differ in a number of important ways from the hard-body *Ixodes* ticks that transmit *B.*

burgdorferi. Generally speaking, *Ornithodoros* ticks are 'hit and run' feeders; their soft bodies are distensible so that they can rapidly ingest a large blood meal and make their escape before being noticed. In contrast, *Ixodes* ticks are slow feeders that 'sit and sip,' remaining attached to their mammalian host for many hours before *B. burgdorferi* transmission occurs. *Borrelia burgdorferi* is maintained in nature through an enzootic cycle involving ticks and rodents. The life cycle of the deer tick takes 2 years to complete. Adult female ticks lay eggs on the ground in early spring. By summer, eggs hatch into larvae. Larvae feed on mice, other small mammals and deer in the late summer and early autumn, moult into nymphs, and then are dormant until the next spring. Nymphs feed on rodents, small mammals, birds and humans in the late spring and summer and moult into adults in the fall. In the fall and early spring, adult ticks feed and mate on large mammals (especially deer) and bite humans. The adult female ticks then drop off these animals and lay eggs in spring, completing a 2-year life cycle. In areas where Lyme borreliosis is highly endemic, a high percentage of the *Ixodes* ticks are infected because they feed on white-footed *Peromyscus leucopus* mice that remain bacteraemic for extended periods. In contrast, most parts of California have a very low incidence of Lyme borreliosis because *Ixodes* ticks in arid regions feed on lizards that are not susceptible to *B. burgdorferi* infection. So to understand the epidemiology of Lyme borreliosis, you have to first understand the ecological context in which they are found.

The intimate relationship between *B. burgdorferi* and its arthropod and vertebrate hosts is reflected in the *Borrelia* genome. At 0.91 Mb, the linear chromosome of *B. burgdorferi* is relatively small compared with most bacteria, containing a limited set of housekeeping genes with surprisingly few genes encoding biosynthetic and metabolic functions. This minimalist genome is consistent with dependence on a rich host (alternatively tick and mammal) environment for the building blocks of life. As a result, *Borrelia* cultivation in the laboratory is challenging and requires complex media containing a rich diversity of essential ingredients. Genomic dwarfism and host dependence appears to be nothing new for *B. burgdorferi*, as it contains one of the smallest fractions of inactivated pseudogenes known among bacteria, indicating that the contraction of its chromosome took place in the distant evolutionary past. That is not to say that the borreliae don't dabble in genetic experimentation. *Borrelia burgdorferi* strain B31 carries at least 12 linear and nine circular plasmids that seem to be in a constant state of rearrangement, recombination, and deletion. Recombination occurs both within and between strains, as some versions of the *ospC* and *ospD* genes are mosaics composed from different *B. burgdorferi* lineages. The linear plasmids in particular are subject to a high level of sequence scrambling through a process of reverse telomere resolution. Telomeres are closed hairpin loop structures at the ends of linear genetic elements, which require special acrobatic skills for the enzymes involved in their replication. Linear chromosomes with telomeric ends are the rule in eukaryotes but are extraordinarily rare in bacteria. The plasticity of *Borrelia* plasmids is reflected in the finding that in at least one case, a *B. hermsii* plasmid is able to convert from a linear to a circular form, apparently without ill effects.

In contrast to the parsimony of the linear chromosome, the linear plasmids carry a high percentage of redundant genes, non-functional 'pseudogenes', and non-coding sequences that appear to be largely genomic baggage. However, amongst the baggage are found some functional genes that are extremely important for the *B. burgdorferi* life cycle. For example, linear plasmid lp54 contains the gene encoding outer surface protein A (OspA). OspA is expressed at high levels on the surface of *B. burgdorferi* in the unfed tick and when cultivated in tick-like conditions (low temperature). Understanding the importance of OspA required the development of methods for genetic manipulation of *B. burgdorferi*, which have been slow in coming for several reasons. Selectable markers encoding antibiotic resistance genes and other genetic tools, such as shuttle vectors, needed to be developed and are now available. Linear plasmids lp25 and lp56 encode restriction-modification systems that modify *B. burgdorferi* DNA and digest foreign DNA lacking these modifications. Strains transformed with recombinant plasmids were frequently found to lack lp25, probably because strains that had lost lp25 and its restriction endonuclease were less likely to destroy foreign DNA. This is a problem for virulence studies because lp25 also contains the *pncA* gene encoding nicotinamidase, a biosynthetic enzyme essential for survival of *B. burgdorferi* in the mammalian host. Knocking out the lp25 restriction enzyme in a strain that retains the rest of the lp25 plasmid created a strain that is more transformable and yet retains virulence. Creating *ospA* knockouts was a challenge because they tend to aggregate and are difficult to isolate on solid agar plates. Researchers got around this problem by transforming *B. burgdorferi* strain 297 with a disrupted version of the *ospA* gene and then selecting for mutants in liquid culture in 96-well plates to isolate mutant clones. Using this approach, they showed that *B. burgdorferi* mutants lacking the *ospA* gene can infect mice but are unable to establish infection in the tick. More recently, researchers have begun to study gene function by replacing a gene's native promoter with inducible promoters that can be regulated by small molecule inducers such as isopropyl β-D-thiogalactopyranoside (IPTG). This elegant approach allows researchers to modulate gene expression and assess the effects of higher or lower expression levels during different stages of the *B. burgdorferi* life cycle.

The role of OspA in tick colonization was eventually determined to be as an adhesin for the TROSPA receptor on the surface of tick midgut epithelial cells. OspA expression ensures that *B. burgdorferi* remains in the midgut until the tick nuzzles up to a mammalian host. As the tick partakes of a nutritious blood meal, *B. burgdorferi* begins to shift the expression of its surface proteins. The decrease in surface OspA releases *B. burgdorferi* from tick epithelial cells in preparation for infection of the mammalian host. Tick feeding exposes *B. burgdorferi* to changes not only in temperature, but also in redox potential, pH, and osmolarity. These changes produce a bacterial response characterized by increases in transcriptional regulators such as RpoS. In *B. burgdorferi*, these regulators activate transcription of a portfolio of key mammalian infection genes, notably *ospC*. Outer surface protein C is encoded on circular plasmid cp26, which contains the genes *guaA* and *guaB* that are essential for *B. burgdorferi* viability. The *gua* genes encode non-redundant purine biosynthesis enzymes required for generating the building blocks of DNA. For this reason, cp26 has been referred to as a 'mini-chromosome'. Plasmid cp26 is characteristic of other *B. burgdorferi* circular plasmids, which contrast with linear plasmids in that the circular plasmids contain closely packed genes, the vast majority of which appear to be intact and presumably functional, without large stretches of intervening non-coding DNA.

As the tick ingests its blood meal and *B. burgdorferi* in the midgut shift from OspA expression to OspC expression, organisms begin their migration to the tick's salivary glands, and eventually to the bite wound and tissues of the mammalian host. Blood entering the tick midgut from the host carries the fibrinolytic enzyme plasmin, which *B. burgdorferi* acquires on its surface to digest its way through the lining of the midgut and into the salivary glands. The migration process is driven by flagellar motors that power the organism's elegant, corkscrew motility. *Borrelia burgdorferi* has a plethora of chemotactic genes to coordinate the rotational direction of the flagella and propel the organism towards its desired destinations. Reflecting the importance and complexity of motility and chemotaxis, these functions require 54 genes or 6% of the organism's chromosomal information. The elongated flat-wave shape of the organism seems designed for breaching tissue barriers. *Borrelia burgdorferi* arriving in the salivary gland express OspC, which prepares the organism for host invasion by binding the tick salivary protein, Salp15. *Borrelia burgdorferi* coated with Salp15 are resistant to killing by borreliacidal antibodies. Salp15 also has immunosuppressive properties that inhibit activation of T-helper lymphocyte at the bite wound where *B. burgdorferi* infection is initiated. Once *B. burgdorferi* enters the bite wound and infects mammalian host tissues, its primary goals are dissemination and persistence. Persistence requires evasion of host defence mechanisms such as complement-mediated killing. Complement evasion appears to again involve acquisition of a host protein on its surface, in this case complement regulatory proteins such as factor H. A number of *B. burgdorferi* complement regulator-acquiring surface proteins (CRASPs) have been identified that are upregulated during mammalian host infection. Persistence also requires adhesion to host tissue proteins. In the case of *B. burgdorferi*, the outer membrane proteins (OMPs) DbpA and DbpB mediate binding to decorin, a protein associated with skin collagen, an ideal location for an organism in waiting for acquisition by skin-biting ticks. Additional OMPs that mediate *Borrelia*–host interactions are the fibronectin-binding protein Bbk32, integrin-binding protein P66, and the glycosaminoglycan binding protein Bgp. Another persistence mechanism is antigenic variation of the prominent surface protein, VlsE. The VlsE expression locus and an elaborate system of *vlsE* silent cassette genes are present on linear plasmid lp28-1. The silent cassette genes provide a source of short sequences that insert through recombination into the *vlsE* gene in the expression locus to alter its antigenic characteristics and evade the humoral immune response of the host. Expression of new VlsE surface proteins presumably allows *B. burgdorferi* clones to escape from the host immune response and form the recurrent skin lesions characteristic of secondary Lyme borreliosis.

Antigenic variation is the hallmark of infections caused by the relapsing fever *Borrelia*, mediated by a genetic recombination system distinct from that of *vlsE* of *B. burgdorferi*. In relapsing fever, patients develop recurrent crises characterized by sudden onset of high fever and shaking chills followed by a drenching sweat, low blood pressure, and defervescence. These crises are associated with extraordinarily high densities of relapsing fever borreliae in the bloodstream of affected patients. Crises are precipitated by formation of antibodies to surface lipoproteins encoded by the gene in the variable protein expression locus. Antibody-mediated clearance of such high numbers of organisms from the bloodstream and delivery of large amounts of bacterial antigens to immune cells triggers a shock-like response presumably mediated by the innate immune system. In louse-borne relapsing fever caused by *B. recurrentis*, relapses are generally few in number (1–5), but the crises can be severe, with mortality rates as high as 40%. Tick-borne relapsing fever caused by organisms such as *B. hermsii* is less severe, but the relapses are more persistent, with as many as 13 relapses having been recorded. Each relapse is associated with the emergence of bacteria that have undergone a recombinational event on the expression plasmid involving insertion of genes encoding any one of 12 variable small proteins (Vsps) or 15 variable large proteins (Vlps). The most likely variable gene insertion events in *B. hermsii* is determined by the degree of homology of sequences upstream of the gene currently in the expression locus with the upstream homology sequence (UHS) of the new gene and the distance from the end of the new gene to its downstream homology sequence (DHS). The result is a remarkable preprogrammed pattern of variable gene expression in the *Borrelia* that emerge with each new relapse. These relapses improve the likelihood that organisms will be present in the bloodstream when an uninfected louse or tick bites the infected host.

A major difference between relapsing fever and Lyme borreliosis is the density and distribution of organisms in the mammalian host. Relapsing fever borreliae can be found in blood at densities as high as 10^8/ml, while *B. burgdorferi* rarely achieves densities greater than 100/ml of tissue. What *B. burgdorferi* lacks in numbers, it makes up for in proinflammatory potential. Histological examination of affected tissues typically reveals an intense inflammatory response that is out of proportion to the number of organisms present. The inflammation is reflected in the characteristic bright red erythema migrans skin lesions that occur at the site of the tick bite in 80% of patients. A systemic inflammatory response also occurs with skin lesions often accompanied by influenza-like symptoms, such as malaise and fatigue, headache, arthralgias, myalgias, and fever. Inflammation is also characteristic of the histological and immunological changes seen at sites of *B. burgdorferi* dissemination including the joints, heart, and nervous system. *Borrelia burgdorferi* is not known to secrete toxins, so the inflammation seen in Lyme borreliosis is largely, if not totally, a result of the host immune response to *B. burgdorferi*. The host innate immune response is driven by recognition of bacterial components by Toll-like receptors (TLRs). Fatty acid modification of abundant *B. burgdorferi* lipoproteins is specifically recognized by TLR2. TLR recognition of *B. burgdorferi* lipoproteins causes various cells, especially macrophages and endothelial cells, to elaborate NF-κB dependent cytokines, adhesion molecules, and chemokines. The inflammatory response is a double-edged sword. On the one hand, inflammation is an effective cellular host defence mechanism that recruits and activates lymphocytes and phagocytes that kill *B. burgdorferi* by producing bactericidal antibodies and engulfing organisms, respectively. On the other hand, inflammation also causes havoc in the joints, heart muscle, and nerve cells, which produces the signs and symptoms we recognize as Lyme borreliosis. Mouse models of Lyme borreliosis have been particularly informative with regard to the pros and cons of inflammation. Infection of C57BL/6 mice with *B. burgdorferi* results in minimal or no arthritis. In contrast, C3H mice develop severe joint swelling and inflammation, even though the numbers of organisms in their joints are similar to those found in the joints of C57BL/6 mice. Comparison of the host transcriptional response in C57BL/6 vs. C3H mice early in the course of *B. burgdorferi* infection revealed much higher levels of interferon-inducible gene transcripts in the C3H mice. Interestingly, C57BL/6 mice lacking functional genes for TLR2 or the inflammatory cytokine IL-10 had increased arthritis and much higher interferon-inducible transcripts, similar to the pattern seen in C3H mice. These results indicate that the outcome of *B. burgdorferi* is strongly determined by the host genetic background, particularly those genes that modulate the nature and intensity of the host inflammatory response.

This book is a comprehensive review of the molecular and cellular biology of the Borreliae, written by many of the same investigators that made the seminal discoveries. Considering the dramatic pace of research since *B. burgdorferi* was discovered less than thirty years ago, the stage is now set for the development of new approaches that will revolutionize the diagnosis and prevention of Lyme borreliosis. Serology is the cornerstone of diagnosis and has relied on a two-tier system of ELISA screening and confirmatory immunoblot. Two-tier testing may eventually be replaced by rapid and relatively inexpensive assays using antigens such as C6, an invariant 26-amino-acid region of VlsE. Serological tests for Lyme borreliosis using the C6 assay have been shown to be highly sensitive and specific. Improved diagnostic tests are critical for disease management, especially when patients do not respond promptly to standard therapy. A minority of Lyme borreliosis patients have fatigue, musculoskeletal pain, and/or difficulties with concentration or short-term memory that persists despite antibiotic therapy. These post-Lyme disease symptoms are usually mild and self-limiting. However, some patients with disabling fatigue, neurocognitive symptoms, and persistent pain are labelled as having 'chronic Lyme disease' despite lacking any objective symptoms or diagnostic evidence of *B. burgdorferi* infection. While it is certainly true that some aspects of Lyme borreliosis remain mysterious, especially regarding its ability to cause persistent infection, the best way to solve these mysteries is to continue the powerful lines of research that have revealed so much already, as described by the authors of this book. Ultimately, the best approach will be to prevent infections from occurring. Although experimental evidence for immunoprotection has been obtained for a number of surface-exposed *B. burgdorferi* proteins, OspA has been the most successful approach. *Borrelia burgdorferi* in the tick midgut are exquisitely sensitive to killing by antibodies to OspA. If an infected tick bites an animal or human with antibodies to OspA, the organism will be rapidly killed before transmission can occur. This elegant strategy was the basis for a recombinant OspA vaccine that was marketed in the USA and administered to thousands of patients. Although concerns that the OspA vaccine could induce autoimmune arthritis were never validated, they did prevent widespread acceptance. Eventually, the OspA vaccine was withdrawn by the manufacturer. Additional outer surface proteins, such as OspC, are being explored as alternative vaccine candidates. And there is hope that OspA may be resurrected as an oral vaccine for wildlife as a means to eradicate *B. burgdorferi* from its natural reservoirs, thereby reducing human exposure and preventing Lyme borreliosis from occurring. This approach, which has been successful for eradication of rabies, would usher in a new age in the management of zoonotic infectious diseases. In the meantime, a simple walk in the woods will not be free from risk, and wisdom will dictate a check for tiny eight-legged passengers.

The Worldwide Saga of Lyme Borreliosis

1

Jorge L. Benach and Juan Carlos García-Moncó

Abstract

The clinical–epidemiological studies of a cluster of arthritis in Lyme, Connecticut, intersected with ongoing studies on tick-borne diseases in Long Island, New York, to set the stage for the discovery of the spirochaete agent of Lyme disease. Studies on the microbiology of spirochaetes in other parts of the country were instrumental for the cultivation of the newly discovered organism. The multinational clinical studies in Europe throughout the twentieth century on the dermatological and neurological manifestations characterized and enlarged the syndrome that we now know as Lyme borreliosis. For reasons that are discussed in this chapter, controversies have arisen among advocate patient groups and the clinicians and scientists that are at the forefront of research in Lyme borreliosis. Polarized opinions on what constitutes this disease and how to treat it have resulted in unprecedented litigation involving medical societies and advocate patient groups with no obvious end in sight.

1.1 The epidemic of oligoarticular arthritis in Connecticut and the north-east

The story of Lyme disease in Connecticut has been told on numerous occasions and has often been cited as an example of a very positive interplay among the lay community, academic medicine, and public health workers. In October 1975, two mothers, *Polly Murray* and *Judith Mensch*, from Lyme, Connecticut, contacted the Connecticut State Health Department to report that a number of children who lived in the same neighbourhood in their town had developed what they had been told was juvenile rheumatoid arthritis. Soon thereafter, a joint investigation by the Connecticut State Health Department and the Yale University School of Medicine was begun to investigate the reports made by both mothers. This fortuitous occurrence developed into a most dramatic epidemiological study that resulted in the identification of a world-wide infection.

The possibility of a focal episode of juvenile rheumatoid arthritis in Lyme kindled the interest of Dr Allen Steere, who had just joined the Division of Rheumatology at Yale, which in turn was headed by Dr Stephen Malawista, a noted rheumatologist with a long track record on inflammatory rheumatic diseases. Dr Steere, joined by Dr David Snydman of the State Health Department, put together an active surveillance system in the affected towns of Lyme, Old Lyme and East Haddam (Connecticut) to identify cases of the disorder reported by the two mothers.

The active community surveillance disclosed 39 children (Steere *et al.*, 1977b) with recurrent attacks of swelling and pain in large joints, notably in the knees, and some for substantial periods of time. In addition, there were twelve adults, who were related to or lived near the arthritic children, with similar signs and symptoms. The surveillance showed an epidemic of oligoarticular arthritis in a focal geographic area. The focality of Lyme disease has remained a constant feature of this disease. In fact, even within these affected towns, clustering of cases was actually the norm in that many of the patients lived in several adjoining streets. Furthermore, there was also clustering within families, with six having more than one affected member, although the manifestations of the disease may have had onsets in different years. In the first study, the prevalence of the disease was 4.3 per 1000 residents but was three times higher among children. The overrepresentation of children in Lyme disease has also remained a constant feature of this disease. This investigation essentially excluded juvenile rheumatoid arthritis as the cause of the epidemic given that the prevalence of arthritis in these Connecticut communities was 100 times greater than that expected prevalence for juvenile rheumatoid arthritis. Thus, with a high degree of certainty that the articular manifestations being reported in the three towns in Connecticut were not juvenile rheumatoid arthritis, the disease was named Lyme arthritis. With the enlarging clinical spectrum, this name was changed to Lyme disease.

The enlarging investigation revealed that approximately one-quarter of the patients had developed an erythematous cutaneous lesion that in some cases appeared to expand into reddish, often annular, concentric rings. Interestingly, the skin lesion appeared weeks before the onset of the articular manifestations. This lesion had been previously described in Sweden in 1909 by Arvid Afzelius (see below) who named it 'erythema migrans' because of its propensity to enlarge. Sometime later, the lesion became known as 'erythema chronicum migrans' to emphasize its protracted duration. Afzelius associated the lesion with the bite of *Ixodes ricinus* ticks. During the summer of 1976, the Yale investigators followed prospectively the patients with the cutaneous lesions (Steere *et al.*, 1977a). Many of these patients developed arthritis, while some also developed neurological or cardiac abnormalities (Steere *et al.*, 1980a). The neurological manifestations among the prospective study group subsequently were defined by Reik and coworkers in 1979 (Reik *et al.*, 1979). The addition of neurological and cardiac involvement to the already known skin and articular manifestations underscored the complex, multi-organ system disease that was being characterized in Connecticut.

The established association of erythema migrans with the bite of *Ixodes ricinus* ticks in Europe led to investigations showing that a local *Ixodes* species was associated with Lyme disease (Steere *et al.*, 1978). Two studies provided indirect evidence for a role of *Ixodes scapularis* in the development of the erythema migrans. One study showed that the incidence of Lyme disease was 2.8 cases per 1000 residents in the three communities of Lyme, Old Lyme, and East Haddam, on the east side of the Connecticut river. This was in marked contrast to the almost 30-fold lower incidence of 0.1 cases per 1000 in several communities on the west side of the river (Steere and Malawista, 1979). The other study by Wallis *et al*. (1978) showed that *I. scapularis* were much more abundant on the east side (with a high incidence of erythema migrans and Lyme arthritis) of the Connecticut River than on the west where the incidence of the disease was significantly less. On the east side of the river, subadult *I. scapularis* were also more abundant on white-footed field mice, *Peromyscus leucopus*, while the adults were more abundant on white-tailed deer. Importantly, attempts to recover a possible aetiological agent from the ticks collected in the affected areas were unsuccessful. Nonetheless, the epidemiological evidence strongly implicated these ixodid ticks in the natural history of Lyme disease.

One final important step was taken that led to successful therapy of Lyme disease with antibiotics. European erythema migrans had long been known to be responsive to antibiotic therapy, while studies demonstrating the efficacy of antibiotic therapy for this skin lesion had also appeared in the USA (Scrimenti, 1970). Therefore, Steere and collaborators gave antibiotics to patients with erythema migrans and compared the outcomes with those in untreated patients from their investigations. The results showed that penicillin, given early in the illness, shortened the duration of erythema migrans and appeared to prevent or lessen subsequent arthritis (Steere *et al.*, 1980b).

Prior to the Connecticut epidemic, erythema migrans in North America was first described in 1969 in a case report of a hunter from Wisconsin. The attending physician treated this patient successfully with penicillin (Scrimenti, 1970). A cluster of four cases of erythema migrans that occurred in the summer of 1975 in southeastern Connecticut was also reported before the Yale University studies (Mast and Burrows, 1976a,b). Although determining when erythema migrans first appeared in the USA is difficult, the lack of any reports prior to 1960 indicate that Lyme disease became epidemic in the 1970s coincident with the reports from Connecticut, Long Island, New York and Massachusetts. In retrospect, the Wisconsin case was a harbinger of the subsequent emergence of Lyme disease in the upper Midwest. Moreover, it is likely that cases of Lyme disease in these endemic areas were not diagnosed as such before it gained widespread recognition. For example, in eastern Long Island, which is a major area endemic for Lyme disease, there are persistent anecdotes of a condition known as the 'Montauk knee' that sounds suspiciously like the monoarticular arthritis of Lyme. This entity evidently has been around for decades, and it is tempting to speculate that the Montauk knee may actually have been the late stage of Lyme disease.

1.2 The role of other tick-borne diseases in the discovery of the agent of Lyme disease

1.2.1 The tick revolution of the 1970s

In the early 1970s, an increase in the number of cases of Rocky Mountain spotted fever (RMSF) occurred in the southeast, south central and northeastern states. This increase was amply documented and sparked a renewed interest in tick-borne diseases in this country (Benach *et al.*, 1977; Burgdorfer, 1975). This was particularly true in Long Island and other coastal New England and the Massachusetts offshore islands that are at the northernmost limit of the geographic distribution of this disease. Parallel with the increase in the number of cases of RMSF were reports of very large populations of *Dermacentor variabilis* ticks, the vector of the causative organism of RMSF, *Rickettsia rickettsii*. Although the reasons for this spike in the prevalence of RMSF have been debated, a general consensus has not developed. Nevertheless, it is generally believed that the increase in cases of RMSF in the early 1970s resulted from a number of environmental changes that may have been occurring during these years, and possibly even earlier. One such change was the banning in the late 1960s of DDT, a chlorinated hydrocarbon pesticide that had been widely used for agricultural, suburban pest, and public health insect control. The widespread use of this pesticide may have affected tick populations as non-target organisms and may have helped keep these populations in check; banning of this pesticide may have allowed for a gradual increase in the tick populations in large geographic areas of the country. Another environmental change reflects the increased suburbanization of rural land leading to a closer relationship of wildlife with new human residents. While this demographic trend is certainly well-documented and can explain the increase in the number of cases of tick-borne diseases, it does not explain the large increases in the tick populations. To explain the latter, increases in the populations of reservoirs have been implicated as the sources of large numbers of vectors. This is particularly true in the case of deer and the *Ixodes scapularis* ticks. As mentioned earlier, the increases in cases of RMSF and its vector, *D. variabilis*, were not an isolated event. In fact, another, possibly larger and of higher public health impact, increase in the numbers of human-biting *I. scapularis* led to the epidemic of Lyme disease in the northeast and upper Midwest.

The entomological literature prior to the 1970s clearly documents the presence in the northeast of an Ixodes species whose subadult stages are ectoparasites of field rodents, notably *Peromyscus leucopus*, and whose females feed on white-tailed deer. Depending on the source, this tick species was identified as *I. muris* or as *I. scapularis*. Stepping into this taxonomic void, Spielman and coworkers described a new species using morphological characteristics of ticks collected in Massachusetts which they named *Ixodes dammini*, for Gustave Dammin, a pathologist at Harvard University (Spielman *et al.*, 1979). This tick, once largely thought to be confined to feral hosts, assumed immediate importance as the vector to humans of *Babesia microti*, and a few years later as the vector of the aetiological agent of Lyme disease.

However, concerns that the newly named *Ixodes dammini* was not a new species prompted researchers to use genetic approaches to address this taxonomic question. Reciprocal crosses between *I. dammini* from Massachusetts and *I. scapularis* from Georgia produced offspring through the F3 generation whereas reciprocal crosses between *I. dammini* and *I. pacificus* from California (now known to be the vector of *B.*

burgdorferi on the west coast (Burgdorfer *et al.*, 1985) and *I. scapularis* and *I. pacificus* crosses produced sterile F_1 progeny. Through assortive mating experiments, morphometric measurements, and chromosome and isoenzyme analyses, led Oliver and coworkers (Oliver *et al.*, 1993) to conclude that *I. dammini* was not a species separate from *I. scapularis*, and that the name *Ixodes scapularis,* which had been used for more than 100 years to identify this species, had priority over the name *Ixodes dammini.* General acceptance of the name change resulted in the now near universal use of the *I. scapularis* nomenclature in the scientific literature.

Was there a 'tick revolution' in the 1970s and possibly even earlier? On the face of the question, it appears improbable that virtually parallel increases in the populations of two separate tick species in such geographic proximity over such a short period of time are due to chance. Convincing arguments have been proposed, but single theories do not explain all of the increases in the tick populations and of the epidemics that developed thereafter. As with other environmental events, the causes are probably multifactorial, and their consequences certainly important enough to be debated for years to come. To be sure, significant multifactorial theories have been tested with an extraordinary degree of insight. One study linked concurrent defoliation of oak trees by gypsy moths (another environmental event in the northeast in the 1970s) with changing acorn production by the defoliated oaks, deer eating acorns, and rodent reservoirs eating the pupae of the gypsy moth (Jones *et al.*, 1998; LoGiudice *et al.*, 2003; Ostfeld *et al.*, 2006). This theory was characterized as an environmental chain reaction using disparate animal and plant players as the basis for the changes in vectors and reservoirs that resulted in the tick-borne epidemics of the last 30 or so years.

1.2.2 Human babesiosis

The finding of human cases of babesiosis on Nantucket Island (Massachusetts) (Ruebush *et al.*, 1977, 1981) and on Long Island (New York) (Benach *et al.*, 1982, 1978, 1979) was a pivotal event in the discovery of the causative organism of Lyme disease. Human babesiosis in the northeastern USA is caused by the intraerythrocytic piroplasm *Babesia microti* (Homer *et al.*, 2000; Rudzinska *et al.*, 1976). Babesiosis is a malaria-like opportunistic infection with a broad spectrum of severity, ranging from subclinical in immunologically intact persons to fatal in elderly, immunocompromised, or cirrhotic patients. This protozoan pathogen is maintained enzootically in the white-footed mouse reservoir (*Peromyscus leucopus*) and is transmitted through the bite of *Ixodes scapularis* (Piesman and Spielman, 1980; Piesman *et al.*, 1979; Spielman *et al.*, 1981). The first cases of human babesiosis appeared in the 1970s in the offshore islands of Massachusetts and New York. The number of cases increased from a handful to more than one dozen in a short time, and more ominously, the severity of the illness in elderly patients, added to the detection of transfusion-transmitted and transplacental cases, required investigation by federal and state public health officials (Esernio-Jenssen *et al.*, 1987; Grabowski *et al.*, 1982). The finding of transfusion-acquired cases resulted in a ban on blood collection from the enzootic areas by the local blood banks. A very clear disease pattern emerged from the increasing number of cases (Dammin *et al.*, 1981; Meldrum *et al.*, 1992; Rosner *et al.*, 1984; Ruebush *et al.*, 1981). The typical patient was 50 years old or older and presented with non-periodic fever and chills, myalgia, fatigue, haemolytic anaemia, and haematuria. A number of patients had been splenectomized with a fatality rate of around 10% in this subset. In fact, studies with experimental animal models also have documented that advanced age is a risk factor for more severe disease (Habicht *et al.*, 1983; Vannier *et al.*, 2004). The parasitaemia varied markedly and did not always correlate with clinical severity (Dammin *et al.*, 1981; Filstein *et al.*, 1980; Meldrum *et al.*, 1992; Rosner *et al.*, 1984; Ruebush *et al.*, 1981). In some immunocompromised patients, the parasitaemia was only reduced with red blood cell exchanges (Cahill *et al.*, 1981), and even in these cases, it persisted at low levels for months. Importantly, in the intervening years since the first cases of babesiosis were discovered, its geographic distribution has not changed markedly. While single cases of *B. microti* infection have been reported from areas other than coastal New England and New York, the large majority of the reported cases remains clustered geographically in the same endemic areas.

A serosurvey conducted in Shelter Island (New York) was carried out to detect asymptomatic individuals that could transmit *B. microti* through blood banking procedures and to determine the prevalence and incidence of serologically reactive but asymptomatic babesiosis (Filstein *et al.*, 1980). Healthy individuals at high risk of exposure to ticks were surveyed during a single transmission season. Paired sera obtained in June and October 1978 were tested for antibodies to *B. microti* using the newly developed indirect immunofluorescence test (Chisholm *et al.*, 1978). Point prevalence values of 4.4% in June and 6.9% in October were obtained. Six of 102 persons tested in both months showed at least a fourfold rise in titre of antibodies to *B. microti* for an incidence of 6% for the season. Control sera from inland areas did not show any significant reactivity to *B. microti*, but of the six persons who seroconverted, four had histories of tick bites during the transmission season. Of note, this serosurvey identified a number of individuals who reported a history of erythema migrans and who also had antibodies to *B. microti*. These individuals did not have clinical babesiosis but retrospective inquiries yielded histories compatible with early Lyme disease. Sera banked from these individuals were used several years later to strengthen the association between *B. burgdorferi* and Lyme disease (Burgdorfer *et al.*, 1982). This association between subclinical exposure to *B. microti* and signs of Lyme disease proved to be a common occurrence that underscored the close natural history of both pathogens. This pattern also underscored the opportunistic nature of babesiosis and the high degree of susceptibility of healthy persons to infection with *B. burgdorferi*.

The sharing of reservoir and vector hosts by *B. microti* and *B. burgdorferi* leads to frequent coinfection in nature (Anderson *et al.*, 1986; Piesman *et al.*, 1987; Piesman *et al.*, 1986) and in patients (Benach and Habicht, 1981; Krause *et al.*, 1992b; Sweeney *et al.*, 1998). Evidence for coinfection in humans was noted before the discovery of *B. burgdorferi*. Clinical histories consistent with Lyme disease were elicited from patients who had antibodies to *B. microti* in the serosurvey in 1978 (Filstein *et al.*, 1980). Subsequently, in 1980 a number of patients with clinical babesiosis reported erythema migrans at the onset of the disease (Benach and Habicht, 1981), Although serological confirmation of Lyme disease was not possible at the time, subsequent serological confirmation of coinfection or coexposure was obtained shortly after *B. burgdorferi* was discovered. Clinical coinfection was further documented in a series of patient studies (Benach *et al.*, 1985; Grunwaldt *et al.*, 1983). The existence of clearly established human cases of coinfection with *B. burgdorferi* and *B. microti* led to clinical and animal model studies to determine whether the course of illness is more severe following coinfection (Hilton *et al.*, 1999; Krause *et al.*, 1996; Wang *et al.*, 2000). In the clinical arena, there are conflicting results. One study reported greater severity of Lyme disease (Krause *et al.*, 1996), while another did not report any synergistic interactions between the two diseases (Wang *et al.*, 2000). Experimental studies of coinfection in mice did not show any synergy leading to greater severity of either infection (Coleman *et al.*, 2005), but another study showed greater arthritis severity in double infections (Moro *et al.*, 2002).

1.2.3 *Borrelia* research in Montana and Minnesota

The medium defined by Kelly for the *in vitro* cultivation of *Borrelia hermsii* greatly facilitated the studies of Stoenner on the biology of relapsing fever and made possible the subsequent discovery of the *Borrelia* agent of Lyme disease (Kelly, 1971). Burgdorfer pioneered relapsing fever research as early as 1954, and his work was instrumental in institutionalizing investigations on the tick–borne *Borrelia* at the Rocky Mountain Laboratory (Burgdorfer, 1954, 1970; Davis and Burgdorfer, 1954). Research on relapsing fever *Borrelia* continued with the studies of Stoenner, and then Barbour on the antigenic variation of these organisms (Barbour *et al.*, 1982; Stoenner, 1974; Stoenner *et al.*, 1982). The early *Borrelia* work at the Rocky Mountain Laboratory was paralleled by the seminal studies, also on relapsing fever, of Russell Johnson at the University of Minnesota. His manuscripts (Johnson, 1977) were required reading for all the investigators that joined this field after the identification of the agent of Lyme disease.

1.3 The discovery of the spirochaete

The discovery of the Lyme disease spirochaete was the result of a convergence of events that were occurring in different geographic areas of the USA although not necessarily in isolation from each other:

1. The collaboration between Burgdorfer and Benach centred around the Long Island, New York, endemic areas of several tick-borne diseases, and led to joint investigations that yielded the isolation of the organism.
2. The state of readiness at the Rocky Mountain Laboratory to culture spirochaetes in Kelly's medium was another factor in the discovery of the organism.
3. The description of an enlarging clinical spectrum for Lyme disease by Steere, Malawista, and coworkers permitted the clinical identification of cases outside of the Lyme, Connecticut, area.

The serosurveys for babesiosis in 1978 and 1979 in Shelter Island, New York, had provided serum specimens from individuals with clinically inapparent babesiosis, but who gave a history of erythema migrans and joint manifestations that had been diagnosed as Lyme disease by Dr Edgar Grunwaldt, the internist on Shelter Island. These sera, stored in our serum bank at Stony Brook University, were used for establishing the aetiology of one of the organisms that were seen in *Ixodes scapularis* adults collected from Shelter Island. One such organism was a microfilaria seen in the haemolymph of the ticks. On dark-field microscopy, spirochaetes were seen and later cultured in Kelly's medium. In addition to the banked sera, we provided the serum of a faculty member, a physician at Stony Brook University, who had suffered from many of the manifestations Lyme disease for some time. These sera reacted with the newly cultured spirochaetes, initially thought to be a species of Treponema, and these findings were reported in Science (Burgdorfer *et al.*, 1982). In short order, spirochaetes were isolated from patients with Lyme disease by two groups, thus fulfilling Koch's postulates; both studies were reported in the same issue of the New England Journal of Medicine in 1983 (Benach *et al.*, 1983; Steere *et al.*, 1983). Spirochaetes were subsequently isolated from the skin of patients with erythema migrans (Berger *et al.*, 1985).

Continuing field investigations headed by Edward Bosler of our laboratory, isolated the spirochaetes from the blood of white-footed mice collected in Shelter Island and from the same areas where the serosurveys of 1978 and 1979 had been carried out (Bosler *et al.*, 1983). The isolation of the *Borrelia* organisms from *Peromyscus leucopus* served to identify the main reservoir of this organism. This finding, also published in Science in 1983, was subsequently confirmed in several studies from other parts of the country (Burgdorfer *et al.*, 1988). Interestingly, deer, long thought to be involved in the zoonotic cycle of the spirochaete, proved to be a relatively incompetent host (Telford *et al.*, 1988). Canine Lyme disease was identified through our continuing studies on Long Island (Lissman *et al.*, 1984). Dr Barry Lissman, a veterinarian in private practice in our area, had been a prior collaborator in our studies on tick-borne disease. A few years earlier, Dr Lissman and our group had characterized for the first time the canine cases of Rocky Mountain Spotted Fever (Lissman and Benach, 1980). Attuned to the problem of high tick burdens in Long Island dogs, Dr Lissman had seen a series of animals with lameness and arthritis that were heavily infested by both *I. scapularis* and *D. variabilis*. The discovery of the spirochaete permitted the testing of canine serum against this agent with strongly positive results. Moreover, we were able to grow the *Borrelia* from blood samples from dogs with this syndrome (Lissman *et al.*, 1984).

The transmission of the organism by the feeding of *I. scapularis* ticks was subsequently verified (Benach *et al.*, 1987; Ribeiro *et al.*, 1987).

Dr Robert Lane in California and Dr Burgdorfer were able to grow the *Borrelia* organism from *Ixodes pacificus* ticks, thus identifying a third endemic area in the western USA (Burgdorfer *et al.*, 1985). In 1983, at the first International Meeting on Lyme Disease at Yale University in New Haven, Connecticut, a group of investigators that had been instrumental in the discovery and characterization of the Lyme disease spirochaete agreed that the new organism would be named *Borrelia burgdorferi*, after Willy

Burgdorfer. Dr Johnson wrote the taxonomic study proposing this name (Johnson *et al.*, 1984). Years later, taxonomists in Europe determined that Lyme disease spirochaetes in Europe were sufficiently diverse to diverse to warrant separate species designations (Belfaiza *et al.*, 1993; Canica *et al.*, 1993). This led to the naming of *B. garinii* and *B. afzelii* as well as the designation of the original isolate as *B. burgdorferi* sensu stricto, with the entire complex being *B. burgdorferi* sensu lato (Baranton *et al.*, 1992) (see Chapter 9). More recently, *B. japonica* and other *Borrelia* species of unknown infectivity were described (Kawabata *et al.*, 1993).

A notable milestone in the therapy of Lyme disease was made by Dattwyler *et al.* (1987, 1988a) with the introduction of ceftriaxone as the choice antibiotic (see Chapter 18). Barthold developed a murine model of Lyme arthritis and carditis that has provided the basis for numerous studies on pathogenesis and host responses (Barthold *et al.*, 1988) (see Chapter 14). The genetic transformation barrier for *B. burgdorferi* was broken by the work of Samuels (Samuels, 1995; Samuels *et al.*, 1994a,b) and Rosa and coworkers (Rosa *et al.*, 1996, 2005) (see Chapter 8). These studies have already permitted important research on gene product function in conjunction with the genome of *B. burgdorferi* (Fraser *et al.*, 1997) (see Chapter 2).

1.4 The European evidence of one hundred years

Although Lyme disease was named for the place where it was first described in the USA, the disorder had been present in Europe for a long time, mainly as a dermatological and neurological disorder, in contrast with the original rheumatological presentation in the USA. Furthermore, many years before, several European scientists had linked this disorder to an infectious agent transmitted by a tick bite. The historical hallmarks are presented in Table 1.1.

1.4.1 Erythema migrans and Azfelius

The first description of the most frequent and worldwide pathognomonic skin lesion of what we now know as Lyme borreliosis or Lyme disease, was given at the beginning of the twentieth century by Arvid Azfelius (1857–1923), one of the founders of the Swedish Society of Dermatology. In 1909 he presented before the society an elderly woman with an expanding annular erythema with central fading that had developed at the site of a bite by the tick *Ixodes reduvius* (Azfelius, 1910). Because of its expansion pattern he named this lesion erythema migrans and suggested that it was due to a tick-transmitted infection. A few years later, the Austrian dermatologist, Lipschütz, described another patient with a similar condition that lasted for 7 months, and suggested naming it erythema chronicum migrans (ECM) on the basis of its duration (Lipschutz, 1918). He did not speculate on its aetiology. In 1921 Azfelius reported six additional patients with a similar skin lesion, and two more were presented before the Swedish Society of Dermatology by other authors (Azfelius, 1921). All of the patients had a previous tick bite, a fact that was held by Azfelius as evidence that a tick bite (*Ixodes ricinus*) was implicated in this entity. In 1930, another Swedish dermatologist, Sven Hellerström, from the Karolinska Institute, reported a patient who developed lymphocytic meningitis three months after an ECM, and suggested that both conditions were related (Hellerström, 1930). In the years to follow, other similar cases were discussed by Hellerström, who suggested a common aetiology for the skin lesion and for the accompanying meningitis. Other European authors reported additional patients with meningitis following eythema migrans (Dalsgaard-Nielsen, 1948; Gelbjerg-Hansen, 1945; Zwelleger, 1946). In 1936, Askani from the Dermatology Clinic at the University of Heidelberg reviewed the significance of ticks as vectors of human and animal pathogens and concluded that ECM was caused either by a toxin or by a living pathogen from the salivary glands of ticks (Askani, 1936). Together with Hellerström's findings, this became the preliminary evidence for the salivary route of transmission during tick feeding.

The cause of ECM remained elusive, and Hellerström mentioned a possible role for spirochaetes, at that time being investigated by Carl Lennhoff, also a member of the Dermatology Clinic at the Karolinska Institute. Lennhoff claimed to have developed a staining technique (based on the interaction of mercuric chloride with spirochaetes) that permitted him to identify

Table 1.1 European hallmarks of Lyme borreliosis

Year	Author, country of origin	Hallmark
1883	Alfred Buchwald, Germany	Description of acrodermatitis chronicum atrophicans (ACA)
1902	Herxheimer and Hartman	Further characterization of ACA
1910	Arvid Azfelius, Sweden	Description of the erythema migrans (EM)
1918	Lipschütz, Austria	Suggested the term erythema chronicum migrans
1922	Garin and Bujadoux, France	Description of meningopolyneuritis after tick bite and probable EM
1930	Hellerström, Sweden	Description of meningitis following EM. Suggested common aetiology for both conditions
1943	Bo Bäfverstedt, Sweden	Description of lymphadenosis benigna cutis
1944	Bannwarth, Germany	Further descriptions of meningopolyneuritis after EM
1948	Carl Lennhoff, Germany	Reported seeing spirochaetes in biopsy specimens (turned out to be artefacts)
1949	Hellerström, Sweden	Suggested that a spirochaete might be involved
1949	Thyresson, Sweden	Successful treatment of ACA with penicillin
1951	Hollström, Sweden	Successful treatment of EM with penicillin
1954	Gotz, Germany	Transmission of ACA among humans
1955	Binder *et al.*, Germany	Transmission of EM among humans
1974	Weber, Germany	Successful response of ECM to antibiotics
1985	Ackermann *et al*, Germany	Description of central nervous system involvement
1984–1985	Ackermann *et al.*, Asbrink *et al.*, Wislke *et al.*, Germany and Sweden	Cultivation of *B. burgdorferi* in skin and fluids

spirochaetal elements in many different skin lesions of unknown aetiology, including psoriasis, zoster, varicella, lymphadenosis benigna cutis, and erythema migrans, among others (Lennhoff, 1948). Although his staining method appeared suitable for demonstrating spirochaetes in culture, it was not suitable for identifying these microorganisms in smears or in tissue sections of ECM lesions. In fact, to Lennhoff's disappointment, the structures that he thought were spirochaetes turned out to be artefacts.

In 1949, Schaltenbrand (1949) from the Neurological Clinic at the University of Würzburg showed that the nervous system disorders associated with ECM were different from those of the central European tick-borne encephalitis. He speculated on the relationship between the anatomical location of the tick bite and the subsequent neurological manifestations and assumed that ECM meningitis was caused by an unidentified viral agent associated with *I. ricinus*. He also described prompt improvement after treatment with corticosteroids and tetracycline. In 1950, Hellerström at the Annual Meeting of the Southern Medical Association in Cincinnati, Ohio, presented a study 'ECM Afzelius with meningitis (Hellestrōm, 1950).' He reviewed several cases in which both erythema migrans and subsequent meningocerebrospinal symptoms occurred after tick bites. He also discussed the still unsolved cause of ECM and wondered whether the sheep tick, *I. ricinus,* could be a vector of spirochaetes. He pointed out that 'spirochaetes have been known to cause circular erythemas that subside in the centre and advance peripherally, and to affect the nervous system'. Hellerström injected tick salivary gland extracts into the skin of patients and healthy people and noted an intense reaction in the former, whereas only a weak response was obtained in the later. In 1951, a colleague of Hellerström, Hollström, reported on the successful use of penicillin in treating ECM (Hollström, 1951). Despite the accumulating evidence, and perhaps because spirochaetes had never been found in association with ticks of the genus *Ixodes,* the search for spirochaetes in ticks was not pursued and the cause of ECM remained obscure. In 1955, Binder and coworkers (1955) succeeded in transmitting ECM from human to human by transplanting pieces of skin taken from the peripheral zone of a lesion of a patient with ECM. In each of the three volunteers, typical lesions developed within 1 to 3 weeks, and a subpassage was equally successful. In each case, treatment with penicillin caused the lesions to disappear within a few days. These experiments clearly suggested that the cause of ECM was an infectious agent susceptible to penicillin.

A rickettsial cause of ECM was suggested in 1962 by Degos *et al.* in France (Degos *et al.*, 1962). Based on results obtained by Giroud's slide agglutination test, they reported six of seven patients with significant titres against the epidemic typhus agent, *Rickettsia prowazekii,* the murine typhus rickettsia, *R. mooseri,* and the boutonneuse fever agent, *R. conorii.* Although these results could not be confirmed, the notion that ECM is caused by a rickettsia was kept alive by the electron microscopic demonstration of rickettsia-like microorganisms in macrophages of two ECM patients (Sandbank and Feuerman, 1979).

In 1974, Klaus Weber, from Munich, successfully treated a patient with ECM and meningitis with antibiotics (penicillin and chloramphenicol) and concluded that this nervous disorder was a bacterial illness (Weber, 1974). He ruled out spirochaetes as the causative agent because 'these microorganisms are transmitted by argasid and not ixodid ticks.' He also referred to the negative results of S. Hard who attempted to infect larval, nymphal, and adult *I. ricinus* by allowing them to feed on the erythematous ring of an ECM lesion (Hard, 1966). Subsequent dark field examination of tick tissues showed no spirochaetes, and transmission experiments with nymphal ticks that had fed on the lesion as larvae proved negative. He pursued further the rickettsial hypothesis and examined the sera of 13 ECM patients for the detection of antibodies to 14 different rickettsial antigens (Sandbank and Feuerman, 1979; Weber, 1981). All tests were negative.

1.4.2 Garin-Bujadoux, Bannwarth and the neurological spectrum

In 1922, two French neurologists, Charles Garin and Antoine Bujadoux, from Lyon, gave the first description of neurological complications after a tick bite and of a skin lesion that, although not recognized as such, was compatible with erythema migrans (Garin and Bujadoux, 1922). They reported on a 58-year-old man who developed a reddish, painful and expanding skin lesion, accompanied by inguinal lymphadenopathy, 3 weeks after having been bitten by a tick on his left buttock. This was followed by excruciating bilateral sciatic pain, intercostal pain, and finally a painful unilateral right brachial plexopathy with amyotrophy of the deltoid muscle. He required morphine, obtaining only partial relief. Four months after disease onset the pain had decreased significantly, and he was left with deltoid amyotrophy and weakness. Serological test for syphilis (Wassermann test) was weakly positive, but the authors were aware that 'in certain tick-transmitted diseases such as relapsing fever and Rocky Mountain Spotted Fever these test can be positive'[sic.]. They ruled out syphilis as the cause, and presented the case as an example of 'paralysie par les tiques' (tick palsy), although they also argued that this disorder could be caused by a tick-borne virus. However, in contrast to their patient, they noted that those cases of tick palsy in the literature were not accompanied by a previous skin lesion. Furthermore, tick palsy is usually caused by *Dermacentor* tick bites, while their patient had been bitten by *Ixodes hexagonus,* the hedgehog tick.

Between 1941 and 1944, Bannwarth in Germany reported several patients who suffered meningitis with evidence of cranial or peripheral neuropathy (Bannwarth, 1941, 1944). In his first report, he described 14 patients with lymphocytic meningitis and severe radicular pain in the trunk and extremities. Pain was migrating, neuralgic and more intense at night. Facial palsy was present in six patients, unilateral in two of them, and bilateral in four. Bannwarth, however, did not make the connection between prior tick bites or ECM, which one of the patient seemed to have had, and the neurological disturbances, and favoured, instead, an allergic or rheumatic aetiology.

In 1985 Ackermann *et al.* reported on eight patients with chronic progressive encephalomyelitis and intrathecal synthesis of specific antibodies against *B. burgdorferi* who improved following parenteral antibiotic therapy (Ackermann *et al.*, 1985). This is the first report of central nervous system involvement in Lyme borreliosis, and was followed by larger series (Ackermann *et al.*, 1987). The protean manifestations of the neurological disorder created some confusion with other neurological diseases of uncertain origin, including multiple sclerosis or Guillain–Barré syndrome, among others (Kristoferitsch, 1991; Schmutzhard *et al.*, 1989). In the 1970s and 1980s, Horstrup and Ackerman reviewed the manifestations of this entity in Europe. They (Hörstrup and Ackermann, 1973), and others (Weber, 1974) insisted on an infectious aetiology even before the spirochaetal origin of Lyme disease was discovered.

1.4.3 Acrodermatitis chronicum atrophicans (ACA) and lymphadenosis benigna cutis

ACA was first described in the late nineteenth century by Buchwald (Buchwald, 1883) and then clearly characterized by Herxheimer and Hartmann in 1902 (Herxheimer and Hartman, 1902). Many patients with this intriguing disorder remembered having been bitten by the sheep tick, *I. ricinus*, and some reported subsequent development of ECM-like skin lesions. In 1942, R. H. Kahie, a German dermatologist, noted that six of seven patients with ACA had a positive serological response against *Treponema pallidum*. Ten years later, his mentor, Dr Grüneberg, confirmed these results and speculated that the reactions must have been caused by a group-specific spirochaete as the causative agent of ACA (Grüneberg, 1954). In the 1940s, the beneficial effect of penicillin on ACA was demonstrated (Thyresson, 1949), and Gotz showed in the 1950s that it was a transmissible condition (Gotz, 1954, 1955). He successfully transplanted affected skin of a patient into four volunteers, including himself, and recorded a clinical picture similar to ACA (i.e. inflammation that lasted for as long as 312 days and promptly disappeared after treatment with penicillin). Gotz speculated the aetiological agent to be a large virus, but, lacking evidence, he concluded that the cause of ACA would continue to be an open question. In 1955, Lohel showed antibodies against spirochaetes in mice inoculated with blood from ACA patients (Lohel, 1955). His claim of ACA being a spirochaetosis, however, was rejected. The geographical distribution in Europe and the observation that most patients had been from rural areas further strengthened the suspicion that the tick *Ixodes ricinus* was the vector.

Lymphadenosis benigna cutis, or borrelial lymphocytoma, was first described by Bäfverstedt, another Swedish dermatologist, in 1943 as a benign skin condition characterized by the appearance of lymphoreticular tissue in the epidermis and subcutaneous layers (Bäfverstedt, 1943). He also mentioned a possible association with cases of ACA. In the ensuing years, it was successfully transmitted and treated with penicillin, again reinforcing the idea of an infectious aetiology.

1.4.4 A missing articular manifestation in Europe?

The rheumatological aspects of Lyme disease became evident after the initial epidemic of arthritis in Lyme, Connecticut was first reported (see above). On the other hand, the European descriptions of this disease consisted primarily of a cutaneous and neurological disorder. This notion, that American Lyme disease primarily affects the joints while European Lyme disease is primarily neurological, may represent an acquisition bias. That dermatologists and neurologists in Europe were well aware of this disease, while in the USA rheumatologists led by Stephen Malawista became involved in the arthritis epidemic may at least partially account for this notion. However, a German patient (Herzer *et al.* in 1976) developed an expanding skin lesion, neurological abnormalities and arthritis following a tick bite (Herzer *et al.*, 1983). This patient insisted on her disease being related to the prior tick bite, but was not diagnosed until she read an article on the lay press about the presentation of Lyme disease in the

USA. She was then told that Lyme arthritis did not exist in Europe.

Although joint manifestations had been observed in European patients with erythema migrans, ACA, and lymphocytic meningitis (Bannwarth, 1941; Benjamowistch and Maschkilleisson, 1933; Erbslöh and Kohlmeyer, 1968; Gans, 1933, 1952; Hoövelborn, 1931; Hopf, 1966; Jessner, 1922; Jessner and Loewenstamm, 1924; Schaltenbrand, 1949), the initial reports on European Lyme disease after the discovery of its aetiology basically neglected the rheumatological aspects to the point that they were regarded as curiosities (Ackermann *et al.*, 1980; Hewitt *et al.*, 1980). In fact, the Second International Symposium on Lyme Disease and Related Disorders, held in Vienna in 1985, did not include any studies by European rheumatologists. In the meantime, an increasing number of reports of Lyme arthritis from several European countries have accrued, suggesting that this joint manifestations may have been overlooked or at least underestimated (Herzer, 1983, 1991; Herzer *et al.*, 1983).

1.4.5 Linking the aetiology of the European syndromes to *Borrelia burgdorferi*

In the early 1980s, the clinical spectrum of a disorder of still unknown aetiology had been clearly delineated, particularly in its cutaneous and neurological manifestations, and investigators had concluded that it was caused by an infectious agent, probably a spirochaete, and that it was transmitted by Ixodid ticks. Soon after the discovery of the Lyme agent in the USA, *B. burgdorferi* was finally isolated from European patients with EM by Swedish and German investigators (Ackermann *et al.*, 1984; Asbrink *et al.*, 1984; Wilske *et al.*, 1985). In 1983, Asbrink *et al.* from Sweden, succeeded for the first time in isolating *B. burgdorferi* from the skin of a patient with ACA (Asbrink *et al.*, 1984), thus demonstrating that this cutaneous disease was also part of the spectrum of Lyme disease.

The availability of serological testing following the discovery of *B. burgdorferi* enabled investigators to obtain evidence for the spirochaetal aetiology of the meningopolyneuritis of Garin-Bujadoux and Bannwarth and other less frequent neurological manifestations. Cultivation of the spirochaete from the cerebrospinal fluid of some patients confirmed the conclusions drawn from serological studies (Allal *et al.*, 1986; Pfister *et al.*, 1989; Preac Mursic *et al.*, 1984).

1.5 The rocky road of Lyme disease

1.5.1 The vaccine

The scientific body of work that led to the development of a commercial vaccine for Lyme disease can be considered a major outcome of the interdisciplinary approach that uncovered the major biological features of *B. burgdorferi* and its natural history in both the vector and the vertebrate host (see Chapter 17). The cationic lipoproteins, OspA and OspB (Benach *et al.*, 1988; Bergström *et al.*, 1989; Brandt *et al.*, 1990), which are prominent in the electrophoretic profile of cultured *B. burgdorferi*, were the first to be studied for their role in eliciting an immune response (Barbour *et al.*, 1983; Johnson *et al.*, 1986). The protective role of antibodies to OspA was revealed by passive immunizations of previously infected *scid* mice with anti-OspA monoclonal antibodies (Barthold and Bockenstedt, 1993; Fikrig *et al.*, 1992b, 1994). Using sera from patients with Lyme disease, passive protection of mice was accomplished but only if the sera contained antibodies to OspA (Zhong *et al.*, 1997). Moreover, two additional studies provided evidence for a role for passive immunization and in the attenuation and resolution of chronic disease (Fikrig *et al.*, 1990; Schaible *et al.*, 1990). The lipid moiety turned out to be critical for the immunogenicity of OspA, since only the full-length lipidated molecule protected mice from challenge (Bockenstedt *et al.*, 1993; Erdile *et al.*, 1993; Weis *et al.*, 1994). Clearly, the lipidation is important biologically but also for conferring immunogenicity to the molecule. In addition, antigens expressed by *B. burgdorferi* in culture could be different from those present in ticks or the mammalian host (Gern *et al.*, 1993; Schaible *et al.*, 1993a). This observation led to studies on the extent and nature of the protection that could be elicited following challenge with antigenically diverse strains of this organism. Furthermore, determining if vaccine preparations could protect from challenge with spirochaetes transmitted by ticks became critical. The antibody response to tick or needle inoculated cultured spirochaetes is dependent on the amount of antigen given. Inocula of 1000 cultured spirochaetes or fewer, or tick exposure, generated antibodies to a diverse group of antigens but not to OspA/B. In contrast, higher doses of cultured organisms developed antibodies to these two antigens (Gern *et al.*, 1997; Schaible *et al.*, 1993b).

There were also geographic differences in some areas of the American continent; protection could be achieved by OspA vaccination followed by challenge with infected ticks (Telford *et al.*, 1993). Although OspA can induce a robust antibody response when administered as a purified protein, Lyme disease patients had long been known not to develop antibodies to OspA in the early stages of the disease. This ostensible paradox was subsequently explained on the basis that expression of OspA is downregulated as the tick begins to feed on a host (Schwan *et al.*, 1995). In this manner, the OspA vaccine would induce an antibody to an antigen that is not present in the host as it is lost during transmission from the tick to the host. If this was the case, how can the OspA subunit vaccine be protective? Attempts to answer this question showed that the vaccine was protective by eliminating the organisms that expressed OspA in the feeding tick (de Silva *et al.*, 1996; Fikrig *et al.*, 1992a). Thus, the mode of action of the OspA vaccine is to block transmission by inducing the formation of antibodies that eliminate spirochaetes in the midgut of the feeding tick (de Silva *et al.*, 1996). That antibodies had an effect in the vector underscores the uniqueness of this vaccine paradigm. Antibodies to OspA have a remote effect on *B. burgdorferi*, affecting the organisms outside of the host where the antibodies were developed. This was a novel approach to a vaccine (Connolly and Benach, 2005).

Based on the effectiveness of the antibody response to OspA, two commercial recombinant lipidated, adjuvant-containing, OspA vaccines were formulated, tested and marketed in the USA. The US Food and Drug Administration (FDA) licensed a vaccine, LYMErix, on 21 December 1998, for persons 15–70 years of age. This licensed vaccine was a subunit formulation containing recombinant OspA (rOspA). Two vaccine studies demonstrated the effectiveness of a triple-dose vaccine regimen that protected 100% of the volunteers after the second year in one trial (Steere *et al.*, 1998) and 92% in another (Sigal *et al.*, 1998). Three doses of 0.5 ml (30 μg) of rOspA vaccine administered by intramuscular injection at 0, 1 and 12 months were required for optimal protection in subjects who were between 15 and 70 years of age. The duration of protective immunity for longer than 1 year after the third dose of immunization was not determined at that time. Unfortunately, safety and efficacy of the rOspA vaccine were not established for persons older than 70 years or younger than 15 years of age, two age groups that have been traditionally overrepresented in this disease (Steigbigel and Benach, 1998). Volunteers and patients that received this vaccine developed a positive serological pattern that distinguished their response from that of patients.

As of February 2002, the LYMErix vaccine was withdrawn from the commercial market in the USA. The reasons for its removal from the market were poor market penetration, and not issues related to the vaccine's safety, immunogenicity, or efficacy. Cross-reactivity of OspA with self-antigens may have caused some doubts about the possibility of an autoimmune response (Gross *et al.*, 1998). IgG antibodies to OspA had been shown to correlate with severe and prolonged arthritis, furthering additional concerns about autoreactivity (Akin *et al.*, 1999; Krause *et al.*, 1992a).

On 22 January 2002, officers from the Lyme Disease Association (LDA) and International Lyme and Associated Disease Society (ILADS) met with both scientists and administrators from the Food and Drug Administration (FDA) to express concerns about the vaccine's track record on safety and effectiveness. This meeting was made possible at a request of a Congressman from New Jersey. The evidence presented included the occurrence of joint and muscle pain after receiving the vaccine. On 25 February, a month or so after meeting, the LDA received written answers to their questions from FDA officials. Coincidentally, on or about the same time, Glaxo SmithKline had withdrawn LYMErix from the market (http://www.lymediseaseassociation.org/Vaccine_LYMERIXMeeting.html).

1.5.2 The Lyme disease treatment controversy

Although the role of the advocacy groups in the withdrawal of the vaccine may be difficult to assess, it underscores the controversy involving public advocate groups and members of the academic medicine establishment as growing bitter and increasingly litigious. The internet is brimming with information that supports the existence of a sharp division of opinion regarding the nature and management of Lyme disease that began almost 20 years ago (Barinaga, 1992).

The Lyme Disease Foundation (LDF) was founded in 1988 by Karen Vanderhoof-Forschner and her husband Thomas E. Forschner as a non-profit corporation to foster education and research on this disease. The LDF was remarkably successful in meeting its goals, and over the years, it was instrumental in educating the public about Lyme disease through its outreach programs, videos, and conferences held in various parts of the country. Researchers in this field were given an opportunity to present their work at their venues, and many did. On the other hand, the LDF also sponsored the alternative views to treatment that ultimately fostered polarization between the research establishment and some of the patients. Other advocacy organizations followed the lead of the LDF. Notably, the Lyme Disease Association (LDA) and International Lyme and Associated Disease Society (ILADS) have become quite active in patient advocacy and education. At the very core of the controversy is the belief by some patient advocates and some of their doctors that Lyme disease is a chronic infection that can only be treated with unusually long courses of antibiotic therapy. From this core, other contentious issues follow such as, for example, the reluctance of insurance companies to pay for long-term antibiotic treatment. The insurance companies base their refusal to cover these lengthy treatments on the lack of scientific evidence showing that long-term antibiotic therapy is beneficial (see Chapter 18).

How did we get to this state of affairs? In approaching this topic, we could have delved deeply into the internet sites that reflect the views of advocacy groups. We chose not to as there is sufficient material in the internet to fill this volume, and not all material is useful to understand the genesis of this social phenomenon. There are obviously some legitimate reasons that may have assisted in the development of the controversy, and these relate to the natural history of Lyme disease and the idiosyncrasies of its agent *B. burgdorferi* (Barbour and Fish, 1993).

- Lyme disease produces non-specific symptoms, often separated in time. Its most compelling sign, the erythema migrans, has a variable morphology, and can be absent or not noticed. The neurological manifestations are protean, and often subtle and non-specific and this results in some confusion for the patients (García-Moncó and Benach, 1995).
- Many patients deny a history of the tick bite, mostly because they never saw or recognized the very small nymph.
- *Borrelia burgdorferi* is fastidious. Even laboratory-adapted strains grow slowly, so that cultures of the spirochaetes are not useful for iron-clad diagnoses that are the norm for other bacterial diseases. Hence, serology has become the only means of laboratory diagnosis of Lyme disease. This, too, presents some confounding problems. The antibodies are long lived and do not provide a means to determine a cure. Although clearly not the intended result, a study showing that there is seronegative Lyme disease alerted patients to the notion that they could also have this disease without serological evidence (Dattwyler *et al.*, 1988b).
- Treatment-refractory Lyme disease has been reported in the academic medicine literature (Drouin *et al.*, 2008; Kannian *et al.*, 2007; Steere *et al.*, 2006). In the mind of some patients, a bona fide treatment failure is the main indication for long-term antibiotic therapy.

Despite some of the problems presented by the diagnostic and clinical features of Lyme disease, many advances have been made to solve the problems outright or mitigate them so that this disease can be diagnosed and treated with a degree of certainty that is equal to that of other infections. The scientific community, under the auspices of the Centers for Disease Control, assembled a panel of experts to standardize the serology of Lyme disease. The resulting guidelines have taken the uncertainty of serological validation out of the laboratory diagnosis of Lyme disease [Proceedings of the Second National Conference on Serologic Diagnosis of Lyme Disease (Dearborn, Michigan, 27–29 October 1994). Washington, DC: Association of State and Territorial Public Health Laboratory Directors 1995. Centers for Disease Control and Prevention. Recommendations for test performance and interpretation from the Second National Conference on Serologic Diagnosis of Lyme Disease. MMWR 1995; 44, 590–591]. Moreover, new, more sensitive and specific tests have been developed that will further solidify the laboratory diagnosis of Lyme disease (Philipp *et al.*, 2001, 2005) (see Chapter 16).

The same is true for treatment. For a number of years, chronic Lyme disease has been studied in patient populations using different clinical research approaches (Marques, 2008). In short, despite the diversity of the studies, their results are markedly similar. Lyme disease is generally treatable with the recommended therapies of antibiotics, making prolonged treatment unnecessary. The efficacy of treatment of chronic Lyme disease was evaluated in a randomized, placebo-controlled clinical trial for both seropositive and seronegative patients with persistent symptoms after having received initial antibiotic therapy for the illness. The study showed that the quality of life did not improve after long-term antibiotic therapy. The implicit recommendation was that long-term antibiotic therapy is not warranted. In fact, the Data Safety Monitoring Board of this project recommended that the study be discontinued because there was no difference in treatment efficacy between groups (Klempner *et al.*, 2001). Post treatment chronic Lyme disease patients who have symptoms show no evidence of objective cognitive impairment. Continuing antibiotics was not more effective than placebo (Kaplan *et al.*, 2003). Halperin *et al* provided evidence-based recommendations on the treatment of neuroborreliosis and post-Lyme syndrome. They found that adults and children respond well to recommended antibiotic therapy. In contrast, there is no compelling evidence that prolonged treatment with antibiotics has any beneficial effect in post-Lyme syndrome (Halperin *et al.*, 2007). A multi-author review tackled the chronic Lyme disease issue and reached the predictable and forceful conclusion that there is no evidence in support of long-term antibiotic therapy (Feder *et al.*, 2007).

An expert panel of the Infectious Diseases Society of America (IDSA) provided evidence-based guidelines for the management of patients with Lyme disease. These guidelines list the doses and durations of recommended antimicrobial therapy for both treatment and prevention of Lyme disease. The guidelines also provide a list of therapies to be avoided and proposed a definition of the post-Lyme disease syndrome (Wormser *et al.*, 2006). As a result of these guidelines, an antitrust investigation against the IDSA was started by the Attorney General of the State of Connecticut alleging that the restrictive recommendations prevent patients from receiving care that is not included in the guidelines (www.lymediseaseassociation.org). The IDSA provided a rebuttal standing by the guidelines (www.idsociety.org). In May 2008, the IDSA reached an agreement with the Connecticut Attorney General that ends the investigation. The agreement allows for the guidelines to remain. However, in this agreement with the Attorney General, the IDSA will call for another special review of the guidelines. To this end, the IDSA will convene a review panel to conduct a comprehensive and up-to-date evaluation of the scientific literature, in order to determine whether the 2006 guidelines need to be changed. At the time of this writing, the IDSA issued a call for applications for scientists to participate in this review panel.

While The Attorney General's investigation of the 2006 IDSA guidelines never questioned the medical evidence, there were concerns with the process leading to the recommendations, and the agreement to revisit the guidelines reflects these concerns. As part of the review process, scientists have been invited to submit information about themselves to meet selection criteria for the review panel. The review panel will consider all new and past evidence and determine whether, and how, the guidelines should be revised. If the panel recommends revisions, they will be carried out as per the IDSA Standards and Practice Guidelines Committee.

The views held by the advocacy groups on treatment and even on the very definition of what constitutes Lyme disease are at odds with the views of the scientific community. The potential for harm resulting from long-term antibiotic treatment, not to mention the harm to the patient caused by treating a disease that they may not have, are the main reasons to double the efforts by the scientific community to have its voice heard by the patients. This is not an easy task as advocacy groups demand complete responsibility for the health care that they want to receive and question the very authority of science (Aronowitz, 1991).

References

Ackermann, R., Gollmer, E., and Rehse-Kupper, B. (1985). Progressive *Borrelia* encephalomyelitis: chronic manifestation of erythema chronicum migrans disease of the nervous system. Deutsche medizinische Wochenschrift (1946) *110*, 1039–1042.

Ackermann, R., Horstrup, P., and Schmidt, R. (1984). Tick-borne meningopolyneuritis (Garin-Bujadoux, Bannwarth). Yale J. Biol. Med. *57*, 485–490.

Ackermann, R., Rehse-Küpper, B., Gollmer, E., and Schmidt, R. (1987). Chronic neurologic manifestations of erythema migrans borreliosis. Ann. N.Y. Acad. Sci. *539*, 16–23.

Ackermann, R., Runne, U., Klenk, W., and Dienst, C. (1980). Erythema chronicum migrans with arthritis. Deutsche medizinische Wochenschrift (1946) *105*, 1779–1781.

Akin, E., McHugh, G.L., Flavell, R.A., Fikrig, E., and Steere, A.C. (1999). The immunoglobulin (IgG) antibody response to OspA and OspB correlates with severe and prolonged Lyme arthritis and the IgG response to P35 correlates with mild and brief arthritis. Infect. Immun. *67*, 173–181.

Allal, J., Thomas, P., and Mazzonelli, J. (1986). *Borrelia* isolated from cerebrospinal fluid in a French case of Lyme disease. Ann. Rheum. Dis. *45*, 789–790.

Anderson, J.F., Johnson, R.C., Magnarelli, L.A., Hyde, F.W., and Myers, J.E. (1986). *Peromyscus leucopus* and *Microtus pennsylvanicus* simultaneously infected with *Borrelia burgdorferi* and *Babesia microti*. J. Clin. Microbiol. *23*, 135–137.

Aronowitz, R.A. (1991). Lyme disease: the social construction of a new disease and its social consequences. Milbank Q. *69*, 79–112.

Asbrink, E., Hederstedt, B., and Hovmark, A. (1984). The spirochetal etiology of erythema chronicum migrans Afzelius. Acta Derm. Venereol. *64*, 291–295.

Askani, H. (1936). Zur Átiologie des Erythema chronicum migrans. Dermatol. Wochenschr. *102*, 125.

Azfelius, A. (1910). Verhandlungen der dermatologischen gesellschaft zu Stockholm, 28 October 1909. Arch. Derm. Syph. *101*, 404.

Azfelius, A. (1921). Erythema chronicum migrans. Acta Derm. Venereol. *2*, 120–125.

Bäfverstedt, B. (1943). Über Lymphadenosis benigna cutis. Eine klinische und patologisch-anatomische Studie (Stockholm).

Bannwarth, A. (1941). Chronische lymphocytare meningitis, entzundliche polyneuritis und 'rheumatismus' eing beitrag zumn problem 'allergic und nervensystem'. Arch. Psychiatr. Nervenkr. *113*, 284–376.

Bannwarth, A. (1944). Zur klinik und patohogenese der 'chronischen lymphocytare meningitis'. Arch. Psychiatr. Nervenkr. *117*, 161–185.

Baranton, G., Postic, D., Saint Girons, I., Boerlin, P., Piffaretti, J.C., Assous, M., and Grimont, P.A. (1992). Delineation of *Borrelia burgdorferi* sensu stricto, *Borrelia garinii* sp. nov., and group VS461 associated with Lyme borreliosis. Int. J. Syst. Bacteriol. *42*, 378–383.

Barbour, A.G., and Fish, D. (1993). The biological and social phenomenon of Lyme disease. Science *260*, 1610–1616.

Barbour, A.G., Tessier, S.L., and Stoenner, H.G. (1982). Variable major proteins of *Borrellia hermsii*. J. Exp. Med. *156*, 1312–1324.

Barbour, A.G., Tessier, S.L., and Todd, W.J. (1983). Lyme disease spirochetes and ixodid tick spirochetes share a common surface antigenic determinant defined by a monoclonal antibody. Infect. Immun. *41*, 795–804.

Barinaga, M. (1992). Furor at Lyme disease conference. Science *256*, 1384–1385.

Barthold, S.W., and Bockenstedt, L.K. (1993). Passive immunizing activity of sera from mice infected with *Borrelia burgdorferi*. Infect. Immun. *61*, 4696–4702.

Barthold, S.W., Moody, K.D., Terwilliger, G.A., Jacoby, R.O., and Steere, A.C. (1988). An animal model for Lyme arthritis. Ann. N.Y. Acad. Sci. *539*, 264–273.

Belfaiza, J., Postic, D., Bellenger, E., Baranton, G., and Girons, I.S. (1993). Genomic fingerprinting of *Borrelia burgdorferi* sensu lato by pulsed-field gel electrophoresis. J. Clin. Microbiol. *31*, 2873–2877.

Benach, J.L., Bosler, E.M., Hanrahan, J.P., Coleman, J.L., Habicht, G.S., Bast, T.F., Cameron, D.J., Ziegler, J.L., Barbour, A.G., Burgdorfer, W., *et al.* (1983). Spirochetes isolated from the blood of two patients with Lyme disease. N. Engl. J. Med. *308*, 740–742.

Benach, J.L., Coleman, J.L., and Golightly, M.G. (1988). A murine IgM monoclonal antibody binds an antigenic determinant in outer surface protein A, an immunodominant basic protein of the Lyme disease spirochete. J. Immunol. *140*, 265–272.

Benach, J.L., Coleman, J.L., Habicht, G.S., MacDonald, A., Grunwaldt, E., and Giron, J.A. (1985). Serological evidence for simultaneous occurrences of Lyme disease and babesiosis. J. Infect. Dis. *152*, 473–477.

Benach, J.L., Coleman, J.L., Skinner, R.A., and Bosler, E.M. (1987). Adult *Ixodes dammini* on rabbits: a hypothesis for the development and transmission of *Borrelia burgdorferi*. J. Infect. Dis. *155*, 1300–1306.

Benach, J.L., and Habicht, G.S. (1981). Clinical characteristics of human babesiosis. J. Infect. Dis. *144*, 481.

Benach, J.L., Habicht, G.S., and Hamburger, M.I. (1982). Immunoresponsiveness in acute babesiosis in humans. J. Infect. Dis. *146*, 369–380.

Benach, J.L., White, D.J., Burgdorfer, W., Keelan, T., Guirgis, S., and Altieri, R.H. (1977). Changing patterns in the incidence of Rocky Mountain Spotted Fever on Long Island (1971–1976). Am. J. Epidemiol. *106*, 380–387.

Benach, J.L., White, D.J., and McGovern, J.P. (1978). Babesiosis in Long Island. Host–parasite relationships of rodent- and human-derived *Babesia microti* isolates in hamsters. Am. J. Trop. Med. Hyg. *27*, 1073–1078.

Benach, J.L., White, D.J., McGovern, J.P., and Jacovina, M.M. (1979). Immunological relationships of Long Island isolates of *Babesia microti*. Am. J. Trop. Med. Hyg. *28*, 643–648.

Benjamowistch, E., and Maschkilleisson, L.N. (1933). Witere Beiträge zur Fräge über die Atrophie der Haut. III. Acrodermatitis cronica atrophicans und ihre Beziehungen zur Sklerodermia. Acta Derm. Venereol. (Stockh.) *14*, 313–341.

Berger, B.W., Kaplan, M.H., Rothenberg, I.R., and Barbour, A.G. (1985). Isolation and characterization of the Lyme disease spirochete from the skin of patients with erythema chronicum migrans. J. Am. Acad. Dermatol. *13*, 444–449.

Bergström, S., Bundoc, V.G., and Barbour, A.G. (1989). Molecular analysis of linear plasmid-encoded major surface proteins, OspA and OspB, of the Lyme disease spirochaete *Borrelia burgdorferi*. Mol. Microbiol. *3*, 479–486.

Binder, E., Doepfmer, R., and Homstein, O. (1955). Experimentelle Über-tragung des Erythema chronicum migrans von Mensch zu Mensch. Hautarzi *6*, 494.

Bockenstedt, L.K., Fikrig, E., Barthold, S.W., Kantor, F.S., and Flavell, R.A. (1993). Inability of truncated recombinant Osp A proteins to elicit protective immunity to *Borrelia burgdorferi* in mice. J. Immunol. *151*, 900–906.

Bosler, E.M., Coleman, J.L., Benach, J.L., Massey, D.A., Hanrahan, J.P., Burgdorfer, W., and Barbour, A.G. (1983). Natural distribution of the *Ixodes dammini* spirochete. Science *220*, 321–322.

Brandt, M.E., Riley, B.S., Radolf, J.D., and Norgard, M.V. (1990). Immunogenic integral membrane proteins of *Borrelia burgdorferi* are lipoproteins. Infect. Immun. *58*, 983–991.

Buchwald, A. (1883). Ein fall von diffuser idiopathischcr Haunt-Atrofie. Arch. Dermatol. Syph. *10*, 553–556.

Burgdorfer, W. (1954). On the occult infection in relapsing fevers. Bull. Soc Pathol. Exot. Filiales *47*, 664–667.

Burgdorfer, W. (1970). Hemolymph test. A technique for detection of rickettsiae in ticks. Am. J. Trop. Med. Hyg. *19*, 1010–1014.

Burgdorfer, W. (1975). A review of Rocky Mountain spotted fever (tick-borne typhus), its agent, and its tick vectors in the United States. J. Med. Entomol. *12*, 269–278.

Burgdorfer, W., Barbour, A.G., Hayes, S.F., Benach, J.L., Grunwaldt, E., and Davis, J.P. (1982). Lyme disease – a tick-borne spirochetosis? Science *216*, 1317–1319.

Burgdorfer, W., Hayes, S.F., and Benach, J.L. (1988). Development of *Borrelia burgdorferi* in ixodid tick vectors. Ann. N.Y. Acad. Sci. *539*, 172–179.

Burgdorfer, W., Lane, R.S., Barbour, A.G., Gresbrink, R.A., and Anderson, J.R. (1985). The western black-legged tick, *Ixodes pacificus*: a vector of *Borrelia burgdorferi*. Am. J. Trop. Med. Hyg. *34*, 925–930.

Cahill, K.M., Benach, J.L., Reich, L.M., Bilmes, E., Zins, J.H., Siegel, F.P., and Hochweis, S. (1981). Red cell exchange: treatment of babesiosis in a splenectomized patient. Transfusion *21*, 193–198.

Canica, M.M., Nato, F., du Merle, L., Mazie, J.C., Baranton, G., and Postic, D. (1993). Monoclonal antibodies for identification of *Borrelia afzelii* sp. nov. associated with late cutaneous manifestations of Lyme borreliosis. Scand J. Infect. Dis. *25*, 441–448.

Chisholm, E.S., Ruebush, T.K., 2nd, Sulzer, A.J., and Healy, G.R. (1978). *Babesia microti* infection in man: evaluation of an indirect immunofluorescent antibody test. Am. J. Trop. Med. Hyg. *27*, 14–19.

Coleman, J.L., LeVine, D., Thill, C., Kuhlow, C., and Benach, J.L. (2005). *Babesia microti* and *Borrelia burgdorferi* follow independent courses of infection in mice. J. Infect. Dis. *192*, 1634–1641.

Connolly, S.E., and Benach, J.L. (2005). The versatile roles of antibodies in *Borrelia* infections. Nat. Rev. Microbiol. *3*, 411–420.

Dalsgaard-Nielsen, T. (1948). Allergic meningitis and chronic erythema migrans afzelii after bite by *Ixodes reduvius*. Acta Allerg. (Kbh.) *1*, 388–393

Dammin, G.J., Spielman, A., Benach, J.L., and Piesman, J. (1981). The rising incidence of clinical *Babesia microti* infection. Hum. Pathol. *12*, 398–400.

Dattwyler, R.J., Halperin, J.J., Pass, H., and Luft, B.J. (1987). Ceftriaxone as effective therapy in refractory Lyme disease. J. Infect. Dis. *155*, 1322–1325.

Dattwyler, R.J., Halperin, J.J., Volkman, D.J., and Luft, B.J. (1988a). Treatment of late Lyme borreliosis – randomised comparison of ceftriaxone and penicillin. Lancet *1*, 1191–1194.

Dattwyler, R.J., Volkman, D.J., Luft, B.J., Halperin, J.J., Thomas, J., and Golightly, M.G. (1988b). Seronegative Lyme disease. Dissociation of specific T- and B-lymphocyte responses to *Borrelia burgdorferi*. N. Engl. J. Med. *319*, 1441–1446.

Davis, G.E., and Burgdorfer, W. (1954). On the susceptibility of the guinea pig to the relapsing fever spirochete *Borrelia duttonii*. Bull. Soc Pathol. Exot. Filiales *47*, 498–501.

de Silva, A.M., Telford, S.R., 3rd, Brunet, L.R., Barthold, S.W., and Fikrig, E. (1996). *Borrelia burgdorferi* OspA is an arthropod-specific transmission-blocking Lyme disease vaccine. J. Exp. Med. *183*, 271–275.

Degos, R., Touraine, R., and Arouete, J. (1962). Chronic erythema migrans. Ann. Dermatol. Syphiligr. (Paris) *89*, 247–260.

Drouin, E.E., Glickstein, L., Kwok, W.W., Nepom, G.T., and Steere, A.C. (2008). Human homologues of a *Borrelia* T-cell epitope associated with antibiotic-refractory Lyme arthritis. Mol. Immunol. *45*, 180–189.

Erbslöh, F., and Kohlmeyer, K. (1968). Über polytope Erkrankungen des peripheren Nervensystems bei lymphozytärer Meningitis. Fortschr. Neurol. Psychiat. *35*, 321–342.

Erdile, L.F., Brandt, M.A., Warakomski, D.J., Westrack, G.J., Sadziene, A., Barbour, A.G., and Mays, J.P.

(1993). Role of attached lipid in immunogenicity of *Borrelia burgdorferi* OspA. Infect. Immun. *61*, 81–90.

Esernio-Jenssen, D., Scimeca, P.G., Benach, J.L., and Tenenbaum, M.J. (1987). Transplacental/perinatal babesiosis. J. Pediatr. *110*, 570–572.

Feder, H.M., Jr., Johnson, B.J., O'Connell, S., Shapiro, E.D., Steere, A.C., and Wormser, G.P. (2007). A critical appraisal of 'chronic Lyme disease'. N. Engl. J. Med. *357*, 1422–1430.

Fikrig, E., Barthold, S.W., Kantor, F.S., and Flavell, R.A. (1990). Protection of mice against the Lyme disease agent by immunizing with recombinant OspA. Science *250*, 553–556.

Fikrig, E., Barthold, S.W., Kantor, F.S., and Flavell, R.A. (1992a). Long-term protection of mice from Lyme disease by vaccination with OspA. Infect. Immun. *60*, 773–777.

Fikrig, E., Barthold, S.W., Marcantonio, N., Deponte, K., Kantor, F.S., and Flavell, R.A. (1992b). Roles of OspA, OspB, and flagellin in protective immunity to Lyme borreliosis in laboratory mice. Infect. Immun. *60*, 657–661.

Fikrig, E., Bockenstedt, L.K., Barthold, S.W., Chen, M., Tao, H., Ali-Salaam, P., Telford, S.R., and Flavell, R.A. (1994). Sera from patients with chronic Lyme disease protect mice from Lyme borreliosis. J. Infect. Dis. *169*, 568–574.

Filstein, M.R., Benach, J.L., White, D.J., Brody, B.A., Goldman, W.D., Bakal, C.W., and Schwartz, R.S. (1980). Serosurvey for human babesiosis in New York. J. Infect. Dis. *141*, 518–521.

Fraser, C.M., Casjens, S., Huang, W.M., Sutton, G.G., Clayton, R., Lathigra, R., White, O., Ketchum, K.A., Dodson, R., Hickey, E.K., *et al.* (1997). Genomic sequence of a Lyme disease spirochaete, *Borrelia burgdorferi*. Nature *390*, 580–586.

Gans, O. (1933). Akrodermatitis atrophicans und Sklerodermie. Zbl Haut Geschl Kr *45*, 678–679.

Gans, O. (1952). Akrodermatitis atrophicans arthropathica. Hautarzt 3, 151–155.

García-Moncó, J.C., and Benach, J.L. (1995). Lyme neuroborreliosis. Ann. Neuro.l 37, 691–702.

Garin, C., and Bujadoux (1922). Paralysie par les tiques. J. Med. Lyon *71*, 765–767.

Gelbjerg-Hansen, G. (1945). Erythema chronicum migrans Azfelii and meningitis after a tick bite. Acta Derm. Venereol. (Stockh) *25*, 458–463.

Gern, L., Hu, C.M., Voet, P., Hauser, P., and Lobet, Y. (1997). Immunization with a polyvalent OspA vaccine protects mice against *Ixodes ricinus* tick bites infected by *Borrelia burgdorferi* ss, *Borrelia garinii* and *Borrelia afzelii*. Vaccine *15*, 1551–1557.

Gern, L., Schaible, U.E., and Simon, M.M. (1993). Mode of inoculation of the Lyme disease agent *Borrelia burgdorferi* influences infection and immune responses in inbred strains of mice. J. Infect. Dis. *167*, 971–975.

Gotz, H. (1954). Die Acrodermatitis chronica atrophicans Herx-heimer ais Infektionskrankheit. Hautarzt 5, 491–504.

Gotz, H. (1955). Die Acrodermatitis chronica atrophicans Herx-heimer ais Infektionskrankheit – Erganzung zur I. Mittei-lung. Hautarzt 6, 249–252.

Grabowski, E.F., Giardina, P.J., Goldberg, D., Masur, H., Read, S.E., Hirsch, R.L., and Benach, J.L. (1982). Babesiosis transmitted by a transfusion of frozen-thawed blood. Ann. Intern. Med. *96*, 466–467.

Gross, D.M., Forsthuber, T., Tary-Lehmann, M., Etling, C., Ito, K., Nagy, Z.A., Field, J.A., Steere, A.C., and Huber, B.T. (1998). Identification of LFA-1 as a candidate autoantigen in treatment-resistant Lyme arthritis. Science *281*, 703–706.

Grüneberg, T. (1954). Auffällige serologische Refunde bei Akro-dermatitis chronica atrophicans Herxheimer. Klin. Wochenschr .*32*, 935–936.

Grunwaldt, E., Barbour, A.G., and Benach, J.L. (1983). Simultaneous occurrence of babesiosis and Lyme disease. N. Engl. J. Med. *308*, 1166.

Habicht, G.S., Benach, J.L., Leichtling, K.D., Gocinski, B.L., and Coleman, J.L. (1983). The effect of age on the infection and immunoresponsiveness of mice to *Babesia microti*. Mech. Ageing Dev. *23*, 357–369.

Halperin, J.J., Shapiro, E.D., Logigian, E., Belman, A.L., Dotevall, L., Wormser, G.P., Krupp, L., Gronseth, G., and Bever, C.T., Jr. (2007). Practice parameter: treatment of nervous system Lyme disease (an evidence-based review): report of the Quality Standards Subcommittee of the American Academy of Neurology. Neurology *69*, 91–102.

Hard, S. (1966). Erythema chronicum migrans (Afzelii) associated with mosquito bite. Acta Derm. Venereol. *46*, 473–476.

Hellerstrom, S. (1930). Erythema chronicum migrans Azfelii. Acta Derm. Venereol. (Stockh.) *11*, 315–321.

Hellström, S. (1950). Erythema chronicum migrans Azfelius with meningitis. South. Med. J. *43*, 330–334.

Herxheimer , K., and Hartman, K. (1902). Über Acrodermatitis chronica atrophicans. Arch. Derm. Syph. *61*, 255–300.

Herzer, P. (1983). Lyme disease. A new or newly recognized clinical entity. MMW Munc.h Med. Wochenschr. *125*, 737–738.

Herzer, P. (1991). Joint manifestations of Lyme borreliosis in Europe. Scand J. Infect. Dis. Suppl. *77*, 55–63.

Herzer, P., Schattenkirchner, M., and Zöllner, N. (1983). Lyme arthritis – eine zu selten bedachte Diagnose? . Verh. Dtsch Ges. In.n Med. *89*, 299–302.

Hewitt, J., Illouz, G., Giroin, D., Dolivo, M., Leibowitch, M., and Gutman, L. (1980). Erythema chronicum migrans suivi d'une polyarthrite séronégative. Premier cas européen d'arthrite de Lyme? In Journées de la Societé francaise de Dermatologie (Paris).

Hilton, E., DeVoti, J., Benach, J.L., Halluska, M.L., White, D.J., Paxton, H., and Dumler, J.S. (1999). Seroprevalence and seroconversion for tick-borne diseases in a high-risk population in the northeast United States. Am. J. Med. *106*, 404–409.

Hollström, E. (1951). Successful treatment of erythema migrans Afzelius. Acta Derm. Venereol. (Stockh) *31*, 235–234.

Homer, M.J., Aguilar-Delfin, I., Telford, S.R., 3rd, Krause, P.J., and Persing, D.H. (2000). Babesiosis. Clin. Microbiol. Rev. *13*, 451–469.

Hoövelborn, C. (1931). Gelenkveränderungen bei Acrodermatitis cronica atrophicans. Arch. Derm. Syph. (Berlin) *164*, 349–356.

Hopf, H.C. (1966). Acrodermatitis chronica atrophicans (Herxheimer) and the nervous system. Analysis of clinical, physiologic, histologic and electromyographic findings. Monographien aus dem Gesamtgebiete der Neurologie und Psychiatrie *114*, 1–130.

Hörstrup, P., and Ackermann, R. (1973). Durch zecken übertragene meningopolyneuritis (Garin-Bujadoux, Bannwarth). Fortschr Neurol Psychiat *41*.

Jessner, M. (1922). Weiterer Beitag zur Kenntnis der Acrodermatitis chronica Atrophican. Arch. Derm. Syph. (Berlin) *164*, 349–356.

Jessner, M., and Loewenstamm, A. (1924). Bericht über 66 Fälle von Acrodermatitis cronica atrophican. Dermatol. Wochenschr. *79*, 1171–1177.

Johnson, R.C. (1977). The spirochetes. Annu. Rev. Microbiol. *31*, 89–106.

Johnson, R.C., Schmid, G.P., Hyde, F.W., Steigerwalt, A.G., and Brenner, D.J. (1984). *Borrelia burgdorferi* sp.nov.: etiological agent of Lyme disease. Int. J. Syst. Bacteriol. *34*, 496–497.

Johnson, R.C., Kodner, C., and Russell, M. (1986). Passive immunization of hamsters against experimental infection with the Lyme disease spirochete. Infect. Immun. *53*, 713–714.

Jones, C.G., Ostfeld, R.S., Richard, M.P., Schauber, E.M., and Wolff, J.O. (1998). Chain reactions linking acorns to gypsy moth outbreaks and Lyme disease risk. Science *279*, 1023–1026.

Kannian, P., McHugh, G., Johnson, B.J., Bacon, R.M., Glickstein, L.J., and Steere, A.C. (2007). Antibody responses to *Borrelia burgdorferi* in patients with antibiotic-refractory, antibiotic-responsive, or non-antibiotic-treated Lyme arthritis. Arthritis Rheum. *56*, 4216–4225.

Kaplan, R.F., Trevino, R.P., Johnson, G.M., Levy, L., Dornbush, R., Hu, L.T., Evans, J., Weinstein, A., Schmid, C.H., and Klempner, M.S. (2003). Cognitive function in post-treatment Lyme disease: do additional antibiotics help? Neurology *60*, 1916–1922.

Kawabata, H., Masuzawa, T., and Yanagihara, Y. (1993). Genomic analysis of *Borrelia japonica* sp. nov. isolated from Ixodes ovatus in Japan. Microbiol. Immunol *37*, 843–848.

Kelly, R. (1971). Cultivation of *Borrelia hermsi*. Science *173*, 443–444.

Klempner, M.S., Hu, L.T., Evans, J., Schmid, C.H., Johnson, G.M., Trevino, R.P., Norton, D., Levy, L., Wall, D., McCall, J., *et al.* (2001). Two controlled trials of antibiotic treatment in patients with persistent symptoms and a history of Lyme disease. N. Engl. J. Med. *345*, 85–92.

Krause, A., Burmester, G.R., Rensing, A., Schoerner, C., Schaible, U.E., Simon, M.M., Herzer, P., Kramer, M.D., and Wallich, R. (1992a). Cellular immune reactivity to recombinant OspA and flagellin from *Borrelia burgdorferi* in patients with Lyme borreliosis. Complexity of humoral and cellular immune responses. J. Clin. Invest. *90*, 1077–1084.

Krause, P.J., Telford, S.R., 3rd, Pollack, R.J., Ryan, R., Brassard, P., Zemel, L., and Spielman, A. (1992b). Babesiosis: an underdiagnosed disease of children. Pediatrics *89*, 1045–1048.

Krause, P.J., Telford, S.R., 3rd, Spielman, A., Sikand, V., Ryan, R., Christianson, D., Burke, G., Brassard, P., Pollack, R., Peck, J., *et al.* (1996). Concurrent Lyme disease and babesiosis: evidence for increased severity and duration of illness. JAMA *275*, 1657–1660.

Kristoferitsch, W. (1991). Neurological manifestations of Lyme borreliosis: Clinical definition and differential diagnosis. Scand. J. Infect. Dis. Suppl. *77*, 64–73.

Lennhoff, C. (1948). Spirochetes in aetiologically obscure diseases. Acta Derm. Venereol. (Stockh.) *28*, 295.

Lipschutz, B. (1918). Uber eine seltene erytheform (erythema chronicum migrans). Arch. Dermatol. Syph. *118*, 349–356.

Lissman, B.A., and Benach, J.L. (1980). Rocky Mountain spotted fever in dogs. J. Am. Vet. Med. Assoc. *176*, 994–995.

Lissman, B.A., Bosler, E.M., Camay, H., Ormiston, B.G., and Benach, J.L. (1984). Spirochete-associated arthritis (Lyme disease) in a dog. J. Am. Vet. Med. Assoc. *185*, 219–220.

LoGiudice, K., Ostfeld, R.S., Schmidt, K.A., and Keesing, F. (2003). The ecology of infectious disease: effects of host diversity and community composition on Lyme disease risk. Proc. Natl. Acad. Sci. U.S.A. *100*, 567–571.

Lohel, H. (1955). Experimental studies on etiology of Herxheimer's acrodermatitis chronica atrophicans. Klin. Wochenschr. *33*, 185–186.

Marques, A. (2008). Chronic Lyme disease: a review. Infect. Dis. Clin. North Am. *22*, 341–360, vii–viii.

Mast, W.E., and Burrows, W.M. (1976a). Erythema chronicum migrans and 'Lyme arthritis'. JAMA *236*, 2392.

Mast, W.E., and Burrows, W.M., Jr. (1976b). Erythema chronicum migrans in the United States. JAMA *236*, 859–860.

Meldrum, S.C., Birkhead, G.S., White, D.J., Benach, J.L., and Morse, D.L. (1992). Human babesiosis in New York State: an epidemiological description of 136 cases. Clin. Infect. Dis. *15*, 1019–1023.

Moro, M.H., Zegarra-Moro, O.L., Bjornsson, J., Hofmeister, E.K., Bruinsma, E., Germer, J.J., and Persing, D.H. (2002). Increased arthritis severity in mice coinfected with *Borrelia burgdorferi* and *Babesia microti*. J. Infect. Dis. *186*, 428–431.

Oliver, J.H., Jr., Owsley, M.R., Hutcheson, H.J., James, A.M., Chen, C., Irby, W.S., Dotson, E.M., and McLain, D.K. (1993). Conspecificity of the ticks *Ixodes scapularis* and *I. dammini* (Acari: Ixodidae). J. Med. Entomol. *30*, 54–63.

Ostfeld, R.S., Canham, C.D., Oggenfuss, K., Winchcombe, R.J., and Keesing, F. (2006). Climate, deer, rodents, and acorns as determinants of variation in Lyme-disease risk. PLoS Biol. *4*, e145.

Pfister, H.W., Preac-Mursic, V., Wilske, B., Einhaupl, K.M., and Weinberger, K. (1989). Latent Lyme neuroborreliosis: presence of *Borrelia burgdorferi* in the cerebrospinal fluid without concurrent inflammatory signs. Neurology *39*, 1118–1120.

Philipp, M.T., Bowers, L.C., Fawcett, P.T., Jacobs, M.B., Liang, F.T., Marques, A.R., Mitchell, P.D., Purcell, J.E., Ratterree, M.S., and Straubinger, R.K. (2001). Antibody response to IR6, a conserved immunodominant region of the VlsE lipoprotein, wanes rapidly after antibiotic treatment of *Borrelia burgdorferi* infection in experimental animals and in humans. J. Infect. Dis. *184*, 870–878.

Philipp, M.T., Wormser, G.P., Marques, A.R., Bittker, S., Martin, D.S., Nowakowski, J., and Dally, L.G. (2005). A decline in C6 antibody titer occurs in successfully treated patients with culture-confirmed early localized or early disseminated Lyme borreliosis. Clin. Diagn. Lab. Immunol. *12*, 1069–1074.

Piesman, J., Hicks, T.C., Sinsky, R.J., and Obiri, G. (1987). Simultaneous transmission of *Borrelia burgdorferi* and *Babesia microti* by individual nymphal Ixodes dammini ticks. J. Clin. Microbiol. *25*, 2012–2013.

Piesman, J., Mather, T.N., Telford, S.R., 3rd, and Spielman, A. (1986). Concurrent *Borrelia burgdorferi* and *Babesia microti* infection in nymphal *Ixodes dammini*. J. Clin. Microbiol. *24*, 446–447.

Piesman, J., and Spielman, A. (1980). Human babesiosis on Nantucket Island: prevalence of *Babesia microti* in ticks. Am. J. Trop. Med. Hyg. *29*, 742–746.

Piesman, J., Spielman, A., Etkind, P., Ruebush, T.K., 2nd, and Juranek, D.D. (1979). Role of deer in the epizootiology of *Babesia microti* in Massachusetts, USA. J. Med. Entomol. *15*, 537–540.

Preac Mursic, V., Wilske, B., Schierz, G., Pfister, H.W., and Einhaupl, K. (1984). Repeated isolation of spirochetes from the cerebrospinal fluid of a patient with meningoradiculitis Bannwarth. Eur J. Clin. Microbiol. 3, 564–565.

Reik, L., Steere, A.C., Bartenhagen, N.H., Shope, R.E., and Malawista, S.E. (1979). Neurologic abnormalities of Lyme disease. Medicine (Baltimore) *58*, 281–294.

Ribeiro, J.M., Mather, T.N., Piesman, J., and Spielman, A. (1987). Dissemination and salivary delivery of Lyme disease spirochetes in vector ticks (Acari: Ixodidae). J. Med. Entomol. *24*, 201–205.

Rosa, P., Samuels, D.S., Hogan, D., Stevenson, B., Casjens, S., and Tilly, K. (1996). Directed insertion of a selectable marker into a circular plasmid of *Borrelia burgdorferi*. J. Bacteriol. *178*, 5946–5953.

Rosa, P.A., Tilly, K., and Stewart, P.E. (2005). The burgeoning molecular genetics of the Lyme disease spirochaete. Nat. Rev. Microbiol. *3*, 129–143.

Rosner, F., Zarrabi, M.H., Benach, J.L., and Habicht, G.S. (1984). Babesiosis in splenectomized adults. Review of 22 reported cases. Am. J. Med. *76*, 696–701.

Rudzinska, M.A., Trager, W., Lewengrub, S.J., and Gubert, E. (1976). An electron microscopic study of *Babesia microti* invading erythrocytes. Cell Tissue Res. *169*, 323–334.

Ruebush, T.K., 2nd, Juranek, D.D., Chisholm, E.S., Snow, P.C., Healy, G.R., and Sulzer, A.J. (1977). Human babesiosis on Nantucket Island: evidence for self-limited and subclinical infections. N. Engl. J. Med. *297*, 825–827.

Ruebush, T.K., 2nd, Juranek, D.D., Spielman, A., Piesman, J., and Healy, G.R. (1981). Epidemiology of human babesiosis on Nantucket Island. Am. J. Trop. Med. Hyg. *30*, 937–941.

Samuels, D.S. (1995). Electrotransformation of the spirochete *Borrelia burgdorferi*. Methods Mol. Biol. *47*, 253–259.

Samuels, D.S., Mach, K.E., and Garon, C.F. (1994a). Genetic transformation of the Lyme disease agent *Borrelia burgdorferi* with coumarin-resistant *gyrB*. J. Bacteriol. *176*, 6045–6049.

Samuels, D.S., Marconi, R.T., Huang, W.M., and Garon, C.F. (1994b). *gyrB* mutations in coumermycin A_1-resistant *Borrelia burgdorferi*. J. Bacteriol. *176*, 3072–3075.

Sandbank, M., and Feuerman, E.J. (1979). Ultrastructural observation of rickettsia-like bodies in erythema cronicum migrans. J. Cutan. Pathol. *6*, 253–264.

Schaible, U.E., Gern, L., Wallich, R., Kramer, M.D., Prester, M., and Simon, M.M. (1993a). Distinct patterns of protective antibodies are generated against *Borrelia burgdorferi* in mice experimentally inoculated with high and low doses of antigen. Immunol. Lett. *36*, 219–226.

Schaible, U.E., Kramer, M.D., Eichmann, K., Modolell, M., Museteanu, C., and Simon, M.M. (1990). Monoclonal antibodies specific for the outer surface protein A (OspA) of *Borrelia burgdorferi* prevent Lyme borreliosis in severe combined immunodeficiency (*scid*) mice. Proc. Natl. Acad. Sci. U.S.A. *87*, 3768–3772.

Schaible, U.E., Wallich, R., Kramer, M.D., Gern, L., Anderson, J.F., Museteanu, C., and Simon, M.M. (1993b). Immune sera to individual *Borrelia burgdorferi* isolates or recombinant OspA thereof protect SCID mice against infection with homologous strains but only partially or not at all against those of different OspA/OspB genotype. Vaccine *11*, 1049–1054.

Schaltenbrand, G. (1949). Chronische aseptische Meningitis. Nervenarzt *20*, 433–442.

Schmutzhard, E., Pohl, P., and Stanek, G. (1989). Failure to demonstrate *B. burgdorferi* as an etiologic agent in multiple sclerosis. Zbl. Bakt. Supp.l *18*, 219–226.

Schwan, T.G., Piesman, J., Golde, W.T., Dolan, M.C., and Rosa, P.A. (1995). Induction of an outer surface protein on *Borrelia burgdorferi* during tick feeding. Proc. Natl. Acad. Sci. U.S.A. *92*, 2909–2913.

Scrimenti, R.J. (1970). Erythema chronicum migrans. Arch. Dermatol. *102*, 104–105.

Sigal, L.H., Zahradnik, J.M., Lavin, P., Patella, S.J., Bryant, G., Haselby, R., Hilton, E., Kunkel, M., Adler-Klein, D., Doherty, T., *et al.* (1998). A vaccine consisting of recombinant *Borrelia burgdorferi* outer-surface protein A to prevent Lyme disease. Recombinant Outer-Surface Protein A Lyme Disease Vaccine Study Consortium. N. Engl. J. Med. *339*, 216–222.

Spielman, A., Clifford, C.M., Piesman, J., and Corwin, M.D. (1979). Human babesiosis on Nantucket Island, USA: description of the vector, *Ixodes dammini*, n. sp. (Acarina: Ixodidae). J. Med. Entomol. *15*, 218–234.

Spielman, A., Etkind, P., Piesman, J., Ruebush, T.K., 2nd, Juranek, D.D., and Jacobs, M.S. (1981). Reservoir hosts of human babesiosis on Nantucket Island. Am. J. Trop. Med. Hyg. *30*, 560–565.

Steere, A.C., Batsford, W.P., Weinberg, M., Alexander, J., Berger, H.J., Wolfson, S., and Malawista, S.E. (1980a). Lyme carditis: cardiac abnormalities of Lyme disease. Ann. Intern. Med. *93*, 8–16.

Steere, A.C., Broderick, T.F., and Malawista, S.E. (1978). Erythema chronicum migrans and Lyme arthritis: epidemiologic evidence for a tick vector. Am. J. Epidemiol. *108*, 312–321.

Steere, A.C., Grodzicki, R.L., Kornblatt, A.N., Craft, J.E., Barbour, A.G., Burgdorfer, W., Schmid, G.P., Johnson, E., and Malawista, S.E. (1983). The spirochetal etiology of Lyme disease. N. Engl. J. Med. *308*, 733–740.

Steere, A.C., Klitz, W., Drouin, E.E., Falk, B.A., Kwok, W.W., Nepom, G.T., and Baxter-Lowe, L.A. (2006). Antibiotic-refractory Lyme arthritis is associated with HLA-DR molecules that bind a *Borrelia burgdorferi* peptide. J. Exp. Med. *203*, 961–971.

Steere, A.C., and Malawista, S.E. (1979). Cases of Lyme disease in the United States: locations correlated with distribution of *Ixodes dammini*. Ann. Intern. Med. *91*, 730–733.

Steere, A.C., Malawista, S.E., Hardin, J.A., Ruddy, S., Askenase, W., and Andiman, W.A. (1977a). Erythema chronicum migrans and Lyme arthritis. The enlarging clinical spectrum. Ann. Intern. Med. *86*, 685–698.

Steere, A.C., Malawista, S.E., Newman, J.H., Spieler, P.N., and Bartenhagen, N.H. (1980b). Antibiotic therapy in Lyme disease. Ann. Intern. Med. *93*, 1–8.

Steere, A.C., Malawista, S.E., Snydman, D.R., Shope, R.E., Andiman, W.A., Ross, M.R., and Steele, F.M. (1977b). Lyme arthritis: an epidemic of oligoarticular arthritis in children and adults in three connecticut communities. Arthritis Rheum. *20*, 7–17.

Steere, A.C., Sikand, V.K., Meurice, F., Parenti, D.L., Fikrig, E., Schoen, R.T., Nowakowski, J., Schmid, C.H., Laukamp, S., Buscarino, C., *et al.* (1998). Vaccination against Lyme disease with recombinant *Borrelia burgdorferi* outer-surface lipoprotein A with adjuvant. Lyme Disease Vaccine Study Group. N. Engl. J. Med. *339*, 209–215.

Steigbigel, R.T., and Benach, J.L. (1998). Immunization against Lyme disease – an important first step. N. Engl. J. Med. *339*, 263–264.

Stoenner, H.G. (1974). Biology of *Borrelia hermsii* in Kelly medium. Appl. Microbiol. *28*, 540–543.

Stoenner, H.G., Dodd, T., and Larsen, C. (1982). Antigenic variation of *Borrelia hermsii*. J. Exp. Med. *156*, 1297–1311.

Sweeney, C.J., Ghassemi, M., Agger, W.A., and Persing, D.H. (1998). Coinfection with *Babesia microti* and *Borrelia burgdorferi* in a western Wisconsin resident. Mayo Clin. Proc. *73*, 338–341.

Telford, S.R., 3rd, Fikrig, E., Barthold, S.W., Brunet, L.R., Spielman, A., and Flavell, R.A. (1993). Protection against antigenically variable *Borrelia burgdorferi* conferred by recombinant vaccines. J. Exp. Med. *178*, 755–758.

Telford, S.R., 3rd, Mather, T.N., Moore, S.I., Wilson, M.L., and Spielman, A. (1988). Incompetence of deer as reservoirs of the Lyme disease spirochete. Am. J. Trop. Med. Hyg. *39*, 105–109.

Thyresson, N. (1949). The penicillin treatment of acrodermatitis chronica atrophicans. Acta Derm. Venereol. (Stockh.) *29*, 572–621.

Vannier, E., Borggraefe, I., Telford, S.R., 3rd, Menon, S., Brauns, T., Spielman, A., Gelfand, J.A., and Wortis, H.H. (2004). Age-associated decline in resistance to *Babesia microti* is genetically determined. J. Infect. Dis. *189*, 1721–1728.

Wallis, R.C., Brown, S.E., Kloter, K.O., and Main, A.J., Jr. (1978). Erythema chronicum migrans and Lyme arthritis: field study of ticks. Am. J. Epidemiol. *108*, 322–327.

Wang, T.J., Liang, M.H., Sangha, O., Phillips, C.B., Lew, R.A., Wright, E.A., Berardi, V., Fossel, A.H., and Shadick, N.A. (2000). Coexposure to *Borrelia burgdorferi* and *Babesia microti* does not worsen the long-term outcome of Lyme disease. Clin. Infect. Dis. *31*, 1149–1154.

Weber, K. (1974). Erythema chronicum migrans meningitis – a bacterial infectious disease? MMW Munch. Med. Wochenschr. *116*, 1993–1998.

Weber, K. (1981). Serological study with rickettsial antigens in erythema chronicum migrans. Dermatologica *163*, 460–467.

Weis, J.J., Ma, Y., and Erdile, L.F. (1994). Biological activities of native and recombinant *Borrelia burgdorferi* outer surface protein A: dependence on lipid modification. Infect. Immun. *62*, 4632–4636.

Wilske, B., Preac-Mursic, V., and Schierz, G. (1985). Antigenic heterogeneity of European *Borrelia burgdorferi* strains isolated from patients and ticks. Lancet *1*, 1099.

Wormser, G.P., Dattwyler, R.J., Shapiro, E.D., Halperin, J.J., Steere, A.C., Klempner, M.S., Krause, P.J., Bakken, J.S., Strle, F., Stanek, G., *et al.* (2006). The clinical assessment, treatment, and prevention of Lyme disease, human granulocytic anaplasmosis, and babesiosis: clinical practice guidelines by the Infectious Diseases Society of America. Clin. Infect. Dis. *43*, 1089–1134.

Zhong, W., Stehle, T., Museteanu, C., Siebers, A., Gern, L., Kramer, M., Wallich, R., and Simon, M.M. (1997). Therapeutic passive vaccination against chronic Lyme disease in mice. Proc. Natl. Acad. Sci. U.S.A. *94*, 12533–12538.

Zwelleger, H. (1946). Uber die chronische allergische meningitis. Helv. Paediatr. Acta *1*, 417–426.

Borrelia Genomics: Chromosome, Plasmids, Bacteriophages and Genetic Variation

2

Sherwood R. Casjens, Christian H. Eggers and Ira Schwartz

Abstract

All members of the *Borrelia* genus that have been examined harbour a linear chromosome that is about 900 kbp in length as well as a plethora of both linear and circular plasmids in the 5–220 kbp size range. Genome sequences have been determined for *Borrelia burgdorferi, Borrelia garinii, Borrelia afzelii, Borrelia duttonii* and *Borrelia recurrentis*. The chromosomes, which carry the vast majority of the housekeeping genes, appear to be very constant in gene content and organization across the genus. The content of the plasmids, which carry most of the genes that encode the differentially expressed surface proteins that interact with *Borrelia*'s arthropod and vertebrate hosts, are much more variable. *Borrelia burgdorferi* strain B31, the *B. burgdorferi* type strain, has been studied in the most detail, and harbours 12 linear and nine circular plasmids that constitute about 612 kbp. The plasmids are unusual, as compared to most bacterial plasmids, in that they contain many paralogous sequences, a large number of pseudogenes, and, in some cases, essential genes. In addition, a number of the plasmids have features suggesting that they are prophages. Some correlations between genome content and pathogenicity have been deduced, and comparative whole genome analyses promise future progress in this arena.

2.1 Introduction

The genus *Borrelia* forms a deeply separated lineage within the *Spirochaetes* branch of the bacterial tree of life (Paster *et al.*, 1991). The organisms are genomically unique and not closely related to any other bacteria, including the other spirochaetes. *Borrelia burgdorferi* and *Borrelia hermsii* represent the two major branches within the *Borrelia* genus, wherein *B. burgdorferi* typifies the 'Lyme disease agent-like' branch and *B. hermsii* typifies the 'relapsing fever agent-like' branch (Paster *et al.*, 1991; Marti Ras *et al.*, 1996; Schwan *et al.*, 2007). (The term '*B. burgdorferi*' denotes *B. burgdorferi* sensu stricto throughout this chapter.) By analysing the genomes of both *B. burgdorferi* and *B. hermsii,* Barbour and coworkers originally showed that both have a topologically unusual genome with linear chromosomes and multiple linear plasmids (Plasterk *et al.* 1985; Barbour, 1988; Ferdows and Barbour, 1989; Hinnebusch *et al.*, 1990; Kitten and Barbour, 1990; Hinnebusch and Barbour, 1991). Subsequent work has shown this to be a property of all members of the genus *Borrelia* that have been examined. Each *Borrelia* species carries a linear chromosome about 900 kbp in length and multiple circular and linear low copy number plasmids that are usually, but not always, in the 9–62 kbp range. These features have been demonstrated directly for eight of the defined Lyme agent-like *Borrelia* species, *B. burgdorferi, Borrelia bissettii, Borrelia garinii, Borrelia afzelii, Borrelia andersonii, Borrelia valaisiana, Borrelia lusitaniae,* and *Borrelia japonica,* as well as the relapsing fever species *Borrelia hermsii, Borrelia turicatae, Borrelia duttonii, Borrelia recurrentis Borrelia coriaceae, Borrelia parkeri,* and *Borrelia anserina* (Ferdows and Barbour, 1989; Davidson *et al.*, 1992; Casjens and Huang, 1993; Kitten *et* al., 1993; Ojaimi et al., 1994; Casjens et al., 1995; Lescot et al., 2008; R. Van Vugt and S. Casjens, unpublished) and all members of the genus probably will have genomes of this type. The chromosomes of these bacteria all have quite similar gene contents, whereas the plasmids are more variable (details below).

In light of their importance for disease pathogenesis (see Chapters 12 and 13), the number and diversity of the plasmids carried by *Borrelia* spp. has been the focus of much research. This enterprise, however, has been hampered by their unusual features. First, individual *Borrelia* cells harbour a more diverse plasmid complement than that of any other bacterium – the sequenced genome of *B. burgdorferi* type strain B31 has 21 plasmids, and several additional plasmids appear to have been lost between isolation of the strain in 1982 and the completion of its genome sequence (Casjens *et al.*, 2000). Twenty of these plasmids have been shown to be carried in a single cloned cell (*i.e.* are members of different compatibility groups), and sequence analysis of the four B31 plasmids that have not yet been shown definitively to reside in the same cell (cp9-1, cp9-2, cp32-2 and cp32-5), indicates that only cp32-2 appears to belong to the same compatibility group with a plasmid (cp32-7) still maintained within B31. Thus, at least 23 plasmids were probably harboured in the original B31 cell when it was isolated (there is no other evidence for non-clonality of that isolate) (Casjens *et al.*, 1997b, 2000; Miller *et al.*, 2000). Second, a number of these plasmids have very similar sizes, and thus cannot be separated by pulsed-field gel electrophoresis. Consequently, mapping a gene to a particular plasmid by Southern hybridization can be difficult. Third, different plasmids in the same cell often carry paralogous sequences that can be extremely similar (Casjens, 2000; Casjens *et al.*, 2000). This can and does confuse both plasmid identification and assembly of plasmid sequences. Thus, *Borrelia* researchers are currently placed in the difficult position of being unable to confidently know the complete plasmid content of a particular strain without some type of complete genomic sequence analysis. Finally, assessing the plasmid content of a *Borrelia* cell is further complicated by the fact that plasmids can be lost during primary isolation; knowing whether native plasmids have been lost from any strain isolated in the laboratory is essentially impossible at this time. Furthermore, many of the *Borrelia* plasmids are readily lost during passage or manipulation *in vitro*, and many clones of any given strain will have lost one or more (Xu *et al.*, 1996; Schwan *et al.*, 1988; Purser and Norris, 2000; Elias *et al.*, 2002; Grimm *et al.*, 2003); as a result, two clones of *B. burgdorferi* strain B31 can potentially, and often do, have significantly different genetic contents.

Overall, the 900-kbp chromosomes (often called the 'large' or 'main' chromosome by workers in this field) carry the great majority of the genes that encode metabolic enzymes, and the 400 to 650 kbp (in different isolates) of plasmid DNAs carry the bulk of the surface lipoprotein encoding genes. With the notable exception of cp26 (see below), the plasmids are not required for growth in culture (Sadziene *et al.*, 1993; Casjens *et al.*, 1997b), but have often been found to be required for mouse infectivity or tick transmittability in the laboratory (Xu *et al.*, 1996; Purser and Norris, 2000; Grimm *et al.*, 2005; Revel *et al.*, 2005; Schwan *et al.*, 1988; Stewart *et al.*, 2005; Strother *et al.*, 2005; Strother and de Silva, 2005). The chromosome carries tightly packed genes, as is typical of bacteria, while many of the linear plasmids have substantially lower gene densities and many apparently decaying pseudogenes (Casjens, 2000; Casjens *et al.*, 2000). Currently, the general impression is that of an evolutionarily quite stable chromosome that encodes the machinery required for existence as a bacterial cell along with a rather large menu of much more evolutionarily variable plasmids that encode most of the proteins that interact with the vertebrate and arthropod host environments encountered by the bacteria (there is a small number of important metabolic enzymes encoded by the plasmids; see below).

We note that references cited in this review were chosen to allow the reader access to the latest literature and not necessarily to credit discoverers, the originators of ideas, or the first to obtain particular kinds of data. Thus, cited articles are sometimes more recent articles instead of those in which the original discoveries were first reported.

2.2 The *Borrelia burgdorferi* strain B31 genome sequence

2.2.1 The chromosome and its genes

The complete 910,725-bp sequence of the *B. burgdorferi* strain B31 linear chromosome was published in 1997 by Fraser *et al.* (1997), and it remains the only *Borrelia* strain whose genome sequence is known to be complete (some plasmid sequences remain unassembled in the others). Its nucleotide composition is 28.6% G+C and is predicted to contain 846 apparently intact protein-encoding genes and ten or fewer genes that appear to be truncated or contain frameshift mutations; the fraction of mutationally inactivated chromosomal genes is among the lowest of the analysed bacterial genomes. Although smaller bacterial genomes are known, this chromosome size is near the small end of the spectrum (*e.g.* Casjens, 1998). The protein-encoding genes occupy 93% of the chromosome, a typical value for a bacterial genome not undergoing current reduction in size (Lynch, 2006). These protein-encoding genes are quite highly clustered according to function, often suggesting transcription as operons; about 67% of the genes are oriented such that they are transcribed away from the centre of the chromosome. GC skew analysis and experimental evidence strongly suggest the presence of a replication origin near the centre of the linear chromosome (Picardeau *et al.*, 1999). The chromosome also carries five rRNA genes (two 23S, two 5S and one 16S) clustered very near the centre of the chromosome, one tmRNA gene, and 31 tRNA genes. The latter are predicted to include tRNAs specific for all 20 amino acids and are scattered across the large chromosome in seven clusters and 13 single genes (Schwartz *et al.*, 1992; Fraser *et al.*, 1997).

About 60% of the strain B31 chromosome's predicted genes have some similarity to a gene in another organism whose role or function is at least partly understood; about 10% are similar to known genes in other organisms whose roles are unknown; and about 30% are unique to *Borrelia* and have unknown functions (Fraser *et al.*, 1997). The chromosome carries what appears to be a rather minimal set of the genes that are required for cell maintenance and replication. These include genes for cell wall biosynthesis (but not synthesis of lipopolysaccharide or phospholipids); protein secretion and lipidation; DNA, RNA and protein biosynthesis; DNA repair; nucleotide metabolism (but not *de novo* synthesis); glycolysis and a few enzymes that provide substrates for the glycolytic pathway, the bacterium's sole mode for generating ATP. A large and complete set of the genes known to be required for motility and chemotaxis is also present, as are a number of genes for transport of small molecules across the cytoplasmic membrane. A major conclusion from this bioinformatic analysis is that the biosynthetic and intermediary metabolic capacity of *Borrelia* is very limited. Genes that encode enzymes that perform functions in respiration, amino acid synthesis, nucleotide synthesis, lipid synthesis and enzyme cofactor synthesis are almost completely lacking, in agreement with their many and fastidious requirements for growth in culture and their evolutionary 'decision' to limit themselves to two highly specific host environments, small vertebrates and arthropods. Curiously, no observed or predicted requirement for iron or iron transporter genes have been found (Posey and Gherardini, 2000; Nguyen *et al.*, 2007). Given the requirement that they scavenge essentially all amino acids, nucleosides, lipids and cofactors from their environment, it is perhaps surprising that 'only' 16 transporters were identified in the initial analysis, and only a few more have been identified since then. Some of these transporters probably have generic specificity and have been hypothesized to bring in multiple related small molecules (Fraser *et al.*, 1997) (see Chapters 5 and 7 for additional discussion of metabolism and motility, respectively). Also of note is the lower than expected number of genes identified as transcriptional regulators. Only two two-component systems and two alternative sigma factors have been identified by sequence homology. Inasmuch as some *Borrelia* genes are known to be regulated by environmental signals, either regulatory genes are fewer than in many other bacteria or much gene regulation is carried out by proteins that are not closely related to known regulatory factors (see Chapter 4 for more details). Given that there are well over 200 genes on the B31 chromosome that have no homologues with a known function, the latter is certainly possible.

Since the genome of *B. burgdorferi* strain B31 was sequenced (Fraser *et al.*, 1997; Casjens *et al.*, 2000), the *Borrelia* research community has usually used the 'locus tags' in its GenBank annotation as names for its genes and their encoded proteins (and their homologues discovered in other isolates). Thus, according to bacterial convention, chromosomal genes are named '*bb0xxx*' (lower case and italicized) in ascending order from *bb0001* through *bb0853* across the chromosome, and chromosoally encoded proteins are named 'BB0xxx' (upper case and not italicized, *e.g.* BB0364). The B31 plasmid locus tag names are similar but have the form 'BB$xx' in which '$' indicates a letter, A through U, that denotes which plasmid carries the gene (*e.g. bba74* encodes protein BBA74 and lies on lp54, *bbs09* lies on cp32-3, *etc.*).

2.2.2 Plasmids – structure, compatibility types and nomenclature

It is well known that all *Borrelia* isolates harbour a large number of plasmids, of which some are linear and some are circular (above). The B31 plasmids were originally named 'lp' for linear plasmid and 'cp' for circular plasmid along with numerical designations based on their apparent sizes by agarose gel electrophoresis. For those which have been measured, the B31 circular and linear plasmids have copy numbers in the range of one to two per chromosome. The relative staining intensities of the other plasmid electrophoresis bands are very similar, suggesting that all are low-copy-number plasmids (Hinnebusch and Barbour, 1992; Casjens and Huang, 1993) (this has also been shown for *B. afzelii* strain Pbi; Glöckner *et al.*, 2004). The genes on the plasmid replicons are largely unrelated to known genes in other organisms; only about six percent of the linear plasmid genes are homologues of genes in other bacteria.

Each strain B31 plasmid carries a complete or partial set of four genes that are predicted on the basis of sequence homologies to be involved in replication and compatibility (shown in Fig. 2.1). These genes encode proteins that are members of the paralogous families named PFam32 (homologous to the well-studied ParA proteins that play a central role in the partitioning of many bacterial plasmids (Shih and Rothfield, 2006)) and three families, PFam49, PFam50 and PFam57/62, that have no strong similarity to genes of known function (PFams were originally defined in Casjens *et al.*, 2000). The smaller plasmids (<19 kbp) carry a subset of these genes and the rest carry all four, usually in a contiguous cluster. Bioinformatic and experimental studies have strongly supported the idea that these proteins determine the plasmid compatibility type (Barbour *et al.*, 1996; Zückert and Meyer, 1996; Casjens *et al.*, 2000; Garcia-Lara *et al.*, 2000; Stewart *et al.*, 2001, 2003; Eggers *et al.*, 2002; Grimm *et al.*, 2003, 2004; Beaurepaire and Chaconas, 2005), and *Borrelia* plasmids have been proposed to be named according to the type of PFam32 protein they encode (Stevenson and Miller, 2003; Casjens *et al.*, 2006). Such a nomenclature scheme has the advantage that plasmids within the same compatibility group have related names, but the disadvantages are that (1) the PFam32 gene within a given plasmid must be sequenced in order for the plasmid to have a finalized designation, and (2) two plasmids with the same maintenance elements can differ substantially in gene content. According to this scheme, 14 PFam32 linear plasmid compatibility types have been identified in the B31 complete genome and draft genome sequences of *B. burgdorferi* strains N40, 297 and JD-1 (Qiu *et al.*, 2004; Casjens *et al.*, 2006). Similarly, twelve circular plasmid PFam32 types have been identified (one cp26 type and 11 cp32 types) (Stevenson and Miller, 2003; Casjens *et al.*, 2006). In addition, the small cp9 plasmids of N40 and B31 and lp5 of B31 do not have PFam32 genes and so appear to define two additional plasmid compatibility types. Thus, at present plasmids with 28 different compatibility types are known in *B. burgdorferi*.

2.2.3 Linear plasmids – gene content and paralogous sequences

Borrelia burgdorferi strain B31 carries 12 linear plasmids that range from 5 to 54 kbp in size. Most linear plasmid genes that are not involved in plasmid replication/partition have no homologues outside *Borrelia*, but a few have been shown to be involved in restriction modification (Lawrenz *et al.*, 2002), nucleotide metabolism (Purser *et al.*, 2003; Zhong *et al.*, 2006; Jewett *et al.*, 2007b), and

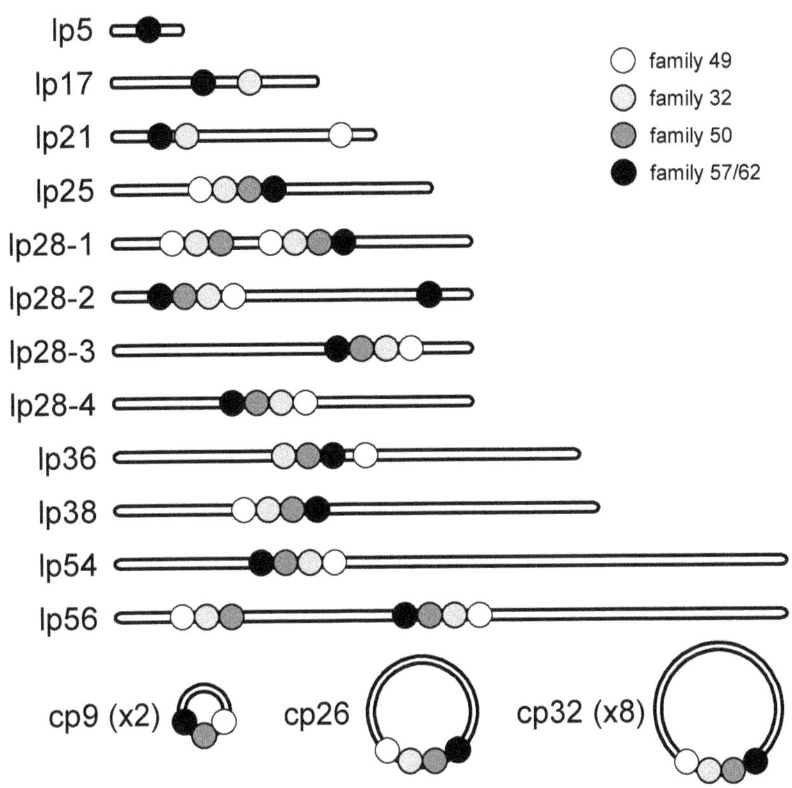

Figure 2.1 Plasmid replication/partitioning genes on the *B. burgdorferi* strain B31 plasmids. Apparently intact genes from the paralogous families thought to be involved in plasmid replication, partitioning and compatibility are indicated by circles with different shading. Paralogous families 57 and 62 are related and so are denoted as one merged family in the figure. The figure is modified from Figure 8 of Casjens *et al.* (2000).

evasion of the host immune response (Purser *et al.*, 2003; Kraiczy *et al.*, 2004, 2008). In addition, a number of linear plasmid genes have been shown to encode surface lipoprotein genes (indicated in Fig. 2.2). Unlike the circular plasmids (below), none of the 12 B31 linear plasmids is homologous throughout its length, so they comprise 12 different plasmids 'types'.

An unusual feature of many of the linear plasmids is that they contain numerous paralogous sequences (Casjens, 2000; Casjens *et al.*, 2000). These are scattered apparently randomly throughout all the linear plasmids. In addition, there is more apparently non-protein coding DNA and a large number of pseudogenes on the linear plasmids (with the exception of lp54 and most of lp28-2). These three features may not be independent of one another. The current best hypothesis for the generation of this state of affairs is that many tandem and non-tandem duplicative rearrangements have taken place on these plasmids. The boundaries of the paralogous sequences thus generated suggest that these events largely involved non-homologous recombination events. Thus, these events often created truncated copies of genes or pseudogenes. In addition, when whole genes were duplicated, the second copy may at times no longer have been under selection. Pairs of these paralogous regions vary from being nearly identical to being only distantly related, suggesting an ongoing process rather than a single massive rearrangement event in the past. Any unselected sequences would be free to accumulate mutations, perhaps in some cases to the point that their sources are no longer recognizable. Such a process could lead to the observed pseudogenes and substantial non-coding regions present on the linear plasmids. The reasons why such rearrangements have been so frequent on these plasmids remain mysterious.

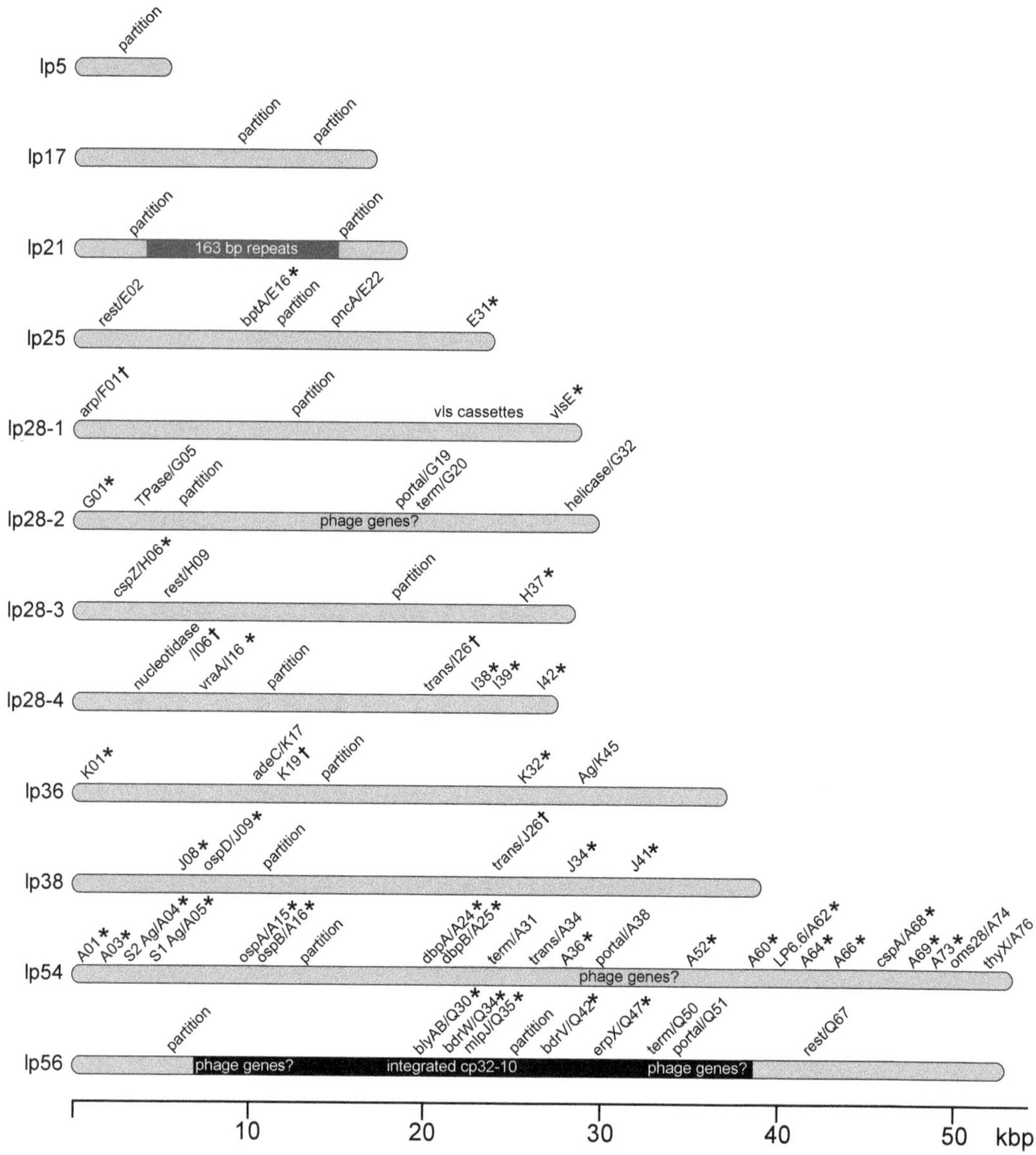

Figure 2.2 The 12 *B. burgdorferi* strain B31 linear plasmids. The B31 linear plasmids are shown approximately to scale. Above each plasmid the genes that have been studied experimentally in some fashion or whose function is robustly predicted are indicated. Black asterisks (*) mark genes for which there is evidence that the encoded protein is membrane bound and/or exposed on the cell surface (marked for each indicated member of the various paralogous families, although this has not always been shown directly for every member); daggers (†) mark genes whose proteins are strongly predicted to be membrane proteins (in most cases N-terminal lipidation prediction); partition, plasmid partitioning (see text); rest, putative restriction/modification enzyme; trans, putative transporter; term, putative phage terminase (DNA packaging enzyme); TPase, putative transposase; Ag, antigenic in infected animals; the first two letters (BB) of GenBank 'locus_tag' gene names have been left out for brevity (*e.g.* H37 instead of BBH037). References for all of the noted genes and their function are too numerous to list here, but see other chapters of this volume, the following articles and references therein (Theisen, 1996; Zhang *et al.*, 1997; Zückert *et al.*, 1999; Skare *et al.*, 1999; Roberts *et al.*, 2000; Porcella *et al.*, 2000; Purser *et al.*, 2003; Feng *et al.*, 2003; Kim *et al.*, 2004; Jacobs *et al.*, 2005; Stewart *et al.*, 2005; Revel *et al.*, 2005; Pinne *et al.*, 2006; Stevenson *et al.*, 2006; Brooks *et al.*, 2006; Jewett *et al.*, 2007b; Antonara *et al.*, 2007; Behera *et al.*, 2007).

2.2.4 Circular plasmids – gene content and highly paralogous sequences

Unlike the strain B31 linear plasmids, which are all substantially different from one another, the B31 strain contains only three 'organizational types' of circular plasmids: cp9, cp26 and seven different cp32s (Fig. 2.3). Also unlike the linear plasmids, the circular plasmids have closely packed genes and relatively few pseudogenes. Cp9 is easily lost and appears not to be required for mouse infectivity (Champion *et al.*, 1994; Purser and Norris, 2000; Elias *et al.*, 2002). Cp26 is the only plasmid, linear or circular, that has been shown to be essential for growth in culture (Byram *et al.*, 2004; Jewett *et al.*, 2007b) and it is almost certainly also essential for maintenance of the spirochaete's enzootic cycle. It carries genes that encode proteins involved in GMP synthesis (Margolis *et al.*, 1994), chitobiose import (Tilly *et al.*, 2004), host integrin binding (Behera *et al.*, 2007), oligopeptide import (Bono *et al.*, 1998),

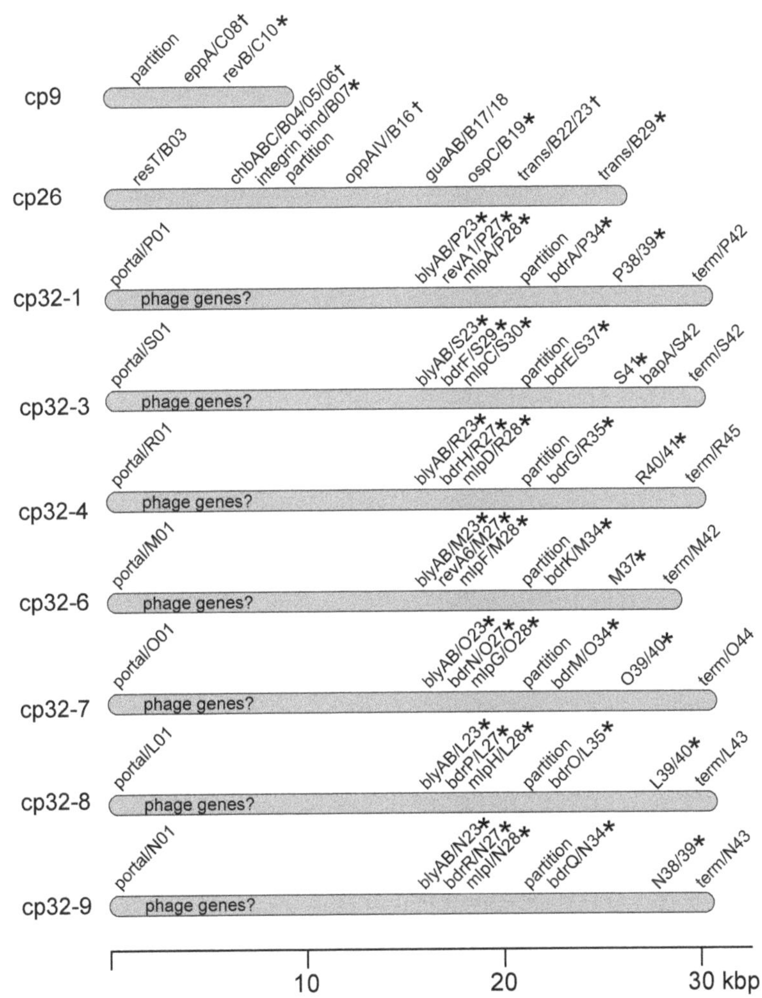

Figure 2.3 The nine *B. burgdorferi* strain B31 circular plasmids. The B31 circular plasmids are shown approximately to scale; they have been opened arbitrarily for linear display as in their GenBank annotation (Casjens *et al.*, 2000). See the legend to Figure 2.2 for abbreviations and references.

and the telomere resolvase (below and Kobryn and Chaconas, 2002). It also encodes one of the important surface antigens that is expressed in the mammalian host, OspC (Wilske *et al.*, 1986; Pal *et al.*, 2004; Grimm *et al.*, 2004; Attie *et al.*, 2007). Apparently very similar cp26 plasmids have been found in all strains examined, which agrees well with its essentiality in B31 (Tilly *et al.*, 1997; Byram *et al.*, 2004; Jewett *et al.*, 2007a). The cp32 plasmids, which appear to be prophages (see below), carry many genes that have no homology to genes outside *Borrelia*, and encode a number of surface exposed proteins, some of which bind the host's factor H complement regulatory protein and may serve as adhesins (Gilmore and Mbow, 1998; Zückert *et al.*, 1999; Porcella *et al.*, 2000; Stevenson *et al.*, 2006; Antonara *et al.*, 2007). The seven strain B31 cp32 plasmids are very similar throughout their lengths – often 99% identical over several kbp (Casjens *et al.*, 1997b, 2000). All isolates that have been carefully examined carry multiple versions of the cp32 plasmids (Stevenson and Miller, 2003; Iyer *et al.*, 2003; Casjens *et al.*, 2006), suggesting that, at least in the aggregate, they provide important functions.

2.2.5 Telomeres of *Borrelia* linear replicons and the complete genome sequence

The 'telomeres' of the chromosome and linear plasmids all have a structure which is quite rare in nature. At each end one DNA strand loops around and becomes the other strand with no break in the phosphodiester backbone (referred to as 'covalently closed hairpin telomeres'). At each dsDNA molecule terminus a number of approximately 21-bp consensus sequences are present (Hinnebusch and Barbour, 1991; Casjens *et al.*, 1997a; Huang *et al.*, 2004b). Hairpin telomeres represent a successful strategy for solving one of the major difficulties associated with replicating the ends of linear DNA molecules (Casjens, 1999). The hairpin ends are created by a DNA resolvase (also called protelomerase) encoded by the *resT* gene on plasmid cp26 (Ravin *et al.*, 2000; Chaconas *et al.*, 2001). This enzyme creates two hairpin ends from the novel sequence joint created when a replication fork passes across a hairpin (Huang *et al.*, 2004a; Deneke *et al.*, 2004; Kobryn *et al.*, 2005; Aihara *et al.*, 2007) (see Chapter 3 for details).

The hairpin telomeres present serious problems in the determination of a rigorously complete *Borrelia* genome sequence. The hairpin ends of the terminal DNA fragments of the linear replicons cannot be ligated to vector plasmids, and so these fragments are not present in a library. Thus, determination of these terminal sequences requires special attention; either (i) these fragments must be isolated and subjected to Maxam–Gilbert type chemical sequencing [as was nobly done for telomeres of B31 lp17 by Hinnebusch *et al.* (1990)], or (ii) the loop at the end must be nicked with a single-strand specific nuclease for ligation to a vector (Zhang *et al.*, 1997) or subjected to ligation-mediated PCR amplification (Casjens *et al.*, 1997a) to obtain sufficient material for sequence determination (PCR reactions around the ends have not been successful, presumably due to the competition between snapback annealing of the two template strands and primer annealing). The former method is ideal, although it is *very* technically demanding, while the latter strategy has the problem of possible terminal nucleotide removal by the enzymes used to nick the loops. In practice, both of these are too labour-intensive to easily allow determination of all terminal sequences in a *Borrelia* genome (strain B31 has 26 such termini). Thus, some nucleotides will be missing from the sequencing of *Borrelia* DNA libraries (the number of missing bp is dependent upon the properties of the DNA library used and the depth of its sequencing). In the B31 genome, the above methods were used to determine the sequences out to the ends of nine telomeres (Hinnebusch *et al.*, 1990; Fraser *et al.*, 1997; Zhang *et al.*, 1997; Huang *et al.*, 2004b; J. Aron, W.M. Huang, and S. Casjens, unpublished). In these cases between 24 and 106 bp were missing from the bulk genome sequence, with the single exception of the left end of B31 plasmid lp28-1, which contains an unclonable sequence about 1300 bp from the telomere. The latter 1300 bp sequence was determined separately by Zhang *et al.* (1997). Restriction mapping using Southern hybridization of electrophoresis gels has shown that none of the other linear plasmids has an unsequenced terminus longer than a few hundred bp (Casjens *et al.*, 2000) (N. Palmer and S. Casjens, unpublished). The current strain B31 whole genome sequence (chromosome and plasmids, without the three 'lost' plasmids above) contains 1,521,318 sequenced bp. If the seventeen unsequenced B31 plasmid termini have on average 100 missing bp, this probable overestimate suggests that there are less than 1700 bp missing from the published 'complete' B31 genome sequence, and that none of the missing terminal sequences is likely to be long enough to contain a protein coding gene.

2.3 *Borrelia burgdorferi* intraspecies genomic variation and virulence

Considerable effort by many laboratories over a number of years has shown that the *Borrelia* species are quite variable. Although it is not the task of this chapter to detail all of these studies, we will note by way of example that in the case of *B. burgdorferi*, the best-studied species, ribosomal RNA and spacer sequences (Liveris *et al.*, 1999; Wormser *et al.*, 1999; Iyer *et al.*, 2001; Bunikis *et al.*, 2004; Attie *et al.*, 2007), macro-restriction pattern variation (Casjens *et al.*, 1995; Mathiesen *et al.*, 1997), multilocus sequence analyses (Qiu *et al.*, 2004; Margos *et al.*, 2008) and a number of other techniques have all documented chromosomal nucleotide sequence variation. Comparison of draft sequences of three *B. burgdorferi* chromosomes shows them to be about 99.4% identical (Qiu *et al.*, 2004; Casjens *et al.*, 2006). These draft sequences also indicate that the three chromosomes are likely to have identical gene orders [with the exception of their right termini, which are in some isolates fused to 7–20 kbp of plasmid-like sequences (Casjens *et al.*, 1997a, 2000; Huang *et al.*, 2004b)]; indeed, we are not aware of any comparative regional sequencing that has demonstrated chromosomal non-synteny. Furthermore, no prophages or insertion sequences have been observed in *B. burgdorferi* chromosomes (Fraser *et al.*, 1997). Microarray-based comparative genome hybridization (CGH) analysis of 15 *B. burgdorferi* isolates (13 human patient and 2 tick) revealed 83–98% whole genome ORF conservation relative to B31 (Terekhova *et al.*, 2006). In this study, sequence divergence was judged by a lack of hybridization to a B31 sequence on the array rather than by determination of DNA sequence, so failure to hybridize could be due to either absence or divergence of the gene. ORF conservation for the chromosome was 93–99% and up to half of this variation was due to the shortening of the chromosome at the right telomere, as already noted. The remaining variable ORFs are distributed along the length of the chromosome (Terekhova *et al.*, 2006). Thus, the chromosomes of *B. burgdorferi* appear to be quite invariant relative to bacterial species, such as *E. coli*, in which differences in gene content and order are common between individual isolates (*e.g.* Perna *et al.*, 2002).

Plasmid content differences have also been noted in numerous studies (Barbour, 1988; Casjens *et al.*, 1995; Palmer *et al.*, 2000; Iyer *et al.*, 2003; and references therein). However, the finding that plasmids lp54, cp26 and cp32 are ubiquitously present in all *B. burgdorferi* isolates analysed has led to the suggestion that these genome elements are, in effect, mini-chromosomes (Barbour, 1993; Casjens *et al.*, 1997b; Tilly *et al.*, 1997; Palmer *et al.*, 2000; Iyer *et al.*, 2003). Of these plasmids, only cp26 has been shown to carry genes required for viability (Byram *et al.*, 2004; Jewett *et al.*, 2007b). All the other B31-like plasmids are either absent from certain isolates, can be lost in culture without loss of viability (Sadziene *et al.*, 1992; Barbour, 1993), or, when present (based on presence of the plasmid-specific partition locus), appear to be substantially different in sequence or gene arrangement from B31 (Iyer *et al.*, 2003; Terekhova *et al.*, 2006). Notably, lp38 and lp56 were not detected in multiple isolates (Norris *et al.*, 1992; Palmer *et al.*, 2000; Iyer *et al.*, 2003). The same isolates also had truncated lp28-1s and the same chromosome macrorestriction pattern, suggesting that a particular plasmid content may be an intrinsic property of particular chromosomal lineages of *B. burgdorferi* (Iyer *et al.*, 2003). Plasmids can be lost on culture passage (Schwan *et al.*, 1988; Xu *et al.*, 1996), so plasmid absence in a strain cultured in the laboratory cannot be definitively considered inherent to a particular isolate. Some of these studies have been performed with low passage isolates (<5), somewhat mitigating this concern.

In addition to presence or absence, a large body of anecdotal comparative sequence evidence (not listed here) indicates that at least some of the plasmids are more variable in sequence than are the chromosomes. For example, genes encoding the immunologically important outer surface protein OspC have been analysed in many isolates and found to be particularly variable (Bunikis *et al.*, 2004; Casjens *et al.*, 2006; Wang *et al.*, 1999; Attie *et al.*, 2007; and references therein). In the only case where complete orthologous plasmid sequences have been reported from two different *B. burgdorferi* isolates, the two cp9 plasmids are about 94% identical overall, and individual cp9 genes vary from 74% to 99% identical in nucleotide sequence (Casjens *et al.*, 2000; Stewart *et al.*, 2001). Comparative genome hybridization (CGH) of *B. burgdorferi* isolates revealed that, in general, plasmid ORFs were 74–98% conserved (Fig. 2.4 and Terekhova *et al.*, 2006); lp54, cp26 and cp32s were the most conserved plasmids by this analysis. Lp54 was 96–100% conserved and, although cp26 was somewhat more divergent (83–100% conservation), this variation was entirely due to variation in the BBB19–22 gene cluster and BBB29. BBB19 encodes OspC and, as noted, is one of the more divergent genes in the *B. burgdorferi* genome (Wang *et al.*, 1999; Qiu *et al.*, 2004; Attie *et al.*, 2007). Virtually all cp32 ORFs are conserved, but the presence of highly paralogous sequences on all these plasmids complicates the analysis by CGH. Linear plasmids represent the most diverse parts of the genome and can be classified into four categories. As noted, lp54 is highly conserved and uniformly present in all isolates (Samuels *et al.*, 1993; Casjens *et al.*, 1995; Terekhova *et al.*, 2006). Plasmids lp38 and lp28-4 were absent from certain isolates, but when present all ORFs were conserved (Fig. 2.4). CGH also revealed that Lp5 and lp21 were absent from almost all isolates studied (Terekhova *et al.*, 2006). Partition loci for the remaining B31-like linear plasmids were detected in almost all isolates, but there is a substantial amount of ORF divergence. The data imply that all *B. burgdorferi* sensu stricto isolates contain most B31-like linear plasmids (based on presence of the corresponding partition locus), but the structure of these plasmids is often different from those of B31. The alterations include deletions and rearrangements that probably result in unique plasmids in many of the isolates.

A relationship between the plasmid content of *B. burgdorferi* and virulence has been well-established (see Chapter 12). In *B. burgdorferi* strain B31, linear plasmids lp25 (where the pncA/BBE22 gene can compensate for lp25 loss) and lp28-1 are necessary for high infectivity in mice (Labandeira-Rey and Skare, 2001; Purser *et al.*, 2003), while lp25 and lp28-4 are important for maintenance in ticks (Grimm *et al.*, 2005; Strother *et al.*, 2005). As noted above, cp26 is found in all isolates and encodes OspC, which is required for establishment of infection in mice (Pal *et al.*, 2004; Grim *et al.* 2004; Stewart *et al.*, 2006; Tilly *et al.*, 2006). Recently, lack of lp36 was shown to correlate with reduced infectivity in mice, and the defect could be partially complemented by the lp36-encoded adenine deaminase (Jewett *et al.*, 2007a). However, *B. burgdorferi* isolates that contain all these plasmids may still be attenuated for virulence in mice (Wang *et al.*, 2002; Iyer *et al.*, 2003) suggesting that additional virulence factors remain to be identified.

Borrelia burgdorferi sensu stricto isolates can be divided into three distinct genotypes (RST1, RST2 and RST3) on the basis of restriction fragment length polymorphism in the 16S-23S rRNA spacer (Liveris *et al.*, 1996; Liveris *et al.*, 1999). RST1 and RST2 isolates are capable of haematogenous dissemination in mice and humans, but certain RST3 isolates are attenuated (Wormser *et al.*, 1999; Wang *et al.*, 2001, 2002). *Borrelia burgdorferi* can be further divided into at least 16 different OspC types, and a correlation between OspC type and pathogenicity has been described (Seinost *et al.*, 1999; Baranton *et al.*, 2001; Lagal *et al.*, 2006). RST and OspC genotypes are in strong linkage disequilibrium (Qiu *et al.*, 1997; Schwartz *et al.*, 2006; Attie *et al.*, 2007). CGH confirms the monophyletic nature of RST1 isolates (all RST1 isolates are either OspC type A or B); 98% of all ORFs were conserved relative to B31 (also an RST1 genotype). RST2 and RST3 isolates had 88% and 83% ORF conservation, respectively (Terekhova *et al.*, 2006). All analysed RST1 isolates harboured all B31-like plasmids, except for a single cp32 in some cases (Iyer *et al.*, 2003; Terekhova *et al.*, 2006). Virtually all RST2 (all are OspC types F, H, K or N) isolates lacked

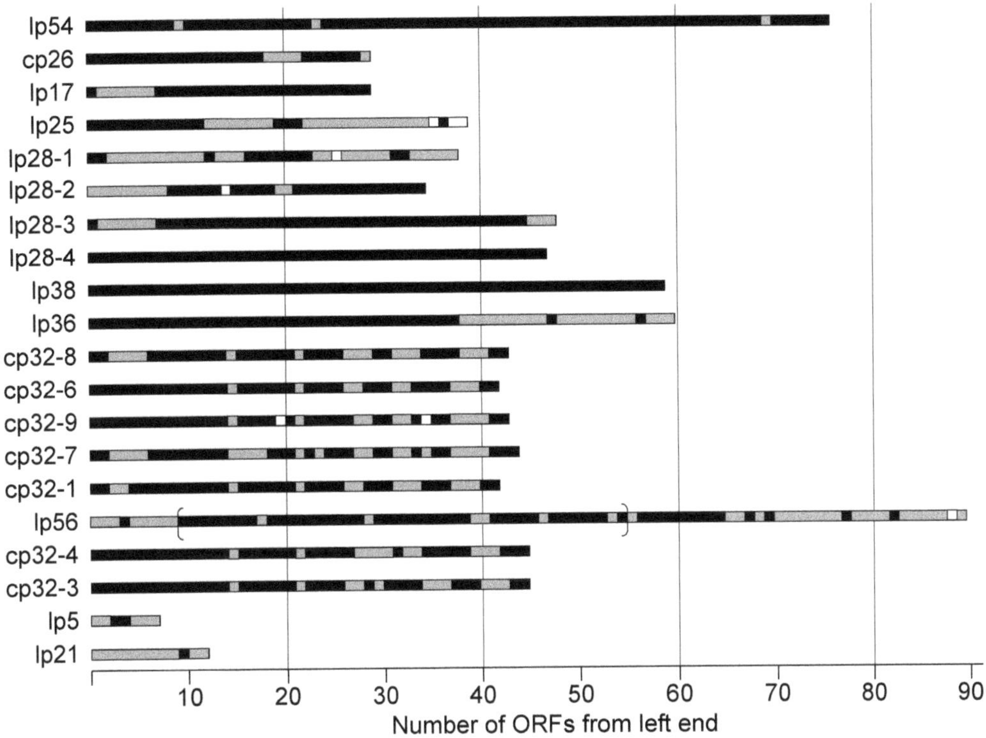

Figure 2.4 Conservation of plasmid ORFs in *B. burgdorferi*. ORF presence or absence was determined by CGH. Black regions represent ORFs present in all of the 17 strains tested, grey regions indicate ORFs not present in all isolates, and white indicates that interpretable data for that region was not available. These data show only whether or not a strain carries a very similar B31 homologue (presumably the orthologue), and do not indicate that the plasmids harbouring these orthologous genes are syntenic in the strains analysed. All plasmids are represented in linear form starting with the gene numbered 01, and plasmid lengths are scaled according to the number of ORFs and pseudogenes in each B31 plasmid (not length in bp) as defined in Casjens *et al.* (2000), which are displayed along each plasmid in the order they are present in the strain B31 plasmids. The bracketed region of lp56 denotes the integrated cp32 in this plasmid. Note that the ORFs carried on plasmids lp38 and lp28-4 were present in all isolates that carried the plasmid; therefore, their ORFs are represented as black even though some isolates (5 and 4 strains, respectively) completely lacked the plasmid. The figure is modified from Figure 2 of Terekhova *et al.* (2006).

plasmids lp38 and lp56 and had truncated lp28-1 (Iyer *et al.*, 2003). Since RST2 isolates are virulent, lp38 and lp56 are probably dispensable for mammalian infection. By contrast, RST3 isolates are much more genomically heterogeneous, with respect to OspC type, ORF conservation and plasmid content (Iyer *et al.*, 2003; Schwartz *et al.*, 2006; Terekhova *et al.*, 2006).

2.4 *Borrelia burgdorferi* interspecies genomic relationships

2.4.1 Chromosomes

Since the publication of the *B. burgdorferi* strain B31 genome sequence, the complete nucleotide sequences of the chromosomes of *B. garinii* strain PBi and *B. afzelii* strain PKo have been published along with the complete sequences of some of their circular and linear plasmids (Glöckner *et al.*, 2004, 2006). In addition, T. Schwan and coworkers have deposited chromosome sequences for relapsing fever isolates *B. hermsii* DAH and *B. turicatae* 91E135 with GenBank (Accession Nos. CP000049 and CP000048, respectively) and Lescot *et al.* (2008) have reported the sequences of *B. duttonii* Ly and *B. recurrentis* A1 and some of their plasmids. These sequences, along with the macro-restriction chromosomal maps that have been constructed for a number of other *B. garinii*, *B. afzelii*, *B. japonica*, *B. andersonii*, *B. valaisiana*, *B. lusitaniae*, and *B. bissettii* strains, strongly suggest that the chromosomes of all the *Borrelia* spp. have highly similar gene orders (Ojaimi *et al.*, 1994; Casjens *et al.*, 1995; Zhong and Barbour, 2004; Bacon *et al.*, 2004; Pettersson *et al.*, 2007; Lescot *et al.*, 2008; R. Van Vugt, and S. Casjens, unpublished).

The analysis of the sequence of the *B. garinii* PBi and *B. afzelii* PKo chromosomes by Glöckner *et al.* (2004, 2006) showed that the two chromosomes are essentially completely co-linear with the *B. burgdorferi* chromosome except for the regions very near the telomeres, and that the common regions of PBi and PKo are about 93% identical in nucleotide sequence. They also indicated the following relationships between the PBi and *burgdorferi* B31 chromosomes. Other than the terminal differences, there are only eight insertions and six deletions in the PBi chromosome relative to the *burgdorferi* B31 chromosome; these vary in size from 138 to 1878 bp in length. Of the 843 genes annotated in the B31 'constant' non-telomeric part of the genomes, the following differences were noted between the B31 and PBi (ignoring a few differences in 'gene calling' by the computer programs used to predict gene locations): (i) PBi has one additional gene that is the apparent result of a duplication of a *bmpA* gene (Aron *et al.*, 1996; Gorbacheva *et al.*, 2000), (ii) one B31 gene is split in PBi and six PBi genes are split in B31 (at least some of these latter differences could be due to frameshifting sequence errors), (iii) seven other genes have substantial length differences in the two strains, due in part to differences in open reading frame (ORF) predictions, and (iv) among the orthologous proteins encoded by the PBi and B31 chromosomes, only 11 have no amino acid differences. Analysis of the *B. afzelii* PKo chromosomes shows a similar picture, with a small number of differences due to insertions and deletions from B31. Although seven locations showed differences from B31 at the same positions in both PKo and PBi (including a *triplication* in the *bmpA* gene of PKo), no ancestral state could be inferred. We deduce from the data of Glöckner *et al.* (2006) that the orthologous sequences of the chromosomes of PKo and PBi are about 93% identical while those of B31 and PKo are 91% identical, indicating that the three species, *B. burgdorferi*, *B. garinii* and *B. afzelii*, are approximately equidistantly related.

The microarray data of Zhong and Barbour (2004) using *B. burgdorferi* B31 sequences as probes found that at least 81% of the B31 chromosomal genes are present in the relapsing fever agent *B. hermsii* strain HS1 (serotype 7). However, since homologous genes might have diverged beyond the threshold for hybridization under the conditions utilized, this value must be taken as a minimum estimate of the number of homologous genes. These authors noted fifteen chromosomal genes that had been sequenced in both *B. burgdorferi* and *B. hermsii*, and that their similarities ranged from 64% to 85% identity (the average is about 74%). Similar comparisons of the homologues of B31 chromosomal genes BB0417, BB0463 and BB0777 (Pettersson *et al.*, 2007) showed that these genes are 89–92% identical between *B. turicatae* and *B. hermsii* and these relapsing fever species genes are 68–77% identical to their *B. burgdorferi* homologues. Analysis of *B. hermsii* strain DAH and *B. turicatae* strain 91E135 chromosomal draft sequences showed that these two strains contain three purine metabolism genes that *B. burgdorferi* lacks and that they lie between genes BB421 and BB422 in the B31 strain (Guyard *et al.*, 2005; Pettersson *et al.*, 2007). In addition, the relapsing fever *Borrelia* chromosomes have two genes that encode enzymes for glycerol-3-phosphate acquisition that are not present in the Lyme disease spirochaetes; these two genes replace *B. burgdorferi*'s gene BB242, whose function is not known (Shang *et al.*, 1997; Schwan *et al.*, 2003; Bacon *et al.*, 2004). Lescot *et al.* (2008) found only 13 genes unique to the Lyme agents and seventeen unique to the relapsing fever agents when their *B. duttonii* and *B. recurrentis* chromosome sequences were compared to the *B. burgdorferi* B31 chromosome sequence. Thus, the relapsing fever and Lyme disease agent chromosomes are quite similar overall. The Lyme agent and relapsing fever groups form two robustly separable groups where intra-group chromosomal sequence differences are less than about 10% (in the Lyme agent group). The Lyme agent and relapsing fever groups have chromosomes that are about 23–32% different from each other (this latter is our value calculated from arbitrarily chosen sections of the sequences mentioned above). A complete and accurate picture awaits the full comparative analyses of the relapsing fever and Lyme agent genomes.

2.4.2 Plasmids

The extensive available data (not listed here) indicate that all the different Lyme agent species have numerous linear and circular plasmids in the same size range as *B. burgdorferi* strain B31. As in *B. burgdorferi*, these vary somewhat between isolates, but in all cases examined there appear to be syntenic homologues of B31 plasmids cp26, the cp32s and lp54. Glöckner *et al.* (2006) recently analysed the sequence of the *B. garinii* PBi and *B. afzelii* PKo genomes showing that they carry eight and six linear plasmids, respectively, as well as at least three and nine circular plasmids, respectively. The circular plasmids are largely homologous to their *B. burgdorferi* relatives; for example, the PBi cp26 has a similar gene order to its B31 homologue (however the B31 ORFs BBB15, BBB21 and BBB22 are not present, suggesting that these very short ORFs may have been falsely predicted in the strain B31 cp26). The *B. garinii* and *B. afzelii* cp32s are also very similar to their B31 counterparts. Among the linear plasmids, both PBi and PKo also carry plasmids that are very similar to *B. burgdorferi* lp54. The *B. afzelii* and *B. garinii* lp54 plasmids have been reported to be systematically several kbp larger than their *B. burgdorferi* counterparts (Samuels *et al.*, 1993; Casjens *et al.*, 1995), and the PBi and PKo sequences indicate additional sequences at their left ends. However, as Fig. 2.5 shows, the vast majority of lp54 plasmids are very similar to that of *B. burgdorferi*, except they have different numbers of tandem Pfam54 genes near their right ends (indicated by the set of parallel lines; B31 has eight such genes in the BBA64-BBA73 gene region).

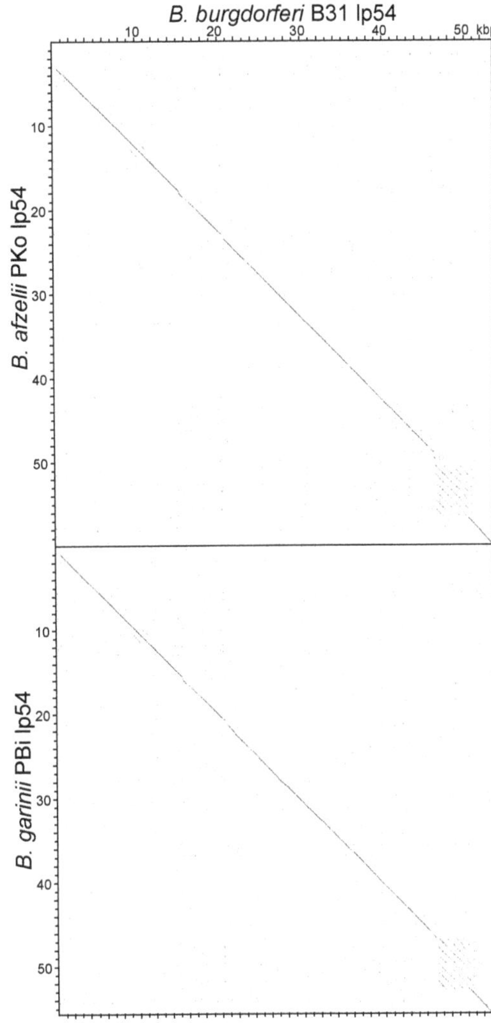

Figure 2.5 Matrix comparison of the lp54 linear plasmids from *B. burgdorferi* strain B31, *B. afzelii* strain PKo and *B. garinii* strain PBi. The B31, PKo and PBi plasmid sequences are from Fraser *et al.* (1997) and Glöckner *et al.* (2004, 2006). The figure was generated with a scanning window of 19 matches out of 23 bp in the computer program DNA STRIDER (Douglas, 1994).

The remaining linear plasmids in PBi and PKo are not complete orthologues of plasmids in B31. However, Dunn *et al.*'s (1994) first complete sequence of a *Borrelia* plasmid, that of cp8.3 carried by *B. afzelii* strain Ip21, turned out to be very similar to *B. burgdorferi* cp9. Fig. 2.6 shows the relationship between these two plasmids. Their predicted orthologous genes vary from 73% to 88% identical, and the *B. burgdorferi* cp9 plasmid has an 1100 bp region of novel sequence (that replaces 600 bp) and a 700 bp insertion relative to the *B. garinii* plasmid. Glöckner *et al.* (2006) noted that there are fragmentary, but in some cases rather extensive, homologous and colinear sequences among the other PBi, PKo and B31 linear plasmids, and that the plasmids of PBi and PKo are overall significantly more similar to each other than to the B31 plasmids. An example of this type of 'mosaic' relationship between related but not non-completely colinear *B. burgdorferi* and *B. afzelii* linear plasmids is shown in Fig. 2.7.

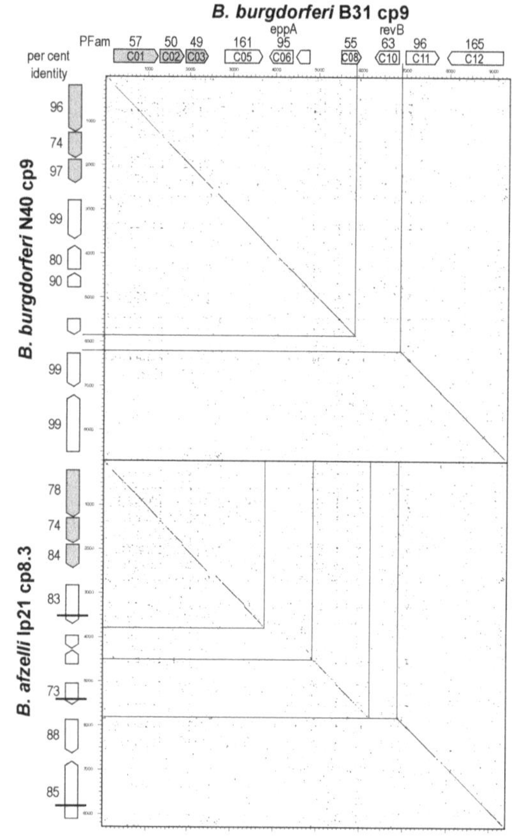

Figure 2.6 Matrix comparison of the small cp9 circular plasmids from *B. burgdorferi* strains B31 and N40 and *B. afzelii* strain lp21. The B31, N40 and lp21 plasmid sequences are from Casjens *et al.* (2000), Stewart *et al.* (2001), and Dunn *et al.* (1994), respectively. The figure was generated with a scanning window of 17 matches out of 23 bp in the computer program DNA STRIDER (Douglas, 1994). The circular sequences were linearized by opening at an arbitrary point; ORFs are indicated as block arrows, pointing in the direction of transcription; and black lines across the genes mark breaks in the reading frames in the lp21 sequence relative to the B31 sequence. Above the B31 ORFs, the paralogous family (PFam) to which the genes belong is indicated (Casjens *et al.*, 2000). Light grey genes are involved in plasmid replication and/or partitioning (see text). Genes *bbc06* and *bbc10* have also been named *eppA* and *revB*, respectively (Champion *et al.*, 1994; Skare *et al.*, 1999; Miller and Stevenson, 2003). On the left, per cent nucleotide sequence identity values are shown for each N40 and lp21 ORF relative to its B31 orthologue.

PKo linear plasmid lp28 contains two separate fairly large blocks of similarity to B31 lp17, one of which is inverted relative to the other. The non-B31 lp17-like parts of PKo lp28 contain mostly homologues of genes on other B31 plasmids, but they are not embedded in larger regions that are obviously orthologous or syntenic to any of those plasmids. For example, a homologue of the BBK32 (fibronectin binding protein; Probert and Johnson, 1998) gene, located on lp36 in B31, is present on PKo lp28. Similar to the B31 linear plasmids, there are substantial regions in PKo lp28 with no apparent ORFs. A detailed comparative analysis of the gene contents of the *B. garinii* and *B. afzelii* plasmids has not yet been reported, but at the current level of understanding, it appears that (1) the plasmids of the different species encode numerous related genes although they are not completely syntenic (and so not obviously orthologous) in the different species, (2) pseudogenes and non-protein-coding regions are over-represented on linear plasmids in each of the species that have been examined, and (3) a few of the plasmids, notably cp26, the cp32s, lp54 and cp9 (when it is present), have nucleotide sequences and gene organizations that appear to be nearly as highly conserved across species lines as the chromosomes.

The relapsing fever *Borrelia* also harbour numerous linear and circular plasmids in the same size ranges as those of the Lyme agents, and at least some isolates have one or two much larger linear plasmids in the 100–220 kbp size range (Kitten and Barbour, 1992; Marconi *et al.*, 1993; Ferdows *et al.*, 1996; Takahashi *et al.*, 2000; Hovis *et al.*, 2004; Schwan *et al.*, 2005; Lescot *et al.* 2008). Stevenson *et al.* (2000a) have shown that *B. hermsii* carries multiple members of the cp32 family and have constructed the sequence of a 'composite' member of this family. It is largely, but not completely, syntenic with the *B. burgdorferi* cp32s. Barbour and coworkers have found that *B. hermsii* and *B. turicatae* have linear plasmids in the 20–40 kbp range that carry putative partitioning protein genes that are similar to those on the Lyme agent plasmids as well as genes that encode major surface antigens (Pennington *et al.*, 1999; Barbour *et al.*, 2000, 2006; Dai *et al.*, 2006). The 180 kbp linear plasmid of *B. hermsii* HS1 carries nucleotide metabolism genes that are not present in the Lyme agent *Borrelia* genomes (Zhong *et al.*, 2006). Interestingly, Ferdows *et al.* (1996) serendipitously isolated a subculture of strain HS1 in which this large plasmid is circular,

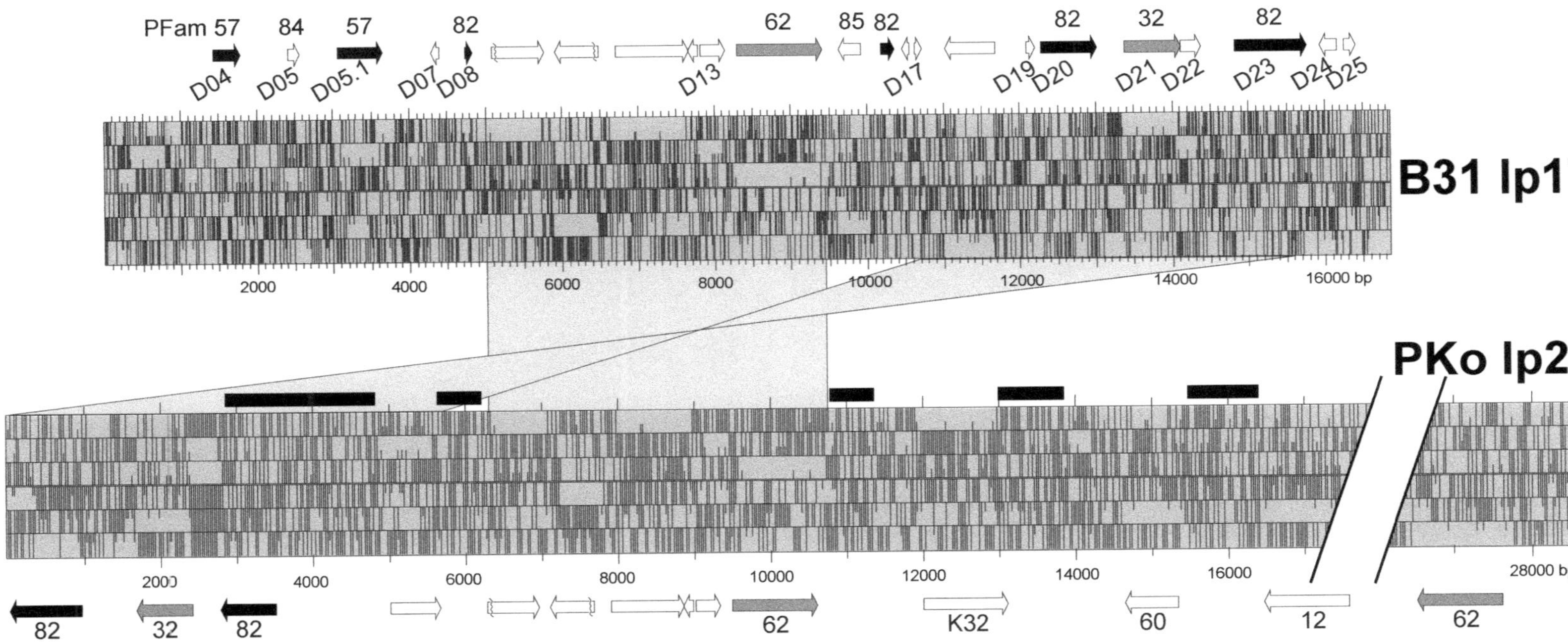

Figure 2.7 Comparison of *B. burgdorferi* strain B31 lp17 and *B. afzelii* PKo lp28. Open reading frame maps are shown for B31 lp17 (Barbour *et al.*, 1996; Fraser *et al.*, 1997) and part of B. afzelii PKo lp28 (Glöckner *et al.*, 2006) in which all six reading frames are shown with full vertical lines within each frame marking nonsense codons and partial vertical lines denote potential ATG start codons (DNA STRIDER; Douglas, 1994). The top three frames are translated rightward and the bottom three leftward in each map. Between the maps, shaded areas mark regions of orthology. At the top of the figure, the predicted genes in B31 lp17 are indicated (Barbour *et al.*, 1996; Fraser *et al.*, 1997); grey arrows indicate putative partitioning genes, and black arrows are truncated or frameshifted pseudogenes. Above the lp17 map, the paralogous families to which the encoded proteins belong are indicated (PFams as in Casjens *et al.*, 2000). Below the PKo lp28 diagram, larger ORFs and pseudogenes are shown by arrows, and the B31 PFam to which they are homologous is given, except for 'K32', which is a homologue of the unique B31 K32 (fibronectin binding) protein located on lp36 in B31. Black bars above the PKo lp28 map point out regions with no significant ORFs in this plasmid.

and bacteria with the linear and circular versions of the plasmid showed no observed differences in gene expression or growth rate. Lescot *et al.* (2008) reported the sequences of seventeen *B. duttonii* and eight *B. recurrentis* plasmid sequence contigs. From this work it is clear that although these carry genes that are similar to *B. burgdorferi* plasmid genes, the organization of genes on the relapsing fever plasmids is often quite different from *B. burgdorferi*. One possible exception to this is organizational similarity between *B. burgdorferi* cp26 and a linear plasmid designated lp23 in *B. duttonii* and *B. recurrentis*, however no detailed comparison of these plasmids has been reported.

2.5 Plasmid–bacteriophage relationships

2.5.1 The ϕBB-1 bacteriophage

A number of studies in which supernatants of *B. burgdorferi* cultures were examined by electron microscopy have reported the presence of tailed bacteriophage-like particles (Hayes *et al.*, 1983; Barbour and Hayes, 1986; Neubert *et al.*, 1993; Schaller and Neubert, 1994; Eggers and Samuels, 1999; Eggers *et al.*, 2000). Sequence analysis also recognized that several *Borrelia* plasmids could be prophages (see below) (Casjens *et al.*, 1995; 1997b). These observations have been borne out by more directed studies in strain CA-11.2A, where small numbers of phage-like particles are spontaneously released, and sublethal concentrations of 1-methyl-3-nitroso-nitroguanidine (MNNG) increase this release significantly; these particles were purified and shown to contain cp32 DNA molecules (Eggers and Samuels, 1999). This phage has been designated φBB-1; it has an apparently contractile tail and is shown in Fig. 2.8 (Eggers and Samuels, 1999; Eggers *et al.*, 2000, 2001a). The presence of cp32 DNA in the particles strongly suggests that these plasmids are prophage DNAs. The demonstration that φBB-1 is capable of transducing a cp32 both between cells of the same isolate and between different *B. burgdorferi* isolates demonstrates that these particles are indeed capable of delivering their DNA into a naive cell (Eggers *et al.*, 2001b). This suggests the exciting possibility that φBB-1 plays a role in enhancing the genetic diversity observed in the complement of cp32 plasmids found in different *B. burgdorferi* isolates (Casjens *et al.*, 1997b; Caimano *et al.*, 2000; Iyer *et al.*, 2003; Stevenson and Miller, 2003). On the other hand, φBB-1 has not yet been propagated by lytic growth in the laboratory, and the reason(s) for this failure – which could theoretically range from a high propensity to form a lysogen, to immunity of strains harbouring other cp32s, to possible genetic defects in DNA replication – are not yet understood. Cp32-like plasmids have been found in most other *Borrelia* species that have been analysed including *B. afzelii, B. garinii, B. bissettii, B. japonica, B. andersonii* and *B. hermsii* (Casjens *et al.*, 1997b; Stevenson *et al.*, 2000a, 2000b; Glöckner *et al.*, 2004, 2006), and cp32-containing bacteriophage have been observed in cultures of *B. bissettii* clone DN127 (Eggers *et al.*, 2000).

The cp32s share organizational properties with the genomes of other tailed bacteriophages. Comparative genomic studies of double-stranded DNA bacteriophage groups, such as the λ-like family of phages, have shown that their genomes are genetic mosaics in which each has an overall conserved gene order, with occasional non-homologous genes inserted into the genomes of

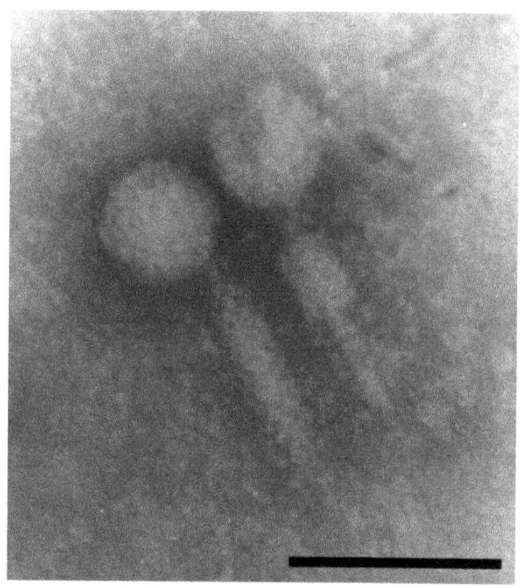

Figure 2.8 *B. burgdorferi* bacteriophage ϕBB-1 virions. Phage particles from the supernatant of an MNNG-treated culture of *B. burgdorferi* strain CA-11.2A were negatively stained for transmission electron microscopy with phosphotungstic acid; bar is 45 nm. Reproduced from Eggers *et al.* (2001b) with permission of the publisher.

different phages (Casjens *et al.*, 1992; Hendrix, 2002; Casjens, 2005; Hendrix and Casjens, 2006); this pattern is similar to that observed among the cp32s (Casjens *et al.*, 1997b, 2000; Caimano *et al.*, 2000; Stevenson *et al.*, 2000b; Stevenson and Miller, 2003). The cp32s have long stretches of nearly identical DNA, broken by three regions of significant variability; these regions correspond to the partitioning/compatibility regions that allow several cp32 isoforms to be stably maintained in a single isolate (see above), and two regions that encode multiple families of differentially expressed OspE/OspF/Elp and Mlp lipoproteins (Zückert and Meyer, 1996; Porcella *et al.*, 1996; Stevenson *et al.*, 1996; Marconi *et al.*, 1996; Casjens *et al.*, 1997b, 2000; Akins *et al.*, 1999; Yang *et al.*, 1999; Caimano *et al.*, 2000; Porcella *et al.*, 2000; Yang *et al.*, 2003).

2.5.2 Potential cp32 phage genes

Two genes that encode proteins with homology to phage virion head assembly proteins are present on the cp32s. Genes L43 (and paralogues on other cp32s; L43 is on cp32–8) and L01 (and paralogues) encode a putative phage large subunit terminase and a portal protein, respectively (Casjens *et al.*, 2000; Eggers *et al.*, 2000, 2001a). Terminases form the ATP-dependent motors that translocate phage DNA into virions, and the portal proteins form the hole through which the DNA is packaged into the preformed phage heads (Catalano, 2005). These two proteins are the most highly conserved tailed-phage proteins, and no homologues have been found that have different functions (Casjens, 2003). The presence of these genes strongly supports the notion that cp32 plasmids are in fact prophages. The positions of genes L43 and L01 (and cp32 paralogues) are perfectly consistent with the typical arrangement of virion head assembly genes in other dsDNA temperate bacteriophages, where the large terminase and portal are often the second and third genes, respectively, in the phage's late operon (see below) (Casjens, 2003). In contrast to L43 and L01, a majority of the predicted proteins encoded on the cp32s have no homologues in current databases outside of *B. burgdorferi*, while the structural proteins for φBB-1 have not yet been identified. This lack of homology may not be surprising, as virion structural proteins are known to be *extremely* variable, and many phage structural proteins have no recognizable homologues, even among proteins with identical functions in other phages (Casjens *et al.*, 1992; Hendrix *et al.*, 1999, 2000; Casjens, 2003, 2005). The known dsDNA phages with genomes in the 20–60 kbp range typically have a single large operon that is expressed late in lytic infection and contains the 20–30 genes required for virion assembly and lysis. The putative terminase and portal protein genes on the cp32s are followed by 25–30 contiguous genes, which are transcribed in the same direction, have no homologues outside the cp32s, and remain uncharacterized (Casjens *et al.*, 2000; Zhang and Marconi, 2005). Based on these ideas, we previously have proposed that this region is probably the 'late operon' of φBB-1 (Eggers *et al.*, 2000, 2001a).

As part of their survival strategy, bacteriophages escape from their host cell by degrading the cell wall and lysing the cell after a cycle of lytic growth. To do this, all tailed bacteriophages encode an endolysin that degrades the cell wall and a holin whose permeabilization of the membrane allows the endolysin access to the cell wall thus controlling the timing of cell lysis. The genes that encode these proteins are sufficiently variable that they are often not recognizable in nucleotide sequence (Wang *et al.*, 2000; Young, 2002). The function of these lysis genes necessitates that their expression be tightly controlled (holins, in particular, are lethal to bacteria; Young and Blasi, 1995), and they most often are clustered at the beginning or the end of the late operon (Casjens *et al.*, 1992; Ford *et al.*, 1998). On the cp32s, *blyA* (L23 and its paralogues) lies near the distal end of the putative phage late operon; the *blyA* protein product has membrane-disruptive activity and has been shown to mediate the release of both the phage lambda encoded endolysin (Damman *et al.*, 2000) and a *E. coli* host-encoded cytolysin (Ludwig *et al.*, 2007) in surrogate systems. These results suggest that *blyA* may encode a holin, although no cp32-encoded endolysin has yet been identified.

Late operons are not expressed from the uninduced prophage DNA and are only expressed during lytic growth of the phage; thus, the above model predicts that these genes will only be expressed during 'inducing' conditions. A second, slightly weaker, prediction of the model is that these genes will in fact be co-transcribed as an operon. Zhang and Marconi (2005) have used RT-PCR across the gene boundaries to demonstrate that the 30 genes from BBL42 to BBL28 (using cp32–8 as a typical cp32) were in fact co-transcribed after *B. burgdorferi* cells were exposed to the φBB-1-inducing agent MNNG, and that these genes are apparently not expressed in uninduced cells (Zhang and Marconi, 2005). Other DNA array transcription studies largely agree that most of these genes are expressed very little if at all without induction (Ojaimi *et al.*, 2003; Anderton *et al.*, 2004; Tokarz *et al.*, 2004). These expression data are completely consistent with the proposal that this block of genes comprise a late phage operon. Curiously, the *blyA* gene's expression pattern is an exception to this rule and not what is expected for a lethal holin. An increase of the *blyA* transcript, in the absence of a concomitant increase in the expression of other genes within the late phage operon, has been observed with an increase in temperature (Ojaimi *et al.*, 2003), after treatment with a bactericidal antibody (Anderton *et al.*, 2004) and in the mammalian host (Zhang and Marconi, 2005). Thus, the *blyA* (and the adjacent *blyB*) gene appears to be under different regulatory control from the rest of the putative cp32 late operon genes. Consequently, a final determination as to whether BlyA is in fact a holin awaits further experimentation. Similarly, the cp32 *bdr* genes (R27 in cp32-4 and paralogues – the most promoter distal member of this gene cluster) appear to be expressed under non-inducing conditions and so may not be part of a late operon (Zückert *et al.*, 1999; Roberts *et al.*, 2000).

There also are three cp32 genes outside of the putative late operon that have homology to phage-related genes: those that encode PFam161, PFam165 and PFam96 genes encode proteins that are weakly similar to phage encoded Erf DNA-binding proteins, exonuclease and recombinases, respectively. The products of these genes could conceivably play a role in the circularization of the linear φBB-1 genome into cp32 or in phage-mediated recombination. Tokarz *et al.* (2004) observed that members of all three of these PFams are upregulated in response to blood, a stimulus that *B. burgdorferi* encounters during the blood meal within the tick vector or during infection of the mammalian host. Regulation of the 'early' genes of even the better studied plasmid prophages remains rather poorly understood (Ravin *et al.*, 2000), so understanding these findings in a larger context is difficult. Additionally, as described above, each cp32 encodes a set of genes that are involved in their replication and stable maintenance. One of these genes (L32 and its PFam32 paralogues) has significant homology to the partitioning gene, *parA* (Zückert and Meyer, 1996; Casjens *et al.*, 1997b, 2000; Stevenson *et al.*, 2000a). Homologues of the *parA* gene are found on a number of well-studied low-copy- number plasmids, including two *E. coli* prophages, P1 and N15, which exist as autonomously replicating plasmids (Austin and Abeles, 1985; Grigoriev and Lobocka, 2001; Surtees and Funnell, 2003).

2.5.3 Lysogenic conversion

Most prophages alter their hosts, often by modifying the bacterial surface, through expression of their 'host conversion' genes (Cheetham *et al.*, 1995; Casjens and Hendrix, 2002; Canchaya *et al.*, 2003; Casjens, 2003; Hendrix and Casjens, 2006), and the cp32s appear to be no exception. The cp32s express several families of proteins that have been well-studied. These are the Mlp, Bdr, Rev, BapA and OspE/OspF/Elp proteins (Gilmore and Mbow, 1998; Zückert *et al.*, 1999; Yang *et al.*, 1999; Skare *et al.*, 1999; Porcella *et al.*, 2000; Bauer *et al.*, 2001; Miller and Stevenson, 2003; Zhang *et al.*, 2005; Stevenson *et al.*, 2006). All of these are membrane lipoproteins and most have been shown to be on the cell surface, but their functional roles are in general not known. An exception to this is OspE and related proteins that are able to bind complement factor H and so appear to play a role in the ability of *B. burgdorferi* to evade the host complement system (Hellwage *et al.*, 2001; Alitalo *et al.*, 2002; Stevenson *et al.*, 2002; Metts *et al.*, 2003; McDowell *et al.*, 2004; Miller and Stevenson, 2004; Hovis *et al.*, 2006). Recently, Antonara *et al.* (2007) have proposed that OspF and related proteins may act as adhesins during *B. burgdorferi* infection. Thus, these genes, as far as we understand them, appear to be involved in the interactions between the pathogenic bacterium and its host, and so can be considered to be lysogenic conversion genes.

2.5.4 A multiplicity of prophages

φBB-1 represents an unusual group of bacteriophages, in that there can be as many as nine quite similar prophages resident within a single cell. The entire complement of cp32s within a given isolate is represented in the extracellular supernatant after phage induction and are all presumably packaged individually within φBB-1 capsids (Eggers *et al.*, 2001b). Additionally, given the remarkable conservation of all potential phage genes identified thus far, the putative structural and functional genes of φBB-1 probably are the same; thus, we anticipate that all cp32s would be packaged within virtually identical capsids. Two B31 plasmids, cp32-4 and cp32-9, have frameshift mutations in one and seven putative late phage operon genes, respectively (Casjens *et al.*, 2000; Eggers *et al.*, 2000), so these two cp32s may no longer be productive prophage genomes by themselves. We have observed, however, that either or both cp32-4 or cp32-9 (the two plasmids were indistinguishable by the methods used) are packaged within φBB-1 capsids released from strain B31 (C. H. Eggers and D. S. Samuels, unpublished), suggesting that these proteins can be cross-utilized among the cp32s (prophages) within the cell.

Several other *B. burgdorferi* plasmids (cp18, lp54, lp56 and lp28–2) contain DNA that is homologous to the cp32 putative virion structural protein genes and so may be phage-related. Cp18 plasmids present in strains 297 and N40 appear to be simple deletions of cp32s that are missing genes from their putative late operons (Stevenson *et al.*, 1997; Caimano *et al.*, 2000); thus they are in effect defective prophages. Strain B31 lp56 contains an integrated cp32 (Casjens *et al.*, 2000). This integration, which appears to be a rather recent event, occurred in the middle of a putative late operon gene, and two of its other late operon genes have frameshift mutations (Casjens *et al.*, 2000), suggesting that the late operon of this cp32 is beginning to decay; such decay is characteristic of prophage genes that are no longer under selection (Canchaya *et al.*, 2003; Casjens, 2003). Lp54 contains 20 genes that are similar (but diverged) in sequence to the putative cp32 late operon genes (including the putative terminase and portal genes). However, in this region of lp54, there are two large insertions, one deletion and one substitution relative to the cp32 sequences. Thus, neither lp56 nor lp54 probably encodes a productive prophage by themselves, and, consistent with these observations, neither intact lp54 nor lp56 have been observed in preparations of phage from induced *B. burgdorferi* cultures. One other strain B31 linear plasmid, lp28-2, is only rather distantly related to the cp32s, but also carries phage-related genes (Eggers *et al.*, 2001a). These genes should encode a potential large subunit terminase (G21), a portal (G20), a tape-measure protein (G08) that is responsible for determining tail length in other bacteriophages (Hendrix, 1988), and several proteins (G10, G22, G23 and G27) with very distant homology to various other phage proteins. Lp28-2 has not yet been found packaged into phage heads, however given that a number of different bacteriophages have been observed in association with *Borrelia,* one of these could potentially be derived from lp28-2 or perhaps a more intact plasmid relative in another strain.

2.6 Summary

The determination of the complete genome sequence of the type strain B31 of *B. burgdorferi* as well as the chromosome and a number of plasmids of *B. garinii* strain PBi and *B. afzelii* strain PKo has led directly to many advances in the study of the *Borrelia* and the molecular basis of Lyme disease. Among the unusual features of *Borrelia* genomes are the following: (1) linear replicons with covalently closed hairpin telomeres; (2) large numbers (up to more than 20) of linear and circular plasmids within individual cells; multiple prophage plasmids within individual cells; (3) the presence of many regions with paralogous sequence in the plasmids; and (4) a low density of protein-coding genes in some of the plasmids as is indicated by many pseudogenes and other regions that do not contain long ORFs. Genes discovered by these sequence analyses have already proven to be very fertile ground for expanding experimental approaches towards the understanding, diagnosis, treatment and prevention of this disease. In addition, the genome sequence is an invaluable basis for understanding the genetic diversity of these bacteria. Such studies have shown some correlations between pathogenicity and genome contents, but these are ongoing and will be augmented considerably by future complete genome sequences of relapsing fever *Borrelia* and additional Lyme agent *Borrelia* isolates.

Acknowledgements

We wish to thank our colleagues for productive interactions as this field of research has grown through the past several decades. Our research was supported by NIH grants 1RO1 AI49003 (SRC) and 1R01 AR42511 and 1R01 AI45801 (IS). CHE was supported by a Quinnipiac University School of Health Sciences Faculty Research Award.

References

Aihara, H., Huang, W.M., and Ellenberger, T. (2007). An interlocked dimer of the protelomerase TelK distorts DNA structure for the formation of hairpin telomeres. Mol. Cell *27*, 901–913.

Akins, D., Caimano, M., Yang, X., Cerna, F., Norgard, M., and Radolf, J. (1999). Molecular and evolutionary analysis of *Borrelia burgdorferi* 297 circular plasmid-encoded lipoproteins with OspE- and OspF-like leader peptides. Infect. Immun. *67*, 1526–1532.

Alitalo, A., Meri, T., Lankinen, H., Seppala, I., Lahdenne, P., Hefty, P., Akins, D., and Meri, S. (2002). Complement inhibitor factor H binding to Lyme disease spirochetes is mediated by inducible expression of multiple plasmid-encoded outer surface protein E paralogs. J. Immunol. *169*, 3847–3853.

Anderton, J., Tokarz, R., Thill, C., Kuhlow, C., Brooks, C., Akins, D., Katona, L., and Benach, J. (2004). Whole-genome DNA array analysis of the response of *Borrelia burgdorferi* to a bactericidal monoclonal antibody. Infect. Immun. *72*, 2035–2044.

Antonara, S., Chafel, R., LaFrance, M., and Coburn, J. (2007). *Borrelia burgdorferi* adhesins identified using *in vivo* phage display. Mol. Microbiol. *66*, 262–276.

Aron, L., Toth, C., Godfrey, H., and Cabello, F. (1996). Identification and mapping of a chromosomal gene cluster of *Borrelia burgdorferi* containing genes expressed *in vivo*. FEMS Microbiol. Lett. *145*, 309–314.

Attie, O., Bruno, J., Xu, Y., Qiu, D., Luft, B., and Qiu, W. (2007). Co-evolution of the outer surface protein C gene (*ospC*) and intraspecific lineages of *Borrelia burgdorferi* sensu stricto in the north-eastern United States. Infect. Genet. Evol. *7*, 1–12.

Austin, S., and Abeles, A. (1985). The partition functions of P1, P7, and F miniplasmids. Basic Life Sci. *30*, 215–226.

Bacon, R., Pilgard, M., Johnson, B., Raffel, S., and Schwan, T. (2004). Glycerophosphodiester phosphodiesterase gene (*glpQ*) of *Borrelia lonestari* identified as a target for differentiating *Borrelia* species associated with hard ticks (Acari: *Ixodidae*). J. Clin. Microbiol. *42*, 2326–2328.

Baranton, G., Seinost, G., Theodore, G., Postic, D., and Dykhuizen, D. (2001). Distinct levels of genetic diversity of *Borrelia burgdorferi* are associated with different aspects of pathogenicity. Res. Microbiol. *152*, 149–156.

Barbour, A. (1988). Plasmid analysis of *Borrelia burgdorferi*, the Lyme disease agent. J. Clin. Microbiol. *26*, 475–478.

Barbour, A. (1993). Linear DNA of *Borrelia* species and antigenic variation. Trends Microbiol. *1*, 236–239.

Barbour, A., Carter, C., Bundoc, V., and Hinnebusch, J. (1996). The nucleotide sequence of a linear plasmid of *Borrelia burgdorferi* reveals similarities to those of circular plasmids of other prokaryotes. J. Bacteriol. *178*, 6635–6639.

Barbour, A., Carter, C., and Sohaskey, C. (2000). Surface protein variation by expression site switching in the relapsing fever agent *Borrelia hermsii*. Infect. Immun. *68*, 7114–7121.

Barbour, A., Dai, Q., Restrepo, B., Stoenner, H., and Frank, S. (2006). Pathogen escape from host immunity by a genome program for antigenic variation. Proc. Natl. Acad. Sci. U.S.A. *103*, 18290–18295.

Barbour, A., and Hayes, S. (1986). Biology of *Borrelia* species. Microbiol. Rev. *50*, 381–400.

Bauer, Y., Hofmann, H., Jahraus, O., Mytilineos, J., Simon, M., and Wallich, R. (2001). Prominent T-cell response to a selectively *in vivo* expressed *Borrelia burgdorferi* outer surface protein (pG) in patients with Lyme disease. Eur J. Immunol. *31*, 767–776.

Beaurepaire, C., and Chaconas, G. (2005). Mapping of essential replication functions of the linear plasmid lp17 of *B. burgdorferi* by targeted deletion walking. Mol. Microbiol. *57*, 132–142.

Behera, A., Durand, E., Cugini, C., Antonara, S., Bourassa, L., Hildebrand, E., Hu, L., and Coburn, J. (2007). *Borrelia burgdorferi* BBB07 interaction with integrin $\alpha_3\beta_1$ stimulates production of proinflammatory mediators in primary human chondrocytes. Cell. Microbiol. *10*, 320–321.

Bono, J., Tilly, K., Stevenson, B., Hogan, D., and Rosa, P. (1998). Oligopeptide permease in *Borrelia burgdorferi*: putative peptide-binding components encoded by both chromosomal and plasmid loci. Microbiology *144*, 1033–1044.

Brooks, C., Vuppala, S., Jett, A., and Akins, D. (2006). Identification of *Borrelia burgdorferi* outer surface proteins. Infect. Immun. *74*, 296–304.

Bunikis, J., Garpmo, U., Tsao, J., Berglund, J., Fish, D., and Barbour, A. (2004). Sequence typing reveals extensive strain diversity of the Lyme borreliosis agents *Borrelia burgdorferi* in North America and *Borrelia afzelii* in Europe. Microbiology *150*, 1741–1755.

Byram, R., Stewart, P., and Rosa, P. (2004). The essential nature of the ubiquitous 26-kilobase circular replicon of *Borrelia burgdorferi*. J. Bacteriol. *186*, 3561–3569.

Caimano, M., Yang, X., Popova, T., Clawson, M., Akins, D., Norgard, M., and Radolf, J. (2000). Molecular and evolutionary characterization of the cp32/18 family of supercoiled plasmids in *Borrelia burgdorferi* 297. Infect. Immun. *68*, 1574–1586.

Canchaya, C., Proux, C., Fournous, G., Bruttin, A., and Brussow, H. (2003). Prophage genomics. Microbiol. Mol. Biol. Rev. *67*, 238–276.

Casjens, S. (1998). The diverse and dynamic structure of bacterial genomes. Annu. Rev. Genet. *32*, 339–377.

Casjens, S. (1999). Evolution of the linear DNA replicons of the *Borrelia* spirochetes. Curr. Opin. Microbiol. *2*, 529–534.

Casjens, S. (2000). Borrelia genomes in the year 2000. J. Mol. Microbiol. Biotechnol. *2*, 401–410.

Casjens, S. (2003). Prophages in bacterial genomics: What have we learned so far? Mol. Microbiol. *249*, 277–300.

Casjens, S., Delange, M., Ley, H., Rosa, P., and Huang, W. (1995). Linear chromosomes of Lyme disease agent spirochetes: genetic diversity and conservation of gene order. J. Bacteriol. *177*, 2769–2780.

Casjens, S., Hatfull, G., and Hendrix, R. (1992). Evolution of dsDNA tailed-bacteriophage genomes. Sem. Virol. *3*, 383–397.

Casjens, S., and Hendrix, R. (2002). Bacteriophage roles in bacterial chromosome evolution. In The Bacterial Chromosome, P. Higgins, ed. (Washington, D.C., ASM Press), pp. 39–52.

Casjens, S., and Huang, W.M. (1993). Linear chromosomal physical and genetic map of *Borrelia burgdorferi*, the Lyme disease agent. Mol. Microbiol. *8*, 967–980.

Casjens, S., Murphy, M., DeLange, M., Sampson, L., van Vugt, R., and Huang, W.M. (1997a). Telomeres of the linear chromosomes of Lyme disease spirochaetes: nucleotide sequence and possible exchange with linear plasmid telomeres. Mol. Microbiol. *26*, 581–596.

Casjens, S., Palmer, N., van Vugt, R., Huang, W., Stevenson, B., Rosa, P., Lathigra, R., Sutton, G., Peterson, J., Dodson, R.J., *et al.* (2000). A bacterial genome in flux: the twelve linear and nine circular extrachromosomal DNAs in an infectious isolate of the Lyme disease spirochete *Borrelia burgdorferi*. Mol. Microbiol. *35*, 490–516.

Casjens, S., van Vugt, R., Tilly, K., Rosa, P., and Stevenson, B. (1997b). Homology throughout the multiple 32-kilobase circular plasmids present in Lyme disease spirochetes. J. Bacteriol. *179*, 217–227.

Casjens, S. (2005). Comparative genomics and evolution of the tailed-bacteriophages. Curr. Opin. Microbiol. *8*, 451–458.

Casjens, S., Huang, W.M., Gilcrease, E., Qiu, W., McCaig, W., Luft, B., and Schutzer, S. (2006). Comparative genomics of *Borrelia burgdorferi*. In Molecular Biology of Spirochetes, F. Cabello, D. Hulinska, and H. Godfrey, eds. (Amsterdam, IOS Press), pp. 79–95.

Catalano, C. (2005). Viral genome packaging machines: genetics, structure, and mechanism (Georgetown, TX, Landes Bioscience).

Chaconas, G., Stewart, P., Tilly, K., Bono, J., and Rosa, P. (2001). Telomere resolution in the Lyme disease spirochete. Embo J. *20*, 3229–3237.

Champion, C., Blanco, D., Skare, J., Haake, D., Giladi, M., Foley, D., Miller, J., and Lovett, M. (1994). A 9.0-kilobase-pair circular plasmid of *Borrelia burgdorferi* encodes an exported protein: evidence for expression only during infection. Infect. Immun. *62*, 2653–2661.

Cheetham, B., Tattersall, D., Bloomfield, G., Rood, J., and Katz, M. (1995). Identification of a gene encoding a bacteriophage-related integrase in a *vap* region of the *Dichelobacter nodosus* genome. Gene *162*, 53–58.

Dai, Q., Restrepo, B., Porcella, S., Raffel, S., Schwan, T., and Barbour, A. (2006). Antigenic variation by *Borrelia hermsii* occurs through recombination between extragenic repetitive elements on linear plasmids. Mol. Microbiol. *60*, 1329–1343.

Damman, C., Eggers, C., Samuels, D.S., and Oliver, D. (2000). Characterization of *Borrelia burgdorferi* BlyA and BlyB proteins: a prophage-encoded holin-like system. J. Bacteriol. *182*, 6791–6797.

Davidson, B., MacDougall, J., and Saint Girons, I. (1992). Physical map of the linear chromosome of the bacterium *Borrelia burgdorferi* 212, a causative agent of Lyme disease, and localization of rRNA genes. J. Bacteriol. *174*, 3766–3774.

Deneke, J., Burgin, A., Wilson, S., and Chaconas, G. (2004). Catalytic residues of the telomere resolvase ResT: a pattern similar to, but distinct from, tyrosine recombinases and type IB topoisomerases. J. Biol. Chem. *279*, 53699–53706.

Douglas, S. (1994). DNA Strider. A Macintosh program for handling protein and nucleic acid sequences. Methods Mol Biol 25, 181–194.

Dunn, J., Buchstein, S., Butler, L., Fisenne, S., Polin, D., Lade, B., and Luft, B. (1994). Complete nucleotide sequence of a circular plasmid from the Lyme disease spirochete, *Borrelia burgdorferi*. J. Bacteriol. *176*, 2706–2717.

Eggers, C., Caimano, M., Clawson, M., Miller, W., Samuels, D.S., and Radolf, J.D. (2002). Identification of loci critical for replication and compatibility of a *Borrelia burgdorferi* cp32 plasmid and use of a cp32-based shuttle vector for the expression of fluorescent reporters in the Lyme disease spirochaete. Mol. Microbiol. *43*, 281–295.

Eggers, C., Casjens, S., Hayes, S., Garon, C., Damman, C., Oliver, D., and Samuels, D.S. (2000). Bacteriophages of spirochetes. J. Mol. Microbiol. Biotechnol. *2*, 365–373.

Eggers, C., Casjens, S., and Samuels, D.S. (2001a). Bacteriophges of *Borrelia burgdorferi* and other spirochetes. In The Spirochetes: Molecular and Cellular Biology, M. Saier, and J. Garcia-Lara, eds. (Norfolk, Horizon Scientific Press), pp. 35–44.

Eggers, C., Kimmel, B., Bono, J., Elias, A., Rosa, P., and Samuels, D.S. (2001b). Transduction by φBB-1, a bacteriophage of *Borrelia burgdorferi*. J. Bacteriol. *183*, 4771–4778.

Eggers, C., and Samuels, D.S. (1999). Molecular evidence for a new bacteriophage of *Borrelia burgdorferi*. J. Bacteriol. *181*, 7308–7313.

Elias, A., Stewart, P., Grimm, D., Caimano, M., Eggers, C., Tilly, K., Bono, J., Akins, D., Radolf, J., Schwan, T., *et al.* (2002). Clonal polymorphism of *Borrelia burgdorferi* strain B31 MI: implications for mutagenesis in an infectious strain background. Infect. Immun. *70*, 2139–2150.

Feng, S., Hodzic, E., Freet, K., and Barthold, S. (2003). Immunogenicity of *Borrelia burgdorferi* arthritis-related protein. Infect. Immun. *71*, 7211–7214.

Ferdows, M., and Barbour, A. (1989). Megabase-sized linear DNA in the bacterium *Borrelia burgdorferi*, the Lyme disease agent. Proc. Natl. Acad. Sci. U.S.A. *86*, 5969–5973.

Ferdows, M., Serwer, P., Griess, G., Norris, S., and Barbour, A. (1996). Conversion of a linear to a circular plasmid in the relapsing fever agent *Borrelia hermsii*. J. Bacteriol. *178*, 793–800.

Ford, M., Sarkis, G., Belanger, A., Hendrix, R., and Hatfull, G. (1998). Genome structure of mycobacteriophage D29: implications for phage evolution. J. Mol. Biol. *279*, 143–164.

Fraser, C.M., Casjens, S., Huang, W.M., Sutton, G., Clayton, R., Lathigra, R., White, O., Ketchum, K., Dodson, R., Hickey, E., *et al.* (1997). Genomic sequence of a Lyme disease spirochaete, *Borrelia burgdorferi*. Nature *390*, 580–586.

Garcia-Lara, J., Picardeau, M., Hinnebusch, B., Huang, W.M., and Casjens, S. (2000). The role of genomics in approaching the study of *Borrelia* DNA replication. J. Mol. Microbiol. Biotechnol 2, 447–454.

Gilmore, R., Jr., and Mbow, M. (1998). A monoclonal antibody generated by antigen inoculation via tick bite is reactive to the *Borrelia burgdorferi* Rev protein, a member of the 2.9 gene family locus. Infect. Immun. *66*, 980–986.

Glöckner, G., Lehmann, R., Romualdi, A., Pradella, S., Schulte-Spechtel, U., Schilhabel, M., Wilske, B., Suhnel, J., and Platzer, M. (2004). Comparative analysis of the *Borrelia garinii* genome. Nucleic Acids Res. *32*, 6038–6046.

Glöckner, G., Schulte-Spechtel, U., Schilhabel, M., Felder, M., Suhnel, J., Wilske, B., and Platzer, M. (2006). Comparative genome analysis: selection pressure on the *Borrelia vls* cassettes is essential for infectivity. BMC Genomics 7, 211.

Gorbacheva, V.Y., Godfrey, H.P., and Cabello, F.C. (2000). Analysis of the *bmp* gene family in *Borrelia burgdorferi* sensu lato. J. Bacteriol. *182*, 2037–2042.

Grigoriev, P.S., and Lobocka, M.B. (2001). Determinants of segregational stability of the linear plasmid-prophage N15 of *Escherichia coli*. Mol. Microbiol. *42*, 355–368.

Grimm, D., Eggers, C.H., Caimano, M.J., Tilly, K., Stewart, P.E., Elias, A.F., Radolf, J.D., and Rosa, P.A. (2004). Experimental assessment of the roles of linear plasmids lp25 and lp28-1 of *Borrelia burgdorferi* throughout the infectious cycle. Infect. Immun. *72*, 5938–5946.

Grimm, D., Elias, A.F., Tilly, K., and Rosa, P.A. (2003). Plasmid stability during in vitro propagation of *Borrelia burgdorferi* assessed at a clonal level. Infect. Immun. *71*, 3138–3145.

Grimm, D., Tilly, K., Byram, R., Stewart, P.E., Krum, J.G., Bueschel, D.M., Schwan, T.G., Policastro, P.F., Elias, A.F., and Rosa, P.A. (2004). Outer-surface protein C of the Lyme disease spirochete: A protein induced in ticks for infection of mammals. Proc. Natl. Acad. Sci. U.S.A. *101*, 3142–3147.

Grimm, D., Tilly, K., Bueschel, D.M., Fisher, M.A., Policastro, P.F., Gherardini, F.C., Schwan, T.G., and Rosa, P.A. (2005). Defining plasmids required by *Borrelia burgdorferi* for colonization of tick vector *Ixodes scapularis* (Acari: Ixodidae). J. Med. Entomol. *42*, 676–684.

Guyard, C., Chester, E.M., Raffel, S.J., Schrumpf, M.E., Policastro, P.F., Porcella, S.F., Leong, J.M., and Schwan, T.G. (2005). Relapsing fever spirochetes contain chromosomal genes with unique direct tandemly repeated sequences. Infect. Immun. *73*, 3025–3037.

Hayes, S.F., Burgdorfer, W., and Barbour, A.G. (1983). Bacteriophage in the *Ixodes dammini* spirochete, etiological agent of Lyme disease. J. Bacteriol. *154*, 1436–1439.

Hellwage, J., Meri, T., Heikkila, T., Alitalo, A., Panelius, J., Lahdenne, P., Seppala, I.J., and Meri, S. (2001). The complement regulator factor H binds to the surface protein OspE of *Borrelia burgdorferi*. J. Biol. Chem. *276*, 8427–8435.

Hendrix, R., and Casjens, S. (2006). Bacteriophage λ and its genetic neighborhood. In The Bacteriophages, 2nd Edition, R. Calendar, ed. (New York City, N.Y., Oxford Press), pp. 409–447.

Hendrix, R. (1988). Tail length determination in double-stranded DNA bacteriophages. Curr. Top. Microbiol. Immunol. *136*, 21–29.

Hendrix, R. (2002). Bacteriophages: evolution of the majority. Theor. Popul. Biol. *61*, 471–480.

Hendrix, R., Lawrence, J., Hatfull, G., and Casjens, S. (2000). The origins and ongoing evolution of viruses. Trends Microbiol. *8*, 504–508.

Hendrix, R., Smith, M., Burns, R., Ford, M., and Hatfull, G. (1999). Evolutionary relationships among diverse bacteriophages and prophages: all the world's a phage. Proc. Natl. Acad. Sci. U.S.A. *96*, 2192–2197.

Hinnebusch, J., and Barbour, A. (1992). Linear- and circular-plasmid copy numbers in *Borrelia burgdorferi*. J. Bacteriol. *174*, 5251–5257.

Hinnebusch, J., and Barbour, A.G. (1991). Linear plasmids of *Borrelia burgdorferi* have a telomeric structure and sequence similar to those of a eukaryotic virus. J. Bacteriol. *173*, 7233–7239.

Hinnebusch, J., Bergström, S., and Barbour, A. (1990). Cloning and sequence analysis of linear plasmid telomeres of the bacterium *Borrelia burgdorferi*. Mol. Microbiol. *4*, 811–820.

Hovis, K., McDowell, J., Griffin, L., and Marconi, R. (2004). Identification and characterization of a linear-plasmid-encoded factor H-binding protein (FhbA) of the relapsing fever spirochete *Borrelia hermsii*. J. Bacteriol. *186*, 2612–2618.

Hovis, K., Tran, E., Sundy, C., Buckles, E., McDowell, J., and Marconi, R. (2006). Selective binding of *Borrelia burgdorferi* OspE paralogs to factor H and serum proteins from diverse animals: possible expansion of the role of OspE in Lyme disease pathogenesis. Infect. Immun. *74*, 1967–1972.

Huang, W.M., Joss, L., Hsieh, T., and Casjens, S. (2004a). Protelomerase uses a topoisomerase IB/Y-recombinase type mechanism to generate DNA hairpin ends. J. Mol. Biol. *337*, 77–92.

Huang, W.M., Robertson, M., Aron, J., and Casjens, S. (2004b). Telomere exchange between linear replicons of *Borrelia burgdorferi*. J. Bacteriol. *186*, 4134–4141.

Iyer, R., Kalu, O., Purser, J., Norris, S., Stevenson, B., and Schwartz, I. (2003). Linear and circular plasmid content in *Borrelia burgdorferi* clinical isolates. Infect. Immun. *71*, 3699–3706.

Iyer, R., Liveris, D., Adams, A., Nowakowski, J., McKenna, D., Bittker, S., Cooper, D., Wormser, G.P., and Schwartz, I. (2001). Characterization of *Borrelia burgdorferi* isolated from erythema migrans lesions: interrelationship of three molecular typing methods. J. Clin. Microbiol. *39*, 2954–2957.

Jacobs, J.M., Yang, X., Luft, B.J., Dunn, J.J., Camp, D.G., 2nd, and Smith, R.D. (2005). Proteomic analysis of Lyme disease: global protein comparison of three strains of *Borrelia burgdorferi*. Proteomics *5*, 1446–1453.

Jewett, M., Byram, R., Bestor, A., Tilly, K., Lawrence, K., Burtnick, M., Gherardini, F., and Rosa, P. (2007a). Genetic basis for retention of a critical virulence plasmid of *Borrelia burgdorferi*. Mol. Microbiol. *66*, 975–990.

Jewett, M., Lawrence, K., Bestor, A.C., Tilly, K., Grimm, D., Shaw, P., VanRaden, M., Gherardini, F., and Rosa, P. (2007b). The critical role of the linear plasmid lp36 in the infectious cycle of *Borrelia burgdorferi*. Mol. Microbiol. *64*, 1358–1374.

Kim, J., Singvall, J., Schwarz-Linek, U., Johnson, B., Potts, J., and Hook, M. (2004). BBK32, a fibronectin-binding MSCRAMM from *Borrelia burgdorferi*, contains a disordered region that undergoes a conformational change on ligand-binding. J. Biol. Chem. *297*, 41706–41714.

Kitten, T., and Barbour, A. (1990). Juxtaposition of expressed variable antigen genes with a conserved telomere in the bacterium *Borrelia hermsii*. Proc. Natl. Acad. Sci. U.S.A. *87*, 6077–6081.

Kitten, T., and Barbour, A. (1992). The relapsing fever agent *Borrelia hermsii* has multiple copies of its chromosome and linear plasmids. Genetics *132*, 311–324.

Kitten, T., Barrera, A., and Barbour, A. (1993). Intragenic recombination and a chimeric outer membrane protein in the relapsing fever agent *Borrelia hermsii*. J. Bacteriol. *175*, 2516–2522.

Kobryn, K., Burgin, A., and Chaconas, G. (2005). Uncoupling the chemical steps of telomere resolution by ResT. J. Biol. Chem. *280*, 26788–26795.

Kobryn, K., and Chaconas, G. (2002). ResT, a telomere resolvase encoded by the Lyme disease spirochete. Mol. Cell *9*, 195–201.

Kraiczy, P., Hellwage, J., Skerka, C., Becker, H., Kirschfink, M., Simon, M., Brade, V., Zipfel, P., and Wallich, R. (2004). Complement resistance of *Borrelia burgdorferi* correlates with the expression of BbCRASP-1, a novel linear plasmid-encoded surface protein that interacts with human factor H and FHL-1 and is unrelated to Erp proteins. J. Biol. Chem. *279*, 2421–2429.

Kraiczy, P., Seling, A., Brissette, C.A., Rossmann, E., Hunfeld, K., Bykowski, T., Burns, L., Troese, M., Cooley, A., Miller, J., *et al.* (2008). *Borrelia burgdorferi* complement regulator-acquiring surface protein 2 (CspZ) as a serological marker of human Lyme disease. Clin Vaccine Immunol *15*, 484–491.

Labandeira-Rey, M., and Skare, J. (2001). Decreased infectivity in *Borrelia burgdorferi* strain B31 is associated with loss of linear plasmid 25 or 28–1. Infect. Immun. *69*, 446–455.

Lagal, V., Portnoi, D., Faure, G., Postic, D., and Baranton, G. (2006). *Borrelia burgdorferi* sensu stricto invasiveness is correlated with OspC-plasminogen affinity. Microbes Infect. *8*, 645–652.

Lawrenz, M., Kawabata, H., Purser, J., and Norris, S. (2002). Decreased electroporation efficiency in *Borrelia burgdorferi* containing linear plasmids lp25 and lp56: impact on transformation of infectious *B. burgdorferi*. Infect. Immun. *70*, 4798–4804.

Lescot, M., Audic, S., Robert, C., Nguyen, T., Blanc, G., Cutler, S., Wincker, P., Couloux, A., Claverie, J., Raoult, D., and Drancourt, N. (2008). The genome of *Borrelia recurrentis*, the agent of deadly louse-bourne relapsing fever, is a degraded subset of tick-borne *Borrelia duttonii*. PLoS Genetics *4*, e100185.

Liveris, D., Varde, S., Iyer, R., Koenig, S., Bittker, S., Cooper, D., McKenna, D., Nowakowski, J., Nadelman, R.B., Wormser, G.P., *et al.* (1999). Genetic diversity of *Borrelia burgdorferi* in Lyme disease patients as determined by culture versus direct PCR with clinical specimens. J. Clin. Microbiol. *37*, 565–569.

Liveris, D., Wormser, G., Nowakowski, J., Nadelman, R., Bittker, S., Cooper, D., Varde, S., Moy, F., Forseter, G., Pavia, C., *et al.* (1996). Molecular typing of *Borrelia burgdorferi* from Lyme disease patients by PCR-restriction fragment length polymorphism analysis. J. Clin. Microbiol. *34*, 1306–1309.

Ludwig, A., von Rhein, C., Mischke, A., and Brade, V. (2007). Release of latent ClyA cytolysin from *Escherichia coli* mediated by a bacteriophage-associated putative holin (BlyA) from *Borrelia burgdorferi*. Int. J. Med. Microbiol. *298*, 473–481.

Lynch, M. (2006). Streamlining and simplification of microbial genome architecture. Annu. Rev. Microbiol. *60*, 327–349.

Marconi, R.T., Samuels, D.S., Schwan, T.G., and Garon, C.F. (1993). Identification of a protein in several *Borrelia* species which is related to OspC of the Lyme disease spirochetes. J. Clin. Microbiol. *31*, 2577–2583.

Marconi, R., Sung, S., Hughes, C., and Carlyon, J. (1996). Molecular and evolutionary analyses of a variable series of genes in *Borrelia burgdorferi* that are related to *ospE* and *ospF*, constitute a gene family, and share a common upstream homology box. J. Bacteriol. *178*, 5615–5626.

Margolis, N., Hogan, D., Tilly, K., and Rosa, P. (1994). Plasmid location of *Borrelia* purine biosynthesis gene homologs. J. Bacteriol. *176*, 6427–6432.

Margos, G., Gatewood, A., Aanensen, D., Hanincova, K., Terekova, D., Vollmer, S., Cornet, M., Piesman, J., Donaghy, M., Bormane, A., *et al.* (2008). MLST of housekeeping genes captures geographic population structure and suggests a European origin of *Borrelia burgdorferi*. Proc. Natl. Acad. Sci. U.S.A. *105*, 8730–8735.

Marti Ras, N., Lascola, B., Postic, D., Cutler, S., Rodhain, F., Baranton, G., and Raoult, D. (1996). Phylogenesis of relapsing fever *Borrelia* spp. Int. J. Syst. Bacteriol. *46*, 859–865.

Mathiesen, D., Oliver, J., Jr., Kolbert, C., Tullson, E., Johnson, B., Campbell, G., Mitchell, P., Reed, K., Telford, S., 3rd, Anderson, J., *et al.* (1997). Genetic heterogeneity of *Borrelia burgdorferi* in the United States. J. Infect. Dis. *175*, 98–107.

McDowell, J., Wolfgang, J., Senty, L., Sundy, C., Noto, M., and Marconi, R. (2004). Demonstration of the involvement of outer surface protein E coiled coil structural domains and higher order structural elements in the binding of infection-induced antibody and the complement-regulatory protein, factor H. J. Immunol. *173*, 7471–7480.

Metts, M., McDowell, J., Theisen, M., Hansen, P., and Marconi, R. (2003). Analysis of the OspE determinants involved in binding of factor H and OspE-targeting antibodies elicited during *Borrelia burgdorferi* infection in mice. Infect. Immun. *71*, 3587–3596.

Miller, J., Bono, J., Babb, K., El-Hage, N., Casjens, S., and Stevenson, B. (2000). Identification of cp9–2, a previously undetected, naturally-occurring circular plasmid of *Borrelia burgdorferi* strain B31. J. Bacteriol. *182*, 6254–6258.

Miller, J., and Stevenson, B. (2003). Immunological and genetic characterization of *Borrelia burgdorferi* BapA and EppA proteins. Microbiology *149*, 1113–1125.

Miller, J., and Stevenson, B. (2004). Increased expression of *Borrelia burgdorferi* factor H-binding surface proteins during transmission from ticks to mice. Int. J. Med. Microbiol. *293 Suppl. 37*, 120–125.

Neubert, U., Schaller, M., Januschke, E., Stolz, W., and Schmieger, H. (1993). Bacteriophages induced by ciprofloxacin in a *Borrelia burgdorferi* skin isolate. Zentralbl. Bakteriol. *279*, 307–315.

Nguyen, K., Wu, J., Boylan, J., Gherardini, F., and Pei, D. (2007). Zinc is the metal cofactor of *Borrelia burgdorferi* peptide deformylase. Arch. Biochem. Biophys. *468*, 217–225.

Norris, S., Carter, C., Howell, J., and Barbour, A. (1992). Low-passage-associated proteins of *Borrelia burgdorferi* B31: characterization and molecular cloning of OspD, a surface-exposed, plasmid-encoded lipoprotein. Infect. Immun. *60*, 4662–4672.

Ojaimi, C., Brooks, C., Casjens, S., Rosa, P., Elias, A., Barbour, A., Jasinskas, A., Benach, J., Katona, L., Radolf, J., *et al.* (2003). Profiling of temperature-induced changes in *Borrelia burgdorferi* gene expression by using whole genome arrays. Infect. Immun. *71*, 1689–1705.

Ojaimi, C., Davidson, B., Saint Girons, I., and Old, I. (1994). Conservation of gene arrangement and an unusual organization of rRNA genes in the linear chromosomes of the Lyme disease spirochaetes *Borrelia burgdorferi, B. garinii* and *B. afzelii*. Microbiology *140*, 2931–2940.

Pal, U., Yang, X., Chen, M., Bockenstedt, L.K., Anderson, J.F., Flavell, R.A., Norgard, M.V., and Fikrig, E. (2004). OspC facilitates *Borrelia burgdorferi* invasion of *Ixodes scapularis* salivary glands. J. Clin. Invest. *113*, 220–230.

Palmer, N., Fraser, C., and Casjens, S. (2000). Distribution of twelve linear extrachromosomal DNAs in natural isolates of the Lyme disease spirochetes. J. Bacteriol. *182*, 2476–2480.

Paster, B., Dewhirst, F., Weisburg, W., Tordoff, L., Fraser, G., Hespell, R., Stanton, T., Zablen, L., Mandelco, L., and Woese, C. (1991). Phylogenetic analysis of the spirochetes. J. Bacteriol. *173*, 6101–6109.

Pennington, P., Cadavid, D., Bunikis, J., Norris, S., and Barbour, A. (1999). Extensive interplasmidic duplications change the virulence phenotype of the relapsing fever agent *Borrelia turicatae*. Mol. Microbiol. *34*, 1120–1132.

Perna, N., Glasner, J., Burland, V., and Plunkett 3rd, G. (2002). The Genomes of *Escherichia coli* K-12 and Pathogenic *E. coli*. In *Escherichia coli*: Virulence mechanisms of a versatile pathogen, M. Donnenberg, ed. (New York, Academic Press, San Diego, CA), pp. 3–53.

Pettersson, J., Schrumpf, M., Raffel, S., Porcella, S., Guyard, C., Lawrence, K., Gherardini, F., and Schwan, T. (2007). Purine salvage pathways among *Borrelia* species. Infect. Immun. *75*, 3877–3884.

Picardeau, M., Lobry, J.R., and Hinnebusch, B.J. (1999). Physical mapping of an origin of bidirectional replication at the centre of the *Borrelia burgdorferi* linear chromosome. Mol. Microbiol. *32*, 437–445.

Pinne, M., Denker, K., Nilsson, E., Benz, R., and Bergström, S. (2006). The BBA01 protein, a member of paralog family 48 from *Borrelia burgdorferi*, is potentially interchangeable with the channel-forming protein P13. J. Bacteriol. *188*, 4207–4217.

Plasterk, R., Simon, M., and Barbour, A. (1985). Transposition of structural genes to an expression sequence on a linear plasmid causes antigenic variation in the bacterium *Borrelia hermsii*. Nature *318*, 257–263.

Porcella, S., Fitzpatrick, C., and Bono, J. (2000). Expression and immunological analysis of the plasmid-borne *mlp* genes of *Borrelia burgdorferi* strain B31. Infect. Immun. *68*, 4992–5001.

Porcella, S., Popova, T., Akins, D., Li, M., Radolf, J., and Norgard, M. (1996). *Borrelia burgdorferi* supercoiled plasmids encode multicopy tandem open reading frames and a lipoprotein gene family. J. Bacteriol. *178*, 3293–3307.

Posey, J., and Gherardini, F. (2000). Ironing out the metal requirements of *Borrelia burgdorferi*, the agent of Lyme disease. Science, *288*, 1651–1653.

Probert, W., and Johnson, B. (1998). Identification of a 47 kd fibronectin-binding protein expressed by *Borrelia burgdorferi* isolate B31. Mol. Microbiol. *30*, 1003–1015.

Purser, J., Lawrenz, M., Caimano, M., Howell, J., Radolf, J., and Norris, S. (2003). A plasmid-encoded nicotinamidase (PncA) is essential for infectivity of *Borrelia burgdorferi* in a mammalian host. Mol. Microbiol. *48*, 753–764.

Purser, J., and Norris, S. (2000). Correlation between plasmid content and infectivity in *Borrelia burgdorferi*. Proc. Natl. Acad. Sci. U.S.A. *97*, 13865–13870.

Qiu, W., Bosler, E., Campbell, J., Ugine, G., Wang, I., Luft, B., and Dykhuizen, D. (1997). A population genetic study of *Borrelia burgdorferi* sensu stricto from eastern Long Island, New York, suggested frequency-dependent selection, gene flow and host adaptation. Hereditas *127*, 203–216.

Qiu, W., Schutzer, S., Bruno, J., Attie, O., Xu, Y., Dunn, J., Fraser, C., Casjens, S., and Luft, B. (2004). Genetic exchange and plasmid transfers in *Borrelia burgdorferi* sensu stricto revealed by three-way genome comparisons and multilocus sequence typing. Proc. Natl. Acad. Sci. U.S.A. *101*, 14150–14155.

Ravin, V., Ravin, N., Casjens, S., Ford, M., Hatfull, G., and Hendrix, R. (2000). Genomic sequence and analysis of the atypical temperate bacteriophage N15. J. Mol. Biol. *299*, 53–73.

Revel, A., Blevins, J., Almazan, C., Neil, L., Kocan, K., de la Fuente, J., Hagman, K., and Norgard, M. (2005). *bptA* (*bbe16*) is essential for the persistence of the Lyme disease spirochete, *Borrelia burgdorferi*, in its natural tick vector. Proc. Natl. Acad. Sci. U.S.A. *102*, 6972–6977.

Roberts, D., Theisen, M., and Marconi, R. (2000). Analysis of the cellular localization of Bdr paralogs in *Borrelia burgdorferi*, a causative agent of Lyme disease: evidence for functional diversity. J. Bacteriol. *182*, 4222–4226.

Sadziene, A., Rosa, P., Thompson, P., Hogan, D., and Barbour, A. (1992). Antibody-resistant mutants of *Borrelia burgdorferi*: *in vitro* selection and characterization. J. Exp. Med. *176*, 799–809.

Sadziene, A., Wilske, B., Ferdows, M., and Barbour, A. (1993). The cryptic *ospC* gene of *Borrelia burgdorferi* B31 is located on a circular plasmid. Infect. Immun. *61*, 2192–2195.

Samuels, D., Marconi, R., and Garon, C. (1993). Variation in the size of the *ospA*-containing linear plasmid, but not the linear chromosome, among the three *Borrelia* species associated with Lyme disease. J. Gen. Microbiol. *139*, 2445–2449.

Schaller, M., and Neubert, U. (1994). Bacteriophages and ultrastructural alterations of *Borrelia burgdorferi* induced by ciprofloxacin. J. Spirochetal Tick-borne Dis. *1*, 37–40.

Schwan, T., Battisti, J., Porcella, S., Raffel, S., Schrumpf, M., Fischer, E., Carroll, J., Stewart, P., Rosa, P., and Somerville, G. (2003). Glycerol-3-phosphate acquisition in spirochetes: distribution and biological activity of glycerophosphodiester phosphodiesterase (GlpQ) among *Borrelia* species. J. Bacteriol. *185*, 1346–1356.

Schwan, T., Burgdorfer, W., and Garon, C. (1988). Changes in infectivity and plasmid profile of the Lyme disease spirochete, *Borrelia burgdorferi*, as a result of *in vitro* cultivation. Infect. Immun. *56*, 1831–1836.

Schwan, T., Raffel, S., Schrumpf, M., Policastro, P., Rawlings, J., Lane, R., Breitschwerdt, E., and Porcella, S. (2005). Phylogenetic analysis of the spirochetes *Borrelia parkeri* and *Borrelia turicatae* and the potential for tick-borne relapsing fever in Florida. J. Clin. Microbiol. *43*, 3851–3859.

Schwan, T., Raffel, S., Schrumpf, M., and Porcella, S. (2007). Diversity and distribution of *Borrelia hermsii*. Emerg. Infect. Dis. *13*, 436–442.

Schwartz, I., Wang, G., Iyer, R., Ojaimi, C., Terekhova, D., Sandigursky, S., Wormser, G., and Liveris, D. (2006). Genotypic variation and *Borrelia burgdorferi* pathogenesis. In Molecular Biology of Spirochetes, F.C. Cabello, D. Hulinska, and H.P. Godfrey, eds. (Amsterdam, IOS Press), pp. 124–134.

Schwartz, J., Gazumyan, A., and Schwartz, I. (1992). rRNA gene organization in the Lyme disease spirochete, *Borrelia burgdorferi*. J. Bacteriol. *174*, 3757–3765.

Seinost, G., Dykhuizen, D., Dattwyler, R., Golde, W., Dunn, J., Wang, I., Wormser, G., Schriefer, M., and Luft, B. (1999). Four clones of *Borrelia burgdorferi* sensu stricto cause invasive infection in humans. Infect. Immun. *67*, 3518–3524.

Shang, E., Skare, J., Erdjument-Bromage, H., Blanco, D., Tempst, P., Miller, J., and Lovett, M. (1997). Sequence analysis and characterization of a 40-kilodalton *Borrelia hermsii* glycerophosphodiester phosphodiesterase homolog. J. Bacteriol. *179*, 2238–2246.

Shih, Y., and Rothfield, L. (2006). The bacterial cytoskeleton. Microbiol. Mol. Biol. Rev. *70*, 729–754.

Skare, J., Foley, D., Hernandez, S., Moore, D., Blanco, D., Miller, J., and Lovett, M. (1999). Cloning and molecular characterization of plasmid-encoded antigens of *Borrelia burgdorferi*. Infect. Immun. *67*, 4407–4417.

Stevenson, B., Bykowski, T., Cooley, A., Babb, K., Miller, J., Woodman, M., von Lackum, K., and Riley, S. (2006). The Lyme disease spirochete Erp protein family: structure, function and regulation of expression. In Molecular Biology of Spirochetes, F.C. Cabello, D. Hulinska, and H.P. Godfrey, eds. (Amsterdam, IOS Press), pp. 354–372.

Stevenson, B., Casjens, S., van Vugt, R., Porcella, S., Tilly, K., Bono, J., and Rosa, P. (1997). Characterization of cp18, a naturally truncated member of the cp32 family of *Borrelia burgdorferi* plasmids. J. Bacteriol. *179*, 4285–4291.

Stevenson, B., El-Hage, N., Hines, M., Miller, J., and Babb, K. (2002). Differential binding of host complement inhibitor factor H by *Borrelia burgdorferi* Erp surface proteins: a possible mechanism underlying the expansive host range of Lyme disease spirochetes. Infect. Immun. *70*, 491–497.

Stevenson, B., and Miller, J. (2003). Intra- and interbacterial genetic exchange of Lyme disease spirochete *erp* genes generates sequence identity amidst diversity. J. Mol. Evol. *57*, 309–324.

Stevenson, B., Porcella, S., Oie, K., Fitzpatrick, C., Raffel, S., Lubke, L., Schrumpf, M., and Schwan, T. (2000a). The relapsing fever spirochete *Borrelia* hermsii contains multiple, antigen-encoding circular plasmids that are homologous to the cp32 plasmids of Lyme disease spirochetes. Infect. Immun. *68*, 3900–3908.

Stevenson, B., Tilly, K., and Rosa, P. (1996). A family of genes located on four separate 32-kilobase circular plasmids in *Borrelia burgdorferi* B31. J. Bacteriol. *178*, 3508–3516.

Stevenson, B., Zückert, W., and Akins, D. (2000b). Repetition, conservation, and variation: the multiple cp32 plasmids of *Borrelia* species. J Molec Microbiol Biotech, 2, 411–422.

Stewart, P., Byram, R., Grimm, D., Tilly, K., and Rosa, P. (2005). The plasmids of *Borrelia burgdorferi*: essential genetic elements of a pathogen. Plasmid *53*, 1–13.

Stewart, P., Chaconas, G., and Rosa, P. (2003). Conservation of plasmid maintenance functions between linear and circular plasmids in *Borrelia burgdorferi*. J. Bacteriol. *185*, 3202–3209.

Stewart, P., Thalken, R., Bono, J., and Rosa, P. (2001). Isolation of a circular plasmid region sufficient for autonomous replication and transformation of infectious *Borrelia burgdorferi*. Mol. Microbiol. *39*, 714–721.

Stewart, P., Wang, X., Bueschel, D., Clifton, D., Grimm, D., Tilly, K., Carroll, J., Weis, J., and Rosa, P. (2006). Delineating the requirement for the *Borrelia burgdorferi* virulence factor OspC in the mammalian host. Infect. Immun. *74*, 3547–3553.

Strother, K., Broadwater, A., and De Silva, A. (2005). Plasmid requirements for infection of ticks by *Borrelia burgdorferi*. Vector Borne Zoonotic Dis *5*, 237–245.

Strother, K., and de Silva, A. (2005). Role of *Borrelia burgdorferi* linear plasmid 25 in infection of *Ixodes scapularis* ticks. J. Bacteriol. *187*, 5776–5781.

Surtees, J., and Funnell, B. (2003). Plasmid and chromosome traffic control: how ParA and ParB drive partition. Curr Top Dev. Biol. *56*, 145–180.

Takahashi, Y., Cutler, S.J., and Fukunaga, M. (2000). Size conversion of a linear plasmid in the relapsing fever agent *Borrelia duttonii*. Microbiol. Immunol. *44*, 1071–1074.

Terekhova, D., Iyer, R., Wormser, G., and Schwartz, I. (2006). Comparative genome hybridization reveals substantial variation among clinical isolates of *Borrelia burgdorferi* sensu stricto with different pathogenic properties. J. Bacteriol. *188*, 6124–6134.

Theisen, M. (1996). Molecular cloning and characterization of *nlpH*, encoding a novel, surface-exposed, polymorphic, plasmid-encoded 33-kilodalton lipoprotein of *Borrelia afzelii*. J. Bacteriol. *178*, 6435–6442.

Tilly, K., Casjens, S., Stevenson, B., Bono, J.L., Samuels, D.S., Hogan, D., and Rosa, P. (1997). The *Borrelia burgdorferi* circular plasmid cp26: conservation of plasmid structure and targeted inactivation of the *ospC* gene. Mol. Microbiol. *25*, 361–373.

Tilly, K., Grimm, D., Bueschel, D., Krum, J., and Rosa, P. (2004). Infectious cycle analysis of a *Borrelia burgdorferi* mutant defective in transport of chitobiose, a tick cuticle component. Vector Borne Zoonotic Dis. *4*, 159–168.

Tilly, K., Krum, J., Bestor, A., Jewett, M., Grimm, D., Bueschel, D., Byram, R., Dorward, D., Vanraden, M., Stewart, P., *et al.* (2006). *Borrelia burgdorferi* OspC protein required exclusively in a crucial early stage of mammalian infection. Infect. Immun. *74*, 3554–3564.

Tokarz, R., Anderton, J., Katona, L., and Benach, J. (2004). Combined effects of blood and temperature shift on *Borrelia burgdorferi* gene expression as determined by whole genome DNA array. Infect. Immun. *72*, 5419–5432.

Wang, G., Ojaimi, C., Iyer, R., Saksenberg, V., McClain, S.A., Wormser, G.P., and Schwartz, I. (2001). Impact of genotypic variation of *Borrelia burgdorferi* sensu stricto on kinetics of dissemination and severity of disease in C3H/HeJ Mice. Infect. Immun. *69*, 4303–4312.

Wang, G., Ojaimi, C., Wu, H., Saksenberg, V., Iyer, R., Liveris, D., McClain, S., Wormser, G., and Schwartz, I. (2002). Disease severity in a murine model of Lyme borreliosis is associated with the genotype of the infecting *Borrelia burgdorferi* sensu stricto strain. J. Infect. Dis. *186*, 782–791.

Wang, I., Dykhuizen, D., Qiu, W., Dunn, J., Bosler, E., and Luft, B. (1999). Genetic diversity of *ospC* in a local population of *Borrelia burgdorferi sensu stricto*. Genetics *151*, 15–30.

Wang, I., Smith, D., and Young, R. (2000). Holins: the protein clocks of bacteriophage infections. Annu. Rev. Microbiol. *54*, 799–825.

Wilske, B., Preac-Mursic, V., Schierz, G., and Busch, K. (1986). Immunochemical and immunological analysis of European *Borrelia burgdorferi* strains. Zentralbl. Bakteriol. Mikrobiol. Hyg. [A] *263*, 92–102.

Wormser, G., Liveris, D., Nowakowski, J., Nadelman, R., Cavaliere, L., McKenna, D., Holmgren, D., and Schwartz, I. (1999). Association of specific subtypes of *Borrelia burgdorferi* with hematogenous dissemination in early Lyme disease. J. Infect. Dis. *180*, 720–725.

Xu, Y., Kodner, C., Coleman, L., and Johnson, R. (1996). Correlation of plasmids with infectivity of *Borrelia burgdorferi* sensu stricto type strain B31. Infect. Immun. *64*, 3870–3876.

Yang, X., Popova, T., Hagman, K., Wikel, S., Schoeler, G., Caimano, M., Radolf, J., and Norgard, M.V. (1999). Identification, characterization, and expression of three new members of the *Borrelia burgdorferi* Mlp (2.9) lipoprotein gene family. Infect. Immun. *67*, 6008–6018.

Yang, X., Hubner, A., Popova, T., Hagman, K., and Norgard, M. (2003). Regulation of expression of the paralogous Mlp family in *Borrelia burgdorferi*. Infect. Immun. *71*, 5012–5020.

Young, R. (2002). Bacteriophage holins: deadly diversity. J. Mol. Microbiol. Biotechnol. *4*, 21–36.

Young, R., and Blasi, U. (1995). Holins: form and function in bacteriophage lysis. FEMS Microbiol. Rev. *17*, 191–205.

Zhang, H., and Marconi, R.T. (2005). Demonstration of cotranscription and 1-methyl-3-nitroso-nitroguanidine induction of a 30-gene operon of *Borrelia burgdorferi*: evidence that the 32-kilobase circular plasmids are prophages. J. Bacteriol. *187*, 7985–7995.

Zhang, H., Raji, A., Theisen, M., Hansen, P., and Marconi, R. (2005). *bdrF2* of Lyme disease spirochetes is coexpressed with a series of cytoplasmic proteins and is produced specifically during early infection. J. Bacteriol. *187*, 175–184.

Zhang, J., Hardham, J., Barbour, A., and Norris, S. (1997). Antigenic variation in Lyme disease borreliae by promiscuous recombination of VMP-like sequence cassettes. Cell *89*, 275–285.

Zhong, J., and Barbour, A. (2004). Cross-species hybridization of a *Borrelia burgdorferi* DNA array reveals infection- and culture-associated genes of the unsequenced genome of the relapsing fever agent *Borrelia hermsii*. Mol. Microbiol. *51*, 729–748.

Zhong, J., Skouloubris, S., Dai, Q., Myllykallio, H., and Barbour, A. (2006). Function and evolution of plasmid-borne genes for pyrimidine biosynthesis in *Borrelia* spp. J. Bacteriol. *188*, 909–918.
Zückert, W., and Meyer, J. (1996). Circular and linear plasmids of Lyme disease spirochetes have extensive homology: characterization of a repeated DNA element. J. Bacteriol. *178*, 2287–2298.
Zückert, W., Meyer, J., and Barbour, A. (1999). Comparative analysis and immunological characterization of the *Borrelia* Bdr protein family. Infect. Immun. *67*, 3257–3266.

Replication of the *B. burgdorferi* Genome and Scrambling of the Linear Replicons through Reverse Telomere Resolution

3

George Chaconas

Abstract

The highly unusual segmented genomes of *Borrelia* species can contain over 20 autonomously replicating DNA molecules. Many of the molecules, including the chromosome, are linear with covalently closed hairpin ends. Current knowledge of the replication and maintenance of DNA molecules will be reviewed, including the process of telomere resolution, whereby the covalently closed hairpin ends are generated from replicative intermediates. Finally, the proposal that reverse telomere resolution is the driving force shaping the ongoing rearrangements and telomere exchanges in the linear replicons of *Borrelia* species will be discussed.

3.1 Introduction

DNA replication in *B. burgdorferi* and other *Borrelia species* must be specifically adapted to the eccentricities of the unusual genomes found in this genus. These include a segmented genome with up to about two dozen pieces of DNA, all peacefully coexisting under one roof (see Chapter 2). This unprecedented number of DNA molecules for a bacterium comes in both circular and linear forms. The linear molecules with their covalently closed hairpin ends require specialized functions for their replication. This chapter will deal with our current state of knowledge on DNA maintenance in *B. burgdorferi*. DNA replication of the chromosome and plasmids in *Borrelia* species has not been extensively studied and a number of mysteries regarding the replication and maintenance of the large cadre of DNA molecules in this organism remain to be investigated.

3.2 Replication of the linear chromosome

The details of the DNA replication process for bacterial chromosomes have been studied in depth in a variety of organisms. The fundamental mechanisms by which replication occurs appear to be highly conserved, although the regulation of the DNA replication process occurs through a variety of strategies in different species (Mott and Berger, 2007). In *B. burgdorferi*, we have only cursory information about the DNA replication process of the linear chromosome, which appears to have a nucleoid structure that differs from that of *E. coli* (Hinnebusch and Bendich, 1997). *Borrelia burgdorferi* has the expected complement of replication proteins (Fraser *et al.*, 1997). A putative region for the initiation of replication was originally identified near the centre of the linear chromosome based upon a very dramatic shift in GC skew (Fraser *et al.*, 1997). Experimental localization of a bidirectional origin by nascent strand analysis (Picardeau *et al.*, 1999) demonstrated that it indeed maps to a 2-kb region in the centre of the linear *B. burgdorferi* chromosome. The 240-bp intergenic stretch between *dnaA* and *dnaN* was predicted as the most likely place for the origin based upon the CG skew polarity shift in this region. Further support for an origin in this region comes from the presence of an Hbb-binding site (Kobryn *et al.*, 2000). Hbb (Tilly *et al.*, 1996) is an architectural protein that severely bends DNA (Kobryn *et al.*, 2000; Mouw and Rice, 2007) and has structural and functional similarity to the IHF and HU family of DNA bending proteins, which are frequently found to bind and bend replication origins.

Sequence gazing through the 240-bp intergenic region did not reveal any perfect matches to consensus DnaA boxes (Old *et al.*, 1993; Picardeau *et al.*, 1999). Moreover, the *B. burgdorferi* DnaA protein has not yet been purified and its recognition sites have not yet been defined. Characterization of the molecular details of the chromosomal *ori*, therefore, awaits further work. Nonetheless, the mapping of a bidirectional origin near the centre of the linear chromosome was a major advance and helped to establish the general pathway for replication of linear DNA with covalently closed hairpin ends (shown in Fig. 3.1), as will be discussed in a later section.

3.3 Replication of the plasmids

One of the many intriguing features of *B. burgdorferi* is the large number of plasmids harboured by this organism. In the type strain B31, there are at least 22 plasmids: 10 circular and 12 linear (Casjens *et al.*, 2000; Fraser *et al.*, 1997; Miller *et al.*, 2000). Most plasmids carry between two and four genes encoding members of paralogous families (PF) 32, 49, 50, 57 and 62, usually clustered in the same region (Casjens *et al.*, 2000). For this reason, it was predicted that these genes were involved in plasmid replication and maintenance. The construction of autonomously replicating *B. burgdorferi* plasmid derivatives carrying only regions encoding the paralogous families, established that these regions indeed carry the essential proteins and DNA sequences required for plasmid replication and maintenance of either circular or linear plasmids (Beaurepaire and Chaconas, 2005; Eggers *et al.*, 2002; Stewart *et al.*, 2003; Stewart *et al.*, 2001).

Each plasmid encodes either PF57 or PF62 suggesting that these proteins, which display a weak but significant similarity, might serve as replication initiators (Casjens *et al.*, 2000). PF57 is found on 17 plasmids in *B. burgdorferi* B31 while PF62 is found on five. However, these proteins show no sequence homology to known replication initiators and contain no identifiable functional motifs. Deletion analysis of a cp32 plasmid revealed that only the PF57 family member and adjacent non-coding sequences were required for replication of a circular cp32 derivative,

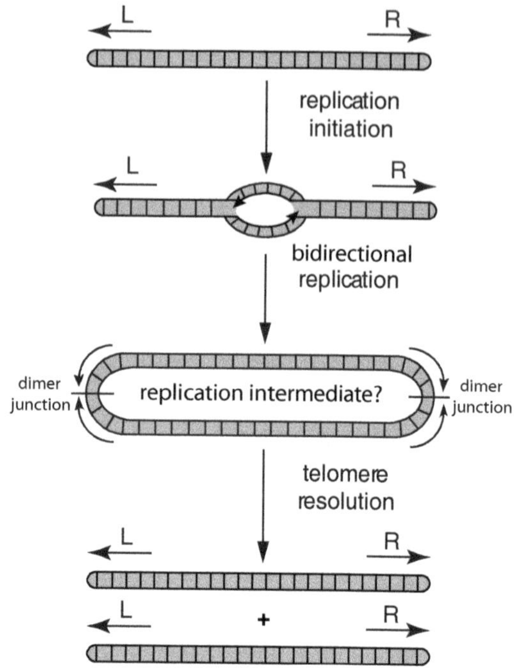

Figure 3.1 Replication pathway for linear DNA from *Borrelia* species. L and R denote the left and right telomeres, respectively. The lines bisecting the dimer junctions are axes of 180 degree rotational symmetry. Whether complete replication precedes telomere resolution or telomere resolution can occur at one end before replication of the other end is not yet known. This figure has been adapted, with permission, from Kobryn and Chaconas (2002).

suggesting PF57 to be the replication initiator (Eggers *et al.*, 2002). Experimental determination of the sequences required for replication of lp17, by progressive deletion of sequences from the left and right side of the linear plasmid through replicated telomere insertion, have shown that only a 1.8-kb region in the centre of the plasmid is essential (Beaurepaire and Chaconas, 2005). This region encodes only BBD14, the PF62 member. Introduction of a stop codon through a single base pair change at either position 20 or 55 resulted in a loss of autonomous replication of the plasmid. These results support the assignment of BBD14 (PF62) as the replication initiator protein for lp17.

Almost all the *B. burgdorferi* plasmids carry a PF32 family member, the only protein showing sequence homology to a known protein. PF32 shows limited sequence similarity to the ParA family of plasmid partition proteins (Casjens *et al.*, 2000; Zückert and Meyer, 1996) suggesting that it might have a role in plasmid partitioning. However, no ParB orthologue has been identified in *Borrelia* species. Moreover, deletion of the PF32 member (BBD21) did not show a significant plasmid stability phenotype (Beaurepaire and Chaconas, 2005) suggesting that perhaps the protein does not play a role in partitioning. However, complementation of the PF32 function on lp17 by a PF32 member encoded by another plasmid cannot be ruled out at this time. The presence of identical PF32 paralogues on different cp32 derivatives has been shown to confer incompatibility upon the two plasmids (Eggers *et al.*, 2002), consistent with a role for PF32 in either partitioning or replication (Funnell and Slavcek, 2004).

Recently, recombinant BBD14 (PF62) and BBD21 (PF32) have been purified from *E. coli* (Deneke and Chaconas, 2008). BBD21 (PF32) was shown to possess an ATPase activity, as predicted by its sequence and the presence of Walker A and B boxes. Surprisingly, PF32 was also found to physically interact with the replication initiator BBD14 (PF62), a property unprecedented for members of the ParA family of proteins. Moreover, there is an absence of any ParB orthologues and a lack of BBD21 oligomerization in the presence of ATP (Deneke and Chaconas, 2008); both of these features are characteristic of the ParA family of proteins. The emerging picture of PF32 members, therefore, suggests that they may be directly involved in DNA replication rather than serving as strict ParA orthologues. A dual role for this family in replication and partitioning cannot be excluded. An *in vitro* activity for BBD14 (PF62) was not observed, and interestingly, this protein required combined codon optimization and co-expression with another *B. burgdorferi* protein for over-expression in *E. coli,* a method that may facilitate overexpression and purification of other *Borrelia* proteins that are difficult to express in *E. coli* (Deneke and Chaconas, 2008).

The other paralogous family members encoded by most *B. burgdorferi* plasmids are PF49 and PF50. These proteins have not yet been purified and their role in plasmid maintenance remains unknown. Perhaps they are involved in regulating the expression of PF57, 62 and 32. Alternatively, they may provide some important function that is not readily discernible because inter-plasmidic complementation of these functions cannot be ruled out. The fact that the genes encoding PF49 and 50 are not readily lost, in spite of the propensity for the linear plasmids to rearrange and delete DNA, suggests that they may play important but as yet unrecognized roles in plasmid maintenance.

The origins of replication on the plasmids also remain uncharacterized at this time. Sequence gazing has suggested a location for the origins near the PF32 member on circular and linear plasmids, based upon changes in CG skew (Picardeau *et al.*, 2000). In a cp32 plasmid, the non-coding region adjacent to PF57 is required for replication and contains six copies of a relaxed DnaA box (Eggers *et al.*, 2002). This region has been proposed as a likely origin. In lp17, in addition to PF62, an adjacent region is also required for replication, suggesting that the origin lies at least partially outside the PF62 coding region (Beaurepaire and Chaconas, 2005). No obvious DnaA-like boxes or iterons were seen in this region. Now with the first report of purification of a *B. burgdorferi* initiator protein, (Deneke and Chaconas, 2008), perhaps establishment of an *in vitro* replication system will be possible, allowing further characterization of the proteins and DNA sites involved in plasmid DNA replication.

3.4 Chromosome and plasmid copy number and topological conversion

Early work on chromosome copy number in *B. hermsii,* by fluorescent dye binding and hybridization, gave values of about six copies of the chromosome in *B. hermsii* grown in culture and 15 copies for spirochaetes recovered from mice (Kitten and Barbour, 1992). In contrast, the *B. burgdorferi* chromosome was estimated to exist as a single copy in culture grown spirochaetes as determined by quantitative PCR of the *recA* gene (Morrison *et al.*, 1999). The discrepancy between these findings may be explained by the pronounced variability observed in the absolute DNA content of different *B. burgdorferi* strains (Beaurepaire and Chaconas, 2007). The copy number of the centre of the *B. burgdorferi* B31-A chromosome has been reported to be four times higher than the right end of the chromosome by quantitative PCR of a kanamycin resistance gene inserted at each location (Beaurepaire and Chaconas, 2007). Hence, multiple initiation events appear to occur as found for *E. coli* grown in rich medium (see Bipatnath *et al.*, 1998). Chromosome copy number appears to be a complex issue and clearly requires further investigation. Early work on native plasmid copy number using DNA hybridization for lp28-1 and lp32-1 in *B. hermsii* (Kitten and Barbour, 1992) and lp17, lp54 and cp26 in *B. burgdorferi* (Hinnebusch and Barbour, 1992) indicated that plasmids were present at about one copy per chromosome. More recently, determination of the relative copy number of lp17 by quantitative PCR gave a value of 3.1 copies per chromosome right end, or 0.8 copies per chromosome centre (Beaurepaire and Chaconas, 2007). This is in good agreement with the previous value of about one copy per chromosome determined by DNA hybridization (Hinnebusch and Barbour, 1992). Relative copy number has also been determined for the cp9-derived shuttle vector pBSV2 (Stewart *et al.*, 2001). Values of 5–10 per chromosome (Tilly *et al.*, 2006) and about 10 per chromosome right end (Beaurepaire and Chaconas, 2007) have been reported by quantitative PCR (Beaurepaire and Chaconas, 2007).

Interestingly, the copy number of the cp9-derived pBSV2 shuttle vector varies with the topological state of the plasmid. When pBSV2 was converted from a circle to a linear plasmid, through the insertion of a telomere resolution site, the copy number decreased five- to tenfold (Beaurepaire and Chaconas, 2007; Chaconas *et al.*, 2001). This corresponded with a similar decrease in the transcription of the PF57 paralogue carried on this plasmid, suggesting that copy number is determined by the level of the replication initiator protein as for *E. coli* (Nordstrom and Dasgupta, 2006). Although changes in copy number for circular versus lp17 derivatives could not be determined in this study, a decrease in the transcription of the replication initiator (PF62) was also observed when the gene was converted from its native linear form to a circular form (Beaurepaire and Chaconas, 2007). Optimization of transcription of the genes encoding the replication initiator proteins for a given topological state would minimize topological interconversions.

3.5 Telomere resolution

As noted earlier, the covalently closed hairpin telomeres found in *Borrelia* species require a specialized mechanism for completing the DNA replication process. Bidirectional replication of a linear DNA molecule with hairpin ends from a central origin will eventually result in a circular dimer intermediate with inverted telomere repeats at the dimer junctions (Fig. 3.1). Some means by which to resolve these dimer junctions in the replicative intermediate is required to generate two daughter molecules, each carrying covalently closed hairpin ends. This process, termed telomere resolution, was demonstrated *in vivo* for *B. burgdorferi* (Chaconas *et al.*, 2001). The dimer junction resulting from replication of a hairpin telomere (see Fig. 3.1) was shown to serve as a telomere resolution site when inserted into lp17. Telomere resolution resulted in the generation of a hairpin telomere at the insertion site. Moreover, insertion of a replicated telomere into a circular *B. burgdorferi* shuttle vector resulted in cleavage of the plasmid at the dimer junction insertion site to generate a linear plasmid carrying hairpin telomeres (Chaconas *et al.*, 2001). This telomere insertion strategy has been used to generate targeted deletions to map the required replication region of lp17 (Beaurepaire and Chaconas, 2005) and to engineer an lp28-1 derivative from which the *vls* locus was deleted (Bankhead and Chaconas, 2007).

At this time, whether both dimer junctions are resolved in a concerted manner or individually processed is not known. In the latter case, resolution of one replicated telomere might occur before replication of the other telomere; hence, the proposed circular dimeric replication intermediate may be resolved before it is formed. Interestingly, a linear head-to-head plasmid dimer has been observed (Marconi *et al.*, 1996), which might represent a partially resolved intermediate, where a mutation has occurred in the telomere resolution site at one of the dimer junctions.

The enzyme that promotes the telomere resolution reaction has been purified and named ResT for Resolvase of Telomeres (Kobryn and Chaconas, 2002). Telomere resolvases, also referred to as protelomerases (see Chaconas, 2005), have thus far also been purified from two phages (Deneke *et al.*, 2000, 2002; Huang *et al.*, 2004a) and a crystal structure for one of the phage enzymes has recently been reported (Aihara *et al.*, 2007).

3.6 The *Borrelia burgdorferi* telomeres

Until recently, only a handful of *B. burgdorferi* hairpin telomeres had been sequenced. Moreover, the ends of lp17 were the only telomeres where sequencing around the covalently closed hairpin ends was performed (Hinnebusch and Barbour, 1991). Other telomeres were sequenced by treatment of the DNA with mung bean nuclease prior to cloning and sequencing, leaving an uncertainty as to whether any nucleotides had been lost by treatment with the mung bean nuclease (Casjens *et al.*, 1997a; Fraser *et al.*, 1997; Hinnebusch *et al.*, 1990; Huang *et al.*, 2004b; Zhang *et al.*, 1997). The sequence of the entire set of *B. burgdorferi* B31 telomeres, except for lp5, has now been reported (Tourand *et al.*, 2009). The telomeres sequences were determined by a genomic sequencing approach using total *B. burgdorferi* plasmid DNA treated with mung bean nuclease as a template. Unique primers near the telomeres were designed based upon the genome sequence. The presence of missing or additional nucleotides at the hairpin termini was tested by generating replicated telomeres substrates corresponding to the determined sequences as well as +1 and −1 derivatives. The set of 19 unique telomeres is shown in Fig. 3.2, in order of decreasing reactivity *in vitro*.

Several interesting features have emerged from the set of sequences. The first is that the previous assignment of three telomere types (Chaconas, 2005; Huang *et al.*, 2004b; Tourand *et al.*, 2003) is still valid, with some updated criteria. All three telomere types carry a now expanded box 3 of eight nucleotides at position 14–21. The box 3 region is recognized by the C-terminal region of ResT (aa 163–449) (Tourand *et al.*, 2007) and this sequence-specific binding is believed to position the active site for cleavage between nucleotides three and four in all telomere types (Tourand *et al.*, 2003).

The type 1 telomeres carry the box 1 sequence TATAAT at position 2–7, the type 2 telomeres harbour a modified box 1 sequence (TATTAT) at position 5–10 and the type 3 telomeres have no box 1 sequence of either kind (Tourand *et al.*, 2009). The only telomere that does not fall cleanly into these three categories is the chromosome left end, which carries an unmodified box 1 (characteristic of type 1 telomeres), but with a type 2 telomere positioning. Because of the box 1 positioning, this telomere has been grouped with the type 2 telomeres.

There appears to be no advantage in reactivity of type 1 versus type 2 telomeres. However, the type 3 telomeres are clearly the least reactive and three of them did not display any *in vitro* activity. Interestingly, the difference in reactivity between the most active and the inactive telomeres *in vitro* was at least 160-fold, raising the possibility that *trans*-acting factors rescue the inactive telomeres *in vivo*.

3.7 The telomere resolvase, ResT

The ResT protein is responsible for generating the covalently closed hairpin telomeres present on the linear replicons of *Borrelia* species. DNA sequencing has revealed a *resT* gene in all species of *Borrelia* sequenced thus far. The *resT* gene, originally referred to as BBB03, is carried on the circular plasmid cp26 in *B. burgdorferi* and is essential for survival of the organism (Byram *et al.*, 2004; Tourand *et al.*, 2006). The essential nature of the enzyme is not surprising since it is required to process replicative intermediates.

Telomere resolvases have also been purified from phage sources (Deneke *et al.*, 2000; Huang *et al.*, 2004a) and have also been referred to as protelomerases (Rybchin and Svarchevsky, 1999), a name that is not preferred because of its scientific ambiguity (see Chaconas, 2005). Telomere resolvases are DNA breakage and reunion enzymes that are mechanistically similar to type IB topoisomerases and the tyrosine recombinase family of site-specific recombinases. These enzymes all perform a two-step transesterification using a 3′-phosphotyrosyl-enzyme intermediate (Aihara *et al.*, 2007; Deneke *et al.*, 2004; Grainge and Jayaram, 1999; Huang *et al.*, 2004a; Kobryn and Chaconas, 2002; Shuman, 1998; Van Duyne, 2002). ResT uses tyrosine 335 as the active site nucleophile that becomes covalently linked to its DNA substrate (Deneke *et al.*, 2004; Kobryn *et al.*, 2005; Kobryn and Chaconas, 2002). Other tyrosine recombinase-like features of the reaction

Type	Telomere	Sequence (box 1, box 3)	Initial Rate (fmol/min)
2	lp28-4R	AATTTATTATCTTTTAGTATAATGA	48.8 ± 2.5
2	lp36R	TATTTATTATCTTTTAGTATAATGA	43.0 ± 7.8
1	ChromR	ATATAATTTTTAATTAGTATAGAAT	41.7 ± 7.3
2	lp28-1R	TATTTATTATCTTTTAGTATATATA	41.4 ± 0.9
1	lp28-1L	ATATAATTTTTAATTAGTATAGATA	40.4 ± 7.4
2	lp28-2L, 36L	AATTTATTATTAATTAGTATAAATA	39.3 ± 13.8
1	lp17L, 28-3L, 28-4L, 21L	ATATAATTTTTTATTAGTATAGAGT	33.7 ± 6.1
2	lp28-3R	ATATTATTATTACTTAGTATAAATT	29.2 ± 6.8
1	lp56R, lp17R	ATATAATATTTATTTAGTACAAAGT	28.4 ± 2.8
2	lp38R	AATTTATTATCTTTTAGTATAATAG	27.4 ± 0.1
2	lp56L	AATTTATTATCTTTTAGTATAATGC	26.5 ± 5.4
3	lp21R	AAATTAGTTTTTTTTAGTATAAAGC	26.5 ± 6.1
2	ChromL	TAAATATAATTTAATAGTATAAAAA	25.6 ± 0.4
3	lp25L	TATAAATTTTTAAATAGTATAGTTA	18.3± 3.1
3	lp25R	AAATATTTTTTTATTAGTATAGAGA	13.6± 2.1
3	lp28-2R	ATACAATTATTAATTAGTATAGAAA	4.7 ± 2.2
3	lp38L	AATTTACAATTTTTTAGTATAAAAA	<0.3
3	lp54L	TGAAGATAATCTATTAGTATACTAA	<0.3
3	lp54R	TAATAAGAGTTTATTAGTATACTAA	<0.3

Figure 3.2 *B. burgdorferi* telomere sequence alignment. Telomere sequences are arranged in descending order, according to the initial rate of telomere resolution. The initial rate, expressed in fmol/min is shown in the right column and telomere sequences are aligned with the hairpins (or symmetry axis in the replicated telomeres) to the left. The telomeres shown are half of the actual replicated telomere substrates used in the telomere resolution reactions. The boxes labelled 1 and 3 refer to previously identified regions of sequence homology (see Chaconas, 2005), with some modifications. The original box 1 sequence, TATAAT (type 1 telomeres) is indicated by a light grey box, while the newly identified box 1 sequence, TATTAT (type 2 telomeres) is shown in darker grey. ChromL has a box 1 characteristic of type 1 telomeres, but its location is in keeping with that of Type 2 telomeres. Because of its spacing, we have grouped ChromL with the type 2 telomeres. The homology box 3 region has been expanded from the five nucleotide sequence TAGTA to the eight nucleotide sequence TTAGTATA shown in the darkest grey. The telomere sequences of lp17L, lp17R, lp21R, lp28-1R, lp56R ChromL and ChromR have been reported previously (Casjens *et al.*, 1997b; Fraser *et al.*, 1997; Hinnebusch and Barbour, 1991; Hinnebusch *et al.*, 1990; Huang *et al.*, 2004b; Zhang *et al.*, 1997). This figure has been adapted, with permission, from Tourand *et al.* (2009).

include a six base-pair overlap region between nicks introduced in the substrate during catalysis, and the lack of a requirement for divalent metal ions or high energy cofactors (Kobryn and Chaconas, 2002). A recent crystal structure of the TelK enzyme from the *Klebsiella oxytoca* phage φK02 in complex with its DNA substrate has revealed structural and catalytic similarities and differences to the tyrosine recombinases and type IB topoisomerases (Aihara *et al.*, 2007).

Borrelia burgdorferi ResT performs a unique reaction in which a phosphodiester bond on each strand in a single DNA molecule is broken and then joined to its partner strand to form a covalently closed hairpin telomere. This is in sharp contrast to the reactions performed by either type I topoisomerases (breakage and reunion of a single DNA strand to change the linking number of a double helix) or site-specific recombinases (generation of a recombinant product through the cleavage and joining of four DNA strands), and thus define telomere resolvases as a new class of DNA breakage and reunion enzyme.

3.8 ResT structure and function

ResT appears to have three structural domains (Fig. 3.3) based upon partial proteolysis and structural fold recognition software (Tourand *et al.*, 2007). Domain 1 (aa 1–163) carries a non-specific DNA-binding region and an amino acid stretch with sequence similarity to the hairpin-binding module found in cut-and-paste transposases (Allingham *et al.*, 2001; Bhasin *et al.*, 1999; Lovell *et al.*, 2002; Rice and Baker, 2001). This module is believed to distort the DNA between the scissile phosphates thereby activating the DNA substrate for cleavage (Bankhead and Chaconas, 2004). The physical nature of the distortion is not currently known, but it is clearly required for cleavage to occur as mutations in this region inhibit the DNA cleavage step. In contrast, a recent crystal structure of the phage φKO2 telomere resolvase, TelK, has not shown a 'hairpin-binding module'; instead, a region in the C-terminal domain of the TelK dimer bends the DNA and disrupts base pairing at the central region of the substrate, between the scissile phosphates, to facilitate hairpin formation and to block reversal of the cleavage reaction (Aihara *et al.*, 2007). Removal of this region has little effect on DNA cleavage by TelK, in contrast to mutagenesis of the hairpin-binding module in ResT, which can have a profound effect upon the cleavage reaction (Bankhead and Chaconas, 2004).

Domain 2 (aa 164–340) carries the catalytic residues (Deneke *et al.*, 2004), including the active site nucleophile, tyrosine 335, which are similar to those of type IB toposiomerases and tyrosine recombinases (Grainge and Jayaram, 1999; Shuman, 1998; Van Duyne, 2002). The crystal structure of TelK, which is believed to have a similar structure to ResT, has shown structural active site features of both tyrosine recombinases and type IB topoisomerases (Aihara *et al.*, 2007). In ResT, the presence of both the hairpin-binding module in domain I and the catalytic residues in domain II argues for a composite active site with a choreographed interaction between these two different components to promote the two-step transesterification by ResT (Fig. 3.4).

A third ResT structural domain (domain 3) has been predicted for residues 341–449 using

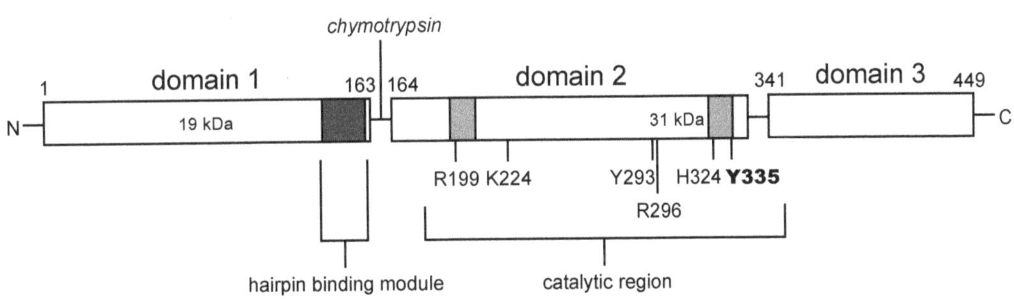

Figure 3.3 Domain structure of the *B. burgdorferi* telomere resolvase ResT. The 449 amino acid protein appears to have three domains. The N-terminal domain displays non-specific DNA binding (Tourand *et al.*, 2007) and carries a 'hairpin-binding module', believed to be similar to that found in cut-and-paste transposases (Bankhead and Chaconas, 2004). Domain 2 carries catalytic residues similar to those found in tyrosine recombinases and type IB topoisomerases (Deneke *et al.*, 2004). The active site tyrosine nucleophile located at position 335 is indicated in bold type. Other residues believed to be involved in catalysis are also labelled. The shaded regions in domain 2 correspond to the GRR(2X)E(6X)F and LGH(4–6X)T(3X) motifs found in all telomere resolvases. Domain 3 has a predicted structural alignment with a transcriptional regulatory protein. This figure has been adapted, with permission from Tourand *et al.* (2007).

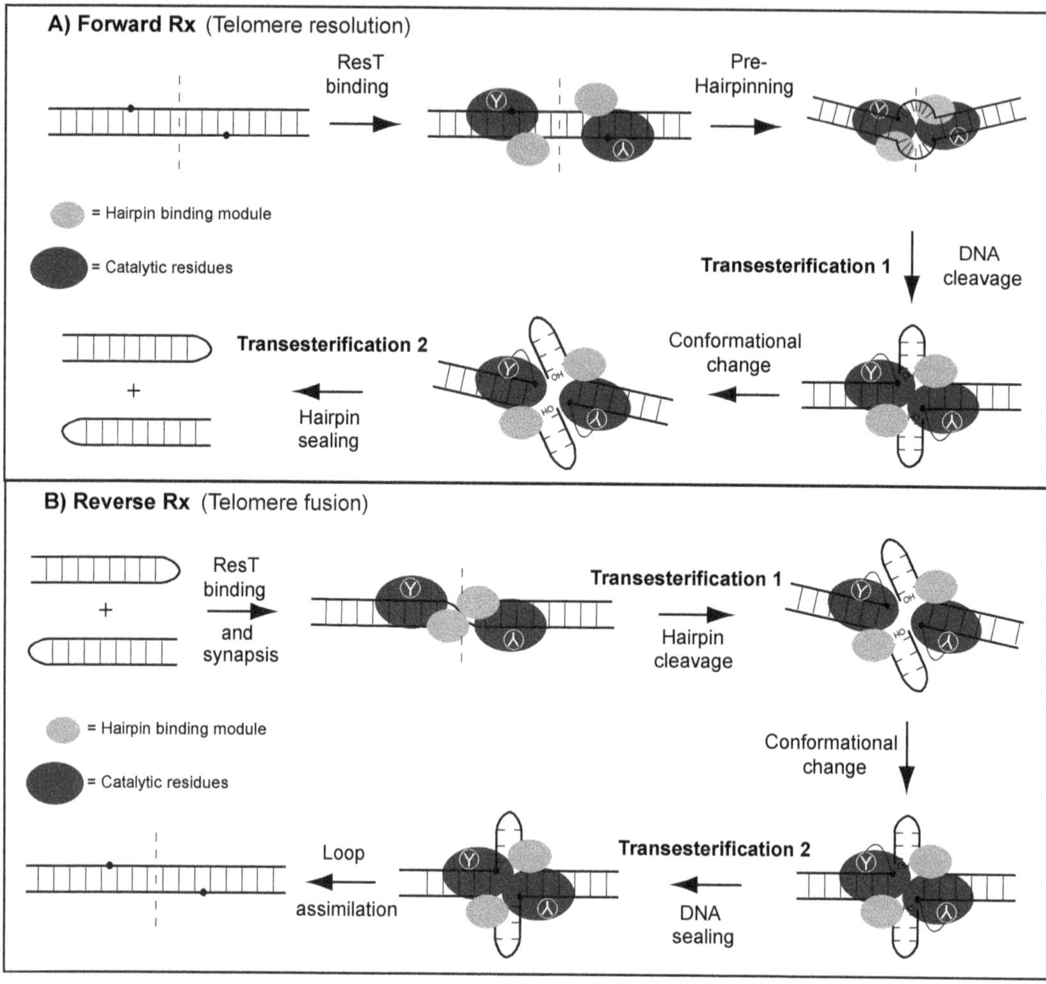

Figure 3.4 Model of the forward and reverse ResT reactions. This figure has been adapted, with permission, from Bankhead and Chaconas (2004) and Chaconas (2005).

fold recognition software (Tourand *et al.*, 2007). This region shows no similarity with the tyrosine recombinase family and aligns with a fold from residues 180–290 of the sex pheromone receptor/transcriptional regulator PrgX from *Enterococcus faecalis* (Shi *et al.*, 2005). Further studies will be required to probe a structural domain boundary in the neighbourhood of residues 340–341. Similarly, further investigation will be required to define the functional role of domain 3 in telomere resolution. A possible role for this domain might include interaction with proteins involved in replication, partitioning and/or cell division.

3.9 Telomere resolution and DNA topology

Most bacterial DNA is found in a negatively supercoiled state inside the cell. As a reflection of the natural toplogical state of the bacterial genome, DNA metabolizing enzymes prefer substrates that are negatively supercoiled (Kanaar and Cozzarelli, 1992). It was, therefore, surprising that a negatively supercoiled plasmid did not serve as a substrate for telomere resolution by ResT (Kobryn and Chaconas, 2002). Because a replicated telomere is an inverted repeat with an axis of 180 degree rotational symmetry (Fig. 3.1), its extrusion as a cruciform can be driven by the free energy of negative supercoiling. The existence of a replicated telomere as a cruciform in negatively supercoiled DNA, coupled with a requirement for a lineform substrate was proposed as the explanation for the lack of reactivity of a supercoiled substrate (Kobryn and Chaconas, 2002). We now know that this is correct because an interaction of ResT monomers across the axis of symmetry is required for the formation of a cross-axis complex to promote the normally concerted reaction (Bankhead *et al.*, 2006; Kobryn *et al.*, 2005).

Even more surprising than the inhibitory effect of negative DNA supercoiling was the finding that positive DNA supercoiling could dramatically stimulate the telomere resolution reaction by a factor of about 20-fold (Bankhead *et al.*, 2006). Positive supercoiling promotes cooperative interactions of ResT across the axis of symmetry in a replicated telomere and facilitates the formation of a cross-axis complex (Bankhead *et al.*, 2006; Kobryn *et al.*, 2005), which is required for reaction chemistry to occur. A likely explanation for the reaction stimulation is the positive twist imparted to the DNA by positive supercoiling. Positive supercoiling in the cell is transient and results from reactions that involve helical unwinding (Liu and Wang, 1987; Wu *et al.*, 1988) or rapid movement of DNA through a fixed translocase (Aussel *et al.*, 2002; Errington *et al.*, 2001; Pease *et al.*, 2005; Saleh *et al.*, 2005). It has been considered a necessary evil requiring removal by the cellular topoisomerases DNA gyrase and topoisomerase IV (Espeli and Marians, 2004; Postow *et al.*, 1999).

Based upon the dramatic stimulation imparted to telomere resolution by positive supercoiling, it is tempting to speculate that the free energy of positive supercoiling might normally be harnessed to promote the reaction *in vivo* and that positive supercoiling might be used to provide spatial and temporal regulation of the reaction. Since both DNA replication and translocation of DNA at the septum by FtsK result in an accumulation of positive supercoils, either of these processes could potentially provide transient positive supercoils to stimulate ResT and promote telomere resolution sometime between the DNA replication and segregation steps. Hence, positive supercoiling would stimulate the enzyme only on replication intermediates, and spurious cleavage of non-replicating molecules would be minimized.

3.10 Reversal of telomere resolution as the driving force for rearrangement of linear DNA molecules in *Borrelia* species

Ongoing rearrangement of the linear DNA molecules in *B. burgdorferi* is an unusual eccentricity whose purpose and mechanism have remained enigmatic (Casjens *et al.*, 2000; Huang *et al.*, 2004b). Dramatic differences in the linear replicons have been reported not only between species (Glockner *et al.*, 2004), but also between clinical isolates within *B. burgdorferi* sensu stricto (Terekhova *et al.*, 2006).

We have demonstrated that the telomere resolvase ResT can promote telomere resolution in reverse (Fig. 3.4B); that is, it can cleave and then join hairpin telomeres on unrelated DNA molecules (Chaconas, 2005; Kobryn and Chaconas, 2005). In most cases, the fused telomeres are resolved back into hairpins. However, when a mutation or deletion that blocks telomere resolution occurs, the fused telomeres would no longer be reactive, resulting in stabilization of the now-linked linear plasmids (Fig. 3.5). The new linked plasmid structure could undergo further rearrangement and fuse to another linear plasmid at a later time. Continual fusion and deletion would be cumulative over evolutionary time and can explain the telomere exchanges that are characteristic of the linear *Borrelia* replicons. We believe that this process may play a prominent role in the evolution and plasticity of the unusual linear DNA molecules in *Borrelia* species. Although our proposal has not been definitively proven *in vivo*, it is the only biochemical mechanism known at this time to explain the telomere exchanges and the data to date support our hypothesis (see also Chaconas, 2005; Kobryn and Chaconas, 2005).

Some specific examples of evolutionary events that can be explained by ResT-mediated stabilized telomere fusion are the telomere exchanges observed at the right end of the linear chromosome of *B. burgdorferi* resulting in progressive addition of DNA (Huang *et al.*, 2004b). The recent genome sequences of *B. recurrentis* and *B. duttonii* (Lescot *et al.*, 2008) provide another interesting example. In these relapsing fever spirochaetes, the *resT* gene is carried on a linear 23-kb plasmid (lp23) that is largely syntenic with

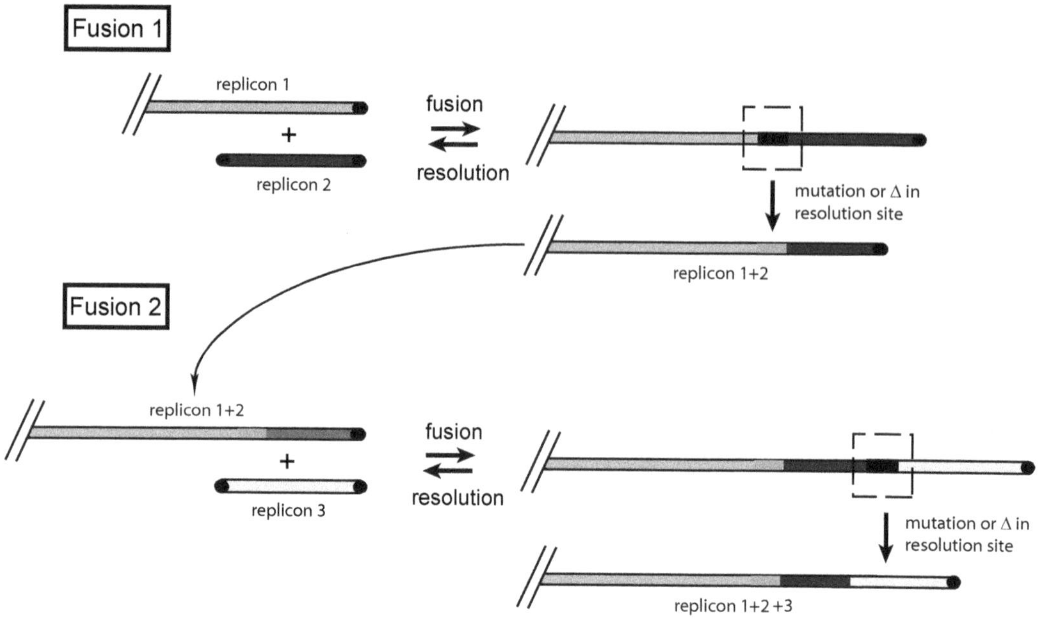

Figure 3.5 Predicted mechanism of telomere exchange in *Borrelia* species. Fusion of different linear DNA molecules can occur through reversal of telomere resolution (Kobryn and Chaconas, 2005), a reaction referred to as telomere fusion. Stabilization of the fused replicons would occur through the mutation of the fused telomeres such that they can no longer be resolved by ResT. This figure has been adapted, with permission, from Kobryn and Chaconas (2005).

circular plasmids encoding ResT in Lyme borreliae (cp26). We have previously proposed that cp26 plasmids were generated through ResT-mediated circularization of a linear form of the plasmid (Chaconas, 2005). The report of a largely syntenic lp23 in *B. recurrentis and B. duttoni* brings us one step closer to validating this hypothesis.

Note added in proof

Since submission of this chapter a new publication regarding the structure and function of ResT (Moriarty and Chaconas, 2009) has appeared.

References

Aihara, H., Huang, W.M., and Ellenberger, T. (2007). An interlocked dimer of the protelomerase TelK distorts DNA structure for the formation of hairpin telomeres. Mol. Cell *27*, 901–913.

Allingham, J.S., Wardle, S.J., and Haniford, D.B. (2001). Determinants for hairpin formation in Tn10 transposition. EMBO J. *20*, 2931–2942.

Aussel, L., Barre, F.X., Aroyo, M., Stasiak, A., Stasiak, A.Z., and Sherratt, D. (2002). FtsK Is a DNA motor protein that activates chromosome dimer resolution by switching the catalytic state of the XerC and XerD recombinases. Cell *108*, 195–205.

Bankhead, T., and Chaconas, G. (2004). Mixing active site components: a recipe for the unique enzymatic activity of a telomere resolvase. Proc. Natl. Acad. Sci. U.S.A. *101*, 13768–13773.

Bankhead, T., and Chaconas, G. (2007). The role of VlsE antigenic variation in the Lyme disease spirochete: persistence through a mechanism that differs from other pathogens. Mol. Microbiol. *65*, 1547–1558.

Bankhead, T., Kobryn, K., and Chaconas, G. (2006). Unexpected twist: harnessing the energy in positive supercoils to control telomere resolution. Mol. Microbiol. *62*, 895–905.

Beaurepaire, C., and Chaconas, G. (2005). Mapping of essential replication functions of the linear plasmid lp17 of *B. burgdorferi* by targeted deletion walking. Mol. Microbiol. *57*, 132–142.

Beaurepaire, C., and Chaconas, G. (2007). Topology-dependent transcription in linear and circular plasmids of the segmented genome of *Borrelia burgdorferi*. Mol. Microbiol. *63*, 443–453.

Bhasin, A., Goryshin, I.Y., and Reznikoff, W.S. (1999). Hairpin formation in Tn5 transposition. J. Biol. Chem. *274*, 37021–37029.

Bipatnath, M., Dennis, P.P., and Bremer, H. (1998). Initiation and velocity of chromosome replication in *Escherichia coli* B/r and K-12. J. Bacteriol. *180*, 265–273.

Byram, R., Stewart, P.E., and Rosa, P. (2004). The essential nature of the ubiquitous 26-kilobase circular replicon of *Borrelia burgdorferi*. J. Bacteriol. *186*, 3561–3569.

Casjens, S., Murphy, M., DeLange, M., Sampson, L., van Vugt, R., and Huang, W.M. (1997a). Telomeres of the linear chromosomes of Lyme disease spirochaetes:

nucleotide sequence and possible exchange with linear plasmid telomeres. Mol. Microbiol. *26*, 581–596.

Casjens, S., Palmer, N., Van Vugt, R., Huang, W.H., Stevenson, B., Rosa, P., Lathigra, R., Sutton, G., Peterson, J., Dodson, R.J., *et al.* (2000). A bacterial genome in flux: the twelve linear and nine circular extrachromosomal DNAs in an infectious isolate of the Lyme disease spirochete *Borrelia burgdorferi*. Mol. Microbiol. *35*, 490–516.

Casjens, S., van Vugt, R., Tilly, K., Rosa, P.A., and Stevenson, B. (1997b). Homology throughout the multiple 32-kilobase circular plasmids present in Lyme disease spirochetes. J. Bacteriol. *179*, 217–227.

Chaconas (2005). Hairpin telomere and genome plasticity in *Borrelia*: all mixed up in the end. Mol. Microbiol. *58*, 625–635.

Chaconas, G., Stewart, P.E., Tilly, K., Bono, J.L., and Rosa, P. (2001). Telomere resolution in the Lyme disease spirochete. EMBO J. *20*, 3229–3237.

Deneke, J., Burgin, A.B., Wilson, S.L., and Chaconas, G. (2004). Catalytic residues of the telomere resolvase ResT: a pattern similar to, but distinct from tyrosine recombinases and type IB topoisomerases. J. Biol. Chem. *279*, 53699–53706.

Deneke, J., and Chaconas, G. (2008). Purification and properties of the plasmid maintenance proteins from the *Borrelia burgdorferi* linear plasmid lp17. J. Bacteriol. *190*, 3992–4000.

Deneke, J., Ziegelin, G., Lurz, R., and Lanka, E. (2000). The protelomerase of temperate *Escherichia coli* phage N15 has cleaving-joining activity. Proc. Natl. Acad. Sci. U.S.A. *97*, 7721–7726.

Deneke, J., Ziegelin, G., Lurz, R., and Lanka, E. (2002). Phage N15 telomere resolution: target requirements for recognition and processing by the protelomerase. J. Biol. Chem. 277, 10410–10419.

Eggers, C.H., Caimano, M.J., Clawson, M.L., Miller, W.G., Samuels, D.S., and Radolf, J.D. (2002). Identification of loci critical for replication and compatibility of a *Borrelia burgdorferi* cp32-based shuttle vector for the expression of fluorescent reporters in the Lyme disease spirochete. Mol. Microbiol. *43*, 281–295.

Errington, J., Bath, J., and Wu, L.J. (2001). DNA transport in bacteria. Nat. Rev. Mol. Cell Biol. *2*, 538–545.

Espeli, O., and Marians, K.J. (2004). Untangling intracellular DNA topology. Mol. Microbiol. *52*, 925–931.

Fraser, C.M., Casjens, S., Huang, W.M., Sutton, G.G., Clayton, R., Lathigra, R., White, O., Ketchum, K.A., Dodson, R., Hickey, E.K., *et al.* (1997). Genomic sequence of a Lyme disease spirochaete, *Borrelia burgdorferi*. Nature *390*, 580–586.

Funnell, B.E., and Slavcek, R.A. (2004). Partition systems of bacterial plasmids. In Plasmid Biology, B.E. Funnell, and P.G.J., eds. (Washington, D.C., ASM Press), pp. 81–104.

Glockner, G., Lehmann, R., Romualdi, A., Pradella, S., Schulte-Spechtel, U., Schilhabel, M., Wilske, B., Suhnel, J., and Platzer, M. (2004). Comparative analysis of the *Borrelia garinii* genome. Nucleic Acids Res. *32*, 6038–6046.

Grainge, I., and Jayaram, M. (1999). The integrase family of recombinase: organization and function of the active site. Mol. Microbiol. 33, 449–456.

Hinnebusch, B.J., and Bendich, A.J. (1997). The bacterial nucleoid visualized by fluorescence microscopy of cells lysed within agarose: comparison of *Escherichia coli* and spirochetes of the genus *Borrelia*. J. Bacteriol. *179*, 2228–2237.

Hinnebusch, J., and Barbour, A.G. (1991). Linear plasmids of *Borrelia burgdorferi* have a telomeric structure and sequence similar to those of a eukaryotic virus. J. Bacteriol. *173*, 7233–7239.

Hinnebusch, J., and Barbour, A.G. (1992). Linear- and circular-plasmid copy numbers in *Borrelia burgdorferi*. J. Bacteriol. *174*, 5251–5257.

Hinnebusch, J., Bergström, S., and Barbour, A.G. (1990). Cloning and sequence analysis of linear plasmid telomeres of the bacterium *Borrelia burgdorferi*. Mol. Microbiol. *4*, 811–820.

Huang, W.M., Joss, L., Hsieh, T., and Casjens, S. (2004a). Protelomerase uses a topoisomerase IB/Y-recombinase type mechanism to generate DNA hairpin ends. J. Mol. Biol. *337*, 77–92.

Huang, W.M., Robertson, M., Aron, J., and Casjens, S. (2004b). Telomere exchange between linear replicons of *Borrelia burgdorferi*. J. Bacteriol. *186*, 4134–4141.

Kanaar, R., and Cozzarelli, N. (1992). Roles of supercoiled DNA structure in DNA transactions. Curr. Opin. Struct. Biol. *2*, 369–379.

Kitten, T., and Barbour, A.G. (1992). The relapsing fever agent *Borrelia hermsii* has multiple copies of its chromosome and linear plasmids. Genetics *132*, 311–324.

Kobryn, K., Burgin, A.B., and Chaconas, G. (2005). Uncoupling the chemical steps of telomere resolution by ResT. J. Biol. Chem. *280*, 26788–26795.

Kobryn, K., and Chaconas, G. (2002). ResT, a telomere resolvase encoded by the Lyme disease spirochete. Mol. Cell 9, 195–201.

Kobryn, K., and Chaconas, G. (2005). Fusion of hairpin telomeres by the *B. burgdorferi* telomere resolvase ResT: implications for shaping a genome in flux. Mol. Cell *17*, 783–791.

Kobryn, K., Naigamwalla, D.Z., and Chaconas, G. (2000). Site-specific DNA binding and bending by the *Borrelia burgdorferi* Hbb protein. Mol. Microbiol. *37*, 145–155.

Lescot, M., Audic, S., Robert, C., Nguyen, T.T., Blanc, G., Cutler, S.J., Wincker, P., Couloux, A., Claverie, J.M., Raoult, D., *et al.* (2008). The genome of *Borrelia recurrentis*, the agent of deadly louse-borne relapsing fever, is a degraded subset of tick-borne *Borrelia duttonii*. PLoS Genet. *4*, e1000185.

Liu, L.F., and Wang, J.C. (1987). Supercoiling of the DNA template during transcription. Proc. Natl. Acad. Sci. U.S.A. *84*, 7024–7027.

Lovell, S., Goryshin, I.Y., Reznikoff, W.R., and Rayment, I. (2002). Two-metal active site binding of a Tn5 transposase synaptic complex. Nat. Struct. Biol. *9*, 278–281.

Marconi, R.T., Casjens, S., Munderloh, U.G., and Samuels, D.S. (1996). Analysis of linear plasmid dimers in *Borrelia burgdorferi* sensu lato isolates: implications concerning the potential mechanism of linear plasmid replication. J. Bacteriol. *178*, 3357–3361.

Miller, J.C., Bono, J.L., Babb, K., El-Hage, N., Casjens, S., and Stevenson, B. (2000). A second allele of *eppA* in *Borrelia burgdorferi* strain B31 is located on the previously undetected circular plasmid cp9-2. J. Bacteriol. *182*, 6254–6258.

Moriarty, T.J and Chaconas, G. (2009). Identification of the determinant conferring permissive substrate usage in the telomere resolvase, ResT. J. Biol. Chem. *284*, 23293–23301.

Morrison, T.B., Ma, Y., Weis, J.H., and Weis, J.J. (1999). Rapid and sensitive quantification of *Borrelia burgdorferi*-infected mouse tissues by continuous fluorescent monitoring of PCR. J. Clin. Microbiol. 37, 987–992.

Mott, M.L., and Berger, J.M. (2007). DNA replication initiation: mechanisms and regulation in bacteria. Nat. Rev. Microbiol. 5, 343–354.

Mouw, K.W., and Rice, P.A. (2007). Shaping the *Borrelia burgdorferi* genome: crystal structure and binding properties of the DNA-bending protein Hbb. Mol. Microbiol. 63, 1319–1330.

Nordstrom, K., and Dasgupta, S. (2006). Copy-number control of the *Escherichia coli* chromosome: a plasmidologist's view. EMBO Rep. 7, 484–489.

Old, I.G., Margarita, D., and Saint Girons, I. (1993). Unique genetic arrangement in the *dnaA* region of the *Borrelia burgdorferi* linear chromosome: nucleotide sequence of the *dnaA* gene. FEMS Microbiol. Lett. *111*, 109–114.

Pease, P.J., Levy, O., Cost, G.J., Gore, J., Ptacin, J.L., Sherratt, D., Bustamante, C., and Cozzarelli, N.R. (2005). Sequence-directed DNA translocation by purified FtsK. Science *307*, 586–590.

Picardeau, M., Lobry, J.R., and Hinnebusch, B.J. (1999). Physical mapping of an origin of bidirectional replication at the centre of the *Borrelia burgdorferi* linear chromosome. Mol. Microbiol. 32, 437–445.

Picardeau, M., Lobry, J.R., and Hinnebusch, B.J. (2000). Analyzing DNA strand compositional asymmetry to identify candidate replication origins of *Borrelia burgdorferi* linear and circular plasmids. Genome Res. *10*, 1594–1604.

Postow, L., Peter, B.J., and Cozzarelli, N.R. (1999). Knot what we thought before: the twisted story of replication. Bioessays *21*, 805–808.

Rice, P.A., and Baker, T.A. (2001). Comparative architecture of transposase and integrase complexes. Nat. Struct. Biol. 8, 302–307.

Rybchin, V.N., and Svarchevsky, A.N. (1999). The plasmid prophage N15: a linear DNA with covalently closed ends. Mol. Microbiol. 33, 895–903.

Saleh, O.A., Bigot, S., Barre, F.X., and Allemand, J.F. (2005). Analysis of DNA supercoil induction by FtsK indicates translocation without groove-tracking. Nat. Struct. Mol. Biol. *12*, 436–440.

Shi, K., Brown, C.K., Gu, Z.Y., Kozlowicz, B.K., Dunny, G.M., Ohlendorf, D.H., and Earhart, C.A. (2005). Structure of peptide sex pheromone receptor PrgX and PrgX/pheromone complexes and regulation of conjugation in *Enterococcus faecalis*. Proc. Natl. Acad. Sci. U.S.A. *102*, 18596–18601.

Shuman, S. (1998). Vaccinia virus DNA topoisomerase: a model eukaryotic type IB enzyme. Biochim. Biophys. Acta *1400*, 321–337.

Stewart, P.E., Chaconas, G., and Rosa, P. (2003). Conservation of plasmid maintenance functions between linear and circular plasmids in *Borrelia burgdorferi*. J. Bacteriol. *185*, 3202–3209.

Stewart, P.E., Thalken, R., Bono, J.L., and Rosa, P. (2001). Isolation of a circular plasmid region sufficient for autonomous replication and transformation of infectious *Borrelia burgdorferi*. Mol. Microbiol. 39, 714–721.

Terekhova, D., Iyer, R., Wormser, G.P., and Schwartz, I. (2006). Comparative genome hybridization reveals substantial variation among clinical isolates of *Borrelia burgdorferi* sensu stricto with different pathogenic properties. J. Bacteriol. *188*, 6124–6134.

Tilly, K., Fuhrman, J., Campbell, J., and Samuels, D.S. (1996). Isolation of *Borrelia burgdorferi* genes encoding homologues of DNA- binding protein HU and ribosomal protein S20. Microbiology *142*, 2471–2479.

Tilly, K., Krum, J.G., Bestor, A., Jewett, M.W., Grimm, D., Bueschel, D., Byram, R., Dorward, D., Vanraden, M.J., Stewart, P., *et al.* (2006). *Borrelia burgdorferi* OspC protein required exclusively in a crucial early stage of mammalian infection. Infect. Immun. *74*, 3554–3564.

Tourand, Y., Bankhead, T., Wilson, S.L., Putteet-Driver, A.D., Barbour, A.G., Byram, R., Rosa, P.A., and Chaconas, G. (2006). Differential telomere processing by *Borrelia* telomere resolvases in vitro but not in vivo. J. Bacteriol. *188*, 7378–7386.

Tourand, Y., Deneke, J., Moriarty, T.J., and Chaconas, G. (2009). Characterization and *in vitro* reaction properties of 19 unique hairpin telomeres from the linear plasmids of the Lyme disease spirochete. J. Biol. Chem. *284*, 7264–7272.

Tourand, Y., Kobryn, K., and Chaconas, G. (2003). Sequence-specific recognition but position-dependent cleavage of two distinct telomeres by the *Borrelia burgdorferi* telomere resolvase, ResT. Mol. Microbiol. *48*, 901–911.

Tourand, Y., Lee, L., and Chaconas, G. (2007). Telomere resolution by *Borrelia burgdorferi* ResT through the collaborative efforts of tethered DNA binding domains. Mol. Microbiol. *64*, 580–590.

Van Duyne, G.D. (2002). A structural view of tyrosine recombinase site-specific recombination. In Mobile DNA II, N.L. Craig, Craigie, R., Gellert, M., and Lambowitz, A.M., ed. (Washington, D.C., ASM Press), pp. 93–117.

Wu, H.-Y., Shyy, S.H., Wang, J.C., and Liu, L.F. (1988). Transcription generates positively and negatively supercoiled domains in the template. Cell 53, 433–440.

Zhang, J.R., Hardham, J.M., Barbour, A.G., and Norris, S.J. (1997). Antigenic variation in Lyme disease borreliae by promiscuous recombination of VMP-like sequence cassettes. Cell *89*, 275–285.

Zückert, W.R., and Meyer, J. (1996). Circular and linear plasmids of Lyme disease spirochetes have extensive homology: characterization of a repeated DNA element. J. Bacteriol. *178*, 2287–2298.

Gene Regulation, Transcriptomics and Proteomics

4

Jon T. Skare, James A. Carroll, X. Frank Yang, D. Scott Samuels and Darrin R. Akins

Abstract

Lyme disease is a tick-borne infection that can lead to chronic, debilitating complications if not recognized or treated appropriately. *Borrelia burgdorferi,* the agent of Lyme disease, is maintained in nature by a complex enzootic cycle involving ticks and various vertebrate hosts. Since the initial finding over a decade ago that outer surface lipoproteins OspA and OspC are differentially expressed in the tick and mammal, many studies have reported that *B. burgdorferi* differentially expresses a vast ensemble of genes and proteins to help it adapt to growth in the unfed tick, feeding tick, and mammalian host environments. Several global microarray studies have utilized various cultivation systems to identify genes regulated by environmental cues and/or mammalian host factors. The combined studies have been directed at identifying genes that are differentially expressed during the unique enzootic life cycle of this pathogen. While the microarray studies have identified hundreds of genes that are differentially expressed under various environmental conditions, recent advances in genetic methodologies have begun providing insight into the mechanisms underlying the differential gene expression observed. Although how most genes are transcriptionally regulated in the tick and/or mammal is still unclear, several regulatory proteins, including alternative sigma factors, DNA-binding proteins, and transcriptional activators/repressors have been identified that either directly or indirectly control the differential expression of many genes in *B. burgdorferi.* Studies on the various transcriptional regulators, alternative sigma factors, and DNA-binding proteins identified thus far have provided important insights into the adaptive strategies employed by *B. burgdorferi* as it cycles between the tick and the mammal. In recent years, functional proteomic-based assays also have become an important tool for examining the complement of proteins synthesized by Lyme disease spirochaetes under experimental conditions that mimic various host environments.

4.1 Differential gene regulation in *Borrelia burgdorferi*

4.1.1 The borrelial enzootic cycle

The enzootic life cycle of *B. burgdorferi* is complex, and the spirochaete is maintained by transmission between ticks of the genus *Ixodes* and various wild rodents, most commonly mice from the genus *Peromyscus* (Lane *et al.,* 1991) (see Chapter 10). Humans can become accidental hosts if they encroach upon this natural life cycle and are fed upon by a *B. burgdorferi*-infected tick (Steere, 2001) (see Chapter 18). Over the past two decades, several laboratories have reported profound changes in global gene transcription and protein expression profiles of *B. burgdorferi* as it migrates from the nymphal tick midgut and salivary glands into the mammalian host during the blood meal. While we are only now starting to understand the molecular mechanisms for pathogenesis of Lyme disease (see Chapter 12) due to advances in genetic manipulation of the bacterium (reviewed in Rosa *et al.,* 2005) (see Chapter 8), tick transmission and the subsequent establishment of infection in the mammalian host is clearly a complex process. Along these lines, several borrelial genes have been shown to impact either the mammalian side of the infectious cycle (*ospC*, *rpoS*, *rpoN*, *rrp2*, *vlsE*, and *dbpBA*; Yang *et al.*, 2003a; Grimm *et al.*, 2004; Pal *et al.*, 2004b; Fisher *et al.*, 2005; Bankhead and Chaconas, 2007; Caimano *et al.*, 2007; Weening *et al.*, 2008; Shi *et al.*, 2008) or the arthropod phase (*ospA*, *bptA*, *napA/dps*, and *bb0365*; Yang *et al.*, 2004; Revel *et al.*, 2005; Li *et al.*, 2007; Pal *et al.*, 2008), providing further support for the contention that *B. burgdorferi* modifies gene expression in a coordinated manner to adapt to different microenvironment(s). *B. burgdorferi* appears to quickly shift its appearance by upregulating genes whose products are needed for transfer from the tick to the mammalian host while at the same time downregulating genes whose products are no longer needed. The virulence factors required for subsequent dissemination by *B. burgdorferi* are probably distinct from those involved in the initial infection and thus may be subject to different regulatory mechanisms. Furthermore, during persistent infection, spirochaetes are exposed to the adaptive host immune response, which may involve yet another set of virulence determinants for survival (see Chapter 15).

4.1.2 Differential gene expression: the *ospA/ospC* paradigm

The seminal report more than a decade ago by Schwan and co-workers (1995), which described the reciprocal expression of the genes encoding the outer surface lipoprotein (Osp) A (downregulated) and OspC (upregulated), has become a paradigm for differential gene expression by the Lyme disease spirochaete during the feeding and transmission cycle. Subsequent to the observation that OspA and OspC are differentially expressed during tick feeding (Schwan *et al.*, 1995), many laboratories identified several differentially expressed *B. burgdorferi* antigens, some of which are upregulated by an increase in temperature (Schwan *et al.*, 1995; Stevenson *et al.*, 1995; Hefty *et al.*, 2001; Revel *et al.*, 2002; Ojaimi *et al.*, 2003) or change in pH (Carroll *et al.*, 1999, 2000, 2001; Yang *et al.*, 2000), while others appear to be expressed exclusively during the mammalian phase of infection (Champion *et al.*, 1994; Akins *et al.*, 1995; Suk *et al.*, 1995; Wallich *et al.*, 1995; Fikrig *et al.*, 1999; Revel *et al.*, 2002; Hefty *et al.*, 2002b). Although there are exceptions, almost all of the differentially expressed *B. burgdorferi* genes and proteins that have been identified to date are encoded by the various plasmids harboured by this organism (Revel *et al.*, 2002; Ojaimi *et al.*, 2003; Brooks *et al.*, 2003; Tokarz *et al.*, 2004) (see Chapter 2). This has led to the conclusion that the extrachromosomal plasmid elements are essential for both *B. burgdorferi* virulence and maintenance of the borrelial enzootic cycle. This notion has been further supported by the observation that changes in plasmid content and/or deletion of specific plasmid-encoded genes correlates with attenuation or loss of *B. burgdorferi* infectivity in the mammalian host (Schwan *et al.*, 1988; Purser and Norris, 2000; McDowell *et al.*, 2001; Labandeira-Rey and Skare, 2001; Byram *et al.*, 2004; Grimm *et al.*, 2004; Jewett *et al.*, 2007) or tick vector (Strother *et al.*, 2005; Strother and de Silva, 2005; Revel *et al.*, 2005; Grimm *et al.*, 2005).

4.1.3 Global gene expression studies

While numerous differentially expressed genes and proteins were identified on a case-by-case basis over the past 15 years, a functional genomic approach to assessing differential expression of borrelial genes began when Norgard and co-workers (Revel *et al.*, 2002) provided the first report of a global microarray analysis of *B. burgdorferi* cultivated in different environments. This was followed closely by other microarray studies that also examined the transcriptional profiles of *B. burgdorferi* genes regulated either by temperature (Ojaimi *et al.*, 2003) or mammalian host factors (Brooks *et al.*, 2003; Tokarz *et al.*, 2004). The overall goal of these studies was to identify changes in the global expression of *B. burgdorferi* genes as the organism adapts from the conditions that model the tick environment (see Chapter 11) to those that recapitulate the mammalian host environment.

Temperature and mammalian host-specific factors result in global changes in gene expression

Microarray analyses were employed to examine: (i) the adaptive response of *B. burgdorferi* to altered temperature and/or pH in culture [which is thought to mimic the tick feeding and transmission process (Revel *et al.*, 2002; Ojaimi *et al.*, 2003)]; (ii) how *B. burgdorferi* alters gene expression under mammalian host-adapted conditions [*B. burgdorferi* cultured in dialysis membrane chambers (DMCs) implanted in rats (Akins *et al.*, 1998; Revel *et al.*, 2002; Brooks *et al.*, 2003; Caimano *et al.*, 2007)]; and (iii) global gene expression changes following cultivation in blood with concomitant temperature shift (Tokarz *et al.*, 2004). When temperature-regulated genes were examined, over 200 different ORFs were differentially expressed at 23°C versus 35°C, with over 130 being expressed at significantly greater levels at 35°C and 82 being upregulated at 23°C. Among the ORFs identified, over 60% are genes of unknown function, indicating that the adaptive response of *B. burgdorferi* is likely to be unique relative to other human pathogens (Ojaimi *et al.*, 2003). Interestingly, on plasmid lp54, 31 of the 76 annotated ORFs exhibited temperature-regulated expression (Ojaimi *et al.*, 2003). These findings suggest that lp54 plays an important role in the adaptive capacity of *B. burgdorferi* as it traffics through its enzootic cycle. A substantial percentage of the differentially regulated genes identified encoded proteins with predicted export signals, an observation indicating that *B. burgdorferi* modifies its surface proteome during the tick feeding and transmission process. The transcriptional profile analysis by Revel *et al.* (2002) also demonstrated that many of the genes upregulated in response to a combination of pH- and temperature-shift were only transiently affected: their expression was tempered as the spirochaete adapted to the mammalian host. This observation led to the conclusion that substantially fewer genes are differentially regulated after infection is established (Revel *et al.*, 2002).

With regard to genes regulated by mammalian host-specific factors, over 120 genes have been identified that are expressed at significantly different levels in the mammalian host compared with temperature-shifted organisms (Brooks *et al.*, 2003), with almost equivalent numbers of genes being upregulated or downregulated. Interestingly, whether using the dialysis membrane implantation model (Akins *et al.*, 1998; Revel *et al.*, 2002; Brooks *et al.*, 2003; Caimano *et al.*, 2007) or a combination of blood and temperature shift for cultivation (Tokarz *et al.*, 2004), a common theme that has emerged is that the vast majority of genes identified on the plasmids (almost 80%) are downregulated, while differentially expressed chromosomal genes are almost entirely upregulated (~90%). Among the ORFs with a known or putative function, the cell envelope category, which would include all outer membrane proteins, was found to contain the most differentially expressed genes. This suggests that there is a dramatic change in the composition of the borrelial surface as it migrates from the tick to mammal. While the mechanism of how all of these genes are regulated is still not clear, some of the upstream promoter regions share a common motif that may represent regulatory sites that are targeted by transcriptional activators and/or repressors during mammalian infection (Brooks *et al.*, 2003). The combined findings of the various microarray experiments have now identified a subset of genes that can be further characterized to help define their role or roles with regard to *B. burgdorferi* virulence and Lyme disease pathogenesis.

4.1.4 Differential expression of chromosome- and plasmid-encoded genes

In microarray experiments that mimicked either the mammalian host environment or the fed tick environment (Revel *et al.*, 2002; Brooks *et al.*, 2003; Tokarz *et al.*, 2004), most of the upregulated genes identified mapped to the chromosome, whereas many of the downregulated genes encoded were observed to be plasmid-encoded. While this observation is consistent with a prior study indicating that a majority of the borrelial surface lipoproteins are downregulated in the mouse model of Lyme disease during chronic infection (Liang *et al.*, 2002c), it does seem to be at odds with previous reports indicating that the plasmids encode proteins that play important and necessary roles in Lyme disease pathogenesis (Simpson *et al.*, 1990). Many of the cell envelope constituents of *B. burgdorferi* are clearly downregulated during the mammalian phase of infection (Revel *et al.*, 2002; Ojaimi *et al.*, 2003; Brooks *et al.*, 2003; Tokarz *et al.*, 2004), although there are a few outer membrane proteins and lipoproteins that are upregulated in the mammalian host. These include *p66*, *bba64*, and *ospC* (Brooks *et al.*,

2003). Outer membrane protein P66 is upregulated in the mammal and binds β_3-chain integrins (Coburn *et al.*, 1999), so this protein may be involved in dissemination of *B. burgdorferi* within the mammalian host (see Chapter 12). OspC and BBA64 had previously been shown to be upregulated during tick transmission and during the acute stage of infection, thus their identification in these 'host-adapted' microarray-based screens was not unexpected (Schwan *et al.*, 1995; Hefty *et al.*, 2001; Gilmore *et al.*, 2001).

4.1.5 Differential gene expression and altered protein localization: multiple levels of regulation may enhance immune evasion

In addition to differential expression of specific genes in either the tick or mammalian host, *B. burgdorferi* also has been shown to alter the cellular location of some surface lipoproteins during the various stages of its enzootic life cycle. Interestingly, the global microarray analyses indicated that there is a dramatic change and concomitant downregulation of many of the *B. burgdorferi* surface lipoproteins as it enters a mammalian host-adapted state (Revel *et al.*, 2002; Brooks *et al.*, 2003; Tokarz *et al.*, 2004). In addition to many surface lipoproteins being downregulated during infection, some surface lipoproteins, when quantified simultaneously for both expression and cellular localization in the mammalian host, display characteristics suggesting that they can be re-localized to the periplasmic space during infection (Hefty *et al.*, 2002b). Taken together, the microarray expression and cellular localization studies strongly suggest that *B. burgdorferi* has evolved a strategy that results in decreasing the overall antigenicity of its outer surface by downregulating and/or altering the cellular location of its immunogenic surface lipoproteins (Hefty *et al.*, 2002b; Liang *et al.*, 2002b,c; Brooks *et al.*, 2003). These combined observations have led to a model that allows for the generation of population heterogeneity and antigenic diversity in *B. burgdorferi* during the mammalian phase of infection and, ultimately, may aid in evasion of the mammalian immune response. The model of surface heterogeneity relies on two major components: a temporal component that describes when a gene/protein is expressed in the tick or mammal and a spatial component that describes where a protein is localized in *B. burgdorferi* (i.e. periplasm or surface). The finding that some surface lipoproteins are downregulated at the transcriptional level while others appear to be sequestered in the periplasm during mammalian infection and not exported to the surface provides for multiple levels of regulation. Therefore, even if a known surface lipoprotein is expressed during infection, it may not be exported to the surface by default in all organisms, which may explain why antibodies against many lipoproteins are not protective against *B. burgdorferi* infection, especially when transmitted via tick infestation (Hagman *et al.*, 2000; Hefty *et al.*, 2002b).

The mechanism(s) by which all of the various chromosomal- and plasmid-encoded genes and proteins are differentially regulated or differentially localized is currently unclear. Unfortunately, progress has been hindered in defining how lipoproteins are localized to the surface of *B. burgdorferi* because the known pathways that are used by other diderm bacteria for localizing lipoproteins to their outer surface are not present in *B. burgdorferi* (Schulze and Zückert, 2006) (see Chapter 6). At the same time, however, much progress has been made over the past several years in the area of gene regulation. Recent characterization of various avirulent and virulent strains that have been mutated in genes encoding alternative sigma factors, sigma factor regulatory proteins, DNA-binding proteins, and transcriptional regulators have started to shed light on the specific pathways and cascades involved in the differential gene expression required for *B. burgdorferi* to successfully move between its arthropod and mammalian hosts.

4.2 Alternative sigma factors in gene regulation

4.2.1 Rationale: environmental signals encountered in the enzootic cycle

As mentioned above, *B. burgdorferi* is maintained in nature via an enzootic cycle alternating between an *Ixodes* tick vector and a vertebrate host, usually small rodents (see Chapter 10). Tick larvae become infected with *B. burgdorferi* when they feed on an infected vertebrate and the spirochaetes are retained in the larvae midgut through the moult to the nymphal stage. At the next blood meal, the spirochaetes migrate from the nymph midgut to the salivary glands and infect the vertebrate host, thus completing the cycle (Lane *et al.*, 1991; Spielman, 1994; Steere, 2006). *B. burgdorferi* synthesizes several proteins that are intimately involved in its enzootic cycle (see Chapters 11 and 12); the most prominent are OspA, OspB, and OspC (Babour *et al.*, 1983, 1984; Howe *et al.*, 1985; Fuchs *et al.*, 1992; Wilske *et al.*, 1993c). OspA binds to a receptor, termed TROSPA, in the tick midgut (Pal *et al.*, 2004a) and is required for *B. burgdorferi* colonization and survival in the tick (Yang *et al.*, 2004). OspB also binds to tick gut and contributes to survival (Neelakanta *et al.*, 2007). OspC binds to both the tick salivary protein SALP15 (Ramamoorthi *et al.*, 2005) and plasminogen (Lagal *et al.*, 2006); it is required for transmission to and/or infection of the vertebrate host (Grimm *et al.*, 2004; Pal *et al.*, 2004b; Stewart *et al.*, 2006; Tilly *et al.*, 2006, 2007; Fingerle *et al.*, 2007; Gilbert *et al.*, 2007). In addition, OspC synthesis correlates with migration from the tick midgut to the salivary gland (Leuba-Garcia *et al.*, 1998; Gilmore, Jr. and Piesman, 2000; Schwan and Piesman, 2000; Fingerle *et al.*, 2002).

4.2.2 *ospC* and *ospAB* regulation

Borrelia burgdorferi senses environmental changes associated with tick feeding and changes its program of gene expression to affect transmission and adapt to the vertebrate host (reviewed in Singh and Girschick, 2004; Rosa *et al.*, 2005; Fikrig and Narasimhan, 2006; Radolf and Caimano, 2008). The regulated production of OspC and OspA has been well documented in *B. burgdorferi* and the underlying molecular mechanisms have recently been described (Hübner *et al.*, 2001; Alverson *et al.*, 2003; Yang *et al.*, 2003a; Caimano *et al.*, 2005; Burtnick *et al.*, 2007; Lybecker and Samuels, 2007; Xu *et al.*, 2007). OspC is almost entirely absent in the unfed tick midgut, but OspC levels increase during the blood meal (Schwan *et al.*, 1995; Montgomery *et al.*, 1996; Leuba-Garcia *et al.*, 1998; Schwan and Piesman, 2000; Ohnishi *et al.*, 2001; Rathinavelu and de Silva, 2001). Schwan *et al.* (1995) initially described that an increase in temperature associated with the blood meal represents an important cue to induce OspC. Overall, several environmental factors affect *ospC* expression, including temperature, pH, host factors, and CO_2 (Schwan *et al.*, 1995; Stevenson *et al.*, 1995; Montgomery *et al.*, 1996; Akins *et al.*, 1998; Carroll *et al.*, 1999; Obonyo *et al.*, 1999; Fingerle *et al.*, 2000; Schwan and Piesman, 2000; Yang *et al.*, 2000, 2001; Babb *et al.*, 2001; Gilmore *et al.*, 2001; Hefty *et al.*, 2001, 2002b; Hübner *et al.*, 2001; Ramamoorthy and Scholl-Meeker, 2001; Rathinavelu and de Silva, 2001; Revel *et al.*, 2002; Brooks *et al.*, 2003; Alverson *et al.*, 2003; Tokarz *et al.*, 2004; Seshu *et al.*, 2004a; Liang *et al.*, 2004b; Caimano *et al.*, 2005; Hyde *et al.*, 2007; Xu *et al.*, 2007).

OspA and OspB are synthesized by *B. burgdorferi* in unfed ticks and in culture (Barbour *et al.*, 1983; Fikrig *et al.*, 1992; Fingerle *et al.*, 1995; Schwan *et al.*, 1995; Montgomery *et al.*, 1996; Schwan and Piesman, 2000; Ohnishi *et al.*, 2001), but these lipoproteins are only rarely expressed in the mammal (Montgomery *et al.*, 1996; Coleman *et al.*, 1997; Hodzic *et al.*, 2003; Crother *et al.*, 2004; Lederer *et al.*, 2005), and when they are, it is typically during the chronic stage of disease (Kalish *et al.*, 1993, 1995; Nowalk *et al.*, 2006a). In fact, ectopic expression of *ospA* during the acute phase of infection in mice elicits an immune response that clears the infection (Strother *et al.*, 2007; Xu *et al.*, 2008b).

Expression of the *ospAB* operon and expression of the *ospC* gene have been considered to be reciprocally regulated (Margolis and Rosa, 1993; Schwan *et al.*, 1995); reduced pH (Yang *et al.*, 2000) or relaxed DNA supercoiling (Alverson *et al.*, 2003) can promote this inverse regulation in culture. However, reciprocal expression of *ospAB* and *ospC* under *in vitro* growth conditions has been controversial since a lack of *ospAB* downregulation upon elevated temperature and reduced pH has also been reported (Carroll *et al.*, 1999, 2003; Caimano *et al.*, 2007). In addition, studies by Ohnishi *et al.* (2001), Hefty *et al.* (2001) and, most recently, Mulay *et al.* (2009) demonstrated that the expression of *ospAB* and *ospC* is not coupled; instead, some spirochaetes produce both outer surface lipoproteins and others produce neither. On the other hand, a recent study by Srivastava and de Silva (2008) suggests that OspA and OspC are reciprocally regulated, although the caveat must be noted that these experiments were performed *in vitro*. In this study, the authors argue that cells have different outer surface lipoprotein profiles due to the random nature associated with *osp* gene regulation and the slow turnover of Osp proteins. Curiously, He *et al.* (2008) recently demonstrated that the loss of OspA and OspB resulted in constitutive expression of OspC, which is consistent with the hypothesis that *B. burgdorferi* compensates for the loss of abundant lipoproteins in its outer membrane by derepressing the expression of other lipoprotein genes (Xu *et al.*, 2008a).

The molecular mechanisms responsible for regulating the *ospAB* operon and the *ospC* gene in response to the different environmental signals are beginning to be elucidated. Host catecholamines have been shown to specifically bind *B. burgdorferi* and induce expression of *ospAB*, which is essential for acquisition during tick feeding (Scheckelhoff *et al.*, 2007). Hübner *et al.* (2001) demonstrated that *ospC* transcription is dependent on the alternative sigma factor RpoS (see below), which was confirmed using an inducible promoter (Gilbert *et al.*, 2007). In addition, Caimano *et al.* (2005) demonstrated that *ospAB* repression is also dependent on RpoS, although full repression is only seen *in vivo* (Caimano *et al.*, 2005; Caimano *et al.*, 2007). A putative repressor binds to the *ospAB* promoter (Margolis and Samuels, 1995), possibly at the upstream T-rich motif (Sohaskey *et al.*, 1999; Caimano *et al.*, 2007). Although this DNA-binding protein is hypothesized to be an RpoS-dependent regulator of *ospAB* transcription, further work is required to identify it and define its function.

4.2.3 The Rrp2–RpoN–RpoS pathway

The sigma factor (σ factor) is an essential but dissociable subunit of bacterial RNA polymerase. Sigma factor binds to RNA polymerase core to form RNA polymerase holoenzyme and provides promoter recognition specificity during transcriptional initiation. In addition to a general housekeeping sigma factor (σ^{70} in *Escherichia coli*), which is responsible for transcription of the majority of genes, most bacterial species have several alternative sigma factors that recognize different promoter sequences and allow transcriptional initiation of a select group of genes in response to certain environmental or developmental signals.

There are two alternative sigma factors encoded in the *B. burgdorferi* genome, RpoN (σ^{54}, encoded by *ntrA* or *rpoN*; *bb0450*) and RpoS (σ^{S} or σ^{38}, encoded by *rpoS*; *bb0771*) (Fraser *et al.*, 1997). Both RpoN and RpoS are widespread in many bacteria and their contributions to differential gene regulation in environmental adaptation have been established (Kazmierczak *et al.*, 2005). However, as initially demonstrated by Hübner *et al.* (2001) and discussed further below, *B. burgdorferi* has evolved a unique σ^{54}–σ^{S} sigma factor cascade (the Rrp2–RpoN–RpoS pathway) that plays a central role in modulating a dramatic switch in the gene expression program as *B. burgdorferi* is transmitted from its tick to vertebrate host.

RpoS

RpoS was first described in *E. coli* as an alternative sigma factor that is essential for cells entering stationary phase (reviewed in Hengge-Aronis, 2002a). Since then, RpoS has been recognized as a global regulator that responds to multiple stresses. The sigma factor accumulates following low or high temperature, low pH, high osmolarity, oxidative stress, UV exposure, and carbon starvation, resulting in the regulation of genes that combat the stressful conditions encountered and prevent cellular damage. RpoS is also involved in regulating expression of virulence factors and was shown to be required for infection by many pathogenic bacteria (Hengge-Aronis, 2000; Kazmierczak *et al.*, 2005). RpoS synthesis and activity are subject to complex regulatory mechanisms, with controls at the level of transcription, translation, and protein stability (reviewed in Hengge-Aronis, 2002b).

One of the significant achievements with recent advances in genetic manipulation of *B. burgdorferi* (see Chapter 8) was the elucidation of the roles of RpoS and RpoN in differential gene expression. Genetic inactivation of *rpoS* was first achieved by Elias *et al.* (2000) in the high-passage, avirulent *B. burgdorferi* strain B31. Hübner *et al.* (2001) generated an *rpoS* mutant in a low-passage infectious strain (*B. burgdorferi* strain 297) that resulted in the loss of OspC and DbpBA, suggesting that RpoS was required for the regulation of the genes encoding these surface lipoproteins. Subsequent studies, including microarray analyses, revealed that RpoS influences expression of more than 100 genes (Fisher *et al.*, 2005; Caimano *et al.*, 2007; Ouyang *et al.*, 2008). Many RpoS-activated genes appeared to be differentially expressed during tick feeding and some, including *ospC, dbpBA, bbk32, oppA-V, bba64,* and *bba66,* have been shown to be required for or associated with mammalian host infection (Fischer *et al.*, 2003; Grimm *et al.*, 2004; Clifton *et al.*, 2006; Fischer *et al.*, 2006; Seshu *et al.*, 2006; Gilmore, Jr. *et al.*, 2007; Medrano *et al.*, 2007; Weening *et al.*, 2008; Gautam *et al.*, 2008; Maruskova *et al.*, 2008; Shi *et al.*, 2008). In addition, 59 genes were upregulated and 34 genes were downregulated by RpoS when spirochaetes were cultivated within dialysis membrane chambers implanted into the peritoneal cavity of rats, which recapitulates the response to mammalian host signals (Akins *et al.*, 1998; Caimano *et al.*, 2007).

RpoN

The second alternative sigma factor present in the *B. burgdorferi* genome is RpoN (σ^{54}). RpoN is the only sigma factor that does not belong to the σ^{70} family (Reitzer and Schneider, 2001; Wigneshweraraj *et al.*, 2008). Accordingly, RpoN-dependent gene activation possesses several unique features. First, activation of RpoN-holoenzyme ($E\sigma^{54}$) exclusively requires a transcriptional activator. Second, RpoN-dependent gene activation requires ATP, a feature often associated with eukaryotic transcriptional activation. Third, RpoN recognizes a highly conserved −24/−12 promoter sequence that differs dramatically from the −35/−10 promoter sequence recognized by the σ^{70} family (Studholme and Buck, 2000). Consequently, in most bacteria, RpoN-dependent gene regulation is tightly controlled and RpoN only involves modulation of a limited number of genes with diverse functions such as nitrogen assimilation, arginine catabolism, and flagellar biosynthesis (Reitzer and Schneider, 2001; Kazmierczak *et al.*, 2005).

Success in generating an *rpoN* mutant by Hübner *et al.* (2001) led to a striking finding that the mutant could no longer produce OspC, DbpA, and DbpB, the same phenotype observed in the *rpoS* mutant. They also showed that abrogation of OspC, DbpA, and DbpB in the *rpoN* mutant was due to the loss of RpoS. Constitutive expression of *rpoS* in the *rpoN* mutant partially rescued expression of OspC. Therefore, RpoN and RpoS constitute a sigma factor cascade (σ^{54}–σ^{S} sigma factor cascade, or the RpoN–RpoS pathway) that governs expression of *ospC* and the *dbpBA* operon. RpoN controlling *rpoS* activation appeared to be direct, as the *rpoS* gene has a consensus −24/−12 RpoN promoter sequence (Hübner *et al.*, 2001; Burtnick *et al.*, 2007; Lybecker and Samuels, 2007; Smith *et al.*, 2007). Further biochemical and genetic analyses indicated that the −24/−12 RpoN-dependent promoter is the major promoter responsible for cell density-dependent *rpoS* expression (Burtnick *et al.*, 2007; Lybecker and Samuels, 2007; Smith *et al.*, 2007). A second, upstream start site exists for a longer *rpoS* transcript that is translated in a temperature-dependent fashion, but lacks a consensus RpoN promoter (see below) (Lybecker and Samuels, 2007). Thus, regulation of *rpoS* expression in *B. burgdorferi* appears to be mainly at the transcriptional level, which is largely controlled by RpoN at high cell density.

In addition to *rpoS*, microarray analysis suggested that RpoN influences the expression of more than 305 borrelial genes (Fisher *et al.*, 2005). Over 80% of these genes were predicted to have RpoN-dependent promoters. This observation is somewhat surprising considering that RpoN usually controls only a few genes in other bacteria (Reitzer and Schneider, 2001). Accordingly, these results have recently been challenged by a study demonstrating that the vast majority of genes influenced by RpoS and RpoN are overlapping, suggesting that RpoN modulates *B. burgdorferi* genes mainly through RpoS (Ouyang *et al.*, 2008). Regardless of this controversy, *rpoS* is the only gene to date whose transcription has been experimentally determined to be directly controlled by RpoN. Nevertheless, further experiments are warranted to validate other RpoN-dependent genes as proposed by the prior microarray analysis (Fisher *et al.*, 2005).

Rrp2

To initiate transcription, RpoN must be activated by a unique group of regulators, the RpoN (σ^{54})-dependent activators. These molecules are also referred to as prokaryotic enhancer-binding proteins (EBPs) since they often bind to a DNA site located 100 to 1000 bp from the −24/−12 RpoN (σ^{54}) promoter (Studholme and Buck, 2000). In *E. coli*, there are 12 RpoN-dependent transcription activators and each activates a set of RpoN-dependent genes in response to a specific signal (Reitzer and Schneider, 2001). In contrast to *E. coli*, *B. burgdorferi* appears to have a single RpoN-dependent activator, Rrp2 (encoded by *rrp2*; *bb0763*). Rrp2 belongs to the well-studied NtrC protein family (Studholme and Buck, 2000). Members of this family contain three putative functional domains: an N-terminal response regulator domain, a central RpoN-activation domain, and a C-terminal helix-turn-helix (HTH) DNA-binding domain (Fig. 4.1, right). The central domain becomes activated upon phosphorylation at a conserved aspartic acid residue (D52) within the N-terminal domain. The activated central domain then contacts the Eσ^{54} holoenzyme through DNA looping, hydrolyses ATP, and promotes open promoter complex formation for transcriptional initiation.

Since Rrp2 is the only annotated RpoN-dependent activator present in the *B. burgdorferi* genome, Yang *et al.* (2003a) predicted that Rrp2 governs activation of RpoN-dependent genes of *B. burgdorferi*, including *rpoS*. Indeed, genetic evidence demonstrated that Rrp2 is the upstream activator for the RpoN–RpoS pathway; a single point mutation of glycine 239 to cysteine (G239C) within one of the ATP-binding motifs in the central activation domain of Rrp2 abolishes expression of *rpoS*, *ospC*, and other RpoN–RpoS pathway-dependent genes (Yang *et al.*, 2003a). More recently, Blevins *et al.* (2009) have shown that recombinant Rrp2 has a phosphorylation-dependent ATPase activity and two studies (Ouyang *et al.*, 2008; Boardman *et al.*, 2008) have demonstrated that Rrp2 is essential for infection of the mammalian host, but not the tick vector, which is consistent with the role of the RpoN–RpoS pathway in regulating genes required for transmission and vertebrate infection.

Interestingly, the mechanism by which Rrp2 activates RpoN-dependant *rpoS* transcription appears to be unique. Unlike NtrC and other genes that require an enhancer-binding site for activation, Rrp2 can activate *rpoS* transcription from only the minimal −24/−12 promoter, suggesting that no enhancer binding site is required for activation (Burtnick *et al.*, 2007; Blevins *et al.*, 2009). The observation that an enhancer-binding protein activates Eσ^{54} without an enhancer-binding site is not unprecedented. Recent genome analyses revealed that there are a few RpoN-dependent activators that do not possess a DNA-binding domain (Beck *et al.*, 2007). One such protein is FlgR in *Helicobacter pylori*, which was shown to be capable of activating transcription of RpoN-dependent genes without DNA binding, presumably by interaction with Eσ^{54}-holoenzyme

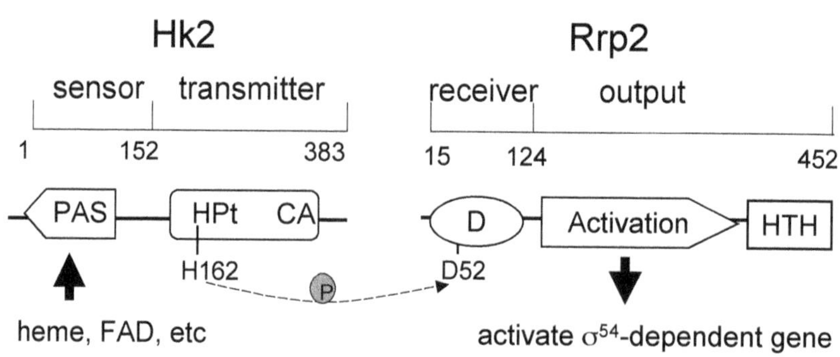

Figure 4.1 The Hk2-Rrp2 two-component system of *B. burgdorferi*. The putative sensor kinase Hk2 is predicted to have a PAS sensor domain and a transmitter (kinase) domain. The putative response regulator Rrp2 contains an N-terminal receiver domain, a central activation domain for σ^{54}-dependent transcriptional activation, and the C-terminal DNA-binding domain with a helix-turn-helix motif (HTH). The presumptive signal transduction is that Hk2 senses a small molecule by its PAS domain and becomes autophosphorylated at H162 and, subsequently, phosphorylates Rrp2 at D52, which leads to activation of RpoN (σ^{54})-dependent genes in *B. burgdorferi*.

in solution (Brahmachary *et al.*, 2004). In this regard, Rrp2 has a predicted HTH-containing C-terminal domain (Yang *et al.*, 2003a) and purified recombinant Rrp2 binds to heparin as well as non-specifically to DNA (Burtnick *et al.*, 2007; Blevins *et al.*, 2009). However, the contribution of this non-specific DNA-binding activity to Rrp2 function remains to be determined (Blevins *et al.*, 2009).

In addition to being the transcriptional activator for the RpoN–RpoS pathway, Rrp2 probably plays an additional role in *B. burgdorferi*. This hypothesis is based on observations that while inactivation of *rpoN* is attainable, disruption of *rrp2* or the region encoding the DNA-binding domain has not been achieved despite exhaustive attempts (Yang *et al.*, 2003a; Burtnick *et al.*, 2007; Blevins *et al.*, 2009), suggesting that *rrp2* is probably essential to *B. burgdorferi* under *in vitro* cultivation condition. One possibility is that although the activation function of Rrp2 is not required for borrelial survival *in vitro*, its DNA-binding activity may be involved in regulating genes, as either a repressor or an activator, in an RpoN-independent fashion. There are no high-affinity Rrp2-binding sites upstream of the *rpoS* gene (Burtnick *et al.*, 2007), but Rrp2 does bind DNA without sequence specificity (Blevins *et al.*, 2009). Further experimental analyses, including identifying putative Rrp2-binding sites in the genome, will be required to examine this hypothesis.

Global microarray analyses of genes influenced by Rrp2 were recently performed in strains B31 and 297 (Boardman *et al.*, 2008; Ouyang *et al.*, 2008). The majority of Rrp2-dependent genes identified in these studies overlapped with RpoN-dependent genes, which further supports the notion that Rrp2 is the activator of RpoN. One caveat of these microarray analyses is using the *rrp2G239C* point mutant rather than a null *rrp2* mutant. Therefore, these studies were limited to identifying genes that are regulated by Rrp2 via activation of RpoN and would not uncover other potential Rrp2-dependent genes that are independent of RpoN, such as Rrp2-repressed genes. Since an Rrp2-deficient mutant has not been achieved, further work using other methodologies will be needed to identify Rrp2-dependent but RpoN-independent genes.

Hk2

Although temperature, pH, CO_2, cell density, and host molecules have been shown to activate *ospC* and other RpoS-dependent genes (see section 4.2.2 above), upstream factors that lead to the activation of the Rrp2–RpoN–RpoS pathway remain undefined. OspA and OspB have been proposed to be involved in a positive feedback loop that activates Rrp2 by an unknown mechanism (He *et al.*, 2008). Since Rrp2 is a two-component response regulator, activation of the RpoN–RpoS pathway is probably modulated by a two-component signalling system (TCS). Such systems are composed of a histidine kinase (HK) that monitors an external signal(s) and a response regulator (RR) that controls gene expression or other physiological activities. Upon signalling, the HK undergoes auto-phosphorylation at the conserved histidine. Then, the histidine-bound phosphoryl group of the HK is transferred onto a conserved aspartic acid residue on the cognate RR, which activates the RR (Stock *et al.*, 2000; Mascher, 2006).

Genome sequences of *B. burgdorferi* revealed that a gene encoding a putative HK, *hk2* (*bb0764*), is immediately upstream of the *rrp2* gene. Therefore, Hk2 has been postulated to be the cognate kinase of Rrp2 that senses signals leading to the activation of the RpoN–RpoS pathway (Yang *et al.*, 2003a; Burtnick *et al.*, 2007). Unlike many HKs that are associated with the membrane, Hk2 in *B. burgdorferi* is predicted to be a cytoplasmic histidine kinase. Therefore, if Hk2 senses a host molecule, then the molecule is probably capable of entering the cytoplasm of the spirochaete. Alternatively, Hk2 could sense an intracellular signal within the cell. Interestingly, Hk2 contains a putative PAS domain (PAS is an acronym for Per-Arnt-Sim), a well-recognized cytoplasmic sensor domain (Fig. 4.1, left). PAS domains comprise a sensing module superfamily found in all kingdoms of life (Taylor and Zhulin, 1999). They recognize stimuli, including light, oxygen, redox potential, and others, by binding cofactors, such as haem, flavin adenine dinucleotide (FAD), and flavin mononucleotide, and modulate global regulation of many biological processes including metabolism, behaviour, development, ion channel function, and hypoxic responses.

Burtnick *et al.* (2007), the only study to date on Hk2, generated and characterized an *hk2* mutant. Surprisingly, disruption of *hk2* did not affect expression of *rpoS* and *ospC* under the *in vitro* growth conditions examined. This finding suggests that, at least *in vitro*, Hk2 is not necessary for the activation of Rrp2 (Burtnick *et al.*, 2007). One caveat of this study is that the *hk2* mutant, despite being in an infectious background (strain B31-A3), lost an endogenous plasmid required for mammalian infection and therefore could not be used *in vivo*. Thus, the role of Hk2 in the activation of the Rrp2–RpoN–RpoS pathway in the enzootic cycle of *B. burgdorferi* remains undetermined. Nevertheless, this study indicates that an additional factor is capable of phosphorylating Rrp2 (Burtnick *et al.*, 2007). Possible candidates include other non-cognate HKs annotated in the *B. burgdorferi* genome (including BB0420, CheA1, or CheA2) or small phosphodonors, such as acetyl phosphate or carbamoyl phosphate (Fig. 4.2) (Lukat *et al.*, 1992; Burtnick *et al.*, 2007).

Rrp1–Hpk1

Recently, a second two-component response regulator system was identified and characterized by Rogers *et al.* (2009). The *rrp1–hpk1*

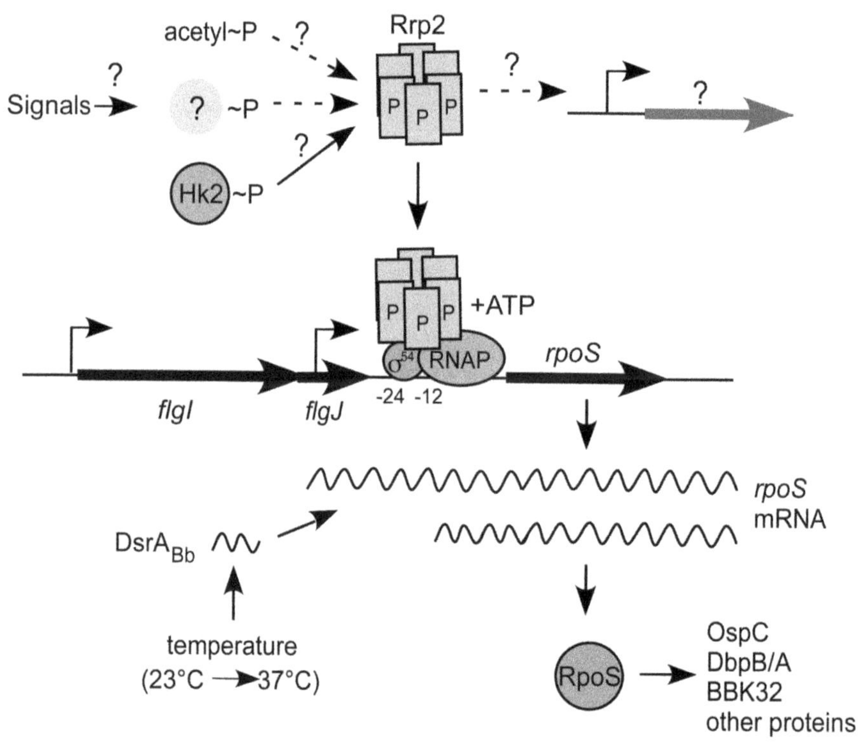

Figure 4.2 The Rrp2–RpoN–RpoS pathway. During tick feeding and mammalian infection, host signals are sensed by the cognate sensor kinase Hk2, small phospho-donors, such as acetyl-phosphate (acetyl~P), or other unknown non-cognate sensor kinases in *B. burgdorferi*. Phosphorylated sensor kinase or accumulated acetyl~P then activates Rrp2 by phosphorylation. Phosphorylated Rrp2, which presumably forms an octamer as does its homologue NtrC in *E. coli*, can then interact with RpoN (σ^{54}) and activate the transcription of genes that have a σ^{54}-type promoter. *rpoS* is the only experimentally demonstrated σ^{54}-dependent gene in *B. burgdorferi* that has a perfect σ^{54}-type promoter (indicated by –24/–12). Rrp2 could function as a repressor for other genes. In addition to the major *rpoS* mRNA species transcribed from the σ^{54}-type promoter, a longer *rpoS* transcript is produced at low cell density from a σ^{70}-type promoter located within the upstream *flgJ* gene. A small RNA molecule ($DsrA_{Bb}$) is proposed to regulate the efficiency of long *rpoS* mRNA species translation in response to temperature (Lybecker and Samuels, 2007). Accumulation of *rpoS* transcript leads to the production of OspC, DbpA, DbpB, BBK32, and other lipoproteins. This figure is modified with permission from Burtnick *et al.* (2007).

two-component system is chromosomally encoded by ORFs *bb0419* and *bb0420*. The genes encoding the *rrp1–hpk1* TCS are co-transcribed from a promoter upstream of *bb0420*. *bb0420* encodes Hpk1, which is a hybrid histidine kinase-response regulator, while *bb0419* encodes the response regulator protein Rrp1. Interestingly, the *hpk1–rrp1* transcript is constitutively expressed during *in vitro* cultivation, but it is significantly upregulated during tick feeding. This observation suggests that the Rrp1–Hpk1 TCS plays an important role during spirochaete transmission from the tick to the mammal. Rrp1 also was shown to contain a conserved GGDEF domain that undergoes phosphorylation and produces the secondary messenger cyclic diguanylate (c-di-GMP) (Rogers *et al.*, 2009), which is an important signalling molecule in many bacterial pathogens (Tamayo *et al.*, 2007; Hengge, 2009). To examine the potential regulatory role of Rrp1–Hpk1 in global gene transcription, the *rrp1* gene was mutated and microarray analyses were performed. Comparison between an avirulent wild-type strain of *B. burgdorferi* and the mutant *rrp1* strain revealed that Rrp1 influences the expression of a large number of functionally diverse genes (Rogers *et al.*, 2009). Surprisingly, the majority of the differentially expressed genes identified in the *rrp1* mutant were shown to encode basic cellular functions. Additionally, the vast majority of genes observed to be regulated by Rrp1 were found to be encoded by what was termed the 'core *Borrelia* genome', which was proposed to be composed of the linear chromosome and plasmids lp54 and cp26. While this recent study did not explore the role of the Rrp1–Hpk1 TCS or the second messenger c-di-GMP in borrelial virulence (Rogers *et al.*, 2009), future studies using virulent strain mutants should help to better define the role of this TCS in the borrelial enzootic life cycle.

4.2.4 Regulation of RpoS by DsrA

The amount of RpoS protein in *E. coli* is regulated at the translational level by three small RNAs (sRNAs) that bind to the upstream region of *rpoS* mRNA, predicted to be in the double-stranded portion of a hairpin stem–loop (Lease and Belfort, 2000; Hengge-Aronis, 2002a; Repoila *et al.*, 2003; Storz *et al.*, 2004; Majdalani *et al.*, 2005; Narberhaus *et al.*, 2006). The small untranslated RNA DsrA induces *rpoS* translation at low temperature (Sledjeski *et al.*, 1996; Hengge-Aronis, 2002a; Repoila *et al.*, 2003). DsrA binds to a region upstream of *rpoS* preventing formation of the hairpin stem–loop that inhibits translation. DsrA transcription and stability are increased at low temperature, resulting in elevated RpoS levels (Repoila and Gottesman, 2001). In fact, DsrA has been called 'a thermometer for the RpoS regulon' (Repoila *et al.*, 2003).

Although *rpoS* mRNA levels in *B. burgdorferi* are regulated by temperature (Caimano *et al.*, 2004), Lybecker and Samuels (2007) identified a chromosomally encoded sRNA in the *B. burgdorferi* genome, not linked to *rpoS*, which was complementary to the *rpoS* promoter sequence. The *B. burgdorferi rpoS* upstream region, like that of *E. coli*, is strongly predicted to assume a hairpin stem–loop structure (Zuker, 2003), consistent with it serving an analogous regulatory role. $DsrA_{Bb}$ exists as four species of 213, 275, 290, and 352 nucleotides and their expression levels are not affected by temperature or cell density.

Analysis of *dsrA*$_{Bb}$ mutant strains demonstrated that $DsrA_{Bb}$ post-transcriptionally regulates the synthesis of RpoS exclusively at low cell density (Lybecker and Samuels, 2007), which is, arguably, most physiologically relevant. Complementation studies suggested that the *dsrA*$_{Bb}$ promoter was in the 3′ region of the hypothetical ORF *bb0577* and that the BB0577 protein was required for $DsrA_{Bb}$ function. $DsrA_{Bb}$ appears to be a molecular thermometer for RpoS regulation and was proposed to function during transmission from the tick vector to the vertebrate host (Lybecker and Samuels, 2007).

There are two different species of *rpoS* mRNAs. One is a novel long mRNA that contains the region complementary to $DsrA_{Bb}$, produced exclusively at low cell density, while the other is a shorter species that lacks the region complementary to $DsrA_{Bb}$, produced primarily at high cell density (Lybecker and Samuels, 2007). $DsrA_{Bb}$-mediated translational control operates only on the long *rpoS* transcript because $DsrA_{Bb}$ is not predicted to bind to the previously described short *rpoS* transcript (Smith *et al.*, 2007). There are no consensus RpoN promoters (Thöny and Hennecke, 1989; Studholme *et al.*, 2000; Studholme and Buck, 2000) upstream of the long *rpoS* 5′ end and the previous studies demonstrating RpoN dependence were performed at high cell density when the short rather than long *rpoS* mRNA is transcribed (Hübner *et al.*, 2001; Smith *et al.*, 2007). This raises crucial questions regarding the requirement for RpoN in the transcription of *rpoS* at low cell density and, of utmost importance, the physiological role of the two *rpoS* mRNA species in the enzootic cycle. Neither the long *rpoS* transcript (Blevins *et al.*, 2009) nor $DsrA_{Bb}$ (M.C. Lybecker and D.S. Samuels, unpublished) are required for vertebrate infection via syringe inoculation, suggesting that the RNA–RNA interaction between $DsrA_{Bb}$ and the long *rpoS* mRNA is only necessary, and possibly sufficient, for RpoS translation during the natural transmission from tick to vertebrate.

4.2.5 RpoS-dependent genes

The ospC *promoter*

Several laboratories have studied the mechanisms by which RpoS controls expression of *ospC* and several other lipoprotein genes. Previously, the two inverted repeats (IRs) located upstream of the –35/–10 *ospC* promoter were hypothesized to function as binding sites for unidentified transactivator(s) that control *ospC* transcription (Fuchs *et al.*, 1992; Marconi *et al.*, 1993; Margolis *et al.*, 1994; Padula *et al.*, 1994; Tilly *et al.*, 1997). Genetic evidence, however, from two independent studies indicated that removal of the IRs did not affect *ospC* induction when spirochaetes were cultivated *in vitro* and *ospC* was expressed *in trans* from a shuttle vector; a minimal –35/–10 sequence of the *ospC* promoter is sufficient for RpoS-dependent expression of *ospC*, suggesting that RpoS governs *ospC* expression by directly binding to the *ospC* promoter (Eggers *et al.*, 2004; Yang *et al.*, 2005). Originally, *ospC* was thought to have a σ^{70} promoter (Marconi *et al.*, 1993) instead of an RpoS (σ^S) promoter, which is not unexpected as these two sigma factors are closely related and recognize similar core promoter elements (Gaal *et al.*, 2001; Hengge-Aronis, 2002b). Promoter selectivity by σ^S and σ^{70} in *E. coli* is complicated, involving minor sequence differences, architectural DNA-binding proteins, or DNA supercoiling (Kusano *et al.*, 1996; Colland *et al.*, 2000; Becker and Hengge-Aronis, 2001; Gaal *et al.*, 2001; Hengge-Aronis, 2002b; Typas *et al.*, 2007). In *B. burgdorferi*, the sequence within the extended –10 region of the promoter contributes to RpoS recognition. A point mutation of C to G at position –13 abolished RpoS-dependent expression of *ospC*. Furthermore, replacing the –10 region of the σ^{70} promoter of the *ospE* gene with the –10 region of the σ^S promoter of *ospF* converted *ospE* to an RpoS-dependent gene (Eggers *et al.*, 2006). Recently, a putative RpoS consensus extended –10 region sequence of TG(G/A)(G/A)ATA(T/A)ATT has been proposed based on analysing the promoter sequences of ten RpoS-dependent genes of *B. burgdorferi* (Caimano *et al.*, 2007).

Although the upstream IRs appeared not to be involved in *ospC* regulation *in vitro*, recent data showed that these IR elements play a vital role in repressing *ospC* expression during mammalian host infection (Xu *et al.*, 2007). OspC is a strong immunogen and an effective target of host protective immunity (Fung *et al.*, 1994; Xu *et al.*, 2006). To develop persistent infection, *B. burgdorferi* must downregulate *ospC* as one of the mechanisms of immune evasion (Liang *et al.*, 2002a; Liang *et al.*, 2002c; Crother *et al.*, 2004; Liang *et al.*, 2004b; Tilly *et al.*, 2006; Xu *et al.*, 2006). Xu *et al.* (2007) reported that spirochaetes carrying an *ospC* gene without the IRs were no longer able to downregulate *ospC* and could not sustain chronic infection in mice. Thus, the IRs may function as binding sites for an unknown repressor(s) or as *cis*-acting sequences responsive to DNA supercoiling.

DNA supercoiling

ospC gene expression is not exclusively regulated by the level of RpoS. As mentioned above, OspC is required for colonizing the mammalian host, but its synthesis must be switched off early in infection for the spirochaetes to persist, while presumably other RpoS-dependent genes, such as *dbpBA* and *ospF* (Hefty *et al.*, 2001; Hefty *et al.*, 2002a; Liang *et al.*, 2004a), must be expressed throughout the course of infection. The *ospC* promoter responds to the level of DNA supercoiling (Alverson *et al.*, 2003) and the region responsible for this *cis*-acting regulatory mechanism has been mapped to the *ospC* promoter (Yang *et al.*, 2005). DNA supercoiling is relaxed in cultured *B. burgdorferi* after a temperature shift that mimics tick feeding, so DNA supercoiling has been hypothesized to serve as a molecular thermometer regulating the tick-to-mammal transmission (Alverson *et al.*, 2003). Increased temperature in other bacteria also results in less negatively supercoiled, or more relaxed, DNA (Tse-Dinh *et al.*, 1997; Hurme and Rhen, 1998; López-García and Forterre, 2000; Rui and Tse-Dinh, 2003). Specifically, DNA gyrase removes, rather than introduces, negative DNA supercoiling following heat shock (Kataoka *et al.*, 1996), until acted upon by the chaperone DnaK (Ogata *et al.*, 1996). The level of DNA supercoiling (Wang and Lynch, 1993; Hurme and Rhen, 1998; Lopez-Garcia and Forterre, 2000; Drolet, 2006) affects the expression of many genes (Peter *et al.*, 2004; Blot *et al.*, 2006; Jeong *et al.*, 2006), including those regulated by RpoS (Cheung *et al.*, 2003) and those encoding virulence factors (Dorman, 1995; Dorman and Corcoran, 2009).

DNA supercoiling is known as a factor in promoter recognition by RpoS. *E. coli* RpoS preferentially recognizes promoters on a relaxed template (Kusano *et al.*, 1996). In addition, DNA relaxation in *E. coli* is required for transcription by RNA polymerase holoenzyme containing RpoS during the cellular response to osmotic stress (Bordes *et al.*, 2003). In fact, DNA supercoiling has been proposed to be a post-translational regulator of RpoS activity (Hengge-Aronis, 2002b; Bordes *et al.*, 2003). Therefore, a role for DNA supercoiling in *ospC* gene regulation is consistent with the effect of DNA supercoiling on RpoS-dependent transcription in *E. coli*. Another role for DNA supercoiling could involve targeting the IRs, identified as part of the *ospC* operator by Xu *et al.* (2007). The IRs are predicted to extrude as stem loops (a cruciform) upon negative supercoiling, potentially blocking *ospC* transcription, which would be relieved by DNA relaxation associated with increased temperature. Accordingly, DNA supercoiling is probably relaxed in *B. burgdorferi* during transmission, thereby enabling RpoS-dependent *ospC* expression. However, negative DNA supercoiling is subsequently restored during the course of mammalian infection, which overrides the ability of RpoS to regulate *ospC* transcription, resulting in the requisite decrease in OspC levels. The corollary is that other RpoS-dependent genes that remain expressed during infection, such as *dbpBA* and *ospF*, would not be sensitive to this DNA supercoiling effect.

Other RpoS-*dependent genes*

In addition to *ospC*, RpoS controls more than 100 *B. burgdorferi* genes (Fisher *et al.*, 2005; Caimano *et al.*, 2007; Ouyang *et al.*, 2008). Most of the RpoS-induced genes have previously been identified by microarray analyses as differentially regulated under various conditions such as temperature, pH, addition of blood, or host-adapted conditions (Revel *et al.*, 2002; Brooks *et al.*, 2003; Ojaimi *et al.*, 2003; Tokarz *et al.*, 2004; Caimano *et al.*, 2007). These findings suggest that RpoS is the major transcriptional regulator controlling differential gene expression in *B. burgdorferi*, particularly during transmission to the mammal. Furthermore, many of these RpoS-dependent genes encode known or putative surface lipoproteins, such as *dbpBA* (Hübner *et al.*, 2001; Eggers *et al.*, 2004), which encodes adhesins DbpB and DbpA that bind the proteoglycan decorin and contribute to virulence in the mammal (Weening *et al.*, 2008; Shi *et al.*, 2008); *bbk32* (Caimano *et al.*, 2007; He *et al.*, 2007), which encodes a fibronectin-binding protein that may be important to mammalian infection (Fischer *et al.*, 2006; Li *et al.*, 2006; Seshu *et al.*, 2006); *mlp* paralogues (Yang *et al.*, 2003b), which encode multicopy lipoproteins (Porcella *et al.*, 1996); *ospF* (Lam *et al.*, 1994; Akins *et al.*, 1999; Caimano *et al.*, 2004; Eggers *et al.*, 2004; Eggers *et al.*, 2006), which is an outer surface lipoprotein shown to be important in transmission of the spirochaete from the tick to mammal (Hefty *et al.*, 2001; Hefty *et al.*, 2002b; Hefty *et al.*, 2002a); *bba64, bba65, bba66,* and *bba73,* which encode a collection of conserved, surface-exposed lipoproteins of unknown function (Hughes *et al.*, 2008; Brooks *et al.*, 2006; Clifton *et al.*, 2006); and *oppA-V* (Medrano *et al.*, 2007), which encodes an oligopeptide permease peptide-binding protein (Lin *et al.*, 2001). The majority of the RpoS-dependent genes are plasmid-borne hypothetical or conserved hypothetical genes. In addition, RpoS controls expression of a limited number of chromosomal genes, including genes involved in chemotaxis and metabolism (Fisher *et al.*, 2005; Caimano *et al.*, 2007; Ouyang *et al.*, 2008).

Mechanisms of RpoS-dependent regulation of some other genes appear more complicated than that of *ospC*. While *ospC* expression is under the direct control of RpoS, other RpoS-dependent genes may be indirectly controlled. For example, electrophoretic mobility shift assay (EMSA) and transcriptional fusion studies using a GFP reporter system demonstrated that a 29-bp inverted repeat (IR) element upstream of the *bba66* promoter is absolutely required for its transcription, suggesting that RpoS may influence *bba66* indirectly via a transactivator that binds to the IR motif (Clifton *et al.*, 2006). Further, this 29-bp IR motif is also found in the putative promoter regions upstream of *bba65, bba71,* and *bba73*, and a synthetic version of the 29-bp IR can competitively inhibit the interaction of the unidentified transacting factor with these promoter elements (D.R. Clifton and J.A. Carroll, unpublished). In addition to genes that are exclusively dependent on RpoS, a subset of RpoS-influenced genes shows only partial control by RpoS, such as *gltP* (encoding a glutamate transporter) and *cdr* (encoding NAD-dependent coenzyme A disulphide reductase). One possibility is that these genes have a promoter that can be recognized by both RpoS and the housekeeping sigma factor σ^{70} (RpoD) (Caimano *et al.*, 2007).

In addition to activating transcription, RpoS was shown to repress as many as 34 genes in mammalian host-adapted spirochaetes cultivated in DMCs (Caimano *et al.*, 2005; Caimano *et al.*, 2007). Several of these repressed genes have been previously demonstrated to be RpoD-dependent and downregulated during tick feeding, including *ospA, lp6.6,* and *bb0365* (de Silva *et al.*, 1996; Schwan and Piesman, 2000; Pal *et al.*, 2008). Some *bdr* paralogues, which encode members of the *Borrelia* direct repeat protein family with no known function (Zückert *et al.*, 1999), are also repressed by RpoS (Roberts *et al.*, 2002). While the mechanisms of RpoS-dependent repression of this group of genes remain unclear, the T-rich track upstream of the promoter was proposed to be involved in such regulation (Sohaskey *et al.*, 1999; Brooks *et al.*, 2003; Caimano *et al.*, 2007). Nevertheless, the findings that RpoS activates synthesis of OspC and other mammalian-associated proteins while repressing production of OspA and other tick-associated molecules demonstrates a molecular mechanism underlying the reciprocal pattern of production of OspA and OspC observed during the borrelial enzootic cycle.

4.2.6 Role of the Rrp2–RpoN–RpoS pathway in Lyme disease pathogenesis

Evidence has accumulated that the Rrp2–RpoN–RpoS pathway activates transcription of genes that are important for transmission or for vertebrate host infection, while repressing genes that are required for spirochaetal survival in the tick vector. Thus, this regulatory system plays a central role in modulating the dramatic switch in gene expression during the borrelial enzootic cycle. Consistent with this notion, Caimano *et al.* (2007) recently showed that *rpoS* expression is upregulated during nymphal tick feeding; in addition, *rrp2, rpoN* and *rpoS* mutants are incapable of establishing infection in the mammalian host (Caimano *et al.*, 2004; Fisher *et al.*, 2005; Boardman *et al.*, 2008; Ouyang *et al.*, 2008). These results are not surprising as this pathway controls OspC, a factor essential for mammalian infection. However, the Rrp2–RpoN–RpoS pathway appears to control additional virulence factors, since complementing the *rrp2* mutant with an *ospC* gene under the *flaB* promoter does not restore infection in mice (Boardman *et al.*, 2008). In addition to the importance to mammalian infection, this pathway appears to regulate expression of genes essential for transmission, since an *rpoN* mutant failed to migrate from the tick gut to the salivary gland during feeding (Fisher *et al.*, 2005).

RpoN and RpoS are widespread among bacteria and each is known to be involved in modulating differential gene expression in response to environmental changes (Kustu *et al.*, 1989; Hengge-Aronis, 2002a). Inactivation of *rpoS* in *Salmonella enterica* serovar Typhimurium significantly reduces virulence, whereas an *rpoN* mutant is much less virulent than the wild-type strain in several bacteria including *Pseudomonas aeruginosa, Vibrio cholerae,* and *Proteus mirabilis* (Kazmierczak *et al.*, 2005). Although a small portion of the genes regulated by RpoN and RpoS may overlap each other in some bacteria, the direct control of RpoS by RpoN to form the sigma factor cascade found in *B. burgdorferi* is unique. Sigma factor cascades are often used to amplify signals and to cross-regulate signalling pathways.

For example, a sigma factor cascade developmentally regulates endospore formation in *Bacillus* (Losick and Stragier, 1992; Kroos, 2007). While the rationale for *B. burgdorferi* employing the Rrp2–RpoN–RpoS cascade has not been fully elucidated, the global response provided by this unique sigma factor cascade is clearly required for *B. burgdorferi* to be transmitted successfully from the arthropod vector to the mammalian host. Although the Rrp2–RpoN–RpoS pathway controls expression of a vast network of genes and operons, there are still an abundant number of differentially expressed genes that are not regulated by this cascade. This observation suggests that other DNA-binding proteins and/or transcriptional regulators must be expressed by *B. burgdorferi* and play important roles in regulating gene transcription.

4.3 Architectural DNA-binding proteins

Only a few DNA-binding proteins potentially involved in gene regulation have been identified in *B. burgdorferi* (Tilly *et al.*, 1996; Fraser *et al.*, 1997; Knight and Samuels, 1999; Casjens *et al.*, 2000; Babb *et al.*, 2006), and only one, BosR/Fur (Boylan *et al.*, 2003; Katona *et al.*, 2004; Seshu *et al.*, 2004b; Boylan *et al.*, 2006; Medrano *et al.*, 2007), has been experimentally shown to regulate transcription (see section 4.4 below). *B. burgdorferi* encodes three different DNA-binding proteins that may play a role in gene regulation: Hbb, Gac, and EbfC/YbaB. Their function has not been defined, but they are likely to be architectural (or nucleoid-associated) DNA-binding proteins that play crucial roles in a variety of molecular processes, including transcription as well as replication, recombination, and chromosome structure maintenance (Drlica and Rouviere-Yaniv, 1987; Schmid, 1990; Travers *et al.*, 1994; Nash, 1996; Rice, 1997; Dame, 2005; Luijsterburg *et al.*, 2008; Saier, Jr., 2008). Most of these macromolecular processes involve multiple interactions between *trans*-acting DNA-binding proteins and *cis*-acting DNA binding sites. Architectural DNA-binding proteins are a class of small DNA-binding proteins, which includes HU, IHF, Fis, and H–NS, that bend DNA and promote the assembly of higher-order nucleoprotein complexes (Drlica and Rouviere-Yaniv, 1987; Schmid, 1990; Travers *et al.*, 1994; Nash, 1996; Rice, 1997; Dame, 2005; Luijsterburg *et al.*, 2008; Saier, 2008).

4.3.1 Hbb

Integration host factor (IHF) was discovered on the basis of its role in the integration of bacteriophage λ into the *E. coli* chromosome (Drlica and Rouviere-Yaniv, 1987; Friedman, 1988; Nash, 1996). IHF binds to DNA at specific sites and induces bending. The origin of replication on the *E. coli* chromosome contains an IHF-binding site, which is conserved in several bacteria, and IHF is essential for replication (Nash, 1996). Another small DNA-binding protein in *E. coli* is HU, which is homologous to IHF (Drlica and Rouviere-Yaniv, 1987; Schmid, 1990; Nash, 1996). HU also bends DNA, but it lacks sequence specificity. Many bacteria have homologues for both IHF and HU, while some species have only a single homologue.

Tilly *et al.* (1996) isolated a gene encoding an IHF/HU homologue, termed Hbb, from *B. burgdorferi* in a genetic screen based on complementation of the λ DNA packaging defect phenotype of *E. coli* IHF/HU mutants. Hbb is most similar to homologues from bacteria that encode a single IHF/HU homologue and is 27–30% identical to IHF and HU from *E. coli* (Tilly *et al.*, 1996). Analysis of the *B. burgdorferi* genome sequence indicates Hbb is the only member in the nucleoproteins category (Fraser *et al.*, 1997; Casjens *et al.*, 2000). Hbb is unable to complement an *E. coli* HU mutant for the defect in bacteriophage Mu growth (Tilly *et al.*, 1996) and it binds DNA in a sequence-specific fashion (Kobryn *et al.*, 2000; Mouw and Rice, 2007), suggesting that Hbb is more like IHF than HU. Kobryn *et al.* (2000) demonstrated that Hbb binds to a specific sequence that is related to the IHF consensus binding site. In addition, IHF binds to this Hbb-binding site with high affinity and Hbb will bind to an IHF consensus site (Mouw and Rice, 2007). However, Hbb affects DNA topology, which is a property of HU, so Hbb has a substantial non-specific DNA-binding activity and is not a straightforward IHF analogue (Kobryn *et al.*, 2000).

Mouw and Rice (2007) solved the structure of an Hbb–DNA complex (Fig. 4.3A). The Hbb structure and mechanism of DNA bending is similar to that of the IHF and HU proteins. Hbb binds DNA as a homodimer with two β-ribbon 'arms' that wrap in the minor groove, with a pair of conserved prolines intercalating between adjacent base pairs (Mouw and Rice, 2007), which results in a DNA bend of more than 180° (Kobryn *et al.*, 2000; Mouw and Rice, 2007). This Hbb-induced bend causes an increase in the sensitivity of the DNA to oxidative damage (Peng *et al.*, 2008).

The binding site identified by Kobryn *et al.* (2000) is near the origin of replication on the chromosome (between *dnaA* and *dnaN*). Thus, Hbb may be involved in the replication initiation (Chaconas, 2005) (see Chapter 3). However, Medrano *et al.* (2010) have recently demonstrated that Hbb binds specifically, and with high affinity, to the promoter region of the *p66* gene and with lower affinity to the promoter regions of the *flaB*, *ospA*, and *ospC* genes, suggesting a putative role in transcriptional regulation. In addition, preliminary data suggest that Hbb specifically binds to the promoter regions of several other genes (C.M. Schwanke, S.W. Knight, C.R. Kuchel, D.Z. Naigamwalla, G. Chaconas, and D.S. Samuels, unpublished), and that an *hbb* mutant displays decreased OspC synthesis and has a growth phenotype (A.C. Ng, M.C. Lybecker, S.F. Bundle, and D.S. Samuels, unpublished). The hypothesis that Hbb is involved in gene expression is supported by results in an *E. coli* surrogate system showing that recombinant Hbb represses transcription of a *p66* promoter fusion (Medrano *et al.*, 2010) and activates transcription of *oppA-II* and *oppA-IV* promoter fusions (Medrano *et al.*, 2007).

4.3.2 Gac

Knight and Samuels (1999) identified Gac in a biochemical screen for proteins that bound to a subtelomeric sequence from linear plasmid lp17. Gac has HU-like non-specific DNA-binding activity (Knight and Samuels, 1999; Corbett *et al.*, 2004), but it is a novel protein that completely lacks sequence similarity to HU or other members of the architectural DNA-binding protein family and has only been identified to date in *B. burgdorferi*. Gac is the C-terminal domain of GyrA (Knight and Samuels, 1999), the A subunit of DNA gyrase, which is the essential type II DNA topoisomerase that introduces negative DNA supercoiling in bacteria (Reece and Maxwell, 1991; Corbett and Berger, 2004). Gac is produced as an independent protein by a mechanism unprecedented in bacteria. *B. burgdorferi* transcribes two mRNAs carrying *gac* sequences: the larger one encodes the entire *gyrBA* operon that presumably generates DNA gyrase, and the

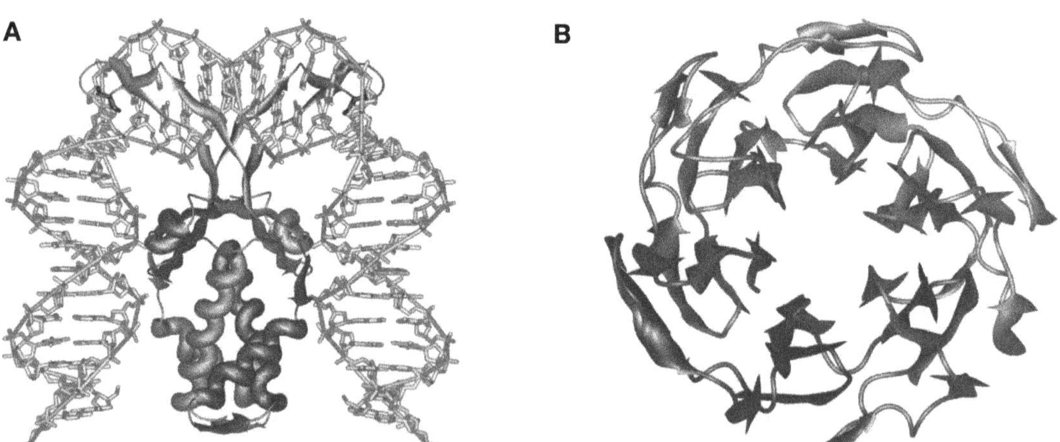

Figure 4.3 Structure of architectural DNA-binding proteins from *B. burgdorferi*. (A) Crystal structure of Hbb with a 35-bp DNA containing an optimized Hbb-binding site as determined by Mouw and Rice (2007); PDB 2NP2. The two Hbb protomers are shown in medium and dark shades of grey with the conserved prolines shown in black and the double helix shown in light grey. (B) Crystal structure of Gac as determined by Corbett *et al.* (2004); PDB 1SUU. Each of the six blades of the β-pinwheel is a different shade of grey. Images were generated using iMol 0.40.

smaller, more abundant one, encodes Gac. The *gac* mRNA is predicted to be transcribed from its own promoter (TTGAAT-N_{17}-TATTAT) embedded in the *gyrA* gene (Knight and Samuels, 1999).

This unusual architectural DNA-binding protein was termed Gac, and its gene *gac*, for GyrA C-terminal domain, because, although it is translated from a portion of the *gyrA* ORF, it is a different gene product of a unique genetic locus (Knight *et al.*, 2000). In other words, mutations can disrupt the synthesis of Gac, but have no phenotypic effect on GyrA. In fact, genetic disruption of Gac was accomplished by mutating the start codon and the next methionine codon (Met 5) to leucine and isoleucine, respectively, which had no detectable phenotype (Knight *et al.*, 2000). Samuels (2006) hypothesized that the phenotype was suppressed by the coumermycin A_1 resistance marker in the upstream *gyrB* gene (see Chapter 8) that was used to introduce the site-specific mutations. *gyrB* mutations suppress HU defects in *E. coli* (Malik *et al.*, 1996; Shanado *et al.*, 1998) and preliminary data using the kanamycin resistance marker inserted downstream of *gyrA*/*gac* suggest that a null *gac* mutant has a profound growth phenotype (D.S. Samuels, E.M. Todd, M.E. Kresge, M.G. Mazzotta, and C.R. Kuchel, unpublished).

Although reverse genetics has so far failed to reveal the function of Gac in *B. burgdorferi*, other experiments suggest that Gac is an architectural DNA-binding protein with HU activity. HU is essential for replicative transposition of bacteriophage Mu in *E. coli*: generating the Type I complex, an early intermediate, requires HU (Craigie *et al.*, 1985). Gac was able to substitute for HU, albeit not as efficiently, in promoting formation of the Type I complex in an *in vitro* donor-cleavage reaction (Knight and Samuels, 1999). Notably, Gac was not able to substitute in the assay for IHF. In addition, Gac, expressed from its own promoter, was able to complement the defect in Mu growth, as assayed by efficiency of plating, in an *E. coli* HU mutant (Knight and Samuels, 1999). Like HU, IHF, and Hbb, Gac bends DNA, which was demonstrated by FRET measurements (Corbett *et al.*, 2004). However, the mechanism by which Gac bends DNA is completely different than the other architectural DNA-binding proteins that involve a β-sheet in the minor groove. Instead, Gac has a novel structure termed a β-pinwheel by Corbett *et al.* (2004), which is similar to a β-propeller (Fig. 4.3B); the DNA wraps around the perimeter of four of the six blades of the pinwheel, resulting in a profound bend. This structure also provided insight into the enzymatic mechanism of DNA gyrase (Corbett *et al.*, 2004), as Gac, and the C-terminal domains of DNA gyrases from other bacteria, impart positive supercoiling on the wrapped DNA, which is converted to negative supercoiling following the strand-passage reaction (Ruthenburg *et al.*, 2005; Huang *et al.*, 2006; Kramlinger and Hiasa, 2006).

4.3.3 EbfC/YbaB

Babb *et al.* (2006) isolated a YbaB homologue, which they termed EbfC (for erp-binding factor, chromosomal), using a biochemical screen for proteins that bound to the upstream regions of the *ospE*, *ospF*, and *elp* genes, and they hypothesized that the protein was involved in regulation of these genes (Babb *et al.*, 2004). Three-dimensional modelling of the *B. burgdorferi* EbfC/YbaB protein against the crystal structure of the *Haemophilus influenzae* YbaB orthologue revealed an unusual structure (Lim *et al.*, 2003) in which two α-helices in the homodimer are predicted to form 'tweezers' that bind the major groove of the DNA helix (Riley *et al.*, 2009). Medrano *et al.* (2007) found that EbfC/YbaB specifically bound to the *oppA-V* promoter and increased transcription of an *oppA-V* promoter fusion in an *E. coli* surrogate system, providing support for the hypothesis that EbfC/YbaB was involved in transcriptional regulation. However, Riley *et al.* (2009) recently demonstrated that EbfC/YbaB not only binds to the sequence GTNAC, which occurs every 650 to 1400 bp in the borrelial genome, but that it also can bind sequences non-specifically that lack the GTNAC motif. This latter observation has led to the conclusion that the EbfC/YbaB DNA-binding protein is not a gene-specific regulatory protein as previously suggested (Babb *et al.*, 2006; Babb *et al.*, 2004), but is more likely an architectural DNA-binding protein that is involved in compacting the highly segmented genome of *B. burgdorferi* (Riley *et al.*, 2009).

While Hbb, Gac, EbfC/YbaB, and BosR/Fur all have DNA-binding activity, Hbb, Gac, and EbfC/YbaB probably have a function in genome structure and nucleation for *B. burgdorferi*. In addition, Hbb and EbfC/YbaB also may be involved in regulating gene transcription, albeit through apparent non-specific DNA–protein interactions. The BosR/Fur DNA-binding protein, however, has been empirically determined to regulate transcription in *B. burgdorferi* and the role of this protein in transcriptional regulation is discussed in further detail in the following section.

4.4 Transcriptional regulation via BosR/Fur protein

4.4.1 The BosR/Fur regulator

The redox environment is probably a physiologically relevant signal *B. burgdorferi* encounters and responds to during its enzootic life cycle. For example, as a *B. burgdorferi*-infected tick takes a blood meal, the spirochaete is exposed to reactive oxygen intermediates as a product of increased tick respiration. Exposure to reactive oxygen intermediates is also likely to be observed when *B. burgdorferi* initiates a mammalian infection and encounters cells of the innate immune response (see Chapter 15). Thus, how *B. burgdorferi* responds to the redox environment it encounters is likely to affect its ability to establish and maintain infection. This response for *B. burgdorferi* is most likely mediated by *trans*-acting factors, including BosR/Fur, that function to regulate genes required for detoxification of oxidizing compounds and may also affect genes that are required for pathogenesis.

A surprising observation from the genome sequence of *B. burgdorferi* was that the spirochaete did not encode many predicted regulatory proteins (Fraser *et al.*, 1997; Casjens *et al.*, 2000). There are only two TCSs and four proteins that share homology with other regulatory protein sequences within the database. One of these four proteins, encoded by *bb0647*, was annotated as Fur by TIGR (Fraser *et al.*, 1997) and contains two separate CXXC motifs involved in either metal binding and/or redox sensing. The observation that *B. burgdorferi* has little to no measurable iron (Fe), and apparently encodes no enzymes that are known to coordinate Fe, suggested that the Fur designation might be erroneous (Posey and Gherardini, 2000). Initial alignments supported the contention that borrelial Fur was actually a homologue of PerR, which is a member of the Fur family of regulators that is known to modulate oxidative stress genes in *Bacillus* spp. (Bsat *et al.*, 1998; Herbig and Helmann, 2001; Fuangthong *et al.*, 2002; Mongkolsuk and Helmann, 2002). Specifically, *B. subtilis* PerR and the *B. burgdorferi* homologue are 27% identical, 49% similar, and encode the two conserved CXXC motifs. Other homologues exhibit up to 35% identity and 55% similarity from a variety of bacterial species, including many firmicutes and a few thermophiles. From a functional perspective, *B. subtilis* PerR represses oxidative stress genes (like catalase) under low oxidizing conditions and disassociates from putative operator sequences after bound Fe is oxidized (Herbig and Helmann, 2001; Lee and Helmann, 2006b). Thus, although there are differences between *Bacillus* PerR and the *B. burgdorferi* homologue in regard to Fe binding, the similarity they share at the amino acid level suggests that they share functional attributes as well.

4.4.2 BosR/Fur as a transcriptional activator

The borrelial Fur/PerR homologue was first characterized by Boylan *et al.* (2003), who determined that this regulatory protein did not repress expression of *napA/dps* (*bb0690*), a gene whose product is involved in the oxidative stress response, but instead was required for transcriptional activation of this locus in response to oxidative stress. Since the *B. burgdorferi* Fur/PerR homologue did not repress *napA/dps*, the Fur/PerR homologue was renamed BosR for <u>b</u>orrelial <u>o</u>xidative <u>s</u>tress <u>r</u>egulator (Boylan *et al.*, 2003). In this study, BosR/Fur required zinc for DNA recognition to sequences upstream of *napA/dps* and the investigators determined that BosR/Fur bound to the *napA/dps* operator only when it was reduced; pre-oxidation was not compatible with binding (Boylan *et al.*, 2003). However, subsequent oxidation was permissible following binding and is presumed to be required for the activation of transcription. The activator function of BosR/Fur for *napA/dps* (*bb0690*) was ascertained through the use of an *E. coli lacZ* heterologous reporter system whereby the expression of *lacZ* from the borrelial *napA/dps* promoter (*napAp*) was dependent on concurrent *bosR/fur* expression (Boylan *et al.*, 2003). Furthermore, the expression of *napAp–lacZ* was accentuated by the addition of reactive oxygen stressors, including H_2O_2 and *t*-butyl hydroperoxide (Boylan *et al.*, 2003). In addition, BosR/Fur was found to protect a 50-bp region that was 137 to 187 bp upstream from the transcriptional start of *napA/dps* (Boylan *et al.*, 2003). Such a large footprint implies that BosR/Fur is capable of forming a large oligomeric structure that binds to a putative regulatory domain for *napA/dps*.

In a separate study, Boylan *et al.* (2006) reported that BosR/Fur also activated the CoA-disulphide reductase gene (*cdr*; *bb0728*) using a similar approach to that used for *napA/dps*. Medrano *et al.* (2007) subsequently reported that BosR/Fur recognizes the putative regulatory domain of *oppA-IV* (*bbb16*), a locus predicted to be involved in peptide transport in *B. burgdorferi*. In this study, transcription of *oppA-IV* was assessed using an *oppA-IVp–lacZ* reporter construct, coupled with *bosR/fur* inducible constructs on a separate plasmid in *E. coli*, comparable to the studies of *napA/dps* and *cdr* transcription (Boylan *et al.*, 2003; Boylan *et al.*, 2006; Medrano *et al.*, 2007). Transcription of *lacZ* was dependent on the induction of *bosR/fur* (Medrano *et al.*, 2007), suggesting that BosR/Fur can transcriptionally activate genes not overtly linked to the oxidative stress response, at least using an *E. coli* surrogate system. More importantly, the expression of *bosR/fur* was activated within the mammalian host, under the same conditions that *oppA-IV* is induced, thereby providing additional support to the idea that BosR/Fur directly activates expression of *oppA-IV* (Medrano *et al.*, 2007).

Based on these results, it appears that BosR/Fur functions more like OxyR than PerR. OxyR is a well-characterized transcriptional activator of the oxidative stress response in *E. coli* where it promotes the transcription of oxidative stress genes in response to H_2O_2 exposure (Zheng *et al.*, 1998; Aslund *et al.*, 1999; Choi *et al.*, 2001). The activation of OxyR requires oxidation of conserved cysteine residues to promote transcription of oxidative stress genes (Zheng *et al.*, 1998; Aslund *et al.*, 1999; Choi *et al.*, 2001). Like BosR/Fur, however, only the reduced form of OxyR is capable of the initial recognition and binding to DNA targets (Zheng *et al.*, 1998; Storz and Imlay, 1999). Additional studies will be needed to further evaluate the extent of a functional link between BosR/Fur and OxyR.

4.4.3 BosR/Fur as a repressor

In contrast to the studies above, an independent study by Benach and co-workers (Katona *et al.*, 2004) referred to BB0647 as Fur and suggested that it functioned as a repressor based on its ability to bind target sequences, including its own putative regulatory region, again only under reduced conditions. Oxidation of borrelial Fur prevented binding of the protein to *napA/dps* operator sequences (Katona *et al.*, 2004). Further studies of borrelial BosR/Fur by Seshu *et al.* (2004b) also suggested that that this transcriptional regulator functioned as a repressor, but these studies were limited to non-infectious *B. burgdorferi* due to the inability to isolate a *bosR/fur* mutant in infectious isolates. Specifically, inactivation of the *bosR/fur* gene resulted in the conversion of a redox-sensitive non-infectious clone of *B. burgdorferi*, CHP100, into a redox-resistant strain, not unlike the redox-resistant infectious *B. burgdorferi* strain B31 derivative MSK5 (Seshu *et al.*, 2004b) (J.T. Skare, unpublished). Complementation with a *bosR/fur* gene from infectious *B. burgdorferi* into strain JS167 (CHP100 *bosR::flgBp-aphI*) resulted in an incomplete complementation of the sensitivity to oxidative stressors. Complementation of JS167 with *bosR/fur* obtained from CHP100, however, did complement the redox-resistant phenotype back to a redox-sensitive one (Seshu *et al.*, 2004b), indicating that the *bosR/fur* alleles from these infectious and non-infectious *B. burgdorferi* strains were distinct. Subsequent sequence analysis indicated that the *bosR/fur* allele from the non-infectious strain contained a mutation that converted an arginine to lysine at residue 39 (R39K) (Seshu *et al.*, 2004b). This seemingly conservative amino acid substitution modified the DNA-binding activity relative to native BosR/Fur resulting in the loss of recognition for the *napA/dps* promoter/operator region and a differential affinity for the putative *sodA*-binding domain (Seshu *et al.*, 2004b). The different oxidative sensitivity phenotypes observed between non-infectious and infectious *B. burgdorferi* was proposed to be due in part to the R39K mutation (Seshu *et al.*, 2004b), but additional uncharacterized factors probably account for the differences observed. Two recent studies in which *bosR/fur* was either genetically disrupted (Hyde *et al.*, 2009) or fused to an inducible promoter (Hyde *et al.*, 2010) in an infectious background suggest that BosR/Fur confers resistance to oxidative stress.

4.4.4 A putative BosR/Fur regulon

The potential role of BosR/Fur in controlling a regulon in *B. burgdorferi* is predicated on the homology that BosR/Fur shares with PerR, a Fur family member, that is a regulatory protein modulating the expression of unlinked genes involved in the oxidative stress response in *B. subtilis* (Helmann *et al.*, 2003). To assess whether such a response is observed for *B. burgdorferi*, the strain containing the *bosRR39K::flgBp-aphI* mutant allele, JS167, was compared relative to its non-infectious parent strain, CHP100, through the use of global transcriptional profiling (Hyde *et al.*, 2006). The results showed that the expression of 88 genes was altered when BosR/Fur(R39K) was missing (Hyde *et al.*, 2006). Most of the genes whose transcripts were differentially regulated encode hypothetical proteins (58%); however, there were several loci whose altered regulation was predicted. Specifically, the sole superoxide dismutase (*sodA*; *bb0153*) was significantly induced when BosR/Fur(R39K) was missing, implying that the mutant BosR/Fur(R39K) functioned as a repressor of *sodA* in the non-infectious strain background analysed (Hyde *et al.*, 2006). The upregulation of *sodA* in JS167 is also consistent with the increased resistance to oxidative stressors (H_2O_2 and *t*-butyl hydroperoxide) observed relative to CHP100, as is the binding of BosR(R39K) to the putative regulatory domain of *sodA* (Seshu *et al.*, 2004b). In contrast, it was somewhat surprising that the transcription of several other genes that are apparently regulated by BosR/Fur, namely *napA/dps*, *cdr*, and *oppA-IV* (Boylan *et al.*, 2003; Boylan *et al.*, 2006; Medrano *et al.*, 2007), were not affected by the loss of BosR/Fur(R39K). Because BosR/Fur activates expression of all three of these genes, the expectation was that these transcripts would be underrepresented in the *bosRR39K::flgBp-aphI* mutant (JS167); however, no significant change was observed at any of these three loci (Hyde *et al.*, 2006). The differences in transcription observed (or lack of a difference) with these non-infectious strains may not be indicative of what occurs in infectious *B. burgdorferi*.

Other genes that were induced in JS167 include several phosphotransfer system (PTS) transporters, the *glp* operon (*glpF*, *glpK*, and *glpA*), *p66*, and *dbpA* (Hyde *et al.*, 2006). Again, as with *sodA*, the induction of these genes in the absence of BosR(R39K) implies that BosR(R39K) is functioning as a repressor for these genes in the non-infectious genetic background. The 7.2-fold induction of *dbpBA* corroborated data from an earlier study demonstrating that strain JS167 produced approximately fourfold more DbpA relative to strain CHP100 (Seshu *et al.*, 2004b). Together with the observed derepression of *p66*, whose gene product binds to host integrins (Coburn *et al.*, 1999; Defoe and Coburn, 2001) (see Chapter 12), BosR/Fur may regulate a subset of borrelial adhesins that are important for host adaptation following mammalian infection.

The induction/derepression of the PTS transporters and *glp* genes in the *bosRR39K::flgBp-aphI* strain implies that BosR/Fur is important for regulating genes involved in energy acquisition. Both the PTS transporters and *glp* gene products could serve to trap transported sugars (via phosphorylation within the cytoplasm following uptake) and concentrate glycerol, respectively, with both ultimately being funnelled into the glycolytic pathway to generate ATP.

The induction of *dbpA* in the absence of BosR/Fur(R39K) suggests that BosR/Fur is involved in adaptation to signals early in mammalian infection since DbpA is known to be important for pathogenic properties during experimental infection (Weening *et al.*, 2008; Shi *et al.*, 2008). In addition to a link to the BosR/Fur regulator, expression of *dbpBA* is Rrp2-RpoN-RpoS-dependent, as indicated above (Yang *et al.*, 2000, 2003a; Hübner *et al.*, 2001). This observation is particularly interesting inasmuch as the gene encoding *rrp2* is significantly downregulated in JS167 relative to CHP100 (Hyde *et al.*, 2006). The exact linkage of BosR/Fur and Rrp2 is not clear at this time. Thus, the regulatory effect observed when strain JS167 and CHP100 are compared may be representative of what happens in infectious *B. burgdorferi*, or it may be specific to the non-infectious CHP100 background that has known genotypic and phenotypic differences (Seshu *et al.*, 2004b; Hyde *et al.*, 2006).

Surprisingly, BosR/Fur was shown to be required for induction of RpoS and its regulon (Hyde *et al.*, 2009, 2010; Ouyang *et al.*, 2009). This observation ties BosR/Fur into the transcriptional regulation of genes involved in the enzootic cycle (see section 4.2 above); it is also consistent with the expression of the *bosR/fur* gene being at its highest during mammalian infection (Medrano *et al.*, 2007) and with BosR/Fur being required for mammalian infection (Hyde *et al.*, 2009; Ouyang *et al.*, 2009). Ouyang *et al.* (2009) also identified 50 genes whose expression was dependent on BosR/Fur, but not RpoS. Of these, eight have consensus Per boxes in their promoter regions that potentially can be recognized by BosR/Fur according to Katona *et al.* (2004). Curiously, transcription of *napA/dps* was not dependent on BosR/Fur in this study (Ouyang *et al.*, 2009), which was discordant with two other recent reports using infectious strains (Hyde *et al.*, 2009, 2010).

4.4.5 Metal and redox requirements for BosR/Fur activity

Despite the recent findings in regard to BosR/Fur function/activity, there are many questions that still remain unanswered. Specifically, which metals does BosR/Fur coordinate? Although metals have been added to BosR/Fur for various assays, no direct measure of metal binding has been tested. This issue is highly relevant because the closest homologue of BosR/Fur, the Fur family member PerR, can coordinate two separate metals simultaneously, either Mn and Zn or Fe and Zn (Herbig and Helmann, 2001). The form of PerR that coordinates Fe is more responsive to oxidative stressors due to metal-catalysed oxidation of the Fe, which oxidizes histidine residues of PerR, and then modifies its binding activity (Herbig and Helmann, 2001). The Mn/Zn form of PerR is also affected by redox but not to the same extent as the Fe/Zn form, as are the CXXC motifs (Lee and Helmann, 2006a). Originally, one of the PerR CXXC domains was believed to serve as the redox sensor but, based on its crystal structure, both of the CXXC motifs coordinate Zn while a separate site in PerR, coordinated via histidine and aspartic acid groups, binds either Fe or Mn and senses the redox microenvironment (Lee and Helmann, 2006a; Traore *et al.*, 2006). Whether BosR/Fur utilizes a similar strategy as PerR to bind different metals that are contingent for optimal function (with the noted exception of Fe) remains to be determined.

4.4.6 Model for BosR/Fur function

A working model can now be generated that takes into account what is currently known or hypothesized to occur regarding BosR/Fur function (Fig. 4.4). BosR/Fur presumably serves to regulate several unlinked genes in response to the redox status of the cell. Although there is some controversy regarding the metal content required for BosR/Fur function (Boylan *et al.*, 2006; Boylan *et al.*, 2003; Katona *et al.*, 2004; Seshu *et al.*, 2004b), the model hypothesizes that Zn is the predominant metal required as a cofactor for activity. Furthermore, the model supposes that metal binding occurs under reduced conditions, but BosR/Fur may coordinate additional metals under both reduced and oxidized conditions.

In the model presented, when cells are exposed to anaerobic (reduced) conditions, BosR/Fur coordinates metal and binds to target sequences and either represses expression or is primed for expression depending on whether BosR/Fur functions as a repressor or activator for a given gene, respectively. What happens next dictates whether BosR/Fur functions as a repressor or activator for a specific gene or operon. Upon oxidation, if BosR/Fur is an activator, the protein remains bound and undergoes a conformational shift that results in transcriptional activation of the target gene. The alternative possibility is that the oxidation results in a conformational change that causes BosR/Fur to disassociate, leading to transcription of the target gene. The possibility that BosR/Fur can function as both a repressor and an activator is supported by the fact that the Fur proteins from other pathogenic bacteria can perform both functions (Grifantini *et al.*, 2003; Lee *et al.*, 2004; Delany *et al.*, 2004; Gao *et al.*, 2008).

The exact signal perceived by *B. burgdorferi* in response to oxidative stress is not known but may be a released alkyl group or a retained alcohol or aldehyde structure as a result of

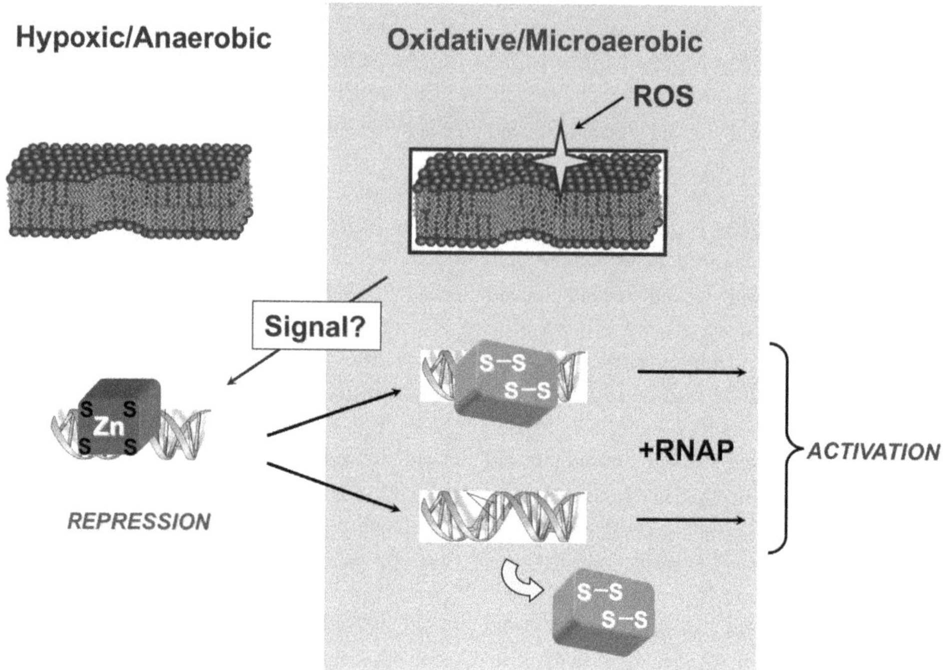

Figure 4.4 Model for BosR/Fur function. In the model proposed, under hypoxic conditions, reduced BosR binds Zn^{2+} (left) and binds to target sequences, repressing their expression. Upon oxidation (via ROS and perhaps an unknown signal or signals), BosR changes conformation (right) that either facilitates its interaction with RNA polymerase (RNAP) (top) or causes it to disassociate, allowing RNAP to bind (bottom). Either way, RNAP initiates transcription.

polyunsaturated lipid peroxidation as recently reported by Boylan *et al.* (2008). Alternatively, the oxidation of carbonyl groups on proteins may be sensed. Furthermore, how the signal is integrated into BosR/Fur is not clear. The oxidation may eliminate the coordinated metal via reduced thiol groups from the CXXC motifs as indicated in the model proposed in Fig. 4.4, but this has not been empirically proven. Alternatively, additional histidine residues may be the 'sensor' of oxidation that results in the change in BosR/Fur activity. As indicated above, this type of response has been observed for PerR from *B. subtilis*, but to date has not been reported for the *B. burgdorferi* BosR/Fur protein (Lee and Helmann, 2006b).

Recently, *bosR/fur* mutants have been generated in an infectious background, which demonstrated that BosR/Fur is required for infection of the mammalian host (Hyde *et al.*, 2009; Ouyang *et al.*, 2009). The finding that BosR/Fur interfaces with the well-characterized Rrp2-RpoN-RpoS pathway (Hyde *et al.*, 2009, 2010; Ouyang *et al.*, 2009) is of great interest (section 4.2; Samuels and Radolf, 2009). However, the regulatory role that BosR/Fur (BB0647) plays in the oxidative stress response, or in transition metal homeostasis, as well as its mechanism of activation and function in the enzootic cycle, remain unclear (Samuels and Radolf, 2009).

4.5 Proteomics

4.5.1 Use of proteomics in tracking differential protein production

Proteomics is the simultaneous analysis of complex protein mixtures, such as cell lysates or tissue extracts, to quantitatively determine changes in protein levels. Since the initial coining of the term, several categories of proteomics have evolved, including expression proteomics, functional proteomics, and cell-mapping proteomics. Expression proteomics, also referred to as global proteomics, classically makes use of two-dimensional (2-D) gels in combination with mass spectrometry to identify protein differences between varying conditions. Recent advancements in expression proteomic methods beyond the use of 2-D-gels have now been employed to compare protein profiles between various bacterial organisms cultivated under different environmental conditions. Similarly, proteomic studies involving *B. burgdorferi* have also focused on identifying proteins differentially produced under various experimental conditions, with the underlying goal being a better understanding of the proteins required for *B. burgdorferi* to successfully transition between its tick and mammalian hosts.

4.5.2 One-dimensional (1-D) and two-dimensional (2-D) proteome comparisons

Prior to the concept of the proteome, investigators compared the protein profiles of various *B. burgdorferi* samples by 1-D and 2-D SDS-PAGE gel electrophoresis to gain insight into the repertoire of virulence determinants expressed by the spirochaete. Early studies using immunoblotting or staining of 1-D- and 2-D-separated proteins demonstrated that there were differences in the proteome profiles of (i) different isolates and genospecies (Wilske *et al.*, 1986, 1992a,b, 1993a–c; Theisen *et al.*, 1993), (ii) repeatedly *in vitro*-cultivated spirochaetes (Norris *et al.*, 1992; Skare *et al.*, 1995, 1999; Carroll and Gherardini, 1996), and (iii) organisms during persistent infection of rodents (Schwan *et al.*, 1988; Schwan *et al.*, 1989; Schwan *et al.*, 1991; Roehrig *et al.*, 1992; Schwan *et al.*, 1993). Many of these differentially expressed proteins were hypothesized to be involved in pathogenesis by allowing the spirochaetes to establish an infection and/or persist in the mammal by evasion of the host immune response. Comparing infectious (low-passage) and non-infectious (high-passage) *B. burgdorferi* isolates has proven advantageous in detecting antigens associated with infectious phenotypes and has assisted in the identification of proteins involved in strain-specific pathogenicity (Schwan *et al.*, 1988; Schwan and Simpson, 1991; Norris *et al.*, 1992; Skare *et al.*, 1995; Carroll and Gherardini, 1996; Skare *et al.*, 1999; Anguita *et al.*, 2000; Schwan and Piesman, 2000). In many instances, the differences observed between low- and high-passage isolates of *B. burgdorferi* can now, in retrospect, be associated with virulence-associated plasmids that were lost during *in vitro* cultivation, particularly lp25 and lp28-1 (Purser and Norris, 2000; Labandeira-Rey and Skare, 2001; McDowell *et al.*, 2001). In regard to differential protein production in infectious isolates, initial investigations by Schwan and co-workers (1995) demonstrated that altering the temperature of the medium could also result in a change in the expression of OspC. This observation led to similar experiments determining that the OspE, OspF, and Elp proteins also were differentially expressed in response to changes in temperature (Akins *et al.*, 1995; Stevenson *et al.*, 1995; Stevenson *et al.*, 1996; Akins *et al.*, 1999; Hefty *et al.*, 2001). These initial observations of the differential regulation of numerous proteins in response to various environmental cues ultimately provided the impetus for subsequent microarray analyses of *B. burgdorferi* (Ojaimi *et al.*, 2002; Revel *et al.*, 2002; Liang *et al.*, 2002b; Ojaimi *et al.*, 2003; Brooks *et al.*, 2003; Narasimhan *et al.*, 2003; Tokarz *et al.*, 2004; Caimano *et al.*, 2007).

4.5.3 Two-dimensional gel electrophoresis and expression proteomics

Two-dimensional SDS-PAGE analyses in combination with immunoblotting have been useful in identifying particular proteins synthesized by *B. burgdorferi* under various environmental conditions. Two-dimensional gel analysis followed by mass spectrometry allowed for the direct identification of proteins (and the genes) that were differentially expressed *in vitro*. Comparisons of *B. burgdorferi* grown at pH 6.0 or pH 7.0 (a pH range similar to that of the fed tick and tissues of the mammalian host) and pH 8.0 (a pH similar to that of the unfed tick midgut) by 2-D non-equilibrium pH-gradient gel electrophoresis (2-D-NEPHGE) (Carroll *et al.*, 1999) revealed over 37 differentially regulated membrane proteins. Most of the changes in the protein profile were observed when samples from pH 7.0 and pH 8.0 were compared. The identity of a majority of these proteins were unknown, but one prominent protein spot whose production increased when cells were grown at pH 7.0 relative to pH 8.0 proved to be the temperature-regulated, RpoS-dependant, outer surface lipoprotein OspC. This finding suggested that OspC, as well as other temperature-regulated proteins, may be under the coordinated control of both temperature and pH (Yang *et al.*, 2000): expression would be upregulated at pH 7.0 and 35°C (mammalian and feeding tick environment) compared to pH 8.0 and 23°C (unfed tick environment), which implies that OspC serves as an indicator for mammalian host adaptation, at least during the early transition from the tick to mammal. In a following study, eleven of the pH-regulated proteins were identified using matrix-assisted laser desorption/ionization-time of flight (MALDI-TOF) mass spectrometry, and Northern analysis of mRNA suggested that these genes were regulated at the level of transcription (Carroll *et al.*, 2000). Using proteomics approaches, over twenty genes regulated, at least in part, in response to pH changes have been identified, including *ospC*, *dbpBA*, *revA*, *bba64*, and *bba66* (Carroll *et al.*, 2000; Nowalk *et al.*, 2006b). Interestingly, many of the proteins identified in these studies are now known to be encoded by genes regulated in an RpoS-dependant manner, which prompted proteome analyses on an RpoS mutant strain. Two-dimensional gel analysis of an *rpoS* mutant in an avirulent high-passage strain as compared to the parental high-passage isolate yielded few differences in the proteomes, although there was an observable difference in their ability to tolerate high salt conditions (Elias *et al.*, 2000). More recently, several investigators have been successful at mutating *rpoS* in infectious strains of *B. burgdorferi*, making these analyses more biologically relevant. As mentioned above, global microarray analyses revealed the expected pleiotropic effect seen following the loss of RpoS expression, which is now known to affect the transcription of multiple unlinked genes (Caimano *et al.*, 2007). Subsequent studies indicated that the disruption of *rpoS* resulted in a decrease in OspC, DbpA, DbpB, BBA65, BBA66, BBA71, and BBA73.

4.5.4 Use of proteomics to assess post-translational modification of *Borrelia burgdorferi* proteins

Proteomics can also be used as an informational tool to define how the loss of a specific protein affects the overall homeostasis of the cell. Examples of this include the loss of either (i) global regulators (i.e. transcriptional regulators or sigma factors) that affect the production of several proteins or (ii) protein-modifying enzymes including proteases and kinases. One example for *B. burgdorferi* involved the effect seen following the loss of CtpA, the C-terminal protease in *B. burgdorferi*. Östberg *et al.* (2004) produced a null mutant in the gene encoding CtpA in *B. burgdorferi* and compared the 2-D-NEPHGE protein profile to that of the parental wild-type strain. This study revealed that along with outer membrane protein P13, the membrane-associated protein BB0323 was also proteolytically cleaved by this protease. BB0323 contains a LysM domain that may allow it to interact with the peptidoglycan layer (Stewart *et al.*, 2004) and is immunoreactive with sera from Lyme patients at various stages of disease (Nowalk *et al.*, 2006a). Moreover, the inactivation of *ctpA* appeared to have pleiotropic effects on the protein profile as seen by the appearance of multiple forms of BBA74 when analysed by 2-D-NEPHGE (Östberg *et al.*, 2004). These multiple forms of BBA74 observed may be the result of either direct or indirect effects that occur in the absence of this protease. One possible explanation involves the role of CtpA in the post-translational regulation of additional proteases that protects BBA74 from proteolytic processing.

4.5.5 Proteomic analysis of host-adapted *Borrelia burgdorferi*

Two techniques have been utilized that allow for a more accurate representation of the proteome of *B. burgdorferi* expressed during infection of mammals. These are the use of (i) the dialysis membrane chamber (DMC) implant system (Akins *et al.*, 1998; Sellati *et al.*, 1999; Stewart *et al.*, 2006) and (ii) Triton X-114 to extract borrelial lipoproteins from the tissues of infected animals, a procedure termed hydrophobic antigen tissue Triton extraction (HATTREX) (Crother *et al.*, 2003; Crother *et al.*, 2004). Analysis of spirochaetes that have been mammalian host-adapted in DMCs compared to cultured spirochaetes indicates an increase in the synthesis of several proteins, including OspC and several OspE and OspF paralogues. Furthermore, the DMC proteome studies also demonstrated that the production of OspA and the small lipoprotein Lp6.6 are dramatically decreased within DMCs, which supported the hypothesis that OspA and

Lp6.6. may be more relevant to *B. burgdorferi* within the tick vector (Lahdenne *et al.*, 1997; Akins *et al.*, 1998). Subsequent studies have corroborated this hypothesis since OspA functions as an adhesin molecule that binds TROSPA in the tick midgut, indicating that it plays a key role in tick colonization and maintenance (see Chapter 11).

The HATTREX method of Crother *et al.* (2004) has been utilized to isolate *B. burgdorferi* in the host-adapted state. Briefly, the technique involves extracting *B. burgdorferi* hydrophobic proteins (membrane proteins and lipoproteins) from infected skin samples of previously infected mice or rabbits at various time points during the course of disease using the non-ionic detergent Triton X-114 (Crother *et al.*, 2003, 2004). Initially HATTREX was combined with separation of extracted proteins by immobilized pH gradient 2-D gel electrophoresis (IPG-2-DGE) and immunoblotted with a small subset of specific antibodies to identify the expression of antigens in various tissues of SCID mice (Crother *et al.*, 2003). Major surface lipoproteins OspA and OspB were absent in all mouse samples tested, as expected, but OspC and DbpA were produced at higher levels in ear, ankle, knee, and heart tissue compared to spirochaetes grown *in vitro*. Production of the VlsE lipoprotein was also observed to be elevated in ear and ankle extracted samples relative to infected heart and *in vitro* grown *B. burgdorferi*. Similar results were obtained with HATTREX samples from the skin of infected New Zealand white rabbits separated by IPG-2-DGE and probed with a similar subset of antibodies (Crother *et al.*, 2004). Specifically, samples from rabbit skin were analysed in a temporal fashion, with tissues harvested at days 5, 7, 9, 11, 14, and 21 post infection. Changes in the protein profile were visualized using infection-derived immune sera, and the subsequent identity of these proteins determined using *in situ* trypsin digestion and micro-liquid chromatography–tandem mass spectrometry (micro-LC-MS/MS). These findings demonstrated differential protein synthesis by *B. burgdorferi* during infection of the skin. Several proteins, including various OspE paralogues, VlsE, BmpA, and DbpA were observed to increase by day seven of infection and remained detectable throughout the study. Other proteins identified using this methodology were OspC, OppA-II, LA7, and the OspE paralogue ErpP. While all of these proteins were identified in the HATTREX method from infected rabbit skin by day seven, none of these proteins were detected at day 21 post-infection. Ultimately, the combined studies by Crother *et al.* (2004) and Akins *et al.* (1998) using mammalian host-adapted spirochaetes have reinforced the notion that the *B. burgdorferi* proteome expressed *in vitro* is dramatically different from the proteome expressed *in vivo* during mammalian infection. These combined global proteome assays, like the global microarray studies, further demonstrate that the borrelial proteome is dynamic and undergoes major changes in regulation during infection.

4.6 Summary

The advent of global transcriptional microarray analyses has provided important insight into how *B. burgdorferi* modulates gene expression to various experimental conditions or environmental stimuli that either individually or collectively simulate a host-adapted condition. The total proteome undergoes major changes during transmission from the tick to mammalian host. The microarray and proteomic analyses are highly correlative, which would suggest that regulation of protein expression in different environments is controlled primarily at the level of transcription. While multiple alternative sigma factors, DNA-binding proteins, and transcriptional regulators have been characterized, much still remains to be elucidated about the transcriptional regulation that occurs during the infectious cycle of *B. burgdorferi*. The studies performed thus far have provided a stable foundation for future studies that can assess other parameters and environmental cues that affect gene expression and protein production in this spirochaete. The recent advances made in genetic manipulation of virulent strains of *B. burgdorferi* also will help to provide new avenues for determining the importance of specific genes and proteins in *B. burgdorferi* virulence and Lyme disease pathogenesis.

Note added in proof

A c-di-GMP-binding protein was recently identified (Freedman *et al.*, 2009).

Acknowledgements

The authors graciously thank the current and former members of their respective laboratories as well as their colleagues for their contributions to research on gene and protein regulation in *B. burgdorferi*. We also thank Melisha Kenedy, Melissa Caimano, and Meghan Lybecker for thoughtful and critical reading of the manuscript. The authors also thank Mary Burtnick for providing the graphics file for Fig. 4.2. Supported by grants AI042345 and AI058086 (JTS), AI083640, AR054023, and AI072144 (XFY), AI051486 (DSS), and AI059373 and AI064629 (DRA) from the National Institutes of Health; JAC was supported by grants 0855452-D (American Heart Association) and CI000181 (Centers for Disease Control and Prevention).

References

Akins, D.R., Bourell, K.W., Caimano, M.J., Norgard, M.V., and Radolf, J.D. (1998). A new animal model for studying Lyme disease spirochetes in a mammalian host-adapted state. J. Clin. Invest. *101*, 2240–2250.

Akins, D.R., Caimano, M.J., Yang, X., Cerna, F., Norgard, M.V., and Radolf, J. (1999). Molecular and evolutionary analysis of *Borrelia burgdorferi* 297 circular plasmid-encoded lipoproteins with OspE- and OspF-like leader peptides. Infect. Immun. *67*, 1526–1532.

Akins, D.R., Porcella, S.F., Popova, T.G., Shevchenko, D., Baker, S.I., Li, M., Norgard, M.V., and Radolf, J.D. (1995). Evidence for *in vivo* but not *in vitro* expression of a *Borrelia burgdorferi* outer surface protein F (OspF) homolog. Mol. Microbiol. *18*, 507–520.

Alverson, J., Bundle, S.F., Sohaskey, C.D., Lybecker, M.C., and Samuels, D.S. (2003). Transcriptional regulation of the *ospAB* and *ospC* promoters from *Borrelia burgdorferi*. Mol. Microbiol. *48*, 1665–1677.

Anguita, J., Samanta, S., Revilla, B., Suk, K., Das, S., Barthold, S.W., and Fikrig, E. (2000). *Borrelia burgdorferi* gene expression *in vivo* and spirochete pathogenicity. Infect. Immun. *68*, 1222–1230.

Aslund, F., Zheng, M., Beckwith, J., and Storz, G. (1999). Regulation of the OxyR transcription factor by hydrogen peroxide and the cellular thiol-disulfide status. Proc. Natl. Acad. Sci. U.S.A. *96*, 6161–6165.

Babb, K., Bykowski, T., Riley, S.P., Miller, M.C., Demoll, E., and Stevenson, B. (2006). *Borrelia burgdorferi* EbfC, a novel, chromosomally encoded protein, binds specific DNA sequences adjacent to *erp* loci on the spirochete's resident cp32 prophages. J. Bacteriol. *188*, 4331–4339.

Babb, K., El-Hage, N., Miller, J.C., Carroll, J.A., and Stevenson, B. (2001). Distinct regulatory pathways control expression of *Borrelia burgdorferi* infection-associated OspC and Erp surface proteins. Infect. Immun. *69*, 4146–4153.

Babb, K., McAlister, J.D., Miller, J.C., and Stevenson, B. (2004). Molecular characterization of *Borrelia burgdorferi erp* promoter/operator elements. J. Bacteriol. *186*, 2745–2756.

Bankhead, T., and Chaconas, G. (2007). The role of VlsE antigenic variation in the Lyme disease spirochete: persistence through a mechanism that differs from other pathogens. Mol. Microbiol. *65*, 1547–1558.

Barbour, A.G., Tessier, S.L., and Hayes, S.F. (1984). Variation in a major surface protein of Lyme disease spirochetes. Infect. Immun. *45*, 94–100.

Barbour, A.G., Tessier, S.L., and Todd, W.J. (1983). Lyme disease spirochetes and ixodid tick spirochetes share a common surface antigenic determinant defined by a monoclonal antibody. Infect. Immun. *41*, 795–804.

Beck, L.L., Smith, T.G., and Hoover, T.R. (2007). Look, no hands! Unconventional transcriptional activators in bacteria. Trends Microbiol. *15*, 530–537.

Becker, G., and Hengge-Aronis, R. (2001). What makes an *Escherichia coli* promoter σ^S dependent? Role of the −13/−14 nucleotide promoter positions and region 2.5 of σ^S. Mol. Microbiol. *39*, 1153–1165.

Blevins, J.S., Xu, H., He, M., Norgard, M.V., Reitzer, L., and Yang, X.F. (2009). Rrp2, a σ^{54}-dependent transcriptional activator of *Borrelia burgdorferi*, activates *rpoS* in an enhancer-independent manner. J. Bacteriol. *191*, 2902–2905.

Blot, N., Mavathur, R., Geertz, M., Travers, A., and Muskhelishvili, G. (2006). Homeostatic regulation of supercoiling sensitivity coordinates transcription of the bacterial genome. EMBO Rep. *7*, 710–715.

Boardman, B.K., He, M., Ouyang, Z., Xu, H., Pang, X., and Yang, X.F. (2008). Essential role of the response regulator Rrp2 in the infectious cycle of *Borrelia burgdorferi*. Infect. Immun. *76*, 3844–3853.

Bordes, P., Conter, A., Morales, V., Bouvier, J., Kolb, A., and Gutierrez, C. (2003). DNA supercoiling contributes to disconnect σ^S accumulation from σ^S-dependent transcription in *Escherichia coli*. Mol. Microbiol. *48*, 561–571.

Boylan, J.A., Hummel, C.S., Benoit, S., Garcia-Lara, J., Treglown-Downey, J., Crane, E.J., III, and Gherardini, F.C. (2006). *Borrelia burgdorferi bb0728* encodes a coenzyme A disulphide reductase whose function suggests a role in intracellular redox and the oxidative stress response. Mol. Microbiol. *59*, 475–486.

Boylan, J.A., Posey, J.E., and Gherardini, F.C. (2003). *Borrelia* oxidative stress response regulator, BosR: a distinctive Zn-dependent transcriptional activator. Proc. Natl. Acad. Sci. U.S.A. *100*, 11684–11689.

Boylan, J.A., Lawrence, K.A., Downey, J.S., and Gherardini, F.C. (2008). *Borrelia burgdorferi* membranes are the primary targets of reactive oxygen species. Mol. Microbiol. *68*, 786–799.

Brahmachary, P., Dashti, M.G., Olson, J.W., and Hoover, T.R. (2004). *Helicobacter pylori* FlgR is an enhancer-independent activator of σ^{54}-RNA polymerase holoenzyme. J. Bacteriol. *186*, 4535–4542.

Brooks, C.S., Hefty, P.S., Jolliff, S.E., and Akins, D.R. (2003). Global analysis of *Borrelia burgdorferi* genes regulated by mammalian host-specific signals. Infect. Immun. *71*, 3371–3383.

Brooks, C.S., Vuppala, S.R., Jett, A.M., and Akins, D.R. (2006). Identification of *Borrelia burgdorferi* outer surface proteins. Infect. Immun. *74*, 296–304.

Bsat, N., Herbig, A., Casillas-Martinez, L., Setlow, P., and Helmann, J.D. (1998). *Bacillus subtilis* contains multiple Fur homologues: identification of the iron uptake (Fur) and peroxide regulon (PerR) repressors. Mol. Microbiol. *29*, 189–198.

Burtnick, M.N., Downey, J.S., Brett, P.J., Boylan, J.A., Frye, J.G., Hoover, T.R., and Gherardini, F.C. (2007). Insights into the complex regulation of *rpoS* in *Borrelia burgdorferi*. Mol. Microbiol. *65*, 277–293.

Byram, R., Stewart, P.E., and Rosa, P. (2004). The essential nature of the ubiquitous 26-kilobase circular replicon of *Borrelia burgdorferi*. J. Bacteriol. *186*, 3561–3569.

Caimano, M.J., Eggers, C.H., Hazlett, K.R., and Radolf, J.D. (2004). RpoS is not central to the general stress response in *Borrelia burgdorferi* but does control expression of one or more essential virulence determinants. Infect. Immun. *72*, 6433–6445.

Caimano, M.J., Iyer, R., Eggers, C.H., Gonzalez, C., Morton, E.A., Gilbert, M.A., Schwartz, I., and Radolf, J.D. (2007). Analysis of the RpoS regulon in *Borrelia burgdorferi* in response to mammalian host signals provides insight into RpoS function during the enzootic cycle. Mol. Microbiol. *65*, 1193–1217.

Caimano, M.J., Eggers, C.H., Gonzalez, C.A., and Radolf, J.D. (2005). Alternate sigma factor RpoS is required for the *in vivo*-specific repression of *Borrelia burgdorferi* plasmid lp54-borne *ospA* and *lp6.6* genes. J. Bacteriol. *187*, 7845–7852.

Carroll, J.A., Cordova, R.M., and Garon, C.F. (2000). Identification of 11 pH-regulated genes in *Borrelia burgdorferi* localizing to linear plasmids. Infect. Immun. *68*, 6677–6684.

Carroll, J.A., El Hage, N., Miller, J.C., Babb, K., and Stevenson, B. (2001). *Borrelia burgdorferi* RevA antigen is a surface-exposed outer membrane protein whose expression is regulated in response to environmental temperature and pH. Infect. Immun. *69*, 5286–5293.

Carroll, J.A., Garon, C.F., and Schwan, T.G. (1999). Effects of environmental pH on membrane proteins in *Borrelia burgdorferi*. Infect. Immun. *67*, 3181–3187.

Carroll, J.A., and Gherardini, F.C. (1996). Membrane protein variations associated with in vitro passage of *Borrelia burgdorferi*. Infect. Immun. *64*, 392–398.

Carroll, J.A., Stewart, P.E., Rosa, P., Elias, A.F., and Garon, C.F. (2003). An enhanced GFP reporter system to monitor gene expression in *Borrelia burgdorferi*. Microbiology *149*, 1819–1828.

Casjens, S., Palmer, N., van Vugt, R., Huang, W.M., Stevenson, B., Rosa, P., Lathigra, R., Sutton, G., Peterson, J., Dodson, R.J., *et al.* (2000). A bacterial genome in flux: the twelve linear and nine circular extrachromosomal DNAs in an infectious isolate of the Lyme disease spirochete *Borrelia burgdorferi*. Mol. Microbiol. *35*, 490–516.

Chaconas, G. (2005). Hairpin telomeres and genome plasticity in *Borrelia*: all mixed up in the end. Mol. Microbiol. *58*, 625–635.

Champion, C.I., Blanco, D.R., Skare, J.T., Haake, D.A., Giladi, M., Foley, D., Miller, J.N., and Lovett, M.A. (1994). A 9.0 kilobase-pair circular plasmid of *Borrelia burgdorferi* encodes an exported protein: evidence for expression only during infection. Infect. Immun. *62*, 2653–2661.

Cheung, K.J., Badarinarayana, V., Selinger, D.W., Janse, D., and Church, G.M. (2003). A microarray-based antibiotic screen identifies a regulatory role for supercoiling in the osmotic stress response of *Escherichia coli*. Genome Res. *13*, 206–215.

Choi, H., Kim, S., Mukhopadhyay, P., Cho, S., Woo, J., Storz, G., and Ryu, S. (2001). Structural basis of the redox switch in the OxyR transcription factor. Cell *105*, 103–113.

Clifton, D.R., Nolder, C.L., Hughes, J.L., Nowalk, A.J., and Carroll, J.A. (2006). Regulation and expression of *bbA66* encoding an immunogenic infection-associated lipoprotein in *Borrelia burgdorferi*. Mol. Microbiol. *61*, 243–258.

Coburn, J., Chege, W., Magoun, L., Bodary, S.C., and Leong, J.M. (1999). Characterization of a candidate *Borrelia burgdorferi* β_3-chain integrin ligand identified using a phage display library. Mol. Microbiol. *34*, 926–940.

Coleman, J.L., Gebbia, J.A., Piesman, J., Degen, J.L., Bugge, T.H., and Benach, J.L. (1997). Plasminogen is required for efficient dissemination of *B. burgdorferi* in ticks and for enhancement of spirochetemia in mice. Cell *89*, 1111–1119.

Colland, F., Barth, M., Hengge-Aronis, R., and Kolb, A. (2000). σ factor selectivity of *Escherichia coli* RNA polymerase: role for CRP, IHF and Lrp transcription factors. EMBO J. *19*, 3028–3037.

Corbett, K.D., and Berger, J.M. (2004). Structure, molecular mechanisms, and evolutionary relationships in DNA topoisomerases. Annu. Rev. Biophys. Biomol. Struct. *33*, 95–118.

Corbett, K.D., Shultzaberger, R.K., and Berger, J.M. (2004). The C-terminal domain of DNA gyrase A adopts a DNA-bending β-pinwheel fold. Proc. Natl. Acad. Sci. U.S.A. *101*, 7293–7298.

Craigie, R., Arndt-Jovin, D.J., and Mizuuchi, K. (1985). A defined system for the DNA strand-transfer reaction at the initiation of bacteriophage Mu transposition: protein and DNA substrate requirements. Proc. Natl. Acad. Sci. U.S.A. *82*, 7570–7574.

Crother, T.R., Champion, C.I., Whitelegge, J.P., Aguilera, R., Wu, X.Y., Blanco, D.R., Miller, J.N., and Lovett, M.A. (2004). Temporal analysis of the antigenic composition of *Borrelia burgdorferi* during infection in rabbit skin. Infect. Immun. *72*, 5063–5072.

Crother, T.R., Champion, C.I., Wu, X.Y., Blanco, D.R., Miller, J.N., and Lovett, M.A. (2003). Antigenic composition of *Borrelia burgdorferi* during infection of SCID mice. Infect. Immun. *71*, 3419–3428.

Dame, R.T. (2005). The role of nucleoid-associated proteins in the organization and compaction of bacterial chromatin. Mol. Microbiol. *56*, 858–870.

de Silva, A.M., Telford, S.R., Brunet, L.R., Barthold, S.W., and Fikrig, E. (1996). *Borrelia burgdorferi* OspA is an arthropod-specific transmission-blocking Lyme disease vaccine. J. Exp. Med. *183*, 271–275.

Defoe, G., and Coburn, J. (2001). Delineation of *Borrelia burgdorferi p66* sequences required for integrin $\alpha_{IIb}\beta_3$ recognition. Infect. Immun. *69*, 3455–3459.

Delany, I., Rappuoli, R., and Scarlato, V. (2004). Fur functions as an activator and as a repressor of putative virulence genes in *Neisseria meningitidis*. Mol. Microbiol. *52*, 1081–1090.

Dorman, C.J. (1995). DNA topology and the global control of bacterial gene expression: implications for the regulation of virulence gene expression. Microbiology *141*, 1271–1280.

Dorman, C.J., and Corcoran, C.P. (2009). Bacterial DNA topology and infectious disease. Nucleic Acids Res. *37*, 672–678.

Drlica, K., and Rouviere-Yaniv, J. (1987). Histonelike proteins of bacteria. Microbiol. Rev. *51*, 301–319.

Drolet, M. (2006). Growth inhibition mediated by excess negative supercoiling: the interplay between transcription elongation, R-loop formation and DNA topology. Mol. Microbiol. *59*, 723–730.

Eggers, C.H., Caimano, M.J., and Radolf, J.D. (2004). Analysis of promoter elements involved in the transcriptional initiation of RpoS-dependent *Borrelia burgdorferi* genes. J. Bacteriol. *186*, 7390–7402.

Eggers, C.H., Caimano, M.J., and Radolf, J.D. (2006). Sigma factor selectivity in *Borrelia burgdorferi*: RpoS recognition of the *ospE/ospF/elp* promoters is dependent on the sequence of the –10 region. Mol. Microbiol. *59*, 1859–1875.

Elias, A.F., Bono, J.L., Carroll, J.A., Stewart, P., Tilly, K., and Rosa, P. (2000). Altered stationary-phase response in a *Borrelia burgdorferi rpoS* mutant. J. Bacteriol. *182*, 2909–2918.

Fikrig, E., Chen, M., Barthold, S.W., Anguita, J., Feng, W., Telford, S.R., III, and Flavell, R.A. (1999). *Borrelia burgdorferi erpT* expression in the arthropod vector and murine host. Mol. Microbiol. *31*, 281–290.

Fikrig, E., Huguenel, E.D., Berland, R., Rahn, D.W., Hardin, J.A., and Flavell, R.A. (1992). Serologic diagnosis of Lyme disease using recombinant outer surface proteins A and B and flagellin. J. Infect. Dis. *165*, 1127–1132.

Fikrig, E., and Narasimhan, S. (2006). *Borrelia burgdorferi*-traveling incognito? Microbes. Infect. *8*, 1390–1399.

Fingerle, V., Goettner, G., Gern, L., Wilske, B., and Schulte-Spechtel, U. (2007). Complementation of a *Borrelia afzelii* OspC mutant highlights the crucial role of OspC for dissemination of *Borrelia afzelii* in *Ixodes ricinus*. Int. J. Med. Microbiol. *297*, 97–107.

Fingerle, V., Hauser, U., Liegl, G., Petko, B., Preac-Mursic, V., and Wilske, B. (1995). Expression of outer surface proteins A and C of *Borrelia burgdorferi* in *Ixodes ricinus*. J. Clin. Microbiol. *33*, 1867–1869.

Fingerle, V., Laux, H., Munderloh, U.G., Schulte-Spechtel, U., and Wilske, B. (2000). Differential expression of outer surface proteins A and C by individual *Borrelia burgdorferi* in different genospecies. Med. Microbiol. Immunol. *189*, 59–66.

Fingerle, V., Rauser, S., Hammer, B., Kahl, O., Heimerl, C., Schulte-Spechtel, U., Gern, L., and Wilske, B. (2002). Dynamics of dissemination and outer surface protein expression of different European *Borrelia burgdorferi* sensu lato strains in artificially infected *Ixodes ricinus* nymphs. J. Clin. Microbiol. *40*, 1456–1463.

Fischer, J.R., LeBlanc, K.T., and Leong, J.M. (2006). Fibronectin binding protein BBK32 of the Lyme disease spirochete promotes bacterial attachment to glycosaminoglycans. Infect. Immun. *74*, 435–441.

Fischer, J.R., Parveen, N., Magoun, L., and Leong, J.M. (2003). Decorin-binding proteins A and B confer distinct mammalian cell type-specific attachment by *Borrelia burgdorferi*, the Lyme disease spirochete. Proc. Natl. Acad. Sci. U.S.A. *100*, 7307–7312.

Fisher, M.A., Grimm, D., Henion, A.K., Elias, A.F., Stewart, P.E., Rosa, P.A., and Gherardini, F.C. (2005). *Borrelia burgdorferi* σ^{54} is required for mammalian infection and vector transmission but not for tick colonization. Proc. Natl. Acad. Sci. U.S.A. *102*, 5162–5167.

Fraser, C.M., Casjens, S., Huang, W.M., Sutton, G.G., Clayton, R., Lathigra, R., White, O., Ketchum, K.A., Dodson, R., Hickey, E.K., *et al.* (1997). Genomic sequence of a Lyme disease spirochaete, *Borrelia burgdorferi*. Nature *390*, 580–586.

Freedman, J.C., Rogers, E.A., Kostick, J.L., Zhang, H., Iyer, R., Schwartz, I., and Marconi, R.T. (2009). Identification and molecular characterization of a cyclic-di-GMP effector protein, PlzA (BB0733): additional evidence for the existence of a functional cyclic-di-GMP regulatory network in the Lyme disease spirochete, *Borrelia burgdorferi*. FEMS Immunol. Med. Microbiol. doi:10.1111/j.1574-695X.2009.00635.x

Friedman, D.I. (1988). Integration host factor: a protein for all reasons. Cell *55*, 545–554.

Fuangthong, M., Herbig, A.F., Bsat, N., and Helmann, J.D. (2002). Regulation of the *Bacillus subtilis fur* and *perR* genes by PerR: not all members of the PerR regulon are peroxide inducible. J. Bacteriol. *184*, 3276–3286.

Fuchs, R., Jauris, S., Lottspeich, F., Preac-Mursic, V., Wilske, B., and Soutschek, E. (1992). Molecular analysis and expression of a *Borrelia burgdorferi* gene encoding a 22 kDa protein (pC) in *Escherichia coli*. Mol. Microbiol. *6*, 503–509.

Fung, B.P., McHugh, G.L., Leong, J.M., and Steere, A.C. (1994). Humoral immune response to outer surface protein C of *Borrelia burgdorferi* in Lyme disease: role of the immunoglobulin M response in the serodiagnosis of early infection. Infect. Immun. *62*, 3213–3221.

Gaal, T., Ross, W., Estrem, S.T., Nguyen, L.H., Burgess, R.R., and Gourse, R.L. (2001). Promoter recognition and discrimination by $E\sigma^S$ RNA polymerase. Mol. Microbiol. *42*, 939–954.

Gao, H., Zhou, D., Li, Y., Guo, Z., Han, Y., Song, Y., Zhai, J., Du, Z., Wang, X., Lu, J., and Yang, R. (2008). The iron-responsive Fur regulon in *Yersinia pestis*. J. Bacteriol. *190*, 3063–3075.

Gautam, A., Hathaway, M., McClain, N., Ramesh, G., and Ramamoorthy, R. (2008). Analysis of the determinants of *bba64* (P35) gene expression in *Borrelia burgdorferi* using a *gfp* reporter. Microbiology *154*, 275–285.

Gilbert, M.A., Morton, E.A., Bundle, S.F., and Samuels, D.S. (2007). Artificial regulation of *ospC* expression in *Borrelia burgdorferi*. Mol. Microbiol. *63*, 1259–1273.

Gilmore, R.D., Jr., Howison, R.R., Schmit, V.L., Nowalk, A.J., Clifton, D.R., Nolder, C., Hughes, J.L., and Carroll, J.A. (2007). Temporal expression analysis of the *Borrelia burgdorferi* paralogous gene family 54 genes BBA64, BBA65, and BBA66 during persistent infection in mice. Infect. Immun. *75*, 2753–2764.

Gilmore, R.D., Mbow, M.L., and Stevenson, B. (2001). Analysis of *Borrelia burgdorferi* gene expression during life cycle phases of the tick vector *Ixodes scapularis*. Microbes and Infection 3, 799–808.

Gilmore, R.D., Jr. and Piesman, J. (2000). Inhibition of *Borrelia burgdorferi* migration from the midgut to the salivary glands following feeding by ticks on OspC-immunized mice. Infect. Immun. *68*, 411–414.

Grifantini, R., Sebastian, S., Frigimelica, E., Draghi, M., Bartolini, E., Muzzi, A., Rappuoli, R., Grandi, G., and Genco, C.A. (2003). Identification of iron-activated and -repressed Fur-dependent genes by transcriptome analysis of *Neisseria meningitidis* group B. Proc. Natl. Acad. Sci. U.S.A. *100*, 9542–9547.

Grimm, D., Tilly, K., Bueschel, D.M., Fisher, M.A., Policastro, P.F., Gherardini, F.C., Schwan, T.G., and Rosa, P.A. (2005). Defining plasmids required by *Borrelia burgdorferi* for colonization of tick vector *Ixodes scapularis* (Acari: *Ixodidae*). J. Med. Entomol. *42*, 676–684.

Grimm, D., Tilly, K., Byram, R., Stewart, P.E., Krum, J.G., Bueschel, D.M., Schwan, T.G., Policastro, P.F., Elias, A.F., and Rosa, P.A. (2004). Outer-surface protein C of the Lyme disease spirochete: a protein induced in ticks for infection of mammals. Proc. Natl. Acad. Sci. U.S.A. *101*, 3142–3147.

Hagman, K.E., Yang, X., Wikel, S.K., Schoeler, G.B., Caimano, M.J., Radolf, J.D., and Norgard, M.V. (2000). Decorin-binding protein A (DbpA) of *Borrelia burgdorferi* is not protective when immunized mice are challenged via tick infestation and correlates with the lack of DbpA expression by *B. burgdorferi* in ticks. Infect. Immun. *68*, 4759–4764.

He, M., Boardman, B.K., Yan, D., and Yang, X.F. (2007). Regulation of expression of the fibronectin-binding protein BBK32 in *Borrelia burgdorferi*. J. Bacteriol. *189*, 8377–8380.

He, M., Oman, T., Xu, H., Blevins, J., Norgard, M.V., and Yang, X.F. (2008). Abrogation of *ospAB* constitutively activates the Rrp2–RpoN–RpoS pathway (σ^N-σ^S cascade) in *Borrelia burgdorferi*. Mol. Microbiol. *70*, 1453–1464.

Hefty, P.S., Brooks, C.S., Jett, A.M., White, G.L., Wikel, S.K., Kennedy, R.C., and Akins, D.R. (2002a). OspE-related, OspF-related, and Elp lipoproteins are immunogenic in baboons experimentally infected with *Borrelia burgdorferi* and in human Lyme disease patients. J. Clin. Microbiol. *40*, 4256–4265.

Hefty, P.S., Jolliff, S.E., Caimano, M.J., Wikel, S.K., and Akins, D.R. (2002b). Changes in temporal and spatial patterns of outer surface lipoprotein expression generate population heterogeneity and antigenic diversity in the Lyme disease spirochete, *Borrelia burgdorferi*. Infect. Immun. *70*, 3468–3478.

Hefty, P.S., Jolliff, S.E., Caimano, M.J., Wikel, S.K., Radolf, J.D., and Akins, D.R. (2001). Regulation of OspE-related, OspF-related, and Elp lipoproteins of *Borrelia burgdorferi* strain 297 by mammalian host-specific signals. Infect. Immun. *69*, 3618–3627.

Helmann, J.D., Wu, M.F., Gaballa, A., Kobel, P.A., Morshedi, M.M., Fawcett, P., and Paddon, C. (2003). The global transcriptional response of *Bacillus subtilis* to peroxide stress is coordinated by three transcription factors. J. Bacteriol. *185*, 243–253.

Hengge, R. (2009). Principles of c-di-GMP signalling in bacteria. Nat. Rev. Microbiol. 7, 263–273.

Hengge-Aronis, R. (2000). A role for the σ^S subunit of RNA polymerase in the regulation of bacterial virulence. Adv. Exp. Med. Biol. *485*, 85–93.

Hengge-Aronis, R. (2002a). Signal transduction and regulatory mechanisms involved in control of the σ^S (RpoS) subunit of RNA polymerase. Microbiol. Mol. Biol. Rev. *66*, 373–395.

Hengge-Aronis, R. (2002b). Stationary phase gene regulation: what makes an *Escherichia coli* promoter σ^S-selective? Curr. Opin. Microbiol. *5*, 591–595.

Herbig, A.F., and Helmann, J.D. (2001). Roles of metal ions and hydrogen peroxide in modulating the interaction of the *Bacillus subtilis* PerR peroxide regulon repressor with operator DNA. Mol. Microbiol. *41*, 849–859.

Hodzic, E., Feng, S., Freet, K.J., and Barthold, S.W. (2003). *Borrelia burgdorferi* population dynamics and prototype gene expression during infection of immunocompetent and immunodeficient mice. Infect. Immun. *71*, 5042–5055.

Howe, T.R., Mayer, L.W., and Barbour, A.G. (1985). A single recombinant plasmid expressing two major outer surface proteins of the Lyme disease spirochete. Science 227, 645–646.

Huang, Y.-Y., Deng, J.-Y., Gu, J., Zhang, Z.-P., Maxwell, A., Bi, L.-J., Chen, Y.-Y., Zhou, Y.-F., Yu, Z.-N., and Zhang, X.-E. (2006). The key DNA-binding residues in the C-terminal domain of *Mycobacterium tuberculosis* DNA gyrase A subunit (GyrA). Nucleic Acids Res. *34*, 5650–5659.

Hübner, A., Yang, X., Nolen, D.M., Popova, T.G., Cabello, F.C., and Norgard, M.V. (2001). Expression of *Borrelia burgdorferi* OspC and DbpA is controlled by a RpoN–RpoS regulatory pathway. Proc. Natl. Acad. Sci. U.S.A. *98*, 12724–12729.

Hughes, J.L., Nolder, C.L., Nowalk, A.J., Clifton, D.R., Howison, R.R., Schmit, V.L., Gilmore, R.D., Jr., and Carroll, J.A. (2008). *Borrelia burgdorferi* surface-localized proteins expressed during persistent murine infection are conserved among diverse *Borrelia* spp. Infect. Immun. *76*, 2498–2511.

Hurme, R., and Rhen, M. (1998). Temperature sensing in bacterial gene regulation-what it all boils down to. Mol. Microbiol. *30*, 1–6.

Hyde, J.A., Seshu, J., and Skare, J.T. (2006). Transcriptional profiling of *Borrelia burgdorferi* containing a unique *bosR* allele identifies a putative oxidative stress regulon. Microbiology *152*, 2599–2609.

Hyde, J.A., Trzeciakowski, J.P., and Skare, J.T. (2007). *Borrelia burgdorferi* alters its gene expression and antigenic profile in response to CO_2 levels. J. Bacteriol. *189*, 437–445.

Hyde, J.A., Shaw, D.K., Smith, R., III, Trzeciakowski, J.P., and Skare, J.T. (2009). The BosR regulatory protein of

Borrelia burgdorferi interfaces with the RpoS regulatory pathway and modulates both homeostatic and pathogenic properties of the Lyme disease spirochete. Mol. Microbiol. *74*, 1344–1355.

Hyde, J.A., Shaw, D.K., Smith, R., III, Trzeciakowski, J.P., and Skare, J.T. (2010). Characterization of a conditional *bosR* mutant in *Borrelia burgdorferi*. Infect. Immun. *78*, 265–274.

Jeong, K.S., Xie, Y., Hiasa, H., and Khodursky, A.B. (2006). Analysis of pleiotropic transcriptional profiles: a case study of DNA gyrase inhibition. PLoS. Genet. *2*, e152.

Jewett, M.W., Lawrence, K., Bestor, A.C., Tilly, K., Grimm, D., Shaw, P., VanRaden, M., Gherardini, F., and Rosa, P.A. (2007). The critical role of the linear plasmid lp36 in the infectious cycle of *Borrelia burgdorferi*. Mol. Microbiol. *64*, 1358–1374.

Kalish, R.A., Leong, J.M., and Steere, A.C. (1993). Association of treatment-resistant chronic Lyme arthritis with HLA-DR4 and antibody reactivity to OspA and OspB of *Borrelia burgdorferi*. Infect. Immun. *61*, 2774–2779.

Kalish, R.A., Leong, J.M., and Steere, A.C. (1995). Early and late antibody responses to full-length and truncated constructs of outer surface protein A of *Borrelia burgdorferi* in Lyme disease. Infect. Immun. *63*, 2228–2235.

Kataoka, K., Mizushima, T., Ogata, Y., Miki, T., and Sekimizu, K. (1996). Heat shock-induced DNA relaxation *in vitro* by DNA gyrase of *Escherichia coli* in the presence of ATP. J. Biol. Chem. *271*, 24806–24810.

Katona, L.I., Tokarz, R., Kuhlow, C.J., Benach, J., and Benach, J.L. (2004). The Fur homologue in *Borrelia burgdorferi*. J. Bacteriol. *186*, 6443–6456.

Kazmierczak, M.J., Wiedmann, M., and Boor, K.J. (2005). Alternative sigma factors and their roles in bacterial virulence. Microbiol. Mol. Biol. Rev. *69*, 527–543.

Knight, S.W., Kimmel, B.J., Eggers, C.H., and Samuels, D.S. (2000). Disruption of the *Borrelia burgdorferi gac* gene, encoding the naturally synthesized GyrA C-terminal domain. J. Bacteriol. *182*, 2048–2051.

Knight, S.W., and Samuels, D.S. (1999). Natural synthesis of a DNA-binding protein from the C-terminal domain of DNA gyrase A in *Borrelia burgdorferi*. EMBO J. *18*, 4875–4881.

Kobryn, K., Naigamwalla, D.Z., and Chaconas, G. (2000). Site-specific DNA binding and bending by the *Borrelia burgdorferi* Hbb protein. Mol. Microbiol. 37, 145–155.

Kramlinger, V.M., and Hiasa, H. (2006). The 'GyrA-box' is required for the ability of DNA gyrase to wrap DNA and catalyze the supercoiling reaction. J. Biol. Chem. *281*, 3738–3742.

Kroos, L. (2007). The *Bacillus* and *Myxococcus* developmental networks and their transcriptional regulators. Annu. Rev. Genet. *41*, 13–39.

Kusano, S., Ding, Q., Fujita, N., and Ishihama, A. (1996). Promoter selectivity of *Escherichia coli* RNA polymerase $E\sigma^{70}$ and $E\sigma^{38}$ holoenzymes: effect of DNA supercoiling. J. Biol. Chem. *271*, 1998–2004.

Kustu, S., Santero, E., Keener, J., Popham, D., and Weiss, D. (1989). Expression of σ^{54} (*ntrA*)-dependent genes is probably united by a common mechanism. Microbiol. Rev. *53*, 367–376.

Labandeira-Rey, M., and Skare, J.T. (2001). Decreased infectivity in *Borrelia burgdorferi* strain B31 is associated with loss of linear plasmid 25 or 28-1. Infect. Immun. *69*, 446–455.

Lagal, V., Portnoi, D., Faure, G., Postic, D., and Baranton, G. (2006). *Borrelia burgdorferi* sensu stricto invasiveness is correlated with OspC-plasminogen affinity. Microbes. Infect. *8*, 645–652.

Lahdenne, P., Porcella, S.F., Hagman, K.E., Akins, D.R., Popova, T.G., Cox, D.L., Radolf, J.D., and Norgard, M.V. (1997). Molecular characterization of a 6.6-kilodalton *Borrelia burgdorferi* outer membrane-associated lipoprotein (lp6.6) which appears to be down-regulated during mammalian infection. Infect. Immun. *65*, 412–421.

Lam, T.T., Nguyen, T.P.K., Montgomery, R.R., Kantor, F.S., Fikrig, E., and Flavell, R.A. (1994). Outer surface proteins E and F of *Borrelia burgdorferi*, the agent of Lyme disease. Infect. Immun. *62*, 290–298.

Lane, R.S., Piesman, J., and Burgdorfer, W. (1991). Lyme borreliosis: relation of its causative agent to its vectors and hosts in North America and Europe. Ann. Rev. Entomol. *36*, 587–609.

Lease, R.A., and Belfort, M. (2000). Riboregulation by DsrA RNA: *trans*-actions for global economy. Mol. Microbiol. *38*, 667–672.

Lederer, S., Brenner, C., Stehle, T., Gern, L., Wallich, R., and Simon, M.M. (2005). Quantitative analysis of *Borrelia burgdorferi* gene expression in naturally (tick) infected mouse strains. Med. Microbiol. Immunol. *194*, 81–90.

Lee, H.W., Choe, Y.H., Kim, D.K., Jung, S.Y., and Lee, N.G. (2004). Proteomic analysis of a ferric uptake regulator mutant of *Helicobacter pylori*: regulation of *Helicobacter pylori* gene expression by ferric uptake regulator and iron. Proteomics. *4*, 2014–2027.

Lee, J.W., and Helmann, J.D. (2006a). Biochemical characterization of the structural Zn^{2+} site in the *Bacillus subtilis* peroxide sensor PerR. J. Biol. Chem. *281*, 23567–23578.

Lee, J.W., and Helmann, J.D. (2006b). The PerR transcription factor senses H_2O_2 by metal-catalysed histidine oxidation. Nature *440*, 363–367.

Leuba-Garcia, S., Martinez, R., and Gern, L. (1998). Expression of outer surface proteins A and C of *Borrelia afzelii* in *Ixodes ricinus* ticks and in the skin of mice. Zentralbl. Bakteriol. *287*, 475–484.

Li, X., Liu, X., Beck, D.S., Kantor, F.S., and Fikrig, E. (2006). *Borrelia burgdorferi* lacking BBK32, a fibronectin-binding protein, retains full pathogenicity. Infect. Immun. *74*, 3305–3313.

Li, X., Pal, U., Ramamoorthi, N., Liu, X., Desrosiers, D.C., Eggers, C.H., Anderson, J.F., Radolf, J.D., and Fikrig, E. (2007). The Lyme disease agent *Borrelia burgdorferi* requires BB0690, a Dps homologue, to persist within ticks. Mol. Microbiol. *63*, 694–710.

Liang, F.T., Brown, E.L., Wang, T., Iozzo, R.V., and Fikrig, E. (2004a). Protective niche for *Borrelia burgdorferi* to evade humoral immunity. Am. J. Pathol. *165*, 977–985.

Liang, F.T., Jacobs, M.B., Bowers, L.C., and Philipp, M.T. (2002a). An immune evasion mechanism for spirochetal persistence in Lyme borreliosis. J. Exp. Med. *195*, 415–422.

Liang, F.T., Nelson, F.K., and Fikrig, E. (2002b). DNA microarray assessment of putative *Borrelia burgdorferi* lipoprotein genes. Infect. Immun. *70*, 3300–3303.

Liang, F.T., Nelson, F.K., and Fikrig, E. (2002c). Molecular adaptation of *Borrelia burgdorferi* in the murine host. J. Exp. Med. *196*, 275–280.

Liang, F.T., Yan, J., Mbow, M.L., Sviat, S.L., Gilmore, R.D., Mamula, M., and Fikrig, E. (2004b). *Borrelia burgdorferi* changes its surface antigenic expression in response to host immune responses. Infect. Immun. *72*, 5759–5767.

Lim, K., Tempczyk, A., Parsons, J.F., Bonander, N., Toedt, J., Kelman, Z., Howard, A., Eisenstein, E., and Herzberg, O. (2003). Crystal structure of YbaB from *Haemophilus influenzae* (HI0442), a protein of unknown function coexpressed with the recombinational DNA repair protein RecR. Proteins *50*, 375–379.

Lin, B., Short, S.A., Eskildsen, M., Klempner, M.S., and Hu, L.T. (2001). Functional testing of putative oligopeptide permease (Opp) proteins of *Borrelia burgdorferi*: a complementation model in *opp*$^-$ *Escherichia coli*. Biochim. Biophys. Acta *1499*, 222–231.

López-García, P., and Forterre, P. (2000). DNA topology and the thermal stress response, a tale from mesophiles and hyperthermophiles. Bioessays *22*, 738–746.

Losick, R., and Stragier, P. (1992). Crisscross regulation of cell-type-specific gene expression during development in *B. subtilis*. Nature *355*, 601–604.

Luijsterburg, M.S., White, M.F., van Driel, R., and Dame, R.T. (2008). The major architects of chromatin: architectural proteins in bacteria, archaea and eukaryotes. Crit Rev. Biochem. Mol. Biol. *43*, 393–418.

Lukat, G.S., McCleary, W.R., Stock, A.M., and Stock, J.B. (1992). Phosphorylation of bacterial response regulator proteins by low molecular weight phospho-donors. Proc. Natl. Acad. Sci. U.S.A. *89*, 718–722.

Lybecker, M.C., and Samuels, D.S. (2007). Temperature-induced regulation of RpoS by a small RNA in *Borrelia burgdorferi*. Mol. Microbiol. *64*, 1075–1089.

Majdalani, N., Vanderpool, C.K., and Gottesman, S. (2005). Bacterial small RNA regulators. Crit Rev. Biochem. Mol. Biol. *40*, 93–113.

Malik, M., Bensaid, A., Rouviere-Yaniv, J., and Drlica, K. (1996). Histone-like protein HU and bacterial DNA topology: suppression of an HU deficiency by gyrase mutations. J. Mol. Biol. *256*, 66–76.

Marconi, R.T., Samuels, D.S., and Garon, C.F. (1993). Transcriptional analyses and mapping of the *ospC* gene in Lyme disease spirochetes. J. Bacteriol. *175*, 926–932.

Margolis, N., Hogan, D., Tilly, K., and Rosa, P.A. (1994). Plasmid location of *Borrelia* purine biosynthesis gene homologs. J. Bacteriol. *176*, 6427–6432.

Margolis, N., and Rosa, P.A. (1993). Regulation of expression of major outer surface proteins in *Borrelia burgdorferi*. Infect. Immun. *61*, 2207–2210.

Margolis, N., and Samuels, D.S. (1995). Proteins binding to the promoter region of the operon encoding the major outer surface proteins OspA and OspB of *Borrelia burgdorferi*. Mol. Biol. Rep. *21*, 159–164.

Maruskova, M., Esteve-Gassent, M.D., Sexton, V.L., and Seshu, J. (2008). Role of the BBA64 locus of *Borrelia burgdorferi* in early stages of infectivity in a murine model of Lyme disease. Infect. Immun. *76*, 391–402.

Mascher, T. (2006). Intramembrane-sensing histidine kinases: a new family of cell envelope stress sensors in Firmicutes bacteria. FEMS Microbiol. Lett. *264*, 133–144.

McDowell, J.V., Sung, S.Y., Labandeira-Rey, M., Skare, J.T., and Marconi, R.T. (2001). Analysis of mechanisms associated with loss of infectivity of clonal populations of *Borrelia burgdorferi* B31MI. Infect. Immun. *69*, 3670–3677.

Medrano, M.S., Ding, Y., Wang, X.-G., Lu, P., Coburn, J., and Hu, L.T. (2007). Regulators of expression of the oligopeptide permease A proteins of *Borrelia burgdorferi*. J. Bacteriol. *189*, 2653–2659.

Medrano, M.S., Policastro, P.F., Schwan, T.G., and Coburn, J. (2010). Interaction of *Borrelia burgdorferi* Hbb with the *p66* promoter. Nucleic Acids Res. in press.

Mongkolsuk, S., and Helmann, J.D. (2002). Regulation of inducible peroxide stress responses. Mol. Microbiol. *45*, 9–15.

Montgomery, R.R., Malawista, S.E., Feen, K.J., and Bockenstedt, L.K. (1996). Direct demonstration of antigenic substitution of *Borrelia burgdorferi ex vivo*: exploration of the paradox of the early immune response to outer surface proteins A and C in Lyme disease. J. Exp. Med. *183*, 261–269.

Mouw, K.W., and Rice, P.A. (2007). Shaping the *Borrelia burgdorferi* genome: crystal structure and binding properties of the DNA-bending protein Hbb. Mol. Microbiol. *63*, 1319–1330.

Mulay, V.B., Caimano, M.J., Iyer, R., Ems, S., Liveris, D., Petzke, M.M., Schwartz, I., and Radolf, J.D. (2009). *Borrelia burgdorferi bba74* is expressed exclusively during tick feeding and is regulated by both arthropod- and mammalian host-specific signals. J. Bacteriol. *191*, 2783–2794.

Narasimhan, S., Caimano, M.J., Liang, F.T., Santiago, F., Laskowski, M., Philipp, M.T., Pachner, A.R., Radolf, J.D., and Fikrig, E. (2003). *Borrelia burgdorferi* transcriptome in the central nervous system of non-human primates. Proc. Natl. Acad. Sci. U.S.A. *100*, 15953–15958.

Narberhaus, F., Waldminghaus, T., and Chowdhury, S. (2006). RNA thermometers. FEMS Microbiol. Rev. *30*, 3–16.

Nash, H.A. (1996). The HU and IHF proteins: accessory factors for complex protein-DNA assemblies. In Regulation of gene expression in *Escherichia coli*, E.C.C. Lin and A.S. Lynch, eds. (Austin, Texas: R.G. Landes Co.), pp. 149–179.

Neelakanta, G., Li, X., Pal, U., Liu, X., Beck, D.S., Deponte, K., Fish, D., Kantor, F.S., and Fikrig, E. (2007). Outer surface protein B is critical for *Borrelia burgdorferi* adherence and survival within *Ixodes* ticks. PLoS. Pathog. 3, e33.

Norris, S.J., Carter, C.J., Howell, J.K., and Barbour, A.G. (1992). Low-passage-associated proteins of *Borrelia burgdorferi* B31: characterization and molecular cloning of OspD, a surface-exposed, plasmid-encoded lipoprotein. Infect. Immun. *60*, 4662–4672.

Nowalk, A.J., Gilmore, R.D., Jr., and Carroll, J.A. (2006a). Serologic proteome analysis of *Borrelia burgdorferi* membrane-associated proteins. Infect. Immun. *74*, 3864–3873.

Nowalk, A.J., Nolder, C., Clifton, D.R., and Carroll, J.A. (2006b). Comparative proteome analysis of subcellular fractions from *Borrelia burgdorferi* by NEPHGE and IPG. Proteomics. *6*, 2121–2134.

Obonyo, M., Munderloh, U.G., Fingerle, V., Wilske, B., and Kurtti, T.J. (1999). *Borrelia burgdorferi* in tick cell culture modulates expression of outer surface proteins A and C in response to temperature. J. Clin. Microbiol. 37, 2137–2141.

Ogata, Y., Mizushima, T., Kataoka, K., Kita, K., Miki, T., and Sekimizu, K. (1996). DnaK heat shock protein of *Escherichia coli* maintains the negative supercoiling of DNA against thermal stress. J. Biol. Chem. *271*, 29407–29414.

Ohnishi, J., Piesman, J., and de Silva, A.M. (2001). Antigenic and genetic heterogeneity of *Borrelia burgdorferi* populations transmitted by ticks. Proc. Natl. Acad. Sci. (USA) *98*, 670–675.

Ojaimi, C., Brooks, C., Akins, D., Casjens, S., Rosa, P., Elias, A., Barbour, A., Jasinskas, A., Benach, J., Katonah, L., Radolf, J., Caimano, M., Skare, J., Swingle, K., Sims, S., and Schwartz, I. (2002). *Borrelia burgdorferi* gene expression profiling with membrane-based arrays. Methods Enzymol. *358*, 165–177.

Ojaimi, C., Brooks, C., Casjens, S., Rosa, P., Elias, A., Barbour, A., Jasinskas, A., Benach, J., Katona, L., Radolf, J., *et al.* (2003). Profiling of temperature-induced changes in *Borrelia burgdorferi* gene expression by using whole genome arrays. Infect. Immun. *71*, 1689–1705.

Östberg, Y., Carroll, J.A., Pinne, M., Krum, J.G., Rosa, P., and Bergström, S. (2004). Pleiotropic effects of inactivating a carboxyl-terminal protease, CtpA, in *Borrelia burgdorferi*. J. Bacteriol. *186*, 2074–2084.

Ouyang, Z., Blevins, J.S., and Norgard, M.V. (2008). Transcriptional interplay among the regulators Rrp2, RpoN and RpoS in *Borrelia burgdorferi*. Microbiology *154*, 2641–2658.

Ouyang, Z., Kumar, M., Kariu, T., Haq, S., Goldberg, M., Pal, U., and Norgard, M.V. (2009). BosR (BB0647) governs virulence expression in *Borrelia burgdorferi*. Mol. Microbiol. *74*, 1331–1343.

Padula, S.J., Sampieri, A., Dias, F., Szczepanski, A., and Ryan, R.W. (1994). Molecular characterization and expression of p23 (OspC) from a North American strain of *Borrelia burgdorferi*. Infect. Immun. *61*, 5097–5105.

Pal, U., Dai, J., Li, X., Neelakanta, G., Luo, P., Kumar, M., Wang, P., Yang, X., Anderson, J.F., and Fikrig, E. (2008). A differential role for BB0365 in the persistence of *Borrelia burgdorferi* in mice and ticks. J. Infect. Dis. *197*, 148–155.

Pal, U., Li, X., Wang, T., Montgomery, R.R., Ramamoorthi, N., Desilva, A.M., Bao, F., Yang, X., Pypaert, M., Pradhan, D., *et al.* (2004a). TROSPA, an *Ixodes scapularis* receptor for *Borrelia burgdorferi*. Cell *119*, 457–468.

Pal, U., Yang, X., Chen, M., Bockenstedt, L.K., Anderson, J.F., Flavell, R.A., Norgard, M.V., and Fikrig, E. (2004b). OspC facilitates *Borrelia burgdorferi* invasion of *Ixodes scapularis* salivary glands. J. Clin. Invest. *113*, 220–230.

Peng, X., Pigli, Y.Z., Rice, P.A., and Greenberg, M.M. (2008). Protein binding has a large effect on radical mediated DNA damage. J. Am. Chem. Soc. *130*, 12890–12891.

Peter, B.J., Arsuaga, J., Breier, A.M., Khodursky, A.B., Brown, P.O., and Cozzarelli, N.R. (2004). Genomic transcriptional response to loss of chromosomal supercoiling in *Escherichia coli*. Genome Biol. *5*, R87.

Porcella, S.F., Popova, T.G., Akins, D.R., Li, M., Radolf, J.D., and Norgard, M.V. (1996). *Borrelia burgdorferi* supercoiled plasmids encode multiple tandem open reading frames and a lipoprotein gene family. J. Bacteriol. *178*, 3293–3307.

Posey, J.E., and Gherardini, F.C. (2000). Lack of a role for iron in the Lyme disease pathogen. Science *288*, 1651–1653.

Purser, J.E., and Norris, S.J. (2000). Correlation between plasmid content and infectivity in *Borrelia burgdorferi*. Proc. Natl. Acad. Sci. U.S.A. *97*, 13865–13870.

Radolf, J.D., and Caimano, M.J. (2008). The long strange trip of *Borrelia burgdorferi* outer-surface protein C. Mol. Microbiol. *69*, 1–4.

Ramamoorthi, N., Narasimhan, S., Pal, U., Bao, F., Yang, X.F., Fish, D., Anguita, J., Norgard, M.V., Kantor, F.S., Anderson, J.F., Koski, R.A., and Fikrig, E. (2005). The Lyme disease agent exploits a tick protein to infect the mammalian host. Nature *436*, 573–577.

Ramamoorthy, R., and Scholl-Meeker, D. (2001). *Borrelia burgdorferi* proteins whose expression is similarly affected by culture temperature and pH. Infect. Immun. *69*, 2739–2742.

Rathinavelu, S., and de Silva, A.M. (2001). Purification and characterization of *Borrelia burgdorferi* from feeding nymphal ticks (*Ixodes scapularis*). Infect. Immun. *69*, 3536–3541.

Reece, R.J., and Maxwell, A. (1991). DNA gyrase: structure and function. Crit Rev. Biochem. Mol. Biol. *26*, 335–375.

Reitzer, L., and Schneider, B.L. (2001). Metabolic context and possible physiological themes of σ^{54}-dependent genes in *Escherichia coli*. Microbiol. Mol. Biol. Rev. *65*, 422–444.

Repoila, F., and Gottesman, S. (2001). Signal transduction cascade for regulation of RpoS: temperature regulation of DsrA. J. Bacteriol. *183*, 4012–4023.

Repoila, F., Majdalani, N., and Gottesman, S. (2003). Small non-coding RNAs, co-ordinators of adaptation processes in *Escherichia coli*: the RpoS paradigm. Mol. Microbiol. *48*, 855–861.

Revel, A.T., Blevins, J.S., Almazán, C., Neil, L., Kocan, K.M., de la Fuente, J., Hagman, K.E., and Norgard, M.V. (2005). *bptA* (*bbe16*) is essential for the persistence of the Lyme disease spirochete, *Borrelia burgdorferi*, in its natural tick vector. Proc. Natl. Acad. Sci. U.S.A. *102*, 6972–6977.

Revel, A.T., Talaat, A.M., and Norgard, M.V. (2002). DNA microarray analysis of differential gene expression in *Borrelia burgdorferi*, the Lyme disease spirochete. Proc. Natl. Acad. Sci. U.S.A. *99*, 1562–1567.

Rice, P.A. (1997). Making DNA do a U-turn: IHF and related proteins. Curr. Opin. Struct. Biol. *7*, 86–93.

Riley, S.P., Bykowski, T., Cooley, A.E., Burns, L.H., Babb, K., Brissette, C.A., Bowman, A., Rotondi, M., Miller, M.C., Demoll, E., Lim, K., Fried, M.G., and Stevenson, B. (2009). *Borrelia burgdorferi* EbfC defines a newly-identified, widespread family of bacterial DNA-binding proteins. Nucleic Acids Res. *37*, 1973–1983.

Roberts, D.M., Caimano, M., McDowell, J., Theisen, M., Holm, A., Orff, E., Nelson, D., Wikel, S., Radolf, J., and Marconi, R.T. (2002). Environmental regulation and differential production of members of the Bdr protein family of *Borrelia burgdorferi*. Infect. Immun. *70*, 7033–7041.

Roehrig, J.T., Piesman, J., Hunt, A.R., Keen, M.G., Happ, C.M., and Johnson, B.J.B. (1992). The hamster immune response to tick-transmitted *Borrelia burgdorferi* differs from the response to needle-innoculated, cultured organisms. J. Immunol. *149*, 3648–3653.

Rogers, E.A., Terekhova, D., Zhang, H.M., Hovis, K.M., Schwartz, I., and Marconi, R.T. (2009). Rrp1, a cyclic-di-GMP-producing response regulator, is an important regulator of *Borrelia burgdorferi* core cellular functions. Mol. Microbiol. *71*, 1551–1573.

Rosa, P.A., Tilly, K., and Stewart, P.E. (2005). The burgeoning molecular genetics of the Lyme disease spirochaete. Nat. Rev. Microbiol. *3*, 129–143.

Rui, S., and Tse-Dinh, Y.-C. (2003). Topoisomerase function during bacterial responses to environmental challenge. Front Biosci. *8*, d256–d263.

Ruthenburg, A.J., Graybosch, D.M., Huetsch, J.C., and Verdine, G.L. (2005). A superhelical spiral in the *Escherichia coli* DNA gyrase A C-terminal domain imparts unidirectional supercoiling bias. J. Biol. Chem. *280*, 26177–26184.

Saier, M.H., Jr. (2008). The bacterial chromosome. Crit Rev. Biochem. Mol. Biol. *43*, 89–134.

Samuels, D.S. (2006). Antibiotic resistance in *Borrelia burgdorferi*: applications for genetic manipulation and implications for evolution. In Molecular Biology of Spirochetes, F.C. Cabello, D. Hulinska, and H.P. Godfrey, eds. (Amsterdam, Netherlands: IOS Press), pp. 56–70.

Samuels, D.S., and Radolf, J.D. (2009). Who is the BosR around here anyway? Mol. Microbiol. *74*, 1295–1299.

Scheckelhoff, M.R., Telford, S.R., Wesley, M., and Hu, L.T. (2007). *Borrelia burgdorferi* intercepts host hormonal signals to regulate expression of outer surface protein A. Proc. Natl. Acad. Sci. U.S.A. *104*, 7247–7252.

Schmid, M.B. (1990). More than just 'histone-like' proteins. Cell *63*, 451–453.

Schulze, R.J., and Zückert, W.R. (2006). *Borrelia burgdorferi* lipoproteins are secreted to the outer surface by default. Mol. Microbiol. *59*, 1473–1484.

Schwan, T.G., Burgdorfer, W., and Garon, C.F. (1988). Changes in infectivity and plasmid profile of the Lyme disease spirochete, *Borrelia burgdorferi*, as a result of in vitro cultivation. Infect. Immun. *56*, 1831–1836.

Schwan, T.G., Karstens, R.H., Schrumpf, M.E., and Simpson, W.J. (1991). Changes in antigenic reactivity of *Borrelia burgdorferi*, the Lyme disease spirochete, during persistant infection in mice. Can. J Microbiol. *37*, 450–454.

Schwan, T.G., Kime, K.K., Schrumpf, M.E., Coe, J.E., and Simpson, W.J. (1989). Antibody response in white-footed mice (*Peromyscus leucopus*) experimentally infected with the Lyme disease spirochete (*Borrelia burgdorferi*). Infect. Immun. *57*, 3445–3451.

Schwan, T.G., and Piesman, J. (2000). Temporal changes in outer surface proteins A and C of the Lyme disease-associated spirochete, *Borrelia burgdorferi*, during the chain of infection in ticks and mice. J. Clin. Microbiol. *38*, 382–388.

Schwan, T.G., Piesman, J., Golde, W.T., Dolan, M.C., and Rosa, P.A. (1995). Induction of an outer surface protein on *Borrelia burgdorferi* during tick feeding. Proc. Natl. Acad. Sci. U.S.A. *92*, 2909–2913.

Schwan, T.G., Schrumpf, M.E., Karstens, R.H., Clover, J.R., Wong, J., Daugherty, M., and Rosa, P.A. (1993). Distribution and molecular analysis of Lyme disease spirochetes, *Borrelia burgdorferi*, isolated from ticks throughout California. J. Clin. Microbiol. *31*, 3096–3108.

Schwan, T.G., and Simpson, W.J. (1991). Factors influencing the antigenic reactivity of *Borrelia burgdorferi*, the Lyme disease spirochete. Scand. J. Infect. Dis. 77 *suppl.*, 94–101.

Sellati, T.J., Bouis, D.A., Caimano, M.J., Feulner, J.A., Ayers, C., Lien, E., and Radolf, J.D. (1999). Activation of human monocytic cells by *Borrelia burgdorferi* and *Treponema pallidum* is facilitated by CD14 and correlates with surface exposure of spirochetal lipoproteins. J. Immunol. *163*, 2049–2056.

Seshu, J., Boylan, J.A., Gherardini, F.C., and Skare, J.T. (2004a). Dissolved oxygen levels alter gene expression and antigen profiles in *Borrelia burgdorferi*. Infect. Immun. *72*, 1580–1586.

Seshu, J., Boylan, J.A., Hyde, J.A., Swingle, K.L., Gherardini, F.C., and Skare, J.T. (2004b). A conservative amino acid change alters the function of BosR, the redox regulator of *Borrelia burgdorferi*. Mol. Microbiol. *54*, 1352–1363.

Seshu, J., Esteve-Gassent, M.D., Labandeira-Rey, M., Kim, J.H., Trzeciakowski, J.P., Hook, M., and Skare, J.T. (2006). Inactivation of the fibronectin-binding adhesin gene *bbk32* significantly attenuates the infectivity potential of *Borrelia burgdorferi*. Mol. Microbiol. *59*, 1591–1601.

Shanado, Y., Kato, J.-i., and Ikeda, H. (1998). *Escherichia coli* HU protein suppresses DNA-gyrase-mediated illegitimate recombination and SOS induction. Genes Cells 3, 511–520.

Shi, Y., Xu, Q., McShan, K., and Liang, F.T. (2008). Both decorin-binding proteins A and B are critical for the overall virulence of *Borrelia burgdorferi*. Infect. Immun. *76*, 1239–1246.

Simpson, W.J., Garon, C.F., and Schwan, T.G. (1990). Analysis of supercoiled circular plasmids in infectious and non-infectious *Borrelia burgdorferi*. Microb. Pathog. *8*, 109–118.

Singh, S.K., and Girschick, H.J. (2004). Molecular survival strategies of the Lyme disease spirochete *Borrelia burgdorferi*. Lancet Infect. Dis. *4*, 575–583.

Skare, J.T., Foley, D.M., Hernandez, S.R., Moore, D.C., Blanco, D.R., Miller, J.N., and Lovett, M.A. (1999). Cloning and molecular characterization of plasmid-encoded antigens of *Borrelia burgdorferi*. Infect. Immun. *67*, 4407–4417.

Skare, J.T., Shang, E.S., Foley, D.M., Blanco, D.R., Champion, C.I., Mirzabekov, T., Sokolov, Y., Kagan, B.L., Miller, J.M., and Lovett, M.A. (1995). Virulent strain associated outer membrane proteins of *Borrelia burgdorferi*. J. Clin. Invest. *96*, 2380–2392.

Sledjeski, D.D., Gupta, A., and Gottesman, S. (1996). The small RNA, DsrA, is essential for the low temperature expression of RpoS during exponential growth in *Escherichia coli*. EMBO J. *15*, 3993–4000.

Smith, A.H., Blevins, J.S., Bachlani, G.N., Yang, X.F., and Norgard, M.V. (2007). Evidence that RpoS (σ^S) in *Borrelia burgdorferi* is controlled directly by RpoN (σ^{54}/σ^N). J. Bacteriol. *189*, 2139–2144.

Sohaskey, C.D., Zückert, W.R., and Barbour, A.G. (1999). The extended promoters for two outer membrane lipoprotein genes of *Borrelia* spp. uniquely include a T-rich region. Mol. Microbiol. *33*, 41–51.

Spielman, A. (1994). The emergence of Lyme disease and human babesiosis in a changing environment. Ann. N. Y. Acad. Sci. *740*, 146–156.

Srivastava, S.Y., and de Silva, A.M. (2008). Reciprocal expression of *ospA* and *ospC* in single cells of *Borrelia burgdorferi*. J. Bacteriol. *190*, 3429–3433.

Steere, A.C. (2001). Lyme disease. N. Engl. J Med *345*, 115–125.

Steere, A.C. (2006). Lyme borreliosis in 2005, 30 years after initial observations in Lyme Connecticut. Wien. Klin. Wochenschr. *118*, 625–633.

Stevenson, B., Schwan, T.G., and Rosa, P.A. (1995). Temperature-related differential expression of antigens in the Lyme disease spirochete, *Borrelia burgdorferi*. Infect. Immun. *63*, 4535–4539.

Stevenson, B., Tilly, K., and Rosa, P.A. (1996). A family of genes located on four separate 32-kilobase circular plasmids in *Borrelia burgdorferi*. J. Bacteriol. *178*, 3508–3516.

Stewart, P.E., Hoff, J., Fischer, E., Krum, J.G., and Rosa, P.A. (2004). Genome-wide transposon mutagenesis of *Borrelia burgdorferi* for identification of phenotypic mutants. Appl. Environ. Microbiol. *70*, 5973–5979.

Stewart, P.E., Wang, X., Bueschel, D.M., Clifton, D.R., Grimm, D., Tilly, K., Carroll, J.A., Weis, J.J., and Rosa, P.A. (2006). Delineating the requirement for the *Borrelia burgdorferi* virulence factor OspC in the mammalian host. Infect. Immun. *74*, 3547–3553.

Stock, A.M., Robinson, V.L., and Goudreau, P.N. (2000). Two-component signal transduction. Annu. Rev. Biochem. *69*, 183–215.

Storz, G., and Imlay, J.A. (1999). Oxidative stress. Curr. Opin. Microbiol. *2*, 188–194.

Storz, G., Opdyke, J.A., and Zhang, A. (2004). Controlling mRNA stability and translation with small, noncoding RNAs. Curr. Opin. Microbiol. *7*, 140–144.

Strother, K.O., and de Silva, A. (2005). Role of *Borrelia burgdorferi* linear plasmid 25 in infection of *Ixodes scapularis* ticks. J. Bacteriol. *187*, 5776–5781.

Strother, K.O., Broadwater, A., and de Silva, A.M. (2005). Plasmid requirements for infection of ticks by *Borrelia burgdorferi*. Vector. Borne. Zoonotic. Dis. *5*, 237–245.

Strother, K.O., Hodzic, E., Barthold, S.W., and de Silva, A.M. (2007). Infection of mice with Lyme disease spirochetes constitutively producing outer surface proteins A and B. Infect. Immun. *75*, 2786–2794.

Studholme, D.J., and Buck, M. (2000). The biology of enhancer-dependent transcriptional regulation in bacteria: insights from genome sequences. FEMS Microbiol. Lett. *186*, 1–9.

Studholme, D.J., Buck, M., and Nixon, T. (2000). Identification of potential σ^N-dependent promoters in bacterial genomes. Microbiology *146*, 3021–3023.

Suk, K., Das, S., Sun, W., Jwang, B., Barthold, S.W., Flavell, R.A., and Fikrig, E. (1995). *Borrelia burgdorferi* genes selectively expressed in the infected host. Proc. Natl. Acad. Sci. (USA) *92*, 4269–4273.

Tamayo, R., Pratt, J.T., and Camilli, A. (2007). Roles of cyclic diguanylate in the regulation of bacterial pathogenesis. Annu. Rev. Microbiol. *61*, 131–148.

Taylor, B.L., and Zhulin, I.B. (1999). PAS domains: internal sensors of oxygen, redox potential, and light. Microbiol. Mol. Biol. Rev. *63*, 479–506.

Theisen, M., Frederiksen, B., Lebech, A.M., Vuust, J., and Hansen, K. (1993). Polymorphism in *ospC* gene of *Borrelia burgdorferi* and immunoreactivity of OspC protein: implications for taxonomy and for use of OspC protein as a diagnostic antigen. J. Clin. Microbiol. *31*, 2570–2576.

Thöny, B., and Hennecke, H. (1989). The –24/–12 promoter comes of age. FEMS Microbiol. Rev. *5*, 341–357.

Tilly, K., Bestor, A., Jewett, M.W., and Rosa, P. (2007). Rapid clearance of Lyme disease spirochetes lacking OspC from skin. Infect. Immun. *75*, 1517–1519.

Tilly, K., Casjens, S., Stevenson, B., Bono, J.L., Samuels, D.S., Hogan, D., and Rosa, P.A. (1997). The *Borrelia burgdorferi* circular plasmid cp26: conservation of plasmid structure and targeted inactivation of the *ospC* gene. Mol. Microbiol. *25*, 361–373.

Tilly, K., Fuhrman, J., Campbell, J., and Samuels, D.S. (1996). Isolation of *Borrelia burgdoreri* genes encoding homologues of DNA-binding protein HU and ribosomal protein S20. Microbiology *142*, 2471–2479.

Tilly, K., Krum, J.G., Bestor, A., Jewett, M.W., Grimm, D., Bueschel, D., Byram, R., Dorward, D., VanRaden, M.J., Stewart, P., and Rosa, P. (2006). *Borrelia burgdorferi* OspC protein required exclusively in a crucial early stage of mammalian infection. Infect. Immun. *74*, 3554–3564.

Tokarz, R., Anderton, J.M., Katona, L.I., and Benach, J.L. (2004). Combined effects of blood and temperature shift on *Borrelia burgdorferi* gene expression as determined by whole genome DNA array. Infect. Immun. *72*, 5419–5432.

Traore, D.A., El, G.A., Ilango, S., Dupuy, J., Jacquamet, L., Ferrer, J.L., Caux-Thang, C., Duarte, V., and Latour, J.M. (2006). Crystal structure of the apo-PerR-Zn protein from *Bacillus subtilis*. Mol. Microbiol. *61*, 1211–1219.

Travers, A.A., Ner, S.S., and Churchill, M.E.A. (1994). DNA chaperones: a solution to a persistence problem? Cell 77, 167–169.

Tse-Dinh, Y.C., Qi, H., and Menzel, R. (1997). DNA supercoiling and bacterial adaptation: thermotolerance and thermoresistance. Trends Microbiol. 5, 323–326.

Typas, A., Becker, G., and Hengge, R. (2007). The molecular basis of selective promoter activation by the σ^S subunit of RNA polymerase. Mol. Microbiol. *63*, 1296–1306.

Wallich, R., Brenner, C., Kramer, M.D., and Simon, M.M. (1995). Molecular cloning and immunological characterization of a novel linear-plasmid-encoded gene, *pG*, of *Borrelia burgdorferi* expressed only *in vivo*. Infect. Immun. 63, 3327–3335.

Wang, J.C., and Lynch, A.S. (1993). Transcription and DNA supercoiling. Curr. Opin. Genet. Dev. *3*, 764–768.

Weening, E.H., Parveen, N., Trzeciakowski, J.P., Leong, J.M., Hook, M., and Skare, J.T. (2008). *Borrelia burgdorferi* lacking DbpBA exhibits an early survival defect during experimental infection. Infect. Immun. *76*, 5694–5705.

Wigneshweraraj, S., Bose, D., Burrows, P.C., Joly, N., Schumacher, J., Rappas, M., Pape, T., Zhang, X., Stockley, P., Severinov, K., and Buck, M. (2008). *Modus operandi* of the bacterial RNA polymerase containing the σ^{54} promoter-specificity factor. Mol. Microbiol. 68, 538–546.

Wilske, B., Barbour, A.G., Bergström, S., Burman, N., Restrepo, B.I., Rosa, P.A., Schwan, T., Soutschek, E., and Wallich, R. (1992a). Antigenic variation and strain heterogeneity in *Borrelia* spp. Res. Microbiol. *143*, 583–596.

Wilske, B., Fingerle, V., Herzer, P., Hofmann, A., Lehnert, G., Peters, H., Preac-Mursic, V., Soutschek, E., and Weber, K. (1993a). Recombinant immunoblot in the serodiagnosis of Lyme borreliosis: comparison with indirect immunofluorescence and enzyme-linked immunosorbent assay. Med. Microbiol. Immunol. *182*, 255–270.

Wilske, B., Luft, B., Schubach, W.H., Zumstein, G., Jauris, S., Preac-Mursic, V., and Kramer, M.D. (1992b). Molecular analysis of the outer surface protein A (OspA) of *Borrelia burgdorferi* for conserved and variable antibody binding domains. Med. Microbiol. Immunol. *181*, 191–207.

Wilske, B., Preac-Mursic, V., Göbel, U.B., Graf, B., Jauris, S., Soutschek, E., Schwab, E., and Zumstein, G. (1993b). An OspA serotyping system for *Borrelia burgdorferi* based on reactivity with monoclonal antibodies and OspA sequence analysis. J. Clin. Microbiol. *31*, 340–350.

Wilske, B., Preac-Mursic, V., Jauris, S., Hofman, A., Pradel, I., Soutschek, E., Schwab, E., Will, G., and Wanner, G. (1993c). Immunological and molecular polymorphisms of OspC, an immunodominant major outer surface protein of *Borrelia burgdorferi*. Infect. Immun. *61*, 2182–2191.

Wilske, B., Preac-Mursic, V., Schierz, G., and Busch, K.V. (1986). Immunochemical and immunological analysis of European *Borrelia burgdorferi* strains. Zentralbl. Bakteriol. Mikrobiol. Hyg. A *263*, 92–102.

Xu, Q., McShan, K., and Liang, F.T. (2007). Identification of an *ospC* operator critical for immune evasion of *Borrelia burgdorferi*. Mol. Microbiol. *64*, 220–231.

Xu, Q., McShan, K., and Liang, F.T. (2008a). Essential protective role attributed to the surface lipoproteins of *Borrelia burgdorferi* against innate defences. Mol. Microbiol. *69*, 15–29.

Xu, Q., McShan, K., and Liang, F.T. (2008b). Modification of *Borrelia burgdorferi* to overproduce OspA or VlsE alters its infectious behaviour. Microbiology *154*, 3420–3429.

Xu, Q., Seemanapalli, S.V., McShan, K., and Liang, F.T. (2006). Constitutive expression of outer surface protein C diminishes the ability of *Borrelia burgdorferi* to evade specific humoral immunity. Infect. Immun. *74*, 5177–5184.

Yang, X., Goldberg, M.S., Popova, T.G., Schoeler, G.B., Wikel, S.K., and Norgard, M.V. (2000). Interdependence of environmental factors influencing reciprocal patterns of gene expression in virulent *Borrelia burgdorferi*. Mol. Microbiol. *37*, 1470–1479.

Yang, X., Popova, T.G., Goldberg, M.S., and Norgard, M.V. (2001). Influence of cultivation media on genetic regulatory patterns in *Borrelia burgdorferi*. Infect. Immun. *69*, 4159–4163.

Yang, X.F., Alani, S.M., and Norgard, M.V. (2003a). The response regulator Rrp2 is essential for the expression of major membrane lipoproteins in *Borrelia burgdorferi*. Proc. Natl. Acad. Sci. U.S.A. *100*, 11001–11006.

Yang, X.F., Hübner, A., Popova, T.G., Hagman, K.E., and Norgard, M.V. (2003b). Regulation of expression of the paralogous Mlp family in *Borrelia burgdorferi*. Infect. Immun. *71*, 5012–5020.

Yang, X.F., Lybecker, M.C., Pal, U., Alani, S.M., Blevins, J., Revel, A.T., Samuels, D.S., and Norgard, M.V. (2005). Analysis of the *ospC* regulatory element controlled by the RpoN–RpoS regulatory pathway in *Borrelia burgdorferi*. J. Bacteriol. *187*, 4822–4829.

Yang, X.F., Pal, U., Alani, S.M., Fikrig, E., and Norgard, M.V. (2004). Essential role for OspA/B in the life cycle of the Lyme disease spirochete. J Exp. Med *199*, 641–648.

Zheng, M., Aslund, F., and Storz, G. (1998). Activation of the OxyR transcription factor by reversible disulfide bond formation. Science *279*, 1718–1721.

Zückert, W.R., Meyer, J., and Barbour, A.G. (1999). Comparative analysis and immunological characterization of the *Borrelia* Bdr protein family. Infect. Immun. *67*, 3257–3266.

Zuker, M. (2003). Mfold web server for nucleic acid folding and hybridization prediction. Nucleic Acids Res. *31*, 3406–3415.

Metabolism and Physiology of *Borrelia* 5

Frank Gherardini, Julie Boylan, Kevin Lawrence and Jon Skare

Abstract

In this chapter, we attempt to define key biochemical pathways and metabolic systems that are the underpinning of the physiology of *Borrelia*. Energy extracted from the fermentation of a few simple sugars fuels these biochemical reactions and also energizes a V-type ATPase (V-ATPase). This establishes a membrane potential that drives motility and the transport of most solutes. After transport, metabolites and biochemical intermediates, such as simple sugars, fatty acids, purines and pyrimidines, peptides and metals ions, are chemically utilized and/or modified to provide an intracellular pool of compounds necessary for protein, nucleic acid, membrane and cell wall biosynthesis. The extremely limited *de novo* biosynthetic capacity of *Borrelia* restricts members of this genus to a host-dependent lifestyle but conserves energy and reflects a reduction of the genome that is an interesting example of adaptive biology.

5.1 Introduction

Like other bacterial pathogens with small genomes, *Borrelia* species have significantly reduced their metabolic capabilities and adapted to their particular hosts and vectors. Therefore, they have taken advantage of metabolites that are provided in excess in the host to eliminate 'expensive' energy-requiring biochemical processes. For example, by utilizing amino acids, nucleotides and fatty acids from the host, energy that is required for the synthesis of these essential cellular building blocks is conserved for the synthesis of macromolecular structures (e.g. membranes, proteins, nucleic acids, flagella) and to fuel transport, motility and cell division. Naturally, this survival 'strategy' does impose limitations on the bacterium. As a consequence of becoming completely host-dependent, it has totally abandoned the ability to exist in a free-living form. While evolutionary limiting, this strategy has been quite successful for a number of symbiotic, commensal, parasitic, and pathogenic bacteria, although it irreversibly links them to their specific host(s).

Analysis of the genome and preliminary biochemical characterization of enzymes and pathways in *Borrelia burgdorferi* provide some insight into its biosynthetic/metabolic capacity. First, its genome lacks the genes encoding enzymes for amino acid, *de novo* nucleotide and fatty acid biosynthesis, as well as those for the TCA cycle (e.g. succinate dehydrogenase, fumarase, aconitase, etc.), respiratory chain (e.g. cytochromes, etc.) and lipopolyaccharide (LPS) synthesis (Fraser *et al.*, 1997). Second, it indicates that *Borrelia* is a homofermentor that degrades a limited number of sugars (e.g. glucose and glucosamine) via the Embden–Myerhof pathway to lactate as the sole electron receptor. Also, some three carbon compounds (e.g. glycerol and glycerol-3-phosphate, G3P) can potentially be used to supplement central metabolism, as well as lipid and lipoprotein biosynthetic pathways (see section 5.6) (Posey and Gherardini, 2000; Schwan *et al.*, 2003). Third, while harbouring the genes encoding enzymes and proteins for DNA, RNA, and protein synthesis as well as those for membrane assembly, cell wall biosynthesis and assembly, cell division, chemotaxis, and motility,

Borrelia lack any *de novo* synthesis of amino acids, nucleotides, vitamin cofactors and fatty acids. Thus, transport of these building blocks becomes critical for the survival of *Borrelia* species (Fraser *et al.*, 1997). Fourth, experimental data indicate that *B. burgdorferi* does not utilize Fe, that the Fe content of the organism is less than 10 Fe atoms per cell (Posey and Gherardini, 2000) and that the bacterium lacks known Fe metalloproteins that are at the core of many biochemical synthetic pathways. This has also had a dramatic effect on the intracellular targets of reactive oxygen species since they do not sustain DNA damage as a result of Fenton chemistry (Boylan *et al.*, 2008).

In the following sections, we will attempt to outline the genes and proteins that impact biochemical pathways that play a role in energy utilization, transport of precursor molecules and compounds (fatty acids, amino acids, nucleotides, etc.), and the modification of those precursors to yield essential components for cell growth, membrane and lipoprotein assembly, nucleic acid synthesis and protein synthesis. The molecular architecture and the assembly of cellular macromolecules in *Borrelia* (proteins, membranes, DNA, and RNAs) are very similar to those described in other bacteria and will not be addressed in these sections.

5.2 Transport

5.2.1 Membrane transport systems

Very little is known about the mechanisms by which solutes are transported across the cell envelopes of spirochaetes. Like other bacteria, however, membrane transport systems presumably provide the necessary exchange of sufficient amounts of essential metabolites from the environment to the appropriate intracellular location for biochemical processes. This is particularly important in *B. burgdorferi* since it has such extremely limited biosynthetic capacity. In fact, ~100 genes of the ~850 genes located on the chromosome encode proteins potentially involved in transport of required biosynthetic intermediates and ions. To date, a limited number of these proteins have been specifically characterized biochemically. However, by comparing genomic, genetic, and uptake/incorporation data combined with biochemical analyses of specific pathways, one can speculate as to the potential role for some of the putative transporters or transport systems. Likewise, the uptake of certain ions or solutes indicates that transport systems are active in *B. burgdorferi* while no specific genes/proteins have been yet identified to explain their accumulation within the cell. For example, Mn has been shown to be actively transported in *B. burgdorferi* in an energy-dependent fashion but no Mn transport system has been identified (Posey and Gherardini, 2000).

Bacteria typically utilize a variety of classes of transport systems:

1. Ion channels (or porins) function as highly selective pores that open and close in response to specific stimuli allowing ions and solutes to rapidly flow down a gradient (Ferguson and Deisenhofer, 2004).
2. Electrochemical transport systems, specifically uniporters, symporters and antiporters, couple electrochemical proton gradients to move ions and solutes across the cytoplasmic membrane. Since electrical gradient potential cannot be maintained across outer membranes, and nucleoside triphosphates are not available in the periplasmic space, transport depends upon the ability of cells to generate electrochemical gradients across cytoplasmic membranes (Bradbeer, 1993; Ferguson and Deisenhofer, 2004). In *B. burgdorferi*, larger solutes, such as fatty acids and nucleotides, appear to follow a similar path (Fraser *et al.*, 1997).
3. Finally, active transporters, including phosphoenolpyruvate sugar phosphotransferase systems (PEP:PTS), the ATP-binding cassette (ABC) transporters and the V-type ATPase both of which utilize the free energy released from PEP or ATP hydrolysis to pump solutes across the inner membrane against concentration gradients. These general mechanisms define the transport of metabolites in *B. burgdorferi*.

The first step in uptake of essential molecules in *B. burgdorferi* is transport across the outer membrane. In Gram-negative bacteria, three different mechanisms are used to pass various substrates into the periplasm:

1. Small molecular weight substrates (<650 Da) can diffuse through non-specific porins such as OmpF or OmpC (Heller and Wilson, 1981; Nakae, 1976).
2. Larger substrates, which exceed the size limits of the general porins, move through specific or specialized porins. Examples of these types of transporters are LamB in *Escherichia coli*, which is essential for the transport of maltodextrans made up of more than three glucose subunits (Wandersman *et al.*, 1979), and ScrY, the sucrose porin from *Klebsiella pneumoniae* (Hardesty *et al.*, 1991). These porins do not bind substrates tightly but rather facilitate transport.
3. Other substrates, which are excluded by size and availability, are transported by high-affinity outer membrane receptors. These receptors bind specific substrates with high affinity and the subsequent release and translocation into the periplasm requires the expenditure of energy. In Gram-negative bacteria, electrochemical energy (proton motive force, PMF) transduction from the cytoplasmic membrane to the transporter is achieved by three inner membrane proteins, ExbB, ExbD and TonB (Eick-Helmerich and Braun, 1989; Eick-Helmerich *et al.*, 1987; Fischer *et al.*, 1989). In *E. coli*, the transport of vitamin B_{12} and Fe–siderophore complexes across the outer membrane is mediated by this energy-dependent mechanism. In *B. burgdorferi*, the absence of homologues for TonB, ExbB or ExbD, suggests that this mechanism does not play a role in transport of substrates across its outer membrane (Fraser *et al.*, 1997).

To date, two outer membrane porins have been identified in the outer leaflet of *B. burgdorferi*. Initially, the approach taken was to isolate the outer membranes, assay the purity of the membrane fraction using inner membrane markers and measure porin activity using a planar lipid bilayer assay to demonstrate porin activity (Skare *et al.*, 1996). Using this methodology, Skare *et al* were able to show that the purified outer membranes of *B. burgdorferi* contained two putative porin activities. The first, designated as Oms28 (encoded by *bba74*), is a 28-kDa, outer membrane protein with a single channel conductance measured at 0.6 nanosiemen (nS) suggesting a relatively small channel (0.64 nm) that could accommodate small molecular weight molecules and ions (Skare *et al.*, 1996), although Mulay *et al* demonstrated that Oms28 does not have porin activity (Mulay *et al.*, 2007). The second, designated Oms66 (also P66, encoded by *bb0603*), is a 66-kDa outer membrane protein also with single channel conductance measuring 12.6 nS (Skare *et al.*, 1997). This would represent a large channel that could accommodate much larger solutes than Oms28. The gene encoding Oms28 resides on plasmid lp54, which has been shown to be lost in certain high-passage, avirulent strains (Skare *et al.*, 1996). Since these isolates have no apparent grow defects, it appears that Oms66 or other unidentified porins are able to compensate for the loss of Oms28 during *in vitro* growth.

Oms66 (P66) is an interesting case. As mentioned in other sections of this chapter, larger molecular weight solutes such as peptides, long chain fatty acids, and deoxynucleotides have no obvious outer membrane transport systems (e.g. fatty acid facilitator, FadL) and are proposed to diffuse across the outer membrane of *B. burgdorferi*. Because of the relative sizes of channels formed by Oms28 and Oms66, it seems likely that Oms66 (P66) may play an important role in the passive transport of some of these solutes. Additionally, using a phage display library of *B. burgdorferi*, Coburn *et al.* (1999) were able to identify P66 as an outer membrane adhesion that bound integrin $\alpha_3\beta_3$. Subsequently, Coburn and Cugini (2003) disrupted *bb0603* and were able to show that the loss of Oms66 (P66) abolished integrin binding activity. Therefore, in addition to porin activity, Oms66 (P66) presumably functions as an integrin binding protein. This same dual function was demonstrated for the 64-kDa and 53-kDa (Msp) proteins from *T. denticola*. Both proteins have porin activity while the 64-kDa protein is also involved in binding of *T. denticola* cells to gingival fibroblasts and Msp is involved in binding to various extracellular matrix proteins (Fenno *et al.*, 1996, 1997). A great deal remains to be determined about transport across the outer membrane of *B. burgdorferi*.

Transport across the inner membrane of bacteria relies on electrochemical transport systems and the ATP-driven import or export systems. *Borrelia burgdorferi* apparently uses similar mechanisms to import and export substrates across its inner membrane; a list of these putative transporters is shown in Table 5.1. The first major category of inner membrane transporters, the electrochemical transporters, can be divided into subgroups that include uniporters, symporters, antiporters, and resistance-nodulation-division (RND) transporters that use electrochemical proton gradients to move substrates across the cytoplasmic membrane. Uniporters transport one solute and are driven directly by the substrate gradient (e.g. BB0074 and BB0042, putative K^+ transporters; BB0380, putative Mg^{2+} transporter; (Fraser *et al.*, 1997). Symporters pump two or more types of solutes in the same direction and are driven by the electrochemical gradient of one solute (e.g. BB0729, BB0401 and BB0814, putative Na^+/dicarboxylate symporters) (Fraser *et al.*, 1997), while antiporters, driven by the single solute gradient, pump substrates in the opposite direction (e.g. BB0447, BB0637 and BB0638, putative Na^+/H^+ antiporters; BB0164, putative Na^+/Ca^+ transporter, and BB0616, putative cation efflux pump) (Fraser *et al.*, 1997).

RND transporters are specialized uniporters that, in Gram-negative bacteria, export toxic substances across both membranes driven by an electrochemical gradient, not by ATP hydrolysis (Li and Nikaido, 2004). The best characterized of these is the AcrAB–TolC drug efflux system in *E. coli* (Nikaido and Zgurskaya, 2001). AcrA and AcrB form an inner membrane complex that is made up of three monomers of each protein. These bind specific substrates and couple export with an inner membrane proton gradient. The TolC component, a membrane fusion protein, is linked to the AcrA/B complex and spans the periplasmic space and the outer membrane. Because of this structural bridge between the inner and outer membrane, substrates can be translocated from the cytoplasm directly to the external medium. Recently, Bunikis *et al.* (2008) identified a RND transport system in *B. burgdorferi* designated BesA (BB0141), BesB (BB0140) and BesC (BB0142; TolC homologue). They showed that the Bes protein complex was involved in resistance of *B. burgdorferi* to various antibiotics and was potentially driven by an electrochemical gradient. In addition, disruption of the Bes operon yielded mutants that were avirulent in experimental animals (Bunikis *et al.*, 2008). Electrochemical transport systems described above are clearly very important to the physiology, biochemistry and survival of *B. burgdorferi*.

Bacteria use group translocation of carbohydrates mediated by PEP:PTS to transport carbohydrates across the cytoplasmic membrane. These systems consist of two cytoplasmic proteins, Enzyme I (EI) and HPr, and a variety of membrane-associated transport complexes, designated as Enzyme II, which are specific for certain sugars (EII^S). These proteins function in concert to transport and phosphorylate incoming sugars for central metabolism. Specifically, EI transfers a phosphoryl group from PEP to HPr, which subsequently transfers a phosphoryl moiety to the various EII transport complexes. EII complexes are made up of three functional units, EIIA, EIIB and EIIC, which occur as three separate protein subunits or as domains within the same polypeptide. EIIA and EIIB proteins or domains receive the phosphoryl group from HPr and transfer it to the incoming sugar, while EIIC, an integral inner membrane protein, functions as the sugar binding protein or domain. The genome of *B. burgdorferi* contains only a single EI homologue, encoded by *bb0558* and two HPr homologues, encoded by *bb0557* and *bb0448* (Fraser *et al.*, 1997). Also, there are homologues for six EIIABC complexes that transport and phosphorylate specific sugars for utilization in central metabolism and RNA synthesis. These include four EIIABC transporters that exist as single polypeptides: (1) BB0116, a putative glucose transporter; (2) BB0408 and BB0629, putative mannose/galactose transporters; and (3) BBB29, a putative glucose transporter. A fifth EIIABC sugar transport system is made up of two proteins encoded by *bb0559* (EIIA) and *bb0645* (EIIBC) where EIIA exists as a single protein and EIIB and EIIC exist as domains within the same polypeptide (Fraser *et al.*, 1997). This PTS system is thought to be involved in glucose transport. The sixth, encoded by *chbC* (*bbb04*; EIIC), *chbA* (*bbb05*; EIIA) and *chbB* (*bbb06*; EIIB), is composed of three distinct proteins and has been implicated in chitobiose uptake and utilization (Tilly *et al.*, 2001). Clearly these PEP:PTS transport systems are important for energy metabolism in *B. burgdorferi*.

ABC transporters are specific systems that transport substrates against a concentration gradient driven by the hydrolysis of ATP. These can function as importers (ABC_{IM}) or exporters (ABC_{EX}), and some have been implicated in the virulence of some bacterial pathogens. ABC_{IM} mediate the transport of nutrients including oligosaccharides, amino acids, peptides, Fe-siderophores and polyamines (Davidson *et al.*, 2008). ABC_{EX} function as secretion systems for variety of molecules including lipids, polysaccharides and protein toxins, such as haemolysins (Davidson *et al.*, 2008). Despite the difference in substrate specificity and the direction of transport, ABC_{IM} and ABC_{EX} are structurally organized in a similar fashion. Generally, most ABC_{IM} are made up of four proteins that form a complex composed of hydrophobic, integral membrane protein(s) (homodimers or heterodimers) and two hydrophilic monomers that are associated with the cytoplasmic side of the membrane complex. The final component of ABC_{IM} are periplasmic binding proteins (PBP) that deliver substrate to the periplasmic side of the cytoplasmic inner membrane complex. This completed complex then mediates the ATP hydrolysis-dependent import of the substrate to the cytoplasm. Conversely, in most ABC_{EX} systems, these functionally distinct protein subunits of the complex are fused into a single polypeptide and export of substrates across the outer membrane is coupled to TolC (Murakami and Yamaguchi, 2003; Poole, 2001; Zgurskaya and Nikaido, 2000).

Several ABC_{IM} systems have been identified in the genome of *B. burgdorferi* (see Table 5.1). The best characterized of these is the oligopeptide transport system, designated as the oligopeptide permease (Opp) operon, which is very similar to the Opp and dipeptide permease (Dpp) systems described in *E. coli* and *Salmonella enterica* (Abouhamad *et al.*, 1991). The homologous Opp system in *E. coli* is composed of a periplasmic binding protein, OppA, two inner membrane transporters, OppB and OppC, and two ATP-binding proteins, OppD and OppF, which hydrolyse ATP to drive peptide transport. The homologous component proteins of the Opp system in *B. burgdorferi* are encoded by *bb0332* (*oppB*), *bb0333* (*oppC*), *bb0334* (*oppD*), *bb0335* (*oppF*) and five periplasmic binding proteins encoded by *bb0328* (*oppA-1*), *bb0329* (*oppA-2*), *bb0330* (*oppA-3*), *bbb16* (*oppA-4*: encoded on plasmid cp26) and *bba34* (*opp-5*: encoded on plasmid lp54) (Fraser *et al.*, 1997). Two additional Opp homologues have been identified in the chromosome of *B. burgdorferi*, encoded by *bb0746* (*oppC*) and *bb0747* (*oppB*), but their role in peptide transport remains to be determined (Fraser *et al.*, 1997). Lin *et al.* (2001) demonstrated that *B. burgdorferi* OppA-1, -2, -4 or -5 with *E. coli* OppB, OppC, OppD and OppF encoded on recombinant plasmids were able to functionally complement an *E. coli oppABCDF* deletion mutant. Further, Wang *et al.* (2004a), using a combinatorial phage display heptapeptide library and competitive binding assays, were able to show that OppA-1 and -2 have a preference for tripeptides over dipeptides or longer peptides. In these assays, OppA-1 had broad substrate specificity, including peptides with bulky hydrophobic residues, while OppA-2 and -3 preferred peptides composed of smaller, non-polar amino acids. Studies of the regulation of the *B. burgdorferi* OppA proteins suggest that these differences in substrate specificity may be important at different stages of the infective cycle. For example, OppA-4 and OppA-5 are upregulated in a mouse while OppA-1 is constitutively expressed *in vivo* or *in vitro* (Wang *et al.*, 2002). More recently, Medrano *et al.* (2007) showed that *oppA-4* was potential regulated by BosR while they and others have shown that *oppA-5* is regulated by RpoN, RpoS and a candidate transcription factor EbfC (Caimano *et al.*, 2007; Medrano *et al.*, 2007). Since *B. burgdorferi* cannot synthesize or transport most amino acids (except possibly glutamate and proline)(Fraser *et al.*, 1997), Opp-mediated peptide transport is critical for amino acid metabolism and survival.

5.2.2 V-type ATPase

Vacuole-type ATPases (V-ATPases) are found in both eukaryotic and prokaryotic cells. In eukaryotes, they are complex structures with at least 13 subunits that function as proton pumps driven by energy derived from the hydrolysis of ATP (Lolkema *et al.*, 2003; Murata *et al.*, 2008; Qi *et al.*,

Table 5.1 Putative transporters from *Borrelia burgdorferi*

Enzyme name	Gene designation	Open reading frame
ABC transporter	–	*bb0753, bb0754, bb0742, bb0466*
ABC transporter	–	*bb0573, bb0574, bbj26, bbj27*
Oligopeptide permease protein	*oppC, oppB*	*bb0746, bb0747*
Oligopeptide transport system	*oppB, oppC, oppD, oppF*	*bb0332, bb0333, bb0334, bb0335*
Oligopeptide binding protein	*oppA-1, -2, -3, -4, -5*	*bb0328, bb0329, bb0330, bbb16, bba34*
Ribose/galactose ABC transporter system	*mglA, mglC1,mglC2*	*bb0677, bb0678, bb0679*
Ribose/galactose ABC transporter system	*bmpB, bmpA, bmpC, bmpD*	*bb0382, bb0383, bb0384, bb0385*
Nucleoside transporter system	–	*bb0316, bb0317, bb0318*
Spermidine/putrescine transport system	–, *potC, potB, potA*	*bb0639, bb0640, bb0641, bb0642*
Mn transporter	*bmtA*	*bb0219*
Phosphate transport system	–, *pstC, pstA, pstB*	*bb0215, bb0216, bb0217, bb0218*
Glycine betaine transport system	–	*bb0144, bb0145, bb0146*
Lipoprotein releasing factor	–	*bb0078, bb0079, bb0080, bb0081*
Phosphate transporter	*phoU*	*bb0042*
Mg transporter	*mgtE*	*bb0380*
Chromate efflux	–	*bb0451, bb0452*
Guanine–hypoxanthine permease	*pbuG*	*bbb22, bbb23*
L-Lactate permease	–	*bb0604*
Potassium transport system	*ktrB, ktrA*	*bb0724, bb0725*
Glycerol uptake facilitator	*glpF*	*bb0240*
Sodium:dicarboxylate symporter	*yhcL*	*bb0729*
Sodium:dicarboxylate symporter	–	*bb0401*
Sodium:pantothenate symporter	–	*bb0814*
PTS system, phosphoenolpyruvate, enzyme I	–	*bb0558*
PTS system, phosphocarrier protein, HPr	–	*bb0557*
PTS system, phosphocarrier protein, HPr	–	*bb0448*
PTS system, fructose-specific enzyme IIABC	–	*bb0629*
PTS system, fructose-specific enzyme IIABC	–	*bb0408*
PTS system, glucose-specific enzyme IIABC	*malX2*	*bbb29*

Enzyme name	Gene designation	Open reading frame
PTS system, glucose-specific enzyme IIABC	*malX1*	*bb0116*
PTS system, glucose-specific enzyme IIAB	*ptsG*	*bb0645*
PTS system, glucose-specific enzyme IIA	*crr*	*bb0559*
PTS system, chitobiose-specific enzyme IIC, A, B	*chbC, chbA, chbB*	*bbb04, bbb05, bbb06*
Multidrug resistance protein B	–	*bbq69, bbi26*
Na^+-driven multidrug efflux pump	–	*bb0583, bb0584*
Na^+-driven multidrug efflux pump	–	*bb0473*
Virulence factor MviN	–	*bb0810*
Arginine/ornithine antiporter	–	*bb0843*
Mechanosensitive ion channel	–	*bb0453*
Sodium–calcium exchanger	–	*bb0164*
Cation efflux	–	*bb0616*
Na^+/H^+ antiporter	–	*bb0638, bb0637, bb0447*
Methicillin resistance factor	–	*bb0586*
Bacitracin resistance factor	–	*bb0258*
60-kDa inner membrane protein, YidC homologue	–	*bb0442*
Borrelia efflux system	*besB, besA, besC*	*bb0140, bb0141, bb0142*
Antibiotic efflux	–	*bb0252*
Protein export (SecDF)	–	*bb0652, bb0653*
V-type ATPase	subunits K, I, D, B, A, E, X, Y	*bb0090, bb0091, bb0092, bb0093, bb0094, bb0096*
Outer membrane porin (oms28)	–	*bba74*
Outer membrane porin (oms66)	–	*bb0603*

2007; Yokoyama and Imamura, 2005). To date, genome comparisons have identified some 37 V-ATPases in 16 bacteria and 21 archaea and more are likely to be found as more genomes are completed. In archaea, these ATPases are common and function in ATP synthesis. In bacteria, V-type ATPases are rare while the vast majority harbour F-type ATPases that function as ATP synthetases. V-Type ATPases are found almost exclusively in low-GC Gram-positive bacteria but some have been identified in members of the phyla Fusobacteria, Chlamydialae and Spirochaetae, but not in Protobacteria. For example, the structure of a V-ATPase from *Enterococcus hirae* has been determined experimentally and, based upon the subunit homology; it seems very similar to those in chlamydial and spirochaetal species in function and structure. More detailed analyses of these bacteria, based upon the organization of the genes in the V-ATPases operons, protein homology and the protein subunits contained therein, show that spirochaetal V-ATPases are most closely related to those found in chlamydial species (Lolkema *et al.*, 2003). *T. pallidum* and *Clostridium tetani* are the only bacteria that contain genes encoding two different V-ATPases.

V-type ATPases from prokaryotes are less complex than their eukaryotic counterparts and contain eight to nine subunits arranged in two heterologous domains designated as V_1 and V_0. V_1 is a hydrophilic, cytosolic domain that is responsible for ATP hydrolysis, while the V_0 is a membrane embedded ion-translocating domain. In *E. hirae*, the V-ATPase contains nine subunits and has been characterized biochemically; the proposed structure is shown in Fig. 5.1B. The V_1 domain is 500-kDa protein complex, made up of subunits A, B, C, D, E, F, and G, functions to hydrolyse ATP and transfer that energy to V_0. Subunits A and B bind ATP, with subunit A functioning as the enzymatic site for hydrolysis. Three monomers each of A and B are arranged around subunits D, G and C, to form the central stem. The hydrolysis of ATP by A causes a rotational shift in the stem complex that 'activates' the V_0 complex to pump ions or protons. The E and F subunits are thought to form an auxiliary stem that might function to stabilize the central stem, however, their exact function is not known. The V_0 is composed of subunits K and I with a stoichiometry of 10 to 1, respectively. Each of the K subunits bind a Na^+ ion and, as the central stem rotates, it drives the rotation of the K ring bringing those bound ions into contact with the ion channels in protein I. The bound Na^+ is 'released' into that channel for translocation into the periplasm.

Borrelia burgdorferi harbours the genes encoding a V-type ATPase that is the most similar to those observed in low-GC Gram-positive bacteria, chlamydial species and other spirochaetes (Fraser *et al.*, 1997). The genes are organized in an operon with the gene order E, *X*, A, B, D, I, K and

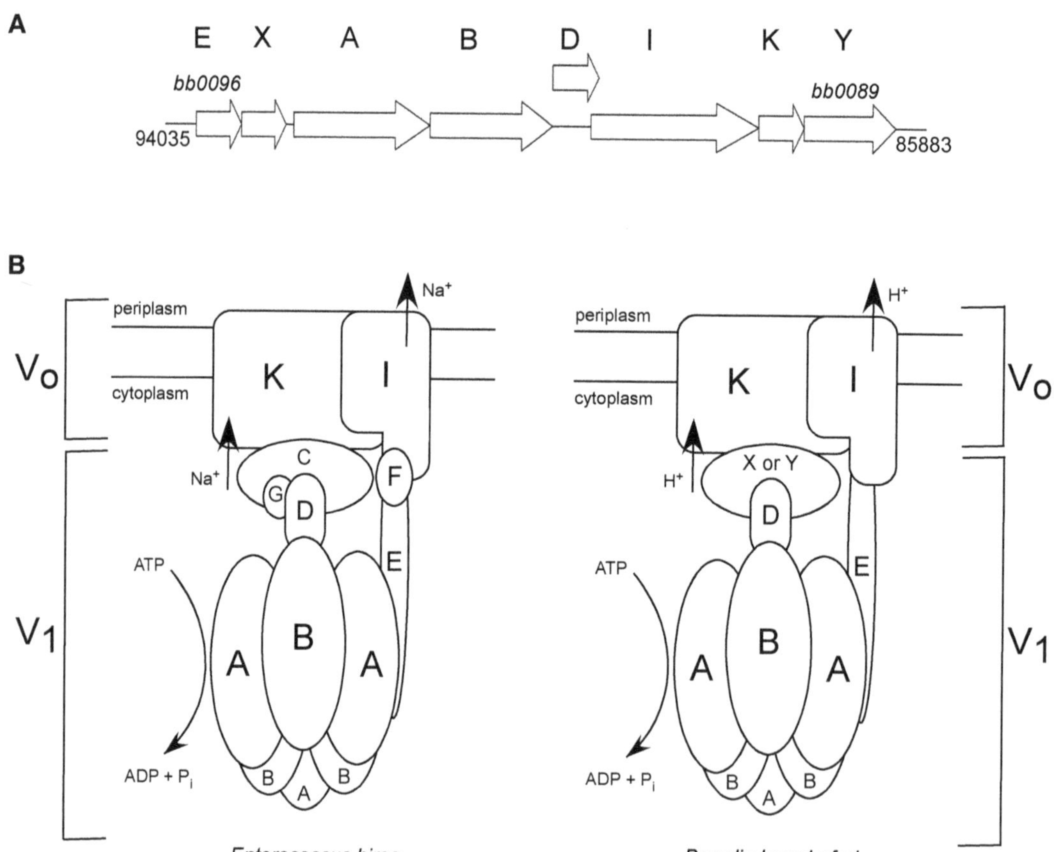

Figure 5.1 (A) Gene organization of the V-ATPase operon from *B. burgdorferi*. (B) Structure model of the V-ATPase transporter in *Enterococcus hirae* compared with the model predicted for *B. burgdorferi*.

Y (Fig. 5.1A). Based upon the well characterized V-ATPase from *E. hirae,* we propose a model for the structure of the V-ATPase from *B. burgdorferi* (Fig. 5.1B). In the V_1 domain, the A and B subunits are proposed to also function in ATP binding and hydrolysis, while D and E function as major and minor components of the central and auxiliary stems, respectively. Conspicuously absent is a gene encoding a C subunit, which is thought to perform an essential function in the interaction between the central stem and V_0 domain. Also absent are homologues for the G and F subunits (Fraser *et al.*, 1997). Lolkema *et al.* (2003) observed that these subunits were missing from some genomes (spirochaetes and chlamydia species) and proposed that other, unidentified genes in their V-ATPase operons might perform the enzymatic functions of C, F, and/or G (see Fig. 5.1, designated X and Y). Homologues of the I and K proteins are contained in the V-ATPase operon of *B. burgdorferi* and presumably perform functions similar to those proposed for the V_0 domain in *E. hirae.* Murata *et al.* (2008) recently showed that binding of Na^+ to the purified K subunit of the V-type ATPase from *E. hirae* was competitively inhibited by Li^+ or H^+, suggesting that those ions could also be transported by this complex. Based upon these observations, we propose that the V-type ATPase in *B. burgdorferi* functions primarily as a H^+ pump that establishes a proton motive force across the inner membrane thereby driving some proton-coupled transport systems (e.g. for Na^+: BB0637, BB0638 and BB0447) and *B. burgdorferi* motility. Importantly, some V-ATPases function as proton transport systems that drive ATP synthesis. Because of the similarities to the V-ATPase from *E. hirae,* this seems unlikely in *B. burgdorferi.* The exact function of this important protein complex remains to be experimentally determined.

5.3 Central metabolism

Borrelia burgdorferi is an obligate bacterial pathogen with no known ability to survive outside of either an arthropod vector or mammalian host. This apparent inability to survive outside of a host implies that *B. burgdorferi* is restricted in terms of its metabolic capabilities and is consistent with its fastidious nature. On the other hand, the ability to survive in such disparate hosts as ticks and warm-blooded mammals implies that *B. burgdorferi* is adept at incorporating metabolites found in either or both of the arthropod and mammalian species they infect. The genome sequence of *B. burgdorferi* indicates that the primary source for energy production is by fermentation via the Embden–Meyerhof pathway (Das *et al.*, 2000; Fraser *et al.*, 1997; von Lackum and Stevenson, 2005). No homologues of TCA enzymes or proteins involved in electron transport and oxidative phosphorylation are encoded in the genome (Das *et al.*, 2000; Fraser *et al.*, 1997; von Lackum and Stevenson, 2005). Despite the limited battery of enzymes involved in energy production, *B. burgdorferi* is able to survive and persist for long periods of time inside the hosts they colonize.

5.3.1 Carbon utilization

One mechanism used by *B. burgdorferi* that aids in ATP generation in glycolysis is their apparent ability to transport several different six-carbon sugars via the phosphotransferase system (PTS). These families of well characterized sugar transporters are predicted to result in the accumulation of glucose, N-acetylglucosamine, chitobiose, and mannose via six distinct PTS transporters (Fig. 5.2) (Fraser *et al.*, 1997; Saier and Paulsen, 2000; von Lackum and Stevenson, 2005). Three of the six PTS transporters are predicted to be competent for glucose uptake implying that, as expected, this is the preferred carbon source of *B. burgdorferi* (Fraser *et al.*, 1997; von Lackum and Stevenson, 2005). In addition to the PTS transporters, *B. burgdorferi* synthesizes an Mgl ABC system that is predicted to transport glucose (Fig. 5.2). Alternative carbon sources, specifically mannose and chitobiose, are predicted to be transported via PTS as well (Fig. 5.2) (Fraser *et al.*, 1997; von Lackum and Stevenson, 2005). Although the genome sequence suggested that additional carbon sources, including fructose, galactose, and ribose, could be utilized, studies have shown that *B. burgdorferi* growth is restricted to glucose, maltose, N-acetylglucosamine, chitobiose, mannose, and glycerol as carbon sources during *in vitro* growth (von Lackum and Stevenson, 2005). Growth limitation in modified BSK media using fructose, ribose and galactose was observed in these studies. Clearly, more studies need to be done to define, experimentally, the specific roles of these

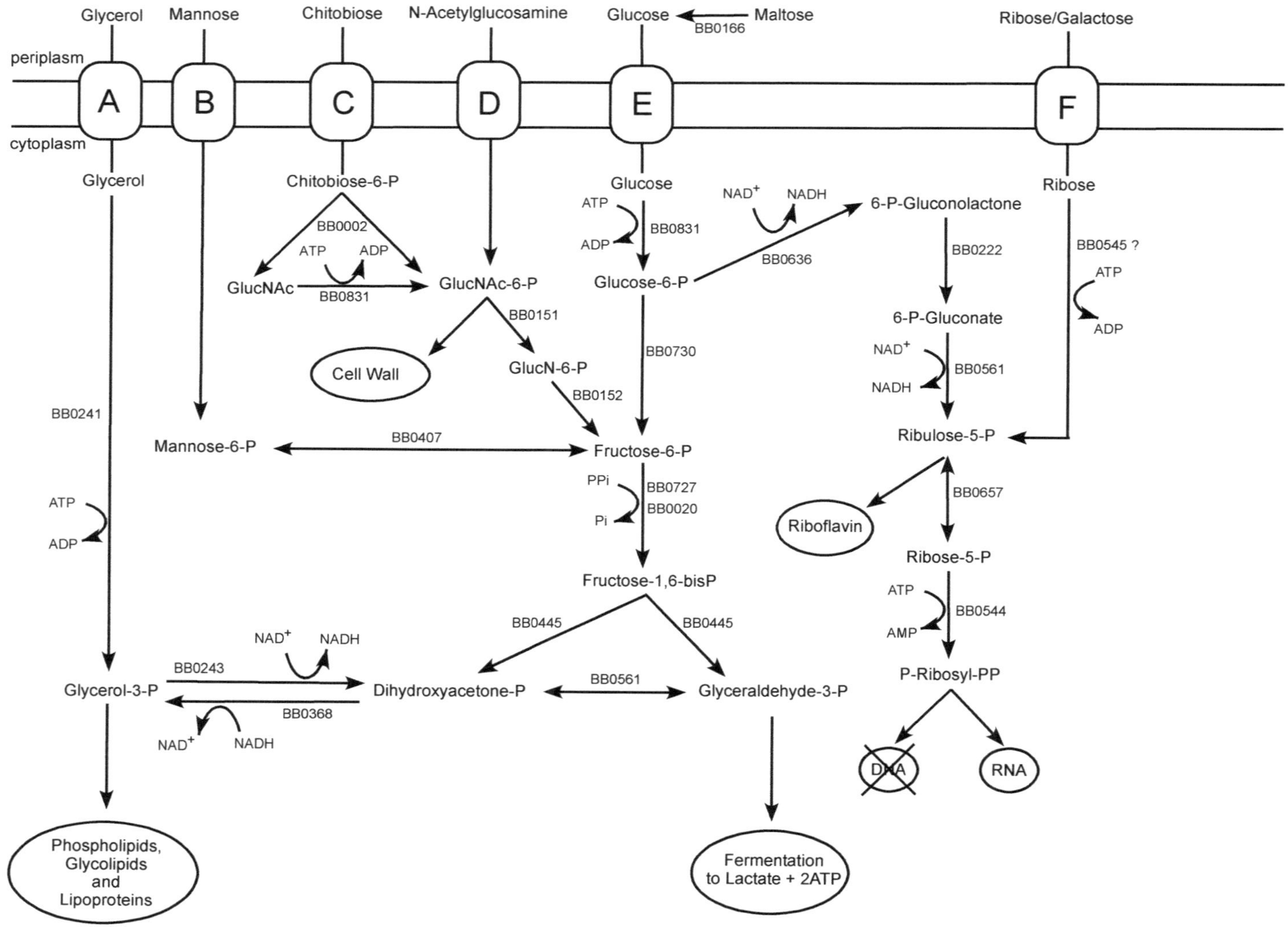
Glycerol
Mannose
Chitobiose
N-Acetylglucosamine
Glucose
Maltose
BB0166
Ribose/Galactose
periplasm
cytoplasm
A
B
C
D
E
F
Glycerol
Chitobiose-6-P
BB0002
ATP
ADP
GlucNAc
BB0831
GlucNAc-6-P
BB0151
Cell Wall
GlucN-6-P
BB0152
Glucose
ATP
ADP
BB0831
Glucose-6-P
NAD+
NADH
BB0636
6-P-Gluconolactone
BB0222
6-P-Gluconate
NAD+
NADH
BB0561
Ribose
BB0545 ?
ATP
ADP
BB0730
BB0241
Mannose-6-P
BB0407
Fructose-6-P
PPi
Pi
BB0727
BB0020
Ribulose-5-P
BB0657
Riboflavin
Ribose-5-P
ATP
AMP
BB0544
P-Ribosyl-PP
DNA
RNA
ATP
ADP
Fructose-1,6-bisP
BB0445
BB0445
Glycerol-3-P
BB0243
NAD+
NADH
BB0368
NAD+
NADH
Dihydroxyacetone-P
BB0561
Glyceraldehyde-3-P
Phospholipids, Glycolipids and Lipoproteins
Fermentation to Lactate + 2ATP

Enzymes Involved in Transport

ORF	Function	Gene Designation
	TRANSPORTER A	
bb0240	Glycerol uptake facilitator	*glpF*
	TRANSPORTER B	
bb0408	PTS, mannose-specific Enzyme IIABC	—
bb0629	PTS, mannose-specific Enzyme IIABC	—
	TRANSPORTER C	
bbb04	PTS, chitobiose-specific Enzyme IIC	*chbC*
bbb05	PTS, chitobiose-specific Enzyme IIA	*chbA*
bbb06	PTS, chitobiose-specific Enzyme IIB	*chbB*
	TRANSPORTER D/E	
bb0559	PTS, glucose-specific Enzyme IIA	*crr*
bb0116	PTS, glucose-specific Enzyme IIABC	*malX1*
bbb29	PTS, glucose-specific Enzyme IIABC	*malX2*
bb0645	PTS, glucose-specific Enzyme IIAB	*ptsG*
	TRANSPORTER F	
bb0382	Ribose/galactose ABC transporter, basic membrane protein B	*bmpB*
bb0383	Ribose/galactose ABC transporter, basic membrane protein A	*bmpA*
bb0384	Ribose/galactose ABC transporter, basic membrane protein C	*bmpC*
bb0385	Ribose/galactose ABC transporter, basic membrane protein D	*bmpD*
bb0677	Ribose/galactose ABC transporter, ATP-binding protein	*mglA*
bb0678	Ribose/galactose ABC transporter, permease protein	*mglC1*
bb0679	Ribose/galactose ABC transporter, permease protein	*mglC2*

Enzymes Involved in Carbon Metabolism

ORF	Function	Gene Designation
bb0002	beta-N-acetylhexosaminidase	*cht*
bb0020	pyrophosphate-fructose-6-P 1-phosphotransferase	*pfkA*
bb0151	N-acetylglucosamine-6-P deacetylase	*nagA*
bb0166	4-alpha-glucanotransferase	*malQ*
bb0222	6-phosphoglucanolactonase	*pgl*
bb0241	glycerol kinase	*glpK*
bb0243	glycerol-3-P dehydrogenase	*glpD*
bb0368	glycerol-3-P dehydrogenase	*gpsA*
bb0407	mannose-6-P isomerase	*manA*
bb0445	fructose-bis-P adolase	*fbaA*
bb0544	ribose phosphate pyrophosphokinase	*prsA*
bb0454	monogalactosyl-1,2-diacylglycerol synthase	-
bb0561	6-phosphogluconate dehydrogenase	*gnd*
bb0636	glucose-6-P 1-dehydrogenase	*zwf*
bb0657	ribose-5-P isomerase	*rpi*
bb0727	pyrophosphate-fructose-6-P 1-phosphotransferase	*pfpB*
bb0730	glucose-6-P isomerase	*pgi*
bb0831	glucokinase	*nagC*

Figure 5.2 Carbohydrate transport and metabolism in *B. burgdorferi*. Enzymes are identified by ORF number in the diagram and correspond to the function and gene designation given in the inserted tables (modified from von Lackum and Stevenson, 2005).

carbon sources on the central metabolism of *B. burgdorferi*.

Analyses of other biochemical pathways may explain why galactose and ribose are transported in *B. burgdorferi*. As shown in section 5.6, galactose is required for the synthesis of the major membrane glycolipid, α-MgalDAG. This glycolipid has been proposed to be required for membrane stability and function (Mannock *et al.*, 2001). In the case of ribose, homologues of all of the enzymes required for the oxidative phase of the pentose phosphate pathway are present within *B. burgdorferi* (Fraser *et al.*, 1997), while several enzymes are missing from the non-oxidative phase of the pentose phosphate pathway (notably ribulose-phosphate 3-epimerase, transketolase, and transaldolase). Given that the oxidative phase of the pentose phosphate pathway pushes ribose towards riboflavin and RNA biosynthesis while the non-oxidative phase yields glyceraldehyde 3-phosphate and fructose 6-phosphate that fuel glycolysis and lipid and lipoprotein biosynthesis (Fig. 5.2), ribose is probably required for RNA (see section 5.4) and riboflavin (see section 5.5) biosynthesis, not for the generating energy. This would explain the presence of a putative ribose transport system in *B. burgdorferi* (Fraser *et al.*, 1997) and the observation of von Lackum and Stevenson that *B. burgdorferi* cannot use ribose as a carbon source during *in vitro* growth (von Lackum and Stevenson, 2005).

The transport of chitobiose is of particular interest as it is a derivative of the chitin exoskeleton of the arthropod vector, specifically a β-1,4-linked disaccharide of N-acetyl glucosamine (NAG) that is not found in mammals. The ability of *B. burgdorferi* to transport chitobiose was originally shown by Tilly *et al.* (2001). A subsequent study demonstrated that genetic inactivation of the *chbC* gene (the IIC component of the chitobiose PTS transporter) did not affect growth in either the tick vector or mammalian hosts, indicating that other sugars present within the tick can compensate for the loss of this transport system (Tilly *et al.*, 2004). Alternatively, a putative beta-glucosidase, encoded by *bb0620*, could function to cleave chitobiose to generate NAG molecules that can then be transported through PTS transporters (Fig. 5.2). Curiously, while NAG is sufficient as a carbon source to promote growth, the non-acetylated form, glucosamine, is not (von Lackum and Stevenson, 2005). This observation suggests that glucosamine cannot be funnelled into the glycolytic pathway because *B. burgdorferi* does not encode a kinase that converts glucosamine to glucosamine-6-phosphate, which, following deamination, forms fructose 6-phosphate, a critical intermediate substrate in the glycolytic pathway (Fig. 5.2). In contrast, after NAG is phosphorylated, it can then either be incorporated into cell wall biosynthesis or deacetylated to generate glucosamine-6-phosphate that is then converted to fructose 6-phosphate for use in glycolysis (Fig. 5.2) (von Lackum and Stevenson, 2005). Again, the observation that NAG can support *in vitro* growth of *B. burgdorferi* suggests that it can serve an alternative role as a carbon source in addition to its more widely accepted function as a building block in cell wall biosynthesis.

Finally, an operon (*bb0240* through *bb0243*) is predicted to encode a glycerol transporter (GlpF; *bb0240*) and enzymes that convert glycerol to glycerol-3-phosphate (GlpK; *bb0241*) and isomerize glycerol 3-phosphate to dihydroxyacetone phosphate (GlpA/GlpD; *bb0243*) (Fraser *et al.*, 1997; Saier and Paulsen, 2000; von Lackum and Stevenson, 2005). The exact role of the protein encoded by *bb0242* is not known. Once dihydroxyacetone phosphate is formed it can isomerize into glyceraldehyde-3-phosphate, which then proceeds to the glycolytic pathway to generate ATP (Fig. 5.2). In addition to its entry into glycolysis, glycerol 3-phosphate can be shuttled into phospholipid, lipoprotein and glycolipid biosynthesis.

5.3.2 ATP synthesis

As indicated above, *B. burgdorferi* lacks genes homologous to those encoding proteins involved in oxidative phosphorylation or the TCA cycle (Das *et al.*, 2000; Fraser *et al.*, 1997; von Lackum and Stevenson, 2005). They also do not appear to make energy via the pentose phosphate pathway, but instead use it to generate building blocks for DNA and RNA (Das *et al.*, 2000; von Lackum and Stevenson, 2005). However, they do have all of the genes required for the glycolytic pathway (Das *et al.*, 2000; Fraser *et al.*, 1997; von Lackum and Stevenson, 2005). The major metabolic pathway for ATP generation in *B. burgdorferi* appears to be substrate-level phosphorylation mediated by the phosphoglycerate kinase and pyruvate kinase enzymatic steps during fermentation (Das *et al.*, 2000; Fraser *et al.*, 1997; von Lackum and Stevenson, 2005). Because oxidative phosphorylation is not operative in *B. burgdorferi*, the ability to harness the energy stored in these reduced substrates is not possible. Thus, reduced $NADH^+$ is oxidized independent of oxidative phosphorylation and instead is dependent on reduction of pyruvate to lactate by lactate dehydrogenase and other redox enzymes that can reduce $NADH^+$, including coenzyme disulphide reductase (i.e. Cdr encoded by *bb0728*; Boylan *et al.*, 2006).

Most bacteria generate a proton gradient across their cytoplasmic membrane due to electron transport systems that result in a greater concentration of protons outside the cytoplasm. The protons then come down their concentration gradient through an F-type ATPase in a process that generates ATP within the cytoplasm of the bacterium. However, *B. burgdorferi* does not encode an F-type ATPase. The maintenance of its proton gradient is probably mediated by the function of the V-type ATPase in a process that hydrolyses ATP (see section 5.2) (Fraser *et al.*, 1997,1998). Effectively, V-type ATPases function in reverse of F-type ATPases and thus utilize energy instead of synthesizing ATP. Previously, V-type ATPases were associated with vacuolar membranes but were more recently found in plasma membranes of most eukaryotes (Beyenbach and Wieczorek, 2006) and select bacteria, including *B. burgdorferi* and *T. pallidum* (Fraser *et al.*, 1997; Fraser *et al.*, 1998). Although, in theory, the pumping of protons in reverse through its V-type ATPase could be used to generate ATP, no evidence exists that this occurs in spirochaetes or any other prokaryote.

5.4 Purine and pyrimidine salvage

Nucleotide metabolism serves an essential function in all living organisms as a means of providing critical precursors for nucleic acid biosynthesis and a variety of cofactors for other pathways. The complexity of these biochemical pathways varies among organisms and spans the range from rudimentary salvage pathways that rely upon exogenous sources of nucleosides, to more complex *de novo* biosynthetic pathways. The degree to which an organism is dependent upon exogenous sources of nucleotide precursors is the direct result of the complexity of the organism's nucleotide biosynthetic pathway and is an important determinant in the niche where this organism can survive. Nucleotide salvage pathways differ among *Borrelia* species (Pettersson *et al.*, 2007). The genomes of Lyme disease spirochaetes (e.g. *B. burgdorferi*, *B. garinii*, and *B. afzelii*) are missing genes encoding enzymes such as ribonucleotide reductase, adenylosuccinate lyase, adenylosuccinate synthase and hypoxanthine phosphoribosyltransferase (Table 5.2). Conversely, these genes are present in the genomes of relapsing fever species (*B. hermsii* and *B. turicatae*). *B. hermsii* encodes a classical nucleotide salvage pathway, similar to those described for other organisms (Barbour *et al.*, 2005; Pettersson *et al.*, 2007). This type of pathway allows for the uptake of purine and pyrimidine bases, which can undergo enzymatic transformation to ribonucleotides for incorporation into RNA or be reduced to deoxyribonucleotides by ribonucleotide reductase (RNR) for incorporation into DNA.

Because *B. hermsii* has a functional RNR, this classical salvage pathway does not require an exogenous source of deoxynucleosides for DNA synthesis and, as a result, does not limit the type of host tissue where this spirochaete can survive (i.e. blood). Hypoxanthine phosphoribosyltransferase in *B. hermsii* is responsible for the production of inosine monophosphate through the transfer of a ribose-5′-monophosphate from PRPP to hypoxanthine (Pettersson *et al.*, 2007). The presence of this enzyme provides for the direct utilization of hypoxanthine for the production of GMP and dGMP through inosine monophosphate dehydrogenase (GuaB) and GMP synthase (GuaA). Furthermore, adenylosuccinate lyase (PurB) and adenylosuccinate synthase (PurA) could provide an additional advantage to *B. hermsii* by performing the conversion of IMP to AMP. This pathway may allow for the utilization of hypoxanthine for the synthesis of AMP and dAMP. Without PurAB, hypoxanthine can only serve to produce GMP and dGMP, thereby imposing a requirement on the bacterium for host-derived sources of adenine nucleosides.

Table 5.2 Enzymes in *Borrelia burgdorferi* nucleotide salvage pathway

Enzyme name	Gene designation	Open reading frame
Adenine deaminase	*adeC*	*bbk17*
Adenine phosphoribosyltransferase	*apt*	*bb0777*
Adenylate kinase	*adk*	*bb0417*
Deoxynucleoside kinase	*dgk*	*bb0239*
Deoxyribosyl transferase	*ptd*	*bb0426*
GMP synthase	*guaA*	*bbb18*
IMP dehydrogenase	*guaB*	*bbb17*
Guanine–hypoxanthine permease	*pbuG*	*bbb22, bbb23*
Nucleoside diphosphate kinase	*ndk*	*bb0463*
Xanthine–guanine phosphoribosyltransferase	*gpt*	*bb0103*
Cytidine deaminase	*cda*	*bb0618*
Thymidine kinase	*tdk*	*bb0791*
CTP synthase	*pyrG*	*bb0575*
Uridylate kinase	*pyrH*	*bb0571*
Uridine kinase	*udk*	*bb0015*
Cytidylate kinase	*cmk*	*bb0819, bb0128*
Thymidylate synthase	*thyX*	*bba76*

5.4.1 Ribonucleotide reductase

In silico analysis of *B. burgdorferi*'s genome failed to identify any sequence with homology to known RNR enzymes (Fraser *et al.*, 1997). RNRs catalyse the reduction of ribonucleotides to their corresponding deoxyribonucleotides as essential precursors for DNA synthesis. RNRs are widely believed to be an absolute requirement for all living organisms due to their role in DNA synthesis (Jordan and Reichard, 1998), although the genome of *Ureaplasma urealyticum* has also been reported to be missing an RNR homologue (Carnrot *et al.*, 2003). Additional evidence for the absence of an RNR in *B. burgdorferi* was illustrated by the lack of cross-hybridization in a Southern blot when *B. burgdorferi* DNA was probed with *nrdF* from *B. hermsii* (Pettersson *et al.*, 2007). Furthermore the absence of RNR enzyme activity was demonstrated in a radiolabelled nucleotide incorporation assay when *B. burgdorferi* was unable to convert ribonucleotides to deoxyribonucleotides (Lawrence *et al.*, 2009). The consequences to an organism deficient in this enzyme are (i) the organism must depend on exogenous sources of deoxynucleosides from host tissue and/or (ii) the organism must have evolved an alternative means of producing and regulating deoxynucleotide pools. *Borrelia burgdorferi* takes advantage of both approaches and has a novel biochemical approach to generate deoxynucleotides for DNA synthesis, which is outlined in Fig. 5.3. Thus, the availability of exogenous deoxynucleosides, which are required to sustain these pathways, is a determining factor in the survival of *B. burgdorferi* in different host tissues and sites (i.e. tick midgut or mammalian tissue).

Deoxynucleotides and various nucleic acid precursors can be found in mammalian tissue due to cell lysis as the result of an immune inflammatory response or cell apoptosis. *Borrelia burgdorferi* can take advantage of this by efficiently transporting and incorporating into DNA free deoxynucleosides in addition to purine bases such as hypoxanthine and adenine that are also

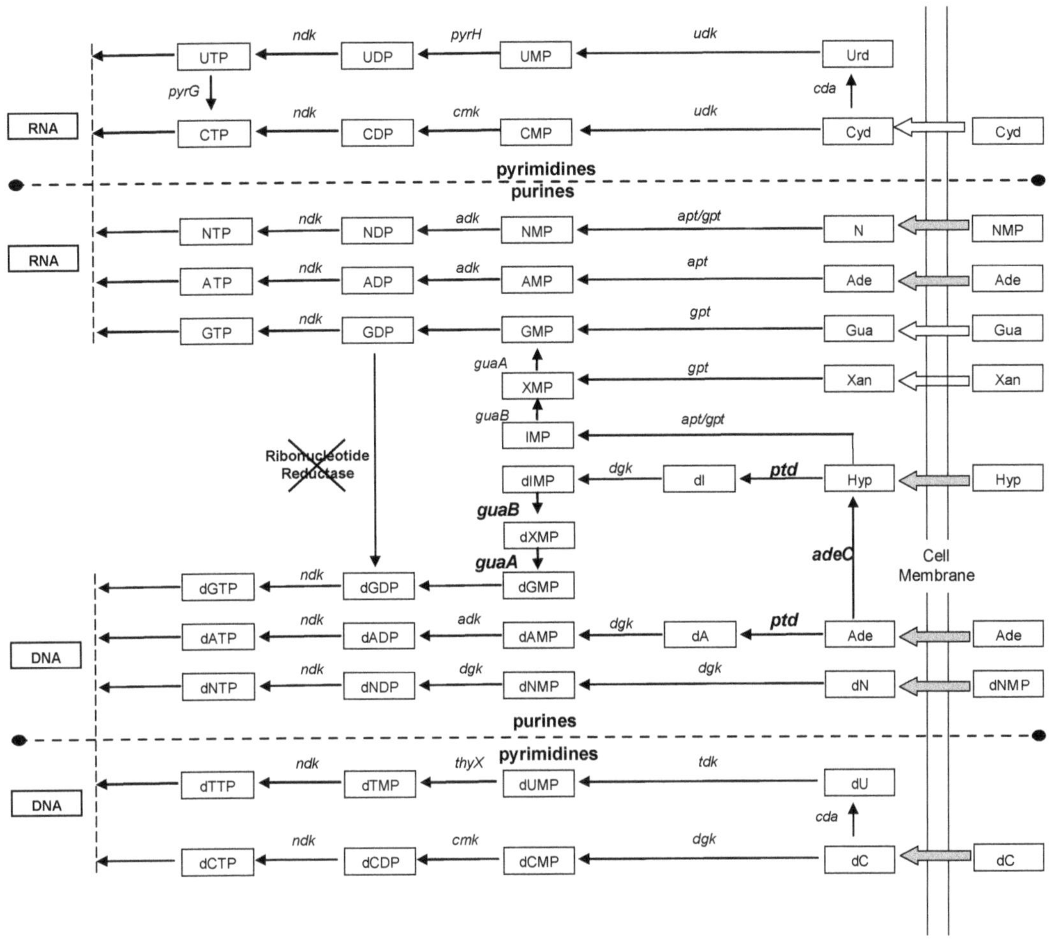

Figure 5.3 Putative purine and pyrimidine salvage pathway in *Borrelia burgdorferi*. Gene designations in bold indicate characterized enzyme activities. Shaded arrows specify membrane transport that was demonstrated experimentally.

available in blood and other tissues (Lawrence *et al.*, 2009). Although the specific transporters that are responsible for this uptake have not been characterized, the genome of *B. burgdorferi* contains two ORFs that are believed to encode guanine-hypoxanthine permeases (PbuG) (Byram *et al.*, 2004) and an operon encoding a putative nucleoside permease system, *bb0677–79* (Overbeek *et al.*, 2003). These transporters may also be responsible for the uptake of several additional nucleotides precursors. Given these few building blocks and a unique purine salvage pathway, *B. burgdorferi* can produce DNA in sufficient quantity and with suitable efficiency to support cell division without the requirement for an RNR.

5.4.2 2′-Deoxyribosyltransferase

In *B. burgdorferi*, a key player that is crucial to its purine salvage pathway is the 2′-deoxyribosyltransferase (DRTase), a 21-kDa protein that is encoded by *bb0426* (Lawrence *et al.*, 2009). This enzyme functions by catalysing the transfer of a 2′-deoxyribose moiety from a donor purine deoxynucleoside to an acceptor purine base. Two types of deoxyribosyltransferase have been described for several *Lactobacillus* species and characterization of both proteins has included elucidation of catalytic mechanisms and crystal structures (Anand *et al.*, 2004; Kaminski, 2002). Biochemical studies using recombinant BB0426 have demonstrated that *B. burgdorferi* harbours a purine-specific 2′-deoxyribosyltransferase designated DRTase I (Lawrence *et al.*, 2009). Enzymatic activity was not detected when pyrimidine pairs or pyrimidine–purine pairs were tested as substrates. *In vivo*, the source of the deoxynucleoside substrates for *B. burgdorferi* is restricted to uptake from host tissues, due to the absence of the reductive capability of an RNR. Once a deoxynucleoside has been transported into the cell, DRTase I can transfer its deoxyribose to another purine base and after phosphorylation, the product can be incorporated into DNA. In particular, this function benefits *B. burgdorferi* in the production of dGMP for DNA synthesis since hypoxanthine, as a deoxyribose recipient, is converted to deoxyinosine (dI). With subsequent phosphorylation to dIMP, and the catalytic action of GuaA and GuaB (discussed below), dGMP is produced. Utilizing DRTase I as an alternative mechanism RNR provides *B. burgdorferi* with the ability to regulate the composition of deoxynucleotide pools when the availability of one or more deoxynucleosides is limited from exogenous sources.

5.4.3 Inosine monophosphate dehydrogenase and GMP synthase

Another critical component to *B. burgdorferi*'s purine salvage pathway is inosine-5′-monophosphate dehydrogenase (GuaB). This enzyme is a 44-kDa gene product of *bbb17* (*guaB*) located on circular plasmid cp26. GuaB has been shown to catalyse the oxidation of IMP to XMP through the corresponding reduction of NAD^+ (Zhou *et al.*, 1997). This activity has been the primary focus of enzymatic studies due to the importance of GuaB in the pathway for GMP production. In *B. burgdorferi*, recombinant GuaB has also been shown to catalyse the conversion of dIMP to dXMP (Lawrence *et al.*, 2009). This activity, along with GMP synthase (GuaA), satisfies the essential role of dGMP production for DNA synthesis in the absence of a functional RNR. The advantage of this mechanism is the utilization of host-derived hypoxanthine for the production of deoxyinosine (dI) through the action of DRTase. Subsequent phosphorylation provides dIMP, which serves as a substrate for GuaB and GuaA, arriving at dGMP for DNA synthesis. In *B. burgdorferi*, the absence of *guaAB* through site-directed mutagenesis was reported to be avirulent *in vivo* using the standard mouse model although this strain did not have a growth defect during *in vitro* cultivation (Jewett *et al.*, 2009).

5.4.4 Adenine deaminase

Adenine deaminase (AdeC) is a 61-kDa protein encoded by *bbk17* in *B. burgdorferi*, whose function has been shown to catalyse the deamination of adenine to produce hypoxanthine (Jewett *et al.*, 2007). This mechanism provides a distinct advantage to *B. burgdorferi*'s purine salvage pathway by allowing for the catabolism of adenine nucleotides when exogenous sources of hypoxanthine are limited in host cell tissue. This secondary source of hypoxanthine can then be utilized in the production of GMP and dGMP as described above. Mutation of *adeC* in *B. burgdorferi* results in an attenuated virulence phenotype *in vivo*, but, like the *guaAB* mutant, the *adeC* mutant did not have a growth defect when cultured *in vitro* (Jewett *et al.*, 2007).

5.5 Cofactor biosynthesis

Enzymes are biological molecules that catalyse chemical reactions and many require the presence of an additional, non-protein cofactor to participate in catalysis. There are two types of cofactors: essential ions (mostly metal ions) and coenzymes (organic compounds) that act as group-transfer reagents. These can be divided into two groups based on their metabolic origin: metabolite coenzymes and vitamin-derived coenzymes. Metabolite coenzymes include a number of nucleoside triphosphates, of which ATP is by far the most abundant. The vitamin-derived coenzymes are derivatives of vitamins, which are compounds that cannot be synthesized by animals and must be obtained as nutrients. This section will focus on the three major vitamin-derived coenzymes of *B. burgdorferi*: coenzyme A (CoA), nicotinamide adenine dinucleotide (NAD), and flavin adenine dinucleotide (FAD).

5.5.1 Coenzyme A

Coenzyme A (CoA) and acyl carrier protein (ACP) are the two predominant acyl group carriers in cells. These compounds are essential cofactors in all cells and participate in numerous reactions in intermediary metabolism. In addition, reduced coenzyme A probably plays other roles in *B. burgdorferi*. The high levels of reduced CoA inside the cell suggest a role in supporting the reduced intracellular environment and the thiol/disulphide distribution of other thiol-containing molecules (Boylan *et al.*, 2006). Although putative genes have been identified in *B. burgdorferi* that may be involved in the synthesis and regulation of CoA, most of what is known about the biochemistry and genetics of the enzymes involved in the formation of pantothenate and CoA has been derived from research with *E. coli* and *S. typhimurium*.

Pantothenate transport

The first step in the synthesis of coenzyme A is the production of pantothenate. Most bacteria are capable of synthesizing pantothenate using intermediates from the amino acid pathways. However, *B. burgdorferi* is missing key enzymes in this biosynthetic pathway and therefore, must acquire pantothenate from its environment via a transport system (Fraser *et al.*, 1997). Pantothenic acid is widely distributed in mammalian tissues (liver, kidney, heart, skin, and muscle) and available for transport into the bacterium. The best-characterized transport system is the high-affinity pantothenate permease of *E. coli* that catalyses the concentrative uptake of pantothenate by a sodium ion cotransport mechanism (Nakamura and Tamura, 1973; Vallari and Rock, 1985). Pantothenate permease belongs to the group of inner membrane permeases that catalyse active cation-dependent symport. This transport activity is required for growth of bacteria that lack a pantothenate biosynthetic pathway (e.g. *Lactobacillus* spp.). *Borrelia burgdorferi* harbours a gene (*bb0814*) encoding a putative pantothenate permease. Without genes encoding pantothenate biosynthetic enzymes, this is the likely pathway for CoA biosynthesis.

CoA biosynthesis

Pantothenate phosphorylation by pantothenate kinase (BB0527) to form 4′-phosphopantothenate is the first step in the biosynthesis of CoA. In *E. coli*, this is the most highly regulated step in the process, and it is regulated through feedback inhibition by CoA and its thioesters (Davis, 1950; Jackowski and Rock, 1981; Maas and Davis, 1950). Non-esterified CoA was significantly more potent than CoA thioesters in inhibiting the enzyme in extracts from *E. coli*. CoA is a competitive inhibitor with respect to ATP, thus providing a mechanism to coordinate CoA production with the energy state of the cell. In *E. coli*, mutants have been isolated in which the pantothenate kinase activity in crude extracts is refractory to feedback inhibition by CoA. These strains have CoA levels that are significantly higher than in strains containing the wild-type kinase. Also, the effects of pantothenate kinase overexpression on the size of the CoA pool have been studied. Strains expressing significantly more wild-type kinase exhibited only a small increase in the steady-state CoA level, corroborating the conclusion that modulation of pantothenate kinase activity by feedback regulation is the critical factor controlling the intracellular CoA concentration. At this time, there is no information on the regulation of the putative pantothenate kinase in *B. burgdorferi*.

Following pantothenate phosphorylation, a cysteine is added to 4′-phosphopantothenate to generate 4′-phosphopantothenoylcysteine, which is then decarboxylated by 4′-phosphopantetheine decarboxylase to generate 4′-phosphopantetheine. This two-step enzymatic process is predicted to be carried out by the phosphopantothenate–cysteine ligase and phosphopantothenoylcysteine decarboxylase activities of the protein encoded by *bb0812* (Fraser *et al.*, 1997). Basically nothing is known about the biochemistry or genetics of this enzyme. The next step requires two enzymes that convert 4′-phosphopantetheine to CoA. First, phosphopantetheine-phosphate adenylyltransferase (BB0702) transfers the AMP moiety to 4′-phosphopantetheine, with the release of pyrophosphate (PP_i) to form dephospho-CoA. Next, a phosphate group is added to the 3′-hydroxyl of dephospho-CoA by dephospho-CoA kinase (BB0547) to yield CoA. Again, little is known about the biochemistry or genetics of these two enzymes. Interestingly, in mammals, these enzymes co-purifiy and exist as a bifunctional protein that is designated CoA synthase (Worrall and Tubbs, 1983). However, these two activities have been separated by chromatographic fractionation of *Brevibacterium ammoniagenes* extracts, suggesting that individual enzymes catalyse the two steps in bacteria (Martin and Drueckhammer, 1993). This has been identified as a secondary regulatory point in CoA biosynthesis (Jackowski and Rock, 1981; Jackowski and Rock, 1984a; Jackowski and Rock, 1984b). Double-label experiments indicate that extracellular 4′-phosphopantetheine is derived from the degradation of ACP, and reutilization of this intermediate is regulated at the adenylyltransferase step. Excretion of 4′-phosphopantetheine is an irreversible mechanism for reduction of the intracellular CoA and ACP pools since *E. coli* is unable to assimilate exogenous 4′-phosphopantetheine into CoA. *Borrelia burgdorferi* may use a similar mechanism to modulate intracellular levels of CoA and ACP. Understanding the biochemical mechanisms that regulate the activity of 4′-phosphopantetheine-phosphate adenylyltransferase will be necessary to establish the role of this enzyme in controlling the intracellular CoA concentration.

CoA composition

The composition of the intracellular CoA pool has been examined in *B. burgdorferi* by using the thiol-specific labelling reagent monobromobimane to label small thiols, which are then separated and identified by their retention time on high-performance liquid chromatography (HPLC) (Boylan *et al.*, 2006). Aerobic cells grown in the presence of pantothenic acid had CoA levels of 3.02 ± 0.19 μmol per gram of cells (dry weight), while those grown without panthothenic acid had 2.77 ± 0.23 μmol per gram of cells. *Borrelia* cells grown in the presence of reduced CoA contained levels of CoA equivalent to cells grown without supplements (2.59 ± 0.48 μmol per gram of cells), indicating that exogenous CoA does not alter the CoA level within the cell. The CoA present in the cell was almost entirely in the thiol form, as the disulphide content was below the detection limit of this assay [determined by calculating: (thiol + disulphide) – thiol]. This thiol form is maintained in the cell by the specific disulphide reductase, CoADR (BB0729).

ACP synthesis

CoA is then used as the 4′-phosphopantetheine donor in the synthesis of ACP from apo-ACP catalysed by ACP synthase (BB0010). In *E. coli*, metabolic labelling following the starvation of pantothenate auxotrophs shows that the level of active ACP is maintained at the expense of CoA (Alberts and Vagelos, 1966; Jackowski and Rock, 1981, 1984a,b). The ACP prosthetic group is removed by the action of ACP phosphodiesterase (Vagelos and Larrabes, 1967), an enzyme not found in *B. burgdorferi*. ACP prosthetic group turnover is rapid compared with the stability of the protein moiety (Jackowski and Rock, 1984b; Powell *et al.*, 1969, 1973), thus generating a cycle of ACP synthesis and degradation. The physiological significance of this cycle remains a mystery, although it seems unlikely to be related to the regulation of fatty acid biosynthesis (Jackowski and Rock, 1983). The 4′-phosphopantetheine released from ACP can be either reincorporated into CoA or excreted from the cell (Jackowski and Rock, 1984b). When the CoA level is low, however, turnover of the ACP prosthetic group is faster and excretion of 4′-phosphopantetheine is minimal, suggesting that the cycle participates in the metabolic maintenance of a basal CoA concentration.

5.5.2 NAD^+

NAD^+ is an essential cofactor that is ubiquitous in nature. It contains a reactive nicotinamide ring that accepts H^+ and electrons during the oxidation of molecules (Westheimer, 1987). In addition to its role in oxidation–reduction reactions, NAD^+ can also be used as a substrate for the modification of proteins by ADP-ribosyltransferases and for the synthesis and repair of DNA by prokaryotic DNA ligase. Therefore, NAD^+ has an impact, directly or indirectly, on most pathways within the cell. In most prokaryotes, NAD^+ can be synthesized *de novo* from aspartate through a series of five reactions (White, 1982). However, of the five enzymes in this pathway, *B. burgdorferi* harbours only one, deamido-NAD^+ pyrophosphorylase encoded by *nadD* (*bb0782*). This enzyme catalyses the conversion of nicotinate mononucleotide (NAMN) to deamido-NAD^+ and the conversion of nicotinamide d-ribonucleotide (NMN) to NAD^+, both at the expense of ATP. Therefore, for *de novo* synthesis to occur, the bacterium must be transporting NMN or nicotinate mononucleotide through an unidentified transporter. *E. coli* and *S. typhimurium* can scavenge from their environments a variety of exogenous pyridines, including NMN, through a membrane-associated NAD^+ pyrophosphatase (PnuE) or other NMN uptake systems (PnuA, PnuB, PnuC, or PnuD). However, the *B. burgdorferi* genome does not contain any orthologues for these transport systems. Given the numerous transporters that have been identified in the *B. burgdorferi* genome with no assigned function, one may function as a specific NMN transporter or a generalized pyridine nucleotide uptake system.

In addition to *de novo* synthesis, prokaryotes such as *E. coli* can scavenge NAD^+ from environmental sources through the pyridine nucleotide cycles (PNCs). NAD^+-dependent oxidation–reduction reactions do not consume NAD. However, NAD^+ is degraded and the half-life during aerobic growth of *Salmonella typhimurium* is approximately 90 min (Park *et al.*, 1989). Therefore, systems functioning to recycle the pyridine moiety are advantageous to the cell. These cycles have been well-characterized in both *S. typhimurium* and *E. coli*. An important feature of the salvage pathways of these organisms is the energetically favourable conservation of the pyridine ring. When recycling is possible, it takes precedence over *de novo* synthesis (McLaren *et al.*, 1973). The existence of various NAD^+ recycling pathways consisting of four-, five-, and six-step cycles is referred to as PNCs IV, V, and VI, respectively. PNCs IV and V function predominantly for intracellular recycling, while PNC VI plays an important role in scavenging preformed pyridines from the environment.

The majority of intracellular NAD^+ recycling begins with hydrolysis of the PP_i bond, releasing NMN and AMP (Hillyard *et al.*, 1981; Manlapaz-Fernandez and Olivera, 1973). One enzyme that can degrade NAD^+ to NMN is prokaryotic DNA ligase, which utilizes the energy released from the PP_i bond to reseal nicks in double-stranded DNA. This reaction is important in the repair, recombination, and replication of DNA. This degradation by DNA ligase was originally thought to be the primary cause of NAD^+ recycling. However, Park *et al.* (1989) demonstrated that *S. typhimurium* cells containing a T4 phage ATP-dependent DNA ligase do not exhibit any significant decrease in NAD^+ turnover. Interestingly, the NAD^+ turnover rate is much higher (three- to four-fold) under aerobic growth conditions than under anaerobic conditions. This may possibly be due to some unknown NAD^+- dependent enzyme that protects membranes or DNA from oxidative damage. The enzyme responsible for initiating most intracellular NAD^+ degradation has yet to be identified. Once NMN has been generated, it can be converted to NAD^+ through the actions of *nadD*.

In addition to NAD^+ recycling, *B. burgdorferi* may be capable of scavenging nicotinamide from the environment to synthesize NAD^+ through the PNC VI pathway. Nicotinamide (NAm) is found in various tissues, including the skin and liver, and, therefore, would be available for transport into the cell. Both *S. typhimurium* and *E. coli* have been shown to transport NAm; however, uptake systems have not been identified in *B. burgdorferi* (Fraser *et al.*, 1997). Nicotinamide is then deamidated to nicotinic acid (NA) by the nicotinamidase PncA (BBE22), encoded on the linear plasmid lp25. Interestingly, numerous studies have demonstrated that lp25 is an essential plasmid for mammalian infection (Labandeira-Rey and Skare, 2001; Purser *et al.*, 2003). *Borrelia burgdorferi* cells lacking lp25 lost infectivity in mice which was restored by introducing a shuttle vector containing *bbe22* (Purser *et al.*, 2003). This indicates that the nicotinamidase encoded by *bbe22* is essential for *B. burgdorferi* infectivity and suggests it is important in the production of NAD^+ in a mammalian host. The next step is the formation of NAMN through action of a NA phosphoribosyltransferase, PncB (BB0635) (Foster *et al.*, 1979), which links the recycling and scavenging systems to the *de novo* pathway. The final series of reactions in the synthesis of NAD, catalysed by NadD and NadE, are essential for cell viability. The NAMN produced by recycling or scavenging is adenylated by NadD (NAMN adenylyltransferase; BB0782), producing NA adenine dinucleotide, which is then amidated by NadE (NAD^+ synthetase; BB0522) to form NAD^+.

5.5.3 Riboflavin

Riboflavin is the central component of the cofactors FAD and FMN, and is therefore required by all flavoproteins. The biosynthesis of one riboflavin molecule requires one molecule of GTP and two molecules of ribulose 5-phosphate as substrates (Bacher *et al.*, 2000). This pathway has been well characterized in *E. coli*. The imidazole ring of GTP is hydrolytically opened, yielding a 4,5-diaminopyrimidine, which is converted to 5-amino-6-ribitylamino-2,4(1H,3H)-pyrimidinedione by deamination, side chain reduction and dephosphorylation. Condensation of 5-amino-6-ribitylamino-2,4(1H,3H)-pyrimidinedione with 3,4-dihydroxy-2-butanone 4-phosphate obtained from ribulose 5-phosphate produces 6,7-dimethyl-8-ribityllumazine. Dismutation of the lumazine derivative yields riboflavin and 5-amino-6-ribitylamino-2,4(1H,3H)-pyrimidinedione, which is recycled in the biosynthetic pathway.

In order to perform its metabolic role, riboflavin must be converted to the cofactors FMN and FAD. This is accomplished through the phosphorylation of the 5′-hydroxyl group of riboflavin generating FMN and is catalysed by riboflavin kinase. FMN is then converted to FAD by FAD synthetase. However, genome analysis indicates that *B. burgdorferi* does not contain any of the enzymes responsible for riboflavin, FMN, or FAD synthesis (Fraser *et al.*, 1997). Yet, the genome does contain predicted flavo-containing enzymes. Also, HPLC analysis of *B. burgdorferi* cells determined the concentration of FAD to be approximately 1 μM, indicating that the bacterium probably transports FAD from the environment via an unidentified uptake system (unpublished data).

5.6 Lipid and lipoprotein biosynthesis

Membrane lipids, glycolipids and lipoproteins are particularly important, not only for cell growth and nutrient acquisition, but also for forming a structural scaffold and creating a physiochemical environment for efficient membrane function. This is important for cells as they adapt to environmental conditions and, in the case of bacterial pathogens, to various hosts and/or different conditions within a particular host.

5.6.1 Phosphatidylcholine (PC), phosphatidylglycerol (PG) and α-monogalactosyl diacylglycerol (α-MgalDAG) synthesis

Lipid and lipoprotein biochemistry in bacteria involves the synthesis of fatty acids and the attachment of these hydrophobic moieties to glycerol-3-phosphate (G3P) for the synthesis of membrane phospholipids and glycolipids, or the modification of lipoproteins. In prokaryotes and eukaryotes, there are a host of enzymes to catalyse reactions that follow a basic biochemical pathway in which either acetyl-ACP or acetyl-CoA are condensed to malonyl-ACP for conversion to acetoacetyl-ACP or condensation with the growing acyl chain. Subsequent reactions catalyse the reduction, dehydration, and final reduction to acyl-ACP for more rounds of elongation. In addition, β-hydroxydecanoyl-dehydrase can catalytically trigger the synthesis of unsaturated fatty acids (Pugh and Wakil, 1965). With the exception of photosynthetic prokaryotes (Tasaka *et al.*, 1996), most bacteria do not synthesize polyunsaturated fatty acids. In *B. burgdorferi*, there are no genes encoding any enzymes involved in fatty acid biosynthesis, thus, these membrane and lipoprotein components must be acquired from the host or growth medium.

Several groups have shown that the membranes of *B. burgdorferi* are primarily composed of phosphatidylcholine (PC; ~11%), phosphatidylglycerol (PG; ~10%) and monogalactosyl diacylglycerol (α-MGalDAG; ~36) as well as an immunogenic sterol galactoside (Belisle *et al.*, 1994; Hossain *et al.*, 2001; Radolf *et al.*, 1995). The acyl subcomponents of these moieties have been shown to be predominantly palmitic acid (~55%) while 35% of the remaining fatty acids were unsaturated (oleic acid; ~22%) and polyunsaturated (linoleic and linolenic acids; 12% and 1% respectively) fatty acids. *Borrelia burgdorferi* selectively incorporates these fatty acids from the growth medium and the polyunsaturated moieties are the primary targets of reactive oxygen species (ROS) in these cells (see section 5.8.1) (Boylan *et al.*, 2008). The effect of growth in the arthropod vector or the various mammalian hosts on the fatty acid composition is not known. Clearly, these different environments may require alterations in the composition of the membranes due to physiochemical effects (such as temperature) or potential changes simply may be the result of availability.

The synthesis of lipids, glycolipids and lipoproteins in *B. burgdorferi* requires; (i) the transport of fatty acids, (ii) the activation of apo-ACP by reduced CoA (see section 5.5.1), (iii) the covalent linkage of fatty acids to CoA or holo-ACP, (iv) the synthesis of CDP-1,2 diacylglycerol (CDP-DAG) from G3P, acyl-ACP or acyl-CoA and CTP, (v) the addition of choline, G3P or galactose to CDP-DAG (or DAG) yielding PC, PG or α-MgalDAG, and, finally, (vi) the addition of PG to proteins to generate lipoproteins. The first step in the process involves the transport of fatty acids, which has been extensively studied in *E. coli*. Fatty acids were originally proposed to passively diffuse across the outer membrane and not require a protein carrier. This seems to be the case for short-chain fatty acids (SCFA: C_4-C_6), since specific transport proteins for these acyl moieties have not been identified. In the absence of specific transport systems, SCFAs are thought to diffuse through porins in the outer membrane. However, in *E. coli*, transport of long chain fatty acids (LCFA: C_{12}-C_{18}) across the outer membrane is facilitated by the protein encoded by *fadL*, which is upregulated by changes in osmolarity (Black, 1991; Higashitani *et al.*, 1993; Nunn and Simons, 1978). FadL binds LCFAs with high affinity and binding opens a LCFA-specific channel that allows the movement of the fatty acid into the periplasmic space. When *fadL* was genetically inactivated in *E. coli*, long-chain fatty acids could still enter the cell but at a rate that could not support growth. No FadL homologue has been identified in *B. burgdorferi* and experimental data suggests that no specific LCFA transport system may be required in these and some other spirochaetes. Cox and Radolf (Cox and Radolf, 2001) showed that FA–aminofluorescein conjugates could readily insert into the outer membrane of *B. burgdorferi* and *T. pallidum*. They hypothesized that the lack of LPS in the spirochaete membranes made them more permeable to lipophilic moieties promoting the uptake of FAs that could not be synthesized by these bacteria. This mechanism very likely explains SCFA and LCFA transport across the outer membrane of *B. burgdorferi*.

Subsequently, LCFAs in *E. coli* are transported across the cytoplasmic membrane by long chain acyl-CoA synthetase (ACAS), which is encoded by *fadD* (Black *et al.*, 1992). This multifunctional protein binds LCFAs, covalent link the acyl moiety to reduced CoA and release the activated form into the cytosol. In addition to CoA, the enzyme requires ATP; a soluble form of the protein has been proposed to bind ATP and this ATP-ACAS-bound form inserts into the inner membrane. Once in the membrane, ATP-ACAS retrieves a LCFA, binds CoA, enzymatically transfers the acyl moiety to CoA with the release of AMP and acyl-CoA into the cell. The *B. burgdorferi* genome harbours two genes that encode putative FadD homologues, *bb0137* and *bb0593* (Fraser *et al.*, 1997). These proteins have ~36% identity to each other and presumable function to transport fatty acids across the inner membrane increasing the intracellular pools of acyl-CoA for lipid, glycolipid and lipoprotein biosynthesis (see below). In *E. coli*, acyl-CoA is the substrate for β-oxidation when fatty acids are utilized as a carbon source; however, *B. burgdorferi* does not contain any of the enzymes required to catalyse these reactions. This suggests that acyl-CoA is only a substrate or precursor for lipid, glycolipid and lipoprotein biosynthetic pathways in *B. burgdorferi*. Acyl-ACP plays a pivotal role in lipid biosynthesis in bacteria and the transfer of the acyl moiety from acyl-CoA to ACP could target fatty acids to membrane lipid biosynthesis. Whether or not this transfer is taking place or required is not clear since acyl-CoA could potentially channel fatty acids into the same biosynthetic pathway(s). The exact role of acyl-CoA or acyl-ACP in lipid and lipoprotein synthesis *B. burgdorferi* remains to be determined experimentally.

The process for the synthesis of membrane lipids from G3P and acyl-ACP is outlined in Fig. 5.4. BB0037 (glycerol-3-phosphate acyltransferase) begins the process by catalysing the addition of two acyl moieties to G3P generating phosphatidic acid (PA), which is subsequently converted to CDP-PA by BB0119 (CDP-diacylglycerol synthase). Either acyl-ACP or acyl-CoA may serve as the acyl donor to G3P since they can potentially be interconverted by the transfer of an acyl moiety between CoA and ACP in a reversible reaction catalysed by CoA acyltransferase (BB0137). Currently, the carrier that serves as the major acyl donor for G3P is not known. At this stage, there are three potential fates for CDP-diacylglycerol (CDP-DAG). In one pathway, phosphatidylcholine synthase (BB0249) converts CDP-DAG to phosphatidylcholine (PC) using choline that has to be transported into the cell (Lopez-Lara and

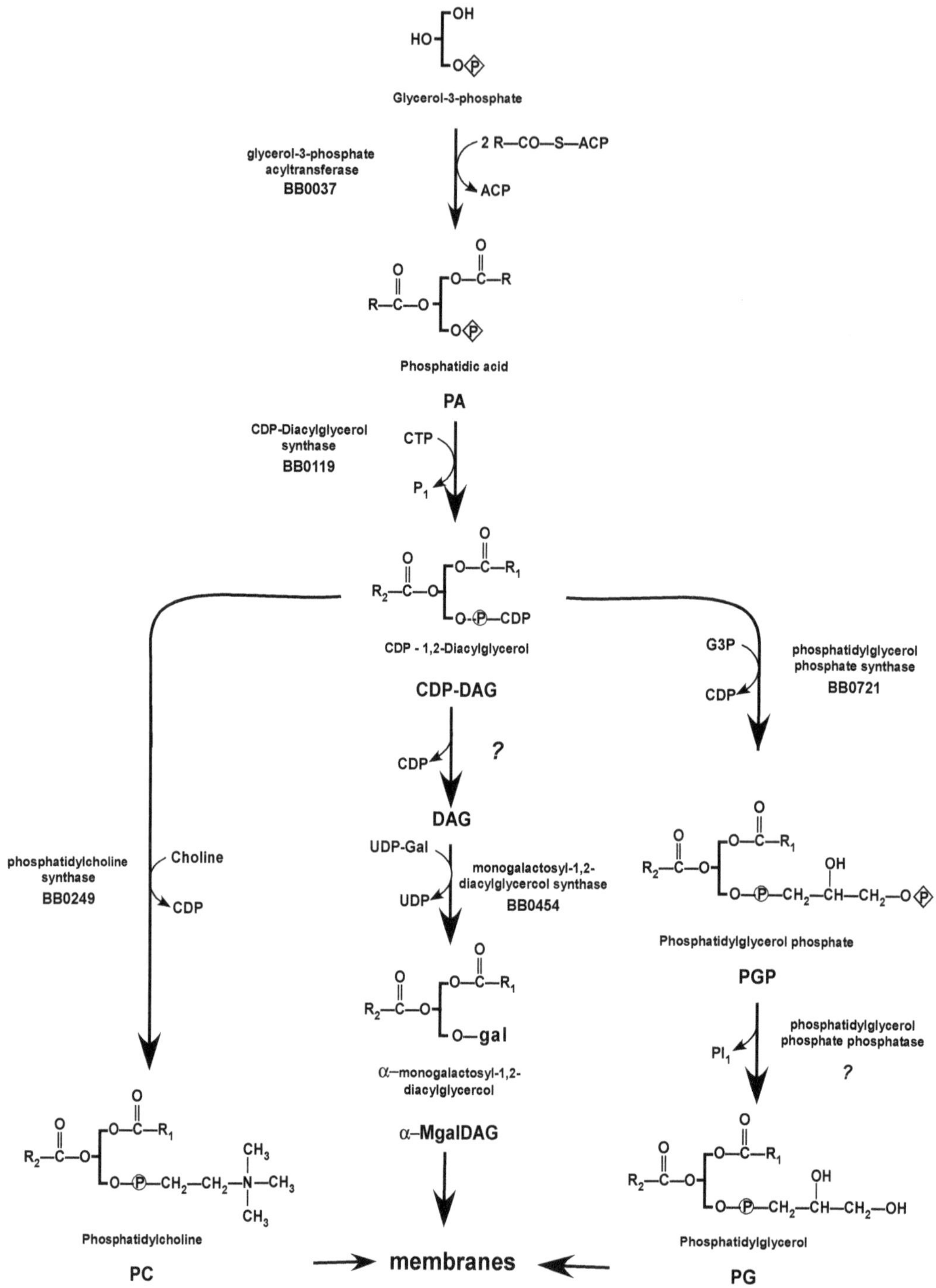

Figure 5.4 Putative lipid and glycolipid biosynthetic pathways in *Borrelia burgdorferi*.

Geiger, 2001; Sohlenkamp *et al.*, 2000; Wang *et al.*, 2004b) (Fig. 5.4, left branch of the pathway). In the second pathway, CDP-DAG is condensed with G3P to phosphatidylglycerol phosphate (PGP) by phosphatidylglycerol phosphate synthase (BB0721) (Wang *et al.*, 2004b). The next step requires the dephosphorylation of PGP. In *E. coli*, this reaction is catalysed by PGP phosphatase

(PgpA), which specifically hydrolyses PGP to PG. A second enzyme, encoded by *pgpB*, has similar activity against PGP but also has activity against PA suggesting that PgpA was the primary enzyme involved in PG synthesis. However, when both *pgpA* and *pgpb* were disrupted, phosphatase activity was reduced but PG synthesis remained at wild-type levels suggesting that one or more additional phosphatases were able to keep the PG biosynthetic pathway functioning at normal levels in *E. coli* (Funk *et al.*, 1992). The enzyme(s) responsible for this activity has not been identified. While no homologues for PgpA or PgpB have been identified in *B. burgdorferi*, the accumulation of high levels of PG in its membranes suggests that other, unidentified phosphatases are catalysing this final step in the biosynthesis of PG (Fig. 5.4, right branch of the pathway).

The major component of membrane lipids in *B. burgdorferi* is the glycolipid α-MgalDAG (Östberg *et al.*, 2007; Radolf *et al.*, 1995). This important component of the outer and inner membranes of *B. burgdorferi* has been shown to be immunogenic (Kinjo *et al.*, 2006). Glycolipids, like α-MgalDAG, are most often found in the membranes of Gram-positive bacteria and rarely in those of Gram-negative bacteria (exceptions include photosynthetic bacteria and some *Pseudomonas* species (Shaw, 1970)) whose most abundant glycolipid is lipopolysaccharide. The synthesis of α-MgalDAG in *B. burgdorferi* from DAG and UDP-galactose (UDP-Gal) is catalysed by α-MgalDAG synthase encoded by *bb0454*. This enzyme has been purified and characterized and was shown to transfer the galactose moiety from UDP-Gal to DAG (Östberg *et al.*, 2007). UDP-Gal is essential for this reaction and is believed to be generated from UDP-Glu by an epimerase encoded by *bb0444*. The second substrate in this reaction is DAG, which was shown to be the only diacyl donor for the synthesis of α-MgalDAG. How DAG is generated in *B. burgdorferi* remains to be determined. Typically, DAG is synthesized from PA by PA phosphatase (PAS) (e.g. in *B. subtilis*); however, no homologue for PAS has been identified in *B. burgdorferi*. An uncharacterized lipid phosphatase may catalyse this reaction or DAG may be generated from CDP-DAG by an unknown mechanism (see Fig. 5.4, central pathway). This remains to be experimentally determined.

5.6.2 Lipoprotein synthesis

The synthesis of lipoproteins is important in *Borrelia* species. Genome analysis of *Borrelia* suggests that at least 150 genes encode lipoproteins (Fraser *et al.*, 1997). While the function of most of these are not known, several are thought to function as transporters (see section 5.2) of essential compounds, such as sugars, peptides, fatty acids, nucleotides, and metals. The potential pathway for synthesis of lipoproteins in *B. burgdorferi* is outlined in Fig. 5.5. This putative pathway is believed to be similar to that in *E. coli* and other Gram-negative bacteria. Diacylglycerol transferase (BB0362) mediates the transfer of the DAG moiety from PG to apolipoproteins (Sankaran *et al.*, 1995). Notably, the acyl side chains in DAG are heterogeneous and contain unsaturated as well as saturated fatty acid moieties. In fact, Belisle *et al.* have demonstrated that the lipoproteins of *B. burgdorferi* contain palmitic, oleic, linoleic and linolenic acid (Belisle *et al.*, 1994) similar to those found in the membrane lipids. By deriving the lipid moieties on lipoproteins from PG, the holo-lipoproteins directly reflect the composition bacterial membrane providing a conducive environment for proper lipoprotein function. After the DAG is added to the apolipoprotein, the signal sequence is removed by signal peptidase II (BB0469) (Regue *et al.*, 1984). The final step in the biosynthesis is the addition of the last acyl moiety by lipoprotein acyltransferase to the amino terminus of the apolipoprotein. The genome of *B. burgdorferi* harbours no apparent homologue to lipoprotein acyltransferase, so if and how the last acyl moiety is added to the apolipoprotein is not known (Fraser *et al.*, 1997). Assembly and processing of lipoproteins in bacteria takes place on the cytosolic membrane and, in Gram negative bacteria, targeting these proteins to the correct intracellular location (e.g. OspC to the outer membrane) is critical for proper function. How this protein trafficking is accomplished in *B. burgdorferi* is described in detail in Chapter 6.

5.7 Metal utilization

5.7.1 Metal utilization

Metalloproteins serve as biocatalysts and electron carriers, so they play an important role in the

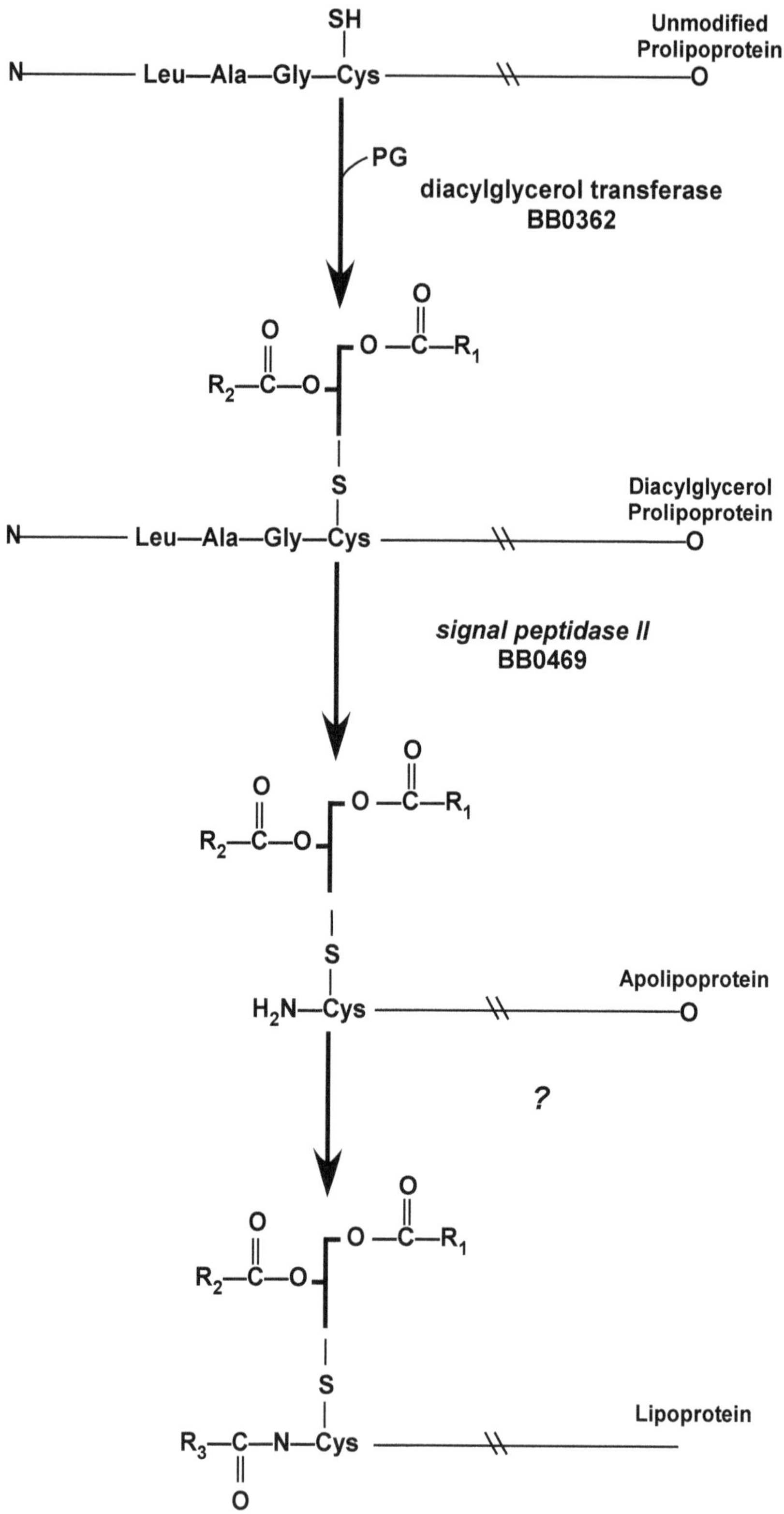

Figure 5.5 Putative lipoprotein biosynthetic pathway in *Borrelia burgdorferi*.

homeostasis of living systems. Approximately a third of all structurally characterized proteins are metalloproteins (Butler, 1998). Several of these metalloproteins function as essential enzymes; therefore, the metal cofactors (usually divalent transition metals) associated with them are absolutely essential. When present in excess, transition metals can be extremely toxic (Finney and O'Halloran, 2003; Rosen, 2002); as such, both eukaryotic and prokaryotic cells must carefully regulate the uptake and efflux of metals across biological membranes to maintain an equilibrium at subtoxic levels (Hall, 2002; Rosen, 2002). Traditionally, essential metals have included the alkali metals potassium (K) and sodium (Na), the alkaline earth metals calcium (Ca) and magnesium (Mg), and the transition metals iron (Fe), manganese (Mn), zinc (Zn), and copper (Cu). Other transition metals that have less well-defined or more obscure biological functions include cobalt (Co), molybdenum (Mo) and nickel (Ni). Metals such as Fe are particularly important to biological systems because of their need as an essential cofactor for critical enzymes (e.g. ribonucleotide reductase and aconitase) as well as their involvement in the damaging side-effects inherent to its redox biology under physiological conditions.

Metals are important for many biological processes including respiration, nitrogen fixation, methanogenesis, nucleic acid synthesis, oxygen transport, oxygen tolerance, proteolysis, and hydrogen production and consumption since the various reaction steps in these processes are catalysed by metalloenzymes (Andrews *et al.*, 2003; Butler, 1998; Klebba, 2003). There are two main classes of metal-requiring enzymes that can be distinguished from one another by the strengths of their metal-binding interactions. Metalloenzymes contain tightly bound metals, most commonly the transition metals Fe, Mn, Zn, Co, or Cu, whereas metal-activated enzymes loosely bind metal ions such as the alkali and alkaline earth metals Ca, K, Na and Mg. Metals are known to participate in the catalytic process in three ways: (i) by binding to substrates so as to orient them properly for reaction, (ii) by mediating redox reactions through reversible changes in the metals oxidation state, and (iii) by electrostatically stabilizing or shielding negative charges.

Although the prior dogma dictated that all living organisms required Fe for survival, a number of studies demonstrated that there were exceptions. Specifically, the lactobacilli have been shown to grow in Fe-depleted media (Archibald, 1983; Bruyneel *et al.*, 1990). Under these conditions, *Lactobacillus plantarum* contains only one to two Fe atoms per cell. This extremely low Fe content is consistent with the ability of lactobacilli to grow in highly Fe-limited environments such as milk (Weinberg, 1997). Under these conditions, *L. plantarum* substitutes Mn for Fe in key metalloproteins. The genome sequence of the *B. burgdorferi* indicated that this pathogen encoded few orthologues of known Fe-requiring proteins (Fraser *et al.*, 1997). In addition, Fe limitation had no effect on growth in defined medium and an independent evaluation of the total metal composition indicated that, while Mg, Mn, Zn, and Ca were readily detected, the organism carried less than 10 Fe atoms per cell (Posey and Gherardini, 2000). Thus, host-imposed Fe restriction has no apparent effect on *B. burgdorferi* growth.

5.7.2 Acquisition and transport

The ability of bacterial pathogens, such as *B. burgdorferi*, to acquire essential elements from their host during an infection is critical for their survival (Blencowe and Morby, 2003; Cornelissen, 2003; Mulrooney and Hausinger, 2003; Outten and O'Halloran, 2001; Rensing and Grass, 2003). Although host tissues, fluids and secretions are seemingly rich in nutrients, transition metals are not readily available. Bacterial pathogens have evolved both passive and active mechanisms for acquiring metals from their host environments (Ferguson and Deisenhofer, 2004). Passive mechanisms are associated with high rates of transport and low metal-binding affinities, while active mechanisms demonstrate slower transport rates with higher binding affinities (Nikaido and Saier, 1992). With the exception of Fe, Gram-negative organisms appear to acquire cations via the use of outer membrane porins (Blencowe and Morby, 2003; Rensing and Grass, 2003). These proteins form barrel-shaped, fluid-filled channels in the outer membrane of cells allowing for the transport of small, hydrophilic solutes including divalent metals into their periplasmic space (Jap and Walian, 1996). Since porin-based mechanisms are a passive means by which to acquire metals, cations transported by this method must be present at concentrations far higher than that of iron. Potential host sources of these cations include ceruloplasmin, albumin, α_2-macroglobulin and β_1-globulin as well as other host serum proteins (de la Calle Guntinas *et al.*,2002).

Very little is known about the mechanisms that *Borrelia* species employ to transport metals across their cell envelopes. Like other organisms, however, membrane transport systems presumably modulate the exchange of metals between the four compartments that define Gram-negative cell envelopes – the outer membrane, the periplasmic space, the cytoplasmic membrane, and the cytoplasm (Ferguson and Deisenhofer, 2004). In order to accomplish this, Gram-negative organisms typically utilize three main classes of bacterial transport systems: (i) ion channels (or porins), which may function as highly selective pores that open and close in response to specific stimuli allowing ions to rapidly flow down a gradient (Ferguson and Deisenhofer, 2004), (ii) secondary active transporters, which couple electrochemical proton gradients to transport, and (iii) primary active transporters, or ATP-binding cassette (ABC) transporters (see section 5.2). Most cations appear to diffuse passively into the periplasmic space via outer membrane porins and then are actively pumped into the cytoplasm by specific transport systems (Jap and Walian, 1996) with the exception of Fe^{3+}, which must be actively transported through the outer and inner membrane of the cell, in most cases, requiring a functional TonB protein as a part of an energy transduction complex (Bradbeer, 1993; Ferguson and Deisenhofer, 2004).

Analysis of the *B. burgdorferi* genome demonstrates the presence of very few predicted metal transporters (Fraser *et al.*, 1997). Interestingly, none of those that were originally annotated by TIGR were predicted to participate in the transport of transition metals. Some time after the publication of the *B. burgdorferi* genome, several investigators noted that it included a gene encoding a member of the ZIP family of metal transporters (encoded by *bb0219*), which are widely distributed among prokaryotes, eukaryotes and archaea (Begum *et al.*, 2002; Guerinot, 2000). Some members of the ZIP family are promiscuous, effectively transporting many different metals including Fe^{2+}, Zn^{2+}, Cd^{2+}, and Mn^{2+}, while others show a preference for specific metals (Grass *et al.*, 2005; Grass *et al.*, 2002). The prokaryotic homologues include the ZupT proteins that are found in several bacteria including *E. coli*, *Campylobacter jejuni* and *B. burgdorferi* (Begum *et al.*, 2002). Based on studies in *E. coli*, ZupT appears to be a divalent metal transporter that transports Zn^{2+}, Mn^{2+}, Fe^{2+}, $Co_2{}^{+}$, Cu^{2+} and Cd^{2+}, although the affinities vary for different metals and the transport of Zn^{2+} seems to be preferred (Grass *et al.*, 2005; Grass *et al.*, 2002). Given that *B. burgdorferi* can transport detectable amounts of Zn^{2+} and Mn^{2+}, but does not accumulate Fe (Posey and Gherardini, 2000), the *B. burgdorferi* ZupT homologue would be likely to be involved in transport of Zn^{2+} and/or Mn^{2+}. In fact, Ouyang *et al.* have recently shown that *bb0219* encodes a Mn specific transporter, designated BmtA (Ouyang *et al.*, 2009).

The requirement of Zn^{2+} in *B. burgdorferi* is supported by several studies indicating that this metal serves as a cofactor for important metalloproteins that are Fe^{2+}-dependent proteins in other bacteria. Specifically, BosR, the redox regulator of *B. burgdorferi* (a Ferric uptake regulator homologue) (Boylan *et al.*, 2006; Boylan *et al.*, 2003; Seshu *et al.*, 2004), and peptide deformylase (Pdf) (Nguyen *et al.*, 2007) both require Zn^{2+} for DNA-binding activity and enzymatic function, respectively. In the case of BosR, the binding of Zn^{2+} is believed to be coordinated by two conserved cysteine residues and is required for efficient DNA binding (Boylan *et al.*, 2006; Boylan *et al.*, 2003; Seshu *et al.*, 2004). The exact configuration of this coordination is not known and the ability to bind additional metals, as is seen for the PerR protein from *Bacillus subtilis*, which can bind either Fe^{2+} and Zn^{2+} or Mn^{2+} and Zn^{2+} (Herbig and Helmann, 2001), remains to be determined. The Pdf from *B. burgdorferi* is a very interesting metalloprotein. All other characterized Pdfs have been shown to require Fe^{2+} for maximum activity *in vivo*. The enzyme, purified from *B. burgdorferi*, contains Zn^{2+} as a cofactor and, in this form, has enzyme activity levels that approach those measured for the Fe^{2+} Pdf purified from *E. coli* (Nguyen *et al.*, 2007). Taken together, these data indicate the *B. burgdorferi* has eliminated the need for Fe^{2+} by substituting other metals in key metalloproteins (e.g. BosR, Pdf, and, see section 5.8, SodA) and by eliminating biosynthetic pathways that normally require Fe^{2+}-containing enzymes (see section 5.4, ribonucleotide reductase).

With regard to other divalent metal transporters, *B. burgdorferi* also encodes both Mg^{2+} (*bb0380*) and Ca^{2+} (*bb0164*) transporters that are homologues of MgtE and CaCA transporters, respectively (Saier and Paulsen, 2000). These same transporters are also found in *Treponema pallidum* (Fraser *et al.*, 1998). Unlike *B. burgdorferi,* however, *T. pallidum* encodes other ABC transporters that are involved in the uptake of divalent metals, including Mn^{2+}, Zn^{2+}, Fe, and Co_2^{+} (Fraser *et al.*, 1998). The best-characterized *T. pallidum* ABC transporter is encoded by the *tro* operon (*troRABCD*) and transports several divalent metals including Mn^{2+}, Zn^{2+} and perhaps Fe^{2+} (Desrosiers *et al.*, 2007; Hazlett *et al.*, 2003; Posey *et al.*, 1999). In contrast to *T. pallidum,* key ABC transporters in *B. burgdorferi* are associated with peptide and ribose/galactose uptake (Fraser *et al.*, 1997); none of the *Borrelia* ABC transporters are predicted to transport metals.

Although there is only one apparent transition metal transporter in *B. burgdorferi,* there are several secondary transporters that appear to transport monovalent cations, i.e. Na^{+} and K^{+}. Briefly, *B. burgdorferi* encodes three Na^{+}/H^{+} antiporters (*bb0447, bb0637,* and *bb0638*), two dicarboxylic acid/Na^{+} (or H^{+}) symporters (*bb0401* and *bb0729*), a pantothenate/Na^{+} symporter (*bb0814*), and a TrkH homologue that transports K^{+} (*bb0724*). When compared with *T. pallidum, B. burgdorferi* encodes three times more monovalent cation transporters (Fraser *et al.*, 1997; Fraser *et al.*, 1998; Saier and Paulsen, 2000); the significance of this observation is not known. *Borrelia burgdorferi* also contains a cation diffusion facilitator (i.e. CDF family member; *bb0616*) that presumably is used to pump out excess metal to promote the proper cytosolic homeostasis and prevent intracellular levels from reaching toxic levels. Finally, genes encoding proteins putatively involved chromate transporters/efflux (*bb0451* and *bb0452*) have been identified in the *Borrelia* genome. Given that chromate usually pairs with Na^{+} or K^{+}, this may serve as another modality for *B. burgdorferi* to obtain monovalent alkali metals. Interestingly, the chromate transporters are genetically linked to a MscS homologue (*bb0453*), a mechanosensitive ion channel protein that helps to maintain the osmotic balance of cells in response to changes in environmental osmolarity. This may be the function of this system in specific infection and colonization sites during the infective cycle (e.g. tick midgut).

5.8 Oxidative stress response

Oxidative stress occurs when reactive oxygen species (ROS) exceeds the detoxification capacity of the cell. Reactive oxygen species are a potentially damaging consequence of life in an oxygen-rich environment and are generated as a byproduct of normal aerobic metabolism (e.g. respiration or fatty acid oxidation) and/or when redox enzymes are oxidized by O_2. However, because *B. burgdorferi* generates ATP via Embden–Meyerhof pathway, the generation of ROS does not come from the metabolism of oxygen. Pathogenic bacteria are exposed to exogenous ROS and/or reactive nitrogen species (RNS) that are used as cytotoxic weapons of the immune system and inflammatory response against invading microorganisms. The key reactive species are superoxide radicals (O_2^{-}), hydrogen peroxide (H_2O_2), hydroxyl radicals ($^{\bullet}OH$), and singlet oxygen, which result from the sequential univalent reductions of molecular oxygen and nitric oxide (NO). The ability to neutralize toxic ROS and RNS constitutes a formidable challenge that the bacterium must overcome to successfully establish and maintain an infection. Presumably *B. burgdorferi* would need to respond to ROS and RNS (i) in the tick following a blood meal and (ii) in the host due to the inflammatory response that occurs early in the infectious process.

5.8.1 Reactive oxygen species

Borrelia burgdorferi is metabolically limited, as its genome encodes no enzymes involved in the TCA cycle or respiration via the electron transport system (Fraser *et al.*, 1997) and oxidative phosphorylation is absent, therefore the source of the majority of endogenous O_2^{-} is absent. However, due to its infective cycle, *B. burgdorferi* is exposed to anaerobic and microaerobic environments, so it may encounter O_2 at levels that could generate biological relevant levels of cytosolic O_2^{-}.

Hydrogen peroxide readily diffuses across biological membranes such that the intracellular concentration is easily influenced by both exogenous and endogenous sources. Intracellular production of peroxides can occur via a non-enzymatic process or by superoxide dismutase (SOD)-catalysed dismutation. Also, *B. burgdorferi* encounters external H_2O_2 due to the inflammatory response induced by the infection, e.g. the erythema migrans rash as well as the inflammation associated with carditis and arthritis. A proposed model for the oxidative stress response is shown in Fig. 5.6.

5.8.2 Cellular targets

The biological targets for ROS are DNA, RNA, proteins, and lipids (Imlay, 2003). Superoxide can inactivate enzymes and proteins containing Fe–S clusters (e.g. aconitase). The most deleterious effects of O_2^- are related to its propensity to be converted to more potent oxidants like H_2O_2 or $^{\bullet}OH$. Hydrogen peroxide efficiently oxidizes protein-thiols, and is likely to inactive enzymes that rely upon active-site cysteine residues for catalytic activity. However, the oxidation of cysteine residues and subsequent inactivation of thiol containing proteins may occur during oxidative stress in bacteria, but does not appear to be a major target of ROS (Imlay, 2002). The most damaging effects of H_2O_2 result from its interactions with reduced Fe (and to a lesser extent, Cu) centres in proteins, generating $^{\bullet}OH$ via the Fenton reaction (Imlay, 2003). $^{\bullet}OH$ reacts with most biomolecules and, because it is so reactive, will oxidize structures within cells depending largely upon proximity. Because Fe can localize along the phosphodiester backbone of nucleic acid, DNA is a major target of $^{\bullet}OH$. Several recent studies suggest that DNA serves as the major target for host toxicity by ROS and RNS assuming that Fe accumulates within the cell and intracellular concentrations of H_2O_2 approach 0.2 to 1 μM (Keyer and Imlay, 1996; Seaver and Imlay, 2001). However, *B. burgdorferi* has been shown to not require Fe for growth and has intracellular Fe concentrations estimated to be <10 atoms per cell (Posey and Gherardini, 2000). At these low levels, *B. burgdorferi* DNA is not likely a target for ROS. In fact, when *B. burgdorferi* cells were exposed to high concentrations of ROS (e.g. H_2O_2), there was no effect on the spontaneous mutation rate or the number of DNA lesions (AP sites or 8-oxoguanine) (Boylan *et al.*, 2008). The lack of damage may indicate that *B. burgdorferi* DNA is not subjected to the same challenge from ROS as is *E. coli* DNA.

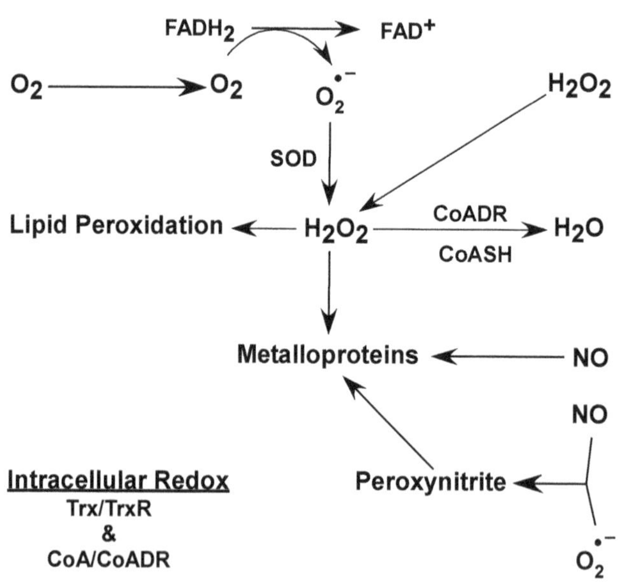

Figure 5.6 Reactive oxygen (ROS) and nitrogen species (RNS) encountered by *Borrelia burgdorferi*. Cellular lipids and proteins are the likely targets of ROS/RNS.

Bacteria that incorporate polyunsaturated fatty acids into lipid structure risk having these structures targeted for peroxidation (Beermann *et al.*, 2000; Ben-Menachem *et al.*, 2003; Imlay, 2003; Tasaka *et al.*, 1996). In this situation, free radicals can attack the polyunsaturated fatty acids and initiate lipid peroxidation. A primary effect of lipid peroxidation is the loss of membrane fluidity that can alter the function of membrane-associated proteins. The operative question is whether bacterial cells contain polyunsaturated fatty acids and, more pointedly, whether *Borrelia* spp. contain these molecules. Since *B. burgdorferi* cannot synthesize its own lipids, it must scavenge them from its host so that their membrane composition reflects the host's lipid profile or their growth medium (Barbour, 1984; Fraser *et al.*, 1997). Several publications demonstrate that *B. burgdorferi* incorporates polyunsaturated fatty acids, specifically linoleic and linolenic acids, into their membranes (Beermann *et al.*, 2000; Ben-Menachem *et al.*, 2003; Boylan and Gherardini, 2008; Boylan *et al.*, 2008), suggesting that their membranes could undergo peroxidation if exposed to ROS (Boylan and Gherardini, 2008; Boylan *et al.*, 2008). Based on the previously mentioned lack of Fe accumulation and the lack of molecules to generate endogenous superoxide, polyunsaturated lipids may serve as the most likely target for oxidative damage in *B. burgdorferi*. In fact, when *Borrelia* cells are treated with oxidants, including H_2O_2, *t*-butyl hydroperoxide, and lipoxidase, the polyunsaturated lipid composition, specifically linoleic acid, decreased in amounts suggesting that they were substrates for oxidation while other fatty acids remain unaffected (Boylan *et al.*, 2008). In addition, as linoleic acid concentrations decreased, aldehydes accumulated. Furthermore, the overall morphology of *B. burgdorferi* was adversely affected by exposure to oxygen and to lipoxidase. Taken together, these observations indicate that the presence of polyunsaturated fatty acids puts *B. burgdorferi* lipids containing these targets at risk for oxidative damage.

5.8.3 Cellular defences

Cellular defences against the damaging effects of ROS involve both enzymatic and non-enzymatic components. The enzymatic components may directly scavenge ROS or may act by producing the non-enzymatic antioxidants. There are five categories of enzymes that provide the bulk of protection against deleterious reactions involving ROS in bacteria: superoxide dismutases (SODs)/superoxide reductases (SORs), hydroperoxidases, catalases, AhpR homologues, and Dps/Dpr proteins. Superoxide dismutases, ubiquitous among aerobic and most aerotolerant organisms, provide a defence against oxidative stress by catalysing the dismutation of superoxide into oxygen and hydrogen peroxide through alternative oxidation and reduction of their catalytic metal ions (McCord and Fridovich, 1969; McCord *et al.*, 1971). To date, there are four types of superoxide dismutase have been identified: manganese-dependent (MnSODs), iron-dependent (FeSODs), copper/zinc-dependent (Cu/ZnSODs), and nickel-dependent (NiSOD) (Palenik *et al.*, 2003; Youn *et al.*, 1996a; Youn *et al.*, 1996b). The MnSOD was the first identified form and is encoded by *sodA* (Touati, 1983). Austin and colleagues cloned and characterized a gene encoding a SOD from *B. burgdorferi* (Austin *et al.*, 1981; Whitehouse *et al.*, 1997). This gene had 68% similarity to *sodA* from *Haemophilus influenzae* and the partially purified enzyme had superoxide dismutase activity. The subsequent genome sequence confirmed that *B. burgdorferi* encoded a SOD gene designated *sodA* (*bb0153*) (Fraser *et al.*, 1997). Based on its homology, the metal ligand associated with this SodA is most likely manganese. The product of SOD is H_2O_2, which is itself a reactive species. Cellular catalases, which destroy H_2O_2 with remarkable rapidity, and peroxidases, which require NADH or NADPH as an electron source, are both capable of ridding the cells of H_2O_2. Since *B. burgdorferi* does not encode either a catalase or peroxidase, how *B. burgdorferi* contended with the H_2O_2 that was generated by the SodA was not clear. A potential explanation for this comes from the study of the CoADR (encoded by *bb0728*): reduced CoA can react to reduce H_2O_2, thereby acting as a weak peroxidase (Boylan *et al.*, 2006).

Borrelia cells contain a large amount of the low molecular weight thiol coenzyme A (CoA) that can function as a glutathione equivalent in *B. burgdorferi* and thus serve as a molecular sink for toxic oxidizing compounds within the cytoplasm (Boylan *et al.*, 2006). Specifically, CoA and

CoADR function as a target for redox recycling and an enzyme to regenerate the reduced form of CoA (CoASH) from oxidized CoA, respectively (Boylan *et al.*, 2006). This, coupled with the absence of any identifiable gene encoding a glutathione synthase or glutathione reductase within the borrelial genome, strongly suggested that CoA and CoADR are responsible for establishing and maintaining the redox balance in *B. burgdorferi* (Boylan *et al.*, 2006; Fraser *et al.*, 1997). The genes encoding thioredoxin (*trxA*; *bb0061*) and thioredoxin reductase (*trxB*; *bb0515*) may represent an additional redox active system in *B. burgdorferi*; however, these proteins have not yet been characterized.

In addition to the few detoxification genes found in the *B. burgdorferi* genome, a homologue of *dps/dpr* is encoded by open reading frame *bb0690*, which is called *napA* (Fraser *et al.*, 1997; Gherardini *et al.*, 2006). Dps/Dpr proteins have been associated with several functions related to oxidative stress in other bacterial systems. One activity is to protect DNA either by non-specific binding (Almiron *et al.*, 1992; Frenkiel-Krispin *et al.*, 2001; Martinez and Kolter, 1997; Morikawa *et al.*, 2006; Wolf *et al.*, 1999) or indirectly without physical interaction (Ceci *et al.*, 2003). The other activity attributed to Dps/Dpr proteins is that of a ferritin-like molecule that can sequester Fe (Ceci *et al.*, 2003; Grant *et al.*, 1998; Ren *et al.*, 2003; Roy *et al.*, 2004) such that the cell is protected from Fenton-based chemistry and subsequent damage to its DNA. Because *B. burgdorferi* does not accumulate and has no apparent requirement for Fe (Posey and Gherardini, 2000), the ferritin-like activity would not likely be relevant to this spirochaete unless it could also sequester other toxic transition metals that *B. burgdorferi* obtained. Yet, the recombinant form of the *B. burgdorferi* NapA protein does attain a dodecyl stoichiometric configuration that binds iron (Li *et al.*, 2007). However, the NapA protein did not bind DNA relative to the *E. coli* homologue (Li *et al.*, 2007). Given the lack of Fe present in *B. burgdorferi* and the inability of NapA to non-specifically protect DNA, how NapA contributes in a detoxification scheme to protect *B. burgdorferi* from oxidative stress is not clear. Other Dps homologues protect DNA in the absence of direct binding (Ceci *et al.*, 2003) and *B. burgdorferi* may utilize a similar strategy.

The most extensively studied organisms for understanding the genetic regulation in response to reactive oxygen/nitrogen species have been the enteric bacteria *E. coli* and *S. typhimurium* as well as the Gram-positive bacterium *B. subtilis*. These bacteria can activate the expression of some 30–40 proteins in response to ROS (Pomposiello *et al.*, 2001; Zheng *et al.*, 2001). The key regulators of these adaptive responses are the SoxRS and OxyR transcription factors of *E. coli* and the peroxide-sensitive repressor PerR from *B. subtilis*. OxyR is a transcription factor, which, when oxidized, forms an intramolecular disulphide bond. It is this oxidized form that is capable of activating transcription of genes in the oxidative stress response (Zheng *et al.*, 1998). SoxR contains a redox active Fe–S cluster that regulates DNA binding (Hidalgo *et al.*, 1997). The PerR protein is a metalloprotein that binds one or more copies of a conserved, 14-bp inverted repeat to repress expression of target genes. The oxidation of the Zn^{2+}–Fe^{2+} form of PerR (the most redox-sensitive form of this regulator) by H_2O_2 abolishes binding, causing derepression of target genes (Herbig and Helmann, 2001).

Borrelia burgdorferi does not contain a SoxRS or OxyR homologue, but it does contain a PerR orthologue (a member of the Fur family of regulatory proteins and annotated as Fur by TIGR) that responds to ROS, albeit in a manner more consistent with functional OxyR, so it has been designated BosR for *Borrelia* oxidative stress regulator (Boylan *et al.*, 2003). Although there is no apparent amino acid homology to OxyR (~15%), BosR is capable of binding DNA under both oxidizing and reducing conditions, requires Zn^{2+} for efficient binding, and has an unusually large (45 bp to 50 bp) target sequence-binding domain based on DNase I footprint analysis (Boylan *et al.*, 2003). BosR binds target sequences upstream of *sodA* (*bb0153*) (Seshu *et al.*, 2004), *napA* (*bb0690*) (Boylan *et al.*, 2003; Seshu *et al.*, 2004), *cdr*, *oppA-4* (Medrano *et al.*, 2007) and *bosR* (*bb0647*) (Katona *et al.*, 2004). BosR acts as a transcriptional activator of oxidative stress genes, including *cdr*, *napA* and *oppA-4* (Boylan *et al.*, 2003; Medrano *et al.*,

2007), although other investigators have suggested that this regulator functions as a repressor (Katona *et al.*, 2004). Several members of the Fur family of regulators can function as both an activator and a repressor (Delany *et al.*, 2004; Gao *et al.*, 2008; Grifantini *et al.*, 2004), therefore BosR may also function in this manner. With the advent of improved genetics (reviewed in Rosa *et al.*, 2005) and the introduction of inducible systems in *B. burgdorferi* (Blevins *et al.*, 2007; Gilbert *et al.*, 2007), a *bosR* inducible system could be used to define the BosR regulon directly in *B. burgdorferi*.

5.9 Conclusion

In studying the metabolism and physiology of *Borrelia,* one sees several features that are quite interesting and most likely related to evolutionary adaptation. This is reflected in the observation the *B. burgdorferi* does not utilize Fe. The lack of Fe containing proteins (e.g. ribonucleotide reductase) is most likely a manifestation of adjustment to a parasitic lifestyle. By colonizing various hosts, *Borrelia* was able to utilize host derived compounds and gradually eliminate energy expensive *de novo* biochemical pathways (e.g. biosynthesis of amino acid, nucleotides, fatty acids, some cofactors etc.) some of which require Fe metalloproteins. This is also reflected in central metabolism which has no TCA cycle or respiratory systems which also have critical Fe containing proteins (e.g. succinate dehydrograse, cytochromes) as key components in their pathways. These adaptions have had subsequent effects on the overall physiology of *Borrelia.* For example, the effects of ROS on *B. burgdorferi* cellular components has shifted from the oxidation of DNA (the major target in most prokaryotes) to the oxidation membrane lipids that are synthesized using host derived fatty acids, such as linoleic acid. Thus one could speculate that utilizing host derived metabolic intermediates, while irreversibly linking *Borrelia* to its various hosts, has gradually led to the elimination metabolic pathways that were expendable and shaped their overall physiology.

5.9.1 Websites

Transporter Protein Analysis Database: http://membranetransport.org/

References

Abouhamad, W.N., Manson, M., Gibson, M.M., and Higgins, C.F. (1991). Peptide transport and chemotaxis in *Escherichia coli* and *Salmonella typhimurium*: characterization of the dipeptide permease (Dpp) and the dipeptide-binding protein. Mol. Microbiol. *5*, 1035–1047.

Alberts, A.W., and Vagelos, P.R. (1966). Acyl carrier protein. 8. Studies of acyl carrier protein and coenzyme A in *Escherichia coli* pantothenate or betaalanine auxotrophs. J. Biol. Chem. *241*, 5201–5204.

Almiron, M., Link, A.J., Furlong, D., and Kolter, R. (1992). A novel DNA-binding protein with regulatory and protective roles in starved *Escherichia coli*. Genes Dev. *6*, 2646–2654.

Anand, R., Kaminski, P.A., and Ealick, S.E. (2004). Structures of purine 2′-deoxyribosyltransferase, substrate complexes, and the ribosylated enzyme intermediate at 2.0 Å resolution. Biochemistry *43*, 2384–2393.

Andrews, S.C., Robinson, A.K., and Rodriguez-Quinones, F. (2003). Bacterial iron homeostasis. FEMS Microbiol. Rev. *27*, 215–237.

Archibald, F.S. (1983). *Lactobacillus plantarum*, an organism not requiring iron. FEMS Microbiol. Lett. *19*, 29–32.

Austin, F.E., Barbieri, J.T., Corin, R.E., Grigas, K.E., and Cox, C.D. (1981). Distribution of superoxide dismutase, catalase and peroxidase activities among *Treponema pallidum* and other spirochetes. Infect. Immun. *33*, 372–379.

Bacher, A., Eberhardt, S., Fischer, M., Kis, K., and Richter, G. (2000). Biosynthesis of vitamin B_2 (riboflavin). Annu Rev Nutr *20*, 153–167.

Barbour, A.G. (1984). Isolation and cultivation of Lyme disease spirochetes. Yale J Biol Med *57*, 521–525.

Barbour, A.G., Putteet-Driver, A.D., and Bunikis, J. (2005). Horizontally acquired genes for purine salvage in *Borrelia* spp. causing relapsing fever. Infect. Immun. *73*, 6165–6168.

Beermann, C., Lochnit, G., Geyer, R., Groscurth, P., and Filgueira, L. (2000). The lipid component of lipoproteins from *Borrelia burgdorferi*: structural analysis, antigenicity, and presentation via human dendritic cells. Biochem. Biophys. Res. Commun. *267*, 897–905.

Begum, N.A., Kobayashi, M., Moriwaki, Y., Matsumoto, M., Toyoshima, K., and Seya, T. (2002). *Mycobacterium bovis* BCG cell wall and lipopolysaccharide induce a novel gene, BIGM103, encoding a 7-TM protein: identification of a new protein family having Zn-transporter and Zn-metalloprotease signatures. Genomics *80*, 630–645.

Belisle, J.T., Brandt, M.E., Radolf, J.D., and Norgard, M.V. (1994). Fatty acids of *Treponema pallidum* and *Borrelia burgdorferi* lipoproteins. J. Bacteriol. *176*, 2151–2157.

Ben-Menachem, G., Kubler- Kielb, J., Coxon, B., Yergey, A., and Schneerson, R. (2003). A newly discovered cholesteryl galactoside from *Borrelia burgdorferi*. Proc. Natl. Acad. Sci. U.S.A. *100*, 7913–7918.

Beyenbach, K.W., and Wieczorek, H. (2006). The V-type H^+ ATPase: molecular structure and function, physiological roles and regulation. J. Exp. Biol. *209*, 577–589.

Black, P.N. (1991). Primary sequence of the *Escherichia coli fadL* gene encoding an outer membrane protein required for long-chain fatty acid transport. J. Bacteriol. *173*, 435–442.

Black, P.N., DiRusso, C.C., Metzger, A.K., and Heimert, T.L. (1992). Cloning, sequencing, and expression of the *fadD* gene of *Escherichia coli* encoding acyl coenzyme A synthetase. J. Biol. Chem. *267*, 25513–25520.

Blencowe, D.K., and Morby, A.P. (2003). Zn(II) metabolism in prokaryotes. FEMS Microbiol. Rev. *27*, 291–311.

Blevins, J.S., Revel, A.T., Smith, A.H., Bachlani, G.N., and Norgard, M.V. (2007). Adaptation of a luciferase gene reporter and lac expression system to *Borrelia burgdorferi*. Appl. Environ. Microbiol. *73*, 1501–1513.

Boylan, J.A., and Gherardini, F.C. (2008). Determining the cellular targets of reactive oxygen species in *Borrelia burgdorferi*. Methods Mol Biol *431*, 213–221.

Boylan, J.A., Hummel, C.S., Benoit, S., Garcia-Lara, J., Treglown-Downey, J., Crane, E.J., 3rd, and Gherardini, F.C. (2006). *Borrelia burgdorferi bb0728* encodes a coenzyme A disulphide reductase whose function suggests a role in intracellular redox and the oxidative stress response. Mol. Microbiol. *59*, 475–486.

Boylan, J.A., Lawrence, K.A., Downey, J.S., and Gherardini, F.C. (2008). *Borrelia burgdorferi* membranes are the primary targets of reactive oxygen species. Mol. Microbiol. *68*, 786–799.

Boylan, J.A., Posey, J.E., and Gherardini, F.C. (2003). Borrelia oxidative stress response regulator, BosR: a distinctive Zn-dependent transcriptional activator. Proc. Natl. Acad. Sci. U.S.A. *100*, 11684–11689.

Bradbeer, C. (1993). The proton motive force drives the outer membrane transport of cobalamin in *Escherichia coli*. J. Bacteriol. *175*, 3146–3150.

Bruyneel, B., Vande Woestyne, M., and Verstraete, W. (1990). Iron complexation as a tool to direct mixed clostridia-lactobacilli fermentations. J. Appl. Bacteriol. *69*, 80–85.

Bunikis, I., Denker, K., Östberg, Y., Andersen, C., Benz, R., and Bergström, S. (2008). An RND-type efflux system in *Borrelia burgdorferi* is involved in virulence and resistance to antimicrobial compounds. PLoS Pathog. *4*, e1000009.

Butler, A. (1998). Acquisition and utilization of transition metal ions by marine organisms. Science *281*, 207–210.

Byram, R., Stewart, P.E., and Rosa, P. (2004). The essential nature of the ubiquitous 26-kilobase circular replicon of *Borrelia burgdorferi*. J. Bacteriol. *186*, 3561–3569.

Caimano, M.J., Iyer, R., Eggers, C.H., Gonzalez, C., Morton, E.A., Gilbert, M.A., Schwartz, I., and Radolf, J.D. (2007). Analysis of the RpoS regulon in *Borrelia burgdorferi* in response to mammalian host signals provides insight into RpoS function during the enzootic cycle. Mol. Microbiol. *65*, 1193–1217.

Carnrot, C., Wehelie, R., Eriksson, S., Bolske, G., and Wang, L. (2003). Molecular characterization of thymidine kinase from *Ureaplasma urealyticum*: nucleoside analogues as potent inhibitors of mycoplasma growth. Mol. Microbiol. *50*, 771–780.

Ceci, P., Ilari, A., Falvo, E., and Chiancone, E. (2003). The Dps protein of *Agrobacterium tumefaciens* does not bind to DNA but protects it toward oxidative cleavage: x-ray crystal structure, iron binding, and hydroxyl-radical scavenging properties. J. Biol. Chem. *278*, 20319–20326.

Coburn, J., Chege, W., Magoun, L., Bodary, S.C., and Leong, J.M. (1999). Characterization of a candidate *Borrelia burgdorferi* β_3-chain integrin ligand identified using a phage display library. Mol. Microbiol. *34*, 926–940.

Coburn, J., and Cugini, C. (2003). Targeted mutation of the outer membrane protein P66 disrupts attachment of the Lyme disease agent, *Borrelia burgdorferi*, to integrin $\alpha_v\beta_3$. Proc. Natl. Acad. Sci. U.S.A. *100*, 7301–7306.

Cornelissen, C.N. (2003). Transferrin-iron uptake by Gram-negative bacteria. Front. Biosci. *8*, d836–847.

Cox, D.L., and Radolf, J.D. (2001). Insertion of fluorescent fatty acid probes into the outer membranes of the pathogenic spirochaetes *Treponema pallidum* and *Borrelia burgdorferi*. Microbiology *147*, 1161–1169.

Das, R., Hegyi, H., and Gerstein, M. (2000). Genome analyses of spirochetes: a study of the protein structures, functions and metabolic pathways in *Treponema pallidum* and *Borrelia burgdorferi*. J. Mol. Microbiol. Biotechnol. *2*, 387–392.

Davidson, A.L., Dassa, E., Orelle, C., and Chen, J. (2008). Structure, function, and evolution of bacterial ATP-binding cassette systems. Microbiol. Mol. Biol. Rev. *72*, 317–364.

Davis, B.D. (1950). Studies on nutritionally deficient bacterial mutants isolated by means of penicillin. Experientia *6*, 41–50.

de la Calle Guntinas, M.B., Bordin, G., and Rodriguez, A.R. (2002). Identification, characterization and determination of metal-binding proteins by liquid chromatography. A review. Anal Bioanal Chem *374*, 369–378.

Delany, I., Rappuoli, R., and Scarlato, V. (2004). Fur functions as an activator and as a repressor of putative virulence genes in *Neisseria meningitidis*. Mol. Microbiol. *52*, 1081–1090.

Desrosiers, D.C., Sun, Y.C., Zaidi, A.A., Eggers, C.H., Cox, D.L., and Radolf, J.D. (2007). The general transition metal (Tro) and Zn^{2+} (Znu) transporters in *Treponema pallidum*: analysis of metal specificities and expression profiles. Mol. Microbiol. *65*, 137–152.

Eick-Helmerich, K., and Braun, V. (1989). Import of biopolymers into *Escherichia coli*: nucleotide sequences of the *exbB* and *exbD* genes are homologous to those of the *tolQ* and *tolR* genes, respectively. J. Bacteriol. *171*, 5117–5126.

Eick-Helmerich, K., Hantke, K., and Braun, V. (1987). Cloning and expression of the *exbB* gene of *Escherichia coli* K-12. Mol Gen Genet *206*, 246–251.

Fenno, J.C., Muller, K.H., and McBride, B.C. (1996). Sequence analysis, expression, and binding activity of recombinant major outer sheath protein (Msp) of *Treponema denticola*. J. Bacteriol. *178*, 2489–2497.

Fenno, J.C., Wong, G.W., Hannam, P.M., Muller, K.H., Leung, W.K., and McBride, B.C. (1997). Conservation of *msp*, the gene encoding the major outer membrane protein of oral *Treponema* spp. J. Bacteriol. *179*, 1082–1089.

Ferguson, A.D., and Deisenhofer, J. (2004). Metal import through microbial membranes. Cell *116*, 15–24.

Finney, L.A., and O'Halloran, T.V. (2003). Transition metal speciation in the cell: insights from the chemistry of metal ion receptors. Science *300*, 931–936.

Fischer, E., Gunter, K., and Braun, V. (1989). Involvement of ExbB and TonB in transport across the outer membrane of *Escherichia coli*: phenotypic complementation of *exb* mutants by overexpressed *tonB* and physical stabilization of TonB by ExbB. J. Bacteriol. *171*, 5127–5134.

Foster, J.W., Kinney, D.M., and Moat, A.G. (1979). Pyridine nucleotide cycle of *Salmonella typhimurium*: regulation of nicotinic acid phosphoribosyltransferase and nicotinamide deamidase. J. Bacteriol. *138*, 957–961.

Fraser, C.M., Casjens, S., Huang, W.M., Sutton, G.G., Clayton, R., Lathigra, R., White, O., Ketchum, K.A., Dodson, R., Hickey, E.K., *et al.* (1997). Genomic sequence of a Lyme disease spirochaete, *Borrelia burgdorferi*. Nature *390*, 580–586.

Fraser, C.M., Norris, S.J., Weinstock, G.M., White, O., Sutton, G.G., Dodson, R., Gwinn, M., Hickey, E.K., Clayton, R., Ketchum, K.A., *et al.* (1998). Complete genome sequence of *Treponema pallidum*, the syphilis spirochete. Science *281*, 375–388.

Frenkiel-Krispin, D., Levin-Zaidman, S., Shimoni, E., Wolf, S.G., Wachtel, E.J., Arad, T., Finkel, S.E., Kolter, R., and Minsky, A. (2001). Regulated phase transitions of bacterial chromatin: a non-enzymatic pathway for generic DNA protection. Embo J. *20*, 1184–1191.

Funk, C.R., Zimniak, L., and Dowhan, W. (1992). The *pgpA* and *pgpB* genes of *Escherichia coli* are not essential: evidence for a third phosphatidylglycerophosphate phosphatase. J. Bacteriol. *174*, 205–213.

Gao, H., Zhou, D., Li, Y., Guo, Z., Han, Y., Song, Y., Zhai, J., Du, Z., Wang, X., Lu, J., and Yang, R. (2008). The iron-responsive Fur regulon in *Yersinia pestis*. J. Bacteriol. *190*, 3063–3075.

Gherardini, F.C., Boylan, J.A., and Brett, P.J. (2006). Metal Utilization and Oxidative Stress, In Treponema: Molecular and Cellular Biology, J.D. Radolf, and L. S, eds. (Norwich: Horizon Scientific/Caister Academic Press), pp. 103–128.

Gilbert, M.A., Morton, E.A., Bundle, S.F., and Samuels, D.S. (2007). Artificial regulation of *ospC* expression in *Borrelia burgdorferi*. Mol. Microbiol. *63*, 1259–1273.

Grant, R.A., Filman, D.J., Finkel, S.E., Kolter, R., and Hogle, J.M. (1998). The crystal structure of Dps, a ferritin homolog that binds and protects DNA. Nat. Struct. Biol. *5*, 294–303.

Grass, G., Franke, S., Taudte, N., Nies, D.H., Kucharski, L.M., Maguire, M.E., and Rensing, C. (2005). The metal permease ZupT from *Escherichia coli* is a transporter with a broad substrate spectrum. J. Bacteriol. *187*, 1604–1611.

Grass, G., Wong, M.D., Rosen, B.P., Smith, R.L., and Rensing, C. (2002). ZupT is a Zn(II) uptake system in *Escherichia coli*. J. Bacteriol. *184*, 864–866.

Grifantini, R., Frigimelica, E., Delany, I., Bartolini, E., Giovinazzi, S., Balloni, S., Agarwal, S., Galli, G., Genco, C., and Grandi, G. (2004). Characterization of a novel *Neisseria meningitidis* Fur and iron-regulated operon required for protection from oxidative stress: utility of DNA microarray in the assignment of the biological role of hypothetical genes. Mol. Microbiol. *54*, 962–979.

Guerinot, M.L. (2000). The ZIP family of metal transporters. Biochim. Biophys. Acta *1465*, 190–198.

Hall, J.L. (2002). Cellular mechanisms for heavy metal detoxification and tolerance. J. Exp. Bot. 53, 1–11.

Hardesty, C., Ferran, C., and DiRienzo, J.M. (1991). Plasmid-mediated sucrose metabolism in *Escherichia coli*: characterization of *scrY*, the structural gene for a phosphoenolpyruvate-dependent sucrose phosphotransferase system outer membrane porin. J. Bacteriol. *173*, 449–456.

Hazlett, K.R., Rusnak, F., Kehres, D.G., Bearden, S.W., La Vake, C.J., La Vake, M.E., Maguire, M.E., Perry, R.D., and Radolf, J.D. (2003). The *Treponema pallidum tro* operon encodes a multiple metal transporter, a zinc-dependent transcriptional repressor, and a semi-autonomously expressed phosphoglycerate mutase. J. Biol. Chem. *278*, 20687–20694.

Heller, K.B., and Wilson, T.H. (1981). Selectivity of the *Escherichia coli* outer membrane porins OmpC and OmpF. FEBS Lett. *129*, 253–255.

Herbig, A.F., and Helmann, J.D. (2001). Roles of metal ions and hydrogen peroxide in modulating the interaction of the *Bacillus subtilis* PerR peroxide regulon repressor with operator DNA. Mol. Microbiol. *41*, 849–859.

Hidalgo, E., Ding, H., and Demple, B. (1997). Redox signal transduction via iron-sulfur clusters in the SoxR transcription activator. Trends Biochem. Sci *22*, 207–210.

Higashitani, A., Nishimura, Y., Hara, H., Aiba, H., Mizuno, T., and Horiuchi, K. (1993). Osmoregulation of the fatty acid receptor gene *fadL* in *Escherichia coli*. Mol Gen Genet *240*, 339–347.

Hillyard, D., Rechsteiner, M., Manlapaz-Ramos, P., Imperial, J.S., Cruz, L.J., and Olivera, B.M. (1981). The pyridine nucleotide cycle. Studies in *Escherichia coli* and the human cell line D98/AH2. J. Biol. Chem. *256*, 8491–8497.

Hossain, H., Wellensiek, H.J., Geyer, R., and Lochnit, G. (2001). Structural analysis of glycolipids from *Borrelia burgdorferi*. Biochimie *83*, 683–692.

Imlay, J.A. (2002). How oxygen damages microbes: oxygen tolerance and obligate anaerobiosis. Adv. Microb. Physiol. *46*, 111–153.

Imlay, J.A. (2003). Pathways of oxidative damage. Annu. Rev. Microbiol. *57*, 395–418.

Jackowski, S., and Rock, C.O. (1981). Regulation of coenzyme A biosynthesis. J. Bacteriol. *148*, 926–932.

Jackowski, S., and Rock, C.O. (1983). Ratio of active to inactive forms of acyl carrier protein in *Escherichia coli*. J. Biol. Chem. *258*, 15186–15191.

Jackowski, S., and Rock, C.O. (1984a). Metabolism of 4′-phosphopantetheine in *Escherichia coli*. J. Bacteriol. *158*, 115–120.

Jackowski, S., and Rock, C.O. (1984b). Turnover of the 4′-phosphopantetheine prosthetic group of acyl carrier protein. J. Biol. Chem. *259*, 1891–1895.

Jap, B.K., and Walian, P.J. (1996). Structure and functional mechanism of porins. Physiol Rev *76*, 1073–1088.

Jewett, M.W., Lawrence, K., Bestor, A.C., Tilly, K., Grimm, D., Shaw, P., VanRaden, M., Gherardini, F., and Rosa, P.A. (2007). The critical role of the linear plasmid lp36 in the infectious cycle of *Borrelia burgdorferi*. Mol. Microbiol. *64*, 1358–1374.

Jewett, M.W., Lawrence, K.A., Bestor, A., Byram, R., Gherardini, F., and Rosa, P.A. (2009). GuaA and GuaB are essential for *Borrelia burgdorferi* survival in the tick–mouse infection cycle. J. Bacteriol. *191*, 6231–6241.

Jordan, A., and Reichard, P. (1998). Ribonucleotide reductases. Annu. Rev. Biochem *67*, 71–98.

Kaminski, P.A. (2002). Functional cloning, heterologous expression, and purification of two different N-deoxyribosyltransferases from *Lactobacillus helveticus*. J. Biol. Chem. 277, 14400–14407.

Katona, L.I., Tokarz, R., Kuhlow, C.J., Benach, J., and Benach, J.L. (2004). The Fur homologue in *Borrelia burgdorferi*. J. Bacteriol. *186*, 6443–6456.

Keyer, K., and Imlay, J.A. (1996). Superoxide accelerates DNA damage by elevating free-iron levels. Proc. Natl. Acad. Sci. U.S.A. *93*, 13635–13640.

Kinjo, Y., Tupin, E., Wu, D., Fujio, M., Garcia-Navarro, R., Benhnia, M.R., Zajonc, D.M., Ben-Menachem, G., Ainge, G.D., Painter, G.F., *et al.* (2006). Natural killer T-cells recognize diacylglycerol antigens from pathogenic bacteria. Nat. Immunol. *7*, 978–986.

Klebba, P.E. (2003). Three paradoxes of ferric enterobactin uptake. Front. Biosci. *8*, s1422–1436.

Labandeira-Rey, M., and Skare, J.T. (2001). Decreased infectivity in *Borrelia burgdorferi* strain B31 is associated with loss of linear plasmid 25 or 28–1. Infect. Immun. *69*, 446–455.

Lawerence, K.A., Jewett, M.W., Rosa, P.A., and Gherardini, F.C. (2009). *Borrelia burgdorferi bb0426* encodes a 2′-deoxyribosyltransferase that plays a central role in purine salavage. Mol. Microbiol. *72*, 1517–1529.

Li, X., Pal, U., Ramamoorthi, N., Liu, X., Desrosiers, D.C., Eggers, C.H., Anderson, J.F., Radolf, J.D., and Fikrig, E. (2007). The Lyme disease agent *Borrelia burgdorferi* requires BB0690, a Dps homologue, to persist within ticks. Mol. Microbiol. *63*, 694–710.

Li, X.Z., and Nikaido, H. (2004). Efflux-mediated drug resistance in bacteria. Drugs *64*, 159–204.

Lin, B., Short, S.A., Eskildsen, M., Klempner, M.S., and Hu, L.T. (2001). Functional testing of putative oligopeptide permease (Opp) proteins of *Borrelia burgdorferi*: a complementation model in *opp⁻ Escherichia coli*. Biochim. Biophys. Acta *1499*, 222–231.

Lolkema, J.S., Chaban, Y., and Boekema, E.J. (2003). Subunit composition, structure, and distribution of bacterial V-type ATPases. J. Bioenerg. Biomembr. *35*, 323–335.

Lopez-Lara, I.M., and Geiger, O. (2001). Novel pathway for phosphatidylcholine biosynthesis in bacteria associated with eukaryotes. J. Biotechnol. *91*, 211–221.

Maas, W.K., and Davis, B.D. (1950). Pantothenate studies. I. Interference by D-serine and L-aspartic acid with pantothenate synthesis in *Escherichia coli*. J. Bacteriol. *60*, 733–745.

Manlapaz-Fernandez, P., and Olivera, B.M. (1973). Pyridine nucleotide metabolism in *Escherichia coli*. IV. Turnover. J. Biol. Chem. *248*, 5150–5155.

Mannock, D.A., Lewis, R.N., McElhaney, R.N., Harper, P.E., Turner, D.C., and Gruner, S.M. (2001). An analysis of the relationship between fatty acid composition and the lamellar gel to liquid-crystalline and the lamellar to inverted nonlamellar phase transition temperatures of phosphatidylethanolamines and diacyl-α-d-glucosyl glycerols. Eur Biophys. *J*. *30*, 537–554.

Martin, D.P., and Drueckhammer, D.G. (1993). Separate enzymes catalyze the final two steps of coenzyme A biosynthesis in *Brevibacterium ammoniagenes*: purification of pantetheine phosphate adenylyltransferase. Biochem. Biophys. Res. Commun. *192*, 1155–1161.

Martinez, A., and Kolter, R. (1997). Protection of DNA during oxidative stress by the nonspecific DNA-binding protein Dps. J. Bacteriol. *179*, 5188–5194.

McCord, J.M., and Fridovich, I. (1969). The utility of superoxide dismutase in studying free radical reactions. I. Radicals generated by the interaction of sulfite, dimethyl sulfoxide, and oxygen. J. Biol. Chem. *244*, 6056–6063.

McCord, J.M., Keele, B.B., Jr., and Fridovich, I. (1971). An enzyme-based theory of obligate anaerobiosis: the physiological function of superoxide dismutase. Proc. Natl. Acad. Sci. U.S.A. *68*, 1024–1027.

McLaren, J., Ngo, D.T., and Olivera, B.M. (1973). Pyridine nucleotide metabolism in *Escherichia coli*. 3. Biosynthesis from alternative precursors *in vivo*. J. Biol. Chem. *248*, 5144–5149.

Medrano, M.S., Ding, Y., Wang, X.G., Lu, P., Coburn, J., and Hu, L.T. (2007). Regulators of expression of the oligopeptide permease A proteins of *Borrelia burgdorferi*. J. Bacteriol. *189*, 2653–2659.

Morikawa, K., Ohniwa, R.L., Kim, J., Maruyama, A., Ohta, T., and Takeyasu, K. (2006). Bacterial nucleoid dynamics: oxidative stress response in *Staphylococcus aureus*. Genes Cells *11*, 409–423.

Mulay, V., Caimano, M.J., Liveris, D., Desrosiers, D.C., Radolf, J.D., and Schwartz, I. (2007). *Borrelia burgdorferi* BBA74, a periplasmic protein associated with the outer membrane, lacks porin-like properties. J. Bacteriol. *189*, 2063–2068.

Mulrooney, S.B., and Hausinger, R.P. (2003). Nickel uptake and utilization by microorganisms. FEMS Microbiol. Rev. *27*, 239–261.

Murakami, S., and Yamaguchi, A. (2003). Multidrug-exporting secondary transporters. Curr. Opin. Struct. Biol. *13*, 443–452.

Murata, T., Yamato, I., Kakinuma, Y., Shirouzu, M., Walker, J.E., Yokoyama, S., and Iwata, S. (2008). Ion binding and selectivity of the rotor ring of the Na^+-transporting V-ATPase. Proc. Natl. Acad. Sci. U.S.A. *105*, 8607–8612.

Nakae, T. (1976). Outer membrane of *Salmonella*. Isolation of protein complex that produces transmembrane channels. J. Biol. Chem. *251*, 2176–2178.

Nakamura, H., and Tamura, Z. (1973). Pantothenate uptake in *Escherichia coli* K-12. J. Nutr. Sci. Vitaminol. (Tokyo) *19*, 389–400.

Nguyen, K.T., Wu, J.C., Boylan, J.A., Gherardini, F.C., and Pei, D. (2007). Zinc is the metal cofactor of *Borrelia burgdorferi* peptide deformylase. Arch. Biochem. Biophys. *468*, 217–225.

Nikaido, H., and Saier, M.H., Jr. (1992). Transport proteins in bacteria: common themes in their design. Science *258*, 936–942.

Nikaido, H., and Zgurskaya, H.I. (2001). AcrAB and related multidrug efflux pumps of *Escherichia coli*. J. Mol. Microbiol. Biotechnol. *3*, 215–218.

Nunn, W.D., and Simons, R.W. (1978). Transport of long-chain fatty acids by *Escherichia coli*: mapping and characterization of mutants in the *fadL* gene. Proc. Natl. Acad. Sci. U.S.A. *75*, 3377–3381.

Östberg, Y., Berg, S., Comstedt, P., Wieslander, A., and Bergström, S. (2007). Functional analysis of a lipid galactosyltransferase synthesizing the major envelope lipid in the Lyme disease spirochete *Borrelia burgdorferi*. FEMS Microbiol. Lett. *272*, 22–29.

Outten, C.E., and O'Halloran, T.V. (2001). Femtomolar sensitivity of metalloregulatory proteins controlling zinc homeostasis. Science *292*, 2488–2492.

Ouyang, Z., He, M., Oman, T., Yang, F., and Norgard, M. (2009). A manganase transporter, BB0219 (BmtA), is required for virulence by the Lyme disease spirochete, *Borrelia burgdorferi*. Proc. Natl. Acad. Sci. U.S.A. *106*, 3449–3454.

Overbeek, R., Larsen, N., Walunas, T., D'Souza, M., Pusch, G., Selkov, E., Jr., Liolios, K., Joukov, V., Kaznadzey, D., Anderson, I., *et al.* (2003). The ERGO genome analysis and discovery system. Nucleic Acids Res. *31*, 164–171.

Palenik, B., Brahamsha, B., Larimer, F.W., Land, M., Hauser, L., Chain, P., Lamerdin, J., Regala, W., Allen, E.E., McCarren, J., *et al.* (2003). The genome of a motile marine *Synechococcus*. Nature *424*, 1037–1042.

Park, U.E., Olivera, B.M., Hughes, K.T., Roth, J.R., and Hillyard, D.R. (1989). DNA ligase and the pyridine nucleotide cycle in *Salmonella typhimurium*. J. Bacteriol. *171*, 2173–2180.

Pettersson, J., Schrumpf, M.E., Raffel, S.J., Porcella, S.F., Guyard, C., Lawrence, K., Gherardini, F.C., and Schwan, T.G. (2007). Purine salvage pathways among *Borrelia* species. Infect. Immun. *75*, 3877–3884.

Pomposiello, P.J., Bennik, M.H., and Demple, B. (2001). Genome-wide transcriptional profiling of the *Escherichia coli* responses to superoxide stress and sodium salicylate. J. Bacteriol. *183*, 3890–3902.

Poole, K. (2001). Multidrug resistance in Gram-negative bacteria. Curr. Opin. Microbiol. *4*, 500–508.

Posey, J.E., and Gherardini, F.C. (2000). Lack of a role for iron in the Lyme disease pathogen. Science *288*, 1651–1653.

Posey, J.E., Hardham, J.M., Norris, S.J., and Gherardini, F.C. (1999). Characterization of a manganese-dependent regulatory protein, TroR, from *Treponema pallidum*. Proc. Natl. Acad. Sci. U.S.A. *96*, 10887–10892.

Powell, G.L., Bauza, M., and Larrabee, A.R. (1973). The stability of acyl carrier protein in *Escherichia coli*. J. Biol. Chem. *248*, 4461–4466.

Powell, G.L., Elovson, J., and Vagelos, P.R. (1969). Acyl carrier protein. XII. Synthesis and turnover of the prosthetic group of acyl carrier protein *in vivo*. J. Biol. Chem. *244*, 5616–5624.

Pugh, E.L., and Wakil, S.J. (1965). Studies on the mechanism of fatty acid synthesis. XIV. The prosthetic group of acyl carrier protein and the mode of its attachment to the protein. J. Biol. Chem. *240*, 4727–4733.

Purser, J.E., Lawrenz, M.B., Caimano, M.J., Howell, J.K., Radolf, J.D., and Norris, S.J. (2003). A plasmid-encoded nicotinamidase (PncA) is essential for infectivity of *Borrelia burgdorferi* in a mammalian host. Mol. Microbiol. *48*, 753–764.

Qi, J., Wang, Y., and Forgac, M. (2007). The vacuolar (H^+)-ATPase: subunit arrangement and *in vivo* regulation. J. Bioenerg. Biomembr. *39*, 423–426.

Radolf, J.D., Goldberg, M.S., Bourell, K., Baker, S.I., Jones, J.D., and Norgard, M.V. (1995). Characterization of outer membranes isolated from *Borrelia burgdorferi*, the Lyme disease spirochete. Infect. Immun. *63*, 2154–2163.

Regue, M., Remenick, J., Tokunaga, M., Mackie, G.A., and Wu, H.C. (1984). Mapping of the lipoprotein signal peptidase gene (*lsp*). J. Bacteriol. *158*, 632–635.

Ren, B., Tibbelin, G., Kajino, T., Asami, O., and Ladenstein, R. (2003). The multi-layered structure of Dps with a novel di-nuclear ferroxidase center. J. Mol. Biol. *329*, 467–477.

Rensing, C., and Grass, G. (2003). *Escherichia coli* mechanisms of copper homeostasis in a changing environment. FEMS Microbiol. Rev. *27*, 197–213.

Rosa, P.A., Tilly, K., and Stewart, P.E. (2005). The burgeoning molecular genetics of the Lyme disease spirochaete. Nat. Rev. Microbiol. *3*, 129–143.

Rosen, B.P. (2002). Transport and detoxification systems for transition metals, heavy metals and metalloids in eukaryotic and prokaryotic microbes. Comp. Biochem. Physiol. A Mol. Integr. Physiol. *133*, 689–693.

Roy, S., Gupta, S., Das, S., Sekar, K., Chatterji, D., and Vijayan, M. (2004). X-ray analysis of *Mycobacterium smegmatis* Dps and a comparative study involving other Dps and Dps-like molecules. J. Mol. Biol. *339*, 1103–1113.

Saier, M.H., Jr., and Paulsen, I.T. (2000). Whole genome analyses of transporters in spirochetes: *Borrelia burgdorferi* and *Treponema pallidum*. J. Mol. Microbiol. Biotechnol. *2*, 393–399.

Sankaran, K., Gupta, S.D., and Wu, H.C. (1995). Modification of bacterial lipoproteins. Methods Enzymol. *250*, 683–697.

Schwan, T.G., Battisti, J.M., Porcella, S.F., Raffel, S.J., Schrumpf, M.E., Fischer, E.R., Carroll, J.A., Stewart, P.E., Rosa, P., and Somerville, G.A. (2003). Glycerol-3-phosphate acquisition in spirochetes: distribution and biological activity of glycerophosphodiester phosphodiesterase (GlpQ) among *Borrelia* species. J. Bacteriol. *185*, 1346–1356.

Seaver, L.C., and Imlay, J.A. (2001). Hydrogen peroxide fluxes and compartmentalization inside growing *Escherichia coli*. J. Bacteriol. *183*, 7182–7189.

Seshu, J., Boylan, J.A., Hyde, J.A., Swingle, K.L., Gherardini, F.C., and Skare, J.T. (2004). A conservative amino acid change alters the function of BosR, the redox regulator of *Borrelia burgdorferi*. Mol. Microbiol. *54*, 1352–1363.

Shaw, N. (1970). Bacterial glycolipids. Bacteriol Rev *34*, 365–377.

Skare, J.T., Champion, C.I., Mirzabekov, T.A., Shang, E.S., Blanco, D.R., Erdjument-Bromage, H., Tempst, P., Kagan, B.L., Miller, J.N., and Lovett, M.A. (1996). Porin activity of the native and recombinant outer membrane protein Oms28 of *Borrelia burgdorferi*. J. Bacteriol. *178*, 4909–4918.

Skare, J.T., Mirzabekov, T.A., Shang, E.S., Blanco, D.R., Erdjument-Bromage, H., Bunikis, J., Bergström, S., Tempst, P., Kagan, B.L., Miller, J.N., and Lovett, M.A. (1997). The Oms66 (p66) protein is a *Borrelia burgdorferi* porin. Infect. Immun. *65*, 3654–3661.

Sohlenkamp, C., de Rudder, K.E., Rohrs, V.V., Lopez-Lara, I.M., and Geiger, O. (2000). Cloning and characterization of the gene for phosphatidylcholine synthase. J. Biol. Chem. *275*, 27500.

Tasaka, Y., Gombos, Z., Nishiyama, Y., Mohanty, P., Ohba, T., Ohki, K., and Murata, N. (1996). Targeted mutagenesis of acyl-lipid desaturases in *Synechocystis*: evidence for the important roles of polyunsaturated membrane lipids in growth, respiration and photosynthesis. EMBO J. *15*, 6416–6425.

Tilly, K., Elias, A.F., Errett, J., Fischer, E., Iyer, R., Schwartz, I., Bono, J.L., and Rosa, P. (2001). Genetics and regulation of chitobiose utilization in *Borrelia burgdorferi*. J. Bacteriol. *183*, 5544–5553.

Tilly, K., Grimm, D., Bueschel, D.M., Krum, J.G., and Rosa, P. (2004). Infectious cycle analysis of a *Borrelia burgdorferi* mutant defective in transport of chitobiose, a tick cuticle component. Vector Borne Zoonotic Dis *4*, 159–168.

Touati, D. (1983). Cloning and mapping of the manganese superoxide dismutase gene (*sodA*) of *Escherichia coli* K-12. J. Bacteriol. *155*, 1078–1087.

Vagelos, P.R., and Larrabes, A.R. (1967). Acyl carrier protein. IX. Acyl carrier protein hydrolase. J. Biol. Chem. *242*, 1776–1781.

Vallari, D.S., and Rock, C.O. (1985). Pantothenate transport in *Escherichia coli*. J. Bacteriol. *162*, 1156–1161.

von Lackum, K., and Stevenson, B. (2005). Carbohydrate utilization by the Lyme borreliosis spirochete, *Borrelia burgdorferi*. FEMS Microbiol. Lett. *243*, 173–179.

Wandersman, C., Schwartz, M., and Ferenci, T. (1979). *Escherichia coli* mutants impaired in maltodextrin transport. J. Bacteriol. *140*, 1–13.

Wang, X.G., Kidder, J.M., Scagliotti, J.P., Klempner, M.S., Noring, R., and Hu, L.T. (2004a). Analysis of differences in the functional properties of the substrate binding proteins of the *Borrelia burgdorferi* oligopeptide permease (Opp) operon. J. Bacteriol. *186*, 51–60.

Wang, X.G., Lin, B., Kidder, J.M., Telford, S., and Hu, L.T. (2002). Effects of environmental changes on expression of the oligopeptide permease (*opp*) genes of *Borrelia burgdorferi*. J. Bacteriol. *184*, 6198–6206.

Wang, X.G., Scagliotti, J.P., and Hu, L.T. (2004b). Phospholipid synthesis in *Borrelia burgdorferi*: *bb0249* and *bb0721* encode functional phosphatidylcholine synthase and phosphatidylglycerolphosphate synthase proteins. Microbiology *150*, 391–397.

Weinberg, E.D. (1997). The *Lactobacillus* anomaly: total iron abstinence. Perspect. Biol. Med. *40*, 578–583.

Westheimer, F.H. (1987). Mechanism of action of the pyridine nucleotides, In Coenzymes and Cofactors, D. Dolphin, R. Poulson, and O. Avramovic, eds. (New York: John Wiley & Sons, Inc), pp. 253–322.

White, B.H. (1982). Biosynthetic and salvage pathways of pyridine nucleotide coenzymes, In the Pyridine Nucleotide Coenzymes, J. Everse, B. Anderson, and K.-S. You, eds. (New York: Academic Press), pp. 225–248.

Whitehouse, C.A., Williams, L.R., and Austin, F.E. (1997). Identification of superoxide dismutase activity in *Borrelia burgdorferi*. Infect. Immun. *65*, 4865–4868.

Wolf, S.G., Frenkiel, D., Arad, T., Finkel, S.E., Kolter, R., and Minsky, A. (1999). DNA protection by stress-induced biocrystallization. Nature *400*, 83–85.

Worrall, D.M., and Tubbs, P.K. (1983). A bifunctional enzyme complex in coenzyme A biosynthesis: purification of pantetheine phosphate adenylyltransferase and dephospho-CoA kinase. Biochem. J. *215*, 153–157.

Yokoyama, K., and Imamura, H. (2005). Rotation, structure, and classification of prokaryotic V-ATPase. J. Bioenerg. Biomembr. *37*, 405–410.

Youn, H.D., Kim, E.J., Roe, J.H., Hah, Y.C., and Kang, S.O. (1996a). A novel nickel-containing superoxide dismutase from *Streptomyces* spp. Biochem. J *318 (Pt 3)*, 889–896.

Youn, H.D., Youn, H., Lee, J.W., Yim, Y.I., Lee, J.K., Hah, Y.C., and Kang, S.O. (1996b). Unique isozymes of superoxide dismutase in *Streptomyces griseus*. Arch. Biochem. Biophys. *334*, 341–348.

Zgurskaya, H.I., and Nikaido, H. (2000). Multidrug resistance mechanisms: drug efflux across two membranes. Mol. Microbiol. *37*, 219–225.

Zheng, M., Aslund, F., and Storz, G. (1998). Activation of the OxyR transcription factor by reversible disulfide bond formation. Science *279*, 1718–1721.

Zheng, M., Wang, X., Templeton, L.J., Smulski, D.R., LaRossa, R.A., and Storz, G. (2001). DNA microarray-mediated transcriptional profiling of the *Escherichia coli* response to hydrogen peroxide. J. Bacteriol. *183*, 4562–4570.

Zhou, X., Cahoon, M., Rosa, P., and Hedstrom, L. (1997). Expression, purification, and characterization of inosine 5′-monophosphate dehydrogenase from *Borrelia burgdorferi*. J. Biol. Chem. *272*, 21977–21981.

Structure, Function and Biogenesis of the *Borrelia* Cell Envelope

6

Sven Bergström and Wolfram R. Zückert

Abstract

Over the past two decades, modern molecular tools have been crucial in shaping our current concepts of the *Borrelia* cell envelope. Although *Borrelia* spirochaetes are often, but mistakenly, described as Gram-negative bacteria due to their diderm, i.e. double-membrane envelopes, a closer examination reveals significant differences in composition and architecture. Probably most striking are the lack of LPS, the presence of major surface lipoproteins at the host–pathogen interface during transmission, persistence and ensuing pathogenic processes, and the additional function of periplasmic flagella in defining cell shape. While surface lipoproteins, such as the Osps, interact with a variety of ligands in different organ tissues, they are also targets of the immune response and several have emerged as vaccine candidates. Some of the identified periplasmic lipoproteins, i.e. the OppAs, are components of substrate transport complexes. Investigations into integral membrane proteins led to the identification of several *Borrelia* porins: P13, whose structure and function is unknown; DipA, which is specific for dicarboxylates; and P66 (Oms66), which serves as a pore-forming outer membrane protein with an extremely high single channel conductance and as an adhesin for β3-integrin. The recently identified Tol homologues BesA, -B, and -C appear to form a type I 'channel' to export exogenous toxic agents such as antibiotics and to maintain infectivity by an unknown mechanism. Initial studies on envelope biogenesis pathways based on diderm proteobacterial model organisms already revealed significant deviations from the norm. This further bolsters the unique status of *Borrelia* among microbial pathogens.

6.1 Introduction

The complex life style of *Borrelia* spirochaetes, i.e. their ability to shuttle between haematophagous arthropods and various vertebrates, exposes them to a variety of niches differing in nutritional content and immunological pressure. Survival, transmission and ultimately pathogenesis, therefore, require a large degree of adaptive biological capacity of these organisms. First, the host–pathogen interface repeatedly needs to be optimized for interaction with environments and tissues as different as the tick midgut, the mammalian skin or joint cartilage. A variety of outer surface-exposed proteins play a critical role in this adaptive process by providing ligands for receptor-mediated adhesion, mechanisms of host immune response avoidance, as well as pathways for the acquisition of nutrients. As they are antigens evoking immune reactions in the mammalian host, they continue to represent preeminent vaccine candidates. The spatiotemporal regulation of borrelial outer surface proteins is being understood in ever increasing depth, and structure–function relationships have begun to emerge. Genomics, proteomics, novel genetic tools and sophisticated visualization techniques now aim to reveal a more complete picture of the structure, function and biogenesis of the *Borrelia* cell envelope. Of particular interest are transport and assembly of its abundant lipoproteins, the uptake of essential nutrients and the efflux of toxic and undesired compounds.

6.2 Structural features of Borrelia membranes

6.2.1 General morphology

Borrelia has a diderm (i.e. two lipid bilayer) ultrastructure that is distinct from that of Gram-negative bacteria, such as *Escherichia coli.* An overview of the *Borrelia* cell envelope is depicted in Fig. 6.1. As in other spirochaetes, a fragile outer membrane surrounds the protoplasmic cylinder, which consists of a peptidoglycan layer, a cytoplasmic (inner) membrane and the enclosed cytoplasmic contents (Barbour *et al.*, 1986; Johnson *et al.*, 1984; Kudryashev *et al.*, 2009). A ribbon of 7–11 flagella is located in the periplasmic space where it is attached to the poles and wraps around the cell cylinder, giving the bacterium its characteristic flat wave shape (Barbour *et al.*, 1986; Charon *et al.*, 2009; Goldstein *et al.*, 1994; Motaleb *et al.*, 2000).

The outer membrane of all bacteria is a selective barrier between the cell and the environment that excludes certain molecules due to its physicochemical properties and simultaneously allows entry of appropriate nutrient substances by diffusion through the outer membrane, either via porins (Benz, 2001) or by receptor-mediated uptake (Nikaido, 2003). Additionally, outer membrane proteins (OMPs) play a critical role in the virulence of bacterial pathogens by responding to different environments, by mediating interactions with the host, or by competing with specific host defence mechanisms, thereby facilitating invasion and colonization of various tissues.

The borrelial outer cell membrane is fluid and consists of 45–62% protein, 23–50% lipid and 3–4% carbohydrate (Barbour *et al.*, 1986). Its composition differs significantly from that of Gram-negative bacteria. A first, particularly distinguishing, feature is the absence of phosphatidylethanolamine (Belisle *et al.*, 1994) and lipopolysaccharide (LPS) (Takayama *et al.*, 1987) and the presence of non-LPS glycolipid antigens (Belisle *et al.*, 1994; Eiffert *et al.*, 1991; Wheeler *et al.*, 1993). These glycolipids represent about 50% of the total lipids and contain only galactose as the monosaccharide constituent (Berg, 2001; Hossain *et al.*, 2001). It is unclear whether these

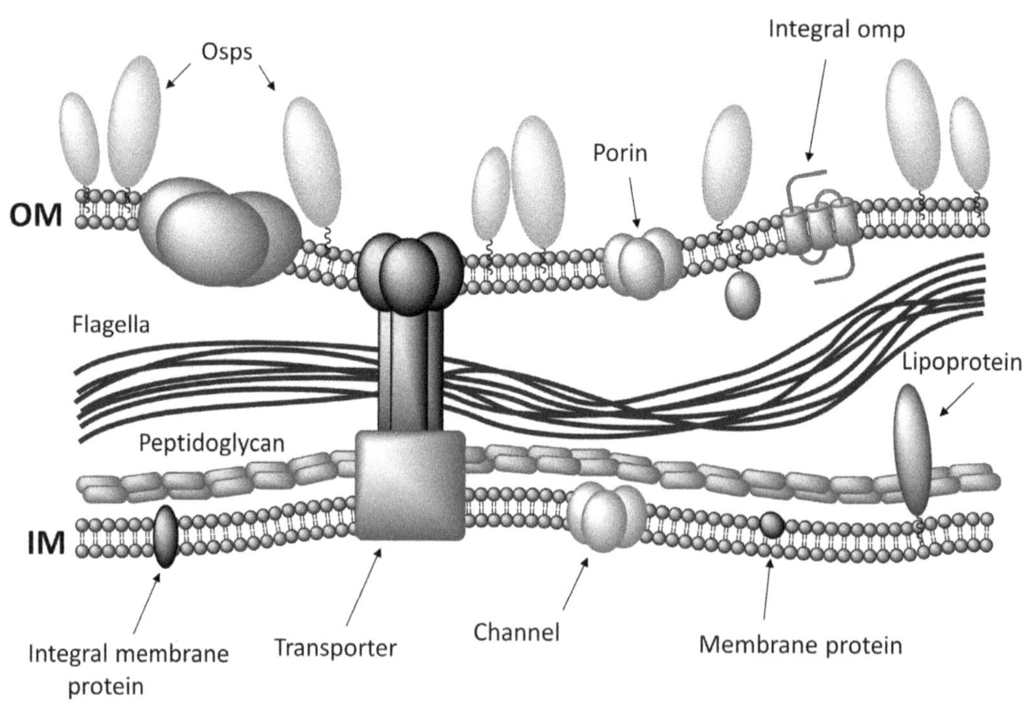

Figure 6.1 *Borrelia* envelope components. OM, outer membrane. IM, inner membrane. Osps, outer surface proteins (lipoproteins). Adapted from Pinne *et al.* (2006b), with permission from IOS Press.

glycolipids are asymmetrically distributed in the borrelial outer membrane as is the case with LPS in Gram-negative bacteria (Kamio and Nikaido, 1976). Secondly, the *B. burgdorferi* outer membrane exhibits a relatively low density of transmembrane-spanning proteins as determined by freeze fracture EM studies (Jones *et al.*, 1995; Radolf *et al.*, 1994; Walker *et al.*, 1991). This may explain why *Borrelia* is more susceptible than Gram-negative bacteria to detergents or to disruption by routine physical manipulations such as centrifugation and resuspension. Thirdly, the outer membrane contains an unusually large number of lipoproteins (Brandt *et al.*, 1990) many of which are on the bacterial surface, the host–pathogen interface, where they act as (i) adhesins, (ii) targets for bactericidal antibodies, or (iii) receptors for various molecules.

6.2.2 Lipid bilayer

Several studies from the pre-genomic era suggested LPS to be a surface component of *Borrelia*. However, genomic sequence provided definitive evidence that the Lyme disease (LD) spirochaete lacks LPS (Fraser *et al.*, 1997). Despite this lack of LPS, other low molecular weight glycolipids are present in borreliae (Honarvar *et al.*, 1994). The phospholipid content of borrelial membranes differs from those of other bacteria, including other spirochaetes, such as *Treponema pallidum*. Mass spectroscopic analysis identified α-monogalactosyl diacylglycerol (α-MGalDAG, 36.1%), phosphatidylcholine (PC, 11.3%), and phosphatidylglycerol (PG, 10.5%) as the major lipids in *B. burgdorferi* (Hossain *et al.*, 2001). Also, a cholesteryl galactoside with an immunogenic motif has been described in *B. burgdorferi* (Ben-Menachem *et al.*, 2003; Schroder *et al.*, 2008). Previous studies also showed that the membranes of the relapsing fever (RF) species *Borrelia hermsii* contain MGalDAG, PC, PC, cholesteryl glucoside, and an acylated cholesteryl glucoside (Livermore *et al.*, 1978). Diglyceride-based glycolipids, like the major MGalDAG of the *B. burgdorferi* envelope, are most widespread in Gram-positive bacteria and mycoplasmas, but essentially absent among Gram-negative bacteria, with the exception of many photosynthetic species (including plant chloroplasts) and some *Pseudomonas* species (Shaw, 1970). Based on the discovery of a novel pathway in which a phosphatidylcholine synthase (Pcs) directly condenses exogenously acquired choline with CDP-DAG to form PC (de Rudder *et al.*, 2000; Lopez-Lara and Geiger, 2001; Sohlenkamp *et al.*, 2004; Sohlenkamp *et al.*, 2000; Wang *et al.*, 2004), the activities of two enzymes involved in synthesis of phosphatidylcholine and phosphatidylglycerol have been characterized (Wang *et al.*, 2004). Recently, the monogalactosyl-1,2-diacylglycerol synthase from *B. burgdorferi* (bbMGS) was identified, cloned, and functionally verified to show that this enzyme catalyses synthesis of the major membrane lipid MGalDAG. Interestingly, the characterized enzyme is closely related to many Gram-positive analogues (Östberg *et al.*, 2007). The finding that mouse natural killer cells were able to recognize the α-MGalDAG lipid in *B. burgdorferi* with concomitant immune activation and antibody production further implies the importance of these glycolipid molecules in the infection biology during Lyme borreliosis (Kinjo *et al.*, 2006; Tupin *et al.*, 2008).

The basic question is if the borrelial glycolipids can functionally replace LPS. This was suggested by Ben-Menachem and coworkers in a study where they thoroughly analysed the properties of the major glycolipids of *B. burgdorferi* and concluded that they have several properties compatible with LPS-like function (Ben-Menachem *et al.*, 2003). They suggested this based on the facts that these two *Borrelia* glycolipids and Gram-negative LPS are main components of their respective outer membranes, surface exposed, and have a three-domain structure. Interestingly, these two surface-exposed glycolipids identified in *B. burgdorferi* react with sera from Lyme disease patients (Cinco *et al.*, 1991; Eiffert *et al.*, 1991; Wheeler *et al.*, 1993).

6.2.3 Lipoproteins

Borrelia spirochaetes contain an unusual abundance of proteins covalently modified by lipids, i.e. lipoproteins (Haake, 2000). An initial analysis of the genome sequence of *B. burgdorferi* sensu stricto predicted 132 open reading frames that contain the signal sequence characteristic for lipoproteins and 32 additional open reading frames that have other similarities to lipoproteins (Casjens *et al.*, 2000; Fraser *et al.*, 1997; Hayashi and Wu, 1990). A recent re-evaluation using a modified prediction algorithm indicated up to 127 lipoproteins in that genome, corresponding to 7.8% of open reading frames (Setubal *et al.*, 2006). This proportion is significantly higher than in other complete bacterial genomes such as *Treponema pallidum* (2.1%) or *Helicobacter pylori* (1.3%) (Casjens *et al.*, 2000; Fraser *et al.*, 1997, 1998). The abundance of surface-localized lipoproteins together with the general importance of outer surface proteins in bacterial adaptation and persistence has led many researchers to focus their investigations on this class of borrelial lipoproteins. Furthermore, the unusual membrane architecture of spirochaetes and the ancient phylogeny of these bacteria suggest that the export, structure and function of spirochaetal lipoproteins are unique to these organisms.

Borrelial lipoproteins are believed to play a major role in host inflammatory response activation and inflammation of tissues at sites of infection during Lyme borreliosis. The immunopotentiating activities of bacterial lipoproteins are related to their N-termini based on two lines of evidence: (i) acylation is essential for their inflammatory activities and (ii) synthetic lipopeptides reproduce the proinflammatory activities of the native molecules (Bessler *et al.*, 1985; Deres *et al.*, 1989; Erdile *et al.*, 1993; Hoffmann *et al.*, 1988; Lex *et al.*, 1986; Radolf *et al.*, 1995). While activation of the innate immune response by lipoproteins is crucial for defence against *B. burgdorferi* infection, surface exposed lipoproteins also play an important role in adaptive responses and pathogenicity of *Borrelia* spirochaetes. Lipoproteins dominate pathogenic mechanisms exploited by *B. burgdorferi*, including those of antigenic variation (Zhang *et al.*, 1997; Zhang and Norris, 1998a,b), evasion of complement killing (Alitalo *et al.*, 2002; Kraiczy *et al.*, 2001a,b; Stevenson *et al.*, 2002) and adherence mechanisms (Coburn *et al.*, 2005). Furthermore, a dynamic interplay between lipoprotein expression and life cycle transitions of *B. burgdorferi* has been envisaged (Revel *et al.*, 2002).

The most abundant and best studied outer membrane proteins of *Borrelia* spirochaetes are lipoproteins called outer surface proteins (Osps). Osps have very little sequence or structural homology to any other known proteins. Characterization of Osps in different *B. burgdorferi* sensu lato strains revealed considerable heterogeneity within and among different species. The most studied Osps are OspA, OspB, OspC, OspE/Erps/Crasps, and the Dbps.

OspA, whose coding sequence was the first one determined for *B. burgdorferi*, remains one of the most studied proteins in these bacteria (Bergström *et al.*, 1989; Howe *et al.*, 1986). It is encoded by the same operon as OspB on the 54-kb linear plasmid lp54 (Barbour and Garon, 1988; Bergström *et al.*, 1989). OspA and OspB have been shown to play important roles in the pathogenesis of Lyme borreliosis in a large number of *in vivo* and *in vitro* experiments. OspA is preferentially expressed by the spirochaetes in the unfed tick midgut (Pal *et al.*, 2000) and mediates adhesion to the midgut epithelium (Pal *et al.*, 2004a).

OspC is encoded on cp26 (Fraser *et al.*, 1997) and forms a genus-wide, structurally conserved protein family with the variable small proteins (Vsps) in RF *Borrelia* (Carter *et al.*, 1994; Eicken *et al.*, 2001; Kumaran *et al.*, 2001; Lawson *et al.*, 2006; Zückert *et al.*, 2001). OspC is very heterogeneous; a large degree of molecular polymorphism of the *ospC* gene has been shown for the different *Borrelia* genospecies (Jauris-Heipke *et al.*, 1995; Livey *et al.*, 1995). So far 19 major OspC groups have been identified (Wang *et al.*, 1999). Molecular analysis of the *ospC* sequences from different *Borrelia* genospecies suggests that recombination between different *ospC* alleles can occur frequently and that this genetic exchange is mediated by lateral transfer of *ospC* sequences within the same genospecies or between different *Borrelia* species (Livey *et al.*, 1995; Wang *et al.*, 1999). Of particular interest is the finding that certain invasive *Borrelia* isolates belongs to defined genotypes or OspC groups. This finding could imply that the OspC protein is an important factor for the invasiveness and cause of dissemination of Lyme borreliosis (Lagal *et al.*, 2003). The cause of this invasive capacity has been suggested to be associated with differences in OspC affinity for plasminogen (Lagal *et al.*, 2006). OspC was initially thought to facilitate the migration of *Borrelia* to tick salivary glands, which is initiated by a blood meal promoting the transmission of spirochaetes to the mammalian host (Gilmore and Mbow, 1999; Pal *et al.*, 2004b). Yet, in other more recent studies, OspC was not necessary to correctly localize or migrate in the tick, but was strictly required to infect mice (Grimm *et al.*, 2004). This might be due to binding of Salp15, an immunosuppressive tick salivary (Ramamoorthi *et al.*, 2005), which would provide an initially immunoprivileged site in the mammalian skin.

The reciprocal expression of OspA/B and OspC during *Borrelia* transition from its tick vector to the mammalian host represents the prototypical paradigm of how stage-specific protein expression can contribute to pathogenesis during the natural cycle of spirochaetal transmission. In unfed ticks, *Borrelia* expresses large amounts of OspA and OspB and almost no OspC. The production of OspA and OspB proteins decreases in the mammalian host and instead the spirochaete promotes production of OspC (Ohnishi *et al.*, 2001; Schwan *et al.*, 1995). OspC expression is induced 36–48 hours into the blood meal. Concomitant downregulation of OspA probably mediates the escape of spirochaete from the tick midgut via the haemolymph to the salivary glands from where it enters the new vertebrate host (Anderson *et al.*, 1989; Ohnishi *et al.*, 2001; Schwan and Piesman, 2000; Schwan *et al.*, 1995; Seshu and Skare, 2000). Antibodies against OspA in the blood of vertebrate hosts kill spirochaetes in the tick midgut before dispersal to the salivary glands and thereby block transmission to the vertebrate host (de Silva *et al.*, 1996; de Silva *et al.*, 1999). Antibodies against OspC are also efficient, probably blocking the transmission by preventing migration of spirochaetes to the salivary glands (Gilmore and Piesman, 2000). Vaccination with a rOspA protects against subsequent *B. burgdorferi* infection (Wallich *et al.*, 1996). Despite its efficacy, it is not used as a vaccine any more, in part because of assumed side-effects (Lathrop *et al.*, 2002). Immunization with OspC also confers protective immunity in animal studies (Gilmore *et al.*, 1996; Sigal *et al.*, 1998; Steere *et al.*, 1998)

The ultrastructures of OspA, OspC and C-terminal part of OspB have been solved by X-ray crystallography. OspA consists of four β-sheets, formed by 21 anti-parallel β-strands and a C-terminal α-helix (Li and Lawson, 1995). The structure depicted was not consistent with that of an enzyme. The structure revealed a hydrophobic cavity buried in a positively charged cleft in the carboxy-terminal domain that might be a binding site for an unknown ligand; the presence of conserved charged groups in a pronounced cleft is typical of proteins that bind small or linear polymeric ligands. OspA could thus recognize either a soluble or host surface molecule and possibly transmit this information to another membrane component, suggesting that OspA could act as a receptor (Li *et al.*, 1997). The structure of OspA originally suggested that the conserved NH_2-terminus could interact with another membrane protein (Li *et al.*, 1997). It is tempting to speculate that this would be the binding site for the *Ixodes scapularis* receptor of *B. burgdorferi* (TROSPA) (Pal *et al.*, 2004a). The crystal structure of the OspB C-terminus (residues 152–296) was identical to that of OspA (Becker *et al.*, 2005). Strikingly, three-dimensional structure analysis of OspC and the related Vsps revealed a totally different structural organization. OspC predominantly consists of α-helices, organized as dimer of five parallel α-helices and two short β-strands (Eicken *et al.*, 2001; Kumaran *et al.*, 2001; Lawson *et al.*, 2006; Zückert *et al.*, 2001).

The Mlp (multicopy lipoproteins) family is one of many paralogue protein families in *B. burgdorferi*, and up to ten *mlp* genes are encoded on cp32 and cp18 (Caimano *et al.*, 2000; Porcella *et al.*, 1996). Mlps are differentially expressed in response to various *in vitro* culture conditions as well as in the host-adapted environment. Thus, Mlp expression is upregulated coordinately by either elevated temperature (37°C versus 23°C), reduced culture pH (6.8 versus 8.0), increased spirochaete cell density, and by some mammalian host factors (Ojaimi *et al.*, 2003; Ojaimi *et al.*, 2005; Porcella *et al.*, 2000; Yang *et al.*, 1999; Yang *et al.*, 2003). These data indirectly suggest that Mlp proteins are also involved in *B. burgdorferi* infection of the mammalian host. In addition, this protein family was shown to be recognized by late Lyme disease patient sera in a microarray assay and is therefore suggested to be a candidate diagnostic antigen (Xu *et al.*, 2008). Another protein proposed to be important for host adaptation is Lp6.6, a small lipoprotein, that is significantly downregulated during mammalian infection, which indicates that its function might be associated with the environment that Borrelia encounter in the tick (Lahdenne *et al.*, 1997). The expression of the Lp6.6 protein has been shown to be downregulated during tick feeding through an RpoS-dependent mechanism (Caimano *et al.*, 2005; Yang *et al.*, 2000). Thus, it is possible that the Lp6.6 is important for the Borrelia spirochaete in the tick environment or during the transition from the tick to the vertebrate host (Caimano *et al.*, 2005).

The RevA protein is synthesized during mammalian infection, and transcription of *revA* is upregulated shortly after the *Borrelia*-infected tick feeds on the mammal (Gilmore and Mbow, 1998; Gilmore *et al.*, 2001). Recently, several new *B. burgdorferi* outer surface exposed lipoproteins have been identified (BB0689, BBA36, BBA66, BBA69, and BBI42) that are actively expressed and immunogenic during *Borrelia* infections and are targets for bactericidal antibody action, making them attractive vaccine candidates for Lyme disease (Brooks *et al.*, 2006).

Attachment to tissues and cellular matrix is an important prerequisite for colonization of the host by pathogenic bacteria. It is not surprising, therefore, that several borrelial lipoproteins have been shown to function as adhesins (see Chapter 12). To facilitate colonization of collagenous tissues by adhering to collagen-associated proteoglycan decorin, the bacteria express two ~20-kDa lipoproteins called decorin-binding proteins A and B (DbpA/B) (Guo *et al.*, 1995, 1998; Hanson *et al.*, 1998). As decorin is found in various tissues and presumably in skin and joints, the sites of high prevalence during infection by *B. burgdorferi*, it is tempting to hypothesize that the decorin binding proteins might be involved in tissue tropism of Lyme disease spirochaetes. Moreover, expression of both *dbpA* and *dbpB* are drastically upregulated at 35°C versus 23°C (Ojaimi *et al.*, 2003) and antibodies against DbpA can prevent an infection (Hagman *et al.*, 1998; Hanson *et al.*, 1998), illustrating the importance of adhesion in the pathogenesis of Lyme disease.

Bgp (*Borrelia* glycosaminoglycan-binding protein) is a *B. burgdorferi* haemagglutinin with the capacity to bind heparan sulphate and dermatan sulphate. Bgp is surface exposed, but its classification as a lipoprotein is still unresolved (Parveen and Leong, 2000). Since the adaptation of spirochaetes to the mammalian host environment leads to enhanced glycosaminoglycan (GAG) binding, Bgp may play a significant role during mammalian infection (Parveen *et al.*, 2003). Moreover, Bgp production was upregulated during conditions that mimicked tick feeding (Ramamoorthy and Philipp, 1998), while secretion of Bgp into extracellular environment has even been observed indicating a possible role as a immune response decoy (Cluss *et al.*, 2004).

P35 has been used to designate two unrelated immunogenic lipoproteins, BBA64 and BBK32. Not only do these serve as important diagnostic tools for early Lyme disease, being recognized by sera from *B. burgdorferi* infection, but they also are expressed *in vivo* and upregulated at 35°C versus 23°C (Fikrig *et al.*, 2000; Indest and Philipp, 2000; Indest *et al.*, 1997; Ojaimi *et al.*, 2003). BBK32 binds fibronectin (Probert and Johnson, 1998) and promotes *Borrelia* attachment to GAGs and is further suggested to be involved in the initiation of microvascular interactions via fibronectin (Fischer *et al.*, 2006; Norman *et al.*, 2008). Furthermore, inactivation of *bbk32* in low-passage, infectious *B. burgdorferi* leads to significantly attenuated phenotype in the mouse model of infection (Seshu *et al.*, 2006). Thus, the borrelial fibronectin–BBK32 interaction appears to be an important pathogenic mechanism of *B. burgdorferi*, particularly for dissemination and persistence (Seshu *et al.*, 2006).

The Bmp lipoprotein family has four members, BmpA-D and are encoded by *bmp* genes of paralogous chromosomal gene family 36 (Aron *et al.*, 1994; Gorbacheva *et al.*, 2000; Ramamoorthy *et al.*, 1996). Bmp lipoproteins are immunogenic and BmpA (P39) is widely used in Lyme disease diagnosis assays (Simpson *et al.*, 1991, 1990). These lipoproteins may be necessary for *in vitro* and *in vivo* growth and they may have a role in virulence (Dobrikova *et al.*, 2001; Gorbacheva *et al.*, 2000; Ramamoorthy *et al.*, 1996). BmpA also activates proinflammatory responses in human synovial cells (Yang *et al.*, 2008).

One family of lipoproteins, the VlsE, is of particular interest because it is believed to contribute to the ability of spirochaetes to cause persistent infection. It has been suggested that the cause of this persistence might be achieved by antigenic variation of Vls (Vmp-like sequence). Vmp stands for variable major proteins of RF *Borrelia* and are the best studied antigenically variable proteins in spirochaetes (Barbour, 1990; Barbour, 1993; Burman *et al.*, 1990) (see Chapter 13). However, the relevance of antigenic variation of VlsE in the pathogenesis of *B. burgdorferi s.l.* remains unclear. Although there is a correlation between the presence of the plasmid that encodes the Vls lipoprotein and infectivity, the Vls proteins appear to be a minor constituent of the bacterial surface. Nevertheless, the general thought is that the antigenic variation accomplished by the *vls* gene locus of LD *Borrelia* is used for this spirochaete to persist in mammals and to evade the host immune responses.

VlsE is encoded on lp28-1 of *B. burgdorferi*, and it undergoes antigenic variation through an elaborate gene conversion mechanism (Coutte *et al.*, 2009; Zhang *et al.*, 1997; Zhang and Norris, 1998a,b). Fifteen non-expressed *vls* cassettes can be recombined into a functional and expressed *vlsE* region resulting in antigenic variation of the expressed lipoprotein (Zhang *et al.*, 1997) and involving partial gene conversion of the expressed *vls* gene by parts of *vls* genes at silent loci in a unidirectional process (Zhang and Norris, 1998b). The mechanism of Vls antigenic variation in LD *Borrelia* spp. was suggested to be similar to the mechanism that is responsible for the recombination between the expressed pilin gene *pilE* and silent copies of *pilS* in *Neisseria gonorrhoeae* (Haas and Meyer, 1986; Hagblom *et al.*, 1985; Koomey, 1997; Zhang *et al.*, 1992, 1997).

VlsE contains two invariable domains and one variable domain that include six variable and six invariable regions. Little is known about the signals that trigger recombination events at the *vls*-locus. Recombination occurs *in vivo* as soon as four days after experimental infection of mice but not *in vitro*, suggesting that the mammalian host provides the signal for Vls recombination (Zhang and Norris, 1998a). The role of the mammalian host is important since the VlsE recombination rate of *B. burgdorferi s.l.* were found to be lower in IFN-γ R-deficient mice than in control animals. These results suggest that the murine immune response can promote the *in vivo* adaptation of *B. burgdorferi s.l.* (Anguita *et al.*, 2001). The solved 3-D structure of VlsE is unlike OspA-C, consisting of 11 α-helices and four short β-strands (Eicken *et al.*, 2002). This is in spite of the fact that VlsE and OspC/Vlps are sharing overall structural features, such as the close proximity of their N and C termini and a hypervariable sequence corresponding to a surface exposed region (Cullen *et al.*, 2004; Lawson *et al.*, 2006; Radolf and Caimano, 2008).

Finally, there is also a class of *Borrelia* lipoproteins of importance for serum resistance. Human and animal sera show bactericidal activity through the action of the complement system (Medzhitov and Janeway, 2000). Complement is a system of plasma proteins that can be activated by antibody or lectins bound to the bacterial surface or by the pathogen surface itself. Gram-negative bacteria have developed multiple strategies to avoid the bactericidal effects of complement by utilizing LPS, capsule or OMPs (Rautemaa and Meri, 1999). Several borrelial lipoproteins promote resistance to complement-mediated killing by preventing complement activation or membrane attack complex formation.

OspE and OspF are the prototypes of a large, functionally diverse group of proteins. OspE, OspF, and Elp (OspE/F-like leader peptides) are found encoded on identical loci on both cp32 and cp18 plasmids, and their genes are transcriptionally linked (Akins *et al.*, 1999; Hefty *et al.*, 2001). It is thought that gene rearrangements within the these families has generated sequence diversity allowing for the parasitic and immunoevasive strategies of Lyme disease *Borrelia* (Akins *et al.*, 1999; Caimano *et al.*, 2000). Members of the these protein families are differentially expressed during *B. burgdorferi* migration in tick and transmission to mammalian host (Akins *et al.*, 1995; Hefty *et al.*, 2001; Stevenson *et al.*, 1998) and host-specific signals were found to alter the expression patterns and final cellular localization of these lipoproteins (Hefty *et al.*, 2002). In the mammalian host, multiple if not all alleles can be expressed simultaneously (Akins *et al.*, 1999; Casjens *et al.*, 2000; Casjens *et al.*, 1997; El Hage *et al.*, 1999; Stevenson *et al.*, 1998; Stevenson *et al.*, 1996). Some members are complement regulator acquiring surface proteins (CRASPs) that bind complement inhibitor factor H, enabling *B. burgdorferi* to avoid host-mediated serum killing (Hellwage *et al.*, 2001; Stevenson *et al.*, 2002). Additionally, *B. burgdorferi* BbCRASP-1 is a homodimer that forms a novel α-helical fold (Cordes *et al.*, 2005). CRASPs of *B. afzelii* also bind complement inhibitor factor H and factor H-like protein to provide complement resistance (Kraiczy *et al.*, 2001). The classification of *Borrelia* as resistant, intermediate-sensitive or sensitive to the bactericidal effects of human serum (Kraiczy *et al.*, 2001) might correlate with possession of one or more of the CRASPs and OspE. *Borrelia burgdorferi* OspE and *B. afzelii* CRASP-II might even be homologous proteins (Cullen *et al.*, 2004). More recently, OspF was found to serve as an adhesin *in vivo* (Antonara *et al.*, 2007) (see Chapter 12).

6.2.4 Porins

The molecular sieving properties of bacterial cell envelopes are the result of channel forming proteins in their outer membranes, called porins. Porins are integral membrane proteins that form large water-filled pores through the outer membrane. Their function is mainly for uptake of substances from the environment and can be subdivided into two classes. The porins of the first class can be described as general diffusion pores, such as OmpF of *E. coli* K12. These porins sort mainly according to the molecular mass of the solutes and show a linear relation between translocation rate and solute concentration gradient. The second class contains pores with a binding site inside the channel. These specific porins are responsible for the rapid uptake of classes of solutes such as carbohydrates or nucleosides.

Borrelia are limited in their metabolic and biosynthetic capacities and therefore highly dependent on nutrients provided by their hosts (Casjens *et al.*, 2000; Fraser *et al.*, 1997; Thein *et al.*, 2008) (see Chapter 5). Consequently, these parasites need to have an efficient regulation of the nutrient uptake across the cell envelope. The first indication of porins in *B. burgdorferi* came through investigation of channel-forming activities in the planar lipid bilayer assay of outer membrane vesicles (OMV). Two porin activities of 0.6 nS and 12.6 nS were found (Skare *et al.*, 1995). Subsequent work has so far characterized three possible porins in *Borrelia burgdorferi*: P13 (Östberg *et al.*, 2002; Pinne *et al.*, 2004), P66 (Skare *et al.*, 1997), and Oms38/Dip A (Thein *et al.*, 2008). Another protein, Oms28 exhibit channel-forming activity but has later been shown not to be an outer membrane pore. Both its localization in the membrane and function has been challenged and this protein, now designated BBA74, was shown to be a periplasmic outer membrane associated protein that lacks typical porin properties. In contrast to LD *Borrelia* species, the knowledge of porin activities in RF *Borrelia* is rather limited. There are indications of several pore-forming activities in outer membrane preparations of RF spirochaetes (Shang *et al.*, 1998; Thein *et al.*, 2008) and genes with a high homology to *B. burgdorferi p13, oms28* and *p66* can be found in the published genomes of the RF agents *B. duttonii, B. recurrentis* and *B. hermsii.* There are also indications of uncharacterized channel-forming activities present in *B. burgdorferi* (Östberg *et al.*, 2002).

The P13 family

Borrelia spirochaetes exhibit a very high level of genomic redundancy, resulting in large numbers of plasmids and paralogous gene families. The biological importance of this feature remains unresolved. In the absence of major Osps (OspA-D and DbpA/B) in *B. burgdorferi* strain B313, a new 13kDa surface antigen (P13) was discovered (Sadziene *et al.*, 1995). Mice infected with B313 were used to generate MAbs against P13, which in turn inhibited the growth of the Osp-deficient strain but not the wild-type strain (Sadziene *et al.*, 1995). This implied that surface exposed lipoproteins must hide the P13-epitope, in a similar manner proposed for another integral outer membrane protein, P66 (Bunikis and Barbour, 1999). Surface exposure of P13 was later confirmed by a variety of techniques, while the presence of a N-terminal signal sequence, signal peptidase I cleavage site and three transmembrane spanning α-helices were predicted by computer analysis (Noppa *et al.*, 2001). Furthermore, C-terminal sequencing revealed that P13 is processed at the C-terminus, but N-terminal sequencing was blocked by a possible modification at the N-terminus (Noppa *et al.*, 2001), which by combined mass-spectrometry was confirmed to be pyroglutamylation: pyroglutamate formation at a N-terminal located glutamine (Nilsson *et al.*, 2002). Pyroglutamate modification is either catalysed by the enzyme glutamine cyclotransferase or can occur spontaneously (Awade *et al.*, 1994; Busby *et al.*, 1987; Khandke *et al.*, 1989).

The latter might occur in P13 since no glutamine cyclotransferase homologue was found in the *B. burgdorferi* genome. Although pyroglutamate formation at N-terminal glutamine residues has been suggested to protect some proteins from proteolytic degradation (Strobl *et al.*, 1997), the role of P13 pyroglutamylation remains to be elucidated. While P13 is chromosomally encoded, another 8 additional members of this paralogous gene family 48 are plasmid encoded (Fraser *et al.*, 1997; Noppa *et al.*, 2001).

The surface exposure and predicted transmembrane spanning domains of P13 indicated that it might be a protein with channel-forming properties. Purified native P13 from an outer membrane protein preparation (Magnarelli *et al.*, 1989) of *B. burgdorferi* B313 was used in a planar lipid bilayer assay. The purity of the OM fraction was verified by the absence of contaminating periplasmic FlaB protein (M. Pinne, unpublished). Purified P13 had a channel-forming activity with an average single-channel conductance of 3.5 nS. The ion selectivity measurements revealed that cations are preferentially transported through the P13 channel ($P_{cation}/P_{anion} = 2.1$), although anions can also penetrate the channel (Östberg *et al.*, 2002) (M. Pinne, unpublished). This might indicate that positively charged molecules are actually transported through the P13 pore. Most porins exist in either open or closed states depending on the transmembrane potential, a phenomenon known as voltage gating. Whether this is physiologically significant is questionable since porin voltage dependence measurements in planar lipid bilayer assay suggest that the applied critical voltage (V_c), above which general porins close, far exceeds the naturally occurring Donnan potential across the outer membrane (Koebnik *et al.*, 2000). However, this strategy is still widely used to characterize porins. The channel formed by P13 was not influenced by voltage increases, reflecting a stably incorporated pore in the lipid bilayer. Since substrate-specific channels tend to be voltage independent (Koebnik *et al.*, 2000), perhaps P13 is a specific porin. While numerous substrates were tested, no substrate specificity for the P13 channel has yet been demonstrated.

The channel-forming activity of P13 was also confirmed by genetic experiments. The *p13* gene was inactivated by allelic-exchange mutagenesis, this being the first reported example of disruption of a gene encoding an integral outer membrane protein of *B. burgdorferi*. The porin activities of OM protein preparations derived from knock-out and wild-type strains were compared using the planar lipid bilayer assay. The 3.5 nS activity corresponding to P13 was eliminated in the *p13* knock-out strain. Moreover, these analyses revealed evidence of additional porin activities present in OM of *B. burgdorferi*, distinct from the activities of Oms28, P66, and the P13.

Several interesting characteristics of P13, such as its small size, unusual amount of subunits in an oligomer, hydrophobicity and predicted membrane-spanning α-helices are all in disagreement with common features of conventional porins (Östberg *et al.*, 2002; Pinne *et al.*, 2006a). The unusual C-terminal cleavage of P13 may also play a role in translocation and/or functionality of the protein *in vivo*. To investigate this, an attempt to identify the protease responsible for C-terminal processing of P13 was performed. Using the amino acid sequence of the known carboxyl-terminal protease A (CtpA) from *Bartonella bacilliformis* (Mitchell and Minnick, 1997) to BLAST search the *B. burgdorferi* B31 MI genome (Casjens *et al.*, 2000; Fraser *et al.*, 1997), a homologous gene (*bb0359*) was found (Östberg *et al.*, 2004).

To elucidate if the activity of P13 and/or other *Borrelia* proteins are influenced by CtpA, the *ctpA* gene was inactivated by allelic exchange mutagenesis (Östberg *et al.*, 2004). Immunoblot analysis revealed that P13 was larger and had a more acidic pI in the *ctpA* knock-out, consistent with the theoretical size and pI of P13 if it was not processed at the C-terminus. The 2-D gel electrophoresis of *B. burgdorferi* total proteins showed that nine proteins were only present in the *ctpA* mutant, six proteins were observed only in wild-type (B31-A). Mass spectroscopy analysis of these potential CtpA substrates revealed that P13 and BB0323 are processed by CtpA, while BBA74 is upregulated and produced in multiple isoforms in the absence of CtpA. These effects were due to loss of CtpA because complementation with a wild type copy of *ctpA* restored wild type-like protein expression profiles (Östberg *et al.*, 2004). However, BBA74 is not a substrate for CtpA, so that its upregulation is probably a secondary effect of *ctpA* inactivation (Östberg *et al.*, 2004) (Y. Östberg, unpublished). Interestingly, the expression level of CtpA in the complemented strain is higher than in wild type, reducing the levels of BBA74 to beyond its detection limit (Östberg *et al.*, 2004). Thus, CtpA has pleiotropic effects, processing P13 and influencing the appearance of several other proteins.

Being extremely hydrophobic (hydropathicity index, 0.47) and spanning the membrane with α-helices, one would have assumed that P13 would stack in the inner membrane by the conventional SRP-dependent protein secretion machinery. Nevertheless, the unusual feature of C-terminal cleavage by the C-terminal processing enzyme A (CtpA) led to the hypothesis that this event initiates translocation to the outer membrane (Östberg *et al.*, 2004). Yet, P13 is present in OM protein preparations from the *ctpA* knock-out mutant. Therefore, it might be that the 28-amino-acid C-terminal extension of P13 that is not required for transportation to the outer membrane, but cleavage may be necessary for correct P13 assembly. This mechanism is supported by the function of CtpA in photosynthetic organisms. A substrate for CtpA in photosystem II, the hydrophobic peptide D1 that spans the thylakoid membranes of chloroplasts (Hankamer *et al.*, 2001), must be cleaved at its C-terminus for correct assembly and translocation and to ensure a functional photosystem II (Ivleva *et al.*, 2000; Nixon *et al.*, 1992).

The C-terminus of P13 is processed immediately after the alanine residue removing the extreme C-terminal 28 aa (Noppa *et al.*, 2001). Interestingly, the C-terminal extension removed in D1 from different organisms varies from between 8 to 16 aa, but it is always cleaved after an alanine (Ivleva *et al.*, 2000). Clearly, similarities between P13 and D1 exist and imply that CtpA might play a critical role in the translocation and assembly of functional P13. Moreover, CtpA of *B. burgdorferi* contains a predicted N-terminal signal sequence reminiscent of the need to transport CtpA across the thylakoid membrane in chloroplasts (Karnauchov *et al.*, 1997). By analogy, we interpret that CtpA is transported through the *Borrelia* inner membrane and that C-terminal processing of P13 occurs in the periplasmic space.

P66 (Oms66)

P66 is a well-described pore-forming outer membrane protein present in Lyme disease *Borrelia* (Bunikis *et al.*, 1995; Pinne *et al.*, 2007; Skare *et al.*, 1997). *Borrelia burgdorferi* P66 was shown by Skare and colleagues to form pores in artificial membranes with an atypical huge single channel conductance of 9.6 nS in 1 M KCl (Skare *et al.*, 1997).

P66, present in both Lyme disease and relapsing fever spirochaetes interestingly exhibits dual functions. It was shown to both act as an adhesin which can bind to β_3-integrin (Coburn *et al.*, 1999; Coburn and Cugini, 2003; Defoe and Coburn, 2001) and it also acts as a porin in the outer membrane as its capability of pore-formation was clearly shown by studies in planar lipid membranes (Pinne *et al.*, 2007; Skare *et al.*, 1997) (M. Thein, I. Bàrcena-Uribarri, A. Sacher, I. Bunikis, S. Bergström, and R. Benz, unpublished). Additionally, P66 contains surface-exposed domains (Bunikis *et al.*, 1995; Bunikis *et al.*, 1996) and exhibits immunogenic potential (Barbour *et al.*, 2008). Given the large channel-forming conductance, the absence of predicted β-sheets, a typical porin secondary structure, and a clear ligand for $\alpha_v\beta_3$ integrins, P66 is not a typical porin. In several studies the P66 protein was identified as a candidate ligand for β_3-chain integrins by using a phage display library of total genomic *B. burgdorferi* DNA (Antonara *et al.*, 2007; Coburn *et al.*, 1999). The binding ability was further inhibited using a predicted synthetic peptide corresponding to P66 amino acids 203–209, which lie within the region selected from the phage library, which inhibited *B. burgdorferi* attachment to integrin $\alpha_{IIb}\beta_3$. The final demonstration that P66 is involved in the β_3-chain integrin binding was shown by a targeted mutagenesis of the integrin-binding domain of P66, thus, demonstrating that the obtained mutants are deficient in binding to $\alpha_v\beta_3$ integrin (Coburn and Cugini, 2003). The pore-forming ability of Oms66 is described below. Further structural or physical function of P66 might be the interaction between P66 and surface located lipoproteins. Osp proteins, in particular OspA, have been shown to limit the access of both antibodies and trypsin to the major surface loop of P66. Thus, the possible close contact between P66 and OspA, and other Osp proteins may hinder the accessibility to the P66 protein that would be a possible vaccine target (Bunikis and Barbour, 1999).

FPLC-purified P66 exhibited voltage-dependent closure of the channels, although at higher voltages than described before by Skare *et al.* (1997). The zero-current membrane potential was −5 mV for an five-fold KCl-gradient, which means that the permeability ratio P_K/P_{Cl} was about 0.77 as calculated from the Goldman–Hodgkin–Katz equation (Benz *et al.*, 1979), suggesting that P66 is permeable to anions and cations. Incubation of purified P66 with corresponding polyclonal rabbit antiserum resulted in total loss of the 11 nS channel-forming activity (Pinne *et al.*, 2007) (M. Pinne, unpublished). Monoclonal antibody H914 or non-P66 specific rabbit serum did not affect the channel-forming ability of P66. In addition, P66 was found to be sensitive to SDS since no single channel-conductance steps were observed for P66 purified by preparative SDS-PAGE (M. Pinne, unpublished).

A search in current published relapsing fever genomes, which includes *B. duttonii*, *B. recurrentis* and *B. hermsii*, reveals the presence of P66 homologues within these species (Lescot *et al.*, 2008).

The deduced P66 amino acid sequences of the Lyme disease and relapsing fever borreliae were compared and showed an inter-species amino acid identity of 41% (M. Thein, unpublished). The P66 homologues of the six species were purified to homogeneity by anion exchange chromatography and their pore-forming capabilities and properties were compared using the black lipid bilayer assay. All purified P66 homologue of RF *Borrelia* were highly active in artificial membranes and formed pores with huge single-channel conductances between 9 and 11 nS. Thus, the question is if this huge single-channel conductance is due to a similarly outstanding huge channel diameter. To answer this question, a channel sizing has been performed (M. Thein, unpublished). The P66 single-channel conductance was analysed in black lipid bilayers in the presence of different non-electrolytes with known hydrodynamic radii. The filling of the channel with those non-electrolytes was calculated and revealed that non-electrolytes with hydrodynamic radii smaller than 0.94 nm enter the pore, whereas non-charged molecules with a greater radius were non-permeable. Thus, the effective diameter of the P66 channel lumen was determined to be ~1.9 nm (M. Thein, unpublished). As derived from single- and multichannel experiments, the P66-induced membrane conductance could be blocked by 80–90% after addition of the non-electrolytes PEG 400, PEG 600 and maltohexaose in the low millimolar range (M. Thein, unpublished). The blockage of one P66 single-channel conductance unit of 11 nS occurred by seven subconducting states, thus indicating a heptameric constitution of P66. The heptameric character of P66 could be confirmed by Blue native PAGE and immunoblot analysis which revealed that P66 forms a complex with a corresponding mass of approximately 440 kDa.

Oms38/DipA

Borrelia spirochaetes are – due to their small genome – metabolically and biosynthetically deficient, thereby making them highly dependent on nutrients provided by their various hosts (see Chapter 5). So far, none of the above-described channel-forming proteins displays any specificity, although several nutrient uptake systems are expected. The first characterized porin in RF *Borrelia*, Oms38 (Thein *et al.*, 2008), was shown to be present in the OM fractions of *B. duttonii*, *B. hermsii* and *B. recurrentis*. These pore-forming proteins were purified by hydroxyapatite chromatography to homogeneity and designated Oms38. Biophysical characterizations of the porins were achieved by using the black lipid bilayer method. Oms38 formed water-filled pores of small conductance (80 pS in 1 M KCl) and are selective for anions. Partial sequencing of the protein and searches of genomic DNA sequences from ongoing projects revealed sequences of *oms38* in *B. duttonii*, *B. hermsii*, *B. recurrentis* and *B. turicatae*. A sequence alignment showed that the deduced amino acid sequences of the different Oms38 are highly conserved among the studied RF species (Thein *et al.*, 2008). Further analyses of analogous proteins in LD borreliae revealed a 36-kDa protein (BB0418) with similar specificity as seen in RF *Borrelia* (M. Thein, unpublished).

This 36-kDa protein of Lyme disease *Borrelia* was further characterized and shown to function as a putative dicarboxylate-specific porin (M. Thein, unpublished). The protein was purified from outer membranes of *B. burgdorferi* by hydroxyapatite chromatography and designated DipA, to denote dicarboxylate-specific porin A. DipA was partially sequenced, and based on the resulting amino acid sequence, the corresponding gene, *dipA*, was identified in the published genome sequences of *B. burgdorferi* B31, *B. garinii* PBi and *B. afzelii* PKo (Fraser *et al.*, 1997; Glockner *et al.*, 2006).

Strikingly, the DipA sequence exhibits high homology to the Oms38 porin of relapsing fever *Borrelia. Borrelia burgdorferi* DipA was biophysically characterized using the black lipid bilayer assay (M. Thein, unpublished). The protein has an average single-channel conductance of 50 pS in 1 M KCl. Preliminary results suggest that DipA does not form a general diffusion pore, but a porin with a binding site specific for dicarboxylates which play important key roles in the deficient metabolic and biosynthetic pathways of *Borrelia* species (M. Thein, unpublished).

6.2.5 Envelope-spanning protein complexes

Efflux system and transporters

Transport proteins catalyse the translocation of solutes by various systems that differ with respect to energy coupling, substrate specificity and transmembrane topology. So far, four distinct types of functionally characterized transport systems have been described. These are (i) bacterial channel proteins, (ii) secondary transporters that utilize chemo-osmotic energy, i.e. proton gradient, (iii) primary transporters utilizing chemical energy, typically in the form of ATP, and (iv) group-translocators that phosphorylate their substrates during transport (PTS). From a comparative genome analysis of transporters from 18 different prokaryotes by Paulsen and co-workers (2000), it appeared that, relative to genome size, the spirochaetes *B. burgdorferi* and *T. pallidum* have few transporters, particularly secondary transporters (Saier and Paulsen, 2000). Interestingly, both spirochaetes contained the lowest numbers of transmembrane segments in their predicted integral membrane protein topologies (Paulsen *et al.*, 2000). In another comparative study of 215 protein-translocation outer membrane porins of Gram-negative bacteria, only the TolC homologue (BB0142), which belongs to the outer membrane factor (OMF) family, was identified in *B. burgdorferi* (Yen *et al.*, 2002). Among the bacteria compared, *B. burgdorferi* has the largest percentage of phosphotransferase system (PTS) permeases. Notably, *B. burgdorferi* also encodes a large number of peptide transporters (Paulsen *et al.*, 2000).

An interesting OMF family is the so called channel-tunnels which are engaged in exporting substances across the membranes of Gram-negative cell envelopes. This family incorporates the efflux pumps which confer resistance to various harmful substances like antibiotics, heavy metals, dyes, bile salts and detergents. Efflux pumps consist of an inner membrane transporter, a periplasmic accessory protein and an outer membrane channel-tunnel. The well-studied *E. coli* channel-tunnel, TolC, is involved in protein secretion and can also be part of diverse efflux pumps (Andersen *et al.*, 2000). The RND (resistance-nodulation division) transporters (Li and Nikaido, 2004) exist in all kingdoms of living organisms but seem to be involved in drug resistance especially in Gram-negative bacteria, in which they export toxic substances across both membranes of the cell envelope in a single energy-coupled step (Koronakis *et al.*, 2000). The TolC crystal structure has been solved and consists of channel, equatorial, and tunnel domains (Koronakis *et al.*, 2000). The channel domain is embedded in the outer membrane and consists of non-polar residues directed towards lipids and polar residues that locate to the interior resulting in typical amphipathic β-strands. The β-barrel of channel-tunnel is different from those of typical porins formed by single peptide chain. To form a TolC channel domain, three protomers are necessary in order to generate a single β-barrel consisting of four β-strands. The tunnel domain of TolC consists entirely of α-helices assembled in an α-barrel, representing a unique structure. In the *B. burgdorferi* outer membrane, a channel-forming activity corresponding to a TolC homologue, BesC, has been identified, which is part of a putative export system comprising the components BesA, BesB and BesC (Bunikis *et al.*, 2008). BesC forms channels in planar lipid bilayers. A structural model of the efflux apparatus was described showing the putative spatial orientation of BesC with respect to the AcrAB homologues BesAB. This structural modelling showed both similarities and differences between the TolC of *E. coli* and BesC of *B. burgdorferi* (Bunikis *et al.*, 2008). For TolC of *E. coli*, it is well known that residues lining the periplasmic entrance are critical for the electrophysiological behaviour. The aspartate ring consists of six aspartate residues at the TolC tunnel entrance viewing from the periplasm is believed to be responsible for its high cation selectivity (Andersen *et al.*, 2002). In BesC, the aspartate ring is replaced by positively charged lysine residues (K366) and negatively charged aspartate residues (D363), resulting in a non-selective BesC channel, in contrast to the cation selective *E. coli* TolC channel (Bunikis *et al.*, 2008). The hypothesis is that the lysine residue (K366) and the aspartate residue (D363) in BesC could neutralize one another, meaning that no net charge exists in the tunnel entrance resulting in a non-selective channel in *B. burgdorferi* (Bunikis *et al.*, 2008). Another difference between BesC and TolC is that the BesC channel has a nearly fourfold higher conductance in 1 M KCl than TolC (80 pS in 1 M KCl) (Benz *et al.*, 1993).

Bunikis and coworkers further demonstrated that the BesC protein was necessary for *B. burgdorferi* to establish infection in mice and also involved in antibiotic resistance. This efflux system might be part of a Type I secretion machinery for maintenance of cellular homeostasis or export of exogenous toxic agents, perhaps also necessary for survival in vastly different host environments (Bunikis *et al.*, 2008).

The genome of *B. burgdorferi* encodes a large amount of homologous carbohydrate transporters that belongs to the PTS group (Fraser *et al.*, 1997). Several studies have attempted to characterize some of these transporters. Chitobiose is a component of the tick cuticle and peritrophic matrix. The products of *chbA*, *chbB*, and *chbC* allow *Borrelia* to transport and utilize chitobiose in the tick environment (Tilly *et al.*, 2001; Tilly *et al.*, 2004). Different *chb* genes may respond differently to various temperatures – *chbB* transcription is upregulated at 23°C, whereas *chbC* transcription increases at 35°C (Ojaimi *et al.*, 2003). Glucose is the major energy source for *B. burgdorferi* involving several PTS glucose transporters (von Lackum *et al.*, 2005). BBB29 is one of the components of a glucose transporter system of *B. burgdorferi* (Byram *et al.*, 2004). BBB29 expression is elevated in a *B. burgdorferi* non-disseminating clinical isolate compared to a disseminating isolate (Ojaimi *et al.*, 2005) and at 23°C versus 35°C (Ojaimi *et al.*, 2003). Mannose-modified host proteins are also a likely carbohydrate source for *B. burgdorferi*, with the BB0408 and BB0629 proteins suggested to be mannose transporters (von Lackum *et al.*, 2005).

The members of major intrinsic protein (MIP) group consists of essential channel-forming proteins that maintain an osmotic cell equilibrium and include aquaporins and glycerol facilitators (GlpF) (Froger *et al.*, 2001; Paulsen *et al.*, 2000). The GlpF homologue in *B. burgdorferi* (BB0240) is involved in glycerol uptake, which can either be used as an energy source or for phospholipid and lipoprotein biosynthesis (Schwan *et al.*, 2003). The process of glycerol uptake (*glpF*) and utilization (*glpK; bb0241* and *glpA; bb0243*) might be required when *Borrelia* resides in its arthropod host since these *glp* genes are significantly upregulated in *B. burgdorferi* grown at 23°C versus 35°C (Ojaimi *et al.*, 2003).

The oligopeptide permeases (Opp)

The oligopeptide permeases (Opps) belong to the family of ATP-binding cassette (ABC) transporters used for transportation of peptides into bacteria to be subsequently used as carbon and nitrogen sources (Lazazzera, 2001). Although the major role of bacterial Opp is in nutrient uptake, they may also be involved in sensing environmental signals in the form of peptides leading to diverse responses, including expression of virulence determinants, quorum sensing, chemotaxis, sporulation, conjugation, and competence (Leonard *et al.*, 1996; Perego *et al.*, 1991; Rudner *et al.*, 1991). It is not known if oligopeptide permease has a sensing capability in *B. burgdorferi s.l.* similar to other bacteria although an interesting feature to note is that Opp-like proteins are involved in expression of virulence determinants in other bacteria (Finlay and Falkow, 1989; Gominet *et al.*, 2001; Goodell and Higgins, 1987; Payne and Gilvarg, 1968). The Opp complex in *B. burgdorferi* consists of peptide-binding lipoprotein OppA, transmembrane proteins OppB and OppC, and the ATPases OppD and OppF (Bono *et al.*, 1998; Das *et al.*, 1996; Kornacki and Oliver, 1998).

Interestingly, *B. burgdorferi* has five *oppA* genes each controlled by promoters which respond differently to various environmental signals (Ojaimi *et al.*, 2003; Wang *et al.*, 2002). However, *oppA* genes can also be co-coordinately upregulated in *B. burgdorferi* clinical isolates with different capacities for haematogenous dissemination (Ojaimi *et al.*, 2005). Perhaps each of OppA possesses a different substrate binding affinity that is active during different stages of *B. burgdorferi* life cycle (Wang *et al.*, 2004), which would attribute to the targeting and penetration of specific tissues in invertebrate and vertebrate hosts (Ojaimi *et al.*, 2005).

The Opp system is particularly interesting in *B. burgdorferi s.l.* since it was evident from the genome sequence that only rudimentary machinery for synthesis or transport of amino acids exists. In fact, only a single intact peptide transporter operon could be found. From the genome analysis of *B. burgdorferi s.l.* it is also apparent that the *oppA* operon differs from that of *E. coli* as it has three separate chromosomally encoded substrate binding proteins (Lin *et al.*, 2001; Wang *et al.*, 2002), OppA-1, 2 and 3 and two additional plasmid encoded homologues Oppa-4 and 5 (Wang *et al.*, 2002). Interestingly, all of the *B. burgdorferi s.l.* OppA proteins were found to complement *E. coli* OppBCDF to form a functional peptide transport system for nutritional purposes (Lin *et al.*, 2001). All five *oppA* genes have their own promoters; the apparent differences in regulation of *oppA* gene expression implies that these genes respond to different environmental signals for the adaptation of the *Borrelia* spirochaetes to various hosts (Wang *et al.*, 2002). There was also a considerable difference in the substrate specificities of the five different OppA proteins, which suggests that they may have separate functions in the spirochaete (Wang *et al.*, 2004). These findings led to an in-depth investigation of the factors that actually regulate the different *oppA* promoters. Thus, the alternative sigma factors, RpoS and RpoN, regulate the expression of *oppA5* (Caimano *et al.*, 2007), the BosR/Fur homologue of *Borrelia* interacts with the *oppA4* promoter, and another candidate transcription factor, EbfC, interacts with the *oppA5* promoter (Medrano *et al.*, 2007) (see Chapter 4). Further studies on the regulation of expression of the different Opp proteins will give insight into the molecular mechanisms by which *Borrelia* spirochaetes survive in a specific host environment.

6.3 Cell envelope biogenesis

The biogenesis of the *Borrelia* cell envelope is still very much a mystery. Inference based on known mechanisms and pathways in other diderm bacteria provides some intriguing clues, but few of these clues have been experimentally tested. The acquisition/synthesis of phosphatidylcholine and phosphatidylglycerol, the two phospholipids making up the *Borrelia* membrane lipid bilayer, is discussed in Chapter 5.. The following paragraphs will, therefore, concentrate on lipoproteins and integral membrane proteins. The working model of *Borrelia burgdorferi* envelope biogenesis that we develop below (see also Fig. 6.2) is, therefore, largely speculative in nature. It should serve as a working model to delineate how these two important components of the borrelial envelope are synthesized and transported to the sites where they are biologically active.

6.3.1 Lipoprotein transport

As evidenced by other chapters in this book, lipoproteins represent a class of membrane-anchored proteins that are crucial for the natural maintenance, transmission and virulence of the spirochaetes. In contrast to integral membrane proteins, lipoproteins are only peripherally tethered to the lipid bilayer leaflets of the inner (IM) or outer membranes (OM) via their acylated amino terminal cysteine residues. This not only mirrors their variable functions as flexible components of the periplasmic space or the host–pathogen interface, but has also implications for their thermodynamically different translocation through both membranes and the aqueous periplasmic space in between. Detailed studies on lipoprotein transport in Gram-negative bacteria have revealed a general three-step process: (i) Sec-dependent transport of prolipoproteins through the IM, (ii) lipid modification on the periplasmic face of the IM and cleavage of the signal peptide by signal peptidase II, and (iii) retention or release of acylated proteins and transport to the inner or outer leaflet of the OM. The latter sorting step is currently known to be mediated by either the Lol pathway (Tokuda and Matsuyama, 2004), a type

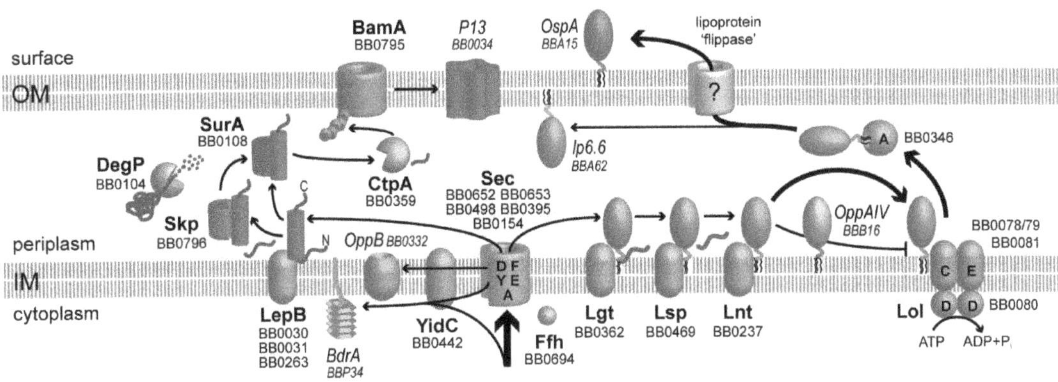

Figure 6.2 Model of *Borrelia burgdorferi* envelope biogenesis pathways. Proposed integral membrane protein (left) and lipoprotein (right) secretion pathways are indicated by arrows. *Borrelia burgdorferi* homologues of known components are indicated by their TIGR ORF number (e.g. BB0154 for SecA). Model substrate proteins (e.g. OspA) are indicated in narrow italic print. See text.

II (T2SS) (Pugsley, 1993) or a type V/autotransporter (T5SS/AT) secretion system (Coutte *et al.*, 2003; van Ulsen *et al.*, 2003).

The enzymes responsible for lipoprotein maturation were first discovered in *E. coli* (Sankaran *et al.*, 1995). The modification of lipoproteins occurs in three steps. First, the signal sequence LXYC specifies the addition of diacyl glycerol in a thioether linkage to the sulphur of the cysteine side chain. Then signal peptidase II cleaves at the amino side of the cysteine residue and a third fatty acid is added in an amide linkage to the new amino terminus (Hayashi and Wu, 1990). Homologues for these three enzymes have been identified in the *B. burgdorferi* genome (Fraser *et al.*, 1997), suggesting that spirochaetal lipoproteins are processed in an identical way.

Lipoproteins entering the *Escherichia coli* Lol pathway (Tokuda and Matsuyama, 2004) first interact with an IM ABC transporter-like complex, LolCDE, which releases OM-targeted lipoproteins from the IM in an ATP-dependent manner (Yakushi *et al.*, 2000). Released lipoproteins then form a water-soluble complex with the periplasmic chaperone LolA (Yokota *et al.*, 1999). After crossing the periplasm, this complex interacts with the OM receptor LolB, which mediates anchoring of lipoproteins to the inner leaflet of the OM (Yokota *et al.*, 1999). The initial '+2 lipoprotein sorting rule' established in *E. coli* indicated that Asp at the penultimate (+2) N-terminal position of the mature lipoprotein leads to IM retention (Yamaguchi *et al.*, 1988) while Ser or other amino acids at the +2 position allow the lipoproteins to be released to the OM. One potential mechanistic explanation is that the interaction of the negatively charged aspartate with the zwitterionic phosphatidylethanolamine blocks an interaction with the LolCDE sortase (Hara *et al.*, 2003; Masuda *et al.*, 2002). Yet, its general applicability remains to be determined as it has become clear that the '+2 rule' is far from universal. In *E. coli*, amino acids at position +3 can also affect localization (Lewenza *et al.*, 2006; Masuda *et al.*, 2002; Seydel *et al.*, 1999; Terada *et al.*, 2001). Furthermore, lipoprotein sorting in other Gram-negative pathogens such as *Pseudomonas aeruginosa* (Lewenza *et al.*, 2008) and *Yersinia pestis* (Silva-Herzog *et al.*, 2008) appears to rely on N-terminal signals at positions +3/+4 and beyond.

While the Lol machinery operates within the periplasm, established pathways of lipoprotein translocation through the OM involve either a T2SS or T5SS/AT mechanism. The prototypical T2SS, the *Klebsiella oxytoca* Pul secretion (Pugsley, 1993), secretes the surface-anchored enzyme PulA. Its secretion signal is located within at least three non-adjacent regions of the structural protein (Francetic and Pugsley, 2005; Sauvonnet and Pugsley, 1996). In the absence of a type II secretion system, PulA's penultimate Asp residue functions as IM retention/Lol avoidance signal (Pugsley *et al.*, 1990). Lipoproteins exported via a T5SS/AT are translocated through the OM via a C-terminal autotransporter domain and include the *Bordetella pertussis* subtilisin SphB1 (Coutte *et al.*, 2003) and the *Neisseria meningitis* serine protease NalP (van Ulsen *et al.*, 2003).

A database homology search for analogous spirochaetal lipoprotein export pathways showed that orthologues of all essential components of the Sec translocase complex and the three enzymes required for lipoprotein modification are present in *Borrelia* (see Fig. 6.2 for orthologues), *Leptospira* and *Treponema* (Fraser *et al.*, 1997; Fraser *et al.*, 1998; Haake, 2000; Nascimento *et al.*, 2004; Seshadri *et al.*, 2004). Yet, further Sec-dependent or -independent bacterial protein secretion pathways are missing in spirochaetes, with the exception of the recently described Type I Bes secretion system in *Borrelia burgdorferi* (Bunikis *et al.*, 2008) or potential twin-arginine translocation (Tat) and type II secretion orthologues in *Leptospira* (Nascimento *et al.*, 2004). We identified orthologues of the IM ABC transporter-like sortase complex LolCDE (Yakushi *et al.*, 2000), the periplasmic lipoprotein-binding chaperone LolA (Yokota *et al.*, 1999), but not the OM lipoprotein receptor LolB (Yokota *et al.*, 1999), in the *B. burgdorferi* genome. Our working model of the *Borrelia* lipoprotein transport pathway therefore consists of LolCDE (BB0078–81) and LolA (BB0346) orthologues, but diverges from that of other diderm bacteria by utilizing a so far unidentified additional OM module to facilitate translocation, or 'flipping', of surface lipoproteins through the OM (Fig. 6.2). Genetically engineered *B. burgdorferi* express and present on their surface biologically active VMPs of RF *Borrelia* (Zückert *et al.*, 2004). This indicates that lipoprotein export machineries in different species of the genus *Borrelia* are functionally conserved.

The proposed transport of *Borrelia* lipoproteins along a Lol-like pathway entails that the sorting signals are most likely at their N-terminus. Indeed, chimeras between the outer surface lipoprotein OspA or the periplasmic oligopeptide-binding lipoprotein OppAIV with mRFP1, a monomeric red fluorescent reporter protein, have supported this hypothesis (Schulze and Zückert, 2006). Localization of OspA and OppAIV point mutants showed that *Borrelia* lipoproteins do not follow the '+2' sorting rule. Fusions of mRFP1 to short N-terminal lipopeptides of OspA, and surprisingly OppAIV, were targeted to the spirochaetal surface. Mutagenesis of the OspA N-terminus defined less than five N-terminal amino acids as the minimal secretion signal. With the exception of negative charges that can act as partial subsurface retention signals in certain peptide contexts, lipoprotein secretion occurred independent of N-terminal sequence. Together, these data indicate that *Borrelia* lipoproteins are exported to the bacterial surface by default, but can be retained in the periplasm by sequence-specific signals. Differential regulation of any pathway component may lead to the accumulation of translocation intermediates and explain the periplasmic localization of Osp proteins observed in some studies (Brusca *et al.*, 1991; Cox *et al.*, 1996; Hefty *et al.*, 2002).

6.3.2 Integral membrane protein transport

While other bacteria commonly ensure correct localization of their membrane proteins by including transmembrane-spanning domains, *Borrelia* seemingly prefer N-terminal lipidation as a membrane anchoring method. Proteins containing transmembrane-spanning domains appear to be rare compared to Gram-negative bacteria. Biosynthetic labelling of *Borrelia* cultures with [^{14}C] amino acids identified only a few amphiphilic polypeptides that did not co-migrate with lipoproteins (Brandt *et al.*, 1990) and freeze-fracture electron microscopy visualized only few outer membrane pore structures (Radolf *et al.*, 1994; Walker *et al.*, 1991).

Integral transmembrane proteins (TMPs) are important components of both the cytoplasmic inner (IM) and outer membrane (OM) lipid bilayers and required for their selective permeability. Studies on the mechanisms of processing, folding and membrane insertion in model systems have identified several IM, periplasmic and OM components involved in the assembly of TMPs in both the IM and OM. Fig. 6.2 shows potential functional *B. burgdorferi* orthologues identified in sequence homology searches. The long-standing textbook dogma on membrane protein structures stated that α-helical transmembrane (TM) domains span the IM and β-barrels span the

OM. Yet, recent data suggest that this might be an oversimplification. The Wza translocon for *E. coli* polysaccharides forms an α-helical barrel in the OM (Dong *et al.*, 2006). P13, a multimeric *B. burgdorferi* OM porin, is predicted to have largely α-helical secondary structure and might form a similar tertiary structure in the OM (Nilsson *et al.*, 2002; Noppa *et al.*, 2001).

TMPs, like LPs, are also predominantly targeted to the Sec translocon, but their signal peptides are cleaved by a separate periplasmic enzyme, signal peptidase I. Intriguingly, the *B. burgdorferi* genome encodes for three potential signal I pepdidases, BB0030, BB0031 and BB0263. It is currently unknown whether all three paralogues are functional. If they are, they might have different specificities and differentially affect protein secretion and virulence, as described in the Gram-positive bacterium *Listeria monocytogenes* (Bonnemain *et al.*, 2004) Subsequent to being threaded through, the TM domains of IM-localizing TMPs exit the Sec channel through a lateral gate into the hydrophobic environment of the lipid bilayer. This process is catalysed by YidC, an IM insertase (Xie and Dalbey, 2008). Interestingly, YidC can also mediate a Sec-independent insertion of IM TMPs, particularly those with only one to two TM domains (Samuelson *et al.*, 2000). The Bdr proteins, a family of bitopic type II membrane proteins with cytoplasmic N-termini and single C-terminal TM domains (Zückert *et al.*, 1999), might therefore insert into the IM in a YidC/BB0442-dependent, but Sec-independent manner.

TMPs destined to the OM somehow avoid IM insertion and enter the periplasmic space where they interact with two chaperones, Skp and SurA. A double *skp surA* mutation is synthetically lethal in *E. coli* (Rizzitello *et al.*, 2001). Yet, it remains unresolved whether Skp and SurA act in two parallel, redundant pathways, or rather sequentially. In the parallel scenario, SurA is the primary chaperone involved in periplasmic transit of OM TMPs, Skp attempts to rescue proteins that fall off the SurA pathway, and DegP degrades proteins abandoned by Skp (Sklar *et al.*, 2007). In the sequential scenario, Skp prevents aggregation of non-native TMPs, and SurA then promotes structural folding (Bos *et al.*, 2007). The biological roles of the respective *B. burgdorferi* paralogues BB0104, BB0108 and BB0796 (Fig. 6.2) remain to be determined.

The insertion of OM TMPs is catalysed by the Bam complex. The OM β-barrel TMP BamA (formerly YaeT or Omp85) (Voulhoux *et al.*, 2003) displays several periplasmic polypeptide-transport-associated (POTRA) domains (Sanchez-Pulido *et al.*, 2003). These N-terminal domains have been shown to recognize C-terminal motifs of OM TMPs and might initiate the structural assembly process of TMPs (Robert *et al.*, 2006). They are also thought to provide a scaffold for the interaction with four associated lipoproteins, BamB (YfgL), BamC (NlpB), BamD (YfiO) and BamE (SmpA) (Kim *et al.*, 2007). The exact roles of these periplasmic OM lipoproteins in OM biogenesis are under investigation. So far, only BamD appears to be essential (Malinverni *et al.*, 2006; Onufryk *et al.*, 2005). Interestingly, a PSI BLAST scouring of the *B. burgdorferi* genome did not reveal any discernible BamB to -E sequence homologues, so Bam complex components will have to be identified experimentally. Yet, like the case of the missing LolB homologue, this again suggests that spirochaetal protein transport have diverged significantly from that in other diderm, e.g. Gram-negative bacteria. It is tempting to speculate that the *Borrelia* BamA homologue BB0795, either directly or via interaction with yet unidentified accessory proteins, has evolved to also catalyse the insertion of outer membrane porins with predicted α-helical TM domains such as the outer membrane porin P13 (Noppa *et al.*, 2001; Sadziene *et al.*, 1995). If this were indeed the case, P13 would have to somehow avoid recognition and sorting to the IM by the Sec/YidC complex, possibly through a 28 amino acid C-terminal domain that is ultimately removed by the periplasmic C-terminal peptidase CtpA (BB0359) (Östberg *et al.* 2004).

6.4 Blebs and cyst formation

B. burgdorferi s.l. is capable of producing extracellular vesicles (blebs) when grown in culture media (Barbour *et al.*, 1986). From earlier studies on the aetiological agent of relapsing fever it is evident that a number of different chemical and physical agents can disturb the bacteria. That is, when the borrelial cells are treated with penicillin, freeze thawing, prolonged cultivation, or the addition of a specific antibody together with a complement source, the spirochaetes form outer membrane blebs which detach from the parent organism (Barbour *et al.*, 1986). In the case of *B. burgdorferi*, the exposure to stress from the mammalian milieu, even in the absence of anti-*B. burgdorferi* antibodies is enough to trigger the production of membrane-bound and possibly secreted blebs (Shoberg and Thomas, 1995). All borrelial blebs appear to contain an outer membrane, while some also contain cytoplasmic membranes (Radolf, 1994). However, all vesicle preparations made to date have no detectable quantities of the periplasmic 41-kDa flagellin investigated by Western blotting (Dorward *et al.*, 1991; Shoberg and Thomas, 1995). These blebs contain plasmid DNA (Garon *et al.*, 1989), including *osp* DNA (Persing *et al.*, 1994), and in their membranes the OspA and OspB proteins and another low molecular weight lipoprotein, Lp6.6, have been found (Dorward *et al.*, 1991; Katona *et al.*, 1992; Lahdenne *et al.*, 1997). The bleb membranes are also thought to contain the borrelial adhesin for human endothelial cells since they are capable of binding to these cells (Shoberg and Thomas, 1993). As an anecdote, in the first half of this century an alternative viewpoint was that these large blebs or 'gemmae' represented another stage of the life-cycle of spirochaetes and this encouraged the early classification of borreliae as protozoans (Barbour *et al.*, 1986).

The recently described phenomenon of cyst formation is another sign of disturbance of cell architecture. Cyst formation involves changes in membrane integrity and cell shape (Alban *et al.*, 2000). It is apparently a starvation response that can be induced by cultivation in serum-free BSKII medium. It is a slow but active process requiring protein synthesis and is reversible in up to 53% of the cells. It is not known whether cyst formation is important for persistence of borreliae in the host and the explanation of survival during antibiotic therapy.

6.5 Future directions

Since the discovery of the first outer membrane components, the borrelial cell envelope has given up many secrets, but many more await unravelling. The completed genome sequences of several species have also aided in the understanding of the common biology of *Borrelia* spirochaetes. In future studies, structure–function relationships of lipoprotein virulence factors need to be determined beyond their antibody epitopes. The probably unusual conformations of *Borrelia* OM porins are to be revealed by recent advances in protein structure determination to further understand the adaptive capacity in various ecological niches of its hosts. The proposed membrane biogenesis pathways have to be tested experimentally. OM blebs or vesicles might represent a means of protein delivery to mammalian cells, as described for other pathogens (Kuehn and Kesty, 2005). If our work with *Borrelia* has taught us one thing, it is to expect the unexpected. So we should not be surprised to discover that this phylogenetic ancient spirochaetal microorganism has developed several solutions of its own and that it does not always conform to the dogmatic structure and function of the cell envelope of Gram-negative bacteria. We therefore anticipate a bright future with many challenges and unsolved mysteries for several generations of *Borrelia* researchers.

Acknowledgements

The authors wish to acknowledge support by the United States National Institutes of Health (NIH/NIAID Grant R01AI63261 to W.R.Z.) and the Swedish Research Council (Grant 07922 to S.B.).

References

Akins, D.R., Caimano, M.J., Yang, X., Cerna, F., Norgard, M.V., and Radolf, J.D. (1999). Molecular and evolutionary analysis of *Borrelia burgdorferi* 297 circular plasmid-encoded lipoproteins with OspE- and OspF-like leader peptides. Infect. Immun. *67*, 1526–1532.

Akins, D.R., Porcella, S.F., Popova, T.G., Shevchenko, D., Baker, S.I., Li, M., Norgard, M.V., and Radolf, J.D. (1995). Evidence for *in vivo* but not *in vitro* expression of a *Borrelia burgdorferi* outer surface protein F (OspF) homologue. Mol. Microbiol. *18*, 507–520.

Alban, P.S., Johnson, P.W., and Nelson, D.R. (2000). Serum-starvation-induced changes in protein synthesis and morphology of *Borrelia burgdorferi*. Microbiology *146 (Pt 1)*, 119–127.

Alitalo, A., Meri, T., Lankinen, H., Seppala, I., Lahdenne, P., Hefty, P.S., Akins, D., and Meri, S. (2002). Complement inhibitor factor H binding to Lyme disease spirochetes is mediated by inducible expression of multiple plasmid-encoded outer surface protein E paralogs. J. Immunol. *169*, 3847–3853.

Andersen, C., Hughes, C., and Koronakis, V. (2000). Chunnel vision. Export and efflux through bacterial channel-tunnels. EMBO Rep. *1*, 313–318.

Andersen, C., Koronakis, E., Hughes, C., and Koronakis, V. (2002). An aspartate ring at the TolC tunnel entrance determines ion selectivity and presents a target for blocking by large cations. Mol. Microbiol. *44*, 1131–1139.

Anderson, J.F., Magnarelli, L.A., LeFebvre, R.B., Andreadis, T.G., McAninch, J.B., Perng, G.C., and Johnson, R.C. (1989). Antigenically variable *Borrelia burgdorferi* isolated from cottontail rabbits and *Ixodes dentatus* in rural and urban areas. J. Clin. Microbiol. *27*, 13–20.

Anguita, J., Thomas, V., Samanta, S., Persinski, R., Hernanz, C., Barthold, S.W., and Fikrig, E. (2001). *Borrelia burgdorferi*-induced inflammation facilitates spirochete adaptation and variable major protein-like sequence locus recombination. J. Immunol. *167*, 3383–3390.

Antonara, S., Chafel, R.M., LaFrance, M., and Coburn, J. (2007). *Borrelia burgdorferi* adhesins identified using *in vivo* phage display. Mol. Microbiol. *66*, 262–276.

Aron, L., Alekshun, M., Perlee, L., Schwartz, I., Godfrey, H.P., and Cabello, F.C. (1994). Cloning and DNA sequence analysis of *bmpC*, a gene encoding a potential membrane lipoprotein of *Borrelia burgdorferi*. FEMS Microbiol. Lett. *123*, 75–82.

Awade, A.C., Cleuziat, P., Gonzales, T., and Robert-Baudouy, J. (1994). Pyrrolidone carboxyl peptidase (Pcp): an enzyme that removes pyroglutamic acid (pGlu) from pGlu-peptides and pGlu-proteins. Proteins *20*, 34–51.

Barbour, A.G. (1990). Antigenic variation of a relapsing fever *Borrelia* species. Annu. Rev. Microbiol. *44*, 155–171.

Barbour, A.G. (1993). Linear DNA of *Borrelia* species and antigenic variation. Trends Microbiol. *1*, 236–239.

Barbour, A.G., and Garon, C.F. (1988). The genes encoding major surface proteins of *Borrelia burgdorferi* are located on a plasmid. Ann N Y Acad Sci *539*, 144–153.

Barbour, A.G., Hayes, S.F., Heiland, R.A., Schrumpf, M.E., and Tessier, S.L. (1986). A *Borrelia*-specific monoclonal antibody binds to a flagellar epitope. Infect. Immun. *52*, 549–554.

Barbour, A.G., Jasinskas, A., Kayala, M.A., Davies, D.H., Steere, A.C., Baldi, P., and Felgner, P.L. (2008). A genome-wide proteome array reveals a limited set of immunogens in natural infections of humans and white-footed mice with *Borrelia burgdorferi*. Infect. Immun. *76*, 3374–3389.

Becker, M., Bunikis, J., Lade, B.D., Dunn, J.J., Barbour, A.G., and Lawson, C.L. (2005). Structural investigation of *Borrelia burgdorferi* OspB, a bactericidal Fab target. J. Biol. Chem. *280*, 17363–17370.

Belisle, J.T., Brandt, M.E., Radolf, J.D., and Norgard, M.V. (1994). Fatty acids of *Treponema pallidum* and *Borrelia burgdorferi* lipoproteins. J. Bacteriol. *176*, 2151–2157.

Ben-Menachem, G., Kubler-Kielb, J., Coxon, B., Yergey, A., and Schneerson, R. (2003). A newly discovered cholesteryl galactoside from *Borrelia burgdorferi*. Proc. Natl. Acad. Sci. U.S.A. *100*, 7913–7918.

Benz, R. (2001). Porins – structure and function. In: Microbial Transport Systems. (Winkelmann, G, ed), WILEY-VCH Verlag GmbH, Weinheim/Germany, pp. 227–246.

Benz, R., Janko, K., and Lauger, P. (1979). Ionic selectivity of pores formed by the matrix protein (porin) of *Escherichia coli*. Biochim. Biophys. Acta *551*, 238–247.

Benz, R., Maier, E., and Gentschev, I. (1993). TolC of *Escherichia coli* functions as an outer membrane channel. Zentralbl Bakteriol *278*, 187–196.

Berg, S. (2001). The glucolipid pathway in *Acholeplasma laidawii*: Functional and structural properties of three consequetively-acting enzymes and similarities to orthologs in prokaryotes. Doctoral thesis, Umeå University, Umeå, Sweden ISBN 91-7305-025-3.

Bergström, S., Bundoc, V.G., and Barbour, A.G. (1989). Molecular analysis of linear plasmid-encoded major surface proteins, OspA and OspB, of the Lyme disease spirochaete *Borrelia burgdorferi*. Mol. Microbiol. *3*, 479–486.

Bessler, W.G., Cox, M., Lex, A., Suhr, B., Wiesmuller, K.H., and Jung, G. (1985). Synthetic lipopeptide analogs of bacterial lipoprotein are potent polyclonal activators for murine B lymphocytes. J. Immunol. *135*, 1900–1905.

Bonnemain, C., Raynaud, C., Reglier-Poupet, H., Dubail, I., Frehel, C., Lety, M.A., Berche, P., and Charbit, A. (2004). Differential roles of multiple signal peptidases in the virulence of *Listeria monocytogenes*. Mol. Microbiol. *51*, 1251–1266.

Bono, J.L., Tilly, K., Stevenson, B., Hogan, D., and Rosa, P. (1998). Oligopeptide permease in *Borrelia burgdorferi*: putative peptide-binding components encoded by both chromosomal and plasmid loci. Microbiology *144 (Pt 4)*, 1033–1044.

Bos, M.P., Robert, V., and Tommassen, J. (2007). Biogenesis of the Gram-negative bacterial outer membrane. Annu. Rev. Microbiol. *61*, 191–214.

Brandt, M.E., Riley, B.S., Radolf, J.D., and Norgard, M.V. (1990). Immunogenic integral membrane proteins of *Borrelia burgdorferi* are lipoproteins. Infect. Immun. *58*, 983–991.

Brooks, C.S., Vuppala, S.R., Jett, A.M., and Akins, D.R. (2006). Identification of *Borrelia burgdorferi* outer surface proteins. Infect. Immun. *74*, 296–304.

Brusca, J.S., McDowall, A.W., Norgard, M.V., and Radolf, J.D. (1991). Localization of outer surface proteins A and B in both the outer membrane and intracellular compartments of *Borrelia burgdorferi*. J. Bacteriol. *173*, 8004–8008.

Bunikis, I., Denker, K., Östberg, Y., Andersen, C., Benz, R., and Bergström, S. (2008). An RND-type efflux system in *Borrelia burgdorferi* is involved in virulence and resistance to antimicrobial compounds. PLoS Pathog. *4*, e1000009.

Bunikis, J., and Barbour, A.G. (1999). Access of antibody or trypsin to an integral outer membrane protein (P66) of *Borrelia burgdorferi* is hindered by Osp lipoproteins. Infect. Immun. *67*, 2874–2883.

Bunikis, J., Noppa, L., and Bergström, S. (1995). Molecular analysis of a 66-kDa protein associated with the outer membrane of Lyme disease *Borrelia*. FEMS Microbiol. Lett. *131*, 139–145.

Bunikis, J., Noppa, L., Östberg, Y., Barbour, A.G., and Bergström, S. (1996). Surface exposure and species specificity of an immunoreactive domain of a 66-kilodalton outer membrane protein (P66) of the *Borrelia* spp. that cause Lyme disease. Infect. Immun. *64*, 5111–5116.

Burman, N., Bergström, S., Restrepo, B.I., and Barbour, A.G. (1990). The variable antigens Vmp7 and Vmp21 of the relapsing fever bacterium *Borrelia hermsii* are structurally analogous to the VSG proteins of the African trypanosome. Mol. Microbiol. *4*, 1715–1726.

Busby, W.H., Jr., Quackenbush, G.E., Humm, J., Youngblood, W.W., and Kizer, J.S. (1987). An enzyme(s) that converts glutaminyl-peptides into pyroglutamyl-peptides: presence in pituitary, brain, adrenal medulla, and lymphocytes. J. Biol. Chem. *262*, 8532–8536.

Byram, R., Stewart, P.E., and Rosa, P. (2004). The essential nature of the ubiquitous 26-kilobase circular replicon of *Borrelia burgdorferi*. J. Bacteriol. *186*, 3561–3569.

Caimano, M.J., Yang, X., Popova, T.G., Clawson, M.L., Akins, D.R., Norgard, M.V., and Radolf, J.D. (2000). Molecular and evolutionary characterization of the cp32/18 family of supercoiled plasmids in *Borrelia burgdorferi* 297. Infect. Immun. *68*, 1574–1586.

Caimano, M.J., Eggers, C.H., Gonzalez, C.A., and Radolf, J.D. (2005). Alternate sigma factor RpoS is required for the in vivo-specific repression of *Borrelia burgdorferi* plasmid lp54-borne *ospA* and *lp6.6* genes. J. Bacteriol. *187*, 7845–7852.

Caimano, M.J., Iyer, R., Eggers, C.H., Gonzalez, C., Morton, E.A., Gilbert, M.A., Schwartz, I., and Radolf, J.D. (2007). Analysis of the RpoS regulon in *Borrelia burgdorferi* in response to mammalian host signals provides insight into RpoS function during the enzootic cycle. Mol. Microbiol. *65*, 1193–1217.

Carter, C.J., Bergström, S., Norris, S.J., and Barbour, A.G. (1994). A family of surface-exposed proteins of 20 kilodaltons in the genus *Borrelia*. Infect. Immun. *62*, 2792–2799.

Casjens, S., Palmer, N., van Vugt, R., Huang, W.M., Stevenson, B., Rosa, P., Lathigra, R., Sutton, G., Peterson, J., Dodson, R.J., *et al.* (2000). A bacterial genome in flux: the twelve linear and nine circular extrachromosomal DNAs in an infectious isolate of the Lyme disease spirochete *Borrelia burgdorferi*. Mol. Microbiol. *35*, 490–516.

Casjens, S., van Vugt, R., Tilly, K., Rosa, P.A., and Stevenson, B. (1997). Homology throughout the multiple 32-kilobase circular plasmids present in Lyme disease spirochetes. J. Bacteriol. *179*, 217–227.

Charon, N.W., Goldstein, S.F., Marko, M., Hsieh, C., Gebhardt, L.L., Motaleb, M.A., Wolgemuth, C.W., Limberger, R.J., and Rowe, N. (2009). The flat-ribbon configuration of the periplasmic flagella of *Borrelia burgdorferi* and its relationship to motility and morphology. J. Bacteriol. *191*, 600–607.

Cinco, M., Banfi, E., Balanzin, D., Godeas, C., and Panfili, E. (1991). Evidence for (lipo) oligosaccharides in *Borrelia burgdorferi* and their serological specificity. FEMS Microbiol. Immunol. *3*, 33–38.

Cluss, R.G., Silverman, D.A., and Stafford, T.R. (2004). Extracellular secretion of the *Borrelia burgdorferi* Oms28 porin and Bgp, a glycosaminoglycan binding protein. Infect. Immun. *72*, 6279–6286.

Coburn, J., Chege, W., Magoun, L., Bodary, S.C., and Leong, J.M. (1999). Characterization of a candidate *Borrelia burgdorferi* β_3-chain integrin ligand identified using a phage display library. Mol. Microbiol. *34*, 926–940.

Coburn, J., and Cugini, C. (2003). Targeted mutation of the outer membrane protein P66 disrupts attachment of the Lyme disease agent, *Borrelia burgdorferi*, to integrin $\alpha_v\beta_3$. Proc. Natl. Acad. Sci. U.S.A. *100*, 7301–7306.

Coburn, J., Fischer, J.R., and Leong, J.M. (2005). Solving a sticky problem: new genetic approaches to host cell adhesion by the Lyme disease spirochete. Mol. Microbiol. *57*, 1182–1195.

Cordes, F.S., Roversi, P., Kraiczy, P., Simon, M.M., Brade, V., Jahraus, O., Wallis, R., Skerka, C., Zipfel, P.F., Wallich, R., *et al.* (2005). A novel fold for the factor H-binding protein BbCRASP-1 of *Borrelia burgdorferi*. Nat. Struct. Mol. Biol. *12*, 276–277.

Coutte, L., Botkin, D.J., Gao, L., and Norris, S.J. (2009). Detailed analysis of sequence changes occurring during *vlsE* antigenic variation in the mouse model of *Borrelia burgdorferi* infection. PLoS Pathog. *5*, e1000293.

Coutte, L., Willery, E., Antoine, R., Drobecq, H., Locht, C., and Jacob-Dubuisson, F. (2003). Surface anchoring of bacterial subtilisin important for maturation function. Mol. Microbiol. *49*, 529–539.

Cox, D.L., Akins, D.R., Bourell, K.W., Lahdenne, P., Norgard, M.V., and Radolf, J.D. (1996). Limited surface exposure of *Borrelia burgdorferi* outer surface lipoproteins. Proc. Natl. Acad. Sci. U.S.A. *93*, 7973–7978.

Cullen, P.A., Haake, D.A., and Adler, B. (2004). Outer membrane proteins of pathogenic spirochetes. FEMS Microbiol. Rev. *28*, 291–318.

Das, S., Shraga, D., Gannon, C., Lam, T.T., Feng, S., Brunet, L.R., Telford, S.R., Barthold, S.W., Flavell, R.A., and Fikrig, E. (1996). Characterization of a 30-kDa *Borrelia burgdorferi* substrate-binding protein homologue. Res. Microbiol *147*, 739–751.

de Rudder, K.E., Lopez-Lara, I.M., and Geiger, O. (2000). Inactivation of the gene for phospholipid N-methyltransferase in *Sinorhizobium meliloti*: phosphatidylcholine is required for normal growth. Mol. Microbiol. *37*, 763–772.

de Silva, A.M., Telford, S.R., 3rd, Brunet, L.R., Barthold, S.W., and Fikrig, E. (1996). *Borrelia burgdorferi* OspA is an arthropod-specific transmission-blocking Lyme disease vaccine. J. Exp. Med. *183*, 271–275.

de Silva, A.M., Zeidner, N.S., Zhang, Y., Dolan, M.C., Piesman, J., and Fikrig, E. (1999). Influence of outer surface protein A antibody on *Borrelia burgdorferi* within feeding ticks. Infect. Immun. *67*, 30–35.

Defoe, G., and Coburn, J. (2001). Delineation of *Borrelia burgdorferi* p66 sequences required for integrin $\alpha_{IIb}\beta_3$ recognition. Infect. Immun. *69*, 3455–3459.

Deres, K., Schild, H., Wiesmuller, K.H., Jung, G., and Rammensee, H.G. (1989). *In vivo* priming of virus-specific cytotoxic T-lymphocytes with synthetic lipopeptide vaccine. Nature *342*, 561–564.

Dobrikova, E.Y., Bugrysheva, J., and Cabello, F.C. (2001). Two independent transcriptional units control the complex and simultaneous expression of the *bmp* paralogous chromosomal gene family in *Borrelia burgdorferi*. Mol. Microbiol. *39*, 370–378.

Dong, C., Beis, K., Nesper, J., Brunkan-Lamontagne, A.L., Clarke, B.R., Whitfield, C., and Naismith, J.H. (2006). Wza the translocon for *E. coli* capsular polysaccharides defines a new class of membrane protein. Nature *444*, 226–229.

Dorward, D.W., Schwan, T.G., and Garon, C.F. (1991). Immune capture and detection of *Borrelia burgdorferi* antigens in urine, blood, or tissues from infected ticks, mice, dogs, and humans. J. Clin. Microbiol. *29*, 1162–1170.

Eicken, C., Sharma, V., Klabunde, T., Lawrenz, M.B., Hardham, J.M., Norris, S.J., and Sacchettini, J.C. (2002). Crystal structure of Lyme disease variable surface antigen VlsE of *Borrelia burgdorferi*. J. Biol. Chem. *277*, 21691–21696.

Eicken, C., Sharma, V., Klabunde, T., Owens, R.T., Pikas, D.S., Hook, M., and Sacchettini, J.C. (2001). Crystal structure of Lyme disease antigen outer surface protein C from *Borrelia burgdorferi*. J. Biol. Chem. *276*, 10010–10015.

Eiffert, H., Lotter, H., Jarecki-Khan, K., and Thomssen, R. (1991). Identification of an immunoreactive non-proteinaleous component in *Borrelia burgdorferi*. Med. Microbiol. Immunol. *180*, 229–237.

El Hage, N., Lieto, L.D., and Stevenson, B. (1999). Stability of *erp* loci during *Borrelia burgdorferi* infection: recombination is not required for chronic infection of immunocompetent mice. Infect. Immun. *67*, 3146–3150.

Erdile, L.F., Brandt, M.A., Warakomski, D.J., Westrack, G.J., Sadziene, A., Barbour, A.G., and Mays, J.P. (1993). Role of attached lipid in immunogenicity of *Borrelia burgdorferi* OspA. Infect. Immun. *61*, 81–90.

Fikrig, E., Feng, W., Barthold, S.W., Telford, S.R., 3rd, and Flavell, R.A. (2000). Arthropod- and host-specific *Borrelia burgdorferi bbk32* expression and the inhibition of spirochete transmission. J. Immunol. *164*, 5344–5351.

Finlay, B.B., and Falkow, S. (1989). Common themes in microbial pathogenicity. Microbiol Rev 53, 210–230.

Fischer, J.R., LeBlanc, K.T., and Leong, J.M. (2006). Fibronectin binding protein BBK32 of the Lyme disease spirochete promotes bacterial attachment to glycosaminoglycans. Infect. Immun. 74, 435–441.

Francetic, O., and Pugsley, A.P. (2005). Towards the identification of type II secretion signals in a nonacylated variant of pullulanase from *Klebsiella oxytoca*. J. Bacteriol. *187*, 7045–7055.

Fraser, C.M., Casjens, S., Huang, W.M., Sutton, G.G., Clayton, R., Lathigra, R., White, O., Ketchum, K.A., Dodson, R., Hickey, E.K., *et al.* (1997). Genomic sequence of a Lyme disease spirochaete, *Borrelia burgdorferi*. Nature 390, 580–586.

Fraser, C.M., Norris, S.J., Weinstock, G.M., White, O., Sutton, G.G., Dodson, R., Gwinn, M., Hickey, E.K., Clayton, R., Ketchum, K.A., *et al.* (1998). Complete genome sequence of *Treponema pallidum*, the syphilis spirochete. Science *281*, 375–388.

Froger, A., Rolland, J.P., Bron, P., Lagree, V., Le Caherec, F., Deschamps, S., Hubert, J.F., Pellerin, I., Thomas, D., and Delamarche, C. (2001). Functional characterization of a microbial aquaglyceroporin. Microbiology *147*, 1129–1135.

Garon, C.F., Dorward, D.W., and Corwin, M.D. (1989). Structural features of *Borrelia burgdorferi* – the Lyme disease spirochete: silver staining for nucleic acids. Scanning Microsc Suppl 3, 109–115.

Gilmore, R.D., Jr., Kappel, K.J., Dolan, M.C., Burkot, T.R., and Johnson, B.J. (1996). Outer surface protein C (OspC), but not P39, is a protective immunogen against a tick-transmitted *Borrelia burgdorferi* challenge: evidence for a conformational protective epitope in OspC. Infect. Immun. *64*, 2234–2239.

Gilmore, R.D., Jr., and Mbow, M.L. (1998). A monoclonal antibody generated by antigen inoculation via tick bite is reactive to the *Borrelia burgdorferi* Rev protein, a member of the 2.9 gene family locus. Infect. Immun. *66*, 980–986.

Gilmore, R.D., Jr., and Mbow, M.L. (1999). Conformational nature of the *Borrelia burgdorferi* B31 outer surface protein C protective epitope. Infect. Immun. *67*, 5463–5469.

Gilmore, R.D., Jr., Mbow, M.L., and Stevenson, B. (2001). Analysis of *Borrelia burgdorferi* gene expression during life cycle phases of the tick vector *Ixodes scapularis*. Microbes Infect. 3, 799–808.

Gilmore, R.D., Jr., and Piesman, J. (2000). Inhibition of *Borrelia burgdorferi* migration from the midgut to the salivary glands following feeding by ticks on OspC-immunized mice. Infect. Immun. *68*, 411–414.

Glockner, G., Schulte-Spechtel, U., Schilhabel, M., Felder, M., Suhnel, J., Wilske, B., and Platzer, M. (2006). Comparative genome analysis: selection pressure on the *Borrelia vls* cassettes is essential for infectivity. BMC Genomics 7, 211.

Goldstein, S.F., Charon, N.W., and Kreiling, J.A. (1994). *Borrelia burgdorferi* swims with a planar waveform similar to that of eukaryotic flagella. Proc. Natl. Acad. Sci. U.S.A. *91*, 3433–3437.

Gominet, M., Slamti, L., Gilois, N., Rose, M., and Lereclus, D. (2001). Oligopeptide permease is required for expression of the *Bacillus thuringiensis plcR* regulon and for virulence. Mol. Microbiol. *40*, 963–975.

Goodell, E.W., and Higgins, C.F. (1987). Uptake of cell wall peptides by *Salmonella typhimurium* and *Escherichia coli*. J. Bacteriol. *169*, 3861–3865.

Gorbacheva, V.Y., Godfrey, H.P., and Cabello, F.C. (2000). Analysis of the *bmp* gene family in *Borrelia burgdorferi* sensu lato. J. Bacteriol. *182*, 2037–2042.

Grimm, D., Tilly, K., Byram, R., Stewart, P.E., Krum, J.G., Bueschel, D.M., Schwan, T.G., Policastro, P.F., Elias, A.F., and Rosa, P.A. (2004). Outer-surface protein C of the Lyme disease spirochete: a protein induced in ticks for infection of mammals. Proc. Natl. Acad. Sci. U.S.A. *101*, 3142–3147.

Guo, B.P., Brown, E.L., Dorward, D.W., Rosenberg, L.C., and Höök, M. (1998). Decorin-binding adhesins from *Borrelia burgdorferi*. Mol. Microbiol. *30*, 711–723.

Guo, B.P., Norris, S.J., Rosenberg, L.C., and Höök, M. (1995). Adherence of *Borrelia burgdorferi* to the proteoglycan decorin. Infect. Immun. *63*, 3467–3472.

Haake, D.A. (2000). Spirochaetal lipoproteins and pathogenesis. Microbiology *146*, 1491–1504.

Haas, R., and Meyer, T.F. (1986). The repertoire of silent pilus genes in *Neisseria gonorrhoeae*: evidence for gene conversion. Cell *44*, 107–115.

Hagblom, P., Segal, E., Billyard, E., and So, M. (1985). Intragenic recombination leads to pilus antigenic variation in *Neisseria gonorrhoeae*. Nature *315*, 156–158.

Hagman, K.E., Lahdenne, P., Popova, T.G., Porcella, S.F., Akins, D.R., Radolf, J.D., and Norgard, M.V. (1998). Decorin-binding protein of *Borrelia burgdorferi* is encoded within a two-gene operon and is protective in the murine model of Lyme borreliosis. Infect. Immun. *66*, 2674–2683.

Hankamer, B., Morris, E., Nield, J., Carne, A., and Barber, J. (2001). Subunit positioning and transmembrane helix organisation in the core dimer of photosystem II. FEBS Lett. *504*, 142–151.

Hanson, M.S., Cassatt, D.R., Guo, B.P., Patel, N.K., McCarthy, M.P., Dorward, D.W., and Hook, M. (1998). Active and passive immunity against *Borrelia burgdorferi* decorin binding protein A (DbpA) protects against infection. Infect. Immun. *66*, 2143–2153.

Hara, T., Matsuyama, S., and Tokuda, H. (2003). Mechanism underlying the inner membrane retention of *Escherichia coli* lipoproteins caused by Lol avoidance signals. J. Biol. Chem. *278*, 40408–40414.

Hayashi, S., and Wu, H.C. (1990). Lipoproteins in bacteria. J. Bioenerg. Biomembr. *22*, 451–471.

Hefty, P.S., Jolliff, S.E., Caimano, M.J., Wikel, S.K., and Akins, D.R. (2002). Changes in temporal and spatial patterns of outer surface lipoprotein expression generate population heterogeneity and antigenic diversity in the Lyme disease spirochete, *Borrelia burgdorferi*. Infect. Immun. *70*, 3468–3478.

Hefty, P.S., Jolliff, S.E., Caimano, M.J., Wikel, S.K., Radolf, J.D., and Akins, D.R. (2001). Regulation of OspE-related, OspF-related, and Elp lipoproteins of *Borrelia burgdorferi* strain 297 by mammalian host-specific signals. Infect. Immun. *69*, 3618–3627.

Hellwage, J., Meri, T., Heikkila, T., Alitalo, A., Panelius, J., Lahdenne, P., Seppala, I.J., and Meri, S. (2001). The complement regulator factor H binds to the surface protein OspE of *Borrelia burgdorferi*. J. Biol. Chem. *276*, 8427–8435.

Hoffmann, P., Heinle, S., Schade, U.F., Loppnow, H., Ulmer, A.J., Flad, H.D., Jung, G., and Bessler, W.G. (1988). Stimulation of human and murine adherent cells by bacterial lipoprotein and synthetic lipopeptide analogues. Immunobiology *177*, 158–170.

Honarvar, N., Schaible, U.E., Galanos, C., Wallich, R., and Simon, M.M. (1994). A *14*,000 MW lipoprotein and a glycolipid-like structure of *Borrelia burgdorferi* induce proliferation and immunoglobulin production in mouse B-cells at high frequencies. Immunology *82*, 389–396.

Hossain, H., Wellensiek, H.J., Geyer, R., and Lochnit, G. (2001). Structural analysis of glycolipids from *Borrelia burgdorferi*. Biochimie *83*, 683–692.

Howe, T.R., LaQuier, F.W., and Barbour, A.G. (1986). Organization of genes encoding two outer membrane proteins of the Lyme disease agent *Borrelia burgdorferi* within a single transcriptional unit. Infect. Immun. *54*, 207–212.

Indest, K.J., and Philipp, M.T. (2000). DNA-Binding proteins possibly involved in regulation of the post-logarithmic-phase expression of lipoprotein P35 in *Borrelia burgdorferi*. J. Bacteriol. *182*, 522–525.

Indest, K.J., Ramamoorthy, R., Sole, M., Gilmore, R.D., Johnson, B.J., and Philipp, M.T. (1997). Cell-density-dependent expression of *Borrelia burgdorferi* lipoproteins in vitro. Infect. Immun. *65*, 1165–1171.

Ivleva, N.B., Shestakov, S.V., and Pakrasi, H.B. (2000). The carboxyl-terminal extension of the precursor D1 protein of photosystem II is required for optimal photosynthetic performance of the cyanobacterium *Synechocystis* sp. PCC 6803. Plant Physiol *124*, 1403–1412.

Jauris-Heipke, S., Liegl, G., Preac-Mursic, V., Rossler, D., Schwab, E., Soutschek, E., Will, G., and Wilske, B. (1995). Molecular analysis of genes encoding outer surface protein C (OspC) of *Borrelia burgdorferi* sensu lato: relationship to *ospA* genotype and evidence of lateral gene exchange of *ospC*. J. Clin. Microbiol. *33*, 1860–1866.

Johnson, R.C., Schmid, G.P., Hyde, F.W., Steigerwalt, A.G., and Brenner, D.J. (1984). *Borrelia burgdorferi* sp. nov.: Etiologic agent of Lyme disease. Int. J. Syst. Bacteriol. *34*, 496–497.

Jones, J.D., Bourell, K.W., Norgard, M.V., and Radolf, J.D. (1995). Membrane topology of *Borrelia burgdorferi* and *Treponema pallidum* lipoproteins. Infect. Immun. *63*, 2424–2434.

Kamio, Y., and Nikaido, H. (1976). Outer membrane of *Salmonella typhimurium*: accessibility of phospholipid head groups to phospholipase C and cyanogen bromide activated dextran in the external medium. Biochemistry *15*, 2561–2570.

Karnauchov, I., Herrmann, R.G., Pakrasi, H.B., and Klosgen, R.B. (1997). Transport of CtpA protein from the cyanobacterium *Synechocystis* 6803 across the thylakoid membrane in chloroplasts. Eur. J. Biochem. *249*, 497–504.

Katona, L.I., Beck, G., and Habicht, G.S. (1992). Purification and immunological characterization of a major low-molecular-weight lipoprotein from *Borrelia burgdorferi*. Infect. Immun. *60*, 4995–5003.

Khandke, K.M., Fairwell, T., Chait, B.T., and Manjula, B.N. (1989). Influence of ions on cyclization of the amino terminal glutamine residues of tryptic peptides of streptococcal PepM49 protein: resolution of cyclized peptides by HPLC and characterization by mass spectrometry. Int J Pept Protein Res. *34*, 118–123.

Kim, S., Malinverni, J.C., Sliz, P., Silhavy, T.J., Harrison, S.C., and Kahne, D. (2007). Structure and function of an essential component of the outer membrane protein assembly machine. Science *317*, 961–964.

Kinjo, Y., Tupin, E., Wu, D., Fujio, M., Garcia-Navarro, R., Benhnia, M.R., Zajonc, D.M., Ben-Menachem, G., Ainge, G.D., Painter, G.F., *et al.* (2006). Natural killer T-cells recognize diacylglycerol antigens from pathogenic bacteria. Nat. Immunol. *7*, 978–986.

Koebnik, R., Locher, K.P., and Van Gelder, P. (2000). Structure and function of bacterial outer membrane proteins: barrels in a nutshell. Mol. Microbiol. *37*, 239–253.

Koomey, M. (1997). Bacterial pathogenesis: a variation on variation in Lyme disease. Curr Biol 7, R538–540.

Kornacki, J.A., and Oliver, D.B. (1998). Lyme disease-causing *Borrelia* species encode multiple lipoproteins homologous to peptide-binding proteins of ABC-type transporters. Infect. Immun. *66*, 4115–4122.

Koronakis, V., Sharff, A., Koronakis, E., Luisi, B., and Hughes, C. (2000). Crystal structure of the bacterial membrane protein TolC central to multidrug efflux and protein export. Nature *405*, 914–919.

Kraiczy, P., Skerka, C., Brade, V., and Zipfel, P.F. (2001a). Further characterization of complement regulator-acquiring surface proteins of *Borrelia burgdorferi*. Infect. Immun. *69*, 7800–7809.

Kraiczy, P., Skerka, C., Kirschfink, M., Zipfel, P.F., and Brade, V. (2001b). Mechanism of complement resistance of pathogenic *Borrelia burgdorferi* isolates. Int. Immunopharmacol. *1*, 393–401.

Kuederling, I., Morrell, J.M., and Nayudu, P.L. (1996). Collection of semen from marmoset monkeys (Callithrix jacchus) for experimental use by vaginal washing. Lab Anim 30, 260–266.

Kuehn, M.J., and Kesty, N.C. (2005). Bacterial outer membrane vesicles and the host–pathogen interaction. Genes Dev. *19*, 2645–2655.

Kumaran, D., Eswaramoorthy, S., Luft, B.J., Koide, S., Dunn, J.J., Lawson, C.L., and Swaminathan, S. (2001). Crystal structure of outer surface protein C (OspC) from the Lyme disease spirochete, *Borrelia burgdorferi*. Embo J. *20*, 971–978.

Lagal, V., Portnoi, D., Faure, G., Postic, D., and Baranton, G. (2006). *Borrelia burgdorferi* sensu stricto invasiveness is correlated with OspC-plasminogen affinity. Microbes Infect. *8*, 645–652.

Lagal, V., Postic, D., Ruzic-Sabljic, E., and Baranton, G. (2003). Genetic diversity among *Borrelia* strains determined by single-strand conformation polymorphism analysis of the *ospC* gene and its association with invasiveness. J. Clin. Microbiol. *41*, 5059–5065.

Lahdenne, P., Porcella, S.F., Hagman, K.E., Akins, D.R., Popova, T.G., Cox, D.L., Katona, L.I., Radolf, J.D., and Norgard, M.V. (1997). Molecular characterization of a 6.6-kilodalton *Borrelia burgdorferi* outer membrane-associated lipoprotein (lp6.6) which appears to be down-regulated during mammalian infection. Infect. Immun. *65*, 412–421.

Lathrop, S.L., Ball, R., Haber, P., Mootrey, G.T., Braun, M.M., Shadomy, S.V., Ellenberg, S.S., Chen, R.T., and Hayes, E.B. (2002). Adverse event reports following vaccination for Lyme disease: December 1998-July 2000. Vaccine *20*, 1603–1608.

Lawson, C.L., Yung, B.H., Barbour, A.G., and Zückert, W.R. (2006). Crystal structure of neurotropism-associated variable surface protein 1 (Vsp1) of *Borrelia turicatae*. J. Bacteriol. *188*, 4522–4530.

Lazazzera, B.A. (2001). The intracellular function of extracellular signaling peptides. Peptides *22*, 1519–1527.

Leonard, B.A., Podbielski, A., Hedberg, P.J., and Dunny, G.M. (1996). *Enterococcus faecalis* pheromone binding protein, PrgZ, recruits a chromosomal oligopeptide permease system to import sex pheromone cCF10 for induction of conjugation. Proc. Natl. Acad. Sci. U.S.A. *93*, 260–264.

Lescot, M., Audic, S., Robert, C., Nguyen, T.T., Blanc, G., Cutler, S.J., Wincker, P., Couloux, A., Claverie, J.M., Raoult, D., and Drancourt, M. (2008). The genome of *Borrelia recurrentis*, the agent of deadly louse-borne relapsing fever, is a degraded subset of tick-borne *Borrelia duttonii*. PLoS Genet *4*, e1000185.

Lewenza, S., Mhlanga, M.M., and Pugsley, A.P. (2008). Novel inner membrane retention signals in *Pseudomonas aeruginosa* lipoproteins. J. Bacteriol. *190*, 6119–6125.

Lewenza, S., Vidal-Ingigliardi, D., and Pugsley, A.P. (2006). Direct visualization of red fluorescent lipoproteins indicates conservation of the membrane sorting rules in the family Enterobacteriaceae. J. Bacteriol. *188*, 3516–3524.

Lex, A., Wiesmuller, K.H., Jung, G., and Bessler, W.G. (1986). A synthetic analogue of *Escherichia coli* lipoprotein, tripalmitoyl pentapeptide, constitutes a potent immune adjuvant. J. Immunol. *137*, 2676–2681.

Li, H., Dunn, J.J., Luft, B.J., and Lawson, C.L. (1997). Crystal structure of Lyme disease antigen outer surface protein A complexed with an Fab. Proc. Natl. Acad. Sci. U.S.A. *94*, 3584–3589.

Li, H., and Lawson, C.L. (1995). Crystallization and preliminary X-ray analysis of *Borrelia burgdorferi* outer surface protein A (OspA) complexed with a murine monoclonal antibody Fab fragment. J. Struct. Biol. *115*, 335–337.

Li, X.Z., and Nikaido, H. (2004). Efflux-mediated drug resistance in bacteria. Drugs *64*, 159–204.

Lin, B., Short, S.A., Eskildsen, M., Klempner, M.S., and Hu, L.T. (2001). Functional testing of putative oligopeptide permease (Opp) proteins of *Borrelia burgdorferi*: a complementation model in *opp*$^-$ *Escherichia coli*. Biochim. Biophys. Acta *1499*, 222–231.

Livermore, B.P., Bey, R.F., and Johnson, .R.C. (1978). Lipid metabolism of *Borrelia hermsii*. Infect. Immun. *20*, 215–220.

Livey, I., Gibbs, C.P., Schuster, R., and Dorner, F. (1995). Evidence for lateral transfer and recombination in OspC variation in Lyme disease *Borrelia*. Mol. Microbiol. *18*, 257–269.

Lopez-Lara, I.M., and Geiger, O. (2001). Novel pathway for phosphatidylcholine biosynthesis in bacteria associated with eukaryotes. J. Biotechnol. *91*, 211–221.

Magnarelli, L.A., Anderson, J.F., and Barbour, A.G. (1989). Enzyme-linked immunosorbent assays for Lyme disease: reactivity of subunits of *Borrelia burgdorferi*. J. Infect. Dis. *159*, 43–49.

Malinverni, J.C., Werner, J., Kim, S., Sklar, J.G., Kahne, D., Misra, R., and Silhavy, T.J. (2006). YfiO stabilizes the

YaeT complex and is essential for outer membrane protein assembly in *Escherichia coli*. Mol. Microbiol. *61*, 151–164.

Masuda, K., Matsuyama, S., and Tokuda, H. (2002). Elucidation of the function of lipoprotein-sorting signals that determine membrane localization. Proc. Natl. Acad. Sci. U.S.A. *99*, 7390–7395.

Medrano, M.S., Ding, Y., Wang, X.G., Lu, P., Coburn, J., and Hu, L.T. (2007). Regulators of expression of the oligopeptide permease A proteins of *Borrelia burgdorferi*. J. Bacteriol. *189*, 2653–2659.

Medzhitov, R., and Janeway, C., Jr. (2000). Innate immune recognition: mechanisms and pathways. Immunol Rev *173*, 89–97.

Mitchell, S.J., and Minnick, M.F. (1997). A carboxy-terminal processing protease gene is located immediately upstream of the invasion-associated locus from *Bartonella bacilliformis*. Microbiology *143 (Pt 4)*, 1221–1233.

Motaleb, M.A., Corum, L., Bono, J.L., Elias, A.F., Rosa, P., Samuels, D.S., and Charon, N.W. (2000). *Borrelia burgdorferi* periplasmic flagella have both skeletal and motility functions. Proc. Natl. Acad. Sci. U.S.A. *97*, 10899–10904.

Nascimento, A.L., Ko, A.I., Martins, E.A., Monteiro-Vitorello, C.B., Ho, P.L., Haake, D.A., Verjovski-Almeida, S., Hartskeerl, R.A., Marques, M.V., Oliveira, M.C., *et al.* (2004). Comparative genomics of two *Leptospira interrogans* serovars reveals novel insights into physiology and pathogenesis. J. Bacteriol. *186*, 2164–2172.

Nikaido, H. (2003). Molecular basis of bacterial outer membrane permeability revisited. Microbiol. Mol. Biol. Rev. *67*, 593–656.

Nilsson, C.L., Cooper, H.J., Hakansson, K., Marshall, A.G., Östberg, Y., Lavrinovicha, M., and Bergström, S. (2002). Characterization of the P13 membrane protein of *Borrelia burgdorferi* by mass spectrometry. J. Am. Soc. Mass. Spectrom. *13*, 295–299.

Nixon, P.J., Trost, J.T., and Diner, B.A. (1992). Role of the carboxy terminus of polypeptide D1 in the assembly of a functional water-oxidizing manganese cluster in photosystem II of the cyanobacterium *Synechocystis* sp. PCC 6803: assembly requires a free carboxyl group at C-terminal position 344. Biochemistry *31*, 10859–10871.

Noppa, L., Östberg, Y., Lavrinovicha, M., and Bergström, S. (2001). P*13*, an integral membrane protein of *Borrelia burgdorferi*, is C-terminally processed and contains surface-exposed domains. Infect. Immun. *69*, 3323–3334.

Norman, M.U., Moriarty, T.J., Dresser, A.R., Millen, B., Kubes, P., and Chaconas, G. (2008). Molecular mechanisms involved in vascular interactions of the Lyme disease pathogen in a living host. PLoS Pathog. *4*, e1000169.

Ohnishi, J., Piesman, J., and de Silva, A.M. (2001). Antigenic and genetic heterogeneity of *Borrelia burgdorferi* populations transmitted by ticks. Proc. Natl. Acad. Sci. U.S.A. *98*, 670–675.

Ojaimi, C., Brooks, C., Casjens, S., Rosa, P., Elias, A., Barbour, A., Jasinskas, A., Benach, J., Katona, L., Radolf, J., *et al.* (2003). Profiling of temperature-induced changes in *Borrelia burgdorferi* gene expression by using whole genome arrays. Infect. Immun. *71*, 1689–1705.

Ojaimi, C., Mulay, V., Liveris, D., Iyer, R., and Schwartz, I. (2005). Comparative transcriptional profiling of *Borrelia burgdorferi* clinical isolates differing in capacities for hematogenous dissemination. Infect. Immun. *73*, 6791–6802.

Onufryk, C., Crouch, M.L., Fang, F.C., and Gross, C.A. (2005). Characterization of six lipoproteins in the σ^E regulon. J. Bacteriol. *187*, 4552–4561.

Östberg, Y., Berg, S., Comstedt, P., Wieslander, A., and Bergström, S. (2007). Functional analysis of a lipid galactosyltransferase synthesizing the major envelope lipid in the Lyme disease spirochete *Borrelia burgdorferi*. FEMS Microbiol. Lett. *272*, 22–29.

Östberg, Y., Carroll, J.A., Pinne, M., Krum, J.G., Rosa, P., and Bergström, S. (2004). Pleiotropic effects of inactivating a carboxyl-terminal protease, CtpA, in *Borrelia burgdorferi*. J. Bacteriol. *186*, 2074–2084.

Östberg, Y., Pinne, M., Benz, R., Rosa, P., and Bergström, S. (2002). Elimination of channel-forming activity by insertional inactivation of the *p13* gene in *Borrelia burgdorferi*. J. Bacteriol. *184*, 6811–6819.

Pal, U., de Silva, A.M., Montgomery, R.R., Fish, D., Anguita, J., Anderson, J.F., Lobet, Y., and Fikrig, E. (2000). Attachment of *Borrelia burgdorferi* within *Ixodes scapularis* mediated by outer surface protein A. J. Clin. Invest. *106*, 561–569.

Pal, U., Li, X., Wang, T., Montgomery, R.R., Ramamoorthi, N., Desilva, A.M., Bao, F., Yang, X., Pypaert, M., Pradhan, D., *et al.* (2004a). TROSPA, an *Ixodes scapularis* receptor for *Borrelia burgdorferi*. Cell *119*, 457–468.

Pal, U., Yang, X., Chen, M., Bockenstedt, L.K., Anderson, J.F., Flavell, R.A., Norgard, M.V., and Fikrig, E. (2004b). OspC facilitates *Borrelia burgdorferi* invasion of *Ixodes scapularis* salivary glands. J. Clin. Invest. *113*, 220–230.

Parveen, N., Caimano, M., Radolf, J.D., and Leong, J.M. (2003). Adaptation of the Lyme disease spirochaete to the mammalian host environment results in enhanced glycosaminoglycan and host cell binding. Mol. Microbiol. *47*, 1433–1444.

Parveen, N., and Leong, J.M. (2000). Identification of a candidate glycosaminoglycan-binding adhesin of the Lyme disease spirochete *Borrelia burgdorferi*. Mol. Microbiol. *35*, 1220–1234.

Paulsen, I.T., Nguyen, L., Sliwinski, M.K., Rabus, R., and Saier, M.H., Jr. (2000). Microbial genome analyses: comparative transport capabilities in eighteen prokaryotes. J. Mol. Biol. *301*, 75–100.

Payne, J.W., and Gilvarg, C. (1968). Size restriction on peptide utilization in *Escherichia coli*. J. Biol. Chem. *243*, 6291–6299.

Perego, M., Higgins, C.F., Pearce, S.R., Gallagher, M.P., and Hoch, J.A. (1991). The oligopeptide transport system of *Bacillus subtilis* plays a role in the initiation of sporulation. Mol. Microbiol. *5*, 173–185.

Persing, D.H., Rutledge, B.J., Rys, P.N., Podzorski, D.S., Mitchell, P.D., Reed, K.D., Liu, B., Fikrig, E., and

Malawista, S.E. (1994). Target imbalance: disparity of *Borrelia burgdorferi* genetic material in synovial fluid from Lyme arthritis patients. J. Infect. Dis. *169*, 668–672.

Pinne, M., Östberg, Y., Comstedt, P., and Bergström, S. (2004). Molecular analysis of the channel-forming protein P13 and its paralogue family 48 from different Lyme disease *Borrelia* species. Microbiology *150*, 549–559.

Pinne, M., Denker, K., Nilsson, E., Benz, R., and Bergström, S. (2006a). The BBA01 protein, a member of paralog family 48 from *Borrelia burgdorferi*, is potentially interchangeable with the channel-forming protein P13. J. Bacteriol. *188*, 4207–4217.

Pinne, M., Östberg, Y., and Bergström, S. (2006b). Porins of Borrelia. In F.C. Cabello, D. Hulinska, and H.P. Godfrey (eds), Molecular Biology of Spirochetes (Amsterdam: IOS Press), pp. 250–263.

Pinne, M., Thein, M., Denker, K., Benz, R., Coburn, J., and Bergström, S. (2007). Elimination of channel-forming activity by insertional inactivation of the *p66* gene in *Borrelia burgdorferi*. FEMS Microbiol. Lett. *266*, 241–249.

Porcella, S.F., Fitzpatrick, C.A., and Bono, J.L. (2000). Expression and immunological analysis of the plasmid-borne *mlp* genes of *Borrelia burgdorferi* strain B31. Infect. Immun. *68*, 4992–5001.

Porcella, S.F., Popova, T.G., Akins, D.R., Li, M., Radolf, J.D., and Norgard, M.V. (1996). *Borrelia burgdorferi* supercoiled plasmids encode multicopy tandem open reading frames and a lipoprotein gene family. J. Bacteriol. *178*, 3293–3307.

Probert, W.S., and Johnson, B.J. (1998). Identification of a 47 kDa fibronectin-binding protein expressed by *Borrelia burgdorferi* isolate B31. Mol. Microbiol. *30*, 1003–1015.

Pugsley, A.P. (1993). The complete general secretory pathway in Gram-negative bacteria. Microbiol. Rev. *57*, 50–108.

Pugsley, A.P., Kornacker, M.G., and Ryter, A. (1990). Analysis of the subcellular location of pullulanase produced by *Escherichia coli* carrying the *pulA* gene from Klebsiella pneumoniae strain UNF5023. Mol. Microbiol. *4*, 59–72.

Radolf, J.D. (1994). Role of outer membrane architecture in immune evasion by *Treponema pallidum* and *Borrelia burgdorferi*. Trends Microbiol. *2*, 307–311.

Radolf, J.D., Arndt, L.L., Akins, D.R., Curetty, L.L., Levi, M.E., Shen, Y., Davis, L.S., and Norgard, M.V. (1995). *Treponema pallidum* and *Borrelia burgdorferi* lipoproteins and synthetic lipopeptides activate monocytes/macrophages. J. Immunol. *154*, 2866–2877.

Radolf, J.D., Bourell, K.W., Akins, D.R., Brusca, J.S., and Norgard, M.V. (1994). Analysis of *Borrelia burgdorferi* membrane architecture by freeze-fracture electron microscopy. J. Bacteriol. *176*, 21–31.

Radolf, J.D., and Caimano, M.J. (2008). The long strange trip of *Borrelia burgdorferi* outer-surface protein C. Mol. Microbiol. *69*, 1–4.

Ramamoorthi, N., Narasimhan, S., Pal, U., Bao, F., Yang, X.F., Fish, D., Anguita, J., Norgard, M.V., Kantor, F.S., Anderson, J.F., *et al.* (2005). The Lyme disease agent exploits a tick protein to infect the mammalian host. Nature *436*, 573–577.

Ramamoorthy, R., and Philipp, M.T. (1998). Differential expression of *Borrelia burgdorferi* proteins during growth in vitro. Infect. Immun. *66*, 5119–5124.

Ramamoorthy, R., Povinelli, L., and Philipp, M.T. (1996). Molecular characterization, genomic arrangement, and expression of *bmpD*, a new member of the *bmp* class of genes encoding membrane proteins of *Borrelia burgdorferi*. Infect. Immun. *64*, 1259–1264.

Rautemaa, R., and Meri, S. (1999). Complement-resistance mechanisms of bacteria. Microbes Infect. *1*, 785–794.

Revel, A.T., Talaat, A.M., and Norgard, M.V. (2002). DNA microarray analysis of differential gene expression in *Borrelia burgdorferi*, the Lyme disease spirochete. Proc. Natl. Acad. Sci. U.S.A. *99*, 1562–1567.

Rizzitello, A.E., Harper, J.R., and Silhavy, T.J. (2001). Genetic evidence for parallel pathways of chaperone activity in the periplasm of *Escherichia coli*. J. Bacteriol. *183*, 6794–6800.

Robert, V., Volokhina, E.B., Senf, F., Bos, M.P., Van Gelder, P., and Tommassen, J. (2006). Assembly factor Omp85 recognizes its outer membrane protein substrates by a species-specific C-terminal motif. PLoS Biol. *4*, e377.

Rudner, D.Z., LeDeaux, J.R., Ireton, K., and Grossman, A.D. (1991). The *spo0K* locus of *Bacillus subtilis* is homologous to the oligopeptide permease locus and is required for sporulation and competence. J. Bacteriol. *173*, 1388–1398.

Sadziene, A., Thomas, D.D., and Barbour, A.G. (1995). *Borrelia burgdorferi* mutant lacking Osp: biological and immunological characterization. Infect. Immun. *63*, 1573–1580.

Saier, M.H., Jr., and Paulsen, I.T. (2000). Whole genome analyses of transporters in spirochetes: *Borrelia burgdorferi* and *Treponema pallidum*. J. Mol. Microbiol. Biotechnol. 2, 393–399.

Samuelson, J.C., Chen, M., Jiang, F., Moller, I., Wiedmann, M., Kuhn, A., Phillips, G.J., and Dalbey, R.E. (2000). YidC mediates membrane protein insertion in bacteria. Nature *406*, 637–641.

Sanchez-Pulido, L., Devos, D., Genevrois, S., Vicente, M., and Valencia, A. (2003). POTRA: a conserved domain in the FtsQ family and a class of β-barrel outer membrane proteins. Trends Biochem. Sci *28*, 523–526.

Sankaran, K., Gupta, S.D., and Wu, H.C. (1995). Modification of bacterial lipoproteins. Methods Enzymol. *250*, 683–697.

Sauvonnet, N., and Pugsley, A.P. (1996). Identification of two regions of *Klebsiella oxytoca* pullulanase that together are capable of promoting β-lactamase secretion by the general secretory pathway. Mol. Microbiol. *22*, 1–7.

Schroder, N.W., Eckert, J., Stubs, G., and Schumann, R.R. (2008). Immune responses induced by spirochetal outer membrane lipoproteins and glycolipids. Immunobiology *213*, 329–340.

Schulze, R.J., and Zückert, W.R. (2006). *Borrelia burgdorferi* lipoproteins are secreted to the outer surface by default. Mol. Microbiol. *59*, 1473–1484.

Schwan, T.G., Battisti, J.M., Porcella, S.F., Raffel, S.J., Schrumpf, M.E., Fischer, E.R., Carroll, J.A., Stewart, P.E., Rosa, P., and Somerville, G.A. (2003). Glycerol-3-phosphate acquisition in spirochetes: distribution and biological activity of glycerophosphodiester phosphodiesterase (GlpQ) among *Borrelia* species. J. Bacteriol. *185*, 1346–1356.

Schwan, T.G., and Piesman, J. (2000). Temporal changes in outer surface proteins A and C of the lyme disease-associated spirochete, *Borrelia burgdorferi*, during the chain of infection in ticks and mice. J. Clin. Microbiol. *38*, 382–388.

Schwan, T.G., Piesman, J., Golde, W.T., Dolan, M.C., and Rosa, P.A. (1995). Induction of an outer surface protein on *Borrelia burgdorferi* during tick feeding. Proc. Natl. Acad. Sci. U.S.A. *92*, 2909–2913.

Seshadri, R., Myers, G.S., Tettelin, H., Eisen, J.A., Heidelberg, J.F., Dodson, R.J., Davidsen, T.M., DeBoy, R.T., Fouts, D.E., Haft, D.H., *et al.* (2004). Comparison of the genome of the oral pathogen *Treponema denticola* with other spirochete genomes. Proc. Natl. Acad. Sci. U.S.A. *101*, 5646–5651.

Seshu, J., Esteve-Gassent, M.D., Labandeira-Rey, M., Kim, J.H., Trzeciakowski, J.P., Hook, M., and Skare, J.T. (2006). Inactivation of the fibronectin-binding adhesin gene *bbk32* significantly attenuates the infectivity potential of *Borrelia burgdorferi*. Mol. Microbiol. *59*, 1591–1601.

Seshu, J., and Skare, J.T. (2000). The many faces of *Borrelia burgdorferi*. J. Mol. Microbiol. Biotechnol. *2*, 463–472.

Setubal, J.C., Reis, M., Matsunaga, J., and Haake, D.A. (2006). Lipoprotein computational prediction in spirochaetal genomes. Microbiology *152*, 113–121.

Seydel, A., Gounon, P., and Pugsley, A.P. (1999). Testing the '+2 rule' for lipoprotein sorting in the *Escherichia coli* cell envelope with a new genetic selection. Mol. Microbiol. *34*, 810–821.

Shang, E.S., Skare, J.T., Exner, M.M., Blanco, D.R., Kagan, B.L., Miller, J.N., and Lovett, M.A. (1998). Isolation and characterization of the outer membrane of *Borrelia hermsii*. Infect. Immun. *66*, 1082–1091.

Shaw, N. (1970). Bacterial glycolipids. Bacteriol. Rev., 365–377.

Shoberg, R.J., and Thomas, D.D. (1993). Specific adherence of *Borrelia burgdorferi* extracellular vesicles to human endothelial cells in culture. Infect. Immun. *61*, 3892–3900.

Shoberg, R.J., and Thomas, D.D. (1995). *Borrelia burgdorferi* vesicle production occurs via a mechanism independent of immunoglobulin M involvement. Infect. Immun. *63*, 4857–4861.

Sigal, L.H., Zahradnik, J.M., Lavin, P., Patella, S.J., Bryant, G., Haselby, R., Hilton, E., Kunkel, M., Adler-Klein, D., Doherty, T., *et al.* (1998). A vaccine consisting of recombinant *Borrelia burgdorferi* outer-surface protein A to prevent Lyme disease. Recombinant Outer-Surface Protein A Lyme Disease Vaccine Study Consortium. N. Engl. J. Med. 339, 216–222.

Silva-Herzog, E., Ferracci, F., Jackson, M.W., Joseph, S.S., and Plano, G.V. (2008). Membrane localization and topology of the *Yersinia pestis* YscJ lipoprotein. Microbiology *154*, 593–607.

Simpson, W.J., Burgdorfer, W., Schrumpf, M.E., Karstens, R.H., and Schwan, T.G. (1991). Antibody to a 39-kilodalton *Borrelia burgdorferi* antigen (P39) as a marker for infection in experimentally and naturally inoculated animals. J. Clin. Microbiol. *29*, 236–243.

Simpson, W.J., Schrumpf, M.E., and Schwan, T.G. (1990). Reactivity of human Lyme borreliosis sera with a 39-kilodalton antigen specific to *Borrelia burgdorferi*. J. Clin. Microbiol. *28*, 1329–1337.

Skare, J.T., Mirzabekov, T.A., Shang, E.S., Blanco, D.R., Erdjument-Bromage, H., Bunikis, J., Bergström, S., Tempst, P., Kagan, B.L., Miller, J.N., and Lovett, M.A. (1997). The Oms66 (p66) protein is a *Borrelia burgdorferi* porin. Infect. Immun. *65*, 3654–3661.

Skare, J.T., Shang, E.S., Foley, D.M., Blanco, D.R., Champion, C.I., Mirzabekov, T., Sokolov, Y., Kagan, B.L., Miller, J.N., and Lovett, M.A. (1995). Virulent strain associated outer membrane proteins of *Borrelia burgdorferi*. J. Clin. Invest. *96*, 2380–2392.

Sklar, J.G., Wu, T., Kahne, D., and Silhavy, T.J. (2007). Defining the roles of the periplasmic chaperones SurA, Skp, and DegP in *Escherichia coli*. Genes Dev. *21*, 2473–2484.

Sohlenkamp, C., de Rudder, K.E., and Geiger, O. (2004). Phosphatidylethanolamine is not essential for growth of *Sinorhizobium meliloti* on complex culture media. J. Bacteriol. *186*, 1667–1677.

Sohlenkamp, C., de Rudder, K.E., Rohrs, V.V., Lopez-Lara, I.M., and Geiger, O. (2000). Cloning and characterization of the gene for phosphatidylcholine synthase. J. Biol. Chem. *275*, 27500.

Steere, A.C., Sikand, V.K., Meurice, F., Parenti, D.L., Fikrig, E., Schoen, R.T., Nowakowski, J., Schmid, C.H., Laukamp, S., Buscarino, C., *et al.* (1998). Vaccination against Lyme disease with recombinant *Borrelia burgdorferi* outer-surface lipoprotein A with adjuvant. Lyme Disease Vaccine Study Group. N. Engl. J. Med. *339*, 209–215.

Stevenson, B., Bono, J.L., Schwan, T.G., and Rosa, P. (1998). *Borrelia burgdorferi* Erp proteins are immunogenic in mammals infected by tick bite, and their synthesis is inducible in cultured bacteria. Infect. Immun. *66*, 2648–2654.

Stevenson, B., El-Hage, N., Hines, M.A., Miller, J.C., and Babb, K. (2002). Differential binding of host complement inhibitor factor H by *Borrelia burgdorferi* Erp surface proteins: a possible mechanism underlying the expansive host range of Lyme disease spirochetes. Infect. Immun. *70*, 491–497.

Stevenson, B., Tilly, K., and Rosa, P.A. (1996). A family of genes located on four separate 32-kilobase circular plasmids in *Borrelia burgdorferi* B31. J. Bacteriol. *178*, 3508–3516.

Strobl, S., Gomis-Ruth, F.X., Maskos, K., Frank, G., Huber, R., and Glockshuber, R. (1997). The α-amylase from the yellow meal worm: complete primary structure, crystallization and preliminary X-ray analysis. FEBS Lett. *409*, 109–114.

Takayama, K., Rothenberg, R.J., and Barbour, A.G. (1987). Absence of lipopolysaccharide in the Lyme disease spirochete, *Borrelia burgdorferi*. Infect. Immun. *55*, 2311–2313.

Terada, M., Kuroda, T., Matsuyama, S.I., and Tokuda, H. (2001). Lipoprotein sorting signals evaluated as the LolA-dependent release of lipoproteins from the cytoplasmic membrane of *Escherichia coli*. J. Biol. Chem. *276*, 47690–47694.

Thein, M., Bunikis, I., Denker, K., Larsson, C., Cutler, S., Drancourt, M., Schwan, T.G., Mentele, R., Lottspeich, F., Bergström, S., and Benz, R. (2008b). Oms38 is the first identified pore-forming protein in the outer membrane of relapsing fever spirochetes. J. Bacteriol. *190*, 7035–7042.

Tilly, K., Elias, A.F., Errett, J., Fischer, E., Iyer, R., Schwartz, I., Bono, J.L., and Rosa, P. (2001). Genetics and regulation of chitobiose utilization in *Borrelia burgdorferi*. J. Bacteriol. *183*, 5544–5553.

Tilly, K., Grimm, D., Bueschel, D.M., Krum, J.G., and Rosa, P. (2004). Infectious cycle analysis of a *Borrelia burgdorferi* mutant defective in transport of chitobiose, a tick cuticle component. Vector Borne Zoonotic Dis *4*, 159–168.

Tokuda, H., and Matsuyama, S. (2004). Sorting of lipoproteins to the outer membrane in E. coli. Biochim. Biophys. Acta *1694*, 1–9.

Tupin, E., Benhnia, M.R., Kinjo, Y., Patsey, R., Lena, C.J., Haller, M.C., Caimano, M.J., Imamura, M., Wong, C.H., Crotty, S., *et al.* (2008). NKT-cells prevent chronic joint inflammation after infection with *Borrelia burgdorferi*. Proc. Natl. Acad. Sci. U.S.A. *105*, 19863–19868.

van Ulsen, P., van Alphen, L., ten Hove, J., Fransen, F., van der Ley, P., and Tommassen, J. (2003). A Neisserial autotransporter NalP modulating the processing of other autotransporters. Mol. Microbiol. *50*, 1017–1030.

von Lackum, K., Miller, J.C., Bykowski, T., Riley, S.P., Woodman, M.E., Brade, V., Kraiczy, P., Stevenson, B., and Wallich, R. (2005). *Borrelia burgdorferi* regulates expression of complement regulator-acquiring surface protein 1 during the mammal-tick infection cycle. Infect. Immun. *73*, 7398–7405.

Voulhoux, R., Bos, M.P., Geurtsen, J., Mols, M., and Tommassen, J. (2003). Role of a highly conserved bacterial protein in outer membrane protein assembly. Science *299*, 262–265.

Walker, E.M., Borenstein, L.A., Blanco, D.R., Miller, J.N., and Lovett, M.A. (1991). Analysis of outer membrane ultrastructure of pathogenic *Treponema* and *Borrelia* species by freeze-fracture electron microscopy. J. Bacteriol. *173*, 5585–5588.

Wallich, R., Kramer, M.D., and Simon, M.M. (1996). The recombinant outer surface protein A (lipOspA) of *Borrelia burgdorferi*: a Lyme disease vaccine. Infection *24*, 396–397.

Wang, G., van Dam, A.P., and Dankert, J. (1999). Evidence for frequent OspC gene transfer between *Borrelia valaisiana* sp. nov. and other Lyme disease spirochetes. FEMS Microbiol. Lett. *177*, 289–296.

Wang, X.G., Kidder, J.M., Scagliotti, J.P., Klempner, M.S., Noring, R., and Hu, L.T. (2004). Analysis of differences in the functional properties of the substrate binding proteins of the *Borrelia burgdorferi* oligopeptide permease (Opp) operon. J. Bacteriol. *186*, 51–60.

Wang, X.G., Lin, B., Kidder, J.M., Telford, S., and Hu, L.T. (2002). Effects of environmental changes on expression of the oligopeptide permease (*opp*) genes of *Borrelia burgdorferi*. J. Bacteriol. *184*, 6198–6206.

Wang, X.G., Scagliotti, J.P., and Hu, L.T. (2004). Phospholipid synthesis in *Borrelia burgdorferi*: BB0249 and BB0721 encode functional phosphatidylcholine synthase and phosphatidylglycerolphosphate synthase proteins. Microbiology *150*, 391–397.

Wheeler, C.M., García-Moncó, J.C., Benach, J.L., Golightly, M.G., Habicht, G.S., and Steere, A.C. (1993). Nonprotein antigens of *Borrelia burgdorferi*. J. Infect. Dis. *167*, 665–674.

Xie, K., and Dalbey, R.E. (2008). Inserting proteins into the bacterial cytoplasmic membrane using the Sec and YidC translocases. Nat. Rev. Microbiol. *6*, 234–244.

Xu, Y., Bruno, J.F., and Luft, B.J. (2008). Profiling the humoral immune response to *Borrelia burgdorferi* infection with protein microarrays. Microb Pathog *45*, 403–407.

Yakushi, T., Masuda, K., Narita, S., Matsuyama, S., and Tokuda, H. (2000). A new ABC transporter mediating the detachment of lipid-modified proteins from membranes. Nat. Cell Biol. *2*, 212–218.

Yamaguchi, K., Yu, F., and Inouye, M. (1988). A single amino acid determinant of the membrane localization of lipoproteins in *E. coli*. Cell *53*, 423–432.

Yang, X., Goldberg, M.S., Popova, T.G., Schoeler, G.B., Wikel, S.K., Hagman, K.E., and Norgard, M.V. (2000). Interdependence of environmental factors influencing reciprocal patterns of gene expression in virulent *Borrelia burgdorferi*. Mol. Microbiol. 37, 1470–1479.

Yang, X., Izadi, H., Coleman, A.S., Wang, P., Ma, Y., Fikrig, E., Anguita, J., and Pal, U. (2008). *Borrelia burgdorferi* lipoprotein BmpA activates proinflammatory responses in human synovial cells through a protein moiety. Microbes Infect. *10*, 1300–1308.

Yang, X., Popova, T.G., Hagman, K.E., Wikel, S.K., Schoeler, G.B., Caimano, M.J., Radolf, J.D., and Norgard, M.V. (1999). Identification, characterization, and expression of three new members of the *Borrelia burgdorferi* Mlp (2.9) lipoprotein gene family. Infect. Immun. *67*, 6008–6018.

Yang, X.F., Hübner, A., Popova, T.G., Hagman, K.E., and Norgard, M.V. (2003). Regulation of expression of the paralogous Mlp family in *Borrelia burgdorferi*. Infect. Immun. *71*, 5012–5020.

Yen, M.R., Peabody, C.R., Partovi, S.M., Zhai, Y., Tseng, Y.H., and Saier, M.H. (2002). Protein-translocating outer membrane porins of Gram-negative bacteria. Biochim. Biophys. Acta *1562*, 6–31.

Yokota, N., Kuroda, T., Matsuyama, S., and Tokuda, H. (1999). Characterization of the LolA-LolB system as the general lipoprotein localization mechanism of *Escherichia coli*. J. Biol. Chem. *274*, 30995–30999.

Zhang, J.R., Hardham, J.M., Barbour, A.G., and Norris, S.J. (1997). Antigenic variation in Lyme disease borreliae by promiscuous recombination of VMP-like sequence cassettes. Cell *89*, 275–285.

Zhang, J.R., and Norris, S.J. (1998a). Kinetics and in vivo induction of genetic variation of *vlsE* in *Borrelia burgdorferi*. Infect. Immun. *66*, 3689–3697.

Zhang, J.R., and Norris, S.J. (1998b). Genetic variation of the *Borrelia burgdorferi* gene *vlsE* involves cassette-specific, segmental gene conversion. Infect. Immun. *66*, 3698–3704.

Zhang, Q.Y., DeRyckere, D., Lauer, P., and Koomey, M. (1992). Gene conversion in *Neisseria gonorrhoeae*: evidence for its role in pilus antigenic variation. Proc. Natl. Acad. Sci. U.S.A. *89*, 5366–5370.

Zückert, W.R., Kerentseva, T.A., Lawson, C.L., and Barbour, A.G. (2001). Structural conservation of neurotropism-associated VspA within the variable *Borrelia* Vsp-OspC lipoprotein family. J. Biol. Chem. *276*, 457–463.

Zückert, W.R., Lloyd, J.E., Stewart, P.E., Rosa, P.A., and Barbour, A.G. (2004). Cross-species surface display of functional spirochetal lipoproteins by recombinant *Borrelia burgdorferi*. Infect. Immun. *72*, 1463–1469.

Zückert, W.R., Meyer, J., and Barbour, A.G. (1999). Comparative analysis and immunological characterization of the *Borrelia* Bdr protein family. Infect. Immun. *67*, 3257–3266.

The Chic Motility and Chemotaxis of *Borrelia burgdorferi*

7

Stuart F. Goldstein, Chunhao Li, Jun Liu, Michael Miller, Md. Abdul Motaleb, Steven J. Norris, Ruth E. Silversmith, Charles W. Wolgemuth and Nyles W. Charon

Abstract

The motility of *Borrelia burgdorferi* is complex, enigmatic, yet quite elegant. Here we address specific and unique aspects of borrelial motility and chemotaxis. For *B. burgdorferi*, in contrast to most spirochaete species, the periplasmic flagella have a skeletal function that affects the entire shape of the cell. These organelles allow the spirochaete to swim by generating backward-moving waves along the length of the cell, and they do so by rotating within the periplasmic space. The well-studied paradigm of *Escherichia coli* and *Salmonella enterica* serovar Typhimurium motility and chemotaxis does not directly apply to these unique bacteria. Recent advances are presented and summarized with respect to cell and periplasmic flagella structure, along with an exciting analysis of the *in situ* hook-basal body–motor complex. In addition, advances in the dynamics of motility, how *B. burgdorferi* swims, their chemotaxis, gene regulation, and the role of motility and chemotaxis in the biology of *B. burgdorferi* are summarized. Now that a foundation in this field has been established, the future can only continue to be exciting as we sort out the structure, function, regulation, and the role of motility and chemotaxis in the life cycle of *B. burgdorferi*.

7.1 General structure of spirochaetes

To really try to begin to understand *B. burgdorferi* motility and chemotaxis, it is important to relate its structure to that of other bacteria and spirochaetes. Spirochaetes constitute a very diverse but monophyletic group of bacteria that have common morphological attributes (Canale-Parola, 1984; Holt, 1978; Paster and Dewhirst, 2000). All species of spirochaetes have periplasmic flagella (PFs), and these organelles have been shown to be essential for motility (see reviews, including some that are quite recent: Charon *et al.*, 1992b; Goldstein and Charon, 1988; Li *et al.*, 2000b; Charon and Goldstein, 2002; Wolgemuth *et al.*, 2006; Canale-Parola, 1978; Lux *et al.*, 2000). One to several hundred PFs are subterminally attached near each end of the protoplasmic cell cylinder and extend towards the centre of the cell where they may or may not overlap. The number of PFs, and whether they overlap at the centre of the cell, is dependent on the species. Surrounding the entire cell is a lipid bilayer membrane. This outer membrane is often referred to as an outer membrane sheath, as it surrounds the entire spirochaete including the PFs. Thus, spirochaetes differ from other flagellated bacteria in that the spirochaete organelles for motility – the PFs – reside within the periplasmic space. There is some variation to this theme. Recent results using electron cryotomography indicate that in the termite gut spirochaete *Treponema primitia*, the PFs reside between two walls; underneath the outer membrane there is one wall, then the PFs, and then the protoplasmic cell cylinder (Murphy *et al.*, 2008).

7.1.1 General structure of *Borrelia burgdorferi*

Early work indicated that, in *B. burgdorferi*, between 7 and 11 PFs are subterminally attached near the cell ends (Hovind Hougen, 1984). These

PFs extend inward along the cell cylinder beneath the outer membrane sheath (Goldstein *et al.*, 1996; Barbour and Hayes, 1986). Occasionally, the PFs protrude, and when they do so they often rotate (Goldstein *et al.*, 1994). The latter observations with *B. burgdorferi* and other spirochaetes were critical, as they put an end to the longstanding question of whether these organelles do in fact rotate in a manner similar to that of other bacterial flagella (Goldstein *et al.*, 1994; Charon *et al.*, 1992a). Studies on swimming cells indicated that *B. burgdorferi* was not helical like other spirochaetes such as *Treponema phagedenis* (Charon *et al.*, 1991), *Treponema denticola* (Ruby *et al.*, 1997), and the Leptospiraceae (Goldstein *et al.*, 1996). In fact, *B. burgdorferi* was found to have a flat wave morphology (Goldstein *et al.*, 1994) (Fig. 7.1). *Treponema pallidum* may also be a flat wave (Cox, 1972). As discussed later in the chapter, these waves propagate from the front to the back end of the cell as the cell swims. Also, as discussed later in the chapter, the PFs have a true skeletal function (Wolgemuth *et al.*, 2006); mutants that lack the PFs are rod-shaped (Sartakova *et al.*, 2001; Motaleb *et al.*, 2000).

How are the PFs arranged within the *B. burgdorferi* cell body? For many years we did not know whether the PFs wrap around the cell body in a right-handed or left-handed sense (a right-handed helix rotates clockwise going away from an observer, and a left and helix rotates counter-clockwise). The early attempts to answer this and related questions using serial sectioning coupled with transmission electron microscopy were unsatisfactory, and this failure was in part due to the thin diameter (see below) of *B. burgdorferi*. However, subsequent work in collaboration with Karolyn Buttle of the high voltage electron microscope facility at the Wadsworth Centre in New York yielded definitive answers. This million volt electron microscope allows for the viewing of thick sections (up to 0.75 μm in thickness) of embedded *B. burgdorferi* cells. By tilting and photographing the sections, three dimensional images can be obtained using stereo-pairs (Goldstein *et al.*, 1996) (Fig. 7.2), thus allowing more of the cell to be visualized in a given section. Several important findings with respect to structure were revealed using this approach, and some of the details will be summarized here. First the cells had a diameter of 0.33 μm and a length of 10 to 20 μm;

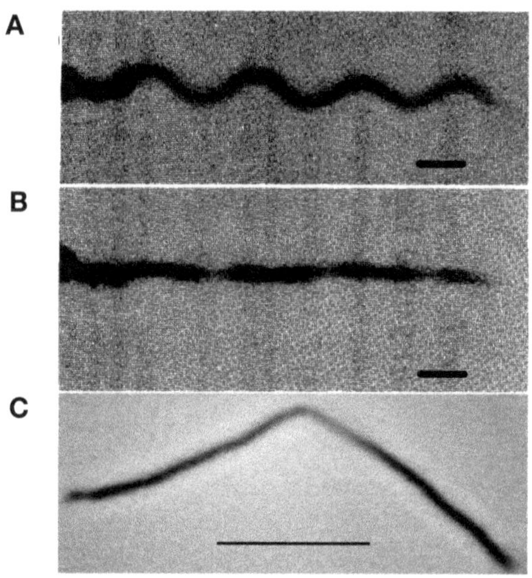

Figure 7.1 The morphology and architecture of *B. burgdorferi*, a spirochaete with a planar, flat-wave morphology. (a) When viewed from one perspective, the cell body appears wave-like. Bar, 1 μm. (b) When rotated by 90 degrees, the cell shape appears straight. Bar, 1 μm. Panels a and b are modified and reprinted with permission from Proc. Natl. Acad. Sci. U.S.A. (Goldstein *et al.*, 1994). (c) Mutants lacking FlaB do not produce periplasmic flagella, and the cells are rod-shaped. The cell is bent in the centre, as it is in the process of division. Bar, 5 μm. Reprinted with permission from *Annu. Rev. Genet.* (Charon and Goldstein, 2002).

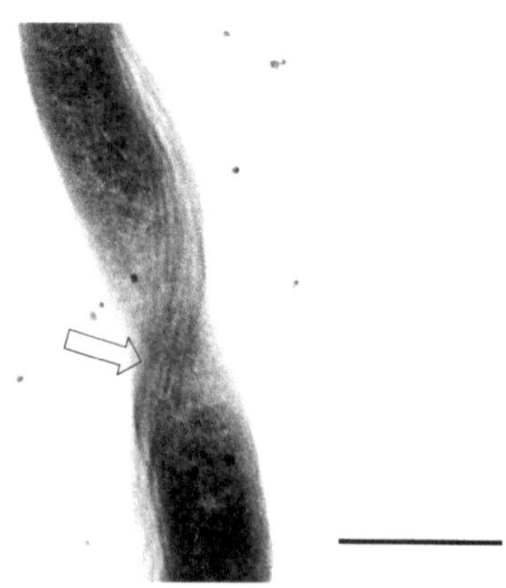

Figure 7.2 High-voltage electron micrograph image of *B. burgdorferi.* Arrow points to a region of the cell where the periplasmic flagella cross over the top of the cell from lower left to upper right. Bar, 0.5 μm.

thus they are 30–60 times longer than they are wide. Second, the PFs grouped together formed a continuous bundle (see discussion below with respect to cryoelectron microscopy) of 0.67 nm in diameter that extended from one end of the cell to the other. Third, the PFs wrap around the axis of the cell cylinder (i.e. take the centre of the cell cylinder and draw a line through it – that's the axis of the cell cylinder) in a right-handed sense. Finally, in space, or along the cell axis (since *B. burgdorferi* is a flat wave, it resembles a sine wave; the cell axis is analogous to the X axis of the sine wave), the PFs wrap around the cell axis in left-handed sense. In addition, the helix pitch and diameter of the PFs *in situ* (2.83 μm and 0.38 μm, respectively) is different from that found with most of the purified PFs *in vitro* (left-handed, 1.48 μm and 0.28, respectively) (Charon *et al.*, 1992a). Most important, the wave length of the cell (2.83 μm) is equivalent to the helix pitch of the PFs (Fig. 7.3). As discussed below in the section on motility, these observations were critical in developing a model of *B. burgdorferi* swimming.

One question that remained an enigma until recently is the following: How can a rod shaped cell cylinder, and two bundles of helical, left-handed periplasmic flagella (see below) lead to a cell with a flat-wave morphology? The main clue from the experiments is that both the cell cylinder and the periplasmic flagella have a different configuration when they are attached to each other than they do when they are separated. Why this happens is easy to understand. Since the helix radius of purified PFs is not equal to the radius of the cell cylinder, the PFs must bend to stay within the periplasmic space. But, to bend the flagella requires that the cell cylinder must exert a force on the PFs. Newton's third law, however, requires that if the cell cylinder exerts a force on the PFs, then the PFs must exert an equal but opposite force back on the cell cylinder. Therefore, the cell cylinder must bend, too. How much the cell

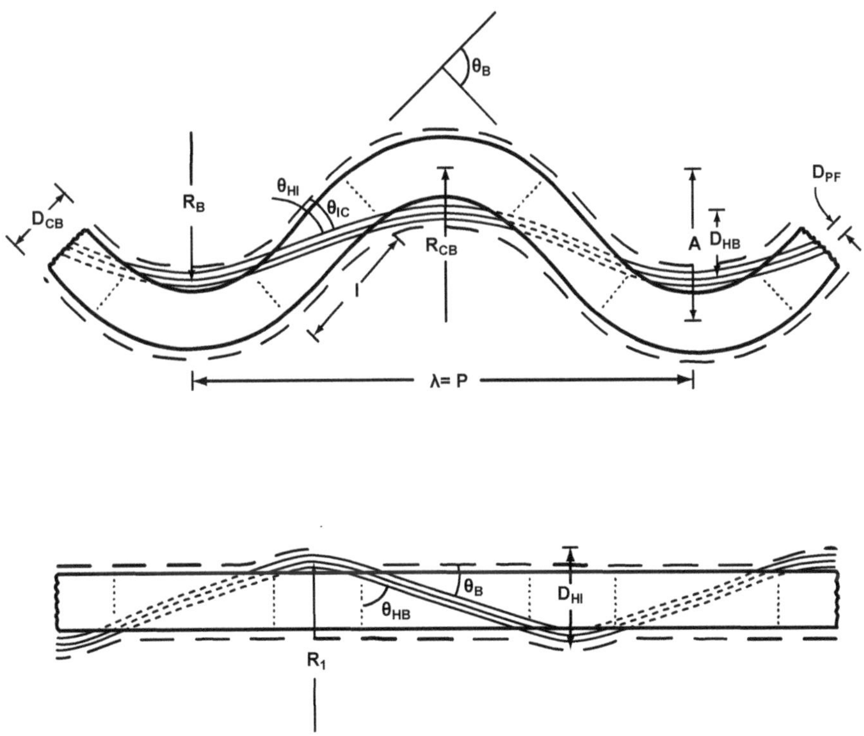

Figure 7.3 Schematic representation of *B. burgdorferi* flat on (top) and edge on (bottom) views. Dashed lines represent the outer membrane. The peak to peak amplitude *(A)* is 0.78 μm and the cell body diameter (D_{CB}) is 0.33 μm. The wave length (λ) and periplasmic flagella helix pitch *(P)* are both 2.83 μm. See Goldstein *et al.* (1996) for details and other measured parameters. Reprinted with permission from *J. Bacteriol.* (Goldstein *et al.*, 1996).

cylinder bends relative to the PFs depends on how stiff the cell cylinder is compared to the PFs. If the cell cylinder were made of steel and the flagella were made out of rubber, then the cell cylinder should deform very little compared to the PFs. Since the observed deformation of the cell cylinder is appreciable, though, we conclude that the PFs must be of at least comparable stiffness to the cell cylinder, if not greater.

To explore whether this conceptual picture is sufficient to explain the flat-wave morphology, Wolgemuth and coworkers developed a mathematical model that explains the final shape of *B. burgdorferi* (Dombrowski *et al.*, 2009). The model treats the cell cylinder as an elastic, rod-shaped filament and the periplasmic flagella as elastic, helical-shaped filaments. The energy required to twist or bend a filament depends on two elastic moduli, known as the twisting and bending moduli, respectively. These parameters are determined by the stiffness of the material (i.e. the Young's modulus or the shear modulus) and the radius of the filament. The model explicitly states that the PFs reside at the cell cylinder's radius and uses the physical concept that the overall shape of the cell cylinder plus the PFs should be such that the total deformation (bending and twisting) energy is minimized. Using reasonable assumptions, there is only one free parameter in this model, the ratio of the bending modulus of the flagella to that of the cell cylinder, $\mathcal{A}$. Dombrowski *et al.* (2009) showed that the morphology predicted by the model compares well with the experimentally observed shape of *B. burgdorferi* when the value of $\mathcal{A}$ is approximately 3–5. The model also predicts that the PFs should wrap about the cell cylinder in the opposite sense of their own handedness; i.e. a left-handed flagellum should wrap about the cell cylinder in a right-handed fashion, which agrees with previous experimental measurements using high voltage electron microscopy (Goldstein *et al.*, 1996).

To test the mathematical model, the stiffness of the cell cylinder and the periplasmic flagella were measured using optical trapping methods. The effective bending modulus found for the cell cylinder plus flagella was 42 ± 24 pN/m^2. The mathematical model predicts that this effective bending modulus is equal to the bending modulus of the cell cylinder plus 0.6 times the bending modulus of the flagella. The stiffness of the purified PFs was also measured. From these experiments, the bending modulus for a single PF was estimated to be 6.7 ± 3.7 pN μm^2, which is of the same order as measurements of the bending modulus of flagellar filaments from *Salmonella enterica* serovar Typhimurium (Kim and Powers, 2005). Using a flagellar diameter of 10 nm, the Young's modulus of the flagellum was estimated to be 700 MPa. Therefore, if there are eight periplasmic flagella along the length of *B. burgdorferi*, the total bending modulus for the PFs should be approximately 53 pN μm^2, and the bending modulus of the cell cylinder is about 10 pN μm^2, which implies that $\mathcal{A} \approx 5$, in good agreement with the results from the mathematical model. For an elastic tube, such as the cell wall, the Young's modulus, E, is related to the bending modulus as $A \approx \pi E a^3 t$. Here t is the thickness of the cell wall, which we estimate to be about 6 nm. Therefore, the Young's modulus of the cell wall of *B. burgdorferi* is about 0.1 MPa, which is comparable to that measured for *Magnetospirillum gryphiswaldense* (Arnoldi *et al.*, 2000) and *Myxococcus xanthus* (Pelling *et al.*, 2006) but substantially lower than what has been estimated for *Escherichia coli* and *Bacilllus subtilis* (Thwaites and Mendelson, 1991; Mendelson *et al.*, 2000). The Young's moduli that are estimated for the cell cylinder and PFs of *B. burgdorferi* imply that the flagellar stiffness compared to that of the cell wall is like steel compared to rubber. This combination of mathematical modelling and micromanipulation experiments show that the mechanical coupling of the helical periplasmic flagella to the rod-shaped cell cylinder is sufficient to determine the flat-wave morphology of *B. burgdorferi*. Interestingly, the physical considerations and mathematics of this model for the flat-wave shape of *B. burgdorferi* has wide applicability. Specifically, it is likely to be useful in describing the complex physics of conjoined elastic filaments such as the axonemes that constitute the cilia and flagella of eukaryotic cells, and the alpha-helical bundles that are found in many proteins.

Cryoelectron tomography has also been critical in sorting out the structure of *B. burgdorferi*. The major advantage of this methodology is that the cells are frozen at very low temperature, are not treated with fixatives or stains, and can be directly observed in their frozen state in the electron microscope (Jensen and Briegel, 2007). Thus, many artefacts associated with classically embedded sectioning are avoided. In addition, because the frozen sections can be tilted, a series of pictures can be taken through 120 degrees. Using sophisticated software programs, these images can then be used to create a tomogram of the cells; it allows for making a CAT scan of a spirochaete. Details that in the past were not previously apparent were now obvious with *B. burgdorferi*. First, in contrast to previous results with embedded cells that indicated that the PFs form a bundle in the periplasmic space (Goldstein *et al.*, 1996), cryoelectron microscopy revealed that the PFs form a tight ribbon in this space as they wrap around the cell cylinder (Charon *et al.*, 2009) (Fig. 7.4). They do so in quite an elegant manner, with very little space between each PF. In addition, in agreement with the high voltage electron microscopy study, the ribbon was found to wrap around the body axis in a right handed sense. This observation has real implications, as it indicates that each PF is tightly associated with the cell cylinder, and thus each can influence cell shape. In addition, as discussed below, the ribbon configuration impacts how *B. burgdorferi* swims. Third, there is a noticeable increase in the periplasmic space in the domain where the PFs reside. The space is approximately 16 nm without the PFs, and 42 nm with the PFs. The size of this increase is slightly larger than the diameter of the PFs (16 nm) (Hovind Hougen, 1984). Thus, the outer membrane fits tightly about the cell cylinder and the periplasmic space widens due to the obstruction produced by the periplasmic flagella. In addition, in the bend regions of the cell, the periplasmic flagella may pull in towards the cell axis, causing the periplasmic space to widen more than the diameter of the filaments.

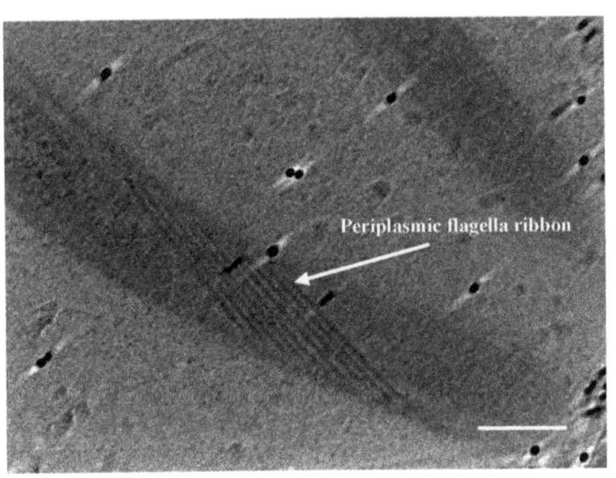

Figure 7.4 Cryo-electron tomograph of *B. burgdorferi* illustrating the ribbon of periplasmic flagella as they wrap around the cell axis in a right handed sense. Bar, 0.21 μm.

One question that is obvious is how the tight ribbon of PFs forms? Two factors may come into play. First, perhaps the spacing of the insertion of the PFs is critical. We find that the PFs are inserted on a line along the subterminal end of the cell approximately 90 nm apart. Second, perhaps the formation of the ribbon depends on having active PFs that rotate. Cells incubated in the cold, or cells treated with the iononophore carbonyl cyanide-m-chlorophenylhydrazone (CCCP) that dissipates the proton gradient, become immotile, but still retain the flat wave morphology (Motaleb *et al.*, 2000). Thus, it is possible that such cells already have their ribbons of PFs in place. The real test to determine if rotation of the PFs is necessary for ribbon formation is to construct a non-motile paralysed mutant (in *motA* or *motB*) that forms but cannot rotate the PFs. Such mutants have been constructed, but they turned out to have polar effects, and more sophisticated in-frame deletion mutants will be needed to answer this question.

7.2 Structure of the periplasmic flagella

There are between 7 and 11 PFs attached at each end of the cell, and they overlap in the centre (Hovind Hougen, 1984). As previously mentioned, the PFs are essential for motility. In addition, these structures are also critical in providing the characteristic flat-wave morphology of the intact spirochaete (Sartakova *et al.*, 2001; Motaleb *et al.*, 2000; Sal *et al.*, 2008). Thus, insertion-deletion mutants in *flaB* (major flagellin filament protein), *flgE* (the flagellar hook protein), and *fliF* (the protein that constitutes the MS ring of the basal body) resulted in mutants that lacked PF filaments and were non-motile. All these mutants were also rod-shaped. Because genetic complementation of both *flaB* and *flgE* restored the wild-type phenotypes of the flat-wave morphology and motility, secondary mutations were not responsible in formation of the mutant phenotypes.

Borrelia burgdorferi PFs are similar in structure to flagella from other bacteria; a given PF is composed of a basal body, hook, and filament containing a single major flagellin FlaB, and a minor flagellin FlaA (Ge *et al.*, 1998; Barbour and Hayes, 1986; Hovind Hougen, 1984; Sal *et al.*, 2008). FlaB is similar in sequence to flagellin of other bacteria at the N- and C- terminal regions (Wilson and Beveridge, 1993; Norris and Treponema pallidum polypeptide research group., 1993). This protein assembles into a polymer, and in the absence of FlaA is still left-handed and helically shaped (Charon *et al.*, 1992a; Motaleb *et al.*, 2004). Its amino acid sequence is similar to FlaB proteins of other spirochaetes (Norris and Treponema pallidum polypeptide research group., 1993). It lacks a signal sequence, and is thus secreted into the periplasmic space by a type III secretion system through the basal body. In other species of spirochaetes, there are multiple FlaB proteins comprising the core of the filament (FlaB1, FlaB2, FlaB3) (Charon and Goldstein, 2002; Li *et al.*, 2000b). *Borrelia burgdorferi* is an exception, as it has only one FlaB protein that comprises the PF core. FlaA of *B. burgdorferi* and other spirochaetes shows no sequence similarity to FlaB (Charon and Goldstein, 2002; Li *et al.*, 2000b; Ge *et al.*, 1998). Because DNA sequence indicates that FlaA has a typical peptidase I cleavage site near its N-terminal sequence and that this sequence is not present in the final protein, FlaA is probably secreted into the periplasmic space by a type II SecA-dependent system (Ge and Charon, 1997a). In other species of spirochaetes, FlaA comprises a flagellar sheath around the FlaB core (Charon and Goldstein, 2002; Li *et al.*, 2000b). In *B. burgdorferi* there is considerably less FlaA than FlaB, which is different than other spirochaetes (Ge *et al.*, 1998; Motaleb *et al.*, 2004). Preliminary evidence indicates that the limited amount of FlaA in the cell resides on the surface of a PF proximal to the basal body (S. Satoshi, M. Motaleb, S. Aizawa, N.W. Charon, unpublished). The precise role of FlaA in motility is unknown, but we know from work on *Brachyspira hydysenteriae* that FlaA influences the helical shape of the PFs (Li *et al.*, 2000a, 2008). It may be that FlaA in *B. burgdorferi* stabilizes the shape of the PF to a specific helical configuration near the basal region. Site-directed mutagenesis of *flaA* in *B. burgdorferi*, and in other spirochaetes, results in cells that still have PFs and are motile, but their motility appears decreased compared to the wild type (M.A. Motaleb and N.W. Charon, unpublished) (Li *et al.*, 2000a, 2008).

The structure of the PF has been studied in some detail, and except for the presence of the FlaA protein, they are similar in structure to flagella of other bacteria. As mentioned, most purified PFs are helical, left-handed, and have a helix pitch of 1.48 μm, and a helix diameter of 0.28 μm at neutral pH (Charon *et al.*, 1992a). However, recent results (S. Satoshi, C. Aizawa, and N.W. Charon, unpublished) suggest that there is a minor population of PFs (10%) with a larger helix pitch of 1.88 μm. In addition, the PFs change their shape on increasing pH, indicating that the PFs of *B. burgdorferi* can undergo helical transformation in a manner similar to the flagella of other bacteria. The presence or absence of FlaA associated with the PFs did not seem to influence the helical shape of the PFs. Furthermore, FlaA was found to readily dissociate from the PFs from the FlaB core at high alkalinity (at a pH greater than 10.8).

The hook structure associated with the PF is unique in spirochaetes in those species so far examined. This structure serves as a flexible coupling between the PF filament and the motor apparatus. In most bacteria, the hook is made up of a polymer of FlgE that readily dissociates into monomers. However, in *T. phagedenis*, *T. denticola*, and *B. burgdorferi*, FlgE is covalently cross-linked (Sal *et al.*, 2008; Chi *et al.*, 2002; Limberger *et al.*, 1994). Thus, Western blots were used to analyse cell lysates and purified hook-basal bodies complexes of *B. burgdorferi* with antibodies to FlgE (Sal *et al.*, 2008). The strongest reactive proteins migrated as high molecular mass complexes greater than 160 kDa, whereas there was slight reactivity for the monomeric form (46 kDa). Reagents such as dithiothreitol, 8 M urea, and 8 M guanidine-HCl did not result in dissolving the complex. Furthermore, the small amount of FlgE detected in a *fliF* mutant was not cross-linked. Because *fliF* encodes the MS ring of the basal body and is essential for secretion of other basal body proteins, evidently FlgE needs to be in the periplasmic space for cross linking to occur. Why is the hook protein cross-linked in spirochaetes? One possibility is that cross-linking of FlgE adds extra strength and flexibility to the hook in a manner similar to what is found in the cross-linking of the head proteins of lamdoid phage HK97 and the pili proteins of *Streptococcus pyogenes* (Kang *et al.*, 2007; Hendrix, 2005). This cross-linking probably results in an increase in structural strength and flexibility. It may be that PFs need to exert a stronger force as they rotate in the periplasmic space as compared to flagella of other bacteria that are exposed to the ambient medium and that a flexible and stable hook structure is necessary to achieve this strength. Most recently, using proteome array methodology, infected humans and mice were found to have a strong immune response to FlgE (Barbour, 2008). Conceivably, the cross-linked nature of FlgE could possibly enhance its immunogenicity during infection.

There are some new exciting results associated with the basal body–motor complex of *B. burgdorferi*. The three dimensional molecular structure of *B. burgdorferi*'s flagellar motor *in situ* was studied by using cryo-electron tomography. As noted by Murphy *et al.* in their study of *T. primitia* (Murphy *et al.*, 2006), the analysis of basal bodies *in situ* in certain spirochaetes has an advantage over other bacteria, as the cell's small diameter allows better visualization than wider bacteria such as *E. coli*. By averaging 2-D images of over 950 flagellar motors from about 250 frozen-hydrated cells, a detailed structure of an intact flagellar motor at 3 nm resolution was obtained, as shown in Fig. 7.5 (Liu *et al.*, 2009). The centre section of the averaged model (Fig. 7.5A) without using any symmetry shows the hook, C-ring (cytoplasmic part of the motor), stator (closely

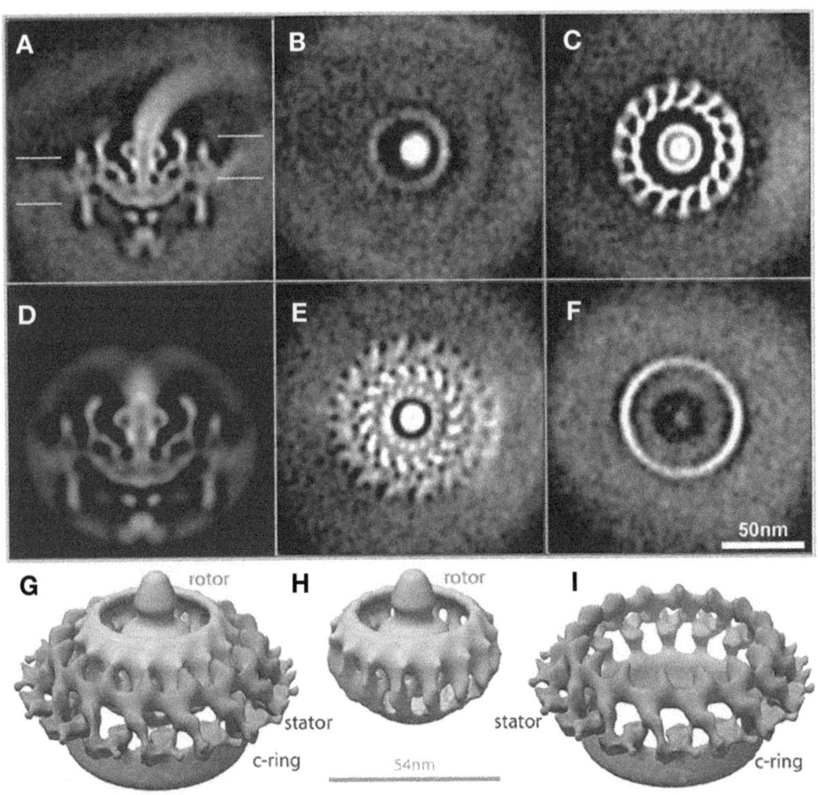

Figure 7.5 Three-dimensional architecture of *B. burgdorferi* flagellar motor *in situ*. The centre sections of 3-D average of the flagellar motor (A) without and (D) with 16-fold symmetry. Locations of the hook, rotor, stator and C-ring are shown. (B, C, E and F) are horizontal cross sections through a flagella motor at hook (B), rotor (c and e), and C-ring (F) as indicated in (A). The map clearly shows the 16-fold symmetry of rotor and its adjacent stators. 3-D maps of the intact flagellar motor (G), rotor (H), C-ring and stator (I) *in situ*. [Please supply a high resolution version of this figure]

associated with the peptidoglycan), rotor, P ring (peptidoglycan bushing ring), and export apparatus (the apparatus involved in secreting flagellar protein components via the type III secretion pathway to the periplasmic space). Fig. 7.5B–F are horizontal cross-sections through a flagella motor, and the level of the sections are indicated in Fig. 7.5A. In Fig. 7.5B, the cross-section skims the top of the motor and the core of the flagella and the uppermost portion of the rotor are visible. Deeper in the inner membrane in Fig. 7.5C, we find the flagella core surrounded by the P ring cuff. The rotor surrounds these structures, and at the level of 7.5C, it interacts with the stator to help drive flagellar rotation. Fig. 7.5E transects the middle of the rotor body. The protein interactions present at this level are exceedingly complex and rotor–rotor-stator (inner to outer) rings surrounding the flagella core take on a geared appearance. The C ring, comprising proteins FliM, FliN and FliG, is visible in Fig. 7.5A and F. A hint of the export apparatus responsible for the growth of the flagella is present in the centre of the C ring formed by FliM and N. There is 16-fold symmetry in the major portion of the flagellar motor as indicated in Fig. 7.5C and E, although there is no visible symmetry in the C-ring as shown in Fig. 7.5F. By imposing 16-fold symmetry along the rod axis, we improved the structure with better signal-to-noise ratio and without much loss of information as shown in Fig. 7.5D and G. This unique structure provides very important information about the architecture (Fig. 7.5G) of the intact flagellar motor *in situ*. The novel reconstruction shows the rotor (Fig. 7.5H) and its surrounding stator (Fig. 7.5I), which has not been seen in the previous electron microscopy structures (Berg, 2003). It is interesting that a structure surrounding the export apparatus that may be a new unreported feature is present in the averaged data. In addition, the C ring complex is also larger compared to other bacteria. Similar results with 7.0 nm resolution were obtained from cryoelectron microscopy tomography analysis of the termite gut spirochaete *T. primetia* (Murphy *et al.*, 2006). These results are consistent with the notion that the flagella in spirochaetes need to generate a higher torque as they reside and rotate in the periplasmic space, and also concomitantly interact with the cell cylinder to impact cell shape.

More recently, distinctive structural changes have been identified from resulting mutations of a single flagellar gene (J. Liu, D. J. Botkin, and S.J. Norris, unpublished). These exciting results indicate that a comprehensive approach will permit a thorough analysis of the structure and assembly of the flagellar motor of *B. burgdorferi* at a molecular level. Moreover, the results obtained should be relevant to achieving a general understanding of the workings of the bacterial flagellar motor.

7.3 Dynamics of motility

Most externally flagellated bacteria have two different swim modalities: runs and tumbles (Macnab, 1996; Armitage, 1999; Berg, 1975; Berg, 2000). For example, in *E. coli*, runs – or translational motility – occur when the flagella rotate counterclockwise (as a frame of reference, the flagella are viewed from the end of the flagella towards its insertion into the cell). During a run species that have peritrichous flagella form a bundle behind the cell and act as a propeller. When one or more of the flagella reverse direction and rotate clockwise, the cell tumbles (Macnab, 1996; Armitage, 1999; Turner *et al.*, 2000). Some of the filaments leave the flagellar bundle and the cell body reorients. It then runs in a new direction when the flagella revert to the counterclockwise direction. In some species, such as *Rhodobacter sphaeroides* and *Sinorhizobium meliloti*, the flagella stop or slow down instead of changing their direction of rotation (Armitage and Schmitt, 1997).

The swimming behaviour of spirochaetes, including *B. burgdorferi*, is more complicated than that of other bacteria as a consequence of their complex geometry. *Borrelia burgdorferi* and other spirochaetes have three motility modes (Fig. 7.6): two modes of runs (with either end leading, top two panels), and one mode of non-translational motility (bottom two panels) (Fosnaugh and Greenberg, 1988; Charon *et al.*, 1984; Goldstein *et al.*, 1994; Li *et al.*, 2000b). During runs, the anterior ribbon of PFs rotates in one direction (counterclockwise), and the posterior ribbon rotates in the opposite direction (clockwise). Thus, the ribbons of PFs rotate asymmetrically. Reversals occur when both ribbons change their direction of rotation. The third mode is a non-translational interval. In this mode, the ribbons rotate in the same direction: both clockwise or

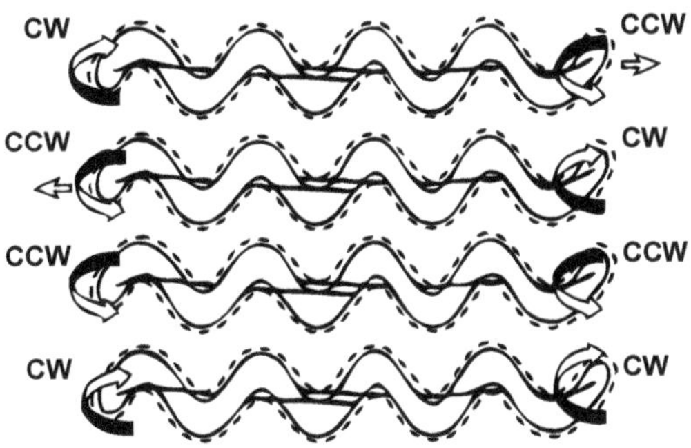

Figure 7.6 Swimming cells of *B. burgdorferi* as a function of direction of rotation of the periplasmic flagella. Open arrows indicate direction of swimming. Dotted lines represent the outer membrane sheath. Grey arrows indicate direction of rotation of the periplasmic flagella (*thin lines*). For simplification, only one periplasmic flagellum is shown attached at each end of the cell cylinder. In *B. burgdorferi*, there are between 7 and 11. The top two panels show translational forms, and the bottom two non-translational forms. Reprinted with permission from the *Annual Review of Genetics* (Charon and Goldstein, 2002).

both counterclockwise. In many species of spirochaetes, including *B. burgdorferi*, the cells become quite distorted during this interval and may bend in the middle; this distorted form is referred to as a flex (Goldstein *et al.*, 1994; Fosnaugh and Greenberg, 1988; Goldstein *et al.*, 1994). The two non-translating forms are morphologically indistinguishable for most spirochaete species except for the Leptospiraceae (Goldstein and Charon, 1988, 1990; Fosnaugh and Greenberg, 1988; Charon *et al.*, 1984; Berg *et al.*, 1978). In *B. burgdorferi*, preliminary evidence suggests that the flex is associated with both groups of PFs rotating clockwise. However, there is a second non-translating form; specifically, the cell seems not distorted and both ends are either gyrating counterclockwise or stopped. This stopped form may be due to the PFs being jammed or the motors not being energized (S.F. Goldstein, unpublished)

The swimming behaviour of translating *B. burgdorferi* has been studied in some detail (Goldstein *et al.*, 1994). Waves propagate from the front to the back at a specific frequency, referred to as the beat frequency, of 5–10 Hz in growth medium at room temperature. In contrast to eukaryotic flagella, propagating waves were full size at the anterior end instead of starting as small bends and increasing in size. The cells swim at a mean speed of 4.25 μm/s. These waves propagate at a speed of 34 μm/sec relative to the cell. The ratio of the swim speed of the cell to wave propagation is 0.12, i.e. in the time it takes a wave to travel the length of the cell, the cell advances 12% of its length through the medium. As with other spirochaetes, the swim speed of *B. burgdorferi* increases markedly with viscosity, with the specific requirement that the agent that increases viscosity causes the medium to be gel-like. Thus, methylcellulose and hyaluronic acid markedly increase the speed of *B. burgdorferi*, whereas Ficoll, which increases viscosity but does not cause the medium to be gel-like, slows the cell down (S.F. Goldstein and N.W. Charon, unpublished) (Berg and Turner, 1979; Kimsey and Spielman, 1990; Goldstein *et al.*, 1994). The current thought is that *B. burgdorferi* and other spirochaetes increase their speed by being able to exert forces normal to the complex matrix of the gel due to their flat wave or helical body. In addition, they can do so with little slippage, i.e. with each wave (or turn of the helix in the helically shaped spirochaetes) propagated, they move that much farther forward. Along these lines, spirochaetes such as the Leptospiraceae and *B. burgdorferi*, when tethered to a glass surface, stationary latex bead, or a eukaryotic cell, rapidly move back and forth across with no slippage but are unable to break free of the attachment (Charon *et al.*, 1981, 1984; Malawista and de Boisfleury, 2008). This motion has been referred to in the literature as staple motion: it is as if the cell is stapled to the stationary solid surface – it moves but can't get away (Cox and Twigg, 1974). To explain this motion, several lines of evidence suggest that part of the outer membrane surface is attached to the solid surface, perhaps by an outer surface protein. This protein is able to rapidly move laterally relative to the cell through the outer membrane as the cell swims; it is analogous to a capping phenomenon seen in eukaryotic cells. One common question concerns whether both ribbons of PFs actively rotate during translation as the description above contends. As an alternative model, perhaps one ribbon rotates while the other is stationary. In this model, cell reversals occur when motor rotation switches from one ribbon to the other. To test this hypothesis, *B. burgdorferi* cells were tethered to a glass surface and their motions video-recorded (Li *et al.*, 2002). Because the PFs have a skeletal function, PF rotation was monitored by tracking the direction of gyration of a given end. Tethered cells were analysed that gyrated their cell ends with little or no motion in the central part of the cell. Both ends actively gyrated, and changes in gyrational direction (clockwise, counterclockwise) of one end relative to the other were readily apparent. Occasionally, the ends would alternate with one end stopping and the other starting. In addition, in cells with both ends simultaneously gyrating, the beat frequency at one end was occasionally seen to vary from that of the other end. These results suggest that the PFs at both cell ends in *B. burgdorferi* are capable of rotation, that the ribbons can both be active at the same time, and that they can change direction of rotation independent of the direction of rotation at the other end. A similar conclusion based on tethered cell analyses was reached with the Leptospiraceae and *T. phagedenis* (N.W. Charon, unpublished; Charon *et al.*, 1984, 1991).

The skeletal function of the PFs forms the basis of motility models for several spirochaete species including that for *B. burgdorferi* (Goldstein and Charon, 1988; Charon *et al.*, 1991; Berg *et al.*, 1978; Goldstein *et al.*, 1994; Li *et al.*, 2000b; Goldstein *et al.*, 1996). All these models rely on evidence that the PFs rotate between the outer membrane sheath and cell cylinder. Recently, a model was described for *B. burgdorferi* translational motility (Goldstein *et al.*, 1994; Li *et al.*, 2000b; Goldstein *et al.*, 1996) (http://stock.cabm.rutgers.edu/blast/video1.mov). This model states that, *in situ*, the relatively rigid PFs are juxtaposed with the highly flexible protoplasmic cell cylinder due to containment by the outer membrane sheath. The PFs have a defined shape that, as mentioned, is a left-handed helix. Because of the close interaction of the PFs with the protoplasmic cell cylinder, the PFs influence the shape of the entire cell. Translational motility occurs as the result of counterclockwise rotation of the PFs as viewed from the back of the cell. This rotation causes backward-moving waves to be propagated down the length of the cell. Thus, the action of the PFs is likened to that of a worm gear in generating wave propagation. Clockwise rotation of the cell about the body axis counterbalances the rotation of the PFs. Mutants that lack the PFs are non-motile because they have lost the ability to propagate backward-moving waves. In addition, these mutants are rod-shaped as a result of their loss of skeletal function associated with the PFs.

The mathematical model that describes the shape of *B. burgdorferi* suggests a slightly different mechanism for motility. As mentioned previously, the work by Dombrowski *et al.* (2009) shows that the PFs are stiffer than the cell body, which is consistent with the motility model just mentioned. However, the PFs are not rigid compared to the cell body and are significantly distorted when embedded in the periplasmic space compared to their shape when purified. Therefore, it is possible that rotation of the flagella leads to further deformation of the PFs, which in turn drives the cell body to undulate as a travelling wave.

Now that the physical parameters (i.e. the bending moduli) of the PF and cell body have been measured, a grounded mathematical model that describes the physical mechanism of motility is possible and should answer the question of the specific dynamic interaction between the PF and cell body that leads to motility of *B. burgdorferi*. When coupled with recent work describing swimming in gel-like media (Lauga, 2007; Fu *et al.*, 2007, 2008), these models could lead to a further understanding of tissue invasion during the disease process.

7.4 Chemotaxis

Bacterial chemotaxis is a dynamic and complex sensory transduction pathway that enables cells to respond to environmental stimuli. Cells swim towards a favourable medium, or away from one that is toxic. A two component phosphorelay system plays an essential role for this response. From the strandpoint of chemotaxis, the best studied organisms are *Escherichia coli* and *Salmonella enterica* serovar Typhimurium; these bacteria serve as the paradigm of this highly complex stimulus response system (Armitage, 1999; Hazelbauer *et al.*, 2008; Bren and Eisenbach, 2000). Briefly, a family of chemoreceptor proteins that span the cytoplasmic *membrane*, referred to as methyl-accepting chemotaxis proteins (MCPs), regulates the autophosphorylation of the sensor histidine kinase CheA, which is associated with the cytoplasmic portion of the MCP. These MCPs bind to attractant ligands in the periplasmic space, and the binding modulates the degree of CheA autophosphorylation. CheA, in turn, transfers the phosphoryl group to the response regulator CheY, producing CheY-P, which then interacts with the switch complex at the flagellar motor to change direction of flagellar rotation from counterclockwise to clockwise. Counterclockwise rotation results in smooth swimming cells that 'run', and clockwise rotation results in tumbling. When MCPs bind attractants, runs are extended, yielding a 'biased random walk' towards attractants (Wadhams and Armitage, 2004); the opposite occurs when MCPs bind repellents. Null mutants in *cheY* or *cheA* are non-chemotactic and continuously rotate their flagella counterclockwise and consequently fail to tumble. For the bacteria to respond within a short period of time, CheZ promotes dephosphorylation of CheY-P. Mutants in CheZ are non-chemotactic and constantly tumble. Bacteria become adapted to the environment that contains the attractant, and with time run and tumble to the same frequency as in the pre-stimulus state. Methylation of the MCPs by CheR results in a dampening of the response that returns the bacterium to its pre-stimulus state. In addition, CheB, which is also activated by phosphotransfer from CheA, promotes demethylation of the MCPs so that they can respond again to a further change in attractant levels.

The chemotactic mechanism of *B. burgdorferi* is enigmatic (Charon and Goldstein, 2002). There are some obvious differences between *B. burgdorferi* and the *E. coli-S. enterica* paradigm. One major difference is gene copy. Whereas *E. coli* and *S. enterica* have only one copy of each of the chemotaxis and motility genes, *B. burgdorferi*, like many bacteria, has several homologues of these genes. Based on sequence identity with other bacteria, it has two *cheA* (*cheA1* and *cheA2*), three *cheY* (*cheY1*, *cheY2*, *cheY3*), two *cheB* (*cheB1*, *cheB2*), two *cheR* (*cheR1*, *cheR2*), and two *fliG* (*fliG1*, *fliG2*) homologues (Fraser *et al.*, 1997). FliG in other bacteria is part of the C-ring complex of the motor that determines the direction of flagellar rotation (Berg, 2003). Second, *B. burgdorferi* lacks *cheZ*, but encodes a *cheX*, which, as discussed below, is a phosphatase active on CheY-P. Finally, *B. burgdorferi* has three modes of motility as already mentioned (run, reverse, and flex – flex is analogous to 'tumble' in other bacteria), not just two as found for *E. coli* and *S. enterica*. *B. burgdorferi* must be able to control the direction of the PF ribbons at each end to carry out the chemotactic response. For example, for the organism to run, there is asymmetrical rotation of the groups of PFs: the anterior end rotates its PFs counterclockwise, and the posterior end rotates its PFs clockwise. Thus, during a run, instead of all the flagella rotating counterclockwise as they do for *E. coli* and *S. enterica*, the ribbons of PFs need to rotate in opposite directions. Indeed, there is a higher level of complexity for *B. burgdorferi* to carry out the chemotactic response. Moreover, *B. burgdorferi* can reverse directions quite fast (less than 0.3 seconds). Thus, both groups of PFs change direction almost simultaneously during a reversal. How they achieve this reversal is not known, but it is possible that a membrane potential mediates coordination of the groups of PFs at each end of the cell, which is essential for chemotaxis [see Charon and Goldstein (2002) for more detailed discussion].

To investigate putative chemotaxis genes in *B. burgdorferi*, it was necessary to develop a rapid, quantitative chemotaxis assay and identify chemoattractants. Traditional capillary tube chemotaxis assays, in which cells entering capillary tubes with attractants are plated and colonies are counted the next day (Adler, 1973), were not amenable to studying *B. burgdorferi*, due to their slow growth rate (8–12 hr doubling time). Initially, rabbit serum and tick saliva were shown to be attractants, by manually counting cells in capillary tubes (Shi *et al.*, 1998; Shih *et al.*, 2002). A modified capillary tube assay was developed, in which cells entering capillary tubes were rapidly enumerated by flow cytometry (Bakker *et al.*, 2007; Motaleb *et al.*, 2007a). Significant progress has recently been made in identifying a number of defined chemicals as chemoattractants. These attractants included glucosamine, N-acetylglucosamine, glucosamine dimers, and glutamate. Glucose was found to be a weak attractant (Motaleb *et al.*, 2007a; Bakker *et al.*, 2007). Having established the modified capillary tube chemotaxis assay, specific genes could be targeted, and their roles in *B. burgdorferi* chemotaxis could be investigated.

cheA1 and *cheA2* were initial putative chemotaxis genes that were targeted to investigate their roles in chemotaxis. Both proteins are expressed in *B. burgdorferi* (Li *et al.*, 2002; Motaleb *et al.*, 2005). Mutants in *cheA2* were found to be non-chemotactic, and they constantly ran in one direction – they failed to flex, stop or reverse (Li *et al.*, 2002). The altered phenotype in *cheA2* mutants did not appear to be the result of any polar effect (Li *et al.*, 2002). In contrast, *cheA1* mutants appeared to exhibit wild-type cell phenotype in that they ran, reversed and flexed; and they were chemotactic (Li *et al.*, 2002). In addition, the double mutant *cheA1cheA2* had the same phenotype as the single *cheA2* mutant. Thus, *cheA2*, but not *cheA1*, was shown to be necessary for chemotaxis. Similar approaches were used to determine the roles of the three *B. burgdorferi cheY*s in chemotaxis and motility. Preliminary evidence indicates that *cheY3* mutants have the same phenotype as *cheA2* (M. Motaleb, R. Bakker, C. Li, and N. Charon, unpublished data). In addition, phosphotransfer studies demonstrated that CheA2 was capable of autophosphorylation using ATP and that CheA2-P was able to transfer phosphate to CheY3 (Motaleb *et al.*, 2005; Motaleb *et al.*, 2007b). It is not clear why *cheA2* but not *cheA1* appears to be required for wild-type motility and chemotaxis. One possibility is that under culture conditions tested (BSK-II medium, 6% rabbit serum, 33°C) only CheA2 is involved in chemotaxis/motility, but under different conditions, such as in the tick, CheA1 may play dominant roles in chemotaxis/motility. Additional studies are required to answer this puzzling question.

Why did the *cheA2* mutant constantly run? In *E. coli* and *S. enterica*, *cheA* and *cheY* mutants constantly run (Parkinson, 1977; Bren and Eisenbach, 2000), so *B. burgdorferi* is similar to these bacteria. However, in these bacteria, their flagella constantly rotate counterclockwise. It follows, then, if all the PFs of *B. burgdorferi* rotated counterclockwise due to the absence of CheA2, the cells should constantly flex, not run. One possible explanation for these results is that in the absence of CheA2 (the default state), the motors at one end rotate counterclockwise and the other clockwise. Conceivably, specific proteins which influence flagellar rotation in the default state localize at the motors at one cell end and not at the other. This possibility (referred to as the asymmetrical motor model) is further strengthened by what is found in other bacteria; some bacteria show differences of specific protein localization at one cell end compared to the other (Losick and Shapiro, 1999; Lybarger and Maddock, 2001).

The role of *cheX* was examined using targeted mutagenesis. As mentioned earlier, *cheZ* mutants of *E. coli* and *S. enterica* constantly tumble. The explanation for this phenomenon is that in a *cheZ* mutant, the pool of CheY-P is relatively high. A high CheY-P pool should increase the probability of CheY-P binding to the motor complex with a resultant clockwise flagellar rotation. The *cheX* mutant was studied in some detail. Purified CheX was prepared and found to dephosphorylate CheY3-P, and did not affect the phosphorylation state of CheA2-P, indicating it exhibited the expected activity to remove phosphate from *B. burgdorferi* CheY-P (Motaleb *et al.*, 2005; Motaleb *et al.*, 2007b). When the phenotype of the *cheX* mutant was analysed, it was found to be similar to that of the *E. coli* and *S. enterica cheZ* mutants: It constantly flexed. This result was unexpected based on the asymmetrical motor model: under low CheY-P conditions as with a *cheA2* mutant, one group of motors rotates clockwise, the other counterclockwise. Under high CheY-P conditions, as with a *cheX* mutant, the cell should constantly run, as a *cheA2* mutant, but in the opposite direction. One possible explanation for the constantly flexing phenotype is that there may be another phosphatase in the *B. burgdorferi* genome that has yet to be identified and that attacks CheY-P. Regardless, it is clear at this time that much has to be sorted out before we can understand the steps involved in deciphering the genetics and physiology of the complex chemotactic response of *B. burgdorferi*.

It is intriguing that CheX appears to be the sole phosphatase that operates in *B. burgdorferi* chemotaxis (Motaleb *et al.*, 2005). This finding suggests that CheX could play a central role in mediating chemotaxis in a variety of the environments experienced in the *B. burgdorferi* lifecycle, thus making it a potential target for future therapeutics. Recently, there have been some notable advances in elucidating the structure and enzymatic mechanism of CheX orthologues. CheX bears sequence homology to both CheC and FliY, chemotaxis phosphatases present in some other bacteria, but the CheX/CheC/FliY family bears no sequence similarity to CheZ, the *E. coli* chemotaxis phosphatase. The X-ray crystal structure of *Thermotoga maritima* CheX (Park *et al.*, 2004) revealed the protein fold that presumably also describes *B. burgdorferi* CheX. Although the overall fold – a globular domain with both alpha helices and beta strands –contrasts the unique long four-helix bundle of CheZ (Zhao *et al.*, 2002), there are some striking structural similarities. CheZ is a homodimer with two identical CheY binding sites; CheX is also a homodimer, both in the X-ray structure (Park *et al.*, 2004) and in solution (Motaleb *et al.*, 2005). Each CheY binding surface on CheZ has two conserved residues – an aspartate and a glutamine – exactly one alpha helical turn apart that insert directly into the CheY active site and are directly involved in catalysis (Zhao *et al.*, 2002). Likewise, there are two conserved residues – again an acidic residue (glutamate) and an amide residue (asparagine)– on CheX (correlating to Glu96 and Asn99 of *B. burgdorferi* CheX) that are located on consecutive turns of an exterior, solvent-exposed alpha helix on the *T. maritima* CheX structure. It is probable that the CheX acid and amide residues interact with CheY in a similar manner to that which occurs in CheZ.

It will be interesting to see if CheX shares functional features of CheZ. CheZ accelerates the intrinsic autodephosphorylation rate of CheY by about 100 fold bringing it within the appropriate time scale (~100–200 ms half-life) to correlate with the ~200 ms chemotactic response time in *E. coli*. The catalytic activity of CheZ is regulated by its CheY-P substrate, such that activity is suppressed at low CheY-P concentrations and enhanced at high CheY-P levels (Silversmith *et al.*, 2008). This type of allosteric regulation may be mediated by the dimeric structure of CheZ, which CheX shares. Finally, in the cell, a portion of the *E. coli* CheZ pool co-localizes with the large polar chemosensory receptor complex (Vaknin and Berg, 2004). Cell localization studies of *B. burgdorferi* CheX will reveal whether association of the chemotaxis phosphatase with chemoreceptors and the sensor kinase is a common theme in chemotaxis. Such studies may also shed light on mechanisms of maintaining asymmetric motor rotation that are fundamental to *B. burgdorferi* swimming behaviour.

7.5 Organization, regulation and expression of motility genes

There are at least 57 putative motility and chemotaxis genes in *B. burgdorferi* (Fig. 7.7). These genes are identified as such by targeted mutagenesis in *B. burgdorferi*, their similarity in sequence to motility and chemotaxis genes in other bacteria, or their clustering with other known motility genes. These putative motility and chemotaxis genes comprise at least 7% of all the deduced chromosomal open reading frames. In addition, the whole genome two-hybrid assays of *T. pallidum* and *C. jejuni* have predicted that the number of proteins that are involved in the network of bacterial motility is more than we have known. For example, at least 18 new motility-related proteins were identified in the genome of *T. pallidum* (Rajagopala *et al.*, 2007). Some of these homologues are also present in the genome of *B. burgdorferi*, indicating that there are more genes involved in motility and chemotaxis than what was annotated. The presence of such a high percentage of motility genes emphasizes the importance of motility for survival in nature. In analysing the motility genes, both the gene order and the deduced amino acid sequences are most similar to those of other spirochaete species, followed by *Bacillus subtilis*. These results again reinforce the common evolutionary ancestry of spirochaetes. Because spirochaete motility genes have such a high homology to the motility genes of other bacteria, the basic apparatus and mechanism required for flagella rotation are also well conserved and present in spirochaetes. In fact, *fliI* (part of the secretion apparatus of the flagella) of *B. burgdorferi* is so similar to those of *E. coli* and *S. typhimurium* that its expression in these bacteria results in negative complementation effects on flagella synthesis and motility (Ge *et al.*, 1997b). A likely explanation for this negative complementation is that the *B. burgdorferi* motility proteins compete and interfere with their counterparts with respect to assembly of flagellar components (Ge *et al.*, 1997b).

The motility genes are clustered very tightly with little intervening sequences between genes.

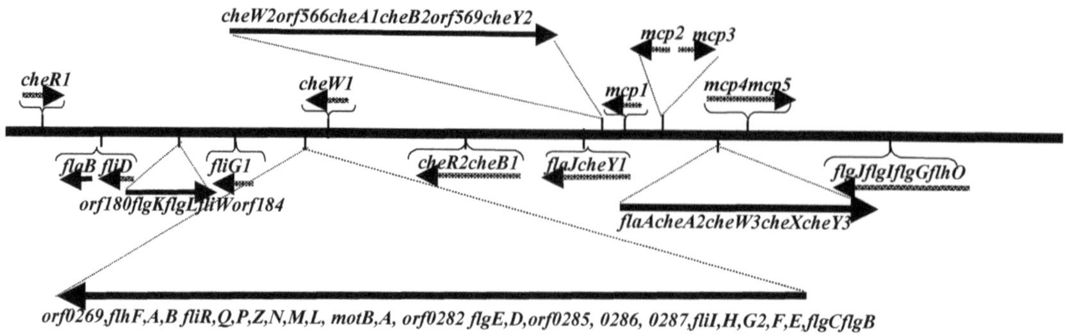

Figure 7.7 Organization of the motility genes of *B. burgdorferi* (Fraser *et al.*, 1997; Ge and Charon, 1997a–c; Ge *et al.*, 1997a,b). Black arrows indicate direction of transcription of genes, and promoters have been identified by primer extension and operons by RT-PCR (Fraser *et al.*, 1997; Ge and Charon, 1997a–c; Ge *et al.*, 1997a,b). Grey stippled arrows indicate presumed promoters, operons, and direction of transcription based solely on sequence analysis.

In *E. coli* there are 13 motility operons (Macnab, 1996). We estimate that there are approximately 16 in *B. burgdorferi*, and some of these genes are co-transcribed with other genes that are probably unrelated to motility and chemotaxis (Fig. 7.7). In *B. burgdorferi*, there is a remarkably large motility gene cluster of 21 kb comprising 26 open reading frames (Ge *et al.*, 1997b). Its gene order is remarkably similar to that of a gene cluster found in *B. subtilis*. Primer extension and RT-PCR analysis indicate that this cluster constitutes an operon with transcription initiated by a single σ^{70}-like promoter that maps upstream of the rod protein *flgB* within the *hslU* gene. Some of the proposed motility genes are unique. For example, there are a number of genes initially, designated as *flb* genes (Ge *et al.*, 1997b), which cluster with other motility genes but have no obvious homologues in other bacteria besides spirochaetes (Fraser *et al.*, 1997). Because *B. burgdorferi* must control the rotation of the PFs at both ends of the cell, perhaps the genes of unknown function that map with the motility gene clusters are involved in this coordination. As already mentioned, there are also multiple copies of genes involved in chemotaxis and motility. For example, there are two *fliG* flagella motor switch gene homologues in both *B. burgdorferi* and in other spirochaetes whereas there is only one *fliG* gene in other bacteria. In *B. burgdorferi*, these two genes that are located within different operons (Fig. 7.7). Mutant analysis indicates that *fliG2* has a similar function as that in *E. coli* and *S. enterica* (C. Li and N.W. Charon, unpublished). Specifically, null mutants in this gene have a non-motile and aflagellated phenotype (Kihara *et al.*, 2000). On the other hand, mutants in *fliG1* are still motile, constantly run, and are unable to swim in highly viscous gel-like medium. Preliminary evidence suggests that FliG1 localizes at one end of the cell (C. Li and N.W. Charon, unpublished). One possibility is that FliG1 associates with the motors at that end of the cell to change the default state from counterclockwise to clockwise, and thus may augment the asymmetrical rotation of the PFs as described earlier.

With respect to the chemotaxis genes, it is not clear why there are multiple *cheA*, *cheY*, *cheB*, and *cheR* homologues. As previously mentioned, among the *cheA* and *cheY* genes, mutant analysis indicated that *cheA2* and *cheY3* were involved with chemotaxis. It may be that these other homologues of chemotaxis genes are involved in another phenomenon besides chemotaxis. Along these lines, other species of bacteria have multiple copies of chemotaxis genes and, in many cases, they do not seem to function in chemotaxis (Bourret *et al.*, 2002). Alternatively, as already mentioned, they may still function in *B. burgdorferi* chemotaxis, but under different conditions than they were tested. Along these lines, a body of microarray analyses indicates that some of these chemotaxis genes have altered their expression under different culture conditions or in response to different stimuli (Brooks *et al.*, 2003; Revel *et al.*, 2002; Tokarz *et al.*, 2004; Ojaimi *et al.*, 2003).

Borrelia burgdorferi is unique with respect to motility gene regulation. Motility gene expression, as described in *E. coli* and *S. enterica,* again serves as a well-studied paradigm. It is a finely orchestrated succession of motility gene expression and protein assembly that requires tight regulation. Motility gene expression and flagella synthesis is regulated by a cascade of transcriptional events involving the ordered expression of class 1, class 2, and class 3 motility genes (Chilcott and Hughes, 2000; Aldridge and Hughes, 2002). Expression of class 1 genes, which comprise the master operon, directs the transcription of class 2 genes encoding the structural proteins involved with basal body synthesis as well as specific regulatory proteins. These regulatory proteins control the expression of the class 3 genes, including those encoding flagellin and chemotaxis proteins. Among the class 2 regulatory genes are *fliA* (encoding the motility specific transcription factor σ^{28}) and *flgM* (encoding the anti-σ^{28} factor FlgM). FlgM and σ^{28} remain bound as a complex in the cytoplasm during basal body and hook synthesis. Upon completion of the hook, the hook-basal body complex provides an export route for FlgM. As FlgM exits the cell, σ^{28} is free to initiate the transcription of the class 3 genes, including the genes encoding flagellin and chemotaxis proteins. Thus, completion of the hook-basal body structure acts as a general assembly checkpoint in the regulation of flagella synthesis whereby flagellin and other genes are transcriptionally controlled in response to the state of hook completion (Hughes *et al.*, 1993). A similar scheme of cascade control of motility gene expression is found in many species of bacteria, and in some species, σ^{54} (RpoN) also directly participates in the cascade control of motility gene regulation (Anderson *et al.*, 1995; Niehus *et al.*, 2004; McCarter, 2001; Dasgupta *et al.*, 2003; Aizawa *et al.*, 2001). However, the regulation of PF synthesis in *B. burgdorferi* is markedly different compared to other motile bacteria. Homologues for the class 2 transcriptional regulatory genes *fliA* and *flgM* are not apparent, and no σ^{28} promoter consensus sequences are evident in the *B. burgdorferi* genome (Fraser *et al.*, 1997; Ge *et al.*, 1997b; Li *et al.*, 2000b). Furthermore, when analysing promoter sequences in *B. burgdorferi,* only σ^{70} promoters are evident in initiating motility and chemotaxis gene expression (Fraser *et al.*, 1997; Ge *et al.*, 1997b; Li *et al.*, 2000b). As discussed below, they lack cascade control of motility gene expression as found in other bacteria.

Mutant analysis has been important in analysing the control of motility genes in *B. burgdorferi.* As previously mentioned, targeted mutants in *flaB* fail to synthesize PFs, are non-motile, and have a rod-shaped morphology (Motaleb *et al.*, 2000). The only putative motility and chemotaxis protein that failed to accumulate to wild-type levels in the *flaB* mutant was the minor filament protein FlaA. However, the decrease of FlaA was found to occur at a post-transcriptional level since the level of *flaA* message in the mutant was equivalent to the wild type (Motaleb *et al.*, 2004). Experiments with the protein synthesis inhibitor spectinomycin indicated that FlaA turned over with a half-life of two hours. These results suggest that the decrease in FlaA could be either due to protein turnover or translational control, or both (Motaleb *et al.*, 2004).

Experiments with a targeted mutant in *flgE* were instructive and illustrate the uniqueness of the *B. burgdorferi* motility control system. As described above in other bacteria, the hook provides a regulatory checkpoint during flagella synthesis. Thus, in these bacteria, mutants that do not synthesize the hook protein fail to synthesize flagellar filament message and the encoded filament protein. When *flgE* was inactivated by insertional mutagenesis in *B. burgdorferi,* the inhibition of flagellin synthesis was similar in one way to these bacteria: Western blot analysis indicated that mutants in *flgE* markedly inhibited both FlaA and FlaB accumulation (Sal *et al.*, 2008). However, in contrast to these bacteria, both *flaA* and *flaB* message in the *flgE* mutants were the same as the wild type using quantitative RT-PCR assays. Further experiments indicated that there was no turnover of FlaB in the *flgE* mutant. As with the *flaB* mutant noted above, FlaA had a half-life of approximately 2 hours. These results strongly indicate that there is translational control of FlaB, and possibly FlaA, in *B. burgdorferi.* Although sigma factor cascade transcriptional control of flagella synthesis is a major mechanism of control in *C. crescentus, B. subtilis* and *S. enterica,* translation control of flagellin also occurs in these species (Aizawa *et al.*, 2001; Anderson and Newton, 1997; Yakhnin *et al.*, 2007; Llewellyn *et al.*, 2005; Aldridge and Hughes, 2002). As suggested by Yakhanin *et al.* for *B. subtilis* (Yakhnin *et al.*, 2007), such control could serve to fine tune flagellin synthesis in response to environmental changes. In the absence of cascade control of motility gene transcription, *B. burgdorferi* appears to rely exclusively on translational control of its motility genes. Finally, in contrast to many other species of bacteria, mutants in *flgE* did not result in the inhibition of chemotaxis gene expression (Sal *et al.*, 2008). These genes are generally the last genes transcribed in the cascade. We found that inactivation of *flgE* had no effect on the synthesis of CheA2, CheX, and CheY3, which are all known to be involved in chemotaxis by targeted mutagenesis (Komeda, 1982; Dasgupta *et al.*, 2003; Kutsukake *et al.*, 1990). These results further reinforce the proposition that *B. burgdorferi* lacks cascade control of motility gene transcription.

7.6 Role of motility and chemotaxis in Lyme disease

Deciphering the roles of motility and chemotaxis in Lyme disease is at its early stage, but significant progress is beginning to be made. Many Lyme disease researchers have long believed that motility and chemotaxis are involved in colonization and/or the disease process. This hypothesis is supported by the following reports: As previously stated, approximately 7 percent of the total open reading frames on the linear chromosome comprise putative chemotaxis and motility genes (Fraser *et al.*, 1997; Casjens *et al.*, 2000), suggesting that motility and chemotaxis is likely to be important for the organism's life cycle. Additionally, *B. burgdorferi* reside in different host environments. Experimental conditions mimicking these disparate environments indicate that the expression of many genes, including those for motility and chemotaxis, are expressed and in some cases modulated during infection (Brooks, 2003;Revel, 2002; Tokarz, 2004; Ojaimi, 2003; Barbour, 2008). *Borrelia burgdorferi* is presumably highly invasive (Szczepanski *et al.*, 1990). The organism can traverse the extracellular matrix, penetrate the endothelial cell lining and colonize the large joints, skin, heart and other tissues in the mammalian host. Most recently, exciting experiments indicate that *B. burgdorferi* labelled with green fluorescent protein can be tracked *in vivo*, and that these cells are seen to penetrate the endothelium (Moriarty *et al.*, 2008; Norman *et al.*, 2008). As stated by the authors, '...this is the first report of high-resolution 2-D and 4-D visualization of dissemination of a bacterial pathogen in a living mammalian host, and provides the first direct insight into spirochaete dissemination *in vivo*.' In addition, motility mutants of *B. burgdorferi* are found to be less able to penetrate tissue cell monolayers and bind to type 1 collagen lattices (Sadziene *et al.*, 1991; Zambrano *et al.*, 2004). Thus, motility, and the ability to swim in highly viscous gel-like media including connective tissue, is thought to play a role in the attachment and invasive process. These hypotheses are supported by the fact that motility and chemotaxis were shown to be critical for the disease process in several species of pathogenic bacteria. Specifically, motility and chemotaxis mutants in *Helicobacter pylori, Vibrio cholerae, S. enterica*, and the spirochaete *B. hyodysenteriae* were all shown to be unable to infect and/or colonize their respective host tissues (McGee *et al.*, 2005; Josenhans and Suerbaum, 2002; Ottemann and Miller, 1997; Kennedy *et al.*, 1997).

Chemotaxis and motility may be critical for *B. burgdorferi* to complete its life cycle, going back and forth from the tick to the mammalian host. One could hypothesize the following: First, within the tick, *B. burgdorferi* motility and chemotaxis is essential for the spirochaetes to migrate from the intestinal gut to the salivary glands and initiate a new infection. Second, once deposited in the skin of the mammalian host, chemotaxis and motility is important for the organisms to migrate from the site of deposition into the circulatory system of the host. Third, localization of the spirochaetes into certain tissues, such as the joints, is dependent on chemotaxis and motility. Finally, when the tick bites a mammalian host, the spirochaetes are attracted to the site (e.g. by recognition of components of the tick saliva in the extracellular matrix), and are taken up during feeding to complete the cycle (Shih *et al.*, 2002). The experiments are beginning to be done to test these hypotheses.

One major concern is that virulence is often unstable, and sorting out whether a given gene is involved in the disease process is complicated. As discussed elsewhere in this book, this is particularly true for *B. burgdorferi*, as infection *in vivo* is associated with plasmids that are often lost during growth *in vitro* (Purser and Norris, 2000; Labandeira-Rey *et al.*, 2003; Elias *et al.*, 2002). Thus, it is critical to use a low passage infectious strain in virulence studies, and to insure that all the natural plasmids are retained before such studies are performed. These experiments are difficult, and constructing the mutants and complemented strains is quite demanding. These technical challenges have slowed progress in genetically dissecting the role of motility and chemotaxis in virulence. However, there have been some recent breakthroughs, including two recent studies. In both of these studies, PCR tests were done to insure that there was no loss of plasmids essential for virulence. First, a mutant in a flagellar motor protein encoded by the *fliG1* was constructed using the *Himar 1* transposon and analysed for its role in a mouse model. Needle inoculation of the *fliG1* mutant into mice was reported to be non-infectious (Botkin *et al.*, 2006). Additionally, a targeted mutant in the *fliG1* was constructed and the mutant was also complemented using a shuttle vector (C. Li and F.T. Liang, unpublished). The targeted mutant was found to be avirulent in a mouse model of Lyme disease whereas the complemented strain restored the infectivity to the wild-type level. Secondly, a targeted mutant in the periplasmic core protein encoded by the *flaB* was found to be avirulent in mice (M. Motaleb, P. Stewart, A. Bestor, P. Rosa, and N. Charon, unpublished). Artificially infected *Ixodes scapularis* ticks were also unable to transmit the organism to mice. Similar results were also obtained when a chemotaxis mutant was tested *in vivo* (M. Motaleb, A. Bestor, P. Stewart, P. Rosa, and N. Charon, unpublished). Complementation studies are in progress to further support their precise role during infection *in vivo*. Although these results are preliminary, it appears that motility and chemotaxis are probably involved in the pathogenesis of Lyme disease. We expect that in the next several years, significant progress will be made in sorting out how motility and chemotaxis enables the bacterium disseminate during Lyme disease. With the *in vivo* tracking system of *B. burgdorferi* available (Moriarty *et al.*, 2008; Norman *et al.*, 2008), it will be interesting to investigate the interaction of motility and chemotaxis mutants with their host tissues or how, and if, these cells penetrate the endothelium.

7.7 Conclusion

The title of this chapter refers to the chic motility and chemotaxis of *B. burgdorferi*. Why chic? Those of us who have observed these treacherous creatures by microscopy have marvelled at their highly sophisticated, elegant, and complex structure, and their fascinating motions as they swim. In addition, as pointed out in this review, the old paradigm of chemotaxis as known in other bacteria just does not fit with *B. burgorferi* – they are enigmatic! Much is known about the motility and chemotaxis of other species of bacteria, but so little is known about these functions in *B. burgdorferi*. As we are beginning to delve into their secrets, we are concomitantly finding new avenues of research and insights relevant not only to Lyme disease, but, as discussed in the review, to other infections and questions relevant to central biological problems as well. If that is not being chic, what is?

Acknowledgements

This research in this chapter was supported by Public Health Service (PHS) grant AR050656 awarded to C.L., PHS grant GM0072004 awarded to C.W.W., PHS grant AR054582 awarded M.A.M., PHS grant RR01219 to J. Frank supporting the Resource for Visualization of Biological Complexity, PHS grant AI-29743 awarded to N.W.C., National Science Foundation grant DMS 0201063 to S.F.G., and American Heart Association grant 0755614B awarded to M.M. R.E.S. was supported by PHS grant GM050860 awarded to Robert B. Bourret and J. L. and S.J.N. were supported by PHS grant AI059048. N.W.C. wishes to express his appreciation for the continued interest in spirochaete motility by both the spirochaete and bacterial motility research communities. We thank S.I. Aizwa, S. Shibata, and D. J. Botkin, T. Lin, and F.T. Liang for sharing unpublished information.

References

Adler, J. (1973). A method for measuring chemotaxis and use of the method to determine optimum conditions for chemotaxis by *Escherichia coli*. J. Gen. Microbiol. *74*, 77–91.

Aizawa, S.-I., Zhulin, I.B., Marquez-Magana, L.M., and Ordal, G.W. (2001). Chemotaxis and motility in *Bacillus subtilis*. In *Bacillus subtilis* and its relatives: from genes to cells, A.L. Sonenshein, R. Losick, and J.A. Hoch, eds. (Washington, D.C.: ASM Press), pp. 437–452.

Aldridge, P., and Hughes, K.T. (2002). Regulation of flagellar assembly. Curr. Opin. Microbiol. *5*, 160–165.

Anderson, D.K., and Newton, A. (1997). Posttranscriptional regulation of *Caulobacter* flagellin genes by a late flagellum assembly checkpoint. J. Bacteriol. *179*, 2281–2288.

Anderson, D.K., Ohta, N., Wu, J., and Newton, A. (1995). Regulation of the *Caulobacter crescentus rpoN* gene and function of the purified σ^{54} in flagellar gene transcription. Mol. Gen. Genet. *246*, 697–706.

Armitage, J.P. (1999). Bacterial tactic responses. Adv. Microb. Physiol *41*, 229–289.

Armitage, J.P., and Schmitt, R. (1997). Bacterial chemotaxis: *Rhodobacter sphaeroides* and *Sinorhizobium meliloti* – variations on a theme? Microbiol. *143*, 3671–3682.

Arnoldi, M., Fritz, M., Bauerlein, E., Radmacher, M., Sackmann, E., and Boulbitch, A. (2000). Bacterial turgor pressure can be measured by atomic force microscopy. Phys. Rev. E. Stat. Phys. Plasmas. Fluids Relat Interdiscip. Topics. *62*, 1034–1044.

Bakker, R.G., Li, C., Miller, M.R., Cunningham, C., and Charon, N.W. (2007). Identification of specific chemoattractants and genetic complementation of a *Borrelia burgdorferi* chemotaxis mutant: A flow cytometry-based capillary tube chemotaxis assay. Appl. Environ. Microbiol 73, 1180–1188.

Barbour, A.G., and Hayes, S.F. (1986). Biology of *Borrelia* species. Microbiol. Rev. *50*, 381–400.

Barbour, A.G., Jasinskas, A., Kayala, M.A., Davies, D.H., Steere, A.C., Baldi, P., and Felgner, P.L. (2008). A genome-wide proteome array reveals a limited set of immunogens in natural infections of humans and white-footed mice with *Borrelia burgdorferi*. Infect. Immun. *76*, 3374–3389.

Berg, H.C. (1975). Chemotaxis in bacteria. Annu. Rev. Biophys. Bioeng. *4*, 119–136.

Berg, H.C. (2000). Motile behavior of bacteria. Phys. Today *53*, 24–29.

Berg, H.C. (2003). The rotary motor of bacterial flagella. Annu. Rev. Biochem. *72*, 19–54.

Berg, H.C., Bromley, D.B., and Charon, N.W. (1978). Leptospiral motility. Symp. Soc. Gen. Microbiol. *28*, 285–294.

Berg, H.C., and Turner, L. (1979). Movement of microorganisms in viscous environments. Nature (London) *278*, 349–351.

Botkin, D.J., Abbott, A.N., Stewart, P.E., Rosa, P.A., Kawabata, H., Watanabe, H., and Norris, S.J. (2006). Identification of potential virulence determinants by *Himar1* transposition of infectious *Borrelia burgdorferi* B31. Infect. Immun. *74*, 6690–6699.

Bourret, R.B., Charon, N.W., Stock, A.M., and West, A.H. (2002). Bright lights, abundant operons–fluorescence and genomic technologies advance studies of bacterial locomotion and signal transduction: review of the BLAST meeting, Cuernavaca, Mexico, 14 to 19 January 2001. J. Bacteriol. *184*, 1–17.

Bren, A., and Eisenbach, M. (2000). How signals are heard during bacterial chemotaxis: protein–protein interactions in sensory signal propagation. J Bacteriol. *182*, 6865–6873.

Brooks, C.S., Hefty, P.S., Jolliff, S.E., and Akins, D.R. (2003). Global analysis of *Borrelia burgdorferi* genes regulated by mammalian host-specific signals. Infect. Immun. *71*, 3371–3383.

Canale-Parola, E. (1978). Motility and chemotaxis of spirochetes. Annu. Rev. Microbiol. *32*, 69–99.

Canale-Parola, E. (1984). The Spirochetes. In Bergey's Manual of Systematic Bacteriology, N.R. Krieg and J.G. Holt, eds. (Baltimore: Williams and Wilkins), pp. 38–70.

Casjens, S., Palmer, N., van Vugt, R., Huang, W.M., Stevenson, B., Rosa, P., Lathigra, R., Sutton, G., Peterson, J., Dodson, R.J., Haft, D., Hickey, E., Gwinn, M., White, O., and Fraser, C.M. (2000). A bacterial genome in flux: the twelve linear and nine circular extrachromosomal DNAs in an infectious isolate of the Lyme disease spirochete *Borrelia burgdorferi*. Mol. Microbiol. 35, 490–516.

Charon, N.W., Daughtry, G.R., McCuskey, R.S., and Franz, G.N. (1984). Microcinematographic analysis of tethered *Leptospira illini*. J. Bacteriol. *160*, 1067–1073.

Charon, N.W., and Goldstein, S.F. (2002). Genetics of motility and chemotaxis of a fascinating group of bacteria: the Spirochetes. Annu. Rev. Genet. *36*, 47–73.

Charon, N.W., Goldstein, S.F., Block, S.M., Curci, K., Ruby, J.D., Kreiling, J.A., and Limberger, R.J. (1992a). Morphology and dynamics of protruding spirochete periplasmic flagella. J. Bacteriol. *174*, 832–840.

Charon, N.W., Goldstein, S.F., Curci, K., and Limberger, R.J. (1991). The bent-end morphology of *Treponema phagedenis* is associated with short, left-handed periplasmic flagella. J. Bacteriol. *173*, 4820–4826.

Charon, N.W., Goldstein, S.F., Marko, M., Hsieh, C., Gebhardt, L.L., Motaleb, M.A., Wolgemuth, C.W., Limberger, R.J., and Rowe, N. (2009). The flat-ribbon configuration of the periplasmic flagella of *Borrelia burgdorferi* and its relationship to motility and morphology. J Bacteriol. *191*, 600–607.

Charon, N.W., Greenberg, E.P., Koopman, M.B.H., and Limberger, R.J. (1992b). Spirochete chemotaxis, motility, and the structure of the spirochetal periplasmic flagella. Res. Microbiol. *143*, 597–603.

Charon, N.W., Lawrence, C.W., and O'Brien, S. (1981). Movement of antibody-coated latex beads attached to the spirochete *Leptospira interrogans*. Proc. Natl. Acad. Sci. U.S.A. *78*, 7166–7170.

Chi, B., Limberger, R.J., and Kuramitsu, H.K. (2002). Complementation of a *Treponema denticola flgE* mutant with a novel coumermycin A_1-resistant *T. denticola* shuttle vector system. Infect. Immun. *70*, 2233–2237.

Chilcott, G.S., and Hughes, K.T. (2000). Coupling of flagellar gene expression to flagellar assembly in *Salmonella enterica* serovar Typhimurium and *Escherichia coli*. Microbiol. Mol. Biol. Rev. *64*, 694–708.

Cox, C.D. (1972). Shape of *Treponema pallidum*. J. Bacteriol. *109*, 943–944.

Cox, P.J., and Twigg, G.I. (1974). Leptospiral motility. Nature (London) *250*, 260–261.

Dasgupta, N., Wolfgang, M.C., Goodman, A.L., Arora, S.K., Jyot, J., Lory, S., and Ramphal, R. (2003). A four-tiered transcriptional regulatory circuit controls flagellar biogenesis in *Pseudomonas aeruginosa*. Mol. Microbiol. *50*, 809–824.

Dombrowski, C., Kan, W., Motaleb, M.A., Charon, N.W., Goldstein, R.E., and Wolgemuth, C.W. (2009). The elastic basis for the shape of *Borrelia burgdorferi*. Biophys. J. *96*, 4409–4417.

Elias, A.F., Stewart, P.E., Grimm, D., Caimano, M.J., Eggers, C.H., Tilly, K., Bono, J.L., Akins, D.R., Radolf, J.D., Schwan, T.G., and Rosa, P. (2002). Clonal polymorphism of *Borrelia burgdorferi* strain B31 MI: implications for mutagenesis in an infectious strain background. Infect. Immun. *70*, 2139–2150.

Fosnaugh, K., and Greenberg, E.P. (1988). Motility and chemotaxis of *Spirochaeta aurantia*: computer-assisted motion analysis. J. Bacteriol. *170*, 1678–1774.

Fraser, C.M., Casjens, S., Huang, W.M., Sutton, G.G., Clayton, R., Lathigra, R., White, O., Ketchum, K.A., Dodson, R., Hickey, E.K., *et al.* (1997). Genomic sequence of a Lyme disease spirochaete, *Borrelia burgdorferi*. Nature *390*, 580–586.

Fu, H.C., Powers, T.R., and Wolgemuth, C.W. (2007). Theory of swimming filaments in viscoelastic media. Phys. Rev. Lett. *99*, 258101.

Fu, H.C., Wolgemuth, C.W., and Powers, T.R. (2008). Beating patterns of filaments in viscoelastic fluids. Phys. Rev. E *78*, 041913.

Ge, Y., and Charon, N.W. (1997a). An unexpected *flaA* homolog is present and expressed in *Borrelia burgdorferi*. J. Bacteriol. *179*, 552–556.

Ge, Y., and Charon, N.W. (1997b). Identification of a large motility operon in *Borrelia burgdorferi* by semi-random PCR chromosome walking. Gene *189*, 195–201.

Ge, Y., and Charon, N.W. (1997c). Molecular characterization of a flagellar/chemotaxis operon in the spirochete *Borrelia burgdorferi*. FEMS Microbiol. Lett. *153*, 425–431.

Ge, Y., Li, C., Corum, L., Slaughter, C.A., and Charon, N.W. (1998). Structure and expression of the FlaA periplasmic flagellar protein of *Borrelia burgdorferi*. J. Bacteriol. *180*, 2418–2425.

Ge, Y., Old, I., Saint Girons, I., and Charon, N.W. (1997a). The *flgK* motility operon of *Borrelia burgdorferi* is initiated by a σ^{70}-like promoter. Microbiol. *143*, 1681–1690.

Ge, Y., Old, I.G., Saint Girons, I., and Charon, N.W. (1997b). Molecular characterization of a large *Borrelia burgdorferi* motility operon which is initiated by a consensus σ^{70} promoter. J. Bacteriol. *179*, 2289–2299.

Goldstein, S.F., Buttle, K.F., and Charon, N.W. (1996). Structural analysis of *Leptospiraceae* and *Borrelia burgdorferi* by high-voltage electron microscopy. J. Bacteriol. *178*, 6539–6545.

Goldstein, S.F., and Charon, N.W. (1988). The motility of the spirochete *Leptospira*. Cell Motil. Cytoskel. *9*, 101–110.

Goldstein, S.F., and Charon, N.W. (1990). Multiple exposure photographic analysis of a motile spirochete. Proc. Natl. Acad. Sci. U.S.A. *87*, 4895–4899.

Goldstein, S.F., Charon, N.W., and Kreiling, J.A. (1994). *Borrelia burgdorferi* swims with a planar waveform similar to that of eukaryotic flagella. Proc. Natl. Acad. Sci. U.S.A. *91*, 3433–3437.

Hazelbauer, G.L., Falke, J.J., and Parkinson, J.S. (2008). Bacterial chemoreceptors: high-performance signaling in networked arrays. Trends Biochem. Sci. *33*, 9–19.

Hendrix, R.W. (2005). Bacteriophage HK97: assembly of the capsid and evolutionary connections. Adv. Virus Res. *64*, 1–14.

Holt, S.C. (1978). Anatomy and chemistry of spirochetes. Microbiol. Rev. *42*, 114–160.

Hovind Hougen, K. (1984). Ultrastructure of spirochetes isolated from *Ixodes ricinus* and *Ixodes dammini*. Yale J. Biol. Med. *57*, 543–548.

Hughes, K.T., Gillen, K.L., Semon, M.J., and Karlinsey, J.E. (1993). Sensing structural intermediates in bacterial flagellar assembly by export of a negative regulator. Science *262*, 1277–1280.

Jensen, G.J., and Briegel, A. (2007). How electron cryotomography is opening a new window onto prokaryotic ultrastructure. Curr. Opin. Struct. Biol. *17*, 260–267.

Josenhans, C., and Suerbaum, S. (2002). The role of motility as a virulence factor in bacteria. Int. J. Med. Microbiol. *291*, 605–614.

Kang, H.J., Coulibaly, F., Clow, F., Proft, T., and Baker, E.N. (2007). Stabilizing isopeptide bonds revealed in gram-positive bacterial pilus structure. Science *318*, 1625–1628.

Kennedy, M.J., Rosey, E.L., and Yancey, R.J., Jr. (1997). Characterization of *flaA*⁻ and *flaB*⁻ mutants of *Serpulina hyodysenteriae*: both flagellin subunits, FlaA and FlaB, are necessary for full motility and intestinal colonization. FEMS Mirobiol. Lett. *153*, 119–128.

Kihara, M., Miller, G.U., and Macnab, R.M. (2000). Deletion analysis of the flagellar switch protein FliG of *Salmonella*. J. Bacteriol. *182*, 3022–3028.

Kim, M., and Powers, T.R. (2005). Deformation of a helical filament by flow and electric or magnetic fields. Phys. Rev. E. Stat. Nonlin. Soft. Matter Phys. *71*, 021914.

Kimsey, R.B., and Spielman, A. (1990). Motility of Lyme disease spirochetes in fluids as viscous as the extracellular matrix. J. Infect. Dis. *162*, 1205–1208.

Komeda, Y. (1982). Fusions of flagellar operons to lactose genes on a Mu *lac* bacteriophage. J. Bacteriol. *150*, 16–26.

Kutsukake, K., Ohya, Y., and Iino, T. (1990). Transcriptional analysis of the flagellar regulon of *Salmonella typhimurium*. J Bacteriol. *172*, 741–747.

Labandeira-Rey, M., Seshu, J., and Skare, J.T. (2003). The absence of linear plasmid 25 or 28-1 of *Borrelia burgdorferi* dramatically alters the kinetics of experimental infection via distinct mechanisms. Infect. Immun. *71*, 4608–4613.

Lauga, E. (2007). Floppy swimming: viscous locomotion of actuated elastica. Phys. Rev. E. Stat. Nonlin. Soft. Matter Phys. *75*, 041916.

Li, C., Bakker, R.G., Motaleb, M.A., Sartakova, M.L., Cabello, F.C., and Charon, N.W. (2002). Asymmetrical flagellar rotation in *Borrelia burgdorferi* nonchemotactic mutants. Proc. Natl. Acad. Sci. U.S.A. *99*, 6169–6174.

Li, C., Corum, L., Morgan, D., Rosey, E.L., Stanton, T.B., and Charon, N.W. (2000a). The spirochete FlaA periplasmic flagellar sheath protein impacts flagellar helicity. J Bacteriol. *182*, 6698–6706.

Li, C., Motaleb, M.A., Sal, M., Goldstein, S.F., and Charon, N.W. (2000b). Spirochete periplasmic flagella and motility. J Mol. Microbiol. Biotechnol. *2*, 345–354.

Li, C., Wolgemuth, C.W., Marko, M., Morgan, D.G., and Charon, N.W. (2008). Genetic analysis of spirochete flagellin proteins and their involvement in motility, filament assembly, and flagellar morphology. J. Bacteriol. *190*, 5607–5615.

Limberger, R.J., Slivienski, L.L., and Samsonoff, W.A. (1994). Genetic and biochemical analysis of the flagellar hook of *Treponema phagedenis*. J. Bacteriol. *176*, 3631–3637.

Liu, J., Lin, T., Botkin, D.J., McCrum, E., Winkler, H., and Norris S.J. (2009). Intact flagellar motor of *Borrelia burgdorferi* revealed by cryo-electron tomography: evidence for stator ring curvature and rotor/C-ring assembly flexion. J. Bacteriol. *191*, 5026–5036.

Llewellyn, M., Dutton, R.J., Easter, J., O'donnol, D., and Gober, J.W. (2005). The conserved *flaF* gene has a critical role in coupling flagellin translation and assembly in *Caulobacter crescentus*. Mol. Microbiol. *57*, 1127–1142.

Losick, R., and Shapiro, L. (1999). Changing views on the nature of the bacterial cell: from biochemistry to cytology. J. Bacteriol. *181*, 4143–4145.

Lux, R., Moter, A., and Shi, W. (2000). Chemotaxis in pathogenic spirochetes: directed movement toward targeting tissues? J Mol. Microbiol. Biotechnol. *2*, 355–364.

Lybarger, S.R., and Maddock, J.R. (2001). Polarity in action: asymmetric protein localization in bacteria. J. Bacteriol. *183*, 3261–3267.

Macnab, R.M. (1996). Flagella and Motility. In *Escherichia coli* and *Salmonella*. Cellular and Molecular Biology, F.C. Neidhardt, R. Curtiss III, J.L. Ingraham, E.C.C. Lin, K.B. Low, B. Magasanik, W.S. Reznikoff, M. Riley, M. Schaechter, and H.E. Umbarger, eds. (Washington, D.C.: ASM Press), pp. 123–145.

Malawista, S.E., and de Boisfleury, C.A. (2008). Clocking the Lyme spirochete. PLoS. ONE. *3*, e1633.

McCarter, L.L. (2001). Polar flagellar motility of the *Vibrionaceae*. Microbiol. Mol. Biol. Rev. *65*, 445–462.

McGee, D.J., Langford, M.L., Watson, E.L., Carter, J.E., Chen, Y.T., and Ottemann, K.M. (2005). Colonization and inflammation deficiencies in Mongolian gerbils infected by *Helicobacter pylori* chemotaxis mutants. Infect. Immun. 73, 1820–1827.

Mendelson, N.H., Sarlls, J.E., Wolgemuth, C.W., and Goldstein, R.E. (2000). Chiral self-propulsion of growing bacterial macrofibers on a solid surface. Phys. Rev. Lett. *84*, 1627–1630.

Moriarty, T.J., Norman, M.U., Colarusso, P., Bankhead, T., Kubes, P., and Chaconas, G. (2008). Real-time high resolution 2-D imaging of the Lyme disease spirochete adhering to and escaping from the vasculature of a living host. PLoS. Pathog. *4*, e1000090.

Motaleb, M.A., Corum, L., Bono, J.L., Elias, A.F., Rosa, P., Samuels, D.S., and Charon, N.W. (2000). *Borrelia burgdorferi* periplasmic flagella have both skeletal and motility functions. Proc. Natl. Acad. Sci. U.S.A. *97*, 10899–10904.

Motaleb, M.A., Miller, M.R., Bakker, R.G., Li, C., and Charon, N.W. (2007a). Isolation and characterization of chemotaxis mutants of the Lyme disease spirochete *Borrelia burgdorferi* using allelic exchange mutagenesis, flow cytometry, and cell tracking. Methods Enzymol. *422*, 419–437.

Motaleb, M.A., Miller, M.R., Li, C., Bakker, R.G., Goldstein, S.F., Silversmith, R.E., Bourret, R.B., and Charon, N.W. (2005). CheX is a phosphorylated CheY phosphatase essential for *Borrelia burgdorferi* chemotaxis. J. Bacteriol. *187*, 7963–7969.

Motaleb, M.A., Miller, M.R., Li, C., and Charon, N.W. (2007b). Phosphorylation assays of chemotaxis two-component system proteins in *Borrelia burgdorferi*. Methods Enzymol. *422*, 438–447.

Motaleb, M.A., Sal, M.S., and Charon, N.W. (2004). The decrease in FlaA observed in a *flaB* mutant of *Borrelia burgdorferi* occurs posttranscriptionally. J. Bacteriol. *186*, 3703–3711.

Murphy, G.E., Leadbetter, J.R., and Jensen, G.J. (2006). *In situ* structure of the complete *Treponema primitia* flagellar motor. Nature *442*, 1062–1064.

Murphy, G.E., Matson, E.G. Leadbetter, J.R., Berg, H.C., and Jensen, G.J. (2008). Novel ultrastructures of *Treponema primitia* and their implications for motility. Mol. Microbiol. *67*, 1184–1195.

Niehus, E., Gressmann, H., Ye, F., Schlapbach, R., Dehio, M., Dehio, C., Stack, A., Meyer, T.F., Suerbaum, S., and Josenhans, C. (2004). Genome-wide analysis of transcriptional hierarchy and feedback regulation in the flagellar system of *Helicobacter pylori*. Mol. Microbiol. *52*, 947–961.

Norman, M.U., Moriarty, T.J., Dresser, A.R., Millen, B., Kubes, P., and Chaconas, G. (2008). Molecular mechanisms involved in vascular interactions of the Lyme disease pathogen in a living host. PLoS Pathog. *4*, e1000169.

Norris, S.J., and *Treponema pallidum* polypeptide research group. (1993). Polypeptides of *Treponema pallidum*: progress toward understanding their structural, functional, and immunologic roles. Microbiol. Rev. *57*, 750–779.

Ojaimi, C., Brooks, C., Casjens, S., Rosa, P., Elias, A., Barbour, A., Jasinskas, A., Benach, J., Katona, L., Radolf, J., Caimano, M., Skare, J., Swingle, K., Akins, D., and Schwartz, I. (2003). Profiling of temperature-induced changes in *Borrelia burgdorferi* gene expression by using whole genome arrays. Infect. Immun. *71*, 1689–1705.

Ottemann, K.M., and Miller, J.F. (1997). Roles for motility in bacterial–host interactions. Mol. Microbiol. *24*, 1109–1117.

Park, S.Y., Chao, X., Gonzalez-Bonet, G., Beel, B.D., Bilwes, A.M., and Crane, B.R. (2004). Structure and function of an unusual family of protein phosphatases: the bacterial chemotaxis proteins CheC and CheX. Mol. Cell *16*, 563–574.

Parkinson, J.S. (1977). Behavioral genetics in bacteria. Annu. Rev. Genet. *11*, 397–414.

Paster, B.J., and Dewhirst, F.E. (2000). Phylogenetic foundation of spirochetes. J. Mol. Microbiol. Biotechnol. *2*, 341–344.

Pelling, A.E., Li, Y., Cross, S.E., Castaneda, S., Shi, W., and Gimzewski, J.K. (2006). Self-organized and highly ordered domain structures within swarms of *Myxococcus xanthus*. Cell Motil. Cytoskeleton *63*, 141–148.

Purser, J.E., and Norris, S.J. (2000). Correlation between plasmid content and infectivity in *Borrelia burgdorferi*. Proc. Natl. Acad. Sci. U.S.A. *97*, 13865–13870.

Rajagopala, S.V., Titz, B., Goll, J., Parrish, J.R., Wohlbold, K., McKevitt, M.T., Palzkill, T., Mori, H., Finley, R.L., Jr., and Uetz, P. (2007). The protein network of bacterial motility. Mol. Syst. Biol. *3*, 128.

Revel, A.T., Talaat, A.M., and Norgard, M.V. (2002). DNA microarray analysis of differential gene expression in *Borrelia burgdorferi*, the Lyme disease spirochete. Proc. Natl. Acad. Sci. U.S.A. *99*, 1562–1567.

Ruby, J.D., Li, H., Kuramitsu, H., Norris, S.J., Goldstein, S.F., Buttle, K.F., and Charon, N.W. (1997). Relationship of *Treponema denticola* periplasmic flagella to irregular cell morphology. J. Bacteriol. *179*, 1628–1635.

Sadziene, A., Thomas, D.D., Bundoc, V.G., Holt, S.C., and Barbour, A.G. (1991). A flagella-less mutant of *Borrelia burgdorferi*. Structural, molecular, and in vitro functional characterization. J. Clin. Invest. *88*, 82–92.

Sal, M.S., Li, C., Motalab, M.A., Shibata, S., Aizawa, S., and Charon, N.W. (2008). *Borrelia burgdorferi* uniquely regulates its motility genes and has an intricate flagellar hook-basal body structure. J. Bacteriol. *190*, 1912–1921.

Sartakova, M.L., Dobrikova, E.Y., Motaleb, M.A., Godfrey, H.P., Charon, N.W., and Cabello, F.C. (2001). Complementation of a nonmotile *flaB* mutant of *Borrelia burgdorferi* by chromosomal integration of a plasmid containing a wild-type *flaB* allele. J. Bacteriol. *183*, 6558–6564.

Shi, W., Yang, Z.M., Geng, Y., Wolinsky, L.E., and Lovett, M.A. (1998). Chemotaxis in *Borrelia burgdorferi*. J Bacteriol. *180*, 231–235.

Shih, C.M., Chao, L.L., and Yu, C.P. (2002). Chemotactic migration of the Lyme disease spirochete (*Borrelia burgdorferi*) to salivary gland extracts of vector ticks. Am. J. Trop. Med. Hyg. *66*, 616–621.

Silversmith, R.E., Levin, M.D., Schilling, E., and Bourret, R.B. (2008). Kinetic characterization of catalysis by the chemotaxis phosphatase CheZ. Modulation of activity by the phosphorylated CheY substrate. J. Biol. Chem. *283*, 756–765.

Szczepanski, A., Furie, M.B., Benach, J.L., Lane, B.P., and Fleit, H.B. (1990). The interaction between *Borrelia burgdorferi* and endothelium in vitro. J. Clin. Invest. *85*, 1637–1647.

Thwaites, J.J., and Mendelson, N.H. (1991). Mechanical behaviour of bacterial cell walls. Adv. Microb. Physiol *32*, 173–222.

Tokarz, R., Anderton, J.M., Katona, L.I., and Benach, J.L. (2004). Combined effects of blood and temperature shift on *Borrelia burgdorferi* gene expression as determined by whole genome DNA array. Infect. Immun. *72*, 5419–5432.

Turner, L., Ryu, W.S., and Berg, H.C. (2000). Real-time imaging of fluorescent flagellar filaments. J. Bacteriol. *182*, 2793–2801.

Vaknin, A., and Berg, H.C. (2004). Single-cell FRET imaging of phosphatase activity in the *Escherichia coli* chemotaxis system. Proc. Natl. Acad. Sci. U.S.A. *101*, 17072–17077.

Wadhams, G.H., and Armitage, J.P. (2004). Making sense of it all: bacterial chemotaxis. Nat. Rev. Mol. Cell Biol. *5*, 1024–1037.

Wilson, D.R., and Beveridge, T.J. (1993). Bacterial flagellar filaments and their component flagellins. Can. J. Microbiol. *39*, 451–472.

Wolgemuth, C.W., Charon, N.W., Goldstein, S.F., and Goldstein, R.E. (2006). The flagellar cytoskeleton of the spirochetes. J Mol. Microbiol. Biotechnol. *11*, 221–227.

Yakhnin, H., Pandit, P., Petty, T.J., Baker, C.S., Romeo, T., and Babitzke, P. (2007). CsrA of *Bacillus subtilis* regulates translation initiation of the gene encoding the flagellin protein (*hag*) by blocking ribosome binding. Mol. Microbiol. *64*, 1605–1620.

Zambrano, M.C., Beklemisheva, A.A., Bryksin, A.V., Newman, S.A., and Cabello, F.C. (2004). *Borrelia burgdorferi* binds to, invades, and colonizes native type I collagen lattices. Infect. Immun. 72, 3138–3146.

Zhao, R., Collins, E.J., Bourret, R.B., and Silversmith, R.E. (2002). Structure and catalytic mechanism of the *E. coli* chemotaxis phosphatase CheZ. Nat. Struct. Biol. *9*, 570–575.

Genetic Manipulation of *Borrelia burgdorferi*

8

Patricia A. Rosa, Felipe Cabello and D. Scott Samuels

Abstract

Genetic studies in *Borrelia burgdorferi* require special consideration of the bacterium's segmented genome, complex growth requirements and evolutionary distance from other genetically tractable bacteria, among other features. Despite these challenges, molecular genetics has been applied with growing success to study the biology and virulence of this important human pathogen. In this chapter we summarize the tools and techniques that are currently available for the genetic manipulation of *B. burgdorferi*, placing them in the context of their utility and shortcomings. Our primary objective is to help researchers discern what is feasible and what is not practical when thinking about potential genetic experiments in *B. burgdorferi*. We have summarized published methods and highlighted their critical elements, but we have not provided detailed protocols. Although a number of advances have been made since *B. burgdorferi* was first transformed nearly 15 years ago, some fairly standard components of a good genetic system remain elusive. We mention these limitations as well, and why they persist, if known. We hope to encourage investigators to explore what might be possible, in addition to recognizing what currently can be achieved, through genetic manipulation of *B. burgdorferi*.

8.1 General comments

Genetic analysis is a valuable tool that has been used widely and effectively to study many aspects of bacterial pathogenesis (Beckwith, 1991). In recent years a molecular genetic approach has been applied with growing success to study the biology and virulence of the tick-borne pathogen *Borrelia burgdorferi* (Tilly *et al.*, 2000; Cabello *et al.*, 2001; Rosa *et al.*, 2005; Samuels, 2006). Both *in vitro* and *in vivo* analyses have thus been able to identify and describe the basic functions of a number of *B. burgdorferi* genes and gene products, and their contributions to the relationship of this important pathogen with its tick vector and mammalian host reservoirs (Rosa *et al.*, 2005; see Chapters 11 and 12). Direct applications of genetic tools developed for Gram-positive and Gram-negative bacteria have encountered limitations, however, due to the genetic distance and distinct codon usage of *B. burgdorferi* relative to other bacteria (Rosa *et al.*, 2005). Hence there is still a significant need to develop additional and alternative genetic tools for *B. burgdorferi* in particular, and for spirochaetes in general.

Twenty years ago, Dr Stanley Falkow restated Koch's postulates as they could be applied to the relatively new field of microbial pathogenesis, utilizing the tools of molecular genetics (Falkow, 1988). Just as the original postulates were developed as a guideline for establishing the connection between an infectious disease and its causative agent, their molecular form extended this instruction to demonstrating the connection between a virulence trait of an infectious microbe and the gene(s) that determine it. In their simplest form, Koch's (Falkow's) molecular postulates entail the ability to create mutations, restore genes and analyse the phenotypes of an otherwise isogenic set of microbes, but their interpretation and application continues to evolve as new technologies emerge with which to investigate the complex biology of virulence (Falkow, 1988, 2004; Bosse *et al.*, 2006; Salama and Manoil, 2006).

Construction of isogenic mutants in *B. burgdorferi* has permitted genetic analysis of regulatory and structural genes, and definition of their functions (Tilly *et al.*, 2000; Cabello *et al.*, 2001; Rosa *et al.*, 2005; Samuels, 2006). Progress along these lines has encountered some limitations, stemming principally from the low frequency of transformation of *B. burgdorferi*, which has impeded construction and screening of comprehensive genomic or mutant libraries. In addition, generation of isogenic wild-type, mutant and complemented clones presents unique challenges in *B. burgdorferi* due to the segmented and unstable nature of the genome and poorly defined, probably multifactorial barriers to the uptake and incorporation of exogenous DNA. Polar effects following insertional inactivation and allelic exchange can arise as a consequence of the compact chromosome of *B. burgdorferi*, which contains many genes with overlapping transcriptional and termination signals (Fraser *et al.*, 1997). The presence of paralogous gene families encoding products with similar functions and with overlapping transcriptional signals in *B. burgdorferi* also can complicate the interpretation of mutant phenotypes (Casjens *et al.*, 2000). Establishing the connection between a mutation and a phenotype through complementation has proven to be essential but challenging. Isolation of nutritional mutants in *B. burgdorferi* is limited by the complex and undefined nature of the growth media and a relatively long doubling time (>5 hours under optimal conditions) (Barbour, 1984). Despite these myriad potential and real challenges, substantial progress has been made in utilizing a genetic approach to study factors involved in the biology and pathogenesis of *B. burgdorferi*, as illustrated by a number of chapters in this book.

8.2 Clone and strain considerations

To date, several *B. burgdorferi* sensu stricto strains and at least one *B. afzelii* strain (Fingerle *et al.*, 2007) have been utilized in genetic studies (Table 8.1). Strain choice generally stems from what has been used previously by individual labs and there are no data indicating that one strain is inherently more suitable for genetic studies than the others. Individual clones derived from the same strain can exhibit substantial heterogeneity in a number of traits, including the ability to be transformed (Elias *et al.*, 2002). Although differences among clones in plasmid content (and putative restriction enzyme genes) clearly contribute to the efficiency with which they are transformed (Lawrenz *et al.*, 2002; Kawabata *et al.*, 2004), additional traits whose genetic underpinnings are unknown also influence transformability. These include surface properties that permit or restrict resuspension in electroporation solution, and survival following electroporation (Tilly *et al.*, 2000). Basic growth properties, such as doubling time and colony formation, also vary among clones and significantly impact the utility of different clones for genetic studies. Finally, an important caveat is that a wild-type 'clone' is neither clonal nor isogenic if passed through a mouse to maintain infectivity, due to rampant recombination at the *vlsE* locus that occurs in the mammalian environment (Zhang *et al.*, 1997; Zhang and Norris, 1998). The re-isolate now comprises a complex population of genetically and phenotypically distinct bacteria expressing different alleles of an abundant and immunogenic surface protein (VlsE) (Zhang *et al.*, 1997; Zhang and Norris, 1998). This mixed population can serve as the source from which to derive a new infectious clone by plating or limiting dilution, but it should not be used directly as the starting material for genetic manipulation.

The availability of the complete and annotated sequences for the chromosome and plasmids of the type strain B31 (Fraser *et al.*, 1997; Casjens *et al.*, 2000) currently offers a significant advantage for genetic studies with respect to generating inactivation and complementation constructs, and monitoring plasmid content. Draught sequences of several other strains are now available (Glöckner *et al.*, 2004; Qiu *et al.*, 2004; Glöckner *et al.*, 2006) and their complete and annotated forms will probably follow, rendering them equally convenient for genetic studies. The high degree of variability in plasmid structure and gene content among strains suggests that the phenotype of a particular mutant (role/contribution of a particular gene) may depend on the context (strain background) in which it is analysed. Ideally, mutations would be introduced and analysed in more than a single clone or strain (Gilbert *et al.*, 2007; Lybecker and Samuels, 2007). However, given the limitations and challenges presented by genetic studies in *B. burgdorferi*, recognizing that experimental outcome could differ among strains and clones, and interpreting results accordingly, is perhaps sufficient.

8.3 Transformation protocol

Borrelia burgdorferi was first genetically transformed in 1994, using a fairly standard methodology (Samuels *et al.*, 1994a). Prior to this study, the effect of buffers and electroporation parameters had been assayed on *B. burgdorferi* (Sambri and Lovett, 1990). Although increasingly more sophisticated genetic tools and techniques have become available, the only significant improvement to the original protocol is the use of large amounts (10 µg or more) of DNA as a substrate (Tilly *et al.*, 2000), which facilitated the transformation of low-passage, infectious strains (Hübner *et al.*, 2001; Stewart *et al.*, 2001; Eggers *et al.*, 2002; Elias *et al.*, 2002). A detailed protocol for electrotransformation of *B. burgdorferi* is presented in Samuels (1995).

Two calculations are typically used to assess how well a transformation has worked. The first, termed transformation frequency, is without units and describes what fraction of the population after electroporation (or other transformation method) has stably acquired the introduced DNA. This is determined by plating aliquots of the post-electroporation culture both with and without selection for the transforming DNA, normalized to the same volume of culture. For example, if 1 ml of a transformation culture yields 50 colonies on a plate with selection and 10^{-5} ml of the same culture yields 100 colonies on a plate without selection, the transformation frequency is 50 transformed bacteria out of 10^7 viable bacteria per ml, or 5×10^{-6}. The other standard calculation is transformation efficiency, which is determined by dividing the total number of transformants obtained in an experiment by the amount of DNA used in the electroporation. For example, if 1 ml of a 10 ml transformation culture yields 50 colonies on a plate with selection and 10 µg of DNA was used in the electroporation, the transformation efficiency is 500 transformants per 10 µg of DNA, or 50 transformants per microgram. Transformation frequency and efficiency can only be determined when selection is imposed and bacteria are plated shortly after recovery from electroporation, because otherwise the number of transformed bacteria in the outgrowth no longer reflects the primary transformation event. Accurate determination of transformation frequency and efficiency also requires verification that bacteria growing under selection are transformants and not spontaneous resistance mutants.

8.3.1 Preparation of competent cells

The preparation of electrocompetent cells involves growing cells to a suitable density, followed by a series of washes to remove culture medium, especially salts, which cause arcing, cell death, and, of course, decreased transformation efficiency (Samuels *et al.*, 1994a; Samuels, 1995). These washes have a profound effect on the ultrastructure of *B. burgdorferi*: the outer membrane appears to be disrupted in transmission electron micrographs of competent and electroporated cells (Fig. 8.1) (Samuels and Garon, 1997), although the cell viability is apparently not affected (Samuels, 1995). Cell density or growth phase has proven to be an important criterion for successful electrotransformation (Samuels, 1995; Tilly *et al.*, 2000; Hübner *et al.*, 2001). Cultures in mid-log phase have the highest transformation frequency, but late-log phase cultures, which have a greater number of cells, yield the most transformants (Samuels, 1995; Tilly *et al.*, 2000). Briefly, cells are harvested sterilely by centrifugation, washed several times, and finally resuspended in cold electroporation solution as a concentrated cell slurry. Some strains aggregate during the washes and cannot be resuspended in electroporation solution (Tilly *et al.*, 2000); this undesirable trait is intensified when bacteria are grown to higher cell density. Competent cell aliquots can be transformed by electroporation immediately or they can be stored at –80°C, which decreases transformation efficiency by about half (Samuels, 1995; Samuels and Garon, 1997).

8.3.2 Electrotransformation

Electroporation conditions to transform *B. burgdorferi* are similar to those of most other bacteria (Samuels *et al.*, 1994a; Samuels, 1995).

Table 8.1 Strains used for genetic manipulation

Strain	Designation	Comments	Reference(s)
B. burgdorferi sensu stricto			
B31	–	High passage, non-infectious, uncloned strain B31; cou^r *gyrB* gene of *B. burgdorferi* used as selectable marker; allelic exchange at *gyrB* with PCR product or genomic DNA	Samuels *et al.* (1994a,b)
		Site-directed mutagenesis of *gyrB* with oligonucleotides	Samuels *et al.* (1997)
	–	High passage, non-infectious, uncloned strain B31; targeted gene inactivation at non-*gyrB* locus with cou^r *gyrB*; allelic exchange construct on non-replicating plasmid DNA	Rosa *et al.* (1996)
	–	High passage, non-infectious, uncloned strain B31; transient transformation with non-replicating plasmid DNA	Sohaskey *et al.* (1997)
	B31-A	High passage, non-infectious, transformable B31 clone; hybrid cassette conferring kan^r as selectable marker; stable integration of a non-replicating plasmid	Bono *et al.* (2000)
	HP-J	High-passage, non-infectious B31 clone lacking many plasmids; highly transformable with shuttle vector	Lawrenz *et al.* (2002)
	B313 and B314	High-passage, non-infectious B31 clones lacking most plasmids; expression of surface protein genes on shuttle vector	Šadžiene *et al.* (1993), Schulze *et al.* (2006)
	5A1 to 5A19 series	Low-passage B31 clones with defined plasmid content and varying infectivity phenotypes	Purser *et al.* (2000)
		Shuttle vector transformation of wild-type clones limited by the presence of lp25 and lp56	Lawrenz *et al.* (2002)
		Infectivity restored to clones lacking lp25 by reintroduction of *pncA* (*bbe22*) gene on a shuttle vector	Purser *et al.* (2003)
	5A4 NP1 and 5A18 NP1	Low-passage, infectious B31 clones with *bbe02* gene inactivated; enhanced shuttle vector transformation	Kawabata *et al.* (2004)
	MSK, ML and MJ series	Low passage B31 clones with defined plasmid content and infectivity phenotypes	Labandeira-Rey *et al.* (2001)
	ML23	Transformable non-infectious clone lacking lp25	Seshu *et al.* (2004)
	B31 A3	Infectious low-passage wild-type B31 clone; low transformation frequency with shuttle vectors; allelic exchange and shuttle vector transformants can retain infectivity	Elias *et al.* (2002)
	A3-LS	Non-infectious clonal derivative of B31 A3 (missing lp28-1); *lacI* cassette and selectable marker inserted in *bbe02* on lp25	Gilbert *et al.* (2007)
297	–	Infectious, low-passage, uncloned strain 297; non-infectious transformants independent of introduced mutation	Hübner *et al.* (2001)
	c155	Infectious, low-passage 297 clone; infectivity of shuttle vector transformants not addressed	Eggers *et al.* (2002)
	BbAH130	Infectious, low passage 297 clone; infectivity of allelic exchange and shuttle vector transformants not addressed	Yang *et al.* (2004)
	297-LK	Non-infectious clonal derivative of BbAH130 (missing lp28-1); *lacI* cassette and selectable marker inserted in *bbe02*	Gilbert *et al.* (2007)

Strain	Designation	Comments	Reference(s)
	CE162	Infectious, low-passage 297 clone; shuttle vector transformants can retain infectivity	Caimano *et al.* (2004)
	PL133	Infectious, low-passage, uncloned strain 297 mouse reisolate; allelic exchange transformants can retain infectivity	Revel *et al.* (2005)
	BDTR630	Infectious clonal derivative of PL133; selectable marker inserted into lp25; infectivity of shuttle vector transformants not addressed	Blevins *et al.*, (2007)
HB19	–	Uncloned strain HB19 (infectious status not reported); allelic exchange at *gyrB* with PCR products and oligonucleotides	Samuels *et al.* (1997)
	clone 1	Non-infectious HB19 clone; allelic exchange transformants	Coburn and Cugini (2003)
	–	Undesignated clonal derivative of strain HB19 (infectious status and passage history not reported); stable and transient plasmid transformants	Alverson *et al.* (2003)
N40	–	Infectious, clonal derivative of strain N40 (passage history not reported); infectivity of shuttle vector transformants not addressed	Stewart *et al.* (2001)
	D10/E9	Low-passage, infectious N40 clone; allelic exchange transformants can retain infectivity	Parveen *et al.* (2006)
CA-11	CA-11.2A	Clonal derivative of strain CA-11 (passage history and infectivity not reported); targeted insertion of marker into a cp32 prophage, ϕBB-1; transduction of marker on ϕBB-1 into strains B31 and Sh2–82; first direct evidence for lateral gene transfer in *B. burgdorferi*	Eggers *et al.* (2001)
B. afzelii			
PKo	cPKo345	Wild-type uncloned strain PK0 and clonal derivative cPKo345; first reported transformation of a *B. afzelii* strain; mouse infectivity of shuttle vector transformants not addressed	Fingerle *et al.* (2007)

A high-voltage electric pulse generates pores in the membrane through which DNA can enter the cell (Shigekawa and Dower, 1988; Trevors *et al.*, 1992; Nickoloff, 1995). Transmission electron microscopy of electroporated *B. burgdorferi* reveals darkly stained regions on the membrane along the length of the cell that are thought to be disruptions in the surface (Fig. 8.1) (Samuels and Garon, 1997). The putative pores are no longer visible after a 30-min incubation in culture medium.

As mentioned above, a significant improvement to the original protocol was increasing the amount of DNA used per electroporation, from the nanogram levels typically used with other bacteria, to the microgram quantities needed to routinely transform *B. burgdorferi* (Bono *et al.*, 2000; Tilly *et al.*, 2000; Hübner *et al.*, 2001; Stewart *et al.*, 2001; Eggers *et al.*, 2002; Elias *et al.*, 2002). The basis for the extremely low transformation efficiency of *B. burgdorferi* remains a mystery. Perhaps very little of the transforming DNA actually gets into the cell, it may be rapidly targeted for degradation once it enters, or some other feature of the spirochaete limits the stable retention of transforming DNA.

Transformation substrate

In addition to the amount of DNA, its source and form have an effect on transformation efficiency.

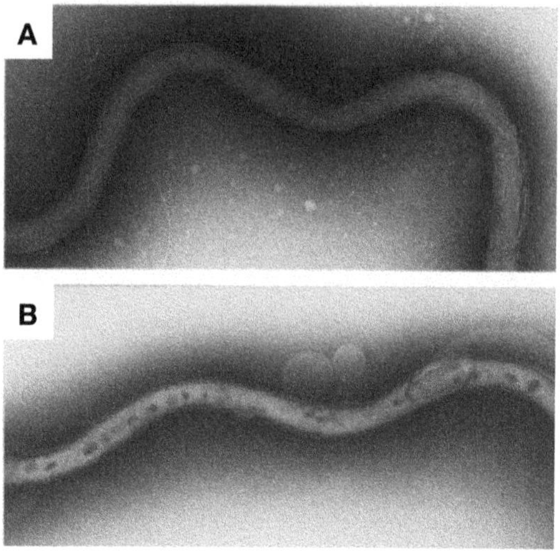

Figure 8.1 Structure of electroporated *B. burgdorferi*. Electron micrographs of competent cells before (A) and after (B) electroporation. Cells were negatively stained and observed by transmission electron microscopy. Dr Claude Garon is acknowledged for the micrographs, which are reprinted with permission from Samuels and Garon, 1997.

Allelic exchange transformation was first demonstrated by introducing site-directed mutations that conferred antibiotic resistance, using PCR-generated linear DNA substrates (Samuels *et al.*, 1994a). Short oligonucleotides also can be used as transformation substrates (Samuels and Garon, 1997), although this has limited utility because the site-directed mutations encoded by the oligonucleotide must confer a selectable advantage (currently limited to antibiotic resistance). Linear DNA is still commonly used in allelic exchange, although almost always using cloned DNA. A circular suicide plasmid was shown to recombine into the genome (Stevenson *et al.*, 1998) at the same frequency as linear DNA and, as described below, circular shuttle vectors have been identified (Sartakova *et al.*, 2000) or constructed (Stewart *et al.*, 2001; Eggers *et al.*, 2002; Stewart *et al.*, 2003; Byram *et al.*, 2004) that replicate autonomously in *B. burgdorferi* as well as *E. coli*. In addition, linear shuttle vectors have been constructed using replication functions from both native circular and linear plasmids (Chaconas *et al.*, 2001; Beaurepaire and Chaconas, 2007). The transformation efficiency of autonomously replicating shuttle vectors into strains lacking lp25 and lp56 (see below) is generally higher than with allelic exchange constructs (Elias *et al.*, 2003), presumably because the additional recombination step is relatively rare and limiting. Remarkably, whole native plasmids up to 28 kb have been successfully transformed into *B. burgdorferi*. (Grimm *et al.*, 2004a).

DNA modification

One of the barriers to efficient transformation of *B. burgdorferi* appears to be putative restriction-modification systems (Lawrenz *et al.*, 2002; Kawabata *et al.*, 2004). Some, but not all, *B. burgdorferi* strains have methylated adenines in the sequence GATC (Hughes and Johnson, 1990; Hughes *et al.*, 1992). Transformation of a set of closely related *B. burgdorferi* clones with distinct plasmid complements indicated that loss of two linear plasmids, lp25 and lp56, was correlated with higher transformation efficiency with a shuttle vector (Lawrenz *et al.*, 2002). These two plasmids carry genes (*bbe02* and *bbq67*) that, based on sequence homology and catalytic motifs, are predicted to encode type IV restriction-modification enzymes. Disruption of *bbe02* increased shuttle vector transformation efficiency approximately 40-fold (Kawabata *et al.*, 2004). The transformation barrier imposed by putative restriction enzymes on lp56 and lp25 applies primarily to shuttle vector transformations (Lawrenz *et al.*, 2002; Kawabata *et al.*, 2004), not gene inactivation by allelic exchange (Elias *et al.*, 2002; Hübner *et al.*, 2003; Tilly *et al.*, 2004). This probably reflects the requirement for intact circular DNA for autonomous plasmid replication, whereas linearized DNA remains an effective template for allelic exchange. Consequently, complementation presents a greater challenge than gene inactivation in clones retaining lp25.

In addition to *bbe02*, lp25 also carries genes required for mouse and tick infectivity (Purser and Norris, 2000; Labandeira-Rey and Skare, 2001; Purser *et al.*, 2003; Grimm *et al.*, 2005; Revel *et al.*, 2005; Strother *et al.*, 2005). The linkage on lp25 of these disparate functions represents one of the major reasons that low-passage infectious strains (containing lp25) are difficult to genetically transform. Transformation efficiency into low-passage infectious strains can be increased by first transforming the DNA into a strain carrying lp56 but not lp25 (Jewett *et al.*, 2007b). The resulting DNA is presumably partially modified and thus presents less of an obstacle when introduced in subsequent transformations into strains containing lp25. Chen *et al.* (2008) recently demonstrated that *in vitro* CpG methylation of foreign shuttle vector DNA prior to transformation of *B. burgdorferi* significantly enhanced its transformation efficiency, thus facilitating complementation of infectious *B. burgdorferi*. CpG methylation appears to overcome the transformation barrier imposed by lp56 (*bbq67*), but not that presented by lp25.

8.4 Selectable markers

Antibiotic resistance markers are one of the most effective selective tools used to single out rare bacterial variants containing new genetic properties, as well as to select for and screen mutants deficient in the expression of specific genes (Pich *et al.*, 2006). *Borrelia burgdorferi* is exquisitely susceptible to many antibiotics (Hunfeld *et al.*, 2002; Ruzic-Sabljic *et al.*, 2005; Hunfeld and Brade, 2006), but the choice of antibiotic resistance genes as genetic markers is somewhat limited by biosafety considerations: markers encoding resistance to ampicillin and tetracycline are prohibited for use in *B. burgdorferi* because these antibiotics are clinically useful in the treatment of Lyme disease (Steere, 2001; Wormser, 2006). The selectable markers currently available for genetic studies in *B. burgdorferi* and their relevant features are listed in Table 8.2.

8.4.1 Endogenous genes

To date, only antibiotic resistance has been used to select transformants (Samuels *et al.*, 1994a; Bono *et al.*, 2000; Sartakova *et al.*, 2000; Elias *et al.*, 2003; Frank *et al.*, 2003). An allele of the endogenous *B. burgdorferi gyrB* gene (Samuels *et al.*, 1994b), which encodes the B subunit of DNA gyrase, was developed into the first selectable marker used in *Borrelia* (reviewed in Samuels, 2006). The allele *gyrB301*, colloquially termed 'NGR', has four point mutations that encode changes in three amino acid residues: Asn-102 to Asp, Gly-104 to Asp, and Arg-133 to Ile, which confer high-level resistance to coumarin antibiotics, including coumermycin A_1. A synthetic version of this allele was generated (Elias *et al.*, 2003), which solved the problem of having to screen large numbers of transformants due to homologous recombination into the native *gyrB* locus (Rosa *et al.*, 1996; Knight *et al.*, 2000). However, the coumermycin A_1-resistant *gyrB* marker is not routinely used because of other problems, such as pleiotropic effects, which are due to relaxed DNA supercoiling, and include increased expression of *groEL* (Alverson and Samuels, 2002) and *ospC* (Alverson *et al.*, 2003). Nonetheless, the use of coumermycin A_1 resistance was seminal for demonstrating the feasibility of transformation and genetic manipulation of *B. burgdorferi* (Samuels *et al.*, 1994a; Rosa *et al.*, 1996). In addition, a number of genes were disrupted with *gyrB* in high-passage non-infectious strains: *ospC* (Tilly *et al.*, 1997), *guaB* (Tilly *et al.*, 1998), *oppAIV* (Bono *et al.*, 1998), *gac* (Knight *et al.*, 2000), *rpoS* (Elias *et al.*, 2000), and *chbC* (Tilly *et al.*, 2001).

8.4.2 Hybrid genes

A major breakthrough in genetic analysis of *B. burgdorferi* was the construction of hybrid genes using native *B. burgdorferi* promoters fused to foreign open reading frames. This was first used to assay transcription by fusing various borrelial promoters to *cat*, encoding chloramphenicol acetyl transferase (Sohaskey *et al.*, 1997), although, unfortunately, *cat* does not confer chloramphenicol resistance in *B. burgdorferi* (Sohaskey and Barbour, 1999). Bono *et al.* (2000) constructed the first efficient selectable marker by fusing the strong constitutive *flgB* and *flaB* promoters of *B. burgdorferi* to the *aphI* gene from Tn*903*, which confers resistance to the aminoglycoside kanamycin; the resulting cassette confers antibiotic resistance in both *E. coli* and *B. burgdorferi*. Several other chimeric genes have been constructed in this fashion, notably selectable markers conferring resistance to gentamicin from the *aacC1* gene from Tn*1696* (Elias *et al.*, 2003) and resistance to spectinomycin and streptomycin from the *aadA* gene from *Shigella flexneri* plasmid R100 (Frank *et al.*, 2003). However, other antibiotic resistance genes have been assayed that do not function in *B. burgdorferi*. The *cat* and *pac* genes, fused to the *flgB* promoter, do not confer resistance to chloramphenicol and puromycin, respectively (Elias *et al.*, 2003). The failure of *cat* may be due to the presence of esterases in the medium, which remove

Table 8.2 Selectable markers

Selectable marker			Antibiotic[a]					
				Selection (μg/ml)[b]				
Gene	Promoter	Source	Resistance	Ec	Bb[d]	Mutation frequency[c]	Comments	Reference(s)
gyrB301	Native (*gyrB*)	*B. burgdorferi* (chromosome)	Coumermycin A_1[e]	NA[f]	0.2–5	10^{-7}	High rate of recombination with genomic *gyrB* and pleiotropic effects	Samuels *et al.* (1994a,b), Rosa *et al.* (1996)
aphI	*flgB*, *flaB*	*E. coli* Tn*903*	Kanamycin	50	200–400	10^{-7} to 10^{-8}		Bono *et al.* (2000)
ermC	Native	*Staphylococcus aureus*	Erythromycin	2–10	0.03		Requires ERM-susceptible Ec mutant; some Bb strains are ERM resistant	Sartakova *et al.* (2000)
aacC1	*flgB*, *flaB*	*Pseudomonas aeruginosa* Tn*1696*	Gentamicin	5–8	40	10^{-8}	Variability in Ec susceptibility (may be strain related)	Elias *et al.* (2002, 2003)
aph(3′)-IIIa	native	*Streptococcus faecalis*	Kanamycin	50	240–350	10^{-7} to 10^{-8}		Sartakova *et al.* (2003) Bugrysheva *et al.* (2005)
gyrB$_{syn}$	native (*gyrB*)	synthetic	Coumermycin A_1[e]	NA[f]	0.2–5	10^{-7}	Nucleotide sequence changed to reduce recombination into genomic *gyrB*	Elias *et al.* (2003)
aadA	*flgB*	*Shigella flexneri* plasmid R100	Spectinomycin; streptomycin	50 SPT	50 STR	6×10^{-6} (SPT) $3–20 \times 10^{-9}$ (STR)	SPT used in Ec and STR used in Bb[g]	Frank *et al.* (2003)
aad9	Native	*S. aureus* Tn*554*	Spectinomycin	50	15–360	6×10^{-6}	Potential[h]	Sartakova *et al.* (2003)
linA′	Native	*S. aureus*	Lincomycin		0.2–0.3		Potential[h]	Sartakova *et al.* (2003)
aac(3)-VIa	Native	*Enterobacter cloacae*	Sisomycin; gentamicin		750	10^{-8} (GEN)	Potential[h]	Sartakova *et al.* (2003)

the acetyl group added by chloramphenicol acetyl transferase (Sohaskey and Barbour, 1999); this may also be the reason why *pac* does not confer puromycin resistance. Currently, hybrid cassettes conferring resistance to kanamycin (*flgBp–aphI* and *flaBp–aphI*), gentamicin (*flgBp–aacC1* and *flaBp–aacC1*), and streptomycin (*flgBp–aadA* and *flaBp–aadA*) are used to disrupt genes in *B. burgdorferi* by allelic exchange, insert site-directed mutations, and select for shuttle vectors and endogenous plasmids (Table 8.2).

8.4.3 Foreign antibiotic resistance markers

Foreign antibiotic resistance genes were thought not to be adequately expressed in *B. burgdorferi* under their own promoters because spirochaetes are genetically distant from both Gram positive and Gram negative bacteria (Paster *et al.*, 1984, 1991; Paster and Dewhirst, 2000), and little was known about the structure of their promoters. This notion was strengthened by the observation that insertion of the *aphI* gene from Tn*903* into the *B. burgdorferi* chromosome conferred little, if any, resistance to kanamycin in the spirochaete when expressed from its native promoter (Stevenson *et al.*, 1998; Bono *et al.*, 2000). As described above, the first antibiotic resistance markers used for genetic selection of *B. burgdorferi* were therefore endogenous to *B. burgdorferi* (Samuels *et al.*, 1994a,b) or were genes from other bacteria placed under the control of strong and constitutive *B. burgdorferi* promoters (Bono *et al.*, 2000). Subsequent demonstration that some foreign antibiotic resistance genes could be adequately expressed in *B. burgdorferi* from their own promoters has expanded the potential range of antibiotic resistance markers that can be used (Sartakova *et al.*, 2000; Sartakova *et al.*, 2003) (Table 8.2).

Sartakova and colleagues discovered that the *ermC* gene from *Staphylococcus aureus* conferred erythromycin resistance in *B. burgdorferi* (Sartakova *et al.*, 2000). However, since erythromycin can be prescribed to treat Lyme disease in certain situations, some Institutional Biosafety Committees have banned or restricted its use. In addition, *B. burgdorferi* strains and clinical isolates vary in their susceptibility to erythromycin (Terekhova *et al.*, 2002). Four other foreign genes expressed from their own promoters confer antibiotic resistance in *B. burgdorferi* (Table 8.2): *aph(3')-IIIa*, *aad9*, *linA'*, and *aac(3)-VIa* confer resistance to kanamycin, spectinomycin, lincomycin, and sisomycin or gentamicin, respectively (Sartakova *et al.*, 2003). Selection with spectinomycin fails because of the high frequency of spontaneous resistance to this antibiotic in *B. burgdorferi* (Criswell *et al.*, 2006), but lincomycin resistance has the potential to be developed into a new selectable marker, which would increase the number of markers available for increasingly complex genetic experiments. These studies clearly indicate that some foreign antibiotic resistance genes can be expressed from their own promoters and used as selective markers in *B. burgdorferi*.

8.4.4 Selection parameters

Transformants were originally isolated (Samuels *et al.*, 1994a) as subsurface colonies in semi-solid medium using a modification (Rosa and Hogan,

---

Notes to Table 8.2

a SPT, spectinomycin; STR, streptomycin; GEN, gentamicin; and ERM, erythromycin.
b Concentration of antibiotic used to select for transformants (or resistant mutants) in *E. coli* (Ec) and *B. burgdorferi* (Bb).
c Frequency of background mutations to antibiotic resistance in *B. burgdorferi* (frequency to spectinomycin, streptomycin, kanamycin, and gentamicin from Criswell *et al.*, 2006).
d Some strains of *B. burgdorferi* can have different susceptibilities to various antibiotics.
e Confers resistance to other coumarin antibiotics, such as novobiocin (used at 5 μg/ml).
f Not available: *E. coli* is relatively resistant to coumarin antibiotics.
g The *flgBp-aadA* cassette confers resistance to both spectinomycin (SPT) and streptomycin (STR) in both *E. coli* (Ec) and *B. burgdorferi* (Bb); however, spectinomycin is used for selection in *E. coli* because many strains are streptomycin-resistant and streptomycin is used for selection in *B. burgdorferi* because the frequency of background mutation to spectinomycin resistance is high (see Criswell *et al.*, 2006).
h These are markers that have been shown to confer resistance in *B. burgdorferi*, but have not yet been used experimentally.

1992) of the plating protocol described by Kurtti *et al.* (1987). Following electroporation, cells are allowed to recover for approximately 18 hours in the absence of selection. The semi-solid medium is formulated by mixing molten agarose with a modified and concentrated medium containing the selective agent (Table 8.2). The molten medium is cooled, mixed with the transformed cells, and poured into plates. The semi-solid medium with cells is allowed to solidify and the plates are incubated in a humidified atmosphere containing CO_2. Transformed bacteria form colonies in the semi-solid medium in five days to two weeks and can be picked by aspirating a plug of agarose containing the colony to liquid medium. A PCR screen for the presence of the selectable marker can be used to distinguish transformants from spontaneous antibiotic resistant mutants (Table 8.2).

An alternative method was developed by Yang *et al.* (2004), who were unable to isolate *ospA* mutants in solid medium. Recovered transformants are diluted into 100 ml of liquid medium with selection and distributed into several 96-well plates. Again, the plates are incubated in a humidified atmosphere containing CO_2 and positive wells are detected by the change in colour of the medium from red to yellow, which is due to the production of lactic acid by the bacteria during growth, as detected by phenol red in the medium. Incubating the 96-well plates at ambient atmosphere for 1 to 4 h increases the colour difference by shifting the bicarbonate–CO_2 equilibrium, which results in more basic (pinker) wells where there is no growth. Putative transformants can be expanded in a larger volume of medium containing antibiotic for confirmation and subsequent analysis. Because of its convenience, this method of selection and cloning in liquid medium has become increasingly popular.

Regardless of the method of isolating transformed clones, there is the possibility that two spirochaetes are intertwined and thus a colony in semi-solid medium or a positive well in liquid medium may not be derived from a single cell; therefore, extensive mixing, vortexing, and/or pipetting is required to separate cells before plating. In addition, the probability that a well was inoculated with a single cell is 0.94 if 10 of 96 wells are positive (Jon Graham, personal communication); however, a typical transformation usually yields fewer positive wells when diluted as described. Note that due to a mathematical error, the formulation of BSK II medium described in the original transformation protocol (Samuels, 1995) is 80% less concentrated than the original and commonly used formulation (Barbour, 1984; Pollack *et al.*, 1993); however, this dilute version of BSK-II has been used in several laboratories without adverse effects on growth or transformation.

8.4.5 Counterselection

Counterselectable markers select for the loss of genetic information (negative selection) and can thus be used to introduce unmarked mutations and cure plasmids (Reyrat *et al.*, 1998). The *rpsL* gene, which encodes ribosomal protein S12, has previously been developed into a counterselectable marker for the spirochete *Leptospira biflexa* (Picardeau *et al.*, 2001) and has recently been described for *B. burgdorferi* (Drecktrah *et al.*, 2010). Single point mutations in *rpsL* confer resistance to the aminoglycoside antibiotic streptomycin (Criswell *et al.*, 2006). Drecktrah *et al.* (2010) demonstrated that streptomycin susceptibility is dominant in an *rpsL* merodiploid. Moreover, streptomycin selects for the loss of a shuttle vector carrying the streptomycin-susceptible *rpsL* allele *in trans* in a streptomycin-resistant background (with the mutant *rpsL* allele on the chromosome). Therefore, *rpsL* functions as a counterselectable marker, although it has not yet been experimentally utilized.

Bactericidal antibodies that recognize outer surface proteins (Osps) of *B. burgdorferi* have been used to select for rare escape variants (Cinco, 1992; Coleman *et al.*, 1992; Šadžiene *et al.*, 1992), so one of these Osps has the potential to be developed into a counterselectable marker. In addition, an antibiotic-resistant allele of *parC* has been described (Galbraith *et al.*, 2005) that can be fashioned into a counterselectable marker (Samuels, 2006).

8.5 Gene inactivation and complementation

8.5.1 Site-directed mutagenesis

Targeted gene inactivation mediated by homologous recombination between transforming DNA

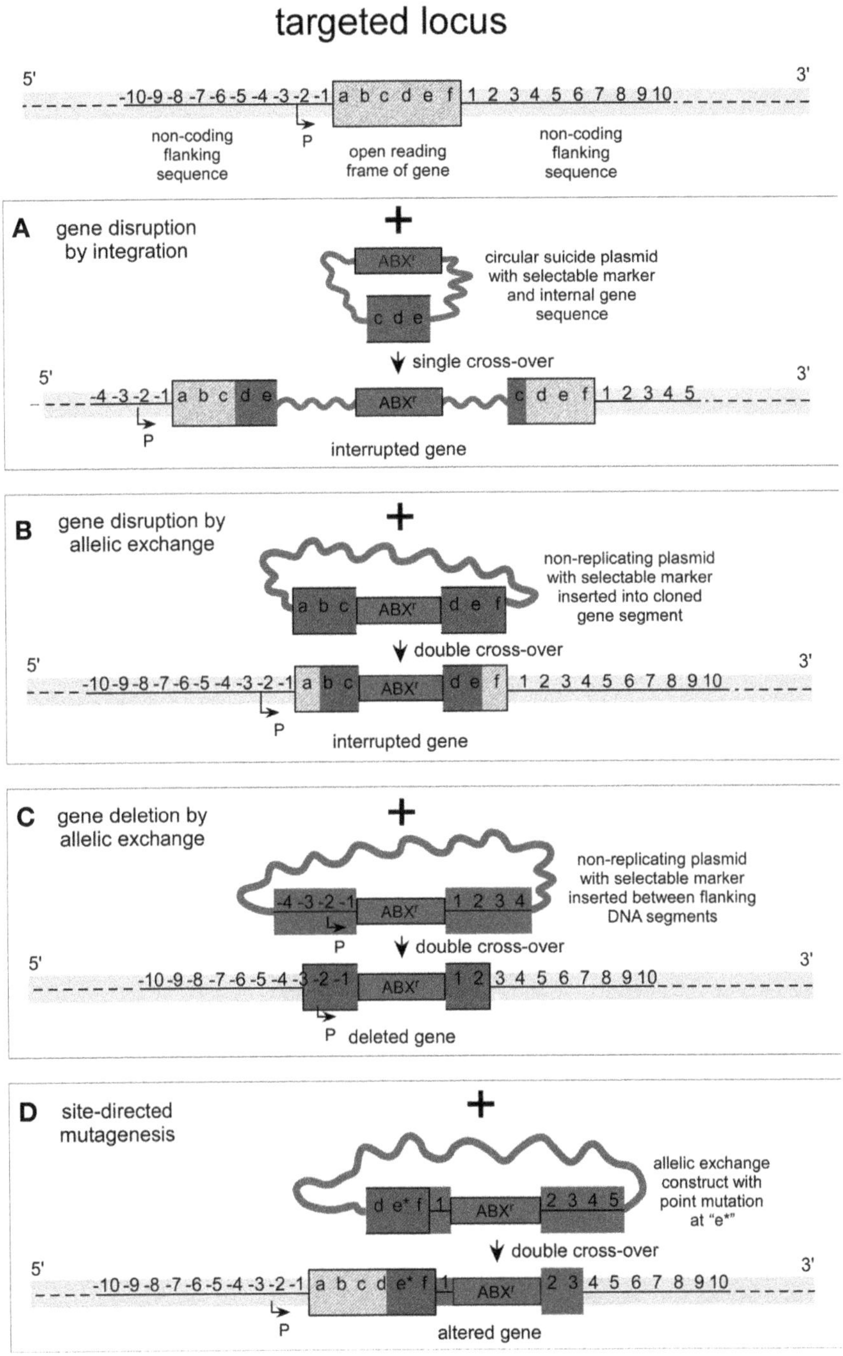

Figure 8.2 Schematic diagram of targeted gene inactivation by homologous recombination. The structure of the cloned inactivation construct determines whether the targeted gene is disrupted (as illustrated in parts A and B), deleted (part C) or altered (part D). In all cases, stable retention of the selectable marker results from homologous recombination between common sequences present at both the targeted locus (shown in light grey) and on the introduced DNA (in dark grey). The precise point of recombination (crossover site) can occur anywhere along these homologous sequences and not only at the positions indicated on these diagrams. If a single rather than double crossover occurs between an allelic exchange construct and the targeted locus, the entire plasmid would integrate and the recovered transformant would carry both wild-type and mutant copies of the targeted gene (not illustrated). This outcome can be minimized by linearizing the suicide plasmid carrying the allelic exchange construct prior to electroporation. A colour version of this figure is located in the plate section at the back of the book.

and endogenous sequences is straightforward but inefficient in *B. burgdorferi*, requiring microgram quantities of DNA and ranging in frequency from 10^{-5} to 10^{-8}, depending upon the recipient organism (Samuels *et al.*, 1994a; Tilly *et al.*, 2000; Elias *et al.*, 2002). Fortunately, illegitimate recombination (insertion of transforming DNA at a non-homologous site) has never been described following transformation of *B. burgdorferi* and thus does not appear to present a significant hurdle to the recovery of intended mutants. Simple gene disruption can be accomplished by integration (through a single crossover event) of a circular suicide vector carrying an appropriate selectable marker and a fragment of DNA homologous with the targeted gene (as depicted in Fig. 8.2A). Alternatively, if homologous sequences flank the selectable marker in the inactivation construct, allelic exchange can occur with the targeted gene through a double crossover event. Depending upon the design of the inactivation construct, allelic exchange can result in simple disruption (Fig. 8.2B) or deletion (Fig. 8.2C) of the targeted gene. To date, a number of plasmid and chromosomal loci in *B. burgdorferi* have been inactivated through allelic exchange. A somewhat more elegant form of allelic exchange entails modifying the coding sequence of the targeted gene (Fig. 8.2D), rather than disrupting or deleting it, to assess the resulting mutant phenotype. This has been used in *B. burgdorferi* to alter the active site of a kinase (Yang *et al.*, 2003), as well as to selectively interrupt a second protein product from the *gyrA* gene (Knight *et al.*, 2000).

A single crossover event can occur with allelic exchange constructs if they are introduced as circular DNA. This can result in the unintentional introduction into *B. burgdorferi* of other antibiotic resistance markers present on the cloning vector. Some selectable markers commonly used in *E. coli*, such as genes conferring resistance to ampicillin and tetracycline, are not permitted for genetic studies in *B. burgdorferi* due to the clinical utility of these antibiotics in treating Lyme disease. At the minimum, cloning vectors carrying these markers should be linearized prior to transformation into *B. burgdorferi*. Preferably, allelic exchange constructs should be cloned into vectors that do not contain such restricted markers or the genes encoding them should be inactivated prior to transformation of *B. burgdorferi*.

8.5.2 Complementation

Complementation plays a crucial role in genetic studies of *B. burgdorferi* due to the inefficiency with which targeted gene inactivation occurs and the inherent instability of the plasmid component of the genome, as described below (in this section and 8.10.1). Although measures can be taken to assure that wild-type and derivative clones contain the same plasmids (Purser and Norris, 2000; Labandeira-Rey and Skare, 2001; Eggers *et al.*, 2002; Elias *et al.*, 2002; Parveen *et al.*, 2006), screening for more subtle genomic alterations, such as point mutations, rearrangements, small deletions, or polar effects on flanking genes, is not feasible. To control for such unforeseen or unidentified second site mutations, the goal of complementation is to leave the original mutation unaltered, but to reintroduce a wild-type copy of the targeted gene, typically *in trans* on an autonomously replicating shuttle vector or elsewhere in the genome. Ideally, such a shuttle vector is stable and present at approximately the same copy number and supercoiling state as the endogenous replicon carrying the mutated gene. As described below, a number of shuttle vectors have been developed for use in *B. burgdorferi*.

Endogenous plasmid shuttle vectors

Several shuttle vectors have been derived from the endogenous plasmids of *B. burgdorferi*. The complete genome sequence of *B. burgdorferi* identified a set of paralogous gene families (PGFs), members of which were present on all plasmids, and suggested that these genes or regions might be involved in plasmid replication and maintenance, although lacking strict sequence homology to such functions in other bacteria (Zückert and Meyer, 1996; Casjens *et al.*, 2000). An early shuttle vector, derived from a small non-essential endogenous circular plasmid (cp9) of *B. burgdorferi*, demonstrated that the region encompassing these PGFs was sufficient for autonomous replication in *B. burgdorferi* (but did not support replication in *E. coli*) (Stewart *et al.*, 2001). This shuttle vector, termed pBSV2, and its derivatives have become the workhorse for complementation and other applications in *B. burgdorferi*. In the absence of strong counterselection against it (Tilly *et al.*, 2006), pBSV2 is relatively stable both *in vitro* and *in vivo* (Stewart *et al.*, 2001) and present at approximately five copies per cell (Tilly *et al.*, 2006). Fluctuation in plasmid copy number during the cell cycle could theoretically result in incomplete complementation, but such changes have not been documented. Similarly, additional shuttle vectors have been derived from other endogenous plasmids (lp25, lp28-1, cp26, and a cp32) by cloning their PGF counterparts together with a functional selectable marker (Eggers *et al.*, 2002; Stewart *et al.*, 2003; Byram *et al.*, 2004; Beaurepaire and Chaconas, 2005). Along with autonomous replication, the PGF regions also mediate incompatibility, resulting in displacement of the endogenous *B. burgdorferi* plasmid from which they were derived (Stewart *et al.*, 2001, 2003; Eggers *et al.*, 2002; Beaurepaire and Chaconas, 2005), unless selection exists for the coexistence of both replicons (Byram *et al.*, 2004). In addition to their utility as genetic tools, this trait of shuttle vectors also permits the experimental manipulation of plasmid content in *B. burgdorferi* (Stewart *et al.*, 2003; Byram *et al.*, 2004; Grimm *et al.*, 2004a; Jewett *et al.*, 2007a) through targeted deletion of a native plasmid by an incompatible shuttle vector.

Broad-host-range plasmids

Many plasmids have extended host range and can multiply in Gram negative and Gram positive bacteria (Kok *et al.*, 1984; Thomas and Smith, 1987). Such broad-host-range plasmids have been used extensively as genetic tools because their wide host range allows them to be used as shuttle cloning vectors in diverse bacterial species (Kok *et al.*, 1984; Thomas and Smith, 1987). This information, and the fact that *B. burgdorferi* has a GC content of 28% (Fraser *et al.*, 1997), suggested that a broad-host-range plasmid of low GC content Gram positive bacteria might replicate in *B. burgdorferi*.

Plasmid pGK12 is a wide-host-range plasmid from *Lactococcus lactis* that is able to replicate in *Staphylococcus*, *Streptococcus*, *Bacillus*, and *Escherichia coli* (Kok *et al.*, 1984; Maguin *et al.*, 1992). This plasmid, containing the antibiotic resistance genes *ermC* and *cat* from *Staphylococcus aureus* plasmids pE194 and pC194, respectively, was introduced by electroporation into *B. burgdorferi* (Sartakova *et al.*, 2000). *Borrelia burgdorferi* containing pGK12 were selected with erythromycin; the plasmid was also used to express green fluorescent protein (Sartakova *et al.*, 2000). The lactococcus plasmid pJRS525 is similar to pGK12 (McIver and Scott, 1997). It can replicate in both *E. coli* and *Streptococcus*, and has been used to complement *B. burgdorferi* mutants (Hübner *et al.*, 2001).

These experiments indicate that *B. burgdorferi* can sustain the replication of broad-host-range plasmids, that these plasmids can be used to manipulate *Borrelia* genetically, and that foreign genes can be expressed in this pathogen from their endogenous promoters (Sartakova *et al.*, 2000, 2003; Hübner *et al.*, 2001). Undoubtedly, the use of broad-host-range plasmids as genetic tools in *B. burgdorferi* has its shortcomings. These include instability in the absence of antibiotics and, in the case of pGK12, low copy number (Sartakova *et al.*, 2000, 2003). However, instability and low copy number may also be advantageous under some genetic experimental conditions that may require plasmid curing (transposition) and balanced expression of borrelial chromosomal and plasmids genes (complementation). Moreover, the long history of use of broad-host-range cloning vectors in microbiology has generated a vast array of genetic resources that may be useful in *B. burgdorferi* genetic studies in the future, including heat- and cold-sensitive replication mutants, copy number mutants, and conjugative transposons.

Genetic reconstitution

An alternative complementation strategy is genetic reconstitution, in which the mutated copy of the gene is replaced with a wild-type copy and closely linked marker (Hübner *et al.*, 2001; Tilly *et al.*, 2001; Li *et al.*, 2007), similar to the strategy used to introduce point mutations (Knight *et al.*, 2000; Yang *et al.*, 2003) (Fig. 8.1D). Allelic exchange in infectious *B. burgdorferi* clones, while inefficient, is not similarly limited by the presence of lp25 and lp56, presumably because double-crossover events do not require intact circular DNA. This method is less preferable than complementation *in trans* on a shuttle vector because it does not leave the original mutation unaltered and hence position effects on flanking genes, either of the mutation or subsequent complementation event, could complicate interpretation of the results. An improvement on this strategy would be to insert the complementing gene somewhere else in the genome, such as at a non-coding site on lp25, lp28-1 or lp36 (Kawabata *et al.*, 2004; Grimm *et al.*, 2005; Blevins *et al.*, 2007; Gilbert *et al.*, 2007; Jewett *et al.*, 2007b), thereby leaving the original mutation undisturbed and providing selection for complemented mutants retaining an essential plasmid.

A potential advantage to complementation by genetic reconstitution, at the site of the mutated locus or elsewhere on the same replicon, is that the reintroduced wild-type gene copy is at the same copy number and similar supercoiling state as the original gene. Some attempts to complement mutations *in trans* on a shuttle vector have only partially or unsuccessfully restored the wild-type phenotype (Sartakova *et al.*, 2001), raising questions about whether the context of the complementing gene has significantly altered its expression. Alternatively, such results could indicate the presence of an unrecognized second site mutation, which is responsible for the defective phenotype of the complemented mutant. Distinguishing among the possible causes of unsuccessful complementation is addressed below.

Limitations and alternatives

Complementation, while critical to demonstrating the link between a mutant phenotype and gene, presents a substantial challenge when working with infectious *B. burgdorferi* clones retaining the full set of endogenous plasmids. As mentioned above, this is primarily due to the activities of putative restriction enzymes encoded by genes on two *B. burgdorferi* plasmids, lp25 and lp56 (Lawrenz *et al.*, 2002; Kawabata *et al.*, 2004; Chen *et al.*, 2008). The sites recognized by, and properties of, these putative restriction enzymes have not been identified, but circumstantial evidence suggests that such sites are present (and unmodified) on the *E. coli*-derived shuttle vectors bearing the complementing genes, thereby leading to the selective recovery of complemented *B. burgdorferi* mutants that have lost lp25 and lp56 (Lawrenz *et al.*, 2002; Kawabata *et al.*, 2004). Loss of lp56 has no discernible impact on infectivity (Purser and Norris, 2000; Jacobs *et al.*, 2006), whereas loss of lp25 renders *B. burgdorferi* non-infectious in mice and seriously attenuated in ticks (Purser and Norris, 2000; Labandeira-Rey and Skare, 2001; Grimm *et al.*, 2004a; Grimm *et al.*, 2005; Revel *et al.*, 2005; Strother *et al.*, 2005). Despite these limitations, persistent effort can usually yield complemented *B. burgdorferi* mutants carrying a wild-type copy of the mutated gene on the shuttle vector and retaining all essential plasmids.

Several approaches have been taken to address the essential but problematic nature of lp25 for genetic studies in infectious *B. burgdorferi*. One solution has been to inactivate the *bbe02* gene (encoding one of the putative restriction enzymes) in a clonal background lacking lp56 (Kawabata *et al.*, 2004; Gilbert *et al.*, 2007). Such clones are highly transformable by the shuttle vector pBSV2 and infectious in both ticks and mice (Kawabata *et al.*, 2004; Jacobs *et al.*, 2006). While no longer available for downstream manipulations, the selectable marker used to inactivate *bbe02* can ensure retention of lp25 in subsequent applications. A variation on this approach entails inserting a marker at a different nonessential site on lp25 to stabilize the plasmid (Grimm *et al.*, 2004a; Blevins *et al.*, 2007). Another approach utilizes a non-infectious, highly transformable clone lacking both lp25 and lp56 for mutant construction (Kawabata *et al.*, 2004). Subsequently, the *bbe22* gene (*pncA*) of lp25 (which is sufficient to restore mouse infectivity to this clone) (Purser *et al.*, 2003), is reintroduced on the shuttle vector used for complementation (Xu *et al.*, 2005; Seshu *et al.*, 2006). This method is attractive because of the ease with which this lp25-minus clone can be manipulated, but infectious cycle analyses of such mutants are limited to the mammalian host because additional genes carried by lp25 are required for colonization and persistence in the tick vector (Grimm *et al.*, 2005; Revel *et al.*, 2005; Strother *et al.*, 2005).

When repeated efforts do not yield a complemented mutant retaining all other essential genes, an argument can be made linking a mutant phenotype and targeted gene if several independently derived mutants are compared with an isogenic wild-type clone. Such mutants must be obtained in separate transformation experiments, to guarantee that they are not siblings that arose from a single primary transformant (Stewart *et al.*, 2008). This argument is based on the low probability that second site mutations resulting in a similar mutant phenotype would have arisen independently in separate transformations. This is not a terribly robust argument, however, and does not address concerns about the possible polar effects of a mutation on flanking genes. All the same, it is stronger than relying upon the analysis of a single uncomplemented mutant.

Failed complementation

Occasionally, a quandary arises when a wild-type copy of the gene has been reintroduced successfully into the mutant strain, yet the defective phenotype persists, leading to the conclusion that complementation has failed. There are a number of possible explanations for this troublesome outcome and identifying the source generally entails a process of elimination through additional experiments.

To start, it is wise to recover the complementation construct from the mutant strain and confirm, by re-sequencing, that the nucleotide sequence of the open reading frame and regulatory region of the complementing gene are intact and correct. Once this has been verified, the expression of the reintroduced gene in the complemented mutant should be compared with that of the same gene in the wild-type clone, ideally at both transcript and protein levels. This addresses the possibility that the complementing gene may not be appropriately expressed *in trans* due to changes in supercoiling, copy number, or absence of key regulatory sequences.

If significant differences in expression between the wild type and complementing genes are found, then a different complementation strategy should be employed, such as genetic reconstitution, in which the genomic location of the mutated gene is recapitulated with the complementing gene. If persistence of the mutant phenotype cannot be explained by faulty gene expression, the next possibility to investigate is polar effects on the transcription of adjacent genes. If expression of adjacent genes appears to be altered by the mutation, then a different gene inactivation strategy should be employed that minimizes potential polar effects. Finally, if the complementing gene is intact and appropriately expressed, and there is no evidence that the mutation impacts adjacent genes, then the most likely explanation for the defective phenotype of the complemented mutant strain is an unlinked and unrecognized mutation elsewhere in the genome. In this case, complementation has failed because the phenotype of the mutant strain does not derive from the introduced mutation, thus illustrating the essential role of complementation in genetic studies.

8.5.3 Surrogate genetics

Limitations on the genetic manipulation of *B. burgdorferi* have stimulated researchers to perform surrogate genetics in *E. coli* because of the well-developed genetic systems of the latter organism and the availability of extensively characterized mutants (Beckwith, 1991). Numerous *B. burgdorferi* genes have been expressed in *E. coli* from either their endogenous promoters or from cloning vector promoters, primarily as a source of purified recombinant protein. A number of *B. burgdorferi* genes have been used to complement *E. coli* mutants deficient in the homologous genes, which has permitted partial functional characterization of these borrelial genes. For example, the GMP synthetase gene *guaA* (Margolis *et al.*, 1994), an oligopetide permease gene *oppA* (Lin *et al.*, 2001), the architectural DNA-binding protein encoded by *gac* (Knight and Samuels, 1999), the DNA recombination gene *recA* (Liveris *et al.*, 2004; Putteet-Driver *et al.*, 2004), the sigma regulatory gene *rpoS* (Eggers *et al.*, 2004), the nicotinamidase gene *pncA* (Purser *et al.*, 2003), and the (p)ppGpp synthase gene rel_{Bbu} (Bugrysheva *et al.*, 2005) have all been able to complement *E. coli* or *Salmonella* mutants deficient in their homologues. Similarly, cloning of *B. burgdorferi* genes encoding phosphatidylcholine and phosphatidylglycerolphosphate synthase in *E. coli* permitted the characterization of these enzymatic activities (Wang *et al.*, 2004). Surrogate genetics in *E. coli* also permitted the characterization of *B. burgdorferi* promoter elements involved in transcription by *B. burgdorferi* sigma factors RpoS and RpoN (Alverson *et al.*, 2003; Caimano *et al.*, 2004; Eggers *et al.*, 2004, 2006; Burtnick *et al.*, 2007) and an oxidative stress response regulator, BosR (Boylan *et al.*, 2003; see Chapter 4). Expression in *E. coli* of a phage display library led to the identification of a putative integrin-binding adhesin of *B.*

burgdorferi (Coburn *et al.*, 1999; Coburn and Cugini, 2003; see Chapter 12). The advances in genetic manipulation of *B. burgdorferi* described in this chapter suggest that the need to perform surrogate genetic experiments in other bacterial hosts may decrease in the future. However, given its complex culture medium and slow growth rate, it is unlikely that *B. burgdorferi* will ever be as easy to manipulate as *E. coli*. Hence, alternative hosts and their genetic systems will probably continue to provide useful information regarding *B. burgdorferi* gene expression and function.

8.6 Random tagged mutagenesis

The use of transposon mutagenesis as a tool for bacterial genetics represents an important advance that has generally superseded chemical and radiation mutagenesis (Hayes, 2003; Bosse *et al.*, 2006; Salama and Manoil, 2006; Burrack and Higgins, 2007). Mutagenized bacteria are easily isolated on the basis of antibiotic resistance or other selectable markers carried on the transposable element, and the mutated gene can be readily identified by routine molecular procedures (Judson and Mekalanos, 2000; Hayes, 2003; Bosse *et al.*, 2006; Salama and Manoil, 2006; Burrack and Higgins, 2007). Transposon mutagenesis can be applied to a range of bacteria, including organisms for which genetic tools are largely unavailable (Lampe *et al.*, 1996; Gueiros-Filho and Beverley, 1997; Lampe *et al.*, 1999). Transposons can also be used to facilitate DNA sequencing and gene fusions with reporter genes (Hayes, 2003; Valdivia *et al.*, 2006). Ideally, transposons used for saturating mutagenesis encode all necessary functions for mobilization, transpose at a relatively high frequency, have a strong selectable marker, insert into a nearly random DNA target sequence, and function in a wide range of bacterial species (Salama and Manoil, 2006).

Genome-wide transposon mutagenesis of *B. burgdorferi* was first demonstrated with a suicide plasmid that contained a modified *himar1* element of the *mariner* family (Stewart *et al.*, 2004). In this system, called pMarGent, the inverted repeats of the transposable element flank a ColE1 origin of replication and a gentamicin resistance cassette that functions in both *E. coli* and *B. burgdorferi*, while the *himar1* transposase is carried on a portion of the suicide plasmid that is not mobilized, and hence only transiently present in *B. burgdorferi* (Stewart *et al.*, 2004; Stewart and Rosa, 2008). Each gentamicin-resistant mutant contains a single, stable transposon insertion that can be rescued in *E. coli* along with flanking *B. burgdorferi* DNA. The transposon insertion site is identified by sequencing the rescued plasmid, using primers that read out from the transposon into the flanking DNA (Stewart *et al.*, 2004; Stewart and Rosa, 2008). This approach yielded a saturated mutant library in a non-infectious *B. burgdorferi* clone and permitted the isolation and characterization of a large number of mutants, with random transposon insertions in the linear chromosome and all linear and circular plasmids (Stewart *et al.*, 2004). These experiments demonstrated the strength and feasibility of transposon mutagenesis in *B. burgdorferi*.

Another approach to generate transposon mutants in *B. burgdorferi* utilized a constitutively expressed *himar1* transposase gene on a replicating plasmid (Morozova *et al.*, 2005). *Borrelia burgdorferi* carrying this plasmid and synthesizing the transposase were subsequently transformed by electroporation with a suicide plasmid containing the *mariner* transposon (Morozova *et al.*, 2005). The site of the transposon insertion in recovered transformants was determined by sequencing the DNA flanking the transposon after amplification by modified inverse PCR. Mutants were identified with insertions in coding and non-coding regions of the chromosome and plasmids. Although in this system the *himar1* transposase is present on a replicating plasmid and thus the *mariner* element can continue to 'hop', a substantial number of the transposon mutants lose the plasmid and are stable (Morozova *et al.*, 2005).

The utility of transposon mutagenesis as a tool to study *B. burgdorferi* virulence was illustrated by using the pMarGent system to generate a limited number of transposon mutants in an infectious *B. burgdorferi* clone and subsequently assessing their ability to infect mice (Botkin *et al.*, 2006a; Botkin *et al.*, 2006b). Further studies, including targeted inactivation and complementation, are required to clarify the roles of the implicated genes in mouse infectivity. Isolation of a random, saturated mutant library requires an insertion every 200 base pairs on average, thus a completely saturated transposon library of *B. burgdorferi* would contain at least 7500 independent mutants, since the *B. burgdorferi* genome is approximately 1,500,000 bp in size (Fraser *et al.*, 1997; Casjens *et al.*, 2000; Salama and Manoil, 2006). The generation of such a large number of independent transposon mutants in an infectious *B. burgdorferi* clone currently is limited by the low frequency of transformation in this background (Stewart *et al.*, 2004; Botkin *et al.*, 2006b). The difficulty in efficiently introducing foreign DNA into infectious *B. burgdorferi* also limits the application of methods successfully developed in other bacterial species to identify *in vivo*-induced virulence genes and essential genes based on *in vitro* and *in vivo* transposition (Bosse *et al.*, 2006). However, as existing and new strategies are devised to enhance genetic manipulation of *B. burgdorferi*, the use of transposon mutagenesis and other approaches to investigate the biology and virulence of this pathogen will be possible.

8.7 Reporter genes

Reporter genes encode proteins whose function can be readily assayed; they have been used extensively in bacteria to monitor variations in gene expression. Three reporter genes have been successfully used in *B. burgdorferi* to study gene expression: chloramphenicol acetyl transferase, *cat*; luciferase, *luc*; and, more extensively, green fluorescent protein, *gfp* (Sohaskey *et al.*, 1997; Sartakova *et al.*, 2000; Carroll *et al.*, 2003; Eggers *et al.*, 2004; Blevins *et al.*, 2007).

Chloramphenicol acetyltransferase was originally used in *B. burgdorferi* to ascertain the structure of *B. burgdorferi* promoters (Sohaskey *et al.*, 1997; Sohaskey *et al.*, 1999) and to study the effect on transcription of DNA supercoiling and temperature (Alverson *et al.*, 2003). The assays to detect chloramphenicol acetyltransferase are not as simple and direct as those to detect fluorescent markers, and they are time- and reagent-intensive. Probably for these reasons, green fluorescent protein (GFP) and its detection by fluorescent microscopy, spectrofluorometers, flow cytometry and immunoblots have replaced the use of chloramphenicol acetyl transferase as the reporter of choice in *B. burgdorferi* studies. Moreover, GFP permits determination of gene transcription in individual living cells (Carroll *et al.*, 2003; Eggers *et al.*, 2004; Falkow, 2004; Valdivia *et al.*, 2006; Srivastava and de Silva, 2008). GFP expression can be influenced by temperature, pH and oxygen concentration, and its levels depend on its synthesis and dilution mediated by cell division; this last factor limits its use when it is necessary to detect rapid changes in gene expression (Valdivia *et al.*, 2006). Alternative fluorescent reporter genes that have been expressed in *B. burgdorferi* include yellow fluorescent protein, *yfp*; cyan fluorescent protein, *cfp* (Eggers *et al.*, 2002); and red fluorescent protein, *mRFP* (Schulze and Zückert, 2006).

Detection of the fluorescence of GFP and its derivatives in *B. burgdorferi* has usually involved the cloning of these genes into multicopy plasmids (Sartakova *et al.*, 2000; Eggers *et al.*, 2002; Carroll *et al.*, 2003). The use of GFP derivatives with increased sensitivity and rapid synthesis (Valdivia *et al.*, 2006) may permit visualization of single copy chromosomal gene fusions, which have not yet been described in *B. burgdorferi*. GFP has permitted study of gene expression of *B. burgdorferi* in ticks (Bykowski *et al.*, 2006; Miller *et al.*, 2006), and recently this reporter was used to visualize the interaction of the spirochaete with the mouse vasculature in real time, high resolution imaging (Moriarty *et al.*, 2008). GFP has not yet been used in more complex genetic endeavours in *B. burgdorferi*, such as the cellular localization of borrelial proteins in complex processes like cell division and signal transduction or the study of engineered regulatory networks (Triccas *et al.*, 2002; Valdivia *et al.*, 2006). However, monomeric red fluorescent protein (mRFP) has been effectively used to investigate the lipoprotein export pathway of *Borrelia* (Schulze and Zückert, 2006). A critical advantage of mRFP over GFP for this purpose is that mRFP continues to fluoresce when localized to the periplasm and the outer surface of bacteria, while GFP only fluoresces within the cytoplasm of the cell (Schulze and Zückert, 2006).

Some of the disadvantages of CAT and GFP discussed above suggest that the genetic toolbox for *B. burgdorferi* would benefit from use of alternative reporter genes. Recently, a codon optimized *luc* gene from firefly luciferase has been adapted for use as a reporter in *B. burgdorferi* (Blevins *et al.*, 2007). Luciferase (measured by

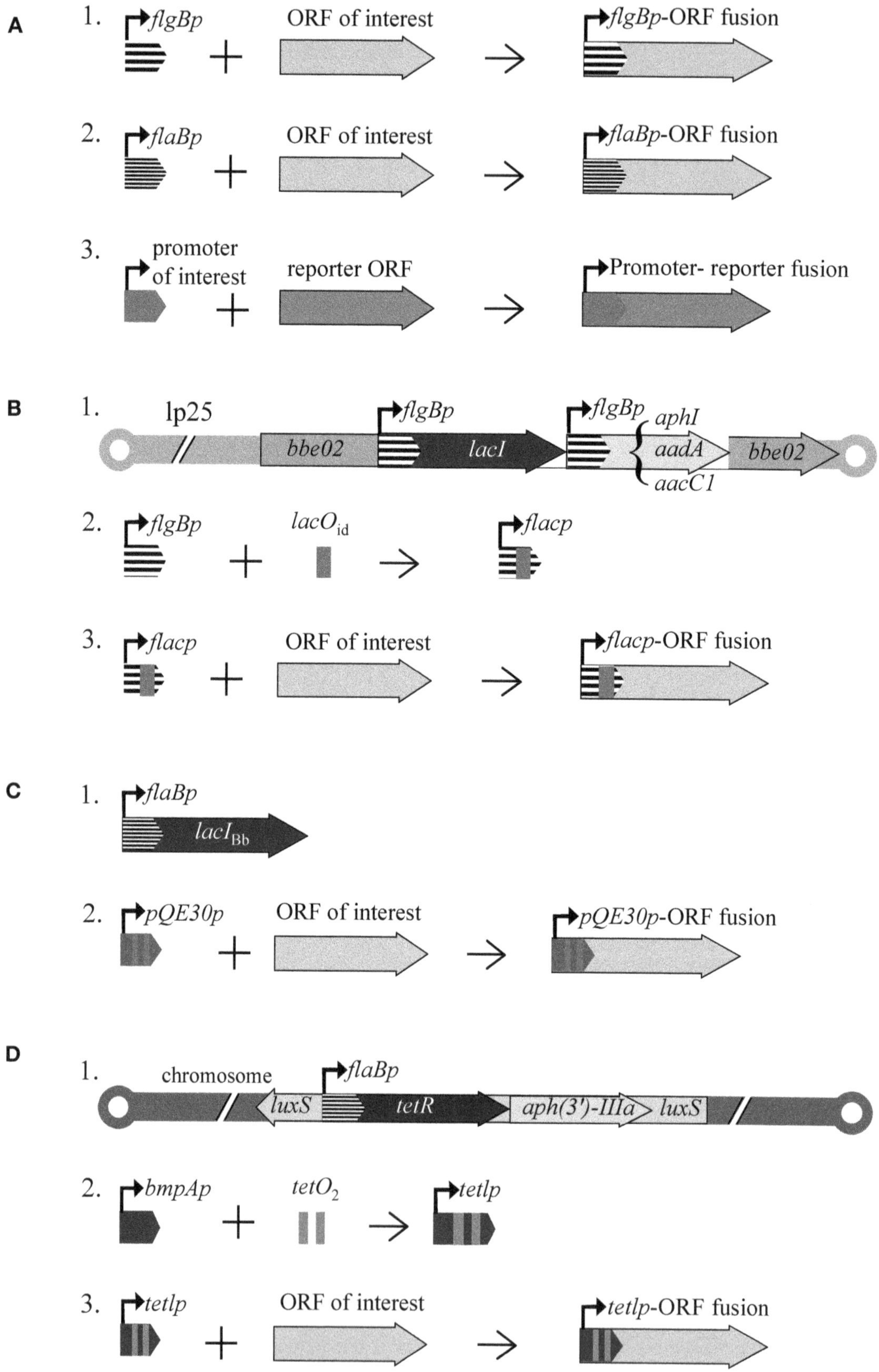
A
1.
flgBp
ORF of interest
flgBp-ORF fusion
2.
flaBp
ORF of interest
flaBp-ORF fusion
3.
promoter of interest
reporter ORF
Promoter- reporter fusion
B
1.
lp25
bbe02
flgBp
lacI
flgBp
aphI
aadA
aacC1
bbe02
2.
flgBp
lacOid
flacp
3.
flacp
ORF of interest
flacp-ORF fusion
C
1.
flaBp
lacIBb
2.
pQE30p
ORF of interest
pQE30p-ORF fusion
D
1.
chromosome
flaBp
luxS
tetR
aph(3')-IIIa
luxS
2.
bmpAp
tetO2
tetlp
3.
tetlp
ORF of interest
tetlp-ORF fusion

luminometer) was used to study variations of transcription mediated by an endogenous Borrelia promoter or by an ITPG-induced *lac* expression system (Blevins *et al.*, 2007). Additional studies will be necessary to determine the usefulness of firefly and other luciferases and their detection systems in the study of *B. burgdorferi in vitro* and *in vivo*. Luciferase genes have been used extensively in other Gram-positive and Gram-negative pathogens during infection of living animals to follow their tissue tropism and to assess the effect of therapy on the evolution of the infection (Contag *et al.*, 1995; Francis *et al.*, 2001). The usefulness of other standard reporter genes, such as β-galactosidase and β-glucuronidase, for studies in *B. burgdorferi* has not been assessed.

8.8 Engineered gene expression

8.8.1 Endogenous promoter fusions

The first tools employed to control gene expression in *B. burgdorferi* were its endogenous *flaB* and *flgB* promoters (Fig. 8.3A) (Sohaskey *et al.*, 1997; Bono *et al.*, 2000; Sartakova *et al.*, 2000). They were used to express a foreign antibiotic resistance gene as a selectable marker (Bono *et al.*, 2000) and either a *cat* or *gfp* gene as a reporter (Sohaskey *et al.*, 1997; Sartakova *et al.*, 2000) in *B. burgdorferi*. Subsequently, these two promoters have been used extensively to constitutively express endogenous and foreign genes in *B. burgdorferi* and to develop new genetic markers and genetic tools, as described throughout this chapter. Constitutive expression mediated by *flaBp* has also been utilized to study the role of OspC downregulation in the ability of *B. burgdorferi* to evade mammalian host defences (Xu *et al.*, 2006). Similarly, *B. burgdorferi* containing *flaBp–ospAB* gene fusions were used to demonstrate that constitutive expression of these proteins led to rapid clearance of *Borrelia* in mice by the immune system (Strother *et al.*, 2007).

8.8.2 Inducible gene expression

Inducible promoter systems that can regulate gene expression in a graded manner offer certain advantages over the total ablation of gene expression generated by null mutants, or full constitutive expression generated by *flaBp* and *flgBp*. For example, information about the function of unknown but essential genes can be obtained

Figure 8.3 (opposite) Schematic diagram of engineered gene expression. (A) Constitutive gene expression (Sohaskey *et al.*, 1997; Bono *et al.*, 2000; Sartakova *et al.*, 2000). The *flgB* promoter (1), *flaB* promoter (2), or other promoters (not shown) can be fused to several open reading frames (or RNA-encoding genes), including antibiotic resistance genes to select for transformants, reporter genes to visualize cells, and endogenous genes to bypass their native regulation or to express them at high levels. Endogenous promoters can be fused to reporter genes to assay transcriptional activity (3). (B) *lac*-based inducible gene expression using the hybrid *flac* promoter (Gilbert *et al.*, 2007). The *flgB* promoter was fused to the *lacI* open reading frame (which encodes the *lac* repressor) and inserted, along with a linked selectable marker, in the *bbe02* gene (a putative barrier to transformation) on lp25 (a plasmid essential for infection); the resulting strains constitutively produce LacI repressor (1). The sequence between the conserved –10 element and the putative ribosome binding site of the *flgB* promoter was replaced with *lac*O_{id} (a synthetic, 'ideal' lac operator), resulting in the inducible *flac* promoter, which is repressed by LacI repressor (2). The *flac* promoter can be fused to the coding region of any gene, either in *cis* in the genome by replacing the native promoter and linking to a selectable marker or in *trans* on a shuttle vector (3). Dr Mike Gilbert contributed to the design and drawing of this panel. (C) *lac*-based inducible gene expression using the *pQE30* promoter (Blevins *et al.*, 2007). The *flaB* promoter was fused to a *lacI* open reading frame that was codon optimized for *B. burgdorferi* (*lacI*$_{Bb}$) and cloned onto a shuttle vector (1). The *pQE30* promoter (from plasmid pQE30) is a T5 promoter with two *lacO* sequences that can be fused to the coding region of any gene (2). (D) *tet*-based inducible gene expression (Cabello *et al.*, 2006). The *flaB* promoter was fused to the *tetR* open reading frame (which encodes the tetracycline repressor) and inserted in the *luxS* gene on the chromosome; the resulting strains constitutively produce TetR repressor (1). Two operator (*tetO*) sequences were inserted in the *bmpA* promoter, one upstream of the conserved –35 element and the other between the –35 and the –10 elements, resulting in the inducible *tetI* promoter, which is repressed by TetR repressor (2). The *tetI* promoter can be fused to the coding region of any gene, either in *cis* in the genome by replacing the native promoter and linking to a selectable marker or *in trans* on a shuttle vector (3). Dr Henry Godfrey contributed to the design of this panel. A colour version of this figure is located in the plate section at the back of the book.

through modulating gene expression and monitoring the phenotypic changes that occur when gene products become limiting. In *B. burgdorferi*, the lack of knowledge regarding the structure and regulatory physiology of a wide range of promoters precludes the use of endogenous genetic switches. A genetic switch that has been used extensively in other bacteria is the lactose operon of *E. coli*, which entails repression by LacI, a promoter containing *lacO* operators that bind LacI, and inducers that relieve repression by binding to LacI (Jacob and Monod, 1961). Another well characterized genetic switch is based on the *E. coli* tetracycline resistance gene of *Tn*10, which encodes an efflux pump (Gossen and Bujard, 1992; Bujard, 1999). In this system, regulation of gene expression involves the TetR repressor, a promoter containing *tet* operators, and tetracycline analogues that bind the repressor and induce the promoter at concentrations 100 times lower than those that inhibit protein synthesis (Gossen and Bujard, 1992; Bujard, 1999). In both the *tet* and *lac* systems, induction of gene expression occurs as a consequence of blocking repression.

lac *system*

Published reports indicate that modifications of the *lac* system can be used to experimentally regulate gene expression in *B. burgdorferi* (Fig. 8.3B and C) (Blevins *et al.*, 2007; Gilbert *et al.*, 2007). Gilbert *et al.* (2007) constructed the *flacp* inducible promoter by integrating the synthetic $lacO_{id}$ operator (Oehler *et al.*, 1994) into the *B. burgdorferi flgB* promoter (Fig. 8.3B) (Ge *et al.*, 1997). This inducible promoter was fused to *ospC* and *rpoS* by directly incorporating *flacp* into the genome and replacing the native promoters of the respective genes (Gilbert *et al.*, 2007). These investigators also generated *B. burgdorferi* strains that produce high levels of the LacI repressor by fusing the *flgB* promoter to *lacI* and inserting the hybrid into lp25, a plasmid required for infectivity (Purser and Norris, 2000; Labandeira-Rey and Skare, 2001). The expression of *flacp*-controlled genes in this background is tightly regulated in response to the inducer isopropyl β-D-thiogalactopyranoside (IPTG); genomic *flacp–ospC* and *flacp–rpoS* fusions bypass the normal regulation by environmental factors (Gilbert *et al.*, 2007). In addition, *flacp-rpoS* was used to demonstrate that RpoS was not sufficient to downregulate OspA *in vitro* (Caimano *et al.*, 2007). This *lac* system was used *in vivo* to further test the hypothesis that OspC was required at an early time point in infection (Tilly *et al.*, 2006; Tilly *et al.*, 2007) and to assess the temporal requirement for OspC (Gilbert *et al.*, 2007). This experiment indicated that the inducible system appears to function *in vivo* and should be useful to test genes required for infectivity (Gilbert *et al.*, 2007).

An alternative system of gene regulation based on the *E. coli lac* operon was also shown to be functional in *B. burgdorferi* (Fig. 8.3C) (Blevins *et al.*, 2007). Gene regulation by the lactose operon was demonstrated using the modified *luc* reporter described previously and by detecting expression of the endogenous *B. burgdorferi* protein BptA. In this case, the *B. burgdorferi lac* expression system consisted of a single plasmid based on the pKFSS1 shuttle vector with the T5/*lac* hybrid promoter *pQE30* from plasmid QE30 (QIAGEN); this promoter contains two *lac* operators (Blevins *et al.*, 2007). On the same plasmid, a codon-optimized *lacI* gene was present under the control of the *flaB* promoter (Blevins *et al.*, 2007). Adding IPTG to *B. burgdorferi* carrying this plasmid with either *luc* or *bptA* cloned under the control of the *pQE30* promoter resulted in increased expression of these gene products. This *lac* system works relatively well in *B. burgdorferi*, but induction was incomplete, which may result from either insufficient levels of IPTG accumulating inside *B. burgdorferi*, *pQE30* being a weak promoter in *Borrelia*, or high levels of LacI.

tet *system*

Preliminary results indicated that an inducible system expressing the TetR repressor from either a plasmid or a chromosomal location in *B. burgdorferi* under the control of the *flaB* promoter, and a hybrid *B. burgdorferi bmpA* promoter containing one or two *tet* operators, can be used to regulate the expression of GFP and BmpA (Fig. 8.2-D) (Cabello *et al.*, 2006). The same experiments showed that induction of expression in this system is linear between anhydrotetracycline concentrations of 0 to 1.5 mg ml^{-1} and that repression is leaky even when using the hybrid promoter containing two *tet* operators (Cabello *et al.*, 2006). A recent report described a single plasmid-based system of inducible gene expression in *B. burgdorferi* (Whetsine *et al.*, 2009).

These recent results demonstrate that gene expression can be artificially regulated in *B. burgdorferi* using either the *lac* system or the *tet* system. Modifications yielding stronger and tighter induction of genes under the control of these inducible promoters should be possible. The availability of these systems for use in *B. burgdorferi* should facilitate the isolation of conditional lethal mutants by permitting regulated repression of essential genes. Use of these genetic switches will also permit manipulation of *B. burgdorferi* gene expression *in vivo* and open a new avenue for studies of pathogenesis in Lyme borreliosis.

8.9 Unavailable or not routinely used

8.9.1 Other mechanisms of DNA exchange

A number of genetic tools used to investigate gene function remain unavailable for *B. burgdorferi.* Neither conjugation nor transduction has been experimentally harnessed for moving DNA into *B. burgdorferi*, although both mechanisms have been observed. The transfer of erythromycin resistance by conjugation between *B. burgdorferi* and *Bacillus subtilis* or *Enterococcus faecalis* has been described at frequencies several orders of magnitude above background, although the genetic basis for resistance and the mechanism of conjugation have not been defined (Jackson *et al.*, 2007). Each *B. burgdorferi* cell contains multiple members of the 32-kb circular plasmid family, cp32 (Stevenson *et al.*, 1996; Casjens *et al.*, 2000). At least one, and possibly all, of the cp32s are prophages of φBB-1 (Eggers and Samuels, 1999; Zhang and Marconi, 2005) (see Chapter 2) and can be transduced between strains (Eggers *et al.*, 2001). Generalized transduction has been adapted for experimental gene transfer in the spirochaete *Brachyspira (Serpulina) hyodysenteriae* (Stanton *et al.*, 2001) and a shuttle vector has been derived from cp32 (Eggers *et al.*, 2002), so this avenue may be fruitful in expanding the experimental armamentarium available to *Borrelia* geneticists. Other mechanisms for the transfer of DNA between *B. burgdorferi* cells, either natural or artificial, have been described. Chemical transformation using a solution containing DMSO and PEG to permeabilize membranes and deliver DNA (Chung *et al.*, 1989) yields transformation efficiencies of *B. burgdorferi* comparable to electroporation (Tilly *et al.*, 2000; Elias *et al.*, 2002), yet the technique has not yet been widely utilized.

8.9.2 Conditional mutants and complete libraries

Conditional vector replication and conditional lethal mutations have not been described yet in *Borrelia,* but are probably feasible given the recent development of inducible gene expression (Cabello *et al.*, 2006; Blevins *et al.*, 2007; Gilbert *et al.*, 2007; Whetstine *et al.*, 2009) and the ever-expanding catalogue of genes responsive to varying environmental (culture) conditions (Revel *et al.*, 2002; Brooks *et al.*, 2003; Ojaimi *et al.*, 2003; Tokarz *et al.*, 2004). The development of conditional lethal mutations has been hindered by the lack of a defined medium and the ability to isolate and manipulate auxotrophs with varying metabolic requirements (see section 8.3). Finally, to date, the poor efficiency of transformation into infectious *B. burgdorferi* clones has limited the feasibility of genetic screens requiring the introduction of a complete genomic library or saturating transposon mutagenesis. Such screens in an infectious clone would be powerful tools for identifying *B. burgdorferi* virulence traits or genes required for *in vivo* growth. The development of highly transformable infectious clones lacking restriction/modification genes (as described above) should facilitate the construction of saturated transposon or genomic libraries in *B. burgdorferi* (Kawabata *et al.*, 2004; Botkin *et al.*, 2006b).

8.9.3 Media and auxotrophic mutants

Essential tools for the genetic manipulation of many bacteria have historically included complete characterization of the nutritional requirements, isolation of auxotrophic mutants in biosynthetic and catabolic pathways, and manipulation of nutritional requirements and media (Beckwith, 1991). These approaches appear to be less relevant to *B. burgdorferi* as its complex nutritional requirements are not yet defined and a well-characterized minimal medium that supports growth does not exist (Kelly, 1971; Barbour,

1984). In addition, the limited biosynthetic capacity of *B. burgdorferi*, with the corresponding reduced number of pathways that can be altered by mutation, potentially limits the usefulness of auxotrophic mutants in this bacterial species (Fraser *et al.*, 1997; see Chapter 5). However, *B. burgdorferi* lacking lp25, which carries *pncA*, is unable to make nicotinamidase (needed for synthesis of nicotinamide adenine dinucleotide) and is an example of an auxotrophic mutant that grows well *in vitro* but not *in vivo* (Purser *et al.*, 2003). Manipulations of the BSKII media have been used to study the metabolism of chitobiose in *B. burgdorferi* and *chbC* mutants, which are unable to take up chitobiose, exhibit an unusual growth pattern *in vitro* but have no growth defect in mice or ticks (Tilly *et al.*, 2001, 2004). The most common nutritional manipulation of *B. burgdorferi* has been serum starvation or supplementation, to study patterns of protein synthesis and gene expression (Alban *et al.*, 2000; Bugrysheva *et al.*, 2003; Concepcion and Nelson, 2003; Tokarz *et al.*, 2004). Progress in understanding the biology of *B. burgdorferi* should enhance efforts to develop a defined growth medium and isolate potential auxotrophic mutants.

8.10 Special considerations

8.10.1 Segmented and unstable genome

Several features of *B. burgdorferi* require special consideration for genetic studies, the most obvious being the segmented nature of the genome. Approximately one third of the cellular DNA comprises more than 20 different DNA molecules (plasmids), some of which are not always faithfully segregated to daughter cells. This scenario mandates ongoing vigilance in monitoring plasmid content throughout an experiment, in order to assure isogenicity between a parental clone and derivative mutants. This applies not only at the initial stage of mutant isolation and characterization, but also during subsequent experiments in which wild-type and mutant organisms are compared. Practically speaking, this is accomplished by a PCR screen, using a complete set of primer pairs specific for every plasmid in the sequenced genome of the type strain B31 (Casjens *et al.*, 2000; Purser and Norris, 2000; Labandeira-Rey and Skare, 2001; McDowell *et al.*, 2001; Elias *et al.*, 2002). Although primer sets have been adapted for strains other than B31 (Eggers *et al.*, 2002; Parveen *et al.*, 2006), the heterogeneity in plasmid structure and content evident from draught sequences of these strains (Qiu *et al.*, 2004) (see Chapter 2) suggests that an accurate assessment of plasmid content in any strain other than B31 awaits its complete sequence.

Of the more than 20 plasmids present in strain B31, almost half do not appear to be required for mouse infectivity, including lp5, cp9, lp21, lp28-2, lp28-3, lp28-4, cp32-3, cp32-7, cp32-9 and lp56 (Purser and Norris, 2000; Labandeira-Rey and Skare, 2001; McDowell *et al.*, 2001; Elias *et al.*, 2002). Some of these plasmids have been found to be dispensable in the tick vector as well (Grimm *et al.*, 2005; Strother *et al.*, 2005) (see Chapter 11). Although not essential, many of the ORFs present on these plasmids are widely conserved among *B. burgdorferi* strains (Terekhova *et al.*, 2006) (see Chapter 2), suggesting they provide a selective advantage during the natural infectious cycle that is not manifest during experimental infections. Whereas retention of these plasmids is not critical for mouse infectivity, comparisons should be made among isogenic clones with identical plasmid contents.

The *in vivo* requirements for a number of other plasmids carried by B31 remain undefined because they have not been lost from clones whose infectivity phenotypes and plasmid content have been correlated, nor are they known to encode essential functions. These include lp17, cp32-1, cp32-4, cp32-6, cp32-8 and lp38. In addition, two highly conserved and stably maintained plasmids, cp26 and lp54 (Tilly *et al.*, 1997; Byram *et al.*, 2004; Qiu *et al.*, 2004; Terekhova *et al.*, 2006), carry genes that are required at different stages of the infectious cycle (Grimm *et al.*, 2004b; Pal *et al.*, 2004; Yang *et al.*, 2004; Blevins *et al.*, 2008; Shi *et al.*, 2008). Loss of lp54 has only been observed *in vitro* as an escape mechanism in the face of artificial selection against some of the major outer surface proteins it encodes (Sădziene *et al.*, 1993). The ubiquitous cp26 represents a unique situation because it also encodes essential housekeeping functions required under all growth conditions (Byram *et al.*, 2004; Jewett *et al.*, 2007a).

Finally, three linear plasmids, lp25, lp28-1 and lp36, have been shown to be critical during the experimental mouse/tick infectious cycle, but non-essential and unstable during *in vitro* growth (Purser and Norris, 2000; Labandeira-Rey and Skare, 2001; Jewett *et al.*, 2007b). These three plasmids require the most attention during genetic manipulations and subsequent experimentation because their loss has a profound impact on the pathogenicity of a clone. A PCR screen of plasmid content using total DNA from a culture yields a qualitative result; a plasmid can be detected by PCR even if present in only a small fraction of the bacteria from which the DNA was prepared. A more quantitative assessment of lp25, lp28-1 and lp36 should be undertaken for inocula used for *in vivo* studies to determine what proportion of the injected bacteria contains these essential but unstable plasmids. This is achieved by plating an aliquot of the inoculum at the same time as initiating a tick or animal infection. Individual colonies (or wells) are subsequently screened by PCR with primer pairs specific for lp25, lp28-1 and lp36. The total number of colonies reflects the number of viable bacteria injected, whereas the infectious dose comprises only those bacteria in the inoculum that retain all essential plasmids.

In summary, although the segmented genome clearly presents unique challenges and concerns for genetic studies in *B. burgdorferi*, only a few of the many plasmids it carries are essential *in vivo* but unstable *in vitro*. Hence the task of maintaining and assuring isogenicity of clones is not as daunting as it might seem. Additionally, for those specific plasmids that do present problems, various strategies have been devised to maintain them, restore them, or get by without them during genetic manipulations of infectious *B. burgdorferi* clones.

8.10.2 *In vivo* phenotype

Another aspect of genetic studies with *B. burgdorferi* that requires recognition (but is not unique to this bacterial pathogen) is the phenotype of the organism and the context in which gene function is addressed. Cultured spirochaetes look and respond quite differently from genetically identical organisms present in infected ticks or mice. Thus *in vitro* assays may not accurately reflect or predict the function that a particular *B. burgdorferi* gene product provides in an *in vivo* setting, because other differentially synthesized bacterial proteins could mask or expose its function. There is no ready solution to this potential problem, but interpretation of results obtained from *in vitro* assays should take this into account. Propagation of spirochaetes in implanted chambers in the peritoneal cavity of rats results in bacteria exhibiting a partially host-adapted state, as assessed by altered patterns of protein synthesis and gene expression (Akins *et al.*, 1998; Revel *et al.*, 2002; Brooks *et al.*, 2003) and present a somewhat more relevant source of bacteria when testing the *in vitro* properties of mutants.

Offsetting the inefficiencies and challenges of genetic studies in *B. burgdorferi* is the availability of relevant animal models in which to analyse mutants. Rodents are natural reservoir hosts, and both ticks and mice can be experimentally infected, making the entire infectious cycle amenable to investigation. Another chapter in this book describes the animal models of Lyme disease in greater detail (see Chapter 14). Several parameters of the *in vivo* model that merit attention with respect to mutant characterization are infectious dose, mode and route of challenge, mouse strain, number of animals and tissues analysed, quantitation of tissue burden and gene expression, and the statistics used to evaluate the data. Experimental design typically is influenced by the mutant phenotype and the focus of the investigation.

Although animal models of *B. burgdorferi* infection have been in place for over 20 years, the actual dynamics of infection are rather murky. If a million spirochaetes are injected into a mouse, will every potentially infectious organism persist or do only a few bacteria initiate the ensuing chronic infection? This is probably not pertinent to the natural infectious cycle where presumably fewer organisms are transmitted during tick feeding, but it is relevant to *in vivo* genetic screens utilizing a highly complex inoculum (such as a saturated library of transposon mutants), in which the ability to be recovered (or not) is a defining trait.

'*In vivo*' for *B. burgdorferi* encompasses an entire infectious cycle, which includes colonization, persistence and transmission of spirochaetes, both into and out of the tick vector and mammalian host. Even when all phases of the infectious cycle are experimentally recapitulated in the lab, a number of pertinent elements of the natural cycle are not present. These include tick co-infection with other strains of *B. burgdorferi* and other bacterial species (Guttman *et al.*, 1996; Wang *et al.*, 1999; Holden *et al.*, 2005; Schulze *et al.*, 2005; Masuzawa *et al.*, 2008), where horizontal gene transfer (Livey *et al.*, 1995; Qiu *et al.*, 2004) and relative fitness could influence survival. The experimental infectious cycle also does not generally take into account the array of mammalian species on which an infected tick might feed (Ryan *et al.*, 1998; Brisson and Dykhuizen, 2004,2006; Brisson *et al.*, 2008; see Chapter 10) or the immune status of the host (Bunikis *et al.*, 2004; see Chapter 15). Finally, experimentally infected mice and ticks are maintained in the laboratory in a physical environment in which temperature, humidity, light, nutrients and material surroundings are optimal and tightly controlled; this differs significantly from the variable conditions to which ticks and mice are subjected in nature. Thus, even the excellent *in vivo* models that exist to assess the phenotype of *B. burgdorferi* mutants only partially mimic the natural scenario and selective forces with which the spirochaete has evolved.

8.11 Conclusion and future directions

Targeted gene inactivation by allelic exchange is now a fairly standard technique in studying *B. burgdorferi*. Stable shuttle vectors have been developed and are widely used for complementation and heterologous gene expression. A system for random, tagged transposon mutagenesis is also available. A limiting feature of most *B. burgdorferi* transformations, however, is that the desired outcome represents an extremely rare event and is recovered at a rate that would be overlooked with many other bacteria for which well-developed genetic systems exist. This undesirable characteristic is most pronounced in infectious clones retaining the full complement of plasmids, thus representing a substantial challenge to genetic studies designed to test or identify potential virulence determinants *in vivo*. Some of the molecular determinants limiting transformation have been identified and circumvented, but additional advances are needed for efficient introduction of foreign DNA into *B. burgdorferi*. On a positive note, the recently described inducible promoter systems represent a significant addition to the genetic toolbox of *B. burgdorferi*. In summary, there remains considerable work to be done, but the future is promising for genetic studies in this fascinating spirochaete.

Acknowledgements

The authors thank the current and former members of their respective laboratories, as well as their colleagues, for their contributions to this topic. They also thank Gary Hettrick for expert graphic assistance and Jon Graham for statistical advice. F.C.C. thanks Dr Henry Godfrey for helpful discussions. Supported by grants AI051486 (D.S.S.) and AI48856 (F.C.C.) from the National Institutes of Health and by the Intramural Research Program of the National Institute of Allergy and Infectious Diseases, National Institutes of Health (P.A.R.).

References

Akins, D.K., Bourell, K.W., Caimano, M.J., Norgard, M.V., and Radolf, J.D. (1998). A new animal model for studying Lyme disease spirochetes in a mammalian host-adapted state. J. Clin. Invest. *101*, 2240–2250.

Alban, P.S., Johnson, P.W., and Nelson, D.R. (2000). Serum-starvation-induced changes in protein synthesis and morphology of *Borrelia burgdorferi*. Microbiology *146*, 119–127.

Alverson, J., Bundle, S.F., Sohaskey, C.D., Lybecker, M.C., and Samuels, D.S. (2003). Transcriptional regulation of the *ospAB* and *ospC* promoters from *Borrelia burgdorferi*. Mol. Microbiol. *48*, 1665–1677.

Alverson, J., and Samuels, D.S. (2002). *groEL* expression in *gyrB* mutants of *Borrelia burgdorferi*. J. Bacteriol. *184*, 6069–6072.

Barbour, A.G. (1984). Isolation and cultivation of Lyme disease spirochetes. Yale J. Biol. Med. *57*, 521–525.

Beaurepaire, C., and Chaconas, G. (2005). Mapping of essential replication functions of the linear plasmid lp17 of *Borrelia burgdorferi* by targeted deletion walking. Mol. Microbiol. *57*, 132–142.

Beaurepaire, C., and Chaconas, G. (2007). Topology-dependent transcription in linear and circular plasmids of the segmented genome of *Borrelia burgdorferi*. Mol. Microbiol. *63*, 443–453.

Beckwith, J. (1991). Strategies for finding mutants. Methods Enzymol. *204*, 3–18.

Blevins, J.S., Hagman, K.E., and Norgard, M.V. (2008). Assessment of decorin-binding protein A to the infectivity of *Borrelia burgdorferi* in the murine models of needle and tick infection. BMC Microbiol. *8*, 82.

Blevins, J.S., Revel, A.T., Smith, A.H., Bachlani, G.N., and Norgard, M.V. (2007). Adaptation of a luciferase gene reporter and *lac* expression system to *Borrelia burgdorferi*. Appl. Environ. Microbiol. 73, 1501–1513.

Bono, J.L., Elias, A.F., Kupko, J.J. III, Stevenson, B., Tilly, K., and Rosa, P. (2000). Efficient targeted mutagenesis in *Borrelia burgdorferi*. J. Bacteriol. *182*, 2445–2452.

Bono, J.L., Tilly, K., Stevenson, B., Hogan, D., and Rosa, P. (1998). Oligopeptide permease in *Borrelia burgdorferi*: putative peptide-binding components encoded by both chromosomal and plasmid loci. Microbiology *144*, 1033–1044.

Bosse, J.T., Zhou, L., Kroll, J.S., and Langford, P.R. (2006). High-throughput identification of conditionally essential genes in bacteria: from STM to TSM. Infect. Disord. Drug Targets 6, 241–262.

Botkin, D.J., Abbott, A., Howell, J.K., Mosher, M., Stewart, P.E., Rosa, P.A., Kawabata, H., Watanabe, H., and Norris, S.J. (2006a). Transposon mutagenesis of infectious *Borrelia burgdorferi* B31: a pilot study, In Molecular Biology of Spirochetes, NATO Advanced Research Workshop Series, F.C. Cabello, and D. Hulinska, eds. (Amsterdam: IOS Press), pp. 13–24.

Botkin, D.J., Abbott, A., Stewart, P.E., Rosa, P.A., Kawabata, H., Watanabe, H., and Norris, S.J. (2006b). Identification of potential virulence determinants by *Himar1* transposition of infectious *Borrelia burgdorferi* B31. Infect. Immun. 74, 6690–6699.

Boylan, J.A., Posey, J.E., and Gherardini, F.C. (2003). *Borrelia* oxidative stress response regulator, BosR: a distinctive Zn-dependent transcriptional activator. Proc. Natl. Acad. Sci. U.S.A. *100*, 11684–11689.

Brisson, D., and Dykhuizen, D.E. (2004). *ospC* diversity in *Borrelia burgdorferi*: different hosts are different niches. Genetics *168*, 713–722.

Brisson, D., and Dykhuizen, D.E. (2006). A modest model explains the distribution and abundance of *Borrelia burgdorferi* strains. Am. J. Trop. Med. Hyg. 74, 615–622.

Brisson, D., Dykhuizen, D.E., and Ostfeld, R.S. (2008). Conspicuous impacts of inconspicuous hosts on the Lyme disease epidemic. Proc. Biol. Sci. 275, 227–235.

Brooks, C.S., Hefty, P.S., Jolliff, S.E., and Akins, D.R. (2003). Global analysis of *Borrelia burgdorferi* genes regulated by mammalian host-specific signals. Infect. Immun. *71*, 3371–3383.

Bugrysheva, J., Dobrikova, E.Y., Sartakova, M.L., Caimano, M.J., Daniels, T., Radolf, J.D., Godfrey, H.P., and Cabello, F. (2003). Characterization of the stringent response and *rel*($_{Bbu}$) expression of *Borrelia burgdorferi*. J. Bacteriol. *185*, 957–965.

Bugrysheva, J.V., Bryksin, A.V., Godfrey, H.P., and Cabello, F.C. (2005). *Borrelia burgdorferi rel* is responsible for generation of guanosine-3′-diphosphate-5′-triphosphate and growth control. Infect. Immun. 73, 4972–4981.

Bujard, H. (1999). Controlling genes with tetracyclines. J. Gene Med. *1*, 372–374.

Bunikis, J., Tsao, J., Luke, C.J., Lunna, M.G., Fish, D., and Barbour, A.G. (2004). *Borrelia burgdorferi* infection in a natural population of *Peromyscus leucopus* mice: a longitudinal study in an area where Lyme Borreliosis is highly endemic. J. Infect. Dis. *189*, 1515–1523.

Burrack, L.S., and Higgins, D.E. (2007). Genomic approaches to understanding bacterial virulence. Curr. Opin. Microbiol. *10*, 4–9.

Burtnick, M.N., Downey, J.S., Brett, P.J., Boylan, J.A., Frye, J.G., Hoover, T.R., and Gherardini, F.C. (2007). Insights into the complex regulation of *rpoS* in *Borrelia burgdorferi*. Mol. Microbiol. *65*, 277–293.

Bykowski, T., Babb, K., von Lackum, K., Riley, S.P., Norris, S.J., and Stevenson, B. (2006). Transcriptional regulation of the *Borrelia burgdorferi* antigenically variable VlsE surface protein. J. Bacteriol. *188*.

Byram, R., Stewart, P.E., and Rosa, P.A. (2004). The essential nature of the ubiquitous 26 kb circular replicon of *Borrelia burgdorferi*. J. Bacteriol. *186*, 3561–3569.

Cabello, F.C., Dubytska, L., Bryksin, A.V., Bugrysheva, J.V., and Godfrey, H.P. (2006). Genetic studies of the *Borrelia burgdorferi bmp* gene family, In NATO Advanced Research Workshop on Molecular Biology of Spirochetes, F.C. Cabello, Hulinska, D and Godfrey, H.P., ed. (Prague, Czech Republic: IOS Press, Amsterdam), pp. 235–249.

Cabello, F.C., Sartakova, M.L., and Dobrikova, E.Y. (2001). Genetic manipulation of spirochetes- light at the end of the tunnel. Trends Microbiol. *9*, 245–248.

Caimano, M.J., Eggers, C.H., Hazlett, K.R., and Radolf, J.D. (2004). RpoS is not central to the general stress response in *Borrelia burgdorferi* but does control expression of one or more essential virulence determinants. Infect. Immun. *72*, 6433–6445.

Caimano, M.J., Iyer, R., Eggers, C.H., Gonzalez, C., Morton, E.A., Gilbert, M.A., Schwartz, I., and Radolf, J.D. (2007). Analysis of the RpoS regulon in *Borrelia burgdorferi* in response to mammalian host signals provides insight into RpoS function during the enzootic cycle. Mol. Microbiol. *65*, 1193–1217.

Carroll, J., Stewart, P., Rosa, P., and Garon, C. (2003). An enhanced GFP reporter system to monitor gene expression in *Borrelia burgdorferi*. Microbiology *149*, 1819–1828.

Casjens, S., Palmer, N., van Vugt, R., Huang, W.M., Stevenson, B., Rosa, P., Lathigra, R., Sutton, G., Peterson, J., Dodson, R.J., *et al.* (2000). A bacterial genome in flux: the twelve linear and nine circular extrachromosomal DNAs in an infectious isolate of the Lyme disease spirochete *Borrelia burgdorferi*. Mol. Microbiol. 35, 490–516.

Chaconas, G., Stewart, P.E., Tilly, K., Bono, J.L., and Rosa, P. (2001). Telomere resolution in the Lyme disease spirochete. EMBO J. *20*, 3229–3237.

Chen, Q., Fischer, J.R., Benoit, V.M., Dufour, N.P., Youderian, P., and Leong, J.M. (2008). *In vitro* CpG methylation increases the transformation efficiency of *Borrelia burgdorferi* strains harboring the endogenous linear plasmid lp56. J. Bacteriol. *190*, 7885–7891.

Chung, C.T., Niemla, S.L., and Miller, R.H. (1989). One-step preparation of competent *Escherichia coli*: transformation and storage of bacterial cells in the same solution. Proc. Natl. Acad. Sci. U.S.A. *86*, 2172–2175.

Cinco, M. (1992). Selection of a *Borrelia burgdorferi* antigenic variant by cultivation in the presence of increasing amounts of homologous immune serum. FEMS Microbiol. Lett. *92*, 15–18.

Coburn, J., Chege, W., Magoun, L., Bodary, S.C., and Leong, J.M. (1999). Characterization of a candidate *Borrelia burgdorferi* β_3-chain integrin ligand identified using a phage display library. Mol. Microbiol. *34*, 926–940.

Coburn, J., and Cugini, C. (2003). Targeted mutation of the outer membrane protein P66 disrupts attachment of the Lyme disease agent, *Borrelia burgdorferi*, to integrin $\alpha_v\beta_3$. Proc. Natl. Acad. Sci. U.S.A. *100*, 7301–7306.

Coleman, J.L., Rogers, R.C., and Benach, J.L. (1992). Selection of an escape variant of *Borrelia burgdorferi* by use of bactericidal monoclonal antibodies to OspB. Infect. Immun. *60*, 3098–3104.

Concepcion, M.B., and Nelson, D.R. (2003). Expression of *spoT* in *Borrelia burgdorferi* during serum starvation. J. Bacteriol. *185*, 444–452.

Contag, C.H., Contag, P.R., Mullins, J.I., Spilman, S.D., Stevenson, D.K., and Benaron, D.A. (1995). Photonic detection of bacterial pathogens in living hosts. Mol. Microbiol. *18*, 593–603.

Criswell, D., Tobiason, V.L., Lodmell, J.S., and Samuels, D.S. (2006). Mutations conferring aminoglycoside and spectinomycin resistance in *Borrelia burgdorferi*. Antimicrob. Agents Chemother. *50*, 445–452.

Drecktrah, D., Douglas, J.M., and Samuels, D.S. (2010). Use of *rpsL* as a counterselectable marker in *Borrelia burgdorferi*. Appl. Environ. Microbiol. *76*, in press.

Eggers, B.J., Caimano, M.J., and Radolf, J.D. (2006). Sigma factor selectivity in *Borrelia burgdorferi:* RpoS recognition of the *ospE/ospF/elp* promoters is dependent on the sequence of the – 10 region. Mol. Microbiol. *59*, 1859–1875.

Eggers, C.H., Caimano, M.J., Clawson, M.L., Miller, W.G., Samuels, D.S., and Radolf, J.D. (2002). Identification of loci critical for replication and compatibility of a *Borrelia burgdorferi* cp32 plasmid and use of a cp32-based shuttle vector for expression of fluorescent reporters in the Lyme disease spirochaete. Mol. Microbiol. *43*, 281–295.

Eggers, C.H., Caimano, M.J., and Radolf, J.D. (2004). Analysis of promoter elements involved in the transcriptional initiation of RpoS-dependent *Borrelia burgdorferi* genes. J. Bacteriol. *186*, 7390–7402.

Eggers, C.H., Kimmel, B.J., Bono, J.L., Elias, A.F., Rosa, P., and Samuels, D.S. (2001). Transduction by φBB-1, a bacteriophage of *Borrelia burgdorferi*. J. Bacteriol. *183*, 4771–4778.

Eggers, C.H., and Samuels, D.S. (1999). Molecular evidence for a new bacteriophage of *Borrelia burgdorferi*. J. Bacteriol. *181*, 7308–7313.

Elias, A.F., Bono, J.L., Carroll, J.A., Stewart, P., Tilly, K., and Rosa, P. (2000). Altered stationary phase response in a *Borrelia burgdorferi rpoS* mutant. J. Bacteriol. *182*, 2909–2918.

Elias, A.F., Bono, J.L., Kupko, J.J., Stewart, P.E., Krum, J.G., and Rosa, P.A. (2003). New antibiotic resistance cassettes suitable for genetic studies in *Borrelia burgdorferi*. J Mol. Microbiol. Biotechnol. *6*, 29–40.

Elias, A.F., Stewart, P.E., Grimm, D., Caimano, M.J., Eggers, C.H., Tilly, K., Bono, J.L., Akins, D.R., Radolf, J.D., Schwan, T.G., and Rosa, P. (2002). Clonal polymorphism of *Borrelia burgdorferi* strain B31 MI: implications for mutagenesis in an infectious strain background. Infect. Immun. *70*, 2139–2150.

Falkow, S. (1988). Molecular Koch's postulates applied to microbial pathogenicity. Rev. Infect. Dis. *10* (suppl. *2*), S274–S276.

Falkow, S. (2004). Molecular Koch's postulates applied to bacterial pathogenicity – a personal recollection 15 years later. Nat. Rev. Microbiol. *2*, 67–72.

Fingerle, V., Goettner, G., Gern, L., Wilske, B., and Schulte-Spechtel, U. (2007). Complementation of a *Borrelia afzelii* OspC mutant highlights the crucial role of OspC for dissemination of *Borrelia afzelii* in *Ixodes ricinus*. Int. J. Med. Microbiol. *297*, 97–107.

Francis, K.P., Yu, J., Bellinger-Kawahara, C., Joh, D., Hawkinson, M.J., Xiao, G., Purchio, T.F., Caparon, M.G., Lipstich, M., and Contag, P.R. (2001). Visualizing pneumococcal infections in the lungs of live mice using bioluminescent *Streptococcus pneumoniae* transformed with a novel Gram-positive *lux* transposon. Infect. Immun. *69*, 3350–3358.

Frank, K.L., Bundle, S.F., Kresge, M.E., Eggers, C.H., and Samuels, D.S. (2003). *aadA* confers streptomycin resistance in *Borrelia burgdorferi*. J. Bacteriol. *185*, 6723–6727.

Fraser, C.M., Casjens, S., Huang, W.M., Sutton, G.G., Clayton, R., Lathigra, R., White, O., Ketchum, K.A., Dodson, R., Hickey, E.K., *et al.* (1997). Genomic sequence of a Lyme disease spirochaete, *Borrelia burgdorferi*. Nature *390*, 580–586.

Galbraith, K.M., Ng, A.C., Eggers, B.J., Kuchel, C.R., Eggers, C.H., and Samuels, D.S. (2005). *parC* mutations in fluoroquinolone-resistant *Borrelia burgdorferi*. Antimicrob. Agents Chemother. *49*, 4354–4357.

Ge, Y., Old, I.G., Saint Girons, I., and Charon, N.W. (1997). Molecular characterization of a large *Borrelia burgdorferi* motility operon which is initiated by a consensus σ^{70} promoter. J. Bacteriol. *179*, 2289–2299.

Gilbert, M.A., Morton, E.A., Bundle, S.F., and Samuels, D.S. (2007). Artificial regulation of *ospC* expression in *Borrelia burgdorferi*. Mol. Microbiol. *63*, 1259–1273.

Glöckner, G., Lehmann, R., Romualdi, A., Pradella, S., Schulte-Spechtel, U., Schilhabel, M., Wilske, B., Suhnel, J., and Platzer, M. (2004). Comparative analysis of the *Borrelia garinii* genome. Nucleic Acids Res. *32*, 6038–6046.

Glöckner, G., Schulte-Spechtel, U., Schilhabel, M., Felder, M., Suhnel, J., Wilske, B., and Platzer, M. (2006). Comparative genome analysis: selection pressure on the *Borrelia vls* cassettes is essential for infectivity. BMC Genomics 7, 211–220.

Gossen, M., and Bujard, H. (1992). Tight control of gene expression in mammalian cells by tetracycline-responsive promoters. Proc. Natl. Acad. Sci. U.S.A. *89*, 5547–5551.

Grimm, D., Eggers, C.H., Caimano, M.J., Tilly, K., Stewart, P.E., Elias, A.F., Radolf, J.D., and Rosa, P.A. (2004a). Experimental assessment of the roles of linear plasmids lp25 and lp28-1 of *Borrelia burgdorferi* throughout the infectious cycle. Infect. Immun. *72*, 5938–5946.

Grimm, D., Tilly, K., Bueschel, D.M., Fisher, M.A., Policastro, P.F., Gherardini, F.C., Schwan, T.G., and Rosa, P.A. (2005). Defining plasmids required by *Borrelia burgdorferi* for colonization of tick vector *Ixodes scapularis* (Acari: Ixodidae). J. Med. Entomol. *42*, 676–684.

Grimm, D., Tilly, K., Byram, R., Stewart, P.E., Krum, J.G., Bueschel, D.M., Schwan, T.G., Policastro, P.F., Elias, A.F., and Rosa, P.A. (2004b). Outer-surface protein C of the Lyme disease spirochete: a protein induced in ticks for infection of mammals. Proc. Natl. Acad. Sci. U.S.A. *101*, 3142–3147.

Gueiros-Filho, F.J., and Beverley, S.M. (1997). Trans-kingdom transposition of the *Drosophila* element *mariner* within the protozoan *Leishmania*. Science *276*, 1716–1719.

Guttman, D.S., Wang, P.W., Wang, I.N., Bosler, E.M., Luft, B.J., and Dykhuizen, D.E. (1996). Multiple infections of *Ixodes scapularis* ticks by *Borrelia burgdorferi* as revealed by single-strand conformation polymorphism analysis. J. Clin. Microbiol. *34*, 652–656.

Hayes, F. (2003). Transposon-based strategies for microbial functional genomics and proteomics. Annu. Rev. Genet. *37*, 3–29.

Holden, K., Hodzic, E., Feng, S., Freet, K.J., Lefebvre, R.B., and Barthold, S.W. (2005). Coinfection with *Anaplasma phagocytophilum* alters *Borrelia burgdorferi* population distribution in C3H/HeN mice. Infect. Immun. *73*, 3440–3444.

Hübner, A., Revel, A.T., Nolen, D.M., Hagman, K.E., and Norgard, M.V. (2003). Expression of a *luxS* gene is not required for *Borrelia burgdorferi* infection of mice via needle inoculation. Infect. Immun. *71*, 2892–2896.

Hübner, A., Wang, X., Nolen, D.M., Popova, T.G., Cabello, F.C., and Norgard, M. (2001). Expression of *Borrelia burgdorferi* OspC and DbpA is controlled by a RpoN–RpoS regulatory pathway. Proc. Natl. Acad. Sci. U.S.A. *98*, 12724–12729.

Hughes, C.A.N., and Johnson, R.C. (1990). Methylated DNA in *Borrelia* species. J. Bacteriol. *172*, 6602–6604.

Hughes, C.A.N., Kodner, C.B., and Johnson, R.C. (1992). DNA analysis of *Borrelia burgdorferi* NCH-1, the first northcentral U.S. human Lyme disease isolate. J. Clin. Microbiol. *30*, 698–703.

Hunfeld, K.P., and Brade, V. (2006). Antimicrobial susceptibility of *Borrelia burgdorferi* sensu lato: what we know, what we don't know, and what we need to know. Wien Klin Wochenschr *118*, 659–668.

Hunfeld, K.P., Kraiczy, P., Kekoukh, K., Schafer, V., and Brade, V. (2002). Standardised *in vitro* susceptibility testing of *Borrelia burgdorferi* against well-known and newly developed antimicrobial agents-possible implications for new therapeutic approaches to Lyme disease. Int. J. Med. Microbiol. *291*, 125–137.

Jackson, C.R., Boylan, J.A., Frye, J.G., and Gherardini, F.C. (2007). Evidence of a conjugal erythromycin resistance element in the Lyme disease spirochete *Borrelia burgdorferii*. Int. J. Antimicrob. Agents *30*, 496–504.

Jacob, F., and Monod, J. (1961). Genetic regulatory mechanisms in the synthesis of proteins. J. Mol. Biol. *3*, 318–356.

Jacobs, M.B., Norris, S.J., Phillippi-Falkenstein, K.M., and Philipp, M.T. (2006). Infectivity of the highly transformable BBE02⁻ lp56⁻ mutant of *Borrelia burgdorferi*, the Lyme disease spirochete, via ticks. Infect. Immun. *74*, 3678–3681.

Jewett, M.W., Byram, R., Bestor, A., Tilly, K., Lawrence, K., Burtnick, M.N., Gherardini, F., and Rosa, P.A. (2007a). Genetic basis for retention of a critical virulence plasmid of *Borrelia burgdorferi*. Mol. Microbiol. *66*, 975–990.

Jewett, M.W., Lawrence, K., Bestor, A.C., Tilly, K., Grimm, D., Shaw, P., VanRaden, M., Gherardini, F., and Rosa, P.A. (2007b). The critical role of the linear plasmid lp36 in the infectious cycle of *Borrelia burgdorferi*. Mol. Microbiol. *64*, 1358–1374.

Judson, N., and Mekalanos, J.J. (2000). Transposon-based approaches to identify essential bacterial genes. Trends Microbiol. *8*, 521–526.

Kawabata, H., Norris, S.J., and Watanabe, H. (2004). BBE02 disruption mutants of *Borrelia burgdorferi* B31 have a highly transformable, infectious phenotype. Infect. Immun. 72, 7147–7154.

Kelly, R. (1971). Cultivation of *Borrelia hermsii*. Science *173*, 443–444.

Knight, S.W., Kimmel, B.J., Eggers, C.H., and Samuels, D.S. (2000). Disruption of the *Borrelia burgdorferi gac* gene, encoding the naturally synthesized GyrA C-terminal domain. J. Bacteriol. *182*, 2048–2051.

Knight, S.W., and Samuels, D.S. (1999). Natural synthesis of a DNA-binding protein from the C-terminal domain of DNA gyrase A *in Borrelia burgdorferi*. EMBO J. *18*, 4875–4881.

Kok, J., van der Vossen, J.M.B.M., and Venema, G. (1984). Construction of plasmid cloning vectors for lactic streptococci which also replicate in *Bacillus subtilis* and *Escherichia coli*. Appl. Environ. Microbiol. *48*, 726–731.

Kurtti, T.J., Munderloh, U.G., Johnson, R.C., and Ahlstrand, G.G. (1987). Colony formation and morphology in *Borrelia burgdorferi*. J. Clin. Microbiol. 25, 2054–2058.

Labandeira-Rey, M., and Skare, J.T. (2001). Decreased infectivity in *Borrelia burgdorferi* strain B31 is associated with loss of linear plasmid 25 or 28-1. Infect. Immun. *69*, 446–455.

Lampe, D., Churchill, M., and Robertson, H. (1996). A purified *mariner* transposase is sufficient to mediate transposition *in vitro*. EMBO J. *15*, 5470–5479.

Lampe, D.J., Akerley, B.J., Rubin, E.J., Mekalanos, J.J., and Robertson, H.M. (1999). Hyperactive transposase mutants of the *Himar1 mariner* transposon. Proc. Natl. Acad. Sci. U.S.A. *96*, 11428–11433.

Lawrenz, M.B., Kawabata, H., Purser, J.E., and Norris, S.J. (2002). Decreased electroporation efficiency in *Borrelia burgdorferi* containing linear plasmids lp25 and lp56: impact on transformation of infectious *Borrelia*. Infect. Immun. *70*, 4851–4858.

Li, X., Pal, U., Ramamoorthi, N., Liu, X., Desrosiers, D.C., Eggers, C.H., Anderson, J.F., Radolf, J.D., and Fikrig, E. (2007). The Lyme disease agent *Borrelia burgdorferi* requires BB0690, a Dps homologue, to persist within ticks. Mol. Microbiol. *63*, 694–710.

Lin, B., Short, S.A., Eskildsen, M., Klempner, M.S., and Hu, L.T. (2001). Functional testing of putative oligopeptide permease (Opp) proteins of

Borrelia burgdorferi: a complementation model in *opp⁻ Escherichia coli.* Biochim. Biophys. Acta *1499,* 222–231.

Liveris, D., Mulay, V., and Schwartz, I. (2004). Functional properties of *Borrelia burgdorferi recA.* J. Bacteriol. *186,* 2275–2280.

Livey, I., Gibbs, C.P., Schuster, R., and Dorner, F. (1995). Evidence for lateral transfer and recombination in OspC variation in Lyme disease *Borrelia.* Mol. Microbiol. *18,* 257–269.

Lybecker, M.C., and Samuels, D.S. (2007). Temperature-induced regulation of RpoS by a small RNA in *Borrelia burgdorferi.* Mol. Microbiol. *64,* 1075–1089.

Maguin, E., Duwat, P., Hege, T., Ehrlich, D., and Gruss, A. (1992). New thermosensitive plasmid for gram-positive bacteria. J. Bacteriol. *174,* 5633–5638.

Margolis, N., Hogan, D., Tilly, K., and Rosa, P.A. (1994). Plasmid location of *Borrelia* purine biosynthesis gene homologs. J. Bacteriol. *176,* 6427–6432.

Masuzawa, T., Kharitonenkov, I.G., Okamoto, Y., Fukui, T., and Ohashi, N. (2008). Prevalence of *Anaplasma phagocytophilum* and its coinfection with *Borrelia afzelii* in *Ixodes ricinus* and *Ixodes persulcatus* ticks inhabiting Tver Province (Russia)–a sympatric region for both tick species. J. Med. Microbiol. *57,* 986–991.

McDowell, J.V., Sung, S.-Y., Labandeira-Rey, M., Skare, J.T., and Marconi, R.T. (2001). Analysis of mechanisms associated with loss of infectivity of clonal populations of *Borrelia burgdorferi* B31 MI. Infect. Immun. *69,* 3670–3677.

McIver, K.S., and Scott, J.R. (1997). Role of *mga* in growth phase regulation of virulence genes of the group A streptococcus. J. Bacteriol. *179,* 5178–5187.

Miller, J.C., von Lackum, K., Woodman, M.E., and Stevenson, B. (2006). Detection of *Borrelia burgdorferi* gene expression during mammalian infection using transcriptional fusions that produce green fluorescent protein. Microb. Pathog. *41,* 43–47.

Moriarty, T.J., Norman, M.U., Colarusso, P., Bankhead, T., Kubes, P., and Chaconas, G. (2008). Real-time high resolution 2-D imaging of the Lyme disease spirochete adhering to and escaping from the vasculature of a living host. PLoS Pathog. *4,* e1000090.

Morozova, O.V., Dubytska, L.P., Ivanova, L.B., Moreno, C.X., Bryksin, A.V., Sartakova, M.L., Dobrikova, E.Y., Godfrey, H.P., and Cabello, F.C. (2005). Genetic and physiological characterization of 23S rRNA and *ftsJ* mutants of *Borrelia burgdorferi* isolated by mariner transposition. Gene 357, 63–72.

Nickoloff, J.A. (1995). Electroporation Protocols for Microorganisms (Totowa, New Jersey: Humana Press).

Oehler, S., Amouyal, M., Kolkhof, P., von Wilcken-Bergmann, B., and Müller-Hill, B. (1994). Quality and position of the three *lac* operators of *E. coli* define efficiency of repression. EMBO J. *13,* 3348–3355.

Ojaimi, C., Brooks, C., Casjens, S., Rosa, P., Elias, A., Barbour, A.G., Jasinskas, A., Benach, J., Katona, L., Radolf, J., *et al.* (2003). Profiling temperature-induced changes in *Borrelia burgdorferi* gene expression using whole genome arrays. Infect. Immun. *71,* 1689–1705.

Pal, U., Yang, X., Chen, M., Bockenstedt, L.K., Anderson, J.F., Flavell, R.A., Norgard, M.V., and Fikrig, E. (2004). OspC facilitates *Borrelia burgdorferi* invasion of *Ixodes scapularis* salivary glands. J. Clin. Invest. *113,* 220–230.

Parveen, N., Cornell, K., Bono, J., Chamberland, C., Rosa, P., and Leong, J. (2006). Bgp, a secreted GAG-binding protein of *Borrelia burgdorferi* strain N40, displays nucleosidase activity and is not essential for infection of immunodeficient mice. Infect. Immun. *74,* 3016–3020.

Paster, B.J., and Dewhirst, F.E. (2000). Phylogenetic foundation of spirochetes. J. Mol. Microbiol. Biotechnol. *2,* 341–344.

Paster, B.J., Dewhirst, F.E., Weisburg, W.G., Tordoff, L.A., Fraser, G.J., Hespell, R.B., Stanton, T.B., Zablen, L., Mandelco, L., and Woese, C.R. (1991). Phylogenetic analysis of the spirochetes. J. Bacteriol. *173,* 6101–6109.

Paster, B.J., Stackebrandt, E., Hespell, R.B., Hahn, C.M., and Woese, C.R. (1984). The phylogeny of the spirochetes. Syst. Appl. Microbiol. *5,* 337–351.

Picardeau, M., Brenot, A., and Saint Girons, I. (2001). First evidence for gene replacement in *Leptospira* spp. Inactivation of *L. biflexa flaB* results in non-motile mutants deficient in endoflagella. Mol. Microbiol. *40,* 189–199.

Pich, O.Q., Burgos, R., Planell, R., Querol, E., and Pinol, J. (2006). Comparative analysis of antibiotic resistance gene markers in *Mycoplasma genitalium*: application to studies of the minimal gene complement. Microbiology *152,* 519–527.

Pollack, R.J., Telford, S.R., and Spielman, A. (1993). Standardization of medium for culturing Lyme disease spirochetes. J. Clin. Microbiol. *31,* 1251–1255.

Purser, J.E., Lawrenz, M.B., Caimano, M.J., Radolf, J.D., and Norris, S.J. (2003). A plasmid-encoded nicotinamidase (PncA) is essential for infectivity of *Borrelia burgdorferi* in a mammalian host. Mol. Microbiol. *48,* 753–764.

Purser, J.E., and Norris, S.J. (2000). Correlation between plasmid content and infectivity in *Borrelia burgdorferi.* Proc. Natl. Acad. Sci. U.S.A. *97,* 13865–13870.

Putteet-Driver, A.D., Zhong, J., and Barbour, A.G. (2004). Transgenic expression of RecA of the spirochetes *Borrelia burgdorferi* and *Borrelia hermsii* in *Escherichia coli* revealed differences in DNA repair and recombination phenotypes. J. Bacteriol. *186,* 2266–2274.

Qiu, W.-G., Schutzer, S.E., Bruno, J.F., Attie, O., Xu, Y., Dunn, J.J., Fraser, C., Casjens, S.R., and Luft, B. (2004). Genetic exchange and plasmid transfers in *Borrelia burgdorferi* sensu stricto revealed by three-way genome comparisons and multilocus sequence typing. Proc. Natl. Acad. Sci. U.S.A. *101,* 14150–14155.

Revel, A.T., Blevins, J.S., Almazán, C., Neil, L., Kocan, K.M., de la Fuente, J., Hagman, K.E., and Norgard, M.V. (2005). *bptA* (*bbe16*) is essential for the persistence of the Lyme disease spirochete, *Borrelia burgdorferi,* in its natural tick vector. Proc. Natl. Acad. Sci. U.S.A. *102,* 6972–6977.

Revel, A.T., Talaat, A.M., and Norgard, M.V. (2002). DNA microarray analysis of differential gene expression in

Borrelia burgdorferi, the Lyme disease spirochete. Proc. Natl. Acad. Sci. U.S.A. *99*, 1562–1567.

Reyrat, J.M., Pelicic, V., Gicquel, B., and Rappuoli, R. (1998). Counterselectable markers: untapped tools for bacterial genetics and pathogenesis. Infect. Immun. *66*, 4011–4017.

Rosa, P., Samuels, D.S., Hogan, D., Stevenson, B., Casjens, S., and Tilly, K. (1996). Directed insertion of a selectable marker into a circular plasmid of *Borrelia burgdorferi*. J. Bacteriol. *178*, 5946–5953.

Rosa, P.A., and Hogan, D. (1992). Colony formation by *Borrelia burgdorferi* in solid medium: clonal analysis of *osp* locus variants, In Proceeding of the First International Conference on Tick Borne Pathogens at the Host-Vector Interface, U.G. Munderloh, and T.J. Kurtti, eds. (University of Minnesota, St. Paul, Minn.), pp. 95–103.

Rosa, P.A., Tilly, K., and Stewart, P.E. (2005). The burgeoning molecular genetics of the Lyme disease spirochaete. Nature Rev. Microbio.l *3*, 129–143.

Ruzic-Sabljic, E., Podreka, T., Maraspin, V., and Strle, F. (2005). Susceptibility of *Borrelia afzelii* strains to antimicrobial agents. Int. J. Antimicrob. Agents *25*, 474–478.

Ryan, J.R., Levine, J.F., Apperson, C.S., Lubke, L., Wirtz, R.A., Spears, P.A., and Orndorff, P.E. (1998). An experimental chain of infection reveals that distinct *Borrelia burgdorferi* populations are selected in arthropod and mammalian hosts. Mol. Microbiol. *30*, 365–379.

Sădziene, A., Rosa, P.A., Thompson, P.A., Hogan, D.M., and Barbour, A.G. (1992). Antibody-resistant mutants of *Borrelia burgdorferi*: in vitro selection and characterization. J. Exp. Med. *176*, 799–809.

Sădziene, A., Wilske, B., Ferdows, M.S., and Barbour, A.G. (1993). The cryptic *ospC* gene of *Borrelia burgdorferi* B31 is located on a circular plasmid. Infect. Immun. *61*, 2192–2195.

Salama, N.R., and Manoil, C. (2006). Seeking completeness in bacterial mutant hunts. Curr. Opin. Microbiol. *9*, 307–311.

Sambri, V., and Lovett, M.A. (1990). Survival of *Borrelia burgdorferi* in different electroporation buffers. Microbiologica *13*, 79–83.

Samuels, D.S. (1995). Electrotransformation of the spirochete *Borrelia burgdorferi*, In Methods in Molecular Biology, J.A. Nickoloff, ed. (Totowa, NJ: Humana Press, Inc.), pp. 253–259.

Samuels, D.S. (2006). Antibiotic resistance in *Borrelia burgdorferi*: applications for genetic manipulation and implications for evolution, In NATO Advanced Research Workshop on Molecular Biology of Spirochetes, F.C. Cabello, D. Hulinska, and H.P. Godfrey, eds. (Prague, Czech Republic: IOS Press, Amsterdam), pp. 56–70.

Samuels, D.S., and Garon, C.F. (1997). Oligonucleotide-mediated genetic transformation of *Borrelia burgdorferi*. Microbiol *143*, 519–522.

Samuels, D.S., Mach, K.E., and Garon, C.F. (1994a). Genetic transformation of the Lyme disease agent *Borrelia burgdorferi* with coumarin-resistant *gyrB*. J. Bacteriol. *176*, 6045–6049.

Samuels, D.S., Marconi, R.T., Huang, W.M., and Garon, C.F. (1994b). *gyrB* mutations in coumermycin A_1-resistant *Borrelia burgdorferi*. J. Bacteriol. *176*, 3072–3075.

Sartakova, M., Dobrikova, E., and Cabello, F.C. (2000). Development of an extrachromosomal cloning vector system for use in *Borrelia burgdorferi*. Proc. Natl. Acad. Sci. U.S.A. *97*, 4850–4855.

Sartakova, M.L., Dobrikova, E.Y., Motaleb, M.A., Godfrey, H.P., Charon, N.W., and Cabello, F.C. (2001). Complementation of a nonmotile *flaB* mutant of *Borrelia burgdorferi* by chromosomal integration of a plasmid containing a wild-type *flaB* allele. J. Bacteriol. *183*, 6558–6564.

Sartakova, M.L., Dobrikova, E.Y., Terekhova, D.A., Devis, R., Bugrysheva, J.V., Morozova, O.V., Godfrey, H.P., and Cabello, F.C. (2003). Novel antibiotic-resistance markers in pGK12-derived vectors for *Borrelia burgdorferi*. Gene *303*, 131–137.

Schulze, R.J., and Zückert, W.R. (2006). *Borrelia burgdorferi* lipoproteins are secreted to the outer surface by default. Mol. Microbiol. *59*, 1473–1484.

Schulze, T.L., Jordan, R.A., Schulze, C.J., Mixson, T., and Papero, M. (2005). Relative encounter frequencies and prevalence of selected *Borrelia, Ehrlichia* and *Anaplasma* infections in *Amblyomma americanum* and *Ixodes* scapularis (Acari: Ixodidae) ticks from central New Jersey. J. Med. Entomol. *42*, 450–456.

Seshu, J., Esteve-Gassent, M.D., Labandeira-Rey, M., Kim, J.H., Trzeciakowski, J.P., Hook, M., and Skare, J.T. (2006). Inactivation of the fibronectin-binding adhesin gene *bbk32* significantly attenuates the infectivity potential of *Borrelia burgdorferi*. Mol. Microbiol. *59*, 1591–1601.

Shi, Y., Xu, Q., McShan, K., and Liang, F.T. (2008). Both decorin-binding proteins A and B are critical for overall virulence of *Borrelia burgdorferi*. Infect. Immun. *76*, 1239–1246.

Shigekawa, K., and Dower, W.J. (1988). Electroporation of eukaryotes and prokaryotes: a general approach to the introduction of macromolecules into cells. BioTechniques *6*, 742–751.

Sohaskey, C.D., Arnold, C., and Barbour, A.G. (1997). Analysis of promoters in *Borrelia burgdorferi* by use of a transiently expressed reporter gene. J. Bacteriol. *179*, 6837–6842.

Sohaskey, C.D., and Barbour, A.G. (1999). Esterases in serum-containing growth media counteract chloramphenicol acetyltransferase activity in vitro. Antimicrob. Agents Chemother. *43*, 655–660.

Sohaskey, C.D., Zückert, W.R., and Barbour, A.G. (1999). The extended promoters for two outer membrane lipoprotein genes of *Borrelia* spp. uniquely include a T-rich region. Mol. Microbiol. *33*, 41–51.

Srivastava, S.Y., and de Silva, A.M. (2008). Reciprocal expression of *ospA* and *ospC* in single cells of *Borrelia burgdorferi*. J. Bacteriol. *190*, 3429–3433.

Stanton, T.B., Matson, E.G. and Humphrey, S.B. (2001). *Brachyspira (Serpulina) hyodysenteriae gyrB* mutants and interstrain transfer of coumermycin A_1 resistance. Appl. Environ. Microbiol. *67*, 2037–2043.

Steere, A.C. (2001). Lyme disease. N. Engl. J. Med. *345*, 115–125.

Stevenson, B., Bono, J.L., Elias, A., Tilly, K., and Rosa, P. (1998). Transformation of the Lyme disease spirochete *Borrelia burgdorferi* with heterologous DNA. J. Bacteriol. *180*, 4850–4855.

Stevenson, B., Tilly, K., and Rosa, P.A. (1996). A family of genes located on four separate 32-kilobase circular plasmids in *Borrelia burgdorferi* B31. J. Bacteriol. *178*, 3508–3516.

Stewart, P.E., Bestor, A., Cullen, J.N., and Rosa, P.A. (2008). Tightly regulated surface protein of *Borrelia burgdorferi* is not essential to the mouse-tick infectious cycle. Infect. Immun. *76*, 1970–1978.

Stewart, P.E., Chaconas, G., and Rosa, P. (2003). Conservation of plasmid maintenance functions between linear and circular plasmids in *Borrelia burgdorferi*. J. Bacteriol. *185*, 3202–3209.

Stewart, P.E., Hoff, J., Fischer, E., Krum, J.G., and Rosa, P.A. (2004). Genome-wide transposon mutagenesis of *Borrelia burgdorferi* for identification of phenotypic mutants. Appl. Environ. Microbiol. *70*, 5973–5979.

Stewart, P.E., and Rosa, P.A. (2008). Transposon mutagenesis of the Lyme disease agent *Borrelia burgdorferi*. Methods Mol Biol *431*, 85–95.

Stewart, P.E., Thalken, R., Bono, J.L., and Rosa, P. (2001). Isolation of a circular plasmid region sufficient for autonomous replication and transformation of infectious *Borrelia burgdorferi*. Mol. Microbiol. *39*, 714–721.

Strother, K.O., Broadwater, A., and de Silva, A. (2005). Plasmid requirements for infection of ticks by *Borrelia burgdorferi*. Vector Borne Zoonotic Dis *5*, 237–245.

Strother, K.O., Hodzic, E., Barthold, S.W., and de Silva, A.M. (2007). Infection of mice with Lyme disease spirochetes constitutively expressing outer surface proteins A and B. Infect. Immun. *75*, 2786–2794.

Terekhova, D., Iyer, R., Wormser, G.P., and Schwartz, I. (2006). Comparative genome hybridization reveals substantial variation among clinical isolates of *Borrelia burgdorferi* sensu stricto with different pathogenic properties. J. Bacteriol. *188*, 6124–6134.

Terekhova, D., Sartakova, M.L., Wormser, G.P., Schwartz, I., and Cabello, F.C. (2002). Erythromycin resistance in *Borrelia burgdorferi*. Antimicrob. Agents Chemother. *46*, 3637–3640.

Thomas, C.M., and Smith, C.A. (1987). Incompatibility group P plasmids: genetics, evolution and use in genetic manipulation. Annu. Rev. Microbiol. *41*, 77–101.

Tilly, K., Bestor, A., Jewett, M.W., and Rosa, P. (2007). Rapid clearance of Lyme disease spirochetes lacking OspC from skin. Infect. Immun. *75*, 1517–1519.

Tilly, K., Casjens, S., Stevenson, B., Bono, J.L., Samuels, D.S., Hogan, D., and Rosa, P. (1997). The *Borrelia burgdorferi* circular plasmid cp26: conservation of plasmid structure and targeted inactivation of the *ospC* gene. Mol. Microbiol. *25*, 361–373.

Tilly, K., Elias, A.F., Bono, J.L., Stewart, P., and Rosa, P. (2000). DNA exchange and insertional inactivation in spirochetes. J. Mol. Microbiol. Biotechnol. *2*, 433–442.

Tilly, K., Elias, A.F., Errett, J., Fischer, E., Iyer, R., Schwartz, I., Bono, J.L., and Rosa, P. (2001). Genetics and regulation of chitobiose utilization in *Borrelia burgdorferi*. J. Bacteriol. *183*, 5544–5553.

Tilly, K., Grimm, D., Bueschel, D.M., Krum, J.G., and Rosa, P. (2004). Infectious cycle analysis of a *Borrelia burgdorferi* mutant defective in transport of chitobiose, a tick cuticle component. Vector Borne Zoonotic Dis. *4*, 159–168.

Tilly, K., Krum, J.G., Bestor, A., Jewett, M.W., Grimm, D., Bueschel, D., Byram, R., Dorward, D., Stewart, P., and Rosa, P. (2006). *Borrelia burgdorferi* OspC protein required exclusively in a crucial early stage of mammalian infection. Infect. Immun. *74*, 3554–3564.

Tilly, K., Lubke, L., and Rosa, P. (1998). Characterization of circular plasmid dimers in *Borrelia burgdorferi*. J. Bacteriol. *180*, 5676–5681.

Tokarz, R., Anderton, J.M., Katona, L.I., and Benach, J.L. (2004). Combined effects of blood and temperature shift on *Borrelia burgdorferi* gene expression as determined by whole genome DNA array. Infect. Immun. *72*, 5419–5432.

Trevors, J.T., Chassy, B.M., Dower, W.J., and Blaschek, H.P. (1992). Electrotransformation of bacteria by plasmid DNA, In Guide to electroporation and electrofusion, D.C. Chang, B.M. Chassy, J.A. Saunders, and A.E. Sowers, eds. (San Diego: Academic Press), pp. 265–290.

Triccas, J.A., Pinto, R., and Britton, W.J. (2002). Destablized green fluorescent protein for monitoring transient changes in mycobacterial gene expression. Res. Microbiol *153*, 379–383.

Valdivia, R.H., Cormack, B.P., and Falkow, S. (2006). The uses of green fluorescent protein in prokaryotes. Methods Biochem. Anal. *47*, 163–178.

Wang, I.N., Dykhuizen, D.E., Qiu, W., Dunn, J.J., Bosler, E.M., and Luft, B.J. (1999). Genetic diversity of *ospC* in a local population of *Borrelia burgdorferi sensu stricto*. Genetics *151*, 15–30.

Wang, X.-G., Scagliotti, J.P., and Hu, L.T. (2004). Phospholipid synthesis in *Borrelia burgdorferi: BB0249* and *BB0721* encode functional phosphatidylcholine synthase and phosphatidylglycerolphosphate synthase proteins. Microbiology *150*, 391–397.

Whetstine, C.R., Slusser, J.G., and Zückert, W.R. (2009). Development of a single-plasmid-based regulatable gene expression system for *Borrelia burgdorferi*. Appl. Environ. Microbiol. *75*, 6553–6558.

Wormser, G.P. (2006). Clinical practice. Early Lyme disease. N. Engl. J. Med. *354*, 2794–2801.

Xu, Q., Seemanapalli, S.V., Lomax, L., McShan, K., Li, X., Fikrig, E., and Liang, F.T. (2005). Association of linear plasmid 28–1 with an arthritic phenotype of *Borrelia burgdorferi*. Infect. Immun. *73*, 7208–7215.

Xu, Q., Seemanapalli, S.V., McShan, K., and Liang, F.T. (2006). Constitutive expression of outer surface protein C diminishes the ability of *Borrelia burgdorferi* to evade specific humoral immunity. Infect. Immun. *74*, 5177–5184.

Yang, X.F., Alani, S.M., and Norgard, M.V. (2003). The response regulator Rrp2 is essential for the expression of major membrane lipoproteins in *Borrelia burgdorferi*. Proc. Natl. Acad. Sci. U.S.A. *100*, 11001–11006.

Yang, X.F., Pal, U., Alani, S.M., Fikrig, E., and Norgard, M.V. (2004). Essential role for OspA/B in the life cycle of the Lyme disease spirochete. J. Exp. Med. *199*, 641–648.

Zhang, H., and Marconi, R.T. (2005). Demonstration of cotranscription and 1-methyl-3-nitrosonitroguanidine induction of a 30-gene operon of *Borrelia burgdorferi*: evidence that the 32-kilobase circular plasmids are prophages. J. Bacteriol. *187*, 7985–7995.

Zhang, J.R., Hardham, J.M., Barbour, A.G., and Norris, S.J. (1997). Antigenic variation in Lyme disease borreliae by promiscuous recombination of VMP-like sequence cassettes. Cell *89*, 275–285.

Zhang, J.R., and Norris, S.J. (1998). Kinetics and in vivo induction of genetic variation of *vlsE* in *Borrelia burgdorferi*. Infect. Immun. *66*, 3689–3697.

Zückert, W.R., and Meyer, J. (1996). Circular and linear plasmids of Lyme disease spirochetes have extensive homology: characterization of a repeated DNA element. J. Bacteriol. *178*, 2287–2298.

Evolutionary Genetics of *Borrelia burgdorferi* sensu lato

9

Daniel Dykhuizen and Dustin Brisson

Abstract

Evolution can inform our understanding of the biology of *Borrelia*. Four forces propel evolutionary change: mutation, genetic drift, migration, and natural selection. Although these microevolutionary forces act primarily within populations, the same forces create macroevolutionary divergence of species. The prevailing data suggest that *Borrelia burgdorferi* sensu lato was once a wide-ranging species in the Northern Hemisphere that rapidly separated into the species present today. The limited neutral divergence within species suggests that the long-term effective population size of each modern species is small. The combined observations that *Borrelia* was once a very wide-ranging species and that the current effective population sizes of the species are very small suggests that *Borrelia* populations can expand rapidly when conditions are favourable and shrink rapidly when conditions are unfavourable. Divergence among and within genospecies has progressed primarily by mutation, as horizontal gene transfers are rare; so rare that most observed recombinants seen in nature are likely to have been selected to maintain genetic diversity. Recent migration, as well as range expansions and contractions, have resulted in overlapping patchy distributions of genospecies across the Northern Hemisphere. Most likely, infected birds are responsible for long distance migration, while infected mammals are important in range expansion and migration between local populations. Natural selection, especially that promoting host specificity, continues to shape the geographic distribution as well as the genetic diversity within and between species.

9.1 Introduction

The evolutionary history of an organism is written in the language of DNA and recorded as within- and between- species genetic variation. The field of evolutionary genetics deciphers the cryptic messages written on the genome to reveal the history of life. Evolutionary genetics is the result of the integration of classical genetics with Darwinian evolution during the Modern Synthesis (Huxley, 1942). The central theme of evolutionary genetics is to characterize the evolutionary forces that shape the patterns of biodiversity observed in nature.

Evolutionary change within populations is the result of changes in gene and genotype frequencies caused by both natural selection and non-selective forces. The four evolutionary forces – mutation, random genetic drift, natural selection, and gene flow – acting within and among populations cause microevolutionary change, which is sufficient to account for the variation between species given adequate time. That is, the microevolutionary forces create the macroevolutionary patterns that characterize higher taxonomic groups.

Genetic mutations are the fundamental source of variation within populations. The majority of mutations are deleterious – causing a reduction in evolutionary fitness – or neutral – causing no effect on fitness. The frequency, persistence, and rate of fixation of neutral genetic variants is governed by chance events and heavily influenced by population size and gene flow. Neutral genetic variation within and between populations and species exposes the phylogenetic relationships among taxa and can illuminate both recent and historical migration events. Gene flow results in cohesion among populations and hinders speciation.

Natural selection plays a central role in evolutionary studies because of its crucial role in adaptation – the fit between organism and environment. Natural selection is the only evolutionary force that produces adaptation or preserves genetic states over long time periods in the face of the dispersive forces of mutation and drift. Advantageous mutations typically have a small positive effect on fitness and provide the raw materials for gradual adaptive evolution. The effect of each evolutionary force on genetic variation within and among populations has been developed in the theory of population genetics. Most empirical investigations have focused on one of two prominent patterns in nature: (a) adaptation, or the 'fit' between organism and environment; or (b) polymorphism, the maintenance of multiple genetic forms in a population.

Evolutionary genetic studies of *Borrelia* combine diverse scientific fields to infer the mutational processes, historical migration events, random drift and population size, and the forces causing selective pressure that have resulted in the genomic structure and the patterns of genetic variation within populations.

The genus *Borrelia* is defined by a linear chromosome with covalently closed telomeres (Barbour and Garon, 1987). *Borrelia burgdorferi* s.s. has a single linear chromosome (~900 kb) that carries the majority of the metabolic 'housekeeping' genes. The majority of the genes on the chromosome display little genetic variation (Qiu *et al.*, 2004). *Borrelia* species typically have multiple linear and circular plasmids ('lp' and 'cp', respectively). In *B. burgdorferi* B31, there are twelve linear plasmids ranging in size from 5 kb to 54 kb and nine circular plasmids ranging from 9 kb to 32 kb (see Chapter 2). There can be extensive genetic variation at plasmid loci (Qiu *et al.* 2004). For example, the highly variable outer surface protein C gene (*ospC*) is located on the 26-kb circular plasmid ('cp26'). Lp54 contains the gene for outer surface protein A (*ospA*). Plasmids up to about 180 kb are found in the relapsing fever Borrelia (Zhong *et al.*, 2006).

Numerous cp32 plasmids are found in all strains. These plasmids resemble prophages that give rise to the φBB-1 bacteriophage (Eggers *et al.*, 2001; Zhang and Marconi, 2005). They contain two highly variable regions, the partition/incompatibility loci and the highly variable *ospE/ospF/elp* (*erp*) locus (Stevenson *et al.*, 1996). Usually the alleles at a locus have a recent common ancestor and the same function. This is not true at the *ospE/ospF/elp* (*erp*) locus. The genes in the *ospE/ospF/elp* locus produce three very different families of membrane lipoproteins, the OspE, OspF and Elp families (Marconi *et al.*, 1996; Akins *et al.*, 1999; Caimano *et al.*, 2000; Alitalo *et al.*, 2002), giving rise to two different naming systems (i.e. Table 9.1). Since the purpose of names in science is to help with identification, we suggest that the OspE/OspF/Elp names be used rather than Erp, because the former names provide more information about the function of the particular allele, but we will continue to use *erp* when referring simply to the site on cp32 where *ospE, ospF* and *elp* are found. The OspE proteins, have been shown to bind with factor H and may help *B. burgdorferi* avoid the vertebrate immune system (Hellwage *et al.*, 2001; Stevenson *et al.*, 2002; McDowell *et al.*, 2003; Metts *et al.*, 2003; Hovis *et al.*, 2006), but the OspF paralogues seem to be adhesins (Antonara *et al.*, 2007). Furthermore, the genes are regulated differently: *ospF* promoters are RpoS dependent, but *ospE* and *elp* promoters are sigma 70 dependent (Eggers *et al.*, 2004; Caimano *et al.*, 2004; Caimano *et al.*, 2005; Eggers *et al.*, 2006). The cp32 plasmids, unlike most of the other plasmids, seem to be found in all species in the genus (see Chapter 2).

9.1.1 Common methods used in evolutionary genetics studies

Evolutionary genetics employs a variety of molecular and statistical techniques to infer the evolutionary history, phylogenetic structure, and dynamics of a population or species. *Borrelia* species and strains can be distinguished by *serotype*. Serotypically different strains react differently to serotype-specific monoclonal antibodies (MAb). There are eight different OspA serotypes of human-infectious *B. burgdorferi* sensu lato (s.l.) species. *Borrelia burgdorferi* sensu stricto (s.s.) is

Table 9.1 Frequency distribution of three OspE, three OspF and four Elp alleles in six OspC serotypes

Strains per serotype	Serotype	OspE			OspF			Elp			
		ErpA	ErpC	ErpP	ErpF	ErpG	ErpY	ErpB	ErpM	ErpQ	ErpX
7[a]	OspC A	2[b]	2	7	1	4	4	5	0	2	4
4	OspC B	2	0	4	3	3	2	4	1	4	1
10	OspC K	1	7	10	10	3	0	9	3	0	4
2	OspC E	0	0	2	2	0	0	0	0	0	1
2	OspC G	1	1	2	2	0	0	0	0	1	1
2	OspC N	0	0	2	1	0	0	0	0	1	1

a The number of strains examined of each OspC serotype.
b The number of strains that tested positive for both the OspC serotype (row) and the *erp* locus (column)

serotype 1, *B. afzelii* is serotype 2, and serotypes 3–7 are found within *B. garinii* (Wilske *et al.*, 1993a). The relative homogeneity of *B. burgdorferi* s.s. and *B. afzelii* and heterogeneity of *B. garinii* is also found in DNA sequences and DNA–DNA hybridization. *B. garinii* serotype 5 represents a recombinant of serotypes 4 and 6 (Wilske *et al.*, 1993a). Serotype 4 is found primarily in rodents where the other *B. garinii* serotypes are specific to birds (Kurtenbach *et al.*, 2002). Serotyping was also used to show that, within any OspA serotype, there are many different OspC serotypes (Wilske *et al.*, 1996).

Restriction fragment length polymorphism (RFLP) involves PCR amplification of DNA followed by digestion with a few restriction enzymes. The sizes of the resulting fragments are determined by gel electrophoresis. RFLP detects 5–10% of the nucleotide differences between strains. Thus, to be useful, RFLP has to be done in regions with moderate genetic variability; too little variability and there is little to no discrimination; too much variability and one cannot determine the sizes of all of the DNA fragments. The intergenic ribosomal spacer region in *B. burgdorferi* s.s. has just the right amount of variation and can be used to distinguish three major clades (RST1, RST2, and RST3), each of which are at a frequency of about a third in *B. burgdorferi* s.s. populations (Liveris *et al.*, 1996; Liveris *et al.*, 1999). The intergenic ribosomal spacer region is a neutrally evolving gene that is appropriate for studies of population dynamics, population structure, and phylogenetic relationships.

Cold single-strand conformational polymorphism (SSCP) can detect single base differences between DNA sequences. PCR amplified fragments are melted and snap chilled to form tertiary structure from single DNA strands. The two strands of the DNA fold into different conformations and have different mobilities on gels resulting in a two-band pattern for every sequence. Four patterns were detected from a 345-bp fragment of the *ospA* gene from a single *B. burgdoreferi* s.s. population (Guttman *et al.*, 1996). DNA sequencing revealed that four of the five polymorphic nucleotides in this region were discriminated by SSCP. SSCP has also been used to distinguish genetic variation within *ospC*, even though the sequence difference between alleles (groups) is on the order of 15% (Wang *et al.*, 1999b). This was possible only because there is very little genetic variation within groups.

Reverse Line Blotting (RLB) is a reverse Southern blot procedure designed to distinguish *ospC* groups (Qiu *et al.*, 2002). Short oligonucleotides (probes) that bind to specific *ospC* alleles are cross-linked to a nylon membrane in parallel rows. Fluorescein-labelled *ospC* DNA is placed in a column of a manifold that allows the DNA to potentially hybridize to any of the specific probes. This method allows easy visualization of the *ospC* allele present in a particular sample. RLB has been used to determine that 16 *ospC* alleles are held in each *B. burgdorferi* s.s. population in the northeastern US by balancing selection (Wang *et al.*, 1999b; Qiu *et al.*, 2002; Brisson and Dykhuizen, 2004).

9.2 Taxonomy

9.2.1 Genospecies

Lyme borreliosis is caused by spirochaetes in the *B. burgdorferi* sensu lato complex (Burgdorfer *et al.*, 1982; van Dam *et al.*, 1993). Thirteen *B. burgdorferi* s.l. genospecies have been properly described by DNA/DNA hybridization, restriction fragment length polymorphism, or DNA sequence analysis (Johnson *et al.*, 1984; Baranton *et al.*, 1992; Canica *et al.*, 1993; Kawabata *et al.*, 1993; Marconi *et al.*, 1995; Fukunaga and Koreki, 1996; Le Fleche *et al.*, 1997; Wang *et al.*, 1997; Postic *et al.*, 1998; Masuzawa *et al.*, 2001; Richter *et al.*, 2004b). All *B. burgdorferi* s.l. genospecies are transmitted among vertebrate wildlife species by ticks in the *Ixodes ricinus* complex. However, only *B. burgdorferi* sensu stricto, *B. garinii*, and *B. afzelii* are zoonoses that commonly cause human disease (van Dam *et al.*, 1993; Busch *et al.*, 1996; Wilske, 2005). The remaining genospecies – *B. bissettii, B. japonica, B. spielmani, B. turdi, B. tanukii, B. valaisiana, B. californiensis, B. lusitaniae, B. andersoni,* and *B. sinica* – also infect multiple vertebrate species and are vectored by *I. ricinus* ticks, but are rarely found in human Lyme disease patients (Derdakova and Lencakova, 2005). Several other *Borrelia* isolates have recently been discovered that may be novel genospecies. These isolates, including genomospecies 1 and genomospecies 2 from California (Postic *et al.*, 2007) and a European group (Qiu *et al.*, 2008; Margos *et al.*, 2008), have not yet been properly described or named. There are two additional species – *B. lonestari* and *B. miyamotoi* (recently this clade has been referred to as *B. miyamotoi* s.l. because of the genetic diversity and wide-spread distribution) – which are found in hard-bodied *Ixodes* ticks, but are more closely related to the relapsing fever group of *Borrelia* than to the species in the *B. burgdorferi* s.l. complex (Bunikis *et al.*, 2004b).

9.2.2 Species evolution

The evolutionary relationships among *B. burgdorferi* s.l. genospecies are of public health as well as academic importance. Despite similar life histories and ecologies, only three of the 13 genospecies cause extensive human disease. The mechanisms that allow some genospecies to cause human disease may result from shared evolutionary histories. Evolutionary relationships are typically investigated by phylogenetic reconstruction using DNA sequence data from housekeeping genes. Unfortunately, no single study has analysed all described genospecies simultaneously. Thus we constructed a consensus gene tree that pieces together the results from all strongly supported relationships reported to date (Fig. 9.1).

Ribosomal RNA data support three major clades: a *B. garinii* clade is separated from a *B. afzelii* clade with *B. andersoni* as an outgroup (Fukunaga *et al.*, 1996; Wang *et al.*, 1999a; Masuzawa *et al.*, 2001; Pecchioli *et al.*, 2007) (Fig. 9.1). Many of the relationships within these major clades are well supported. Within the *B. garinii* clade, *B. burgdorferi* s.s. is closely related to genomospecies 1 and genomospecies 2 (De Michelis *et al.*, 2000; Pecchioli *et al.*, 2007), while *B. bissettii* is close to *B. japonica* (Wang *et al.*, 1999a). In the *B. afzelii* clade, *B. valaisiana* is closely related to *B. tanukii* with *B. turdi* as their outgroup (Fukunaga *et al.*, 1996; Masuzawa *et al.*, 2001). *B. lusitaniae* is also closely related to *B. valaisiana*, but its relationship with *B. tanukii* and *B. turdi* were not investigated (Pecchioli *et al.*, 2007). Thus, the ribosomal RNA data suggest that *B. garinii* and *B. afzelii* are the basal taxa of the two major clades of *B. burgdorferi* s.l. with *B. andersoni* falling outside both clades (Fig. 9.1).

Analysis of the flagellin gene (*flaB*) disagrees with analyses of ribosomal RNA data. Using the entire length of the *flaB* gene, Masuzawa *et al.* (2001) reports that *B. garinii* and *B. turdi* are related, as are *B. burgdorferi* s.s. and *B. andersoni*. *B. sinica* is associated with the *B. garinii/B. turdi* clade by *flaB*, but placed with *B. japonica* by 16S rRNA (Masuzawa *et al.*, 2001). None of the between-species nodes for the *flaB* tree are strongly supported except the relationship of *B. valasisiana* with *B. tanukii* and their association with *B afzelii*.

flaB and rRNA are on the major chromosome (Fraser *et al.*, 1997), a genetic element that has little evidence of recombination within or between genes (Qiu *et al.*, 2004; Attie *et al.*, 2007). The discrepancies between the *flaB* and the rRNA

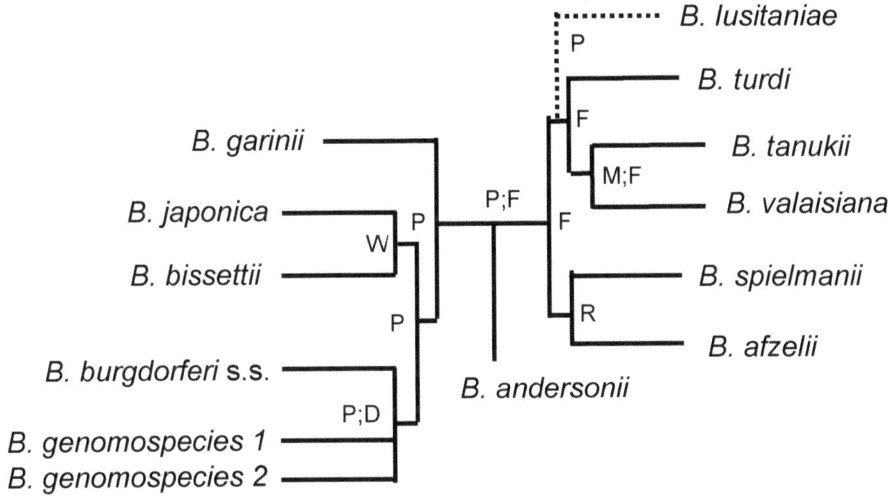

Figure 9.1 Consensus genospecies species tree using ribosomal RNA sequences. *B. lusitaniae* forms a tight group with *B. valaisiana* but its relationship with *B. turdi* and *B. tanukii* has not been investigated. Clades are strongly supported by the following studies: P, Pecchioli *et al.* (2007); D, De Michelis *et al.* (2000); W, Wang *et al.* (1999a); F, Fukunaga *et al.* (1996); R, Richter *et al.* (2006); M, Masuzawa *et al.* (2001).

phylogenetic reconstructions are probably not caused by recombination. Disagreement in the pattern of relationships can occur if several speciation events occur within a short time, resulting in little or no phylogenetic information between speciation events. This could happen if a widely distributed species suddenly fragmented into many species. The species fragmentation hypothesis is supported by recent multi-locus sequence typing (MLST) analyses of *B. burgdorferi* s.l. genospecies (Richter *et al.*, 2006; Postic *et al.*, 2007). The phylogenetic analyses of *B. burgdorferi* s.l. based on concatenated sequences of seven housekeeping loci – *hbb, groEL, ospA, fla, rrs* (16S rRNA), and *rrf* (5S rRNA)-*rrl* (23S rRNA) intergenic spacer region (IGS) – provided little discriminatory power. None of the relationships among *B. garinii, B. burgdorferi* s.s., *B. afzelii, B. valaisiana, B. lusitaniae,* and *B. spielmanii* were statistically supported (Richter *et al.*, 2006). However, the newly described *B. spielmanii* is probably the sister taxa of *B. afzelii* and the newly described *B. californensis* is the likely sister of *B. bissettii* (Richter *et al.*, 2006; Postic *et al.*, 2007). The reason for the low discriminatory power is that each individual gene results in a different phylogenetic tree topology. The internal branches between genospecies on gene trees are usually very short (for example, see Masuzawa *et al.*, 2001); while all the isolates of a genospecies cluster together tightly. A different pattern is expected if horizontal transfer between species is causing the discrepancies between gene trees. The gene trees will have a highly variable structure, often with a long internal branches. The observation that the concatenated gene tree has little statistical resolution while particular species can be associated using a particular gene can be explained by the fact that a few homoplastic base changes in a particular gene can group one set of species together, whereas homoplastic changes in other genes would group a different set of species together.

Phylogenetic analyses of surface-exposed proteins which are encoded on plasmids like outer surface protein A (OspA) and outer surface protein C (OspC), often show evidence of lateral gene transfers between genospecies. For example, the *ospA* gene in *B. valaisiana* has three distinct morphs, two of which cluster with *B. burgdorferi* s.s. and one that clusters with one of the four *B. garinii ospA* morphs (Godfroid *et al.*, 2003). *ospC* shows even greater evidence of lateral gene transfers. Very similar sequences are found in multiple genospecies including the three human-infectious species (Baranton *et al.*, 1998). Both of these genes interact with their environments and may be subject to strong selection (Wang *et al.*, 1999b; Pal *et al.*, 2000; Qiu *et al.*, 2002).

9.3 Clonality and recombination

Unlike plants and animals, reproduction and sex (genetic recombination between lineages) are separate functions in bacteria. Reproduction by binary fission results in a clone of genetically identical individuals that differ only by rare chance mutations. Daughter cells can further diverge by sex, or horizontal gene transfer (HGT), independent of reproduction. The amount of sex differs greatly among bacterial species, presumably depending on the life history of the species. Within a clone, cells that differ by random mutations of single nucleotides can often be distinguished from cells that changed because of HGT. Random mutations are distributed throughout the genome, while nucleotide changes are clustered as a result of HGT.

Normally in bacteria, only a short piece of DNA is transferred. In *Borrelia*, the transferred piece during HGT is usually less than a thousand base pairs (Dykhuizen and Baranton, 2001; Attie *et al.*, 2007). We have already expanded the definition of a clone by including within the clone individuals that contain differences in sequence due to mutation and HGT. As the clone continues to expand, multiple mutations and regions of non-homology because of HGT will accumulate, making individuals quite different. However, individuals in this group still share considerable identity because they arose from a recent common ancestor. This shared identity can be used to infer the sequence of the ancestor of the clone. This sequence is called a clonal frame and all the strains descended from this ancestor are part of a clonal complex. Eventually, the clonal complex concept can become otiose as it becomes difficult to distinguish the clonal frame from derived mutations and horizontally transferred DNA. However, the clonal complex concept remains a serviceable stratagem in *B. burgdorferi* as HGT is rare and mutations are often easily identified.

Unlike other species of Eubacteria, *B. burgdorferi* shows little evidence of HGT despite a genome extensively fragmented into many smaller, more easily transferred units. Alleles within a clonal complex are almost always found linked to alleles at other loci, even when the loci are on different plasmids. By comparing gene trees, Dykhuizen *et al.* (1993) showed that the evolutionary history of two chromosomal genes (*p93* and *flaB*) are identical within and between *Borrelia* species and that the history of *ospA*, a plasmid-borne gene, is nearly identical with the evolutionary history of the chromosomal genes. The *ospA* phylogeny differs by only one small DNA fragment (~ 200 bp) that has been transferred into *B. garinii* strain PHei from a distantly related strain. Unlike most *Borrelia* genes, intragenic recombination is extensive in *ospC* both within and between species (Livey *et al.*, 1995; Wang *et al.*, 1999b; Dykhuizen and Baranton, 2001). Most HGT described to date involves short DNA fragments such that *ospC* sequences are combinations of multiple ancestors.

Recent MLST analyses confirm the low rate of HGT in *Borrelia* genospecies. MLST combines six to eight housekeeping gene sequences to provide high discriminatory power while retaining signatures of evolutionary relationships within and between clonal complexes (Maiden *et al.*, 1998; Enright and Spratt, 1999; Feil and Enright, 2004). Housekeeping genes are ideal for among-species phylogenetic analysis because they are found in all species, evolve slowly, are rarely under directional selection and are rarely horizontally transferred (Maiden *et al.*, 1998). Thus, these genes often indicate the true evolutionary history of a species. Analyses of multiple loci can also safeguard against accepting biased evolutionary relationships obtained in a single-locus analysis (Maiden *et al.*, 1998; Urwin and Maiden, 2003). One of the advantages of MLST is that the data from different laboratories can be combined. However, the three groups that have done MLST (Richter *et al.*, 2006; Postic *et al.*, 2007; Margos *et al.*, 2008; Qiu *et al.*, 2008) have all used different loci.

Limited HGT can arise due to lack of opportunity or lack of a recombination system. Multiple *B. burgdorferi* s.l. strains and species are often found in the same host and in the same tick (Guttman *et al.*, 1996), giving *B. burgdorferi* ample opportunity for HGT. Further, *Borrelia* species have a complete set of genes to carry out homologous recombination (Fraser *et al.*, 1997). Within lineages, there is strong evidence for homologous recombination within genomes (Stevenson *et al.*, 1998; Stevenson and Miller, 2003) and homologous recombination happens in the laboratory during insertion of a selectable marker (Samuels *et al.*, 1994; Rosa *et al.*, 1996). Nevertheless, the rate of HGT is very low in nature and transformation in the lab is difficult, suggesting that natural transformation is rare (Samuels, 1995). Transduction of cp32 plasmids has been shown (Eggers and Samuels, 1999; Eggers *et al.*, 2001), however transduction of other plasmids or genes has not been shown. The primary mechanism of HGT is probably a gene transfer agent restricted to small DNA fragments (Dykhuizen and Baranton, 2001). Limited HGT that is restricted to small DNA fragments guards against DNA parasites such as transposable elements, restriction modification systems, and parasitic plasmids like the F plasmid in *E. coli*. Simultaneously, this system prevents large evolutionary advances enjoyed by other bacterial species, such as the incorporation of antibiotic resistance or pathogenicity islands.

The rate and extent of HGT is the inverse of the degree of linkage disequilibrium observed between genes. In a purely clonal species, alleles at one locus will always be observed in the same genomes with specific alleles at other loci as a result of common descent. Recombination breaks up these associations. Within geographic regions, identical alleles are nearly always associated within each *ospC* clone. For example, no recombination between four genes (p66, IGS, *ospA*, and *ospC*) was found in 61 *B. burgdorferi* s.s. isolates from southern Connecticut nor between IGS and *ospC* in 73 *B. afzelii* isolates from southern Sweden (Bunikis *et al.*, 2004a). However, the same *ospC* group can be found in different backgrounds in different geographic regions (Fig. 9.2A). For example, the *ospC* type B clonal complex in the Midwest differs from the *ospC* type B clonal complex in the Northeast; the *ospC* type L clonal complex in the Midwest differs from the *ospC* type L clonal complex in Europe; and one Californian *ospC* type A allele is in a very different genomic background than the other *ospC* type A alleles from the Northeast and from California. In Europe, one clonal background includes *ospC* types B, 'ND', 'M1' and Q, suggestive of multiple recombinations into an advantageous background.

9.3.1 Realized recombination rate

The realized, or observed, HGT rate is the product of the combined effects of recombination and natural selection. That is, the observed HGT rate may be low for a particular gene region, even if genes are regularly transferred among strains, if the recombinant bacteria are at a selective disadvantage. Likewise, if the HGT rate is very low, then gene regions showing recombination are likely to be selected for diversity.

Northeastern *B. burgdorferi* experienced an extreme population bottleneck resulting in decreased genetic diversity at most loci (Qiu *et al.*, 2002). The average divergence among *B. burgdorferi* s.s. alleles is only 0.7% (Qiu *et al.*, 2004). However, a few genetic regions have retained high diversity.

We can use regions of high diversity remaining after a bottleneck to estimate the realized HGT rate. Qiu *et al.*, (2004) found 12 regions containing alleles with high sequence divergence; only one of which was on the main chromosome. The chromosomal region contained only two alleles. There are twelve *ospC* alleles and the remaining plasmid-borne loci have two to five alleles. Thus, a maximum of one HGT event occurred on the chromosome while as many as 31 occurred on plasmid loci. The gene trees of the plasmid-borne loci are inconsistent with the ribosomal spacer region sequence on the chromosome – generally used to form a strain phylogeny – strongly suggesting additional HGT. Unexpectedly, the *ospC* gene tree was consistent with the ribosomal spacer region sequence (Attie *et al.*, 2007). The other gene trees showed 28 examples of homoplasy (Attie *et al.*, 2007). This suggests that there were an additional 28 recombinations represented by the homoplasy. Thus, there is one recombination for the chromosome of over 900 kb compared with 59 (31 + 28) on the plasmids, which totals about 500 kb, or about a hundred-fold higher rate of recombination on plasmids than on the chromosome. Although *B. burgdorferi* is not strictly clonal, HGT events are rare in general and nearly non-existent on the chromosome. The low rate of HGT in *Borrelia* strongly implies

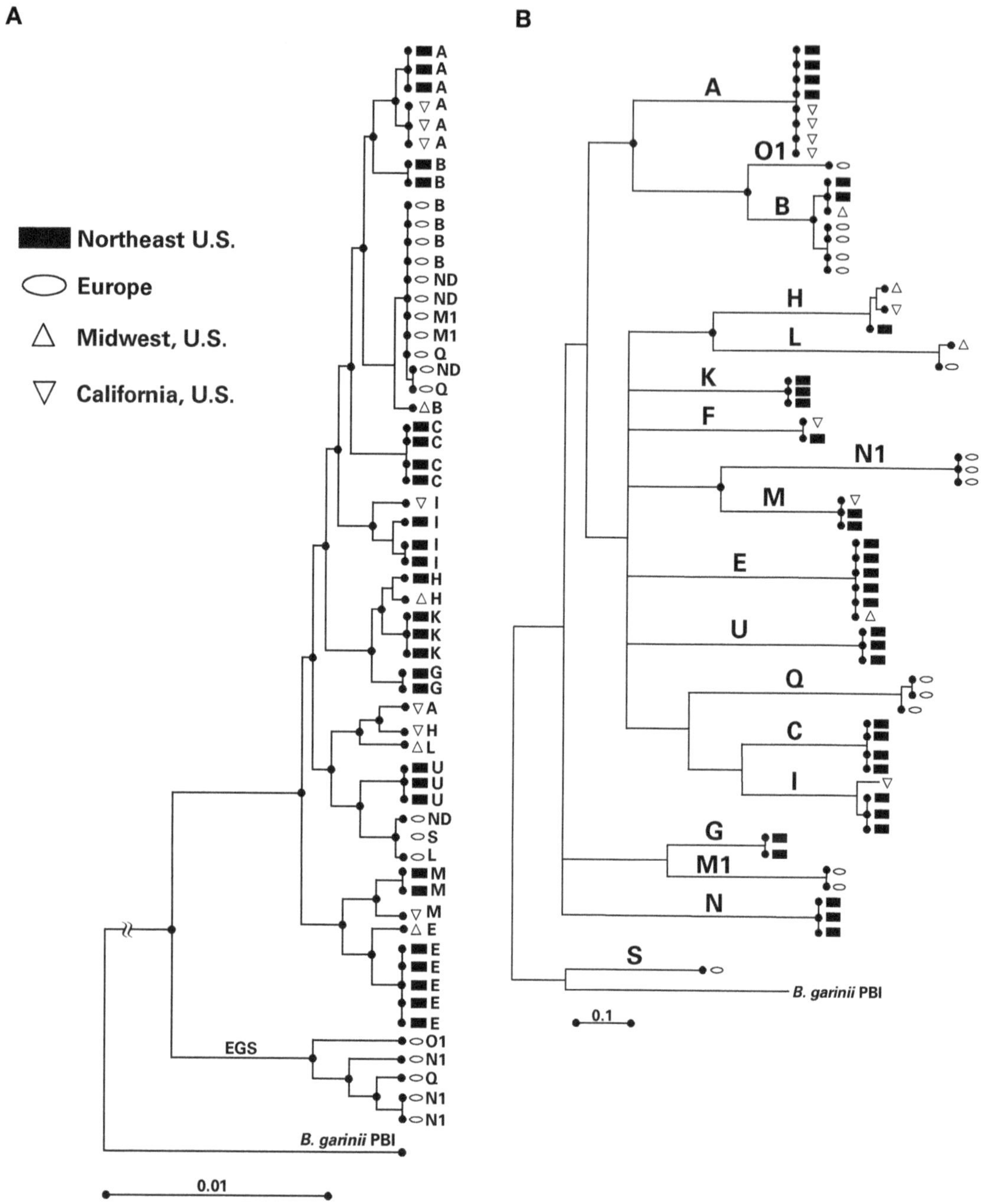

Figure 9.2 Phylogenetic trees of *Borrelia burgdorferi s.s.* (A) The tree created using MLST data, giving an idea of clonal background. (B) A tree created using the OspC sequence. These trees are redrawn from Margos *et al.* (2008). All strains are *B. burgdorferi* s.s., except for a single *B. garinii* strain used as an outgroup. EGS stands for a group of *B. burgdorferi* s.s. strains that could be a separate genospecies (Margos *et al.*, 2008). The symbols represent the origin of the strain: open oval, Europe; solid rectangle, northeast USA, open triangle, Midwest USA; upside-down open triangle, California. The OspC groups are the letters with M1, N1 and O1 as temporary names of newly discovered groups. ND is a group or groups that could not be amplified. Given our experience with these OspC groups, we expect this is a single group and have treated it as such on the figure and in the text. (ND is, of course, not present in B because the sequence of OspC was not determined). OspC group S is linked to *B. garinii* as expected since group S contains a fragment from *B. garinii* H13 (Dykhuizen and Baranton, 2001).

that most or all the genes showing these diverse alleles are likely to be genetic systems under balancing selection where recombinants have a selective advantage.

The transferred DNA fragments are small, generally smaller than a single gene. However, the transferred fragment size was not invariant; fragments incorporated into plasmids range from 444 bp to 936 bp (average 562 bp) while the transferred chromosomal fragment is 1128 bp. There was some evidence of whole plasmid transfer in this data set (Qiu *et al.*, 2004), but this was not followed up. There is as yet no firm evidence of plasmid transfer except maybe cp32.

9.3.2 cp32 plasmids

The cp32 plasmids are prophages (Eggers *et al.* 2001; Zhang and Marconi, 2005) and, therefore, could transfer at a high rate independently of other HGT systems. We examined the association between *ospC* alleles and cp32-borne *ospE*/*ospF*/*elp* (*erp*) alleles in 27 *B. burgdorferi* s.s. isolates cultured from human patients (M.R. Turner, unpublished data). We designed a PCR procedure to discriminate among eleven published *ospE*/*ospF*/*elp* sequences from B31, N40 and Sh-2–82 by constructing primer pairs specific to each allele. Except for *erpP*, an *ospE* paralogue that was found in every strain tested, all the other *ospE*/*ospF*/*elp* genes were found in a subset of strains and were not consistently associated with *ospC* (Table 9.1). The overwhelming impression is that the cp32 plasmids transfer between strains at a much faster rate than other genes and, consequently, cp32 genetic variants are not tightly associated with OspC serotypes.

9.4 Phylogeography

Geographic proximity can indicate phylogenetic proximity as 'Every species has come into existence coincident both in space and time with a pre-existing closely allied species' (Wallace, 1855). The study of the historical processes that are responsible for contemporary geographic distributions is called phylogeography (Avise *et al.*, 1987). Phylogeography considers the geographic distribution of individuals in light of the patterns associated with a gene genealogy (Avise, 2000). Phylogeography is a guiding tenet of evolution and is built on the ideas of an originator of the theory of natural selection, Alfred Russel Wallace.

Lyme *Borrelia* are distributed throughout Eurasia, North America, and northern Africa (Kurtenbach *et al.* 2006). However, each individual genospecies occupies only limited regions of the total area in which the *B. burgdorferi* s.l. complex is endemic. *Borrelia burgdorferi* genospecies have overlapping but not identical distributions such that they can be found in the same forests and sometimes in the same animals (Humair *et al.*, 1995; Humair and Gern, 1998; Kurtenbach *et al.*, 1998b; Humair *et al.*, 1999; Richter *et al.*, 2004a). However, the phylogenetic relationships among genospecies do not recapitulate geographic relationships. That is, geographic proximity does not indicate phylogenetic proximity.

9.4.1 Long-distance migration

Borrelia burgdorferi s.s. is endemic to all three North American Lyme disease foci – the northeast, the Midwest, and California – as well as in Europe (Fig. 9.3). Its closest relatives are genomospecies 1 and genomospecies 2, which are located in California (De Michelis *et al.*, 2000; Pecchioli *et al.*, 2007; Postic *et al.*, 2007). This may suggest a Californian origin of *B. burgdorferi* s.s. with subsequent colonization of the Northeast and Europe (Postic *et al.*, 2007). However, *B. burgdorferi* s.s. probably did arise in the American northeast (Foretz *et al.*, 1997). Invoking Vavilov's theory of the centres of origin – the centre of origin of a species is in the region of highest genetic and/or phenotypic diversity – *B. burgdorferi* s.s. originated in the northeast of North America. Of the 22 major *ospC* types (Seinost *et al.*, 1999; Wang *et al.*, 1999b; Qiu *et al.*, 2002), five types are found both in Europe and North America, 12 in only North America, and five in only Europe. Three of the five *ospC* types found only in Europe show clear homology with *ospC* in *B. garinii* or *B. afzelii*, suggesting lateral gene transfer into *B. burgdorferi* s.s. (Marti Ras *et al.*, 1997; Dykhuizen and Baranton, 2001). The depauperate *ospC* variation in European *B. burgdorferi* s.s. is compatible with the hypothesis that relatively few strains, and thus limited *ospC* diversity, migrated from North America. Gene trees support this hypothesis, indicating that North American genospecies like *B. bissettii* and the California genospecies are the

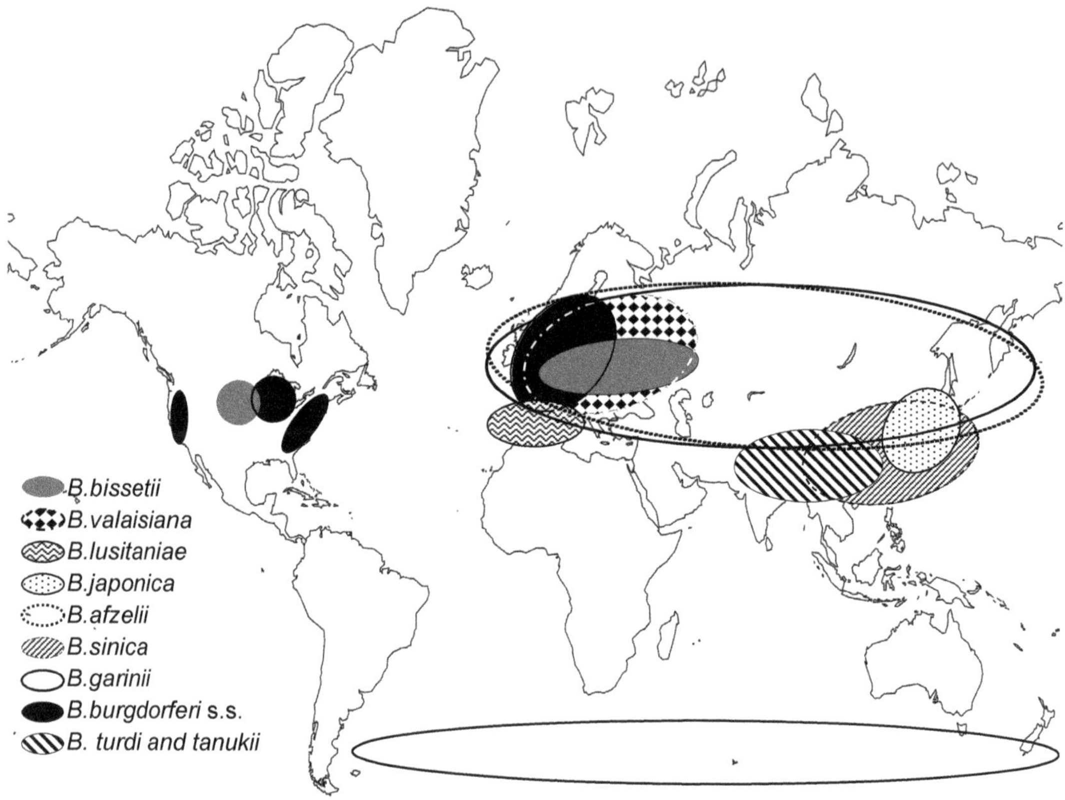

Figure 9.3 *Borrelia* species distributions. The geographical distribution of *B. burgdorferi s.l.* is shown on this map, which is redrawn from Kurtenbach *et al.* (2006) with additional data on *B. sinica* from Masuzawa *et al.* (2001).

closest relatives of *B. burgdorferi* s.s. (Dykhuizen *et al.*, 1993; Marti Ras *et al.*, 1997; Dykhuizen and Baranton, 2001).

More recent analyses suggest a more complicated picture (Margos *et al.*, 2008; Qiu *et al.*, 2008). Recent migration of *ospC* major group E and two of the most important human infectious *ospC* types – A and K – from North America to Europe is likely. The support for the idea that the migration is very recent is that strains from North America and Europe within each of these three groups are identical over 11,167 bp of six chromosomal genes (Qiu *et al.*, 2008). Other *B. burgdorferi* s.s. strains in Europe are likely to have been present for a considerable amount of time. The *ospC* type B shows divergent evolutionary lineages on the two continents (Fig. 9.2A), suggesting isolation by distance (Qiu *et al.*, 2008). Other *ospC* types unique to Europe are also found in this background (Fig. 9.2A), suggesting recombination of these types into the background of type B, a background that migrated from North America.

The branch on Fig. 9.2A labelled EGS is a divergent group of strains found in Europe (Margos *et al.*, 2008) (Fig. 9.2A). These strains have been isolated in Finland, France and Germany, and represent either a highly divergent *B. burgdorferi* s.s. lineage (Margos *et al.*, 2008) or a new species (Qiu *et al.*, 2008). These strains are more closely related to the other *B. burgdorferi* s.s. strains than they are related to *B. bissettii*. Margos *et al.* (2008) used this observation to suggest that the ancestor of *B. burgdorferi* s.s. could be from Europe. The same argument can be used to suggest that *B. burgdorferi* s.s. arose in California because genospecies 1 and 2 were isolated there. Both these statements cannot be true, and both are probably false.

To see why neither Europe nor California are the probable the centres of origin of *B. burgdorferi* s.s. complex consider the pattern of sequence divergence of the Californian *B. burgdorferi* s.s. isolates from the northeastern North American *B. burgdorferi* s.s. isolates. The *ospC* type A isolates from California are in a different genomic background than *ospC* type A isolates from northeastern North America or from Europe. The Californian *ospC* type I isolate shows both genomic and *ospC* sequence divergence (Fig. 9.2) (Foretz *et al.*, 1997). These sequence differences suggest that this strain has been in California for quite a while as the divergence in *ospC* type B shows that the strains in Europe have resided in Europe for a considerable amount of time. There are no highly divergent clones in the northeastern US. All the present *ospC* clones in the Northeast cluster tightly, suggesting there is no migration from California or Europe into the northeastern population. These observations suggest that *B. burgdorferi* s.s. originated in the northeastern part of North America with regular out migrations into Europe and California.

The data presented here are explained by the hypothesis that *B. burgdorferi* s.s. was once a wide-ranging species that subsequently separated into isolated populations. The populations in Europe, northeastern North America, and California then evolved independently. Northeastern North American strains have continuously migrated into the European and Californian populations as is apparent in the phylogenetic reconstructions (Fig. 9.2).

9.4.2 Species distributions

B. andersonii and *B. bissettii* are the only other genospecies in the *B. burgdorferi* s.l. complex endemic to North America. *B. andersoni* and *B. burgdorferi* have largely overlapping distributions in the eastern half of the US, although they appear to be distantly related. rRNA loci tend to place *B. andersoni* as the most distant relative to all other genospecies in the complex. However, the placement of *B. andersoni* is uncertain as it is rarely included in phylogenetic analyses, probably because it is not infectious in humans.

B. bissettii is endemic to the middle of North America and parts of Europe. It is most closely related to *B. japonica*, which is only found on the Japanese islands. Further, the *B. bissettii*/*B. japonica* clade is closely related to *B. burgdorferi* s.s. and genomospecies 1 and 2. While *B. burgdorferi* s.s. and *B. bissettii* have overlapping distributions in Europe, there is little evidence suggesting that phylogenetic relationships recapitulate geographic relationships.

B. afzelii, *B. garinii*, and *B. valaisiana* are widely distributed in Eurasia. *B. garinii*, however, is closely related to the North American clade (which includes *B. japonica*). *B. afzelii* appears to give rise to the remaining Eurasian genospecies. *B. turdi*, *B. tanukii*, and *B. valaisiana* form a tight group and are all found in Asia. Thus, there probably were at least two introductions of *B. burgdorferi* s.l. species into east Asia. The origin of *B. sinica* is unknown, but may have come from its geographic neighbour, *B. turdi*, on mainland China. *B. lusitaniae* is found around the Mediterranean and east Asia and is part of the *B. afzelii* clade.

9.4.3 Vector distribution

The current distribution of each genospecies is affected not only by its phylogenetic history but also by migration and the current and historical distributions of suitable tick vectors and vertebrate hosts. Several tick species in the *Ixodes ricinus* complex populate the North American and Eurasian continents. *I. dentatus* (*B. andersonii*), *I. spinipalpis* (*B. bissettii*), *I. pacificus* (*B. burgdorferi* s.s. and *B. bissettii*) and *I. scapularis* (*B. burgdorferi* s.s.) are the dominant Lyme *Borrelia* vectors in North America. *I. ricinus* (*B. burgdorferi* s.s., *B. lusitaniae*, *B. garinii*, *B. afzelii*, and *B. valaisiana*) is the primary human-biting vector in western Europe while *I. persulcatus* (*B. garinii* and *B. afzelii*) is the primary human-biting vector in eastern Europe, China, and Japan. East Asia has several other competent Lyme borreliosis vectors including *I. columnae* (*B. valaisiana*) and *I. turdus* (*B. turdi*) in Japan, *I. granulatus* (*B. valaisiana*) in China and Japan; *I. ovatus* (*B. sinica* and *B. japonica*) in China, Nepal, and Japan; and *I. tanuki* (*B. tanukii*) in Nepal and Japan (Masuzawa, 2004). *I. uriae* (*B. garinii*) is found around the Atlantic ring and is primarily associated with sea birds (Olsen *et al.*, 1993). The majority of *I. ricinus* species tested have been shown to be capable vectors for most of the *B. burgdorferi* s.l. species tested (Mather and Mather, 1990; Gern *et al.*, 1991; Lane *et al.*, 1994; Dolan *et al.*, 1998; Lane *et al.*, 1999; Dolan *et al.*, 2000; Peavey *et al.*, 2000; Eisen *et al.*, 2003;

Sun and Xu, 2003; Sun *et al.*, 2003). Thus, factors other than vector competence probably restrict the distributions of *B. burgdorferi* genospecies. In fact, the current distribution of *B. lusitaniae* is restricted to a subregion of the distribution of its principal vector, *I. ricinus* (Hubalek and Halouzka, 1997).

9.4.4 Host distribution and migration

Each *B. burgdorferi* genospecies uses a different set of vertebrate species as hosts (Kurtenbach *et al.*, 1998c; Kurtenbach *et al.*, 2002). Thus, different distributions of vertebrate host species can have a dramatic effect on the distributions of *B. burgdorferi* s.l. genospecies. For example, the preponderance of pheasants in the British Isles has resulted in an overrepresentation of *B. garinii* and very few *B. afzelii* (Kurtenbach *et al.*, 1998a; Kurtenbach *et al.*, 1998b). *Borrelia burgdorferi* genospecies that utilize birds as hosts, especially migratory birds, should have broader distributions than those that primarily use mammals (Kurtenbach *et al.*, 2002). Birds can move *B. burgdorferi* and ticks into areas that can then propagate a new population (Ogden *et al.*, 2008). The discovery of identical *B. garinii flaB* sequences in *I. uriae* from the Northern and Southern hemispheres strongly suggests transequatorial transport of *B. garinii* by migrating seabirds either as infections of birds or within infected ticks carried by birds (Olsen *et al.*, 1995). However, northern and southern *I. uriue* are reproductively isolated, suggesting that *B. garinii* is not dispersed between hemispheres by ticks attached to seabirds (Gylfe *et al.*, 2001). Thus, long-distance migration of *B. garinii*, if it occurs, is mediated through active infections of birds and is then transmitted to an established vector population. Long distance migration via infected ticks is probably a rare event because for *B. burgdorferi* to successfully establish a new population, the tick would have to have become infected as a larva, then attach to the bird as a nymph and infect a susceptible host as an adult. This is the only way a genotype that does not infect a bird could move large distances. If the uninfected larvae attach to the bird and move with the bird, they would simply carry the same strains as the migrant bird.

Mammalian hosts have very limited migration, limiting the potential movement of *B. burgdorferi*. Evidence from the northeastern US suggests that migration between *I. scapularis* populations is severely restricted (Qiu *et al.*, 2002). This may be one reason that *B. lusitaniae* has not expanded to areas of Europe beyond the Mediterranean region. However, recent reports suggest that migratory birds appear to be reservoirs for *B. lusitaniae* (Marie-Angele *et al.*, 2006). In contrast to the evidence for *B. lusitaniae*, genospecies associated with birds should have broad distributions and be relatively homogeneous throughout the regions inhabited by *I. ricinus* ticks. *Borrelia burgdorferi* genospecies associated with mammals should be more heterogeneous over their range due to limited migration. Additionally, *B. burgdorferi* genospecies associated with birds would be better able to disperse to areas opened by climatic events such as coming out of an ice age (Kurtenbach *et al.*, 2002).

9.4.5 Glaciation and range expansion

Both *B. burgdorferi* s.s. and the northern lineage of *I. scapularis* lack substantial neutral genetic variation; the variation that does exist forms a star-shaped phylogeny (Qiu *et al.*, 2002). Star phylogenies are characteristic of recently derived evolutionary lineages that have undergone a rapid population and geographic expansion (Slatkin and Hudson, 1991; Harpending *et al.*, 1998; Qiu *et al.*, 2002;). In contrast, the southern lineage of *I. scapularis* shows considerable population differentiation with a coalescence pattern typical of long-standing populations (Qiu *et al.*, 2002). Qiu *et al.* (2002) proposed that the shared evolutionary history of *B. burgdorferi* s.s. and the northern *I. scapularis* was caused by range contraction during the Pleistocene glaciation and subsequent range expansion approximately 18,000 years ago. This geological event decreased the genetic diversity of many plant and animal species compared to southern populations not affected by the glaciation (Avise, 1992; Merila *et al.*, 1996). Higher genetic diversity in tick populations in Long Island and southern Connecticut suggest an interglacial refugium existed in the region (Qiu *et al.*, 2002). This refugium probably supplied modern populations in New England with *B. burgdorferi* migrants after the glacial maximum.

9.5 Population genetics and natural selection

Population genetics, the theoretical foundation of modern Darwinism, explores evolutionary change via mutation, genetic drift, gene flow, natural selection, population subdivision, and population structure in space. Adaptation and speciation are well within the purview of population genetics. Population geneticists take a statistical approach to integrating Mendelian genetics with evolution. Abstract mathematical models provide a framework for understanding the patterns of genetic variation in actual populations and the evolutionary forces that have caused that pattern (Crow and Kimura, 1970; Hartl and Clark, 2007). Random genetic drift tends to purify populations of genetic variation through the statistical effect of sampling error. That is, some alleles will be carried into future generations while others tend towards extinction through purely chance events. This gives a typical pattern of gene frequencies, the Ewens–Watterson neutral distribution (Ewens, 1972; Watterson, 1978). Statistical tests can determine if the observed allele frequencies are significantly different from those expected if genetic drift is the primary force changing gene frequency. Gene frequencies more even than expected under a neutral drift hypothesis suggest that balancing selection is preserving genetic variation.

9.5.1 Balancing selection

The patterns of genetic variation within a *Borrelia* species have been extensively studied in the northeastern *B. burgdorferi* s.s., particularly the surface-exposed proteins encoded by *ospA* and *ospC*. Four *ospA* alleles are found in northeastern tick populations at frequencies that are more even than expected by chance events alone, suggesting balancing selection is maintaining the diversity (Qiu *et al.*, 1997). This finding is interesting as most of the *ospA* alleles differ at synonymous sites of codons that are unlikely to have a significant selective effect on evolutionary fitness.

Unlike *ospA*, the 16 *ospC* major groups of alleles found in the northeastern US differ by an average of 14% at the amino acid level. Wang *et al.* (1999b) defined an *ospC* major group as alleles that are less than 2% different and more than 8% from alleles in other major groups. The lower end of this *ospC* variation (8–11%) represents the variation between two *ospC* groups that share a segment homogenized by recombination (Livey *et al.*, 1995; Wang *et al.*, 1999b). For example, *ospC* group C in *B. burgdorferi* s.s. is a recombinant composed of parts of *ospC* major groups I, B, and E (Qiu *et al.*, 2002; Attie *et al.*, 2007), so a pairwise comparison between group C and I would be about a third lower than expected. Short sequence motifs are often shared among *ospC* groups (Attie *et al.*, 2007). Even though there is tremendous variability between *ospC* major groups, the N-terminal region includes a leader sequence that shows very little variation within species but differs between species (Livey *et al.*, 1995). In addition to the 16 *ospC* major groups found in northeastern *B. burgdorferi* s.s. (A to N, T, and U) (Seinost *et al.* 1999; Wang *et al.* 1999b; Qui *et al.*, 2002; Brisson and Dykhuizen, 2004), there are five major groups found only in Europe (P, Q, R, S, and V) and additional groups that have been given provisional names (Qiu *et al.*, 2008; Margos *et al.*, 2008, unpublished data). The naming of these groups should be systematized. Similar *ospC* variation found in other *B. burgdorferi* genospecies (Baranton *et al.*, 2001) should be similarly systematized.

OspC major groups are serotypically different. The serological response to OspC by the mammalian immune system occurs early in infection and is very strong (Padula *et al.* 1993; Jauris-Heipke *et al.*, 1993; Wilske *et al.*, 1993b; Margolis *et al.*,1994; Stevenson and Barthold, 1994; Wilske *et al.*, 1995; Pohl-Koppe *et al.*, 2001; Earnhart *et al.*, 2005; Lovrich *et al.*, 2005). OspC is one of the most immunodominant antigens of early Lyme disease, second only to an unexposed flagellar protein (Schwan *et al.*, 1991; Fung *et al.*, 1994; Gilmore *et al.*, 1996). Despite an ample antibody response, *B. burgdorferi* s.s. lives in *Peromyscus leucopus* for the mouse's lifetime without changing the *ospC* sequence (Hodzic *et al.*, 2000), contrary to the expected action of an immune-escape gene (Liang *et al.*, 2002). Instead, *B. burgdorferi* avoids antibody-mediated death by downregulating *ospC* once established in a vertebrate (Schwan *et al.*, 1995; Gilmore *et al.*, 2001; Fingerle *et al.*, 2002; Hodzic *et al.*, 2002). Once infected, a mouse cannot be infected by another strain with the same *ospC* major group allele but can be infected by strains with alternative alleles (Gilmore *et al.*,1996). Thus, the mouse immune system discriminates among the *ospC* major groups. The surface-exposed regions of OspC display the highest amino acid variability, further supporting the serotypic difference hypothesis (Eicken *et al.*, 2001; Kumaran *et al.*, 2001).

Nearly every OspC serotype is found in every *B. burgdorferi* s.s. population at frequencies that are more even than expected by chance, similar to the *ospA* surveys (Qiu *et al.*, 1997; Wang *et al.*, 1999b; Qiu *et al.*, 2002). Both *ospA* and *ospC* allele frequencies are not significantly different among populations (Qiu *et al.*, 1997, 2002), suggesting that either very high migration rates between populations or balancing selection maintains allele frequencies across a wide geographical area. The migration rate of the tick vector appears to be limited, even among geographically proximate populations (Qiu *et al.*, 2002). The major source of *B. burgdorferi* s.s. migrants would then have to be infected hosts with limited tick burdens. However, as noted above, most small mammals – the main vertebrate hosts for *B. burgdorferi* s.s. – have very limited migration patterns.

9.5.2 Negative frequency-dependent selection

Balancing selection maintains multiple alleles at intermediate frequencies with high sequence diversity in a population (Vekemans and Slatkin, 1994), as seen at the *ospC* locus (Wang *et al.*, 1999b; Lin *et al.*, 2002; Qiu *et al.*, 2002). The most likely forms of balancing selection in bacteria are negative frequency-dependent selection and multiple niche polymorphism. Negative frequency-dependent selection occurs when rare genotypes have a selective advantage over common genotypes (Gromko, 1977). The strong immune response generated against OspC could give rare serotypes a selective advantage and maintain the *ospC* polymorphism. Immune memory to specific OspC serotypes protects vertebrates from future infections of the same serotype (Barthold, 1999), but does not prevent infection with alternative OspC serotypes (Preac-Mursic *et al.*, 1992; Gilmore *et al.*, 1996; Probert *et al.*, 1997; Barthold, 1999). If a secondary immune response to OspC is the only factor preventing *B. burgdorferi* s.s. from infecting an animal, every host would be infected by all serotypes that they had been challenged with during the nymphal season. The number of susceptible hosts is much greater for rare OspC serotypes, such that they should increase in frequency. With simple negative frequency-dependent selection, all serotypes would be at approximately the same frequency at equilibrium. In natural populations, however, serotypes vary in frequency by 20-fold (Qiu *et al.*, 1997; Wang *et al.*, 1999b), suggesting that negative frequency-dependent selection by immune exclusion is not the major cause of balancing selection.

9.5.3 Multiple-niche polymorphism

Multiple-niche polymorphism, also referred to as diversifying selection, can maintain diversity within populations when the environment is heterogeneous and no one serotype has the highest fitness in all environments. The environments *B. burgdorferi* s.s. encounter – different species of vertebrate hosts – are heterogeneous with respect to serotype fitness. The fitness of serotypes differs dramatically among four vertebrate species sampled from a natural northeastern forest (Brisson and Dykhuizen, 2004). In fact, the majority of serotypes had fitnesses of nearly zero in most vertebrate species. That is, only a subset of serotypes can both infect and be transmitted to ticks by each species, and the subsets differ among species (Brisson and Dykhuizen, 2004). Within a species, there is no variation in the serotypes transmitted to feeding ticks. The correlation between OspC serotype and vertebrate species suggests that the host-species range of *B. burgdorferi* s.s. is defined by the OspC serotype or a tightly linked locus.

We developed a model whereby the frequencies of the OspC serotypes in questing nymphal ticks are determined by the frequency of the various species of hosts, the larval tick load on each species, and the transmission probability of each OspC serotype to feeding larvae (Brisson and Dykhuizen, 2006). This model predicted that at least one major host species was not sampled in the original study. This host was determined to be a Sorex shrew, which infects about 25% of the ticks infected as larvae (Brisson *et al.*, 2008). The success of this model in predicting that there was another major host supports the contention that multiple niche polymorphism is the major cause of balancing selection in *B. burgdorferi* s.s. The model predicts that mice infect only 25% of the infected ticks. The prevalence of infected ticks was reduced by 25% when mice were vaccinated against *B. burgdorferi* s.s. (Tsao *et al.*, 2004), which is an interesting and perhaps not simple coincidence. The model suggests that another important host, perhaps a bird, infects at most 20% of the infected ticks.

9.5.4 Human infection by OspC serotypes

Similar to evolutionarily relevant host species, only five OspC serotypes (A, B, K, I, and N) infect humans in the Northeast (Seinost *et al.*, 1999; Dykhuizen *et al.*, 2008). OspC serotype Q – which is restricted to Europe – also causes chronic disease in humans (Lagal *et al.*, 2003). Only a subset of the OspC serotypes in *B. garinii* and *B. afzelii* are found in disseminated human infections as well (Baranton *et al.*, 2001; Lagal *et al.*, 2003). Seinost *et al.* (1999) defined invasive strains as any strain found in the blood. This definition should be revised (Dykhuizen *et al.*, 2008). All 'highly invasive' serotypes have been isolated from the human central nervous system (Seinost *et al.*, 1999; Lagal *et al.*, 2003; Earnhart *et al.*, 2007), but other OspC serotypes have been isolated from human blood at very low rates (Alghaferi *et al.*, 2005; Earnhart *et al.*, 2007; Dykhuizen *et al.*, 2008). If relative invasiveness (the probability of an OspC serotype being found in human blood divided by the probability of exposure to that OspC serotype) is estimated for all serotypes, the five 'highly invasive' serotypes have substantially higher relative invasiveness. OspC serotype H shows an intermediate relative invasiveness score potentially due to its rarity in questing ticks. All other serotypes show a very low relative invasiveness and are rarely found in blood. Human invasiveness, like most biological phenomena, is not simply true/false but a bimodally distributed continuum.

A recent experiment may explain why the infection continuum found in humans is replaced with an all-or-none ability to cause chronic infection in other animal hosts (Derdakova *et al.*, 2004). In the field, OspC serotype A produces a persistent infection in *P. leucopus*, while OspC serotype E does not (Brisson and Dykhuizen, 2004). Experimentally, both serotypes can disseminate into the blood of a mouse that has never been exposed to *B. burgdorferi* s.s. Only serotype A results in a persistent infection while serotype E loses its ability to infect larvae over time. Serotype A strains can also infect mice that have been previously exposed to serotype E; the reverse is not true (Derdakova *et al.*, 2004). Since most humans that contract Lyme disease are immunologically naïve, they are initially permissive hosts for a wider range of serotypes. By analogy, we expect that if immunologically naïve humans are infected by a non-persistent strain, the disease will resolve after a few months.

9.5.5 OspC function

The molecular mechanism of OspC is currently unknown; however, through genetic studies, there is information about the function of OspC (see Chapter 12). OspC is induced when a tick starts feeding on a mammal (Schwann *et al.*, 1995) and is repressed two weeks after infection (Liang *et al.*, 2002), suggesting that OspC is important for the initiation of infection in mammals. Ohnishi *et al* (2001) have shown by immunofluorescence that spirochaetes without OspC on the cell surface can penetrate the tick salivary glands and are found in the skin shortly after inoculation. Mutants without an expressed OspC cannot establish infection in mammals (either by tick bites or by needle inoculation), but can maintain an infection once established (Grimm *et al.*, 2004; Tilly *et al.*, 2006). The innate immune system seems to clear the spirochaetes lacking OspC on the outer surface of the membrane within 48 hours (Tilly *et al.* 2007). This function of OspC can be replaced by any number of other outer surface proteins (Xu *et al.* 2008). However, these other proteins cannot replace OspC in speed and extent of dissemination from the tick bite (Xu *et al.*, 2008; Radolf and Caimano, 2008). Thus, the function of OspC seems to be to promote rapid dissemination through tissue.

The majority of population genetic studies have used *ospC* variation as a marker to distinguish various clones in order to study the population dynamics and evolutionary history of *Borrelia*. There was an implicit assumption that the differences in infectivity could be caused by *ospC* or any number of tightly linked loci. Several lines of evidence suggest that an action of OspC causes the differential infectiveness.

1 *Population genetics theory*. Balancing selection preserves multiple alleles at a locus within populations at intermediate frequencies and with high sequence diversity (Vekemans and Slatkin, 1994), giving the alleles a deeper coalescence than neutrally evolving alleles (May *et al.*, 1999; Schierup *et al.*, 2000). This is the pattern seen at the *ospC* locus (Jauris-Heipke *et al.*, 1993; Wilske *et al.*, 1993b; Marconi *et al.*, 1993; Livey *et al.*, 1995; Theisen *et al.*, 1995; Masuzawa *et al.*, 1997b; Wang *et al.* 1999b; Seinost *et al.*, 1999; Qiu *et al.*, 2002; Lin *et al.*, 2002; Brisson and Dykhuizen, 2004). If another locus were the focus of the strong balancing selection, OspC serotypes should be lost by genetic drift when linkage disequilibrium breaks down (as the B, Q, M1 and ND types in Fig. 9.2). This does not happen. Rather, OspC shows the greatest number of types of any variable region and the deepest coalescence (Qiu *et al.*, 2004), logically supposing that *ospC* is the locus under balancing selection. If another locus caused differential infectiveness, recombination would shuffle *ospC* alleles into different backgrounds, distorting the correlation between OspC serotypes and invasiveness in different species.
2 *High variation at* ospC *is selected*. Variation at *ospC* must be selectively important as populations with low *ospC* diversity increase variation by recombining in *ospC* from other genospecies (Dykhuizen and Baranton, 2001). OspC serotype S in *B. burgdorferi* s.s. is derived from an *ospC* in *B. garinii* (Fig. 9.2B).
3 ospE/ospF/elp *variation*. *ospE/ospF/elp* loci are sufficiently variable to cause the host-species specificity but are transferred among *ospC* backgrounds. This rapid transfer breaks down the linkage disequilibrium required to explain the correlation between OspC serotypes and host species due to a linked gene. Further, new evidence suggests that *ospE/ospF/elp* do not provide the discrimination required (Woodman *et al.*, 2007).
4 *Multiple niches*. The explanatory power of the multiple niche hypothesis – different species of hosts transmit different sets of serotypes – suggests that this is the major mechanism producing balancing selection. The pattern of strains infecting mice, chipmunks, squirrels, humans, and two species of shrew requires ten classes. No other variable locus has this many classes except *ospC* (Qiu *et al.*, 2004). Ten classes can be created from more than one locus, but the pattern of major variation seen at other sites does not combine in such a way to make the ten required classes (Qiu *et al.*, 2004).
5 *Hypothetical host protein*. If OspC binds to and uses a host protein to migrate rapidly through tissue, and if host protein is genetically variable between species then species specificity for *ospC* clones could be explained.

9.6 Host species specificity

Like the majority of zoonotic pathogens, each of the *B. burgdorferi* s.l. genospecies is maintained in nature by a broad spectrum of vertebrate species including birds, mammals, and reptiles (Richter *et al.*, 2000; Gern and Humair, 2002; Korenberg *et al.*, 2002; Miyamoto and Masuzawa, 2002; Piesman, 2002; LoGiudice *et al.*, 2003). The primary tick vectors of *B. burgdorferi* s.l. are generalist ectoparisites that feed on many different vertebrate species (Anderson, 1989a,b; Gern *et al.*, 1998; Randolph, 1998). The *B. burgdorferi* s.l. genospecies, however, tend to specialize on a specific vertebrate species and are not truly generalists (Donahue *et al.*, 1987; Anderson, 1988; Humair *et al.*, 1995; Kurtenbach *et al.*, 1998b; Humair *et al.*, 1999; Brisson and Dykhuizen, 2006; Brisson *et al.*, 2008), despite the between-species transmission opportunity provided by the generalist nature of the tick vectors. As many as six genospecies coexist in the same habitats in Eurasia in overlapping transmission cycles that include the same tick vector species but different sets of vertebrate hosts (Olsen *et al.*, 1993; Humair *et al.*, 1995; Kurtenbach *et al.*, 1998b; Humair *et al.*, 1999). *Borrelia burgdorferi* genospecies may parse the environment giving the appearance of overlapping distributions while limiting competition. In this way, *B. burgdorferi* s.l. genospecies can be regarded as different ecological species or ecotypes (Cohan, 2002; Kurtenbach *et al.*, 2006).

9.6.1 Europe

The host species specificity is of particular importance to the global and local distributions of genospecies. Each genospecies is sustained in habitats consisting of suitable vertebrate host species and suitable vector species. Seven *B. burgdorferi* s.l. genospecies are endemic to Europe (Hubalek and Halouzka, 1997; Masuzawa, 2004; Kurtenbach *et al.*, 2006). These can be artificially separated into those that rarely cause human disease (*B. valaisiana, B. lusitaniae, B. bissettii, B garinii* serotypes 1, 2, 3, 5 and 6, and the newly named *B. spielmanii*), and those that regularly cause disseminated human disease (*B. afzelii, B. garinii* serotype 4, and *B. burgdorferi* s.s.). European genospecies are maintained in nature primarily by the sheep tick, *Ixodes ricinus*, an extreme host-species generalist. *B. afzelii* primarily infects rodents and insectivores across Asia and the more northern climes in Europe (Humair *et al.*, 1995; Humair *et al.*, 1999). Sea birds and song birds maintain *B. garinii* but not *B. afzelii* populations (Olsen *et al.*, 1993; Olsen *et al.*, 1995; Gylfe *et al.*, 2001). However, *B. garinii* is serotypically and ecologically diverse in Eurasia, with five of the six European OspA serotypes maintained by birds while serotype 4 specializes on rodents and insectivores (Hu *et al.*, 2001). Interestingly, OspA serotype 4 is the most common *B. garinii* serotype found in disseminated infections in humans (Huegli *et al.*, 2002). Further, genetically diverse *B. garinii* from Asia are associated with rodents, not birds (Nakao *et al.*, 1994; Masuzawa *et al.*, 1997a, 2001). None of the *B. garinii* serotypes uses either dormice or red squirrels as reservoir hosts, despite being sympatric with these rodents in nature (Humair and Gern, 1998). *Borrelia burgdorferi* s.s. is the most general of the European genospecies, infecting insectivores, rodents, and birds (Nakao *et al.* 1994; Humair *et al.*, 1995; Kurtenbach *et al.*, 1998b; Michalik *et al.*, 2005; Duneau *et al.*, 2008). The *Borrelia* genospecies that are not human infectious are commonly sympatric with the more invasive and more studied genospecies. Considerably less is known about the basic biology, evolution, and ecology of these species than the medically relevant genospecies. *B. valaisiana* is a pan-Eurasian species that readily infects ground-foraging and migratory sea-birds (Saint Girons *et al.*, 1998; Kipp *et al.*, 2006; Michalik *et al.*, 2008). *B. bissettii* is found in Europe and North America and is maintained primarily by rodents (Burkot *et al.*, 2000; DeNatale *et al.*, 2002; Lane *et al.*, 2005; Brown *et al.*, 2006). *B. lusitaniae* has been found in rodents, birds, and reptiles and *B. spielmanii* may specialize on dormice (Richter *et al.*, 2004b; Marie-Angele *et al.* 2006; Amore *et al.* 2007; Duneau *et al.* 2008).

9.6.2 Causes of host species specificity

Host-species specificity is probably due to differences in infectivity. Genospecies infectiousness has been proposed to be mediated by sensitivity to host serum components (Kurtenbach *et al.*, 1998c; Hellwage *et al.*, 2001; Kurtenbach *et al.*, 2002; Stevenson *et al.*, 2002; Kraiczy *et al.*, 2003;; Metts *et al.*, 2003; Miller and Stevenson, 2004; Haupt *et al.*, 2007). Genospecies differ in their ability to bind to host factor H and factor H-like protein-1/reconectin. Factor H and factor H-like proteins protect host cells from lysis by modulating the membrane attack complex of the alternative complement pathway (Rodriguez de Cordoba *et al.*, 2004). *Borrelia burgdorferi* prevents complement-mediated lysis by binding these control proteins. Eurasian genospecies are resistant to lysis *in vitro* by the complement of species they use as reservoir hosts, but not from vertebrate species they cannot infect (Kurtenbach *et al.*, 1998c; Lane and Quistad, 1998; Kuo *et al.*, 2000; Nelson *et al.*, 2000; Ullmann *et al.*, 2003). Thus, complement sensitivity could prevent genospecies from infecting certain host species. Deflection of the complement system may be a key determinant of host-species adaptation of *B. burgdorferi* s.l. species. However, Woodman *et al.* (2007) found no difference in the ability of *B. burgdorferi* s.s. to infect wild-type mice and those missing complement factors B or C3. Also, BbCRASP-2, which binds factor H and factor H-like protein and is regulated such that it is expected to be important in host immune evasion (Hartmann *et al.*, 2006), does not to contribute to complement resistance (Coleman *et al.*, 2008). These results suggest that differences in the innate immune system between species may not cause different *B. burgdorferi* genospecies to use different host species.

9.6.3 North America

Although the western USA, southeastern USA and Asian *B. burgdorferi* s.l. foci are more impoverished in the number of genospecies compared to European foci (Hubalek and Halouzka, 1997; Masuzawa, 2004; Kurtenbach *et al.*, 2006), the transmission dynamics are substantially more complex for two reasons. First, each of these regions has at least two vector-competent *Ixodes* tick species that parasitize different but overlapping sets of vertebrate species (James and Oliver, 1990; Miyamoto *et al.*, 1992; Kawabata *et al.*, 1993; Anderson *et al.*, 1996; Fukunaga *et al.*, 1996; Dennis *et al.*, 1998; Eisen *et al.*, 2003). This complex transmission 'network', as opposed to the transmission 'cycle' in Europe, maintains *B. burgdorferi* in multiple sets of species that are commonly parasitized by different tick species. If *B. burgdorferi* becomes extinct in one set of vertebrates, the population can be re-established by occasional vector transmission from vertebrate hosts that still maintain *B. burgdorferi*. For example, *B. burgdorferi* s.s. is maintained by least two tick species in the western US – *I. pacificus* and *I. jellisoni* – while *I. spinipalpis* appears to be a major vector for *B. bissettii* but also vectors *B. burgdorferi* s.s. in nature (Brown and Lane, 1992; Brown *et al.*, 2006). Ecological and evolutionary models, as well as public health efforts, are intrinsically more difficult due to this complexity. Second, each of these regions has at least two *B. burgdorferi* s.l. genospecies with overlapping but distinct transmission networks of ticks and vertebrate hosts. For example, both *B. burgdorferi* s.s. and *B. bissettii* are endemic in California and present in dusky-footed woodrat (*Neotoma fuscipes*), while only *B. burgdorferi* s.s. has been isolated from the California kangaroo rat (*Dipodomys californicus*) and *B. bissettii* from two *Peromyscus* species (Lane *et al.*, 2005; Brown *et al.*, 2006). Nevertheless, all *B. burgdorferi* s.l., aside from *B. burgdorferi* s.s., appear to predominantly use a specific set of vertebrate host species as reservoirs. Genospecies-specific enzootic transmission cycles may exist with unique combinations of vectors and vertebrate host species, simplifying the transmission networks of these infectious bacteria species.

The ecology of *B. burgdorferi* s.s. in the northeastern and Midwestern US is in many aspects the least complex. There is one *B. burgdorferi* genospecies, *B. burgdorferi* s.s., and one tick vector, *I. scapularis* (Burgdorfer and Gage, 1986; Dennis *et al.*, 1998). Additionally, *B. burgdorferi* s.s. has been heavily studied in the northeastern US due to human Lyme disease incidence rates reported in these states. The seasonal activity of the life-stages of *I. scapularis* ticks makes for very efficient transmission of *B. burgdorferi* s.s. from infected ticks to hosts to naïve ticks (Wilson and Spielman, 1985; Sonenshine, 1992; Daniels *et al.*, 1996). Larval ticks, which hatch uninfected from eggs (Piesman *et al.*, 1986; Lane and Burgdorfer, 1987; Magnarelli *et al.*, 1987; Mather *et al.*, 1991; Patrican, 1997), actively seek a blood-meal in late summer when they can acquire *B. burgdorferi* before overwintering and moulting to the nymphal stage. The nymphs are active in early summer, infecting vertebrates such that the maximum proportion of animals will be infected during the following larval activity season (Sonenshine, 1992; Goodwin *et al.*, 2001). The prevalence of *B. burgdorferi* in ticks in the western and southeastern USA is much lower because of changes in tick feeding and host species. In these regions, the ticks feed throughout the year, not seasonally, thus lowering the probability a larvae will feed on an infected vertebrate (Oliver, 1996; Durden *et al.*, 2002). Additionally, the host species commonly parasitized by ticks in the southeastern and western US are often poor reservoirs for *B. burgdorferi* (Oliver *et al.*, 1993; Oliver, 1996; Slowik and Lane, 2001; Durden *et al.*, 2002; Eisen *et al.*, 2004). In fact, some species, such as the western fence lizard (*Sceloporus occidentalis*) or alligator lizard (*Elgaria multicarinata*), kill *Borrelia* in ticks that feed upon them (Lane and Loye, 1989; Lane, 1990; Lane and Quistad, 1998; Wright *et al.*, 1998; Lane *et al.*, 2006). These alternative tick cycles act to reduce the prevalence of *Borrelia* by deflecting infectious tick bites from species that amplify and transmit the bacteria well to species that rarely or never transfer *B. burgdorferi* to ticks (Ostfeld and Keesing, 2000; LoGiudice *et al.*, 2003).

The transmission dynamics and ecology of the northeastern and Midwestern *B. burgdorferi* s.s. populations are complicated by the extreme generalist behaviour of both *B. burgdorferi* and *I. scapularis*. *I. scapularis* subadults feed on at least 31 species of mammals from seven orders and 49 species of birds from 17 families (Anderson, 1988; Sonenshine, 1992). *Borrelia burgdorferi* s.s. have been found in more than 20 vertebrate species including several species of rodents, insectivores, and birds (Callister *et al.*, 1988; Battaly and Fish, 1993; Brown and Lane, 1994; Stafford *et al.*, 1995; Oliver, 1996; LoGiudice *et al.*, 2003; Giery and Ostfeld, 2007). The white-footed mouse (*Peromyscus leucopus*) has long been posited to be the primary reservoir species maintaining *B. burgdorferi* s.s. in nature (Schwan *et al.*, 1989; Mather and Mather, 1990; Rand *et al.*, 1993; Porco, 1999; Ostfeld *et al.*, 2001; Derdakova *et al.*, 2004; Anderson and Norris, 2006). This tenet has recently been questioned. Recent evidence suggests that four mammalian species including two rodents – white-footed mice and eastern chipmunks (*T. striatus*) – and two insectivorous shrews – short-tailed shrews (*Blarina brevicauda*) and masked shrews (*Sorex cinereus*) – provide the blood meal and *B. burgdorferi* s.s. to at least 75% of the infected tick population in one hyper-endemic locale in the northeastern US (Brisson and Dykhuizen, 2006; Brisson *et al.*, 2008). The white-footed mouse is an important reservoir species of *B. burgdorferi* s.s., infecting about one-quarter of the infected nymphs, but is neither the primary nor principal host for *B. burgdorferi* s.s. Birds, especially the ground foraging robin (*Turdus migratorius*), may be significant reservoirs for *B. burgdorferi* s.s. responsible for infecting up to 20% of the infected nymph population.

9.6.4 Effects of within-species diversity

The *B. burgdorferi* s.s. transmission cycle is further complicated by genotypic diversity within the genospecies (Guttman *et al.*, 1996; Liveris *et al.*, 1996; Wang *et al.*, 1999b). Lineages of *B. burgdorferi* s.s., differentiated by the allele at the *ospC* locus, are found predominantly in overlapping but distinct sets of vertebrate species (Brisson and Dykhuizen, 2004; Hanincova *et al.*, 2006). Thus, host-species specificity in *B. burgdorferi* s.s. *ospC* genotypes may be similar to host-species specificity in genospecies in Europe. However, *B. burgdorferi* s.s. genotypes do not specialize on a related set of vertebrate species, as genospecies do in Europe. The mechanism promoting host specificity may be associated with complement resistance (Stevenson *et al.*, 2002). However, this is unlikely to be correct since *B. burgdorferi* s.s. strain ZS7 is equally resistant to serum derived from 11 of the 14 vertebrate species against which it was tested (Kurtenbach *et al.*, 1998c). A molecular mechanism of OspC leading to host specificity is equally uncertain. *ospC* is in linkage disequilibrium with many other *B. burgdorferi* s.s. loci and no molecular mechanism based on OspC has been validated (Dykhuizen *et al.*, 1993, 2008; Bunikis *et al.*, 2004a).

Similar to other animal species, humans are infected primarily by only a subset of *ospC* genotypes, despite the commonness of many others in ticks in the surrounding habitats (Seinost *et al.*, 1999; Baranton *et al.*, 2001; Dykhuizen *et al.*, 2008). This pattern is not restricted to *B. burgdorferi* s.s. Across the three human infectious *B. burgdorferi* s.l., only 10 of the more than 60 *ospC* genotypes are commonly found in disseminated human infections (Baranton *et al.*, 2001).

9.7 Hard bodies versus soft bodies

While hard versus soft bodies are meaningful at the gym, this difference is not of any evolutionary importance for humans. Not so for ticks and, consequently, not so for *Borrelia*. Soft-bodied argasid ticks feed quickly. Thus, the *Borrelia* that are vectored by soft-bodied ticks must be at high concentration in the vertebrate blood during tick feeding to be passively transferred to the tick. However, *Borrelia* at high concentrations in the blood are under constant duress by the vertebrate immune system. Relapsing fever *Borrelia* have overcome these opposing selection pressures by changing there surface antigens to avoid the immune system and reach high blood concentrations for short periods of time. The periods of high *Borrelia* concentration lead to periods of acute illness. Thus, the life strategy of the tick directly influences the virulence of the pathogen in vertebrates (see Chapters 10 and 11).

Hard-bodied ixodid ticks feed for several days or even weeks. While feeding in leisure, the injected tick saliva proteins prevent blood coagulation and suppress the innate immune system. *Borrelia* that are vectored by hard-bodied ticks sense and migrate towards tick saliva during tick feeding (Motaleb *et al.*, 2005; Bakker *et al.*, 2007). These Lyme *Borrelia* have time to swim through tissue towards feeding ticks. Thus, high blood concentrations of Lyme *Borrelia* are not necessary for transfer to feeding ticks. Hard-bodied ticks in northern climes often have seasonal activity patterns such that the different tick life stages feed in different seasons. As a result, *Borrelia* must establish chronic infections in the vertebrate host that must avoid the immune system. This can be accomplished by remaining at low densities, spread out in the infected tissue. The putative quorum sensing system in *Borrelia* (Stevenson *et al.*, 2003) may serve to spread out bacteria, rather than help them congregate, thus, achieving a density too low at any one spot for the immune system to recognize the clone. The classifications of acute and chronic are not external categories that bacterial infections are fixed in, but rather represent life history strategies that can be changed during evolution as the ecology changes.

9.8 Future directions

We have proposed a number of ideas throughout this chapter that could be followed up. These include the following:

1. Pathogens like *B. burgdorferi* will expand rapidly in numbers and range during favourable periods and rapidly contract to small isolated populations during unfavourable periods. The small isolated populations will differentiate, forming separate species. This pattern will lead to a species tree that is like a bush. This cycle can happen repeatedly giving a nested pattern of bursts of speciation.
2. The within species variation at OspC is the cause of the host species specificity. Thus, determining the molecular mechanism of OspC action is important and could lead to practical results.
3. The host generality of the vector and the host specificity of the *Borrelia* imply that there is a strong diluting effect inhibiting the growth of *Borrelia* populations. This could be used to cause extinction of particularly pathogenic clones or species by targeted vaccination.

Finally we suggest that evolutionary biology and population genetics should be incorporated as a regular part of training and research in infectious disease biology. These fields can complement the molecular biology, biochemistry and cell biology that are the main focus of infectious disease research and help create directions of research that will lead to improved treatment and prevention of infectious diseases.

Acknowledgements

This work was partially funded by Public Health Service grant GM060731 to DED. We thank Nicholas Friedenberg for a careful editing of the manuscript.

References

Akins, D.R., Caimano, M.J., Yang, X., Cerna, F., Norgard, M.V., and Radolf, J.D. (1999). Molecular and evolutionary analysis of *Borrelia burgdorferi* 297 circular plasmid-encoded lipoproteins with OspE- and OspF-like leader peptides. Infect. Immun. *67*, 1526–1532.

Alghaferi, M., Anderson, J.M., Park, J., Auwaerter, P.G., Aucott, J.N., Norris, D.E., and Dumler, J.S. (2005). *Borrelia burgdorferi ospC* heterogeneity among human and murine isolates from a defined region of northern Maryland and southern Pennsylvania: Lack of correlation with invasive and noninvasive genotypes. J. Clin. Microbiol. *43*, 1879–1884.

Alitalo, A., Meri, T., Lankinen, H., Seppala, I., Lahdenna, P., Hefty, P.S., Akins, D., and Meri, S. (2002). Complement inhibitor factor H binding to Lyme disease spirochetes is mediated by inductable expression of multiple plasmid-encoded outer surface protein E paralogs. J. Immunol. *169*, 3847–3853.

Amore, G., Tomassone, L., Grego, E., Ragagli, C., Bertolotti, L., Nebbia, P., Rosati, S., and Mannelli, A. (2007). *Borrelia lusitaniae* in immature *Ixodes ricinus* (Acari: Ixodidae) feeding on common wall lizards in Tuscany, central Italy. J. Med. Entomol. *44*, 303–307.

Anderson, J.F. (1988). Mammalian and avian reservoirs for *Borrelia burgdorferi*. Ann. N.Y. Acad. Sci. *539*, 180–191.

Anderson, J.F. (1989a). Ecology of Lyme disease. Conn. Med. *53*, 343–346.

Anderson, J.F. (1989b). Epizootiology of *Borrelia* in *Ixodes* tick vectors and reservoir hosts. Rev. Infect. Dis. *11 Suppl. 6*, S1451–1459.

Anderson, J.F., Flavell, R.A., Magnarelli, L.A., Barthold, S.W., Kantor, F.S., Wallich, R., Persing, D.H., Mathiesen, D., and Fikrig, E. (1996). Novel *Borrelia burgdorferi* isolates from *Ixodes scapularis* and *Ixodes dentatus* ticks feeding on humans. J. Clin. Microbiol. *34*, 524–529.

Anderson, J.M., and Norris, D.E. (2006). Genetic diversity of *Borrelia burgdorferi* sensu stricto in *Peromyscus leucopus*, the primary reservoir of Lyme disease in a region of endemicity in southern Maryland. Appl. Environ. Microbiol. 72, 5331–5341.

Antonara, S., Chafel, R.M., LaFrance, M., and Coburn, J. (2007). *Borrelia burgdorferi* adhesins identified using *in vivo* phage display. Mol. Microbiol. *66*, 262–276.

Attie, O., Bruno, J.F., Xu, Y., Qiu, D., Luft, B.J., and Qiu, W.G. (2007). Co-evolution of the outer surface protein C gene (*ospC*) and intraspecific lineages of *Borrelia burgdorferi* sensu stricto in the north-eastern United States. Infect. Genet. Evol. 7, 1–12.

Avise, J.C. (1992). Molecular population-structure and the biogeographic history of a regional fauna – a case-history with lessons for conservation biology. Oikos 63, 62–76.

Avise, J.C. (2000). Phylogeography: The History and Formation of Species (Cambridge, MA: Harvard University Press).

Avise, J.C., Arnold, J., Ball, R.M., Bermingham, E., Lamb, T., Neigel, J.E., Reeb, C.A., and Saunders, N.C. (1987). Intraspecific phylogeography – the mitochondrial-DNA bridge between population-genetics and systematics. Ann. Rev. Ecol. Syst. *18*, 489–522.

Bakker, R.G., Li, C., Miller, M.R., Cunningham, C., and Charon, N.W. (2007). Identification of specific chemoattractants and genetic complementation of a *Borrelia burgdorferi* chemotaxis mutant: flow cytometry-based capillary tube chemotaxis assay. Appl. Environ. Microbiol. 73, 1180–1188.

Baranton, G., Marti Ras, N., and Postic, D. (1998). Molecular epidemiology of the aetiological agents of Lyme borreliosis. Wien. Klin. Wochenschr. *110*, 850–855.

Baranton, G., Postic, D., Saint Girons, I., Boerlin, P., Piffaretti, J.C., Assous, M., and Grimont, P.A. (1992). Delineation of *Borrelia burgdorferi* sensu stricto, *Borrelia garinii* sp. nov., and group VS461 associated with Lyme borreliosis. Int. J. Syst. Bacteriol. *42*, 378–383.

Baranton, G., Seinost, G., Theodore, G., Postic, D., and Dykhuizen, D. (2001). Distinct levels of genetic diversity of *Borrelia burgdorferi* are associated with different aspects of pathogenicity. Res. Microbiol. *152*, 149–156.

Barbour, A.G., and Garon, C.F. (1987). Linear plasmids of the bacterium *Borrelia burgdorferi* have covalently closed ends. Science 237, 409–411.

Barthold, S.W. (1999). Specificity of infection-induced immunity among *Borrelia burgdorferi* sensu lato species. Infect. Immun. *67*, 36–42.

Battaly, G.R., and Fish, D. (1993). Relative importance of bird species as hosts for immature *Ixodes dammini* (Acari: Ixodidae) in a suburban residential landscape of southern New York State. J. Med. Entomol. *30*, 740–747.

Brisson, D., and Dykhuizen, D.E. (2004). *ospC* diversity in *Borrelia burgdorferi*: different hosts are different niches. Genetics *168*, 713–722.

Brisson, D., and Dykhuizen, D.E. (2006). A modest model explains the distribution and abundance of *Borrelia burgdorferi* strains. Am. J. Trop. Med. Hyg. *74*, 615–622.

Brisson, D., Dykhuizen, D.E., and Ostfeld, R.S. (2008). Conspicuous impacts of inconspicuous hosts on the Lyme disease epidemic. Proc. R. Soc. B *275*, 227–235.

Brown, R.N., and Lane, R.S. (1992). Lyme disease in California: a novel enzootic transmission cycle of *Borrelia burgdorferi*. Science *256*, 1439–1442.

Brown, R.N., and Lane, R.S. (1994). Natural and experimental *Borrelia burgdorferi* infections in woodrats and deer mice from California. J. Wildl. Dis. *30*, 389–398.

Brown, R.N., Peot, M.A., and Lane, R.S. (2006). Sylvatic maintenance of *Borrelia burgdorferi* (Spirochaetales) in Northern California: untangling the web of transmission. J. Med. Entomol. *43*, 743–751.

Bunikis, J., Garpmo, U., Tsao, J., Berglund, J., Fish, D., and Barbour, A.G. (2004a). Sequence typing reveals extensive strain diversity of the Lyme borreliosis agents *Borrelia burgdorferi* in North America and *Borrelia afzelii* in Europe. Microbiology *150*, 1741–1755.

Bunikis, J., Tsao, J., Garpmo, U., Bergland, J., Fish, D., and Barbour, A.G. (2004b). Typing of *Borrelia* relapsing fever group strains. Emerging. Infec. Dis. *10*, 1661–1664.

Burgdorfer, W., Barbour, A.G., Hayes, S.F., Benach, J.L., Grunwaldt, E., and Davis, J.P. (1982). Lyme disease – a tick-borne spirochetosis. Science *216*, 1317–1319.

Burgdorfer, W., and Gage, K.L. (1986). Susceptibility of the black-legged tick, *Ixodes scapularis*, to the Lyme disease spirochete, *Borrelia burgdorferi*. Zentralbl. Bakteriol. Mikrobiol. Hyg. A *263*, 15–20.

Burkot, T.R., Schneider, B.S., Pieniazek, N.J., Happ, C.M., Rutherford, J.S., Slemenda, S.B., Hoffmeister, E., Maupin, G.O., and Zeidner, N.S. (2000). *Babesia microti* and *Borrelia bissettii* transmission by *Ixodes spinipalpis* ticks among prairie voles, *Microtus ochrogaster*, in Colorado. Parasitology *121*, 595–599.

Busch, U., Hizo-Teufel, C., Boehmer, R., Fingerle, V., Nitschko, H., Wilske, B., and Preac-Mursic, V. (1996). Three species of *Borrelia burgdorferi* sensu lato (*B. burgdorferi* sensu stricto, *B. afzelii*, and *B. garinii*) identified from cerebrospinal fluid isolates by pulsed-field gel electrophoresis and PCR. J. Clinic. Microbiol. *34*, 1072–1078.

Caimano, M.J., Eggers, C.H., Gonzalez, C., and Radolf, J.D. (2005). Alternate sigma factor, RpoS, is required for the in vivo-specific repression of the *Borrelia burgdorferi* lp54-borne *ospA* and *lp6.6* genes. J. Bacteriol. *187*, 7845–7852.

Caimano, M.J., Eggers, C.H., Hazlett, K.R., and Radolf, J.D. (2004). RpoS is not central to the general stress response in *Borrelia burgdorferi*, but does control expression of one or more virulence determinants. Infect. Immun. 72, 6433–6445.

Caimano, M.J., Yang, X., Popova, T.G., Clawson, M.L., Akins, D.R., Norgard, M.V., and Radolf, J.D. (2000) Molecular and evolutionary characterization of the cp32/18 family of supercoiled plasmids in *Borrelia burgdorferi* 297. Infect. Immun. *68*, 1574–1586.

Callister, S.M., Agger, W.A., Schell, R.F., and Ellingson, J.L. (1988). *Borrelia burgdorferi* infection surrounding La Crosse. Wis. J. Clinic. Microbiol. *26*, 2632–2636.

Canica, M.M., Nato, F., du Merle, L., Mazie, J.C., Baranton, G., and Postic, D. (1993). Monoclonal antibodies for identification of *Borrelia afzelii* sp. nov. associated with late cutaneous manifestations of Lyme borreliosis. Scand. J. Infect. Dis. 25, 441–448.

Cohan, F.M. (2002). What are bacterial species? Ann. Rev. Microbiol. 56, 457–487.

Coleman, A.S., Yang, X., Kumar, M., Zhang, X., Promnares, K., Shroder, D., Kenedy, M.R. Anderson, J.F., and Pal. U. (2008). *Borrelia burgdorferi* complement regulator-acquiring surface protein 2 does not contribute to complement resistance or host infectivity. PLoS ONE 3, 3010e.

Crow, J.F., and Kimura M. (1970). An Introduction to Population Genetics Theory (New York: Harper & Row).

Daniels, T.J., Falco, R.C., Curran, K.L., and Fish, D. (1996). Timing of *Ixodes scapularis* (Acari: Ixodidae) oviposition and larval activity in southern New York. J. Med. Entomol. 33, 140–147.

De Michelis, S., Sewell, H.S., Collares-Pereira, M., Santos-Reis, M., Schouls, L.M., Benes, V., Holmes, E.C., and Kurtenbach, K. (2000). Genetic diversity of *Borrelia burgdorferi* sensu lato in ticks from mainland Portugal. J. Clinic. Microbiol. 38, 2128–2133.

DeNatale, C.E., Burkot, T.R., Schneider, B.S., and Zeidner, N.S. (2002). Novel potential reservoirs for *Borrelia* sp. and the agent of human granulocytic ehrlichiosis in Colorado. J. Wildl. Dis. 38, 478–482.

Dennis, D.T., Nekomoto, T.S., Victor, J.C., Paul, W.S., and Piesman, J. (1998). Reported distribution of *Ixodes scapularis* and *Ixodes pacificus* (Acari: Ixodidae) in the United States. J. Med. Entomol. 35, 629–638.

Derdakova, M., Dudioak, V., Brei, B., Brownstein, J.S., Schwartz, I., and Fish, D. (2004). Interaction and transmission of two *Borrelia burgdorferi* sensu stricto strains in a tick-rodent maintenance system. Appl. Environ. Microbiol. 70, 6783–6788.

Derdakova, M., and Lencakova, D. (2005). Association of genetic variability within the *Borrelia burgdorferi* sensu lato with the ecology, epidemiology of Lyme borreliosis in Europe. Ann. Agric. Environ. Med. 12, 165–172.

Dolan, M.C., Lacombe, E.H., and Piesman, J. (2000). Vector competence of *Ixodes muris* (Acari: Ixodidae) for *Borrelia burgdorferi*. J. Med. Entomol. 37, 766–768.

Dolan, M.C., Piesman, J., Mbow, M.L., Maupin, G.O., Peter, O., Brossard, M., and Golde, W.T. (1998). Vector competence of *Ixodes scapularis* and *Ixodes ricinus* (Acari: Ixodidae) for three genospecies of *Borrelia burgdorferi*. J. Med. Entomol. 35, 465–470.

Donahue, J.G., Piesman, J., and Spielman, A. (1987). Reservoir competence of white-footed mice for Lyme disease spirochetes. Am. J. Trop. Med. Hyg. 36, 92–96.

Duneau, D., Boulinier, T., Gomez-Diaz, E., Petersen, A., Tveraa, T., Barrett, R.T., and McCoy, K.D. (2008). Prevalence and diversity of Lyme borreliosis bacteria in marine birds. Infect. Genet. Evol, 8, 352–359.

Durden, L.A., Oliver, J.H., Jr., Banks, C.W., and Vogel, G.N. (2002). Parasitism of lizards by immature stages of the blacklegged tick, *Ixodes scapularis* (Acari, Ixodidae). Exp. Appl. Acarol. 26, 257–266.

Dykhuizen, D.E., and Baranton, G. (2001). The implications of a low rate of horizontal transfer in *Borrelia*. Trends Microbiol. 9, 344–350.

Dykhuizen, D.E., Brisson, D., Sandigursky, S., Wormser, G.P., Nowakowski, J., Nadelman, R., and Schwartz, I. (2008). The propensity of different *Borrelia burgdorferi* sensu stricto genotypes to cause disseminated infections in humans. Am. J. Trop. Med. Hyg. 78, 806–810.

Dykhuizen, D.E., Polin, D.S., Dunn, J.J., Wilske, B., Preacmursic, V., Dattwyler, R.J., and Luft, B.J. (1993). *Borrelia burgdorferi* is clonal – Implications for taxonomy and vaccine development. Proc. Natl. Acad. Sci. U.S.A. 90, 10163–10167.

Earnhart, C.G., Buckles, E.L., Dumler, J.S., and Marconi, R.T. (2005). Demonstration of OspC type diversity in invasive human Lyme disease isolates and identification of previously uncharacterized epitopes that define the specificity of the OspC murine antibody response. Infect. Immun. 73, 7869–7877.

Eggers, C.H., Caimano, M.J., and Radolf, J.D. (2004). Analysis of promoter elements involved in the transcriptional initiation of RpoS-dependent *Borrelia burgdorferi* genes. J. Bacteriol. 186, 7390–7402.

Eggers, C.H., Caimano, M.J., and Radolf, J.D. (2006). Sigma factor selectivity in *Borrelia burgdorferi*: RpoS recognition of the *ospE/ospF/elp* promoters is dependent on the sequence of the −10 region. Mol Microbiol. 59, 1859–1875.

Eggers, C.H., Kimmel, B.J., Bono, J.L., Elias, A.F., Rosa, P., and Samuels, D.S. (2001). Transduction by φBB-1, a bacteriophage of *Borrelia burgdorferi*. J. Bacteriol. 183, 4771–4778.

Eggers C.H., and Samuels, S.D. (1999). Molecular evidence for a new bacteriophage of *Borrelia burgdorferi*. J. Bacteriol. 181, 7308–7313.

Eicken, C., Sharma, V., Klabunde, T., Owens, R.T., Pikas, D.S., Hook, M., and Sacchettini, J.C. (2001). Crystal structure of Lyme disease antigen outer surface protein C from *Borrelia burgdorferi*. J. Biol. Chem. 276, 10010–10015.

Eisen, L., Dolan, M.C., Piesman, J., and Lane, R.S. (2003). Vector competence of *Ixodes pacificus* and *I. spinipalpis* (Acari: Ixodidae), and reservoir competence of the dusky-footed woodrat (*Neotoma fuscipes*) and the deer mouse (*Peromyscus maniculatus*), for *Borrelia bissettii*. J. Med. Entomol. 40, 311–320.

Eisen, L., Eisen, R.J., and Lane, R.S. (2004). The roles of birds, lizards, and rodents as hosts for the western black-legged tick *Ixodes pacificus*. J. Vector Ecol. 29, 295–308.

Enright, M.C., and Spratt, B.G. (1999). Multilocus sequence typing. Trends Microbiol. 7, 482–487.

Ewens, W.J. (1972). The sampling theory of selectively neutral alleles. Theoret. Pop. Biol. 3, 87–112.

Feil, E.J., and Enright, M.C. (2004). Analyses of clonality and the evolution of bacterial pathogens. Curr. Opin. Microbiol. 7, 308–313.

Fingerle, V., Rauser, S., Hammer, B., Kahl, O., Heimerl, C., Schulte-Spechtel, U., Gern, L., and Wilske, B. (2002). Dynamics of dissemination and outer surface protein expression of different European *Borrelia burgdorferi* sensu lato strains in artificially infected *Ixodes ricinus* nymphs. J. Clinic Microbiol. 40, 1456–1463.

Foretz, M., Postic, D., and Baranton, G. (1997). Phylogenetic analysis of *Borrelia burgdorferi* sensu stricto by arbitrarily primed PCR and pulsed-field gel electrophoresis. Internat. J. Sys. Bacteriol. *47*, 11–18.

Fraser, C.M., Casjens, S., Huang, W.M., Sutton, G.G., Clayton, R., Lathigra, R., White, O., Ketchum, K.A., Dodson, R., Hickey, E.K., *et al.* (1997). Genomic sequence of a Lyme disease spirochaete, *Borrelia burgdorferi*. Nature *390*, 580–586.

Fukunaga, M., Hamase, A., Okada, K., Inoue, H., Tsuruta, Y., Miyamoto, K., and Nakao, M. (1996). Characterization of spirochetes isolated from ticks (*Ixodes tanuki, Ixodes turdus,* and *Ixodes columnae*) and comparison of the sequences with those of *Borrelia burgdorferi* sensu lato strains. Appl. Environ. Microbiol. *62*, 2338–2344.

Fukunaga, M., and Koreki, Y. (1996). A phylogenetic analysis of *Borrelia burgdorferi* sensu lato isolates associated with Lyme disease in Japan by flagellin gene sequence determination. Int. J. Syst. Bacteriol. *46*, 416–421.

Fung, B.P., McHugh, G.L., Leong, J.M., and Steere, A.C. (1994). Humoral immune response to outer surface protein C of *Borrelia burgdorferi* in Lyme disease: role of the immunoglobulin M response in the serodiagnosis of early infection. Infect. Immun. *62*, 3213–3221.

Gern, L., Estrada-Pena, A., Frandsen, F., Gray, J.S., Jaenson, T.G., Jongejan, F., Kahl, O., Korenberg, E., Mehl, R., and Nuttall, P.A. (1998). European reservoir hosts of *Borrelia burgdorferi* sensu lato. Zentralbl. Bakteriol. *287*, 196–204.

Gern, L., and Humair, P.F. (2002). Ecology of *Borrelia burgdorferi* sensu lato in Europe. Lyme Borreliosis: Biology, Epidemiology, and Control (Wallingford, UK: CABI Publishing).

Gern, L., Toutoungi, L.N., Hu, C.M., and Aeschlimann, A. (1991). *Ixodes (Pholeoixodes) hexagonus,* an efficient vector of *Borrelia burgdorferi* in the laboratory. Med. Vet. Entomol. *5*, 431–435.

Giery, S.T., and Ostfeld, R.S. (2007). The role of lizards in the ecology of Lyme disease in two endemic zones of the north-eastern United States. J. Parasitol. *93*, 511–517.

Gilmore, R.D., Kappel, K.J., Dolan, M.C., Burkot, T.R., and Johnson, B.J.B. (1996). Outer surface protein C (OspC), but not P39, is a protective immunogen against a tick-transmitted *Borrelia burgdorferi* challenge: Evidence for a conformational protective epitope in OspC. Infect. Immun. *64*, 2234–2239.

Gilmore, R.D., Mbow, M.L., and Stevenson, B. (2001). Analysis of *Borrelia burgdorferi* gene expression during life cycle phases of the tick vector *Ixodes scapularis*. Microb. Infect. *3*, 799–808.

Godfroid, E., Min Hu, C., Humair, P.F., Bollen, A., and Gern, L. (2003). PCR-reverse line blot typing method underscores the genomic heterogeneity of *Borrelia valaisiana* species and suggests its potential involvement in Lyme disease. J. Clin. Microbiol. *41*, 3690–3698.

Goodwin, B.J., Ostfeld, R.S., and Schauber, E.M. (2001). Spatiotemporal variation in a Lyme disease host and vector: Black-legged ticks on white-footed mice. Vector Borne and Zoonotic Diseases *1*, 129–138.

Grimm, D., Tilly, K., Bryam, R., Stewart, P.E., Krum, J.G., Bueschel, D.M., Schwan, T.G., Policastro, P.F., Elias, A.F., and Rosa, P.A. (2004) Outer-surface protein C of the Lyme disease spirochete: A protein induced in ticks for infection in mammals. Proc. Natl. Acad. Sci. U.S.A. *101*, 3142–3147.

Gromko, M.H. (1977). What is frequency-dependent selection. Evolution *31*, 438–442.

Guttman, D.S., Wang, P.W., Wang, I.N., Bosler, E.M., Luft, B.J., and Dykhuizen, D.E. (1996). Multiple infections of *Ixodes scapularis* ticks by *Borrelia burgdorferi* as revealed by single-strand conformation polymorphism analysis. J. Clinic. Microbiol. *34*, 652–656.

Gylfe, A., Yabuki, M., Drotz, M., Bergström, S., Fukunaga, M., and Olsen, B. (2001). Phylogeographic relationships of *Ixodes uriae* (Acari: Ixodidae) and their significance to transequatorial dispersal of *Borrelia garinii*. Hereditas *134*, 195–199.

Hartl, D.L., and Clark A.G. (2007). Principles of Population Genetics (Sunderland, MA: Sinauer Associates).

Hartmann, K., Corvey, C., Skerka, C., Kirschfink, M., Karas, M., Brade, A., Miller, J.C., Stevenson, B., Wallich, R., Zipfel, P.F., and Kraiczy, P. (2006). Functional characterization of BbCRASP-2, a distinct outer membrane protein of *Borrelia burgdorferi* that binds host complement regulators factor H and FHL-1. Mol. Microbiol. *61*, 1220–1236.

Hanincova, K., Kurtenbach, K., Diuk-Wasser, M., Brei, B., and Fish, D. (2006). Epidemic spread of Lyme borreliosis, north-eastern United States. Emerg. Infect. Dis. *12*, 604–611.

Harpending, H.C., Batzer, M.A., Gurven, M., Jorde, L.B., Rogers, A.R., and Sherry, S.T. (1998). Genetic traces of ancient demography. Proc. Natl. Acad. Sci. U.S.A. *95*, 1961–1967.

Haupt, K., Kraiczy, P., Wallich, R., Brade, V., Skerka, C., and Zipfel, P.F. (2007). Binding of human factor H-related protein 1 to serum-resistant *Borrelia burgdorferi* is mediated by borrelial complement regulator-acquiring surface proteins. J. Infect. Dis. *196*, 124–133.

Hellwage, J., Meri, T., Heikkila, T., Alitalo, A., Panelius, J., Lahdenne, P., Seppala, I.J., and Meri, S. (2001). The complement regulator factor H binds to the surface protein OspE of *Borrelia burgdorferi*. J. Biol. Chem. *276*, 8427–8435.

Hodzic, E., Feng, S., Freet, K.J., Borjesson, D.L., and Barthold, S.W. (2002). *Borrelia burgdorferi* population kinetics and selected gene expression at the host–vector interface. Infect. Immun. *70*, 3382–3388.

Hodzic, E., Feng, S.L., and Barthold, S.W. (2000). Stability of *Borrelia burgdorferi* outer surface protein C under immune selection pressure. J. Infect. Dis. *181*, 750–753.

Hovis, K.M., Tran, E., Sundy, C.M., Buckles, E., McDowell, J.V., and Marconi, R.T. (2006). Selective binding of *Borrelia burgdorferi* OspE paralogs to factor H and serum proteins from diverse animals: possible expansion of the role of OspE in Lyme disease pathogenesis. Infect. Immun. *74*, 1967–1972.

Hu, C.M., Wilske, B., Fingerle, V., Lobet, Y., and Gern, L. (2001). Transmission of *Borrelia garinii* OspA serotype 4 to BALB/c mice by *Ixodes ricinus* ticks collected in the field. J. Clin. Microbiol. *39*, 1169–1171.

Hubalek, Z., and Halouzka, J. (1997). Distribution of *Borrelia burgdorferi* sensu lato genomic groups in Europe, a review. Eur. J. Epidemiol. *13*, 951–957.

Huegli, D., Hu, C.M., Humair, P.F., Wilske, B., and Gern, L. (2002). Apodemus species mice are reservoir hosts of *Borrelia garinii* OspA serotype 4 in Switzerland. J. Clin. Microbiol. *40*, 4735–4737.

Humair, P.F., and Gern, L. (1998). Relationship between *Borrelia burgdorferi* sensu lato species, red squirrels (*Sciurus vulgaris*) and *Ixodes ricinus* in enzootic areas in Switzerland. Acta Trop. *69*, 213–227.

Humair, P.F., Peter, O., Wallich, R., and Gern, L. (1995). Strain variation of Lyme disease spirochetes isolated from *Ixodes ricinus* ticks and rodents collected in two endemic areas in Switzerland. J. Med. Entomol. *32*, 433–438.

Humair, P.F., Rais, O., and Gern, L. (1999). Transmission of *Borrelia afzelii* from Apodemus mice and Clethrionomys voles to *Ixodes ricinus* ticks: differential transmission pattern and overwintering maintenance. Parasitology *118* (pt. *1*), 33–42.

Huxley, J. (1942). Evolution: The Modern Synthesis (London: Allen & Unwin).

James, A.M., and Oliver, J.H., Jr. (1990). Feeding and host preference of immature *Ixodes dammini, I. scapularis,* and *I. pacificus* (Acari: Ixodidae). J. Med. Entomol. *27*, 324–330.

Jauris-Heipke, S., Fuchs, R., Motz, M., Preacmursic, V., Schwab, E., Soutschek, E., Will, G., and Wilske, B. (1993). Genetic heterogenity of the genes coding for the outer surface protein C (*ospC*) and the flagellin of *Borrelia burgdorferi*. Med. Microbiol. Immunol. *182*, 37–50.

Johnson, R.C., Hyde, F.W., and Rumpel, C.M. (1984). Taxonomy of the Lyme disease spirochetes. Yale J. Biol. Med. *57*, 529–537.

Kawabata, H., Masuzawa, T., and Yanagihara, Y. (1993). Genomic analysis of *Borrelia japonica* sp. nov. isolated from *Ixodes ovatus* in Japan. Microbiol. Immunol. *37*, 843–848.

Kipp, S., Goedecke, A., Dorn, W., Wilske, B., and Fingerle, V. (2006). Role of birds in Thuringia, Germany, in the natural cycle of *Borrelia burgdorferi* sensu lato, the Lyme disease spirochaete. Int. J. Med. Microbiol. *296 Suppl. 40*, 125–128.

Korenberg, E., Gorelova, N., and Kovalevskii, Y. (2002). Ecology of *Borrelia burgdorferi* sensu lato in Russia. In Lyme Borreliosis: Biology, Epidemiology and Control, J.S. Gray, O. Kahl,, R.S.Lane, and G. Stanek, ed., (Wallingford, Oxfordshire, UK: CABI Publishing). pp. 175–200.

Kraiczy, P., Hellwage, J., Skerka, C., Kirschfink, M., Brade, V., Zipfel, P.F., and Wallich, R. (2003). Immune evasion of *Borrelia burgdorferi*: mapping of a complement-inhibitor factor H-binding site of BbCRASP-3, a novel member of the Erp protein family. Eur. J. Immunol. *33*, 697–707.

Kumaran, D., Eswaramoorthy, S., Luft, B.J., Koide, S., Dunn, J.J., Lawson, C.L., and Swaminathan, S. (2001). Crystal structure of outer surface protein C (OspC) from the Lyme disease spirochete, *Borrelia burgdorferi*. EMBO J. *20*, 971–978.

Kuo, M.M., Lane, R.S., and Giclas, P.C. (2000). A comparative study of mammalian and reptilian alternative pathway of complement-mediated killing of the Lyme disease spirochete (*Borrelia burgdorferi*). J. Parasitol. *86*, 1223–1228.

Kurtenbach, K., Carey, D., Hoodless, A.N., Nuttall, P.A., and Randolph, S.E. (1998a). Competence of pheasants as reservoirs for Lyme disease spirochetes. J. Med. Entomol. *35*, 77–81.

Kurtenbach, K., De Michelis, S., Etti, S., Schafer, S.M., Sewell, H.S., Brade, V., and Kraiczy, P. (2002). Host association of *Borrelia burgdorferi* sensu lato – the key role of host complement. Trends Microbiol. *10*, 74–79.

Kurtenbach, K., Hanincova, K., Tsao, J.I., Margos, G., Fish, D., and Ogden, N.H. (2006). Fundamental processes in the evolutionary ecology of Lyme borreliosis. Nat. Rev. Microbiol. *4*, 660–669.

Kurtenbach, K., Peacey, M., Rijpkema, S.G., Hoodless, A.N., Nuttall, P.A., and Randolph, S.E. (1998b). Differential transmission of the genospecies of *Borrelia burgdorferi* sensu lato by game birds and small rodents in England. Appl. Environ. Microbiol. *64*, 1169–1174.

Kurtenbach, K., Sewell, H.S., Ogden, N.H., Randolph, S.E., and Nuttall, P.A. (1998c). Serum complement sensitivity as a key factor in Lyme disease ecology. Infect. Immun. *66*, 1248–1251.

Lagal, V., Postic, D., Ruzic-Sabljic, E., and Baranton, G. (2003). Genetic diversity among *Borrelia* strains determined by single-strand conformation polymorphism analysis of the *ospC* gene and its association with invasiveness. J. Clin. Microbiol. *41*, 5059–5065.

Lane, R.S. (1990). Susceptibility of the western fence lizard (*Sceloporus occidentalis*) to the Lyme borreliosis spirochete (*Borrelia burgdorferi*). Am. J. Trop. Med. Hyg. *42*, 75–82.

Lane, R.S., Brown, R.N., Piesman, J., and Peavey, C.A. (1994). Vector competence of *Ixodes pacificus* and *Dermacentor occidentalis* (Acari: Ixodidae) for various isolates of Lyme disease spirochetes. J. Med. Entomol. *31*, 417–424.

Lane, R.S., and Burgdorfer, W. (1987). Transovarial and transstadial passage of *Borrelia burgdorferi* in the western black-legged tick, *Ixodes pacificus* (Acari: Ixodidae). Am. J. Trop. Med. Hyg. *37*, 188–192.

Lane, R.S., and Loye, J.E. (1989). Lyme disease in California: interrelationship of *Ixodes pacificus* (Acari: Ixodidae), the western fence lizard (*Sceloporus occidentalis*), and *Borrelia burgdorferi*. J. Med. Entomol. *26*, 272–278.

Lane, R.S., Mun, J., Eisen, L., and Eisen, R.J. (2006). Refractoriness of the western fence lizard (*Sceloporus occidentalis*) to the Lyme disease group spirochete *Borrelia bissettii*. J. Parasitol. *92*, 691–696.

Lane, R.S., Mun, J., Eisen, R.J., and Eisen, L. (2005). Western gray squirrel (Rodentia: Sciuridae): a primary reservoir host of *Borrelia burgdorferi* in Californian oak woodlands? J. Med. Entomol. *42*, 388–396.

Lane, R.S., Peavey, C.A., Padgett, K.A., and Hendson, M. (1999). Life history of *Ixodes (Ixodes) jellisoni* (Acari: Ixodidae) and its vector competence for *Borrelia burgdorferi* sensu lato. J. Med. Entomol. *36*, 329–340.

Lane, R.S., and Quistad, G.B. (1998). Borreliacidal factor in the blood of the western fence lizard (*Sceloporus occidentalis*). J. Parasitol. *84*, 29–34.

Le Fleche, A., Postic, D., Girardet, K., Peter, O., and Baranton, G. (1997). Characterization of *Borrelia lusitaniae* sp. nov. by 16S ribosomal DNA sequence analysis. Int. J. Syst. Bacteriol. *47*, 921–925.

Liang, F.T., Jacobs, M.B., Bowers, L.C., and Philipp, M.T. (2002). An immune evasion mechanism for spirochetal persistence in Lyme borreliosis. J. Exper. Med. *195*, 415–422.

Lin, T., Oliver, J.H., Jr. and Gao, L. (2002). Genetic diversity of the outer surface protein C gene of southern *Borrelia* isolates and its possible epidemiological, clinical, and pathogenetic implications. J. Clin. Microbiol. *40*, 2572–2583.

Liveris, D., Varde, S., Iyer, R., Koening, S., Bittker, S., Cooper, D., McKenna, D., Nowakowski, J., Nadleman, R.B., Wormser, G.P., and Schwartz, I. (1999). Genetic diversity of *Borrelia burgdorferi* in Lyme disease patients as determined by culture versus direct PCR with clinical specimens. J. Clin. Microbiol. *37*, 565–569.

Liveris, D., Wormser, G.P., Nowakowski, J., Nadelman, R., Bittker, S., Cooper, D., Varde, S., Moy, F.H., Forseter, G., Pavia, C.S., and Schwartz, I. (1996). Molecular typing of *Borrelia burgdorferi* from Lyme disease patients by PCR-restriction fragment length polymorphism analysis. J. Clin. Microbiol. *34*, 1306–1309.

Livey, I., Gibbs, C.P., Schuster, R., and Dorner, F. (1995). Evidence for lateral transfer and recombination in OspC variation in Lyme disease *Borrelia*. Mol. Microbiol. *18*, 257–269.

LoGiudice, K., Ostfeld, R.S., Schmidt, K.A., and Keesing, F. (2003). The ecology of infectious disease: Effects of host diversity and community composition on Lyme disease risk. Proc. Natl. Acad. Sci. U.S.A. *100*, 567–571.

Lovrich, S.D., Jobe, D.A., Schell, R.F., and Callister, S.M. (2005). Borreliacidal OspC antibodies specific for a highly conserved epitope are immunodominant in human Lyme disease and do not occur in mice or hamsters. Clin. Diagn. Lab. Immunol. *12*, 746–751.

Magnarelli, L.A., Anderson, J.F., and Fish, D. (1987). Transovarial transmission of *Borrelia burgdorferi* in *Ixodes dammini* (Acari:Ixodidae). J. Infect. Dis. *156*, 234–236.

Maiden, M.C., Bygraves, J.A., Feil, E., Morelli, G., Russell, J.E., Urwin, R. Zhang, Q., Zhou, J. Zurth, K., Caugant, D.A., Feavers, I.M., Achtman, M., and Spratt, B.G. (1998). Multilocus sequence typing: a portable approach to the identification of clones within populations of pathogenic microorganisms. Proc. Natl. Acad. Sci. U.S.A. *95*, 3140–3145.

Marconi, R.T., Liveris, D., and Schwartz, I. (1995). Identification of novel insertion elements, restriction fragment length polymorphism patterns, and discontinuous 23S rRNA in Lyme disease spirochetes: phylogenetic analyses of rRNA genes and their intergenic spacers in *Borrelia japonica* sp. nov. and genomic group 21038 (*Borrelia andersonii* sp. nov.) isolates. J. Clin. Microbiol. 33, 2427–2434.

Marconi, R.T., Sung, S.Y., Hughes, C.A.N., and Carlyon, J.A. (1996) Molecular and evolutionary analysis of a variable series of genes in *Borrelia burgdorferi* that are related to *ospE* and *ospF*, consititute a gene family, and share a common upstream homology box. J. Bacteriol. *178*, 5615–5626.

Marconi, R.T., Samuels, D.S., and Garon, C.F. (1993). Transcriptional analyses and mapping of the *ospC* gene in Lyme disease spirochetes. J. Bacteriol. *175*, 926–932.

Margolis, N., Hogan, D., Cieplak, W., Schwan, T.G., and Rosa, P.A. (1994). Homology between *Borrelia burgdorferi* OspC and members of the family of *Borrelia hermsii* variable major proteins. Gene *143*, 105–110.

Margos, G., Gatewood, A.G., Aanensen, D.M., Hanincova, K., Terekhova, D., Vollmer, S.A., Cornet, M., Piesman, J., Donaghy, M., Bormane, A., Hurn, M.A., Feil, E.J., Fish, D., Casjens, S., Wormser, G.P., Schwartz, I., and Kurtenbach, K. (2008). MLST of housekeeping genes captures geographic population structure and suggests a European origin of *Borrelia burgdorferi*. Proc. Natl. Acad. Sci. U.S.A. *105*, 8730–8735.

Marie-Angele, P., Lommano, E., Humair, P.F., Douet, V., Rais, O., Schaad, M., Jenni, L., and Gern, L. (2006). Prevalence of *Borrelia burgdorferi* sensu lato in ticks collected from migratory birds in Switzerland. Appl. Environ. Microbiol. *72*, 976–979.

Marti Ras, N., Postic, D., Foretz, M., and Baranton, G. (1997). *Borrelia burgdorferi* sensu stricto, a bacterial species 'made in the U.S.A.'? Int. J. Syst. Bacteriol. *47*, 1112–1117.

Masuzawa, T. (2004). Terrestrial distribution of the Lyme borreliosis agent *Borrelia burgdorferi* sensu lato in East Asia. Jpn. J. Infect. Dis. *57*, 229–235.

Masuzawa, T., Iwaki, A., Sato, Y., Miyamoto, K., Korenberg, E.I., and Yanagihara, Y. (1997a). Genetic diversity of *Borrelia burgdorferi* sensu lato isolated in far eastern Russia. Microbiol. Immunol. *41*, 595–600.

Masuzawa, T., Komikado, T., and Yanagihara, Y. (1997b). PCR-restriction fragment length polymorphism analysis of the *ospC* gene for detection of mixed culture and for epidemiological typing of *Borrelia burgdorferi* sensu stricto. Clin. Diagn. Lab. Immunol. *4*, 60–63.

Masuzawa, T., Takada, N., Kudeken, M., Fukui, T., Yano, Y., Ishiguro, F., Kawamura, Y., Imai, Y., and Ezaki, T. (2001). *Borrelia sinica* sp. nov., a Lyme disease-related *Borrelia* species isolated in China. Int. J. Syst. Evol. Microbiol. *51*, 1817–1824.

Mather, T.N., and Mather, M.E. (1990). Intrinsic competence of three ixodid ticks (Acari) as vectors of the Lyme disease spirochete. J. Med. Entomol. *27*, 646–650.

Mather, T.N., Telford, S.R., 3rd and Adler, G.H. (1991). Absence of transplacental transmission of Lyme disease spirochetes from reservoir mice (*Peromyscus leucopus*) to their offspring. J. Infect. Dis. *164*, 564–567.

May, G., Shaw, F., Badrane, H., and Vekemans, X. (1999). The signature of balancing selection: Fungal mating compatibility gene evolution. Proc. Natl. Acad. Sci. U.S.A. *96*, 9172–9177.

McDowell, J.V., Wolfgang, J., Tran, E., Metts, M.S., Hamilton, D., and Marconi, R.T. (2003). Compprehensive analysis of the factor H binding capabilities of *Borrelia* species associated with Lyme disease: delineation of two distinct classes of Factor H binding proteins. Infect. Immun. *71*, 3597–3602.

Merila, J., Bjorklund, M., and Baker, A.J. (1996). Genetic population structure and gradual northward decline of genetic variability in the greenfinch (*Carduelis chloris*). Evolution *50*, 2548–2557.

Metts, M.S., McDowell, J.V., Theisen, M., Hansen, P.R., and Marconi, R.T. (2003). Analysis of the OspE determinants involved in binding of factor H and OspE-targeting antibodies elicited during *Borrelia burgdorferi* infection in mice. Infect. Immun. *71*, 3587–3596.

Michalik, J., Skotarczak, B., Skoracki, M., Wodecka, B., Sikora, B., Hofman, T., Rymaszewska, A., and Sawczuk, M. (2005). *Borrelia burgdorferi* sensu stricto in yellow-necked mice and feeding *Ixodes ricinus* ticks in a forest habitat of west central Poland. J. Med. Entomol. *42*, 850–856.

Michalik, J., Wodecka, B., Skoracki, M., Sikora, B., and Stanczak, J. (2008). Prevalence of avian-associated *Borrelia burgdorferi* s.l. genospecies in *Ixodes ricinus* ticks collected from blackbirds (*Turdus merula*) and song thrushes (*T. philomelos*). Int. J. Med. Microbiol. *298 suppl.1*, 129–138.

Miller, J.C., and Stevenson, B. (2004). Increased expression of *Borrelia burgdorferi* factor H-binding surface proteins during transmission from ticks to mice. Int. J. Med. Microbiol. *293 Suppl. 37*, 120–125.

Miyamoto, K., and Masuzawa, T. (2002). Ecology of *Borrelia burgdorferi* sensu lato in Japan and East Asia. In Lyme Borreliosis: Biology, Epidemiology and Control (ed. J.S. Gray, O. Kahl, R.S. Lane, and G. Stanek, ed. (Wallingford, Oxfordshire, UK: CABI Publishing), pp. 201–222.

Miyamoto, K., Nakao, M., Uchikawa, K., and Fujita, H. (1992). Prevalence of Lyme borreliosis spirochetes in ixodid ticks of Japan, with special reference to a new potential vector, *Ixodes ovatus* (Acari: Ixodidae). J. Med. Entomol. *29*, 216–220.

Motaleb, M.A., Miller, M.R., Li, C., Bakker, R.G., Goldstein, S.F., Silversmith, R.E., Bourret, R.B., and Charon, N.W. (2005). CheX is a phosphorylated CheY phosphatase essential for *Borrelia burgdorferi* chemotaxis. J. Bacteriol. *187*, 7963–7969.

Nakao, M., Miyamoto, K., and Fukunaga, M. (1994). Lyme disease spirochetes in Japan: enzootic transmission cycles in birds, rodents, and *Ixodes persulcatus* ticks. J. Infect. Dis. *170*, 878–882.

Nelson, D.R., Rooney, S., Miller, N.J., and Mather, T.N. (2000). Complement-mediated killing of *Borrelia burgdorferi* by nonimmune sera from sika deer. J. Parasitol. *86*, 1232–1238.

Ogden, N.H., Lindsay, L.R., Hanincova, K., Barker, I.K., Bigras-Poulin, M., Charron, D.F., Heagy, A., Francis, C.M., O'Callaghan, C.J., Schwartz, I., and Thompson, R.A. (2008). Role of migratory birds in introduction and range expansion of *Ixodes scapularis* ticks and of *Borrelia burgdorferi* and *Anaplasma phagocytophilum* in Canada. Appl. Environ. Microbiol. *74*, 1780–1790.

Ohnishi, J., Piesman, J., and de Silva, A.M. (2001). Antigenic and genetic heterogeneity of *Borrelia burgdorferi* populations transmitted by ticks. Proc. Natl. Acad. Sci. U.S.A. *98*, 670–675.

Oliver, J.H., Jr. (1996). Lyme borreliosis in the southern United States: a review. J. Parasitol. *82*, 926–935.

Oliver, J.H., Jr., Cummins, G.A., and Joiner, M.S. (1993). Immature *Ixodes scapularis* (Acari: Ixodidae) parasitizing lizards from the southeastern U.S.A. J. Parasitol. *79*, 684–689.

Olsen, B., Duffy, D.C., Jaenson, T.G., Gylfe, A., Bonnedahl, J., and Bergström, S. (1995). Transhemispheric exchange of Lyme disease spirochetes by seabirds. J. Clin. Microbiol. *33*, 3270–3274.

Olsen, B., Jaenson, T.G., Noppa, L., Bunikis, J., and Bergström, S. (1993). A Lyme borreliosis cycle in seabirds and *Ixodes uriae* ticks. Nature *362*, 340–342.

Ostfeld, R., and Keesing, F. (2000). The function of biodiversity in the ecology of vector-borne zoonotic diseases. Canadian Journal of Zoology-Revue Canadienne De Zoologie *78*, 2061–2078.

Ostfeld, R.S., Schauber, E.M., Canham, C.D., Keesing, K., Jones, C.G., and Wolff, J.O. (2001). Effects of acorn production and mouse abundance on abundance and *Borrelia burgdorferi* infection prevalence of nymphal *Ixodes scapularis* ticks. Vect. Born. Zoon. Dis. *1*, 55–64.

Padula, S.J., Sampieri, A., Dias, F., Szczepanski, A., and Ryan, R.W. (1993). Molecular characterization and expression of p23 (OspC) from a North American strain of *Borrelia burgdorferi*. Infect. Immun. *61*, 5097–5105.

Pal, U., de Silva, A.M., Montgomery, R.R., Fish, D., Anguita, J., Anderson, J.F., Lobet, Y., and Fikrig, E. (2000). Attachment of *Borrelia burgdorferi* within *Ixodes scapularis* mediated by outer surface protein A. J. Clinic. Invest. *106*, 561–569.

Patrican, L.A. (1997). Acquisition of Lyme disease spirochetes by cofeeding *Ixodes scapularis* ticks. Am. J. Trop. Med. Hyg. *57*, 589–593.

Peavey, C.A., Lane, R.S., and Damrow, T. (2000). Vector competence of *Ixodes angustus* (Acari: Ixodidae) for *Borrelia burgdorferi* sensu stricto. Exp. Appl. Acarol. *24*, 77–84.

Pecchioli, E., Hauffe, H.C., Tagliapietra, V., Bandi, C., Genchi, C., and Rizzoli, A. (2007). Genospecies of *Borrelia burgdorferi* sensu lato in *Ixodes ricinus* ticks from the Autonomous Province of Trento, Italy. Int. J. Med. Microbiol. *297*, 53–59.

Piesman, J. (2002). Ecology of *Borrelia burgdorferi* sensu lato in North America. In Lyme Borreliosis: Biology, Epidemiology and Control, J.S. Gray, O. Kahl, R.S. Lane, and G. Stanek, ed. (Wallingford, Oxfordshire, UK: CABI Publishing), pp. 223–250.

Piesman, J., Donahue, J.G., Mather, T.N., and Spielman, A. (1986). Transovarially acquired Lyme disease spirochetes (*Borrelia burgdorferi*) in field-collected larval *Ixodes dammini* (Acari, Ixodidae). J. Med. Entomol. *23*, 219–219.

Pohl-Koppe, A., Kaunicnik, A., and Wilske, B. (2001). Characterization of the cellular and humoral immune response to outer surface protein C and outer surface protein 17 in children with early disseminated Lyme borreliosis. Med. Microbiol. Immunol. (Berl.) *189*, 193–200.

Porco, T.C. (1999). A mathematical model of the ecology of Lyme disease. Math. Med. Biol. *16*, 261–296.

Postic, D., Garnier, M., and Baranton, G. (2007). Multilocus sequence analysis of atypical *Borrelia burgdorferi* sensu lato isolates – description of *Borrelia californiensis* sp. nov., and genomospecies 1 and 2. Int. J. Med. Microbiol. *297*, 263–271.

Postic, D., Ras, N.M., Lane, R.S., Hendson, M., and Baranton, G. (1998). Expanded diversity among Californian borrelia isolates and description of *Borrelia bissettii* sp. nov. (formerly *Borrelia* group DN127). J. Clinic. Microbiol. *36*, 3497–3504.

Preac-Mursic, V., Wilske, B., Patsouris, E., Jauris, S., Will, G., Soutschek, E., Rainhardt, S., Lehnert, G., Klockmann, U., and Mehraein, P. (1992). Active immunization with pC protein of *Borrelia burgdorferi* protects gerbils against *B. burgdorferi* infection. Infection *20*, 342–349.

Probert, W.S., Crawford, M., Cadiz, R.B., and LeFebvre, R.B. 1997. Immunization with outer surface protein (Osp) A, but not OspC, provides cross-protection of mice challenged with North American isolates of *Borrelia burgdorferi*. J. Infect. Dis. *175*, 400–405.

Qiu, W.G., Bosler, E.M., Campbell, J.R., Ugine, G.D., Wang, I.N., Luft, B.J., and Dykhuizen, D.E. (1997). A population genetic study of *Borrelia burgdorferi* sensu stricto from eastern Long Island, New York, suggested frequency-dependent selection, gene flow and host adaptation. Hereditas *127*, 203–216.

Qiu, W.G., Bruno, J.F., McCaig, W.D., Xu, Y., Livey, I., Schriefer, M.E., and Luft, B.J. (2008). Wide distribution of a high-virulence *Borrelia burgdorferi* clone in Europe and North America. Emerg. Infect. Dis. *14*, 1097–1104.

Qiu, W.G., Dykhuizen, D.E., Acosta, M.S., and Luft, B.J. (2002). Geographic uniformity of the Lyme disease spirochete (*Borrelia burgdorferi*) and its shared history with tick vector (*Ixodes scapularis*) in the northeastern United States. Genetics *160*, 833–849.

Qiu, W.G., Schutzer, S.E., Bruno, J.F., Attie, O., Xu, Y., Dunn, J.J., Fraser, C.M., Casjens, S.R., and Luft, B.J. (2004). Genetic exchange and plasmid transfers in *Borrelia burgdorferi* sensu stricto revealed by three-way genome comparisons and multilocus sequence typing. Proc. Natl. Acad. Sci. U.S.A. *101*, 14150–14155.

Radolf, J.D., and Caimano, M.J. (2008). The long strange trip of *Borrelia burgdorferi* outer-surface protein C. Mol. Microbiol. *69*, 1–4.

Rand, P.W., Lacombe, E.H., Smith, R.P., Jr., Rich, S.M., Kilpatrick, C.W., Dragoni, C.A., and Caporale, D. (1993). Competence of *Peromyscus maniculatus* (Rodentia: Cricetidae) as a reservoir host for *Borrelia burgdorferi* (Spirochaetares: Spirochaetaceae) in the wild. J. Med. Entomol. *30*, 614–618.

Randolph, S.E. (1998). Ticks are not insects: Consequences of contrasting vector biology for transmission potential. Parasit. Today *14*, 186–192.

Richter, D., Klug, B., Spielman, A., and Matuschka, F.R. (2004a). Adaptation of diverse Lyme disease spirochetes in a natural rodent reservoir host. Infect. Immun. *72*, 2442–2444.

Richter, D., Postic, D., Sertour, N., Livey, I., Matuschka, F.-R., and Baranton, G. (2006). Delineation of *Borrelia burgdorferi* sensu lato species by multilocus sequence analysis and confirmation of the delineation of *Borrelia spielmanii* sp. nov. Int. J. Syst. Evol. Microbiol. *56*, 873–881.

Richter, D., Schlee, D.B., Allgower, R., and Matuschka, F.R. (2004b). Relationships of a novel Lyme disease spirochete, *Borrelia spielmani* sp. nov., with its hosts in Central Europe. Appl. Environ. Microbiol. *70*, 6414–6419.

Richter, D., Spielman, A., Komar, N., and Matuschka, F.R. (2000). Competence of American robins as reservoir hosts for Lyme disease spirochetes. Emerg. Infect. Dis. *6*, 133–138.

Rodriguez de Cordoba, S., Esparza-Gordillo, J., Goicoechea de Jorge, E., Lopez-Trascasa, M., and Sanchez-Corral, P. (2004). The human complement factor H: functional roles, genetic variations and disease associations. Mol. Immunol. *41*, 355–367.

Rosa, P., Samuels, D.S., Hogan, D., Stevenson, B., Casjens, S., and Tilly, K. (1996). Directed insertion of a selectable marker into a circular plasmid of *Borrelia burgdorferi*. J. Bacteriol. *178*, 5946–5953.

Saint Girons, I., Gern, L., Gray, J.S., Guy, E.C., Korenberg, E., Nuttall, P.A., Rijpkema, S.G., Schonberg, A., Stanek, G., and Postic, D. (1998). Identification of *Borrelia burgdorferi* sensu lato species in Europe. Zentralbl. Bakteriol. *287*, 190–195.

Samuels, D.S. (1995). Electrotransformation of the spirochete *Borrelia burgdorferi*. In Methods in Molecular Biology, J.A. Nickoloff, ed. (Totowa, New Jersey: Humana Press), pp. 253–259.

Samuels, D.S., Mach, K.E., and Garon, C.F. (1994). Genetic transformation of the Lyme disease agent *Borrelia burgdorferi* with coumarin-resistant *gyrB*. J. Bacteriol. *176*, 6045–6049.

Schierup, M.H., Vekemans, X., and Charlesworth, D. (2000). The effect of subdivision on variation at multiallelic loci under balancing selection. Genet. Res. *76*, 51–62.

Schwan, T.G., Karstens, R.H., Schrumpf, M.E., and Simpson, W.J. (1991). Changes in antigenic reactivity of *Borrelia burgdorferi*, the Lyme disease spirochete, during persistent infection in mice. Canad. J. Microbiol. 37, 450–454.

Schwan, T.G., Kime, K.K., Schrumpf, M.E., Coe, J.E., and Simpson, W.J. (1989). Antibody response in white-footed mice (*Peromyscus leucopus*) experimentally infected with the Lyme disease spirochete (*Borrelia burgdorferi*). Infect. Immunol. 57, 3445–3451.

Schwan, T.G., Piesman, J., Golde, W.T., Dolan, M.C., and Rosa, P.A. (1995). Induction of an outer surface protein on *Borrelia burgdorferi* during tick feeding. Proc. Natl. Acad. Sci. U.S.A. 92, 2909–2913.

Seinost, G., Dykhuizen, D.E., Dattwyler, R.J., Golde, W.T., Dunn, J.J., Wang, I.N., Wormser, G.P., Schriefer, M.E., and Luft, B.J. (1999). Four clones of *Borrelia burgdorferi* sensu stricto cause invasive infection in humans. Infect. Immun. 67, 3518–3524.

Slatkin, M., and Hudson, R.R. (1991). Pairwise comparisons of mitochondrial-DNA sequences in stable and exponentially growing populations. Genetics *129*, 555–562.

Slowik, T.J., and Lane, R.S. (2001). Birds and their ticks in northwestern California: minimal contribution to *Borrelia burgdorferi* enzootiology. J. Parasitol. *87*, 755–761.

Sonenshine, D.E. (1992). The biology of ticks (New York, USA: Oxford University Press).

Stafford, K.C., Bladen, V.C., and Magnarelli, L.A. (1995). Ticks (Acari, Ixodidae) infesting wild birds (Aves) and white-footed mice in Lyme, Ct. J. Med. Entomol. *32*, 453–466.

Stevenson, B., and Barthold, S.W. (1994). Expression and sequence of outer surface protein C among North American isolates of *Borrelia burgdorferi*. FEMS Microbiol. Lett. *124*, 367–372.

Stevenson, B., Casjens, S., and Rosa P.A. (1998). Evidence of past recombination events among the genes encoding the Erp antigens of *Borrelia burgdorferi*. Microbiology *144*, 1869–1879.

Stevenson, B., El-Hage, N., Hines, M.A., Miller, J.C., and Babb, K. (2002). Differential binding of host complement inhibitor factor H by *Borrelia burgdorferi* Erp surface proteins: a possible mechanism underlying the expansive host range of Lyme disease spirochetes. Infect. Immun. *70*, 491–497.

Stevenson, B., and Miller., J.C. (2003). Intra- and interbacterial genetic exchange of Lyme disease spirochete *erp* genes generates sequence identity amidst diversity. J. Mol. Evol. *57*, 309–324.

Stevenson, B., Tilly, K., and Rosa, P.A. (1996). A family of genes located on four seperate 32-kilobase circular plasmids in *Borrelia burgdorferi*. J. Bacteriol. *178*, 3508–3516.

Stevenson, B., von Lackum, K., Wattier, R.L., McAlister, J.D., Miller, J.C., and Babb, K. (2003). Quorum sensing by the Lyme disease spirochete. Microbes Infect. *5*, 991–997.

Sun, Y., and Xu, R. (2003). Ability of *Ixodes persulcatus, Haemaphysalis concinna* and *Dermacentor silvarum* ticks to acquire and transstadially transmit *Borrelia garinii*. Exp. Appl. Acarol. *31*, 151–160.

Sun, Y., Xu, R., and Cao, W. (2003). *Ixodes sinensis*: competence as a vector to transmit the Lyme disease spirochete *Borrelia garinii*. Vect. Born. Zoon. Dis. *3*, 39–44.

Theisen, M., Borre, M., Mathiesen, M.J., Mikkelsen, B., Lebech, A.M., and Hansen, K. (1995). Evolution of the *Borrelia burgdorferi* outer surface protein OspC. J. Bacteriol. *177*, 3036–3044.

Tilly, K., Bestor, A., Jewett, M.W., and Rosa, P. (2007). Rapid clearence of Lyme disease spirochetes lacking OspC from the skin. Infect. Immun. *75*, 1517–1519.

Tilly, K., Krum, J.G., Bestor, A., Jewett, M.W., Grimm, D., Bueschel, D., Byram, R., Doward, D., VanRaden, M.J., Stewart, P., and Rosa, P. (2006). *Borrelia burgdorferi* OspC protein required exclusively in a crucial stage of mammalian infection. Infect. Immun. *74*, 3554–3564.

Tsao, J.I., Wootton, J.T., Bunikis, J., Luna, M.G., Fish, D., Barbour, A.G., and Gotschlich, E.C. (2004). An ecological approach to preventing human infection: vaccinating wild mouse reservoirs intervenes in the Lyme disease cycle. Proc. Natl. Acad. Sci. U.S.A. *52*, 18159–18164.

Ullmann, A.J., Lane, R.S., Kurtenbach, K., Miller, M., Schriefer, M.E., Zeldner, N., and Piesman, J. (2003). Bacteriolytic activity of selected vertebrate sera for *Borrelia burgdorferi* sensu stricto and *Borrelia bissettii*. J. Parasitol. *89*, 1256–1257.

Urwin, R., and Maiden, M.C. (2003). Multi-locus sequence typing: a tool for global epidemiology. Trends Microbiol. *11*, 479–487.

van Dam, A.P., Kuiper, H., Vos, K., Widjojokusumo, A., de Jongh, B.M., Spanjaard, L., Ramselaar, A.C., Kramer, M.D., and Dankert, J. (1993). Different genospecies of *Borrelia burgdorferi* are associated with distinct clinical manifestations of Lyme borreliosis. Clin. Infect. Dis. *17*, 708–717.

Vekemans, X., and Slatkin, M. (1994). Gene and allelic genealogies at a gametophytic self-incompatibility locus. Genetics *137*, 1157–1165.

Wallace, A.R. (1855). On the law which has regulated the introduction of new species. Annals and Magazine of Natural History *16*, S20.

Wang, G., van Dam, A.P., Le Fleche, A., Postic, D., Peter, O., Baranton, G., de Boer, R., Spanjaard, L., and Dankert, J. (1997). Genetic and phenotypic analysis of *Borrelia valaisiana* sp. nov. (*Borrelia* genomic groups VS116 and M19). Int. J. Syst. Bacteriol. *47*, 926–932.

Wang, G., van Dam, A.P., Schwartz, I., and Dankert, J. (1999a). Molecular typing of *Borrelia burgdorferi* sensu lato: taxonomic, epidemiological, and clinical implications. Clin. Microbiol. Rev. *12*, 633–653.

Wang, I.N., Dykhuizen, D.E., Qiu, W.G., Dunn, J.J., Bosler, E.M., and Luft, B.J. (1999b). Genetic diversity of *ospC* in a local population of *Borrelia burgdorferi sensu stricto*. Genetics *151*, 15–30.

Watterson, G.A. (1978). The homozygosity test of neutrality. Genetics *88*, 405–417.

Wilske, B. (2005). Epidemiology and diagnosis of Lyme borreliosis. Ann. Med. *37*, 568–579.

Wilske, B., Jauris-Heipke, S., Lobentanzer, R., Pradel, I., Preac-Mursic, V., Rossler, D., Soutschek, E., and Johnson, R.C. (1995). Phenotypic analysis of outer surface protein C (OspC) of *Borrelia burgdorferi* sensu lato by monoclonal antibodies: relationship to genospecies and OspA serotype. J. Clin. Microbiol. *33*, 103–109.

Wilske, B., Busch, U., Fingerie, V., Jauris-Heipke, U., Preac-Mursic, S.V., Rossler, D., and Will, G. (1996). Immunological and molecular variability of OspA and OspC. Implications for *Borrelia* vaccine development. Infection *24*, 208–212.

Wilske, B., Preac-Mursic, V., Goebel, U.B., Graf, B., Jauris, S., Soutschek, E., Schwab, E., and Zumstein, G. (1993a). An OspA serotyping system for *Borrelia burgdorferi* based on reactivity with monoclonal antibodies and OspA sequence analysis. J. Clin. Microbiol. *31*, 340–350.

Wilske, B., Preac-Mursic, V., Jauris, S., Hofmann, A., Pradel, I., Soutschek, E., Schwab, E., Will, G., and Wanner, G. (1993b). Immunological and molecular polymorphisms of OspC, an immunodominant major outer surface protein of *Borrelia burgdorferi*. Infect. Immun. *61*, 2182–2191.

Wilson, M.L., and Spielman, A. (1985). Seasonal activity of immature *Ixodes dammini* (Acari: Ixodidae). J. Med. Entomol. *22*, 408–414.

Woodman, M.E., Cooley, A.E., Miller, J.C., Lazarus, J.J., Tucker, K., Bykowski, T., Botto, M., Hellwage, J., Wooten, R.M., and Stevenson, B. (2007). *Borrelia burgdorferi* binding of host complement factor H is not required for efficient mammalian infection. Infec. Immun. *75*, 3131–3139.

Wright, S.A., Lane, R.S., and Clover, J.R. (1998). Infestation of the southern alligator lizard (Squamata: Anguidae) by *Ixodes pacificus* (Acari: Ixodidae) and its susceptibility to *Borrelia burgdorferi*. J. Med. Entomol. *35*, 1044–1049.

Xu, Q., McShan, K., and Liang, F.T. (2008). Essential protective role attributed to the surface lipoproteins of *Borrelia burgdorferi* against innate defences. Mol. Microbiol. *69*, 15–29.

Zhang, H., and Marconi, R.T. (2005). Demonstration of cotranscription and 1-methyl-3-nitroso-nitroguanidine induction of a 30-gene operon of *Borrelia burgdorferi*: evidence that the 32-kilobase circular plasmids are prophages. J. Bacteriol. *187*, 7985–7995.

Zhong, J., Skouloubris, S., Dai, Q., Myllykallo, H., and Barbour, A.G. (2006). Function and evolution of plasmid-borne genes for pyrimidine biosysnthesis in *Borrelia* spp. J. Bacteriol. *188*, 909–918.

Ecology of Borreliae and Their Arthropod Vectors

10

Joseph Piesman and Tom G. Schwan

Abstract

Borreliae are unique among the pathogenic spirochaetes by requiring obligate blood-feeding arthropods for their transmission and maintenance in susceptible vertebrate host populations. With one exception, ticks transmit all borreliae and nearly all species are maintained in enzootic foci with humans being only accidental victims to infection. The true relapsing fever spirochaetes are transmitted by many species of soft (Argasidae) ticks in the genus *Ornithodoros* and in many geographic areas, specific species of ticks transmit only one species of spirochaete. Specific tick-spirochaete associations range from near-desert conditions to high elevation coniferous forests.

The four principal hard tick (Ixodidae) species that transmit Lyme disease spirochaetes (*Borrelia burgdorferi* sensu lato) include *Ixodes scapularis* in eastern North America, *I. pacificus* in western North America, *I. persulcatus* in Asia, and *I. ricinus* in Europe. These ticks are basically forest dwellers, spending most of their time hiding in the leaf litter of the forest floor, where humidity is high and the risk of desiccation is low.

Borrelia burgdorferi sensu stricto, *B. garinii*, and *B. afzelii* are the principal genospecies causing human disease. Of these three genospecies, *B. burgdorferi* sensu stricto is unique in that it has a Holarctic distribution, spanning both the Old World (Palearctic Region) and the New World (Nearctic Region) in the northern hemisphere. Specific rodents were originally identified as the principal reservoirs of *B. burgdorferi* sensu stricto, including the white footed mouse (*Peromyscus leucopus*) in North America, and the wood mouse (*Apodemus sylvaticus*), yellow-necked mouse (*Apodemus flavicollis*), and bank vole (*Clethrionomys glareolus*) in Europe. Birds also have been identified as reservoirs of *B. burgdorferi* sensu stricto on both sides of the Atlantic.

Since the discovery of the Lyme borreliosis spirochaete in 1982, a plethora of knowledge concerning the ecology of the spirochaete, its tick vectors, and vertebrate hosts has been brought forth through an incredible focus of research efforts. The task before the public health research community is not solely producing more ecological knowledge regarding the Lyme borreliosis spirochaetes cycle in nature. Clearly, our immediate challenge is to synthesize and harness available knowledge of spirochaete–tick–host interactions to develop ecologically based interventions to combat Lyme borreliosis.

10.1 Introduction

Borreliae are unique among the pathogenic spirochaetes by requiring obligate blood-feeding arthropods for their transmission and maintenance in susceptible vertebrate host populations (Canale-Parola, 1984). This dependency on two very different animals adds many levels of ecological complexity that do not exist for other pathogenic spirochaetes. For example, the syphilis spirochaete *Treponema pallidum*, which infects only humans, is transmitted by direct intimate contact from one person to another. For borrelioses, humans are, with few exceptions, only accidental victims when fed upon by infected ticks that normally feed upon other animals in the wild. Ecological parameters necessary to maintain foci that lead to human infection include the bringing together of wild vertebrate hosts, tick vectors, borreliae, appropriate habitats and climates, and the activities of the accidental human hosts.

Two ecological divisions of borreliae exist, which include those enzootic species that are maintained by wild vertebrate populations versus those species that are maintained only by human populations (Felsenfeld, 1971). Another division between these spirochaetes that impacts the ability of scientists to gather ecological data is rooted in the two phylogenetic groups, which include the Lyme borreliosis and related spirochaetes (transmitted by ixodid ticks) as opposed to the true relapsing fever spirochaetes (transmitted by argasid ticks) (Fukunaga *et al.*, 1996a,b).

As detailed below, Lyme borreliosis spirochaetes are maintained in natural cycles utilizing small vertebrates, primarily rodents and birds, and ixodid ticks (Lane *et al.*, 1991). The small mammals, once infected, probably remain infected for life; isolating spirochaetes from various tissues, such as the skin and urinary bladder, from these persistently infected animals is relatively efficient (Schwan *et al.*, 1988; Sinsky and Piesman, 1989). In addition, the immature *Ixodes* ticks that serve as the primary vectors feed for several days, thus collecting these ticks directly from their hosts, including humans, is easily accomplished. Furthermore, ixodid ticks quest for their hosts, by positioning themselves on vegetation with their front legs reaching out to facilitate their being swept up onto the hair and feathers of a passing host (Sonenshine, 1991). This behaviour greatly facilitates collection of these ticks by flagging, a procedure in which a one-metre square cotton cloth is dragged over the vegetation to pick up ticks waiting for hosts. Additionally, Lyme borreliosis spirochaetes are relatively easy to grow and isolate in BSK medium from tick and vertebrate tissues.

The different species of relapsing fever spirochaetes vary in their ecological requirements (Felsenfeld, 1971), yet the general theme is strikingly different compared to the Lyme borreliosis spirochaetes. North American relapsing fever spirochaetes (discussed below), are maintained in cycles involving squirrels, chipmunks and other relatively small mammals as the primary vertebrate hosts and soft ticks in the genus *Ornithodoros* as the vector (Dworkin *et al.*, 2002b). However, the vertebrate hosts are infected for a relatively short period of time because the recurrent spirochaetaemias eventually cease. This greatly reduces the opportunity for investigators to detect and isolate spirochaetes from their natural hosts. In addition, the *Ornithodoros* ticks that transmit these spirochaetes are nocturnal and feed quickly, in 15 to 90 minutes, and leave the host immediately after engorgement. This feeding behaviour makes collecting these ticks from their host nearly impossible. Furthermore, *Ornithodoros* ticks do not quest for their hosts in the same manner, as do the ixodid ticks. They live in close association with the nests of their hosts so they rarely travel far to feed and, therefore, cannot be collected by flagging vegetation. Their refuges in tree stumps or snags, host nests, burrows and other structures must be found and examined. Lastly, the relapsing fever spirochaetes are more difficult to isolate in BSK medium than are the Lyme borreliosis spirochaetes. These basic differences in the duration of infection of vertebrate hosts, the feeding behaviours of ixodid versus argasid ticks, and the efficiency of culturing spirochaetes have made delineating natural enzootic foci of Lyme borreliosis significantly easier than for relapsing fever.

10.2 Louse-borne (epidemic) relapsing fever

Louse-borne (epidemic) relapsing fever is unique among this group of infections. The disease is caused by only one species of spirochaete, *Borrelia recurrentis*, which is transmitted solely by the human body louse, *Pediculus humanus humanus* (Fig. 10.1). Transmission from lice to humans occurs primarily when these ectoparasites are crushed and the spirochaete-infected haemolymph from the body cavity of the louse contaminates abrasions in the skin or mucous membranes (Buxton, 1946). These spirochaetes also may be shed in louse faeces, allowing for contaminative or posterior station transmission (Houhamdi and Raoult, 2005). This disease is not a zoonosis, and infections are maintained only in human populations subject to conditions in which body louse infestations are prevalent.

Spirochaetes, subsequently named *B. recurrentis*, were identified as the probable cause of

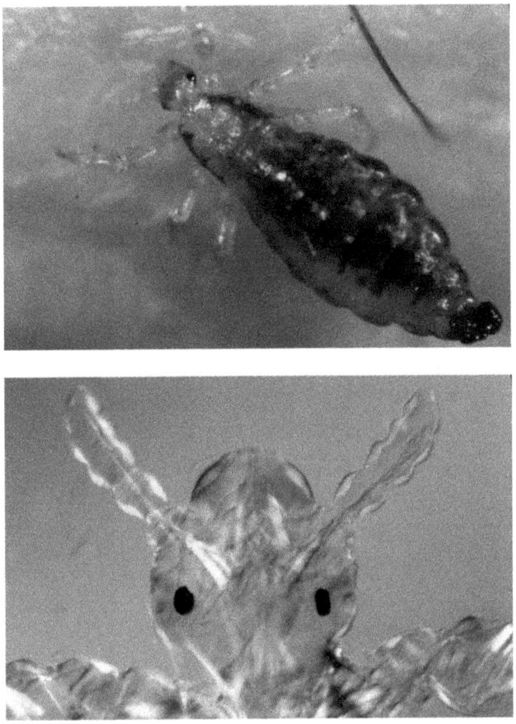

Figure 10.1 Human body louse (*Pediculus humanus*). A colour version of this figure is located in the plate section at the back of the book.

relapsing fever (*recurrenskranken*) in 1868 and 1873 during epidemics in Berlin. The young physician, Otto Obermeier, detected spirochaetes microscopically in the blood of many relapsing fever patients during their acute illness, but not when the patients felt better and their body temperature had returned to normal (Birkhaug, 1942; Obermeier, 1873). In 1907, Mackie demonstrated in India that the human body louse was probably involved with the human-to-human transmission of these bacteria when he found many spirochaete-infected lice infesting boys with relapsing fever (Mackie, 1907). During the next several years, many studies confirmed the role of these obligate blood-feeding, ectoparasitic insects as the primary biological vector (Buxton, 1946). These lice are exquisitely host specific and feed only on humans. The life cycle includes the egg or nit, three larval instars, and adults. Although called body lice, these insects spend a significant amount of time in the clothing, where the eggs are laid and glued onto the fibres. The entire life cycle after emerging from the egg may last up to about 80 days if clothes are removed at night, which lowers the ambient temperature for the lice. Alternatively, continual wearing of clothes may shorten the survival of the lice to only a few weeks.

Body lice become infected when they feed on a human host who is spirochaetaemic with *B. recurrentis*. Spirochaetes enter the midgut with the blood meal and disseminate into the louse's haemocoele. During this period, spirochaetes become increasingly difficult to detect microscopically; before this was recognized, it was thought that the newly acquired spirochaetes altered their shape and became unrecognizable during early infection in lice. No evidence supports such a change in spirochaete morphology. Rather, the difficulty in detecting spirochaetes by microscopy is due to their decreasing number in the louse midgut before they infect and multiply within the body cavity. Spirochaetes generally infect no other tissues of the louse's body, although Mackie reported differently in his original demonstration (Mackie, 1907). Thus, with no infection of the salivary glands, spirochaetes are not transmitted by bite. With no infection of reproductive tissues, there is no vertical (transovarial) transmission to the next generation of lice. Transmission occurs when infected lice are crushed to release spirochaete-infected haemolymph onto the human skin. This crude mechanism for transmission, which results in the death of the louse, suggests that the association of *B. recurrentis* with human body lice evolved recently compared to the other relapsing fever spirochaetes transmitted by argasid ticks (Nicolle and Anderson, 1927; Weyer, 1960). Buxton's classic review of the biology of human lice and the disease agents they transmit, including *B. recurrentis*, should be read for additional information (Buxton, 1946).

Louse-borne relapsing fever has become much less prevalent and more geographically restricted over the last century. Regions of Africa remain 'hotspots' where there has been a long history of large epidemics (Kirk, 1939). Ethiopia and Sudan remain as the two countries continually plagued with human infections (de Jong *et al.*, 1995), although the incidence has declined in Ethiopia in recent years (Ramos *et al.*, 2008). The largest and most recent epidemic occurred in the Darfur region of southwestern Sudan in 1999–2000. This outbreak was initially thought to be a viral haemorrhagic fever as the patients experienced severe bleeding. However, thrombocytopenia and bleeding are also associated with louse-borne relapsing fever (Dennis *et al.*, 1976; Southern and Sanford, 1968). An on-site epidemiological investigation by CDC and WHO in Sudan, however, proved that the outbreak was caused by *B. recurrentis* with approximately 20,000 human cases and a 10% mortality rate (Dennis, 1999).

10.3 Tick-borne (endemic) relapsing fever

In contrast to louse-borne relapsing fever, tick-borne or endemic relapsing fever is caused by numerous species of spirochaetes distributed throughout the world (Felsenfeld, 1971), although many of these bacteria named in the older literature await characterization with newer molecular techniques. These spirochaetes are transmitted by argasid ('soft') ticks in the genus *Ornithodoros*. Few specific spirochaete-tick associations have been studied in detail, however spirochaete transmission occurs when ticks feed on their vertebrate hosts (Burgdorfer, 1951; Dutton and Todd, 1905; Koch, 1905, 1906; Schwan and Hinnebusch, 1998; Varma, 1962; Wheeler, 1938b). Spirochaetes are delivered in the saliva directly into the dermis, or via the contamination of infected coxal fluid that is excreted while some species of soft ticks feed (Burgdorfer, 1951; Varma, 1962). All tick-borne relapsing fever spirochaetes, with the exception of *B. duttonii*, are maintained in enzootic cycles that involve many species of smaller mammals and ticks in different regions of the world. *B. duttonii*, which causes relapsing fever in many areas of sub-Saharan Africa, infects only humans and is transmitted by *Ornithodoros moubata*. Most species of *Ornithodoros* ticks are relatively long-lived and may retain infectious spirochaetes for many years (Francis, 1938), making these arthropods the true reservoirs of the spirochaetes rather than the vertebrate hosts.

David Livingstone, the legendary Scottish-born missionary, doctor and explorer, is credited with the first written account of what was later to be named tick-borne relapsing fever. In his 1857 book on his explorations across Southern Africa (Livingstone, 1857) he wrote the following based on events in what is now Angola: 'When sleeping in the house of the Commandant, I was bitten on the foot by a kind of tick, known in the southern country by the name tampan, and common in all the native huts in this country.' Later, when in what is now Mozambique, Livingstone added:

> *He [the commandant] also generously presented my men with abundant provisions of millet; and, by giving them lodgings in a house of his own until they could erect their own huts, he preserved them from the bite of the tampans, here named carapotos. We had heard frightful accounts of this insect while among the Banyai; and Major Sicard assured me that to strangers its bite is more especially dangerous, as it sometimes causes fatal fever.*

These accounts were recorded in 1855, 13 years before Obermeier detected spirochaetes in his Berlin patients, so the spirochaetal aetiology of what was called at first tick fever was yet to be determined. In 1877, the tick referred to as tampan, carapotos, and many other local tribal names was cursorily described by Murray and named *Argas moubata* (Murray, 1877), later to be reassigned to the genus *Ornithodoros*.

The events surrounding the discovery that tick fever in Africa was relapsing fever caused by a tick-transmitted spirochaete are fascinating. Alfred Cook, a British missionary doctor working in Uganda, observed numerous patients with relapsing fever between 1899 to 1903, and in January 1904 he was the first to report the presence of spirochaetes in the blood of these individuals (Cook, 1904). Cook called the spirochaetes *Spirillum Obermeieri*, but did not implicate ticks with transmission. On 26 November 1904, Philip Ross and AD Milne published their discovery of 'spirilla' in the blood of 8 patients, also in Uganda, who claimed they had been recently bitten by ticks (Ross and Milne, 1904). These investigators implicated ticks as the vector of the spirochaetes and concluded that the disease long known as tick fever in Africa was relapsing fever. Yet they neither examined ticks for infection nor attempted transmission experiments, nor did they cite Cook's findings. Curiously, on the very same date (26 November 1904), J. Everett Dutton and John L. Todd communicated to their sponsors at the Liverpool School of Tropical Medicine that they had discovered a spirochaete to be the pathogenic agent of tick fever in the eastern Congo Free State (now the Democratic Republic of Congo), yet their seminal report was not published until November 1905 (Dutton and Todd, 1905). Both Dutton and Todd accidentally infected themselves, most likely during patient autopsies, and fever and spirochaetaemia curves for both of them were presented in their publication. By 15 January 1905, Todd had recovered with no subsequent problems. Dutton, however, relapsed three times through mid January and appeared to have recovered yet died February 27, 1905 (Dutton and Todd, 1905). When Dutton and Todd's work was published in 1905, Todd wrote the following in the preface: 'Four months ago we were in possession of the main facts stated in this paper. Illness and death retarded the gathering of information and the recording of our observations.' And later, 'For almost a year and a half Dr Dutton and I have worked together in the Congo. This communication represents only a small part of the results of our observations. Other reports, based on our common work, must soon be written by me alone. May any weakness in all be recognized as mine, and in nowise due to my absent comrade.' Dutton and Todd's efforts were the first to prove that the tick *Ornithodoros moubata* was the vector for spirochaetes causing tick-borne relapsing fever (Dutton and Todd, 1905), which was soon confirmed by Robert Koch while working in German East Africa (now Tanzania) (Koch, 1905, 1906).

Tick-borne relapsing fever is now known to be endemic in many regions of the world excluding Australia, New Zealand and Oceania (Felsenfeld, 1971). Distinguishing tick-borne from louse-borne relapsing fever may be difficult in areas where both infections occur (Scott *et al.*, 2005), especially if patients are few, sporadic and isolated. Additionally, in Africa both zoonotic and non-zoonotic tick-associated infections exist, which confuses the true epidemiological significance of specific ticks and the spirochaetes they transmit.

As recounted above, the *O. moubata* and *B. duttonii* association has no proven vertebrates other than humans that maintain the transmission cycle in nature. Domestic animals such as chickens and pigs that live in or close to tick-infested huts in Tanzania have long been known to be potential alternative hosts for the ticks (Hoogstraal, 1981; Nuttall *et al.*, 1908). Such animals may also be exposed to *B. duttonii* if fed on by infected ticks (McCall *et al.*, 2007), yet no proof of any animal, wild or domestic, being a reservoir for *B. duttonii* and a source for human infections exists. Also, the ecology of relapsing fever in East and Central Africa is complicated by the complex taxonomic status of *O. moubata* and *O. porcinus*, and differences in their behaviour and host preferences (Walton, 1962). Regardless, tick-borne relapsing fever caused by *B. duttonii* transmitted by *O. moubata* remains a significant public health problem in Africa (Barclay and Coulter, 1990; Cutler, 2006; Dupont *et al.*, 1997; Jongen *et al.*, 1997).

In West Africa, André Leger first found relapsing fever spirochaetes in the musk shrew, *Crocidura stämpflii*, near Dakar, Senegal, in 1917 and proposed the name *Spirochaeta crocidurae* (Leger, 1917). These spirochaetes are now known as *Borrelia crocidurae* and are transmitted by *Ornithodoros sonrai*, the only known vector of *B. crocidurae* in West Africa (Vial *et al.*, 2006b), and were previously referred to as the small form of *Ornithodoros erraticus* (Hoogstraal *et al.*, 1954). These ticks parasitize many species of small burrowing mammals, but feed primarily on rodents (i.e. *Arvicanthis* species, *Mastomys* species) and insectivores (*Crocidura* species) (Trape *et al.*, 1991). This association is a true enzootic cycle although many people become infected because of the proximity of tick-infested rodent burrows to the stick-and-mud huts in which the people reside. Some burrow systems may open directly into the huts, where ticks may quest for hosts at night and feed on people while sleeping (Vial *et al.*, 2006a).

The most extensive epidemiological and ecological studies performed in recent years on any endemic focus of tick-borne relapsing fever have been done in Senegal and neighbouring countries in West Africa (Godeluck *et al.*, 1994; Logan *et al.*, 1993; Trape *et al.*, 1991, 1996; Vial *et al.*, 2006a). The *O. sonrai* ticks are widespread, parasitize many species of burrowing rodents and shrews, and in many areas have a high prevalence of infection with *B. crocidurae*. As a result, in some villages of Senegal, tick-borne relapsing fever is the second most common cause of outpatient visits to hospitals and clinics with only malaria more prevalent (Vial *et al.*, 2006a). Additionally, in West Africa as in many other areas south of the Sahara Desert, patients with recurrent febrile illnesses are usually thought to have malaria, and may be treated as such without laboratory confirmation to eliminate other possible infections. A recent serological survey in Togo identified nearly 10% of the patients to be seropositive for the relapsing fever antigen GlpQ (Nordstrand *et al.*, 2007). *O. sonrai* may be expanding its range to the south across West Africa with the continued sub-Saharan drought (Trape *et al.*, 1996). Thus tick-borne relapsing fever is most likely currently under-diagnosed in this part of Africa and may probably expand its distribution with climate changes that promote the spread and increased abundance of the tick vector.

Across the African continent in Egypt, *Ornithodoros erraticus* ticks are abundant in the burrows of many species of small mammals, especially the Nile grass rat (*Arvicanthis niloticus*), as well as reptiles, amphibians and birds (Hoogstraal *et al.*, 1954). When the Egyptian surveys were done 50 years ago, the small and large form designations for *O. erraticus* were still in use. However, most of the ticks were attributed to the small form, and would now be considered *O. sonrai*. The Egyptian ticks had a high prevalence of infection with *B. crocidurae*, which gave rise to some interesting experiments in the laboratory under the direction of the late Harry Hoogstraal. *B. crocidurae* quickly disseminated to numerous tick tissues after *O. sonrai* fed on spirochaetaemic hamsters, and spirochaetes were vertically maintained in ticks via transovarial and transstadial transmission (Gaber *et al.*, 1984). More interesting was the finding that male *O. sonrai* venereally transmitted *B. crocidurae* to females, which later transmitted spirochaetes when they fed (Gaber *et al.*, 1982). Also, unfed ticks were capable of feeding on those ticks that recently engorged on spirochaetemic hamsters. This form of hyperparasitism resulted in those ticks that fed on infected ticks to also acquire the infection and subsequently transmit spirochaetes during feeding (Helmy *et al.*, 1983). Thus, one could argue as to whether or not vertebrate hosts are necessary to maintain *B. crocidurae* in these ticks, given the efficient vertical transmission through the tick's life cycle, and the horizontal transmission associated with the ticks mating and feeding on each other. We suspect that the frequency of hyperparasitism may be magnified in dense laboratory colonies of these ticks and may be only infrequent events in nature, which would maintain a prominent role for the small mammals in these Egyptian foci. Given the rather widespread distribution and abundance of *B. crocidurae*-infected ticks (Davis and Hoogstraal, 1954; Khalil *et al.*, 1984), the lack of any demonstrated human disease in Egypt is intriguing.

Across parts of North Africa from Libya and Egypt, to the Middle East and beyond, the tick *Ornithodoros tholozani* is the vector of *Borrelia persica* (Felsenfeld, 1971). This spirochaete has been characterized from tick and human blood samples by PCR and DNA sequence analysis (Assous *et al.*, 2006; Halperin *et al.*, 2006). The species clearly segregates into a clade distinct from other species of relapsing fever spirochaetes for which molecular data are available. The vector is sometimes referred to as the cave tick, as these parasites prefer dark, moist areas such as caves, abandoned buildings, and animal burrows (Felsenfeld, 1971; Sidi *et al.*, 2005). In Israel, where the most recent epidemiological studies have been done, the wild vertebrate hosts that support the *O. tholozani* ticks are not known (Sidi *et al.*, 2005). However, many humans acquire the infection while occupying caves, and a significant incidence of infection continues in military recruits while occupying caves during training manoeuvres (Hasin *et al.*, 2006a; Sidi *et al.*, 2005).

10.4 Tick-borne relapsing fever in North America

The first documented outbreak of tick-borne relapsing fever in North America occurred in Colorado in 1915 (Meader, 1915). Spirochaetes were detected microscopically in two of five patients, but ticks were not implicated in spite of the discoveries made in Africa 10 years previously. Instead, Meader suggested that the patients were infected as a result of 'body vermin' disseminating from gypsies that visited the camp where the patients had slept. Six years later, two patients contracted a recurrent illness after staying in a cabin in the Sierra Nevada Mountains near the Truckee River in California (Briggs, 1922). While malaria was initially suspected, microscopic examination of the blood for plasmodia-infected erythrocytes, instead, found spirochaetes in both patients. Bedbugs and ticks were suspected as possible vectors for infection, but elucidation of the true vector associated with relapsing fever in the mountains of western North America would have to wait until the next decade.

Today, tick-borne relapsing fever in humans occurs in scattered foci throughout much of western North America from southern British Columbia, Canada, through the mountains to southern California, and to the east from Montana to Texas (Dworkin *et al.*, 1998; Dworkin *et al.*, 2002a; Dworkin *et al.*, 2002b). Because of the difficulty in finding *Ornithodoros* ticks and infected vertebrate hosts, much of our knowledge about the geographic distribution of relapsing fever spirochaetes is based on human infections and their probable site of exposure. When such cases are identified, restricted and brief epidemiological investigations at the putative site of exposure may occur, which may or may not result in finding infected ticks or rodents.

In North America, three species of spirochaetes, each transmitted by a different species of soft tick, cause relapsing fever: *Borrelia hermsii* transmitted by *Ornithodoros hermsi* (Fig. 10.2); *Borrelia turicatae* transmitted by *Ornithodoros turicata*; *Borrelia parkeri* transmitted by *Ornithodoros parkeri* (Davis, 1942a; Kemp *et al.*, 1934). In a series of exhaustive studies performed in the 1940s, Gordon Davis and his colleagues at the Rocky Mountain Laboratory demonstrated the extraordinary species-specific association between spirochaete and tick (Davis, 1942a,b). Field crews would collect these ticks from throughout the western USA, return them to the laboratory, separate them by species, and feed them on mice (Fig. 10.3). Those mice that became spirochaetaemic were then fed upon by all three species of colony-reared, uninfected ticks, and held until they fed again on normal mice. These studies demonstrated that spirochaetes that originated from *O. hermsi* were subsequently transmitted only by *O. hermsi* and not by *O. turicatae* or *O. parkeri*, and such was the case for the spirochaetes originating from the other two species of ticks. The basis for this spirochaete-tick specificity is not yet understood, but may lie in the interactions that enable spirochaetes to persist in the salivary glands of only one species of tick (Schwan and Piesman, 2002), or it could reflect a more global physiological mismatch that completely prevents survival of spirochaetes in the 'wrong' ticks (Kemp *et al.*, 1934).

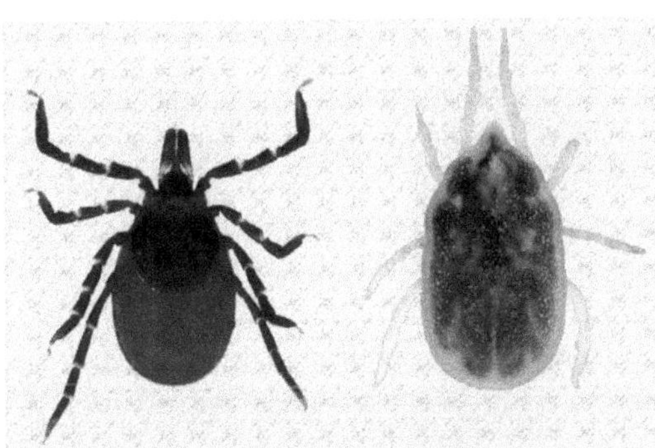

Figure 10.2 *Ixodes scapularis* (left) and *Ornithodoros hermsi* (right). A colour version of this figure is located in the plate section at the back of the book.

Most human cases of tick-borne relapsing fever in North America are caused by *B. hermsii* (Dworkin *et al.*, 2002a). Pioneering work on the ecology of *B. hermsii*-caused relapsing fever was done during the 1930s in two mountain areas of California, including the Lake Tahoe region in the Sierra Nevada Mountains in east-central California and the San Bernardino Mountains of southern California (Beck, 1937, 1942; Coleman, 1934; Herms and Wheeler, 1936; Wynns, 1936; Wynns and Beck, 1935). Such studies elucidated the enzootic nature of the infection with a previously unidentified tick transmitting spirochaetes to tree squirrels (*Tamiasciurus* species) and chipmunks (*Tamias* species) and to people that slept in mountain cabins infested with the spirochaete-infected ticks (Briggs, 1922; Coleman, 1934; Wynns and Beck, 1935). The 'new' tick was named *Ornithodoros hermsi*, in honour of Dr William B. Herms, a medical entomologist at the University of California, Berkeley (Wheeler, 1935; Wheeler *et al.*, 1935).

The identification of *B. hermsii* as a cause of relapsing fever was based primarily on 'guilt by association'. Spirochaetes in the blood of patients were called *B. hermsii* if the patients had been exposed in mountain cabins, from which *O. hermsi*

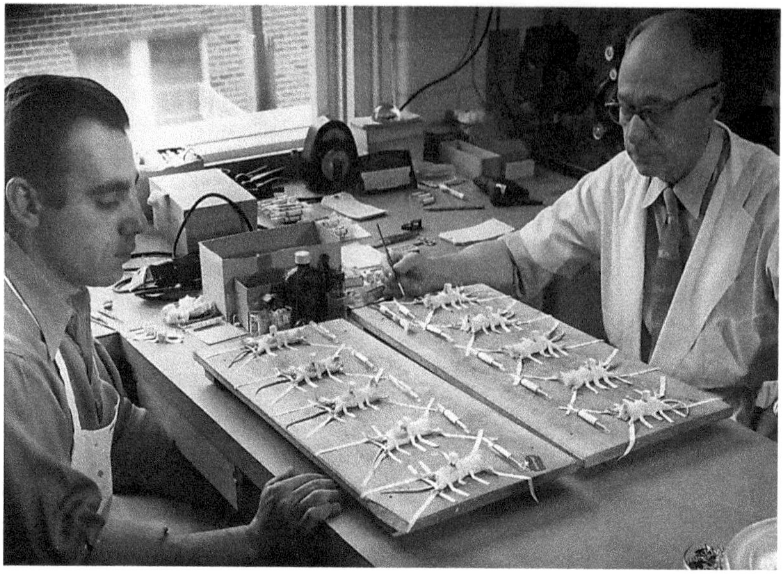

Figure 10.3 Dr Gordon E. Davis (right) performing experimental tick transmission studies at the Rocky Mountain Laboratory to demonstrate the species specificity of relapsing fever spirochaetes and *Ornithodoros* ticks.

may or may not have been found; if the ecological setting fit, the identity of the spirochaete was assumed. With the advent of a culture medium that supported the *in vitro* growth of *B. hermsii* (Kelly, 1971), and immunological and molecular techniques to identify and characterize these spirochaetes (Bunikis *et al.*, 2004; Hovis *et al.*, 2006; Ras *et al.*, 1996; Schwan *et al.*, 1992, 2007; Scott *et al.*, 2005; Uhlmann *et al.*, 2007), we now have a much better understanding of the distribution and genetic relationship of *B. hermsii* to other species of spirochaetes (Porcella *et al.*, 2005; Schwan *et al.*, 2007). Nevertheless, our understanding of the ecology of *B. hermsii* has not changed or expanded much since the early investigations during the 1930s. This is in striking contrast to the efforts that have been directed at the ecology of Lyme borreliosis since its emergence in the mid-1970s (Steere *et al.*, 1977a,b).

B. hermsii is maintained in endemic cycles including its tick vector *O. hermsi* and a variety of small vertebrate hosts in higher elevation coniferous forests (Wynns, 1936; Wynns and Beck, 1935) (Fig. 10.4). Because vertical or transovarial transmission of *B. hermsii* in *O. hermsi* is rare (Longanecker, 1951; Wheeler, 1938a,b), susceptible vertebrate hosts are essential for the spirochaetes to infect new cohorts of ticks and maintain the transmission cycle in nature. Few studies have investigated endemic foci; chipmunks and tree squirrels, however, have long been considered primary hosts for both *O. hermsi* and *B. hermsii* (Coleman, 1934; Fritz *et al.*, 2004; Trevejo *et al.*, 1998) and pine squirrels (*Tamiasciurus hudsonicus*) experience high spirochaetaemias when infected (Burgdorfer and Mavros, 1970). These diurnal rodents return to their nests at night, where the fast-feeding, nocturnal ticks parasitize them. Woodrats (*Neotoma* species) may play a role, and birds, which also are susceptible to *B. hermsii* infection, may serve as hosts for *O. hermsi* (Gregson, 1949; Schwan *et al.*, 2007; Thomas *et al.*, 2002).

Vector-borne diseases in the temperate latitudes are frequently more prevalent during warmer seasons of the year when cold-blooded arthropods are more active and abundant, resulting in higher rates of feeding on the blood of vertebrates. Such is true for relapsing fever associated with the bite of *O. hermsi*. Most human cases caused by *B. hermsii* occur during May through September, although people have been infected during all months of the year (Dworkin *et al.*, 1998). Human exposure to infected *O. hermsi* occurs most often when sleeping in recreational cabins in higher elevation forests (Dworkin *et* al., 1998), which has led to numerous, multiple case outbreaks (Boyer *et al.*, 1977; CDC, 1990; Schwan *et al.*, 2003b; Thompson *et al.*, 1969; Trevejo *et al.*, 1998). Wooden cabins serve as sturdy alternative refuges from the usual hollow trees and fallen logs in the forest for squirrels and chipmunks to nest in and overwinter. During the warmer months of the year, infected *O. hermsi*, like other ticks, become active and can use a gradient of carbon dioxide emanating from respiring vertebrates to find their hosts at night (Adeyeye and Butler, 1991; Garcia, 1962), feed quickly and return to whence they came. Human daytime activities, such as cleaning out rodent and tick infested structures in forests, can also result in tick bites and infection (Dworkin *et al.*, 1998; Schwan *et al.*, 2009a).

B. turicatae and *B. parkeri* are the two other geographically widespread species of relapsing fever spirochaetes in North America (Breitschwerdt *et al.*, 1994; Brumpt and Brumpt, 1939; Gage *et al.*, 2001; Schwan *et al.*, 2005). Again, both species of spirochaetes are restricted to their own species of tick for their maintenance and transmission in nature, *O. turicata* and *O. parkeri*, respectively (Davis, 1942a). The ecological settings where these ticks and spirochaetes live are quite different compared to the mountain forests where *O. hermsi* is found. Given the very limited knowledge of the true public health potential of *B. parkeri* (California, 1983), we will not discuss it further.

O. turicata occurs in appropriate habitats across the southern half of USA from California to Kansas, Oklahoma and Texas, south into Mexico, with a widely disjunct population in Florida (Cooley and Kohls, 1944; Davis, 1936, 1941a). These ticks occur at lower elevations in relatively dry landscapes, and are parasites of a variety of ground dwelling and burrowing animals (Cooley and Kohls, 1944). However, the few ecological studies directed at this tick have been done in Florida where the ticks occupy the sandy burrows occupied by gopher tortoises (*Gopherus polyphemus*) and other small animals that may share their lairs (Adeyeye and Butler, 1989, 1991). Large infestations of *O. turicata* larvae and nymphs have also been found on desert tortoises (*Gopherus agassizi*) in the southern

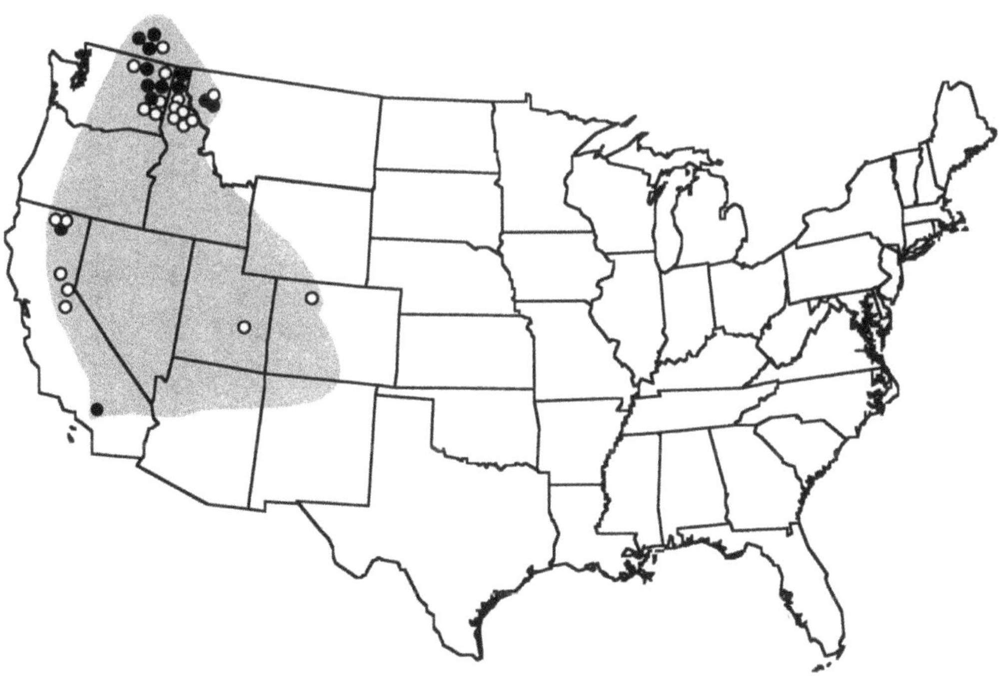

Figure 10.4 Distribution of *Borrelia hermsii* in the USA and southern British Columbia, Canada. The open circles and solid dots represent source localities for Genomic Group I and Genomic Group II spirochaetes, respectively.

California deserts (Ryckman and Kohls, 1962). One female tortoise had 110 ticks attached to the skin between the plates of the carapace, which is most striking given that these immature ticks feed for only 15 to 75 minutes on suckling mice in the laboratory (Beck *et al.*, 1986). Future laboratory efforts investigating the life cycle of *O. turicata* with tortoises as the host would be helpful to better understand this host-parasite association.

Tick-borne relapsing fever associated with the bite of *O. turicata* was first reported in Texas in 1927 (Coleman and Wright, 1942) and Kansas in 1934 (Closson, 1934). During the 1930s and 1940s, central Texas became a well defined enzootic focus with approximately 425 human cases reported for 1930 through 1941 (numbers vary with source) (Coleman and Wright, 1942; Kemp *et al.*, 1935; Rawlings, 1995). Persons most frequently infected were farmers and ranchers (Kemp *et al.*, 1935), while today most cases are associated with spelunking or occupying tick-infested caves in Texas (Davis *et al.*, 2002; Dworkin *et al.*, 2002a; Rawlings, 1995).

No isolates of *B. turicatae* have yet been established for analysis from infected humans or wild animals, yet the identity of this spirochaete as the cause of human disease is still being based solely on the association with spirochaete-infected *O. turicata* (Davis *et al.*, 2002). In contrast, several well-characterized isolates of *B. turicatae* have been cultured from the blood of clinically ill, domestic dogs in Florida and Texas (Breitschwerdt *et al.*, 1994; Schwan *et al.*, 2005; Whitney *et al.*, 2007). One case of equine abortion, in which spirochaetes were detected histologically but not isolated, was reported from southern California (Walker *et al.*, 2002). However, PCR and DNA sequence analysis of infected liver tissues demonstrated that the spirochaetes were very closely related to *B. turicatae* and *B. parkeri*. Clearly, more attention is needed to better understand the true significance of *B. turicatae* as a human and veterinary health problem in North America. Beyond a few site investigations associated with human cases, little is known about the ecology of this potentially interesting and widespread pathogen. Recent discoveries of relapsing fever spirochaetes closely related to, but distinct from, *B. turicatae* and *B. parkeri* in the bat tick *Carios kelleyi* in the central USA are quite intriguing and also await in-depth ecological and experimental studies (Gill *et al.*, 2008; Loftis *et al.*, 2005; Schwan *et al.*, 2009b).

Early studies investigating the vector competency of *O. turicata* infected with *B. turicatae* illustrate the dual role of this tick as both the vector and reservoir for this spirochaete (Davis, 1941b, 1943; Francis, 1938). First recall that the life cycle and feeding behaviour of this and other argasid ticks are vastly different compared to those of ixodid ticks, including those that transmit Lyme borreliosis spirochaetes (Balashov, 1972; Sonenshine, 1991). *O. turicata* has one larval stage followed by up to six successive nymphal stages, each requiring a blood meal, before it differentiates into male and female (Beck *et al.*, 1986). As adults, a male may feed and copulate up to 30 times over a 36-month period (Davis, 1941b). Females also feed repeatedly over many months to years, producing eggs after each blood meal. Thus this tick has tremendous longevity compared to *Ixodes scapularis* that has only three stages, larva, one nymph, and adults, which then reproduce and die after a single blood meal at the end of a 2-year life cycle (Dammin, 1986; Spielman *et al.*, 1979).

O. turicata also vertically transmits *B. turicatae* efficiently through both transovarial and transstadial transmission (Davis, 1943; Francis, 1938). In an impressive series of tick feeding experiments, Davis documented the maintenance of *B. turicatae* through five successive generations of ticks spanning five years while feeding them only on uninfected mice; the prevalence of larvae infected transovarially varied from 35% to 100% with each new generation (Davis, 1943). Additionally, *O. turicata* can survive long periods of time under the right conditions without a blood meal, as can many other species of argasid ticks (Balashov, 1972; Sonenshine, 1991). Francis (Francis, 1938) reported that infected, adult *O. turicata* that fasted for 3, 4 and 5 years were able to transmit *B. turicatae* when fed on rhesus monkeys. Additionally, one group of ticks that had transmitted spirochaetes after fasting for 4 years, transmitted spirochaetes again when fed 2.5 years later. The ability of these ticks to repeatedly transmit spirochaetes is also quite remarkable. Davis followed 12 infected male *O. turicata* during successive blood-meals for 1 to 3 years. During this time one tick transmitted spirochaetes during all 23 times it fed on mice (Davis, 1943).

10.5 Ecology of Lyme disease spirochaetes

10.5.1 The tick vectors

The four principal hard tick (Ixodidae) species that transmit Lyme disease spirochaetes (*Borrelia burgdorferi* sensu lato) include *Ixodes scapularis* in eastern North America, *I. pacificus* in western North America (Fig. 10.2), *I. persulcatus* in Asia, and *I. ricinus* in Europe. This group of closely related ticks is in the genus *Ixodes,* subgenus *Ixodes*. Taxonomically, they have been placed within a species complex originally called the *I. persulcatus* complex (Filippova, 1991), but now generally called the *I. ricinus* complex (Filippova, 1999; Keirans *et al.*, 1999; Xu *et al.*, 2003). These ticks are basically forest dwellers, spending most of their time hiding in the leaf litter of the forest floor, where humidity is high and the risk of desiccation is low. The immature stages of the tick, the six-legged larvae and the eight-legged nymphs, quest on low lying vegetation and tend to infest smaller hosts to obtain their blood meals, especially rodents, insectivores, and birds. The dimorphic adult ticks, tend to quest higher up on vegetation and generally infest larger hosts, especially ungulates like deer; females take a full blood meal, drop off the hosts, and oviposit into vegetation and die. Male *Ixodes* transfer sperm to female ticks either on or off the host, but generally do not ingest blood. In many ways, these ticks are ideally suited as vectors of zoonotic pathogens, since they feed on a wide variety of animals but also include humans within the range of hosts they willing feed upon.

In eastern North America, *I. scapularis*, commonly called the blacklegged tick, is found over a wide area in forests east of the 100th meridian (Dennis *et al.*, 1998; Diuk-Wasser *et al.*, 2006; Ogden *et al.*, 2008) (Fig. 10.5). The historical distribution records of this tick species suggests

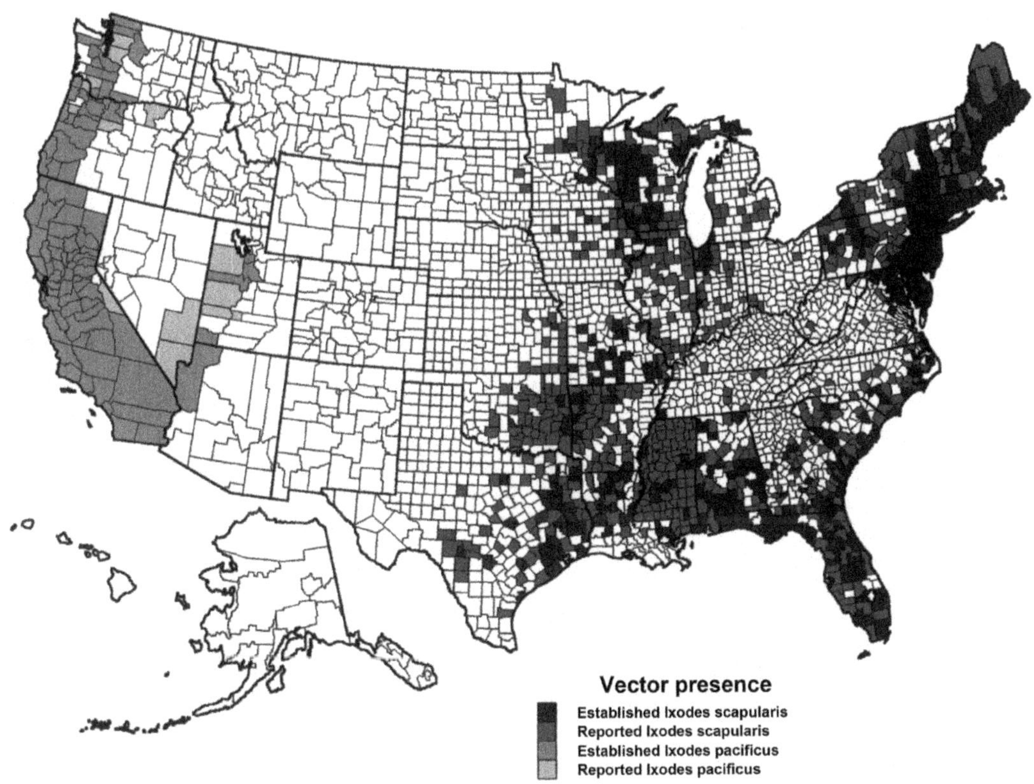

Figure 10.5 Distribution of the tick vectors associated with Lyme disease in the USA. A colour version of this figure is located in the plate section at the back of the book.

that it was once plentiful wherever deer and forests were found in eastern North America, but, around 1900, widespread deforestation for agriculture and aggressive hunting of white-tailed deer (*Odocoileus virginianus*) reduced the population of *I. scapularis* to a few isolated refugia. After World War II ended (1945), changes in lifestyle allowed dense forests to bloom in suburban settings and deer to flourish in the eastern USA. Accordingly, during this period, the deer population in eastern North America has dramatically increased from perhaps hundreds to millions (Kilpatrick and LaBonte, 2007). With reforestation and the resurgence of the deer population, the increased density of *I. scapularis* caught the attention of various scientists in the 1970s right around the time the first case of erythema migrans was recognized in North America (Scrimenti, 1970) (see Chapter 1).

The current overall distribution of *I. scapularis* covers much of the eastern half of North America from the Atlantic Coast up to the 100th meridian; the 100th meridian is where the relatively dry and open Great Plains begins, with conditions inhospitable for this moisture- and forest-loving tick. There are discrete gaps in the distribution of *I. scapularis* in the Appalachian Mountain region as well as parts of the heavily agricultural Midwestern USA (Dennis *et al.*, 1998). In far northern Canada, temperatures may be too frigid for *I. scapularis* (Lindsay *et al.*, 1995). Although *I. scapularis* is widely distributed as a species, there are only a few unique foci where the nymphal stage of this tick (the key stage for transmission, as detailed below) openly quests on the forest floor and frequently encounters people. The three general areas where these conditions occur are: (1) the northeastern USA from coastal Maryland through coastal Maine, (2) the Midwestern USA, especially along the Minnesota–Wisconsin border, and (3) southern Ontario and Quebec (Dennis *et al.*, 1998; Diuk-Wasser *et al.*, 2006; Lindsay *et al.*, 1999; Ogden *et al.*, 2006). The implications for transmission of the Lyme disease spirochaete to humans as a consequence of nymphal *I. scapularis* questing habits are, indeed, profound. In regions hyperendemic for Lyme disease in the northeastern USA, hundreds of nymphal *I. scapularis* can be collected in just a few hours of flagging vegetation, a process that involves dragging a cloth metre square flag through forested habitat (Piesman *et al.*, 1987a; Rodgers *et al.*, 2007; Stafford 3rd *et al.*, 1998a); human tick-bites by nymphal *I. scapularis* are extremely common in this region from mid-May to mid-July and the risk of Lyme disease is extremely high (CDC, 2002; Falco and Fish, 1988; Falco *et al.*, 1996). In contrast, nymphal *I. scapularis* in the southern USA are incredibly difficult to collect by flagging and rarely bite humans. During a week of flagging in areas of Georgia and Mississippi, where adult *I. scapularis* were known to be abundant, less than a dozen nymphal *I. scapularis* where collected (Piesman, 2002; Goddard and Piesman, 2006). A comparable flagging effort in the northeastern USA in the Lyme disease endemic regions might have produced thousands of nymphs (JP, personal observations). Moreover, nymphal *I. scapularis* rarely bite humans in the southern USA; in a review of human tick-bites, only a single nymphal *I. scapularis* was reported from this region (Merten and Durden, 2000).

Why are the biting habits of the nymphal stage of *I. scapularis* so important to the ecology and epidemiology of Lyme disease? Laboratory research using murine models has demonstrated that the longer *I. scapularis* is attached, the greater the risk that these ticks will transmit an infectious dose of *B. burgdorferi* sensu stricto; transmission does not take place before 24 hours of attachment, is rare from 24–48 hours and increases during the third day of attachment (des Vignes *et al.*, 2001; Hojgaard *et al.*, 2008; Piesman *et al.*, 1987b). This delay is due to an alteration in the outer surface proteins of the spirochaetes from OspA to OspC during tick feeding, which allows the spirochaetes to move from the midgut of the tick to the salivary glands (see Chapters 4 and 11). A key to this migration is that when OspA is downregulated spirochaetes are no longer bound to a midgut receptor protein called TROSPA. Once the spirochaetes get to the salivary glands the OspC they are expressing binds a protein called Salp15, which may protect the spirochaetes from some aspects of the host response when they are transferred to vertebrates (Hovius *et al.*, 2007; Pal *et al.*, 2004; Ramamoorthi *et al.*, 2005; Rosa, 2005). Since nymphs are quite small (<2 mm) compared to the much larger adult females, nymphs are able to feed much longer than adults before detection and removal by the tick-bite victim (Falco *et al.*, 1996). Consequently, human Lyme disease cases in the northeastern and Midwestern USA occur mainly during the summer months (CDC, 2002, 2007), coinciding with the period of biting activity of nymphal *I. scapularis.* This is also a time when adult *I. scapularis,* which feed mainly during the winter months, are inactive. A high proportion of nymphal *I. scapularis* in the northeastern and Midwestern USA are infected with *B. burgdorferi* sensu stricto. The proportion infected in endemic regions varies from 15% to 30%, with an average of about 25% infection in large samples of nymphal *I. scapularis* (Maupin *et al.*, 1991; Wang *et al.*, 2003). In summary, nymphal *I. scapularis* in the northeastern USA and Midwestern USA are commonly infected with Lyme disease spirochaetes, frequently bite humans, often escape detection by the tick-bite victim, and feed long enough to transmit infectious spirochaetes. They are exquisite vectors of Lyme disease spirochaetes. In contrast, in the southern USA, nymphal *I. scapularis* rarely quest on vegetation, rarely infest humans, and although infection with *B. burgdorferi* sensu stricto has been reported from southern populations of *I. scapularis* (Lin *et al.*, 2004; Oliver Jr. *et al.*, 2003) such infection is exceedingly rare (Piesman *et al.*, 1999).

There has been taxonomic controversy concerning the northern and southern populations of *I. scapularis.* Based on morphological and behavioural differences, the northern populations of *I. scapularis* were named as a separate species, *Ixodes dammini* (Spielman *et al.*, 1979), but this species determination subsequently was challenged on morphometric grounds (Hutcheson *et al.*, 1995). Molecular studies have generally shown that there are two clades of *I. scapularis*: the extremely genetically diverse southern clade and the less diverse but more widespread 'northern' clade, also called the 'All-American' clade (Norris *et al.*, 1996; Qiu *et al.*, 2002; Rich *et al.*, 1995). Although differences in transmission risk of *B. burgdorferi* in the northern and southern USA has been suggested to be due to the genetic differences between the two clades of *I. scapularis* (Telford 3rd, 1998), doubt is thrown on this hypothesis since the 'northern' or 'all-American' clade is distributed widely from Mississippi to Maine yet the cycle of *B. burgdorferi* is inefficient in southern clines. The distribution of suitable hosts for immature ticks, including the abundance of lizards in the southern USA (a host refractory to *B. burgdorferi* sensu stricto) may have more to do with the potential for the Lyme disease spirochaete cycle to be maintained efficiently, than the genetic background of the tick populations. More research on this question is certainly required.

Ixodes pacificus, the western blacklegged tick, is the principal vector of Lyme disease spirochaetes in the western USA and British Columbia, Canada (Banerjee *et al.*, 1994; Lane *et al.*, 1991) (Fig. 10.5). Although this tick can be found in isolated deciduous forests in the drier parts of North America, including Arizona and Utah, they are mainly found along the humid areas of the Pacific Coast (Dennis *et al.*, 1998; Kain *et al.*, 1999; Kain *et al.*, 1997). Nymphal *I. pacificus* were originally thought to be difficult to collect from vegetation, but isolated foci of questing nymphal *I. pacificus* infected with *B. burgdorferi* sensu lato were subsequently found in forested habitats (Clover and Lane, 1995). A model based on the ideal woodland habitat where questing nymphal *I. pacificus* can be collected in abundance, based on observations in Mendicino County, California, has been extended to predict the focal risk of Lyme disease spirochaete transmission in the state of California (Eisen *et al.*, 2006). Interestingly, nymphal *I. pacificus* seem to be more easily collected from downed logs and tree trunks, as compared to the forest floor (Lane *et al.*, 2007). In general, people encounter *I. pacificus* far from their homes, as a result of recreational rather than domestic pursuits. Nevertheless, instances of peridomestic transmission of *B. burgdorferi* by *I. pacificus* have been reported (Lane *et al.*, 1991).

Ixodes persulcatus, the taiga tick, is widely distributed over a huge part of the globe in Eurasia, from Japan in the Pacific Ocean through China and Russia up to the Baltic Sea (Korenberg *et al.*, 1987, 2001, 2002). Within the boundaries of the Russian Federation, *I. persulcatus* has been extensively studied as the principal tick vector of the extremely pathogenic Tick-Borne Encephalitis Virus (TBEV) (Korenberg and Kovalevskii, 1999). This tick can be found in broad-leaved, mixed or southern taiga forests (Filippova, 1991). The taiga is a unique habitat present in much of Siberia that is characterized by predominantly coniferous forests overlying formerly glaciated or permafrost areas. In the Russian Federation, *I. persulcatus* appears to be a more efficient vector of Lyme disease spirochaetes than does the European species, *I. ricinus* (Korenberg *et al.*, 2002; Korenberg *et al.*, 2001). Interestingly, most cases of Lyme disease in the Russian Federation within the range of *I. persulcatus* that can be traced to specific tick bites are associated with bites by the adult stage of the tick rather than the nymphal stage (Korenberg *et al.*, 2002). Whether this is due to the strains of *B. burgdorferi* sensu lato present within the range of *I. persulcatus* or some aspect of *I. persulcatus* feeding dynamics is presently unknown.

Ixodes ricinus, known commonly as the castor bean tick or sheep tick, is the principal vector of Lyme disease spirochaetes in Europe. It is distributed along a northsouth axis extending from Northern Scandinavia (Lindgren *et al.*, 2000) to the Kroumiry Mountains in Tunisia (Zhioua *et al.*, 1999), and along an east–west axis extending from Ireland (Grey *et al.*, 1999) to the European section of the Russian Federation (Korenberg *et al.*, 2002). *Ixodes ricinus* infests heavily populated areas of Europe and may possibly transmit more infections with *B. burgdorferi* than does any other tick species. At least 60,000 cases of Lyme disease were estimated to occur per year within the distribution of *I. ricinus* (O'Connell *et al.*, 1998). In the majority of its range, *I. ricinus* is found in heavily forested areas, including montane regions up to an altitude of 1350 m (Stünzner *et al.*, 2006). Although *I. ricinus*, like the other members of the *I. ricinus* complex is mainly a forest dwelling creature, it can be found on open pasture in areas that are subject to heavy rainfall. A series of classic papers was published around 1950 that described in exquisite detail the ecology of *I. ricinus* in sheep pastures (Lees and Milne, 1951; Milne, 1948, 1950a, 1950b). Immature *I. ricinus* have been reported to infest an extremely wide range of hosts including rodents, birds, insectivores, and lagomorphs, as well as medium sized mammals; adult *I. ricinus* feed on larger hosts, including roe deer (*Capreolus capreolus*), sheep, and also wild boar (*Sus scrofa*) (Humair *et al.*, 2007; Moran Cadenas *et al.*, 2007). Although birds and mammals serve as the primary hosts for immature *I. ricinus* over much of its range, in the southern extremes of its range, it can be found primarily infesting lizards (Amore *et al.*, 2007; Dsouli *et al.*, 2006). Both the nymphal and adult female stages of *I. ricinus* aggressively bite humans (Fingerle *et al.*, 1998). Unlike nymphal *I. scapularis*, nymphal *I. ricinus* can transmit an infectious dose of *B. burgdorferi* sensu lato, or more specifically *B. afzelii*, during the first day of tick attachment; but, as with *I. scapularis*, the risk of transmission of Lyme disease spirochaetes increases with the duration of nymphal *I. ricinus* attachment (Crippa *et al.*, 2002; Kahl *et al.*, 1998).

10.5.2 The spirochaetes and hosts

Of the many genospecies of *B. burgdorferi* sensu lato, *B. burgdorferi* sensu stricto, *B. garinii*, and *B. afzelii* are the principal genospecies causing human disease. Of these three genospecies, *B. burgdorferi* sensu stricto is unique in that it has a Holarctic distribution, meaning it spans both the Old World (Palearctic Region) and the New World (Nearctic Region) in the northern hemisphere. Specific rodents were originally identified as the principal reservoirs of *B. burgdorferi* sensu stricto, including the white footed mouse (*Peromyscus leucopus*) in North America (Anderson *et al.*, 1987; Donahue *et al.*, 1987), and the wood mouse (*Apodemus sylvaticus*), yellow-necked mouse (*Apodemus flavicollis*), and bank vole (*Clethrionomys glareolus*) in Europe (De Boer *et al.*, 1993; Kurtenbach *et al.*, 1994; Tälleklint *et al.*, 1993). In some areas of Europe, where *I. ricinus* is the predominant vector, tree squirrels were shown to be important reservoirs of *B. burgdorferi* sensu stricto (Humair and Gern, 1998); a similar observation has been made in California, where *I. pacificus* is the principal vector (Lane *et al.*, 2005). Chipmunks also play a role as important reservoirs on both sides of the Atlantic (Schulze *et al.*, 2005; Slajchert *et al.*, 1997; Vourc'h *et al.*, 2007). However, *B. burgdorferi* sensu stricto can clearly infect an extremely wide range of hosts, besides rodents. On islands off the coast of Sweden, lagomorphs alone, in the absence of rodents, can efficiently maintain the *B. burgdorferi* sensu stricto cycle (Tälleklint and Jaenson, 1993).

Birds also have been identified as reservoirs of *B. burgdorferi* sensu stricto on both sides of the Atlantic. In North America, a wide variety of birds are infected with *B. burgdorferi* sensu stricto, and these birds have been found to be competent reservoirs infecting larval *I. scapularis* ticks feeding on them (Ginsberg *et al.*, 2005; Richter *et al.*, 2000; Stafford 3rd *et al.*, 1995). In general, birds that spend most of their time on the ground, either ground feeding or ground nesting or both play the greatest role as reservoirs for *B. burgdorferi* sensu stricto in North America. The American robin (*Turdus migratorius*) in particular may be an important avian reservoir of *B. burgdorferi* sensu stricto in North America in suburban habitats where Lyme disease is endemic. In addition to playing a role as reservoirs of *B. burgdorferi* sensu stricto, birds may play an extremely important role in bringing the spirochaete to new regions, acting as 'phoretic' hosts. An extremely comprehensive study in Canada, along the border with the USA, demonstrated that birds introduce spirochaete-infected ticks into areas where the *B. burgdorferi* enzootic cycle has not been previously established (Ogden *et al.*, 2008). In Europe, birds also play a role as reservoirs of *B. burgdorferi* sensu stricto, but other genospecies are more common in European birds than is *B. burgdorferi* sensu stricto (Comstedt *et al.*, 2006; Hanincova *et al.*, 2003a; Humair *et al.*, 1998).

Besides rodents, lagomorphs, and birds, *B. burgdorferi* sensu stricto has been shown to utilize insectivores heavily to infect ticks. Indeed, in certain areas shrews may infect more ticks with *B. burgdorferi* sensu stricto than all the other potential reservoir hosts combined (Brisson *et al.*, 2008). Comprehensive studies on the spectrum of reservoirs serving to infect ticks with *B. burgdorferi* sensu stricto have been conducted in several locations (Brisson and Dykhuizen, 2006; Hanincova *et al.*, 2006; LoGiudice *et al.*, 2003). In summary, the breadth of hosts that *B. burgdorferi* can utilize as reservoirs is extremely diverse, and varies from locale to locale.

In Europe, *B. garinii* and *B. afzelii* are more common in patients, ticks, and hosts than is *B. burgdorferi* sensu stricto (Cadenas *et al.*, 2007; Hanincova *et al.*, 2003a; Ruzić-Sabljić *et al.*, 2001; Taragel'ová *et al.*, 2008). Generally, *B. garinii* utilizes birds as reservoir hosts (Comstedt *et al.*, 2006; Taragel'ová *et al.*, 2008), whereas, in sharp contrast, *B. afzelii* utilizes rodents as reservoir hosts (Hanincova *et al.*, 2003b; Humair *et al.*, 1999). This dichotomy in the host distribution of *B. garinii* and *B. afzelii* is thought to be due to a difference in the sensitivity of these genospecies to host serum complement. In an elegant series of experiments, Kurtenbach *et al.* (1998) demonstrated that rodent sera were borreliacidal for *B. garinii*; in contrast, avian sera were borreliacidal for *B. afzelii*. Subsequent research has demonstrated that molecules on the surface of *Borrelia* called CRASPs (complement regulator-acquiring surface proteins) bind host immune regulators (FHL-1 and factor H) that protect spirochaetes from complement lysis (Haupt *et al.*, 2007; Kraiczy *et al.*, 2006; Kraiczy and Wurzner, 2006; Wallich *et al.*, 2005). Essentially, most *B. garinii* strains have the proper repertoire of CRASPs to bind avian immune regulators to resist lysis by avian sera, whereas *B. afzelii* have a repertoire of CRASPs that allow them to resist lysis by rodent sera. The one exception to this rule is that *B. garinii* serotype 4 is able to resist lysis in rodent sera and can be found utilizing rodents as reservoir hosts (Hu *et al.*, 2001; Huegli *et al.*, 2002).

In addition to the three major genospecies commonly causing human Lyme disease (*B. burgdorferi* sensu stricto, *B. garinii*, and *B. afzelii*), other less common genospecies are often detected in ticks and occasionally cause human disease. The recently described genospecies *B. spielmanii* (Richter *et al.*, 2006) was originally reported as *B. burgdorferi* sensu lato strain A14S isolated from a patient in the Netherlands (van Dam *et al.*, 1993; Wang *et al.*, 1999). Apparently, *B. spielmanii* has been associated with erythema migrans in discrete foci (Fingerle *et al.*, 2008); the key reservoir host for this genospecies is the garden dormice (*Eliomys quercinus*) (Richter *et al.*, 2004). Mapping how closely human disease caused by *B. spielmanii* reflects the distribution of the rodent reservoir host *E. quercinus* should be interesting.

The genospecies *B. lusitaniae* was originally isolated from ticks in Portugal (Le Fleche *et al.*, 1997). Although *B. lusitaniae* is widely distributed in Europe, it is particularly common in ticks from the Mediterranean Basin in southern Europe and northern Africa (De Michelis *et al.*, 2000; Grego *et al.*, 2007; Younsi *et al.*, 2005). Moreover, *B. lusitaniae* has been demonstrated to cause human disease (Collares-Pereira *et al.*, 2004; da Franca *et al.*, 2005). The reservoirs of *B. lusitaniae* were a mystery for some time, but lizards were recently shown to competently infect *I. ricinus* with *B. lusitaniae* both in the laboratory and in the field (Amore *et al.*, 2007; Dsouli *et al.*, 2006; Majlathova *et al.*, 2006; Richter and Matuschka, 2006). This is especially surprising since lizard sera have been found to be lytic to a variety of *Borrelia* genospecies (Kuo *et al.*, 2000; Ullmann *et al.*, 2003). One of the reasons why Lyme disease is thought to be rare in the southern USA is due to the abundance of zooprophylactic lizards serving as hosts for immature *I. scapularis* (Spielman *et al.*, 1985). The manner in which *B. lusitaniae* resists lysis by reptilian sera has not yet been determined, but characterization of the CRASPs present in *B. lusitaniae* should be interesting.

The importance of the various other genospecies of *B. burgdorferi* sensu lato to public health is questionable but several of them have interesting ecological niches. In western North America, *B. bissettii* has been frequently identified infecting woodrats (*Neotoma* spp.) and the tick *I. spinipalpis* in both Colorado (Schneider *et al.*, 2000) and California (Brown *et al.*, 2006). Although there have been isolated reports of *B. bissettii* in ticks or patients from eastern North America and Europe (Strle *et al.*, 1997), a definitive enzootic cycle for *B. bissettii* outside of western North America has not been established. The genospecies *B. califoriensis* has recently been described circulating in kangaroo rats (*Dipodomys californicus*) in arid climates of the western USA, with the tick *I. jellisonii* as a suspect vector (Postic *et al.*, 2007). In the eastern USA, the genospecies *B. andersonii* has been closely associated with cottontail rabbits (*Sylvilagus floridanus*) and the rabbit feeding tick, *I. dentatus* (Anderson *et al.*, 1989; Marconi *et al.*, 1995).

In Europe and Asia, *B. valaisiana* is associated with birds and ticks feeding on birds, and often infects the same ticks and hosts as *B. garinii* (Hanincova *et al.*, 2003a; Taragel'ová *et al.*, 2008). Interestingly, a human case of infection with *B. valaisiana* was reported in a Japanese man with a complicated travel history (Saito *et al.*, 2007). Furthermore, in Japan isolates of three additional genospecies have been reported: *B. tanukii* from *I. tanuki* ticks, *B. turdae* from *I. turdus*, and *B. japonicum* from *I. ovatus* (Fukunaga *et al.*, 1996a,b). In general, *I. tanuki* feeds on rodents and medium-sized mammals, *I. turdus* feeds exclusively on birds, and *I. ovatus* has been associated with shrews in Japan. Probably the least known genospecies ecologically is the genospecies *B. sinica*, isolated from *Niviventer confucianus*, the Chinese white-bellied rat captured in Sichuan province, China, and a single *I. ovatus* tick captured in Anhui province, China (Masuzawa *et al.*, 2001). Clearly, more ecological studies are needed to further understand these relatively rare Asian genospecies of *B. burgdorferi* sensu lato.

10.6 Ecology of hard tick relapsing fever spirochaetes

Although the majority of relapsing fever spirochaetes occur in soft ticks (*Argasidae*), several relapsing fever group spirochaetes have been associated with hard ticks (*Ixodidae*). The first hard tick relapsing fever spirochaete studied was an organism seen in blood smears of cattle by Sir Arnold Theiler in South Africa in 1902 (Callow, 1967). It was named *Borrelia theileri* and described from cattle and ticks in Africa, Australia, and South America. The ticks that transmit this spirochaete were originally placed in two separate genera (*Rhipicephalus* and *Boophilus*) but all these tick species are now placed within the genus *Rhipicephalus*, including the known vectors of *B. theileri*: *R. eversti*, *R.* (*Boophilus*) *microplus*, *R.* (*Boophilus*) *annulatus*, and *R.* (*Boophilus*) *decoloratus*. Experimental studies demonstrated that *B. theileri* caused a fever of short duration in cattle that coincided with the presence of spirochaetes in blood smears; the spirochaetes passed from the female ticks of *R.* (*Boophilus*) *microplus* through the ovaries to larval ticks that subsequently transmitted infection to other cattle (Smith *et al.*, 1978). Unfortunately, *B. theileri* has not been grown in culture media to date.

Another hard tick relapsing fever spirochaete was discovered as a result of the search to determine whether a microorganism transmitted by the Lone star tick (*Amblyomma americanum*) caused the condition known as southern tick-associated rash illness, an erythema migrans-like condition that results from *A. americanum* bites. Many researchers observed spirochaetes in *A. americanum*, but noticed that these spirochaetes differed from *B. burgdorferi*. Molecular characterization of the *A. americanum* spirochaetes

(Barbour *et al.*, 1996) established that these spirochaetes were quite distinct from *B. burgdorferi* sensu lato and more closely related to relapsing fever spirochaetes; the spirochaete was tentatively named *B. lonestari* (Barbour *et al.*, 1996). In rapid succession, *B. lonestari* was characterized by molecular phylogenetic studies as being closely related to *B. theileri* (Rich *et al.*, 2001), cultured *in vitro* in the presence of tick cell lines (Varela *et al.*, 2004), demonstrated to be infective to white-tailed deer (Moore 4th *et al.*, 2003), and found to pass transovarially from female ticks to larval offspring (Stromdahl *et al.*, 2003). Ironically, despite this rapid progress on our knowledge of the ecology and transmission of *B. lonestari*, we are still uncertain as to whether *B. lonestari* actually causes the erythema migrans-like rash associated with *A. americanum* bites in humans.

An additional hard tick relapsing fever spirochaete was isolated on the northern island of Japan (Hokkaido), from midguts from four *I. persulcatus* ticks and from the blood of a single rodent (*Apodemus argenteus*) (Fukunaga and Koreki, 1995). All of these strains originally grew in BSK culture medium and several of them including the type strain HT31 is widely available for study. A series of studies attempting to characterize the molecular profile of spirochaetes found in ticks of the *I. ricinus* 'complex', soon discovered that spirochaetes extremely similar to *B. miyamotoi* could be found in *I. scapularis* (Scoles et al., 2001), *I. ricinus* (Fraenkel *et al.*, 2002), and *I. pacificus* (Mun *et al.*, 2006). An in-depth molecular study of *B. miyamotoi* spirochaetes from *I. ricinus* in Sweden and *I. scapularis* in Connecticut, USA, demonstrated extremely limited diversity between these widely geographically separated populations (Bunikis *et al.*, 2004). Accordingly, this group of spirochaetes has been called *B. miyamotoi* sensu lato wherever it is found in *Ixodes* ticks (Bunikis *et al.*, 2004).

Interestingly, hard tick relapsing fever spirochaetes share the pathway in which they hydrolyse phospholipids through the use of glycerophosphodiester phosphodiesterase (GlpQ) with soft tick relapsing fever spirochaetes (Bacon *et al.*, 2004; Schwan *et al.*, 2003a; Ullmann *et al.*, 2005; Yparraguirre *et al.*, 2007); the use of GlpQ may allow these spirochaetes to achieve high cell densities in vertebrate blood (Schwan *et al.*, 2003a). However, whether hard tick relapsing fever spirochaetes actually must amplify in a vertebrate host or rely exclusively on transovarial passage in the tick, passing from female to larval tick, to survive is not currently known. Larval *A. americanum* are transovarially infected with *B. lonestari* prior to feeding on a vertebrate host (Stromdahl *et al.*, 2003). Indeed, many reports of transovarial transmission of spirochaetes in the *B. burgdorferi* sensu lato group were performed prior to the identification of *B. miyamotoi* sensu lato in *I. ricinus* and *I. scapularis*. Studies to determine transovarial passage of *Borrelia* in *Ixodes* ticks should now be performed using molecular tools to discriminate *B. burgdorferi* sensu lato from *B. miyamotoi* sensu lato.

10.7 Prevention of diseases caused by *Borrelia*

More prevention-related research has been directed towards the prevention of Lyme borreliosis than any other tick-borne disease. With the introduction in 1999 of an OspA-based recombinant vaccine against *B. burgdorferi* in the USA (Steere *et al.*, 1998), hopes were rapidly raised in the campaign to prevent Lyme borreliosis. These hopes were dashed, however, when the vaccine was withdrawn from the market in 2002. Personal protection and tick control remain as the two primary pillars in the fight against Lyme borreliosis in Asia, Europe and North America.

Several guides to an integrated pest management (IPM) or integrated tick management (ITM) approach to tick control for the prevention of Lyme disease are available (Schulze and Jordan, 2006; Stafford 3rd, 2007). Inherent in a discussion of tick control is the basic paradox that although application of traditional chemical acaricides, like deltamethrin, is extremely effective in killing *Ixodes* ticks (Schulze *et al.*, 2001), residents of Lyme disease-endemic regions in both North America and Europe are extremely reluctant to use area-wide acaricides (Piesman, 2006; Piesman and Eisen, 2008). This reluctance has spurred research on alternative tick control methodologies. Landscape modifications through vegetation management such as burning, clearing brush and mowing, removing leaf litter, or placing wood chip barriers between lawns and forest edge have been tested (Piesman, 2006; Schulze *et al.*, 1995; Stafford 3rd *et al.*, 1998b; Wilson, 1986).

Host targeted applications of acaricides directed towards rodents or deer have also been employed (Carroll *et al.*, 2002; Dolan *et al.*, 2004; Mather *et al.*, 1987; Pound *et al.*, 2000), as have least toxic pesticides including desiccants and soaps (Patrican and Allan, 1995). An especially promising avenue towards increasing community participation in tick control efforts is the utilization of biological control methods, including the use of the entomopathogenic fungus, *Metarhizium anisopliae* to kill *Ixodes* ticks (Benjamin *et al.*, 2002). Rather than using a single method to control ticks, the combination of several methods into an ITM campaign, has been shown to be effective (Schulze *et al.*, 2007). The future of tick control for Lyme borreliosis prevention may brighten if residents are offered a variety of choices, with each homeowner or property manager selecting their most acceptable choice.

Personal protection against Lyme borreliosis requires multiple measures: application of tick repellents to skin or clothing, avoidance of tick infested habitats, daily tick checks during the height of the nymphal tick season, and prompt removal of attached ticks. Residents and visitors to Lyme borreliosis-endemic regions such as Martha's Vineyard Island will employ personal protection measures if they believe that the potential benefit outweighs the perceived inconvenience (Shadick *et al.*, 1997). If residents of endemic regions fail to prevent tick bite, but detect and remove an attached tick, antibiotic prophylaxis may be an option (Nadelman *et al.*, 2001; Wormser *et al.*, 2006).

Although much less prevention-based research has been directed towards relapsing fever than towards Lyme borreliosis, some basic preventative strategies have been employed. In the western USA, where tick-borne borne relapsing fever is associated with indoor rodent reservoirs and rodent associated ticks (*O. hermsi*), rodent proofing rustic cabins or houses can be the first line of defence against tick-borne relapsing fever (Boyer *et al.*, 1977; Paul *et al.*, 2002). Spraying of infested cabins with insecticides need only be performed if the rodent proofing measures have failed. In Israel, prompt antibiotic prophylaxis of persons exposed to *B. persica* through the bite of *O. tholozani* was 100% effective in preventing tick-borne relapsing fever (Hasin *et al.*, 2006b). In Africa, measures to intervene and stop epidemics of louse-borne relapsing fever include aggressive antibiotic treatment of cases combined with delousing in refugee transit camps (Sundnes and Haimanot, 1993).

10.8 Conclusions

In spite of over a century of work on borrelioses, we still know relatively little about the diverse physical, biological and ecological factors that are necessary to maintain endemic foci of tick-borne relapsing fever. This is due, in part, to the relatively few human cases that are recognized each year, and as a result, the lack of adequate funding to support such research. Earlier investigations were also hampered by the inability to culture these bacteria and characterize them at the molecular level, well beyond just what species of tick the spirochaetes were associated with in nature. Even today, as more microbiologists expand their research to include borreliae other than those that cause Lyme borreliosis, few investigators are working either in the field or with the ticks that transmit these spirochaetes. In many endemic areas, we still do not have an understanding of what species of wild animals are responsible for maintaining these spirochaetes, and are thus an essential link in the chain of infection to humans. Until we have a much better understanding of the distribution of these pathogens in nature, and what are the critical enzootic hosts, humans will continue to be infected, misdiagnosed and suffer prolonged, needless and costly illnesses.

Since the discovery of the Lyme borreliosis spirochaete in 1982, a plethora of knowledge concerning the ecology of the spirochaete, its tick vectors, and vertebrate hosts has been brought forth through an incredible focus of research efforts. We now understand much about the ecology of Lyme borreliosis from islands off the coast of Massachusetts to the taiga forests of Siberia. The flexibility and plasticity of the Lyme borreliosis spirochaete cycle in nature makes research in the field of Lyme borreliosis ecology fascinating, but requires caution in extrapolating results obtained in one ecological setting to different habitats. The task before the public health research community is not solely producing more ecological knowledge regarding the Lyme borreliosis spirochaetes cycle in nature. Clearly, our immediate challenge is to synthesize and harness available knowledge of spirochaete–tick–host interactions to develop ecologically based intervention techniques that will be widely accepted by the public, cost-effective, and implemented on a large enough scale to reduce the human risk of Lyme borreliosis. The challenge is great, but the potential public health benefit of reducing Lyme borreliosis spirochaete transmission risk is substantial.

References

Adeyeye, O.A., and Butler, J.F. (1989). Population structure and seasonal intra-burrow movement of *Ornithodoros turicata* (Acari: Argasidae) in gopher tortoise burrows. J. Med. Entomol. *26*, 279–283.

Adeyeye, O.A., and Butler, J.F. (1991). Field evaluation of carbon dioxide baits for sampling *Ornithodoros turicata* (Acari: Argasidae) in gopher tortoise burrows. J. Med. Entomol. *28*, 45–48.

Amore, G., Tomassone, L., Grego, E., Ragagli, C., Bertolotti, L., Nebbia, P., Rosati, S., and Mannelli, A. (2007). *Borrelia lusitaniae*in immature *Ixodes ricinus* (Acari: Ixodidae) feeding on common wall lizards in Tuscany, central Italy. J. Med. Entomol. *44*, 303–307.

Anderson, J.F., Johnson, R.C., and Magnarelli, L.A. (1987). Seasonal prevalence of *Borrelia burgdorferi* in natural populations of white-footed mice, *Peromyscus leucopus*. J. Clin. Microbiol. *25*, 1564–1566.

Anderson, J.F., Magnarelli, L.A., LeFebvre, R.B., Andreadis, T.G., McAninch, J.B., Perng, G.-C., and Johnson, R.C. (1989). Antigenically variable *Borrelia burgdorferi* isolated from cottontail rabbits and *Ixodes dentatus* in rural and urban areas. J. Clin. Microbiol. *27*, 13–20.

Assous, M.V., Wilamowski, A., Bercovier, H., and Marva, E. (2006). Molecular characterization of tickborne relapsing fever *Borrelia*, Israel. Emerg. Infect. Dis. *12*, 1740–1743.

Bacon, R.M., Pilgard, M.A., Johnson, B.J.B., Raffel, S.J., and Schwan, T.G. (2004). Glycerophosphodiester phosphodiesterase gene (*glpQ*) of *Borrelia lonestari* identified as a target for differentiating *Borrelia* species associated with hard ticks (Acari: Ixodidae). J. Clin. Microbiol. *42*, 2326–2328.

Balashov, Y.S. (1972). Bloodsucking ticks (Ixodoidea)-vectors of diseases of man and animals. Misc. Publ. Entomol. Soc. Am. *8*, 161–376.

Banerjee, S.N., Smith, J.A., and Fernando, K. (1994). Lyme disease in British Columbia-An update. British Columbia Med. J. *36*, 540–541.

Barbour, A.G., Maupin, G.O., Teltow, G.J., Carter, C.J., and Piesman, J. (1996). Identification of an uncultivable *Borrelia* species in the hard tick *Amblyomma americanum*: possible agent of a Lyme disease-like illness. J. Infect. Dis. *173*, 403–409.

Barclay, A.J.G., and Coulter, J.B.S. (1990). Tick-borne relapsing fever in central Tanzania. Trans. Roy. Soc. Trop. Med. Hyg. *84*, 852–856.

Beck, A.F., Holscher, K.H., and Butler, J.F. (1986). Life cycle of *Ornithodoros turicata americanus* (Acari: Argasidae) in the laboratory. J. Med. Entomol. *23*, 313–319.

Beck, M.D. (1937). California field and laboratory studies on relapsing fever. J. Infect. Dis. *60*, 64–80.

Beck, M.D. (1942). Present distribution of relapsing fever in California. In: A symposium on relapsing fever in the Americas, F.R. Moulton, ed. (Washington, D.C.: Am. Assoc. Adv. Sci.), pp. 20–25.

Benjamin, M.A., Zhioua, E., and Ostfeld, R.S. (2002). Laboratory and field evaluation of the entomopathogenic fungus *Metarhizium anisopliae* (Deuteromycetes) for controlling questing adult *Ixodes scapularis* (Acari: Ixodidae). J. Med. Entomol. *39*, 723–728.

Birkhaug, K. (1942). Otto H.F. Obermeier: February *13*, 1843 – August *20*, 1873. In: A symposium of relapsing fever in the Americas, F.R. Moulton, ed. (Washington, DC: Am. Assoc. Adv. Sci.), pp. 7–14.

Boyer, K.M., Munford, R.S., Maupin, G.O., Pattison, C.P., Fox, M.D., Barnes, A.M., Jones, W.L., and Maynard, J.E. (1977). Tick-borne relapsing fever: an interstate outbreak originating at Grand Canyon National Park. Am. J. Epidemiol. *105*, 469–479.

Breitschwerdt, E.B., Nicholson, W.L., Kiehl, A.R., Steers, C., Meuten, D.J., and Levine, J.F. (1994). Natural infections with *Borrelia* spirochetes in two dogs from Florida. J. Clin. Microbiol. *32*, 352–357.

Briggs, L.H. (1922). Relapsing fever in California. JAMA *79*, 941–944.

Brisson, D., and Dykhuizen, D.E. (2006). A modest model explains the distribution and abundance of *Borrelia burgdorferi* strains. Am. J. Trop. Med. Hyg. *74*, 615–622.

Brisson, D., Dykhuizen, D.E., and Ostfeld, R.S. (2008). Conspicuous impacts of inconspicuous hosts on the Lyme disease epidemic. Proc. Biol. Sci. *275*, 227–235.

Brown, R.N., Peot, M.A., and Lane, R.S. (2006). Sylvatic maintenance of *Borrelia burgdorferi* (Spirochaetales) in Northern California: untangling the web of transmission. J. Med. Entomol. *43*, 743–751.

Brumpt, E., and Brumpt, L.C. (1939). Identite du spirochete des fievres recurrentes a tiques des plateaux mexicains et du *Spirochaeta turicatae* agent de la fievre recurrente sporadique des Etats-Unis. Ann. Parasitol. Hum. Compar. *17*, 287–298.

Bunikis, J., Tsao, J., Garpmo, U., Berglund, J., Fish, D., and Barbour, A.G. (2004). Typing of *Borrelia* relapsing fever group strains. Emerg. Infect. Dis. *10*, 1661–1664.

Burgdorfer, W. (1951). Analyse des Infektionsverlaufes bei *Ornithodorus moubata* (Murray) und der naturlichen Uebertragung von *Spirochaeta duttoni*. Acta Trop *8*, 194–262.

Burgdorfer, W., and Mavros, A.J. (1970). Susceptibility of various species of rodents to the relapsing fever spirochete, *Borrelia hermsii*. Infect. Immun. *2*, 256–259.

Buxton, P.A. (1946). The Louse. An Account of the Lice Which Infest Man, Their Medical Importance and Control, 2nd edn (Baltimore, Williams & Wilkins).

Cadenas, F.M., Rais, O., Jouda, F., Douet, V., Humair, P.F., Moret, J., and Gern, L. (2007). Phenology of *Ixodes ricinus* and infection with *Borrelia burgdorferi* sensu lato along a north- and south-facing altitudinal gradient on Chaumont Mountain, Switzerland. J. Med. Entomol. *44*, 683–693.

California, D.H.S. (1983). Relapsing fever. In Control of communicable diseases in California (Sacramento, American Public Health Association), pp. 374–377.

Callow, L.L. (1967). Observations on tick-transmitted spirochaetes of cattle in Austrialia and South Africa. Br Vet. J. *123*, 492–497.

Canale-Parola, E. (1984). Order 1. Spirochaetales Buchanan 1917, 163AL. In Bergey's Manual of Sytematic Bacteriology, N.R. Krieg, ed. (Baltimore, Williams & Williams), pp. 38–39.

Carroll, J.F., Allen, P.C., Hill, D.E., Pound, J.M., Miller, J.A., and George, J.E. (2002). Control of *Ixodes scapularis* and *Amblyomma americanum* through use of the '4-poster' treatment device on deer in Maryland. Exp Appl Acarol *28*, 289–296.

CDC (1990). Common source outbreak of relapsing fever – California. MMWR *39*, *579*, 585–586.

CDC (2002). Lyme disease – United States, 2000. MMWR *51*, 29–31.

CDC (2007). Lyme disease-United States, 2003–2005. MMWR *56*, 573–576.

Closson, H.O. (1934). Relapsing Fever. J Kansas Med Soc *35*, 53–54.

Clover, J.R., and Lane, R.S. (1995). Evidence implicating nymphal *Ixodes pacificus* (Acari: Ixodidae) in the epidemiology of Lyme disease in California. Am. J. Trop. Med. Hyg. *53*, 237–240.

Coleman, G.E. (1934). Relapsing fever in California. III. The carrier condition; epidemiology. J. Infect. Dis. *54*, 282–294.

Coleman, J.M., and Wright, H.E. (1942). The present distribution of relapsing fever in Texas. Am. Assoc. Adv. Sci. 26–28.

Collares-Pereira, M., Couceiro, S., Franca, I., Kurtenbach, K., Schafer, S.M., Vitorino, L., Goncalves, L., Baptista, S., Vieira, M.L., and Cunha, C. (2004). First isolation of *Borrelia lusitaniae* from a human patient. J. Clin. Microbiol. *42*, 1316–1318.

Comstedt, P., Bergström, S., Olsen, B., Garpmo, U., Marjavaara, L., Mejlon, H., Barbour, A.G., and Bunikis, J. (2006). Migratory passerine birds as reservoirs of Lyme borreliosis in Europe. Emerg. Infect. Dis. *12*, 1087–1095.

Cook, A.R. (1904). Relapsing fever in Uganda. J. Trop. Med. Hyg. *7*, 24–26.

Cooley, R.A., and Kohls, G.M. (1944). The Argasidae of North America, Central America and Cuba. American Midland Naturalist *Monograph* No. *1*, 1–152.

Crippa, M., Rais, O., and Gern, L. (2002). Investigations on the mode and dynamics of transmission and infectivity of *Borrelia burgdorferi* sensu stricto and *Borrelia afzelii* in *Ixodes ricinus* ticks. Vector Borne Zoonotic Dis. *2*, 3–9.

Cutler, S.J. (2006). Possibilities for relapsing fever reemergence. Emerg. Infect. Dis. *12*, 369–374.

da Franca, I., Santos, L., Mesquita, T., Collares-Pereira, M., Baptista, S., Vieira, L., Viana, I., Vale, E., and Prates, C. (2005). Lyme borreliosis in Portugal caused by *Borrelia lusitaniae*? Clinical report on the first patient with a positive skin isotate.. Wien Klin. Wochenschr. *117*, 429–432.

Dammin, G.J. (1986). Lyme disease: its transmission and diagnostic features. Lab. Manag. *24*, 33–38.

Davis, G.E. (1936). *Ornithodoros turicata*: the possible vector of relapsing fever in southwestern Kansas. Publ. Health Rep. *51*, 1719.

Davis, G.E. (1941a). *Ornithodoros turicata* and relapsing fever spirochetes in New Mexico. Public Health Rep. *56*, 2258–2261.

Davis, G.E. (1941b). *Ornithodoros turicata;* The male; feeding and copulation habits, fertility, span of life, and transmission of relapsing fever spirochetes. Publ. Health Rep. *56*, 1799–1802.

Davis, G.E. (1942a). Species unity or plurality of the relapsing fever spirochetes. In: A symposium of relapsing fever in the Americas, F.R. Moulton, ed. (Washington, DC: Am. Assoc. Adv. Sci.), pp. 41–47.

Davis, G.E. (1942b). Tick vectors and life cycles of ticks. In: A symposium of relapsing fever in the Americas, F.R. Moulton, ed. (Washington, DC: Am. Assoc. Adv. Sci.), pp. 67–76.

Davis, G.E. (1943). Relapsing fever: the tick *Ornithodoros turicata* as a spirochetal reservoir. Public Health Rep. *58*, 839–842.

Davis, G.E., and Hoogstraal, H. (1954). The relapsing fevers: a survey of the tick-borne spirochetes of Egypt. J. Egyptian Publ. Health Assoc. *29*, 139–143.

Davis, H., Vincent, J.M., and Lynch, J. (2002). Tick-borne relapsing fever caused by *Borrelia turicatae*. Pediatr. Infect. Dis. J. *21*, 703–705.

De Boer, R., Hovius, K.E., Nohlmans, M.K., and Gray, J.S. (1993). The woodmouse (*Apodemus sylvaticus*) as a reservior of tick-transmitted spirochetes (*Borrelia burgdorferi*) in The Netherlands. Zentrabl. Bakteriol. *279*, 404–416.

de Jong, J., Wilkinson, R.J., Schaeffers, P., Sondorp, H.E., and Davidson, R.N. (1995). Louse-borne relapsing fever in southern Sudan. Trans. Roy. Soc. Trop. Med. Hyg. *89*, 621.

De Michelis, S., Sewell, H.S., Collares-Pereira, M., Santos-Reis, M., Schouls, L.M., Benes, V., Holmes, E.C., and Kurtenbach, K. (2000). Genetic diversity of *Borrelia burgdorferi* sensu lato in ticks from mainland Portugal. J. Clin. Microbiol. *38*, 2128–2133.

Dennis, D. (1999). Epidemic (louse-borne) relapsing fever, Southern Sudan, 1999: Investigation and control of an outbreak with unique features. Abstract from the 39th Interscience Conference on Antimicrobial Agents and Chemotherapy (American Society for Microbiology).

Dennis, D.T., Awoke, S., Doberstyn, E.B., and Fresh, J.W. (1976). Bleeding in louse-borne relapsing fever: clinical and laboratory features in 29 patients. E. Afr. Med. J. *53*, 220–225.

Dennis, D.T., Nekomoto, T.S., Victor, J.C., Paul, W.S., and Piesman, J. (1998). Reported distribution of *Ixodes scapularis* and *Ixodes pacificus* (Acari: Ixodidae) in the United States. J. Med. Entomol. *35*, 629–638.

des Vignes, F., Piesman, J., Heffernan, R., Schulze, T.L., Stafford 3rd, K.C., and Fish, D. (2001). Effect of tick removal on transmission of *Borrelia burgdorferi* and

Ehrlichia phagocytophila by *Ixodes scapularis* nymphs. J. Infect. Dis. *183*, 773–778.

Diuk-Wasser, M.A., Gatewood, A.G., Cortinas, M.R., Yaremych-Hamer, S., Tsao, J., Kitron, U., Hickling, G., Brownstein, J.S., Walker, E., Piesman, J., *et al.* (2006). Spatiotemporal patterns of host-seeking *Ixodes scapularis* nymphs (Acari: Ixodidae) in the United States. J. Med. Entomol. *43*, 166–176.

Dolan, M.C., Maupin, G.O., Schneider, B.S., Denatale, C., Hamon, N., Cole, C., Zeidner, N.S., and Stafford 3rd, K.C. (2004). Control of immature *Ixodes scapularis* (Acari: Ixodidae) on rodent reservoirs of *Borrelia burgdorferi* in a residential community of southeastern Connecticut. J. Med. Entomol. *41*, 1043–1054.

Donahue, J.G., Piesman, J., and Spielman, A. (1987). Reservoir competence of white-footed mice for Lyme disease spirochetes. Am. J. Trop. Med. Hyg. *36*, 92–96.

Dsouli, N., Younsi-Kabachii, H., Postic, D., Nouira, S., Gern, L., and Bouattour, A. (2006). Reservoir role of lizard *Psammodromus algirus* in transmission cycle of *Borrelia burgdorferi* sensu lato (Spirochaetaceae) in Tunisia. J. Med. Entomol. *43*, 737–742.

Dupont, H.T., La Scola, B., Williams, R., and Raoult, D. (1997). A focus of tick-borne relapsing fever in southern Zaire. Clin. Infect. Dis. *25*, 139–144.

Dutton, J.E., and Todd, J.L. (1905). The nature of human tick-fever in the eastern part of the Congo Free State with notes on the distribution and bionomics of the tick. Liverpool School Trop Med Mem. *17*, 1–18.

Dworkin, M.S., Anderson Jr., D.E., Schwan, T.G., Shoemaker, P.C., Banerjee, S.N., Kassen, B.O., and Burgdorfer, W. (1998). Tick-borne relapsing fever in the northwestern United States and southwestern Canada. Clin. Infect. Dis. *26*, 122–131.

Dworkin, M.S., Schwan, T.G., and Anderson, D.E. (2002a). Tick-borne relapsing fever in North America. Med. Clin. North Am. *86*, 417–433.

Dworkin, M.S., Shoemaker, P.C., Fritz, C.L., Dowell, M.E., and Anderson Jr., D.E. (2002b). The epidemiology of tick-borne relapsing fever in the United States. Am. J. Trop. Med. Hyg. *66*, 753–758.

Eisen, R.J., Lane, R.S., Fritz, C.L., and Eisen, L. (2006). Spatial patterns of Lyme disease risk in California based on disease incidence data and modeling of vector-tick exposure. Am. J. Trop. Med. Hyg. *75*, 669–676.

Falco, R.C., and Fish, D. (1988). Ticks parasitizing humans in a Lyme disease endemic area of southern New York State. Am. J. Epidemiol. *128*, 1146–1152.

Falco, R.C., Fish, D., and Piesman, J. (1996). Duration of tick bites in a Lyme disease-endemic area. Am. J. Epidemiol. *143*, 192–197.

Felsenfeld, O. (1971). Borrelia. Strains, Vectors, Human and Animal Borreliosis (St. Louis, MO, Warren H. Green, Inc.).

Filippova, N.A., ed. (1991). A hypothesis for the palaeogenesis of the distribution of the main vectors for Lyme disease. (Prague, Academia, and The Hague, SPB Academic Publishing bv).

Filippova, N.A., ed. (1999). Systematic relationships of the *Ixodes ricinus* complex in the palearctic faunal region. (Columbus, Ohio: Ohio Biological Survey).

Fingerle, V., Liegl, G., Munderloh, U., and Wilske, B. (1998). Expression of outer surface proteins A and C of *Borrelia burgdorferi* in *Ixodes ricinus* ticks removed from humans. Med Microbiol. Immunol. *187*, 121–126.

Fingerle, V., Schulte-Spechtel, U.C., Ruzic-Sabljic, E., Leonhard, S., Hofmann, H., Weber, K., Pfister, K., Strle, F., and Wilske, B. (2008). Epidemiological aspects and molecular characterization of *Borrelia burgdorferi* s.l. from southern Germany with special respect to the new species *Borrelia spielmanii* sp.nov. Int. J. Med. Microbiol. *298*, 279–290.

Fraenkel, C.J., Garpmo, U., and Berglund, J. (2002). Determination of novel *Borrelia* genospecies in Swedish *Ixodes ricinus* ticks. J. Clin. Microbiol. *40*, 3308–3312.

Francis, E. (1938). Longevity of the tick *Ornithodoros turicata* and of *Spirochaeta recurrentis* with this tick. Publ. Health. Rep. *53*, 2220–2241.

Fritz, C.L., Bronson, L.R., Smith, C.R., Schriefer, M.E., Tucker, J.R., and Schwan, T.G. (2004). Isolation and characterization of *Borrelia hermsii* associated with two foci of tick-borne relapsing fever in California. J. Clin. Microbiol. *42*, 1123–1128.

Fukunaga, M., and Koreki, Y. (1995). The flagellin gene of *Borrelia miyamotoi* sp. nov. and its phylogenetic relationship among *Borrelia* species. FEMS Microbiol. Lett. *134*, 255–258.

Fukunaga, M., Hamase, A., Okada, K., Inoue, H., Tsuruta, Y., Miyamoto, K., and Nakao, M. (1996a). Characterization of spirochetes isolated from ticks (*Ixodes tanuki, Ixodes turdus,* and *Ixodes columnae*) and comparison of the sequences with those of *Borrelia burgdorferi* sensu lato strains. Appl. Environ. Microbiol. *62*, 2338–2344.

Fukunaga, M., Okada, K., Nakao, M., Konishi, T., and Sato, Y. (1996b). Phylogenetic analysis of *Borrelia* species based on flagellin gene sequences and its application for molecular typing of Lyme disease borreliae. Int. J. Syst. Bacteriol. *46*, 898–905.

Gaber, M.S., Khalil, G.M., and Hoogstraal, H. (1982). *Borrelia crocidurae*: venereal transfer in Egyptian *Ornithodoros erraticus* ticks. Exp Parasitol *54*, 182–184.

Gaber, M.S., Khalil, G.M., Hoogstraal, H., and Aboul-Nasr, A.E. (1984). *Borrelia crocidurae* localization and transmission in *Ornithodoros erraticus* and *O. savignyi*. Parasitol *88*, 403–413.

Gage, K.L., Eggleston, M.E., Gilmore Jr., R.D., Dolan, M.C., Montenieri, J.A., Tanda, D.T., and Piesman, J. (2001). Isolation and characterization of *Borrelia parkeri* in *Ornithodoros parkeri* (Ixodida: Argasidae) collected in Colorado. J. Med. Entomol. *38*, 665–674.

Garcia, R. (1962). Carbon dioxide as an attractant for certain ticks (Acarina: Argasidae and Ixodidae). Ann. Entomol. Soc. Am. *55*, 605.

Gill, J.S., Ullmann, A.J., Loftis, A.D., Schwan, T.G., Raffel, S.J., Schrumpf, M.E., and Piesman, J. (2008). Novel relapsing fever spirochete in bat tick. Emerg. Inf. Dis. *14*, 522–523.

Ginsberg, H.S., Buckley, P.A., Balmforth, M.G., Zhioua, E., Mitra, S., and Buckley, F.G. (2005). Reservoir competence of native North American birds for the

Lyme disease spirochete, *Borrelia burgdorferi*. J. Med. Entomol. *42*, 445–449.

Godeluck, B., Duplantier, J.M., Ba, K., and Trape, J.F. (1994). A longitudinal survey of *Borrelia crocidurae* prevalence in rodents and insectivores in Senegal. Am. J. Trop. Med. Hyg. *50*, 165–168.

Gray, J.S., Kirstein, F., Robertson, J.N., Stein, J., and Kahl, O. (1999). *Borrelia burgdorferi* sensu lato in *Ixodes ricinus* ticks and rodents in a recreational park in southwestern Ireland. Exp. Appl. Acarol. *23*, 717–729.

Grego, E., Bertolotti, L., Peletto, S., Amore, G., Tomassone, L., and Mannelli, A. (2007). *Borrelia lusitaniae* OspA gene heterogeneity in Mediterranean basin area. J. Mol. Evol. *65*, 511–518.

Gregson, J.D. (1949). Notes on the occurrence of *Ornithodoros hermsi* in British Columbia, and its probable relation to relapsing fever. Argasidae, Ixodoidea. Proc. Entomol. Soc. British Columbia *45*, 15–16.

Halperin, T., Orr, N., Cohen, R., Hasin, T., Davidovitch, N., Klement, E., Kayouf, R., Baneth, G., Cohen, D., and Yavzori, M. (2006). Detection of relapsing fever in human blood samples from Israel using PCR targeting the glycerophosphodiester phosphodiesterase (GlpQ) gene. Acta Trop. *98*, 189–195.

Hanincova, K., Kurtenbach, K., Diuk-Wasser, M., Brei, B., and Fish, D. (2006). Epidemic spread of Lyme borreliosis, north-eastern United States. Emerg. Infect. Dis. *12*, 604–611.

Hanincova, K., Schafer, S.M., Etti, S., Sewell, H.S., Taragelova, V., Ziak, D., Labuda, M., and Kurtenbach, K. (2003b). Association of *Borrelia afzelii* with rodents in Europe. Parasitology *126*, 11–20.

Hanincova, K., Taragelova, V., Koci, J., Schafer, S.M., Hails, R., Ullmann, A.J., Piesman, J., Labuda, M., and Kurtenbach, K. (2003a). Association of *Borrelia garinii* and *B. valaisiana* with songbirds in Slovakia. Appl. Environ. Microbiol. *69*, 2825–2830.

Hasin, T., Davidovitch, N., Cohen, R., Dagan, T., Romem, A., Orr, N., Klement, E., Lubezky, N., Kayouf, R., Sela, T., *et al.* (2006b). Postexposure treatment with doxycycline for the prevention of tick-borne relapsing fever. N. Engl. J. Med. *355*, 148–155.

Haupt, K., Kraiczy, P., Wallich, R., Brade, V., Skerka, C., and Zipfel, P.F. (2007). Binding of human factor H-related protein 1 to serum-resistant *Borrelia burdgorferi* is mediated by borrelial complement regulator-acquiring surface proteins. J. Infect. Dis. *196*, 124–133.

Helmy, N., Khalil, G.M., and Hoogstraal, H. (1983). Hyperparasitism in *Ornithodoros erraticus*. J. Parasitol. *69*, 229–233.

Herms, W.B., and Wheeler, C.M. (1936). *Ornithodoros hermsi* Wheeler as a vector of relapsing fever in California. J. Parasitol. *22*, 276–282.

Hojgaard, A., Eisen, R.J., and Piesman, J. (2008). Transmission dynamics of *Borrelia burgdorferi* s.s. during the key third day of feeding by nymphal *Ixodes scapularis* (Acari: Ixodidae). J. Med. Entomol. *45*, 732–736.

Hoogstraal, H. (1981). Changing patterns of tickborne diseases in modern society. Annu. Rev. Entomol. *26*, 75–99.

Hoogstraal, H., Salah, A.A., and Kaiser, M.N. (1954). Summary of the known distribution and biology of *Ornithodoros erraticus* (Lucas, 1849) (Ixodoidea, Argasidae) in Egypt. J. Egyptian Publ. Health Assoc. *29*, 127–138.

Houhamdi, L., and Raoult, D. (2005). Excretion of living *Borrelia recurrentis* in feces of infected human body lice. J. Infect. Dis. *191*, 1898–1906.

Hovis, K.M., Schriefer, M.E., Bahlani, S., and Marconi, R.T. (2006). Immunological and molecular analyses of the *Borrelia hermsii* factor H and factor-H like protein 1 binding protein, FhbA: demonstration of its utility as a diagnostic marker and epidemiological tool for tick-borne relapsing fever. Infect. Immun. *74*, 4519–4529.

Hovius, J.W., van Dam, A.P., and Fikrig, E. (2007). Tick–host–pathogen interactions in Lyme borreliosis. Trends Parasitol. *23*, 434–438.

Hu, C.M., Wilske, B., Fingerle, V., Lobet, Y., and Gern, L. (2001). Transmission of *Borrelia garinii* OspA serotype 4 to BALB/c mice by *Ixodes ricinus* ticks collected in the field. J. Clin. Microbiol. *39*, 1169–1171.

Huegli, D., Hu, C.M., Humair, P.F., Wilske, B., and Gern, L. (2002). *Apodemus* species mice are reservoir hosts of *Borrelia garinii* OspA serotype 4 in Switzerland. J. Clin. Microbiol. *40*, 4735–4737.

Humair, P.F., Douet, V., Cadenas, F.M., Schouls, L.M., Van De Pol, I., and Gern, L. (2007). Molecular identification of blood meal source in *Ixodes ricinus* ticks using 12S rDNA as a genetic marker. J. Med. Entomol. *44*, 869–880.

Humair, P.F., and Gern, L. (1998). Relationship between *Borrelia burgdorferi* sensu lato species, red squirrels (*Sciurus vulgaris*) and *Ixodes ricinus* in enzootic areas in Switzerland. Acta Trop. *69*, 213–227.

Humair, P.F., Postic, D., Wallich, R., and Gern, L. (1998). An avian reservoir (*Turdus merula*) of Lyme borreliosis spirochetes. Zentrabl. Bakteriol. *287*, 521–538.

Humair, P.F., Rais, O., and Gern, L. (1999). Transmission of *Borrelia afzelii* from *Apodemus* mice and *Clethrionomys* voles to *Ixodes ricinus* ticks; differential transmission pattern and overwintering maintenance. Parasitology *118*, 33–42.

Hutcheson, H.J., Oliver Jr., J.H., Houck, M.A., and Strauss, R.E. (1995). Multivariate morphometric discrimination of nymphal and adult forms of the blacklegged tick (Acari: Ixodidae), a principal vector of the agent of Lyme disease in eastern North America. J. Med. Entomol. *32*, 827–842.

Jongen, V.H.W.M., van Roosmalen, J., Tiems, J., van Holten, J., and Wetsteyn, J.C.F.M. (1997). Tick-borne relapsing fever and pregnancy outcome in rural Tanzania. Acta Obstet Gynecol Scand *76*, 834–838.

Kahl, O., Janetzki-Mittmann, C., Gray, J.S., Jonas, R., Stein, J., and de Boer, R. (1998). Risk of infection with *Borrelia burgdorferi* sensu lato for a host in relation to the duration of nymphal *Ixodes ricinus* feeding and the method of tick removal. Zentralbl. Bakteriol. *287*, 41–52.

Kain, D.E., Sperling, F.A., Daly, H.V., and Lane, R.S. (1999). Mitochondrial DNA sequence variation in *Ixodes pacificus* (Acari: Ixodidae). Heredity *83*, 378–386.

Kain, D.E., Sperling, F.A., and Lane, R.S. (1997). Population genetic structure of *Ixodes pacificus* (Acari: Ixodidae) using allozymes. J. Med. Entomol. *34*, 441–450.

Keirans, J.E., Needham, G.R., and Oliver Jr., J.H., eds. (1999). The *Ixodes (Ixodes) ricinus* complex worldwide: Diagnosis of the species in the complex, host, and distribution. (Columbus, Ohio: Ohio Biological Survey).

Kelly, R. (1971). Cultivation of *Borrelia hermsi*. Science *173*, 443–444.

Kemp, H.A., Moursund, W.H., and Wright, H.E. (1934). Relapsing fever in Texas. II. The specificity of the vector, *Ornithodorus turicata*, for the spirochete. Am. J. Trop. Med. *14*, 159–162.

Kemp, H.A., Moursund, W.H., and Wright, H.E. (1935). Relapsing fever in Texas. V. A survey of the epidemiology and clinical manifestations of the disease as it occurs in Texas. Am. J. Trop. Med. *15*, 495–506.

Khalil, G.M., Helmy, N., Hoogstraal, H., and El-Said, A. (1984). Seasonal dynamics of *Ornithodoros (Pavlovskyella) erraticus* (Acari: Ixodoidea: Argasidae) and the spirochete *Borrelia crocidurae* in Egypt. J. Med. Entomol. *21*, 536–539.

Kilpatrick, H.J., and LaBonte, A.M. (2007). Managing Urban Deer in Connecticut. A Guide for Residents and Communities (Hartford, Connecticut: Bureau of Natural Resources/Wildlife Division).

Kirk, R. (1939). The epidemiology of relapsing fever in the Anglo-Egyptian Sudan. Ann. Trop. Med. Parasitol. 33, 125–140.

Koch, R. (1905). Vorlaufige Mitteilungen uber die Ergebnisse einer Forchungreise nach Ostafrika. Deutsche Medizinishce Wochenschrift *31*, 1865–1869.

Koch, R. (1906). Ueber afrikanischen Recurrens. Berliner Klinische Wochenschrift *43*, 185–194.

Korenberg, E.I., Gorelova, N.B., and Kovalevskii, Y.V. (2002). Ecology of *Borrelia burgdorferi* sensu lato in Russia. In Lyme Borreliosis: Biology, Epidemiology, and Contol. (Wallingford, Oxon, UK, CAB International).

Korenberg, E.I., and Kovalevskii, Y.V. (1999). Main features of Tick-Borne Encephalitis eco-epidemiology in Russia. Zentrabl Bakteriol *289*, 525–539.

Korenberg, E.I., Kovalevskii, Y.V., Levin, M.L., and Shchyogoleva, T.V. (2001). The prevalence of *Borrelia burgdorferi* sensu lato in *Ixodes persulcatus* and *I. ricinus* ticks in the zone of their sympatry. Folia Parasitol (Praha) *48*, 63–68.

Korenberg, E.I., Kryuchechnikov, V.N., Ananyina, Y.V., and Chernukha, Y.G. (1987). Prerequisities of the existence of Lyme disease in the USSR. Zentrabl. Bakteriol. Mikrobiol. Hyg. *263*, 471–472.

Kraiczy, P., Rossman, E., Brade, V., Simon, M.M., Skerka, C., Zipfel, P.F., and Wallich, R. (2006). Binding of human complement regulators FHL-1 and factor H to CRASP-1 orthologs of *Borrelia burgdorferi*. Wien Klin. Wochenschr. *118*, 669–676.

Kraiczy, P., and Wurzner, R. (2006). Complement escape of human pathogenic bacteria by acquisition of complement regulators. Mol. Immunol. *43*, 31–44.

Kuo, M.M., Lane, R.S., and Giclas, P.C. (2000). A comparative study of mammalian and reptilian alternative pathway of complement-mediated killing of the Lyme disease spirochetes (*Borrelia burgdorferi*). J. Parasitol. *86*, 1223–1228.

Kurtenbach, K., Dizij, A., Seitz, H.M., Margos, G., Moter, S.E., Kramer, M.D., Wallich, R., Schaible, U.E., and Simon, M.M. (1994). Differential immune responses to *Borrelia burgdorferi* in European wild rodent species influence spirochete transmission to *Ixodes ricinus* L. (Acari: Ixodidae). Infect. Immun. *62*, 5344–5352.

Kurtenbach, K., Sewell, H.-S., Ogden, N.H., Randolph, S.E., and Nuttall, P.A. (1998). Serum complement sensitivity as a key factor in Lyme disease ecology. Infect. Immun. *66*, 1248–1251.

Lane, R.S., Mun, J., Eisen, R.J., and Eisen, L. (2005). Western gray squirrel (Rodentia: Sciuridae): a primary reservoir host of *Borrelia burgdorfer* in California oak woodlands? J. Med. Entomol. *42*, 388–396.

Lane, R.S., Mun, J., Peribanez, M.A., and Stubbs, H.A. (2007). Host-seeking behavior of *Ixodes pacificus* (Acari: Ixodidae) nymphs in relation to environmental parameters in dense-woodland and woodland-grass habitats. J. Vector Ecol. *32*, 342–357.

Lane, R.S., Piesman, J., and Burgdorfer, W. (1991). Lyme borreliosis: relation of its causative agent to its vectors and hosts in North America and Europe. Annu. Rev. Entomol. *36*, 587–609.

Le Fleche, A., Postic, D., Girardet, K., Peter, O., and Baranton, G. (1997). Characterization of *Borrelia lusitaniae* sp. nov. by 16S ribosomal DNA sequence analysis. Int. J. Syst. Bacteriol. *47*, 921–925.

Lees, A.D., and Milne, A. (1951). The seasonal and diurnal activities of individual sheep ticks (*Ixodes ricinus* L.). Parasitology *41*, 189–208.

Leger, A. (1917). Spirochete de la musaraigne (*Crocidura stampflii* Jentink). Bull. Soc. Path. Ex. *10*, 280–281.

Lin, T., Oliver, J.H., Jr., and Gao, L. (2004). Molecular characterization of *Borrelia* isolates from ticks and mammals from the southern United States. J. Parasitol. *90*, 1298–1307.

Lindgren, E., Talleklint, L., and Polfeldt, T. (2000). Impact of climatic change on the northern latitude limit and population density of the disease-transmitting European tick *Ixodes ricinus*. Environ. Health Perspect. *108*, 119–123.

Lindsay, L.R., Barker, I.K., Surgeoner, G.A., McEwen, S.A., Gillespie, T.J., and Robinson, J.T. (1995). Survival and development of *Ixodes scapularis* (Acari: Ixodidae) under various climatic conditions in Ontario, Canada. J. Med. Entomol. 32, 143–152.

Lindsay, L.R., Mathison, S.W., Barker, I.K., McEwen, S.A., and Surgeoner, G.A. (1999). Abundance of *Ixodes scapularis* (Acari: Ixodidae) larvae and nymphs in relation to host density and habitat on Long Point, Ontario. J. Med. Entomol. *36*, 243–254.

Livingstone, D. (1857). Missionary travels and researches in South Africa (London, John Murray).

Loftis, A.D., Gill, J.S., Schriefer, M.E., Levin, M.L., Eremeeva, M.E., Gilchrist, M.J.R., and Dasch, G.A. (2005). Detection of *Rickettsia*, *Borrelia*, and *Bartonella* in *Carios kelleyi* (Acari: Argasidae). J. Med. Entomol. *42*, 473–480.

Logan, T.M., Wilson, M.L., and Cornet, J.P. (1993). Association of ticks (Acari: Ixodoidea) with rodent burrows in northern Senegal. J. Med. Entomol. *30*, 799–801.

LoGiudice, K., Ostfeld, R.S., Schmidt, K.A., and Keesing, F. (2003). The ecology of infectious disease: effects of host diversity and community composition on Lyme disease risk. Proc. Natl. Acad. Sci. U.S.A. *100*, 567–571.

Longanecker, D.S. (1951). Laboratory and field studies on the biology of the relapsing fever tick vector (*Ornithodoros hermsi* Wheeler) in the high mountains of California. Am. J. Trop. Med. *31*, 373–380.

Mackie, F.P. (1907). The part played by *Pediculus corporis* in the transmission of relapsing fever. Br. Med. J. *2*, 1706–1709.

Majlathova, V., Majlath, I., Derdakova, M., Vichova, B., and Pet'ko, B. (2006). *Borrelia lusitaniae* and green lizards (*Lacerta viridis*) Karst Region, Slovakia. Emerg. Infect. Dis. *12*, 1895–1901.

Marconi, R.T., Liveris, D., and Schwartz, I. (1995). Identification of novel insertion elements, restriction fragment length polymorphism patterns, and discontinuous 23S rRNA in Lyme disease spirochetes: phylogenetic analyses of rRNA genes and their intergenic spacers in *Borrelia japonica* sp. nov. and genomic group 21038 (*Borrelia andersonii* sp. nov.) isolates. J. Clin. Microbiol. 33, 2427–2434.

Masuzawa, T., Takada, N., Kudeken, M., Fukui, T., Yano, Y., Ishiguro, F., Kawamura, Y., Imai, Y., and Ezaki, T. (2001). *Borrelia sinica* sp. nov., a Lyme disease-related *Borrelia* species isolated in China. Int. J. Syst. Evol. Microbiol *51*, 1817–1824.

Mather, T.N., Ribeiro, J.M., and Spielman, A. (1987). Lyme disease and babesiosis: acaricide focused on potentially infected ticks. Am. J. Trop. Med. Hyg. *36*, 609–614.

Maupin, G.O., Fish, D., Zultowsky, J., Campos, E.G. and Piesman, J. (1991). Landscape ecology of Lyme disease in a residential area of Weschester County, New York. Am. J. Epidemiol. *133*, 1105–1113.

McCall, P.J., Hume, J.C.C., Motshegwa, K., Pignatelli, P., Talbert, A., and Kisinza, W. (2007). Does tick-borne relapsing fever have an animal reservoir in East Africa? Vector Borne Zoonotic Dis. *7*, 659–666.

Meader, C.N. (1915). Five cases of relapsing fever originating in Colorado, with positive blood findings in two. Colorado Med. *12*, 365–369.

Merten, H.A., and Durden, L.A. (2000). A state-by-state survey of ticks recorded from humans in the United States.. J. Vector Ecol. *25*, 102–113.

Milne, A. (1948). Pasture improvement and the control of sheep tick, *Ixodes ricinus* L. Ann. Appl. Biol. *35*, 369–378.

Milne, A. (1950a). The ecology of the sheep tick, *Ixodes ricinus* L.; microhabitat economy of the adult tick. Parasitology *40*, 14–34.

Milne, A. (1950b). The ecology of the sheep tick, *Ixodes ricinus* L.; spatial distribution. Parasitology *40*.

Moore 4th, V.A., Varela, A.S., Yabsley, M.J., Davidson, W.R., and Little, S.E. (2003). Detection of *Borrelia lonestari*, putative agent of southern tick-associated rash illness, in white-tailed deer (*Odocoileus virginianus*) from the southeastern United States. J. Clin. Microbiol. *41*, 424–427.

Moran Cadenas, F., Rais, O., Humair, P.F., Douet, V., Moret, J., and Gern, L. (2007). Identification of host blood meal source and *Borrelia burgdorferi* sensu lato in field-collected *Ixodes ricinus* ticks in Chaumont (Switzerland). J. Med. Entomol. *44*, 1109–1117.

Mun, J., Eisen, R.J., Eisen, L., and Lane, R.S. (2006). Detection of a *Borrelia miyamotoi* sensu lato relapsing-fever group spirochete from *Ixodes pacificus* in California. J. Med. Entomol. *43*, 120–123.

Murray, A. (1877). Economic entomology. Aptera., Vol 1. (London, Chapman & Hall).

Nadelman, R.B., Nowakoski, J., Fish, D., Falco, R.C., Freeman, K., McKenna, D., Welch, P., Marcus, R., Aguero-Rosenfeld, M.E., Dennis, D.T., *et al.* (2001). Prophylaxis with single-dose doxycycline for the prevention of Lyme disease after an *Ixodes scapularis* tick bite. N. Engl. J. Med. *345*, 79–84.

Nicolle, C., and Anderson, C. (1927). Sur l'origine des fievres recurrentes humaines. Bull. Inst. Pasteur (Paris) *25*.

Nordstrand, A., Bunikis, I., Larsson, C., Tsogbe, K., Schwan, T.G., Nilsson, M., and Bergström, S. (2007). Tickborne relapsing fever diagnosis obscured by malaria, Togo. Emerg. Inf. Dis. *13*, 117–123.

Norris, D.E., Klompen, J.S., Keirans, J.E., and Black 4th, W.C. (1996). Population genetics of *Ixodes scapularis* (Acari: Ixodidae) based on mitochondrial 16S and 12S genes. J. Med. Entomol. 33, 78–89.

Nuttall, G.H.F., Warburton, C., Cooper, W.F., and Robinson, L.E. (1908). Ticks. A monograph of the Ixodoidea. Part I. Argasidae (Cambridge, Cambridge Univ. Press).

O'Connell, S., Granstrom, M., Gray, J.S., and Stanek, G. (1998). Epidemiology of European Lyme borreliosis. Zentrabl. Bakteriol. *287*, 229–240.

Obermeier, O. (1873). Vorkommen feinster eine Eigenbewegung zeigender Faden im Blute von Recurrenskranken. Centralblatt medicinischen Wissenschaften *11*, 145–147.

Ogden, N.H., Lindsay, L.R., Hanincova, K., Barker, I.K., Bigras-Poulin, M., Charron, D.F., Heagy, A., Francis, C.M., O'Callaghan, C.J., Schwartz, I., *et al.* (2008). Role of migratory birds in introduction and range expansion of *Ixodes scapularis* ticks and of *Borrelia burgdorferi* and *Anaplasma phagocytophilum* in Canada. Appl. Environ. Microbiol. *74*, 1780–1790.

Ogden, N.H., Trudel, L., Artsob, H., Barker, I.K., Beauchamp, G., Charron, D.F., Drebot, M.A., Galloway, T.D., O'Handley, R., Thompson, R.A., *et al.* (2006). *Ixodes scapularis* ticks collected by passive surveillance in Canada: analysis of geographic distribution and infection with Lyme borreliosis agent *Borrelia burgdorferi*. J. Med. Entomol. *43*, 600–609.

Oliver Jr., J.H., Lin, T., Gao, L., Clark, K.L., Banks, C.W., Durden, L.A., James, A.M., and Chandler Jr., F.W. (2003). An enzootic transmission cycle of Lyme borreliosis spirochetes in the southeastern United States. Proc. Natl. Acad. Sci. U.S.A. *100*, 11642–11645.

Pal, U., Li, X., Wang, T., Montgomery, R.R., Ramamoorthi, N., de Silva, A.M., Bao, F., Yang, X., Pypaert, M., Pradhan, D., *et al.* (2004). TROSPA, an

Ixodes scapularis receptor for *Borrelia burgdorferi*. Cell *119*, 457–468.

Patrican, L.A., and Allan, S.A. (1995). Application of desiccant and insecticidal soap treatments to control *Ixodes scapularis* (Acari: Ixodidae) nymphs and adults in a hyperendemic woodland site. J. Med. Entomol. *32*, 859–863.

Paul, W.S., Maupin, G., Scott-Wright, A.O., Craven, R.B., and Dennis, D.T. (2002). Outbreak of tick-borne relapsing fever at the North Rim of the Grand Canyon: evidence for effectiveness of preventive measures. Am. J. Trop. Med. Hyg. *66*, 71–75.

Piesman, J. (2002). Ecology of *Borrelia burgdorferi* sensu lato in North America. In Lyme borreliosis: Biology, Epidemiology, and Control, J.S. Gray, O. Kahl, R.S. Lane, and G. Stanek, eds (Wallingford, UK: CABI Publishing), pp. 223–249.

Piesman, J. (2006). Strategies for reducing the risk of Lyme borreliosis in North America. Int. J. Med. Microbiol. *296*, S17–S22.

Piesman, J., and Eisen, L. (2008). Prevention of tick-borne diseases. Annu. Rev. Entomol. *53*, 323–343.

Piesman, J., Clark, K.L., Dolan, M.C., Happ, C.M., and Burkot, T.R. (1999). Geographic survey of vector ticks (*Ixodes scapularis* and *Ixodes pacificus*) for infection with the Lyme disease spirochete, *Borrelia burgdorferi*. J. Vector Ecol. *24*, 91–98.

Piesman, J., Mather, T.N., Dammin, G.J., Telford 3rd, S.R., Lastavica, C.C., and Spielman, A. (1987a). Seasonal variation of transmission risk: Lyme disease and human babesiosis. Am. J. Epidemiol. *126*, 1187–1189.

Piesman, J., Mather, T.N., Sinsky, R.J., and Spielman, A. (1987b). Duration of tick attachment and *Borrelia burgdorferi* transmission. J. Clin. Microbiol. *25*, 557–558.

Porcella, S.F., Raffel, S.J., Anderson Jr., D.E., Gilk, S.D., Bono, J.L., Schrumpf, M.E., and Schwan, T.G. (2005). Variable tick protein in two genomic groups of the relapsing fever spirochete *Borrelia hermsii* in western North America. Infect. Immun. *73*, 6647–6658.

Postic, D., Garnier, M., and Baranton, G. (2007). Multilocus sequence analysis of atypical *Borrelia burgdorferi* sensu lato isolates – description of *Borrelia californiensis* sp. nov., and genomospecies 1 and 2. Int. J. Med. Microbiol. *297*, 263–271.

Pound, J.M., Miller, J.A., George, J.E., and Lemeilleur, C.A. (2000). The '4-poster' passive topical treatment device to apply acaricide for controlling ticks (Acari: Ixodidae) feeding on white-tailed deer. J. Med. Entomol. *37*, 588–594.

Qiu, W.G., Dykhuizen, D.E., Acosta, M.S., and Luft, B.J. (2002). Geographic uniformity of the Lyme disease spirochete (*Borrelia burgdorferi*) and its shared history with tick vector (*Ixodes scapularis*) in the Northeastern United States. Genetics *160*, 833–849.

Ramamoorthi, N., Narasimhan, S., Pal, U., Bao, F., Yang, X.F., Fish, D., Anguita, J., Norgard, M.V., Kantor, F.S., Anderson, J.F., *et al.* (2005). The Lyme disease agent exploits a tick protein to infect the mammalian host. Nature *436*, 573–577.

Ramos, J.M., Malmierca, E., Reyes, F., and Tesfamariam, A. (2008). Results of a 10-year survey of louse-borne relapsing fever in southern Ethiopia: a decline in endemicity. Ann. Trop. Med. Parasitol. *102*, 467–469.

Ras, N.M., Lascola, B., Postic, D., Cutler, S.J., Rodhain, F., Baranton, G., and Raoult, D. (1996). Phylogenesis of relapsing fever *Borrelia* spp. Int. J. Syst. Bacteriol. *46*, 859–865.

Rawlings, J.A. (1995). An overview of tick-borne relapsing fever with emphasis on outbreaks in Texas. Tex. Med. *91*, 56–59.

Rich, S.M., Armstrong, P.M., Smith, R.D., and Telford 3rd, S.R. (2001). Lone star tick-infecting borreliae are most closely related to the agent of bovine borreliosis. J. Clin. Microbiol. *39*, 494–497.

Rich, S.M., Caporale, D.A., Telford 3rd, S.R., Kocher, T.D., Hartl, D.L., and Spielman, A. (1995). Distribution of the *Ixodes ricinus*-like ticks of eastern North America. Proc. Natl. Acad. Sci. U.S.A. *92*, 6284–6288.

Richter, D., and Matuschka, F.R. (2006). Perpetuation of the Lyme disease spirochete *Borrelia lusitaniae* by lizards. Appl. Environ. Microbiol. *72*, 4627–4632.

Richter, D., Postic, D., Sertour, N., Livey, I., Matuschka, F.R., and Baranton, G. (2006). Delineation of *Borrelia burgdorferi* sensu lato species by multilocus sequence analysis and confirmation of the delineation of *Borrelia spielmanii* sp. nov. Int. J. Syst. Evol. Microbio.l *56*, 873–881.

Richter, D., Schlee, D.B., Allgöwer, R., and Matuschka, F.R. (2004). Relationships of a novel Lyme disease spirochete, *Borrelia spielmani* sp. nov., with its hosts in Central Europe. Appl. Environ. Microbiol. *70*, 6414–6419.

Richter, D., Spielman, A., Komar, N., and Matuschka, F.R. (2000). Competence of American robins as reservoir hosts for Lyme disease spirochetes. Emerg. Inf. Dis. *6*, 133–138.

Rodgers, S.E., Miller, N.J., and Mather, T.N. (2007). Seasonal variation in nymphal blacklegged tick abundance in southern New England forests. J. Med. Entomol. *44*, 898–900.

Rosa, P. (2005). Lyme disease agent borrows a practical coat. Nat. Med. *11*, 831–832.

Ross, P.H., and Milne, A.D. (1904). Tick fever. Br. Med. J. *2*, 1453–1454.

Ruzić-Sabljić, E., Arnez, M., Lotric-Furlan, S., Maraspin, V., Cimperman, J., and Strle, F. (2001). Genotypic and phenotypic characterization of *Borrelia burgdorferi* sensu lato strains isolated from human blood. J Med Microbiol *50*, 896–901.

Ryckman, R.E., and Kohls, G.M. (1962). The desert tortoise, *Gopherus agassizi*, a host for the tick *Ornithodoros turicata* in California. J. Parasitol. *48*, 502–503.

Saito, K., Ito, T., Asashima, N., Ohno, M., Nagai, R., Fujita, H., Koizumi, N., Takano, A., Watanabe, H., and Kawabata, H. (2007). Case report: *Borrelia valaisiana* infection in a Japanese man associated with traveling to foreign countries. Am. J. Trop. Med. Hyg. *77*, 1124–1127.

Schneider, B.S., Zeidner, N.S., Burkot, T.R., Maupin, G.O., and Piesman, J. (2000). *Borrelia* isolates in northern Colorado identified as *Borrelia bissettii*. J. Clin. Microbiol. *38*, 3103–3105.

Schulze, T.L., and Jordan, R.A. (2006). Assessment and Management of Vector Tick Populations in New

Jersey. A Guide for Pest Management Professionals, Land Managers, and Public Health Officials. Freehold Township, New Jersey: Freehold Township Health Department.

Schulze, T.L., Jordan, R.A., and Hung, R.W. (1995). Suppression of subadult *Ixodes scapularis* (Acari: Ixodidae) following removal of leaf litter. J. Med. Entomol. *32*, 730–733.

Schulze, T.L., Jordan, R.A., Hung, R.W., Taylor, R.C., Markowski, D., and Chomsky, M.S. (2001). Efficacy of granular deltamethrin against *Ixodes scapularis* and *Amblyomma americanum* (Acari: Ixodidade) nymphs. J. Med. Entomol. *38*, 344–346.

Schulze, T.L., Jordan, R.A., and Schulze, C.J. (2005). Host associations of *Ixodes scapularis* (Acari: Ixodidae) in residential and natural settings in a Lyme disease-endemic area in New Jersey. J. Med. Entomol. *42*, 966–973.

Schulze, T.L., Jordan, R.A., Schulze, C.J., Healy, S.P., Jahn, M.B., and Piesman, J. (2007). Integrated use of 4-poster passive topical treatment devices for deer, targeted acaricide applications, and Maxforce TMS bait boxes to rapidly suppress populations of *Ixodes scapularis* (Acari: Ixodidae) in a residential landscape. J. Med. Entomol. *44*, 830–835.

Schwan, T.G., Battisti, J.M., Porcella, S.F., Raffel, S.J., Schrumpf, M.E., Fischer, E.R., Carroll, J.A., Stewart, P.E., Rosa, P., and Somerville, G.A. (2003a). Glycerol-3-phosphate acquisition in spirochetes: distribution and biological activity of glycerophosphodiester phosphodiesterase (GlpQ) among *Borrelia* spirochetes. J. Bacteriol. *185*, 1346–1356.

Schwan, T.G., Burgdorfer, W., Schrumpf, M.E., and Karstens, R.H. (1988). The urinary bladder, a consistent source of *Borrelia burgdorferi* in experimentally infected white-footed mice (*Peromyscus leucopus*). J. Clin. Microbiol. *26*, 893–895.

Schwan, T.G., Gage, K.L., Karstens, R.H., Schrumpf, M.E., Hayes, S.F., and Barbour, A.G. (1992). Identification of the tick-borne relapsing fever spirochete *Borrelia hermsii* by using a species-specific monoclonal antibody. J. Clin. Microbiol. *30*, 790–795.

Schwan, T.G., and Hinnebusch, B.J. (1998). Bloodstream-versus tick-associated variants of a relapsing fever bacterium. Science *280*, 1938–1940.

Schwan, T.G., and Piesman, J. (2002). Vector interactions and molecular adaptations of Lyme disease and relapsing fever spirochetes associated with transmission by ticks. Emerg. Infect. Dis. *8*, 115–121.

Schwan, T.G., Policastro, P.F., Miller, Z., Thompson, R.L., Damrow, T., and Keirans, J.E. (2003b). Tick-borne relapsing fever caused by *Borrelia hermsii*, Montana. Emerg. Infect. Dis. *9*, 1151–1154.

Schwan, T.G., Raffel, S.J., Schrumpf, M.E., Policastro, P.F., Rawlings, J.A., Lane, R.S., Breitschwerdt, E.B., and Porcella, S.F. (2005). Phylogenetic analysis of the spirochetes *Borrelia parkeri* and *Borrelia turicatae* and the potential for tick-borne relapsing fever in Florida. J. Clin. Microbiol. *43*, 3851–3859.

Schwan, T.G., Raffel, S.J., Schrumpf, M.E., and Porcella, S.F. (2007). Diversity and distribution of *Borrelia hermsii*. Emerg. Infect. Dis. *13*, 436–442.

Schwan T.G., Raffel, S.J., Schrumpf, M.E., Schrumpf, M.E., Webster, L.S., Marques, A.R., Spano, R., Rood, M., Burns, J., and Hu, R. (2009a). Tick-borne relapsing fever and *Borrelia hermsii*, Los Angeles County, California, USA. Emerg. Infect. Dis. *15*, 1026–1031.

Schwan, T.G., Raffel, S.J., Schrumpf, M.E., Gill, J.S., and Piesman, J. (2009b). Characterization of a novel relapsing fever spirochete in the midgut, coxal fluid, and salivary glands of the bat tick *Carios kelleyi*. Vector Borne Zoonotic Dis. *9*, 643–647.

Scoles, G.A., Papero, M., Beati, L., and Fish, D. (2001). A relapsing fever group spirochete transmitted by *Ixodes scapularis* ticks. Vector Borne Zoonotic Dis. *1*, 21–34.

Scott, J.C., Wright, D.J.M., and Cutler, S.J. (2005). Typing African relapsing fever spirochetes. Emerg. Infect. Dis. *11*, 1722–1729.

Scrimenti, R.J. (1970). Erythema chronicum migrans. Arch. Dermatol. *102*, 104–105.

Shadick, N.A., Daltroy, L.H., Phillips, C.B., Liang, U.S., and Liang, M.H. (1997). Determinants of tick-avoidance behaviors in an endemic area for Lyme disease. Am. J. Prev. Med. *13*, 265–270.

Sidi, G., Davidovitch, N., Balicer, R.D., Anis, E., Grotto, I., and Schwartz, E. (2005). Tickborne relapsing fever in Israel. Emerg. Infect. Dis. *11*, 1784–1786.

Sinsky, R.J., and Piesman, J. (1989). Ear punch biopsy method for detection and isolation of *Borrelia burgdorferi* from rodents. J. Clin. Microbiol. *27*, 1723–1727.

Slajchert, T., Kitron, U.D., Jones, C.J., and Mannelli, A. (1997). Role of the eastern chipmunk (*Tamias striatus*) in the epizootiology of Lyme borreliosis in northwestern Illinois, USA. J. Wildl. Dis. *33*, 40–46.

Smith, R.D., Brener, J., Osorno, M., and Ristic, M. (1978). Pathobiology of *Borrelia theileri* in the tropical cattle tick, *Boophilus microplus*. J. Invert. Pathol. *32*, 182–190.

Sonenshine, D.E. (1991). Biology of Ticks, Vol. 1 (New York, Oxford University Press).

Southern, P.M., and Sanford, J.P. (1968). Relapsing fever: a clinical and microbiological review. Medicine *48*, 129–149.

Spielman, A., Clifford, C.M., Piesman, J., and Corwin, M.D. (1979). Human babesiosis on Nantucket Island, USA: description of the vector, *Ixodes (Ixodes) dammini*, n. sp. (Acarina: Ixodidae). J. Med. Entomol. *15*, 218–234.

Spielman, A., Wilson, M.L., Levine, J.F., and Piesman, J. (1985). Ecology of *Ixodes dammini*-borne babesiosis and Lyme disease. Ann. Rev. Entomol. *30*, 439–460.

Stafford 3rd, K.C. (2007). Tick Management Handbook. An integrated guide for homeowners, pest control operators, and public health officials for the prevention of tck-associated diseases. (New Haven, CT: The Connecticut Agricultural Experiment Station).

Stafford 3rd, K.C., Bladen, V.C., and Magnarelli, L.A. (1995). Ticks (Acari: Ixodidae) infesting wild birds (Aves) and white-footed mice in Lyme, CT. J. Med. Entomol. *32*, 453–466.

Stafford 3rd, K.C., Cartter, M.L., Magnarelli, L.A., Ertel, S.H., and Mshar, P.A. (1998a). Temporal correlations between tick abundance and prevalence of ticks infected with *Borrelia burgdorferi* and increasing incidence of Lyme disease. J. Clin. Microbiol. *36*, 1240–1244.

Stafford 3rd, K.C., Ward, J.S., and Magnarelli, L.A. (1998b). Impact of controlled burns on the abundance of *Ixodes scapularis* (Acari: Ixodidae). J. Med. Entomol. *35*, 510–513.

Steere, A.C., Malawista, S.E., Hardin, J.A., Ruddy, S., Askenase, P.W., and Andiman, W.A. (1977a). Erythema chronicum migrans and Lyme arthritis. The enlarging clinical spectrum. Ann. Intern. Med. *86*, 685–698.

Steere, A.C., Malawista, S.E., Snydman, D.R., Shope, R.E., Andiman, W.A., Ross, M.R., and Steele, F.M. (1977b). Lyme arthritis: an epidemic of oligoarticular arthritis in children and adults in three Connecticut communities. Arthritis Rheum. *20*, 7–17.

Steere, A.C., Sikand, V.K., Meurice, F., Parenti, D.L., Fikrig, E., Schoen, R.T., Nowakowski, J., Schmid, C.H., Laukamp, S., Buscarino, C., *et al.* (1998). Vaccination against Lyme disease with recombinant *Borrelia burgdorferi* outer-surface lipoprotein A with adjuvant. N. Engl. J. Med. *339*, 209–215.

Strle, F., Picken, R.N., Cheng, Y., Cimperman, J., Maraspin, V., Lotric-Furlan, S., Ruzic-Sabljic, E., and Picken, M.M. (1997). Clinical findings for patients with Lyme borreliosis caused by *Borrelia burgdorferi* sensu lato with genotypic and phenotypic similarities to strain 25015. Clin. Infect. Dis. *25*, 273–280.

Stromdahl, E.Y., Williamson, P.C., Kollars Jr., T.M., Evans, S.R., Barry, R.K., Vince, M.A., and Dobbs, N.A. (2003). Evidence of *Borrelia lonestari* DNA in *Amblyomma americanum* (Acari: Ixodidae) removed from humans. J. Clin. Microbiol. *41*, 5557–5562.

Stünzner, D., Hubálek, Z., Halouzka, J., Wendelin, I., Sixl, W., and Marth, E. (2006). Prevalence of *Borrelia burgdorferi* sensu lato in the tick *Ixodes ricinus* in the Styrian mountains of Austria. Wien Klin. Wochenschr. *118*, 682–685.

Sundnes, K.O., and Haimanot, A.T. (1993). Epidemic of louse-borne relapsing fever in Ethiopia. Lancet *343*, 1213–1215.

Tälleklint, L., and Jaenson, T.G. (1993). Maintenance by hares of European *Borrelia burgdorferi* in ecosystems without rodents. J. Med. Entomol. *30*, 273–276.

Tälleklint, L., Jaenson, T.G., and Mather, T.N. (1993). Seasonal variation in the capacity of the bank vole to infect larval ticks (Acari: Ixodidae) with the Lyme disease spirochete, *Borrelia burgdorferi*. J. Med. Entomol. *30*, 812–815.

Taragel'ová, V., Koci, J., Hanincová, K., Kurtenbach, K., Derdáková, M., Ogden, N.H., Literák, I., Kocianová, E., and Labuda, M. (2008). Blackbirds and song thrushes constitute a key reservoir of *Borrelia garinii*, the causative agent of borreliosis in Central Europe. Appl. Environ. Microbiol. *74*, 1289–1293.

Telford 3rd, S.R. (1998). The name *Ixodes dammini* epidemiologically justified. Emerg. Infect. Dis. *4*, 132–134.

Thomas, N.J., Bunikis, J., Barbour, A.G., and Wolcott, M.J. (2002). Fatal spirochetosis due to a relapsing fever-like *Borrelia* sp. in a northern spotted owl. J Wildl Dis *38*, 187–193.

Thompson, R.S., Burgdorfer, W., Russell, R., and Francis, B.J. (1969). Outbreak of tick-borne relapsing fever in Spokane County, Washington. JAMA *210*, 1045–1050.

Trape, J.F., Duplantier, J.M., Bouganali, H., Godeluck, B., Legros, F., J.P., C., and Camicas, J.L. (1991). Tick-borne borreliosis in West Africa. Lancet *337*, 473–475.

Trape, J.F., Godeluck, B., Diatta, G., Rogier, C., Legros, F., Albergel, J., Pepin, Y., and Duplantier, J.M. (1996). The spread of tick-borne borreliosis in West Africa and its relationship to sub-saharan drought. Am. J. Trop. Med. Hyg. *54*, 289–293.

Trevejo, R.T., Schriefer, M.E., Gage, K.L., Safranek, T.J., Orloski, K.A., Pape, W.J., Montenieri, J.A., and Campbell, G.L. (1998). An interstate outbreak of tick-borne relapsing fever among vacationers at a Rocky Mountain cabin. Am. J. Trop. Med. Hyg. *58*, 743–747.

Uhlmann, E.J., Seed, P.C., Schwan, T.G., and Storch, G.A. (2007). Polymerase chaing reaction of tick-borne relapsing fever caused by *Borrelia hermsii*. Pediatr. Infect. Dis. J. *26*, 267–269.

Ullmann, A.J., Gabitzsch, E.S., Schulze, T.L., Zeidner, N.S., and Piesman, J. (2005). Three multiplex assays for detection of *Borrelia burgdorferi* sensu lato and *Borrelia miyamotoi* sensu lato in field-collected *Ixodes* nymphs in North America. J. Med. Entomol. *42*, 1057–1062.

Ullmann, A.J., Lane, R.S., Kurtenbach, K., Miller, M., Schriefer, M.E., Zeidner, N., and Piesman, J. (2003). Bacteriolytic activity of selected vertebrate sera for *Borrelia burgdorferi* sensu stricto and *Borrelia bissettii*. J. Parasitol. *89*, 1256–1257.

van Dam, A.P., Kuiper, H., Vos, K., Widjojokusumo, A., de Jongh, B.M., Spanjaard, L., Ramselaar, A.C.P., Kramer, M.D., and Dankert, J. (1993). Different genospecies of *Borrelia burgdorferi* are associated with distinct clinical manifestations of Lyme borreliosis. Clin. Infect. Dis. *17*, 708–717.

Varela, A.S., Luttrell, M.P., Howerth, E.W., Moore, V.A., Davidson, W.R., Stallknecht, D.E., and Little, S.E. (2004). First culture isolation of *Borrelia lonestari*, putative agent of southern tick-associated rash illness. J. Clin. Microbiol. *42*, 1163–1169.

Varma, M.G.R. (1962). Transmission of relapsing fever spirochaetes by ticks. Symp. Zool. Soc. Lond. *6*, 61–82.

Vial, L., Diatta, G., Tall, A., Ba, E.H., Bouganali, H., Durand, P., Sokhna, C., Rogier, C., Renaud, F., and Trape, J.F. (2006a). Incidence of tick-borne relapsing fever in west Africa: longitudinal study. Lancet *368*, 37–43.

Vial, L., Durand, P., Arnathau, C., Halos, L., Diatta, G., Trape, J.F., and Renaud, F. (2006b). Molecular divergences of the *Ornithodoros sonrai* soft tick species, a vector of human relapsing fever in West Africa. Microbes Infect. *8*, 2605–2611.

Vourc'h, G., Marmet, J., Chassagne, M., Bord, S., and Chapuis, J.L. (2007). *Borrelia burgdorferi* sensu lato in Siberian chipmunks (*Tamias sibiricus*) introduced in suburban forests in France. Vector Borne Zoonotic Dis. *7*, 637–641.

Walker, R.L., Read, D.H., Hayes, D.C., and Nordhausen, R.W. (2002). Equine abortion associated with the *Borrelia parkeri-B. turicatae* tick-borne relapsing fever spirochete group. J. Clin. Microbiol. *40*, 1558–1562.

Wallich, R., Pattathu, J., Kitiratschky, V., Brenner, C., Zipfel, P.F., Brade, V., Simon, M.M., and Kraiczy, P.

(2005). Identification and functional characterization of complement regulator-acquiring surface protein 1 of the Lyme disease spirochetes *Borrelia afzelii* and *Borrelia garinii*. Infect. Immun. *73*, 2351–2359.

Walton, G.A. (1962). The *Ornithodoros moubata* superspecies problem in relation to human relapsing fever epidemiology. Symp. Zool. Soc. Lond. *6*.

Wang, G., Liveris, D., Brei, B., Wu, H., Falco, R.C., Fish, D., and Schwartz, I. (2003). Real-time PCR for simultaneous detection and quantification of *Borrelia burgdorferi* in field-collected *Ixodes scapularis* ticks from the North-eastern United States. Appl. Environ. Microbiol. *69*, 4561–4565.

Wang, G., van Dam, A.P., Schwartz, I., and Dankert, J. (1999). Molecular typing of *Borrelia burgdorferi* sensu lato: taxonomic, epidemiological, and clinical implications. Clin. Microbiol. Rev. *12*, 633–653.

Weyer, F. (1960). Biological relationships between lice (Anoplura) and microbial agents. Ann. Rev. Entomol. *5*, 405–420.

Wheeler, C.M. (1935). A new species of tick which is a vector of relapsing fever in California. Am. J. Trop. Med. *15*, 435–438.

Wheeler, C.M. (1938a). Progress of spirochaete infection in the developmental stages of the host tick, *Ornithodoros hermsi*, Wheeler. Am. J. Trop. Med. Hyg. *18*, 413–419.

Wheeler, C.M. (1938b). Relapsing fever in California: attempts to transmit spirochaetes of California relapsing fever to human subjects by means of the bite of the vector, *Ornithodoros hermsi* Wheeler. Am. J. Trop. Med. *18*, 641–659.

Wheeler, C.M., Herms, W.B., and Meyer, K.F. (1935). A new tick vector of relapsing fever in California. Proc. Soc. Exp. Biol. Med. 32, 1290–1292.

Whitney, M.S., Schwan, T.G., Sultemeier, K.B., McDonald, P.S., and Brillhart, M.N. (2007). Spirochetemia caused by *Borrelia turicatae* infection in 3 dogs in Texas. Vet. Clin. Pathol. *36*, 212–216.

Wilson, M.L. (1986). Reduced abundance of adult *Ixodes dammini* (Acari: Ixodidae) following destruction of vegetation. J. Econ. Entomol. *79*, 693–696.

Wormser, G.P., Dattwyler, R.J., Shapiro, E.D., Halperin, J.J., Steere, A.C., Klempner, M.S., Krause, P.J., Bakken, J.S., Strle, F., Stanek, G., *et al.* (2006). The clinical assessment, treatment, and prevention of Lyme disease, human granulocytic anaplasmosis, and babesiosis: clinical practice guidelines by the Infectious Diseases Society of America. Clin. Infect. Dis. *43*, 1089–1134.

Wynns, H.L. (1936). Epidemiology. Special Bulletin 61. Relapsing Fever. Calif State Dept Publ Health, 13–18.

Wynns, H.L., and Beck, M.D. (1935). Epidemiological studies on relapsing fever in California. Am. J. Publ. Hlth*7*, 270–276.

Xu, G., Fang, Q.Q., Keirans, J.E., and Durden, L.A. (2003). Molecular phylogenetic analyses indicate that the *Ixodes ricinus* complex is a paraphyletic group. J. Parasitol. *89*, 452–457.

Younsi, H., Sarih, M., Jouda, F., Godfroid, E., Gern, L., Bouattour, A., Baranton, G., and D., P. (2005). Characterization of *Borrelia lusitaniae* isolates collected in Tunisia and Morocco. J. Clin. Microbiol. *43*, 1587–1593.

Yparraguirre, L.A., Machado-Ferreira, E., Ullmann, A.J., Piesman, J., Zeidner, N.S., and Soares, C.A.G. (2007). A hard tick relapsing fever group spirochete in a Brazilian *Rhipicephalus (Boophilus) microplus.* Vector Borne Zoonotic Dis. *7*, 717–721.

Zhioua, E., Bouattour, A., Hu, C.M., Gharbi, M., Aeschliman, A., Ginsberg, H.S., and Gern, L. (1999). Infection of *Ixodes ricinus* (Acari: Ixodidae) by *Borrelia burgdorferi* sensu lato in North Africa. J. Med. Entomol. *36*, 216–218.

Tick Interactions

11

Utpal Pal and Erol Fikrig

Abstract

Borrelia burgdorferi causes Lyme disease in a variety of animals and humans. The spirochaete is maintained in an enzootic cycle that primarily involves small mammals and *Ixodes spp.* ticks. In the northeastern USA, *I. scapularis* is the main vector. As only *I. scapularis* and closely related ticks transmit *B. burgdorferi*, the spirochaete–tick interactions are thought to be highly specific. Several spirochaete and tick proteins that directly or indirectly contribute to the natural cycle of *B. burgdorferi* infection have been identified. This chapter will focus on past discoveries and future challenges that are relevant to our understanding of *B. burgdorferi*–tick interactions and on the development of novel preventative measures that interfere with the *B. burgdorferi* life cycle.

11.1 Introduction

Ixodes ticks transmit a wide range of infections to animals and humans (Nadelman and Wormser, 1998; Steere *et al.*, 2004). Lyme disease is a prominent tick-borne human illness (Barbour and Fish, 1993) caused by a group of spirochaetes belonging to the *Borrelia burgdorferi* sensu lato complex (Burgdorfer *et al.*, 1982). *Borrelia* species within this complex commonly associated with human infections include *B. burgdorferi* sensu stricto, prevalent throughout the USA and Europe, and *B. afzelii* and *B. garinii*, which are primarily distributed in Eurasia (Nadelman and Wormser, 1998; Piesman and Gern, 2004). In the northeastern USA, Lyme disease spirochaetes are maintained in nature through a complex tick–rodent infection cycle. Larval ticks acquire the pathogen while feeding on infected wild hosts, such as white-footed mice (Anderson, 1989; Anderson *et al.*, 1987), and then transstadially maintain *B. burgdorferi* before transmitting the pathogen back to a naïve mouse during the subsequent blood meal (Steere, 2001; Steere *et al.*, 2004). Humans and domestic animals that develop Lyme disease are infected as incidental hosts and do not play a role in the natural transmission cycle.

Borrelia burgdorferi does not readily infect most tick species, suggesting that the interactions between *I. scapularis* and *B. burgdorferi* are highly specific, and limited studies have shed important light on the complex interactions between *B. burgdorferi* and ticks (Fikrig and Narasimhan, 2006; de Silva *et al.*, 2009; Munderloh and Kurtti, 1995; Tsao, 2009). With the advent of genetic tools for creating *Borrelia* mutants (Rosa *et al.*, 2005) and current efforts to sequence the *I. scapularis* genome (Hill and Wikel, 2005), investigators will have increasingly robust tools for studying *Borrelia*–tick interactions during the arthropod phase of the spirochaete life cycle and at the interface with the mammalian host. A few spirochaete proteins already have been shown to be required in the vector and to interact with tick receptors in selected organs, such as the gut or salivary gland (Pal *et al.*, 2004a; Ramamoorthi *et al.*, 2005; Rosa, 2005). *Borrelia* adaptation to the vector or host also requires tightly controlled regulation of bacterial gene expression (Caimano *et al.*, 2005, 2007; Fisher *et al.*, 2005; Hubner *et al.*, 2001). Exciting new studies have shed light on the complex gene regulation of *B. burgdorferi* in ticks. The RpoN/RpoS pathway has been shown to be active in feeding nymphs (Caimano *et al.*, 2007) and to induce a set of proteins required for *B. burgdorferi* transmission from ticks as well as for host infection (Fisher *et al.*, 2005). Future studies on *Borrelia*–tick interaction will allow us to understand unique adaptive strategies of the bacterium to survive in an extremely diverse range of host tissues. This information will contribute to the development of novel preventative measures against *B. burgdorferi* infection. The following sections will highlight the *B. burgdorferi* infection cycle through the vector and will review how molecular interactions of selected *B. burgdorferi* and *I. scapularis* gene products support the pathogen life cycle in nature. We will also comment on recent advances in *Ixodes* genomics and on how tick interaction studies may contribute to the development of a transmission-blocking Lyme disease vaccine.

11.2 The *Borrelia burgdorferi* life cycle in ticks

Borrelia burgdorferi survives in nature in a tick–mammal infection cycle (Anguita *et al.*, 2003; Pal and Fikrig, 2003). Given the lack of transovarial transmission, *B. burgdorferi* needs to be acquired during one of the life stages of its vector (Anderson, 1989; Munderloh and Kurtti, 1995; Pal and Fikrig, 2003) at the time of engorgement on infected mammals, primarily wild rodents. Within the first 12 hours of tick attachment to its host, as the arthropod prepares for ingestion of a blood meal from the feeding lesion, the transfer of spirochaetes to the vector has already begun. Details of how *B. burgdorferi* transits between the arthropod and the vertebrate host, either by active migration or via passive transfer along with host fluid, are uncertain. Spirochaetes enter the tick gut from the host dermis and, presumably, continue to migrate until the ticks are fully engorged, which usually takes 72–96 hours (de Silva and Fikrig, 1995; Piesman *et al.*, 1990). *Borrelia burgdorferi* intimately associates with the gut tissues, which is recorded within 48 hours of feeding (Yang *et al.*, 2004), and can persist in the gut throughout the life span of the arthropod (Pal *et al.*, 2001). Thus, *B. burgdorferi* spends a significant part of its life cycle confined to the tick gut. During a subsequent blood meal, the spirochaetes multiply in the gut. An unknown fraction of the *B. burgdorferi* population exits the gut, enters the haemocoele and disseminates towards the salivary glands, through which the bacteria can be transmitted to a new mammalian host (Coleman *et al.*, 1997; de Silva and Fikrig, 1995; Piesman *et al.*, 2001; Schwan and Piesman, 2002). Thus, a larval tick acquires the pathogen from the infected host, moults into a nymph, which transmits the infection to the next host. Maintenance of the spirochaete in the enzootic cycle involves successful persistence of *B. burgdorferi* through multiple developmental stages of the arthropod and coordinated dissemination through tick tissues to the new host. The following section will highlight the varied environments and formidable challenges that *B. burgdorferi* encounters during its entry, long-term persistence and subsequent transmission through ticks.

11.2.1 *Borrelia burgdorferi* encounters diverse tick environments

The feeding gut during acquisition

When transmitted from mammals, *B. burgdorferi* adapts to an entirely new environment in the vector. Little is known about the immune system of the tick. Nevertheless, it can be presumed that the feeding tick gut, which *B. burgdorferi* encounters immediately following acquisition, presents a hostile environment. As spirochaetes migrate into the luminal spaces of the gut and colonize the tissue, the bacterium must avoid being digested with the blood meal and must also bypass tick innate immune defence mechanisms (Sonenshine *et al.*, 2005). Although gastric digestion in ticks is primarily intracellular (Sonenshine, 1993), with intraluminal proteases virtually absent, degradation of blood meal components, such as haemoglobin, could lead to peptides that can have antimicrobial activities (Fogaca *et al.*, 1999; Nakajima *et al.*, 2003). Ticks may also produce conventional antimicrobial peptides in the gut (Hynes *et al.*, 2005; Sonenshine *et al.*, 2002). An *I. ricinus* defensin-like gene is upregulated in the gut after infection with *B. burgdorferi* (Rudenko *et al.*, 2005); however, the mechanism by which spirochaetes evade this innate immune offence is unknown. Phagocytosis allows tick gut cells to capture components of blood meals, such as cellular remnants and haemoglobin crystals (Sonenshine, 1993), but ingestion of *B. burgdorferi* is rarely observed. *Ixodes* ticks also form a transient peritrophic membrane that surrounds the blood meal (elaborated below). How the acellular peritrophic barrier membrane influences *B. burgdorferi* acquisition and colonization of gut epithelium in ticks is not clear. Invading spirochaetes could rapidly colonize the gut epithelial cells prior to the formation of a peritrophic barrier or, alternatively, spirochaetes could cross the peritrophic membrane after its formation and colonize the underlying gut epithelium. In either case, the peritrophic barrier could afford protection of tissue-bound bacteria against the toxic contents of the lumen.

The unfed gut

In its natural cycle, *B. burgdorferi* persists in ticks for months, mostly confined in the gut of unfed ticks. *Ixodes* ticks have a relatively long life cycle, spanning at least 18 months (Ullmann *et al.*, 2005), a significant part of which is dominated by phases of nutrient deprivation and limited metabolic activity (Sonenshine, 1993). Little is known about the mechanisms by which *B. burgdorferi* persists in unfed ticks. Living within the gut of the poikilotherm, spirochaetes must also survive through temperature extremes caused by seasonal changes or even daily alteration of temperatures that can fluctuate by many degrees. The lumen of the unfed gut (Fig. 11.1) is poor in residual nutrients, primarily because gut epithelia store nutrient macromolecules within cellular endosomes, leaving the lumen devoid of nutrients (Sonenshine, 1993). Genes that are expressed by *B. burgdorferi* in the unfed gut or in 'tick-like' growth conditions *in vitro* could be associated with maintenance of the most elemental functions for persistence in nutrient-depleted anoxic arthropod environments (Ojaimi *et al.*, 2002; Revel *et al.*, 2002). Spirochaetes are known to upregulate specific gene products that protect DNA against starvation or oxidative stress-induced damage during the long inter-moult period between blood meals (described below). Cystic forms of *B. burgdorferi* have also been reported, especially under unfavourable conditions *in vitro* (Alban *et al.*, 2000; Brorson and Brorson, 1998,

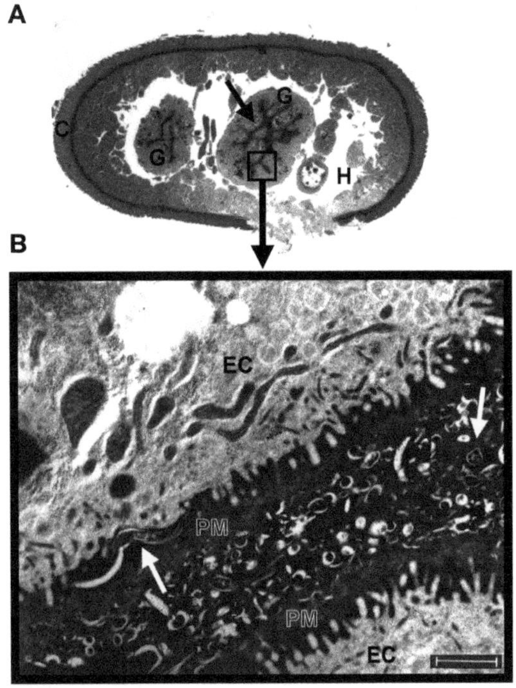

Figure 11.1 Gut of nymphal *Ixodes scapularis*. (A) Cross-sectional view of the whole tick near the abdominal area in an unfed *I. scapularis*. Sections (5 µm) were stained with toluidine blue, which reveals orientation of the gut diverticula with conspicuous luminal spaces, as indicated by an arrow. (B) Transmission electron micrograph of feeding gut lumen (similar areas as indicated by the box in panel A) from a 36-hour fed and infected *I. scapularis* nymph. The sample was processed for electron microscopy as detailed before (Pal *et al.*, 2004a). Image shows formation of a conspicuous peritrophic membrane (PM) that surrounds the luminal contents. *Borrelia burgdorferi* in cross- or longitudinal sections (white arrows) are on either side of the PM. C, cuticle; G, gut tissue; H, haemocoelic spaces, EC, epithelial cells. Bar = 1 µm.

2004) and in incidental hosts *in vivo* (Brorson and Brorson, 1998; Duray *et al.*, 2005); however, the occurrence of such *B. burgdorferi* cysts in the natural infection cycle, such as in dormant ticks, is not known.

Feeding gut during transmission

Tick attachment to a host initiates metabolic changes in the gut (Sonenshine, 1993) that probably prepare *B. burgdorferi* for the transmission event. During the first 36 hours of host attachment, blood enters the tick slowly and there is little digestive activity (Sauer, 1986; Sonenshine, 1993). Spirochaetes in the gut must sense the appropriate physiochemical stimuli and begin multiplying. Following 24 hours of host attachment, *B. burgdorferi* is principally confined within the lumen of the tick gut; whereas, by 72 hours, they are dispersed throughout the entire gut tissue (Pal and Fikrig, 2003). Although the density of *B. burgdorferi* in the gut increases exponentially as feeding progresses (de Silva and Fikrig, 1995; Piesman *et al.*, 1990), generating a phenotypically diverse population of spirochaetes (Ohnishi *et al.*, 2001; Srivastava and de Silva, 2008), the majority of the spirochaetes do not exit the gut. A fraction of spirochaetes, however, crosses the gut epithelial barrier and disseminates to the haemocoele for transmission to the host via the salivary gland (Coleman *et al.*, 1997). The molecular determinants that influence *B. burgdorferi* dissemination from the gut and the mechanism for spirochaete invasion across the gut epithelia, including the apical and basal membranes (Sauer, 1986; Sonenshine, 1993), are unknown. Host-derived plasminogen has been implicated in efficient dissemination of *B. burgdorferi* within the tick and also for enhancement of spirochaetaemia in the vertebrate hosts (Coleman *et al.*, 1997; Coleman *et al.*, 1995). The mechanism by which *B. burgdorferi* disseminates through the infected gut tissue is also not known; however, in the feeding gut, spirochaetes are detected both intra- and extracellularly across the gut epithelia (Zung *et al.*, 1989). As previously mentioned, an ephemeral, but solid, peritrophic membrane is formed in the feeding infected tick gut, clearly visible within 36 hours of onset of feeding (Fig. 11.1), which engulfs the blood meal and *B. burgdorferi* present in the lumen (Rudzinska *et al.*, 1982; Sonenshine, 1993). This membrane is a solid mechanical barrier and even small particles, such as ribosomes derived from ingested blood meal, could be retained in the lumen proper (Rudzinska *et al.*, 1982). *Borrelia burgdorferi* can be detected in the lumen of the feeding gut (Pal and Fikrig, 2003; Pal *et al.*, 2004a); however, how the physical division of gut luminal spaces and epithelial tissues imposed by the peritrophic membrane influences *B. burgdorferi* transmission is not known. *Babesia microti*, a tick-borne pathogen is reported to develop a highly specialized organelle, the arrowhead, which helps the microbe to cross the peritrophic barrier of *Ixodes* ticks (Rudzinska *et al.*, 1982). The peritrophic membrane has been shown to limit and strongly affect pathogen transmission in many other arthropod-borne pathogens (Beerntsen *et al.*, 2000; Pimenta *et al.*, 1997).

The haemocoele

During transmission, *B. burgdorferi* needs to travel through tick haemocoelic spaces to migrate to the salivary gland. Cellular constituents of the haemocoele in arthropods are dominated by actively phagocytic blood cells or haemocytes that are the primary mediators of arthropod innate immune responses (Sauer, 1986; Sonenshine, 1993). Little is known about how *B. burgdorferi* travels through the tick haemolymph, avoids phagocytosis and the innate immune defence of haemocytes, selectively recognizes and enters the salivary gland. Studies indicate that a significant percentage ($13 \pm 5\%$) of isolated haemocytes in feeding *I. scapularis* could be associated with disseminated spirochaetes (Coleman *et al.*, 1997). Although, occasionally, haemocytes with ingested *B. burgdorferi* have been noted (Coleman *et al.*, 1997), the biological significance of *Ixodes* haemocyte–spirochaete association is not clear. In another study, when *B. burgdorferi* was experimentally injected into the haemocoele of *I. scapularis* and *Dermacentor variabilis*, a non-competent tick vector of *Borrelia*, the spirochaetes readily survived in the *Ixodes* haemolymph; whereas, they were rapidly killed by haemocytes in the *D. variabilis* haemolymph, suggesting that *B. burgdorferi* is resistant to killing by *Ixodes* haemocytes (Johns *et al.*, 2001). Although the number of *B. burgdorferi* in the infected tick gut increases exponentially (Piesman *et al.*, 1990; Piesman *et al.*, 2001), the exact proportion of spirochaetes that cross the gut epithelial barrier and systemically disseminate to the salivary gland through the haemocoele is unknown. Studies in feeding infected *I. scapularis* indicate that approximately 5% and 0.7% of the gut spirochaetes can be detected in haemolymph and salivary gland, respectively (Coleman *et al.*, 1997). Future studies on the kinetics and mechanism of *B. burgdorferi* dissemination through the *Ixodes* haemocoele should provide important insights for the development of novel preventative approaches to interfere with the spirochaete transmission from ticks to mammals.

The salivary gland

The salivary gland is the only other known tick organ (Sauer, 1986; Sonenshine, 1993) that spirochaetes must traverse during transmission to a mammalian host. *Borrelia burgdorferi,* however, do not infect salivary glands of feeding *Ixodes* ticks upon their acquisition from infected mammals or during persistence in unfed ticks (Fikrig and Narasimhan, 2006; Pal and Fikrig, 2003; Piesman, 1993; Piesman *et al.*, 2001). Spirochaetes migrate from the infected tick gut to the gland several hours after the initiation of feeding, which lasts for 3–7 days (Fikrig and Narasimhan, 2006; Pal and Fikrig, 2003; Piesman, 1993; Piesman *et al.*, 2001). *B. hermsii,* the relapsing fever spirochaetes, on the other hand, readily colonizes the salivary glands of unfed soft ticks, *Ornithodoros hermsi* (Schwan and Piesman, 2002; Sonenshine, 1993). Soft ticks are fast-feeders, however, relapsing fever spirochaetes are quickly transmitted within minutes of their attachment to a mammalian host (Schwan and Piesman, 2002). Irrespective of the kinetics of transmission, tick salivary glands potentially act as the second distinct barrier to efficient transmission of most of the arthropod-borne pathogens. Other tick-borne bacterial pathogens, such as *Anaplasma marginale,* require active replication within the salivary gland for effective transmission (Ueti *et al.*, 2007), yet there is no evidence for *B. burgdorferi* replication in tick salivary glands. Spirochaetes invade the salivary gland by an undefined mechanism, and, once inside the gland, may be passively carried to the host dermis along with tick saliva. As several salivary gland genes are upregulated in infected *I. scapularis* nymphs as compared to uninfected nymphs (Ribeiro *et al.*, 2006), salivary gland proteins may play an important role in *B. burgdorferi* transmission and infection.

11.2.2 Regulation of gene expression in ticks

Borrelia burgdorferi alters its gene expression profile as it cycles between arthropods and mammals (Anguita *et al.*, 2003; de Silva and Fikrig, 1997; Rosa *et al.*, 2005; Schwan, 2003), and specific genes maintain the spirochaete infection cycle in nature (Coburn and Cugini, 2003; Grimm *et al.*, 2004b; Li *et al.*, 2007b; Pal *et al.*, 2004b, 2008b; Purser *et al.*, 2003; Revel *et al.*, 2005; Rosa *et al.*, 2005; Yang *et al.*, 2004). The diverse microenvironments within vector tissues probably induce the changes in pathogen gene expression required for survival. In contrast to many pathogenic bacteria, *B. burgdorferi* devotes large portions of its genome (more than 8% of coding genes) towards producing approximately 150 lipoproteins (Casjens *et al.*, 2000; Fraser *et al.*, 1997), many of which are likely to be exposed on the bacterial surface and be differentially expressed. Although the host or vector environment is likely to be too complex and dynamic to be accurately modelled artificially, a number of microarray studies provided indirect but important clues about the regulation and function of *B. burgdorferi* genes in tick-specific environments (Brooks *et al.*, 2003; Fisher *et al.*, 2005; Narasimhan *et al.*, 2002; Ojaimi *et al.*, 2002; Revel *et al.*, 2002; Tokarz *et al.*, 2004). As temperature is a critical environmental factor known to affect *B. burgdorferi* gene expression, microarray studies attempted to identify *B. burgdorferi* genes that are responsive to temperature fluctuations *in vitro* (Ojaimi et *al.*, 2002). Such studies have important implications as the temperature in the tick gut changes rapidly during tick feeding from the ambient arthropod temperature (represented by 23°C) to mammalian temperature (approximately 37°C). Transcriptome analysis of cultured spirochaetes grown in two different temperatures, 23°C and 35°C, indicates that a significant number of spirochaete genes were indeed differentially expressed at the two temperatures; 133 genes were expressed at significantly greater levels at 35°C, and 82 were more significantly expressed at 23°C (Ojaimi *et al.*, 2002). Unfortunately, the function of the vast majority of these temperature-influenced genes is currently unknown; however, these results underscore their potential role in *B. burgdorferi* survival under diverse environmental conditions. *Borrelia burgdorferi* gene expression that undergoes alteration upon growth within the dialysis membrane chamber (DMC) implanted in a mammalian host (Brooks *et al.*, 2003) or addition of blood (Tokarz *et al.*, 2004), also provides important information that can be extrapolated to speculate regulation and function of *B. burgdorferi* genes *in vivo*. During

DMC cultivation, 125 spirochaete genes display dramatic alterations in expression. Interestingly, the vast majority of these differentially expressed genes that are downregulated (79%) are encoded on plasmids, whereas almost all the upregulated genes (93%) are encoded on the chromosome (Brooks *et al.*, 2003). Another recent study identified 34 genes that are downregulated by RpoS when spirochaetes are cultured within the DMC (Caimano *et al.*, 2007). The identified genes that show repression in experimental conditions like DMC, might not have important roles for *B. burgdorferi* infectivity in mammalian dermis, but could be relevant in other phases of the life cycle including *B. burgdorferi* persistence in ticks. The majority of these RpoS-repressed genes in DMC-grown spirochaetes are localized in plasmids, and almost one-third are known or predicted to encode lipoproteins, such as OspA. Other microarray studies indicated that the addition of blood to the spirochaete culture maintained at 35°C for 48 hours, simulating conditions in infected ticks prior to transmission, had a dramatic effect on the expression of 154 genes, including a significant downregulation of 79 genes, such as those relevant to chemotaxis and sensory regulons, oligopeptide transporters, proteases, and of the recombination genes of cp32 (Tokarz *et al.*, 2004). These dynamic changes in the *B. burgdorferi* transcriptome could influence adaptation of spirochaetes *in vivo* following a blood meal in the tick.

Limited studies have attempted to identify *B. burgdorferi* genes that are selectively upregulated in the environments of either unfed or fed ticks *in vivo* (Narasimhan *et al.*, 2002). Functional roles of these vector-induced genes or the mechanism of their selective upregulation in ticks, is unknown. Regulatory pathways involving Rrp1 (Rogers *et al.*, 2009) or the *rpoN* and *rpoS* alternative sigma factor pathways, where RpoN controls the expression of RpoS (Hübner *et al.*, 2001), are regarded as the key genetic regulatory networks in *B. burgdorferi*. These pathways may activate many spirochaete genes in feeding ticks, implying a role for the RpoN/RpoS pathways in *B. burgdorferi* transmission through ticks. A *rpoN* null mutant successfully colonizes tick guts, but is unable to invade salivary glands or infect the vertebrate host (Fisher *et al.*, 2005). These studies indicate that RpoN-regulated genes are involved in transmission from the vector to the host. A more recent study indicates that expression of the alternative sigma factor, *rpoS*, is selectively induced during the nymphal blood meal, suggesting a potential role of RpoS as an activator of genes relevant to spirochaete transmission and establishment of early infection in a mammalian host (Caimano *et al.*, 2004; Caimano *et al.*, 2007). Individual spirochaete genes are also identified that show dramatic expression in ticks, such as *ospA, ospB, ospC, ospD, ospE, ospF, elp, bptA, dps, cspA, la7* and *bba74*, that may play important roles for tick tissue colonization, transmission, and infection of a new host (Brooks *et al.*, 2003; Bykowski *et al.*, 2007, 2008; Caimano *et al.*, 2007; Grimm *et al.*, 2004b; Hefty *et al.*, 2001; Li *et al.*, 2007b; Mulay *et al.*, 2009; Neelakanta *et al.*, 2007; Ojaimi *et al.*, 2002; Pal *et al.*, 2008a; Revel *et al.*, 2005; Schwan and Piesman, 2000; Schwan *et al.*, 1995; Tokarz *et al.*, 2004; Yang *et al.*, 2004). Although many of these genes belong to *B. burgdorferi* paralogous gene families, such as Erps (OspE/F-related proteins), individual members have been shown to have distinct expression patterns in the infected tick during feeding and transmission to the mammalian host. For example, while *ospE, ospF* and *Bbk2.11* display progressive induction in the gut while the arthropod is being engorged on the blood meal, certain members of *elp* family, such as *Bbk2.10* and *ospE/F* homologue *p21* are only expressed at some later points of feeding and dissemination (Hefty *et al.*, 2001). Although the details of environmental signals or borrelial regulatory networks that influence induction of selected *B. burgdorferi* genes in ticks are largely unknown, two genes, *ospA* and *ospC*, have been the most extensively studied. In fact, the principles developed with these proteins provided important clues for understanding the regulation of bacterial gene expression in ticks. *ospA* is prominently expressed in vector-specific *B. burgdorferi* life cycle including the unfed gut; whereas, *ospC* is induced only during arthropod feeding (Schwan, 2003; Schwan *et al.*, 1995). A simple reciprocal relationship between expression of *ospA* and *ospC* has been proposed where the similar signals and pathways that induce *ospC* decrease the production of *ospA* (Yang *et al.*, 2000; He *et al.*, 2008), which was also recently demonstrated at the single cell level (Srivastava and de Silva, 2008). These contentions have been supported by additional recent studies that show a reciprocally regulated expression of *ospA* and *ospC*-like genes by *B. burgdorferi* (Caimano *et al.*, 2007). However, as spirochaete populations producing both OspA and OspC, or neither protein, have been reported in ticks (Ohnishi *et al.*, 2001; Srivastava and de Silva, 2008), expression of these genes *in vivo* might also be regulated through unknown complex and independent regulatory networks. Although not clear for *ospA* regulation, physiochemical signals, such as temperature, pH and cell density, probably play important roles in regulating the expression of *ospC* in the feeding tick (Carroll *et al.*, 1999; Indest *et al.*, 1997; Stevenson *et al.*, 1995; Yang *et al.*, 2000). Expression of *ospC*-like genes in *B. burgdorferi* is shown to be influenced by a two-component system consisting of a putative membrane sensor kinase (HK2) and a cytoplasmic response regulator protein (Rrp2) (Revel *et al.*, 2002; Yang *et al.*, 2003). Rrp2, together with the alternative sigma factor RpoN, induces a second sigma factor, RpoS (Blevins *et al.*, 2009), which activates the transcription of *ospC* and *ospC*-like genes associated with tick transmission and host infection (Fisher *et al.*, 2005; Hubner *et al.*, 2001; Yang *et al.*, 2003). In contrast, relatively little is known about the pathways and signals regulating *ospA* and *ospA*-like gene expression in ticks. RpoS is required for the downregulation of the tick phase-specific lipoproteins OspA and lp6.6 in response to mammalian host-derived signals (Caimano *et al.*, 2005, 2007). Unlike *ospC*, *ospA* is relatively unresponsive to physiochemical signals; however, *ospA* expression may be influenced by specific biochemical factors in the blood meal. In fact, host immunoglobulin has recently been shown to influence *B. burgdorferi* OspA production (Hodzic *et al.*, 2005).

Overall, our current knowledge of *borrelial* gene regulation is mainly supported by the microarray studies that significantly contributed to our understanding of differentially expressed *B. burgdorferi* genes that might function in the adaptation of spirochaetes through diverse environments *in vivo*. Exciting molecular tools have recently been developed (Gilbert *et al.*, 2007) that should foster new studies delineating the regulatory mechanisms of *ospC* or additional *B. burgdorferi* genes (Alverson *et al.*, 2003; Caimano *et al.*, 2005; Caimano *et al.*, 2007; Hübner *et al.*, 2001; Lybecker and Samuels, 2007; Xu *et al.*, 2007). Additional work is certainly needed to understand the intrinsic signal transduction pathways in spirochaetes and the details of environmental factors that influence spirochaete gene expression in ticks.

11.3 *Borrelia burgdorferi* genes known to function in ticks

The *B. burgdorferi* genome encodes approximately 1780 genes that are distributed in a single chromosome and 21 plasmids (Casjens *et al.*, 2000; Fraser *et al.*, 1997). The vast majority of these genes have been annotated as 'hypothetical proteins' as they have little resemblance to known proteins in the databases, and thus, could potentially carry out specialized functions in the spirochaete's enzootic life cycle (Casjens *et al.*, 2000; Fraser *et al.*, 1997). Previous efforts have focused on identifying *B. burgdorferi* genes that are selectively upregulated at specific phases of the spirochaete life cycle. As discussed above, microarray studies of transcriptome profiling of cultured *B. burgdorferi* may identify genes that are differentially expressed in response to environmental changes, and are relevant to the spirochaete natural life cycle and transmission through ticks. Indeed, recent mutagenesis studies indicate that preferentially regulated *B. burgdorferi* gene products, (Grimm *et al.*, 2004b; Li *et al.*, 2007b; Neelakanta *et al.*, 2007; Pal *et al.*, 2004b; Yang *et al.*, 2004) could serve important functions in supporting the natural cycle of the spirochaete. The following sections describe *B. burgdorferi* genes known to be highly expressed in the arthropod vector, many of which are thought to be important in ticks (Grimm *et al.*, 2004b; Li *et al.*, 2007a,b;; Munderloh and Kurtti, 2005; Pal *et al.*, 2004b, 2008a;; Revel *et al.*, 2005; Rosa *et al.*, 2005; Yang *et al.*, 2004). These genes are encoded on plasmids (*ospAB*, *ospC*, *ospD*, and *bbe16*) or the chromosome (*dps* and *la7*).

11.3.1 *Borrelia burgdorferi* plasmids

The *B. burgdorferi* genome contains nine circular (cp) and 12 linear (lp) plasmids (Casjens *et al.*, 2000; Fraser *et al.*, 1997). Several studies have indicated that specific plasmids are indispensable for *B. burgdorferi* survival in ticks (Grimm *et al.*, 2005; Purser and Norris, 2000; Rosa *et al.*, 2005; Stewart *et al.*, 2005). As noted earlier, microarray studies indicate that, compared to chromosomal genes, expression of plasmid-encoded genes is more responsive to environmental changes, indicating that they have a pivotal role in spirochaete survival in diverse environments (Brooks *et al.*, 2003; Caimano *et al.*, 2007; Ojaimi *et al.*, 2002; Tokarz *et al.*, 2004). *Borrelia burgdorferi* isolates missing selected plasmids reinforce the role of plasmid-encoded genes in the spirochaete life cycle in nature. While two linear plasmids, lp25 and lp28-1, were required for spirochaete persistence in the mammalian host, only lp25 has been found to be essential for tick infection (Labandeira-Rey and Skare, 2001; Grimm *et al.*, 2004a; Purser *et al.*, 2003; Strother *et al.*, 2005; Strother and de Silva, 2005). *bbe22* on lp25 encodes for a nicotinamidase that is most likely to be required for the biosynthesis of nicotinamide adenine dinucleotide. Complementation of *bbe22* alone was sufficient to restore the ability of spirochaetes lacking lp25 to infect mice (Purser *et al.*, 2003) whereas *bbe22* complementation only partially restored tick infectivity (Strother and de Silva, 2005). *Borrelia burgdorferi* missing lp28-4 also displayed a reduced ability to persist in ticks and to transmit to mice (Strother *et al.*, 2005). The same study also determined that the plasmids lp5, lp28-1, and cp9 were not required for spirochaete infectivity of tick guts. A recent study also conclusively established that *B. burgdorferi* missing the plasmid lp36 showed no defect in survival in ticks, while murine infectivity was significantly compromised (Jewett *et al.*, 2007). Individual genes carried on the same plasmid may have selective host- or vector-specific roles. For example, genes on plasmid lp54 include *ospA*, which is essential in ticks (Yang *et al.*, 2004) and *dbpAB* is important in mice (Liang *et al.*, 2004; Shi *et al.*, 2008; Shi *et al.*, 2006). Microarray studies also indicated differential expression of genes from many of the *B. burgdorferi* plasmids, most notably lp54, when spirochaetes are cultured in the presence or absence of mammal-specific factors, such as blood, or in *B. burgdorferi* cultured within DMC (Brooks *et al.*, 2003; Caimano *et al.*, 2007; Tokarz *et al.*, 2004).

11.3.2 OspA and OspB

Borrelia burgdorferi genes *ospA* and *ospB* constitute a single operon and are among the most abundant outer membrane lipoproteins of *B. burgdorferi*. Both proteins show vector-specific expression in the life cycle of spirochaetes. *ospA* and *ospB* are upregulated when spirochaetes exit the infected mammalian host and migrate to feeding *Ixodes* ticks (de Silva and Fikrig, 1997; de Silva *et al.*, 1996; Fikrig *et al.*, 2004; Neelakanta *et al.*, 2007; Schwan, 2003; Schwan and Piesman, 2000). Recombinant OspA and OspB specifically adhere to the tick gut, indicating that these outer surface lipoproteins are important for spirochaete adaptation in the tick gut (Fikrig *et al.*, 2004; Pal *et al.*, 2000). Experimental transfer of non-borreliacidal OspA and OspB antibodies to the murine host prevents tissue adherence of spirochaetes in feeding ticks, further implicating these proteins in spirochaete colonization of the tick gut (Pal *et al.*, 2001). An *ospAB* null mutant has been generated and used in tick–mouse infection studies (Yang *et al.*, 2004). The *ospAB* mutant retains full murine infectivity and pathogenicity, but is unable to infect ticks. OspA/B-deficient *B. burgdorferi* migrates from infected mice into ticks, but fails to survive in the feeding tick gut, reinforcing the important role of the *ospAB* operon in the maintenance of *Borrelia* in ticks. An OspB-deficient *B. burgdorferi* isolate was generated, and infection studies further indicated that OspB is also involved in colonization and persistence of *B. burgdorferi* in the arthropod vector (Neelakanta *et al.*, 2007). A recent study indicated that while OspA is important for spirochaete survival in ticks and necessary for *B. burgdorferi* transmission through the vector, neither OspA nor OspB gene products are essential for microbial persistence through larval–nymphal moult (Battisti *et al.*, 2008). In addition to the biological role of gut adhesin, OspA also binds plasminogen (Fuchs *et al.*, 1994), potentially contributing to the dissemination of *B. burgdorferi* in the vector (Coleman *et al.*, 1997). Furthermore, OspA was recently shown to protect spirochaetes from host acquired immunity in the vector (Battisti *et al.*, 2008), hypothetically by shielding the access of host-derived borreliacidal antibodies to the conserved spirochaete antigens, such as P66 (Bunikis *et al.*, 1999), although P66 is immunogenic in murine reservoir hosts (Brunet *et al.*, 1995) where *ospA* is not expressed. Overall, these experiments demonstrate the importance of gene products of OspA/B operon in the *B. burgdorferi* adaptation in ticks.

11.3.3 OspC

OspC is a prominent outer membrane lipoprotein of *B. burgdorferi* and is induced during transmission of spirochaetes from the feeding tick to the host (Schwan *et al.*, 1995). *ospC* is the archetype of *rpoS*-dependent genes associated with tick transmission and host infection (Fisher *et al.*, 2005; Hubner *et al.*, 2001; Yang *et al.*, 2003). Studies with *ospC* deletion mutants have demonstrated that OspC is essential for infecting the murine host (Grimm *et al.*, 2004b; Pal *et al.*, 2004b). However, the exact role of OspC in the *B. burgdorferi* transmission cycle through ticks warrants further investigation, as there are conflicting results about the phenotype of *ospC* mutants in ticks. In one study, the *ospC* null mutant failed to invade *I. scapularis* salivary glands in significant numbers and was unable to establish infection in the host (Pal *et al.*, 2004b). Subsequent studies showed that OspC binds a tick salivary gland protein, Salp15, in the salivary gland, and that OspC–tick protein interaction facilitated establishment of spirochaete infection of the murine host (further discussed below). A recent report with *B. afzelii*, which is prevalent in Europe and Asia and is transmitted by *Ixodes* ticks, also supports the notion that OspC is critical for spirochaete dissemination from the tick gut to the salivary gland (Fingerle *et al.*, 2007). Complementation of a naturally occurring *ospC* mutant (Fingerle *et al.*, 2000) with a wild-type copy of the *ospC* gene resulted in more efficient migration to the salivary glands. Another group independently created a series of *ospC* mutants and assayed them in tick-mouse infection studies (Grimm *et al.*, 2004b; Stewart *et al.*, 2006; Tilly *et al.*, 2006, 2007). Their results show that OspC is required for establishing infection in the murine host, at least at the early stages (Tilly *et al.*, 2009); however, *ospC* deletion had no influence on the migration of *B. burgdorferi* from the tick gut to the salivary gland. Differences in experimental approaches to tick infection (microinjection versus immersion infection) and the use of different isolates of *B. burgdorferi* (strain 297 versus B31) may have contributed to the different results. Kinetics of *B. burgdorferi* dissemination from tick gut to the salivary gland could also vary among isolates. Finally, host-derived plasminogen in ticks (Coleman *et al.*, 1997; Coleman *et al.*, 1995) is also known to bind *B. burgdorferi* gene products, such as *ospC* (Lagal *et al.*, 2006). However, how specificity of the interactions between the same spirochaete antigen, such as OspC, with multiple host or vector molecules, such as plasminogen or Salp15 contributes to *B. burgdorferi* infections remains to be studied.

11.3.4 Dps

One of the *B. burgdorferi* chromosomal genes that display regulated expression in the *B. burgdorferi* enzootic cycle is *bb0690,* which encodes a *B. burgdorferi* protein similar to the DNA-binding protein of the starved bacteria (Dps) family of proteins (Li *et al.*, 2007b). *dps* expression is low throughout *B. burgdorferi* infection in mice, but high during tick intermoult periods. Dps-deficient spirochaetes retain full murine infectivity, readily migrate to the feeding ticks, and persist in repleted ticks. The mutant, however, fails to survive throughout the extended intermoult periods of resting *I. scapularis,* indicating that Dps is critical for persistence of spirochaetes in unfed ticks (Li *et al.*, 2007b). Dps is known to protect bacterial DNA against starvation or oxidative stress-induced damage. Therefore, the *B. burgdorferi* Dps homologue may protect the spirochaete against DNA damage during the long intermoult periods between tick blood meals and dormancy.

11.3.5 OspD

OspD is an outer membrane protein of *B. burgdorferi,* which shows a tightly regulated expression in spirochaete life cycle. *ospD,* in general, shows an elevated expression in ticks as compared to mice (Li *et al.*, 2007a). Two recent studies examined *ospD* expression in select stages of tick-mouse infection cycle of *B. burgdorferi* (Li *et al.*, 2007a; Stewart *et al.*, 2008). Although results from these studies differ in the detail kinetics of *ospD* induction, both studies show that *ospD* transcripts are significantly expressed in a narrow time window that corresponds to the phases of tick detachment from the host following feeding. An

OspD-deficient isolate of *B. burgdorferi* was created and examination of the mutant phenotype in mice indicated that OspD has no obvious influence on *B. burgdorferi* survival in the murine host or on the acquisition of spirochaetes by ticks (Li *et al.*, 2007a; Stewart *et al.*, 2008). Recombinant OspD binds to tick gut lysates *in vitro,* suggesting a role for this protein in spirochaete adherence of ticks (Li *et al.*, 2007a). However, while the *ospD* mutant has a decreased capacity to attach to the tick gut *in vivo,* this decrease does not influence subsequent spirochaete transmission during a second blood meal. Taken together, these data suggest that OspD may have a secondary or non-essential role in *B. burgdorferi* persistence within ticks (Li *et al.*, 2007a; Stewart *et al.*, 2008).

11.3.6 BptA

BptA, annotated as *bbe16,* is a putative lipoprotein that is possibly surface-exposed on the outer membrane of *B. burgdorferi*. *bbe16* is located on the plasmid lp25, which has been closely associated with *B. burgdorferi* infectivity (Purser *et al.*, 2003; Purser and Norris, 2000; Strother *et al.*, 2005; Strother and de Silva, 2005). A microarray analysis first identified that *bbe16* is a differentially expressed gene and was upregulated when *B. burgdorferi* was cultivated in DMC, compared with spirochaetes grown in culture mimicking the conditions of fed ticks (Revel *et al.*, 2002). Later studies show that *btpA* is selectively expressed in culture conditions mimicking the tick, suggesting a functional role for spirochaete persistence in ticks (Revel *et al.*, 2005). *In vivo* analyses of generated infectious isolates of *btpA* mutant and complemented isolates conclusively demonstrated a requirement for BtpA protein in persistent infection of ticks (Revel *et al.*, 2005). The exact function of the protein in the *B. burgdorferi* natural cycle, however, remains to be elucidated.

11.3.7 La7

Borrelia burgdorferi gene *la7,* annotated as *bb0365,* encodes P22, which is also known as lipoprotein LA7 (Grewe and Nuske, 1996; Lam *et al.*, 1994). *la7* is produced in the mammal (von Lackum *et al.*, 2007) and is highly upregulated in spirochaetes entering feeding ticks (Pal *et al.*, 2008a). Mutant spirochaetes lacking *la7* do not differ from the parental isolates in the establishment of murine infection (Pal *et al.*, 2008a). When naïve ticks engorge on spirochaete-infected mice, the *la7* mutant enters ticks, but has a markedly decreased survival within feeding ticks when compared to wild-type *B. burgdorferi*. LA7 may, therefore, be important for spirochaete survival in feeding ticks.

Specific sets of genes, located on either plasmids or on the chromosome, show induced expression in ticks, and a few of these have already proven to be of functional significance. Additional *B. burgdorferi* genes have also been identified as being expressed *in vivo*, but lack an essential role. *Borrelia burgdorferi* mutants lacking either *luxS* (Hubner *et al.*, 2003) or *chbC* (Tilly *et al.*, 2004), the gene products implicated in the quorum sensing pathway or the transport of chitobiose – a component of tick cuticle, respectively, were generated. *luxS* or *chbC* functions are, however, not essential for *B. burgdorferi* persistence in ticks nor during their natural infection cycle (Hubner *et al.*, 2003; Tilly *et al.*, 2004). The recent development of genetic tools has allowed researchers to create specific mutants of *B. burgdorferi* and to identify *Borrelia* proteins that function at different stages in the vector phase of the life cycle. As *B. burgdorferi* spends a significant part of its life inside the arthropod, additional genes are likely to be identified that are important in ticks. High-throughput expression analysis, such as quantitative RT-PCR and RNA-amplification-based microarray assays, have the potential to catalogue additional genes that are highly expressed in ticks and may perform a specialized function in the complex enzootic life cycle of *B. burgdorferi.* Recently, function of a manganese transporter, termed as BmtA, found to be essential to the infectious life cycle of *B. burgdorferi* in ticks and mammals (Ouyang *et al.*, 2009).

11.4 Molecular interfaces: interaction of microbial and vector gene products

Regardless of the nature of the ecological relationship between a microbe and its host, development of specific molecular interactions is an essential prerequisite for the survival of the microbe *in vivo* (Dale and Moran, 2006; Finlay and Cossart, 1997). Predominantly based on the vector-specific expression of limited gene products of *B. burgdorferi,* scientists have uncovered a handful of discrete interactions between the spirochaete and tick vector that are summarized below. These studies have established that certain *B. burgdorferi* proteins, either directly or via molecular interactions with specific vector receptors, contribute to the maintenance of the spirochaete infection cycle through ticks.

11.4.1 *Borrelia burgdorferi* interaction with tick gut proteins

Recent studies have identified a receptor in the tick gut that specifically interacts with *B. burgdorferi* OspA (Pal *et al.,* 2004a). OspA binds the tick receptor for OspA (TROSPA), which is predominantly expressed in the gut of *I. scapularis* ticks. This receptor–ligand interaction is supported by the findings that TROSPA co-localizes with OspA in the tick gut *in vivo* and that OspA specifically bound recombinant TROSPA in a series of *in vitro* assays (Pal *et al.,* 2004a). Interestingly, *TROSPA* expression within the tick was upregulated during infection with *B. burgdorferi* and decreased during engorgement, correlating with *ospA* expression by *B. burgdorferi* in ticks. When uninfected ticks were allowed to feed on infected mice passively administered with TROSPA antisera, acquisition of *B. burgdorferi* by the ticks was significantly reduced, compared to infected mice treated with pre-immune sera. TROSPA-immunized ticks also have a reduced capacity for *B. burgdorferi* transmission. Similarly, RNA interference of *TROSPA* in nymphal ticks significantly diminished the ability of the ticks to acquire *B. burgdorferi* from infected mice (Pal *et al.,* 2004a). *Borrelia burgdorferi* appears to colonize the tick gut by adhering to TROSPA through OspA; the OspA–TROSPA interaction is important for supporting the *B. burgdorferi* infection cycle. Although TROSPA supports the *B. burgdorferi* life cycle in ticks, the native function of TROSPA in the arthropod is currently unknown. *TROSPA* is not consistently expressed in all stages of *I. scapularis* ticks, and its expression is highly influenced by environmental signals, such as tick engorgement or *B. burgdorferi* infection. These observations indicate a developmental rather than a housekeeping function of TROSPA in tick biology.

11.4.2 *Borrelia burgdorferi* interaction with tick salivary gland proteins

Since the identification of immunodominant antigens that are deposited in the host dermis by feeding ticks, *I. scapularis* salivary gland proteins (Salps) became an intense focus of *Borrelia* research (Das *et al.,* 2000, 2001; Francischetti *et al.,* 2005; Ribeiro *et al.,* 2006). Indeed, novel salivary gland vaccine targets have been identified for other vector-borne diseases, such as leishmaniasis (Morris *et al.,* 2001; Valenzuela *et al.,* 2001). Since many of the Salps are soluble, feeding-induced, and co-transmitted to the host along with *B. burgdorferi,* these proteins have an opportunity to influence, either directly or indirectly, the spirochaete migration between vector and host (Ramamoorthi *et al.,* 2005; Rosa, 2005). The generation of host immunity against *I. scapularis* salivary gland proteins, those expressed as early as 24 hours after feeding, is sufficient to impair *B. burgdorferi* transmission between ticks and host (Narasimhan *et al.,* 2007a). Following are representative examples of tick Salps that are shown to influence spirochaete infection:

Salp15

As discussed earlier, *ospC* is a prominent borrelial outer membrane protein that is swiftly upregulated during tick transmission (Schwan *et al.,* 1995) and binds a tick salivary protein, salp15 (Anguita *et al.,* 2002; Ramamoorthi *et al.,* 2005). The OspC–Salp15 interaction has important functional consequences that impact *B. burgdorferi* transmission and pathogenesis in a multipartite manner involving the pathogen, the vector and the host (Ramamoorthi *et al.,* 2005). *Borrelia burgdorferi* carries Salp15 (via OspC binding) during its transmission from the tick salivary gland to the murine host, and this interaction facilitates survival of spirochaetes within the infected host (Anguita *et al.,* 2002). Salp15 specifically binds *B. burgdorferi* OspC and this interaction prevents killing of spirochaetes by borreliacidal antibodies *in vitro.* Thus, coating of the spirochaete via binding to Salp15 in the tick salivary gland could have important implications for spirochaete survival in the host. Salp15 binding to spirochaetes could potentially mask the exposure of OspC to components of the host immune system, as

OspC is shown to be highly immunogenic and a target of host borreliacidal antibodies (Buckles *et al.*, 2006; Earnhart *et al.*, 2007; Earnhart and Marconi, 2007a,b; Fung *et al.*, 1994; Liang *et al.*, 2002; Mbow *et al.*, 1999; Wilske, 2005; Wilske *et al.*, 1993, 1996). Additionally, by itself, Salp15 carries host immunosuppressive properties by inhibiting $CD4^+$ T-cell activation (Anguita *et al.*, 2002). Thus, OspC-bound Salp15 provides additional benefits to spirochaetes, potentially affording protection against immune cells that infiltrate the tick-feeding site. Interestingly, OspC–Salp15 interaction could be beneficial to both vector and pathogen (Ramamoorthi *et al.*, 2005). *Borrelia burgdorferi* induces *salp15* expression in feeding salivary glands. Therefore, *B. burgdorferi* infection in ticks may provide survival advantages to the arthropod that produces this protein for effective feeding at the host site via binding to Salp15 and plasminogen (Coleman *et al.*, 1995, 1997; Lagal *et al.*, 2006; Ramamoorthi *et al.*, 2005).

Salp25D

A recent study indicates that one of the abundant tick Salps, Salp25D (Das *et al.*, 2001), plays an important role in the mammalian host for acquisition of *B. burgdorferi* by the vector (Narasimhan *et al.*, 2007b). Silencing of *salp25D* in tick salivary glands blocks spirochaete acquisition by ticks from infected mice. Salp25D functions as an antioxidant, thus Salp25D-mediated detoxification of reactive oxygen radicals at the tick–pathogen–host interface potentially provides a survival advantage to *B. burgdorferi* at the tick-feeding site in mice.

Salp20

In addition to Salp15, which binds to *B. burgdorferi* and contributes to the protection of spirochaetes against host immunity, Salp20 has been shown to protect serum-sensitive strains of *Borrelia* from lysis in *in vitro* assays (Tyson *et al.*, 2007). Thus, Salp20 may also protect *Borrelia* from host complement within the vector blood meal and during transmission. Salp20 is related to *Ixodes scapularis* anti-complement *(Isac)* proteins that may block the host alternative complement pathway (Ribeiro *et al.*, 2006; Soares *et al.*, 2005). RNA interference studies have established that *Isac* and related family members of tick proteins aid efficient tick feeding and growth of spirochaetes in ticks (Soares *et al.*, 2005). The characterized tick salivary gland transcriptome (Francischetti *et al.*, 2005; Ribeiro *et al.*, 2006; Ribeiro and Francischetti, 2003) will foster research identifying the roles of additional tick salivary gland proteins that support *B. burgdorferi* infection and transmission to the host. The annotated *Ixodes* salivary gland transcriptome harbours proteins with signature functional domains including basic-tail, Kunitz, proline-rich peptides, metalloproteases (class PIII-like), as well as novel proteins with similarities to disintegrins, histamine-binding proteins, anticomplements and a neuropeptide-like protein (nlp-31) with GGY repeats that may have antimicrobial activity (Francischetti *et al.*, 2005). Further characterization of these protein functions will greatly contribute to our understanding of host immunity and vaccine development against both ticks and tick-borne pathogens.

Collectively, all of the above studies indicate the existence of highly discrete molecular interfaces where specific microbial and vector gene products interact, contributing to *B. burgdorferi* survival in a natural enzootic cycle.

11.5 *Ixodes* genomics

Exploration of the *Ixodes* genome will offer new avenues for scientific investigation and reveal novel features of the microbe–vector interaction. Studies of *Ixodes* genomics, however, largely await the major development of gene manipulation tools and completion of whole genome sequencing of *I. scapularis,* which are summarized below.

11.5.1 Gene manipulation

Stable methods of gene manipulation in prominent arthropod vectors, such as mosquitoes, are well developed (Catteruccia, 2007; Catteruccia *et al.*, 2000) and are most likely to be influenced by the development of advanced genetic tools in the fly, *Drosophila* sp. (Venken and Bellen, 2005). Gene manipulation in ticks is poorly developed, possibly because few of the available genetic tools are readily applicable to ticks. RNA interference (RNAi) is now viewed as the only possible gene-silencing technique in ticks (de la Fuente *et al.*, 2007a; Hajdusek *et al.*, 2009), although the knockdown effects are transient. Unfortunately, unlike many other arthropods (Bucher *et al.*,

2002; He *et al.*, 2006; Khila and Grbic, 2007), the use of parental RNAi to introduce stable RNAi effects in offspring has had limited success in ticks. Additional formidable challenges are the lack of tick embryology and genome information, and further progress in this area would stimulate gene manipulation research in ticks. The long life span and slow post-embryonic development of *Ixodes* ticks would pose serious problems for phenotypic analysis of the mutants once we are able to manipulate the tick genome. These efforts will be complicated by the fact that *I. scapularis* embryonic development, including preoviposition, pre-ecolosion and premoult development, is shown to be highly sensitive to environmental conditions, such as temperature (Heath, 1979; Ogden *et al.*, 2004). Nevertheless, cell lines from *I. scapularis* (Munderloh *et al.*, 1994) are available to the research community and the use of *Ixodes* cell lines should allow us to optimize the manipulation of the tick genome in cell culture.

11.5.2 Whole genome sequence

Recently, the *I. scapularis* Genome Project (IGP), a collaborative effort between the international community of tick researchers and genome sequencing centres, was initiated and is supported by the National Institutes of Health (Hill and Wikel, 2005; Pagel Van Zee *et al.*, 2007). The IGP aims to perform whole genome sequencing of *I. scapularis*, and presently, more than 18 million trace reads, representing approximately a five-fold coverage of the genome, have been deposited at the National Centre for Biotechnology Information (NCBI) trace-archives (Pagel Van Zee *et al.*, 2007). In addition, 20 *Ixodes* BAC (bacterial artificial chromosome) clones, 370,000 BAC-end reads and more than 80,000 expressed sequence tags (ESTs) have been sequenced. By 2008, sequencing, assembly and annotation of six-fold coverage of the *Ixodes* genome are expected, but could take longer owing to the large genome size, genomic complexity and presence of extensive repetitive DNA in *I. scapularis* (Geraci *et al.*, 2007; Ullmann *et al.*, 2005). The tick genome will impact development of gene manipulation studies and also contribute to our understanding of the tick biology relevant to pathogen survival and transmission through ticks.

11.6 Interaction-blocking Lyme disease vaccine

Understanding the molecular details of the *B. burgdorferi* life cycle in the arthropod is likely to contribute to the development of novel vaccines against Lyme disease. This notion is supported by the success and mechanism of protection afforded by the former human Lyme disease vaccine, based on the *B. burgdorferi* antigen OspA (Fikrig *et al.*, 1990; Sigal *et al.*, 1998; Steere *et al.*, 1998). The recombinant OspA vaccine induces antibodies in the host that enter the guts of infected feeding ticks and prevent the transmission of spirochaetes from the vector to the host (de Silva *et al.*, 1996). Multi-host pathogens, such as *B. burgdorferi*, can survive in mammals and cause persistent infections. Thus, the pathogen may have evolved defensive mechanisms to elude the sophisticated immune systems of mammals, including antigenic variation via variable recombination and differential gene expression (de Silva and Fikrig, 1997; Zhang *et al.*, 1997; Zhang and Norris, 1998). Therefore, antigens expressed in the host are, often, not effective vaccine candidates. The vector, on the other hand, lacks an adaptive immune system and *Borrelia* antigens expressed in the vector are mostly conserved. Additionally, antigenic variation mechanisms appear to be minimally active in the vector (Indest *et al.*, 2001; Nosbisch and de Silva, 2007). Further identification of *B. burgdorferi* antigens that are essential for spirochaete persistence and transmission through the vector could potentially contribute to the development of effective vaccines to prevent the incidence of Lyme disease. Along with studies on laboratory animal models to contribute to the development of a transmission-blocking vaccine, field experiments are also being attempted to assess if these vaccines can be used in the wild reservoir hosts to reduce the proportion of infected ticks in nature. Immunizing wild mice with OspA vaccine was shown to reduce the *B. burgdorferi* prevalence in ticks in the confined geographic area under study (Tsao *et al.*, 2004). In addition to bacterial antigens, proteins from the vector have a potential for vaccine development (de la Fuente *et al.*, 2007b). Vaccines targeting the tick gut or salivary proteins that interact with *Borrelia* could inhibit pathogen development in ticks and block pathogen transmission.

11.7 Concluding remarks

Recent studies have indicated that the interaction of a distinct set of ligand receptors contributes to the successful persistence of human pathogens, such as *B. burgdorferi*, through a complex life cycle in nature. These studies have also established the clear evolution of an intimate relationship between the pathogen and its arthropod vector. *Borrelia burgdorferi* can induce the expression of specific tick genes, such as *TROSPA* and *salp15*, and experimental evidence suggests that at least one of them (Salp15) can potentially modulate host immunity that favours the tick feeding process. Therefore, *B. burgdorferi* infection may provide survival advantages to ticks. Further work is needed to determine the exact ecological relationship between ticks and *B. burgdorferi*, to determine if it is parasitic or mutualistic. Identification of additional microbial and vector gene products actively engaged in the acquisition, maintenance and transmission of the microbe through the arthropod is needed. The long lifespan and slow post-embryonic development of *Ixodes* ticks, coupled with limited genomic information, hinders the application of gene manipulation tools to tick research and the effective analysis of the tick transcriptome and proteome. Future research on tick embryology, optimization of gene manipulation in *Ixodes* cells lines, and the NIH-funded tick genome project will certainly expand experimental opportunities. These efforts will also contribute to our understanding of tick biology and help unravel the details of the molecular interactions of *Borrelia*-tick gene products. Specific questions to be addressed are the molecular details how *B. burgdorferi* evades tick immunity and intense gut digestive activities upon its arrival in feeding ticks, and how it continues to persist in ticks throughout the intermoult periods dominated by temperature extremities and starvation. Studies should also determine how *B. burgdorferi* crosses solid physical barriers, such as the chitinous gut peritrophic membrane, and then selectively migrates through the gut, haemocoele and salivary gland epithelia during their transmission to mammals. *Ixodes* ticks and spirochaetes have probably evolved numerous but distinct molecular interfaces that aid in their coexistence and coevolution. Globally, incidences of tick-borne diseases are increasing, including *B. burgdorferi* infections (Piesman and Eisen, 2008; Vorou *et al.*, 2007). Understanding the biology of the tick-*B. burgdorferi* interface will provide further information that may contribute to the development of preventative measures, such as novel transmission-blocking vaccines, to control *B. burgdorferi* infection.

Acknowledgements

We are grateful to Aravinda M. de Silva for his critical comments and valuable suggestions in preparation of this manuscript. We are also thankful to Deborah Shroder and Ireen Dryburgh-Barry for assistance with the preparation of the manuscript. Lyme disease research in our laboratories is supported by the National Institutes of Health, Arthritis Foundation and American Heart Association. Utpal Pal is the recipient of a Scientist Development Grant from American Heart Association. Erol Fikrig is an Investigator of the Howard Hughes Medical Institute.

References

Alban, P.S., Johnson, P.W., and Nelson, D.R. (2000). Serum-starvation-induced changes in protein synthesis and morphology of *Borrelia burgdorferi*. Microbiology *146 (Pt 1)*, 119–127.

Alverson, J., Bundle, S.F., Sohaskey, C.D., Lybecker, M.C., and Samuels, D.S. (2003). Transcriptional regulation of the *ospAB* and *ospC* promoters from *Borrelia burgdorferi*. Mol. Microbiol. *48*, 1665–1677.

Anderson, J.F. (1989). Epizootiology of *Borrelia* in *Ixodes* tick vectors and reservoir hosts. Rev Infect Dis *11 Suppl 6*, S1451–1459.

Anderson, J.F., Duray, P.H., and Magnarelli, L.A. (1987). Prevalence of *Borrelia burgdorferi* in white-footed mice and *Ixodes dammini* at Fort McCoy, Wis. J. Clin. Microbiol. *25*, 1495–1497.

Anguita, J., Hedrick, M.N., and Fikrig, E. (2003). Adaptation of *Borrelia burgdorferi* in the tick and the mammalian host. FEMS Microbiol. Rev. *27*, 493–504.

Anguita, J., Ramamoorthi, N., Das, S., Conze, D., Askenase, P., Rincón, M., Kantor, F.S., and Fikrig, E. (2002). Salp15, an *Ixodes scapularis* saliva protein, inhibits $CD4^+$ T-cell activation. Immunity *16*, 849–859.

Barbour, A.G., and Fish, D. (1993). The biological and social phenomenon of Lyme disease. Science *260*, 1610–1616.

Battisti, J.M., Bono, J.L., Rosa, P.A., Schrumpf, M.E., Schwan, T.G., and Policastro, P.F. (2008). Outer surface protein A protects Lyme disease spirochetes from acquired host immunity in the tick vector. Infect. Immun. *76*, 5228–5237.

Blevins, J.S., Xu, H., He, M., Norgard, M.V., Reitzer, L., and Yang, X.F. (2009). Rrp2, a sigma54-dependent transcriptional activator of *Borrelia burgdorferi*, activates *rpoS* in an enhancer-independent manner. J. Bacteriol. *191*, 2902–2905.

Beerntsen, B.T., James, A.A., and Christensen, B.M. (2000). Genetics of mosquito vector competence. Microbiol. Mol. Biol. Rev. *64*, 115–137.

Brooks, C.S., Hefty, P.S., Jolliff, S.E., and Akins, D.R. (2003). Global analysis of *Borrelia burgdorferi* genes regulated by mammalian host-specific signals. Infect. Immun. *71*, 3371–3383.

Brorson, Ø., and Brorson, S.H. (1998). *In vitro* conversion of *Borrelia burgdorferi* to cystic forms in spinal fluid, and transformation to mobile spirochetes by incubation in BSK-H medium. Infection *26*, 144–150.

Brorson, Ø., and Brorson, S.H. (2004). An in vitro study of the susceptibility of mobile and cystic forms of *Borrelia burgdorferi* to tinidazole. Int Microbiol *7*, 139–142.

Brunet, L.R., Sellitto, C., Spielman, A., and Telford, S.R., 3rd (1995). Antibody response of the mouse reservoir of *Borrelia burgdorferi* in nature. Infect. Immun. *63*, 3030–3036.

Bucher, G., Scholten, J., and Klingler, M. (2002). Parental RNAi in Tribolium (Coleoptera). Curr Biol *12*, R85–86.

Buckles, E.L., Earnhart, C.G., and Marconi, R.T. (2006). Analysis of antibody response in humans to the type A OspC loop 5 domain and assessment of the potential utility of the loop 5 epitope in Lyme disease vaccine development. Clin Vaccine Immunol *13*, 1162–1165.

Bunikis, J., and Barbour, A.G. (1999). Access of antibody or trypsin to an integral outer membrane protein (P66) of *Borrelia burgdorferi* is hindered by Osp lipoproteins. Infect. Immun. *67*, 2874–2883.

Burgdorfer, W., Barbour, A.G., Hayes, S.F., Benach, J.L., Grunwalt, E., and Davis, J.P. (1982). Lyme disease-A tickborne spirochetosis? Science *216*, 1317–1319.

Bykowski, T., Woodman, M.E., Cooley, A.E., Brissette, C.A., Brade, V., Wallich, R., Kraiczy, P., and Stevenson, B. (2007). Coordinated expression of *Borrelia burgdorferi* complement regulator-acquiring surface proteins during the Lyme disease spirochete's mammal-tick infection cycle. Infect. Immun. *75*, 4227–4236.

Bykowski, T., Woodman, M.E., Cooley, A.E., Brissette, C.A., Wallich, R., Brade, V., Kraiczy, P., and Stevenson, B. (2008). *Borrelia burgdorferi* complement regulator-acquiring surface proteins (BbCRASPs): Expression patterns during the mammal-tick infection cycle. Int. J. Med. Microbiol. *298*, 249–256.

Caimano, M.J., Eggers, C.H., Gonzalez, C.A., and Radolf, J.D. (2005). Alternate sigma factor RpoS is required for the in vivo-specific repression of *Borrelia burgdorferi* plasmid lp54-borne *ospA* and *lp6.6* genes. J. Bacteriol. *187*, 7845–7852.

Caimano, M.J., Eggers, C.H., Hazlett, K.R., and Radolf, J.D. (2004). RpoS is not central to the general stress response in *Borrelia burgdorferi* but does control expression of one or more essential virulence determinants. Infect. Immun. *72*, 6433–6445.

Caimano, M.J., Iyer, R., Eggers, C.H., Gonzalez, C., Morton, E.A., Gilbert, M.A., Schwartz, I., and Radolf, J.D. (2007). Analysis of the RpoS regulon in *Borrelia burgdorferi* in response to mammalian host signals provides insight into RpoS function during the enzootic cycle. Mol. Microbiol. *65*, 1193–1217.

Carroll, J.A., Garon, C.F., and Schwan, T.G. (1999). Effects of environmental pH on membrane proteins in *Borrelia burgdorferi*. Infect. Immun. *67*, 3181–3187.

Casjens, S., Palmer, N., van Vugt, R., Huang, W.M., Stevenson, B., Rosa, P., Lathigra, R., Sutton, G., Peterson, J., Dodson, R.J., *et al.* (2000). A bacterial genome in flux: the twelve linear and nine circular extrachromosomal DNAs in an infectious isolate of the Lyme disease spirochete *Borrelia burgdorferi*. Mol. Microbiol. *35*, 490–516.

Catteruccia, F. (2007). Malaria vector control in the third millennium: progress and perspectives of molecular approaches. Pest Manag Sci *63*, 634–640.

Catteruccia, F., Nolan, T., Loukeris, T.G., Blass, C., Savakis, C., Kafatos, F.C., and Crisanti, A. (2000). Stable germline transformation of the malaria mosquito *Anopheles stephensi*. Nature *405*, 959–962.

Coburn, J., and Cugini, C. (2003). Targeted mutation of the outer membrane protein P66 disrupts attachment of the Lyme disease agent, *Borrelia burgdorferi*, to integrin alpha-v-beta-3. Proc. Natl. Acad. Sci. U.S.A. *100*, 7301–7306.

Coleman, J.L., Gebbia, J.A., Piesman, J., Degen, J.L., Bugge, T.H., and Benach, J.L. (1997). Plasminogen is required for efficient dissemination of *Borrelia burgdorferi* in ticks and for enhancement of spirochetemia in mice. Cell *89*, 1111–1119.

Coleman, J.L., Sellati, T.J., Testa, J.E., Kew, R.R., Furie, M.B., and Benach, J.L. (1995). *Borrelia burgdorferi* binds plasminogen, resulting in enhanced penetration of endothelial monolayers. Infect. Immun. *63*, 2478–2484.

Dale, C., and Moran, N.A. (2006). Molecular interactions between bacterial symbionts and their hosts. Cell *126*, 453–465.

Das, S., Banerjee, G., DePonte, K., Marcantonio, N., Kantor, F.S., and Fikrig, E. (2001). Salp25D, an *Ixodes scapularis* antioxidant, is 1 of 14 immunodominant antigens in engorged tick salivary glands. J. Infect. Dis. *184*, 1056–1064.

Das, S., Marcantonio, N., Deponte, K., Telford, S.R., 3rd, Anderson, J.F., Kantor, F.S., and Fikrig, E. (2000). Salp16, a gene induced in *Ixodes scapularis* salivary glands during tick feeding. Am. J. Trop. Med. Hyg. *62*, 99–105.

de la Fuente, J., Kocan, K.M., Almazan, C., and Blouin, E.F. (2007a). RNA interference for the study and genetic manipulation of ticks. Trends Parasitol 23, 427–433.

de la Fuente, J., Kocan, K.M., and Blouin, E.F. (2007b). Tick vaccines and the transmission of tick-borne pathogens. Vet. Res. Commun. *31 Suppl. 1*, 85–90.

de Silva, A.M., and Fikrig, E. (1995). Growth and migration of *Borrelia burgdorferi* in Ixodes ticks during blood feeding. Am. J. Trop. Med. Hyg. *53*, 397–404.

de Silva, A.M., and Fikrig, E. (1997). Arthropod- and host-specific gene expression by *Borrelia burgdorferi*. J. Clin. Invest. *99*, 377–379.

de Silva, A.M., Telford, S.R., Brunet, L.R., Barthold, S.W., and Fikrig, E. (1996). *Borrelia burgdorferi* OspA is an arthropod-specific transmission-blocking Lyme disease vaccine. J. Exp. Med. *183*, 271–275.

de Silva, A.M., Tyson, K.R., and Pal, U. (2009). Molecular characterization of the tick–Borrelia interface. Front. Biosci. *14*, 3051–3063.

Duray, P.H., Yin, S.R., Ito, Y., Bezrukov, L., Cox, C., Cho, M.S., Fitzgerald, W., Dorward, D., Zimmerberg, J., and Margolis, L. (2005). Invasion of human tissue *ex vivo* by *Borrelia burgdorferi*. J. Infect. Dis. *191*, 1747–1754.

Earnhart, C.G., Buckles, E.L., and Marconi, R.T. (2007). Development of an OspC-based tetravalent, recombinant, chimeric vaccinogen that elicits bactericidal antibody against diverse Lyme disease spirochete strains. Vaccine 25, 466–480.

Earnhart, C.G., and Marconi, R.T. (2007a). An octavalent lyme disease vaccine induces antibodies that recognize all incorporated OspC type-specific sequences. Hum. Vaccin. *3*, 281–289.

Earnhart, C.G., and Marconi, R.T. (2007b). OspC phylogenetic analyses support the feasibility of a broadly protective polyvalent chimeric Lyme disease vaccine. Clin. Vaccine Immunol *14*, 628–634.

Fikrig, E., Barthold, S.W., Kantor, F.S., and Flavell, R.A. (1990). Protection of mice against the Lyme disease agent by immunizing with recombinant OspA. Science *250*, 553–556.

Fikrig, E., and Narasimhan, S. (2006). *Borrelia burgdorferi* – traveling incognito? Microbes Infect. *8*, 1390–1399.

Fikrig, E., Pal, U., Chen, M., Anderson, J.F., and Flavell, R.A. (2004). OspB antibody prevents *Borrelia burgdorferi* colonization of *Ixodes scapularis*. Infect. Immun. *72*, 1755–1759.

Fingerle, V., Laux, H., Munderloh, U.G., Schulte-Spechtel, U., and Wilske, B. (2000). Differential expression of outer surface proteins A and C by individual *Borrelia burgdorferi* in different genospecies, Med. Microbiol. Immunol. (Berlin) *189*, 59–66.

Fingerle, V., Goettner, G., Gern, L., Wilske, B., and Schulte-Spechtel, U. (2007). Complementation of a *Borrelia afzelii* OspC mutant highlights the crucial role of OspC for dissemination of *Borrelia afzelii* in *Ixodes ricinus*. Int. J. Med. Microbiol. *297*, 97–107.

Finlay, B.B., and Cossart, P. (1997). Exploitation of mammalian host cell functions by bacterial pathogens. Science *276*, 718–725.

Fisher, M.A., Grimm, D., Henion, A.K., Elias, A.F., Stewart, P.E., Rosa, P.A., and Gherardini, F.C. (2005). *Borrelia burgdorferi* σ^{54} is required for mammalian infection and vector transmission but not for tick colonization. Proc. Natl. Acad. Sci. U.S.A. *102*, 5162–5167.

Fogaca, A.C., da Silva, P.I., Jr., Miranda, M.T., Bianchi, A.G., Miranda, A., Ribolla, P.E., and Daffre, S. (1999). Antimicrobial activity of a bovine hemoglobin fragment in the tick *Boophilus microplus*. J. Biol. Chem. *274*, 25330–25334.

Francischetti, I.M., My Pham, V., Mans, B.J., Andersen, J.F., Mather, T.N., Lane, R.S., and Ribeiro, J.M. (2005). The transcriptome of the salivary glands of the female western black-legged tick *Ixodes pacificus* (Acari: Ixodidae). Insect Biochem. Mol. Biol. 35, 1142–1161.

Fraser, C.M., Casjens, S., Huang, W.M., Sutton, G.G., Clayton, R., Lathigra, R., White, O., Ketchum, K.A., Dodson, R., Hickey, E.K., *et al.* (1997). Genomic sequence of a Lyme disease spirochaete, *Borrelia burgdorferi*. Nature *390*, 580–586.

Fuchs, H., Wallich, R., Simon, M.M., and Kramer, M.D. (1994). The outer surface protein A of the spirochete *Borrelia burgdorferi* is a plasmin(ogen) receptor. Proc. Natl. Acad. Sci. (U. S. A.) *91*, 12594–12598.

Fung, B.P., McHugh, G.L., Leong, J.M., and Steere, A.C. (1994). Humoral immune response to outer surface protein C of *Borrelia burgdorferi* in Lyme disease: role of the immunoglobulin M response in the serodiagnosis of early infection. Infect. Immun. *62*, 3213–3221.

Geraci, N.S., Spencer Johnston, J., Paul Robinson, J., Wikel, S.K., and Hill, C.A. (2007). Variation in genome size of argasid and ixodid ticks. Insect Biochem. Mol. Biol. *37*, 399–408.

Gilbert, M.A., Morton, E.A., Bundle, S.F., and Samuels, D.S. (2007). Artificial regulation of *ospC* expression in *Borrelia burgdorferi*. Mol. Microbiol. *63*, 1259–1273.

Grewe, C., and Nuske, J.H. (1996). Immunolocalization of a 22 kDa protein (IPLA7, P22) of *Borrelia burgdorferi*. FEMS Microbiol. Lett. *138*, 215–219.

Grimm, D., Eggers, C.H., Caimano, M.J., Tilly, K., Stewart, P.E., Elias, A.F., Radolf, J.D., and Rosa, P.A. (2004a). Experimental assessment of the roles of linear plasmids lp25 and lp28-1 of *Borrelia burgdorferi* throughout the infectious cycle. Infect. Immun. *72*, 5938–5946.

Grimm, D., Tilly, K., Bueschel, D.M., Fisher, M.A., Policastro, P.F., Gherardini, F.C., Schwan, T.G., and Rosa, P.A. (2005). Defining plasmids required by *Borrelia burgdorferi* for colonization of tick vector *Ixodes scapularis* (Acari: Ixodidae). J. Med. Entomol. *42*, 676–684.

Grimm, D., Tilly, K., Byram, R., Stewart, P.E., Krum, J.G., Bueschel, D.M., Schwan, T.G., Policastro, P.F., Elias, A.F., and Rosa, P.A. (2004b). Outer-surface protein C of the Lyme disease spirochete: a protein induced in ticks for infection of mammals. Proc. Natl. Acad. Sci. U.S.A. *101*, 3142–3147.

Hajdusek, O., Sojka, D., Kopacek, P., Buresova, V., Franta, Z., Sauman, I., Winzerling, J., and Grubhoffer, L. (2009). Knockdown of proteins involved in iron metabolism limits tick reproduction and development. Proc. Natl. Acad. Sci. U.S.A. *106*, 1033–1038.

He, Z.B., Cao, Y.Q., Yin, Y.P., Wang, Z.K., Chen, B., Peng, G.X., and Xia, Y.X. (2006). Role of *hunchback* in segment patterning of *Locusta migratoria manilensis* revealed by parental RNAi. Dev. Growth Differ. *48*, 439–445.

He, M., Oman, T., Xu, H., Blevins, J., Norgard, M.V., and Yang, X.F. (2008). Abrogation of ospAB constitutively activates the Rrp2–RpoN–RpoS pathway (sigmaN-sigmaS cascade) in *Borrelia burgdorferi*. Mol. Microbiol. *70*, 1453–1464.

Heath, A.C. (1979). The temperature and humidity preferences of *Haemaphysalis longicornis*, *Ixodes Holocyclus* and *Rhipicephalus sanguineus* (Ixodidae): studies on eggs. Int J. Parasitol. *9*, 33–39.

Hefty P.S., Jolliff, S.E., Caimano, M.J., Wikel, S.K., Radolf, J.D., and Akins, D.R. (2001). Regulation of OspE-related, OspF-related, and Elp lipoproteins of *Borrelia burgdorferi* strain 297 by mammalian host-specific signals. Infect. Immun. *69*, 3618–3627.

Hill, C.A., and Wikel, S.K. (2005). The *Ixodes scapularis* Genome Project: an opportunity for advancing tick research. Trends Parasitol *21*, 151–153.

Hodzic, E., Tunev, S., Feng, S., Freet, K.J., and Barthold, S.W. (2005). Immunoglobulin-regulated expression of *Borrelia burgdorferi* outer surface protein A *in vivo*. Infect. Immun. *73*, 3313–3321.

Hübner, A., Revel, A.T., Nolen, D.M., Hagman, K.E., and Norgard, M.V. (2003). Expression of a *luxS* gene is not required for *Borrelia burgdorferi* infection of mice via needle inoculation. Infect. Immun. *71*, 2892–2896.

Hübner, A., Yang, X., Nolen, D.M., Popova, T.G., Cabello, F.C., and Norgard, M.V. (2001). Expression of *Borrelia burgdorferi* OspC and DbpA is controlled by a RpoN–RpoS regulatory pathway. Proc. Natl. Acad. Sci. U.S.A. *98*, 12724–12729.

Hynes, W.L., Ceraul, S.M., Todd, S.M., Seguin, K.C., and Sonenshine, D.E. (2005). A defensin-like gene expressed in the black-legged tick, *Ixodes scapularis*. Med Vet Entomol *19*, 339–344.

Indest, K.J., Howell, J.K., Jacobs, M.B., Scholl-Meeker, D., Norris, S.J., and Philipp, M.T. (2001). Analysis of *Borrelia burgdorferi vlsE* gene expression and recombination in the tick vector. Infect. Immun. *69*, 7083–7090.

Indest, K.J., Ramamoorthy, R., Solé, M., Gilmore, R.D., Johnson, B.J., and Philipp, M.T. (1997). Cell-density-dependent expression of *Borrelia burgdorferi* lipoproteins *in vitro*. Infect. Immun. *65*, 1165–1171.

Jewett, M.W., Lawrence, K., Bestor, A.C., Tilly, K., Grimm, D., Shaw, P., VanRaden, M., Gherardini, F., and Rosa, P.A. (2007). The critical role of the linear plasmid lp36 in the infectious cycle of *Borrelia burgdorferi*. Mol. Microbiol. *64*, 1358–1374.

Johns, R., Ohnishi, J., Broadwater, A., Sonenshine, D.E., de Silva, A.M., and Hynes, W.L. (2001). Contrasts in tick innate immune responses to *Borrelia burgdorferi* challenge: immunotolerance in *Ixodes scapularis* versus immunocompetence in *Dermacentor variabilis* (Acari: Ixodidae). J. Med. Entomol. *38*, 99–107.

Khila, A., and Grbic, M. (2007). Gene silencing in the spider mite *Tetranychus urticae*: dsRNA and siRNA parental silencing of the Distal-less gene. Dev. Genes Evol. *217*, 241–251.

Labandeira-Rey, M., and Skare, J.T. (2001). Decreased infectivity in *Borrelia burgdorferi* strain B31 is associated with loss of linear plasmid 25 or 28-1. Infect. Immun. *69*, 446–455.

Lagal, V., Portnoi, D., Faure, G., Postic, D., and Baranton, G. (2006). *Borrelia burgdorferi* sensu stricto invasiveness is correlated with OspC-plasminogen affinity. Microbes Infect. *8*, 645–652.

Lam, T.T., Nguyen, T.P., Fikrig, E., and Flavell, R.A. (1994). A chromosomal *Borrelia burgdorferi* gene encodes a 22-kilodalton lipoprotein, P22, that is serologically recognized in Lyme disease. J. Clin. Microbiol. *32*, 876–883.

Li, X., Neelakanta, G., Liu, X., Beck, D.S., Kantor, F.S., Fish, D., Anderson, J.F., and Fikrig, E. (2007a). Role of outer surface protein D in the *Borrelia burgdorferi* life cycle. Infect. Immun. *75*, 4237–4244.

Li, X., Pal, U., Ramamoorthi, N., Liu, X., Desrosiers, D.C., Eggers, C.H., Anderson, J.F., Radolf, J.D., and Fikrig, E. (2007b). The Lyme disease agent *Borrelia burgdorferi* requires BB0690, a Dps homologue, to persist within ticks. Mol. Microbiol. *63*, 694–710.

Liang, F.T., Brown, E.L., Wang, T., Iozzo, R.V., and Fikrig, E. (2004). Protective niche for *Borrelia burgdorferi* to evade humoral immunity. Am. J. Pathol. *165*, 977–985.

Liang, F.T., Jacobs, M.B., Bowers, L.C., and Philipp, M.T. (2002). An immune evasion mechanism for spirochetal persistence in Lyme borreliosis. J. Exp. Med. *195*, 415–422.

Lybecker, M.C., and Samuels, D.S. (2007). Temperature-induced regulation of RpoS by a small RNA in *Borrelia burgdorferi*. Mol. Microbiol. *64*, 1075–1089.

Mbow, M.L., Gilmore, R.D., Jr., and Titus, R.G. (1999). An OspC-specific monoclonal antibody passively protects mice from tick-transmitted infection by *Borrelia burgdorferi* B31. Infect. Immun. *67*, 5470–5472.

Morris, R.V., Shoemaker, C.B., David, J.R., Lanzaro, G.C., and Titus, R.G. (2001). Sandfly maxadilan exacerbates infection with *Leishmania major* and vaccinating against it protects against *L. major* infection. J. Immunol. *167*, 5226–5230.

Mulay, V.B., Caimano, M.J., Iyer, R., Dunham-Ems, S., Liveris, D., Petzke, M.M., Schwartz, I., and Radolf, J.D. (2009). *Borrelia burgdorferi bba74* is expressed exclusively during tick feeding and is regulated by both arthropod- and mammalian host-specific signals. J. Bacteriol. *191*, 2783–2794.

Munderloh, U.G., and Kurtti, T.J. (1995). Cellular and molecular interrelationships between ticks and prokaryotic tick-borne pathogens. Annu. Rev. Entomol. *40*, 221–243.

Munderloh, U.G., and Kurtti, T.J. (2005). The ABCs of Lyme disease spirochaetes in ticks. Lancet *366*, 962–964.

Munderloh, U.G., Liu, Y., Wang, M., Chen, C., and Kurtti, T.J. (1994). Establishment, maintenance and description of cell lines from the tick *Ixodes scapularis*. J. Parasitol. *80*, 533–543.

Nadelman, R.B., and Wormser, G.P. (1998). Lyme borreliosis. Lancet 352, 557–565.

Nakajima, Y., Ogihara, K., Taylor, D., and Yamakawa, M. (2003). Antibacterial hemoglobin fragments from the midgut of the soft tick, *Ornithodoros moubata* (Acari: Argasidae). J. Med. Entomol. *40*, 78–81.

Narasimhan, S., Deponte, K., Marcantonio, N., Liang, X., Royce, T.E., Nelson, K.F., Booth, C.J., Koski, B., Anderson, J.F., Kantor, F., and Fikrig, E. (2007a). Immunity against *Ixodes scapularis* salivary proteins expressed within 24 hours of attachment thwarts tick feeding and impairs *Borrelia* transmission. PLoS ONE *2*, e451.

Narasimhan, S., Santiago, F., Koski, R.A., Brei, B., Anderson, J.F., Fish, D., and Fikrig, E. (2002). Examination of the *Borrelia burgdorferi* transcriptome in *Ixodes scapularis* during feeding. J. Bacteriol. *184*, 3122–3125.

Narasimhan, S., Sukumaran, B., Bozdogan, U., Thomas, V., Liang, X., DePonte, K., Marcantonio, N., Koski, R.A., Anderson, J.F., Kantor, F., and Fikrig, E. (2007b). A tick antioxidant facilitates the Lyme disease agent's successful migration from the mammalian host to the arthropod vector. Cell Host Microbe *2*, 7–18.

Neelakanta, G., Li, X., Pal, U., Liu, X., Beck, D.S., Deponte, K., Fish, D., Kantor, F.S., and Fikrig, E. (2007). Outer surface protein B Is critical for *Borrelia burgdorferi* adherence and survival within *Ixodes* ticks. PLoS Pathog. 3, e33.

Nosbisch, L.K., and de Silva, A.M. (2007). Lack of detectable variation at *Borrelia burgdorferi vlsE* locus in ticks. J. Med. Entomol. *44*, 168–170.

Ogden, N.H., Lindsay, L.R., Beauchamp, G., Charron, D., Maarouf, A., O'Callaghan, C.J., Waltner-Toews, D., and Barker, I.K. (2004). Investigation of relationships between temperature and developmental rates of tick *Ixodes scapularis* (Acari: Ixodidae) in the laboratory and field. J. Med. Entomol. *41*, 622–633.

Ohnishi, J., Piesman, J., and de Silva, A.M. (2001). Antigenic and genetic heterogeneity of *Borrelia burgdorferi* populations transmitted by ticks. Proc. Natl. Acad. Sci. U.S.A. *98*, 670–675.

Ojaimi, C., Brooks, C., Akins, D., Casjens, S., Rosa, P., Elias, A., Barbour, A., Jasinskas, A., Benach, J., Katonah, L., *et al.* (2002). Borrelia burgdorferi gene expression profiling with membrane-based arrays. Methods Enzymol. *358*, 165–177.

Ouyang, Z., He, M., Oman, T., Yang, X.F., and Norgard, M.V. (2009). A manganese transporter, BB0219 (BmtA), is required for virulence by the Lyme disease spirochete, *Borrelia burgdorferi*. Proc. Natl. Acad. Sci. U.S.A. *106*, 3449–3454.

Pagel Van Zee, J., Geraci, N.S., Guerrero, F.D., Wikel, S.K., Stuart, J.J., Nene, V.M., and Hill, C.A. (2007). Tick genomics: the *Ixodes* genome project and beyond. Int J. Parasitol. *37*, 1297–1305.

Pal, U., Dai, J., Li, X., Neelakanta, G., Luo, P., Kumar, M., Wang, P., Yang, X., Anderson, J.F., and Fikrig, E. (2008a). A differential role for BB0365 in the persistence of *Borrelia burgdorferi* in mice and ticks. J. Infect. Dis. *197*, 148–155.

Pal, U., de Silva, A.M., Montgomery, R.R., Fish, D., Anguita, J., Anderson, J.F., Lobet, Y., and Fikrig, E. (2000). Attachment of *Borrelia burgdorferi* within *Ixodes scapularis* mediated by outer surface protein A. J. Clin. Invest. *106*, 561–569.

Pal, U., and Fikrig, E. (2003). Adaptation of *Borrelia burgdorferi* in the vector and vertebrate host. Microbes Infect. *5*, 659–666.

Pal, U., Li, X., Wang, T., Montgomery, R.R., Ramamoorthi, N., de silva, A.M., Bao, F., Yang, X., Pypaert, M., Pradhan, D., *et al.* (2004a). TROSPA, an *Ixodes scapularis* receptor for *Borrelia burgdorferi*. Cell *119*, 457–468.

Pal, U., Montgomery, R.R., Lusitani, D., Voet, P., Weynants, V., Malawista, S.E., Lobet, Y., and Fikrig, E. (2001). Inhibition of *Borrelia burgdorferi*–tick interactions in vivo by outer surface protein A antibody. J. Immunol. *166*, 7398–7403.

Pal, U., Wang, P., Bao, F., Yang, X., Samanta, S., Schoen, R., Wormser, G.P., Schwartz, I., and Fikrig, E. (2008b). *Borrelia burgdorferi* basic membrane proteins A and B participate in the genesis of Lyme arthritis. J. Exp. Med. *205*, 133–141.

Pal, U., Yang, X., Chen, M., Bockenstedt, L.K., Anderson, J.F., Flavell, R.A., Norgard, M.V., and Fikrig, E. (2004b). OspC facilitates *Borrelia burgdorferi* invasion of *Ixodes scapularis* salivary glands. J. Clin. Invest. *113*, 220–230.

Piesman, J. (1993). Dynamics of *Borrelia burgdorferi* transmission by nymphal Ixodes dammini ticks. J. Infect. Dis. *167*, 1082–1085.

Piesman, J., and Eisen, L. (2008). Prevention of tick-borne diseases. Annu. Rev. Entomol. *53*, 323–343.

Piesman, J., and Gern, L. (2004). Lyme borreliosis in Europe and North America. Parasitology *129 Suppl.*, S191–220.

Piesman, J., Oliver, J., and Sinsky, R. (1990). Growth kinetics of the Lyme disease spirochete (*Borrelia burgdorferi*) in vector ticks (*Ixodes dammini*). Am. J. Trop. Med. Hyg. *42*, 352–357.

Piesman, J., Schneider, B.S., and Zeidner, N.S. (2001). Use of quantitative PCR to measure density of *Borrelia burgdorferi* in the midgut and salivary glands of feeding tick vectors. J. Clin. Microbiol. *39*, 4145–4148.

Pimenta, P.F., Modi, G.B., Pereira, S.T., Shahabuddin, M., and Sacks, D.L. (1997). A novel role for the peritrophic matrix in protecting *Leishmania* from the hydrolytic activities of the sand fly midgut. Parasitology *115 (Pt 4)*, 359–369.

Purser, J.E., Lawrenz, M.B., Caimano, M.J., Howell, J.K., Radolf, J.D., and Norris, S.J. (2003). A plasmid-encoded nicotinamidase (PncA) is essential for infectivity of *Borrelia burgdorferi* in a mammalian host. Mol. Microbiol. *48*, 753–764.

Purser, J.E., and Norris, S.J. (2000). Correlation between plasmid content and infectivity in *Borrelia burgdorferi*. Proc. Natl. Acad. Sci. U.S.A. *97*, 13865–13870.

Ramamoorthi, N., Narasimhan, S., Pal, U., Bao, F., Yang, X.F., Fish, D., Anguita, J., Norgard, M.V., Kantor, F.S., Anderson, J.F., *et al.* (2005). The Lyme disease agent exploits a tick protein to infect the mammalian host. Nature *436*, 573–577.

Revel, A.T., Blevins, J.S., Almazan, C., Neil, L., Kocan, K.M., de la Fuente, J., Hagman, K.E., and Norgard, M.V. (2005). *bptA* (*bbe16*) is essential for the persistence of the Lyme disease spirochete, *Borrelia burgdorferi*, in its natural tick vector. Proc. Natl. Acad. Sci. U.S.A. *102*, 6972–6977.

Revel, A.T., Talaat, A.M., and Norgard, M.V. (2002). DNA microarray analysis of differential gene expression in *Borrelia burgdorferi*, the Lyme disease spirochete. Proc. Natl. Acad. Sci. U.S.A. *99*, 1562–1567.

Ribeiro, J.M., Alarcon-Chaidez, F., Francischetti, I.M., Mans, B.J., Mather, T.N., Valenzuela, J.G., and Wikel, S.K. (2006). An annotated catalog of salivary gland transcripts from *Ixodes scapularis* ticks. Insect Biochem. Mol. Biol. *36*, 111–129.

Ribeiro, J.M., and Francischetti, I.M. (2003). Role of arthropod saliva in blood feeding: sialome and post-sialome perspectives. Annu. Rev. Entomol. *48*, 73–88.

Rogers, E.A., Terekhova, D., Zhang, H.M., Hovis, K.M., Schwartz, I., and Marconi, R.T. (2009). Rrp1, a cyclic-di-GMP-producing response regulator, is an important regulator of *Borrelia burgdorferi* core cellular functions. Mol. Microbiol. *71*, 1551–1573.

Rosa, P. (2005). Lyme disease agent borrows a practical coat. Nat. Med. *11*, 831–832.

Rosa, P.A., Tilly, K., and Stewart, P.E. (2005). The burgeoning molecular genetics of the Lyme disease spirochaete. Nat. Rev. Microbiol. *3*, 129–143.

Rudenko, N., Golovchenko, M., Edwards, M.J., and Grubhoffer, L. (2005). Differential expression of *Ixodes ricinus* tick genes induced by blood feeding or *Borrelia burgdorferi* infection. J. Med. Entomol. *42*, 36–41.

Rudzinska, M.A., Spielman, A., Lewengrub, S., Piesman, J., and Karakashian, S. (1982). Penetration of the peritrophic membrane of the tick by *Babesia microti*. Cell Tissue Res. *221*, 471–481.

Sauer, J.R., and Hair, A.J (1986). Morphology, Physiology and Behavioral Biology of Ticks, (Chichester: Ellis Horwood Limited).

Schwan, T.G. (2003). Temporal regulation of outer surface proteins of the Lyme-disease spirochaete *Borrelia burgdorferi*. Biochem. Soc. Trans. *31*, 108–112.

Schwan, T.G., and Piesman, J. (2000). Temporal changes in outer surface proteins A and C of the Lyme disease-associated spirochete, *Borrelia burgdorferi*, during the chain of infection in ticks and mice. J. Clin. Microbiol. *38*, 382–388.

Schwan, T.G., and Piesman, J. (2002). Vector interactions and molecular adaptations of Lyme disease and relapsing fever spirochetes associated with transmission by ticks. Emerg. Infect. Dis. *8*, 115–121.

Schwan, T.G., Piesman, J., Golde, W.T., Dolan, M.C., and Rosa, P.A. (1995). Induction of an outer surface protein on *Borrelia burgdorferi* during tick feeding. Proc. Natl. Acad. Sci. U.S.A. *92*, 2909–2913.

Shi, Y., Xu, Q., McShan, K., and Liang, F.T. (2008). Both decorin-binding proteins A and B are critical for the overall virulence of *Borrelia burgdorferi*. Infect. Immun. *76*, 1239–1246.

Shi, Y., Xu, Q., Seemanapalli, S.V., McShan, K., and Liang, F.T. (2006). The *dbpBA* locus of *Borrelia burgdorferi* is not essential for infection of mice. Infect. Immun. *74*, 6509–6512.

Sigal, L.H., Zahradnik, J.M., Lavin, P., Patella, S.J., Bryant, G., Haselby, R., Hilton, E., Kunkel, M., Adler-Klein, D., Doherty, T., *et al.* (1998). A vaccine consisting of recombinant *Borrelia burgdorferi* outer-surface protein A to prevent Lyme disease. N. Engl. J. Med. *339*, 216–222.

Soares, C.A., Lima, C.M., Dolan, M.C., Piesman, J., Beard, C.B., and Zeidner, N.S. (2005). Capillary feeding of specific dsRNA induces silencing of the *isac* gene in nymphal *Ixodes scapularis* ticks. Insect Mol. Biol. *14*, 443–452.

Sonenshine, D.E. (1993). Biology of Ticks (New York: Oxford University Press).

Sonenshine, D.E., Ceraul, S.M., Hynes, W.E., Macaluso, K.R., and Azad, A.F. (2002). Expression of defensin-like peptides in tick hemolymph and midgut in response to challenge with *Borrelia burgdorferi*, *Escherichia coli* and *Bacillus subtilis*. Exp. Appl. Acarol. *28*, 127–134.

Sonenshine, D.E., Hynes, W.L., Ceraul, S.M., Mitchell, R., and Benzine, T. (2005). Host blood proteins and peptides in the midgut of the tick *Dermacentor variabilis* contribute to bacterial control. Exp. Appl. Acarol. *36*, 207–223.

Srivastava, S.Y., and de Silva, A.M. (2008). Reciprocal expression of *ospA* and *ospC* in single cells of *Borrelia burgdorferi*. J. Bacteriol. *190*, 3429–3433.

Steere, A.C. (2001). Lyme disease. N. Engl. J. Med. *345*, 115–125.

Steere, A.C., Coburn, J., and Glickstein, L. (2004). The emergence of Lyme disease. J. Clin. Invest. *113*, 1093–1101.

Steere, A.C., Sikand, V.K., Meurice, F., Parenti, D.L., Fikrig, E., Schoen, R.T., Nowakowski, J., Schmid, C.H., Laukamp, S., Buscarino, C., and Krause, D.S. (1998). Vaccination against Lyme disease with recombinant *Borrelia burgdorferi* outer-surface lipoprotein A with adjuvant. N. Engl. J. Med. 339, 209–215.

Stevenson, B., Schwan, T.G., and Rosa, P.A. (1995). Temperature-related differential expression of antigens in the Lyme disease spirochete, *Borrelia burgdorferi*. Infect. Immun. *63*, 4535–4539.

Stewart, P.E., Bestor, A., Cullen, J.N., and Rosa, P.A. (2008). A tightly regulated surface protein of *Borrelia burgdorferi* is not essential to the mouse-tick infectious cycle. Infect. Immun. *76*, 1970–1978.

Stewart, P.E., Byram, R., Grimm, D., Tilly, K., and Rosa, P.A. (2005). The plasmids of *Borrelia burgdorferi*: essential genetic elements of a pathogen. Plasmid *53*, 1–13.

Stewart, P.E., Wang, X., Bueschel, D.M., Clifton, D.R., Grimm, D., Tilly, K., Carroll, J.A., Weis, J.J., and Rosa, P.A. (2006). Delineating the requirement for the *Borrelia burgdorferi* virulence factor OspC in the mammalian host. Infect. Immun. 74, 3547–3553.

Strother, K.O., Broadwater, A., and De Silva, A. (2005). Plasmid requirements for infection of ticks by *Borrelia burgdorferi*. Vector Borne Zoonotic Dis. *5*, 237–245.

Strother, K.O., and de Silva, A. (2005). Role of *Borrelia burgdorferi* linear plasmid 25 in infection of *Ixodes scapularis* ticks. J. Bacteriol. *187*, 5776–5781.

Tilly, K., Bestor, A., Dulebohn, D.P., and Rosa, P. (2009). OspC-independent infection and dissemination by host-adapted *Borrelia burgdorferi*. Infect. Immun. *77*, 2672–2682.

Tilly, K., Bestor, A., Jewett, M.W., and Rosa, P. (2007). Rapid clearance of Lyme disease spirochetes lacking OspC from skin. Infect. Immun. *75*, 1517–1519.

Tilly, K., Grimm, D., Bueschel, D.M., Krum, J.G., and Rosa, P. (2004). Infectious cycle analysis of a *Borrelia burgdorferi* mutant defective in transport of chitobiose, a tick cuticle component. Vector Borne Zoonotic Dis. *4*, 159–168.

Tilly, K., Krum, J.G., Bestor, A., Jewett, M.W., Grimm, D., Bueschel, D., Byram, R., Dorward, D., Vanraden, M.J., Stewart, P., and Rosa, P. (2006). *Borrelia burgdorferi* OspC protein required exclusively in a crucial early stage of mammalian infection. Infect. Immun. *74*, 3554–3564.

Tokarz, R., Anderton, J.M., Katona, L.I., and Benach, J.L. (2004). Combined effects of blood and temperature shift on *Borrelia burgdorferi* gene expression as determined by whole genome DNA array. Infect. Immun. *72*, 5419–5432.

Tsao, J.I. (2009). Reviewing molecular adaptations of Lyme borreliosis spirochetes in the context of reproductive fitness in natural transmission cycles. Vet Res. *40*, 36.

Tsao, J.I., Wootton, J.T., Bunikis, J., Luna, M.G., Fish, D., and Barbour, A.G. (2004). An ecological approach to preventing human infection: vaccinating wild mouse reservoirs intervenes in the Lyme disease cycle. Proc. Natl. Acad. Sci. U.S.A. *101*, 18159–18164.

Tyson, K., Elkins, C., Patterson, H., Fikrig, E., and de Silva, A. (2007). Biochemical and functional characterization of Salp20, an *Ixodes scapularis* tick salivary protein that inhibits the complement pathway. Insect Mol. Biol. *16*, 469–479.

Ueti, M.W., Reagan, J.O., Jr., Knowles, D.P., Jr., Scoles, G.A., Shkap, V., and Palmer, G.H. (2007). Identification of midgut and salivary glands as specific and distinct barriers to efficient tick-borne transmission of *Anaplasma marginale*. Infect. Immun. 75, 2959–2964.

Ullmann, A.J., Lima, C.M., Guerrero, F.D., Piesman, J., and Black, W.C. IV. (2005). Genome size and organization in the blacklegged tick, *Ixodes scapularis* and the Southern cattle tick, *Boophilus microplus*. Insect Mol. Biol. *14*, 217–222.

Valenzuela, J.G., Belkaid, Y., Garfield, M.K., Mendez, S., Kamhawi, S., Rowton, E.D., Sacks, D.L., and Ribeiro, J.M. (2001). Toward a defined anti-*Leishmania* vaccine targeting vector antigens: characterization of a protective salivary protein. J. Exp. Med. *194*, 331–342.

Venken, K.J., and Bellen, H.J. (2005). Emerging technologies for gene manipulation in *Drosophila melanogaster*. Nat. Rev. Genet. 6, 167–178.

von Lackum, K., Ollison, K.M., Bykowski, T., Nowalk, A.J., Hughes, J.L., Carroll, J.A., Zückert, W.R., and Stevenson, B. (2007). Regulated synthesis of the *Borrelia burgdorferi* inner-membrane lipoprotein IpLA7 (P22, P22-A) during the Lyme disease spirochaete's mammal-tick infectious cycle. Microbiology *153*, 1361–1371.

Vorou, R.M., Papavassiliou, V.G., and Tsiodras, S. (2007). Emerging zoonoses and vector-borne infections affecting humans in Europe. Epidemiol. Infect. *135*, 1231–1247.

Wilske, B. (2005). Epidemiology and diagnosis of Lyme borreliosis. Ann. Med. 37, 568–579.

Wilske, B., Busch, U., Fingerle, V., Jauris-Heipke, S., Preac Mursic, V., Rossler, D., and Will, G. (1996). Immunological and molecular variability of OspA and OspC. Implications for *Borrelia* vaccine development. Infection 24, 208–212.

Wilske, B., Fingerle, V., Herzer, P., Hofmann, A., Lehnert, G., Peters, H., Pfister, H.W., Preac-Mursic, V., Soutschek, E., and Weber, K. (1993). Recombinant immunoblot in the serodiagnosis of Lyme borreliosis. Comparison with indirect immunofluorescence and enzyme-linked immunosorbent assay. Med. Microbiol. Immunol. *182*, 255–270.

Xu, Q., McShan, K., and Liang, F.T. (2007). Identification of an *ospC* operator critical for immune evasion of *Borrelia burgdorferi*. Mol. Microbiol. *64*, 220–231.

Yang, X., Goldberg, M.S., Popova, T.G., Schoeler, G.B., Wikel, S.K., Hagman, K.E., and Norgard, M.V. (2000). Interdependence of environmental factors influencing reciprocal patterns of gene expression in virulent *Borrelia burgdorferi*. Mol. Microbiol. 37, 1470–1479.

Yang, X.F., Alani, S.M., and Norgard, M.V. (2003). The response regulator Rrp2 is essential for the expression of major membrane lipoproteins in *Borrelia burgdorferi*. Proc. Natl. Acad. Sci. U.S.A. *100*, 11001–11006.

Yang, X.F., Pal, U., Alani, S.M., Fikrig, E., and Norgard, M.V. (2004). Essential role for OspA/B in the life cycle of the Lyme disease spirochete. J. Exp. Med. *199*, 641–648.

Zhang, J.R., Hardham, J.M., Barbour, A.G., and Norris, S.J. (1997). Antigenic variation in Lyme disease borreliae by promiscuous recombination of Vmp-like sequence cassettes. Cell *89*, 275–285.

Zhang, J.R., and Norris, S.J. (1998). Genetic variation of the *Borrelia burgdorferi* gene *vlsE* involves cassette-specific, segmental gene conversion. Infect. Immun. *66*, 3698–3704.

Zung, J., Lewengrub, S., Rudzinska, M., Spielman, A., Telford, S., and Piesman, J. (1989). Fine structural evidence for the penetration of the Lyme disease spirochete *Borrelia burgdorferi* through the gut and salivary tissues of *Ixodes dammini*. Can. J. Zool. *67*, 1717–1748.

Pathobiology of Lyme Disease *Borrelia* 12

Steven J. Norris, Jenifer Coburn, John M. Leong, Linden T. Hu and Magnus Höök

Abstract

Lyme disease *Borrelia* are host-dependent, tick-transmitted, invasive, non-toxigenic, persistent pathogens that cause disease in humans and other mammals primarily through the induction of inflammatory reactions. During transmission from the infected tick, the bacteria undergo dramatic changes in gene expression, resulting in adaptation to the mammalian environment. Expression of outer surface protein C (OspC) is essential during these early stages of colonization, although the mechanism by which OspC promotes spirochaetal infectivity is unknown. Organisms multiply and spread locally and induce an inflammatory response that in humans results in an erythema migrans, the hallmark lesion of localized infection. The spirochaetes enter the bloodstream during the primary infection and colonize multiple tissue sites, leading to the disseminated stage of infection. Motility and chemotaxis mechanisms undoubtedly play a role in this process. Several borrelial proteins have been implicated in adherence to host cell surface proteins and extracellular matrix components and are likely to be involved in the homing of *Borrelia* to histological compartments within each tissue, penetration of blood vessels, and adherence to and migration through endothelial cells and tissue strata at distant sites. Activation of plasmin on the bacterial surface and induction of host proteases are thought to facilitate dissemination and/or inflammation. Most tissue damage appears to result from host inflammatory reactions. Although the mechanisms are not entirely understood, induction of cytokine/chemokine expression by bacterial lipoproteins and the resulting recruitment and activation of lymphocytes, macrophages, and granulocytes play a major role in both local histopathology and constitutional symptoms. Despite their relatively low densities in tissues, *Borrelia* cause neurological, cardiovascular, arthritic, and dermatological manifestations during the disseminated and persistent stages of infection by mechanisms that remain largely a mystery. Immune evasion mechanisms, including the *vls* antigenic variation system, complement-regulator acquiring surface proteins (CRASPs), downregulation of highly antigenic surface proteins (such as OspC), and invasion of protective niches, permit the survival of the pathogens for months to years following infection despite robust antibody and cellular responses.

12.1 Introduction

Prokaryotes evolved approximately 3.5 billion years ago (Schopf, 2000), and by definition were initially free-living. The evolution of microbial communities and symbiotic relationships eventually resulted in the development of multicellular eukaryotic organisms about 1 billion years ago, creating new ecological niches. Bacteria colonized these ecological niches as commensal, symbiotic and pathogenic organisms. Development of host-associated pathogens resulted from genetic adaptation to protect themselves, parasitize, and become harmful to other organisms. This adaptation involved modification, acquisition, and loss of a variety of genes, thereby acquiring, creating, and preserving those functions that assure their survival and losing those that are nonessential.

Lyme disease *Borrelia* have developed an unusual lifestyle in that they alternate between vertebrate and arthropod hosts; i.e. they never 'hit the ground'. In addition, they belong to a group of organisms that produce no known toxins yet are capable of invading virtually any mammalian tissue and causing infection and disease manifestations for months to years (Fig. 12.1). In this manner, infection with Lyme disease *Borrelia* parallels that of another spirochaete, *Treponema pallidum* subsp. *pallidum*, the causative agent of syphilis (Fig. 12.2). Both syphilis and Lyme disease exhibit local, disseminated and persistent manifestations in which tissue pathology appears to be due primarily to host inflammatory reactions (see Chapter 15). However, there are important differences between the two in addition to their modes of transmission. For example, the primary lesion of Lyme disease, erythema migrans, consists of a diffuse, expanding rash with central clearing, resulting in the classic bull's eye pattern (see Chapter 18). In contrast, the chancre of syphilis is a well-demarcated, ulcerative lesion that results from tissue necrosis, erosion of the epithelial layer, and haemorrhage, pathogenic mechanisms that are not typically operative in Lyme borreliosis. Both Lyme disease and syphilis have neurological manifestations and can have meningovascular syndromes that occur during the early disseminated stages. In Lyme disease, however, Bell's palsy and other peripheral neuropathies are common, whereas in syphilis demyelination of the dorsal root ganglia (resulting in tabes dorsalis) and frank infection of the brain parenchyma (paresis) can occur during the tertiary phase. Cardiovascular manifestations of Lyme disease consist of cardiac block (arrhythmia), while long-term syphilitic infection can result in weakening of the lamina media of the aorta and other elastic arteries, leading to aortitis and aneurysm formation. Arthritis is a prominent persistent symptom of Lyme disease in North America. Bone involvement is common in both the sexually transmitted and congenital forms of syphilis, and erosion of the nose can occur in endemic syphilis. Finally, congenital transmission of syphilis is common in infected women, whereas there is no convincing evidence for transplacental transmission of spirochaetes in Lyme disease. Overall, Lyme disease and syphilis are both persistent infections with pathology predominantly caused by inflammatory reactions, yet the diseases have distinct manifestations. Overlaid on this observation is the fact that the pathogenesis of Lyme disease is different in Europe and the USA, most likely due to predominance of *B. garinii* and *B. afzelii* among European patients. (Even more extreme distinctions are observed among *T. pallidum* subspecies

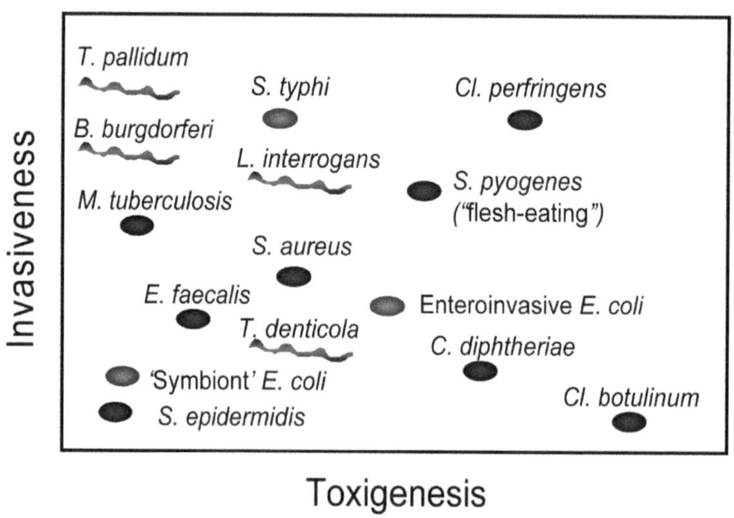

Figure 12.1 Characterization of pathogenic bacteria according to toxigenesis (plotted in the *x* dimension) and invasiveness (plotted in the *y* dimension). *Borrelia burgdorferi*, like *T. pallidum*, is a highly invasive non-toxigenic organism. From Botkin *et al.* (2006a) with permission.

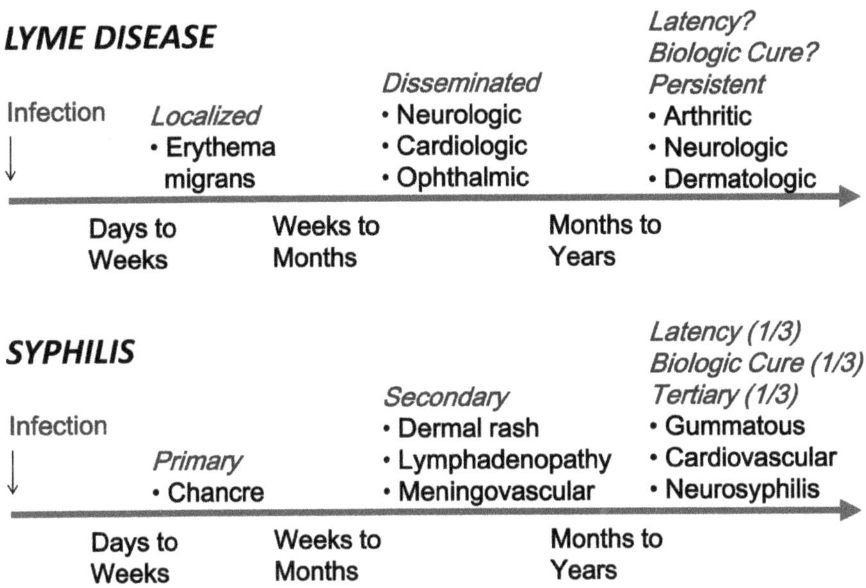

Figure 12.2 Comparison of the time course of Lyme disease and syphilis in humans, showing similarities in the pattern of local, disseminated, and persistent stages of infection.

and the diseases they cause, despite a high degree of genetic similarity.) The dissimilarities observed in these spirochaetal infections reflect important differences in infection patterns, tissue tropisms, and extracellular matrix/host cell interactions that in turn must be related to gene content and expression.

In this chapter, we examine the factors involved in the pathogenesis of Lyme disease. Studying the pathogenesis of invasive, non-toxigenic organisms is relatively complex; purification and characterization of toxins is straightforward compared to the analysis of gene products (e.g. adhesins and other surface molecules) that are best studied in the context of intact organisms. In addition, low passage, infectious strains of *B. burgdorferi* have low transformation rates, associated with the presence of putative restriction-modification systems encoded on plasmids lost during *in vitro* culture (Kawabata *et al.*, 2004; Lawrenz *et al.*, 2002). Thus our knowledge regarding the genes important in disease pathogenesis is still limited but is expanding rapidly (Rosa *et al.*, 2005). Because of the complexity of host–pathogen interactions in Lyme disease, this chapter can include only a portion of applicable studies; the reader is asked to refer to the other chapters for additional detail regarding gene regulation (see Chapter 4), motility (see Chapter 7), and other aspects of the spirochaete that contribute to pathogenesis.

12.2 Virulence determinants versus virulence factors

Current knowledge regarding the virulence determinants of Lyme disease *Borrelia* is summarized in Table 12.1. Virulence determinants include all genes that affect the infection process, and thereby encompass genes involved in basic metabolic processes. Virulence factors can be considered specialized mechanisms or molecules that promote survival or pathogenesis in the infected host and typically are not found in non-pathogenic organisms. A classic example of a virulence determinant that is not a virulence factor is AroD of *Salmonella* species (Miller *et al.*, 1989). AroD is a 3-dehydroquinate dehydratase involved in the synthesis of a chorismate, an intermediate in aromatic amino acid biosynthesis. *Salmonella* Typhi *aroD* mutants are attenuated and have been utilized as live vaccine strains; however, AroD fulfils a basic biosynthetic function and, thus, is not a virulence factor. In *B. burgdorferi*, virulence determinants involved in basic cellular functions include PncA, GuaAB, and AdeC (Table 12.1). Other genes listed in Table 12.1 provide true

Table 12.1 Information regarding potential virulence determinants of *Borrelia burgdorferi* (partial listing). Genes are sorted according to their apparent major functions during the infectious cycle

Gene(s)	Location	Evidence
Host adaptation		
rpoN (*bb0450*), *rpoS* (*bb0771*)	Chromosome	Sigma factors that regulate temperature- and pH-dependent gene expression (Hübner *et al.*, 2001; Ouyang *et al.*, 2008); RpoS interacts with the *ospC* promoter (Eggers *et al.*, 2004; Yang *et al.*, 2005); *rpoN* or *rpoS* inactivation results in loss of mouse infectivity and mammal-arthropod transmission, but does not affect survival in ticks (Fisher *et al.*, 2005; Caimano *et al.*, 2007; Ouyang *et al.*, 2008)
rrp2 (*bb0673*)	Chromosome	Two component response regulator required for activation of the RpoN–RpoS pathway. Inactivation by point mutation abrogates mouse infection (Yang *et al.*, 2003; Boardman *et al.*, 2008; Ouyang *et al.*, 2008)
luxS (*bb0377*)	Chromosome	Potential involvement in quorum sensing (Stevenson and Babb, 2002; von Lackum *et al.*, 2006; Riley *et al.*, 2007); inactivation does not affect mouse infection by needle inoculation (Hübner *et al.*, 2003) or tick transmission (Blevins *et al.*, 2004)
ospC (*bbb19*)	cp26	Increased expression under mammalian tissue conditions (37°C, pH 7.6) (Schwan *et al.*, 1995; Schwan and Piesman, 2000); binds Salp15 (Ramamoorthi *et al.*, 2005) and plasminogen (Lagal *et al.*, 2006); inactivation results in loss of infectivity in mice, but conflicting data exist on the effect of inactivation on migration of *B. burgdorferi* in ticks (Grimm *et al.*, 2004b; Pal *et al.*, 2004b; Stewart *et al.*, 2006; Tilly *et al.*, 2006)
Survival and other basic cellular functions		
pncA (*bbe22*)	lp25	Nicotinamidase involved in NAD metabolism; inactivation results in loss of infectivity in mice (Purser *et al.*, 2003)
guaAB (*bbb17*, *bbb18*)	cp26	Involved in purine interconversion (Margolis *et al.*, 1994); transposon mutagenesis suggests *guaB* is required for mouse infection (Botkin *et al.*, 2006b)
bbb26, *bbb27*	cp26	Periplasmic membrane proteins of unknown function. Expression of either *bbb26* or *bbb27* is essential for *B. burgdorferi* survival *in vitro*. (Jewett *et al.*, 2007a)
resT (*bbb03*)	cp26	Telomere resolvase required for DNA replication and survival of *Borrelia* (Kobryn and Chaconas, 2002)
adeC (*bbk17*)	lp36	Encodes adenine deaminase, which converts adenine to hypoxanthine. Involved in purine interconversion; required for full infectivity in mice, but not in ticks (Jewett *et al.*, 2007b)
fliG-1 (*bb0221*)	Chromosome	Flagellar motor switch protein homologue; *fliG-1* transposon mutants remain motile, but are non-infectious in mice (Botkin *et al.*, 2006b)
recA (*bb131*)	Chromosome	DNA-binding protein involved in homologous recombination; not essential for mouse infection or *vlsE* sequence variation (Liveris *et al.*, 2008)
bmpAB, *bmpCD* (*bb0382*–*bb0385*)	Chromosome	Expressed during mammalian infection (Aron *et al.*, 1994; Simpson *et al.*, 1994; Ramamoorthy and Philipp, 1998; Gorbacheva *et al.*, 2000)

Gene(s)	Location	Evidence
Adherence		
dbpAB (*bba24*, *bba25*)	lp54	Binding to extracellular matrix (ECM) component decorin, GAGs (Guo *et al.*, 1998); lack of expression of either *dbpA* or *dbpB* results in reduced virulence (Shi *et al.*, 2006; Blevins *et al.*, 2008; Shi *et al.*, 2008a; Shi *et al.*, 2008b; Weening *et al.*, 2008)
bbk32	lp36	Binding to fibronectin and GAGs (Fischer *et al.*, 2006; Probert and Johnson, 1998); conflicting results exist on the effects of *bbk32* inactivation on pathogenesis in mice (Li *et al.*, 2006; Seshu *et al.*, 2006)
p66 (*bb0603*)	Chromosome	Porin protein, involved in binding of *B. burgdorferi* to integrins (Coburn *et al.*, 1999; Coburn and Cugini, 2003); inactivation in non-infectious *B. burgdorferi* reduces binding to integrins and eliminates specific channel forming activity (Coburn and Cugini, 2003; Pinne *et al.*, 2007)
Immune evasion		
vlsE (*bbf33*)	lp28-1	Antigenic variation system (Zhang *et al.*, 1997; Coutte *et al.*, 2009); VlsE protein expressed in high quantities in mouse ear and joint, but not heart, tissues (Crother *et al.*, 2003; Crother *et al.*, 2004); *vls* locus required for full infectivity in mice (Bankhead and Chaconas, 2007); complementation of lp28-1$^-$ strain with *vlsE* did not restore full infectivity, suggesting that antigenic variation is needed for immune evasion via VlsE expression (Lawrenz *et al.*, 2004)
crasp1 (*bba68*)	lp54	Complement regulator-acquiring surface protein-1, binds to complement inhibitory factors H and FHL-1; inactivation in a non-infectious strain decreases resistance to killing by normal human serum (Brooks *et al.*, 2005)
crasp2 (*bbH06*)	lp28-3	Complement regulator-acquiring surface protein-2, binds to complement inhibitory factors H and FHL-1 (Hartmann *et al.*, 2006); *crasp2* inactivation does not detectably reduce pathogenesis in mice (Coleman *et al.*, 2008)
ospE/*ospF*/*elp* (*erpA-Q*) (*17 paralogues*)	cp32s, lp28-1	Increased expression in mammalian tissue (Fikrig *et al.*, 1998; Wallich *et al.*, 1995); differential expression in joint, heart, and spleen but not skin (Fikrig *et al.*, 1999); binding to complement inhibitory factor H (Alitalo *et al.*, 2001; Hellwage *et al.*, 2001; Kraiczy *et al.*, 2001a; Stevenson *et al.*, 2002)
Tick infection		
ospAB (*bba15*, *bba16*)	lp54	Highly expressed in tick midgut; involved in adherence to midgut epithelium (Bergström *et al.*, 1989; Schwan and Piesman, 2000; Pal *et al.*, 2004a); required for wild-type levels of tick colonization, but not for mouse infection (Yang *et al.*, 2004; Neelakanta *et al.*, 2007)
bptA (*bbe16*)	lp28-1	*B*orrelia persistence in ticks A (*bptA*); function unknown; inactivation results in loss of persistence in *Ixodes* ticks, decreased infectivity in mice (Revel *et al.*, 2005)
chbC (*bbb04*)	cp26	Involved in transport of chitobiose, an arthropod cytoskeleton component. Not required for mouse or tick infection (Tilly *et al.*, 2004)
ospD (*bbj09*)	lp38	Lipoprotein expressed in ticks at the time of tick detachment and blood meal digestion. Not essential during tick phase or for mammalian infection (Li *et al.*, 2007; Stewart *et al.*, 2008)
bb0365	Chromosome	Unknown function; inactivation results in loss of persistence in *Ixodes* ticks, but has no effect on infectivity in mice (Pal *et al.*, 2008)

pathogenicity-specific functions, as described in greater detail below.

12.3 The infectious process

Fig. 12.3 provides a graphical representation of the basic steps of the infectious process in Lyme borreliosis. The diagram is simplified with regard to both the pathogenic processes of Lyme disease *Borrelia* and the host responses. During transmission and local infection of the dermis, the bacterium first undergoes changes in gene expression that permit it to adapt to the mammalian environment (see Chapter 4). Survival and replication in the dermis following transmission involves adaptation to the increased temperature, altered pH, and differences in nutrient composition, as well as 'fending off' of the host innate immune mechanisms (including the alternative complement pathway and phagocytosis). Binding to a variety of host tissue components involves activities such as adherence and the acquisition of plasmin, which may aid in the tissue invasion process. Motility and chemotaxis also are likely to be important for local dissemination, invasion of the circulatory system, extravasation at distant sites, and penetration into a wide variety of tissue types. Immune responses, both innate and adaptive, contribute to clearance of the organisms, but also are thought to be primarily responsible for tissue pathology and constitutional manifestations (e.g. low grade fever, fatigue, and malaise). Lyme disease *Borrelia* have established mechanisms (including immune evasion and invasion of protective niches) that permit survival for months to years in the mammalian host; indeed, experimentally infected mice appear to be infected for

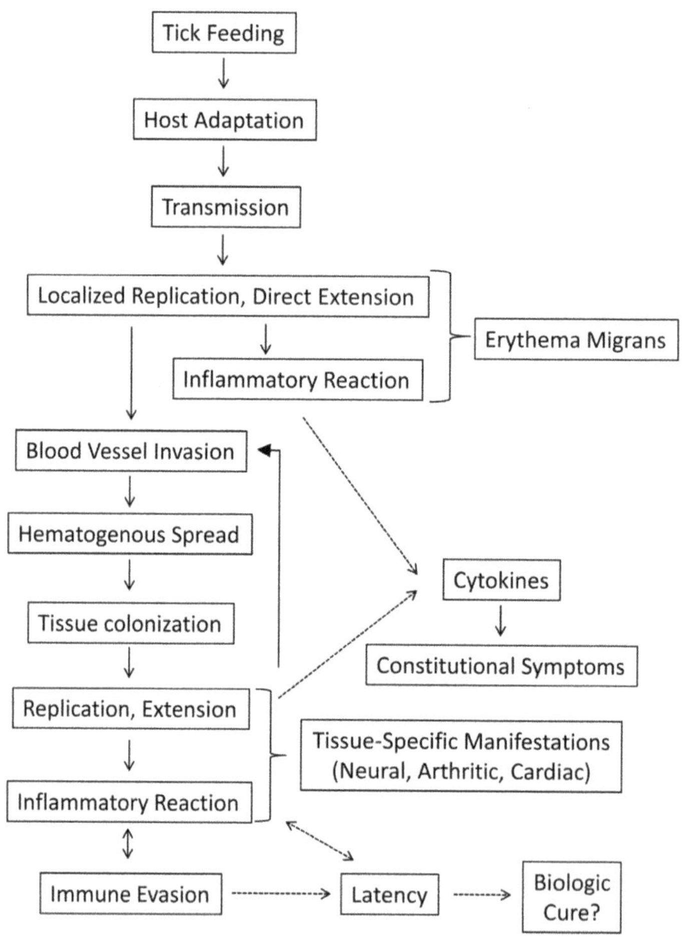

Figure 12.3 Simplified view of the process of Lyme disease *Borrelia* infection in mammalian hosts.

life. Whether biological cure (complete clearance of the organism without antimicrobial therapy) is possible is not clear; there is considerable serological evidence that many individuals in endemic areas have been infected previously but are asymptomatic despite the lack of treatment. These mechanisms are addressed in the remainder of this chapter, and some aspects (e.g. gene regulation, motility, and immune responses) are covered more thoroughly in other chapters (see Chapters 4, 7 and 15.

12.4 Adaptation to the mammalian host

When an *Ixodes* tick begins feeding on a mammalian host, there is a dramatic alteration in the tick midgut environment (reviewed by Schwan and Piesman, 2002; and Pal and Fikrig, 2003) (see Chapter 10). The blood meal and proximity to the mammal result in changes in temperature, pH, availability of nutrients, and exposure to antibodies, complement and other host protective factors. *Borrelia burgdorferi* replicates rapidly in the midgut during and following tick feeding at both the larval and nymphal stages (Piesman *et al.*, 1990). In infected nymphs, the number of *Borrelia* were found to increase from fewer than 500 prior to feeding to about 1.7×10^5 per tick after 72 hours of feeding, a greater than 300-fold increase (de Silva and Fikrig, 1995). The number of spirochaetes decreases about fivefold during the moult phase between larvae and nymph, and approximately tenfold during the nymph-to-adult moult phase, indicating that the unfed 'flat' tick environment is relatively unsupportive to *B. burgdorferi* growth and survival (Piesman *et al.*, 1990).

In addition to rapid multiplication, *B. burgdorferi* also undergoes dramatic changes in gene expression during tick feeding (Piesman *et al.*, 2003) (see Chapter 4). Similar alterations in mRNA levels are observed *in vitro* with changes in temperature and pH, as well as incubation in dialysis membrane chambers or in the presence of blood (Brooks *et al.*, 2003; Ojaimi *et al.*, 2003; Revel *et al.*, 2002; Tokarz *et al.*, 2004); therefore, these environments have been used as models to study gene regulation during the tick to mammal transition. 'Mammal-like' environments (high temperature, presence of blood, and lower pH) result in the upregulation of surface proteins such as OspC, DbpA, DbpB, and P66, and downregulation of tick-related gene products, including OspA and OspB. In addition, genes involved in chemotaxis and transport, and many hypothetical genes on lp54, are upregulated in the presence of blood and in dialysis membrane chambers. Many of these genes are regulated through the Rrp2–RpoN–RpoS sigma factor pathway (Boardman *et al.*, 2008; Caimano *et al.*, 2007; Ouyang *et al.*, 2008), as described in detail in Chapter 4.

Beginning about 36 hours after the initiation of feeding, spirochaetes have migrated to the tick salivary glands and are capable of transmission via saliva into the dermis of the mammalian host (Schwan and Piesman, 2002). The tick saliva protein Salp15 (Ramamoorthi *et al.*, 2005) may contribute to the initial survival of Lyme disease *Borrelia*, perhaps by binding to OspC, protecting against anti-OspC immunoglobulin binding, or interfering with $CD4^+$ T-cell activation (reviewed by Hovius *et al.*, 2007).

Although there are conflicting data regarding the role of OspC in promoting migration to the salivary gland (Grimm *et al.*, 2004b; Pal *et al.*, 2004b), OspC expression is clearly necessary for the initial colonization of mammalian tissue in the mouse model (reviewed by Radolf and Caimano, 2008). The mechanism of the OspC requirement is not as yet known. *ospC* mutants are unable to initiate infection in either immunocompetent or SCID mice (Stewart *et al.*, 2006; Tilly *et al.*, 2007; Tilly *et al.*, 2006). Expression is only required during the initial infection; Δ*ospC* organisms complemented with an *ospC* expressing plasmid consistently lose the complementing plasmid six weeks post infection (Tilly *et al.*, 2006). However, overexpression of other lipoproteins (OspA, OspB, VlsE, or DbpA) can permit survival in SCID mice, leading the authors to conclude that OspC provides a non-specific protective role against innate immune mechanisms (Xu *et al.*, 2008); another possibility is that OspC stabilizes the outer membrane under *in vivo* conditions. Constitutive expression of OspC results in elimination of *B. burgdorferi* in immunocompetent mice, but not SCID mice (Xu *et al.*, 2006). Evidence to date indicates that OspC is expressed transiently early in infection, but is downregulated by an unknown mechanism involving the production of anti-OspC antibodies, possibly being 'replaced' by VlsE as a stabilizing surface protein (Crother *et al.*, 2003; Tilly *et al.*, 2007).

Once colonization occurs, Lyme disease *Borrelia* multiply locally in the dermis, reaching high numbers. Liveris *et al.* (2002) utilized quantitative PCR to determine the number of *B. burgdorferi* in 2 mm biopsies from the leading edges of 50 erythema migrans lesions. Forty of the specimens (80%) were PCR positive; in these, the numbers of organisms detected varied from 10 to 11,000, with a median of 1450. Assuming a biopsy volume of ~5 μl, the median density would be >10^5 per gram tissue. Thus, the initial stages represent a fairly robust infection.

The availability of nutrients in mammalian tissue must differ greatly from those of the tick midgut and, based on the rapid growth rate, must represent a relatively rich environment. The metabolic capabilities of Lyme disease *Borrelia* are quite limited, so the spirochaetes must be highly dependent upon acquisition of fatty acids, carbohydrates, nucleic acid bases, and most amino acids from the host milieu (see Chapter 5). Several enzymes required for survival in mammalian tissue have been identified (Table 12.1). PncA, a nicotinamidase that converts nicotinamide (niacin) to nicotinic acid, is needed to provide a precursor for NAD biosynthesis (Purser *et al.*, 2003). AdeC, an adenine deaminase, converts adenine to hypoxanthine and thus facilitates purine interconversion (Jewett *et al.*, 2007b). Preliminary data from transposon mutagenesis indicates that GuaB, an inosine monophosphate dehydrogenase, is required for mouse infection and may promote the interconversion of inosine and guanine (Botkin *et al.*, 2006b). *Borrelia burgdorferi* lacking the plasmids encoding these genes grow normally *in vitro*, indicating that BSK medium provides compounds that are downstream of these activities. The genes encoding these enzymes are located on plasmids, reinforcing the essential roles of these replicons.

Unlike most pathogens, Lyme disease *Borrelia* produce no known toxins. Thus, the pathogenesis of this disease is due to the ability of spirochaetes to multiply, invade and persist in mammalian tissues, and to induce innate and adaptive inflammatory reactions, as described below and in Chapter 15.

12.5 Dissemination – direct extension and haematogenous spread

Motility plays a major role in the spread of Lyme disease *Borrelia* from the initial site of infection (Charon and Goldstein, 2002) (see Chapter 7), although definitive evidence for this association has been lacking until recently. Remarkable videomicrography studies by Moriarty *et al.* (2008) and Norman *et al.* (2008) have demonstrated directly the movement of *B. burgdorferi* within infected tissue, and the interaction between *Borrelia* organisms and the vascular endothelium. In these experiments, infectious clones derived from *B. burgdorferi* B31 were transformed with a plasmid expressing high levels of green fluorescent protein (GFP). C3H/HeN or C57Bl mice were inoculated with these strains, and organisms were directly observed in living mouse ear and flank skin using a technique called intravital microscopy (IVM). Briefly, mice were anaesthetized and regions of the ear or flank skin were mounted on a microscopic stage and visualized by either spinning disk confocal microscopy or by conventional epifluorescence. Counterstaining with an Alexa Fluor 555-labelled anti-PECAM-1 monoclonal antibody permitted visualization of endothelial cells and hence the vasculature. Fluorescent spirochaetes could be easily found in mouse ear tissue up to 28 days post infection. Organisms were located in the extracellular matrix, predominantly in the perivascular regions. They were observed to undergo the same translational, flexing, and reversal motions that are typically displayed by spirochaetes *in vitro*. Moreover, translational speeds of up to 4 μm/s were witnessed; although 'runs' as long as 40 μm were observed, the bacteria commonly exhibited 'reciprocal translation', i.e. a back and forth motion resulting from frequent reversal of the direction of flagellar rotation and thus spirochaete movement. These observations document the vital role of motility in the direct extension of spirochaetes from a site of infection, as occurs during the erythema migrans phase of pathogenesis.

These studies further revealed the interaction of *B. burgdorferi* with the host vasculature. Prior studies using static cultures of human umbilical vein endothelial cells (HUVECs) showed that Lyme disease *Borrelia* bind to endothelial cells and can transmigrate across HUVEC monolayers. In these reports, *B. burgdorferi* was found to migrate either through intercellular junctions (between neighbouring endothelial cells), or by an intracellular route (through endothelial cells) (Comstock and Thomas, 1989; Comstock and Thomas, 1991; Ma, 1991; Szczepanski *et al.*, 1990; Thomas and Comstock, 1989). Moriarty *et al.* (2008) were able to directly observe *Borrelia*–endothelial interactions *in vivo* by injecting GFP-expressing *B. burgdorferi* into mice in sufficient numbers to yield a concentration of ~10^7/ml of blood, permitting frequent detection of bacterial adherence to blood vessels by IVM. Infectious *B. burgdorferi* clones were found to frequently associate with the endothelia of capillaries and post-capillary venules, but were not observed to adhere to arteriole endothelia; this finding is most likely related to the slower blood flow rate and lower shear force found in the capillary beds and the post-capillary circulatory system. Both short-term and stationary interactions of *Borrelia* with the endothelial cell luminal surface were observed, with the short-term interactions being further subdivided into transient (<1 s to traverse 100 μm of blood vessel) and dragging (1–20 s/100 μm) interactions. The spirochaetes were frequently observed to migrate up and down the vessel surface, indicating an ability to swim 'upstream' against the blood flow. Stationary interactions occurred at endothelial cell junctions (as visualized by anti-PECAM-1 staining) 83% of the time. In these cases, the organisms partially traversed the vessel wall in most instances (79%), whereas transient interactions were nearly always superficial (93%). 'Escape' of *Borrelia* across the blood vessel wall (extravasation) was also observed, and required on average 10.8 min. Further studies by Norman *et al.* (2008) demonstrated that preincubation of the spirochaetes with fibronectin, dalteparin (a therapeutic heparin compound), or dextran sulphate dramatically reduced both transient and stationary interactions with blood vessel endothelia. Whereas non-infectious *B. burgdorferi* lacking several plasmids did not bind to endothelia to a significant extent, transformation with a plasmid expressing the fibronectin- and glycosoaminoglycan (GAG)-binding protein BBK32 (described below) increased transient interactions to near wild-type levels, and greatly expanded the proportion of stationary interactions (Norman *et al.*, 2008). These observations provide credence to the importance of BBK32 and fibronectin- and GAG-mediated adherence in the interactions of *B. burgdorferi* with vascular endothelial cells and hence in dissemination. In addition, these studies suggest that the different types of bacterial interactions with the vasculature are likely to be mediated by different *B. burgdorferi* molecules, and that the dragging interactions may be prerequisites to the stationary interactions. These and other adherence-related properties of Lyme disease *Borrelia* are addressed in depth in the subsequent sections.

12.6 Interaction of the Lyme disease spirochaete with host cells

Adhesion to host cells and extracellular matrix components by bacterial pathogens is thought to be an important first step in the establishment of infection (Finlay and Falkow, 1997), and cell adhesion by the Lyme disease spirochaete was one of the first (putative) pathogenic phenotypes investigated. Several studies showed that the spirochaete binds to a variety of cultured cells of both mammalian and tick origin (Comstock and Thomas, 1989, 1991; García-Moncó *et al.*, 1989, 1992; Hechemy *et al.*, 1989, 1992; Kurtti *et al.*, 1988, 1993; Ma, 1991; Szczepanski and Benach, 1991; Szczepanski *et al.*, 1990; Thomas *et al.*, 1994; Thomas and Comstock, 1989); for recent reviews, see Coburn *et al.* (2005) and Steere *et al.* (2004). Over the years, much effort has gone into identifying molecular host targets for spirochaete adhesion and colonization, as well as bacterial adhesins, i.e. surface proteins that mediate attachment to host cells or extracellular matrix. It is now clear that the spirochaete expresses a plethora of adherence determinants. This complexity presumably reflects the central role that adhesion plays in the pathogenesis of infection but also results in (at least partial) functional redundancy that greatly complicates the analysis of any single pathway. In this context, considering the molecular properties usually associated with *Borrelia* adhesins is useful.

The adhesin, which often is a protein or a carbohydrate structure, should be exposed on the bacterial cell surface and be anchored to or have the ability to associate with the outer membrane. Proteinase K digestion of intact bacteria leading to the removal or a reduction of a putative adhesin protein is often taken as a proof of surface exposure of the adhesin.

In studying a possible adhesin-ligand pair, determining the affinity of the interaction is important. K_D values below 10 μM suggest a very high affinity interaction. Most adhesin–ligand interactions have reported K_D values of 10 μM to 10 nM. Since bacterial attachment to a substrate molecule probably involves multiple pairs of adhesin–ligand interactions, a relatively high K_D (10–50 μM) between individual adhesin-ligand pairs can still result in physiologically relevant adherence. However, the relevance of these low affinity interactions should be demonstrated with *in vitro* experiments.

The specificity of the adhesin ligand interaction is also of importance. For example, most predicted *Borrelia* lipoproteins have basic isoelectric points and contain clusters of positively charged residues. These positively charged proteins will bind to strong polyanions such as heparin. However, heparin is normally not exposed in an animal and is therefore not a likely target for bacterial adherence and colonization. One needs to demonstrate that a glycosaminoglycan (GAG) binding adhesin interacts with a relevant GAG target under physiological conditions. Recent technological advancements make it possible to now study the biological relevance of a putative adhesin–ligand interaction. Spirochaete mutant strains can be generated with gain or loss of a particular adhesin gene. These strains can be examined in both *in vitro* models of bacterial attachment and *in vivo* models of Lyme disease. Furthermore, cell lines can be generated where particular ligand genes are suppressed or knocked out and genetically manipulated mice are being used to examine specific host components in the disease process. These studies are in their infancy, but these types of experiments clearly will be critical for us to understand the molecular pathogenesis of Lyme disease.

12.6.1 Adhesion in the tick: a tissue-specific function of OspA

The first *B. burgdorferi* antigen identified was outer surface protein A (OspA). Although its abundance on the surface of cultured spirochaetes made it a natural candidate adhesin, and preliminary studies suggested that OspA may have a role in *B. burgdorferi* interaction with mammalian cells (Comstock *et al.*, 1993), further analysis did not reveal a significant role for the molecule in binding to mammalian cells (Grab *et al.*, 1999) (J. Leong and E. Fikrig, unpublished). A critical observation that shed light on the possible role of OspA was that its expression is generally restricted to spirochaetes in the tick midgut (and during *in vitro* culture), and that it is not typically expressed in the mammalian host (Schwan and Piesman, 2000; Schwan *et al.*, 1995). The OspA vaccine against *B. burgdorferi* resulted in antibodies that eliminated the bacteria from ticks, suggesting that the function of OspA is critical while the bacteria are in the tick host (de Silva *et al.*, 1996). In fact, OspA binds to extracts of tick gut better than to extracts of tick salivary gland, and mediates spirochaetal attachment to midgut epithelial cells by recognizing the tick protein TROSPA (Pal *et al.*, 2000; Pal *et al.*, 2004a). The importance of this function in the tick was genetically confirmed by demonstration that a *B. burgdorferi* mutant incapable of producing OspA was competent to establish infection in mice and was taken up in the tick blood meal at levels similar to those of the wild type, but was not able to colonize the tick midgut (Yang *et al.*, 2004). The role of TROSPA as the receptor for OspA was also demonstrated genetically, as RNAi knockdown of TROSPA expression decreased (but did not abrogate) *B. burgdorferi* attachment (Pal *et al.*, 2004a).

12.6.2 Adhesion in the mammal: promiscuous binding by the Lyme disease agent

OspA expression by *B. burgdorferi* decreases when the tick starts acquiring its blood meal, an event that may release the bacterium from the midgut and allow efficient transmission into the host. Indeed, *Leishmania donovani*, another arthropod-borne pathogen, utilizes its ability to adhere to midgut epithelial cells in the sandfly to establish a stable infection prior to transmission, and then diminishes this adhesive activity upon differentiation to facilitate transmission to the mammalian host upon sandfly feeding (Sacks, 1992). In addition to diminished expression of

OspA, the expression of outer surface protein C (OspC) increases, suggesting that the bacterium may undergo a developmental shift as it prepares to enter the mammalian host. Consistent with this, binding of the spirochaete to cultured mammalian cells increases after growth in dialysis membrane chambers implanted in the rat peritoneum (Parveen *et al.*, 2003). While only a single pathway, mediated by host cell glycosaminoglycans, has been demonstrated to increase upon host adaptation, current knowledge of expression patterns of *B. burgdorferi* adhesins suggests that several pathways may be induced during mammalian infection (Fikrig *et al.*, 2000; Hagman *et al.*, 2000; Suk *et al.*, 1995).

12.6.3 P66 and BBB07: integrin-binding proteins

Given the diversity of mammalian cells (or components) recognized by *B. burgdorferi*, it is perhaps not surprising that the spirochaete encodes multiple adhesins. One of these, the *B. burgdorferi* protein P66, was originally identified as an antigen to which human Lyme disease patient sera frequently react in immunoblots (Dressler *et al.*, 1993; Ntchobo *et al.*, 2001). Cloning of the *p66* gene (*bb0603*) was achieved by two independent groups (Bunikis *et al.*, 1995; Probert *et al.*, 1995), but the sequence neither revealed any structural motifs aside from the secretion signal nor any similarities to previously identified proteins. In fact, the *p66* gene appears to be unique to the genus *Borrelia*. The protein is at least partially surface-exposed, as determined by proteolysis and antibody reactivity (Bunikis *et al.*, 1995; Bunikis *et al.*, 1996; Probert *et al.*, 1995). In addition, P66 appears to serve as a channel-forming protein in the *B. burgdorferi* outer membrane (Pinne *et al.*, 2007; Skare *et al.*, 1997). Although the P66 primary amino acid sequence lacks similarity to those of β barrel porins of Gram-negative organisms, several potential β-pleated sheet regions are present based on secondary structure predictions.

P66 was later identified as a ligand for the β_3 chain integrins using phage display by selecting for binding to integrin $\alpha_{IIb}\beta_3$ *in vitro* (Coburn *et al.*, 1999). *Borrelia burgdorferi* has been shown to bind to the β_3 chain integrins $\alpha_{IIb}\beta_3$ and $\alpha_v\beta_3$, and to the β_1 chain integrins $\alpha_3\beta_1$ and $\alpha_5\beta_1$ (Behera *et al.*, 2008; Coburn *et al.*, 1999). P66 does not appear to bind to $\alpha_5\beta_1$ with high affinity. It is certainly possible that P66 binds to additional integrins, such as $\alpha_v\beta_1$ and $\alpha_v\beta_5$, but this has not yet been demonstrated experimentally. The phage display approach applied *in vivo* also identified P66 as an adhesin for endothelial cells (Antonara *et al.*, 2007), and the fragment identified *in vivo* overlaps with the fragment identified in the *in vitro* selection for binding to $\alpha_{IIb}\beta_3$. These results support the hypothesis that P66 interaction with integrins can occur *in vivo*.

The potential consequences of P66 interaction *in vivo* go beyond simple adhesion, because integrins are transmembrane signalling molecules that modulate how mammalian cells interact with and respond to their environments. Therefore, P66 may influence host cell biology to promote the survival and dissemination of *B. burgdorferi*. The *in vitro* or *in vivo* significance of this possibility has not yet been tested. However, P66 does cause upregulation of particular inflammatory mediators in primary human chondrocytes (Behera *et al.*, 2008). The receptor(s) that mediate(s) this response is/are unknown. A role for P66 in mammalian infection is further supported by the observations that P66 is expressed during mammalian infection (Dressler *et al.*, 1993; Ntchobo *et al.*, 2001) and during bacterial acquisition or transmission by *Ixodes scapularis* ticks, but not in the unfed tick (Cugini *et al.*, 2003). Although the gene encoding P66 (*bb0603*) has been inactivated in a non-infectious strain (Coburn and Cugini, 2003), testing of the role of P66 *in vivo* awaits analysis of a set of isogenic strains generated in an infectious strain background that do or do not express P66.

P66 is not the only integrin-binding protein expressed by these bacteria. The cp26-encoded protein BBB07 binds to a similar profile of integrins, but signals through integrin $\alpha_3\beta_1$ to induce production of several inflammatory mediators in primary human chondrocytes (Behera *et al.*, 2008). Since sera from *B. burgdorferi*-infected, but not control, mice recognize recombinant BBB07, this protein probably may also influence *B. burgdorferi* interactions with its host through binding to integrins. Interestingly, BBB07 has full-length homology (up to 49% identity/68% similarity) with a large family of predicted proteins encoded by diverse Gram-positive organisms, including

Clostridium and *Bacillus* species. The function(s) and structural locations of this protein superfamily (DUF871; http://www.ncbi.nlm.nih.gov) are unknown; however, given their association with Gram-positive organisms, these proteins would not be expected to have typical outer membrane protein properties. Therefore, the structural location of BBB07 should be analysed to determine whether the protein is surface-exposed in intact *Borrelia* and could thus interact with host cells. Clearly, further investigation of P66 and BBB07 is needed to delineate the roles of these proteins during mammalian infection, to understand their interactions with host receptors at the molecular level, and to identify the consequences of the interactions of these proteins with integrins *in vivo*.

12.6.4 DbpA and DbpB: related lipoproteins that bind the proteoglycan decorin and glycosaminoglycans

The association of *B. burgdorferi* with collagen fibres (Barthold *et al.*, 1993) led investigators to search for spirochaetal proteins that promote binding to collagen or to collagen-associated molecules. In fact, the spirochaete binds to decorin, a collagen-binding proteoglycan that is expressed as a component of the connective tissue in multiple tissues (Guo *et al.*, 1995). Decorin consists of a 36-kDa core protein usually substituted with one GAG chain of the chondroitin dermatan sulphate type. GAGs are long, unbranched, disaccharide repeats that are extremely heterogeneous due to variation in the extent of sulphation and glucuronic acid L-5 epimerization. Dermatan sulphate, heparin, heparan sulphate, and the chondroitin sulphates are different classes of GAGs distinguished by their particular disaccharide repeats and/or differences in their degree of modification. GAGs and proteoglycans are ubiquitous and essential components of mammalian tissues; they have been co-opted as adhesion targets for a large number of bacterial pathogens (reviewed in Rostand and Esko, 1997).

DbpA and DbpB are surface lipoproteins of 20 kDa and 18 kDa. They are encoded by an operon on lp54 and were identified as decorin-binding proteins in gel overlay assays (Guo *et al.*, 1998; Guo *et al.*, 1995; Hagman *et al.*, 1998). Although 56% similar in sequence, DbpA is distinguished from DbpB by its (1) ability to bind decorin more efficiently, (2) capacity to inhibit attachment of *B. burgdorferi* to immobilized decorin, (3) ability to elicit a protective immune response against challenge with *in vitro*-cultivated *B. burgdorferi*, and (4) high degree of polymorphism, with sequence similarities as low as 58% between alleles (Feng *et al.*, 1998; Guo *et al.*, 1998; Hagman *et al.*, 1998; Hanson *et al.*, 1998; Roberts *et al.*, 1998). The decorin-binding region of DbpA has been localized to a 15-residue sequence containing a critical lysine (K82) and DbpB has a related sequence that may also be involved in decorin binding (Pikas *et al.*, 2003).

Neither the isolated core protein nor GAG of decorin are recognized by *B. burgdorferi*, suggesting that both components are required for optimal bacterial binding to decorin (Guo *et al.*, 1995). Nevertheless, *B. burgdorferi* recognizes GAGs on the surface of cultured cells, in extracellular matrix (ECM), and in purified form; fragments of heparin (a highly sulphated analogue of heparan sulphate) as short as 16 monosaccharide residues can interfere with GAG-binding by *B. burgdorferi* (Leong *et al.*, 1998a). Among a diverse collection of Lyme disease spirochaete strains, all of the infectious isolates expressed GAG-binding activity (Isaacs, 1994; Leong *et al.*, 1998b). Due to the high negative charge of GAGs, a potential difficulty in the study of bacterial–GAG interactions is differentiating specific from non-specific binding. Regardless, *B. burgdorferi* strains demonstrate specificity for subsets of GAGs (e.g. binding to dermatan sulphate but not chondroitin sulphate) and the particular GAG-binding specificity varies with the strain (Leong *et al.*, 1998b). In addition, the particular class of GAGs that mediate spirochaetal attachment varies with mammalian cell type (Leong *et al.*, 1998b), such that *B. burgdorferi* strains with different GAG-binding specificities display corresponding differences in the spectrum of cell types recognized (Parveen *et al.*, 1999). The simplest hypothesis to explain these results is that *B. burgdorferi* expresses multiple GAG-binding adhesins, and that the particular repertoire of adhesins expressed by a particular strain is responsible for its particular GAG binding specificity.

To better understand the function of DbpA and DbpA in their native outer membrane location, a shuttle vector harbouring *dbpA* and *dbpB* was introduced into an otherwise nonadherent strain. *Borrelia burgdorferi* B314 is a high-passage strain that has lost numerous plasmids (Sadziene *et al.*, 1995), including lp54, which encodes *dbpA* and *dbpB* (Hagman *et al.*, 1998), and binds cultured mammalian cells very poorly; it provided a 'null' strain background with which to assess the function of exogenously expressed adhesins. Ectopic expression of *dbpA* and *dbpB* conferred onto strain B314 the ability to efficiently bind to purified GAGs and cultured epithelial cells (Fischer *et al.*, 2003). Furthermore, in spite of their sequence relatedness, only DbpB promoted attachment to chondroitin-6-sulphate and to cultured glial cells. Thus, these studies definitively demonstrated that DbpA and DbpB function as adhesins, revealed GAG-binding activities in addition to their known decorin-binding activities, and confirmed the previous hypothesis that *B. burgdorferi* expresses multiple GAG-binding adhesins with varying binding specificities.

DbpA and DbpB are apparently not expressed in the tick vector and appear to be more highly expressed during infection than during *in vitro* culture, suggesting an important function during infection (Hagman *et al.*, 2000). The spirochaetes spread somewhat less efficiently in decorin-deficient mice than in wild-type controls and survive less well in the joints and skin, two tissues that express relatively high levels of decorin, indicating that decorin binding may promote dissemination and/or survival in the host (Brown *et al.*, 2001; Liang *et al.*, 2004a). The incomplete nature of the defects in tissue colonization in the decorin-deficient mouse might reflect other potential mammalian substrates, such as GAGs, recognized by DbpA and/or DbpB. Other studies involving either *B. burgdorferi* strains that are deficient in DbpA or DbpB or that overexpress these proteins also suggest participation in colonization of particular tissues, although definitive conclusions are difficult to derive from these studies, in part due to lack of complementation of mutants (Shi *et al.*, 2006; Xu *et al.*, 2007). More rigorous recent studies from three different laboratories have demonstrated that DbpA and DbpB facilitate *B. burgdorferi* infection, in terms of dissemination to different tissues (kinetics and localization) and resistance to clearance by the immune system (Blevins *et al.*, 2008; Shi *et al.*, 2008a; Shi *et al.*, 2008b; Weening *et al.*, 2008). These studies employed independently generated mutant strains and complemented derivatives, two different mouse strains, and two distinct parental *B. burgdorferi* strain backgrounds. Nevertheless, all three groups showed that the mutants were similarly attenuated in the mouse infection model, requiring approximately 10^4-fold more bacteria to infect 50% of the mice by inoculation with laboratory-grown bacteria. Mutant bacteria that had been acquired from infected mice by larval ticks were no different from the wild-type bacteria when transmitted to mice by nymphal ticks, a striking contrast to the defects seen with the bacteria grown in culture. While there are many possible explanations for this result, the conclusion of all these studies is that, while *dbpAB* is not essential to the ability of *B. burgdorferi* to cause infection in laboratory mice, the encoded proteins do play roles that require further investigation.

12.6.5 BBK32, a fibronectin- and GAG-binding lipoprotein

Fibronectin (Fn) is a complex glycoprotein, present both in plasma and in extracellular matrix, which contains distinct domains that interact with heparin, collagen, gelatin, integrins, and other plasma and tissue components. Benach and coworkers showed that adhesion of *B. burgdorferi* to the ECM was inhibited by antibody directed against Fn (Szczepanski *et al.*, 1990). While it is possible that *B. burgdorferi* encodes more than one Fn-binding protein (Grab *et al.*, 1998; Kopp *et al.*, 1995), the best-characterized Fn-binding protein is the 47-kDa surface lipoprotein BBK32 encoded on lp36 and identified by gel overlay assays and Fn affinity chromatography (Probert and Johnson, 1998). Ectopic expression of BBK32 was sufficient to promote binding of the otherwise non-adhesive strain B314, which lacks lp36, to purified fibronectin and to mammalian cells. The N-terminal domain of BBK32, which contains the minimal Fn binding sequence (Kim *et al.*, 2004; Probert *et al.*, 2001), is largely unstructured and undergoes a significant conformational shift upon Fn-binding, similar to Fn-binding proteins of *Staphylococcus* and *Streptococcus*.

Expression of BBK32 in the non-adherent strain B314 also revealed that, in addition to conferring binding to Fn, BBK32 mediates bacterial binding to GAGs. Purified recombinant BBK32 was capable of binding to heparin, indicating that BBK32 promotes spirochaetal attachment to GAGs through direct binding (Fischer *et al.*,

2006). Heparin was incapable of inhibiting the BBK32–Fn interaction, suggesting that the two binding activities are discrete. Thus, like DbpA and DbpB, BBK32 promotes bacterial attachment to GAGs as well as to at least one more substrate (Fn in the case of BBK32 and decorin in the case of DbpA and DbpB).

Consistent with having a role during infection of the mammalian host, BBK32 is expressed in ticks after initiation of the blood meal and in the mammalian host (Fikrig *et al.*, 2000; Suk *et al.*, 1995), where it elicits a humoral immune response (Fikrig *et al.*, 1997). In addition, expression is RpoS-dependent (Caimano *et al.*, 2007; He *et al.*, 2007; Ouyang *et al.*, 2008) (see Chapter 4). Furthermore, as mentioned above, intravital microscopic studies revealed that BBK32 promotes both transient and stable interactions with the vascular endothelium (Norman *et al.*, 2008). Skare and coworkers reported that the ID_{50} of a *B. burgdorferi* strain defective for the expression of BBK32 is 22-fold higher than that for wild type B31 (Seshu *et al.*, 2006), whereas Fikrig and coworkers, using an independently derived *B. burgdorferi bbk32* mutant, found no role for BBK32 during experimental infection by tick inoculation or needle inoculation with a single (relatively large) dose (Li *et al.*, 2006). Further studies will be required to fully characterize the role of BBK32 in mammals.

12.6.6 Bgp: a candidate glycosaminoglycan binding adhesin with enzymatic activity

Before the identification of DbpA, DbpB, and BBK32 as GAG-binding adhesins, Bgp (*Borrelia* GAG-binding protein) was purified from bacterial lysates based on its heparin-sensitive haemagglutination activity, which correlated with GAG binding by intact spirochaetes (Leong *et al.*, 1995; Parveen and Leong, 2000). Bgp is a secreted protein that is found in the culture media as well as on the surface of the spirochaete (Cluss *et al.*, 2004; Parveen and Leong, 2000). In addition to its GAG-binding properties, Bgp is a 5′-methylthioadenosine/*S*-adenosyl homocysteine nucleosidase (Pfs), which in other bacteria is important in elimination of toxic metabolic products, in nucleotide salvage, and in the production of the quorum-sensing molecule, autoinducer 2 (Winzer and Williams, 2001). Recombinant Bgp partially inhibits *B. burgdorferi* attachment to purified GAGs and to cultured mammalian cells, supporting it as a candidate adhesin, but the non-adherent *B. burgdorferi* strain B314 expresses Bgp, indicating that Bgp expression is not sufficient to promote bacterial binding. A *B. burgdorferi bgp* mutant retains the ability to infect immunodeficient mice (Parveen *et al.*, 2006); its role during infection of immunocompetent mice remains to be evaluated.

12.6.7 Collagen: an 'orphan' host ligand with an unknown bacterial adhesin

Because *B. burgdorferi* has a predilection for connective tissue in mammalian hosts, several laboratories have searched for the ECM components and the corresponding bacterial molecules that might be involved in this interaction. Although previous reports suggested that *B. burgdorferi* does not bind to the collagen, a more recent report demonstrates that collagen binding depends on how the collagen is prepared (Zambrano *et al.*, 2004). Typically, since the extracellular matrix is difficult to disrupt, preparations of collagen are generated by treatment with pepsin, which partially disrupts the fibrillar matrix. However, extracts of rat tendon made with acid instead of pepsin supported *B. burgdorferi* attachment and invasion of the matrix, while controls such as gelatin (hydrolysed collagen) did not (Zambrano *et al.*, 2004). Interaction of the bacteria with the collagen matrix was not affected by removal of GAGs or decorin, supporting a direct *B. burgdorferi*–collagen interaction. While the *B. burgdorferi* molecule(s) responsible have not yet been identified, the observed reduction in attachment after digestion of the intact bacteria with proteinase K is consistent with a role for one or more proteins in this interaction (Zambrano *et al.*, 2004).

12.6.8 'Orphan' candidate *Borrelia burgdorferi* adhesins with unknown host receptors

While several *B. burgdorferi* adhesins and their mammalian receptors have been well characterized *in vitro*, what is occurring *in vivo* has been much more difficult to determine. An *in vivo* filamentous phage selection was therefore used to identify several *B. burgdorferi* proteins as candidate adhesins responsible for interactions with the endothelial cells in the vasculature of living mice (Antonara *et al.*, 2007).

In the *in vivo* phage display selection for candidate adhesins, OspC was the most frequently selected protein. As mentioned above, upon tick feeding, spirochaetes in the midgut decrease expression of OspA and increase expression of OspC (Schwan *et al.*, 1995). While experimental evidence for a role of OspC in colonization of the tick salivary glands is somewhat controversial (Pal *et al.*, 2004b; Tilly *et al.*, 2006), OspC clearly is essential for establishment of *B. burgdorferi* infection in mice (Grimm *et al.*, 2004b; Tilly *et al.*, 2006). Its function, however, remains unknown. The essential nature of OspC is possibly related to a potential role as an adhesin and/or invasin (Lagal *et al.*, 2003; Radolf and Caimano, 2008).

Several other interesting candidates were identified. The OspF family of proteins was frequently selected, suggesting a role for these proteins as adhesins. While the OspE family of proteins has been shown to bind factor H, a mammalian protein that inhibits complement-mediated lysis, the OspF subfamily has not. The *in vivo* relevance of factor H binding is yet to be definitively determined: although factor H-deficient mice are productively infected by *B. burgdorferi* (Woodman *et al.*, 2007), complement killing in these mice may be ineffective due to secondary defects, as the majority (>80%) of the $cfh^{-/-}$ mice have C3 levels that are below the detection limit (Pickering *et al.*, 2002).

Another protein of interest in the pathogenesis of *B. burgdorferi* infection that was selected *in vivo* was Vls-related (Zhang and Norris, 1998a; Zhang *et al.*, 1997; Zhang and Norris, 1998b). Since the fragments of proteins expressed in filamentous phage display libraries preclude the isolation of intact proteins (or the corresponding genes), whether the selected clones derived from the expression site, *vlsE*, or from the silent cassettes is not clear, but VlsE could serve a function in adhesion (Antonara *et al.*, 2007). The mammalian receptors for OspC, OspF proteins, and VlsE have not yet been identified, if indeed they exist, but P66, the integrin ligand, was also selected using this approach. Interestingly, the Dbp proteins, Bgp, and BBK32 were not selected, although the phage particles bearing these adhesins may have been removed from the pool available to bind the vessel wall due to binding of serum proteins or other molecules in solution, or the receptors to which they bind may not be available in the lumen of the vasculature.

12.6.9 How do we make sense of the multiple pathways?

Borrelia burgdorferi certainly encodes many adherence pathways. One might ask the questions: do all of the putative adhesins actually function as adhesins, and if so, why would *B. burgdorferi* need so many? This apparent complexity may reflect a central importance for binding to the host cell surface or extracellular matrix components in the survival of the bacterium in either the mammal or the tick. Like other pathogenic bacteria, *B. burgdorferi* may express several distinct adhesins that play different roles in infection and the development of disease. Adhesins may mediate attachment to host substrates in different tissues or might trigger responses in the host cell. In addition, some binding pathways may be, at least partially, redundant. For example, several pathways involve the recognition of ECM components, such as Fn, collagen, or GAGs. These components consist of repeating subunits of either protein or carbohydrate origin and the corresponding *B. burgdorferi* adhesins might be recognizing these diverse repeat structures with limited specificity. However, even interactions of relatively low specificity may provide an important function by either promoting (relatively non-specific) bacterial binding or facilitating the formation of highly specific interactions mediated by other bacterial and host molecules. Experimentally, the plethora of attachment pathways raises the possibility that inactivation of a single adhesin will have little effect on infectivity in the mouse model or that the defect will be apparent only under particular circumstances or in particular animals, which has been observed already in several instances (Blevins *et al.*, 2008; Li *et al.*, 2006; Parveen *et al.*, 2006; Seshu *et al.*, 2006; Shi *et al.*, 2006, 2008a,b; Weening *et al.*, 2008). A difficult challenge, particularly in an experimental system with limited genetic tools, will be to systematically characterize the adherence pathways of *B. burgdorferi*, with respect to their biochemical functions, their effects on host cell biology, and their roles during infection of mammals and ticks.

12.7 Role of host proteases in dissemination and inflammation

Bacteria that enter through the skin and subsequently cause disseminated disease must be able to break down tissue barriers in order to spread. In contrast to bacteria such as *Staphylococcus aureus* and *Streptococcus pyogenes* that produce multiple exported proteases, a *B. burgdorferi* virulence-related protease capable of digesting extracellular matrix proteins has not been identified. While there has been one report of a high-molecular-weight collagenase found in borrelial lysates (Grab *et al.*, 1996), this finding was not replicated by other laboratories (Coleman *et al.*, 1995; Klempner *et al.*, 1995). In lieu of producing its own proteases, *B. burgdorferi* may utilize host proteases to break tissue barriers. *Borrelia burgdorferi* has been shown to bind and stabilize plasmin, an important protease in the host fibrinolytic cascade (Coleman *et al.*, 1995; Fuchs *et al.*, 1994; Klempner *et al.*, 1995). Plasmin is a broad-spectrum, serine protease that, in addition to degrading fibrin, is capable of degrading extracellular matrix components such as elastin and laminin. Plasmin is also an important activator of other extracellular-matrix-degrading proteases such as matrix metalloproteinases (MMPs). Plasmin is produced by activation of its circulating zymogen form, plasminogen, through the action of activating proteases such as urokinase-type plasminogen activator (u-PA) and tissue-type plasminogen activator (t-PA). The presence of *B. burgdorferi* induces production of u-PA and its receptor (CD87) from circulating monocytes (Coleman *et al.*, 2001; Fuchs *et al.*, 1996; Haile *et al.*, 2006). CD87 can stabilize u-PA against degradation and inactivation by circulating inhibitors (e.g. plasminogen activator inhibitor-2, PAI-2) allowing it to efficiently convert plasminogen to plasmin. Urokinase may also bind directly to *B. burgdorferi*, which also protects u-PA against inactivation (Klempner *et al.*, 1995). Several putative borrelial plasmin(ogen)-binding proteins have been proposed, but the contribution and functional relevance of each remains to be determined (Fuchs *et al.*, 1994; Hu *et al.*, 1997; Lagal *et al.*, 2006; Rossmann *et al.*, 2007). The binding of plasmin to *B. burgdorferi* has been shown to increase the ability of the organism to pass through cell culture monolayers and to avoid inhibition by circulating inhibitors such as α_2-antiplasmin (Coleman *et al.*, 1995; Perides *et al.*, 1996). In studies of plasminogen-deficient mice, plasmin was required for efficient dissemination of *B. burgdorferi* within the tick (i.e. from the midgut to the salivary glands) (Coleman *et al.*, 1997). The presence of plasmin also enhanced spirochaetaemia following feeding of infected ticks on uninfected mice.

The observation that plasmin-binding did not enhance the spread of spirochaetes to distant sites suggests that other proteases may be involved in the process of extracellular matrix degradation and dissemination of *B. burgdorferi.* MMPs are the major host enzymes in degradation of extracellular matrix proteins. They are involved in normal host processes such as tissue remodelling and wound healing and, as a family, can degrade every single component of the extracellular matrix. Infection with *B. burgdorferi* has been shown to induce production of numerous MMPs (including MMP-1, MMP-3, MMP-9, MMP-13, and MMP-19) from different host cells (keratinocytes, fibroblasts, astrocytes, PBMCs, and chondrocytes) in culture (Behera *et al.*, 2005; Gebbia *et al.*, 2001; Lin *et al.*, 2001; Perides *et al.*, 1999; Zhao *et al.*, 2003). MMP-9 is upregulated in erythema migrans skin lesions from infected patients (Zhao *et al.*, 2003). *In vitro* studies showed that MMP-9 can be activated by plasmin bound to *B. burgdorferi* and that MMP-9 activity results in increased penetration through artificial barriers, including artificial blood–brain barrier models (Gebbia *et al.*, 2001; Grab *et al.*, 2005). MMP-9 was also found to increase in the cerebrospinal fluid of patients with neuroborreliosis (Kirchner *et al.*, 2000). *In vivo* studies using MMP-9 knockout mice have shown that MMP-9 plays an important role in penetration of *Streptococcus pneumoniae* through the blood–brain barrier (Meli *et al.*, 2003). However, studies of MMP-9 knockout mice in Lyme disease models have not shown any difference in dissemination of either *B. burgdorferi* (joints and heart) or *B. hermsii* (brain) to distant sites (L.T. Hu, unpublished). Although, because the MMPs show considerable overlap in their substrate specificities and because *B. burgdorferi* induces multiple MMPs from host tissues, the deficiency of any single MMP may be compensated for by other induced enzymes.

While *B. burgdorferi* may co-opt host proteases to assist with dissemination, the induction and activation of MMPs resulting from the presence of *B. burgdorferi* is also responsible for some of the pathology associated with Lyme disease. In particular, MMPs have been closely associated with bone and cartilage degradation in other forms of arthritis (e.g. rheumatoid arthritis). The collagenases MMP-1 and MMP-13 as well as stromelysins (MMP-3 and MMP-19) were found to be elevated in the synovial fluids of patients with Lyme arthritis and/or induced from cultured chondrocytes by the presence of *B. burgdorferi* (Hu *et al.*, 2001; Lin *et al.*, 2001). Studies with monkey cartilage explants have shown that MMPs and a related metalloproteinase from the ADAMTS family (a disintegrin-like and metalloprotease domain with thrombospondin type I motifs), aggrecanase-1/ADAMTS-4, play a key role in the degradation of articular cartilage after *B. burgdorferi* infection (Behera *et al.*, 2006b; Hu *et al.*, 2001). The specific MMPs induced by *B. burgdorferi* differ significantly between animals as well as between tissues and may be responsible for some of the differences that are seen in tissue-specific pathology. For example, collagenases (MMP-1, MMP-8 and MMP-13) are increased in human Lyme arthritis, but are not induced in the joints of C3H/HeN, Balb/c or C57BL/6 mice infected with *B. burgdorferi* (Behera *et al.*, 2005; Crandall *et al.*, 2006). Irreparable cartilage damage occurs only with the degradation of interstitial collagens by collagenases and not with degradation of other cartilage components (e.g. aggrecans and hyaluronans) that is mediated through MMPs belonging to the non-collagenase families. This difference may explain the lack of permanent joint erosions in mouse models of Lyme arthritis compared with that seen in prolonged human disease.

12.8 Induction of cytokines and other mediators of inflammation

Because of the paucity of expressed proteases, toxins and other tissue damaging products by *B. burgdorferi*, the majority of the pathology seen secondary to *B. burgdorferi* infection is thought to be due to the effects of the host response to the organism. *Borrelia burgdorferi* is recognized by pattern recognition receptors on cells participating in the initial innate immune response. Because of the high number of lipoproteins expressed by *B. burgdorferi*, most attention has been focused on toll-like receptor 2 (TLR2), which heterodimerizes with TLR1 to recognize triacyl-lipoproteins. Recognition of borrelial lipoproteins by TLR2 activates many inflammatory mediators including, but not limited to, cytokines and chemokines such as TNF-α, IL-1, IL-6, monocyte chemoattractant-1, CXCL-1 (Hirschfeld *et al.*, 1999). It also induces production of some, but not all MMPs (Gebbia *et al.*, 2004). While it was initially felt that activation of TLR2 occurred through ligation of lipoproteins on the surface of live *B. burgdorferi* with membrane-bound TLR2, several lines of evidence have now suggested that recognition may occur primarily after processing of the organism in phagolysosomes (Behera *et al.*, 2006c; Moore *et al.*, 2007). Despite the important role of TLRs in inflammatory signalling, studies in knockout mice deficient in TLR2, CD14 and MyD88 (a shared adapter molecule) have all shown increased rather than decreased inflammation after infection with *B. burgdorferi* (Behera *et al.*, 2006a; Benhnia *et al.*, 2005; Bolz *et al.*, 2004; Liu *et al.*, 2004; Wooten *et al.*, 2002). Some of the increase may be due to the much larger bacterial burden in these knockout mice as the TLR system appears to be critical in the early control of infection (Bolz *et al.*, 2004; Liu *et al.*, 2004; Wooten *et al.*, 2002). However, this finding clearly suggests that other receptors recognizing different *B. burgdorferi* products can play a role in induction of inflammatory mediators. Among the proposed receptors are integrin receptors, which have been shown to bind *B. burgdorferi* and can activate many of the same signalling pathways as TLRs (Behera *et al.*, 2006c). While the likelihood that *B. burgdorferi* is recognized by multiple receptors complicates studies implicating specific receptors as critical for the inflammatory response to the organism, downstream signalling pathways that can be activated by multiple receptors have clearly been shown to be important in mediating the inflammatory response. Among the signalling pathway components that have been linked to the inflammatory response to *B. burgdorferi* using animal models are: mitogen-activated protein kinase p38 (Anguita *et al.*, 2002a), cyclooxygenase-2 (Anguita *et al.*, 2002b), protein kinase C (Shin *et al.*, 2007), and STAT-1 (Brown *et al.*, 2006). Other inflammatory signalling pathway components that have been shown to play a role *in vitro*, but have not yet been confirmed to play a role *in vivo*, including c-Jun N-terminal kinase (JNK), STAT-3, and extracellular signal-regulated kinase (ERK) (Shin *et al.*, 2008).

12.9 Immune evasion mechanisms

Most pathogenic microorganisms have developed mechanisms for evading the innate and adaptive immune responses, permitting their survival and proliferation in a hostile host environment (van der Woude and Bäumler, 2004). Immune evasion mechanisms are particularly important in pathogens that cause persistent infections that last months, years, and even decades. These infectious agents include Lyme disease and relapsing fever *Borrelia*, *T. pallidum* (syphilis and related illnesses), *M. tuberculosis*, protozoa such as *Plasmodium* and *Babesia spp.* (malaria and related diseases), *Leishmania spp.* (leishmaniasis), and *Trypanosoma* (sleeping sickness and Chagas' disease), the Herpesvirus family (herpes virus I and II, varicella zoster virus, Epstein–Barr virus, and others), and human immunodeficiency virus. Mechanisms of immune evasion are myriad but can be grouped into antigenic variation, inhibition of immune responses, antigen modulation, sequestration, or masking, and protective niches, including intracellularity. Each of these will be considered in the context of Lyme disease *Borrelia*.

12.9.1 Antigenic variation (VlsE)

In 1997, Zhang *et al.* described a 10-kb locus in lp28-1 of *B. burgdorferi* B31 that includes a gene expressing a 35-kDa surface-exposed lipoprotein (*vlsE*) and a series of 15 silent cassettes (*vls2–16*) with a high degree of sequence similarity (76.9–91.4 % at the amino acid level) to the central cassette region of the expression site (Fig. 12.4) (Zhang *et al.*, 1997). This locus was named Vmp-like sequence (*vls*) due to sequence homology with the variable large proteins of the antigenic variation system of *Borrelia hermsii* and other relapsing fever spirochaetes. However, unlike the relapsing fever system (Dai *et al.*, 2006), *vlsE* undergoes random, segmental gene conversion in which portions of any of the silent cassettes can be substituted for the pre-existing sequence within the cassette region of the expression site

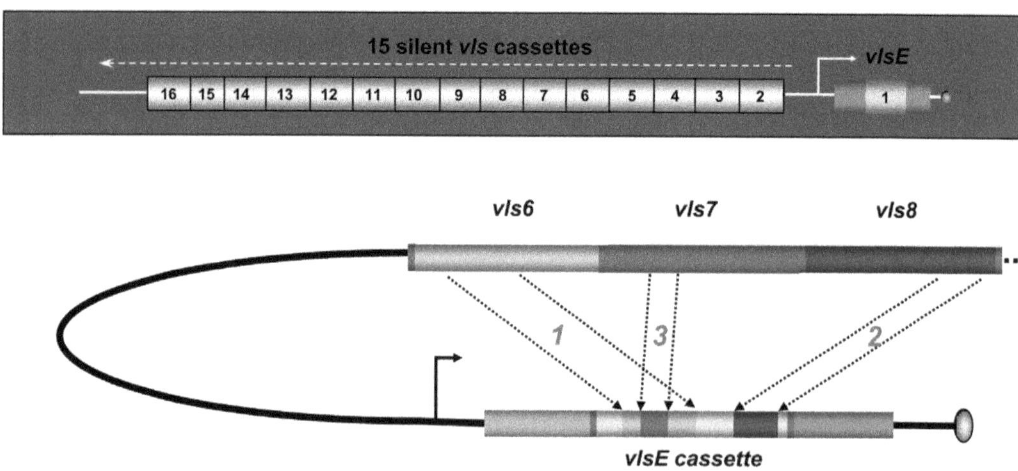

Figure 12.4 The *vls* antigenic variation system of *B. burgdorferi* B31. The top portion of the figure shows the arrangement of the *vls* silent cassettes 1 through 15 and the *vlsE* expression site in plasmid lp28-1. A hypothetical example of 3 sequential gene conversion events involving silent cassettes 6 (event 1), 8 (event 2), and 7 (event 3) is shown. From Norris (2006) with permission.

(Fig. 12.4) (Zhang and Norris, 1998a; Zhang and Norris, 1998b). Although point mutations may also occur (Sung *et al.*, 2001), most sequence changes can be attributed directly to substitution of silent cassette sequences of varying lengths. These recombination events can be detected as early as four days post infection in mice, and by 28 days each clone reisolated from mice exhibits a different sequence containing an estimated 8–13 separate recombination events (Zhang and Norris, 1998b). Sequence variation also occurs in SCID mice; however, the elimination of clones expressing the parental *vlsE* takes longer, presumably because of the lack of immune clearance of invariant clones (Zhang *et al.*, 1998; Coutte *et al.*, 2009). The gene conversion events have not been detected during *in vitro* culture (Zhang and Norris, 1998b) or in infected ticks (Indest *et al.*, 2001), indicating that as yet unidentified environmental signals in mammals regulate the recombination process. Clones lacking lp28-1 survive in the absence of immune pressure (in SCID mice and dialysis membrane chambers) (Labandeira-Rey *et al.*, 2003; Purser *et al.*, 2003), consistent with a role in protection against adaptive immunity. As expected, transformation with lp28-1 does render lp28-1$^-$ strains fully infectious (Grimm *et al.*, 2004a).

Recent data by Bankhead and Chaconas (2007) have demonstrated that the *vls* antigenic variation system is the major element of lp28-1 that is required for mammalian infection. In these studies, lp28-1 was truncated through allelic exchange of a region containing a duplex lp17 telomere sequence into loci either upstream or downstream of the origin of replication. Cleavage of the telomere sequence by the ResT system resulted in the creation of modified lp28-1 replicons lacking either the *vls* locus or genes *bbf01* to *bbf19* from the left end of the plasmid. The Δ*vls* clones exhibited decreased infectivity in C3H/HeN mice similar to that seen with a lp28-1 negative clone, with no positive cultures obtained at 21 or 28 days post infection. In contrast, a clone lacking the open reading frames *bbf01*–*bbf19* were fully infectious in C3H/HeN mice by needle inoculation. The Δ*vlsE* clones were recovered from every tissue site tested 28 days after infection of SCID mice; this result paralleled the infectivity of lp28-1 minus clones in SCID mice observed previously (Labandeira-Rey *et al.*, 2003; Lawrenz *et al.*, 2004; Purser *et al.*, 2003). Shuttle vectors with *vlsE* do not exhibit recombination even if lp28-1 is present (Lawrenz *et al.*, 2004). In addition, complementation with *vlsE* alone does not restore infectivity to clones lacking lp28-1 (Lawrenz *et al.*, 2004) or containing lp28-1::Δ*vls* (Bankhead and Chaconas, 2007). RecA expression is not required for *vlsE* recombination during mouse infection (Liveris *et al.*, 2008), indicating that the process is not dependent upon 'standard' homologous recombination. However, recent studies have shown that the RuvAB Holliday junction helicase/resolvase is required for efficient *vlsE* recombination (Dresser *et al.*, 2009; Lin *et al.*, 2009). Up-regulated expression of an as-yet unidentified specific mediator of *vlsE* recombination during mammalian infection is probably responsible for the occurrence of the variation; downregulation of an inhibitor is a less likely mechanism, since *vlsE* sequence variation is not observed in heterologous system such as *E. coli* (S. J. Norris and J. K. Howell, unpublished). Overall, active recombination of *vlsE* appears to be required for evasion of host adaptive immune responses, most likely by changing variable region epitopes that are targets of antibody reactivity.

No function other than immune evasion has been ascribed to VlsE. VlsE, like the variable surface glycoproteins (VSGs) of *Trypanosoma cruzi* and other pathogenic protozoa, may coat the surface of the organism and thereby prevent the interaction of antibodies with hidden, non-variant antigens. Other possibilities include adhesion to host components, acquisition of nutrients, or inhibition of host clearance mechanisms (like the factor H-binding activity of CRASPs; see below). *Trans*-complementation with an invariant copy of *vlsE* on a shuttle plasmid does not restore full infectivity (Bankhead and Chaconas, 2007; Lawrenz *et al.*, 2004), indicating that sequence variation of VlsE is intimately associated with its function.

The *vls* system has been detected in all strains and genospecies of Lyme disease *Borrelia* examined. The sequence of the complete *vls* system (*vlsE* and the silent cassettes) has been determined only for strain B31 of *B. burgdorferi* (Fraser *et al.*, 1997; Zhang *et al.*, 1997). However, partial or complete sequences of *vlsE* have been obtained for several US *Borrelia burgdorferi* strains (Iyer *et al.*, 2000; H. Kawabata and H. Watanabe, GenBank entry AB041949; M. Caimano and J. D. Radolf, Genbank entry AY052626) and European *B. garinii*, *B. afzelii*, and *B. burgdorferi* strains (Göettner *et al.*, GenBank entries AJ630106–AJ630112 and AJ850088–AJ850089). In addition, the *vls* silent cassette loci of Lyme disease *Borrelia* have been characterized in several publications (Kawabata *et al.*, 1998; Wang *et al.*, 2003; Wang *et al.*, 2001) and in multiple draught genomic sequences. *vlsE* is typically not detected in genomic shotgun sequences, most likely because of its proximity to the telomere and/or a large inverted repeat (Bykowski *et al.*, 2006; Hudson *et al.*, 2001; Zhang *et al.*, 1997).

The structure of VlsE reveals an amazing coordination between sequence variation and structure (Eicken *et al.*, 2002) (Fig. 12.4). Alignment of the cassette regions within *vlsE* and the *vls* silent cassettes shows that most of the sequence differences occur in six variable regions (VRs) interspersed among six invariant regions (IRs) (Fig. 12.4). In the structure, these variable regions correspond precisely with random coil segments that cover nearly the entire membrane distal surface of the protein. This region of the protein is presumably in direct contact with the external milieu, which during mammalian infection consists of body fluids containing immunoglobulins. The lateral surfaces, composed almost entirely of α helices, may be protected from exposure to antibodies by interaction with neighbouring molecules of either VlsE or other proteins (Eicken *et al.*, 2002).

Evidence is accumulating that expression of *vlsE* is upregulated during mammalian infection. Hodzic *et al.* (2003) showed by real time reverse transcription PCR (RT-PCR) that *vlsE* is expressed in multiple tissues, but the transcript levels were not compared with *in vitro* levels. In a novel approach utilizing Triton X-114 extraction to recover *B. burgdorferi* lipoproteins from infected tissue, Crother *et al.* (2004; 2003) found that VlsE was one of the most prominent surface antigens expressed by *B. burgdorferi* during infection of SCID mice or rabbit skin. RT-PCR studies by Liang *et al.* (2004b) showed that *B. burgdorferi* changes its outer surface protein transcript levels in response to host immune responses. In particular, *vlsE* mRNA levels were increased in heart and skin when SCID mice were injected with anti-*B. burgdorferi* antiserum or anti-OspC antiserum. Interestingly, Anguita *et al.* (2001) found that *vlsE* expression levels were decreased in IFN-γ-deficient mice, and also provided data indicating that *vlsE* sequence variation is decreased in these animals compared to wild-type mice. They speculate that the decreased inflammatory response in the IFN-γ-deficient mice affects *B. burgdorferi* gene expression by an as yet unknown mechanism. In a comparison of *B. burgdorferi* gene expression in the salivary glands and midguts of *I. scapularis* ticks during feeding, Piesman *et al.* (2003) noted an increase in *vlsE* transcript levels after 72 to 96 hours of feeding but not at earlier timepoints.

At present, the mechanisms of transcriptional regulation and control of sequence variation in *vlsE* are not clear, although some information is available regarding the effects of environmental conditions. Hudson *et al.* (2001) mapped the transcriptional start site of *vlsE* and also determined that incubation of *B. burgdorferi* with human endothelial cells or isolated membranes preferentially increased the expression of *vlsE* mRNA. Furthermore, they were able to sequence a large, 51-bp inverted repeat between the *vlsE* expression site and the silent cassettes, which also encompasses the –35 region of the *vlsE* promoter. More recent studies have shown that this inverted repeat is somewhat larger (54 bp) and, at least in some B31 clones, has perfect complementarity throughout its length (Bykowski *et al.*, 2006) (J. K. Howell and S. J. Norris, unpublished). This inverted repeat and its predicted stem loop structure may be involved in the regulation of *vlsE* expression and recombination, although Bykowski *et al.* (2006) found that removal of one side of the inverted repeat did not affect expression of a reporter construct. Whole genome array analyses of *B. burgdorferi* B31 gene expression have yielded disparate results (Ojaimi *et al.*, 2002; Ojaimi *et al.*, 2003; Revel *et al.*, 2002; Tokarz *et al.*, 2004), but, in one study, *vlsE* transcript levels were found to increase slightly (1.78-fold) at 35°C as compared with 23°C (Ojaimi *et al.*, 2002). Addition of human blood during a temperature shift from 23°C to 35°C (to mimic conditions during a blood meal) resulted in a 3.3-fold decrease in cDNA hybridizing to *vls* silent cassette sequences, which presumably represents *vlsE* transcripts (Tokarz *et al.*, 2004). Seshu *et al.* (2004) found that removal of dissolved oxygen using Oxyrase increased VlsE, OspC and DbpA expression at the protein and mRNA levels, whereas expression of OspA and NapA (an oxygen-sensitive regulator) were reduced under these conditions. Using a GFP reporter construct and electrophoretic mobility shift assays (EMSAs), Bykowski *et al.* (2006) confirmed that the canonical –35 and –10 sequences found upstream of *vlsE* are required for expression, and showed that both expression levels and complexes detected by EMSA are increased at pH 8.0 and higher temperatures. Thus, regulation of *vlsE* may be due in part to proteins that bind the *vlsE* promoter region.

VlsE is highly antigenic despite its apparent role as an antigenic variation protein. Indeed, nearly all Lyme disease patients beyond the erythema migrans stage express anti-VlsE antibodies that can be readily detected by immunoblot or ELISA techniques (Bacon *et al.*, 2003; Lawrenz *et al.*, 1999; Schulte-Spechtel *et al.*, 2003). Liang and colleagues performed detailed analyses of the invariant regions within the VlsE central cassette and found that IR6, a 23-amino acid segment forming an α-helix, is highly reactive with sera from humans and animals infected with Lyme disease *Borrelia* (Liang *et al.*, 1999; Liang and Philipp, 1999, 2000). Diagnostic tests utilizing either full-length VlsE or IR6 (called the C6 peptide) have been developed, and have been shown to have sensitivity and specificity similar to the existing 'two tier' testing system involving ELISA and immunoblot analyses with whole cell *Borrelia* antigens (Bacon *et al.*, 2003; Schulte-Spechtel *et al.*, 2003).

Although antibodies against IR6 and the C-terminal invariant region can be detected consistently in infected humans and animals, antisera against these regions do not kill *B. burgdorferi in vitro* and do not have demonstrable reactivity with the surface of intact *B. burgdorferi* organisms (Liang *et al.*, 1999). The three-dimensional structure of VlsE demonstrates that IR6 is covered by the variable regions in the membrane distal portion, while the N- and C-terminal regions form α helices within the membrane proximal portion (Eicken *et al.*, 2002). Thus, these antigenic regions are probably inaccessible to antibody in intact *Borrelia*. In contrast, the variable region loops form a 'cap' on the membrane distal surface of VlsE and thus would be readily accessible to antibodies. McDowell *et al.* (2002) performed elaborate cross-absorption studies demonstrating variant VlsE proteins contain both shared and unique epitopes. Absorption of antiserum from *B. burgdorferi*-infected mice with the same VlsE variant expressed by the infecting strain removed all reactive antibodies, whereas absorption with a different VlsE variant left a subset of antibodies that reacted specifically with the variant present in the infecting strain. Because the only sequence differences between the two variant proteins are in the Variable Regions 1–6, these 'variant-specific' antibodies are most likely directed to these six variable regions. This study provides additional evidence that VlsE is a true antigenic variation protein involved in immune evasion.

12.9.2 Complement cascade inhibition (CRASPs)

Another mechanism of Lyme disease (and relapsing fever) *Borrelia* immune evasion involves complement resistance. It has been known for some time that strains and genospecies of Lyme disease *Borrelia* vary in their susceptibility to killing by human and other mammalian sera, and the susceptibility of this killing activity to heat inactivation indicated that the complement system was responsible (reviewed by Kurtenbach *et al.*, 2002). Studies of this phenomenon eventually led to the identification of *Borrelia* proteins that bind factor H, factor H-like protein 1 (FHL-1), and/or FH-related protein 1 (FHR-1) (Kraiczy *et al.*, 2001a; Kraiczy *et al.*, 2001b). These *Borrelia* proteins are collectively called complement regulator-acquiring surface proteins or CRASPs. A recent study by Woodman *et al.* (2007) showed that infection with *B. burgdorferi* progresses in a very similar manner in Factor H-deficient mice as in wild-type C57BL/6 mice, leading the authors to conclude that Factor H may not be essential for *B. burgdorferi* infection and that other mechanisms of complement resistance may exist. However, factor H-deficient mice have extremely low levels of C3 due to the high rate of spontaneous C3 proteolysis (Pickering *et al.*, 2002), presenting a confounding variable in these studies. Therefore, a discussion of CRASPs and their potential role in *Borrelia* pathogenesis is warranted.

CRASPs are members of hypothetical surface protein families and bind Factor H-related proteins, as determined by far-Western immunoblotting. Factor H, FHL-1, and FHR-1 are splice variants of the same gene that bind to the key complement component C3 and facilitate the binding and activation of factor I, which in turn cleaves C3 into an inactive form. This activity is important in the regulation of the complement cascade, helping to prevent 'runaway' complement activation and hence tissue damage. In the CRASP model, *Borrelia* organisms usurp this host system by binding proteins in the Factor H-related group via CRASPs, thereby inhibiting the accumulation of active C3b on the bacterial cell surface. The spirochaetes would thus have less C3b associated with their surface, and would have decreased opsonization (reducing phagocytosis) and less formation of the complement membrane attack complex (decreasing complement-dependent lysis).

Most data accumulated to date support the importance of CRASP-1 in inhibiting the complement cascade. Disruption of *bba68* (encoding CRASP-1) in the complement-resistant *B. burgdorferi* strain B31cF resulted in sensitivity to killing by human complement, and complementation with the shuttle vector pKFSS-1::CRASP-1 restored resistance (Brooks *et al.*, 2005). Similarly, transformation of the serum-sensitive *B. garinii* strain Bg50 with pKFSS-1::CRASP-1 rendered the strain resistant to killing by human complement; in addition, immunofluorescence of intact organisms indicated that CRASP-1 is surface exposed (Brooks *et al.*, 2005). McDowell *et al.* (2006) provided RT-PCR and Western blot analysis data indicating that CRASP-1 is not expressed during experimental infection of mice with different North American *B. burgdorferi* strains, and that human Lyme disease patients do not produce antibodies against CRASP-1; this information brought into question whether CRASP-1 is present during mammalian infection. However, Rossman *et al.* (2006) found that antibodies from Lyme disease patients react specifically with non-denatured, recombinant CRASP-1 in a line blot assay, but not with the denatured protein in Western blots. von Lackum *et al.* (2005) also provided immunofluorescence evidence that CRASP-1 is expressed following transmission of *B. burgdorferi* from ticks to mice. These results indicate that CRASP-1 (or members of its paralogous family) is expressed during mammalian infection and induces the production of antibodies in humans against conformational CRASP-1 epitope(s). The effects of CRASP-1 inactivation on infectivity have not been reported, so further analysis is needed to more firmly establish the role of this protein in pathogenesis.

CRASP-2 is not required for experimental infection of mice and most likely is not involved in the binding of Factor H and related proteins, according to a recent report by Coleman *et al.* (2008). In studies conducted with the infectious clone *B. burgdorferi* B31-A3, they determined that the CRASP-2 gene (*bbh06*) is expressed in all tissues examined during mouse infection, but not in infected ticks. In addition, there is a robust antibody response against CRASP-2 in human Lyme disease patients and in infected mice. However, proteinase K digestion of intact organisms indicates that CRASP-2 is not surface exposed (contrary to a prior report (Hartmann *et al.*, 2006)), and vaccination with recombinant CRASP-2 does not protect mice against *B. burgdorferi* infection. Deletion of *bbh06* did not affect mouse infectivity or pathogenesis, and did not reduce resistance to complement-mediated killing *in vitro*. Unlike the other CRASPs, CRASP-2 has no closely related paralogues in *B. burgdorferi* that could potentially fulfil its function. These results indicate that CRASP-2 is not essential for infectivity or protection against complement activity, despite the fact that it binds Factor H and FHL-1 in *in vitro* far-Western assays. Consistent with its apparent lack of surface exposure, BBH06 probably fulfils another function in *Borrelia* cells.

OspE, OspF and Elp proteins (also known as Erps) represent paralogous lipoprotein families encoded by ~16 mono- and bi-cistronic operons on the cp32 circular plasmids (Brissette *et al.*, 2008; Stevenson *et al.*, 2001). Three OspE proteins, encoded by *erpP*, *erpC*, and *erpA*, have been shown to bind Factor H and FHR-1 in far-Western assays and are thereby called BbCRASP-3, BbCRASP-4, and BbCRASP-5, respectively (Kraiczy *et al.*, 2003). The OspE,

OspF, and Elp paralogues examined appear to be lipidated and surface-exposed (El-Hage *et al.*, 2001). Analysis of this gene family is complicated by the high degree of redundancy; for example, *erpA* and *erpN* are identical copies of the same gene on two different replicons, cp32-1 and cp32-8. Representative OspE, OspF, and Elp family proteins have been shown to be expressed at high levels in acutely infected mice and macaque monkeys, in nymphal ticks during feeding, and in larval ticks during and after feeding (Miller *et al.*, 2005; Miller *et al.*, 2003). The levels of expression were found to differ during chronic infection of mammals (Miller and Stevenson, 2006), consistent with *ospF* promoters being RpoS dependent while *ospE* and *elp* promoters are σ^{70}-dependent (Eggers *et al.*, 2004; Eggers *et al.*, 2006) (see Chapter 4). The *B. burgdorferi* B31 OspE paralogues differ in their ability to bind to Factor H from different mammalian species (Kurtenbach *et al.*, 2002; Stevenson *et al.*, 2002), leading to speculation that expression of multiple OspE proteins may provide complement resistance in a broad range of mammals. To date, however, there are no reports on the relationship of *ospE*, *ospF*, or *elp* gene inactivation and infectivity.

12.9.3 Protective niches and other mechanisms of persistence

Borrelia burgdorferi is able to survive for at least one year in multiple tissues of experimentally infected mice (Barthold *et al.*, 1993; Zhang and Norris, 1998b). Because of their motility and invasive nature, Lyme disease *Borrelia* can penetrate deep into connective tissue, including the dense irregular connective tissue of the dermis and dense regular connective tissue of tendons (Barthold *et al.*, 2006; Cadavid *et al.*, 2003). *Borrelia burgdorferi* was recently shown to bind to type I collagen (Zambrano *et al.*, 2004), which may play a role in tissue tropism, the ability to penetrate connective tissue, and hence the long-term survival of organisms. Cabello *et al.* (2007) recently summarized this information and concluded that Lyme disease *Borrelia* are able to 'hide in plain sight', i.e. in protective niches in mammalian tissue. DbpA and DbpB have also been implicated in long-term colonization of collagenous tissue (Liang *et al.*, 2004a). *T. pallidum* is likewise capable of surviving in host tissue for years. Re-inoculation of the skin of so-called chancre immune rabbits (infected with *T. pallidum* three months previously) showed that most spirochaetes were rapidly destroyed, but were still present in dense collagen bundles, hair follicles, arrector pili muscles, and small nerves in the dermis (Sell and Salman, 1992; Sell *et al.*, 1985). Lyme disease *Borrelia* may similarly become inaccessible to antibody by penetrating protective niches. This mechanism may also protect *Borrelia* against elimination by antimicrobial agents (Bockenstedt *et al.*, 2002; Hodzic *et al.*, 2008).

As described previously, *B. burgdorferi* appears to modulate gene expression during the course of mammalian infection, including downregulation of OspC expression following the initial stages of infection (Liang *et al.*, 2004b) and upregulation of VlsE (Crother *et al.*, 2004; Crother *et al.*, 2003; Liang *et al.*, 2004b). Liang *et al.* (2004b) found that *ospC* mRNA levels were 446-fold lower in immunocompetent mice than in SCID mice, whereas message levels of *vlsE* and another lp28-1 encoded gene, *bbf01*, are 32- and 20-fold higher, respectively, in immunocompetent mice. These results indicate that the adaptive immune response in some way affects gene expression in *B. burgdorferi*. Anguita *et al.* (2001) found that a derivative of *B. burgdorferi* N40 that had been passaged *in vitro* 75 times (N40–75) induced lower levels of IFN-γ and IL-12 relative to low-passage N40. In addition, N40–75 was more readily eliminated by passive transfer of anti-*B. burgdorferi* antiserum. Administration of IFN-γ or IL-12 actually enhanced the level of infection. *Borrelia burgdorferi* B31 also exhibited less *vlsE* recombination in IFN-γ receptor-deficient mice four days post infection (Anguita *et al.*, 2001), which the authors suggest may indicate an immune response-related activation of *vlsE* recombination. Thus far, no mechanism of *B. burgdorferi* gene regulation related to immune responses has been delineated. However, *Escherichia coli* O157:H7 and other pathogenic bacteria sense catecholamines through a two-component regulator and exhibit enhanced growth in the presence of noradrenaline (Clarke *et al.*, 2006). *Borrelia burgdorferi* may 'sense' host components and thereby respond by changes in gene expression.

12.10 Summary

It is striking that nearly all of *B. burgdorferi* replication occurs during or immediately after the blood meal in the tick, or during mammalian infection. The number of *Borrelia* decreases five- to tenfold after the blood meal is depleted and moulting from larva to nymph or nymph to adult has occurred. In a sense, the replication cycle of Lyme disease *Borrelia* occurs primarily in the vertebrate host and during digestion of the blood meal; *B. burgdorferi* is 'maintained' the unfed tick stage, but does not increase in numbers. This scenario is not unlike the vegetative cell/spore cycle of some Gram-positive bacteria, in which the sole purpose of the spore phase is long-term survival and dispersal. In this light, one possibility is that Lyme disease *Borrelia* first evolved from a free-living organism to colonize and become pathogenic in vertebrate hosts, and later developed the ability to survive in and be transmitted by arthropod hosts. In comparison, *Leptospira* appear to have retained the ability to survive and multiply in an aqueous environment after becoming host-adapted, whereas *Treponema* species are only found associated with host species and are incapable of living in the environment. We can only speculate on whether each of these ventures into host colonization and pathogenesis occurred from a common, pathogenic spirochaete or by three independent pathways. Given the genetic and pathogenic differences in these organisms, the most likely scenario is that each evolved separately as vertebrate colonizers and pathogens. Conversely, the closely related Lyme borreliosis and relapsing fever organisms most likely evolved from a common, vertebrate-infecting *Borrelia* ancestor. Perhaps further analysis of the expanding genomic information will also reveal whether this common ancestor was also tick-transmitted prior to bifurcation into hard tick-transmitted and soft tick-transmitted genotypes.

References

Alitalo, A., Meri, T., Ramo, L., Jokiranta, T.S., Heikkila, T., Seppala, I.J., Oksi, J., Viljanen, M., and Meri, S. (2001). Complement evasion by *Borrelia burgdorferi*: serum-resistant strains promote C3b inactivation. Infect. Immun. *69*, 3685–3691.

Anguita, J., Barthold, S.W., Persinski, R., Hedrick, M.N., Huy, C.A., Davis, R.J., Flavell, R.A., and Fikrig, E. (2002a). Murine Lyme arthritis development mediated by p38 mitogen-activated protein kinase activity. J. Immunol. *168*, 6352–6357.

Anguita, J., Samanta, S., Ananthanarayanan, S.K., Revilla, B., Geba, G.P., Barthold, S.W., and Fikrig, E. (2002b). Cyclooxygenase 2 activity modulates the severity of murine Lyme arthritis. FEMS Immunol. Med. Microbiol. *34*, 187–191.

Anguita, J., Thomas, V., Samanta, S., Persinski, R., Hernanz, C., Barthold, S.W., and Fikrig, E. (2001). *Borrelia burgdorferi*-induced inflammation facilitates spirochete adaptation and variable major protein-like sequence locus recombination. J. Immunol. *167*, 3383–3390.

Antonara, S., Chafel, R.M., Lafrance, M., and Coburn, J. (2007). *Borrelia burgdorferi* adhesins identified using *in vivo* phage display. Mol. Microbiol. *66*, 262–276.

Aron, L., Alekshun, M., Perlee, L., Schwartz, I., Godfrey, H.P., and Cabello, F.C. (1994). Cloning and DNA sequence analysis of *bmpC*, a gene encoding a potential membrane lipoprotein of *Borrelia burgdorferi*. FEMS Microbiol. Lett. *123*, 75–82.

Bacon, R.M., Biggerstaff, B.J., Schriefer, M.E., Gilmore, R.D., Jr., Philipp, M.T., Steere, A.C., Wormser, G.P., Marques, A.R., and Johnson, B.J. (2003). Serodiagnosis of Lyme disease by kinetic enzyme-linked immunosorbent assay using recombinant VlsE1 or peptide antigens of *Borrelia burgdorferi* compared with 2-tiered testing using whole-cell lysates. J. Infect. Dis. *187*, 1187–1199.

Bankhead, T., and Chaconas, G. (2007). The role of VlsE antigenic variation in the Lyme disease spirochete: persistence through a mechanism that differs from other pathogens. Mol. Microbiol. *65*, 1547–1558.

Barthold, S.W., de Souza, M.S., Janotka, J.L., Smith, A.L., and Persing, D.H. (1993). Chronic Lyme borreliosis in the laboratory mouse. Am. J. Pathol. *143*, 959–971.

Barthold, S.W., Hodzic, E., Tunev, S., and Feng, S. (2006). Antibody-mediated disease remission in the mouse model of Lyme borreliosis. Infect. Immun. *74*, 4817–4825.

Behera, A.K., Durand, E., Cugini, C., Antonara, S., Bourassa, L., Hildebrand, E., Hu, L.T., and Coburn, J. (2008). *Borrelia burgdorferi* BBB07 interaction with integrin $\alpha_3\beta_1$ stimulates production of proinflammatory mediators in primary human chondrocytes. Cell. Microbiol. *10*, 320–331.

Behera, A.K., Hildebrand, E., Bronson, R.T., Perides, G., Uematsu, S., Akira, S., and Hu, L.T. (2006a). MyD88 deficiency results in tissue-specific changes in cytokine induction and inflammation in interleukin-18-independent mice infected with *Borrelia burgdorferi*. Infect. Immun. 74, 1462–1470.

Behera, A.K., Hildebrand, E., Scagliotti, J., Steere, A.C., and Hu, L.T. (2005). Induction of host matrix metalloproteinases by *Borrelia burgdorferi* differs in human and murine Lyme arthritis. Infect. Immun. 73, 126–134.

Behera, A.K., Hildebrand, E., Szafranski, J., Hung, H.H., Grodzinsky, A.J., Lafyatis, R., Koch, A.E., Kalish, R., Perides, G., Steere, A.C., and Hu, L.T. (2006b). Role of aggrecanase 1 in Lyme arthritis. Arthritis Rheum 54, 3319–3329.

Behera, A.K., Hildebrand, E., Uematsu, S., Akira, S., Coburn, J., and Hu, L.T. (2006c). Identification of a TLR-independent pathway for *Borrelia burgdorferi*-induced expression of matrix metalloproteinases and inflammatory mediators through binding to integrin $\alpha_3\beta_1$. J. Immunol. *177*, 657–664.

Benhnia, M.R., Wroblewski, D., Akhtar, M.N., Patel, R.A., Lavezzi, W., Gangloff, S.C., Goyert, S.M., Caimano, M.J., Radolf, J.D., and Sellati, T.J. (2005). Signaling through CD14 attenuates the inflammatory response to *Borrelia burgdorferi*, the agent of Lyme disease. J. Immunol. *174*, 1539–1548.

Bergström, S., Bundoc, V.G., and Barbour, A.G. (1989). Molecular analysis of linear plasmid-encoded major surface proteins, OspA and OspB, of the Lyme disease spirochaete *Borrelia burgdorferi*. Mol. Microbiol. *3*, 479–486.

Blevins, J.S., Hagman, K.E., and Norgard, M.V. (2008). Assessment of decorin-binding protein A to the infectivity of *Borrelia burgdorferi* in the murine models of needle and tick infection. BMC Microbiol. *8*, 82.

Blevins, J.S., Revel, A.T., Caimano, M.J., Yang, X.F., Richardson, J.A., Hagman, K.E., and Norgard, M.V. (2004). The *luxS* gene is not required for *Borrelia burgdorferi* tick colonization, transmission to a mammalian host, or induction of disease. Infect. Immun. *72*, 4864–4867.

Boardman, B.K., He, M., Ouyang, Z., Xu, H., Pang, X., and Yang, X.F. (2008). Essential role of the response regulator Rrp2 in the infectious cycle of *Borrelia burgdorferi*. Infect. Immun. *76*, 3844–3853.

Bockenstedt, L.K., Mao, J., Hodzic, E., Barthold, S.W., and Fish, D. (2002). Detection of attenuated, noninfectious spirochetes in *Borrelia burgdorferi*-infected mice after antibiotic treatment. J. Infect. Dis. *186*, 1430–1437.

Bolz, D.D., Sundsbak, R.S., Ma, Y., Akira, S., Kirschning, C.J., Zachary, J.F., Weis, J.H., and Weis, J.J. (2004). MyD88 plays a unique role in host defense but not arthritis development in Lyme disease. J. Immunol. *173*, 2003–2010.

Botkin, D.J., Abbott, A., Howell, J.K., Mosher, M., Stewart, P.E., Rosa, P.A., Kawabata, H., Watanabe, H., and Norris, S.J. (2006a). Transposon mutagenesis of infectious *Borrelia burgdorferi* B31: a pilot study. In Molecular Biology of Spirochetes, F.C. Cabello, D. Hulinska, and H.P. Godfrey, eds. (Amsterdam: IOS Press), pp. 13–23.

Botkin, D.J., Abbott, A.N., Stewart, P.E., Rosa, P.A., Kawabata, H., Watanabe, H., and Norris, S.J. (2006b). Identification of potential virulence determinants by *Himar1* transposition of infectious *Borrelia burgdorferi* B31. Infect. Immun. *74*, 6690–6699.

Brissette, C.A., Cooley, A.E., Burns, L.H., Riley, S.P., Verma, A., Woodman, M.E., Bykowski, T., and Stevenson, B. (2008). Lyme borreliosis spirochete Erp proteins, their known host ligands, and potential roles in mammalian infection. Int. J. Med. Microbiol. *298*, 257–267.

Brooks, C.S., Hefty, P.S., Jolliff, S.E., and Akins, D.R. (2003). Global analysis of *Borrelia burgdorferi* genes regulated by mammalian host-specific signals. Infect. Immun. *71*, 3371–3383.

Brooks, C.S., Vuppala, S.R., Jett, A.M., Alitalo, A., Meri, S., and Akins, D.R. (2005). Complement regulator-acquiring surface protein 1 imparts resistance to human serum in *Borrelia burgdorferi*. J. Immunol. *175*, 3299–3308.

Brown, C.R., Blaho, V.A., Fritsche, K.L., and Loiacono, C.M. (2006). Stat1 deficiency exacerbates carditis but not arthritis during experimental Lyme borreliosis. J. Interferon Cytokine Res. *26*, 390–399.

Brown, E.L., Wooten, R.M., Johnson, B.J. B., Iozzo, R.V., Smith, A., Dolan, M.C., Guo, B.P., Weis, J.J., and Höök, M. (2001). Resistance to Lyme disease in decorin-deficient mice. J. Clin. Invest. *107*, 845–852.

Bunikis, J., Noppa, L., and Bergström, S. (1995). Molecular analysis of a 66-kDa protein associated with the outer membrane of Lyme disease *Borrelia*. FEMS Microbiol. Lett. *131*, 139–145.

Bunikis, J., Noppa, L., Östberg, Y., Barbour, A.G., and Bergström, S. (1996). Surface exposure and species specificity of an immunoreactive domain of a 66-kilodalton outer membrane protein (P66) of the *Borrelia spp*. that cause Lyme disease. Infect. Immun. *64*, 5111–5116.

Bykowski, T., Babb, K., von Lackum, K., Riley, S.P., Norris, S.J., and Stevenson, B. (2006). Transcriptional regulation of the *Borrelia burgdorferi* antigenically variable VlsE surface protein. J. Bacteriol. *188*, 4879–4889.

Cabello, F.C., Godfrey, H.P., and Newman, S.A. (2007). Hidden in plain sight: *Borrelia burgdorferi* and the extracellular matrix. Trends Microbiol. *15*, 350–354.

Cadavid, D., Bai, Y., Dail, D., Hurd, M., Narayan, K., Hodzic, E., Barthold, S.W., and Pachner, A.R. (2003). Infection and inflammation in skeletal muscle from nonhuman primates infected with different genospecies of the Lyme disease spirochete *Borrelia burgdorferi*. Infect. Immun. *71*, 7087–7098.

Caimano, M.J., Iyer, R., Eggers, C.H., Gonzalez, C., Morton, E.A., Gilbert, M.A., Schwartz, I., and Radolf, J.D. (2007). Analysis of the RpoS regulon in *Borrelia burgdorferi* in response to mammalian host signals provides insight into RpoS function during the enzootic cycle. Mol. Microbiol. *65*, 1193–1217.

Charon, N.W., and Goldstein, S.F. (2002). Genetics of motility and chemotaxis of a fascinating group of bacteria: the spirochetes. Annu. Rev. Genet. *36*, 47–73.

Clarke, M.B., Hughes, D.T., Zhu, C., Boedeker, E.C., and Sperandio, V. (2006). The QseC sensor kinase: A bacterial adrenergic receptor. Proc. Natl. Acad. Sci. U.S.A. *103*, 10420–10425.

Cluss, R.G., Silverman, D.A., and Stafford, T.R. (2004). Extracellular secretion of the *Borrelia burgdorferi* Oms28 porin and Bgp, a glycosaminoglycan binding protein. Infect. Immun. *72*, 6279–6286.

Coburn, J., Chege, W., Magoun, L., Bodary, S.C., and Leong, J.M. (1999). Characterization of a candidate *Borrelia burgdorferi* β_3-chain integrin ligand identified using a phage display library. Mol. Microbiol. *34*, 926–940.

Coburn, J., and Cugini, C. (2003). Targeted mutation of the outer membrane protein P66 disrupts attachment of the Lyme disease spirochete, *Borrelia burgdorferi*, to integrin $\alpha_v\beta_3$. Proc. Natl. Acad. Sci. U.S.A. *100*, 7301–7306.

Coburn, J., Fischer, J.R., and Leong, J.M. (2005). Solving a sticky problem: new genetic approaches to host cell adhesion by the Lyme disease spirochete. Mol. Microbiol. *57*, 1182–1195.

Coleman, A.S., Yang, X., Kumar, M., Zhang, X., Promnares, K., Shroder, D., Kenedy, M.R., Anderson, J.F., Akins, D.R., and Pal, U. (2008). *Borrelia burgdorferi* complement regulator-acquiring surface protein 2 does not contribute to complement resistance or host infectivity. PLoS ONE 3, 3010e.

Coleman, J.L., Gebbia, J.A., and Benach, J.L. (2001). *Borrelia burgdorferi* and other bacterial products induce expression and release of the urokinase receptor (CD87). J. Immunol. *166*, 473–480.

Coleman, J.L., Gebbia, J.A., Piesman, J., Degen, J.L., Bugge, T.H., and Benach, J.L. (1997). Plasminogen is required for efficient dissemination of *B. burgdorferi* in ticks and for enhancement of spirochetemia in mice. Cell *89*, 1111–1119.

Coleman, J.L., Sellati, T.J., Testa, J.E., Kew, R.R., Furie, M.B., and Benach, J.L. (1995). *Borrelia burgdorferi* binds plasminogen, resulting in enhanced penetration of endothelial monolayers. Infect. Immun. *63*, 2478–2484.

Comstock, L., Fikrig, E., Shoberg, R., Flavell, R., and Thomas, D. (1993). A monoclonal antibody to OspA inhibits association of *Borrelia burgdorferi* with human endothelial cells. Infect. Immun. *61*, 423–431.

Comstock, L.E., and Thomas, D.D. (1989). Penetration of endothelial cell monolayers by *Borrelia burgdorferi*. Infect. Immun. *57*, 1626–1628.

Comstock, L.E., and Thomas, D.D. (1991). Characterization of *Borrelia burgdorferi* invasion of cultured endothelial cells. Microb Pathog *10*, 137–148.

Coutte, L., Botkin, D.J., Gao, L., and Norris, S.J. (2009). Detailed analysis of sequence changes occurring during *vlsE* antigenic variation in the mouse model of *Borrelia burgdorferi* infection. PLoS Pathog. *5*, e1000293.

Crandall, H., Dunn, D.M., Ma, Y., Wooten, R.M., Zachary, J.F., Weis, J.H., Weiss, R.B., and Weis, J.J. (2006). Gene expression profiling reveals unique pathways associated with differential severity of Lyme arthritis. J. Immunol. *177*, 7930–7942.

Crother, T.R., Champion, C.I., Whitelegge, J.P., Aguilera, R., Wu, X.Y., Blanco, D.R., Miller, J.N., and Lovett, M.A. (2004). Temporal analysis of the antigenic composition of *Borrelia burgdorferi* during infection in rabbit skin. Infect. Immun. *72*, 5063–5072.

Crother, T.R., Champion, C.I., Wu, X.Y., Blanco, D.R., Miller, J.N., and Lovett, M.A. (2003). Antigenic composition of *Borrelia burgdorferi* during infection of SCID mice. Infect. Immun. *71*, 3419–3428.

Cugini, C., Medrano, M., Schwan, T.G., and Coburn, J. (2003). Regulation of expression of the *Borrelia burgdorferi* β_3-chain integrin ligand, P66, in ticks and in culture. Infect. Immun. *71*, 1001–1007.

Dai, Q., Restrepo, B.I., Porcella, S.F., Raffel, S.J., Schwan, T.G., and Barbour, A.G. (2006). Antigenic variation by *Borrelia hermsii* occurs through recombination between extragenic repetitive elements on linear plasmids. Mol. Microbiol. *60*, 1329–1343.

de Silva, A.M., and Fikrig, E. (1995). Growth and migration of *Borrelia burgdorferi* in *Ixodes* ticks during blood feeding. Am. J. Trop. Med. Hyg. *53*, 397–404.

de Silva, A.M., Telford, S.R., 3rd, Brunet, L.R., Barthold, S.W., and Fikrig, E. (1996). *Borrelia burgdorferi* OspA is an arthropod-specific transmission-blocking Lyme disease vaccine. J. Exp. Med. *183*, 271–275.

Dresser, A.R., Hardy, P.O., and Chaconas, G. (2009). Investigation of the genes involved in antigenic switching at the *vlsE* locus in *Borrelia burgdorferi*: an essential role for the RuvAB branch migrase. PLoS Pathog. *5*, e1000680.

Dressler, F., Whalen, J.A., Reinhardt, B.N., and Steere, A.C. (1993). Western blotting in the serodiagnosis of Lyme disease. J. Infect. Dis. *167*, 392–400.

Eggers, C.H., Caimano, M.J., and Radolf, J.D. (2004). Analysis of promoter elements involved in the transcriptional initiation of RpoS-dependent *Borrelia burgdorferi* genes. J. Bacteriol. *186*, 7390–7402.

Eggers, C.H., Caimano, M.J., and Radolf, J.D. (2006). Sigma factor selectivity in *Borrelia burgdorferi*: RpoS recognition of the *ospE*/*ospF*/*elp* promoters is dependent on the sequence of the –10 region. Mol. Microbiol. *59*, 1859–1875.

Eicken, C., Sharma, V., Klabunde, T., Lawrenz, M.B., Hardham, J.M., Norris, S.J., and Sacchettini, J.C. (2002). Crystal structure of Lyme disease variable surface antigen VlsE of *Borrelia burgdorferi*. J. Biol. Chem. *277*, 21691–21696.

El-Hage, N., Babb, K., Carroll, J.A., Lindstrom, N., Fischer, E.R., Miller, J.C., Gilmore, R.D., Jr., Mbow, M.L., and Stevenson, B. (2001). Surface exposure and protease insensitivity of *Borrelia burgdorferi* Erp (OspEF-related) lipoproteins. Microbiology *147*, 821–830.

Feng, S., Hodzic, E., Stevenson, B., and Barthold, S.W. (1998). Humoral immunity to *Borrelia burgdorferi* N40 decorin binding proteins during infection of laboratory mice. Infect. Immun. *66*, 2827–2835.

Fikrig, E., Barthold, S.W., Sun, W., Feng, W., Telford, S.R., 3rd, and Flavell, R.A. (1997). *Borrelia burgdorferi* P35 and P37 proteins, expressed in vivo, elicit protective immunity. Immunity *6*, 531–539.

Fikrig, E., Chen, M., Barthold, S.W., Anguita, J., Feng, W., Telford, S.R., 3rd, and Flavell, R.A. (1999). *Borrelia burgdorferi erpT* expression in the arthropod vector and murine host. Mol. Microbiol. *31*, 281–290.

Fikrig, E., Feng, W., Aversa, J., Schoen, R.T., and Flavell, R.A. (1998). Differential expression of *Borrelia burgdorferi* genes during erythema migrans and Lyme arthritis. J. Infect. Dis. *178*, 1198–1201.

Fikrig, E., Feng, W., Barthold, S.W., Telford, S.R., 3rd, and Flavell, R.A. (2000). Arthropod- and host-specific *Borrelia burgdorferi bbk32* expression and the inhibition of spirochete transmission. J. Immunol. *164*, 5344–5351.

Finlay, B.B., and Falkow, S. (1997). Common themes in microbial pathogenicity revisited. Microbiol. Mol. Biol. Rev. *61*, 136–169.

Fischer, J.R., LeBlanc, K.T., and Leong, J.M. (2006). Fibronectin binding protein BBK32 of the Lyme disease spirochete promotes bacterial attachment to glycosaminoglycans. Infect. Immun. *74*, 435–441.

Fischer, J.R., Parveen, N., Magoun, L., and Leong, J.M. (2003). Decorin-binding proteins A and B confer distinct mammalian cell type-specific attachment by *Borrelia burgdorferi*, the Lyme disease spirochete. Proc. Natl. Acad. Sci. U.S.A. *100*, 7307–7312.

Fisher, M.A., Grimm, D., Henion, A.K., Elias, A.F., Stewart, P.E., Rosa, P.A., and Gherardini, F.C. (2005). *Borrelia burgdorferi* σ^{54} is required for mammalian infection and vector transmission but not for tick colonization. Proc. Natl. Acad. Sci. U.S.A. *102*, 5162–5167.

Fraser, C.M., Casjens, S., Huang, W.M., Sutton, G.G., Clayton, R., Lathigra, R., White, O., Ketchum, K.A., Dodson, R., Hickey, E.K., *et al.* (1997). Genomic sequence of a Lyme disease spirochaete, *Borrelia burgdorferi*. Nature *390*, 580–586.

Fuchs, H., Simon, M.M., Wallich, R., Bechtel, M., and Kramer, M.D. (1996). *Borrelia burgdorferi* induces secretion of pro-urokinase-type plasminogen activator by human monocytes. Infect. Immun. *64*, 4307–4312.

Fuchs, H., Wallich, R., Simon, M.M., and Kramer, M.D. (1994). The outer surface protein A of the spirochete *Borrelia burgdorferi* is a plasmin(ogen) receptor. Proc. Natl. Acad. Sci. U.S.A. *91*, 12594–12598.

García-Moncó, J.C., Fernandez-Villar, B., and Benach, J.L. (1989). Adherence of the Lyme disease spirochete to glial cells and cells of glial origin. J. Infect. Dis. *160*, 497–506.

García-Monco, J.C., Fernandez-Villar, B., Rogers, R.C., Szczepanski, A., Wheeler, C.M., and Benach, J.L. (1992). *Borrelia burgdorferi* and other related spirochetes bind to galactocerebroside. Neurology *42*, 1341–1348.

Gebbia, J.A., Coleman, J.L., and Benach, J.L. (2001). *Borrelia* spirochetes up-regulate release and activation of matrix metalloproteinase gelatinase B (MMP-9) and collagenase 1 (MMP-1) in human cells. Infect. Immun. *69*, 456–462.

Gebbia, J.A., Coleman, J.L., and Benach, J.L. (2004). Selective induction of matrix metalloproteinases by *Borrelia burgdorferi* via Toll-like receptor 2 in monocytes. J. Infect. Dis. *189*, 113–119.

Gorbacheva, V.Y., Godfrey, H.P., and Cabello, F.C. (2000). Analysis of the *bmp* gene family in *Borrelia burgdorferi* sensu lato. J. Bacteriol. *182*, 2037–2042.

Grab, D.J., Givens, C., and Kennedy, R. (1998). Fibronectin-binding activity in *Borrelia burgdorferi*. Biochim. Biophys. Acta *1407*, 135–145.

Grab, D.J., Kennedy, R., and Philipp, M.T. (1996). *Borrelia burgdorferi* possesses a collagenolytic activity. FEMS Microbiol. Lett. *144*, 39–45.

Grab, D.J., Perides, G., Dumler, J.S., Kim, K.J., Park, J., Kim, Y.V., Nikolskaia, O., Choi, K.S., Stins, M.F., and Kim, K.S. (2005). *Borrelia burgdorferi*, host-derived proteases, and the blood–brain barrier. Infect. Immun. *73*, 1014–1022.

Grimm, D., Eggers, C.H., Caimano, M.J., Tilly, K., Stewart, P.E., Elias, A.F., Radolf, J.D., and Rosa, P.A. (2004a). Experimental assessment of the roles of linear plasmids lp25 and lp28-1 of *Borrelia burgdorferi* throughout the infectious cycle. Infect. Immun. *72*, 5938–5946.

Grimm, D., Tilly, K., Byram, R., Stewart, P.E., Krum, J.G., Bueschel, D.M., Schwan, T.G., Policastro, P.F., Elias, A.F., and Rosa, P.A. (2004b). Outer-surface protein C of the Lyme disease spirochete: a protein induced in ticks for infection of mammals. Proc. Natl. Acad. Sci. U.S.A. *101*, 3142–3147.

Guo, B.P., Brown, E.L., Dorward, D.W., Rosenberg, L.C., and Höök, M. (1998). Decorin-binding adhesins from *Borrelia burgdorferi*. Mol. Microbiol. *30*, 711–723.

Guo, B.P., Norris, S.J., Rosenberg, L.C., and Höök, M. (1995). Adherence of *Borrelia burgdorferi* to the proteoglycan decorin. Infect. Immun. *63*, 3467–3472.

Hagman, K.E., Lahdenne, P., Popova, T.G., Porcella, S.F., Akins, D.R., Radolf, J.D., and Norgard, M.V. (1998). Decorin-binding protein of *Borrelia burgdorferi* is encoded within a two-gene operon and is protective in the murine model of Lyme borreliosis. Infect. Immun. *66*, 2674–2683.

Hagman, K.E., Yang, X., Wikel, S.K., Schoeler, G.B., Caimano, M.J., Radolf, J.D., and Norgard, M.V. (2000). Decorin-binding protein A (DbpA) of *Borrelia burgdorferi* is not protective when immunized mice are challenged via tick infestation and correlates with the lack of DbpA expression by *B. burgdorferi* in ticks. Infect. Immun. *68*, 4759–4764.

Haile, W.B., Coleman, J.L., and Benach, J.L. (2006). Reciprocal up-regulation of urokinase plasminogen activator and its inhibitor, PAI-2, by *Borrelia burgdorferi* affects bacterial penetration and host-inflammatory response. Cell. Microbiol. *8*, 1349–1360.

Hanson, M.S., Cassatt, D.R., Guo, B.P., Patel, N.K., McCarthy, M.P., Dorward, D.W., and Höök, M. (1998). Active and passive immunity against *Borrelia burgdorferi* decorin binding protein A (DbpA) protects against infection. Infect. Immun. *66*, 2143–2153.

Hartmann, K., Corvey, C., Skerka, C., Kirschfink, M., Karas, M., Brade, V., Miller, J.C., Stevenson, B., Wallich, R., Zipfel, P.F., and Kraiczy, P. (2006). Functional characterization of BbCRASP-2, a distinct outer membrane protein of *Borrelia burgdorferi* that binds host complement regulators factor H and FHL-1. Mol. Microbiol. *61*, 1220–1236.

He, M., Boardman, B.K., Yan, D., and Yang, X.F. (2007). Regulation of expression of the fibronectin-binding protein BBK32 in *Borrelia burgdorferi*. J. Bacteriol. *189*, 8377–8380.

Hechemy, K.E., Samsonoff, W.A., Harris, H.L., and McKee, M. (1992). Adherence and entry of *Borrelia burgdorferi* in Vero cells. J Med Microbiol *36*, 229–238.

Hechemy, K.E., Samsonoff, W.A., McKee, M., and Guttman, J.M. (1989). *Borrelia burgdorferi* attachment to mammalian cells. J. Infect. Dis. *159*, 805–806.

Hellwage, J., Meri, T., Heikkila, T., Alitalo, A., Panelius, J., Lahdenne, P., Seppala, I.J., and Meri, S. (2001). The complement regulator factor H binds to the surface protein OspE of *Borrelia burgdorferi*. J. Biol. Chem. *276*, 8427–8435.

Hirschfeld, M., Kirschning, C.J., Schwandner, R., Wesche, H., Weis, J.H., Wooten, R.M., and Weis, J.J. (1999). Cutting edge: inflammatory signaling by *Borrelia burgdorferi* lipoproteins is mediated by toll-like receptor 2. J. Immunol. *163*, 2382–2386.

Hodzic, E., Feng, S., Freet, K.J., and Barthold, S.W. (2003). *Borrelia burgdorferi* population dynamics and prototype gene expression during infection of immunocompetent and immunodeficient mice. Infect. Immun. *71*, 5042–5055.

Hodzic, E., Feng, S., Holden, K., Freet, K.J., and Barthold, S.W. (2008). Persistence of *Borrelia burgdorferi* following antibiotic treatment in mice. Antimicrob. Agents Chemother. *52*, 1728–1736.

Hovius, J.W. R., van Dam, A.P., and Fikrig, E. (2007). Tick–host–pathogen interactions in Lyme borreliosis. Trends Parasitol. *23*, 434–438.

Hu, L.T., Eskildsen, M.A., Masgala, C., Steere, A.C., Arner, E.C., Pratta, M.A., Grodzinsky, A.J., Loening, A., and Perides, G. (2001). Host metalloproteinases in Lyme arthritis. Arthritis Rheum. *44*, 1401–1410.

Hu, L.T., Pratt, S.D., Perides, G., Katz, L., Rogers, R.A., and Klempner, M.S. (1997). Isolation, cloning, and expression of a 70-kilodalton plasminogen binding protein of *Borrelia burgdorferi*. Infect. Immun. *65*, 4989–4995.

Hübner, A., Revel, A.T., Nolen, D.M., Hagman, K.E., and Norgard, M.V. (2003). Expression of a *luxS* gene is not required for *Borrelia burgdorferi* infection of mice via needle inoculation. Infect. Immun. *71*, 2892–2896.

Hübner, A., Yang, X., Nolen, D.M., Popova, T.G., Cabello, F.C., and Norgard, M.V. (2001). Expression of *Borrelia burgdorferi* OspC and DbpA is controlled by a RpoN–RpoS regulatory pathway. Proc. Natl. Acad. Sci. U.S.A. *98*, 12724–12729.

Hudson, C.R., Frye, J.G., Quinn, F.D., and Gherardini, F.C. (2001). Increased expression of *Borrelia burgdorferi vlsE* in response to human endothelial cell membranes. Mol. Microbiol. *41*, 229–239.

Indest, K.J., Howell, J.K., Jacobs, M.B., Scholl-Meeker, D., Norris, S.J., and Philipp, M.T. (2001). Analysis of *Borrelia burgdorferi vlsE* gene expression and recombination in the tick vector. Infect. Immun. *69*, 7083–7090.

Isaacs, R.D. (1994). *Borrelia burgdorferi* bind to epithelial cell proteoglycan. J. Clin. Invest. *93*, 809–819.

Jewett, M.W., Byram, R., Bestor, A., Tilly, K., Lawrence, K., Burtnick, M.N., Gherardini, F., and Rosa, P.A. (2007a). Genetic basis for retention of a critical virulence plasmid of *Borrelia burgdorferi*. Mol. Microbiol. *66*, 975–990.

Jewett, M.W., Lawrence, K., Bestor, A.C., Tilly, K., Grimm, D., Shaw, P., VanRaden, M., Gherardini, F., and Rosa, P.A. (2007b). The critical role of the linear plasmid lp36 in the infectious cycle of *Borrelia burgdorferi*. Mol. Microbiol. *64*, 1358–1374.

Kawabata, H., Myouga, F., Inagaki, Y., Murai, N., and Watanabe, H. (1998). Genetic and immunological analyses of Vls (VMP-like sequences) of *Borrelia burgdorferi*. Microb. Pathog. *24*, 155–166.

Kawabata, H., Norris, S.J., and Watanabe, H. (2004). BBE02 disruption mutants of *Borrelia burgdorferi* B31 have a highly transformable, infectious phenotype. Infect. Immun. *72*, 7147–7154.

Kim, J.H., Singvall, J., Schwarz-Linek, U., Johnson, B.J., Potts, J.R., and Höök, M. (2004). BBK32, a fibronectin binding MSCRAMM from *Borrelia burgdorferi*, contains a disordered region that undergoes a conformational change on ligand binding. J. Biol. Chem. *279*, 41706–41714.

Kirchner, A., Koedel, U., Fingerle, V., Paul, R., Wilske, B., and Pfister, H.W. (2000). Upregulation of matrix metalloproteinase-9 in the cerebrospinal fluid of patients with acute Lyme neuroborreliosis. J. Neurol. Neurosurg. Psychiatr. *68*, 368–371.

Klempner, M.S., Noring, R., Epstein, M.P., McCloud, B., Hu, R., Limentani, S.A., and Rogers, R.A. (1995). Binding of human plasminogen and urokinase-type plasminogen activator to the Lyme disease spirochete, *Borrelia burgdorferi*. J. Infect. Dis. *171*, 1258–1265.

Kobryn, K., and Chaconas, G. (2002). ResT, a telomere resolvase encoded by the Lyme disease spirochete. Mol. Cell *9*, 195–201.

Kopp, P.A., Schmitt, M., Wellensiek, H.J., and Blobel, H. (1995). Isolation and characterization of fibronectin-binding sites of *Borrelia garinii* N34. Infect. Immun. *63*, 3804–3808.

Kraiczy, P., Hellwage, J., Skerka, C., Kirschfink, M., Brade, V., Zipfel, P.F., and Wallich, R. (2003). Immune evasion of *Borrelia burgdorferi*: mapping of a complement-inhibitor factor H-binding site of BbCRASP-3, a novel member of the Erp protein family. Eur J. Immunol. *33*, 697–707.

Kraiczy, P., Skerka, C., Brade, V., and Zipfel, P.F. (2001a). Further characterization of complement regulator-acquiring surface proteins of *Borrelia burgdorferi*. Infect. Immun. *69*, 7800–7809.

Kraiczy, P., Skerka, C., Kirschfink, M., Brade, V., and Zipfel, P.F. (2001b). Immune evasion of *Borrelia burgdorferi* by acquisition of human complement regulators FHL-1/reconectin and Factor H. Eur J. Immunol. *31*, 1674–1684.

Kurtenbach, K., De Michelis, S., Etti, S., Schafer, S.M., Sewell, H.S., Brade, V., and Kraiczy, P. (2002). Host association of *Borrelia burgdorferi* sensu lato – the key role of host complement. Trends Microbiol. *10*, 74–79.

Kurtti, T.J., Munderloh, U.G., Ahlstrand, G.G., and Johnson, R.C. (1988). *Borrelia burgdoferi* in tick cell culture: Growth and cellular adherence. Ent Soc Am *25*, 256–261.

Kurtti, T.J., Munderloh, U.G., Krueger, D.E., Johnson, R.C., and Schwan, T.G. (1993). Adhesion to and invasion of cultured tick (Acarina: Ixodidae) cells by *Borrelia burgdorferi* (Spirochaetales: *Spirochaetaceae*) and maintenance of infectivity. J. Med. Entomol. *30*, 586–596.

Labandeira-Rey, M., Seshu, J., and Skare, J.T. (2003). The absence of linear plasmid 25 or 28–1 of *Borrelia burgdorferi* dramatically alters the kinetics of experimental infection via distinct mechanisms. Infect. Immun. *71*, 4608–4613.

Lagal, V., Portnoi, D., Faure, G., Postic, D., and Baranton, G. (2006). *Borrelia burgdorferi* sensu stricto invasiveness is correlated with OspC-plasminogen affinity. Microbes Infect. *8*, 645–652.

Lagal, V., Postic, D., Ruzic-Sabljic, E., and Baranton, G. (2003). Genetic diversity among *Borrelia* strains determined by single-strand conformation polymorphism analysis of the *ospC* gene and its association with invasiveness. J. Clin. Microbiol. *41*, 5059–5065.

Lawrenz, M.B., Hardham, J.M., Owens, R.T., Nowakowski, J., Steere, A.C., Wormser, G.P., and Norris, S.J. (1999). Human antibody responses to VlsE antigenic variation protein of *Borrelia burgdorferi*. J. Clin. Microbiol. *37*, 3997–4004.

Lawrenz, M.B., Kawabata, H., Purser, J.E., and Norris, S.J. (2002). Decreased electroporation efficiency in *Borrelia burgdorferi* containing linear plasmids lp25 and lp56: impact on transformation of infectious *B. burgdorferi*. Infect. Immun. *70*, 4798–4804.

Lawrenz, M.B., Wooten, R.M., and Norris, S.J. (2004). Effects of *vlsE* complementation on the infectivity of *Borrelia burgdorferi* lacking the linear plasmid lp28-1. Infect. Immun. *72*, 6577–6585.

Leong, J.M., Morrissey, P.E., Ortega-Barria, E., Pereira, M.E., and Coburn, J. (1995). Hemagglutination and proteoglycan binding by the Lyme disease spirochete, *Borrelia burgdorferi*. Infect. Immun. *63*, 874–883.

Leong, J.M., Robbins, D., Rosenfeld, L., Lahiri, B., and Parveen, N. (1998a). Structural requirements for glycosaminoglycan recognition by the Lyme disease spirochete, *Borrelia burgdorferi*. Infect. Immun. *66*, 6045–6048.

Leong, J.M., Wang, H., Magoun, L., Field, J.A., Morrissey, P.E., Robbins, D., Tatro, J.B., Coburn, J., and Parveen, N. (1998b). Different classes of proteoglycans contribute to the attachment of *Borrelia burgdorferi* to cultured endothelial and brain cells. Infect. Immun. *66*, 994–999.

Li, X., Liu, X., Beck, D.S., Kantor, F.S., and Fikrig, E. (2006). *Borrelia burgdorferi* lacking BBK32, a fibronectin-binding protein, retains full pathogenicity. Infect. Immun. *74*, 3305–3313.

Li, X., Neelakanta, G., Liu, X., Beck, D.S., Kantor, F.S., Fish, D., Anderson, J.F., and Fikrig, E. (2007). Role of outer surface protein D in the *Borrelia burgdorferi* life cycle. Infect. Immun. *75*, 4237–4244.

Liang, F.T., Alvarez, A.L., Gu, Y., Nowling, J.M., Ramamoorthy, R., and Philipp, M.T. (1999). An immunodominant conserved region within the variable domain of VlsE, the variable surface antigen of *Borrelia burgdorferi*. J. Immunol. *163*, 5566–5573.

Liang, F.T., Brown, E.L., Wang, T., Iozzo, R.V., and Fikrig, E. (2004a). Protective niche for *Borrelia burgdorferi* to evade humoral immunity. Am. J. Pathol. *165*, 977–985.

Liang, F.T., and Philipp, M.T. (1999). Analysis of antibody response to invariable regions of VlsE, the variable surface antigen of *Borrelia burgdorferi*. Infect. Immun. *67*, 6702–6706.

Liang, F.T., and Philipp, M.T. (2000). Epitope mapping of the immunodominant invariable region of Borrelia burgdorferi VlsE in three host species. Infect. Immun. *68*, 2349–2352.

Liang, F.T., Yan, J., Mbow, M.L., Sviat, S.L., Gilmore, R.D., Mamula, M., and Fikrig, E. (2004b). *Borrelia burgdorferi* changes its surface antigenic expression in response to host immune responses. Infect. Immun. *72*, 5759–5767.

Lin, B., Kidder, J.M., Noring, R., Steere, A.C., Klempner, M.S., and Hu, L.T. (2001). Differences in synovial fluid levels of matrix metalloproteinases suggest separate mechanisms of pathogenesis in Lyme arthritis before and after antibiotic treatment. J. Infect. Dis. *184*, 174–180.

Lin, T., Gao, L., Edmondson, D.G., Jacobs, M.B., Philipp, M.T., and Norris, S.J. (2009). Central role of the holliday junction helicase RuvAB in *vlsE* recombination and infectivity of *Borrelia burgdorferi*. PLoS Pathog. *5*, e1000679.

Liu, N., Montgomery, R.R., Barthold, S.W., and Bockenstedt, L.K. (2004). Myeloid differentiation antigen 88 deficiency impairs pathogen clearance but does not alter inflammation in *Borrelia burgdorferi*-infected mice. Infect. Immun. *72*, 3195–3203.

Liveris, D., Mulay, V., Sandigursky, S., and Schwartz, I. (2008). *Borrelia burgdorferi vlsE* antigenic variation is not mediated by RecA. Infect. Immun. *76*, 4009–4018.

Liveris, D., Wang, G., Girao, G., Byrne, D.W., Nowakowski, J., McKenna, D., Nadelman, R., Wormser, G.P., and Schwartz, I. (2002). Quantitative detection of *Borrelia burgdorferi* in 2-millimeter skin samples of erythema migrans lesions: correlation of results with clinical and laboratory findings. J. Clin. Microbiol. *40*, 1249–1253.

Ma, Y., A. Sturrock, and J.J. Weis. (1991). Intracellular localization of *Borrelia burgdorferi* within human endothelial cells.. Infect. Immun. *59*, 671–678.

Margolis, N., Hogan, D., Tilly, K., and Rosa, P.A. (1994). Plasmid location of *Borrelia* purine biosynthesis gene homologs. J. Bacteriol. *176*, 6427–6432.

McDowell, J.V., Hovis, K.M., Zhang, H., Tran, E., Lankford, J., and Marconi, R.T. (2006). Evidence that the BBA68 protein (BbCRASP-1) of the Lyme disease spirochetes does not contribute to factor H-mediated immune evasion in humans and other animals. Infect. Immun. *74*, 3030–3034.

McDowell, J.V., Sung, S.Y., Hu, L.T., and Marconi, R.T. (2002). Evidence that the variable regions of the central domain of VlsE are antigenic during infection with Lyme disease spirochetes. Infect. Immun. *70*, 4196–4203.

Meli, D.N., Christen, S., and Leib, S.L. (2003). Matrix metalloproteinase-9 in pneumococcal meningitis: activation via an oxidative pathway. J. Infect. Dis. *187*, 1411–1415.

Miller, I., Chatfield, S., Dougan, G., Desilva, L., Joysey, H., and Hormaeche, C. (1989). Bacteriophage P22 as a vehicle for transducing cosmid gene banks between smooth strains of *Salmonella typhimurium*: use in identifying a role for *aroD* in attenuating virulent *Salmonella* strains. Mol. Gen. Genet. *215*, 312–316.

Miller, J.C., Narayan, K., Stevenson, B., and Pachner, A.R. (2005). Expression of *Borrelia burgdorferi erp* genes during infection of non-human primates. Microb. Pathog. *39*, 27–33.

Miller, J.C., and Stevenson, B. (2006). *Borrelia burgdorferi erp* genes are expressed at different levels within tissues of chronically infected mammalian hosts. Int. J. Med. Microbiol. *296 Suppl 40*, 185–194.

Miller, J.C., von Lackum, K., Babb, K., McAlister, J.D., and Stevenson, B. (2003). Temporal analysis of *Borrelia burgdorferi* Erp protein expression throughout the mammal-tick infectious cycle. Infect. Immun. *71*, 6943–6952.

Moore, M.W., Cruz, A.R., LaVake, C.J., Marzo, A.L., Eggers, C.H., Salazar, J.C., and Radolf, J.D. (2007). Phagocytosis of *Borrelia burgdorferi* and *Treponema pallidum* potentiates innate immune activation and induces gamma interferon production. Infect. Immun. *75*, 2046–2062.

Moriarty, T.J., Norman, M.U., Colarusso, P., Bankhead, T., Kubes, P., and Chaconas, G. (2008). Real-time high resolution 2-D imaging of the Lyme disease spirochete adhering to and escaping from the vasculature of a living host. PLoS Pathog. *4*, e1000090.

Neelakanta, G., Li, X., Pal, U., Liu, X., Beck, D.S., DePonte, K., Fish, D., Kantor, F.S., and Fikrig, E. (2007). Outer surface protein B Is critical for *Borrelia burgdorferi* adherence and survival within *Ixodes* ticks. PLoS Pathog. *3*, e33.

Norman, M.U., Moriarty, T.J., Dresser, A.R., Millen, B., Kubes, P., and Chaconas, G. (2008). Molecular mechanisms involved in vascular interactions of the Lyme disease pathogen in a living host. PLoS Pathog. *4*, e1000169.

Norris, S.J. (2006). Antigenic variation with a twist – the *Borrelia* story. Mol. Microbiol. *60*, 1319–1322.

Ntchobo, H., Rothermel, H., Chege, W., Steere, A.C., and Coburn, J. (2001). Recognition of multiple antibody epitopes throughout *Borrelia burgdorferi* P66, a candidate adhesin, in patients with early or late manifestations of Lyme disease. Infect. Immun. *69*, 1953–1956.

Ojaimi, C., Brooks, C., Akins, D., Casjens, S., Rosa, P., Elias, A., Barbour, A., Jasinskas, A., Benach, J., Katonah, L., *et al.* (2002). *Borrelia burgdorferi* gene expression profiling with membrane-based arrays. Meth Enzymol *358*, 165–177.

Ojaimi, C., Brooks, C., Casjens, S., Rosa, P., Elias, A., Barbour, A., Jasinskas, A., Benach, J., Katona, L., Radolf, J., *et al.* (2003). Profiling of temperature-induced changes in *Borrelia burgdorferi* gene expression by using whole genome arrays. Infect. Immun. *71*, 1689–1705.

Ouyang, Z., Blevins, J.S., and Norgard, M.V. (2008). Transcriptional interplay among the regulators Rrp2, RpoN and RpoS in *Borrelia burgdorferi*. Microbiology *154*, 2641–2658.

Pal, U., Dai, J., Li, X., Neelakanta, G., Luo, P., Kumar, M., Wang, P., Yang, X., Anderson, J.F., and Fikrig, E. (2008). A differential role for BB0365 in the persistence of *Borrelia burgdorferi* in mice and ticks. J. Infect. Dis. *197*, 148–155.

Pal, U., de Silva, A.M., Montgomery, R.R., Fish, D., Anguita, J., Anderson, J.F., Lobet, Y., and Fikrig, E. (2000). Attachment of *Borrelia burgdorferi* within *Ixodes scapularis* mediated by outer surface protein A. J. Clin. Invest. *106*, 561–569.

Pal, U., and Fikrig, E. (2003). Adaptation of *Borrelia burgdorferi* in the vector and vertebrate host. Microbes Infect. *5*, 659–666.

Pal, U., Li, X., Wang, T., Montgomery, R.R., Ramamoorthi, N., Desilva, A.M., Bao, F., Yang, X., Pypaert, M., Pradhan, D., *et al.* (2004a). TROSPA, an *Ixodes scapularis* receptor for *Borrelia burgdorferi*. Cell *119*, 457–468.

Pal, U., Yang, X., Chen, M., Bockenstedt, L.K., Anderson, J.F., Flavell, R.A., Norgard, M.V., and Fikrig, E. (2004b). OspC facilitates *Borrelia burgdorferi* invasion of *Ixodes scapularis* salivary glands. J. Clin. Invest. *113*, 220–230.

Parveen, N., Caimano, M., Radolf, J.D., and Leong, J.M. (2003). Adaptation of the Lyme disease spirochaete to the mammalian host environment results in enhanced glycosaminoglycan and host cell binding. Mol. Microbiol. *47*, 1433–1444.

Parveen, N., Cornell, K.A., Bono, J.L., Chamberland, C., Rosa, P., and Leong, J.M. (2006). Bgp, a secreted glycosaminoglycan-binding protein of *Borrelia burgdorferi* strain N40, displays nucleosidase activity and is not essential for infection of immunodeficient mice. Infect. Immun. *74*, 3016–3020.

Parveen, N., and Leong, J.M. (2000). Identification of a candidate glycosaminoglycan-binding adhesin of the Lyme disease spirochete *Borrelia burgdorferi*. Mol. Microbiol. *35*, 1220–1234.

Parveen, N., Robbins, D., and Leong, J.M. (1999). Strain variation in glycosaminoglycan recognition influences cell-type-specific binding by Lyme disease spirochetes. Infect. Immun. *67*, 1743–1749.

Perides, G., Noring, R., and Klempner, M.S. (1996). Inhibition of *Borrelia burgdorferi*-bound fibrinolytic enzymes by α_2-antiplasmin, PAI-1 and PAI-2. Biochem. Biophys. Res. Commun. *219*, 690–695.

Perides, G., Tanner-Brown, L.M., Eskildsen, M.A., and Klempner, M.S. (1999). *Borrelia burgdorferi* induces matrix metalloproteinases by neural cultures. J. Neurosci. Res. *58*, 779–790.

Pickering, M.C., Cook, H.T., Warren, J., Bygrave, A.E., Moss, J., Walport, M.J., and Botto, M. (2002). Uncontrolled C3 activation causes membranoproliferative glomerulonephritis in mice deficient in complement factor H. Nat. Genet. *31*, 424–428.

Piesman, J., Oliver, J.R., and Sinsky, R.J. (1990). Growth kinetics of the Lyme disease spirochete (*Borrelia burgdorferi*) in vector ticks (*Ixodes dammini*). Am. J. Trop. Med. Hyg. *42*, 352–357.

Piesman, J., Zeidner, N.S., and Schneider, B.S. (2003). Dynamic changes in *Borrelia burgdorferi* populations in *Ixodes scapularis* (Acari: Ixodidae) during transmission: studies at the mRNA level. Vector Borne Zoonotic Dis 3, 125–132.

Pikas, D.S., Brown, E.L., Gurusiddappa, S., Lee, L.Y., Xu, Y., and Höök, M. (2003). Decorin-binding sites in the adhesin DbpA from *Borrelia burgdorferi*: a synthetic peptide approach. J. Biol. Chem. *278*, 30920–30926.

Pinne, M., Thein, M., Denker, K., Benz, R., Coburn, J., and Bergström, S. (2007). Elimination of channel-forming activity by insertional inactivation of the *p66* gene in *Borrelia burgdorferi*. FEMS Microbiol. Lett. *266*, 241–249.

Probert, W.S., Allsup, K.M., and LeFebvre, R.B. (1995). Identification and characterization of a surface-exposed, 66-kilodalton protein from *Borrelia burgdorferi*. Infect. Immun. *63*, 1933–1939.

Probert, W.S., and Johnson, B.J. (1998). Identification of a 47 kDa fibronectin-binding protein expressed by

Borrelia burgdorferi isolate B31. Mol. Microbiol. *30*, 1003–1015.

Probert, W.S., Kim, J.H., Höök, M., and Johnson, B.J. B. (2001). Mapping the ligand-binding region of the *Borrelia burgdorferi* fibronectin-binding protein BBK32. Infect. Immun. *69*, 4129–4133.

Purser, J.E., Lawrenz, M.B., Caimano, M.J., Howell, J.K., Radolf, J.D., and Norris, S.J. (2003). A plasmid-encoded nicotinamidase (PncA) is essential for infectivity of *Borrelia burgdorferi* in a mammalian host. Mol. Microbiol. *48*, 753–764.

Radolf, J.D., and Caimano, M.J. (2008). The long strange trip of *Borrelia burgdorferi* outer-surface protein C. Mol. Microbiol. *69*, 1–4.

Ramamoorthi, N., Narasimhan, S., Pal, U., Bao, F., Yang, X.F., Fish, D., Anguita, J., Norgard, M.V., Kantor, F.S., Anderson, J.F., *et al.* (2005). The Lyme disease agent exploits a tick protein to infect the mammalian host. Nature *436*, 573–577.

Ramamoorthy, R., and Philipp, M.T. (1998). Differential expression of *Borrelia burgdorferi* proteins during growth *in vitro*. Infect. Immun. *66*, 5119–5124.

Revel, A.T., Blevins, J.S., Almazan, C., Neil, L., Kocan, K.M., de la Fuente, J., Hagman, K.E., and Norgard, M.V. (2005). *bptA* (*bbe16*) is essential for the persistence of the Lyme disease spirochete, *Borrelia burgdorferi*, in its natural tick vector. Proc. Natl. Acad. Sci. U.S.A. *102*, 6972–6977.

Revel, A.T., Talaat, A.M., and Norgard, M.V. (2002). DNA microarray analysis of differential gene expression in *Borrelia burgdorferi*, the Lyme disease spirochete. Proc. Natl. Acad. Sci. U.S.A. *99*, 1562–1567.

Riley, S.P., Bykowski, T., Babb, K., von Lackum, K., and Stevenson, B. (2007). Genetic and physiological characterization of the *Borrelia burgdorferi* ORF *BB0374-pfs-metK-luxS* operon. Microbiology *153*, 2304–2311.

Roberts, W.C., Mullikin, B.A., Lathigra, R., and Hanson, M.S. (1998). Molecular analysis of sequence heterogeneity among genes encoding decorin binding proteins A and B of *Borrelia burgdorferi* sensu lato. Infect. Immun. *66*, 5275–5285.

Rosa, P.A., Tilly, K., and Stewart, P.E. (2005). The burgeoning molecular genetics of the Lyme disease spirochaete. Nat. Rev. Microbiol. 3, 129–143.

Rossmann, E., Kitiratschky, V., Hofmann, H., Kraiczy, P., Simon, M.M., and Wallich, R. (2006). *Borrelia burgdorferi* complement regulator-acquiring surface protein 1 of the Lyme disease spirochetes is expressed in humans and induces antibody responses restricted to nondenatured structural determinants. Infect. Immun. *74*, 7024–7028.

Rossmann, E., Kraiczy, P., Herzberger, P., Skerka, C., Kirschfink, M., Simon, M.M., Zipfel, P.F., and Wallich, R. (2007). Dual binding specificity of a *Borrelia hermsii*-associated complement regulator-acquiring surface protein for factor H and plasminogen discloses a putative virulence factor of relapsing fever spirochetes. J. Immunol. *178*, 7292–7301.

Rostand, K.S., and Esko, J.D. (1997). Microbial adherence to and invasion through proteoglycans. Infect. Immun. *65*, 1–8.

Sacks, D.L. (1992). The structure and function of the surface lipophosphoglycan on different developmental stages of *Leishmania* promastigotes. Infect Agents Dis *1*, 200–206.

Sadziene, A., Thomas, D.D., and Barbour, A.G. (1995). *Borrelia burgdorferi* mutant lacking Osp: biological and immunological characterization. Infect. Immun. *63*, 1573–1580.

Schopf, J.W. (2000). Solution to Darwin's dilemma: discovery of the missing Precambrian record of life. Proc. Natl. Acad. Sci. U.S.A. *97*, 6947–6953.

Schulte-Spechtel, U., Lehnert, G., Liegl, G., Fingerle, V., Heimerl, C., Johnson, B.J., and Wilske, B. (2003). Significant improvement of the recombinant *Borrelia*-specific immunoglobulin G immunoblot test by addition of VlsE and a DbpA homologue derived from *Borrelia garinii* for diagnosis of early neuroborreliosis. J. Clin. Microbiol. *41*, 1299–1303.

Schwan, T.G., and Piesman, J. (2000). Temporal changes in outer surface proteins A and C of the Lyme disease-associated spirochete, *Borrelia burgdorferi*, during the chain of infection in ticks and mice. J. Clin. Microbiol. *38*, 382–388.

Schwan, T.G., and Piesman, J. (2002). Vector interactions and molecular adaptations of Lyme disease and relapsing fever spirochetes associated with transmission by ticks. Emerg. Infect. Dis. *8*, 115–121.

Schwan, T.G., Piesman, J., Golde, W.T., Dolan, M.C., and Rosa, P.A. (1995). Induction of an outer surface protein on *Borrelia burgdorferi* during tick feeding. Proc. Natl. Acad. Sci. U.S.A. *92*, 2909–2913.

Sell, S., and Salman, J. (1992). Demonstration of *Treponema pallidum* in axons of cutaneous nerves in experimental chancres of rabbits. Sex Transm Dis *19*, 1–6.

Sell, S., Salman, J., and Norris, S.J. (1985). Reinfection of chancre-immune rabbits with *Treponema pallidum*. I. Light and immunofluorescence studies. Am. J. Pathol. *118*, 248–255.

Seshu, J., Boylan, J.A., Gherardini, F.C., and Skare, J.T. (2004). Dissolved oxygen levels alter gene expression and antigen profiles in *Borrelia burgdorferi*. Infect. Immun. 72, 1580–1586.

Seshu, J., Esteve-Gassent, M.D., Labandeira-Rey, M., Kim, J.H., Trzeciakowski, J.P., Höök, M., and Skare, J.T. (2006). Inactivation of the fibronectin-binding adhesin gene *bbk32* significantly attenuates the infectivity potential of *Borrelia burgdorferi*. Mol. Microbiol. *59*, 1591–1601.

Shi, Y., Xu, Q., McShan, K., and Liang, F.T. (2008a). Both decorin-binding proteins A and B are critical for the overall virulence of *Borrelia burgdorferi*. Infect. Immun. *76*, 1239–1246.

Shi, Y., Xu, Q., Seemanapalli, S.V., McShan, K., and Liang, F.T. (2006). The *dbpBA* locus of *Borrelia burgdorferi* is not essential for infection of mice. Infect. Immun. *74*, 6509–6512.

Shi, Y., Xu, Q., Seemanaplli, S.V., McShan, K., and Liang, F.T. (2008b). Common and unique contributions of decorin-binding proteins A and B to the overall virulence of *Borrelia burgdorferi*. PLoS ONE 3, e3340.

Shin, O.S., Behera, A.K., Bronson, R.T., and Hu, L.T. (2007). Role of novel protein kinase C isoforms in Lyme arthritis. Cell. Microbiol. *9*, 1987–1996.

Shin, O.S., Isberg, R.R., Akira, S., Uematsu, S., Behera, A.K., and Hu, L.T. (2008). Distinct roles for MyD88 and Toll-like receptors 2, 5, and 9 in phagocytosis of *Borrelia burgdorferi* and cytokine induction. Infect. Immun. *76*, 2341–2351.

Simpson, W.J., Cieplak, W., Schrumpf, M.E., Barbour, A.G., and Schwan, T.G. (1994). Nucleotide sequence and analysis of the gene in *Borrelia burgdorferi* encoding the immunogenic P39 antigen. FEMS Microbiol. Lett. *119*, 381–387.

Skare, J.T., Mirzabekov, T.A., Shang, E.S., Blanco, D.R., Erdjument-Bromage, H., Bunikis, J., Bergström, S., Tempst, P., Kagan, B.L., Miller, J.N., and Lovett, M.A. (1997). The Oms66 (p66) protein is a *Borrelia burgdorferi* porin. Infect. Immun. *65*, 3654–3661.

Steere, A.C., Coburn, J., and Glickstein, L. (2004). The emergence of Lyme disease. J. Clin. Invest. *113*, 1093–1101.

Stevenson, B., and Babb, K. (2002). LuxS-mediated quorum sensing in *Borrelia burgdorferi*, the Lyme disease spirochete. Infect. Immun. *70*, 4099–4105.

Stevenson, B., El-Hage, N., Hines, M.A., Miller, J.C., and Babb, K. (2002). Differential binding of host complement inhibitor factor H by *Borrelia burgdorferi* Erp surface proteins: a possible mechanism underlying the expansive host range of Lyme disease spirochetes. Infect. Immun. *70*, 491–497.

Stevenson, B., Zückert, W.R., and Akins, D.R. (2001). Chapter 10. Repetition, conservation, and variation: the multiple cp32 plasmids of *Borrelia* species, In The Spirochetes: Molecular and Cellular Biology, M.H. Saier, Jr., and J. Garcia-Lara, eds. (Wymondham, Norfolk, U.K.: Horizon Press), pp. 87–100.

Stewart, P.E., Bestor, A., Cullen, J.N., and Rosa, P.A. (2008). A tightly regulated surface protein of *Borrelia burgdorferi* is not essential to the mouse-tick infectious cycle. Infect. Immun. *76*, 1970–1978.

Stewart, P.E., Wang, X., Bueschel, D.M., Clifton, D.R., Grimm, D., Tilly, K., Carroll, J.A., Weis, J.J., and Rosa, P.A. (2006). Delineating the requirement for the *Borrelia burgdorferi* virulence factor OspC in the mammalian host. Infect. Immun. *74*, 3547–3553.

Suk, K., Das, S., Sun, W., Jwang, B., Barthold, S.W., Flavell, R.A., and Fikrig, E. (1995). *Borrelia burgdorferi* genes selectively expressed in the infected host. Proc. Natl. Acad. Sci. U.S.A. *92*, 4269–4273.

Sung, S., McDowell, J., and Marconi, R. (2001). Evidence for the contribution of point mutations to *vlsE* variation and for apparent constraints on the net accumulation of sequence changes in *vlsE* during infection with Lyme disease spirochetes. J. Bacteriol. *183*, 5855–5861.

Szczepanski, A., and Benach, J.L. (1991). Lyme borreliosis: host responses to *Borrelia burgdorferi*. Microbiol. Rev. *55*, 21–34.

Szczepanski, A., Furie, M.B., Benach, J.L., Lane, B.P., and Fleit, H.B. (1990). Interaction between *Borrelia burgdorferi* and endothelium *in vitro*. J. Clin. Invest. *85*, 1637–1647.

Thomas, D.D., Cadavid, D., and Barbour, A.G. (1994). Differential association of *Borrelia* species with cultured neural cells. J. Infect. Dis. *169*, 445–448.

Thomas, D.D., and Comstock, L.E. (1989). Interaction of Lyme disease spirochetes with cultured eucaryotic cells. Infect. Immun. *57*, 1324–1326.

Tilly, K., Bestor, A., Jewett, M.W., and Rosa, P. (2007). Rapid clearance of Lyme disease spirochetes lacking OspC from skin. Infect. Immun. *75*, 1517–1519.

Tilly, K., Grimm, D., Bueschel, D.M., Krum, J.G., and Rosa, P. (2004). Infectious cycle analysis of a *Borrelia burgdorferi* mutant defective in transport of chitobiose, a tick cuticle component. Vector Borne Zoonotic Dis. *4*, 159–168.

Tilly, K., Krum, J.G., Bestor, A., Jewett, M.W., Grimm, D., Bueschel, D., Byram, R., Dorward, D., Vanraden, M.J., Stewart, P., and Rosa, P. (2006). *Borrelia burgdorferi* OspC protein required exclusively in a crucial early stage of mammalian infection. Infect. Immun. *74*, 3554–3564.

Tokarz, R., Anderton, J.M., Katona, L.I., and Benach, J.L. (2004). Combined effects of blood and temperature shift on *Borrelia burgdorferi* gene expression as determined by whole genome DNA array. Infect. Immun. *72*, 5419–5432.

van der Woude, M.W., and Bäumler, A.J. (2004). Phase and antigenic variation in bacteria. Clin. Microbiol. Rev. *17*, 581–611.

von Lackum, K., Babb, K., Riley, S.P., Wattier, R.L., Bykowski, T., and Stevenson, B. (2006). Functionality of *Borrelia burgdorferi* LuxS: the Lyme disease spirochete produces and responds to the pheromone autoinducer-2 and lacks a complete activated-methyl cycle. Int. J. Med. Microbiol. *296 Suppl. 40*, 92–102.

von Lackum, K., Miller, J.C., Bykowski, T., Riley, S.P., Woodman, M.E., Brade, V., Kraiczy, P., Stevenson, B., and Wallich, R. (2005). *Borrelia burgdorferi* regulates expression of complement regulator-acquiring surface protein 1 during the mammal-tick infection cycle. Infect. Immun. *73*, 7398–7405.

Wallich, R., Brenner, C., Kramer, M.D., and Simon, M.M. (1995). Molecular cloning and immunological characterization of a novel linear- plasmid-encoded gene, *pG*, of *Borrelia burgdorferi* expressed only in vivo. Infect. Immun. *63*, 3327–3335.

Wang, D., Botkin, D.J., and Norris, S.J. (2003). Characterization of the *vls* antigenic variation loci of the Lyme disease spirochaetes *Borrelia garinii* Ip90 and *Borrelia afzelii* ACAI. Mol. Microbiol. *47*, 1407–1417.

Wang, G., van Dam, A.P., and Dankert, J. (2001). Analysis of a VMP-like sequence (*vls*) locus in *Borrelia garinii* and Vls homologues among four *Borrelia burgdorferi* sensu lato species. FEMS Microbiol. Lett. *199*, 39–45.

Weening, E.H., Parveen, N., Trzeciakowski, J.P., Leong, J.M., Höök, M., and Skare, J.T. (2008). *Borrelia burgdorferi* lacking DbpBA exhibit an early survival defect during experimental infection. Infect Immun, IAI.00690–00608.

Winzer, K., and Williams, P. (2001). Quorum sensing and the regulation of virulence gene expression in pathogenic bacteria. Int. J. Med. Microbiol. *291*, 131–143.

Woodman, M.E., Cooley, A.E., Miller, J.C., Lazarus, J.J., Tucker, K., Bykowski, T., Botto, M., Hellwage, J., Wooten, R.M., and Stevenson, B. (2007). *Borrelia burgdorferi* binding of host complement regulator factor H is not required for efficient mammalian infection. Infect. Immun. 75, 3131–3139.

Wooten, R.M., Ma, Y., Yoder, R.A., Brown, J.P., Weis, J.H., Zachary, J.F., Kirschning, C.J., and Weis, J.J. (2002). Toll-like receptor 2 is required for innate, but not acquired, host defense to *Borrelia burgdorferi*. J. Immunol. *168*, 348–355.

Xu, Q., McShan, K., and Liang, F.T. (2008). Essential protective role attributed to the surface lipoproteins of *Borrelia burgdorferi* against innate defences. Mol. Microbiol. *69*, 15–29.

Xu, Q., Seemanapalli, S.V., McShan, K., and Liang, F.T. (2006). Constitutive expression of outer surface protein C diminishes the ability of *Borrelia burgdorferi* to evade specific humoral immunity. Infect. Immun. *74*, 5177–5184.

Xu, Q., Seemanaplli, S.V., McShan, K., and Liang, F.T. (2007). Increasing the interaction of *Borrelia burgdorferi* with decorin significantly reduces the 50 percent infectious dose and severely impairs dissemination. Infect. Immun. 75, 4272–4281.

Yang, X.F., Alani, S.M., and Norgard, M.V. (2003). The response regulator Rrp2 is essential for the expression of major membrane lipoproteins in *Borrelia burgdorferi*. Proc. Natl. Acad. Sci. U.S.A. *100*, 11001–11006.

Yang, X.F., Lybecker, M.C., Pal, U., Alani, S.M., Blevins, J., Revel, A.T., Samuels, D.S., and Norgard, M.V. (2005). Analysis of the *ospC* regulatory element controlled by the RpoN–RpoS regulatory pathway in *Borrelia burgdorferi*. J. Bacteriol. *187*, 4822–4829.

Yang, X.F., Pal, U., Alani, S.M., Fikrig, E., and Norgard, M.V. (2004). Essential Role for OspA/B in the life cycle of the Lyme disease spirochete. J. Exp. Med. *199*, 641–648.

Zambrano, M.C., Beklemisheva, A.A., Bryksin, A.V., Newman, S.A., and Cabello, F.C. (2004). *Borrelia burgdorferi* binds to, invades, and colonizes native type I collagen lattices. Infect. Immun. 72, 3138–3146.

Zhang, J.-R., and Norris, S.J. (1998a). Genetic variation of the *Borrelia burgdorferi* gene *vlsE* involves cassette-specific, segmental gene conversion. Infect. Immun. *66*, 3698–3704.

Zhang, J.R., Hardham, J.M., Barbour, A.G., and Norris, S.J. (1997). Antigenic variation in Lyme disease borreliae by promiscuous recombination of VMP-like sequence cassettes. Cell *89*, 275–285.

Zhang, J.R., and Norris, S.J. (1998b). Kinetics and *in vivo* induction of genetic variation of *vlsE* in *Borrelia burgdorferi*. Infect. Immun. *66*, 3689–3697.

Zhao, Z., Chang, H., Trevino, R.P., Whren, K., Bhawan, J., and Klempner, M.S. (2003). Selective up-regulation of matrix metalloproteinase-9 expression in human erythema migrans skin lesions of acute Lyme disease. J. Infect. Dis. *188*, 1098–1104.

Pathogenesis of Relapsing Fever

13

Alan G. Barbour and Betty P. Guo

Abstract

Relapsing fever is caused by several species of *Borrelia*, all but one of which are transmitted from reservoir animals to humans by soft (argasid) ticks. The exception is *B. recurrentis*, which is transmitted from one human to another by the body louse and, under certain conditions, may cause large epidemics. Relapsing fever *Borrelia* species have a number of properties that facilitate invasion and dissemination in the mammalian host; these include neurotropism and adhesiveness for erythrocytes and platelets. But their most characteristic pathogenetic feature is multiphasic antigenic variation. Surface-exposed lipoproteins largely determine the antigenic identities of the spirochaetes. Switching between genes encoding these highly polymorphic proteins is accomplished by a recombination in which the active gene at a single expression site is replaced by a copy of an archived, silent gene. The order of appearance of serotypes during infection is partially the function of the location of the archived gene and the surrounding sequences.

13.1 Introduction

> *The latter part of July and the months of August, September, and October, were infested with a fever, which was very frequent during this period, not unlike that of the autumn of the preceding year; which compare also the years 1741, 1745, 1748. It was attended with an intense pain in the head. It terminated in four, for the most part in five or six days, sometimes nine, and commonly in a critical sweat: it was far from being mortal. I was assured of seventy of the poorer sort at the same time in this fever; abandoned to the use of whey and God's good providence, who all recovered. The crisis was very imperfect, for they were subject to relapse, even sometimes to the third time.*
>
> John Rutty, *A Chronological History of the Weather and Seasons and of the Prevailing Diseases in Dublin*, London, Robinson and Roberts, 1770, p. 75.

John Rutty was born in Wiltshire, England, in 1698. After he received his medical degree from the University of Leiden in 1723, he moved to Dublin and practiced medicine there until his death in 1775. He kept a detailed diary, which contained, among the more personal recordings and musings, detailed accounts of the weather and the illnesses he witnessed in the city. The latter observations were assembled for a 1770 publication, from which the epigraph is excerpted.

While recognizing features of relapsing fever is possible in disease descriptions from Greece at the time of Hippocrates (Felsenfeld, 1971), Dr Rutty's description of outbreaks in Dublin in the 18th century is one of the first entries on what is believed to be relapsing fever in the literature of the modern era of empiricism and the scientific method. Obermeier's discovery of spirochaetes in the blood of relapsing fever patients would not occur for another century (Barbour and Hayes, 1986), and thus, we cannot now be certain of the aetiology of the epidemic that Rutty recorded. Yet, the above description is highly characteristic of the clinical course of relapsing fever (reviewed in Barbour, 2006a,b), in particular the occurrence of three or more episodes of fever, the ending of each episode in a 'crisis', the low mortality, and the illness's eventual resolution without specific therapy. While some vector-borne viral infections, such as dengue and Colorado Tick Fever, can present with headache, constitutional symptoms, and 'saddleback fever' that may suggest a relapse, the crisis that marks the end of a febrile period during relapsing fever is not typical of viral infection. During a crisis, which lasts about half an hour, there are shaking chills or rigors, a further rise in the temperature and tachycardia. The crisis is followed by profuse diaphoresis, falling temperature, and hypotension that usually persist for several hours. Mortality is highest during the crisis and its immediate aftermath.

Relapsing fever (RF) is either tick-borne (TBRF) or louse-borne (LBRF) (Felsenfeld, 1971). LBRF is caused by *Borrelia recurrentis*, and TBRF is caused by one of several different *Borrelia* species (see Chapter 9). What Rutty observed among his urban clientele in a northern European climate was almost surely the louse-borne form of relapsing fever. The outbreaks occurred when the weather was turning cooler and Dublin's residents probably were adding more layers of clothing, the preferred niche for the human body louse (see Chapter 10). Most of Rutty's cases were among the 'poorer sort' who lived in close quarters, bathed infrequently, and possessed few changes of seldom-laundered clothes, all risk factors for lousiness.

Over the millennia there have been millions of LBRF infections of humans in large and small epidemics (Bryceson *et al.*, 1970). Endemic transmission of the agent persists in the Horn of Africa and adjacent areas (Borgnolo *et al.*, 1993; Mitiku and Mengistu, 2002; Sundnes and Haimanot, 1993). As expected for a disease with this degree of public health importance, there have been numerous clinical, epidemiological, and entomological studies of LBRF, its agent, and vector. Many of these reports date to the first half of the last century and, with the detail characteristic of publications of that time, provide a wealth of information about the clinical aspects of LBRF. But *B. recurrentis*'s reservoir appears to be restricted to humans, and reports of experimental animal infections with *B. recurrentis* have been neither convincing nor reproducible (Barbour, 1986). *B. recurrentis* may cause a spirochaetaemia in weanling mice or newborn rabbits, but only transiently.

TBRF is a zoonosis that is documented in all regions of the world except Oceania and Antarctica. TBRF consistently accounts for thousands of cases of relapsing fever globally each year. Most of the studies on pathogenesis of and immunity to relapsing fever have been carried out on TBRF species, whose natural reservoirs are rodents and other mammals. For the TBRF *Borrelia* species, infection of mouse and rat provide suitable experimental models (see Chapter 14). This chapter by necessity emphasizes studies of TBRF, especially those of two Nearctic species, *B. hermsii* and *B. turicatae*, and two Afro-Tropical/Palearctic species, *B. duttonii* and *B. crocidurae*, but the review also covers important aspects of LBRF infection. Most studies of the pathophysiology of the 'crisis' during relapsing fever and the related phenomenon of the Jarisch–Herxheimer reaction have been conducted on human disease, because these features of RF have not been well reproduced in an animal model. *B. recurrentis* is descended from the tick-borne species *B. duttonii* or a close relative (Lescot *et al.*, 2008) and, thus, the findings of experimental studies of tick-borne species may be applicable to *B. recurrentis* and LBRF.

13.2 Transmission

Harry Hoogstral, the medical acarologist and chronicler of tick-borne diseases, proposed that the *Borrelia* lineage began as symbionts of ticks rather than as parasites of vertebrates (Hoogstraal, 1979). Spirochaetes are known endosymbionts in a number of invertebrates, notably in termites and molluscs. From that perspective, infection of vertebrates is but the means for transmission of the spirochaetes to another arthropod host, in this case soft-bodied (argasid) ticks of the genus *Ornithodoros* (Cooley and Kohls, 1944; Felsenfeld, 1971; Hoogstraal, 1979). One piece of evidence favouring an arthropod origin for the genus *Borrelia* is the high frequency of transovarial transmission observed for TBRF species, up to 100% of offspring, in most but not all species (Burgdorfer and Varma, 1967; Davis, 1943, 1952). The borreliae in the ovary invade the developing oocyte–yolk complex before the impervious shell forms around the egg (Aeschlimann, 1958). Some species and strains of RF borreliae can be serially transmitted vertically in ticks for several generations, sometimes with loss of infectivity for mammals (Balashov, 1972; Geigy and Aeschlimann, 1964).

The vertebrate reservoir hosts of the TBRF *Borrelia* spp. are usually rodents (Nicolle and Anderson, 1927), but may include pigs, goats, sheep, rabbits, bats, opossums, armadillos, foxes, cats, dogs, and birds (Felsenfeld, 1971; Hoogstraal, 1985; Pavlovsky, 1963; Thomas *et al.*, 2002). These animals usually become infected by the bite of an *Ornithodoros* or related soft-bodied tick (see Chapter 10). There are reports of peroral transmission through consumption of livers and brains of infected animals (reviewed in Barbour and Hayes, 1986). Transplacental transmission occurs in mammals and may be accompanied by placental damage and stillbirths (Larsson *et al.*, 2006a). Peroral and transplacental routes of infection are rare means of transmission in nature and probably insufficient for maintenance of the pathogen. Likewise, for humans, relapsing fever is almost always acquired through contact with a tick or louse carrying the spirochaetes. Rarely the disease is acquired (i) through accidental inoculation of infected blood, (ii) by contact of abraded or lacerated skin, mucous membranes, or conjunctiva with infected patient or animal blood (Herms and Wheeler, 1935), or (iii) by transplacental or perinatal transmission from mother to fetus (Fuchs and Oyama, 1969; Yagupsky and Moses, 1985). Transmission by aerosols, fomites, saliva, urine, faeces, or sexual contact does not occur.

Most *Ornithodoros* ticks that transmit relapsing fever feed for less than 30 minutes, usually at night. The argasid tick's usual duration of feeding is too short to allow for migration of the spirochaetes from midgut to salivary glands while embedded. The TBRF species spirochaetes have already migrated to the salivary glands by the time an *Ornithodoros* tick begins to feed and, consequently, the spirochaetes enter the host's skin and blood within minutes of the start of feeding (Burgdorfer, 1951). In some species of *Ornithodoros*, such as *O. moubata*, there may be transmission at the bite site through entry of infected coxal fluid (Varma, 1956).

In contrast to most TBRF species, transovarial transmission of *B. recurrentis* does not occur (Bryceson *et al.*, 1970; Felsenfeld, 1965). LBRF is not acquired from the bite or the saliva of the louse. After entering the midgut of the louse, the spirochaetes move to the haemolymph where they may persist for the approximately 3-week life-span of the louse. The spirochaetes do not migrate to the salivary glands or appear in faeces of the louse, nor is there regurgitation of spirochaetes from the midgut. Humans become infected with *B. recurrentis* when they crush an infected louse with fingers or 'pop' one between their teeth. The organism is introduced at the bite site, into the skin of the crushing fingers or the conjunctivae when eyes are rubbed, or through the mucous membranes of the mouth.

13.3 *Borrelia*–arthropod interactions

Little is known about *Borrelia* factors important for establishing infection in the arthropod, persistence and growth of the spirochaetes in the arthropod host, or transmission from the tick or louse to a vertebrate. In the early and mid-twentieth century there were many studies of borrelias in ticks and lice (reviewed in Barbour and Hayes, 1986; Burgdorfer and Varma, 1967; Felsenfeld, 1965; Felsenfeld, 1971). An example was Burgdorfer's comprehensive investigations of *B. duttonii* in its tick vector, *O. moubata* (Burgdorfer, 1951). For instance, a high density of spirochaetes were found around ganglia and in the salivary glands of the tick (Burgdorfer, 1959). The steps in the movement of spirochaetes from midgut through the haemolymph to the salivary glands, the staging area for infection of the vertebrate, were known. But the molecular correlates of these events and processes, for the most part, remain to be discovered.

An exception is the Vtp protein, which before its current designation went by 'pI$_C$', 'VmpC', and 'Vsp33' (Barbour *et al.*, 1982; Barstad *et al.*, 1985; Carter *et al.*, 1994). Earlier designations included 'C' because expression of the protein was only observed in populations growing in broth medium culture (Stoenner *et al.*, 1982). Vtp's function was unknown until Schwan and Hinnebusch (1998) demonstrated that Vtp was also preferentially expressed by spirochaetes in the salivary glands of unfed *Ornithodoros hermsi* ticks and that its production could be induced *in vitro* by lowering the incubation temperature to 23° C. Although this tick-associated variant can circulate in the blood of experimentally infected mice (Barbour *et al.*, 2000), it was not noted in a prospective study of relapses in infected mice using a battery of variant-specific antisera, including the tick-associated variant (Stoenner *et al.*, 1982). When spirochaetes expressing Vtp enter a new mouse host from the tick, they change back to the mammal-associated variant they last manifested in the mouse host (Schwan and Hinnebusch, 1998).

The Vtp proteins are homologous to the Vsp proteins of relapsing fever *Borrelia* spp. (see below) and OspC of Lyme borreliosis (LB) species (Carter *et al.*, 1994; Marconi *et al.*, 1993) (see Chapters 6 and 12). Further studies revealed that the Vtp proteins of different strains of *B. hermsii* collected in the western USA are highly polymorphic in sequence (Porcella *et al.*, 2005). This finding suggests the action of positive selection, such as the selection by an adaptive immune system or for survival in different niches, such as different vertebrate reservoirs.

13.4 Mammalian infection

A feeding period for soft ticks, measured in minutes, contrasts with feedings measured in days for the hard tick vectors of Lyme borreliosis species. Under these greatly restricted opportunities for transmission of RF species, the fact that a single spirochaete of the TBRF species *B. duttonii*, *B. hermsii*, or *B. turicatae* is sufficient to initiate an RF infection in experimental animals is not surprising (Beunders and Van Thiel, 1932; Schuhardt and Wilkerson, 1951; Stoenner *et al.*, 1982). The bacteria multiply in the blood at an estimated rate of one cell division every ~6 hours until they number between 10^6 to 10^7, sometimes as many as 10^8, per millilitre of blood at their peak (Fig. 13.1) (Cadavid *et al.*, 1994; Schuhardt and Wilkerson, 1951; Stoenner *et al.*, 1982) (C. Crowder, A.D. Putteet-Driver, A. Phillips, and A.G. Barbour, unpublished). Cultivable RF species achieve similar generation times in broth medium, such as BSK II (Barbour, 1986). The interval between relapses in experimental infections is usually 2 to 7 days. Untreated LBRF tends to have fewer relapses than does TBRF (Bodman and Stewart, 1948; Borgnolo *et al.*, 1993; Brown *et al.*, 1988; Bryceson *et al.*, 1970; Garnham *et al.*, 1947; Salih *et al.*, 1977; Sundnes and Haimanot, 1993).

The incubation period between exposure and onset of fever or other signs of illness is from 3 to 12 days. During relapses of disease the number of spirochaetes in the blood increases again but usually not to the same density as during the first episode of spirochaetaemia. The lower concentration of cells in a relapse spirochaetaemia does not appear to be due to attenuated virulence of the relapse serotypes. Russell (1936) described the relapse populations as being as virulent as the 'attack' serotype in naïve animals. Crowder *et al.* (C. Crowder, A.D. Putteet-Driver, A. Phillips and A.G. Barbour, unpublished) observed that the relapse population of *B. hermsii* increased more slowly than the serotype that initiated the infection in mice, but when each serotype was used to infect naïve mice, their growth rates were indistinguishable. These findings suggest that changes in host factors are also determinants of growth rates during infection. Evidence for this was the observation that during co-infection of experimental animals with a relapsing fever *Borrelia* species and an African trypanosome species the population growth of the parasite was slower than during single infections (Felsenfeld and Wolf, 1973).

The growth rates and maximum cell densities of *Borrelia* species vary by species and environmental conditions. Most notably, RF species and Lyme borreliosis species differ by a factor of 1000 or more in the cell densities they achieve in the blood of immunocompetent as well as immunodeficient mice (Barbour and Bundoc, 2001; Sadziene *et al.*, 1993a) (C. Crowder, A.D. Putteet-Driver, A. Phillips, and A.G. Barbour, unpublished). On the other hand, during *in vitro* cultivation in the same type of medium, representative RF and LB species have similar growth rates and reach stationary cell densities as high as ~10^8 per millilitre (Barbour, 1988; Cadavid *et al.*, 1994; Pennington *et al.*, 1997) (C. Crowder, A.D. Putteet-Driver, A. Phillips, and A.G. Barbour, unpublished). Moreover, expression profiles analysed by genome-wide arrays revealed little difference between cells of *B. hermsii* harvested from BSK II broth medium or in the blood of mice, when the temperature and pH of the broth

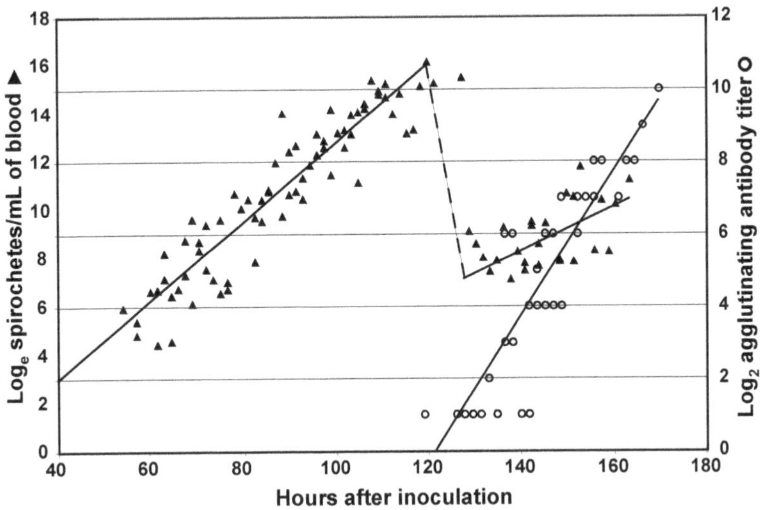

Figure 13.1 Experimental infection of mice with *Borrelia hermsii*. Numbers of spirochaetes and titres of agglutinating antibodies in the blood were determined during the infection of adult BALB/c mice (n = 95) inoculated intraperitoneally with 1–10 cells of serotype 7 of *Borrelia hermsii* at hour 0. Samples were collected every 3 hours beginning 12 hours after inoculation; second samples from each mouse were obtained 7.5 hours after the first. The number of spirochaetes was determined by quantitative polymerase chain reaction assay for the 16S ribosomal RNA gene as described (Bunikis and Barbour, 2005). For the macroagglutination assay twofold dilutions of plasma were incubated with serotype 7 cells as described (Sadziene *et al.*, 1993b). The left *y*-axis gives the $\log_e$ of spirochaetes per millilitre and the right *y*-axis gives the $\log_2$ of the titre of agglutinating antibody. Each data point is a sampling from a single mouse. Linear regression determinations of spirochaete population growth were made separately for hours 50–120 and >120 hours. The mean generation time of the spirochaetes during the first bacteraemia was ~4.5 hours. During the second bacteraemia, which was predominantly serotype 26, the mean generation time was ~7.5 hours. In naive mice the generation times of serotypes 7 and 26 are not significantly different (Restrepo and Barbour, 1994). The data are taken from C. Crowder, A.D. Putteet-Driver, A. Phillips and A.G. Barbour (unpublished).

approximated that of blood, and cell densities at the time of collection were comparable (Zhong and Barbour, 2004). Given these observations, genetic differences in metabolic pathways or other inherent capabilities plausibly account for observed differences between RF and LB species in peak burdens in the blood. Four pathways that differ between the two groups of species are in the acquisition of glycerol-3-phosphate (Bacon *et al.*, 2004; Schwan *et al.*, 2003; Schwan *et al.*, 1996), pyrimidine biosynthesis (Zhong *et al.*, 2006), purine salvage (Barbour *et al.*, 2005; Lescot *et al.*, 2008; Pettersson *et al.*, 2007), and glutamate-arginine-proline biosynthesis (Lescot *et al.*, 2008).

Greater persistence in the blood presumably increases the RF borrelia's fitness in terms of transmission to an infected vertebrate to a naïve set of tick vectors (Barbour and Restrepo, 2000). But RF spirochaetes also leave the vascular compartment through the endothelium to invade the central nervous system, the eye, the liver, testes, and other organs and tissues (Cadavid and Barbour, 1998; Cadavid *et al.*, 2001; Nordstrand *et al.*, 2001; Shamaei-Tousi *et al.*, 2001). The borrelias survive in these extravascular sites when they have been eliminated from the blood (Cadavid *et al.*, 1993; Cadavid *et al.*, 2006; Kazragis *et al.*, 1996; Larsson *et al.*, 2008). Persistence in extravascular niches may also serve to enhance transmissibility, because invasion of the blood from these sites can recur under conditions of immune suppression (Larsson *et al.*, 2006b).

In animal models, tissue invasion appears to be facilitated by recruitment of plasminogen to the borrelial cell surface. Once converted to enzymatically active plasmin, local degradation of extracellular matrix barriers in conjunction with bacterial motility apparently enable spread to distant organs. In plasminogen-deficient mice, *B. crocidurae* invasion of brain and kidney was delayed

(Nordstrand *et al.*, 2001). Another mechanism of tissue invasion occurs indirectly as a consequence of erythrocyte rosettes and microemboli-mediated haemorrhage (Fig. 13.2). In experimental rats, pressure-induced capillary rupture leads to leakage of aggregates of spirochaetes and erythrocytes into adjacent tissue (Shamaei-Tousi *et al.*, 2001).

In tissues the spirochaetes are generally found in the interstitial spaces between cells except when they have been engulfed by phagocytes. There is no compelling evidence that RF borrelias are facultative let alone obligatory intracellular pathogens. The pathological findings of relapsing fever are known mainly from autopsies of fatal human cases and from histological examinations of experimentally infected animals (Bryceson *et al.*, 1970; Cadavid and Barbour, 1998; Southern and Sanford, 1969). Extracellular spirochaetes were demonstrated in tissues, often in perivascular locations (Cadavid *et al.*, 2001; Duray, 1987; Pennington *et al.*, 1997). The liver and spleen were often enlarged and showed inflammation, with microabscesses in the spleen (Anderson and Zimmerman, 1955; Judge *et al.*, 1974b; Thomas *et al.*, 2002). The microabscesses, areas of necrosis, and petechiae may be the consequence, in part, of microemboli of spirochaetes and blood cells, and transendothelial migration of neutrophils (Shamaei-Tousi *et al.*, 2000; Shamaei-Tousi *et al.*, 1999).

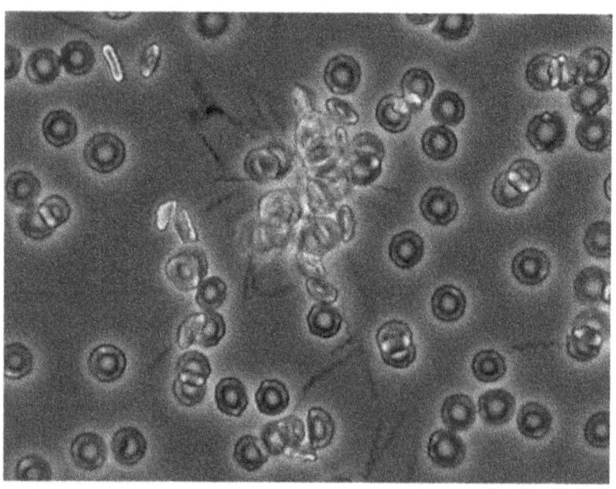

Figure 13.2 Phase contrast micrograph of a rosette of spirochaetes and erythrocytes formed during experimental relapsing fever. Blood was examined four days after inoculation of an adult C3H/HeN mouse with *Borrelia duttonii*. Magnification 400×. Photo courtesy of Marie Andersson and Sven Bergström, University of Umeå.

Intracranial haemorrhage has been observed in both LBRF and TBRF (Anderson and Zimmerman, 1955; Babes, 1916; Dewar and Walmsley, 1945; Judge *et al.*, 1974a; Thomas *et al.*, 2002). When spirochaetes in the brain were noted, they were more common around vessels and adjacent to leptomeninges (Anderson and Zimmerman, 1955; Cadavid *et al.*, 2001; García-Moncó *et al.*, 1997). Although cardiac abnormalities are not usually associated with relapsing fever, myocarditis with the presence of numerous spirochaetes has been demonstrated in experimental infections of both TBRF and LBRF (Breitschwerdt *et al.*, 1996; Cadavid *et al.*, 1994; Judge *et al.*, 1974b; Londono *et al.*, 2005; Wengrower *et al.*, 1984). In experimental *B. turicatae* infection, the severity of illness correlated with the numbers of spirochaetes in the blood (Pennington *et al.*, 1997).

In the blood of infected patients, leucocyte counts are usually in the normal range or only slightly elevated (Parsons, 1947). Leucopoenia and platelet counts below 50,000/mm^3 may occur during the crisis. Haemorrhagic complications, such as epistaxis, are common in LBRF (Dennis *et al.*, 1976; Perine *et al.*, 1971). The erythrocyte sedimentation rate often is moderately elevated. There may be laboratory evidence of hepatitis, such as elevated serum concentrations of aminotransferases and prolonged prothrombin and partial thromboplastin times (Barbour, 2006b). In the cerebrospinal fluid there may be mildly elevated numbers of mononuclear cells and protein levels but glucose levels in the CSF are typically normal.

13.5 The crisis phenomenon and the Jarisch–Herxheimer reaction

The mortality rates for untreated LBRF and TBRF are reported in the ranges of 10–70% and 4–10%, respectively (Barbour, 2006b). The higher mortality rates observed in some series may be attributable in part to the association of LBRF with wars, famine, and refugees, situations in which malnutrition and inadequate health care are common. If death occurs during RF, it is most common during a crisis episode or within hours after the start of antibiotic therapy. The crisis phenomenon corresponds to the rapid clearance of spirochaetes from the blood once a threshold level of neutralizing antibodies has been reached.

Administration of bactericidal antibiotics provides an analogous effect, which was termed the Jarisch–Herxheimer reaction (J–HR) after it was first observed during therapy of patients with secondary syphilis. In 1911 Balfour (Balfour, 1911) observed in chicks experimentally infected with *Borrelia anserina* that treatment with the arsenical '606' (neosalvarsan) produced an 'artificial crisis' in the animals. Subsequently, treatment of both LBRF and TBRF with antibiotics was shown to commonly result in the J–HR (Bryceson, 1976; Bryceson *et al.*, 1972; Warrell *et al.*, 1983; Webster *et al.*, 2002). J–HR is effectively the same as the crisis that follows the appearance of neutralizing antibodies in the blood. In both situations spirochaetes are lysed – by antibodies during the crisis and by antibiotics during J–HR; spirochaetal components are thus released into the circulation. Within 1–2 hours of the start of treatment the patient experiences intense rigors and becomes restless and apprehensive. This is accompanied by an elevation of temperature by 1–2°C, tachycardia, and a rise in blood pressure over 1 to 2 hours. This phase is followed over the next few hours by diaphoresis, exhaustion, a decline in temperature, hypotension, and leucopenia. The shock-like state, in which there is dilated vasculature and increased oxygen demands with fever (Warrell *et al.*, 1983), may be exacerbated by myocardial dysfunction if spirochaetes have invaded the heart.

Because of the similarity of J–HR to septic shock, the phenomenon was first thought to be the consequence of an endotoxin (Bryceson *et al.*, 1972). Borrelias have abundant glycolipids (Ben-Menachem *et al.*, 2003; Hossain *et al.*, 2001; Schroder *et al.*, 2008), but they do not have an endotoxin-like lipopolysaccharide (Ben-Menachem *et al.*, 2003; Takayama *et al.*, 1987). Another hypothesis about the crisis was that it was attributable to antigen–antibody complexes (Galloway *et al.*, 1977), but changes in complement levels were not detected in patients with LBRF (Warrell *et al.*, 1983). A more likely mechanism for J–HR and the crisis is the action of the abundant lipoproteins of spirochaetes on the toll-like 2 (TLR2) receptors of macrophages and other cells (Scragg *et al.*, 2000; Vidal *et al.*, 1998). Levels of tumour necrosis factor (TNFα), interleukin (IL) 6, and IL-8 rose several fold over pre-treatment levels during the J–HR in patients treated for LBRF (Negussie *et al.*, 1992). Infusion of antibodies recognizing human TNFα prior to administration of penicillin reduced the severity of J–HR in a clinical trial (Fekade *et al.*, 1996). In mice experimentally infected with *B. turicatae* or *B. hermsii*, serum concentrations of the anti-inflammatory cytokine IL-10 reach high levels (Londono *et al.*, 2008) (C. Crowder, A.D. Putteet-Driver, A. Phillips, and A.G. Barbour, unpublished).

13.6 Neurotropism

LBRF tends to exhibit a more severe course, but TBRF caused by some species of *Borrelia*, such as *B. duttonii* and *B. turicatae*, results in a higher frequency of neurological complications, which may include facial palsy, weakness, radiculopathy, and, occasionally, stupor or coma (Cadavid and Barbour, 1998). Neurotropism has also been observed experimentally in mammals (Cadavid and Barbour, 1998; Cadavid *et al.*, 1993; Dubois and Pearson, 1939; Kazragis *et al.*, 1996; Larsson *et al.*, 2006b; Shamaei-Tousi *et al.*, 1999; Velu *et al.*, 1930). RF borrelias appear in the brain of mice as early as the second day after inoculation (Andersson *et al.*, 2007). *B. crocidurae* cells remain actively growing in the brain and persist there in numbers of ~1000 per mouse brain (Larsson *et al.*, 2006b; Larsson *et al.*, 2008). An analogous phenomenon may occur in the tick vectors. In *Ornithodoros moubata* ticks, *B. duttonii* spirochaetes accumulated around the central ganglion cells (Burgdorfer, 1951).

In both human cases and experimental infections of rodents, neurotropism is more common following infection by some species than by others (Cadavid and Barbour, 1998). For example, the greater incidence of neurological involvement in cases of relapsing fever due to *B. duttonii* in Africa and *B. turicatae* in North America has long been noted (reviewed by Cadavid and Barbour, 1998). A newly described species of RF *Borrelia* in Spain appears to be particularly neurotropic (García-Moncó *et al.*, 1997).

Neurotropism also varies between variants of a given strain. Cadavid *et al.* (2001; 1997; 1994) demonstrated in an experimental animal model that some variants of a *B. turicatae* strain were present in higher numbers in the brain compared to other variants of the same strain. However, neurotropism is not necessarily associated with greater virulence. A neuroinvasive variant of *B. turicatae* had a lower fatality rate among experimentally infected mice than an isogenic variant with a lower potential for invasion of the brain (Cadavid *et al.*, 1994; Pennington *et al.*, 1997).

13.7 Adhesins

Adherence to host cells enables microbes to gain a foothold on surfaces and resist mechanical removal. Binding of RF *Borrelia* to cellular and soluble blood components, such as integrins and glycoaminoglycans, has been described. In the absence of knock-out mutants, the functional roles of these associations for infectivity and pathogenesis remain to be confirmed, but roles for these host ligands in mediating adherence of borrelias to host tissues and cells are plausible and would be consistent with similar, more established pathogenetic mechanisms of other pathogens, including *B. burgdorferi* (see Chapter 12).

Some species of relapsing fever spirochaetes have been noted to form aggregates or rosettes with platelets or erythrocytes in the blood of infected animals (Alugupalli *et al.*, 2003b; Burman *et al.*, 1998; Mooser, 1958). Fig. 13.2 shows a rosette of spirochaetes and erythrocytes. Rosettes are best viewed when spirochaete density in the blood is at its highest. For *B. crocidurae*, aggregation of spirochaetes and erythrocytes into rosettes appears to delay immune clearance; antibodies appear and rise in the blood more slowly than is observed in experimental infection with *Borrelia* spp. that do not aggregate erythrocytes (Burman *et al.*, 1998). Erythrocyte and spirochaete rosettes also activate endothelial cells and cause microvascular thrombi and emboli that lead in turn to pathological changes in tissues (Shamaei-Tousi *et al.*, 1999, 2000, 2001). Erythrocyte aggregation has also been noted in experimental infections with *B. duttonii* (Takahashi *et al.*, 2000).

B. hermsii attachment to platelets occurs via the integrin $\alpha_{IIb}\beta_3$ (Alugupalli *et al.*, 2001, 2003b), and *B. crocidurae* binds to erythrocytes via the neolactoglycan lacto-N-neotetraose (Guo *et al.*, 2009). The bacterial ligands for these interactions have not yet been identified for RF species. One possible spirochaete ligand is P66, or Oms66, an outer membrane protein porin of both LB and RF *Borrelia* species (Bunikis *et al.*, 1998; Bunikis *et al.*, 1995; Shang *et al.*, 1998; Skare *et al.*, 1997). P66 of *B. burgdorferi* has been reported to be an adhesin for the β_3-integrin (Coburn *et al.*, 1999; Coburn and Cugini, 2003).

Two other possible adhesins are products of open reading frames BH0512 and BH0553 of *B. hermsii*. Both proteins are characterized by multiple repeating domains. Indirect evidence for the adhesin function of these proteins was the finding that antisera to these proteins diminished attachment of *B. hermsii* to erythrocytes (Guyard *et al.*, 2005). One of the surface lipoproteins of *B. turicatae*, Vsp1, binds to glycosaminoglycans to a greater degree than other Vsp proteins of the same species or of *B. hermsii* (Magoun *et al.*, 2000; Zückert *et al.*, 2004).

13.8 Serum, complement and phagocyte resistance

Early investigators noted that RF *Borrelia* was susceptible to non-immune sera from some species of animals but resistant to the sera from other species. For example, *B. duttonii* survived in fresh sera from mice, guinea pigs, rabbits, pigeons, chickens, and horses but not in fresh sera from cow, goat, pig, or sheep (Cuboni, 1929). If the latter sera were first heat-inactivated, viability of the spirochaetes was retained, an indication that the alternative complement pathway was involved. This is similar to the differential serum resistance demonstrated by Lyme borreliosis species (see Chapters 14 and 15). The determinants of serum resistance among RF species have yet to be confirmed, but some bacterial components have been proposed to mediate these effects.

Factor H is a major complement regulator in the serum that prevents C3b formation by accelerating decay of the C3 convertase and by facilitating the irreversible inactivation of C3b. Bacterial acquisition of factor H could enable complement evasion by inactivating C3b or preventing its deposition and, thus, phagocytosis. Cells of some RF *Borrelia* species, including *B. hermsii*, but not cells of *B. anserina* or *B. coriaceae* displayed factor H binding activity (McDowell *et al.*, 2003; Meri *et al.*, 2006). This activity was attributed to plasmid-encoded proteins of ~20 kDa and called FhbA proteins (Hovis *et al.*, 2004). *B. hermsii* also binds to factor H-like protein 1, an alternative splice variant of factor H (Hovis *et al.*, 2006; Lopez *et al.*, 2008). FhbA elicits IgM and IgG responses during experimental infection of mice (Colombo and Alugupalli, 2008; R. Marcsisin and A.G. Barbour, submitted).

The BhCRASP-1 protein of *B. hermsii* is a 20-kDa outer membrane lipoprotein that is orthologous to the CRASP protein BBA68 of *B. burgdorferi*. BhCRASP-1 appears to bind factor H as well as plasminogen (Rossmann *et al.*, 2007). Finally, *B. duttonii* and *B. recurrentis* are reported to associate with C4b-binding protein in the serum (Meri *et al.*, 2006).

13.9 Immunity

Immunity in relapsing fever has been the subject of study since the late nineteenth century (reviewed in Barbour, 1987; Barbour, 2000). The relapse phenomena observed in experimental animals and in patients with neurosyphilis undergoing fever therapy with relapsing fever borrelias provided evidence for early immunologists, such as Paul Ehrlich, of the fine specificity of the adaptive immune response. The role of immunity in bringing about clearance of spirochaetes from the blood was demonstrated by findings that the initial spirochaetaemia of mice or rats can be prolonged by gamma irradiation (Barbour *et al.*, 1983; Newman and Johnson, 1984), treatment with cyclophosphamide (Newman and Johnson, 1984; Stoenner *et al.*, 1982), the severe combined immunodeficiency (SCID) phenotype (Cadavid *et al.*, 1994), or splenectomy (Meleney, 1928).

Fig. 13.1 shows the appearance of anti-*Borrelia* antibodies in response to infection and the proliferation of borrelias in a group of mice infected with <10 cells of *B. hermsii* (C. Crowder, A.D. Putteet-Driver, A. Phillips, and A.G. Barbour, unpublished). Clearance coincides with the rise of variant-specific agglutinating antibodies. This first immune response does not prevent the growth of a second wave of spirochaetes.

As the figure suggests, adaptive immunity to RF species is largely if not exclusively variant-specific. If there is an immune response that provides protection against several antigenic variants of a strain or species of RF *Borrelia*, it has not to our knowledge been reported. This may be one reason why there has not been an effective vaccine against relapsing fever, although the agent has been known for over 120 years. Aristowsky and Wainstein (1929) immunized humans with one variant of a strain; this provided protection against challenge with that particular variant but not variants isolated from relapse populations. The variant-specificity of immune response may also explain the paucity of useful serological assays for confirmation of the diagnosis of RF or for seroepidemiological surveys. The most promising subunit serological assay for RF is based on the GlpQ protein, which is found in relapsing fever *Borrelia* spp. but not in Lyme disease *Borrelia* spp. (Porcella *et al.*, 2000; Schwan *et al.*, 1996).

In 1896, Gabritchewsky showed the importance of humoral immunity for clearance of spirochaetes during relapsing fever when he demonstrated the lytic action of immune serum on borrelias (Gabritchewsky, 1896). Novy and Knapp reported in 1906 that they passively protected animals against infection with immune serum (Novy and Knapp, 1906). Calabi (1959) subsequently found that serum taken at the time of clearance reduced the viable count of spirochaetes in suspension by up to one million-fold. The lesser importance of cellular immunity for resolution of infection was indicated by the findings that T-cell deficient nude mice cleared spirochaetaemia as effectively as wild-type animals and that infection was cleared following reconstitution of lethally irradiated mice with B-cells alone but not T-cells (Barbour and Bundoc, 2001; Newman and Johnson, 1984).

As the foregoing experiments would suggest, IgM antibodies from infected animals are alone sufficient to limit or eliminate infection in the blood (Alugupalli *et al.*, 2003a; Arimitsu and Akama, 1973; Barbour and Bundoc, 2001; Connolly and Benach, 2001; Newman and Johnson, 1981, 1984). Within 24 to 48 hours after the spirochaetes reach a peak of 10^6 to 10^7 organisms per millilitre of blood, they are undetectable by light microscopy in the blood of an immunocompetent animal. When antibodies are administered intravenously, the spirochaetes are cleared from the blood within one hour (Barbour and Bundoc, 2001; Cadavid *et al.*, 1993). As earlier empirical studies had suggested, the neutralizing antibody response is serotype-specific: the animal remains susceptible to other serotypes of the same strain or of different strains (Barbour, 1987; Barbour, 2000). Up to 10 different relapses have been recorded during experimental infections (reviewed in Felsenfeld, 1965). Whether recovery is due to the development of antibodies to all or most serotypes during the infection's course or to the eventual rise of antibodies directed against antigens expressed by all serotypes is unknown. In endemic areas, newcomers or visitors appear to be at greater risk of symptomatic TBRF than long-term residents, an indication that long-lasting immunity to the local strain(s) may occur among humans.

Specific antibodies, even F_{ab} monomers, recognizing outer membrane proteins of *B. hermsii* and other RF species can kill borrelias in the absence of complement or phagocytes (Connolly and Benach, 2001; Connolly *et al.*, 2004; Sadziene *et al.*, 1994). The target for bactericidal antibody lies in the variable region of the *B. hermsii* protein (LaRocca *et al.*, 2008), as it does for bactericidal antibodies against a *B. burgdorferi* surface protein (Sadziene *et al.*, 1994). Incubation of the antibody with *Borrelia* cells leads to bacterial lysis following outer membrane disruption and excessive bleb-formation (LaRocca *et al.*, 2008).

B-cell memory commonly is dependent on stimulation by helper T-cells. The fact that *Borrelia* clearance is not dependent on T-cell function indicated that a different B-cell type was responsible for generating bactericidal antibodies. B1b lymphocytes were shown to confer long-lasting immunity independent of a T-cell response via rapid generation of *Borrelia*-specific IgM (Alugupalli *et al.*, 2003a, 2004). Induction of this response occurs by way of toll-like receptor stimulation and B-cell antigen receptor activation (Alugupalli *et al.*, 2007).

During infection both experimental animals and human patients develop antibody responses against a number of different spirochaetal components, including the periplasmic flagella and heat-shock proteins. However, only antibodies directed against variant-specific proteins appear to be effective in controlling infection (Barbour and Bundoc, 2001; Cadavid *et al.*, 1993; Stoenner *et al.*, 1982). If there are other, less variable antigens to which the infected animal or person eventually responds with neutralizing antibody or, less likely, cell-mediated immunity, they have not yet been identified. The reduced numbers and slower growth of fully virulent spirochaetes in the blood during relapses in comparison to the first fever indicates that there may be other factors that limit infection after the initial attack. Whatever its basis, a host response that eradicates the spirochaetes of the predominant serotype from the blood is often not successful in clearing borrelias from the brain, cerebrospinal fluid, or eye (Barbour, 1987; Cadavid and Barbour, 1998; Larsson *et al.*, 2006b).

13.10 Antigenic variation

13.10.1 Vsp and Vlp proteins

The steady increase of spirochaetes in the blood, followed by a rapid and deep fall in their number, and then their reappearance in the blood (Fig. 13.1) are the direct result of the bacterium's unique capacity for antigenic variation. Investigators as early as 1918 noted that each new crop of spirochaetes differed antigenically from the population it succeeded and the population to succeed it (Jancso, 1918; Meleney, 1928; Russell, 1936). The individual variants of a given strain are called serotypes. By definition, there is little or no antigenic cross-reactivity between serotypes with polyclonal antisera obtained either from infected patients or immunized animal (Coffey and Eveland, 1967; Stoenner *et al.*, 1982). A serotype that had disappeared from an RF *Borrelia* population under the force of immunity could reappear again if even a single spirochaete of the surviving population was transmitted to a naïve animal (Barbour, 1987); thus the hereditary information conferring serotype identity is not lost from a cell. This feature of relapsing fever indicated that antigenic variation was not simply an accelerated evolutionary process with selection for a succession of mutants, as typified by influenza A virus.

B. hermsii, a North American species that is most commonly encountered in the mountainous areas of the West (see Chapter 10), has been the subject of the most extensive studies of antigenic variation (Barbour, 2002). Several serotypes of the HS1 strain of this species were derived from a single cell injected into a mouse (Barbour and Stoenner, 1985; Restrepo *et al.*, 1992; Stoenner *et al.*, 1982). Stoenner raised specific antisera in mice to each of 24 *B. hermsii* serotypes and used these reagents to study the relapses that occurred during experimental infections (Stoenner *et al.*, 1982). With this battery of antisera in immunofluorescence assays of blood smears from infected animals, Stoenner identified the serotypes of ~80–90% of the variants that appeared during relapses in hundreds of mice. Examination of the bulk of *B. hermsii* chromosomal sequences and plasmids revealed additional *vsp* and *vlp* genes for 59 different serotypes (Dai *et al.*, 2006).

Serotype identity is determined by expression of one of a repertoire of polymorphic lipoproteins (Barbour *et al.*, 1982) (see Chapter 6). The processed N-terminus is acylated similarly to other bacterial lipoproteins and anchors the protein in the outer membrane (Carter *et al.*, 1994; Scragg *et al.*, 2000). The known set of about 40 proteins, for which expression is documented, is divided about equally between two different families: variable large proteins (Vlp) of about 36 kDa and variable small proteins (Vsp) of about 20 kDa (Barbour *et al.*, 1983; Burman *et al.*, 1990; Carter *et al.*, 1994; Restrepo *et al.*, 1992). These are encoded by *vlp* and *vsp* genes, respectively. Full-length Vsp and Vlp proteins may have near-identical N-terminal sequences, but after the conserved signal peptide has been cleaved off, the processed proteins are highly divergent in sequence. Vsp and Vlp proteins are orthologues of the OspC proteins and VlsE proteins of LB *Borrelia* species, respectively (Carter *et al.*, 1994; Marconi *et al.*, 1993; Zhang *et al.*, 1997). Another orthologue of Vsp is the Vtp protein that is expressed in ticks (see above).

Fig. 13.3 represents the structure of a Vsp protein dimer. The Vsp and Vlp proteins are not obviously homologous at the primary sequence level, but they resemble each other at the conformational level, being predominantly α-helical in secondary structure (Burman *et al.*, 1990; Zückert *et al.*, 2004) and having similar folds (Lawson *et al.*, 2006). Vsp proteins exist as dimers (Zückert *et al.*, 2001, 2004) and in this configuration resemble the deduced structure of a single Vlp protein (Eicken *et al.*, 2002). The C-terminus is not anchored but is also close to the N-terminus near the cell's surface for both Vsp and Vlp proteins. The loops between the four α-helices of each monomer are exposed to the environment as ligands or epitopes. The first

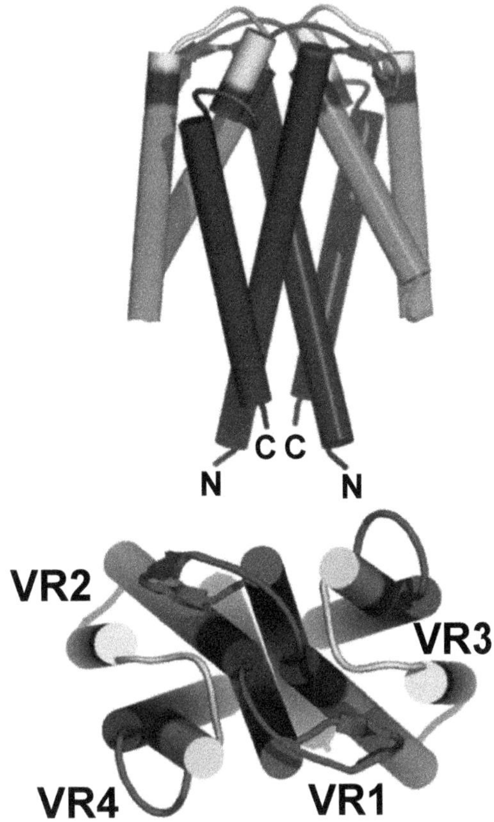

Figure 13.3 Schematic representation of structure of a Vsp lipoprotein dimer of B. turicatae. The four major alpha-helical chains in each monomer are depicted as cylinders. The first and fourth alpha-helix (dark grey) in each Vsp are relatively conserved in sequence between different Vsp proteins of a given species (Dai *et al.*, 2006; Restrepo *et al.*, 1992). The second and third alpha-helix (light grey) and loops between all the helical chains are variable in sequence and are divided into variable regions (VR) 1, 2, 3, and 4. The greatest variability is at the top or dome of the protein, as visualized from above in the lower figure. Adapted from Lawson *et al.* (2006).

and fourth alpha helical chains are conserved in sequence, while the second and third chains and loops between the chains comprise four variable regions of the molecule.

The Vlp family of proteins is furthered divided into four subfamilies with less than 70% sequence identity among them: α, β, γ, and δ (Burman *et al.*, 1990; Dai *et al.*, 2006; Hinnebusch *et al.*, 1998; Restrepo *et al.*, 1992). Different strains of *B. hermsii* from distant geographic origins in western North America had the same four subfamilies (Hinnebusch *et al.*, 1998). Fig. 13.4A shows a phylogram of the genes encoding several Vlp proteins of the HS1 strain of *B. hermsii* and the distances between and within the Vlp subfamilies. Serotype-specific epitopes of Vlp proteins are distributed throughout the protein (Barstad *et al.*, 1985).

The considerable diversity among the sets of *vsp* and *vlp* genes contained in each *Borrelia* cell is the consequence of multiple rounds of intragenic recombination among the alleles within each cell (Rich *et al.*, 2001b). But this process was probably preceded by horizontal transfer of alleles between strains, for otherwise there would not be a fund of heterogeneous sequences upon which recombinatorial processes could operate. Horizontal

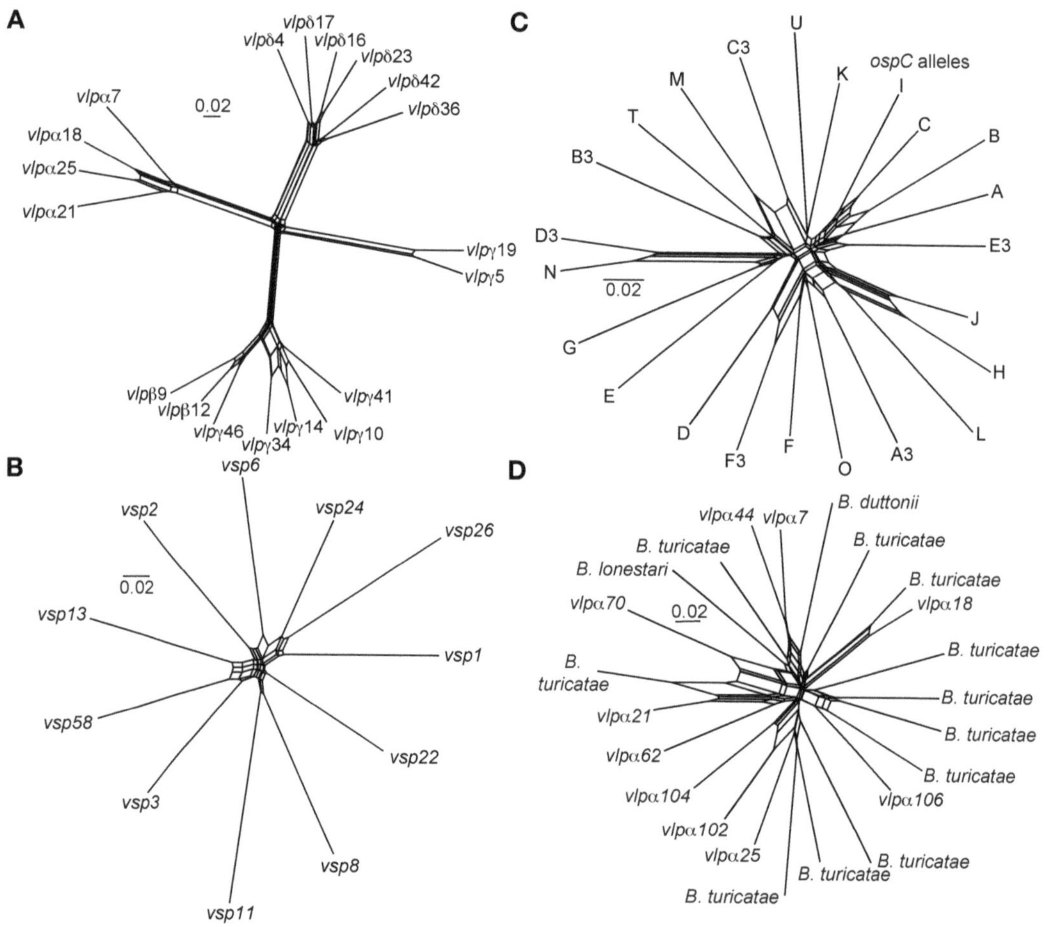

Figure 13.4 Evolution of variable antigen genes of relapsing fever *Borrelia* species. The four panels are unrooted distance phylograms of codon-aligned nucleotide sequences of *vlp* (A and D) and *vsp* (panel B) genes of one strain each of *B. hermsii* (A, B and D), *B. turicatae* (D), *B. duttonii* (D), and *B. lonestari* (D), and the *ospC* genes of each of several strains of the Lyme borreliosis agent *B. burgdorferi* in the northeastern USA (C). Phylogenetic relationships were inferred using the default values for the NeighborNet algorithm of version 4.10 of the SplitsTree4 suite (Huson and Bryant, 2006); the scale of nucleotide distance between the sequences is given in each panel. The star-like patterns of the phylograms and the networks of connections between branches in C and D are indicative of a history of recombination between the alleles. The relapsing fever *Borrelia* species sequences are from Dai *et al.* (2006) and Q. Dai, S. Campeau, S.F. Porcella, T.G. Schwan, and A.G. Barbour (unpublished), and the *B. burgdorferi* sequences are from B. Travinsky, J. Bunikis, and A.G. Barbour (unpublished).

transfer could be accomplished, for example, by the bacteriophage that exist in *B. hermsii* (Barbour and Hayes, 1986) and which may be synonymous with the 30-kb circular plasmids of this species (Eggers and Samuels, 1999; Stevenson *et al.*, 2000; Zhang and Marconi, 2005).

In contrast to the multiple *vsp* alleles in the genomes of RF *Borrelia* species, there is only one *ospC* allele in the genome of cells of LB species. The degree of divergence between the several *vsp* genes of a single strain of *B. hermsii* is comparable to that of *ospC* alleles representing several different strains of *B. burgdorferi* (Fig. 13.4B and C). Both sets of genes show evidence of being under positive selection and a history of multiple rounds of recombination (Livey *et al.*, 1995; Rich *et al.*, 2001b; Wang *et al.*, 1999). Since there is only one type of *ospC* allele per cell, recombination among these genes presumably occurred through genetic transfer between strains. If an ancestor of RF species, which possessed a single proto-*ospC/vsp* gene, became merodiploid through acquisition of a second orthologue, then with subsequent rounds of recombination and gene duplications, the number of alleles might increase from two to several within a single lineage.

A phylogenetic analysis of a selection of *vlp* alleles of the α subfamily from different species in the RF group, including *B. duttonii*, *B. turicatae*, and *B. lonestari* (see below), as well as the full set from *B. hermsii*, indicates that a basic repertoire of variable antigens was in-hand before speciation of the RF group further advanced. Fig. 13.4D shows a phylogram of the codon-aligned sequences and the underlying network of recombination. There was the same evidence of recombination within and perhaps between species in these genes as was observed in *vsp* genes, thus obscuring the evolutionary descent. But note that the *vlpα18* gene of *B. hermsii* and a *vlp* gene of *B. turicatae* are more similar to each other than they are to alleles of their own species, an indication that the common ancestor of *B. hermsii* and *B. turicatae* already had a differentiated *vlpα18* gene.

13.10.2 Genetic mechanisms for antigenic variation

Recognizing the cumulative studies of the biology, biochemistry, and immunology of relapsing fever, a modeller of the genetic mechanism of serotype-switching should account for the following features of the phenomenon (reviewed in Barbour, 2002, 2003): (1) The variation is reversible; a serotype once cleared by an animal can appear again in another animal to whom the pathogen, even a single cell, is passed. (2) It is multiphasic, that is, involving three or more different antigens. (3) Serotype identity is conferred by a set of proteins, only one of which is expressed at a time. (4) Serotype-specific epitopes of the protein are located in different parts of the protein, not just in one region. (5) When there is an antigenic switch in a cell during infection, all or nearly all of the protein changes, not just a limited region of the protein, as in VlsE variation in *B. burgdorferi* (see Chapter 12). (6) Antigen switches are effectively random and spontaneous. In other words, they are not an effect of a change in the environment, such as inoculation into a mouse from culture medium or the appearance of neutralizing antibody in the blood (Stoenner *et al.*, 1982).

Reversibility is achieved by archiving the antigen genes at locations where they are silent and, thus, presumably not under direct selection by immune systems and other host factors (Kitten and Barbour, 1990; Meier *et al.*, 1985; Plasterk *et al.*, 1985). For an antigen gene to be expressed, it must be copied and moved to a site where transcription can occur (Barbour *et al.*, 1991a; Burman *et al.*, 1990; Sohaskey *et al.*, 1999). This site may be altered by subsequent DNA rearrangements, leading to loss of the DNA that was located there, but a copy of that gene remains in the archive for future activation. The archived genes and the expression site for the duplicated gene are located on linear plasmids of 28–32 kb in *B. hermsii* (Ferdows *et al.*, 1996; Kitten and Barbour, 1990; Plasterk *et al.*, 1985) and on plasmids of other sizes in other RF species (Lescot *et al.*, 2008; Penningon *et al.*, 1999; Tabuchi *et al.*, 2002). There are multiple copies of each type of plasmid in each cell (Kitten and Barbour, 1992).

Sequencing of the linear plasmids that bear *vsp* and *vlp* genes in *B. hermsii* further defined the organization and arrangement of the expression site and the several archival sites (Burman *et al.*, 1990; Dai *et al.*, 2006; Restrepo *et al.*, 1992). There appears to be a complete or near-complete *vsp* or *vlp* gene for each Vsp or Vlp protein that a spirochaete is capable of expressing. The report by Dai *et al.* (2006) provides full physical maps of large regions of the plasmids. In brief, the archived *vsp* and *vlp* genes are, with a few exceptions, found in clusters, usually in the same orientation, i.e. head-to-tail, but head-to-head at some locations. Several of the silent genes are full-length, that is, with a start codon and the signal peptide. Most of these complete genes at silent sites also have consensus ribosomal binding sequences and transcriptional start sites but not discernible promoters. Other archived genes are lacking some sequence at the 5′ end. Some *vsp*- and *vlp*-like sequences are probably pseudogenes; they are missing 3′ ends and/or have indels with frameshifts. The expression site for *vsp* and *vlp* genes is located near the telomere of a 28-kb plasmid (Kitten and Barbour, 1990); it has a σ^{70}-type promoter (Barbour *et al.*, 1991a) and an upstream T-rich region that appears to enhance transcription (Barbour *et al.*, 1991b; Sohaskey *et al.*, 1999). Similar arrangements of the repertoires of *vsp* and *vlp* genes have been found on linear plasmids of various sizes in the RF species *B. duttonii* and *B. recurrentis* (Lescot *et al.*, 2008; Tabuchi *et al.*, 2002; Vidal *et al.*, 2002).

In *B. hermsii* three mechanisms can account for all the observed features of antigenic variation (Fig. 13.5). The first and most common mechanism is a non-reciprocal recombination between a linear plasmid with an array of silent, archived *vsp* and *vlp* genes and another linear plasmid with a transcribed *vsp* or *vlp* gene (Kitten and Barbour, 1990; Meier *et al.*, 1985; Plasterk *et al.*, 1985). In *B. hermsii*, as well as in *B. turicatae*, what is effectively a gene conversion results in the replacement of the expressed gene for a variable major protein at a site downstream from a promoter at an expression site (Barbour *et al.*, 1991a; Cadavid *et al.*, 1997; Penningon *et al.*, 1999; Pennington *et al.*, 1999). An archived *vsp* gene can replace an expressed *vlp* gene or vice versa. The boundaries for the recombination are regions of sequence identity between silent and expression sites around and flanking the extreme 5′ and 3′ ends of the expressed and silent *vsp* or *vlp* genes (Barbour *et al.*, 2000; Dai *et al.*, 2006). In some cases, recombination at the expression site occurs within a *vsp* or *vlp* gene (Dai *et al.*, 2006; Kitten *et al.*, 1993). However, if the resultant chimeric protein retained one or more specific epitopes of the original protein, it presumably would still be recognized by circulating antibodies elicited during the preceding spirochaetaemia (Barstad *et al.*, 1985).

In *B. hermsii*, the longest length of converted sequence that has been observed is about 2 kb (Kitten and Barbour, 1990; Restrepo *et al.*, 1992). In *B. turicatae*, a more extensive gene conversion involving 10 or more kb downstream of the promoter on the expression-site plasmid occurs (Penningon *et al.*, 1999). The mechanism of intermolecular recombination between plasmids first demonstrated in Nearctic species, like *B. hermsii*, also seems to occur in Afro-Tropical and Palearctic species. Rearrangements incorporating long lengths of plasmids have also been noted in *B. duttonii* (Takahashi *et al.*, 2000).

The second mechanism for expressing a different variable major protein is an intramolecular DNA rearrangement of the plasmid with the expression site (Fig. 13.5). This mechanism has only been noted when there is another variable major protein gene just downstream of an expressed *vlp* gene (Restrepo *et al.*, 1994). In such instances, the expressed gene and the untranscribed downstream gene are in the same orientation. A deletion between short direct repeats at the 5′ ends of each of the tandemly arrayed genes effectively excises the sitting *vlp* gene, and the downstream gene effectively takes the place of the deleted gene next to the promoter. After this deletion, the newly expressed gene is susceptible to small gene conversions that provide further diversity of variable membrane proteins (Restrepo and Barbour, 1994), but the significance of this phenomenon for further immune avoidance is not known.

The activation of the *vtp* gene when *B. hermsii* is in the tick occurs by a mechanism that does not require recombination or transposition. This is the third mechanism: modification of transcripts by expression site switching (Fig. 13.5). The *vtp* gene is located at another locus on a different plasmid than the expression site. At that location, *vtp* has its own promoter (Barbour *et al.*, 2000; Carter *et al.*, 1994). There is only one copy of *vtp* per genome. The mechanism of differential transcription between the *vtp* locus and the expression site locus is unknown.

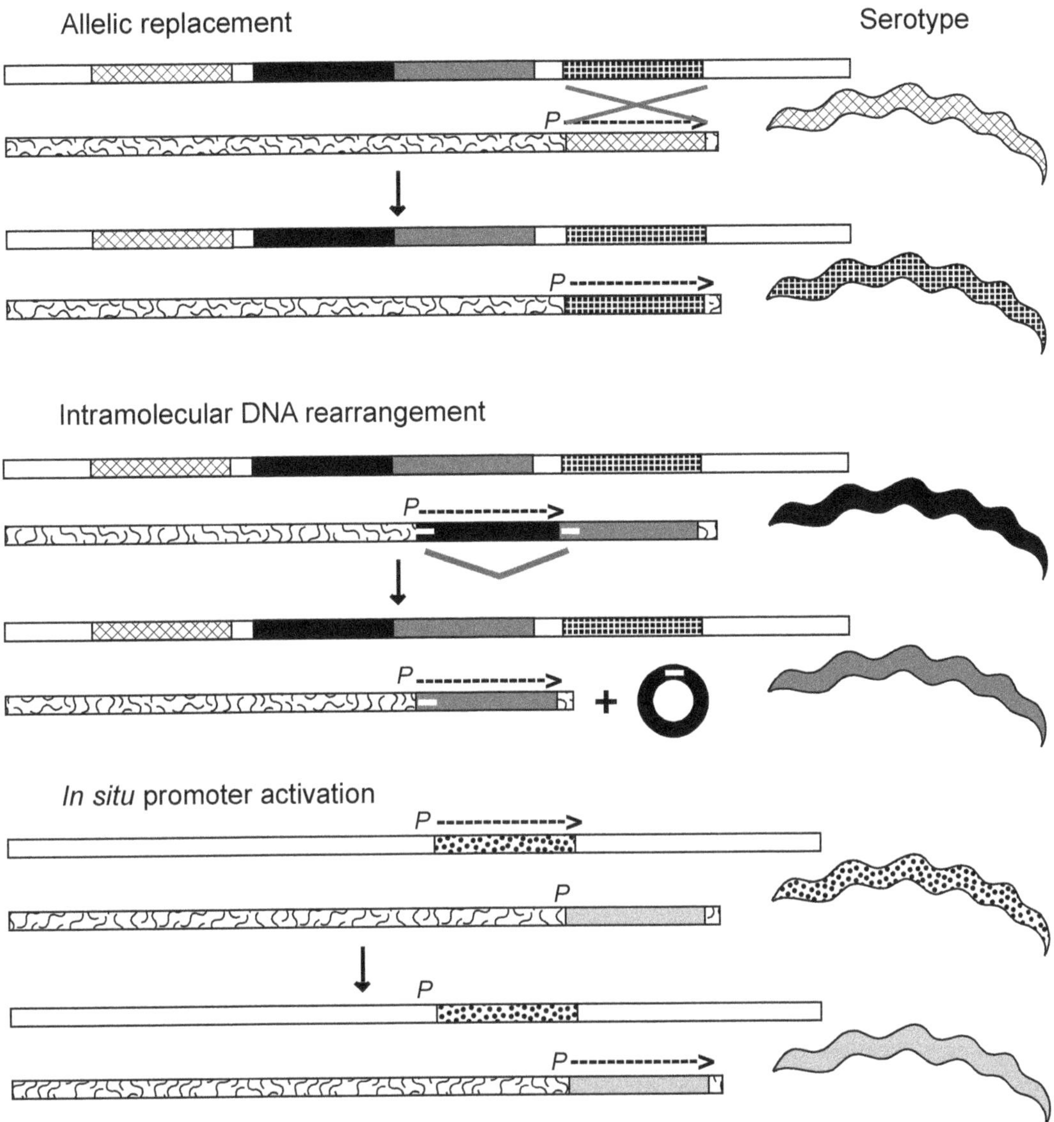

Figure 13.5 Three mechanisms for serotype switches by *B. hermsii*: allelic replacement, intramolecular DNA rearrangement, and *in situ* promoter activation. The different mechanisms are discussed in the text. *P* indicates the promoter and the location of the expression site near the right telomere of a linear plasmid; an overhead arrow shows whether the gene is active and the direction of transcription. *vsp* and *vlp* genes at the expression site location or in archival locations on the same (squiggly line background) or another linear plasmid (white background) are schematically represented by different patterns and greyscale values. The serotype of the cell, similarly indicated by pattern or greyscale, corresponds to *vsp* or *vlp* gene, at the expression site. In the second mechanism a deletion by direct repeats (small white horizontal bars) occurs and results in a non-replicative circle.

13.10.3 Programming of antigenic variation

The overall serotype switch rate for *B. hermsii* is an estimated 10^{-4} to 10^{-3} per cell per generation. That is, by the time a given serotype reaches a population size of 1000 to 10,000 cells in an animal, at least one cell of a different serotype has probably appeared in the population (Barbour and Stoenner, 1985; Stoenner *et al.*, 1982). However, the rates for individual switches, e.g.

from serotype 3 to serotype 7 or from serotype 17 to serotype 19, vary over about a thousand-fold range (Barbour *et al.*, 2006; Barbour and Stoenner, 1985). The rates in each direction are not equivalent. For instance, switches from serotype 7 to serotype 19 are much less frequent than a switch from serotype 19 to serotype 7. These differences in switch rates among all possible antigenic variation events could account for the rough ordering of the appearance serotypes during individual infections that has been observed (Coffey and Eveland, 1967; Frank and Barbour, 2006; Stoenner *et al.*, 1982).

The organization of the *vsp* and *vlp* genes on the plasmids provided clues to the possible mechanisms for the observed differences in frequencies of serotypes during the first relapse (Dai *et al.*, 2006). The more frequent *B. hermsii* serotypes had Vsp or Vlp proteins that were encoded by genes with certain characteristics in their silent or archival locations (Barbour *et al.*, 2006). One important characteristic was the extent of identity between the 5′ end of archived *vsp* or *vlp* sequence and the 5′ end of the expression site (Fig. 13.6). This region of identity of approximately 50 nucleotides is called the UHS region and is the usual site for the upstream crossover during recombination (Dai *et al.*, 2006). The second major variable was the spatial relationship of a given archived gene to an extragenic repetitive sequence of ~200 nucleotides called the DHS element, which directly follows some but not all archived genes and, thus, is at various distances downstream of a silent allele. The DHS contains the downstream crossover site for the recombination (Dai *et al.*, 2006).

The serotypes that were most frequently observed during the first relapse of infection in laboratory mice were specified by *vsp* or *vlp* genes that in their archived locations had the following properties: (1) a UHS sequence that was identical or near-identical to the UHS at the expression site, and (2) a DHS element 50–200 nucleotides downstream. Archived *vsp* or *vlp* alleles that lacked both properties were rarely expressed during the first or early relapses. Silent genes with either property were intermediate in frequency (Barbour *et al.*, 2006). Presumably, possession of both (1) and (2) fostered recombination between the archived sequence and the expression site.

13.11 Other *Borrelia* species

As described in Chapter 9, *Borrelia* species sort into two major clades: one comprises the species that cause LB as well as related species, like *B. bissettii* and *B. andersoni*, and a second clade includes the species that cause relapsing fever. The second clade additionally includes species that do not by definition cause 'relapsing fever' but that cause vertebrate infections sharing with relapsing fever the characteristic of levels of spirochaetaemia that are microscopically

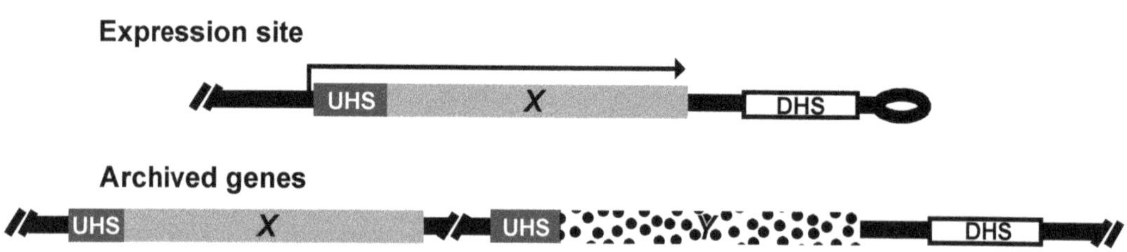

Figure 13.6 Organization of the expression plasmid (top) and one archival plasmid (bottom) in *B. hermsii*. Two examples of *vsp* or *vlp* genes are denoted by 'X' and 'Y' and different fill patterns. There is a single expression site on one plasmid, and silent variants of *vsp* and *vlp* genes are found on the same and other plasmids. The direction and extent of transcription of the duplicate gene at the expression site is indicated by the arrow. The UHS element at the expression site comprises 61 nt around the start codon of the variant gene (Barbour *et al.*, 1991a; Dai *et al.*, 2006; Kitten *et al.*, 1993). Silent genes vary in the extent to which their UHS regions are identical to the expression site UHS; this is represented here by the relative length of the UHS block. There is a 214 nt non-coding DHS element downstream from the expression site and adjacent to the plasmid telomere and at various locations on the plasmids. The lengths of *vsp*/*vlp* gene examples and the UHS and DHS elements are not to scale.

detectable, i.e. cell concentrations of $\geq 10^4$ per ml of blood.

B. anserina and *B. coriaceae* are two species transmitted by soft ticks but not known to cause relapsing fever in humans. *B. anserina*, the agent of avian or fowl spirochaetosis, is transmitted by *Argas persicus* and related species, and is global in distribution in poultry flocks (DaMassa and Adler, 1979; Diab and Soliman, 1977; Zaher *et al.*, 1977). The disease causes severe anaemia, weight loss, and high mortality in affected poultry flocks. Heavy spirochaetaemia of the first episode of illness may extend for 4–6 days in chickens and is prolonged for weeks in bursectomized chickens (Bandopadhyay and Vegad, 1983; DaMassa and Adler, 1979). There are antigenic differences between strains of *B. anserina* and evidence of strain-specific immunity (DaMassa and Adler, 1979; Pachanian and Smith, 1970; Uppal and Rao, 1966). *B. anserina* has not been known for displaying antigenic variation during infection. The mortality of the early infection is often high, and many experimental infections were terminated after the first spirochaetaemia (Bandopadhyay and Vegad, 1983). Nevertheless, the occurrence of immune evasion is suggested by Balfour's report on 'Spirochaetosis of Sudanese fowls' (Balfour, 1911). He wrote that 'in chicks relapses are common, and the disease therefore approaches mammalian spirochaetosis in type.' In experimental infections of chicks, Balfour noted that 2- to 3-day periods of heavy spirochaetaemia culminated in a 'crisis.' This was followed by disappearance of spirochaetes from the blood, then reappearance of spirochaetes in the blood, sometimes in 'large numbers' after an interval of ~4 days. *B. anserina* has a Vsp-like lipoprotein of 22 kDa (Sambri *et al.*, 1999). Chicks that were passively immunized with antisera to this protein were protected against subsequent homologous challenge with ~10^5 borrelias, but antisera to a Vsp-like protein from another strain of *B. anserina* did not provide protection against heterologous challenge.

B. coriaceae occurs in *O. coriaceus* soft ticks in California (Hendson and Lane, 2000; Lane and Manweiler, 1988). These ticks feed on deer and cattle, but the role of these large mammals in maintenance of the infection remains to be determined. *B. coriaceae* was identified by PCR in the blood of a mule deer and a Columbian black-tailed deer in California (Lane *et al.*, 2005; Yabsley *et al.*, 2005). *O. coriaceus* has been associated with epizootic bovine abortion in California and Nevada (Schmidtmann *et al.*, 1976). *B. coriaceae*'s aetiological role in this disease has been suggested but not proven (Lane *et al.*, 1985).

B. theileri is an RF-like *Borrelia* species that is transmitted by hard ticks of the metastriate genus *Rhipicephalus* rather than by argasid ticks (see Chapter 10). *B. theileri* produces a microscopically detectable spirochaetaemia and weakness, loss of appetite, and anaemia in cattle in Africa, Australia, and Europe (Brocklesby *et al.*, 1963; Sharma *et al.*, 2000; Smith *et al.*, 1978). According to Theiler's original description of the infection, there may be two attacks of fever, suggesting the occurrence of antigenic variation (Theiler, 1904). Transovarial transmission of *B. theileri* in *Rhipicephalus (Boophilus) microplus* occurs (Smith *et al.*, 1978). Another *Borrelia* sp. that clusters with *B. theileri* was identified in *Rhipicephalus (Boophilus) microplus* in Brazil (Yparraguirre *et al.*, 2007).

In the last decade, other RF group species that are transmitted by hard rather than soft ticks have been identified. These all cluster with *B. theileri* in a monophyletic group distinct from species transmitted by soft ticks. One of the newly discovered species is *B. lonestari*, which was found in the USA in *Amblyomma americanum*, another metastriate tick (Barbour *et al.*, 1996; Rich *et al.*, 2001a). A similar organism was identified in *A. cajennense* ticks, a species whose range extends to Central America, the Caribbean, and South America (A.G. Barbour, unpublished). The distribution in the southern and eastern USA of *A. americanum* corresponds with the distribution of an illness characterized by a rash resembling erythema migrans of Lyme borreliosis, which has been called southern tick-associated rash illness or STARI (Campbell *et al.*, 1995). Moyer *et al.* (2006) used a low-passage culture isolate of *B. lonestari* to infect white-tailed deer (*Odocoileus virginianus*), a common host of *A. americanum* ticks, as well as dogs, mice, and cattle. Only deer developed spirochaetaemia, as evidenced by blood smear and PCR; the spirochaetaemia lasted for 6–9 days. Yabsley *et al.* (2008) subsequently found that raccoons were refractory to infection with *B. lonestari*. The failure of *B. lonestari* to infect mice is attributable to the bactericidal activity of naïve mouse serum for this species (Grigery *et al.*, 2005).

Another RF-like species of hard ticks, *B. miyamotoi*, was first discovered in *Ixodes persulcatus* ticks in Japan (Fukunaga and Koreki, 1995), and similar organisms have been found in other *Ixodes* species: *I. scapularis* in the northeastern USA (Bunikis *et al.*, 2004; Scoles *et al.*, 2001), *I. pacificus* in California (Mun *et al.*, 2006), and *I. ricinus* in Europe (Bunikis *et al.*, 2004; Fraenkel *et al.*, 2002). *I. scapularis* transovarially transmits the spirochaete (Scoles *et al.*, 2001). Laboratory mice can be infected with *B. miyamotoi* from *I. scapularis* (Scoles *et al.*, 2001), and the levels of spirochaetaemia are comparable to that observed for TBRF *Borrelia* spp. (Schwan *et al.*, 2003). But whether *B. miyamotoi* can cause infections of humans has not been established.

13.12 Future directions

Several other species of *Borrelia* probably have reservoir host-vector cycles that rarely if ever involve human infections and, for this reason, have escaped attention. These may include organisms that, like *B. miyamotoi* and *B. lonestari*, share features with both the relapsing fever and Lyme borreliosis groups of species. These postulated new species may not include a 'missing link' per se, but they could be useful for further defining the differences and similarities in the pathogenesis of relapsing fever and Lyme borreliosis. Facilitating this and other lines of investigation are the increasing number of tools for genetic manipulation of relapsing fever borrelias.

References

Aeschlimann, A. (1958). Développement embryonnaire d'*Ornithodorus moubata* (Murray) et transmission transovarienne de *Borrelia duttoni*. Acta Tropica *15*, 15–64.

Alugupalli, K.R., Akira, S., Lien, E., and Leong, J.M. (2007). MyD88- and Bruton's tyrosine kinase-mediated signals are essential for T-cell-independent pathogen-specific IgM responses. J. Immunol. *178*, 3740–3749.

Alugupalli, K.R., Gerstein, R.M., Chen, J., Szomolanyi-Tsuda, E., Woodland, R.T., and Leong, J.M. (2003a). The resolution of relapsing fever borreliosis requires IgM and is concurrent with expansion of B1b lymphocytes. J. Immunol. *170*, 3819–3827.

Alugupalli, K.R., Leong, J.M., Woodland, R.T., Muramatsu, M., Honjo, T., and Gerstein, R.M. (2004). B1b lymphocytes confer T-cell-independent long-lasting immunity. Immunity *21*, 379–390.

Alugupalli, K.R., Michelson, A.D., Barnard, M.R., Robbins, D., Coburn, J., Baker, E.K., Ginsberg, M.H., Schwan, T.G., and Leong, J.M. (2001). Platelet activation by a relapsing fever spirochaete results in enhanced bacterium–platelet interaction via integrin $\alpha_{IIb}\beta_3$ activation. Mol. Microbiol. *39*, 330–340.

Alugupalli, K.R., Michelson, A.D., Joris, I., Schwan, T.G., Hodivala-Dilke, K., Hynes, R.O., and Leong, J.M. (2003b). Spirochete-platelet attachment and thrombocytopenia in murine relapsing fever borreliosis. Blood *102*, 2843–2850.

Anderson, T.R., and Zimmerman, L.E. (1955). Relapsing fever in Korea: a clinicopathologic study of eleven fatal cases with special attention to association with salmonella infections. Am. J. Pathol. *31*, 1083–1109.

Andersson, M., Nordstrand, A., Shamaei-Tousi, A., Jansson, A., Bergström, S., and Guo, B.P. (2007). In situ immune response in brain and kidney during early relapsing fever borreliosis. J. Neuroimmunol. *183*, 26–32.

Arimitsu, Y., and Akama, K. (1973). Characterization of protective antibodies produced in mice infected with *Borrelia duttonii*. Jpn. J. Med. Sci. Biol. *26*, 229–237.

Aristowsky, W.M., and Wainstein, A.B. (1929). Rekurrens-Schutzimpfungsversuche am Menschen. I. Zeitschrift für Immunitaetsforschung und Experimetelle Therapie *61*, 296–308.

Babes, V. (1916). Hemorragies meningees et autres manifestations hemorragiques dans la fievre recurrente. Comptes Rendus des Seances Societe de Biologie (Paris) *79*, 855–857.

Bacon, R.M., Pilgard, M.A., Johnson, B.J., Raffel, S.J., and Schwan, T.G. (2004). Glycerophosphodiester phosphodiesterase gene (*glpQ*) of *Borrelia lonestari* identified as a target for differentiating *Borrelia* species associated with hard ticks (Acari:Ixodidae). J. Clin. Microbiol. *42*, 2326–2328.

Balashov, Iu.S. (1972). Infestation of natural populations of *Ornithodoros tartakovskyi* 01., 1931 ticks with *Borrelia latyshevi* Sof., 1941 spirochetes in different sections of the geographic area of the vector. Parazitologiia *6*, 97–102.

Balfour, A. (1911). Fourth Report of the Wellcome Tropical Research Laboratories at the Gordon Memorial College Khartoum (London, Wellcome Tropical Research Laboratories), pp. 76–107.

Bandopadhyay, A.C., and Vegad, J.L. (1983). Observations on the pathology of experimental avian spirochaetosis. Res. Vet. Sci. *35*, 138–144.

Barbour, A.G. (1986). Cultivation of *Borrelia*: a historical overview. Zentralbl. Bakteriol. Mikrobiol. Hyg. [A] *263*, 11–14.

Barbour, A.G. (1987). Immunobiology of relapsing fever. Contrib. Microbiol. Immunol. *8*, 125–137.

Barbour, A.G. (1988). Laboratory aspects of Lyme borreliosis. Clin. Microbiol. Rev. *1*, 399–414.

Barbour, A.G. (2000). *Borrelia* infections: relapsing fever and Lyme disease. In Effects of Microbes on the Immune System, M.W. Cunningham, and R.S. Fujinami, eds. (Philadelphia, Lippincott, Williams, and Wilkins), pp. 57–70.

Barbour, A.G. (2002). Antigenic variation by relapsing fever *Borrelia* species and other bacterial pathogens. In Mobile DNA II, N.L. Craig, R. Craigie, M. Gellert, and A. Lambowitz, eds. (Washington, D.C., American Society for Microbiology), pp. 972–994.

Barbour, A.G. (2003). Antigenic variation in *Borrelia*: relapsing fever and Lyme borreliosis. In Antigenic Variation, A. Craig, and A. Scherf, eds. (London, Academic Press), pp. 319–356.

Barbour, A.G. (2006a). Relapsing fever. In Tick-borne Diseases of Humans, J. Goodman, D.T. Dennis, and D. Sonenshine, eds. (Washington, D.C., American Society for Microbiology), pp. 268–291.

Barbour, A.G. (2006b). Relapsing fever and other *Borrelia* diseases. In Tropical Infectious Diseases Principles, Pathogens, & Practice, R.L. Guerrant, D.H. Walker, and P.F. Weller, eds. (Philadelphia, Churchill Livingstone), pp. 499–510.

Barbour, A.G., Barrera, O., and Judd, R.C. (1983). Structural analysis of the variable major proteins of *Borrelia hermsii*. J. Exp. Med. *158*, 2127–2140.

Barbour, A.G., and Bundoc, V. (2001). *In vitro* and *in vivo* neutralization of the relapsing fever agent *Borrelia hermsii* with serotype-specific immunoglobulin M antibodies. Infect. Immun. *69*, 1009–1015.

Barbour, A.G., Burman, N., Carter, C.J., Kitten, T., and Bergström, S. (1991a). Variable antigen genes of the relapsing fever agent *Borrelia hermsii* are activated by promoter addition. Mol. Microbiol. *5*, 489–493.

Barbour, A.G., Carter, C.J., Burman, N., Freitag, C.S., Garon, C.F., and Bergström, S. (1991b). Tandem insertion sequence-like elements define the expression site for variable antigen genes of *Borrelia hermsii*. Infect. Immun. *59*, 390–397.

Barbour, A.G., Carter, C.J., and Sohaskey, C.D. (2000). Surface protein variation by expression site switching in the relapsing fever agent *Borrelia hermsii*. Infect. Immun. *78*, 7114–7121.

Barbour, A.G., Dai, Q., Restrepo, B.I., Stoenner, H.G., and Frank, S.A. (2006). Pathogen escape from host immunity by a genome program for antigenic variation. Proc. Natl. Acad. Sci. U.S.A. *103*, 18290–18295.

Barbour, A.G., and Hayes, S.F. (1986). Biology of *Borrelia* species. Microbiol. Rev. *50*, 381–400.

Barbour, A.G., Maupin, G.O., Teltow, G.J., Carter, C.J., and Piesman, J. (1996). Identification of an uncultivable *Borrelia* species in the hard tick *Amblyomma americanum*: possible agent of a Lyme disease-like illness. J. Infect. Dis. *173*, 403–409.

Barbour, A.G., Putteet-Driver, A.D., and Bunikis, J. (2005). Horizontally acquired genes for purine salvage in *Borrelia* spp. causing relapsing fever. Infect. Immun. *73*, 6165–6168.

Barbour, A.G., and Restrepo, B.I. (2000). Antigenic variation in vector-borne pathogens. Emerg. Infect. Dis. *6*, 449–457.

Barbour, A.G., and Stoenner, H.G. (1985). Antigenic variation of *Borrelia hermsii*. In Genome Rearrangement, M.I. Simon, and I. Herskowitz, eds. (New York, Alan R. Liss, Inc.), pp. 123–135.

Barbour, A.G., Tessier, S.L., and Stoenner, H.G. (1982). Variable major proteins of *Borrellia hermsii*. J. Exp. Med. *156*, 1312–1324.

Barstad, P.A., Coligan, J.E., Raum, M.G., and Barbour, A.G. (1985). Variable major proteins of *Borrelia hermsii*. Epitope mapping and partial sequence analysis of CNBr peptides. J. Exp. Med. *161*, 1302–1314.

Ben-Menachem, G., Kubler-Kielb, J., Coxon, B., Yergey, A., and Schneerson, R. (2003). A newly discovered cholesteryl galactoside from *Borrelia burgdorferi*. Proc. Natl. Acad. Sci. U.S.A. *100*, 7913–7918.

Beunders, B., and Van Thiel, P. (1932). Untersuchungen ueber die persistenz von *Spirochaeta duttoni* in mausegehirnen bei experimenteller febris recurrens. Zeitschrift fuer Hygiene und Infektionskrankheiten *114*, 568–583.

Bodman, R.I., and Stewart, I.S. (1948). Louse-borne relapsing fever in Persia. Br. Med. J., 291–293.

Borgnolo, G., Hailu, B., Ciancarelli, A., Alaviva, M., and Woldemariam, T. (1993). Louse-borne relapsing fever: a clinical and an epidemiological study of 389 patients in Asella Hospital, Ethiopia. Trop. Med. Geograph. Med. *45*, 66–69.

Breitschwerdt, E.B., Geoly, F.J., Meuten, D.J., Levine, J.F., Howard, P., Hegarty, B.C., and Stafford, L.C. (1996). Myocarditis in mice and guinea pigs experimentally infected with a canine-origin *Borrelia* isolate from Florida. Am. J. Vet. *Res. 57*, 505–511.

Brocklesby, D.W., Scott, G.R., and Rampton, C.S. (1963). *Borrelia theileri* and transient fevers in cattle. Vet. Rec. *75*, 103–104.

Brown, V., Larouze, B., Desve, G., Rousset, J.J., Thibon, M., Fourrier, A., and Schwoebel, V. (1988). Clinical presentation of louse-borne relapsing fever among Ethiopian refugees in northern Somalia. Ann. Trop. Med. Parasitol. *82*, 499–502.

Bryceson, A. (1976). Clinical pathology of the Jarisch-Herxheimer reaction. J. Infect. Dis. *133*, 696–704.

Bryceson, A., Parry, E., Perine, P., Warrel, D., Vukotich, D., and Leithead, C. (1970). Louse-borne relapsing fever: A clinical and laboratory study of 62 cases in Ethiopia and a reconsideration of the literature. Q. J. Med. *153*, 129–170.

Bryceson, A.D., Cooper, K.E., Warrell, D.A., Perine, P.L., and Parry, E.H. (1972). Studies of the mechanism of the Jarisch-Herxheimer reaction in louse-borne relapsing fever: evidence for the presence of circulating Borrelia endotoxin. Clin. Sci. *43*, 343–354.

Bunikis, J., and Barbour, A.G. (2005). Third *Borrelia* species in white-footed mice. Emerg. Infect. Dis. *11*, 1150–1151.

Bunikis, J., Garpmo, U., Tsao, J., Berglund, J., Fish, D., and Barbour, A.G. (2004). Sequence typing reveals extensive strain diversity of the Lyme borreliosis agents *Borrelia burgdorferi* in North America and *Borrelia afzelii* in Europe. Microbiology *150*, 1741–1755.

Bunikis, J., Luke, C.J., Bunikienc, E., Bergström, S., and Barbour, A.G. (1998). A surface-exposed region of a novel outer membrane protein (P66) of *Borrelia* spp. is variable in size and sequence. J. Bacteriol. *180*, 1618–1623.

Bunikis, J., Noppa, L., and Bergström, S. (1995). Molecular analysis of a 66-kDa protein associated with the outer membrane of Lyme disease *Borrelia*. FEMS Microbiol. Lett. *131*, 139–145.

Burgdorfer, W. (1951). Analyse des Infektionsverlaufes bei *Ornithodorus moubata* und der natürlichen Uebertragung von *Spirochaeta duttoni*. Acta Tropica *8*, 196–262.

Burgdorfer, W. (1959). Zum organotropismus der Spirochaete *B. duttoni* gegenuber der ubertragenden zecke. Acta Tropica *16*, 242–243.

Burgdorfer, W., and Varma, M.G. (1967). Trans-stadial and transovarial development of disease agents in arthropods. Annu. Rev. Entomol. *12*, 347–376.

Burman, N., Bergström, S., Restrepo, B.I., and Barbour, A.G. (1990). The variable antigens Vmp7 and Vmp21 of the relapsing fever bacterium *Borrelia hermsii* are structurally analogous to the VSG proteins of the African trypanosome. Mol. Microbiol. *4*, 1715–1726.

Burman, N., Shamaei-Tousi, A., and Bergström, S. (1998). The spirochete *Borrelia crocidurae* causes erythrocyte rosetting during relapsing fever. Infect. Immun. *66*, 815–819.

Cadavid, D., and Barbour, A.G. (1998). Neuroborreliosis during relapsing fever: review of the clinical manifestations, pathology, and treatment of infections in humans and experimental animals. Clin. Infect. Dis. *26*, 151–164.

Cadavid, D., Bundoc, V., and Barbour, A.G. (1993). Experimental infection of the mouse brain by a relapsing fever *Borrelia* species: a molecular analysis. J. Infect. Dis. *168*, 143–151.

Cadavid, D., Pachner, A.R., Estanislao, L., Patalapati, R., and Barbour, A.G. (2001). Isogenic serotypes of *Borrelia turicatae* show different localization in the brain and skin of mice. Infect. Immun. *69*, 3389–3397.

Cadavid, D., Pennington, P.M., Kerentseva, T.A., Bergström, S., and Barbour, A.G. (1997). Immunologic and genetic analyses of VmpA of a neurotropic strain of *Borrelia turicatae*. Infect. Immun. *65*, 3352–3360.

Cadavid, D., Sondey, M., Garcia, E., and Lawson, C.L. (2006). Residual brain infection in relapsing-fever borreliosis. J. Infect. Dis. *193*, 1451–1458.

Cadavid, D., Thomas, D.D., Crawley, R., and Barbour, A.G. (1994). Variability of a bacterial surface protein and disease expression in a possible mouse model of systemic Lyme borreliosis. J. Exp. Med. *179*, 631–642.

Calabi, O. (1959). The presence of plasma inhibitors during the crisis phenomenon in experimental relapsing fever (*Borrelia novyi*). J. Exp. Med. *110*, 811–825.

Campbell, G.L., Paul, W.S., Schriefer, M.E., Craven, R.B., Robbins, K.E., and Dennis, D.T. (1995). Epidemiologic and diagnostic studies of patients with suspected early Lyme disease, Missouri, 1990–1993. J. Infect. Dis. *172*, 470–480.

Carter, C.J., Bergström, S., Norris, S.J., and Barbour, A.G. (1994). A family of surface-exposed proteins of 20 kilodaltons in the genus *Borrelia*. Infect. Immun. *62*, 2792–2799.

Coburn, J., Chege, W., Magoun, L., Bodary, S.C., and Leong, J.M. (1999). Characterization of a candidate *Borrelia burgdorferi* β_3-chain integrin ligand identified using a phage display library. Mol. Microbiol. *34*, 926–940.

Coburn, J., and Cugini, C. (2003). Targeted mutation of the outer membrane protein P66 disrupts attachment of the Lyme disease agent, *Borrelia burgdorferi*, to integrin $\alpha_v \beta_3$. Proc. Natl. Acad. Sci. U.S.A. *100*, 7301–7306.

Coffey, E.M., and Eveland, W.C. (1967). Experimental relapsing fever initiated by *Borrelia hermsi*. II. Sequential appearance of major serotypes in the rat. J. Infect. Dis. *117*, 29–34.

Colombo, M.J., and Alugupalli, K.R. (2008). Complement factor H-binding protein, a putative virulence determinant of *Borrelia hermsii*, is an antigenic target for protective B1b lymphocytes. J. Immunol. *180*, 4858–4864.

Connolly, S.E., and Benach, J.L. (2001). Cutting edge: the spirochetemia of murine relapsing fever is cleared by complement-independent bactericidal antibodies. J. Immunol. *167*, 3029–3032.

Connolly, S.E., Thanassi, D.G., and Benach, J.L. (2004). Generation of a complement-independent bactericidal IgM against a relapsing fever *Borrelia*. J. Immunol. *172*, 1191–1197.

Cooley, R.A., and Kohls, G.M. (1944). The Argasidae of North America, Central America, and Cuba (South Bend, Notre Dame University Press).

Cuboni, E. (1929). Sul potere spirocheticida del siero di sangue di alcuni animali. Bol. Inst. Sieroterap Milan *8*, 813–817.

Dai, Q., Restrepo, B.I., Porcella, S.F., Raffel, S.J., Schwan, T.G., and Barbour, A.G. (2006). Antigenic variation by *Borrelia hermsii* occurs through recombination between extragenic repetitive elements on linear plasmids. Mol. Microbiol. *60*, 1329–1343.

DaMassa, A.J., and Adler, H.E. (1979). Avian spirochetosis: natural transmission by *Argas* (*Persicargas*) sanchezi (Ixodoidea: argasidae) and existence of different serologic and immunologic types of *Borrelia anserina* in the United States. Am. J. Vet. *Res. 40*, 154–157.

Davis, G.E. (1943). Relapsing fever: the tick *Ornithodoros turicata* as a spirochetal reservoir. Publ. Health Rep. *58*, 839–842.

Davis, G.E. (1952). Biology as an aid to the identification of two closely related species of ticks of the genus *Ornithodoros*. J. Parasitol. 38, 477–480.

Dennis, D., Awoke, S., Doberstein, E., and Fresh, J. (1976). Bleeding in louse-borne relapsing fever in Ethiopia. East Afr. Med. J. 53, 220–225.

Dewar, H., and Walmsley, R. (1945). Relapsing fever with nephritis and subarachnoid haemorrhage. Lancet Nov. *17*, 630–631.

Diab, F.M., and Soliman, Z.R. (1977). An experimental study of *Borrelia anserina* in four species of Argas ticks. 1. Spirochete localization and densities. Z Parasitenkd *53*, 201–212.

Dubois, A., and Pearson, Y. (1939). Variabilite du neurotropisme d'une souche de *Spirochaeta dutoni*, immunite en l'abscence de neurotropisme. Comptes Rendus des Seances Societe de Biologie (Paris) *130*, 1379–1381.

Duray, P.H. (1987). The surgical pathology of human Lyme disease: an enlarging picture. Am. J. Surg. Pathol. *11*, 47–60.

Eggers, C.H., and Samuels, D.S. (1999). Molecular evidence for a new bacteriophage of *Borrelia burgdorferi*. J. Bacteriol. *181*, 7308–7313.

Eicken, C., Sharma, V., Klabunde, T., Lawrenz, M.B., Hardham, J.M., Norris, S.J., and Sacchettini, J.C. (2002). Crystal structure of Lyme disease variable surface antigen VlsE of *Borrelia burgdorferi*. J. Biol. Chem. *277*, 21691–21696.

Fekade, D., Knox, K., Hussein, K., Melka, A., Lalloo, D.G., Coxon, R.E., and Warrell, D.A. (1996). Prevention of Jarisch-Herxheimer reactions by treatment with antibodies against tumor necrosis factor α. N. Engl. J. Med. *335*, 311–315.

Felsenfeld, O. (1965). Borreliae, human relapsing fever, and parasite–vector–host relationships. Bacteriol.l Rev. *29*, 46–74.

Felsenfeld, O. (1971). Borrelia: Strains, Vectors, Human and Animal Borreliosis (St. Louis, MO, Warren H. Greene, Inc.).

Felsenfeld, O., and Wolf, R.H. (1973). Interference between borreliae and trypanosomes. I. Antigenic analysis, fluorescent antibody and immobilisine studies of *Borrelia turicatae* and *Trypanosoma cruzi*. Ann. Trop. Med. Parasitol. *67*, 335–340.

Ferdows, M.S., Serwer, P., Griess, G.A., Norris, S.J., and Barbour, A.G. (1996). Conversion of a linear to a circular plasmid in the relapsing fever agent *Borrelia hermsii*. J. Bacteriol. *178*, 793–800.

Fraenkel, C.J., Garpmo, U., and Berglund, J. (2002). Determination of novel *Borrelia* genospecies in Swedish *Ixodes ricinus* ticks. J. Clin. Microbiol. *40*, 3308–3312.

Frank, S.A., and Barbour, A.G. (2006). Within-host dynamics of antigenic variation. Infect. Genet. Evol. *6*, 141–146.

Fuchs, P.C., and Oyama, A.A. (1969). Neonatal relapsing fever due to transplacental transmission of *Borrelia*. JAMA *208*, 690–692.

Fukunaga, M., and Koreki, Y. (1995). The flagellin gene of *Borrelia miyamotoi* sp. nov. and its phylogenetic relationship among *Borrelia* species. FEMS Microbiol. Lett. *134*, 255–258.

Gabritchewsky, G.N. (1896). Les bases de la sérothérapie de la fievre récurrente. Annales Institut Pasteur *10*, 630–653.

Galloway, R.E., Levin, J., Butler, T., Naff, G.B., Goldsmith, G.H., Saito, H., Awoke, S., and Wallace, C.K. (1977). Activation of protein mediators of inflammation and evidence for endotoxemia in *Borrelia recurrentis* infection. Am. J. Med. *63*, 933–938.

García-Moncó, J.C., Miller, N.S., Backenson, P.B., Anda, P., and Benach, J.L. (1997). A mouse model of *Borrelia* meningitis after intradermal injection. J. Infect. Dis. *175*, 1243–1245.

Garnham, P., Davis, C., Heisch, R., and Timms, G. (1947). An epidemic of louse-borne relapsing fever in Kenya. Trans. Roy. Soc. Trop. Med. Hyg.*41*, 141–170.

Geigy, R., and Aeschlimann, A. (1964). Langfristige Beobachtungen über transoverielle Übertragung von *Borrelia duttoni* durch *Ornithodorus moubata*. Acta Trop *21*, 87–91.

Grigery, C.N., Moyer, P., Little, S.E., and Masters, E.J. (2005). Bacteriocidal activity of lizard and mouse serum for *Borrelia lonestari*, putative agent of a Lyme-like illness (AKA STARI or Masters disease) in Missouri. Mo Med *102*, 442–446.

Guo, B.P., Teneberg, S., Münch, R., Terunuma, D., Hatano, K., Matsuoka, K., Angström, J., Borén, T., and Bergström, S. (2009). Relapsing fever *Borrelia* binds to neolacto glycans and mediates rosetting of human erythrocytes. Proc. Natl. Acad. Sci. U.S.A. *106*, 19280–19285.

Guyard, C., Chester, E.M., Raffel, S.J., Schrumpf, M.E., Policastro, P.F., Porcella, S.F., Leong, J.M., and Schwan, T.G. (2005). Relapsing fever spirochetes contain chromosomal genes with unique direct tandemly repeated sequences. Infect. Immun. *73*, 3025–3037.

Hendson, M., and Lane, R.S. (2000). Genetic characteristics of *Borrelia coriaceae* isolates from the soft tick *Ornithodoros coriaceus* (Acari: Argasidae). J. Clin. Microbiol. *38*, 2678–2682.

Herms, W.B., and Wheeler, C.M. (1935). Tick transmission of California relapsing fever. J Econ Entom *28*, 846–855.

Hinnebusch, B.J., Barbour, A.G., Restrepo, B.I., and Schwan, T.G. (1998). Population structure of the relapsing fever spirochete *Borrelia hermsii* as indicated by polymorphism of two multigene families that encode immunogenic outer surface lipoproteins. Infect. Immun. *66*, 432–440.

Hoogstraal, H. (1979). Ticks and spirochetes. Acta Trop *36*, 133–136.

Hoogstraal, H. (1985). Argasid and nuttalliellid ticks as parasites and vectors. Adv. Parasitol. *24*, 135–238.

Hossain, H., Wellensiek, H.J., Geyer, R., and Lochnit, G. (2001). Structural analysis of glycolipids from *Borrelia burgdorferi*. Biochimie *83*, 683–692.

Hovis, K.M., McDowell, J.V., Griffin, L., and Marconi, R.T. (2004). Identification and characterization of a linear-plasmid-encoded factor H-binding protein (FhbA) of the relapsing fever spirochete *Borrelia hermsii*. J. Bacteriol. *186*, 2612–2618.

Hovis, K.M., Schriefer, M.E., Bahlani, S., and Marconi, R.T. (2006). Immunological and molecular analyses of the *Borrelia hermsii* factor H and factor H-like protein 1 binding protein, FhbA: demonstration of its utility as a diagnostic marker and epidemiological tool for tick-borne relapsing fever. Infect. Immun. *74*, 4519–4529.

Huson, D.H., and Bryant, D. (2006). Application of phylogenetic networks in evolutionary studies. Mol. Biol. Evol. *23*, 254–267.

Jancso, N. (1918). Experimentelle Untersuchung bezüglich der Pathogenese der Rezidive des Rückfallfiebers. Zentbl. Bakt. Orig. *81*, 457–474.

Judge, D., Samuel, I., Perine, P., and Vukotic, D. (1974a). Louse-borne relapsing fever in man. Arch. Pathol. *97*, 136–140.

Judge, D.M., La Croix, J.T., and Perine, P.L. (1974b). Experimental louse-borne relapsing fever in the grivet monkey, *Cercopithecus aethiops*. II. Pathology. Am. J. Trop. Med. Hyg. *23*, 962–968.

Kazragis, R.J., Dever, L.L., Jorgensen, J.H., and Barbour, A.G. (1996). In vivo activities of ceftriaxone and vancomycin against *Borrelia* spp. in the mouse brain and other sites. Antimicrob. Agents Chemother. *40*, 2632–2636.

Kitten, T., and Barbour, A.G. (1990). Juxtaposition of expressed variable antigen genes with a conserved telomere in the bacterium *Borrelia hermsii*. Proc. Natl. Acad. Sci. U.S.A. *87*, 6077–6081.

Kitten, T., and Barbour, A.G. (1992). The relapsing fever agent *Borrelia hermsii* has multiple copies of its chromosome and linear plasmids. Genetics *132*, 311–324.

Kitten, T., Barrera, A.V., and Barbour, A.G. (1993). Intragenic recombination and a chimeric outer membrane protein in the relapsing fever agent *Borrelia hermsii*. J. Bacteriol. *175*, 2516–2522.

Lane, R.S., Burgdorfer, W., Hayes, S.F., and Barbour, A.G. (1985). Isolation of a spirochete from the soft tick, *Ornithodoros coriaceus*: a possible agent of epizootic bovine abortion. Science *230*, 85–87.

Lane, R.S., and Manweiler, S.A. (1988). *Borrelia coriaceae* in its tick vector, *Ornithodoros coriaceus* (Acari: Argasidae), with emphasis on transstadial and transovarial infection. J. Med. Entomol. *25*, 172–177.

Lane, R.S., Mun, J., Parker, J.M., and White, M. (2005). Columbian black-tailed deer (*Odocoileus hemionus columbianus*) as hosts for *Borrelia* spp. in northern California. J. Wildl. Dis. *41*, 115–125.

LaRocca, T.J., Katona, L.I., Thanassi, D.G., and Benach, J.L. (2008). Bactericidal action of a complement-independent antibody against relapsing fever *Borrelia* resides in its variable region. J. Immunol. *180*, 6222–6228.

Larsson, C., Andersson, M., Guo, B.P., Nordstrand, A., Hagerstrand, I., Carlsson, S., and Bergström, S. (2006a). Complications of pregnancy and transplacental transmission of relapsing-fever borreliosis. J. Infect. Dis. *194*, 1367–1374.

Larsson, C., Andersson, M., Pelkonen, J., Guo, B.P., Nordstrand, A., and Bergström, S. (2006b). Persistent brain infection and disease reactivation in relapsing fever borreliosis. Microbes Infect. *8*, 2213–2219.

Larsson, C., Lundqvist, J., and Bergström, S. (2008). Residual brain infection in murine relapsing fever borreliosis can be successfully treated with ceftriaxone. Microb. Pathog. *44*, 262–264.

Lawson, C.L., Yung, B.H., Barbour, A.G., and Zückert, W.R. (2006). Crystal structure of neurotropism-associated variable surface protein 1 (Vsp1) of *Borrelia turicatae*. J. Bacteriol. *188*, 4522–4530.

Lescot, M., Audic, S., Robert, C., Nguyen, T.T., Blanc, G., Cutler, S.J., Wincker, P., Couloux, A., Claverie, J.M., Raoult, D., *et al.* (2008). The genome of *Borrelia recurrentis*, the agent of deadly louse-borne relapsing fever, is a degraded subset of tick-borne *Borrelia duttonii*. PLoS Genet. *4*, e1000185.

Livey, I., Gibbs, C.P., Schuster, R., and Dorner, F. (1995). Evidence for lateral transfer and recombination in OspC variation in Lyme disease *Borrelia*. Mol. Microbiol. *18*, 257–269.

Londono, D., Bai, Y., Zückert, W.R., Gelderblom, H., and Cadavid, D. (2005). Cardiac apoptosis in severe relapsing fever borreliosis. Infect. Immun. *73*, 7669–7676.

Londono, D., Marques, A., Hornung, R.L., and Cadavid, D. (2008). IL-10 helps control pathogen load during high-level bacteremia. J. Immunol. *181*, 2076–2083.

Lopez, J.E., Schrumpf, M.E., Raffel, S.J., Policastro, P.F., Porcella, S.F., and Schwan, T.G. (2008). Relapsing fever spirochetes retain infectivity after prolonged *in vitro* cultivation. Vector Borne Zoonotic Dis. *8*, 813–820.

Magoun, L., Zückert, W.R., Robbins, D., Parveen, N., Alugupalli, K.R., Schwan, T.G., Barbour, A.G., and Leong, J.M. (2000). Variable small protein (Vsp)-dependent and Vsp-independent pathways for glycosaminoglycan recognition by relapsing fever spirochaetes. Mol. Microbiol. *36*, 886–897.

Marconi, R.T., Samuels, D.S., Schwan, T.G., and Garon, C.F. (1993). Identification of a protein in several *Borrelia* species which is related to OspC of the Lyme disease spirochetes. J. Clin. Microbiol. *31*, 2577–2583.

McDowell, J.V., Tran, E., Hamilton, D., Wolfgang, J., Miller, K., and Marconi, R.T. (2003). Analysis of the ability of spirochete species associated with relapsing fever, avian borreliosis, and epizootic bovine abortion to bind factor H and cleave C3b. J. Clin. Microbiol. *41*, 3905–3910.

Meier, J.T., Simon, M.I., and Barbour, A.G. (1985). Antigenic variation is associated with DNA rearrangements in a relapsing fever *Borrelia*. Cell *41*, 403–409.

Meleney, H.E. (1928). Relapse phenomena of *Spironema recurrentis*. J. Exp. Med. *48*, 65–82.

Meri, T., Cutler, S.J., Blom, A.M., Meri, S., and Jokiranta, T.S. (2006). Relapsing fever spirochetes *Borrelia recurrentis* and *B. duttonii* acquire complement regulators C4b-binding protein and factor H. Infect. Immun. *74*, 4157–4163.

Mitiku, K., and Mengistu, G. (2002). Relapsing fever in Gondar, Ethiopia. East Afr. Med. J. *79*, 85–87.

Mooser, H. (1958). Erytrozyten-adhäsion und hämagglomeration durch rückfallfieber-spirochäten. Tropenmedizin und Parasitologie *9*, 93–111.

Moyer, P.L., Varela, A.S., Luttrell, M.P., Moore, V.A.t., Stallknecht, D.E., and Little, S.E. (2006). White-tailed deer (*Odocoileus virginianus*) develop spirochetemia following experimental infection with *Borrelia lonestari*. Vet. Microbiol. *115*, 229–236.

Mun, J., Eisen, R.J., Eisen, L., and Lane, R.S. (2006). Detection of a *Borrelia miyamotoi* sensu lato relapsing-fever group spirochete from *Ixodes pacificus* in California. J. Med. Entomol. *43*, 120–123.

Negussie, Y., Remick, D.G., DeForge, L.E., Kunkel, S.L., Eynon, A., and Griffin, G.E. (1992). Detection of plasma tumor necrosis factor, interleukins 6, and 8 during the Jarisch-Herxheimer Reaction of relapsing fever. J. Exp. Med. *175*, 1207–1212.

Newman, K., Jr., and Johnson, R.C. (1981). *In vivo* evidence that an intact lytic complement pathway is not essential for successful removal of circulating *Borrelia turicatae* from mouse blood. Infect. Immun. *31*, 465–469.

Newman, K., Jr., and Johnson, R.C. (1984). T-cell-independent elimination of *Borrelia turicatae*. Infect. Immun. *45*, 572–576.

Nicolle, C., and Anderson, C. (1927). Etude compartive de quelques viru recurrents, pathogenes pour l'homme. Arch. Inst. Past. de Tunis *16*, 123–206.

Nordstrand, A., Shamaei-Tousi, A., Ny, A., and Bergström, S. (2001). Delayed invasion of the kidney and brain by *Borrelia crocidurae* in plasminogen-deficient mice. Infect. Immun. *69*, 5832–5839.

Novy, F.G., and Knapp, R.E. (1906). Studies on *Spirillum obermeieri* and related organisms. J. Infect. Dis. *3*, 291–293.

Pachanian, A., and Smith, J.B. (1970). Immunization of chicks against avian spirochetosis with vaccines of *Borrelia anserina*. Tex. Rep. Biol. Med. *28*, 287–301.

Parsons, L. (1947). Relapsing fever at Lake Tahoe, California-Nevada. Am. J. Clin. Pathol. *17*, 388–392.

Pavlovsky, Y.N. (1963). Tick-borne relapsing fever. In Human Disease with Natural Foci, Y.N. Pavlovsky, ed. (Moscow, Foreign Languages Publishing House), pp. 138–184.

Penningon, P.M., Cadavid, D., Bunikis, J., Norris, S.J., and Barbour, A.G. (1999). Extensive interplasmidic duplications change the virulence phenotype of the relapsing fever agent *Borrelia turicatae*. Mol. Microbiol. *34*, 1120–1132.

Pennington, P.M., Allred, C.D., West, C.S., Alvarez, R., and Barbour, A.G. (1997). Arthritis severity and spirochete burden are determined by serotype in the *Borrelia turicatae*-mouse model of Lyme disease. Infect. Immun. *65*, 285–292.

Pennington, P.M., Cadavid, D., and Barbour, A.G. (1999). Characterization of VspB of *Borrelia turicatae*, a major outer membrane protein expressed in blood and tissues of mice. Infect. Immun. *67*, 4637–4645.

Perine, P., Parry, E., Vukotich, D., and al, e. (1971). Bleeding in louse-borne relapsing fever. Trans. Roy. Soc. Trop. Med. Hyg. *65*, 776–781.

Pettersson, J., Schrumpf, M.E., Raffel, S.J., Porcella, S.F., Guyard, C., Lawrence, K., Gherardini, F.C., and Schwan, T.G. (2007). Purine salvage pathways among *Borrelia* species. Infect. Immun. *75*, 3877–3884.

Plasterk, R.H., Simon, M.I., and Barbour, A.G. (1985). Transposition of structural genes to an expression sequence on a linear plasmid causes antigenic variation in the bacterium *Borrelia hermsii*. Nature *318*, 257–263.

Porcella, S.F., Raffel, S.J., Anderson, D.E., Jr., Gilk, S.D., Bono, J.L., Schrumpf, M.E., and Schwan, T.G. (2005). Variable tick protein in two genomic groups of the relapsing fever spirochete *Borrelia hermsii* in western North America. Infect. Immun. *73*, 6647–6658.

Porcella, S.F., Raffel, S.J., Schrumpf, M.E., Schriefer, M.E., Dennis, D.T., and Schwan, T.G. (2000). Serodiagnosis of Louse-borne relapsing fever with glycerophosphodiester phosphodiesterase (GlpQ) from *Borrelia recurrentis*. J. Clin. Microbiol. *38*, 3561–3571.

Restrepo, B.I., and Barbour, A.G. (1994). Antigen diversity in the bacterium *B. hermsii* through 'somatic' mutations in rearranged *vmp* genes. Cell *78*, 867–876.

Restrepo, B.I., Carter, C.J., and Barbour, A.G. (1994). Activation of a vmp pseudogene in *Borrelia hermsii*: an alternate mechanism of antigenic variation during relapsing fever. Mol. Microbiol. *13*, 287–299.

Restrepo, B.I., Kitten, T., Carter, C.J., Infante, D., and Barbour, A.G. (1992). Subtelomeric expression regions of *Borrelia hermsii* linear plasmids are highly polymorphic. Mol. Microbiol. *6*, 3299–3311.

Rich, S.M., Armstrong, P.M., Smith, R.D., and Telford, S.R., 3rd (2001a). Lone star tick-infecting borreliae are most closely related to the agent of bovine borreliosis. J. Clin. Microbiol. *39*, 494–497.

Rich, S.M., Sawyer, S.A., and Barbour, A.G. (2001b). Antigen polymorphism in *Borrelia hermsii*, a clonal pathogenic bacterium. Proc. Natl. Acad. Sci. U.S.A. *98*, 15038–15043.

Rossmann, E., Kraiczy, P., Herzberger, P., Skerka, C., Kirschfink, M., Simon, M.M., Zipfel, P.F., and Wallich, R. (2007). Dual binding specificity of a *Borrelia hermsii*-associated complement regulator-acquiring surface protein for factor H and plasminogen discloses a putative virulence factor of relapsing fever spirochetes. J. Immunol. *178*, 7292–7301.

Russell, H. (1936). Observations on immunity in relapsing fever and trypanosomiasis. Trans. Roy. Soc. Trop. Med. Hyg. *30*, 179–190.

Sadziene, A., Barbour, A.G., Rosa, P.A., and Thomas, D.D. (1993a). An OspB mutant of *Borrelia burgdorferi* has reduced invasiveness in vitro and reduced infectivity in vivo. Infect. Immun. *61*, 3590–3596.

Sadziene, A., Jonsson, M., Bergström, S., Bright, R.K., Kennedy, R.C., and Barbour, A.G. (1994). A bactericidal antibody to *Borrelia burgdorferi* is directed against a variable region of the OspB protein. Infect. Immun. *62*, 2037–2045.

Sadziene, A., Thompson, P.A., and Barbour, A.G. (1993b). In vitro inhibition of *Borrelia burgdorferi* growth by antibodies. J. Infect. Dis. *167*, 165–172.

Salih, S., Mustafa, D., Abdel-Wahab, S., Ahmed, M., and Omer, A. (1977). Louse-borne relapsing fever: I. A clinical and laboratory study of 363 cases in the Sudan. Trans. Roy. Soc. Trop. Med. Hyg. *71*, 43–48.

Sambri, V., Marangoni, A., Olmo, A., Storni, E., Montagnani, M., Fabbi, M., and Cevenini, R. (1999). Specific antibodies reactive with the 22-kilodalton major outer surface protein of *Borrelia anserina* Ni-NL protect chicks from infection. Infect. Immun. *67*, 2633–2637.

Schmidtmann, E.T., Bushnell, R.B., Loomis, E.C., Oliver, M.N., and Theis, J.H. (1976). Experimental and epizootiologic evidence associating *Ornithodoros coriaceus* (Acari:Agrasidae) with the exposure of cattle to epizootic bovine abortion in California. J. Med. Entomol. *13*, 292–299.

Schroder, N.W., Eckert, J., Stubs, G., and Schumann, R.R. (2008). Immune responses induced by spirochetal outer membrane lipoproteins and glycolipids. Immunobiology *213*, 329–340.

Schuhardt, V.T., and Wilkerson, M. (1951). Relapse phenomena in rats infected with single spirochetes (*Borrelia recurrenis* var. turicatae). J. Bacteriol. *62*, 215–219.

Schwan, T.G., Battisti, J.M., Porcella, S.F., Raffel, S.J., Schrumpf, M.E., Fischer, E.R., Carroll, J.A., Stewart, P.E., Rosa, P., and Somerville, G.A. (2003). Glycerol-3-phosphate acquisition in spirochetes: distribution and biological activity of glycerophosphodiester phosphodiesterase (GlpQ) among *Borrelia* species. J. Bacteriol. *185*, 1346–1356.

Schwan, T.G., and Hinnebusch, B.J. (1998). Bloodstream- versus tick-associated variants of a relapsing fever bacterium. Science *280*, 1938–1940.

Schwan, T.G., Schrumpf, M.E., Hinnebusch, B.J., Anderson, D.E., Jr., and Konkel, M.E. (1996). GlpQ: an antigen for serological discrimination between relapsing fever and Lyme borreliosis. J. Clin. Microbiol. *34*, 2483–2492.

Scoles, G.A., Papero, M., Beati, L., and Fish, D. (2001). A relapsing fever group spirochete transmitted by *Ixodes scapularis* ticks. Vector Borne Zoonotic Dis. *1*, 21–34.

Scragg, I.G., Kwiatkowski, D., Vidal, V., Reason, A., Paxton, T., Panico, M., Dell, A., and Morris, H. (2000). Structural characterization of the inflammatory moiety of a variable major lipoprotein of *Borrelia recurrentis*. J. Biol. Chem. *275*, 937–941.

Shamaei-Tousi, A., Burns, M.J., Benach, J.L., Furie, M.B., Gergel, E.I., and Bergström, S. (2000). The relapsing fever spirochaete, *Borrelia crocidurae*, activates human endothelial cells and promotes the transendothelial migration of neutrophils. Cell. Microbiol. *2*, 591–599.

Shamaei-Tousi, A., Collin, O., Bergh, A., and Bergström, S. (2001). Testicular damage by microcirculatory disruption and colonization of an immune-privileged site during *Borrelia crocidurae* infection. J. Exp. Med. *193*, 995–1004.

Shamaei-Tousi, A., Martin, P., Bergh, A., Burman, N., Brännström, T., and Bergström, S. (1999). Erythrocyte-aggregating relapsing fever spirochete *Borrelia crocidurae* induces formation of microemboli. J. Infect. Dis. *180*, 1929–1938.

Shang, E.S., Skare, J.T., Exner, M.M., Blanco, D.R., Kagan, B.L., Miller, J.N., and Lovett, M.A. (1998). Isolation and characterization of the outer membrane of *Borrelia hermsii*. Infect. Immun. *66*, 1082–1091.

Sharma, S.P., Amanfu, W., and Losho, T.C. (2000). Bovine borreliosis in Botswana. Onderstepoort J. Vet. Res. *67*, 221–223.

Skare, J.T., Mirzabekov, T.A., Shang, E.S., Blanco, D.R., Erdjument-Bromage, H., Bunikis, J., Bergström, S., Tempst, P., Kagan, B.L., Miller, J.N., *et al.* (1997). The Oms66 (p66) protein is a *Borrelia burgdorferi* porin. Infect. Immun. *65*, 3654–3661.

Smith, R.D., Brener, J., Osorno, M., and Ristic, M. (1978). Pathobiology of *Borrelia theileri* in the tropical cattle tick, *Boophilus microplus*. J. Invertebr. Pathol. 32, 182–190.

Sohaskey, C.D., Zückert, W.R., and Barbour, A.G. (1999). The extended promoters for two outer membrane lipoprotein genes of *Borrelia* spp. uniquely include a T-rich region. Mol. Microbiol. 33, 41–51.

Southern, P., and Sanford, J. (1969). Relapsing fever. A clinical and microbiological review. Medicine *48*, 129–149.

Stevenson, B., Porcella, S.F., Oie, K.L., Fitzpatrick, C.A., Raffel, S.J., Lubke, L., Schrumpf, M.E., and Schwan, T.G. (2000). The relapsing fever spirochete *Borrelia hermsii* contains multiple, antigen-encoding circular plasmids that are homologous to the cp32 plasmids of Lyme disease spirochetes. Infect. Immun. *68*, 3900–3908.

Stoenner, H.G., Dodd, T., and Larsen, C. (1982). Antigenic variation of *Borrelia hermsii*. J. Exp. Med. *156*, 1297–1311.

Sundnes, K., and Haimanot, A. (1993). Epidemic of louse-borne relapsing fever in Ethiopia. Lancet *342*, 1213.

Tabuchi, N., Mitani, H., Seino, S., and Fukunaga, M. (2002). The 44-kb linear plasmid molecule in the relapsing fever agent *Borrelia duttonii* strain Ly serve as a preservation of *vmp* genes. Microbiol. Immunol. *46*, 159–165.

Takahashi, Y., Cutler, S.J., and Fukunaga, M. (2000). Size conversion of a linear plasmid in the relapsing fever agent *Borrelia duttonii*. Microbiol. Immunol. *44*, 1071–1074.

Takayama, K., Rothenberg, R.J., and Barbour, A.G. (1987). Absence of lipopolysaccharide in the Lyme disease spirochete, *Borrelia burgdorferi*. Infect. Immun. *55*, 2311–2313.

Theiler, A.J. (1904). Spirillosis of cattle. J. Comp. Pathol. Ther. *17*, 47–55.

Thomas, N.J., Bunikis, J., Barbour, A.G., and Wolcott, M.J. (2002). Fatal spirochetosis due to a relapsing fever-like *Borrelia* sp. in a northern spotted owl. J. Wildl. Dis. *38*, 187–193.

Uppal, D.R., and Rao, S.B. (1966). Studies on the preparation of a spirochaete vaccine. I. Indian Vet. J. *43*, 191–195.

Varma, M.G.R. (1956). Infections of *Ornithodoros* ticks with relapsing fever spirochaetes, and the mechanisms of their transmission. Ann. Trop. Med. Parasitol. *50*, 18–31.

Velu, H., Balozet, L., and Zottner, G. (1930). Neurotropisme de *Spirochaeta hispanicum* (souche de Mansouriah) pour le lapin' forme nouvelle de l'infection inapparente. Comptes Rendus des Seances Societe de Biologie (Paris) *103*, 381–383.

Vidal, V., Cutler, S., Scragg, I.G., Wright, D.J., and Kwiatkowski, D. (2002). Characterisation of silent and active genes for a variable large protein of *Borrelia recurrentis*. BMC Infect. Dis. *2*, 25.

Vidal, V., Scragg, I.G., Cutler, S.J., Rockett, K.A., Fekade, D., Warrell, D.A., Wright, D.J., and Kwiatkowski, D. (1998). Variable major lipoprotein is a principal TNF-inducing factor of louse- borne relapsing fever. Nat. Med. *4*, 1416–1420.

Wang, G., van Dam, A.P., and Dankert, J. (1999). Evidence for frequent OspC gene transfer between *Borrelia valaisiana* sp. nov. and other Lyme disease spirochetes. FEMS Microbiol. Lett. *177*, 289–296.

Warrell, D.A., Perine, P.L., Krause, D.W., Bing, D.H., and MacDougal, S.J. (1983). Pathophysiology and immunology of the Jarisch-Herxheimer-like reaction in louse-borne relapsing fever: comparison of tetracycline and slow-release penicillin. J. Infect. Dis. *147*, 898–909.

Webster, G., Schiffman, J.D., Dosanjh, A.S., Amieva, M.R., Gans, H.A., and Sectish, T.C. (2002). Jarisch-Herxheimer reaction associated with ciprofloxacin administration for tick-borne relapsing fever. Pediatr. Infect. Dis. J *21*, 571–573.

Wengrower, D., Knobler, H., Gillis, S., and Chajek-Shaul, T. (1984). Myocarditis in tick-borne relapsing fever. J. Infect. Dis. *149*, 1033.

Yabsley, M.J., Davidson, W.R., Stallknecht, D.E., Varela, A.S., Swift, P.K., Devos, J.C., Jr., and Dubay, S.A. (2005). Evidence of tick-borne organisms in mule deer (*Odocoileus hemionus*) from the western United States. Vector Borne Zoonotic Dis. *5*, 351–362.

Yabsley, M.J., Murphy, S.M., Luttrell, M.P., Little, S.E., Massung, R.F., Stallknecht, D.E., Conti, L.A., Blackmore, C.G., and Durden, L.A. (2008). Experimental and field studies on the suitability of raccoons (*Procyon lotor*) as hosts for tick-borne pathogens. Vector Borne Zoonotic Dis. *8*, 491–503.

Yagupsky, P., and Moses, S. (1985). Neonatal *Borrelia* species infection (relapsing fever). Am. J. Dis. Child. *139*, 74–76.

Yparraguirre, L.A., Machado-Ferreira, E., Ullmann, A.J., Piesman, J., Zeidner, N.S., and Soares, C.A. (2007). A hard tick relapsing fever group spirochete in a Brazilian *Rhipicephalus (Boophilus) microplus*. Vector Borne Zoonotic Dis. *7*, 717–721.

Zaher, M.A., Soliman, Z.R., and Diab, F.M. (1977). An experimental study of *Borrelia anserina* in four species of Argas ticks. 2. Transstadial survival and transovarial transmission. Z. Parasitenkd. *53*, 213–223.

Zhang, H., and Marconi, R.T. (2005). Demonstration of cotranscription and 1-methyl-3-nitroso-nitroguanidine induction of a 30-gene operon of *Borrelia burgdorferi*: evidence that the 32-kilobase circular plasmids are prophages. J. Bacteriol. *187*, 7985–7995.

Zhang, J.R., Hardham, J.M., Barbour, A.G., and Norris, S.J. (1997). Antigenic variation in Lyme disease borreliae by promiscuous recombination of VMP-like sequence cassettes. Cell *89*, 275–285.

Zhong, J., and Barbour, A.G. (2004). Cross-species hybridization of a *Borrelia burgdorferi* DNA array reveals infection- and culture associated genes of the unsequenced genome of the relapsing fever agent *Borrelia hermsii*. Mol. Microbiol. *51*, 729–748.

Zhong, J., Skouloubris, S., Dai, Q., Myllykallio, H., and Barbour, A.G. (2006). Function and evolution of plasmid-borne genes for pyrimidine biosynthesis in *Borrelia* spp. J. Bacteriol. *188*, 909–918.

Zückert, W.R., Kerentseva, T.A., Lawson, C.L., and Barbour, A.G. (2001). Structural conservation of neurotropism-associated VspA within the variable *Borrelia* Vsp-OspC lipoprotein family. J. Biol. Chem. *276*, 457–463.

Zückert, W.R., Lloyd, J.E., Stewart, P.E., Rosa, P.A., and Barbour, A.G. (2004). Cross-species surface display of functional spirochetal lipoproteins by recombinant *Borrelia burgdorferi*. Infect. Immun. *72*, 1463–1469.

Animal Models of Borreliosis

14

Stephen W. Barthold, Diego Cadavid and Mario T. Philipp

Abstract

Animal models have been extremely useful for understanding many features of Lyme and relapsing fever borrelioses, including pathogenesis, host response, host–vector interactions, *in vivo* bacterial gene regulation, vaccinology and chemotherapy. Some degree of overlap is necessary with these subjects that are covered in other chapters, but this chapter attempts to focus on descriptive and practical aspects of various animal models used for borreliosis research.

14.1 Animal models of Lyme borreliosis

Ticks of the *Ixodes persulcatus* complex utilize a remarkable array of reservoir host species, in fact, more than any other group of ticks (Anderson, 1991). Incidental hosts, aside from humans, include domestic cattle, horses, and dogs, among others, as natural victims of this vector–pathogen–host non-selectivity (Barthold *et al.*, 1995b; Chou *et al.*, 2006; Lischer *et al.*, 2000; Magnarelli *et al.*, 2005; Parker and White, 1992; Skotarczak *et al.*, 2005). These indiscriminant features lend themselves well to experimental infection of a wide variety of animal species that are referred to as animal models. Several rodent species are susceptible to experimental *B. burgdorferi* s.l. infection, including hispid cotton rats (*Sigmodon hispidus*) (Burgdorfer and Gage, 1987), Mongolian gerbils (*Meriones unguiculatus*) (Preac-Mursic *et al.*, 1990), white-footed mice (*Peromyscus leucopus*) (reviewed below), grasshopper mice (*Onychomys leucogaster*) (Czub *et al.*, 1992), Syrian hamsters (*Mesocricetus auratus*) (reviewed below), guinea pigs (*Cavia porcellus*) (Krinsky *et al.*, 1982; Sonnesyn *et al.*, 1993), laboratory rats (*Rattus norvegicus*) (reviewed below), and laboratory mice (*Mus musculus*) (reviewed below). In addition, laboratory rabbits (*Oryctolagus cuniculus*) are susceptible to experimental infection (reviewed below). Larger animal models include non-human primates and dogs (reviewed below). Lyme borreliosis is suspected to cause illness in horses and cattle, with variable degrees of success at experimental infection (Burgess *et al.*, 1987; Chang *et al.*, 2000; Chang, 2005; Lischer *et al.*, 2000; Parker and White, 1992; Tuomi *et al.*, 1998).

The natural history of *B. burgdorferi* is to establish persistent infections with minimal harm to its natural hosts. Humans and domestic animals are aberrant hosts, and clinical disease is an aberration from the norm. Thus, a major challenge to modelling human disease has been an inability to experimentally reproduce some aspects of the human disease in animal models. Generally, adult animals are susceptible to infection, but resistant to development of disease. However, infant animals have immature innate and acquired immune responses that favour the evolution of disease, as does immunosuppression and X-irradiation. The various strains and mutants of laboratory mouse also favour susceptibility to disease (see below).

14.1.1 Non-human primate models

The multiorgan nature of *B. burgdorferi* infection in rhesus macaques (*Macaca mulatta*), including the nervous system, the similarity between the signs of Lyme borreliosis to those noted in humans, and the rhesus macaque taxonomic proximity with humans favour this model. Rhesus macaques exhibit signs that are characteristic of all three phases of human Lyme borreliosis, namely early localized, early disseminated and late-stage disease. Neuroborreliosis is an important aspect of this animal model, as neurological lesions are minimal or absent in rodents and lagomorphs.

Multi-organ character of Lyme borreliosis

Spirochaetes disseminate to multiple organs, including the central nervous system (cerebrum, brainstem and cerebellum, spinal cord, and dura mater) (Cadavid *et al.*, 2004; Cadavid *et al.*, 2000; Roberts *et al.*, 1995). Early localized infection is characterized by erythema migrans (EM) in 70–80% of human cases (Steere and Sikand, 2003), and in a similar proportion of rhesus macaques (Pachner *et al.*, 1995b; Philipp *et al.*, 1993). Following tick-borne infection, EM was intense with defined borders in the form of a ring or circular patch of about 8 to 10 cm in diameter; biopsy cultures were almost always positive for *B. burgdorferi*. EM lesions featured monocytic and lymphocytic infiltrates of the superficial perivascular areas as well as the deep dermal adipose tissues and perivascular spaces within muscle (Philipp *et al.*, 1993), similar to human EM (Berger, 1989; Mullegger *et al.*, 2000). Signs of early-disseminated disease included secondary EM (Philipp *et al.*, 1993), splenomegaly (Philipp *et al.*, 1993; Roberts *et al.*, 1995), fever (Roberts *et al.*, 1995), conjunctivitis (Philipp *et al.*, 1993) (a rare manifestation of human Lyme borreliosis) (Lesser, 1995), and cerebrospinal fluid (CSF) pleocytosis (with evidence of spirochaetal DNA) (Pachner *et al.*, 1993, 1995c).

Arthritis and carditis, which are signs of late Lyme borreliosis in humans, are also found in rhesus macaques (Cadavid *et al.*, 2004; Roberts *et al.*, 1995). In a study that involved five animals infected with *B. burgdorferi* JD1 by exposure to *I. scapularis* ticks, and one uninfected control, the joints most commonly affected were the knee and elbow joints, as with humans (Steere *et al.*, 2004). Lesions included synovial hyperplasia and mild periarticular fibrosis. In severely affected joints, articular cartilage was marginally eroded, and in some cases there was pannus formation. Synovial hyperplasia was accompanied by perisynovial fibrosis and perivascular accumulation of both B and T (primarily CD4) lymphocytes, plasma cells, and macrophages. The leucocyte subsets are similar to those described for human Lyme borreliosis (Steere *et al.*, 1988). As with humans (Johnston *et al.*, 1985; Steere *et al.*, 2004), evidence of spirochaetal presence in synovial tissue was obtained by culture from joint tissue fragments (Roberts *et al.*, 1995).

Varying degrees of atrioventricular block, presumed to be due to inflammation, occurs in up to 8% of Lyme disease patients (Steere, 1989; Steere, 2001). In a study that involved 14 immune competent and 11 immune suppressed macaques that received a needle inoculum with *B. burgdorferi* N40, all but one of the animals developed carditis (Cadavid *et al.*, 2004) (Fig. 14.1). In addition, carditis was evident in three of five rhesus macaques that were infected by tick-borne *B. burgdorferi* JD1 (Roberts *et al.*, 1995). Increased numbers of organisms and inflammatory cell infiltrates were observed in animals that were immune suppressed with dexamethasone (2 mg/kg, i.m., q.i.d.) either transiently, from days 7 to 28 post inoculation, or for the duration of the experiment (Bai *et al.*, 2004; Cadavid *et al.*, 2004). Leucocytic infiltrates were composed primarily of plasma cells (Fig. 14.2), monocytes and T-lymphocytes. In addition, there was increased deposition of IgG, IgM and complement membrane attack complex (MAC) (Cadavid *et al.*, 2004). Double-immunostaining revealed that many spirochaetes in the heart of immunosuppressed animals

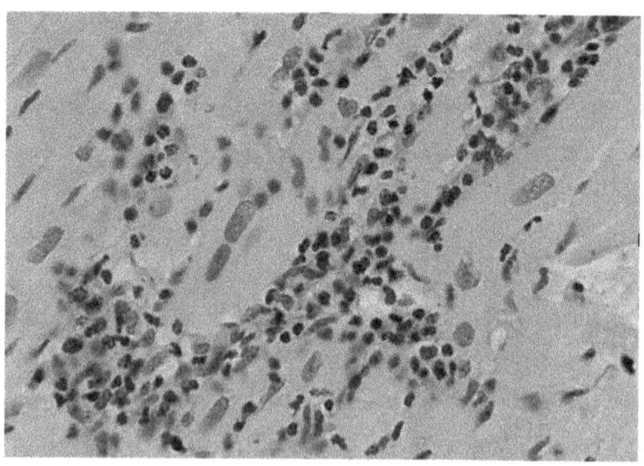

Figure 14.1 Myocarditis in a rhesus macaque infected with *B. burgdorferi*.

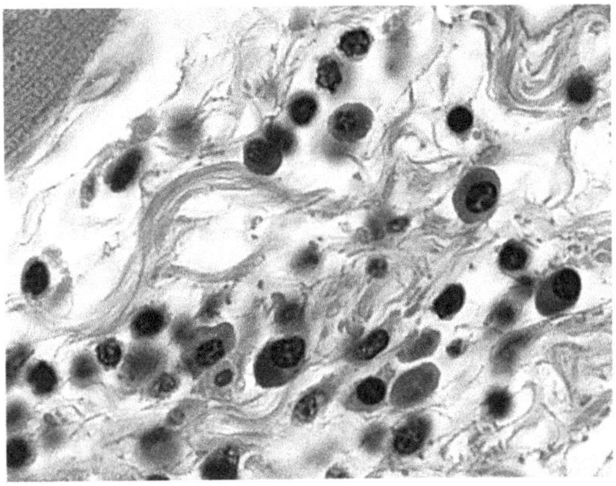

Figure 14.2 Abundance of plasma cells in an area of myocardial injury in a rhesus macaque infected with *B. burgdorferi*.

infected with *B. burgdorferi* N40 had MAC on their membranes, with prominent loss of cardiomyocytes (Cadavid *et al.*, 2004) by apoptosis (Londono *et al.*, 2005). Production of the B-cell chemokine CXCL13 was increased in hearts of infected animals (Cadavid *et al.*, 2004). The possible correlation between Lyme carditis and heart block has not been studied in rhesus macaques but continuous telemetric monitoring of heart function, a technique that would be required for this purpose, is available for rhesus (Chaves *et al.*, 2006) and could be implemented.

Neuroborreliosis of the peripheral nervous system

Peripheral neuroborreliosis has been evaluated both functionally and histologically in rhesus macaques. Nerve conduction studies (NCS) were performed on 11 animals: eight had been inoculated with *B. burgdorferi* JD1 (five via tick bite, and three via needle and syringe) and three were sham-inoculated controls. An additional seven normal animals established a normative database (England *et al.*, 1997). Motor NCS and conduction velocities were recorded on median, ulnar, and tibial nerves. Sensory NCS were performed on median and ulnar nerves. Complete electrophysiological studies were performed on all animals at study entry and sequentially thereafter at 3 months, 8 months, and 13 months. The first observations in the infected group of monkeys were performed 29–32 months after inoculation. Abnormal NCS values were further analysed for evidence of axonal loss or demyelination. Three of the infected monkeys had normal NCS values. The remaining five infected animals had a mononeuropathy or mononeuropathy multiplex nerve lesion pattern that was consistent with neuronal loss. Neuropathies involved five ulnar, two median and one tibial nerve. Concentric needle electromyographic examinations performed on selected muscles showed confirmatory evidence of denervation, concordant with the distribution of nerve lesions in each affected monkey. As reinnervation occurred, abnormal spontaneous activity (fibrillation potentials and positive sharp waves) decreased, and large polyphasic motor unit action potentials appeared. Over the 13-month period of study, all nerve conduction parameters improved, providing evidence of axonal regeneration.

Serial electrophysiological and neuropathological findings were clearly correlative, consistent with axonal loss variety multifocal neuropathies, most of which also exhibited some degree of axonal regeneration. By the time of necropsy, nerve lesions were mostly of a chronic nature, but inflammatory infiltrates observed in some nerves suggested active disease (England *et al.*, 1997). In a study in five animals that had been infected by tick-borne *B. burgdorferi* JD1 for only 3 months, the most prominent and consistent neurological manifestation observed was neuritis (Roberts *et al.*, 1995). The latter was characterized by intraneural perivascular infiltration of mononuclear cells, including macrophages and B lymphocytes, involving multiple nerves. T-lymphocytes were not detected by immunohistochemistry: no $CD4^+$ or $CD8^+$ T-cells were observed and staining with an anti-alpha-beta T-cell receptor antibody and with an anti-$CD3^+$ antibody also yielded negative results (Roberts *et al.*, 1998).

Direct evidence of the presence of *B. burgdorferi* in the PNS was obtained in a study that involved 10 rhesus macaques, seven infected and three controls. The infected animals, of which three were immune competent and four were immune suppressed, were given a needle inoculum of *B. burgdorferi* N40. PCR was used to amplify the *ospA* gene to detect evidence of spirochaetes in tissues. Spirochaetes were present in PNS tissues both in normal and immune suppressed macaques (Cadavid *et al.*, 2000).

Neuroborreliosis of the central nervous system

Evidence that *B. burgdorferi* accesses the CNS of rhesus macaques was first provided by Pachner and co-workers (Pachner *et al.*, 1995b). Their study included five animals that were given intradermal inoculations of *B. burgdorferi* N40. One of these animals was immune compromised by dexamethasone injections. CNS invasion was documented by both culture and PCR of CSF specimens. Pleocytosis and spirochaetal DNA (by PCR) were detected in the CSF of all of the animals, and two of the animals, one of which was the immune compromised one, were CSF-culture positive. This latter animal also showed evidence of CNS inflammation by magnetic resonance imaging (MRI). An important feature of the model is the finding that both PCR positivity and pleocytosis were intermittently detected for as long as 18 weeks post inoculation in all of the immune competent animals (Pachner *et al.*, 1995b). This is an indication that a *B. burgdorferi* infection may persist in the CNS for an extended time. Spirochaetes localize primarily to the meninges in the CNS of rhesus macaques. In a study designed to investigate spirochaetal localization in the CNS and in other organs, animals were immune suppressed to increase spirochaetal burdens in the tissues and thus facilitate the identification of predominant tissue tropisms. Spirochaetes were detected immunohistochemically chiefly in the leptomeninges, nerve roots, and dorsal root ganglia. By *ospA* PCR, spirochaetal DNA was detected in the cerebrum, brainstem and cerebellum, spinal cord and dura mater, not only in immune suppressed animals but also in immune competent rhesus, albeit in much lower amounts (Cadavid *et al.*, 2000). In another study that included eight transiently immune suppressed animals, half of which were needle- or tick-inoculated with *B. burgdorferi* N40, spirochaetal DNA was frequently detected in the cerebrum and brainstem (Pachner, 2001). Spirochaetal DNA also has been detected in spinal leptomeninges, spinal roots, and dorsal root ganglia of cervical, thoracic, and lumbosacral spinal cord tissues, but only minimally in the spinal cord parenchyma of immunosuppressed animals.

Inflammation related to the presence of *B. burgdorferi* occurs in the rhesus CNS, not only in the meninges (Fig. 14.3), with associated pleocytosis, but also in the brain, albeit at a much lower frequency than in organs such as skeletal (Cadavid *et al.*, 2003) (Fig. 14.4) and cardiac muscles (Fig. 14.1) (Cadavid *et al.*, 2004; Pachner, 2001). One inflammatory mediator that was found to be consistently upregulated in skeletal muscle tissue is the B-cell chemokine CXCL13 (Pachner *et al.*, 2002), which was also detected by immunohistochemistry in the brain and spinal cord of *B. burgdorferi*-infected rhesus macaques (Bai *et al.*, 2004; Narayan *et al.*, 2005). CXCL13 was found to be upregulated in the CSF of patients with neuroborreliosis and other CNS inflammatory diseases, but not in patients with non-inflammatory neurological diseases (Rupprecht *et al.*, 2005).

Preclinical assessment of vaccines

While initial experimental assessment of candidate vaccines is usually done in rodents, it is generally agreed that preclinical testing of vaccines for humans may be assessed more reliably in non-human primates (Staprans and Feinberg, 2004). Safety, immunogenicity, and efficacy of OspA vaccines were examined in rhesus macaques (Philipp *et al.*, 1996; Philipp *et al.*, 1997). The vaccines that were tested included one with the same formulation as that subsequently used in humans, namely, lipidated OspA combined with aluminium hydroxide, as well as formulations that included non-lipidated OspA with and without the adjuvant monophosphoryl lipid A (Philipp *et al.*, 1996). The adjuvanted vaccine was given not

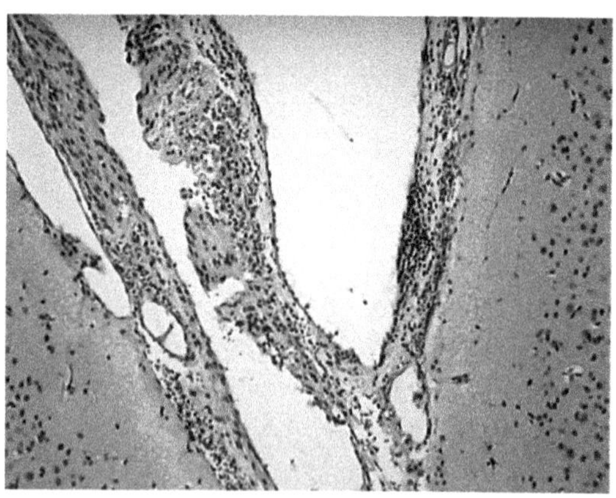

Figure 14.3 Non-suppurative leptomeningitis in a rhesus macaque infected with *B. burgdorferi*.

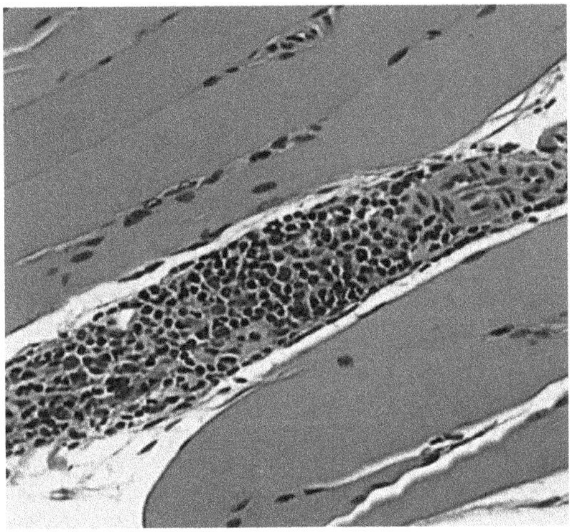

Figure 14.4 Skeletal myositis in a rhesus macaque infected with *B. burgdorferi*. There is infiltration of the endomesial connective tissue with mononuclear leucocytes, which are primarily macrophages. Note similarity of lesion to that found in the mouse model (Fig. 14.11).

only to naïve animals but also to macaques that had previously been infected with *B. burgdorferi*. Safety was assessed in both groups of animals, by physical examination, clinical laboratory analyses of blood and urine samples, and quantification of inflammation in the anterior eye chamber.

No local vaccine site reaction or obvious signs of arthritis were seen in any animal during the duration of the study. All urinalyses, serum chemistries and complete blood count results were either normal or, when not, not attributable to vaccination. No significant intraocular inflammatory responses were found, thus indicating that the powerful adjuvant properties of lipidated OspA do not cause uveitis. Moreover, putative pre-sensitization to the vaccine by a pre-existing *B. burgdorferi* infection also had no deleterious effect. All formulations elicited strong IgG anti-OspA responses, as measured by ELISA titres at 4 weeks after the last vaccine dose, but were significantly lower 9 weeks thereafter and were thus short lived. When the same vaccine formulations and vaccination regimens were used in humans to evaluate the vaccines' safety and immunogenicity, the results of the preclinical assessment in rhesus macaques were essentially confirmed, thereby validating the usefulness of the rhesus model (VanHoecke *et al.*, 1996).

Efficacy of the three OspA vaccine formulations was investigated after challenging the animals with tick-borne *B. burgdorferi* B31. Protection was assessed by a variety of procedures, including evaluation of residual spirochaetes in ticks that had fed on vaccinated compared with control animals, differential post-challenge seroconversion (by immunoblot), and PCR to detect *B. burgdorferi* DNA in skin biopsies collected from the site of inoculation. Post mortem, several tissues were evaluated by PCR and immunohistochemistry with a specific monoclonal antibody. As many as 41 of the 43 ticks that had fed on control animals harboured spirochaetes, an infection rate that was similar to that of unfed nymphs. In contrast, less than 1% of the ticks that fed on vaccinated animals (1/121) had detectable spirochaetes, and at a much lesser density than in the control group. Moreover, none of the vaccinated animals seroconverted or had spirochaetal DNA detectable by PCR in skin biopsy specimens obtained from the site on inoculation, whereas all of the control animals did. However, spirochaetal DNA and antigen were found in organs of some of the vaccinated animals, suggesting a low-level of infection.

To evaluate whether this putative infection persisted or had been transient, a subgroup of the vaccinated animals was radically immune suppressed by surgical splenectomy and administration, one week thereafter, of cyclophosphamide and prednisone sodium phosphate (Bohm Jr *et al.*, 1999). In spite of the immune suppression regimen, the frequency of post-mortem detection of spirochaetal DNA by PCR was not significantly different from that observed in the vaccinated animals that were not immune suppressed. Xenodiagnosis did not yield spirochaetes in ticks applied to the immune suppressed animals, whereas a recently infected and otherwise normal animal was successfully xenodiagnosed. Finally, cultivation of bronchoalveolar lavage and urine samples yielded no *B. burgdorferi* spirochaetes. These results substantiated, on the one hand, the conclusion that the apparent low-level infection detected in some of the vaccinated animals had been transient (Philipp *et al.*, 1997), and underscored, on the other, the versatility of the rhesus model for evaluation of vaccine efficacy.

Diagnostic serology

Serology is a crucial complement to the clinical diagnosis of Lyme borreliosis. A significant advance in Lyme disease serology was brought about by the development of the C6 test (Liang *et al.*, 1999a, 2000a,b). This ELISA-type test is based on the detection of serum antibodies to a peptide (C6) that reproduces the amino acid sequence of a small fragment of VlsE, the antigenic variation protein of *B. burgdorferi*. Invariable region (IR) 6, as the fragment is called, is an invariant, antigenically conserved, and immunodominant segment of the VlsE molecule (Liang *et al.*, 1999a).

Three important features of the test were first noted in rhesus macaques infected with *B. burgdorferi*. The ability of the test to detect antibody early in the infection was precisely determined using serial serum samples. Nine of the 10 macaques used in this study had been infected by ticks, and one by needle and syringe. Antibody was detected in seven animals as early as 3 weeks post inoculation. At week 5 post inoculation, nine animals had responded, and all had responded at week 6 (Liang *et al.*, 2000a). Rhesus immunoglobulin readily cross-reacts with reagents used to quantify human antibodies in enzyme immunoassays. Such an assay design could therefore be applied to human serum specimens. When this was done, the C6 test performed with significantly greater sensitivity than traditional whole-cell antigen-based ELISAs, or with a comparable sensitivity but greater specificity (Cinco and Murgia, 2006; Philipp *et al.*, 2005; Smismans *et al.*, 2006; Tjernberg *et al.*, 2007).

Absence of reactivity of the C6 ELISA with antibodies elicited by the OspA vaccine also was first uncovered with serum specimens from rhesus monkeys that had been vaccinated with the OspA vaccine used in humans (Liang *et al.*, 1999b). This feature of the test was later confirmed both with human and canine serum specimens (Marques *et al.*, 2002; O'Connor *et al.*, 2004). The OspA vaccine for humans is no longer commercially available, but dogs continue to be extensively vaccinated with OspA-based vaccines or spirochaete bacterins that contain OspA among other antigens (Littman *et al.*, 2006).

The marked decline in C6 antibody titre that follows antibiotic treatment of patients with early Lyme borreliosis is the third feature of the C6 test that was first revealed with the rhesus model (Philipp *et al.*, 2001). Seven animals that had been infected with *B. burgdorferi* via tick feeding were treated for 60 days with oral doxycycline (2 mg/kg, bid), starting at week 12 post-inoculation. A longitudinal analysis of serum antibody levels revealed that C6 antibody declined at a much faster rate after treatment than did antibody to *B. burgdorferi* whole-cell antigen (standard Lyme ELISA) (Philipp *et al.*, 2001). By 6 months after the initiation of antibiotic treatment, the level of C6 antibody in all animals had reached baseline level, whereas the mean level of antibody as detected by standard Lyme ELISA was 2.5 times the corresponding baseline value. These data suggest that pre- and post-treatment quantification of C6 antibody titre could be used to evaluate response to treatment in Lyme disease patients, as is done with the VDRL antibody test in the treatment of syphilis (Larsen *et al.*, 1995). Several studies performed subsequently confirmed that a 4-fold decline, or a decline to baseline values, in C6 antibody titre occurs in a significant proportion of successfully treated patients with early localized or early disseminated Lyme borreliosis (Philipp *et al.*, 2003; Philipp *et al.*, 2005) but not in patients with post-treatment Lyme disease syndrome (Fleming *et al.*, 2004) or late disease (Peltomaa *et al.*, 2003). The possibility of adapting the C6 test to serve as a diagnostic device to monitor progression of treatment is tantalizing and has already been realized for the management of canine Lyme disease (http://www.idexx.com/animalhealth/laboratory/c6/).

Chemotherapy

Lyme disease is usually treated with antimicrobial therapy, but some patients who are treated in the early or early-disseminated disease phases may continue to have persistent subjective symptoms (Asch *et al.*, 1994; Shadick *et al.*, 1994). A few patients fail to respond to antibiotic therapy, with signs of persistent infection (Steere *et al.*, 1994; Wormser *et al.*, 2006). The response to treatment in patients with late manifestations is typically slower (Wormser *et al.*, 2006). Post-treatment Lyme disease syndrome is a condition that occurs in some patients after treatment for Lyme borreliosis. It is characterized by non-specific symptoms of fatigue, sleep disorders, headache, memory and concentration difficulties, myalgias and arthralgias. The cause of this syndrome is currently unknown but antibiotic therapy does not seem to be helpful (Klempner *et al.*, 2001).

Decreased efficacy of antibiotic treatment in late disease or ineffectiveness of treatment of chronic Lyme disease may be attributed to several causes, including (1) spirochaetes that persist in the tissues, probably in small numbers, inaccessible or impervious to antibiotic; (2) inflammatory responses to residual antigens from dead organisms; (3) residual tissue damage following pathogen clearance; (4) autoimmune responses, possibly elicited by antigenic mimicry (Bolz and Weis, 2004). In an effort to gain insight into these hypotheses, an experiment was conducted to assess the efficacy of a prolonged regimen of antibiotic treatment in rhesus monkeys that were inoculated with *B. burgdorferi*. The treatment regimen included the administration of ceftriaxone for 30 days followed by 60 days of doxycycline. This regimen is similar to that utilized in two controlled trials of antibiotic treatment in patients with persistent symptoms and a history of Lyme disease (Klempner *et al.* 1997). Doxycycline is used primarily to treat early localized and disseminated disease, whereas ceftriaxone is used to treat late disease (Wormser *et al.*, 2006). Treatment was initiated at 6 months post inoculation, and animals were kept for a minimum of 6 months after treatment termination. Post mortem, tissue samples from multiple organs were collected for (1) *in vitro* culture of *B. burgdorferi*; (2) quantitative real-time PCR and reverse transcriptase (RT) PCR, for detection of *B. burgdorferi* DNA and RNA, respectively; and (3) histopathology and immunohistochemistry, to localize inflammatory lesions and spirochaetes in tissues. An initial analysis of the results indicates that spirochaetal DNA and antigen is still detectable in treated animals (M.T. Philipp, S.W. Barthold, and others, *unpublished*), similar to findings in the dog and mouse models.

Other non-human primate models

Hamadryas baboons (*Papio hamadryas*) have been inoculated with *B. burgdorferi* both via tick and syringe, and infectivity and antibody responses were evaluated. Animals became infected by both inoculation procedures; one of the two tick-inoculated animals exhibited an erythematous lesion surrounding the site of tick attachment (Hefty *et al.*, 2002).

14.1.2 Canine model

Dogs manifest many of the signs seen in human Lyme borreliosis, but many dogs that are naturally infected in endemic areas are seropositive and clinically normal (Levy and Magnarelli, 1992), In contrast, some dogs develop fatal kidney disease (Littman *et al.*, 2006). The pathogenesis of canine Lyme nephropathy is poorly understood, but it has features of an immune-complex glomerulonephritis. It is not associated with the direct presence of spirochaetes in the kidney, and is tragically refractory to palliative clinical intervention. Nearly 30% of nephropathy cases are associated with prior history of immunization with a commercial *B. burgdorferi* bacterin (Dambach *et al.*, 1997). Dogs are at greater risk of infection than humans in endemic areas, and serve as a sentinel species for the presence of *B. burgdorferi* in a region (Lindenmayer *et al.*, 1991).

Although dogs are susceptible to infection by sub-adult ticks, they tend to be preferentially and more efficiently infected by adult female ticks (Anderson and Magnarelli, 1993; Appel *et al.*, 1993; Korshus *et al.*, 2003). Exposure of pregnant bitches to infected ticks resulted in infection of the dams, but not their fetuses or pups (Appel *et al.*, 1993), but in a separate study, intrauterine transmission was documented (Gustafson *et al.*, 1993).

Dogs develop disseminated, persistent infections, with a course that closely mimics human disease, including intermittent bouts of oligoarthritis that arise several weeks after infection. Unlike humans, erythema migrans is inconsistently observed in dogs (or at least is difficult to discern). Because of their physical size, dogs offer the opportunity to collect clinical samples, including blood, synovial fluid, skin biopsies, and CSF. All ages of dogs are susceptible to infection, but 6-week old puppies were more susceptible to development of clinical disease (lameness), compared with older dogs. Clinical illness is silent in some dogs, but in one study, over 60% developed one to six episodes of transient (3 or 4 days) lameness, with reluctance to move and fever commencing at 2 to 5 months after infection. Lameness was sudden in onset, and most often involved the elbow or shoulder joint on the side where ticks were placed. The episodes of lameness resolved spontaneously without therapy. Affected joints had slight swelling with marked exudation of fibrin and neutrophils into the joint space. Spirochaetes were detectable in multiple tissues for 1 year or more, based upon culture and PCR, and could be isolated from joint capsules of both lame and clinically normal dogs. Bacteraemia was variably detected, and transient. As in humans, total IgG levels were increased for up to 17 months. Immunoblot profiles resembled natural infection, with initial reactivity to 39- and 41-kDa antigens (Appel *et al.*, 1993; Barthold *et al.*, 1995b; Burgess, 1986; Summers *et al.*, 2005).

Histopathology from dogs at several months after tick-borne infection revealed multifocal non-suppurative meningitis, encephalitis, perineuritis, periarteritis, polyarthritis, and perivascular infiltrates in skin near the site of initial tick attachment. All of these lesions were characterized by infiltration of lymphocytes, plasma cells, and macrophages. Most dogs, regardless of clinical status, had some joints with focal aggregation of lymphocytes, plasma cells, neutrophils, and fewer numbers of macrophages and eosinophils. Unlike humans, lymphoid follicles with germinal centres were not found, and synovial villous proliferation was rare. Furthermore, the dogs developed enlarged lymph nodes, particularly those near the site of tick attachment, with hyperplasia dominated by numerous plasma cells (Chang *et al.*, 2001; Summers *et al.*, 2005). Corticosteroid treatment at 420 days of infection appeared to reactivate arthritis (Straubinger *et al.*, 1998b). Infected dogs may manifest bouts of increased aggressive behaviour (Evans *et al.*, 1995a; Evans *et al.*, 1995b). The neurological aspects of canine Lyme borreliosis have not been well characterized or compared to the human condition, nor has cardiac disease. A naturally occurring case of complete heart block and myocarditis has been documented in a dog (Levy and Duray, 1988). Dogs would lend themselves well to electrocardiographic studies.

When dogs were infected by tick, spirochaetes disseminated locally within tissues. Detection of spirochaetes by culture and PCR was most apparent in tissues within the somatic quadrant of the site of tick attachment. Likewise, lymph node enlargement was greatest in the gland nearest to the tick site. Spirochaetes were most apt to be detected in collagen-rich tissues, with synovia, fascia, and lymph nodes closest to the tick bite most often infected. Quantitative PCR of skin biopsies of infected dogs have indicated a significant drop in spirochaete numbers that correlated with rising antibody levels (Chang *et al.*, 1996; Straubinger *et al.*, 1997a; Straubinger, 2000). Synovial fluids from dogs during acute lameness had elevated volume, cell count, cell viability, higher proportion of neutrophils, and chemotactic activity for neutrophils. At necropsy, IL-8 was found to be expressed in synovia, pericardium, and peritoneum, and TNF was elevated in enlarged lymph nodes draining affected joints. These indices were lower or absent during clinical silence or in joints with mild inflammation (Straubinger *et al.*, 1997a). Canine synovial explants exposed to *B. burgdorferi* have also revealed upregulation of proinflammatory cytokines, including IL-1α, IL-1β, TNF and IL-8 in response to *B. burgdorferi* (Straubinger *et al.*, 1998a).

As a species of veterinary importance, as well as for modelling the human condition, dogs have been utilized to demonstrate efficacy of vaccination and antibiotic treatment in a number of trials. Vaccination has been shown to be effective against tick-borne and syringe challenge using various formulations of whole-cell bacterins and recombinant OspA/B proteins (Chang *et al.*, 1995; Chu *et al.*, 1992; Coughlin *et al.*, 1995; Keller *et al.*, 1994; Straubinger *et al.*, 1995, 2002; Topfer and Straubinger, 2007). Perhaps the most intriguing findings with the dog model have been the demonstration of persisting spirochaetes after antibiotic treatment. Treatment of tick-infected dogs for 30 days with amoxicillin or doxycycline eliminated clinical disease, and reduced, but did not entirely eliminate infection, based upon PCR of tissues. Antibody titres to *B. burgdorferi* declined following treatment; in some cases to undetectable levels. When dogs were maintained for 6 months after treatment was completed, antibody levels rose again (Straubinger *et al.*, 1997b). Studies with antibiotic-treated dogs have shown a rapid decline in antibody response to C6 of VlsE, in contrast to persistence or a slower decline in overall antibody response to *B. burgdorferi* lysates, despite incomplete clearance of infection

(Philipp *et al.*, 2001). Evidence of persistence was found in dogs treated with amoxicillin, azithromycin, ceftriaxone, or doxycycline (Straubinger *et al.*, 1998b; Straubinger *et al.*, 2000). Notably, evidence of persistent infection in these studies has been based upon PCR, with no success at culturing viable spirochaetes. In untreated dogs, PCR and culture have equal sensitivity. Similar results have been confirmed in the rhesus and mouse models.

14.1.3 Rabbit model

In a seminal publication published in 1982, Willy Burgdorfer and colleagues demonstrated that laboratory rabbits could be infected with treponema-like spirochaetes (later to be named *B. burgdorferi*) following feeding by adult *Ixodes dammini* (later to be renamed *I. scapularis*) (Burgdorfer *et al.*, 1982). Furthermore, the rabbits developed skin lesions reminiscent of EM. Other groups attempted to infect rabbits with inconsistent results (Benach *et al.*, 1984; Krinsky *et al.*, 1982; Pachner *et al.*, 1994). In one early study, an EM-like lesion was shown to contain aggregates of lymphoid cells, and spirochaetes were visualized with silver stain (Kornblatt *et al.*, 1984a). These early efforts also confirmed bacteraemia in rabbits (Benach *et al.*, 1984; Kornblatt *et al.*, 1984a,b), and acquisition of infection by ticks following feeding upon experimentally infected rabbits (Burgdorfer, 1984). Intradermal inoculation with high doses (10^6) of cultured spirochaetes resulted in a high rate of PCR reactivity in the CNS and visceral organs, but tissues were culture negative. Following high dose (10^8) intraperitoneal inoculation of infant Dutch rabbits, spirochaetes could be cultured from multiple organs and the rabbits developed carditis and arthritis at 30 days after inoculation (Moody *et al.*, 1990a).

The rabbit model came into its own with the reproducible development of EM lesions following intradermal inoculation of adult New Zealand White rabbits with as few as 10^3 *B. burgdorferi* B31 spirochaetes, or by tick-borne infection. Lesions arose within 3–8 days, expanded to 7.5–79 mm diameter, and remained visible for 3–7 days, and then regressed. Similar lesions could be induced with a number of *B. burgdorferi* s.s. isolates. EM lesions were similar to EM in humans, with dermal aggregates of lymphocytes and plasma cells around adnexae, vessels and small nerves. In addition, spirochaetes could be cultured from multiple organs, including spinal cord but not cerebral cortex or CSF, for up to 8 weeks, but not thereafter (Foley *et al.*, 1995). The EM lesions are associated with the presence of extracellular spirochaetes in the dermal collagen by immunohistochemistry. Spirochaetes remained in the skin following regression of the lesions, but spirochaetes were eventually cleared (Chong-Cerrillo *et al.*, 2001). The reproducibility of EM induction has been used for demonstrating resistance to re-infection with high challenge doses of the homologous *B. burgdorferi* strain (Foley *et al.*, 1995. 1997; Shang *et al.*, 2000), but not against challenge with a heterologous *B. burgdorferi* s.s. strain (Shang *et al.*, 2001). Similar results occurred when rabbits were challenged with host-adapted spirochaetes, introduced by implantation of infected skin biopsies (Shang *et al.*, 2000, 2001). As with rodent models, rabbits can be protected against infection (and development of EM) when immune rabbit serum was administered before and after (–18, 0, +24, +48, +96 and +216 hours) challenge, but not after (+24, +48, +96, and +216 hours) challenge inoculation (Chong-Cerillo *et al.*, 2000). The rabbit model has also been used to demonstrate efficacy of vaccination with recombinant OspA, outer membrane vesicles, and outer membrane protein Oms66 (P66) (Exner *et al.*, 2000; Shang *et al.*, 2000). The rabbit skin model, which contains (albeit transiently) high numbers of spirochaetes, has also been used to examine expression of hydrophobic proteins of host-adapted spirochaetes between 5 and 21 days after infection, with maximal numbers of spirochaetes at day 11. Using a method to extract and measure hydrophobic antigens from tissues, significant changes in the relative expression of *B. burgdorferi* proteins were found, including loss of detectable OspC by day 9, and increases in VlsE and other proteins (BmpA, OppA-2, DbpA, etc.) from days 7 through 21 (Crowley *et al.*, 2004).

Multiple studies indicate that rabbits, unlike other species, are able to clear infection by 8 weeks. Unlike rodents, dogs, monkeys, and humans, which develop strong and early antibody responses to VlsE invariable region IR6, rabbits mount early antibody to the carboxyl-terminal domain of VlsE, but a very low or no response to IR6 during the first 6–11 weeks. Although *vlsE* gene sequence recombination occurs during infection of rabbits, a significant number of *B. burgdorferi* isolates (25 of 50 clones) obtained from rabbit skin were found to be devoid of lp-28, the B31 plasmid that encodes VlsE (among other genes). This rate of loss has not been observed in mice. Despite these findings, a few rabbits were found to seroconvert with high titres of antibody to IR6 after several weeks of infection, and spirochaetes were cultured from one rabbit at 98 weeks (Embers *et al.*, 2007). Thus, some rabbits, like other species, can be persistently infected, but most rabbits seem to mount an effective sterilizing immune response, which may reflect the unique immunophysiology of lagomorphs. The comparative nuances of the rabbit immune response are well worth further investigation.

14.1.4 Rodent models of Lyme borreliosis

White-footed mouse model

Peromyscus leucopus received early attention as a model, because of its importance as a natural reservoir host in North America and its availability as a laboratory animal. *Peromyscus* mice are readily susceptible to infection by tick, which transmit infection between 48 and 72 hours after attachment (Piesman *et al.*, 1987), but do not become infected transplacentally from mothers (Mather *et al.*, 1991). Studies in experimentally infected hamsters and naturally exposed *Peromyscus* mice revealed that ear tissue could be sampled as a non-invasive method for evaluating infection status (Sinsky and Piesman, 1989). This observation was soon capitalized upon with the advent of PCR for detection of *B. burgdorferi* DNA in museum specimens of rodents, confirming that *B. burgdorferi* was present in suitable reservoir hosts on Cape Cod in the 1890s (Persing *et al.*, 1990). Likewise, this species revealed the high rate of infection of the urinary bladder, which is now generally tested to confirm disseminated infection in animal model studies, regardless of species (Callister *et al.*, 1989; Schwan *et al.*, 1988b). The widely used method of xenodiagnosis, in which infection of the host can be evaluated with a high degree of sensitivity by feeding ticks upon infected hosts was also introduced in this model (Donahue *et al.*, 1987). Microscopic studies in experimentally infected *Peromyscus* mice revealed no evidence of pathology in tissues in persistently infected animals, with spirochaetes observed extracellularly within the interstitium of various tissues (Wright and Nielsen, 1990). However, when *Peromyscus* mice were infected as infants, they developed arthritis and carditis (Moody *et al.*, 1994b).

Hamster model

In a remarkably insightful early study, Russell Johnson and colleagues not only documented the susceptibility of Syrian hamsters to infection with *B. burgdorferi*, but also demonstrated that the animals developed early bacteraemia, which was cleared after day 6, and then developed persistent disseminated infection of multiple organs (up to 52 days) with no overt clinical signs (Johnson *et al.*, 1984). Furthermore, hamsters of both sexes and all ages (4, 10 and 52 weeks of age) were susceptible to disseminated infection. The important concept of loss of spirochaete infectivity following multiple *in vitro* passage was also initially documented with the hamster model. Persistent infection was later confirmed for up to 14 months, with periodic recurrence of bacteraemia in some animals after months of infection (Johnson *et al.*, 1988).

The hamster model was used for examining efficacy of passive immunization, using antiserum from rabbits that had been hyperimmunized intravenously weekly with 10^8 viable spirochaetes. Protective activity was effective when antiserum was administered 17–18 hours before, but not 17–18 hours after challenge. As little as 0.0125 ml of rabbit antiserum was effective against 10^8 intraperitoneal challenge (1000 ID_{50}). Furthermore, passive immunization was effective against challenge with homologous isolates, but not against geographically heterologous isolates (Johnson *et al.*, 1986a; Johnson *et al.*, 1988). These pivotal passive immunization studies confirmed that *B. burgdorferi* was exquisitely susceptible to antibody, but only when antibody was present at the time of challenge, but not once infection had been established. This concept of host adaptation of *B. burgdorferi* has been a significant subject of research interest, and probably underlies the critical mechanisms of persistence.

Inoculation of adult hamsters intraperitoneally or cutaneously with *B. burgdorferi* results in disseminated, persistent infection, but without lesions in internal organs (Duray and Johnson, 1986; Goodman *et al.*, 1991; Johnson *et al.*, 1988). However, when inbred LSH/SsLak hamsters were inoculated in the hind paws with high doses ($\simeq 10^6$) of spirochaetes, acute transient synovitis was evident microscopically. Whole-body non-lethal irradiation prior to inoculation exacerbated the synovitis, with earlier onset, grossly evident swelling, and delayed resolution. Induction of joint swelling could also be induced in irradiated hamsters following intravenous and intraperitoneal inoculation, and with hind paw inoculation of fewer spirochaetes. At 1 week, the tibiotarsal and intertarsal synovia of non-irradiated hamsters became hypertrophied and hyperplastic, with infiltration of neutrophils and fibrin. Numerous spirochaetes were visible in the synovial and periarticular tissues. Inflammation subsided by 3 weeks, which coincided with seroconversion, and at 15 weeks, there was residual fibrosis, focal perineurovascular accumulations of lymphocytes and plasma cells, and spirochaetes were no longer visible. Joints of irradiated hamsters had similar, but more severe inflammatory changes, with persistence of inflammation and spirochaetes at 5 weeks. Inflammation of distant joints (fore paws) arose in irradiated hamsters, but lesions were less severe (Schmitz *et al.*, 1989). A sequential study of arthritis in irradiated and non-irradiated hamsters reiterated the transient nature of the inflammation in non-irradiated hamsters, which was prolonged and more severe with irradiation. The abundance of extracellular spirochaetes within the periarticular oedematous tissue, which disappeared as inflammation resolved, and the tendency of spirochaetes to align themselves along collagen bundles in tendons and fibrous connective tissue was noted (Hejka *et al.*, 1989). Others have shown that hamsters inoculated as 3-day old infants intraperitoneally with 10^6 *B. burgdorferi* developed disseminated infections with polyarthritis and carditis at 30 days (Moody *et al.*, 1990a).

Hamsters appear to require relatively high doses of spirochaetes to establish infection following intraperitoneal or intradermal inoculation, but this may be a reflection of culture conditions or the strain of the inoculum (Obonyo *et al.*, 2002; Roehrig *et al.*, 1992). Despite the apparent need to use high doses of spirochaetes when syringe inoculating, hamsters are readily susceptible to tick-borne infection and can transmit infection to feeding ticks (Kurti *et al.*, 1996; Piesman *et al.*, 1987; Woodrum and Oliver, 1999). Infected nymphal ticks feeding upon hamsters or *Peromyscus* mice transmitted infection inefficiently at 24 and 48 hours, but most animals became infected after 72 hours of attachment (Piesman *et al.*, 1987). Studies with tick-infected hamsters have shown no evidence of venereal, transplacental, or contact transmission among infected hamsters (Woodrum and Oliver, 1999).

The hamster model has been used to elucidate important concepts about biologically relevant host antibody responses during active infection. Serum from actively infected hamsters (immune serum) was protective when passively transferred to naïve irradiated hamsters prior to high dose foot challenge. Using serial dilutions of immune serum to titrate protective activity, protective activity was present within 7 days of infection, peaked between 3 and 5 weeks, and then declined, despite persistent infection. Protection was achieved with as little as 0.0125 ml of immune serum. The dynamics of the protective antibody response differed from the overall antibody reactivity against *B. burgdorferi* lysates, as measured by ELISA, which remained elevated throughout the course of infection (Schmitz *et al.*, 1990, 1991b). The unusual dynamics of protective activity in immune serum were associated with complement (Schmitz *et al.*, 1991a), IgG2 (Schmitz *et al.*, 1992) and *in vitro* borreliacidal activity (Lovrich *et al.*, 1991). Passive immunization with immune sera and challenge with homologous and heterologous isolates has also been used to demonstrate different seroprotective groups among isolates of *B. burgdorferi* and *B. garinii* (Lovrich *et al.*, 1993, 1994).

The hamster model has been utilized for immunization studies, using inactivated *B. burgdorferi*, mutant *B. burgdorferi* and recombinant proteins (Hughes *et al.*, 1993; Jobe *et al.*, 1994; Johnson *et al.*, 1985, 1986b,c; Kurti *et al.*, 1996; Lovrich *et al.*, 1995; Padilla *et al.*, 1996; Rousselle *et al.*, 1998). Immunization with a non-infectious *B. burgdorferi* mutant lacking OspA and OspB was shown to be protective against challenge with infectious wild-type spirochaetes, demonstrating that host protective immunity can be elicited against antigens other than OspA/B (Hughes *et al.*, 1993). Hyperimmunization with formalin-inactivated whole *B. burgdorferi* prior to challenge with a homologous isolate in the hind paw may result in severe arthritis, but only if challenge took place prior to development of borreliacidal antibodies. Borreliacidal antibodies could be circumvented by challenge with a heterologous isolate (different seroprotective group) of *B. burgdorferi* (Lim *et al.*, 1994). Subsequent studies demonstrated that sensitized $CD4^+$ T-cells and macrophages were responsible for this effect (DuChateau *et al.*, 1996, 1997;, 1999; Lim *et al.*, 1995a,b). This phenomenon of destructive arthritis was later shown to occur when hamsters were pre-immunized with recombinant OspA (Croke *et al.*, 2000). This observation provoked controversy, as it was published at a time when the human OspA vaccine (LYMErix) was being marketed (Parenti, 2000).

Rat model

Several tick isolates of *B. burgdorferi* (N8, N13, and N40) were shown to be infectious in 3-week-old inbred Lewis (LEW/SSNcr) rats. The rats developed transient bacteraemia, disseminated infection and mild polyarthritis. Inoculation of infant (3-day- old) Lewis, Fischer 344 (F344/Ncr) and outbred Sprague–Dawley (CR:RAR[SD]) rats revealed similar results, but arthritis was more severe in the younger rats, with grossly evident swelling of tibiotarsal joints and lameness. Spirochaetes were visualized by silver stain in the periarticular oedema, synovium and occasionally in the joint spaces (Barthold *et al.*, 1988a,b). Infant rats developed periarticular tibiotarsal swelling by day 14, which subsided by day 60. Caliper measurements indicated significant persistent joint enlargement beyond 90 days. Bacteraemia was evident at 14 days, but disappeared thereafter, and multiple tissues were infected for up to 1 year. Acute arthritis, with exudation of fibrin and neutrophils, subsided by 90 days, with subsequent fibrosis and infiltration of macrophages, lymphocytes and plasma cells in the perisynovial tissue. When rats were examined at 180, 270 and 360 days, some rats had acute oligoarticular arthritis in addition to chronic changes in other joints, suggesting disease exacerbation. In addition, infiltrates of plasma cells and lymphocytes were noted in the connective tissue at the base of the heart and aortic wall, as well as myocardium (Moody *et al.*, 1990b). Rats were shown to be susceptible to infection with numerous different *B. burgdorferi* isolates from the northeast, Midwest and Pacific US, as well as Europe, with no apparent differences in patterns or ability to disseminate and induce disease (Barthold *et al.*, 1990b). An exception was New York tick *Borrelia* sp. 25015 (*B. bissettii*), which was infectious but non-arthritogenic (Anderson *et al.*, 1990).

Rats could be infected with significantly lower doses of spirochaetes following intradermal inoculation, compared with intraperitoneal inoculation, but could not be infected through contact, oronasal inoculation, in utero transmission, or venereal transmission (Moody and Barthold, 1991). Intravenous inoculation of live *B. burgdorferi*, but not killed spirochaetes, was found to increase the permeability of the blood–brain barrier, with mild pleocytosis and spirochaetes noted in cerebrospinal fluid. These changes were less marked when rats were inoculated with long-term, *in vitro*-passaged organisms (García-Moncó *et al.*, 1990). Intravenously inoculated radiolabelled spirochaetes were cleared from the vasculature within 1 hr of injection, with random invasion of multiple organs (Galbe *et al.*, 1993).

The larger body size of rats compared with other laboratory rodents has facilitated implantation of dialysis membrane chambers containing cultivated *B. burgdorferi* into the peritoneal cavity (see discussion in the section on mouse). The laboratory rat model also lends itself to intra-articular injection for the study of arthritogenesis, using an approach known as antigen-induced allergic arthritis in Wistar rats. Injection of butanol-extracted, water-soluble outer surface proteins into the knee joint was initially shown to be arthritogenic (Gondolf *et al.*, 1994). More recent work demonstrated that lipidated recombinant OspA and OspC, but not OspB, were arthritogenic, whereas non-lipidated proteins were not, and synthetic lipid alone or coupled to bovine serum albumen was not. Thus, results indicated that the lipid and protein moieties were both needed for arthritogenesis (Batsford *et al.*, 2004).

Laboratory mouse model
Effective use of the mouse model requires an awareness that infection may be influenced by mouse genotype, immune competence, and age, as well as *B. burgdorferi* isolate, passage history, adaptive form, dose, route of inoculation, and site of inoculation. A recent addition to this list of variables is genetic mutation of *B. burgdorferi*, and the non-genetic variables involved in creating such mutants. The mouse model shares many features in common with acute disseminated human Lyme borreliosis, including arthritis, carditis, vasculitis, myositis, peripheral neuritis, an intermittent course of disease remission and exacerbation, persistent infection, and immune response. Mice are too small to manifest EM (the whole mouse is an EM, with similar foci of perivasculitis and perineuritis in multiple tissues). Mice do not develop neuroborreliosis, which is most likely due to the paucity of collagenous tissue in their meninges and brain. The most commonly studied feature of the mouse model is arthritis, which is similar in character to that found in other rodents, dogs, non-human primates and humans.

Basic features of the mouse model Infection has two distinct phases in immunocompetent mice: preimmune and immune phases. In the preimmune phase, spirochaetes utilize their full armamentarium to evade host innate responses, proliferate, and disseminate broadly throughout all tissues. Spirochaetes disseminate directly through tissue, as well as in the blood (bacteraemia), and population dynamics reach their zenith in various tissues during this period. Arrival of spirochaetes into joint and heart tissue results in an acute inflammatory response that is directly related to the presence of spirochaetes, but inflammation is curiously minimal or absent in other tissues that are also infected during this phase (Barthold *et al.*, 1991).

As the acquired immune response evolves in the immune phase of infection, there is reduction (but incomplete clearance) of spirochaetes in all tissues, with resolution of acute inflammatory lesions in joints and hearts. Disease resolution coincides with clearance of spirochaetes from much of the heart tissue and the synovium, but persistence of spirochaetes intercalated within collagen of connective tissue of the aortic wall, ligaments, tendons, joint capsules and dermis (Barthold *et al.*, 1991, 1992a, 1993, 2006; Hodzic *et al.*, 2003; Pachner *et al.*, 1995a; Zimmer *et al.*, 1990). Over the course of several months of infection, mice experience periodic exacerbations of acute oligoarticular arthritis, acute carditis, and bacteraemia, associated with return of spirochaetes into the synovium. During the immune phase of infection, random foci of segmental perivasculitis, perineuritis and other chronic lesions are present (Barthold *et al.*, 1991, 1992b, 1993).

Kinetics of spirochaete dissemination and tissue invasion In a comprehensive study utilizing C3H mice inoculated intradermally by syringe on the dorsal thorax, dissemination of spirochaetes to various tissues was evaluated by culture, histochemistry (silver stain), immunohistochemistry, in situ DNA hybridization, PCR and histology. Although that study involved syringe-inoculation of cultured non-clonal *B. burgdorferi* N40, similar events take place following infection with other *B. burgdorferi* isolates, by other means of infection, and with other adaptive states of spirochaetes (see below). Tissues were analysed on days 1, 2, 3, 4, 7, 10, 15, 21, and 30 after inoculation. Blood and spleen of most mice were positive by day 3, and ear by day 10. Inflammation in joints and hearts coincided with invasion of spirochaetes on days 4 through 7 in joints and days 7 through 10 in hearts. Spirochaetes had a general predilection for collagenous tissues of joints, heart, arteries, nerves, muscle, skin and other tissues. Spirochaete populations peaked at 15 days, after which bacterial burdens declined globally in all tissues, but persisted (Barthold *et al.*, 1991). Examination of mice at 30, 90, 180 and 360 days after inoculation confirmed persistence of spirochaetes in multiple collagenous tissues for the duration of the study. At 30 days, hearts and many joints of all limbs were acutely inflamed, whereas at later intervals, both carditis and arthritis had generally resolved. However, some mice sampled at later intervals had acute carditis and oligoarticular arthritis, suggesting recurrent acute disease (Barthold *et al.*, 1993). Quantitative PCR analysis of spirochaete populations in various tissues during the course of infection has demonstrated the early rise, peak at 2 weeks, and then decline but persistence of spirochaetes at the inoculation site, distant skin, ear pinna, heart base, ventricular myocardium, tibiotarsal joint, and quadriceps muscle. The decline in spirochaete populations during the immune phase of the infection is global in all tissues, including bacteraemia, but spirochaetes persist in all tissues in significant numbers. In contrast, spirochaete populations continue to rise in various tissues and bacteraemia persists in C3H-*scid* mice at 4 and 6 weeks. However, spirochaete populations plateau without massive overgrowth in immunodeficient mice, suggesting mediation by innate immunity or quorum sensing by spirochaetes (Hodzic *et al.*, 2003).

Musculoskeletal pathology The 'arthritis' in mice infected with *B. burgdorferi* consists of several different components. Joint disease arises in many joints, but is most evident as swelling of the tibiotarsal joint, and is particularly obvious in immunodeficient mice (Fig. 14.5). It must be emphasized that calliper measurements of joints do not necessarily reflect arthritis. Grossly measurable joint swelling may be due to periarticular oedema (with or without underlying synovitis/arthritis), proliferative synovitis, and/or fibrosis. In the initial pre-immune and early immune phases of infection, there is often grossly visible swelling of joints, particularly the tibiotarsal (ankle) joints, due to periarticular oedema (Fig. 14.6). The oedema arises in the subcutis, and features sparse infiltration of neutrophils and macrophages in response to the presence of large numbers of spirochaetes in the subcutaneous tissue (Barthold *et al.*, 1993). This reaction may or may not be associated with underlying inflammation of joints. Notably, it has been shown that joint swelling and joint inflammation are differentially affected by separate genetic loci in the mouse (Roper *et al.*, 2001). A major component of joint disease is polysynovitis (Fig. 14.7), with synovial proliferation, infiltration of neutrophils, and exudation of neutrophils and fibrin into synovial lumina, involving many bursae, sheaths, and joints. The tibiotarsal region is optimal for examining this response, which peaks at 2–3 weeks, but many joints of both the axial and appendicular skeleton may be involved. Spirochaetes are present within the proliferating synovium and surrounding connective tissue, but

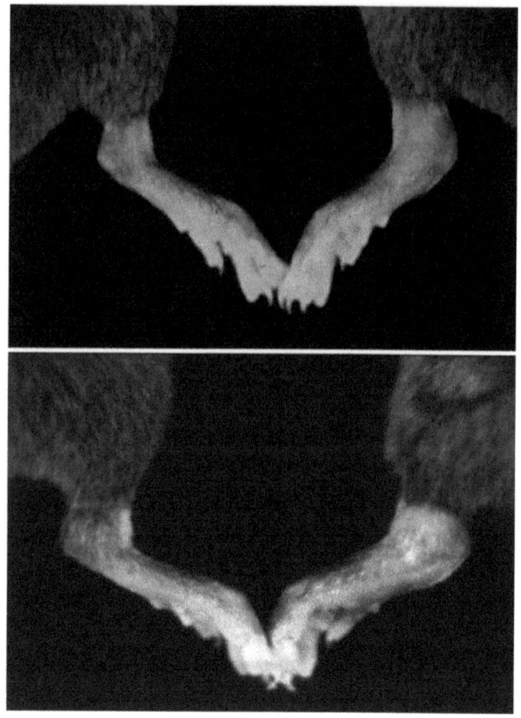

Figure 14.5 Tibiotarsal swelling in C3H mice infected with *B. burgdorferi*. The upper panel depicts an uninfected mouse (left), compared to an infected mouse (right) at around 2–3 weeks of infection. The lower panel depicts an immunocompetent mouse (left) compared to a SCID mouse at around 60 days of infection.

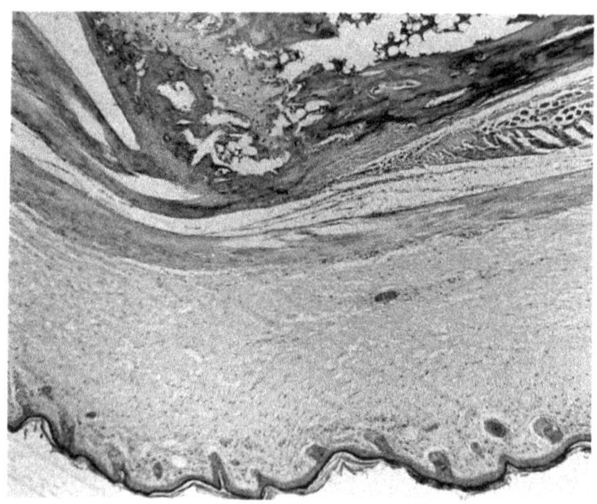

Figure 14.6 Tibiotarsal region of a C3H mouse during acute infection with *B. burgdorferi*. The dermis and subcutis are markedly oedematous in the absence of inflammation of the underlying joints and tendon sheaths. This illustrates the concept that gross tibiotarsal swelling is not an accurate measure of arthritis.

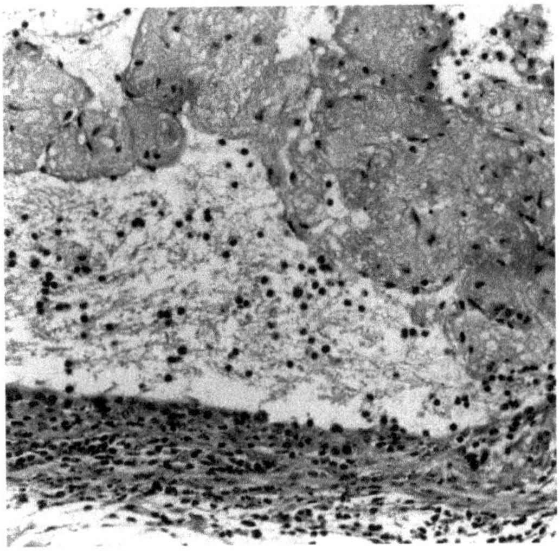

Figure 14.7 Synovium of a C3H mouse during acute infection with *B. burgdorferi*. The synovium is hypercellular due to synovial proliferation and leucocytic infiltration, and the lumen contains exudate consisting of leucocytes and fibrin.

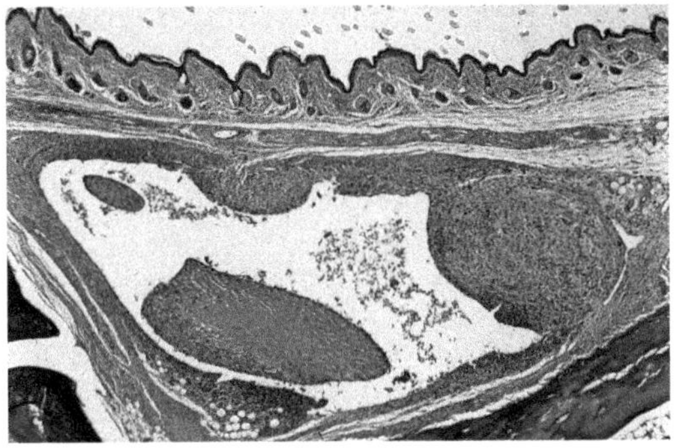

Figure 14.8 Tibiotarsal flexor tendon of a C3H mouse during acute infection with *B. burgdorferi*. The tendon sheath is thickened due to synovial hyperplasia and inflammation and the lumen contains exudate. This is a consistent site for examining synovial pathology in the mouse model.

not in exudate within joint lumina. In addition, inflammation of attachment sites of ligaments and tendons to bone is a frequent finding (enthesopathy) that is associated with spirochaetes among collagen fibres. The tibial crest, Achilles tendon, and patellar ligaments are frequent sites for these lesions. Marked proliferative changes are consistently present around the tibiotarsal flexor tendon, its bursal sheath, and its fibrocartilaginous cushion on the dorsal aspect of the tibiotarsus (Barthold *et al.*, 1990a, 1992a,b, 1993; Ma *et al.*, 1998) (Fig. 14.8). If infection is allowed to progress in SCID mice, there is progressive pannus of joint spaces, with destruction of bone and articular cartilage (Barthold *et al.*, 1992b; Schaible *et al.*, 1990) (Figs. 14.9 and 14.10). During the immune phase of infection in immunocompetent mice, the inflammatory and proliferative changes resolve, with residual capsular fibrosis and sparse mononuclear leucocyte infiltrates (Barthold *et al.*, 1992a, 1993). Tissue-specific analysis of spirochaete burdens during the immune phase of infection has shown that spirochaetes are eliminated from the proliferative synovium, but persist in adjacent collagenous tissues (capsule, tendons, ligaments) (Barthold *et al.*, 2006; Hodzic *et al.*, 2003).

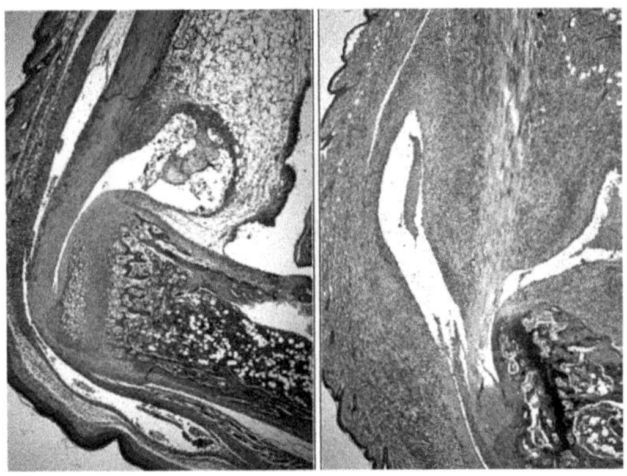

Figure 14.9 Synovitis of the tibiotarsus in a C3H mouse during acute infection with *B. burgdorferi* (left). The inflammation will completely regress and resolve in immunocompetent mice, but synovial disease becomes progressively severe in SCID mice (right) at 60 days of infection.

Myositis of skeletal muscle is also a frequent feature of *B. burgdorferi* infection in the mouse, and is quite apparent in immunodeficient mice. Lesions consist of infiltration of endomesial and perimesial connective tissue with macrophages and lymphocytes, particularly in perivascular regions (Barthold *et al.*, 1991, 1992a; Defosse *et al.*, 1992; Museteanu *et al.*, 1991; Schaible *et al.*, 1990) (Fig. 14.11).

Cardiovascular pathology Evaluation of carditis requires optimal sampling and processing to include both heart base (including aortic root,

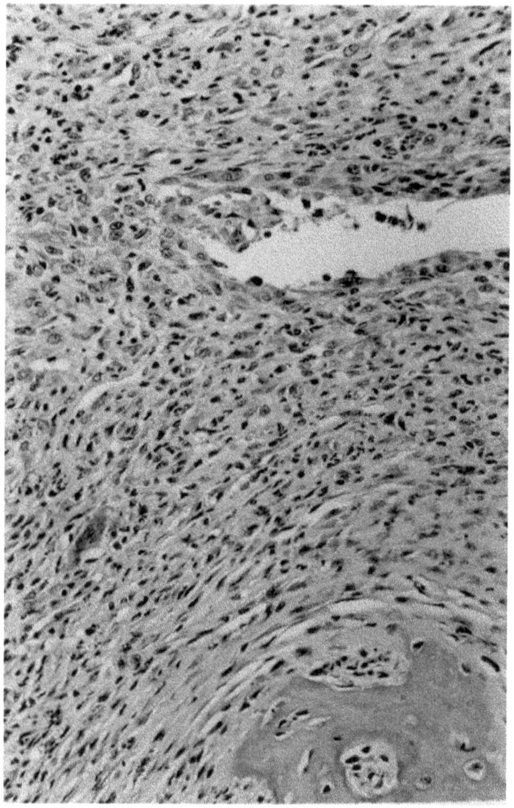

Figure 14.10 Proliferative synovium in a SCID mouse infected with *B. burgdorferi*. The joint lumen is being progressively occluded by proliferating synovium, which is replacing the articular cartilage (note bone at lower right).

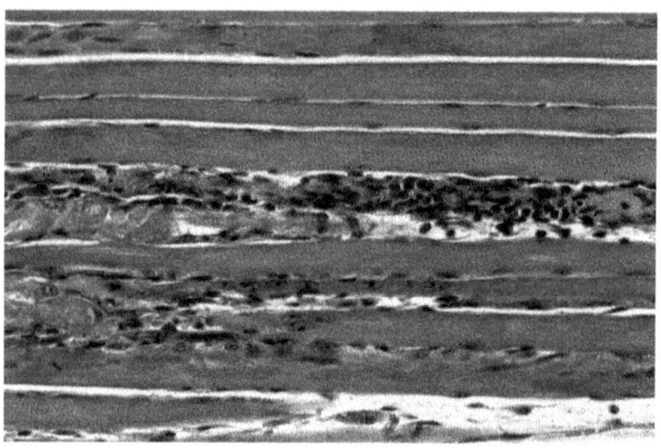

Figure 14.11 Myositis in a C3H mouse chronically infected with *B. burgdorferi*. There is infiltration of the endomesial connective tissue with mononuclear leucocytes, particularly in association with small vessels.

auricles, and ventricular myocardium). Syringe inoculation in the dorsal thoracic midline, in contrast to distant sites, such as tail base, results in more consistent cardiac involvement in immunocompetent mice. Tick-borne infection, which generally results from ticks feeding in the head and neck region, also favours development of carditis. Immunodeficient mice develop carditis, regardless of inoculation site. Although not as extensively studied as arthritis-susceptibility, mouse genetic background is a major determinant of carditis susceptibility, which parallels arthritis susceptibility (Armstrong *et al.*, 1992).

The temporal kinetics of cardiac infection have been studied in the carditis-susceptible C3H mouse following syringe inoculation. Tissues were examined by culture, PCR and immunohistochemistry for spirochaete invasion at 2, 3, 4, 7, 10, 15, 21 and 30 days after inoculation. Spirochaetes were first detectable in heart at day 7, which coincided with onset of inflammation (Barthold *et al.*, 1991). With the exception of the aortitis that occurs during acute infection, the inflammatory response of cardiac tissues to infection with *B. burgdorferi* is dominated by mononuclear leucocytes, and particularly macrophages (Ruderman *et al.*, 1995). During the early, pre-immune phase of infection, there is acute inflammation of the aortic root above the aortic valve (Fig. 14.12). The aortitis is characterized initially as segmental transmural arteritis, with inflammation extending from the endothelium through to the periarterial adventitia. The arteritis features infiltration of neutrophils throughout the vessel wall, and margination of neutrophils, thrombocytes, and fibrin on the endothelium (Fig. 14.13). In addition, there is generalized infiltration of the atrial and ventricular myocardium with macrophages (Fig. 14.14). Many of these lesions undergo resolution with the advent of host immunity, and are therefore most apparent in immunodeficient mice (Armstrong *et al.*, 1992; Barthold *et al.*, 1993; Barthold *et al.*, 1991; Barthold *et al.*, 1992b; Defosse *et al.*, 1992; Ruderman *et al.*, 1995; Schaible *et al.*, 1989a; Zimmer *et al.*, 1990). During the pre-immune phase, lesions feature the presence of predominantly extracellular spirochaetes, but occasional intracellular spirochaetes have been visualized by electron microscopy within cardiac myofibres (Armstrong *et al.*, 1992; Barthold *et al.*, 1991; Pachner *et al.*, 1995a; Zimmer *et al.*, 1990). Following the evolution of host immunity, carditis

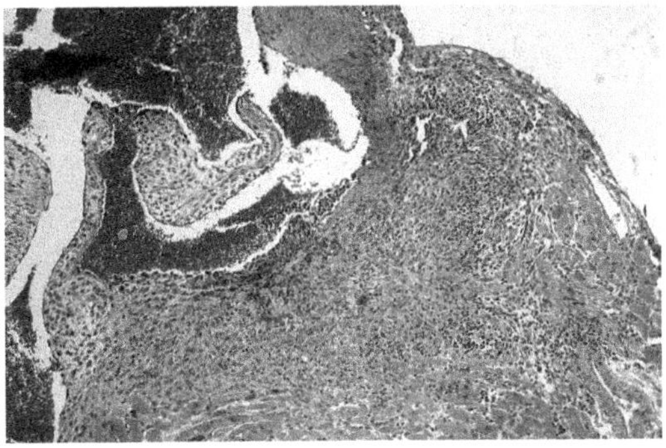

Figure 14.12 Acute carditis in a C3H mouse infected with *B. burgdorferi*. There is marked infiltration of the connective tissue at the heart base with macrophages, which extend into the adjacent myocardium (bottom of figure). The aortic root near the aortic valve has endothelial margination of neutrophils, with extension of inflammation into the vessel wall.

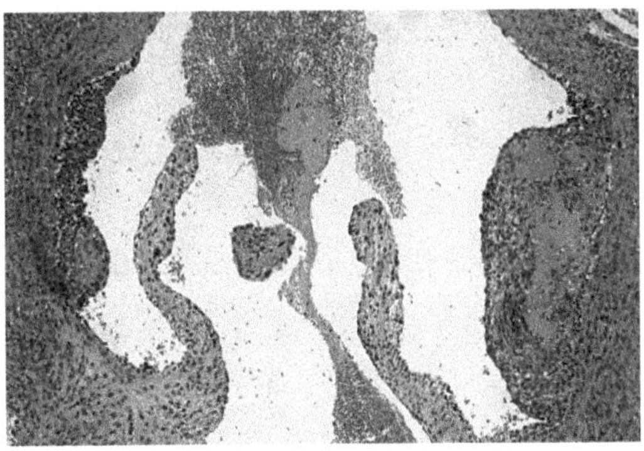

Figure 14.13 Acute thrombosis of aortic root during acute *B. burgdorferi* infection of a C3H mouse. Note masses of marginated fibrin and leucocytes on endothelial surface of the aorta.

wanes, with residual infiltration of epicardium with lymphocytes (Fig. 14.15), and segmental periaortic infiltration of lymphocytes and plasma cells (Fig. 14.16). Spirochaetes are present within all regions of the heart during the pre-immune phase (or in immunodeficient mice), but as the host immune response evolves, spirochaetes are relatively cleared from myocardium and heart base tissue, but persist in the aortic wall at the base of the heart (Barthold *et al.*, 2006; Hodzic *et al.*, 2003; Hodzic *et al.*, 2002). Similar clearance occurs in SCID mice when induced by adoptive lymphocyte transfer, passive transfer of immune

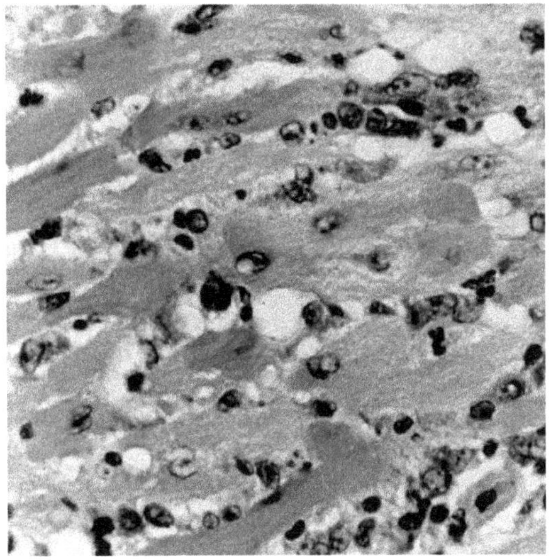

Figure 14.14 Ventricular myocarditis in a C3H mouse during acute *B. burgdorferi* infection. The myocardial interstitium is infiltrated with mononuclear leucocytes.

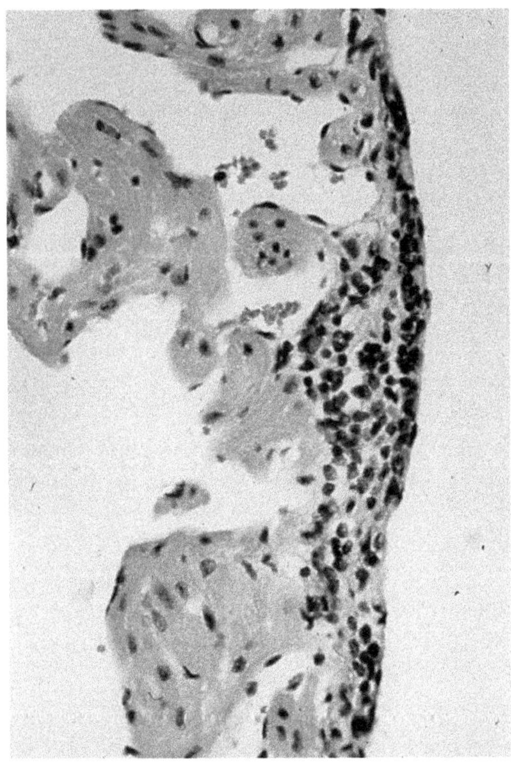

Figure 14.15 Epicardial infiltration of mononuclear leucocytes in a C3H mouse during chronic infection with *B. burgdorferi*.

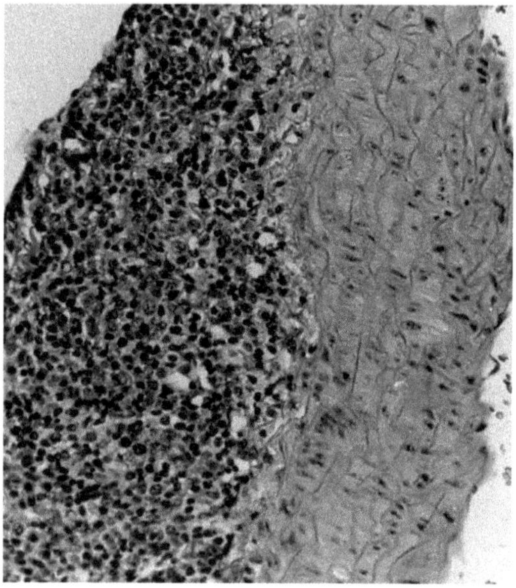

Figure 14.16 Segmental infiltration of periaortic adventitia with mononuclear leucocytes during chronic infection of a C3H mouse with *B. burgdorferi*. These lesions are commonly present around the base of the aorta during chronic infection.

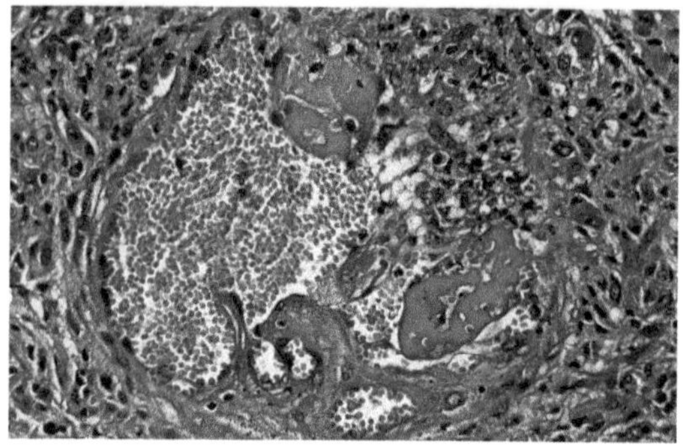

Figure 14.17 Acute endarteritis and thrombosis of the coronary artery during acute *B. burgdorferi* infection of a C3H mouse.

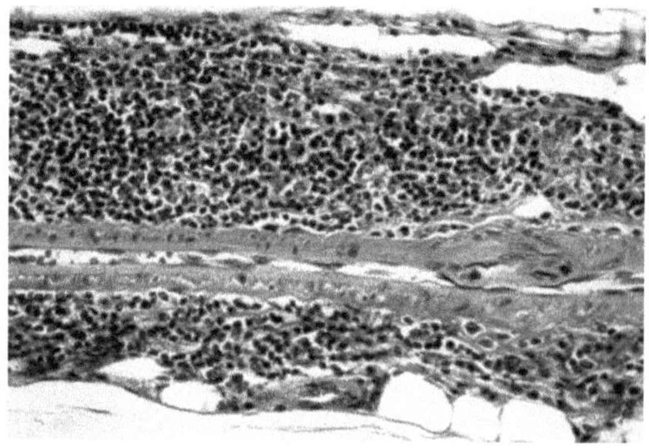

Figure 14.18 Segmental infiltration of periarterial (plantar artery) adventitia in a C3H mouse during chronic infection with *B. burgdorferi*. These lesions are randomly distributed in many tissues, but are most commonly found around major vessels of the extremities.

serum, or passive transfer of DbpA or Arp antisera (Barthold *et al.*, 2006).

In addition to carditis, segmental arteritis of the coronary, femoral, plantar, saphenous, tibial, renal, pulmonary and other arteries may be present. During acute infection, these foci of arteritis are similar in character to the transmural aortitis that occurs at the base of the heart (Fig. 14.17). During chronic infection, vessels frequently feature segmental periarterial lymphoplasmacytic infiltrates in the adventitia (Fig. 14.18), similar to chronic lesions of carditis. These lesions are associated with the presence of visible spirochaetes in silver-stained histosections (Barthold *et al.*, 1993; Barthold *et al.*, 1992b). Although vascular lesions are common in infected mice, their multifocal and variable locations are a challenge to experimental study.

Despite the challenge of a heart rate of over 300 beats per minute, a limited number of studies have confirmed that experimentally infected mice develop cardiac conduction defects, which arise during the acute phase of infection in association with inflammation, and resolve with resolution of inflammation during the immune phase of infection (Armstrong *et al.*, 1992; Saba *et al.*, 2001).

Other pathology Urinary bladder lesions consist of lymphoid aggregates in the lamina propria, increased vascularity, and thickening of vessel walls with perivascular infiltrates. Spirochaetes can be visualized by silver staining in the lamina propria and subserosa (Barthold *et al.*, 1992b, 1993; Czub *et al.*, 1992). Similar lymphoid infiltrates may arise in the renal pelvis and around the ureters (Barthold *et al.*, 1992b). In addition to segmental periarteritis, segmental lymphoplasmacytic cuffs may be found in the perineurium of peripheral nerves (Fig. 14.19), and mononuclear cell infiltrates may also be found in association with connective tissue of dorsal root nuclei (radiculoneuritis). Interstitial inflammatory changes and perivascular infiltrates have been noted in lungs of SCID mice (Barthold *et al.*, 1993; Schaible *et al.*, 1989a; Schaible *et al.*,

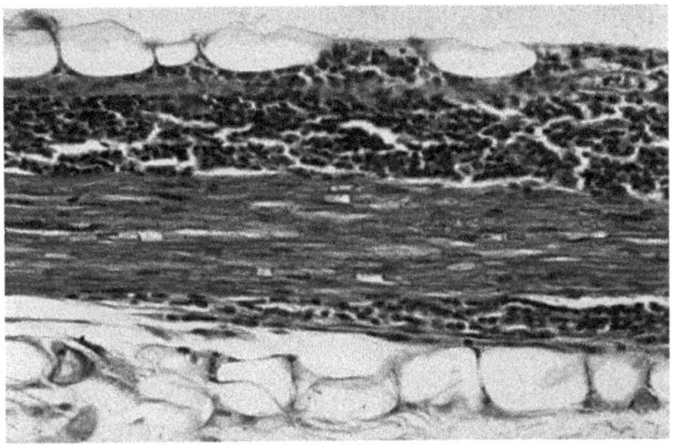

Figure 14.19 Segmental infiltration of perineural tissue around a peripheral nerve in a C3H mouse during chronic infection with *B. burgdorferi*.

1989b; Schaible *et al.*, 1990). The involvement of lung is underscored by development of severe inflammation in infected RAG-null mice reconstituted by adoptive T-cell transfer (McKisic and Barthold, 2000). Non-suppurative perivascular infiltrates have also been observed in the brain of SCID mice (Schaible *et al.*, 1989b). Hepatic inflammation and fibrosis have been reported in SCID mice (Schaible *et al.*, 1990), but hepatic lesions have not been found in other studies. Multinucleate giant cells have been described in the spleens of infected SCID mice (Schaible *et al.*, 1990), but these elements represent megakaryocytes, which increase with the myelopoiesis that is stimulated by inflammatory lesions in distant tissues.

Mouse genotype and age All inbred strains and outbred stocks of laboratory mice are susceptible to *B. burgdorferi* infection. Strains of mice that have been shown to be susceptible to infection include inbred A/J, A/WySn, AKR, BALB/c, BALB/cByJ, C.B-17, C3H/HeJ, C3H/HeN, C3H.OH, C3H.SW, C57BL/6, C.B-17, CAL.20, C57BL/10, DBA/2, PL, B10.A, B10.AQR, B10.A(5R), B10.BR, B10.D2, B10.WB, B10R111, B10.S, NZB, SWR, SJL, SWR, CRL:SKH(hr/hr) Br (hairless), and outbred ddY, NMRI, CD-1, ICR, and RML mice (Barthold, 1991; Barthold *et al.*, 1990a; Bockenstedt *et al.*, 1993; Golde *et al.*, 1993; Grimm *et al.*, 2004b; Masuzawa *et al.*, 1992; Raveche *et al.*, 2005; Schaible *et al.*, 1991; Schaible *et al.*, 1989b; Schwan *et al.*, 1991; Shih *et al.*, 1992; Simon *et al.*, 1996; Tuomi *et al.*, 2002; Yang *et al.*, 1992). In addition, numerous mouse genetic mutants, both spontaneous and induced, have been used for Lyme disease studies (Table 14.1).

Susceptibility to arthritis (and probably carditis) is genetically determined, but is also age dependent. Infection of several diverse genotypes of mice, representing major MHC haplotypes, including BALB/cByJ (H-2d), C3H/HeJ (H-2k), C57BL/6J (H-2b), SJL/J (H-2s), and SWR/J (H-2q) developed uniformly severe disease (based upon histopathology) at 30 days after intraperitoneal inoculation with 10^6 or 10^7 *B. burgdorferi* N40 at 3 days of age, but displayed distinctly different patterns of susceptibility when inoculated at 3 weeks of age. Among mice infected at 3 weeks of age, C3H and SWR mice were most susceptible to induction of arthritis, and C57BL/6, SJL and BALB/c were significantly more resistant to disease. C3H/He mice inoculated at 12 weeks of age were more resistant to disease compared with C3H mice inoculated at 3 weeks. Susceptible strains developed higher IgG ELISA titres to *B. burgdorferi* compared to resistant genotypes (Barthold *et al.*, 1990a). Disease-susceptible C3H mice weigh less than their age-matched control peers, particularly when infected as young mice, but body weight returns to normal during late, persistent infection (Saba *et al.*, 2001).

A subsequent study examined arthritis (tibiotarsal enlargement) and antibody responses following *B. burgdorferi* ZS7 (10^8 tail base) inoculation of mice representing varying H-2 haplotypes, Igh allotypes, and genetic strain backgrounds, including C57BL/6 (H-2b, Igh-b), BALB/c (H-2d, Igh-a), C.B-17 (H-2d, Igh-b), Cal-20 (H-2d, Igh-d), DBA/2 (H-2d, Igh-c), C3H/HeJ and C3H/HeN (H-2k, Igh-a), C3H.OH (H-2 dddk, Igh-a), AKR (H-2k, Igh-d), a number of C57BL/10 (B10) congenic lines with the same Igh allotype (b) but differing H-2 haplotypes, including B10.D2 (H-2d), B10.WB (H-2j), B10.R111 (H-2r), B10.S (H-2s), B10.A (H-2kkkd), B10.AQR (H-2qkkd), and B10.A(5R) (H-2 bbkd) mice. It was concluded that H-2k haplotype mice (AKR, C3H, B10.BR) developed chronic progressive arthritis, H-2 haplotypes H-2b (C57BL/6), H-2j (B10.WB), H-2r (B10.R111 and H-2s (B10.S) developed arthritis of variable duration and intensity that was not progressive, and mice of the H-2d haplotype (BALB/c, DBA/2, C.B-17, B10.D2 and Cal-20)

Table 14.1 Mouse genetic mutations utilized in various Lyme borreliosis studies[a]

Mouse strain – gene mutation	Reference
Natural mutants	
BALB/c-*nude*	Moody *et al.* (1990a
C.B-17-*scid*	Schaible *et al.* (1989a), Zimmer *et al.* (1990), Schaible *et al.* (1990), Schaible *et al.* (1991), Barthold *et al.* (1992b), Simon *et al.* (1996)
C3H-*scid*	Barthold *et al.* (1992b)
B6-*scid*	KeaneMyers *et al.* (1996), Mbow *et al.* (2001)
B6-*beige* (NK deficient)	Barthold and deSouza (1995)
NIH-3 (*scid*, *beige*)	Defosse *et al.* (1992), Simon *et al.* (1996)
C3H/HeJ TLR4 deficient	Barthold *et al.* (1991), Ma *et al.* (1998), Yang *et al.* (1994)
A, AKR, B10.D2, DBA/2, SWR – complement 5 deficient	Bockenstedt *et al.* (1993)
C3H/HeJ-$Tnfsf6^{gld}$ (FasL deficient)	Shi *et al.* (2006)
Targeted mutants[b]	
B6-*IgM* null (μ chain membrane exon)	KeaneMyers *et al.* (1996)
B6-*Igh6* null	McKisic and Barthold (2000), McKisic *et al.* (2000), Mbow *et al.* (2001)
B6-*Rag1*, B6-*Rag2* null	Schaible *et al.* (1989a), Schaible *et al.* (1991), Simon *et al.* (1996), McKisic and Barthold (2000), McKisic *et al.* (2000)
B6-*Ncf1* null (NADPH oxidase *p47phox*)	Crandall *et al.* (2005)
B6-*Cybb* null (NADPH oxidase *gp91phox*)	Crandall *et al.* (2005)
B6-*Tcrβ* null	Mbow *et al.* (2001)
B6-*Tcrδ* null	Mbow *et al.* (2001), Bockenstedt *et al.* (2003)
B6-*Tcrβ/δ* null	Mbow *et al.* (2001), McKisic and Barthold (2000), McKisic *et al.* (2000)
B6-*CIITA* null (MHC II transactivator)	Fikrig *et al.* (1997a)
B6-*MHC II* null	Mbow *et al.* (2001)
B6129-*CD40L* null	Fikrig *et al.* (1996a)
MRL/MpJ-*Fcgrt* null (FcRn binds IgG Fc)	Crowley *et al.* (2006)
B6-*CD1d* null	Kumar *et al.* (2000)
C3H-*CD14* null	Benhnia *et al.* (2005)
DBA-*IL4* null	Brown *et al.* (1999)
BALB/c-*IL4* null	Potter *et al.* (2000), Satoskar *et al.* (2000)
B6-*IL4* null	Potter *et al.* (2000)
BALB/c-*IL4R* null	Potter *et al.* (2000)
BALB/c-*IL4-IL4R* null	Potter *et al.* (2000)
B6-*IL6* null	Anguita *et al.* (1998)

Mouse strain – gene mutation	Reference
B6-*IL10* null	Brown *et al.* (1999), Lazarus *et al.* (2006)
C3H-*IFNγ* null	Brown *et al.* (1999)
B6129-*IFNγ* null	Christopherson *et al.* (2003)
129-*IFNγ Rα* null	Anguita *et al.* (2001), Glickstein *et al.* (2001)
C3H/HeJ-*Cmkar2* null (CXCR2)	Brown *et al.* (2003)
DBA/2- *Cmkar2* null (CXCR2)	Brown *et al.* (2003)
B6-CCR2 null	Brown *et al.* (2003), Montgomery *et al.* (2007)
C3H/HeJ-CCR2 null	Brown *et al.* (2003), Montgomery *et al.* (2007)
B6-*C3* null	Woodman *et al.* (2007), Lawrenz *et al.* (2003)
B6-*Bf* null (complement factor B)	Woodman *et al.* (2007)
B6-*Cfh* null (complement factor H)	Woodman *et al.* (2007)
B6-CD11a null	Guerau-de-Arellano *et al.* (2005a)
B6-CD11c null	Guerau-de-Arellano *et al.* (2005a)
B6-*Itgam* null (CD11b)	Guerau-de-Arellano *et al.* (2005a)
B6-*Igb2* hypomorph (CD18)	Guerau-de-Arellano *et al.* (2005b)
BALB/c-*Cr2* null (CD21/35 complement receptors)	Jacobson *et al.* (2007)
B6-*Cr2* null (CD21/35 complement receptors)	Jacobson *et al.* (2007)
B6129-*TLR1* null	Alexopoulou *et al.* (2002)
B6129-*TLR2* null	Alexopoulou *et al.* (2002)
C3H-*TLR2* null	Bolz *et al.* (2004), Wang *et al.* (2004), Wooten *et al.* (2002)
B6129-*MyD88* null	Liu *et al.* (2004), Bockenstedt *et al.* (2006)
B6-*MyD88* null	Bolz *et al.* (2004), Behera *et al.* (2006)
B6129-*MKK3* null (regulator of p38 MAP kinase)	Anguita *et al.* (2002)
C3H-*iNOS* null	Brown and Reiner (1999a)
DBA/2-*iNOS* null	Brown and Reiner (1999a)
C3H-*Stat1* null	Brown *et al.* (2006)
DBA/2-*Stat1* null	Brown *et al.* (2006)
B6129-*Plg* null (plasminogen)	Coleman *et al.* (1997)
BALB/c-*Dcn* null (decorin)	Liang *et al.* (2004a)
B6-*Dcn* null (decorin)	Brown *et al.* (2001)
C3H/HeN-*Dcn* null (decorin)	Brown *et al.* (2001)
B6129-*E- P-selectin* null	Seiler *et al.* (1998)
B6129-*Spp1* null (osteopontin)	Potter *et al.* (2002)
129-*Spp1* null (osteopontin)	Potter *et al.* (2002)
B6-*Spp1* null (osteopontin)	Craig-Mylius *et al.* (2005)

Table 14.1 Continued

Mouse strain – gene mutation	Reference
Transgenic mutants	
B10.BR-dominant negative p38 MAP kinase	Hedrick *et al.* (2006)
B6C3H-*SAF-1* (serum amyloid A-activating factor-1)	Ray *et al.* (2004)
B10-*DR4Dw4* transgenic	Feng *et al.* (1995)
B6-*ospA* transgenic	Fikrig *et al.* (1996b)
C3H-*ospA* transgenic	Fikrig *et al.* (1996b), Fikrig *et al.* (1997b)
C3H-*flagellin* transgenic	Fikrig and Barthold (1997)

a Effects of mutations (phenotype) are varied and dependent upon experimental design, outcome measured, and methods of assessment.

b Mutations created by homologous recombination in 129 ES cells (129 strain not specified). Parental mouse strains are congenic to 129 (or insipient congenic at N5 or higher). In some cases, mutations were originally maintained on another parental strain, but fully backcrossed (N10) onto another parental strain, in which case the original parental strain is not indicated. If hybrid, the parental strain indicates such.

developed minimal or no arthritis. All types of mice developed similar patterns of antibody reactivity to *B. burgdorferi* antigens on immunoblots (Schaible *et al.*, 1991). The argument for H-2 linkage to arthritis susceptibility, however, was weakened by a study using C3H (susceptible) and B10 (resistant) congenic mice of the same genotype but differing haplotypes, including C3H/HeN (H-2k) compared with C3H/SW (H-2b), B10.BR (H-2k) and B10 (H-2b) mice. Arthritis was equally severe among C3H mice and equally mild among B10 mice, regardless of haplotype (Yang *et al.*, 1992). In other studies, C3H congenic mice were equally susceptible to arthritis and DBA congenic mice were equally resistant, regardless of their H-2 haplotype (H-2k or H-2d) (Brown and Reiner, 2000). The differing conclusions from these MHC studies may reflect different *B. burgdorferi* isolates, doses, routes of inoculation, mouse suppliers, mouse genotypes, and methods of assessing arthritis. In another study using outbred ICR, BALB/c, B10, B10.BR and B10.D2 infected by tick, differences in immunoblot response to various *B. burgdorferi* proteins were noted to be H-2 restricted (Golde *et al.*, 1993).

The C3H/He mouse has emerged as a prototype disease-susceptible genotype, and BALB/c or B6 (the standard abbreviation for C57BL/6) as resistant genotypes for disease modelling, and the genetic basis of arthritis susceptibility has been further explored. Following intradermal inoculation of C3H/HeN and BALB/cAnN mice, C3H mice harboured 5- to 10-fold higher *ospA* and *flagellin* DNA copy numbers in tissues at intervals up to 5 weeks, compared to BALB mice. F1 hybrids were disease-susceptible with high copy numbers of tissue DNA (Yang *et al.*, 1994). The genetic resistance of BALB/c mice can be overcome with higher doses or relative virulence of inocula, resulting in severe disease and spirochaete DNA levels equivalent to C3H mice (Ma *et al.*, 1998; Weis *et al.*, 1997). Thus, the resistance of BALB/c mice was determined to be conditional, whereas B6 mice were resistant to disease at all doses of *B. burgdorferi*. Remarkably, levels of spirochaetal DNA in tissues of B6 mice were equivalent to C3H (C3) mice. B6C3F1 hybrids developed intermediate disease severity (Ma *et al.*, 1998). Susceptibility was evaluated in F2 intercross generation mice derived from C3H/HeN and B6 mice, revealing nine distinct quantitative trait loci (QTL) that were genetically linked to ankle swelling (loci on chromosomes 4 and 5), arthritis histopathology (loci on chromosomes 5 and 11), and humoral responses (loci on chromosomes 6,

9, 11, 12, and 17) (Weis *et al.*, 1999). Reciprocal backcross populations derived from C3H/HeN and B6, and C3H/HeJ and BALB/cAnN mice have been used for composite interval mapping, which revealed 12 QTL, including seven previously reported loci. Further analysis revealed multiple QTL on chromosome 5 that affected disease phenotypes (ankle swelling and histopathological criteria of inflammation) (Roper *et al.*, 2001). Comparative studies between arthritis-resistant DBA/2 and arthritis-susceptible C3H/HeJ mice (Barthold *et al.*, 1990a) confirmed studies with B6 and C3H mice, with equivalent levels of spirochaetes in joint tissues, but markedly different disease severity (Brown and Reiner, 1998). These findings clearly illustrate that various aspects of Lyme borreliosis in the mouse model are under distinctly different and complex genetic control.

The genetic basis of arthritis susceptibility among inbred mouse strains remains apparent in globally immunodeficient mice. Arthritis is more severe in C3H-*scid* mice compared to BALB/c-*scid* mice (Barthold *et al.*, 1992b). Infection of RAG-null mice on C3H, BALB/c, B6 or DBA/2 genetic backgrounds resulted in severe arthritis among C3H- and BALB-*Rag* null mice, whereas arthritis was mild or absent in the B6- or DBA-*Rag* null mice (Brown and Reiner, 1999b).

Infectivity of B. burgdorferi species, strains and isolates Laboratory mice are susceptible to infection with numerous species, strains and isolates of *B. burgdorferi* s.s., *B. afzelii*, *B. garinii*, *B. japonica*, *B. bissetti* and *B. lusitaniae* (Anderson *et al.*, 1990; Kaneda *et al.*, 1998; Masuzawa *et al.*, 1994a; Pachner *et al.*, 2004; Tuomi *et al.*, 2002; Yrjänäinen *et al.*, 2006; Zeidner *et al.*, 2001). Because different patterns of organotropism and disease are suspected in humans infected with different *B. burgdorferi* s.l. and different ribosomal DNA genotypes of *B. burgdorferi* s.s., it is tempting to compare patterns of tissue distribution and disease induction by various isolates.

With few exceptions, if *B. burgdorferi* s.l. isolates or strains are infectious to mice, they will induce disseminated and persistent infection, and will induce similar patterns of disease, but disease severity may vary. One study examined mice infected with *B. burgdorferi* s.s., *B. afzelii*, and *B. garinii* isolates. *Borrelia afzelii* isolates tended to be cultured less often from urinary bladder, but more often from heart and ear, compared to other species and isolates (Tuomi *et al.*, 2002). A similar study suggested that *B. burgdorferi* s.s. isolates were more apt to infect heart, more apt to cause inflammation, and to induce higher antibody responses, compared to *B. afzelii* and *B. garinii* isolates, but levels of *B. afzelii* and *B. garinii* were higher in skin compared to *B. burgdorferi* s.s. isolates (Pachner *et al.*, 2004). The pathogenicity of ten clinical *B. burgdorferi* s.s. isolates, representing 2 distinct ribosomal DNA spacer restriction fragment length polymorphism genotypes (RST1 and RST3) revealed that RST1 isolates were more apt to be infectious in C3H mice following inoculation with high doses of cultured spirochaetes, induced greater ankle swelling and arthritis scores, and were associated with higher levels of spirochaetal *recA* DNA in tissues, corresponding to a higher percentage of human patients with Lyme disease infected with RST1 isolates (Wang *et al.*, 2001; Wang *et al.*, 2002). A similar study was performed with tick-borne exposure, revealing that bacteraemia lasted longer in mice infected with RST1 isolates, but there were no differences in spirochaetal *flagellin* DNA levels in tissues, or disease severity between RST1 and RST3 groups (Dolan *et al.*, 2004).

Infectious dose, adaptive state and route of inoculation The infectious dose of *B. burgdorferi* is lowest following intradermal inoculation. Three-week old C3H and BALB mice were equally susceptible to intradermal inoculation with 10 or fewer *B. burgdorferi* N40 spirochaetes, compared to one hundred-fold more spirochaetes to establish infection following intraperitoneal inoculation (Barthold, 1991; Barthold *et al.*, 1995a). The preferential infectivity by the intradermal route has been confirmed with other *B. burgdorferi* s.s., *B. afzelii*, and *B. garinii* isolates (Tuomi *et al.*, 2002). Mice are also susceptible to infection following intragastric inoculation of very high doses of *B. burgdorferi* N40, but there is no evidence for contact transmission (Barthold, 1991), or in utero (placental) transmission, although maternal infection may cause fetal death (Silver *et al.*, 1995; Weis *et al.*, 1997).

The site of intradermal inoculation can significantly influence the pattern of infection, disease, and host response. Inoculation in the dorsal midline thoracic region results in bacteraemia, antibody response, and evolution of arthritis and carditis more rapidly than when mice are inoculated in the foot (Barthold, 1996), observations that have recently been rediscovered (Motameni *et al.*, 2005). Depending upon site of intradermal inoculation, peripheral blood and splenic lymphocytes manifest different proliferative responses to mitogens. Responses were impaired following inoculation in the shoulder region, whereas no effect occurred following inoculation of the foot (deSouza *et al.*, 1993). Others (KeaneMyers *et al.*, 1996; Matyniak and Reiner, 1995) have shown similar effects with cytokine responses, which could only be detected in regional lymph nodes draining sites of inoculation, but not the spleen. This phenomenon is related to direct tissue invasion of spirochaetes. Mice inoculated at the base of the tail develop more severe arthritis in the rear legs, but mild or no carditis. Likewise, enlargement of regional lymph nodes is greatest near the site of inoculation, despite subsequent disseminated infection. Thus, purposeful targeting of specific sites can influence experimental outcome. Direct footpad inoculation with cultured spirochaetes has also been used to favour induction of local arthritis (Christopherson *et al.*, 2003; Masuzawa *et al.*, 1994b).

Infecting dose and adaptive state of spirochaetes used for inocula can significantly influence the host immune response. The OspA-OspC dichotomous expression serves as an example. Inoculation of high doses of culture-grown spirochaetes stimulates an immune response to antigens within the input inoculum (reflecting *in vitro* conditions), as well as to the spirochaetes that adapt and proliferate in the host (reflecting *in vivo* conditions). Seroconversion to OspA occurs with high-dose cultured inocula, but does not occur when the inoculating dose is low (generally less that 10^4) or introduced by tick (Barthold and Bockenstedt, 1993; Barthold *et al.*, 1995a; Gern *et al.*, 1993; Golde *et al.*, 1994; Pachner *et al.*, 1993; Schaible *et al.*, 1993). Seroconversion following low-dose or tick-borne infection reflects the antigenic stimulation that arises to host-adapted spirochaetes during active infection, which includes reactivity to antigens that are expressed exclusively or preferentially *in vivo* (Barthold *et al.*, 1995a; Gern *et al.*, 1993; Golde *et al.*, 1993, 1994; Schaible *et al.*, 1993; Simpson *et al.*, 1991; Suk *et al.*, 1995; Wallich *et al.*, 1995).

In addition to infection with cultured spirochaetes or by tick-borne exposure, mice can be infected with host-adapted spirochaetes. One means of inoculation with host-adapted spirochaetes is subcutaneous transplantation of allografts from infected mice (Barthold, 1993; Barthold *et al.*, 1995a). Spirochaetes can be purified from peripheral blood of mice during the acute (spirochaetaemic) phase of infection, especially in SCID mice (Cassatt *et al.*, 1998). Subcutaneously implanted metal net chambers (Jonsson *et al.*, 1995) and perforated Teflon chambers (Hurtenbach *et al.*, 1995), which allow spirochaetes to enter from surrounding tissues, have successfully yielded spirochaetes for study. In addition, these approaches have allowed harvest of tissue fluid for cytokine analysis (Hurtenbach *et al.*, 1995). Subcutaneous or intraperitoneal placement of chambers equipped with 0.1 μm pore membrane filters has also been shown to support growth of host-adapted spirochaetes (deSilva *et al.*, 1998). Larger yields of host-adapted spirochaetes have been harvested from dialysis membrane chambers implanted into the peritoneal cavities of rats (Akins *et al.*, 1998).

These various host-adaptive forms of spirochaetes, however, can reflect potentially important differences. Allografts derived from SCID mice contain spirochaetes that express OspA and OspC, which are both influenced by antibody and therefore not expressed (or only transiently so with OspC) in tissues from immunocompetent mice (Hodzic *et al.*, 2003; Hodzic *et al.*, 2005; Liang *et al.*, 2002, 2004b). Spirochaetes that are grown within dialysis membrane chambers undergo antigenic changes that reflect those that occur during infection, but not those mediated by antibody, such as OspC (Akins *et al.*, 1998; Liang *et al.*, 2004b; Parveen *et al.*, 2003; Revel *et al.*, 2002; Stewart *et al.*, 2006). Antigenic expression of spirochaetes during tick feeding has been evaluated by examining the dermal tissue that remains attached to the tick hypostome when the feeding ticks are removed at selected intervals. Spirochaetes that enter the host prior to 53 hours of tick attachment expressed primarily OspA and few expressed OspC, but generally failed to establish infection. In contrast, spirochaetes that entered the host at later intervals of feeding that led to successful infections tended to produce less OspA and more OspC (Ohnishi *et al.*, 2001). It is important to note that passive immunization with immune serum will protect mice from tick-borne infection if administered during 24–48 hours after tick attachment, but not thereafter, whereas immune serum is ineffective at preventing infection by partially fed ticks, which contain spirochaetes that have undergone antigenic modulation (Shih and Liu, 1998). Furthermore, the relative expression of OspC by *B. afzelii* at the tick feeding site has been shown to be influenced by the mouse strain (Leuba-Garcia *et al.*, 1998). Collectively, these various studies emphasize that input spirochaetes may vary in their antigenic profiles, which may influence immunological studies of early infection.

More rapid dissemination, with earlier invasion of joint tissue and onset of disease, occurs when mice are infected with host-adapted spirochaetes through allografts, compared with syringe or tick-borne infection, or with high doses of cultured spirochaetes. In the case of tick-borne infection, it has been shown that spirochaetes entering the mouse must undergo a brief period of adaptation before dissemination (Hodzic *et al.*, 2002; Shih *et al.*, 1992). Notably, dissemination following tick-borne infection is more rapid in genetically disease-susceptible C3H mice compared to resistant CD-1 mice (Shih *et al.*, 1993). Cultured spirochaetes must also undergo a period of local adaptation prior to dissemination, explaining their vulnerability to immune serum for up to 4 days after inoculation, whereas host adapted spirochaetes are not vulnerable (Barthold *et al.*, 1997; deSilva *et al.*, 1998). High-passage infectious spirochaetes remained vulnerable to passive immunization with immune serum for longer intervals after inoculation compared to low-passaged spirochaetes, and were defective in upregulation of genes that are typically preferentially expressed during infection (Anguita *et al.*, 2000). In contrast, OspA antiserum is protective against tick- or syringe-borne spirochaetes, but only if administered prior to or at the time of challenge, and is ineffective in preventing infection with host-adapted spirochaetes (Barthold *et al.*, 1995a).

Tick-borne inoculation and xenodiagnosis Both larval and nymphal stages of *Ixodes spp.* ticks readily feed upon laboratory mice. Larval ticks, which are not transovarially infected, can be fed upon laboratory mice that have been inoculated with any infectious *B. burgdorferi* species, strain, isolate or mutant of interest. Nymphs, infected as larvae with the *B. burgdorferi* isolate of interest, can then be used as a means of infection of experimental mice. Placement of ticks into the pelage of the dorsal neck and shoulder region will result in rapid attachment in the head and neck region (notably, ticks are more apt to attach locally when placed on pigmented strains of mice, and paradoxically tend to wander more when placed on albino mice). These sites are generally not accessible to grooming by the host mouse, but it is important that the mouse be maintained singly within the cage until the ticks have fed to repletion and detached to avoid inter-mouse grooming. Placement of mice in wire-bottomed cages over pans with water allows recovery of replete ticks for further analysis.

Xenodiagnosis is used as a highly sensitive method of detecting infection in rodents (Shih *et al.*, 1995), including detection of low numbers of replication-suppressed spirochaetes following antibiotic treatment of the host (Bockenstedt *et al.*, 2002; Hodzic *et al.*, 2008). Xenodiagnosis involves the placement of larval or nymphal ticks upon a host, allowing them to feed to repletion, moult and harden, and then testing the ticks for infection by culture, PCR, or transmission to other hosts. There is a burst of spirochaete replication within the tick during feeding, which amplifies the ability to detect spirochaetes (deSilva and Fikrig, 1995; Hodzic *et al.*, 2002). Co-feeding of uninfected ticks with infected ticks has been shown to be an effective means of detecting very early infection in mice (Hua *et al.*, 2003). Events (gene expression, population dynamics, etc.) that take place at the tick attachment/feeding site can be assessed by biopsy of the tick attachment site (Hodzic *et al.*, 2002; Leuba-Garcia *et al.*, 1998; Ohnishi *et al.*, 2001; Shih *et al.*, 1992). Forceful removal of ticks results in attachment of a small fragment of mouse dermis to the tick hypostome, which has been successfully examined for spirochaetes, gene, and protein expression (Ohnishi *et al.*, 2001).

Passage, plasmid loss, and infectivity of B. burgdorferi The plasmid content of *B. burgdorferi* isolates, strains or clones is a critical variable that must be taken into consideration with animal model phenotyping. Alan Barbour noted that early passage spirochaetes had a tendency to grow in aggregates or clumps, but lost this property after several *in vitro* subpassages (Barbour, 1984). Russell Johnson and colleagues found that *B. burgdorferi* infectivity for hamsters was lost after 3–30 passages (Johnson *et al.*, 1984). It is now well established that *in vitro* passage can result in loss of plasmids or selective outgrowth of plasmid-deficient subclones in a variety of *B. burgdorferi* s.l. isolates (Anguita *et al.*, 2000; Moody *et al.*, 1990a; Norris *et al.*, 1992, 1995; Schwan *et al.*, 1988a; Siebers *et al.*, 1999; Thomas *et al.*, 2001; Tuomi *et al.*, 2002; Xu *et al.*, 1996). Low-infectivity B31 clones are associated with absence of lp25, whereas intermediate-infectivity phenotypes retained lp25, but were missing lp28-1 (Labandeira-Rey and Skare, 2001; Purser and Norris, 2000; Xu *et al.*, 1996). The critical infectivity determinant on the B31 lp25 plasmid has been defined as a nicotinamidase (PncA) (Purser *et al.*, 2003). The lp28-1 determinant for reduced infectivity has not been defined, but is widely believed to be related to the VlsE locus. Loss of this plasmid results in a phenotype that requires a higher infectious dose. The spirochaetes remain capable of disseminated infection, but are more vulnerable to host-mediated clearance, or may have altered tissue tropism (Elias *et al.*, 2002; Grimm *et al.*, 2004a; Labandeira-Rey and Skare, 2001; Labandeira-Rey *et al.*, 2003; Lawrenz *et al.*, 2004; Norris *et al.*, 1995; Purser and Norris, 2000; Xu *et al.*, 1996; Zhang *et al.*, 1997). Recently, lp36 has been incriminated as necessary for syringe-borne infectivity in C3H mice. Notably, a low level of mouse infectivity could be demonstrated by tick feeding or with very high doses of syringe-borne lp36-deficient spirochaetes, but spirochaete numbers in tissues were very low, and limited to ear tissue (Jewett *et al.*, 2007). Others have shown loss of infectivity among B31 clones that possess lp25 and lp28-1, with no other correlation between specific plasmid loss and infectivity, suggesting that additional mechanisms may contribute to loss of infectivity (McDowell *et al.*, 2001).

Plasmids may be lost from clonal isolates after only a few passages, and freezing and thawing of frozen stocks may contribute to plasmid heterogeneity. In contrast, virulent clones can persist through repeated passage (Grimm *et al.*, 2003). Clonal *B. burgdorferi* N40 retained its infectivity for mice up to passage 119, but infectivity was lost thereafter (Thomas *et al.*, 2001). Furthermore, *B. burgdorferi* clonality is a relative concept, in that spirochaetes containing variable plasmid content can arise from a single colony, and genetic mosaics also arise *in vivo* following inoculation of clonal populations of spirochaetes, with isolates derived from infected tissues displaying different plasmid content, infectivity and tissue tropic phenotypes (Botkin *et al.*, 2006; Elias *et al.*, 2002; Labandeira-Rey and Skare, 2001; McDowell *et al.*, 2001).

Genetic transformation of *B. burgdorferi* through electroporation, selective growth, and *in vitro* passage can readily accentuate vulnerability to plasmid loss and polymorphism, not to mention the non-genetic effects upon biological behaviour (Botkin *et al.*, 2006; Elias *et al.*, 2002). Different *B. burgdorferi* isolates may have significantly different plasmid content and gene locations among the linear and circular plasmid genome (Qiu *et al.*, 2004). Until the genome has been fully sequenced of other *B. burgdorferi* isolates, the B31 strain remains the reference strain for sequence, plasmid content, and gene location. Several low passage, infectious derivatives of B31 are used for genetic modification, including B31 MI-A3, a clonal isolate of B31 MI that is lacking cp9 and has a truncated *ospB* (Elias *et al.*, 2002). B31 5A4 and 5A11 contain all 20 plasmids. Highly transformable variants of B31 have been developed which have increased transformability due to loss of lp56, which contains a putative restriction-modification gene, and targeted inactivation of BBE02, which encodes another putative restriction-modification gene (Jacobs *et al.*, 2006; Kawabata *et al.*, 2004). These variants are infectious to mice via tick feeding, but their phenotype in mice (tissue distribution, arthritogenic activity, persistence, etc.) has yet to be fully characterized. The absence of lp56 has been shown to correlate with lower number of spirochaetes in ticks following feeding on mice (Jacobs *et al.*, 2006). B31 5A13 and B31 ML23 are low-infectivity clones that lack lp25, but infectivity has been restored by transformation with BBE22, encoding a nicotinamidase (Purser *et al.*, 2003; Seshu *et al.*, 2006). Notably, the B31 5A13/BBE22 variant, although infectious to mice, induces less severe arthritis and has a higher ID50 relative to wild-type B31 5A4 (Purser *et al.*, 2003). These factors can therefore influence interpretation of mouse model studies. Several studies have utilized stain 297 for genetic manipulation (Blevins *et al.*, 2004; Hubner *et al.*, 2003; Pal *et al.*, 2004; Revel *et al.*, 2005; Yang *et al.*, 2004). Use of non-B31 clones for genomic studies is handicapped by limited information on genome sequence and plasmid configuration, with significant differences among isolates (Qiu *et al.*, 2004).

Phenotypic analysis of B. burgdorferi *genetic mutants*

Targeted gene disruption, constitutive expression, and complementation are being increasingly utilized to examine phenotypic effects on *B. burgdorferi* (see Chapter 8). Mutants have been studied in mice with targeted disruption, complementation or constitutive expression of *ospA* and/or *ospB* (Yang *et al.* 2004; Strother *et al.* 2007), *ospC* (Grimm *et al.*, 2004b; Pal *et al.* 2004; Stewart *et al.* 2006; Tilly *et al.*, 2006, 2007; Xu *et al.*, 2006,. 2008), *ospD* (Li *et al.*, 2006, 2007), *bbk32* (Li *et al.*, 2006, 2007; Seshu *et al.* 2006), *bmpA/B* (Pal *et al.* 2008b), *dbpA* and/or *dbpB* (Shi *et al.*, 2008; Xu *et al.*, 2007), *vlsE* (Bankhead and Chaconas, 2007), *luxS* (Blevins *et al.*, 2004; Huber *et al.*, 2003), *ntrA* (sigma factor 54) (Fisher *et al.*, 2005), *rpoS* (Caimano *et al.*, 2004), *bba36* (Revel *et al.*, 2005), *bbe16* (Revel *et al.*, 2005), *bb0365* (Pal *et al.*, 2008a), *bbk17* (Jewett *et al.*, 2007), *bb0588* (Parveen *et al.*, 2006), and *besC* (Bunikis *et al.*, 2008). Random transposon mutagenesis, using a spirochaete-adapted *Himar1* delivery vector, has also been used to identify critical genes involved in infectivity and pathogenicity in the mouse (Botkin *et al.*, 2006). Because of the tendency of *B. burgdorferi* to lose plasmids during passage and infection, it is critically important that the spirochaete genotype is confirmed, not only prior to infection, but also following infection (at the time of phenotypic evaluation). This is especially important in genetic mutants created with shuttle vectors, which may be lost during *in vivo* passage. Outgrowth of wild-type spirochaetes during infection may also occur if the original genetic mutant inoculum is not clonal. These issues are particularly germane when assessing disease phenotype or spirochaete persistence, since such experiments tend to take weeks or months. A number of reporters are being utilized to assess gene regulation in *B. burgdorferi*, and these are beginning to be utilized for *in vivo* studies. A *B. burgdorferi* (B31-5A-14) mutant that expresses GFP under control of the *erpAB* promoter has been utilized to visualize live extracellular spirochaetes in ear tissues of mice at 6 months of infection (Miller *et al.*, 2006). A recent study used a *B. burgdorferi* B31 5A4 NP1that expressed GFP under control of the *flaB* promoter to examine dissemination of spirochaetes in the microvasculature of mice following intravenous inoculation (Moriarty *et al.*, 2008). Use of reporters for examination of gene expression during mammalian infection requires awareness of the additional antigenic variable introduced by the reporter itself.

Reported phenotypes vary considerably and in some cases results from different laboratories, such as with *bbk32*, appear to be contradictory. These various studies lack consistency of approach, including different *B. burgdorferi* isolates and clones, infectivity in immunocompetent vs. immunodeficient mice, infecting dose, tick transmission, tick acquisition, tissue distribution, tissue burden, disease severity, measurement of disease, and persistence (beyond 1 month) in immunocompetent hosts. In addition to understanding the nuances of genetic manipulation of *B. burgdorferi*, an overall understanding of the mouse model is becoming increasingly important in approaching the *in vivo* characterization of *B. burgdorferi* genetic mutants.

Humoral response to infection Although T-cells influence the host immune response to infection, key events in the host response are mediated by B-cells or antibodies in the absence of cognate T-cell help (see Chapter 15). Infection of immunodeficient SCID and NIH-3 (triple deficient) mice (Barthold *et al.*, 1992b; Defosse *et al.*, 1992; Hodzic *et al.*, 2003; Schaible, 1989; Schaible *et al.*, 1990; Zimmer *et al.*, 1990) or B6-*Rag1* (or *Rag2*) null mice (McKisic and Barthold, 2000; McKisic *et al.*, 2000) results in progressive disease. A similar course occurs in B-cell deficient (B6-*Igh6* null) mice, whereas T-cell-deficient BALB athymic nude or B6-*Tcr* β/δ null mice develop a disease course that parallels fully immunocompetent mice, with disease remission (Kang *et al.*, 1995; McKisic and Barthold, 2000; McKisic *et al.*, 2000; Moody *et al.*, 1990a). Adoptive transfer of naïve T-cells into infected B6-*Rag1* null mice exacerbated disease, whereas transfer of B-cells induced disease remission. Immune serum from T-cell-deficient mice, but not B-cell-deficient mice, was protective against syringe challenge (McKisic *et al.*, 2000) and induced disease remission (Barthold *et al.*, 2006). In addition, adoptive transfer of B-cells, but not T-cells, elicited protective immunity in SCID mice (Schaible *et al.*, 1994). Collectively, these studies strongly implicate humoral immunity as a critical determinant of the host response to *B. burgdorferi* infection.

The host response to infection is more thoroughly covered in the Chapter 15, but discussion of the mouse model is not complete without an overview of the role of the acquired humoral response in three major biological events: protective immunity, disease-resolving immunity, and global reduction of spirochaetes in tissues (with cessation of bacteraemia) during the immune phase of infection. These critical events can be differentially dissected for experimental analysis. Protective and disease-resolving antibodies develop during active infection of mice with *B. burgdorferi* s.s., *B. afzelii* PKo or *B. garinii* PBi, indicating universality of the phenomena, but are species specific (Barthold, 1999).

- *Protective immunity* Serum from infected immunocompetent mice during the immune phase of infection (immune serum) is protective against tick- and syringe-borne cultured spirochaetes, but not against challenge with host-adapted spirochaetes (Barthold, 1996; Barthold *et al.*, 1997; deSilva *et al.*, 1998). When immune serum is passively transferred to naïve mice before or 1 day after tick attachment, relatively large, but physiological amounts of immune serum (100–500 μl) are protective (Bockenstedt *et al.*, 2003; Shih *et al.*, 1995; Shih and Liu, 1998). In contrast, as little as 0.0125 ml of immune serum is protective against challenge with low doses of syringe-inoculated cultured spirochaetes if administered prior to, but not after inoculation (Barthold and Bockenstedt, 1993; Barthold *et al.*, 1997). Immune sera from human patients and naturally infected dogs are also protective against syringe challenge in mice and hamsters (Barthold *et al.*, 1995b; Callister *et al.*, 1993; Fikrig *et al.*, 1994). Although cultured spirochaetes do not reflect the natural state of spirochaetes, the exquisite susceptibility of cultured spirochaetes to minute amounts of immune serum containing antibody to *in vivo*-expressed antigens can be utilized in a passive protection assay of functional immunity that is amenable to titration. Using serial dilutions of immune serum and a standardized inoculum in such an assay, the passive protective activity in immune serum rises rapidly during early infection, peaks at 30 to 60 days, and then declines during the course of persistent infection. These kinetics do not correspond with the rising titres and expanding reactivity of antibodies that evolve during infection when measured by Western blots and ELISAs using antigens expressed by cultured spirochaetes (Barthold *et al.*, 1993; Barthold *et al.*, 1997).
- *Disease-resolving immunity* When mice are challenged with host-adapted spirochaetes, immune serum is incapable of preventing dissemination, but prevents development of disease (Barthold, 1996; Barthold *et al.*, 1997; deSilva *et al.*, 1998). Transfer of immune serum during ongoing arthritis induces resolution of arthritis in infected SCID mice, but does not eliminate infection (Barthold *et al.*, 1997; Barthold *et al.*, 1996; Barthold *et al.*, 2006; deSilva *et al.*, 1998). These effects require passive immunization with larger volumes of immune serum (150 μl or more) than those required to induce protective immunity against syringe challenge. Using serial dilutions of immune serum to passively immunize infected SCID mice, disease-resolving activity in immune serum follows similar kinetics as protective activity, with peak activity during early infection, followed by decline, despite persisting infection. These

events may explain the recurrent bouts of acute disease during chronic infection.
Somewhat discordant observations have been reported regarding the effect of immune serum on carditis resolution. Early studies indicated that passive immunization of infected SCID mice with immune serum induced only partial resolution of carditis or had no effect (Barthold *et al.*, 1996; Barthold *et al.*, 1997; McKisic and Barthold, 2000). Other studies demonstrated that B-cells were not necessary for carditis resolution, and that CD4+ helper T-cells facilitated carditis resolution (Bockenstedt *et al.*, 2001). However, recent studies have confirmed that immune serum alone is capable of inducing carditis resolution (Barthold *et al.*, 2006). These discrepancies are probably due to the complex features of carditis, the fact that carditis is less sensitive to the effects of antibody, compared to arthritis, and may be impacted by the quality and quantity of antibody in immune sera, including the effects of T-cell help.

- *Global reduction of spirochaete burdens in tissue* Another important effect of immune serum is the reduction of spirochaete numbers in all tissues following passive immunization of infected C3H-*scid* mice with immune serum from infected immunocompetent mice. This parallels the reduction of spirochaete burdens that occurs during the evolution of host immunity in immunocompetent mice (Barthold *et al.*, 1993; Hodzic *et al.*, 2003), and also occurs following adoptive transfer of lymphocytes to infected SCID mice from immunocompetent mice (Barthold *et al.*, 2006). Notably, IL-10 deficient mice exhibit up to 10-fold greater clearance of spirochaetes from tissues compared to wild-type during the immune phase of infection, but their immune serum has similar spirochaete-reducing properties as wild type mice when passively transferred to infected SCID mice (Lazarus *et al.*, 2006).

Passive immunization with prototype antisera Based upon its protective, disease-resolving, and spirochaete-reducing properties, immune serum has been used to screen *B. burgdorferi* genomic expression libraries to identify the responsible antigens. Antiserum against recombinant non-lipidated DbpA elicits protective immunity against syringe-borne challenge (Cassatt *et al.*, 1998; Feng *et al.*, 1998; Hanson *et al.*, 1998) and induces disease resolution (Barthold *et al.*, 2006; Feng *et al.*, 2000), but DbpA immunity does not seem to protect against tick-borne infection (Hagman *et al.*, 2000). Antiserum to recombinant non-lipidated Arp is not protective, but induces disease resolution (Barthold *et al.*, 2006; Feng *et al.*, 2000; Feng *et al.*, 2003). Despite their protective and disease-resolving activity, neither DbpA nor Arp antisera induced reductions in spirochaete burdens when infected SCID mice were passively immunized (Barthold *et al.*, 2006). Reduction of spirochaete burdens in tissues and reduction in joint swelling (arthritis resolution was not confirmed by histopathology) can be induced by passive administration of antisera to recombinant BmpA and BmpB (Pal *et al.*, 2008b). Although these antigenic targets do not explain the whole story, they offer the opportunity to precisely dissect the pathogenesis of Lyme borreliosis in the mouse model, using prototype antigens and antibodies with known biological effect.

Using the ZS7 strain of *B. burgdorferi* s.s., it has been shown that passive immunization of infected SCID mice with polyclonal and monospecific antibody to recombinant OspC is capable of resolving carditis and arthritis, as well as clear infection, based upon culture of ear biopsies (Zhong *et al.*, 1997; Zhong *et al.*, 1999). In contrast, passive immunization of SCID mice infected with *B. burgdorferi* N40 with OspC antibody had no effect upon disease or infection status (Barthold *et al.*, 1997). These results underscore the potential differences in OspC expression during infection among *B. burgdorferi* isolates, strains or species. It has been shown that passive immunization of *B. burgdorferi* B31-infected SCID mice with OspC antibody eliminated OspC-expressing spirochaetes, but failed to clear infection (Liang *et al.*, 2002), but passive immunization of SCID mice infected with *B. burgdorferi* that constitutively expressed OspC resulted in clearance of infection (Xu *et al.*, 2006). Results of these studies are influenced by the tissues examined and methods for detecting spirochaetes.

Chemotherapy

The mouse model has been utilized in several studies of antibiotic efficacy for treatment of early and established *B. burgdorferi* infections. Based upon culture of tissues after antibiotic treatment of infected mice, it has been shown that amoxicillin, ceftriaxone, and penicillin G could eliminate infection and disease, whereas oxytetracycline, doxycyline, chloramphenicol, erythromycin and azithromycin failed to cure infected mice (Kazragis *et al.*, 1996; Konishi and Nakao, 1997; Malawista *et al.*, 1994; Moody *et al.*, 1994a; Pavia *et al.*, 2002; Yrjänäinen *et al.*, 2007). Doxycycline and vancomycin have been shown to be effective at preventing early infection (Kazragis *et al.*, 1996; Zeidner *et al.*, 2004).

The propensity of *B. burgdorferi* to persist in immunologically competent hosts, a concept that has been irrefutably proven with various animal models, presents a challenge to complete eradication of infection with antibiotic therapy. It is becoming increasingly clear that standard infection, treatment and culture experimental designs have proven to be inaccurate for assessing survival of spirochaetes in animal models treated with antibiotics. 'Persisters,' a small subpopulation of bacteria that are dormant and non-dividing, have been documented since 1944 during early studies on the action of penicillin on *Staphylococcus* spp., and have subsequently been found among a wide variety of other bacteria. Persisters are tolerant (in contrast to resistant) to antibiotics because they have little or no cell wall synthesis, translation or topoisomerase activity (reviewed in Lewis, 2007). Although not proven with *B. burgdorferi*, this may explain the reduced effectiveness of antibiotic therapy during late Lyme disease, and results from several studies in animal models (dogs, non-human primates and mice) in which spirochaetes cannot be cultured after antibiotic treatment, but can be detected in tissues by PCR, suggest persister forms. Furthermore, live and non-dividing populations of persisting spirochaetes cannot be detected by culture, but can be detected in antibiotic-treated mice by xenodiagnosis. These forms of spirochaetes are acquired by feeding ticks, but they do not undergo the usual burst of replication in the ticks. The spirochaetes, nevertheless, are transmissible by feeding infected ticks on naïve SCID mice. Although the spirochaetes that infect these mice disseminate, they do not replicate to the levels of spirochaetes from ticks that have fed upon antibiotic-untreated mice (Hodzic *et al.*, 2008). An earlier study in mice suggested that the spirochaetes that were acquired by xenodiagnosis on antibiotic-treated mice were genetically attenuated and non-infectious (Bockenstedt *et al.*, 2002), but this was not found in a subsequent study (Hodzic *et al.*, 2008).

Under normal circumstances, culture and PCR are equally sensitive for detection of spirochaetes in tissues, but this is not the case when spirochaetes are in a non-dividing state. Another caveat that confounds most studies is that following antibiotic treatment, spirochaetes are cleared from several of the tissues that are routinely cultured as read-outs, such as ear and urinary bladder. It is important to examine and test highly collagenous tissue of joints and heart base for maximal detection sensitivity of persisting spirochaetes (Hodzic *et al.*, 2008). These sites are preferential sites of spirochaete persistence during the immune phase of chronic infection (Barthold *et al.*, 2006) (Fig. 14.20). The highly collagenous dermis is also a common site for persistence during chronic infection (Barthold *et al.*, 1993), which fits with the spirochaete's relationship to its tick vector. Despite persisting infection following antibiotic treatment, the serum antibody response declines as if the animals were cured of infection (Bockenstedt *et al.*, 2002; Hodzic *et al.*, 2008).

Although evidence for non-dividing persister forms of *B. burgdorferi* is still incomplete, support for this concept has arisen from other studies. Both penicillin G and doxycycline have been shown to significantly alter the expression pattern of *B. burgdorferi* s.s. LW2 structural proteins, and most of the affected genes are chromosomal and involved in bacterial cell metabolism (Hunfeld *et al.*, 2008). Other studies have shown that calprotectin, an abundant neutrophil cytoplasmic protein, is a potent bacteriostat, and physiological levels of this protein induce *B. burgdorferi* to enter a non-dividing state without cell death. Calprotectin's effects are mediated through chelation of Zn, which is essential to growth of spirochaetes, as well as activation of NADPH oxidase (Lusitani *et al.*, 2003). A potential consequence of this suppression of division would

Figure 14.20 Immunohistochemical staining of a single *B. burgdorferi* spirochaete in collagenous tendon tissue of a C3H mouse. Persisting spirochaetes during chronic infection are typically associated with collagen bundles in various tissues, including tendons, ligaments, vessel walls, and dermis.

be antibiotic tolerance, which indeed has been shown with *B. burgdorferi* exposed to penicillin G, but not doxycycline (Montgomery *et al.*, 2006). It should also be noted that functional levels of calprotectin are present in synovial fluids of Lyme disease and other arthritides (Montgomery *et al.*, 2006). Collectively, these results pose challenges to existing dogma about antibiotic treatment of Lyme borreliosis, and the mouse model offers opportunities for further incisive investigation, but the model must be utilized in increasingly precise ways with full cognizance of these biological features.

Vaccination and heterologous footpad challenge model of arthritis

The mouse has been extensively utilized as a model for vaccination with various recombinant proteins, and challenge by syringe- or tick-borne inoculation. These studies are straight-forward, and generally involve necropsy of mice at 2 or more weeks after challenge. Vaccination strategies are beyond the scope of this chapter, and discussed in Chapter 17.

Apropos to this chapter, however, is a distinctly different mouse arthritis model that has evolved from studies in the hamster involving vaccination with formalin-inactivated whole *B. burgdorferi* 297, followed by challenge in the footpad of IFN-gamma deficient B6 mice with 10^6 live heterologous *B. bissettii.* (Amlong *et al.*, 2006; Burchill *et al.*, 2003; Christopherson *et al.*, 2003; Nardelli *et al.*, 2004, 2005, 2006; Peterson *et al.*, 2007). IFN-gamma-deficient mice are used in this model because neutralization of IFN-gamma augments the arthritis (Munson *et al.*, 2004). The mice in this model system develop severe arthritis in feet that have been inoculated with live *B. bissettii*. Arthritis is exacerbated by the preimmunization with killed *B. burgdorferi* 297. Arthritis will not occur if the mice are challenged with the homologous isolate, as they develop protective borreliacidal immunity that prevents infection. The mechanism for the ensuing severe arthritis in this model is a combination of direct *B. burgdorferi*-induced inflammation, modulation of arthritis by the heterologous borreliacidal antibody response, plus immune-mediated exacerbation of the inflammation, in contrast to the arthritis that evolves in the preimmune phase of infected mice in the standard mouse model, which undergoes antibody-mediated resolution. This model is complex, mechanistically multifactorial, and experimentally artificial.

14.2 Animal models of relapsing fever borreliosis

Relapsing fever (RF) is a multisystemic spirochaetal infection that occurs in two forms, epidemic and endemic, and caused by a variety of Borrelia species. Epidemic RF is louse borne and caused by *B. recurrentis,* whereas endemic RF is tick-borne and caused by different Borrelia species in endemic areas throughout the world. Among the several agents of endemic RF, the best studied have been the New World species *B. turicatae* and *B. hermsii* and the Old World species *B. duttoni* and *B. crocidurae* (Barbour and Hayes, 1986). The clinical and pathological manifestations of RF borreliosis are diverse, depending on the Borrelia species and the genetic background of the host. The relapsing course is mediated by antigenic variation of outer membrane lipoproteins, which allows infection to persist for several weeks with alternating periods of illness at times of high bacteraemia and periods of relative well-being during low bacteraemia (see Chapter 13). In mice, IgM antibodies produced by B1b cells have been shown to be important for clearance of the variant serotypes (Alugupalli *et al.*, 2003a, 2004; Belperron *et al.*, 2005; Connolly and Benach, 2001).

Untreated infection can result in multisystemic complications and even death. However, because of the paucity of human cases since antibiotics were discovered, there has been little modern research on the pathology of RF in humans. In contrast, much has been learned on the pathogenesis of RF borreliosis in experimental animal models. This section will present an overview of RF pathology in experimental animals. For the epidemic form, the only available animal model is the non-human primate. In contrast, mice and rats have proven to be excellent models to study the endemic form (Cadavid and Barbour, 1998; Schuhardt and O'Bryan, 1945; Trape *et al.*, 1991).

14.2.1 Relapsing fever borreliosis in immunocompetent animals

Clinical manifestations

There is great heterogeneity of clinical manifestations of RF borreliosis in experimental animals. As a general rule, higher pathogen loads result in more severe disease. Therefore, animals inoculated with large numbers of organisms tend to fall ill and even die early (Wright and Woodrow, 1980). *Borrelia recurrentis* infection in grivet monkeys (*Cercopithecus aethiops*) uniformly caused fever as high as 40.9°C and killed one in five animals, including two during a severe relapse (Judge *et al.*, 1974a). The clinical manifestations of infection in immunocompetent mice vary with their genetic background, in contrast to antibody-deficient mice that uniformly develop clinical disease independent of genetic background (Cadavid *et al.*, 2006). Old World RF Borrelia species appear to be more pathogenic than the New World species: inoculation of BALB/c mice with *B. crocidurae* resulted in clinical disease, whereas *B. hermsii* did not (Cadavid *et al.*, 1994; Shamaei-Tousi *et al.*, 1999). While inoculation with *B. turicatae* did not result in any clinical disease in any of several inbred mouse strains tested, including C3H/HeJ, SWR/J, or B6, severe vestibular dysfunction was observed in 1 in 4 outbred mice (Cadavid *et al.*, 2006). Deficiency of TLR2, which plays a major role in recognition of borrelia lipoproteins, did not result in observable clinical disease during infection with *B. turicatae* in mice of the B6 background, despite increased bacteraemia (Alugupalli *et al.*, 2007; Cadavid *et al.*, 2006). The strong and transient production of IL-10 observed in immunocompetent mice during peak bacteraemia probably explains the paucity of clinical manifestations (Londono *et al.*, 2008c).

Infection

All outbred stocks and inbred strains of mice that have been tested (B6, BALB/c, C3H/HeJ, SWR) develop relapsing bacteraemia upon inoculation with endemic RF spirochaetes including *B. hermsii* (Cadavid *et al.*, 1993), *B. turicatae* (Cadavid *et al.*, 2006), *B. crocidurae* (Shamaei-Tousi *et al*,, 1999), and an uncultivable strain from Spain (Gebbia *et al.*, 1999). Peak bacteraemia varies from 10^6/ml to 10^8/ml with higher counts usually observed during the first peak and with higher initial inocula (Alugupallli *et al.*, 2003a; Cadavid *et al.*, 1993,2006; Larsson *et al.*, 2006; Stoenner *et al.*, 1982). Spirochaetes were observed free in the blood with the exception of erythrocyte-associated *B. crocidurae* and extravascularly in the kidney interstitium, lung and testis (Shamaei-Tousi *et al.*, 1999). Dissemination to the kidneys was facilitated by plasminogen (Nordstrand *et al.*, 2001; Shamaei-Tousi *et al.*, 1999). Unlike *B. burgdorferi* (Coleman *et al.*, 1997), plasminogen plays no role in the bacteraemia of RF borreliosis (Gebbia *et al.*, 1999; Nordstand *et al.*, 2001).

The subject of brain infection with RF spirochaetes was of particular interest prior to the discovery of penicillin, notably the finding that brain homogenates were infectious after the blood had ceased to be infectious; a phenomenon called residual brain infection (Cadavid and Barbour, 1998). Cadavid *et al.* (2006) found residual brain infection after 1 month in 20% of mice inoculated with a neurotropic strain of *B. turicatae*. TLR2 deficiency did not increase susceptibility to residual brain infection. In the majority of cases, residual brain infection was caused by new serotypes, but in at least one case the originally inoculated serotype was responsible (Cadavid *et al.*, 2006). Larsson *et al.* (2006) also studied residual brain infection with four different RF species and found that it was more prevalent with *B. duttoni*. Different serotypes of *B. duttoni* were capable of causing residual brain infection and the average pathogen load in the brain was around 2×10^3 spirochaetes/gram of brain. In some mice they observed recurrence of bacteraemia with steroid treatment (Larsson *et al.*, 2006). At least two studies have shown that host plasmin facilitated brain infection (Gebbia *et al.*, 1999; Nordstand *et al.*, 2001). There have been several studies to precisely localize RF spirochaetes in the brain. Rare spirochaetes were observed in the brain parenchyma by silver staining techniques in C3H/HeN but not in 129/Black Swiss mice infected with a Spanish strain (García-Moncó *et al.*, 1997; Gebbia *et al.*, 1999). In contrast, *B. crocidurae* was readily observed in the brain parenchyma early after inoculation, although extensive intravascular congestion, rosetting, and microthrombi made it difficult to establish whether these spirochaetes were truly intrathecal (Andersson *et al.*, 2007; Nordstand *et al.*, 2001; Shamaei-Tousi *et al.*, 1999). Others failed to find any spirochaetes microscopically in the brain during the residual infection phase with *B. duttoni* (Larsson *et al.*, 2006).

Inflammation and pathology

During peak bacteraemia, increased numbers of neutrophils, lymphocytes, and monocytes were observed in the blood of most rodent models, in addition to anaemia, which worsened after the peak (Gebbia *et al.*, 1999; Shamaei-Tousi *et al.*, 1999). However, inoculation with large numbers of *B. duttoni* spirochaetes resulted in a progressive decrease in the number of leucocytes that corresponded to the increase in bacteraemia (Wright and Woodrow, 1980). There is a strong correlation between the magnitude of bacteraemia and thrombocytopenia (Alugupalli *et al.*, 2003b). Early and persistent thrombocytopenia was observed in mice infected with *B. duttoni* (Wright and Woodrow, 1980), *B. crocidurae* (Shamaei-Tousi *et al.*, 1999), and an RF strain from Spain (Gebbia *et al.*, 1999). RF spirochaetes bind platelets *in vivo* and *in vitro* (Alugupalli *et al.*, 2001a,b, 2003b; Shamaei-Tousi *et al.*, 1999). Inoculation with 10^6 *B. hermsii* spirochaetes lowered platelet counts 5- to 20-fold and prolonged bleeding time 5 fold due to accelerated platelet clearance but without increasing D-dimers (Alugupalli *et al.*, 2003b). RF infection increased fibrin deposition (Gebbia *et al.*, 1999; Wright and Woodrow, 1980).

Marked enlargement of the spleen and liver is a consistent feature in small animal models of RF (Gebbia *et al.*, 1999; Shamaei-Tousi *et al.*, 1999; Wright and Woodrow, 1980). In grivet monkeys *B. recurrentis* caused multiple microabscesses that replaced the white pulp in the spleen, and hepatitis with foci of midzonal necrosis (Judge *et al.*, 1974b). Accumulation of neutrophils in and around microvessels was also observed in the kidneys with *B. crocidurae* but not with a Spanish strain (Gebbia *et al.*, 1999; Shamaei-Tousi *et al.*, 1999). BALB/c mice and rats inoculated with 1 million *B. crocidurae* spirochaetes rapidly developed haemorrhages in the surface of the lungs, liver, and kidneys. Microscopic examination revealed intravascular erythrocyte aggregation with secondary extravasation and haemorrhage (Shamaei-Tousi *et al.*, 1999; Shamaei-Tousi *et al.*, 2001).

Prominent vascular pathology occurred in mice inoculated with 5×10^7 *B. duttoni* spirochaetes. This very large inoculum resulted in high mortality of 50% after only 4 days (Wright and Woodrow, 1980). The most conspicuous feature in infected tissues was the presence of blood vessel fibrin thrombi, most commonly in the kidney, that distended and occluded glomerular capillary loops. Intravascular fibrin thrombi were also seen in the liver, spleen, and lung. One of the mice showed a cortical infarct in the brain. Infection with *B. crocidurae* resulted in spirochaetes associated with blood vessels and endothelial cell activation (Nordstrand *et al.*, 2001; Shamaei-Tousi *et al*, 2000). Blood vessel thrombosis in the spleen and liver with spirochaetes in the blood vessels of many organs was documented in fatal *B. hermsii* infection of a wild owl (Thomas *et al.*, 2002).

Borrelia recurrentis causes histiocytic myocarditis in grivet monkeys (Judge *et al.*, 1974b). Myocarditis was also observed in rodents infected with Old World (Gebbia *et al.*, 1999; Nordstrand *et al.* 2001), but not with New World RF species (Londono *et al.*, 2005). Infected hearts showed perivascular mononuclear leucocyte infiltrates in the pericardium and focal myocarditis with mononuclear leucocyte infiltrates, interstitial oedema, and isolated cardiomyocyte necrosis (Gebbia *et al.*, 1999).

Animal models that utilize large inocula have prominent tissue injury. Mice inoculated with 5×10^7 *B. duttoni* spirochaetes developed focal areas of endothelial cell swelling in the kidneys and nuclear pyknosis and karyorrhexis in hepatocytes and Kupffer cells in the liver. Spleens were enlarged with large areas of necrosis and macrophages in the red pulp were markedly vacuolated (Wright and Woodrow, 1980). BALB/c mice inoculated with large numbers of *B. crocidurae* spirochaetes developed changes suggestive of cell death in kidneys and lungs. These changes were thought to be the result of vascular occlusion from microthrombi formed by a combination of erythrocytes, spirochaetes, and leucocytes (Shamaei-Tousi *et al.*, 1999). TUNEL staining revealed the presence of apoptotic germ cells during peak bacteraemia in rats infected with *B. crocidurae* (Shamaei-Tousi *et al.*, 2001). In contrast, tissue damage is not a feature in mice inoculated with *B. hermsii* (Shamaei-Tousi *et al.*, 2001), *B. turicatae* (Cadavid *et al.*, 2006), a Spanish strain (Gebbia *et al.*, 1999), or lower doses of *B. duttoni* (Andersson *et al.*, 2007; Larsson *et al.*, 2006). No alterations in gene expression were found in the brain of mice with residual infection by *B. duttoni* (Larsson *et al.*, 2006).

There have been several studies of inflammation in the brain. Wild-type or TLR2-deficient B6 mice inoculated with *B. turicatae* developed leptomeningeal inflammation and brain parenchymal macrophage infiltration/activation with little if any lymphocytes (Cadavid *et al.*, 2006). Inoculation of BALB/c mice with *B. crocidurae* resulted in increased perivascular macrophages/microgliosis as well as small assemblies of lymphocytes and granulocytes in the brain (Andersson *et al.*, 2007). The inflammatory reaction was more pronounced in the choroid plexus and in the leptomeninges, and consisted of lymphocytes and neutrophils. The meninges were congested and occasional petechial haemorrhages in the brain parenchyma were also found; activated macrophages/microglia were by far the most abundant inflammatory cells in the brain, followed by neutrophils with little if any increase in B- or T-cells (Andersson *et al.*, 2007). Brain inflammation during *B. crocidurae* infection decreased in the absence of plasminogen (Nordstand *et al.*, 2001). 129/Black Swiss, BALB/c, and C3H/HeN mice inoculated with a RF strain from Spain also developed perivascular mononuclear cell infiltration of the leptomeninges, accumulation of activated macrophages perivascularly, and plexitis (García-Moncó *et al.*, 1997; Gebbia *et al.*, 1999). In this study, the inflammatory infiltrate was strongly CD45R (B220) positive and there was submeningeal fibrin deposition that was not seen in the absence of plasminogen (Gebbia *et al.*, 1999). No evidence of inflammation was found in the brain of mice experiencing residual brain infection with *B. duttoni* (Larsson *et al.*, 2006) or during any stage of infection with *B. hermsii* (Cadavid *et al.*, 1993; Shamaei-Tousi *et al.*, 1999). One fatal case of *B. hermsii* infection in a wild owl had spirochaetes in the cerebral ventricles and haemorrhage in the choroid plexus (Thomas *et al.*, 2002).

14.2.2 Relapsing fever borreliosis in immune-deficient mice

Since serotype-specific antibodies rapidly clear RF spirochaetes from blood and tissues, mice deficient in B-cells cannot clear the infection, making it possible to compare disease by specific serotypes over time. In the early 1990s there was an outbreak of endemic RF in Texas with prominent neurological complications (Rawlings, 1995). Inoculation of spirochaetes cultured from whole tick homogenates into SCID mice allowed the identification of two *B. turicatae* serotypes that caused distinct differences in pathogen load in blood and brain (Cadavid *et al.*, 1994). These two serotypes were originally named A and B, and subsequently renamed 1 and 2 (or Bt1 and Bt2). The only discernible difference between them is the variable major protein they express, 23 kDa Vsp1 in serotype 1 and 20 kDa Vsp2 in serotype 2 (Cadavid *et al.*, 1997; Pennington *et al.*, 1999a,b). Infection of B-cell-deficient mice with serotypes 1 and 2 is being utilized to understand the basis of their different virulence and tissue infectivity.

Clinical manifestations

Since their original isolation it became apparent that Bt2 was more virulent than Bt1. Both serotypes have similar abilities to establish infection with as little as 1 spirochaete per mouse, but Bt2 causes more severe clinical manifestations. Clinical disease began in the first week by purulent discharge from the eyes and ruffled skin, was followed by tibiotarsal joint swelling and redness by the second week, and signs of vestibular dysfunction arose after 3 weeks (Cadavid *et al.*, 1994). Tibiotarsal and metatarsal joint reddening and swelling began earlier and were more significantly more severe with Bt2 that also resulted in more functional impairment on an equilibrium bar. Despite the absence of specific antibody to control pathogen load, all the clinical manifestations either completely resolved (purulent eye discharge) or significantly decreased (skin and joints) over days to weeks (Cadavid *et al.*, 1994). Since signs of vestibular dysfunction were a delayed manifestation of infection, new groups of SCID mice were infected with 10^4 Bt1 or Bt2 spirochaetes and examined for up to 2 months. Mice co-infected with both Bt1 and Bt2 were compared with mice infected with either serotype alone. Although vestibular dysfunction began earlier with Bt2, only Bt1 caused severe vestibular dysfunction associated with early mortality. Interestingly, this was prevented by co-infection with Bt2. Unlike vestibular dysfunction, the severe arthritis and weight loss caused by Bt2 alone was not influenced by co-infection with Bt1 (Cadavid *et al.*, 2007). In a series of recent studies that compared infection in *Rag1* null (T and B-cell-deficient) and *Igh6* null (B-cell-deficient) mice, it was shown that *Rag1* null mice developed less clinical disease, despite higher pathogen load due to higher production of IL-10 (Gelderblom *et al.*, 2007b; Londono *et al.*, 2008a,b).

Infection

The high number of spirochaetes found in blood and tissues of B-cell-deficient mice inoculated with *B. turicatae* provided an opportunity to study their localization after systemic inoculation. Whole decalcified heads were examined to study multiple tissues simultaneously without disrupting important anatomical relationships. Abundant spirochaetes were observed microscopically in multiple tissues as early as 6 days after inoculation. In the brain, the localization was primarily leptomeningeal, although rarely spirochaetes were also found in the brain parenchyma itself (Cadavid *et al.*, 2001). The site of entry into the brain appeared to be the leptomeningeal microcirculation, where spirochaetes were observed attached to and in the process of crossing brain-microvascular endothelial cells (Sethi *et al.*, 2006). Spirochaetes were also observed in the dura mater, middle and inner ear, endoneurium and perineurium of cranial and peripheral nerves, skin, and bone marrow (Cadavid *et al.*, 2001). Clumps of spirochaetes were observed only in the subarachnoid space and inner ear. Abundant spirochaetes were present in the extracellular matrix of the skin, skeletal and cardiac muscle. No spirochaetes were found in salivary glands or in the choroid plexus and none were intracellular (Cadavid *et al.*, 2001; Sethi *et al.*, 2006). The first tissue where spirochaetes were observed in the head outside of the vasculature was the dura mater (Cadavid *et al.*, 2001). Bt2 was more abundant in the skin (Cadavid *et al.*, 2001), the heart (Londono *et al.*, 2005), and the tibiotarsal joints (Pennington *et al.*, 1997).

Bt2 causes higher bacteraemia than Bt1. Pathogen load during co-infection was studied indirectly by comparing circulating levels of the B-cell chemokine CXCL13 in SCID mice persistently infected with Bt2 alone or in combination with Bt1 (Cadavid *et al.*, 2007; Gelderbom *et al.*, 2007a). This approach was used as an alternative to directly counting spirochaetes because the levels of this chemokine strongly correlate with pathogen load (Narayan *et al.*, 2005). Results revealed that co-infected mice had significantly higher CXCL13 levels: the mean (SD) amount in pg/ml was 72,625 (4750) compared with 42,625 (5406) in mice infected with Bt2 alone ($P < 0.001$).

The greater neurotropism of Bt1 compared with Bt2 had been apparent since the original isolation of these two serotypes. When brain infection was studied by culture of whole brain homogenates, Bt1 was found in the brain much more frequently than Bt2. Furthermore, only Bt1 was apparent by Western blot on brain cultures from SCID mice that were co-infected with Bt1 and Bt2 (Cadavid *et al.*, 1994). Since Bt2 readily infected the brain after intracerebral inoculation (Cadavid *et al.*, 2001), this suggested that Bt1 is better than Bt2 at crossing the blood–brain barrier (BBB). Using immunohistochemistry, five times more Bt1 than Bt2 spirochaetes were found in the leptomeninges in SCID mice examined 18 days after inoculation. The pathogen load in the brain of mice coinfected with Bt1 and Bt2 was intermediate between that of mice infected with either serotype (Cadavid, 2006; Sethi *et al.*, 2006), suggesting that coinfection with Bt2 may decrease the entry of Bt1 into the brain.

Pathology

Examination of tissues from SCID mice persistently infected with Bt1 or Bt2 has confirmed significant differences in their pathogenicity. No inflammation was apparent for at least 5 days after inoculation. However, by 2 weeks arthritis was present and more severe with Bt2 than with Bt1: 19/20 joints from Bt2 infected mice had arthritis compared with 12/20 from Bt1 infected mice ($P < 0.01$); it was severe only with Bt2. H&E-stained sections revealed mixed inflammation of the synovial lining of both joints and tendon sheaths with mononuclear and polymorphonuclear cells, with prominent thickening of the synovium and leucocyte exudate into the joint space. Infected SCID mice also showed prominent inflammation in the heart. Cardiac inflammation involved both the myocardium and pericardium and was characterized by prominent mononuclear leucocytic infiltrate between myocardial fibres with exudate in the pericardium. Cardiac inflammation was more severe in the atria and around the great vessels in the mediastinum (Londono *et al.*, 2005) similar to Lyme borreliosis (Barthold *et al.*, 1993). Masson's trichrome staining showed loss of cardiomyocytes that was more prominent in the atrium; immunohistochemistry showed infiltration by activated macrophages in the atrium and periphery of the heart; and *in situ* hybridization revealed prominent upregulation of expression of IL-6 also more prominent in the heart base. Digital image analysis revealed that the loss of cardiomyocytes in the heart base was significantly worse with Bt2 than with Bt1. Morphological features of apoptosis, including fragmented nuclei and condensed chromatin, were readily apparent in areas of cardiomyocyte injury. The TUNEL assay and immunostaining for BAX confirmed the presence of cardiomyocyte apoptosis that was more prominent near inflamed areas. Analysis of caspase activation by RNAse protection assay showed significant upregulation of several caspases, most notably caspase 1 and caspase 11. The upregulation of caspase 1 was confirmed at the protein level (Londono *et al.*, 2005), with strong expression found in areas of injury that co-localized with activated macrophages. Cardiomyocyte apoptosis was also found in Rag1 and Igh6-deficient mice (Gelderblom *et al.*, 2007b), with strong expression in areas of injury that co-localized with activated macrophages. Cardiomyocyte apoptosis was also found in Rag1 or Igh6 null mice. Molecular analysis of cardiac tissue from Rag1 or Igh6 null mice showed prominent upregulation of both TNF and IL-10 of similar degree with either serotype (Cadavid *et al.*, 2001).

Inflammation in the skin of SCID mice was equally severe with either serotype despite the marked differences in pathogen load (Cadavid *et al.*, 2001). Most inflammatory cells in the skin were activated macrophages (Sethi *et al.*, 2006). Significant inflammation was also found in the vestibular and auditory labyrinth, which probably explained the prominent vestibular dysfunction that characterizes this model (Gelderblom *et al.*, 2007b). Mild inflammation was also found in the leptomeninges with no obvious difference between Bt1 and Bt2 (Fig. 14.21). Although H&E-staining did not reveal any apparent abnormalities in the brain parenchyma, immunostaining for F4/80, a marker of activated microglia/macrophages revealed widespread infiltration by activated macrophages/microglia with a 30-fold increase in Bt1-infected mice compared to uninfected controls (Fig. 14.22). Microgliosis was less apparent in the brain of Bt2 infected mice. The morphology of F4/80-stained cells was heterogeneous with some amoeboid and some ramified. Examination of mice deficient in Rag1 or Igh6 confirmed that persistent infection with Bt1 results in severe cerebral microgliosis. Molecular analysis of the brain from Igh6-deficient mice showed that Bt1 infection resulted in significant

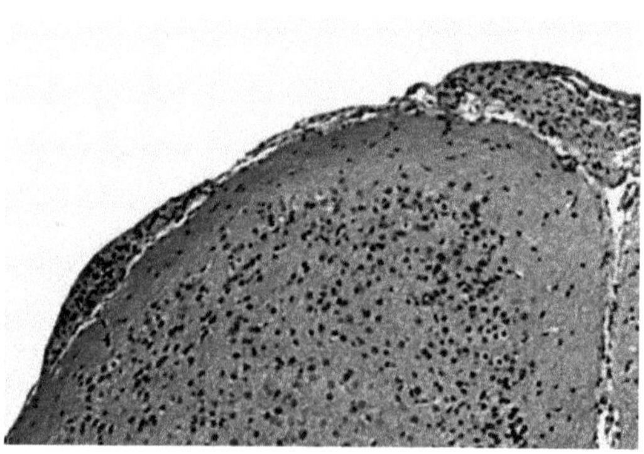

Figure 14.21 Mild leptomeningitis in a RAG2-deficient mouse persistently infected with serotype 1 of *B. turicatae*.

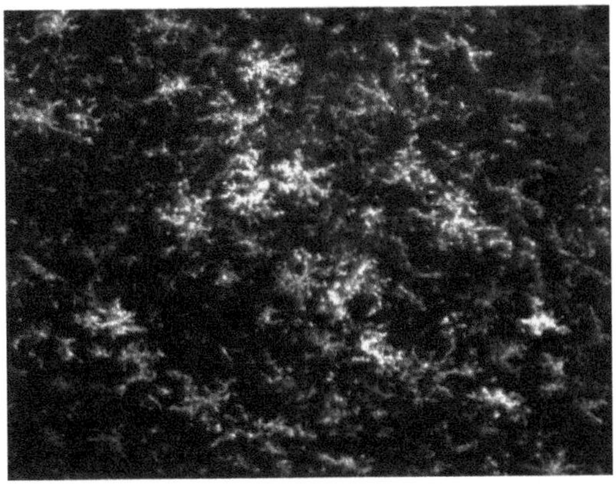

Figure 14.22 Severe microgliosis in the brain of a C.B-17-*scid* mouse persistently infected with serotype 1 of *B. turicatae*, depicted by immunohistochemical staining of F4/80. Note the prominence of ramified microglia.

upregulation of TNF and IL-10 by 2 weeks but, interestingly, by 4 weeks only upregulation of IL-10 persisted. Immunofluorescence showed that fully activated microglia were the main source of IL-10 in Bt1-infected brains. Oligoarray gene expression analysis in Igh6-deficient mice persistently infected with Bt1 compared with uninfected controls revealed 116 out of 14,109 well-documented genes present on the Affymetrix 430A mouse gene chip were upregulated at least 2-fold at $P < 0.01$ (Gelderblom *et al.*, 2007a). Many of them are genes known to be upregulated during microglia activation. CXCL13 was the most highly upregulated gene in infected brains and the most abundant of the chemokines in the blood (Gelderblom *et al.*, 2007b). An analysis of functional categories within the gene ontology database showed that the majority of the 20 categories with the highest EASE scores refer to functions of the immune response, including proteins involved in antigen processing and presentation; chemotaxis, adhesion and migration; LPS/lipoprotein toll-like receptor ligation and signalling; and pathogen control and host defence (Gelderblom *et al.*, 2007b).

The observation by Cooper *et al.* (2000) that humans infected with *Borrelia recurrentis* have extraordinarily high levels of circulating IL-10 at times when they appear to be doing relatively well prompted studies on its possible protective role in experimental animals. High levels of IL-10 characterized persistent infection of B-cell-deficient mice with *B. turicatae* (Gelderblom *et al.*, 2007b). In fact, IL-10 had the highest levels among all the cytokines measured with a strong positive correlation with pathogen load (Londono *et al.*, 2008a,c). The increase in IL-10 levels was associated with a concomitant decrease in circulating levels of TNF and many other proinflammatory cytokines and chemokines The upregulation of IL-10 was also found in tissues, including heart and brain. Exogenous IL-10 given to Igh6 null mice reduced bacteraemia, clinical disease, cerebral microgliosis, and CXCL13. Bt2-infected Rag2 null mice also deficient in IL-10 showed early mortality due to increased bacteraemia, prominent apoptosis of leptomeningeal microvascular endothelial cells (Fig. 14.23), and severe intracerebral haemorrhage (Fig. 14.24). Bt1-infected IL-10- and Rag2-deficient mice did not die and only showed smaller areas of haemorrhage in the brain and other organs when necropsied 2 weeks after inoculation (Londono *et al.*, 2008a; Londono *et al.*, 2008). These results revealed that a functional protective mechanism during RF borreliosis is the production of IL-10 systemically and locally.

14.3 Concluding remarks

Each of the models discussed in this chapter has its strengths and weaknesses, and analogy to human disease is not always possible. The fact

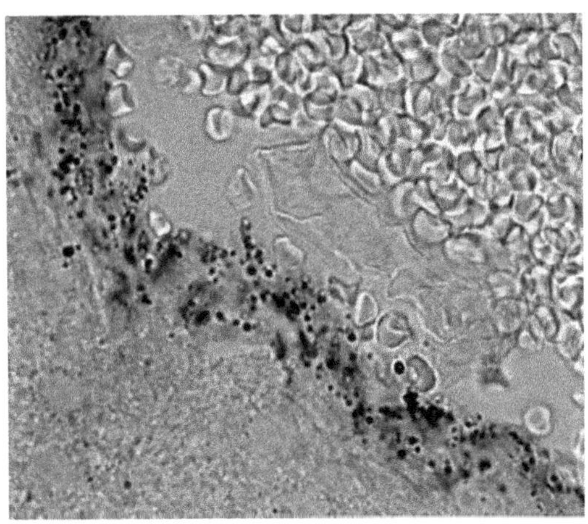

Figure 14.23 TUNEL staining of apoptotic leptomeningeal microvascular endothelial cells next to an area of subarachnoid haemorrhage in a RAG2/IL-10 null mouse infected with serotype 2 of *B. turicatae*.

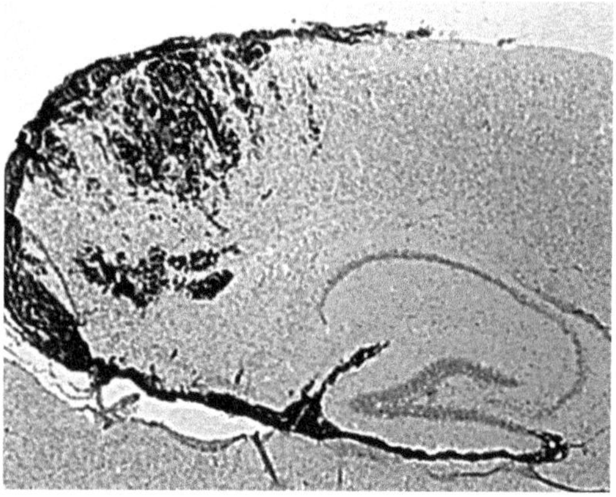

Figure 14.24 Sagittal section of brain from a RAG2/IL-10-deficient mouse at 4 days after inoculation with serotype 2 of *B. turicatae*. Note the extensive subarachnoid and parenchymal haemorrhage.

that a single model does not recapitulate the entire human condition should not negate its overall value for understanding concepts of pathogenesis. It is readily apparent that a comparative commonality pervades all models of each type of borreliosis discussed, and to a certain extent, between different types of borreliosis, which allows reasonable assumptions for extrapolation to the human disease. Sir William Osler, the Father of Modern Medicine, who is also credited as a founder of discipline of veterinary pathology, referred to this commonality among species as 'One Medicine' (Cardiff *et al.*, 2008).

Acknowledgements

Supported by grants R01 AI26815 and R01 AI073911 (Barthold); R21 NS053997 and R21 NS057545-02 (Cadavid); and U01 CI000153, R01 NS048952 and P51 RR00164 (Philipp).

References

Akins, D.R., Bourell, K.W., Caimano, M.J., Norgard, M.V., and Radolf, J.D. (1998). A new animal model for studying Lyme disease spirochetes in a mammalian host-adapted state. J. Clin. Invest. *101*, 2240–2250.

Alexopoulou, L., Thomas, V., Schnare, M., Lobet, Y., Anguita, J., Schoen, R.T., Medzhitov, R., Fikrig, E., and Flavell, R.A. (2002). Hyporesponsiveness to vaccination with *Borrelia burgdorferi* OspA in humans and in TLR1- and TLR2-deficient mice. Nat. Med. *8*, 878–884.

Alugupalli, K.R., Akira, S., Lien, E., and Leong, J.M. (2007). MyD-88- and Bruton's tyrosine kinase-mediated signals are essential for T-cell-independent pathogen-specific IgM responses. J. Immunol. *178*, 3740–3749.

Alugupallli, K.R., Gerstein, R.M., Chen, J., Szomolanyi-Tsuda, E., Woodland, R.T., and Leong, J.M. (2003a). The resolution of relapsing fever borreliosis requires IgM and is concurrent with expansion of B1b lymphocytes. J. Immunol. *170*, 3819–3827.

Alugupalli, K.R., Leong, J.M., Woodland, R.T., Muramatsu, M., Honjo, T., and Gerstein, R.M. (2004). B1b lymphocytes confer T-cell-independent long-lasting immunity. Immunity *21*, 379–390.

Alugapalli, K.R., Michelson, A.D., Barnard, M.R., and Leong, J.M. (2001a). Serial determinations of platelet counts in mice by flow cytometry. Thromb. Haemost. *86*:668–671.

Alugupalli, K.R., Michelson, A.D., Barnard, M.R., Robbins, D., Coburn, J., Baker, E.K., Ginsburg, M.H., Schwan, T.G., and Leong, J.M. (2001b). Platelet activation by a relapsing fever spirochaete results in enhanced bacterium–platelet interaction via integrin $\alpha_{IIb}\beta_3$ activation. Mol. Microbiol. *39*, 330–340.

Alugupalli, K.R., Michelson, A.D., Joris, I., Schwan, T.G., Hodivala-Dilke, K., Hynes, R.O., and Leong, J.M. (2003b). Spirochete-platelet attachment and thrombocytopenia in murine relapsing fever borreliosis. Blood *102*, 2843–2850.

Amlong, C.A., Nardelli, D.T., Peterson, S.H., Warner, T.F., Callister, S.M., and Schell, R.F. (2006). Anti-interleukin 15 prevents arthrtis in *Borrelia*-vaccinated and -infected mice. Clin. Vaccine Immunol. *13*, 289–296.

Anderson, J.F. (1991). Epizootiology of Lyme borreliosis. Scand. J. Infect. Dis. *77*, 23–34.

Anderson, J.F., Barthold, S.W., and Magnarelli, L.A. (1990). Infectious but nonpathogenic isolate of *Borrelia burgdorferi*. J. Clin. Microbiol. *28*, 2693–2699.

Anderson, J.F., and Magnarelli, L.A. (1993). Epizootiology of Lyme disease-causing *Borreliae*. Clin. Dermatol. *11*, 339–351.

Andersson, M., Nordstrand, A., Shamaei-Tousi, A., Jansson, A., Bergström, S., and Guo, B.P. (2007). In situ immune response in brain and kidney during early relapsing fever borreliosis. J. Neuroimmunol. *183*, 26–32.

Anguita, J., Barthold, S.W., Persinski, R., Hedrick, M.N., Huy, C.A., Davis, R.J., Flavell, R.A., and Fikrig, E. (2002). Murine Lyme arthritis development mediated by p38 mitogen-activated protein kinase activity. J. Immunol. *168*, 6352–6357.

Anguita, J., Rincaon, M., Samanta, S., Barthold, S.W., Flavell, R.A., and Fikrig, E. (1998). *Borrelia burgdorferi*-infected, interleukin-6-deficient mice have decreased Th2 responses and increased Lyme arthritis. J. Infect. Dis. *178*, 1512–1515.

Anguita, J., Samanta, S., Revilla, B., Suk, K., Das, S., Barthold, S.W., and Fikrig, E. (2000). *Borrelia burgdorferi* gene expression *in vivo* and spirochete pathogenicity. Infect. Immun. *68*, 1222–1230.

Anguita, J., Thomas, V., Samanta, S., Persinski, R., Hernanz, C., Barthold, S.W., and Fikrig, E. (2001). *Borrelia burgdorferi*-induced inflammation facilitates spirochete adaptation and variable major protein-like sequence locus recombination. J. Immunol. *167*, 3383–3390.

Appel, M.J.G., Allan, S., Jacobson, R.H., Lauderdale, T.L., Chang, Y.F., Shin, S.J., Thomford, J.W., Todhunter, R.J., and Summers, B.A. (1993). Experimental Lyme disease in dogs produces arthritis and persistent infection. J. Infect. Dis. *167*, 651–664.

Armstrong, A.L., Barthold, S.W., Persing, D.H., and Beck, D.S. (1992). Carditis in Lyme disease susceptible and resistant strains of laboratory mice infected with *Borrelia burgdorferi*. Am. J. Trop. Med. Hyg. *47*, 249–258.

Asch, E.S., Bujak, D.I., Weiss, M., Peterson, M.G.E., and Weinstein, A. (1994). Lyme disease: an infectious and postinfectious syndrome. J Rheumatol *21*, 454–461.

Bai, Y., Narayan, K., Dail, D., Sondey, M., Hodzic, E., Barthold, S.W., Pachner, A.R., and Cadavid, D. (2004). Spinal cord involvement in the nonhuman primate model of Lyme disease. Lab. Invest. *84*, 160–172.

Bankhead, T., and Chaconas, G. (2007). The role of VlsE antigenic variation in the Lyme disease spirochete: persistence through a mechanism that differs from other pathogens.. Mol. Microbiol. *65*, 1547–1558.

Barbour, A.G. (1984). Isolation and cultivation of Lyme disease spirochetes. Yale J. Bio.l Med. *57*, 521–525.

Barbour, A.G., and Hayes, S.F. (1986). Biology of *Borrelia* species. Microbiol. Rev. *50*, 381–400.

Barthold, S.W. (1991). Infectivity of *Borrelia burgdorferi* relative to route of inoculation and genotype in laboratory mice. J. Infect. Dis. *163*, 419–420.

Barthold, S.W. (1993). Antigenic stability of *Borrelia burgdorferi* during chronic infections of immunocompetent mice. Infect. Immun. *61*, 4955–4961.

Barthold, S.W. (1996). Lyme borreliosis in the laboratory mouse. J. Tick-Borne Spirochetal Dis. *3*, 22–44.

Barthold, S.W. (1999). Specificity of infection-induced immunity among *Borrelia burgdorferi* sensu lato species. Infect. Immun. *67*, 36–42.

Barthold, S.W., Beck, D.S., Hansen, G.M., Terwilliger, G.A., and Moody, K.D. (1990a). Lyme borreliosis in selected strains and ages of laboratory mice. J. Infect. Dis. *162*, 133–138.

Barthold, S.W., and Bockenstedt, L.K. (1993). Passive immunizing activity of sera from mice infected with *Borrelia burgdorferi*. Infect. Immun. *61*, 4696–4702.

Barthold, S.W., deSouza, M., and Feng, S. (1996). Serum-mediated resolution of Lyme arthritis in mice. Lab. Invest. *74*, 57–67.

Barthold, S.W., and deSouza, M.S. (1995). Exacerbation of Lyme arthritis in beige mice. J. Infect. Dis. *172*, 778–784.

Barthold, S.W., deSouza, M.S., Fikrig, E., and Persing, D.H. (1992a). Lyme borreliosis in the laboratory mouse. In Lyme Disease: Molecular and Immunological Approaches, S.E. Schutzer, ed. (Cold Spring Harbor, NY, Cold Spring Harbor Laboratory Press), pp. 223–242.

Barthold, S.W., deSouza, M.S., Janotka, J.L., Smith, A.L., and Persing, D.H. (1993). Chronic Lyme borreliosis in the laboratory mouse. Am. J. Pathol. *143*, 951–971.

Barthold, S.W., Feng, S., Bockenstedt, L.K., Fikrig, E., and Feen, K. (1997). Protective and arthritis-resolving activity in serum from mice infected with *Borrelia burgdorferi*. Clin. Infect. Dis. *25*, S9–S17.

Barthold, S.W., Fikrig, E., Bockenstedt, L.K., and Persing, D.H. (1995a). Circumvention of outer surface protein A immunity by host-adapted *Borrelia burgdorferi*. Infect. Immun. *63*, 2255–2261.

Barthold, S.W., Hodzic, E., Tunev, S., and Feng, S. (2006). Antibody-mediated disease remission in the mouse model of Lyme borreliosis. Infect. Immun. *74*, 4817–4825.

Barthold, S.W., Levy, S.A., Fikrig, E., Bockenstedt, L.K., and Smith, A.L. (1995b). Serologic response of naturally exposed or vaccinated dogs to *Borrelia burgdorferi*, the agent of Lyme borreliosis. J. Am. Vet. Med. Assoc. *207*, 1435–1440.

Barthold, S.W., Moody, K.D., and Beck, D.S. (1990b). Susceptibility of laboratory rats to isolates of *Borrelia burgdorferi* from different geographic areas. Am. J. Trop. Med. Hyg. *42*, 596–600.

Barthold, S.W., Moody, K.D., Terwilliger, G.A., Duray, P.H., Jacoby, R.O., and Steere, A.C. (1988a). Experimental Lyme arthritis in rats infected with *Borrelia burgdorferi*. J. Infect. Dis. *157*, 842–846.

Barthold, S.W., Moody, K.D., Terwilliger, G.A., Jacoby, R.O., and Steere, A.C. (1988b). An animal model for Lyme borreliosis. Ann. N.Y. Acad. Sci. *539*, 264–273.

Barthold, S.W., Persing, D.H., Armstrong, A.L., and Peeples, R.A. (1991). Kinetics of *Borrelia burgdorferi* dissemination and evolution of disease following intradermal inoculation of mice. Am. J. Pathol. *139*, 263–273.

Barthold, S.W., Sidman, C.L., and Smith, A.L. (1992b). Lyme borreliosis in genetically resistant and susceptible mice with severe combined immunodeficiency. Am. J. Trop. Med. Hyg. *47*, 605–613.

Batsford, S., Dunn, J., and Mihatsch, M. (2004). Outer surface lipoproteins of *Borrelia burgdorferi* vary in their ability to induce experimental joint injury. Arthritis Rheum *50*, 2360–2369.

Behera, A.K., Hildebrand, E., Bronson, R.T., Perides, G., Uematsu, S., Akira, S., and Hu, L.T. (2006). MyD88 deficiency results in tissue-specific changes in cytokine induction and inflammation in interleukin-18-independent mice infected with *Borrelia burgdorferi*. Infect. Immun. *74*, 1462–1470.

Belperron, A.A., Dailey, C.M., and Bockenstedt, L.K. (2005). Infection-induced marginal zone B-cell production of *Borrelia hermsii*-specific antibody is impaired in the absence of CD1d. J. Immunol. *174*, 5681–5686.

Benach, J.L., Bosler, E.M., Coleman, J.L., and Habicht, G.S. (1984). Experimental transmission of the Lyme disease spirochete to rabbits. J. Infect. Dis. *150*, 786–787.

Benhnia, M.R., Wroblewski, D., Akhtar, M.N., Patel, R.A., Lavezzi, W., Gangloff, S.C., Goyert, S.M., Caimano, M.J., Radolf, J.D., and Sellati, T.J. (2005). Signaling through CD14 attenuates the inflammatory response to *Borrelia burgdorferi*, the agent of Lyme disease. J. Immunol. *174*, 1539–1548.

Berger, B.W. (1989). Dermatologic manifestations of Lyme disease. Rev. Infect. Dis. *6*, S1475–1481.

Blevins, J.S., Revel, A.T., Caimano, M.J., Yang, X.F., Richardson, J.A., Hagman, K.E., and Norgard, M.V. (2004). The *luxS* gene is not required for *Borrelia burgdorferi* tick colonization, transmission to a mammalian host, or induction of disease. Infect. Immun. *72*, 4864–4867.

Bockenstedt, L.K., Barthold, S.W., Deponte, K., Marcantonio, N., and Kantor, F.S. (1993). *Borrelia burgdorferi* infection and immunity in mice deficient in the fifth component of complement. Infect. Immun. *61*, 2104–2107.

Bockenstedt, L.K., Kang, I., Shanafelt, M.-C., Chang, C., Persing, D., Hayday, A., and Barthold, S.W. (2001). $CD4^+$ Th1 cells facilitate regression of murine Lyme carditis. Infect. Immun. *69*, 5264–5269.

Bockenstedt, L.K., Liu, N., Schwartz, I., and Fish, D. (2006). MyD88 defciiency enhances acquisition and transmission of *Borrelia burgdorferi* by *Ixodes scapularis* ticks. Infect. Immun. *74*, 2154–2160.

Bockenstedt, L.K., Mao, J., Hodzic, E., Barthold, S.W., and Fish, D. (2002). Detection of attenuated, non-infectious spirochetes after antibiotic treatment of *Borrelia burgdorferi*-infected mice. J. Infect. Dis. *186*, 1430–1437.

Bockenstedt, L.K., Shanafelt, M.C., Belperron, A., Mao, J., and Barthold, S.W. (2003). Humoral immunity reflects altered T helper cell bias in *Borrelia burgdorferi*-infected γ/δ T-cell-deficient mice.. Infect. Immun. *71*, 2938–2940.

Bohm Jr, R.P., Dennis, V., Blanchard, J.L., Philipp, M.T., and Gerber, M.A. (1999). Clinical outcome of a protocol to produce immunosuppression in rhesus macaques (*Macaca mulatta*): application to infectious disease and gene therapy studies. J. Med. Primatol. *28*, 344–352.

Bolz, D.D., Sundsbak, R.S., Ma, Y., Akira, S., Kirschning, C.J., Zachary, J.F., Weis, J.H., and Weis, J.J. (2004). MyD88 plays a unique role in host defense but not arthritis development in Lyme disease. J. Immunol. *173*, 2003–2010.

Bolz, D.D., and Weis, J.J. (2004). Molecular mimicry to *Borrelia burgdorferi*: pathway to autoimmunity? Autoimmunity *37*, 387–392.

Botkin, D.J., Abbott, A.N., Stewart, P.E., Rosa, P.A., Kawabata, H., Watanabe, H., and Norris, S.J. (2006). Identification of potential virulence determinants by *Himar1* transposition of infectious *Borrelia burgdorferi* B31.. Infect. Immun. *74*, 6690–6699.

Brown, C.R., Blaho, V.A., Fritsche, K.L., and Loiacono, C.M. (2006). Stat1 deficiency exacerbates carditis but not arthritis during experimental Lyme borreliosis. J Interferon Cytokine Res. *26*, 390–399.

Brown, C.R., Blaho, V.A., and Loiacono, C.M. (2003). Susceptibility to experimental Lyme arthritis correlates with KC and monocyte chemoattractant protein-1 production in joints and requires neutrophil recruitment via CXCR2. J. Immunol. *171*, 893–901.

Brown, C.R., and Reiner, S.L. (1998). Clearance of *Borrelia burgdorferi* may not be required for resistance to experimental Lyme arthritis. Infect. Immun. *66*, 2065–2071.

Brown, C.R., and Reiner, S.L. (1999a). Development of Lyme arthritis in mice deficient in inducible nitric oxide synthase. J. Infect. Dis. *179*, 1573–1576.

Brown, C.R., and Reiner, S.L. (1999b). Genetic control of experimental Lyme arthritis in the absence of specific immunity. Infect. Immun. *67*, 1967–1973.

Brown, C.R., and Reiner, S.L. (2000). Genes outside the major histocompatibility complex control resistance and susceptibility to experimental Lyme arthritis. Med. Microbiol. Immunol *189*, 85–90.

Brown, E.L., Wooten, R.M., Johnson, B.J.B., Iozzo, R.V., Smith, A., Dolan, M.C., Guo, B.P., Weis, J.J., and Hook, M. (2001). Resistance to Lyme disease in decorin-deficient mice. J. Clin. Invest. *107*, 845–852.

Brown, J.P., Zachary, J.F., Teuscher, C., Weis, J.J., and Wooten, R.M. (1999). Dual role of interleukin-10 in murine Lyme disease: regulation of arthritis severity and host defense. Infect. Immun. *67*, 5142–5150.

Bunikis, I., Denker, K., Östberg, Y., Andersen, C., Benz, R., and Bergström, S. (2008). An RND-type efflux system in *Borrelia burgdorferi* is involved in virulence and resistance to antimicrobial compounds. PLoS Pathog. *4*, e1000009.

Burchill, M.A., Nardelli, D.T., England, D.M., DeCoster, D.J., Christopherson, J.A., Callister, S.M., and Schell, R.F. (2003). Inhibition of interleukin-17 prevents the development of arthritis in vaccinated mice challenged with *Borrelia burgdorferi*. Infect. Immun. *71*, 3437–3442.

Burgdorfer, W. (1984). The New Zealand white rabbit: an experimental host for infecting ticks with Lyme disease spirochetes. Yale J Biol Med *57*, 609–612.

Burgdorfer, W., Barbour, A.G., Hayes, S.F., Benach, J.L., Grunwalt, E., and Davis, J.P. (1982). Lyme disease-A tickborne spirochetosis? Science *216*, 1317–1319.

Burgdorfer, W., and Gage, K.L. (1987). Susceptibility of the hispid cotton rat (*Sigmodon hispidus*) to the Lyme disease spirochete (*Borrelia burgdorferi*). Am. J. Trop. Med. Hyg. *37*, 624–628.

Burgess, E.C. (1986). Experimental inoculation of dogs with *Borrelia burgdorferi*. Zentralbl. Bakt. Mikrobiol. Hyg. *263*, 49–54.

Burgess, E.C., Gendron-Fitzpatrick, A., and Wright, W.O. (1987). Arthritis and systemic disease caused by *Borrelia burgdorferi* infection in a cow. J. Am. Vet. Med. Assoc. *191*, 1468–1470.

Cadavid, D. (2006). The mammalian host response to borrelia infection. Wien Klin. Wochenschr. *118*, 653–658.

Cadavid, D., Bai, Y., Dail, D., Hurd, M., Narayan, K., Hodzic, E., Barthold, S.W., and Pachner, A.R. (2003). Infection and inflammation in skeletal muscle from nonhuman primates infected with different genospecies of the Lyme disease spirochete *Borrelia burgdorferi*. Infect. Immun. *71*, 7087–7098.

Cadavid, D., Bai, Y., Hodzic, E., Narayan, K., Barthold, S.W., and Pachner, A.R. (2004). Cardiac involvement in non-human primates infection with the Lyme disease spirochete *Borrelia burgdorferi*.. Lab. Invest. *84*, 1439–1450.

Cadavid, D., and Barbour, A.G. (1998). Neuroborreliosis during relapsing fever: review of the clinical manifestations, pathology, and treatment of infections in humans and experimental animals. Clin. Infect. Dis. *26*, 151–164.

Cadavid, D., Bundoc, V., and Barbour, A.G. (1993). Experimental infection of the mouse brain by a relapsing fever *Borrelia* species – a molecular analysis. J. Infect. Dis. *168*, 143–151.

Cadavid, D., Garcia, E., and Gelderbom, H. (2007). Coinfection with *Borrelia turicatae* serotype 2 prevents the severe vestibular dysfunction and earlier mortality caused by serotype 1. J. Infect. Dis. *195*, 1686–1693.

Cadavid, D., O'Neill, T., Schaefer, H., and Pachner, A.R. (2000). Localization of *Borrelia burgdorferi* in the nervous system and other organs in a nonhuman primate model of Lyme disease. Lab. Invest. *80*, 1043–1054.

Cadavid, D., Pachner, A.R., Estanislao, L., Patalapati, R., and Barbour, A.G. (2001). Isogenic serotypes of *Borrelia turicatae* show different localization in the brain and skin of mice. Infect. Immun. *69*, 3389–3397.

Cadavid, D., Pennington, P.M., Kerentseva, T.A., Bergström, S., and Barbour, A.G. (1997). Immunologic and genetic analyses of VmpA of a neurotropic strain of *Borrelia turicatae*. Infect. Immun. *65*, 3352–3360.

Cadavid, D., Sondey, M., Garcia, E., and Lawson, C.L. (2006). Residual brain infection in relapsing-fever borreliosis. J. Infect. Dis. *193*, 1451–1458.

Cadavid, D., Thomas, D.D., Crawley, R., and Barbour, A.G. (1994). Variability of a bacterial surface protein and disease expression in a possible mouse model of systemic Lyme borreliosis. J. Exp. Med. *179*, 631–642.

Caimano, M.J., Eggers, C.H., Hazlett, K.R., and Radolf, J.D. (2004). RpoS is not central to the general stress response in *Borrelia burgdorferi* but does control expression of one or more essential virulence determinants. Infect. Immun. *72*, 6433–6445.

Callister, S.M., Agger, W.A., Schell, R.F., and Brand, K.M. (1989). Efficacy of the urinary bladder for isolation of *Borrelia burgdorferi* from naturally infected, wild *Peromyscus leucopus*. J. Clin. Microbiol. *27*, 773–774.

Callister, S.M., Schell, R.F., Case, K.L., Lovrich, S.D., and Day, S.P. (1993). Characterization of the borreliacidal antibody response to *Borrelia burgdorferi* in humans: a serodiagnostic test. J. Infect. Dis. *167*, 158–164.

Cardiff, R.D., Ward, J.M., and Barthold, S.W. (2008). 'One medicine'–'one pathology': are veterinary and human pathology prepared? Lab. Invest. *88*,18–16.

Cassatt, D.R., Patel, N.K., Ulbrandt, N.D., and Hanson, M.S. (1998). DbpA, but not OspA, is expressed by *Borrelia burgdorferi* during spirochemia and is a target for protective antibodies. Infect. Immun. *66*, 5379–5387.

Chang, Y.F., Appel, M.J.G., Jacobson, R.H., shin, S.J., Harpending, P., Straubinger, R., Patrician, L.A., Mohammed, H., and Summers, B.A. (1995). Recombinant OspA protects dogs against infection and disease caused by *Borrelia burgdorferi*. Infect. Immun. *63*, 3543–3549.

Chang, Y.F., Ku, Y.W., Chang, C.F., Chang, C.D., McDonough, S.P., Divers, T., Pough, M., and Torres, A. (2005). Antibiotic treatment of experimentally *Borrelia burgdorferi*-infected ponies. Vet. Microbiol. *107*, 285–294.

Chang, Y.F., Novosel, V., Chang, C.F., Summers, B.A., Ma, D.P., Chiang, Y.W., Acree, W.M., Chu, H.J., Shin, S., and Lien, D.H. (2001). Experimental induction of chronic borreliosis in adult dogs exposed to *Borrelia burgdorferi*-infected ticks and treated with dexamethasone. Am. J. Vet. *Res. 62*, 1104–1112.

Chang, Y.F., Novosol, V., McDonough, S.P., Chang, C.F., Jacobson, R.H., Divers, T., Quimby, F.W., Shin, S., and Lein, D.H. (2000). Experimental infection of ponies with *Borrelia burgdorferi* by exposure to Ixodid ticks. Ve.t Pathol. *37*, 68–76.

Chang, Y.F., Straubinger, R.K., Jacobson, R.H., Kim, J.B., Kim, D., Shin, S.J., and Appel, M.J.G. (1996). Dissemination of *Borrelia burgdorferi* after experimental infection in dogs. J. Spirochetal Tick Borne Dis. 3, 80–86.

Chaves, A.A., Keller, W.J., O'Sullivan, S., Williams, M.A., Fitzgerald, L.E., McPherson, H.E., Goykhman, D., Ward, P.D., Hoe, C.M., Mixson, L., *et al.* (2006). Cardiovascular monkey telemetry: sensitivity to detect QT interval prolongation. J. Pharmacol. Toxicol. Methods *54*, 150–158.

Chong-Cerillo, C., Wu, S.-Y., Wang, Y.-P., Blanco, D.R., and Lovett, M.A. (2000). Immune serum from rabbits infected with *Borrelia burgdorferi* B31 confers complete passive protection against homologous challenge. J. Spirochetal Tick-Borne Dis. *7*, 3–9.

Chong-Cerrillo, C., Shang, E.S., Blanco, D.R., Lovett, M.A., and Miller, J.N. (2001). Immunohistochemical analysis of Lyme disease in the skin of naive and infection-immune rabbits following challenge. Infect. Immun. *69*, 4094–4102.

Chou, J., Wunschmann, A., Hodzic, E., and Borjesson, D.L. (2006). Detection of *Borrelia burgdorferi* DNA in tissues from dogs with presumptive Lyme borreliosis. J. Am. Vet. Med. Assoc. *229*, 1260–1265.

Christopherson, J.A., Munson, E.L., England, D.M., Croke, C.L., Remington, M.C., Molitor, M.L., DeCoster, D.J., Callister, S.M., and Schell, R.F. (2003). Destructive arthritis in vaccinated interferon gamma-deficient mice challenged with *Borrelia burgdorferi*: modulation by tumor necrosis factor alpha. Clin. Diag. Lab. Immunol. *10*, 44–52.

Chu, H.J., Chavez Jr, L.G., Blumer, B.M., Sebring, R.W., Wasmoen, T.L., and Acree, W.M. (1992). Immmunogenicity and efficacy study of a commercial *Borrelia burgdorferi* bacterin.. J. Am. Vet. Med. Assoc. *201*, 403–411.

Cinco, M., and Murgia, R. (2006). Evaluation of the C6 enzyme-linked immunosorbent assay for the serodiagnosis of Lyme borreliosis in north-eastern Italy. New Microbiol. *29*, 139–141.

Coleman, J.L., Gebbia, J.A., Piesman, J., Degen, J.L., Bugge, T.H., and Benach, J.L. (1997). Plasminogen is required for efficient dissemination of *Borrelia burgdorferi* in ticks and for enhancement of spirochetemia in mice. Cell *89*, 1111–1119.

Connolly, S.E., and Benach, J.L. (2001). Cutting edge: the spirochetemia of murine relapsing fever is cleared by complement-independent bacteriocidal antibodies. J. Immunol. *167*, 3029–3032.

Cooper, P.J., Fekade, D., Remick, D.G., Grint, P., Wherry, J., and Griffin, G.E. (2000). Recombinant human interleukin-10 fails to alter proinflammatory cytokine production or physiologic changes associated with the Jarisch-Herxheimer reaction. J. Infect. Dis. *181*, 203–209.

Coughlin, R.T., Fish, D., Mather, T.N., Ma, J.N., Pavia, C., and Bulger, P. (1995). Protection of dogs from Lyme disease with a vaccine containing outer surface protein (Osp) A, OspB and the saponin adjuvant QS21. J. Infect. Dis. *171*, 1049–1052.

Craig-Mylius, K., Weber, G.F., Coburn, J., and Glickstein, L. (2005). *Borrelia burgdorferi*, an extracellular pathogen, circumvents osteopontin in inducing an inflammatory cytokine response. J. Leukoc. Biol. *77*, 710–718.

Crandall, H., Ma, Y., Dunn, D.M., Sundsbak, R.S., Zachary, J.F., Olofsson, P., Holmdahl, R., Weis, J.H., Weiss, R.B., Teucher, C., and Weis, J.J. (2005). *Bb2Bb3* regulation of murine Lyme arthritis is distinct from *Ncf1* and independent of the phagocyte nicotinamide adenine dinucleotide phosphate oxidase. Am. J. Pathol. *167*, 775–785.

Croke, C.L., Munson, E.L., Lovrich, S.D., Christopherson, J.A., Remington, M.C., England, D.M., Callister, S.M., and Schell, R.F. (2000). Occurrence of severe destructive Lyme arthritis in hamsters vaccinated with outer surface protein A and challenged with *Borrelia burgdorferi*.. Infect. Immun. *68*, 658–663.

Crowley, H., Alroy, J., Sproule, T.J., Roopenian, D., and Huber, B.T. (2006). The MHC class I-related FcRn ameliorates murine Lyme arthritis. International Immunol *18*, 409–414.

Crowley, T.R., Champion, C.I., Whitelegge, J.P., Aguilera, R., Wu, X.Y., Blanco, D.R., Miller, J.N., and Lovett, M.A. (2004). Temporal analysis of the antigenic composition of *Borrelia burgdorferi* during infection in rabbit skin. Infect. Immun. *72*, 5063–5072.

Czub, S., Duray, P.H., Thomas, R.E., and Schwan, T.G. (1992). Cystitis induced by infection with the Lyme disease spirochete, *Borrelia burgdorferi*, in mice. Am J. Pathol. *141*, 1173–1179.

Dambach, D.M., Smith, C.A., Lewis, R.M., and VanWinkle, T.J. (1997). Morphologic, immunohistochemical, and ultrastructural characterization of a distinctive renal lesion in dogs putatively associated with *Borrelia burgdorferi* infection: 49 cases (1987–1992). Vet. Pathol. *34*, 85–96.

Defosse, D.L., Duray, P.H., and Johnson, R.C. (1992). The NIH-3 immunodeficient mouse is a model for Lyme borreliosis myositis and carditis. Am. J. Pathol. *141*, 3–10.

deSilva, A., and Fikrig, E. (1995). Growth and migration of *Borrelia burgdorferi* in *Ixodes* ticks during blood feeding. Am. J. Trop. Med. Hyg. 53, 397–404.

deSilva, A., Fikrig, E., Hodzic, E., Telford, S.R. III, and Barthold, S.W. (1998). Immune evasion by tick-borne and host-adapted *Borrelia burgdorferi*. J. Infect. Dis. *177*, 395–400.

deSouza, M.S., Smith, A.L., Beck, D.S., Kim, L.J., Hansen, G.M., Jr., and Barthold, S.W. (1993). Variant responses of mice to *Borrelia burgdorferi* depending on the site of intradermal inoculation. Infect. Immun. *61*, 4493–4497.

Dolan, M.C., Piesman, J., Schneider, B.S., Schriefer, M., Brandt, K., and Zeidner, N.S. (2004). Comparison of disseminated and nondisseminated strains of *Borrelia burgdorferi* sensu stricto in mice naturally infected by tick bite. Infect. Immun. *72*, 5262–5266.

Donahue, J.G., Piesman, J., and Spielman, A. (1987). Reservoir competence of white-footed mice for Lyme disease spirochetes. Am. J. Trop. Med. Hyg. *36*, 92–96.

DuChateau, B.K., England, D.M., Callister, S.M., Lim, L.C., Lovrich, S.D., and Schell, R.F. (1996). Macrophages exposed to *Borrelia burgdorferi* induce Lyme arthritis in hamsters. Infect. Immun. *64*, 2540–2547.

DuChateau, B.K., Jensen, J.R., England, D.M., Callister, S.M., Lovrich, S.D., and Schell, R.F. (1997). Macrophages and enriched populations of T-lymphocytes interact synergistically for the induction of severe, destructive Lyme arthritis. Infect. Immun. *65*, 2829–2836.

DuChateau, B.K., Munson, E.L., England, D.M., Lovrich, S.D., Callister, S.M., Jensen, J.R., and Schell, R.F. (1999). Macrophages interact with enriched population of distinct T-lymphocyte subsets for the induction of severe destructive Lyme arthritis. J. Leukoc. Biol. *65*, 162–170.

Duray, P.H., and Johnson, R.C. (1986). The histopathology of experimentally infected hamsters with the Lyme disease spirochete, *Borrelia burgdorferi*. Proc. Soc. Exp. Biol. Med. *181*, 263–269.

Elias, A.F., Stewart, P.E., Grimm, D., Caimano, M.J., Eggers, C.H., Tilly, K., Bono, J.L., Akins, D.R., Radolph, J.D., Schwan, T.G., *et al.* (2002). Clonal polymorphism of *Borrelia burgdorferi* strain B31 MI: implications for mutagenesis in an infectious strain background. Infect. Immun. *70*, 2139–2150.

Embers, M.E., Liang, F.T., Howell, J.K., Jacobs, M.B., Purcell, J.E., Norris, S.J., Johnsonn, B.J.B., and Philipp, M.T. (2007). Antigenicity and recombination of VlsE, the antigenic variation protein of *Borrelia burgdorferi*, in rabbits, a host putatively resistannt to long-term infection with this spirochete. FEMS Immunol. Med. Microbiol. *50*, 421–429.

England, J.D., Bohm, R.P., Roberts, E.D., and Philipp, M.T. (1997). Mononeuropathy multiplex in rhesus monkeys with chronic Lyme disease. Ann. Neurol. *41*, 375–384.

Evans, R.D., Bosler, E.M., Orthel, F., Robertson, J.L., Schneider, E.M., LeFebvre, R.B., and Graham, M.D. (1995a). Canine Lyme borreliosis. I. Gross clinical observations of laboratory beagles following exposure to ticks infected with *Borrelia burgdorferi*. J. Spirochetal Tick Borne Dis. *2*, 28–32.

Evans, R.D., Robertson, J.L., Graham, M.D., Bosler, E.M., LeFebvre, R.B., and Scheider, E.M. (1995b). Canine Lyme borreliosis. II. Minimal lesions in tissues of laboratory beagles following infection by exposure to Ixodid ticks infected with *Borrelia burgdorferi*. J Spirochetal Tick Borne Dis *2*, 33–36.

Exner, M.M., Wu, X., Blanco, D.R., and Miller, J.N. (2000). Protection elicited by native outer membrane protein Oms66 (P66) against host-adapted *Borrelia burgdorferi*: Conformational nature of bactericidal epitopes. Infect. Immun. *68*, 2647–2654.

Feng, S., Barthold, S.W., Bockenstedt, L.K., Zaller, D.M., and Fikrig, E. (1995). Lyme disease in human DR4Dw4-transgenic mice. J. Infect. Dis. *172*, 286–289.

Feng, S., Hodzic, E., and Barthold, S.W. (2000). Lyme arthritis resolution with antiserum to a 37-kilodalton *Borrelia burgdorferi* protein. Infect. Immun. *68*, 4169–4173.

Feng, S., Hodzic, E., Freet, K., and Barthold, S.W. (2003). Immunogenicity of *Borrelia burgdorferi* arthritis-related protein. Infect. Immun. *71*, 7211–7214.

Feng, S., Hodzic, E., Stevenson, B., and Barthold, S.W. (1998). Humoral immunity to *Borrelia burgdorferi* N40 decorin binding proteins during infection of laboratory mice. Infect. Immun. *66*, 2827–2835.

Fikrig, E., and Barthold, S.W. (1997). Lyme disease in transgenic mice expression the *Borrelia burgdorferi* flagellin epitope implicated in human neuroborreliosis. FEMS Microbiol. Lett. *148*, 137–143.

Fikrig, E., Barthold, S.W., Chen, M., Chang, C.-H., and Flavell, R.A. (1997a). Protective antibodies develop, and murine Lyme arthritis regresses, in the absence of MHC class II and CD4+ T-cells. J. Immunol. *159*, 5682–5686.

Fikrig, E., Barthold, S.W., Chen, M., Grewal, I.S., Craft, J., and Flavell, R.A. (1996a). Protective antibodies in murine Lyme disease arise independently of CD40 ligand. J. Immunol. *157*, 1–3.

Fikrig, E., Barthold, S.W., Chen, M., Tao, M., Ali-Salaam, P., Bockenstedt, L.K., and Flavell, R.A. (1997b). Lyme borreliosis in transgenic mice tolerant to OspA from *Borrelia burgdorferi* 25015. J. Infect. Dis. *175*, 1000–1003.

Fikrig, E., Bockenstedt, L.K., Barthold, S.W., Chen, M., Tao, H., Ali-Salaam, P., Telford, S.R., and Flavell, R.A. (1994). Sera from patients with chronic Lyme disease protect mice from Lyme borreliosis. J. Infect. Dis. *169*, 568–574.

Fikrig, E., Tao, H., Chen, M., Barthold, S.W., and Flavell, R.A. (1996b). Lyme arthritis in transgenic mice tolerant to *Borrelia burgdorferi* OspA or OspB. J. Clin. Invest. *96*, 1706–1714.

Fisher, M.A., Grimm, D., Henion, A.K., Elias, A.F., Stewart, P.E., Rosa, P.A., and Gherardini, F.C. (2005). *Borrelia burgdorferi* σ_{54} is required for mammalian infection and vector transmission but not for tick colonization. Proc. Natl. Acad. Sci. U.S.A. *102*, 5162–5167.

Fleming, R.V., Marques, A.R., Klempner, M.S., Schmid, C.H., Dally, L.G., Martin, D.S., and Philipp, M.T. (2004). Pre-treatment and post-treatment assessment of the C(6) test in patients with persistent symptoms and a history of Lyme borreliosis. Eur J. Clin. Microbiol. Infect Dis 23, 615–618.

Foley, D.M., Gayek, R.J., Skare, J.T., Wagar, E.A., Champion, C.I., Blanco, D.R., Lovett, M.A., and Miller, J.N. (1995). Rabbit model of Lyme borreliosis: Erythema migrans, infection-derived immunity, and identification of *Borrelia burgdorferi* proteins associated with virulence and protective immunity. J. Clin. Invest. *96*, 965–975.

Foley, D.M., Wang, Y.P., Wu, X.Y., Blanco, D.R., Lovett, M.A., and Miller, J.M. (1997). Acquired resistance to *Borrelia burgdorferi* infection in the rabbit. Comparison between outer surface protein A vaccine- and infection-derived immunity. J. Clin. Invest. *99*, 2030–2035.

Galbe, J.L., Guy, E., Zapatero, J.M., Peerschke, E.I.B., and Benach, J.L. (1993). Vascular clearance of *Borrelia burgdorferi* in rats. Microb. Pathogen. *14*, 187–201.

García-Moncó, J.C., FernandezVillar, B., CalvoAlen, J., and Benach, J.L. (1990). *Borrelia burgdorferi* in the central nervous system: experimental and clinical evidence for early invasion. J. Infect. Dis. *161*, 1187–1193.

García-Moncó, J.C., Miller, N.S., Backenson, P.B., Anda, P., and Benach, J.L. (1997). A mouse model of *Borrelia* meningitis after intradermal injection. J. Infect. Dis. *175*, 1243–1245.

Gebbia, J.A., Monco, J.C., Degen, J.L., Bugge, T.H., and Benach, J.L. (1999). The plasminogen activation system enhances brain and heart invasion in murine relapsing fever borreliosis. J. Clin. Invest. *103*, 81–87.

Gelderblom, H., Londono, D., Bai, Y., Cabral, E.S., Quandt, J., Hornung, R., Martin, R., Marques, A., and Cadavid, D. (2007a). High production of CXCL13 in blood and brain during persistent infection with the relapsing fever spirochete *Borrelia turicatae*.. J. Neuropath. Exp. Neurol. *66*, 208–217.

Gelderblom, H., Schmidt, J., Londono, D., Bai, Y., Quandt, J., Hornung, R., Marques, A., Martin, R., and Cadavid, D. (2007b). Role of interleukin 10 during presistent infection with the relapsing fever spirochete *Borrelia turicatae*. Am. J. Pathol. *170*, 251–262.

Gern, L., Schaible, U.E., and Simon, M.M. (1993). Mode of inoculation of the Lyme disease agent *Borrelia burgdorferi* influences infection and immune responses in inbred strains of mice. J. Infect. Dis. *167*, 971–975.

Glickstein, L., Edelstein, M., and Dong, J.Z. (2001). Gamma interferon is not required for arthritis resistance in the murine Lyme disease model. Infect. Immun. *69*, 3737–3743.

Golde, W.T., Burkot, T.R., Sviat, S., Keen, M.G., Mayer, L.W., Johnson, B.J.B., and Piesman, J. (1993). The major histocompatibility complex-restricted response of recombinant inbred strains of mice to natural and tick transmission of *Borrelia burgdorferi*. J. Exp. Med. *177*, 9–17.

Golde, W.T., Kappel, K.J., Dequesne, G., Feron, C., Plainchamp, D., Capiau, C., and Lobet, Y. (1994). Tick transmission of *Borrelia burgdorferi* to inbred strains of mice induces an antibody response to P39 but not to outer surface protein A. Infect. Immun. *62*, 2625–2627.

Gondolf, K.B., Mihatsch, M., Curscellas, E., Dunn, J.J., and Batsford, S.R. (1994). Induction of experimental allergic arthritis with outer surface proteins of *Borrelia burgdorferi*. Arthritis Rheum *37*, 1070–1077.

Goodman, J.L., Jurkovich, P., Kodner, C., and Johnson, R.C. (1991). Persistent cardiac and urinary tract infections with *Borrelia burgdorferi* in experimentally infected Syrian hamsters. J. Clin. Microbiol. *29*, 894–896.

Grimm, D., Eggers, C.H., Caimano, M.J., Tilly, K., Stewart, P.E., Elias, A.F., Radolf, J.D., and Rosa, P.A. (2004a). Experimental assessment of the roles of linear plasmids lp25 and lp28-1 of *Borrelia burgdorferi* throughout the infectious cycle. Infect. Immun. *72*, 5938–5946.

Grimm, D., Elias, A.F., Tilly, K., and Rosa, P.A. (2003). Plasmid stability during *in vitro* propagation of *Borrelia burgdorferi* assessed at a clonal level. Infect. Immun. *71*, 3138–3145.

Grimm, D., Tilly, K., Byram, R., Stewart, P.E., Krum, J.G., Bueschel, D.M., Schwan, T.G., Policastro, P.F., Elias, A.F., and Rosa, P.A. (2004b). Outer-surface protein C of the Lyme disease spirochete: a protein induced in ticks for infection of mammals. Proc. Natl. Acad. Sci. U.S.A. *101*, 3142–3147.

Guerau-de-Arellano, M., Alroy, J., Bullard, D., and Huber, B.T. (2005a). Aggravated Lyme carditis in CD11a$^{-/-}$ and CD11c$^{-/-}$ mice. Infect. Immun. *73*, 7637–7643.

Guerau-de-Arellano, M., Alroy, J., and Huber, B.T. (2005b). β2 integrins control the severity of murine Lyme carditis. Infect. Immun. *73*, 3242–3250.

Gustafson, J.M., Burgess, E.C., Wachel, M.D., and Steinberg, H. (1993). Intrauterine transmission of *Borrelia burgdorferi* in dogs. Am. J. Vet. *Res. 54*, 882–890.

Hagman, K.E., Yang, X., Wikel, S.K., Schoeler, G.B., Caimano, M.J., Radolph, J.D., and Norgard, M.V. (2000). Decorin-binding protein A (DbpA) of *Borrelia burgdorferi* is not protective when immunized mice are challenged via tick infestation and correlates with the lack of DbpA expression by *B. burgdorferi* in ticks. Infect. Immun. *68*, 4759–4764.

Hanson, M.S., Cassatt, D.R., Guo, B.P., Patel, N.K., McCarthy, M.P., Dorward, D.W., and Hook, M. (1998). Active and passive immunity against *Borrelia burgdorferi* decorin binding protein A (DbpA) protects against infection. Infect. Immun. *66*, 2143–2153.

Hedrick, M.N., Olson, C., Conze, D.B., Bates, T.C., Rincon, M., and Anguita, J. (2006). Control of *Borrelia burgdorferi*-specific CD4^{+}-T-cell effector function by interleukin-12- and T-cell receptor-induced p38 mitogen-activated protein kinase activity. Infect. Immun. *74*, 5713–5717.

Hefty, P.S., Brooks, C.S., Jett, A.M., White, G.L., Wikel, S.K., Kennedy, R.C., and Akins, D.R. (2002). OspE-related, OspF-related, and Elp lipoproteins are immunogenic in baboons experimentally infected with *Borrelia burgdorferi* and in human Lyme disease patients.. J. Clin. Microbiol. *40*, 4256–4265.

Hejka, A., Schmitz, J.L., England, D.M., Callister, S.M., and Schell, R.F. (1989). Histopathology of Lyme arthritis in LSH hamsters. Am. J. Pathol. *134*, 1113–1123.

Hodzic, E., Feng, S., Freet, K., and Barthold, S.W. (2003). *Borrelia burgdorferi* population dynamics and prototype gene expression during infection of immunocompetent and immunodeficient mice. Infect. Immun. *71*, 5042–5055.

Hodzic, E., Feng, S., Freet, K.J., Borjesson, D.L., and Barthold, S.W. (2002). *Borrelia burgdorferi* population kinetics and selected gene expression at the host–vector interface. Infect. Immun. *70*, 3382–3388.

Hodzic, E., Feng, S., Holden, K., Freet, K.J., and Barthold, S.W. (2008). Persistence of *Borrelia burgdorferi* following antibiotic treatment in mice. Antimicrob. Agents Chemother. *52*, 1728–1736.

Hodzic, E., Tunev, S., Feng, S., Freet, K., and Barthold, S. (2005). Immunoglobulin-regulated expression of *Borrelia burgdorferi* outer surface protein A *in vivo*. Infect. Immun. *73*, 3313–3321.

Hua, C.M., Cheminade, Y., Perret, J.L., Weynants, V., Lobet, Y., and Gern, L. (2003). Early detection of *Borrelia burgdorferi* sensu lato infection of Balb/c mice by co-feeding *Ixodes ricinus* ticks.. Int. J. Med. Microbiol. *293*, 421–426.

Hubner, A., Revel, A.T., Nolen, D.M., Hagman, K.E., and Norgard, M.V. (2003). Expression of a *luxS* gene is not required for *Borrelia burgdorferi* infection of mice via needle inoculation. Infect. Immun. *71*, 2892–2896.

Hughes, C.A.N., Engstrom, S.M., Coleman, L.A., Kodner, C.B., and Johnson, R.C. (1993). Protective immunity is induced by a *Borrelia burgdorferi* mutant that lacks OspA and OspB. Infect. Immun. *61*, 5115–5122.

Hunfeld, K.P., Burg, S., Hanssen-Huber, C., Karas, M., Brade, V., and Kraiczy, P. (2008). Changes in the expression pattern of structural proteins after exposure of *Borrelia burgdorferi* to penicillin G and doxycycline. Int. J. Med. Microbiol. *298 (S1)*, 325–332.

Hurtenbach, U., Museteanu, C., Gasser, J., Schaible, U.E., and Simon, M.M. (1995). Studies on early events of *Borrelia burgdorferi*-induced cytokine production in immunodeficient SCID mice by using a tissue chamber model for acute inflammation. Int. J. Exp. Pathol. *76*, 111–123.

Jacobs, M.B., Norris, S.J., Phillippi-Falkenstein, K.M., and Philipp, M.T. (2006). Infectivity of the highly transformable BBE02^{-} lp56-mutant of *Borrelia burgdorferi*, the Lyme disease spirochete, via ticks. Infect. Immun. *74*, 3678–3681.

Jacobson, A.C., Ma, Y., Zachary, J.F., Weis, J.J., and Weis, J.H. (2007). Mice lacking CD21 and CD35 proteins mount effective immune responses against *Borrelia burgdorferi* infection. Infect. Immun. *75*, 2075–2078.

Jewett, M.W., Lawrence, K., Bestor, A.C., Tilly, K., Grimm, D., Shaw, P., VanRaden, M., Gherardini, F., and Rosa, P.A. (2007). The critical role of the linear plasmid lp36 in the infectious cycle of *Borrelia burgdorferi*. Mol. Microbiol. *64*, 1358–1374.

Jobe, D.A., Callister, S.M., Limm, L.C.L., Lovrich, S.D., and Schell, R.F. (1994). Ability of canine Lyme disease vaccine to protect hamsters against infection with several isolates of *Borrelia burgdorferi*. J. Clin. Microbiol. *32*, 618–622.

Johnson, B.J.B., Sviat, S.L., Happ, C.M., Dunn, J.J., Frantz, J.C., Mayer, L.W., and Piesman, J. (1995). Incomplete protection of hamsters vaccinated with unlipidated OspA from *Borrelia burgdorferi* infection is associated with low levels of antibody to an epitope defined by mAb La-2. Vaccine *13*, 1086–1094.

Johnson, R.C., Kodner, C., and Russell, M. (1986a). Passive immunization of hamsters against experimental infection with *Borrelia burgdorferi*. Infect. Immun. *53*, 713–716.

Johnson, R.C., Kodner, C., Russell, M., and Duray, P.H. (1986b). Active immunization of hamsters against experimental infection with *Borrelia burgdorferi*. Inf Imm *54*, 897–898.

Johnson, R.C., Kodner, C., Russell, M., and Duray, P.H. (1988). Experimental infection of hamsters with *Borrelia burgdorferi*. Ann. N.Y. Acad. Sci. *539*, 258–263.

Johnson, R.C., Kodner, C.L., and Russell, M.E. (1986c). Vaccination of hamsters against experimental infection with *Borrelia burgdorferi*. Zbl. Bakt. Hyg. A *263*, 45–48.

Johnson, R.C., Marek, N., and Kodner, C. (1984). Infection of Syrian hamsters with Lyme disease spirochetes. J. Clin. Microbiol. *20*, 1099–1101.

Johnston, Y.E., Duray, P.H., Steere, A.C., Kashgarian, M., Buza, J., Malawista, S.E., and Askenase, P.W. (1985). Lyme arthritis: spirochetes found in synovial microangiopathic lesions. Am J. Pathol. *118*, 26–34.

Jonsson, M., Elmros, T., and Berstrom, S. (1995). Subcutaneous implanted chambers in different mouse strains as an animal model to study genetic stability during infection with Lyme disease *Borrelia*. Microbial Pathogenesis *18*, 109–114.

Judge, D.M., LaCroix, J.T., and Perine, P.L. (1974a) Experimental louse-borne relapsing fever in the grivet monkey, *Cercopithecus aeithiops*. I. Clinical course. Am. J. Trop. Med. Hyg. *23*, 957–961.

Judge, D.M., LaCroix, J.T., and Perine, P.L. (1974b). Experimental louse-borne relapsing fever in the grivet monkey, *Cercopithecus aethiops*. II. Pathology. Am. J. Trop. Med. Hyg. *23*, 962–968.

Kaneda, K., Masuzawa, T., Simon, M.M., Isogai, E., Isogai, H., Yasugami, K., suzuki, T., Suzuki, Y., and Yanagihara, Y. (1998). Infectivity and arthritis induction of *Borrelia japonica* in SCID mice and immune competent mice: possible role of galactosylceramide binding activity on initiation of infection. Microbiol. Immunol. *42*, 171–175.

Kang, I., Barthold, S.W., and Bockenstedt, L.K. (1995). Lyme borreliosis in B-cell deficient mice. Arthritis Rheum. *38*, S345.

Kawabata, H., Norris, S.J., and Watanabe, H. (2004). BBE02 disruption mutants of *Borrelia burgdorferi* B31 have a highly transformable, infectious phenotype. Infect. Immun. *72*, 7147–7154.

Kazragis, R.J., Dever, L.L., Jorgensen, J.H., and Barbour, A.G. (1996). *In vivo* activities of ceftriaxone and vancomycin against *Borrelia* spp. in the mouse brain and other sites. Antimicrob. Agents Chemother. *40*, 2632–2636.

KeaneMyers, A., Maliszewski, C.R., Finkelman, F.D., and Nickell, S.P. (1996). Recombinant IL-4 treatment augments resisitance to *Borrelia burgdorferi* infections in both normal susceptible and antibody-deficient susceptible mice. J. Immunol. *156*, 2488–2494.

Keller, D., Koster, F.T., Marks, D.H., Hosbach, P., Erdile, L.F., and Mays, J.P. (1994). Safety and immunogenicity of a recombinant outer surface protein A Lyme vaccine. J. Am. Med. Ass. *27*, 1764–1768.

Klempner, M.S., Hu, L.T., Evans, J., Schmid, C.H., Johnson, G.M., Trevino, R.P., Norton, D., Levy, L., Wall, D., McCall, J., *et al.* (2001). Two controlled trials of antibiotic treatment in patients with persistent symptoms and a history of Lyme disease.. N. Engl. J. Med. *345*, 85–92.

Konishi, T., and Nakao, M. (1997). Antibiotic treatment in mice infected with Japanese *Borrelia garinii*: efficacy of ceftriaxone for eradicating the infection induced by *Ixodes persulcatus* ticks. Microbio.l Immunol. *41*, 165–168.

Kornblatt, A.N., Steere, A.C., and Brownstein, D.G. (1984a). Experimental Lyme disease in rabbits: spirochetes found in erythema migrans and blood. Infect. Immun. *46*, 220–223.

Kornblatt, A.N., Steere, A.C., and Brownstein, D.G. (1984b). Infection in rabbits with the Lyme disease spirochete. Yale J Biol Med *57*, 613–614.

Korshus, J.B., Munderloh, U.G., Bey, R.F., and Kurti, T.J. (2003). Experimental infection of dogs with *Borrelia burgdorferi* sensu stricto using *Ixodes scapularis* ticks artificially infected by capillary feeding. Med Microbiol. Immunol. *193*, 27–34.

Krinsky, W.L., Brown, S.J., and Askenase, P.W. (1982). *Ixodes dammini* skin lesions in guinea pigs and rabbits compared to erythema chronicum migrans in patients with Lyme arthritis.. Exp Parasitol *53*, 381–395.

Kumar, H., Belperron, A., Barthold, S.W., Bendelac, B., and Bockenstedt, L.K. (2000). CD1d deficiency impairs murine host defense against the spirochete, *Borrelia burgdorferi*. J. Immunol. *165*, 4797–4801.

Kurti, T.J., Munderloh, U.G., Hughes, C.A., Engstrom, S.M., and Johnson, R.C. (1996). Resistance to tick-borne spirochete challenge induced by *Borrelia burgdorferi* strains that differ in expression of outer surface proteins. Infect. Immun. *64*, 48–53.

Labandeira-Rey, M., Seshu, J., and Skare, J. (2003). The absence of linear plasmid 25 or 28-1 of *Borrelia burgdorferi* dramatically alters the kinetics of experimental infection via distinct mechanisms. Infect. Immun. *71*, 4608–4613.

Labandeira-Rey, M., and Skare, J.T. (2001). Decreased infectivity in *Borrelia burgdorferi* strain B31 is associated with loss of linear plasmid 25 or 28–1. Infect. Immun. *69*, 446–455.

Larsen, S.A., Steiner, B.M., and Rudolph, A.H. (1995). Laboratory diagnosis and interpretation of tests for syphilis. Clin. Microbiol. *8*, 1–21.

Larsson, C., Andersson, M., Pelkonen, J., Guo, B.P., Nordstrand, A., and Bergström, S. (2006). Persistent brain infection and disease reactivation in relapsing fever borreliosis. Microbes Infect. *8*, 2213–2219.

Lawrenz, M.B., Wooten, R.M., and Norris, S.J. (2004). Effects of *vlsE* complementation on the infectivity of *Borrelia burgdorferi* lacking the linear plasmid lp28-1. Infect. Immun. 72, 6577–6585..

Lawrenz, M.B., Wooten, R.M., Zachary, J.F., Drouin, S.M., Weis, J.J., Wetsel, R.A., and Norris, S.J. (2003). Effect of complement component 3 deficiency on experimental Lyme borreliosis in mice. Infect. Immun. *71*, 4432–4440.

Lazarus, J.J., Meadows, M.J., Lintner, R.E., and Wooten, R.M. (2006). IL-10 deficiency promotes increased *Borrelia burgdorferi* clearance predominantly through enhanced innate immune responses. J. Immunol. *177*.

Lesser, R.L. (1995). Ocular manifestations of Lyme disease. Am. J. Med. *98*, 605–625.

Leuba-Garcia, S., Martinez, R., and Gern, L. (1998). Expression of outer surface proteins A and C of *Borrelia afzelii* in *Ixodes ricinus* ticks and in the skin of mice. Zentralblatt fur Bakteriologe *287*, 475–484.

Levy, S.A., and Duray, P.H. (1988). Complete heart block in a dog seropositive for *Borrelia burgdorferi*. Similarity to human Lyme carditis. J Vet Intern Med *2*, 138–144.

Levy, S.A., and Magnarelli, L.A. (1992). Relationship between development of antibodies to *Borrelia burgdorferi* in dogs and the subsequent development of limb/joint borreliosis. J. Am. Vet. Med. Assoc. *200*, 344–347.

Lewis, K. (2007). Persister cells, dormancy and infectious disease. Nat. Rev. Microbiol. *5*, 48–56.

Li, X., Liu, X., Beck, D.S., Kantor, F.S., and Fikrig, E. (2006). *Borrelia burgdorferi* lacking BBK32, a fibronectin-binding protein, retains full pathogenicity. Infect. Immun. *74*, 3305–3313.

Li, X., Neelakanta, G., Liu, X., Beck, D.S., Kantor, F.S., Fish, D., Anderson, J.F., and Fikrig, E. (2007). Role of outer surface protein D in the *Borrelia burgdorferi* life cycle. Infect. Immun. *75*, 4237–4244.

Liang, F.T., Aberer, E., Cinco, M., Gern, L., Hu, C.M., Lobet, Y.N., Ruscio, M., VoetJr, P.E., Weynants, V.E., and Philipp, M.T. (2000a). Antigenic conservation of an immunodominant invariable region of the VlsE lipoprotein among European pathogenic genospecies of *Borrelia burgdorferi* s.l. J. Infect. Dis. *182*, 1455–1462.

Liang, F.T., Alvarez, A.L., Gu, Y., Nowling, J.M., Ramamoorthy, R., and Philipp, M.T. (1999a). An immunodominant conserved region within the variable domain of VlsE, the variable surface antigen of *Borrelia burgdorferi*. J. Immunol. *163*, 5566–5573.

Liang, F.T., Brown, E.L., Wang, T., Iozzo, R.V., and Fikrig, E. (2004a). Protective niche for *Borrelia burgdorferi* to evade humoral immunity. Am J. Pathol. *165*, 977–985.

Liang, F.T., Jacobs, M.B., Bowers, L.C., and Philipp, M.T. (2002). An immune evasion mechanism for spirochetal persistence in Lyme borreliosis. J. Exp. Med. *195*, 415–422.

Liang, F.T., Jacobson, R.H., Straubinger, R.K., Grooters, A., and Philipp, M.T. (2000b). Characterization of a *Borrelia burgdorferi* VlsE invariable region useful in canine Lyme disease serodiagnosis by enzyme-linked immunosorbent assay. J. Clin. Microbiol. *38*, 4160–4166.

Liang, F.T., Steere, A.C., Marques, A.R., Johnson, B.J.B., Miller, J.N., and Philipp, M.T. (1999b). Sensitive and specific serodiagnosis of Lyme disease by enzyme-linked immunosorbent assay with a peptide based on an immunodominant conserved region of *Borrelia burgdorferi* VlsE. J. Clin. Microbiol. *37*, 3990–3996.

Liang, F.T., Yan, J., Mbow, M.L., Sviat, S.L., Gilmore, R.D., Mamula, M., and Fikrig, E. (2004b). *Borrelia burgdorferi* changes its surface antigenic expression in response to host immune responses. Infect. Immun. *72*, 5759–5767.

Lim, L., England, D., DuChateau, B., Glowacki, N., and Schell, R. (1995a). *Borrelia burgdorferi*-specific T-lymphocytes induce severe destructive Lyme arthritis. Infect. Immun. *63*, 1400–1408.

Lim, L.C.L., England, D.M., DuChateau, B.K., Glowacki, N.J., Creson, J.R., Lovrich, S.D., Callister, S.M., Jobe, D.A., and Schell, R.F. (1994). Development of destructive arthritis in vaccinated hamsters challenged with *Borrelia burgdorferi*. Infect. Immun. *62*, 2825–2833.

Lim, L.C.L., England, D.M., Glowacki, N.J., DuChateau, B.K., and Schell, R.F. (1995b). Involvement of CD4$^+$ T-lymphocytes in induction of severe destructive Lyme arthritis in inbred LSH hamsters. Infect. Immun. *63*, 4818–4825.

Lindenmayer, J.M., Marshall, D., and Onderdonk, A.B. (1991). Dogs as sentinels for Lyme disease in Massachusetts. Am. J. Publ. Health *81*, 1448–1455.

Lischer, C.J., Leutenegger, C.M., Braun, U., and Lutz, H. (2000). Diagnosis of Lyme disease in two cows by the detection of *Borrelia burgdorferi* DNA. Vet. Rec. *146*, 497–499.

Littman, M.P., Goldstein, R.E., Labato, M.A., Lappin, M.R., and Moore, G.E. (2006). ACVIM small animal consensus statement on Lyme disease in dogs: diagnosis, treatment, and prevention. J. Vet. Intern. Med. *20*, 422–434.

Liu, N., Montgomery, R.R., Barthold, S.W., and Bockenstedt, L.K. (2004). Myeloid differentiation antigen 88 deficiency impairs pathogen clearance but does not alter inflammation in *Borrelia burgdorferi*-infected mice. Infect. Immun. 72, 3195–3203.

Londono, D., Bai, Y., Zückert, W.R., Gelderblom, H., and Cadavid, D. (2005). Cardiac apoptosis in severe relapsing fever borreliosis. Infect. Immun. *73*, 7669–7676.

Londono, D., Carvajal, J., Arguelles-Grande, C., Marques, A., and Cadavid, D. (2008a). Interleukin 10 protects the brain microcirulation from spirochetal injury. J. Neuropath. Exp. Neurol. *67*, 976–983.

Londono, D., Marques, A. Hornung, R., and Cadavid, D. (2008b). IL-10 helps control pathogen load during high-level bacteremia. J. Immunol. *181*, 2076–2083.

Londono, D., Marques, A., Hornung, R., and Cadavid, D. (2008c). Relapsing fever borreliosis in interleukin-10-deficient mice. Infect. Immun. *12*, 508–513.

Lovrich, S.D., Callister, S.M., DuChateau, B.K., Lim, L.C.L., Winfrey, J., Day, S.P., and Schell, R.F. (1995). Abilities of OspA proteins from different seroprotective groups of *Borrelia burgdorferi* to protect hamsters from infection. Infect. Immun. *63*, 2113–2119.

Lovrich, S.D., Callister, S.M., Lim, L.C.L., DuChateau, B.K., and Schell, R.F. (1994). Seroprotective groups of Lyme borreliosis spirochetes from North American and Europe. J. Infect. Dis. *170*, 115–121.

Lovrich, S.D., Callister, S.M., Lim, L.C.L., and Schell, R.F. (1993). Seroprotective groups among isolates of *Borrelia burgdorferi*. Infect. Immun. *61*, 4367–4374.

Lovrich, S.D., Callister, S.M., Schmitz, J.L., Alder, J.F., and Schell, R.F. (1991). *Borrelia*cidal activity of sera from hamsters infected with the Lyme disease spirochete. Infect. Immun. *59*, 2522–2528.

Lusitani, D., Malawista, S.E., and Montgomery, R.R. (2003). Calprotectin, an abundant cytosolic protein from human polymorphonuclear leukocytes, inhibits the growth of *Borrelia burgdorferi*. Infect. Immun. *71*, 4711–4716.

Ma, Y., Seiler, K.P., Eichwald, E.J., Weis, J.H., Teuscher, C., and Weis, J.J. (1998). Distinct characteristics of resistance to *Borrelia burgdorferi*-induced arthritis in C57BL/6N mice. Infect. Immun. *66*, 161–168.

Magnarelli, L.A., Bushmich, S.L., IJdo, J.W., and Fikrig, E. (2005). Seroprevalence of antibodies against *Borrelia burgdorferi* and *Anaplasma phagocytophilum* in cats. Am. J. Vet. Res. *66*, 1895–1899.

Malawista, S.M., Barthold, S.W., and Persing, D.H. (1994). Fate of *Borrelia burgdorferi* DNA in tissues of infected mice after antibiotic treatment. J. Infect. Dis. *170*, 1312–1316.

Marques, A.R., Martin, D.S., and Philipp, M.T. (2002). Evaluation of the C6 peptide enzyme-linked immunosorbent assay for individuals vaccinated with the recombinant OspA vaccine. Clin Microbiol *40*, 2591–2593.

Masuzawa, T., Beppu, Y., Kawabata, H., Yangihara, Y., Iwamoto, Y., Shimizu, T., and Johnson, R.C. (1992). Experimental *Borrelia burgdorferi* infection of outbred mice. J. Clin. Microbiol. *30*, 3016–3018.

Masuzawa, T., Kurita, T., Kawabata, H., Muramatsu, K., and Uchikawa, K. (1994a). Infectivity and early antibody response to *Borrelia burgdorferi* sensu lato isolated in Japan in outbred mice. Microbiol. Immunol. *38*, 621–627.

Masuzawa, T., Kurita, T., Kawabata, H., Suzuki, H., and Yanagihara, Y. (1994b). Experimental induction of Lyme arthritis in outbred mice. Microbiol. Immunol. *38*, 263–268.

Mather, T.N., Telford, S.R. III, and Adler, G.H. (1991). Absence of transplacental transmission of Lyme disease spirochetes from reservoir mice (*Peromyscus leucopus*) to their offspring. J. Infect. Dis. *164*, 564–567.

Matyniak, J.E., and Reiner, S.L. (1995). T helper phenotype and genetic susceptibility in experimental Lyme disease. J. Exp. Med. *181*, 1251–1254.

Mbow, M.L., Zeidner, N., Gilmore, R.D., Dolan, M., Piesman, J., and Titus, R.G. (2001). Major histocompatibility complex class II-independent generation of neutralizing antibodies against T-cell-dependent *Borrelia burgdorferi* antigens presented by dendritic cells: regulation by NK and γ/δ T-cells. Infect. Immun. *69*, 2407–2415.

McDowell, J.V., Sung, S.Y., Labandeira-Rey, M., Skare, J.T., and Marconi, R.T. (2001). Analysis of mechanisms associated with loss of infectivity of clonal populations of *Borrelia burgdorferi* B31MI. Infect. Immun. *69*, 3670–3677.

McKisic, M.D., and Barthold, S.W. (2000). T-cell-independent responses to *Borrelia burgdorferi* are critical for protective immunity and resolution of Lyme disease. Infect. Immun. *68*, 5190–5197.

McKisic, M.D., Redmond, W.L., and Barthold, S.W. (2000). Cutting edge: T-cell-mediated pathology in murine Lyme borreliosis. J. Immunol. *164*, 6096–6099.

Miller, J.C., von Lackum, K., Woodman, M.E., and Stevenson, B. (2006). Detection of *Borrelia burgdorferi* gene expression during mammalian infection using transcriptional fusions that produce green flourescence protein. Microbial Pathogen. *41*, 43–47.

Montgomery, R.R., Booth, C.J., Wang, X., Blaho, V.A., Malawista, S.E., and Brown, C.R. (2007). Recruitment of macrophages and polymorphonuclear leukocytes in Lyme carditis. Infect. Immun. *75*, 613–620.

Montgomery, R.R., Schreck, K., Wang, X., and Malawista, S.E. (2006). Human neutrophil calprotectin reduces the susceptibility of *Borrelia burgdorferi* to penicillin.. Infect. Immun. *74*, 2468–2472.

Moody, K.D., Adams, R.L., and Barthold, S.W. (1994a). Effectiveness of antimicrobial treatment against *Borrelia burgdorferi* infection in mice. Antimicrob. Agents Chemother. *38*, 1567–1572.

Moody, K.D., and Barthold, S.W. (1991). Relative infectivity of *Borrelia burgdorferi* in Lewis rats by various routes of inoculation. Am. J. Trop. Med. Hyg. *44*, 135–139.

Moody, K.D., Barthold, S.W., and Terwilliger, G.A. (1990a). Lyme borreliosis in laboratory animals: effect of host species and *in vitro* passage of *Borrelia burgdorferi*. Am. J. Trop. Med. Hyg. *43*, 87–92.

Moody, K.D., Barthold, S.W., Terwilliger], G.A., Beck, D.S., Hansen, G.M., and Jacoby, R.O. (1990b). Experimental chronic Lyme borreliosis in Lewis rats. Am. J. Trop. Med. Hyg. *42*, 65–74.

Moody, K.D., Terwilliger, G.A., Hansen, G.M., and Barthold, S.W. (1994b). Experimental *Borrelia burgdorferi* infection in *Peromyscus leucopus*. J. Wildl. Dis. *30*, 155–161.

Moriarty, T.J., Norman, M.U., Colarusso, P., Bankhead, T., Kubes, P., and Chaconas, G. (2008). Real-time high resolution 2-D imaging of the Lyme disease spirochete adhering to and escaping from the vasculature of a living host. PLos Pathogen. 4:e1000090.

Motameni, A.R., Bates, T.C., Juncadella, I.J., Petty, C., Hedrick, M.N., and Anguita, J. (2005). Distinct bacterial dissemination and disease outcome in mice subcutaneously infected with *Borrelia burgdorferi* in the midline of the back and the footpad. FEMS Immunol. Med. Microbiol. *45*, 279–284.

Mullegger, R.R., McHugh, G., Ruthazer, R., Binder, B., Kerl, H., and Steere, A.C. (2000). Differential expression of cytokine mRNA in skin specimens from patients with erythema migrans or acrodermatitis chronica atrophicans.. J. Invest. Dermatol. *115*, 1115–1123.

Munson, E.L., DeCoster, D.J., Nardelli, D.T., England, D.M., Callister, S.M., and Schell, R.F. (2004). Neutralization of gamma interferon augments borreliacidal antibody production and severe destructive arthritis in C3H/HeJ mice. Clin. Diag. Lab. Immunol. *11*, 35–41.

Museteanu, C., Schaible, U.E., Stehle, T., Kramer, M.D., and Simon, M.M. (1991). Myositis in mice inoculated with *Borrelia burgdorferi*. Am. J. Pathol. *139*, 1267–1271.

Narayan, K., Dail, D., Li, L., Cadavid, D., Amrute, S., Fitzgerald-Bocarsly, P., and Pachner, A.R. (2005). The nervous system as ectopic germinal center: CXC13 and IgG in Lyme neuroborreliosis. Ann. Neurol. *57*, 813–823.

Nardelli, D.T., Burchill, M.A., England, D.M., Torrealba, J., Callister, S.M., and Schell, R.F. (2004). Association of $CD4^+$ $CD25^+$ cells with prevention of severe destructive arthritis in *Borrelia-burgdorferi*-vaccinated and challenged gamma interferon-deficient mice treated with anti-interleukin-17 antibody. Clin. Diag. Lab. Immunol. *11*, 1075–1084.

Nardelli, D.T., Cloute, J.P., Luk, K.H., Torrealba, J., Warner, T.F., Callister, S.M., and Schell, R.F. (2005). CD4(+)CD25(+) T-cells prevent arthritis associated with *Borrelia* vaccination and infection. Clin. Diag. Lab. Immunol. *12*, 786–792.

Nardelli, D.T., Warner, T.F., Callister, S.M., and Schell, R.F. (2006). Anti-CD25 antibody treatment of mice vaccinated and challenged with *Borrelia spp*. does not exacerbate arthritis but inhibits borreliacidal antibody production. Clin Vaccine Immunol *13*, 884–891.

Nordstand, A., Shamaei-Tousi, A., Ny, A., and Bergström, S. (2001). Delayed invasion of the kidney and brain by *Borrelia crocidurae* in plasminogen-deficient mice. Infect. Immun. *69*, 5832–5839.

Norris, S.J., Carter, C.J., Howell, J.K., and Barbour, A.G. (1992). Low-passage associated proteins of *Borrelia burgdorferi* B31: characterization and molecular cloning of OspD, a surface-exposed, plasmid-encoded lipoprotein. Infect. Immun. *60*, 4662–4672.

Norris, S.J., Howell, J.K., Garza, S.A., Ferdows, M.S., and Barbour, A.G. (1995). High- and low-infectivity phenotypes of clonal populations of *in vitro*-cultured *Borrelia burgdorferi*. Infect. Immun. *63*, 2206–2212.

O'Connor, T.P., Esty, K.J., Hanscom, J.L., Shields, P., and Philipp, M.T. (2004). Dogs vaccinated with common Lyme disease vaccines do not respond to IR6, the conserved immunodominant region of the VlsE surface protein of *Borrelia burgdorferi*.. Clin. Diagn. Lab. Immunol. *11*, 458–462.

Obonyo, M., Munderloh, U.G., Sam, T.N., and Kurti, T.J. (2002). Cultivation at 37 degrees C enhances *Borrelia burgdorferi* sensu stricto infectivity for hamsters. Med. Microbiol. Immunol. *191*, 33–39.

Ohnishi, J., Piesman, J., and deSilva, A. (2001). Antigenic and genetic heterogeneity of *Borrelia burgdorferi* populations transmitted by ticks. Proc. Natl. Acad. Sci. U.S.A. *98*, 670–675.

Pachner, A.R. (2001). The rhesus model of Lyme neuroborreliosis. Immunol Rev *183*, 186–204.

Pachner, A.R., Basta, J., Delaney, E., and Hulinska, D. (1995a). Localization of *Borrelia burgdorferi* in murine Lyme borreliosis by electron microscopy. Am. J. Trop. Med. Hyg. *52*, 128–133.

Pachner, A.R., Braswell, S.T., Delaney, E., Amemiya, K., and Major, E. (1994). A rabbit model of Lyme neuroborreliosis: characterization by PCR, serology, and sequencing of the OspA gene from the brain. Neurology *44*, 1938–1943.

Pachner, A.R., Dail, D., Bai, Y., Sondey, M., Pak, L., Narayan, K., and Cadavid, D. (2004). Genotype determines phenotype in experimental Lyme borreliosis. Ann. Neurol. *56*, 361–370.

Pachner, A.R., Dail, D., Narayan, K., Dutta, K., and Cadavid, D. (2002). Increased expression of B-lymphocyte chemoattractant, but not proinflammatory cytokines, in muscle tissue in rhesus chronic Lyme borreliosis. Cytokine *19*, 297–307.

Pachner, A.R., Delaney, E., O'Neill, T., and Major, E. (1995b). Inoculation of nonhuman primates with the N40 strain of *Borrelia burgdorferi* leads to a model of Lyme neuroborreliosis faithful to the human disease. Neurology *45*, 165–172.

Pachner, A.R., Delaney, E., and ONeill, T. (1995c). Neuroborreliosis in the nonhuman primate: *Borrelia burgdorferi* persists in the central nervous system. Ann. Neurol. *38*, 667–669.

Pachner, A.R., Delaney, E., and Ricalton, N.S. (1993). Murine Lyme borreliosis: route of inoculation determines immune response and infectivity. Regional Immunol. *4*, 345–351.

Padilla, M.L., Callister, S.M., Schell, R.F., Bryant, G.L., Jobe, D.A., Lovrich, S.D., DuChateau, B.K., and Jensen, J.R. (1996). Characterization of the protective borreliacidal antibody response in humans and hamsters after vaccination with a *Borrelia burgdorferi* outer surface protein A vaccine. J. Infect. Dis. *174*, 739–746.

Pal, U., Dai, J., Li, X., Neelakanta, G., Luo, P., Kumar, M., Wang, P., Yang, X., Anderson, J.F., and Fikrig, E. (2008a). A differential role for BB0365 in the persistence of *Borrelia burgdorferi* in mice and ticks. J. Infect. Dis. *197*, 148–155.

Pal, U., Wang, P., Bao, F., Yang, X., Samanta, S., Schoen, R., Wormser, G.P., Schwartz, I., and Fikrig, E. (2008b). *Borrelia burgdorferi* basic membrane proteins A and B participate in the genesis of Lyme arthritis. J. Exp. Med. *205*, 133–141.

Pal, U., Yang, X., Chen, M., Bockenstedt, L.K., Anderson, J.F., Flavell, R.A., Norgard, M.V., and Fikrig, E. (2004). OspC facilitates *Borrelia burgdorferi* invasion of *Ixodes scapularis* salivary glands. J. Clin. Invest. *113*, 220–230.

Parenti, D.L. (2000). Outer surface protein A and arthritis in hamsters. Infect. Immun. *68*, 7212.

Parker, J.L., and White, K.K. (1992). Lyme borreliosis in cattle and horses: a review of the literature. Cornell Vet. *82*, 253–274.

Parveen, N., Caimano, M., Radolf, J.D., and Leong, J.M. (2003). Adaptation of the Lyme disease spirochaete to the mammalian host environment results in enhanced glycosaminoglycan and host cell binding. Mol. Microbiol. *47*, 1433–1444.

Parveen, N., Cornell, K.A., Bono, J.L., Chamberland, C., Rosa, P., and Leong, J.M. (2006). Bgp, a secreted glycosaminoglycan-binding protein of *Borrelia burgdorferi* strain N40, displays nucleosidas activity and is not essential for infection of immunodeficient mice.. Infect. Immun. *74*, 3016–3020.

Patterson, J.L., and Carrion Jr, R. (2005). Demand for nonhuman primate resources in the age of biodefense. ILAR J. *46*, 15–22.

Pavia, C., Inchiosa, M.A., and Wormser, G.P. (2002). Efficacy of short-course ceftriaxone therapy for *Borrelia burgdorferi* infection in C3H mice. Antimicrob. Agents Chemother. *46*, 132–134.

Peltomaa, M., McHugh, G., and Steere, A.C. (2003). Persistence of the antibody response to VlsE sixth invariant (IR6) peptide of *Borrelia burgdorferi* after successful antibiotic treatment of Lyme disease.. J. Infect. Dis. *187*, 1178–1186.

Pennington, P.M., Allerd, C.D., West, C.S., Alvarez, R., and Barbour, A.G. (1997). Arthritis severity and spirochete burden are determined by serotype in the *Borrelia turicatae*-mouse model of Lyme disease. Infect. Immun. *65*, 285–292.

Pennington, P.M., Cadavid, D., and Barbour, A.G. (1999a). Characterization of VspB of *Borrelia turicatae*, a major outer surface membrane protein expressed in blood and tissues of mice.. Infect. Immun. *67*, 4637–4645.

Pennington, P.M., Cadavid, D., Bunikis, J., Norris, S.J., and Barbour, A.G. (1999b). Extensive interplasmidic duplications change the virulence phenotype of relapsing fever agent *Borrelia turicatae*.. Mol. Microbiol. *34*, 1120–1132.

Persing, D.H., Telford, S.R. III, Rys, P.N., Dodge, D.E., White, T.J., Malawista, S.E., and Spielman, A. (1990). Detection of *Borrelia burgdorferi* DNA in museum specimens of *Ixodes dammini* ticks. Science *249*, 1420–1423.

Peterson, S.H., Nardelli, D.T., Warner, T.F., Callister, S.M., Torrealba, J.R., and Schell, R.F. (2007). Anit-p19 antibody treatment exacerbates Lyme arthritis and enhances borreliacidal activity. Clin. Vaccine Immunol. *14*, 510–517.

Philipp, M.T., Aydintug, M.K., Bohm Jr., R.P., Cogswell, F.B., Dennis, V.A., Lanners, H.N., Lowrie Jr, R.C., Roberts, E.D., Conway, M.D., Karacorlu, M., *et al.* (1993). Early and early disseminated phases of Lyme disease in the rhesus monkey: a model for infection in humans. Infect. Immun. *61*, 3047–3059.

Philipp, M.T., Bowers, L.C., Fawcett, P.T., Jacobs, M.B., Liang, F.T., Marques, A.R., Mitchell, P.D., Purcell, J.E., Ratterree, M.S., and Straubinger, R.K. (2001). Antibody response to IR6, a conserved immunodominant region of the VlsE lipoprotein, wanes rapidly after antibiotic treatment of *Borrelia burgdorferi* infection in experimental animals and humans. J. Infect. Dis. *184*, 870–878.

Philipp, M.T., Lobet, Y., Bohm Jr, R.P., Conway, M.D., Dennis, V.A., Desmons, P., Gu, Y., Hauser, P., LowrieJr, R.C., and Roberts, E.D. (1996). Safety and immunogenicity of recombinant outer surface protein A (OspA) vaccine formulations in the rhesus monkey. J. Spirochetal Tick-Borne Dis. 3, 67–69.

Philipp, M.T., Lobet, Y., Bohm Jr, R.P., Roberts, E.D., Dennis, V.A., Gu, Y., Lowrie Jr, R.C., Desmons, P., Duray, P.H., Englands, J.D., *et al.* (1997). The outer surface protein A (OspA) vaccine against Lyme disease: efficacy in the rhesus monkey. Vaccine *15*, 1872–1887.

Philipp, M.T., Marques, A.R., Fawcett, P.T., Dally, L.G., and Martin, D.S. (2003). C6 test as an indicator of therapy outcome for patients with localized or disseminated Lyme borreliosis. J. Clin. Microbiol. *41*, 4955–4960.

Philipp, M.T., Wormser, G.P., Marques, A.R., Bittker, S., Martin, D.S., Nowakowski, J., and Dally, L.G. (2005). A decline in C6 antibody titer occurs in successfully treated patients with culture-confirmed early localized or disseminated Lyme borreliosis. Clin. Diagn. Lab. Immunol. *12*, 1069–1974.

Piesman, J., Mather, T.N., Sinsky, R.J., and Spielman, A. (1987). Duration of tick attachment and *Borrelia burgdorferi* transmission. J. Clin. Microbiol. *25*, 557–558.

Potter, M.R., Noben-Trauth, N., Weis, J.H., Teurscher, C., and Weis, J.J. (2000). Interleukin-4 (IL-4) and IL-13 signaling pathways do not regulate *Borrelia burgdorferi*-induced arthritis in mice: IgG1 is not required for host control of tissue spirochetes. Infect. Immun. *68*, 5603–5609.

Potter, M.R., Rittling, S.R., Denhardt, D.T., Roper, R.J., Weis, J.H., Teuscher, C., and Weis, J.J. (2002). Role of osteopontin in murine Lyme arthritis and host defense against *Borrelia burgdorferi*. Infect. Immun. *70*, 1372–1381.

Preac-Mursic, V., Patsouris, E., Wilske, B., Reinhardt, S., Gos, B., and Mehraein, P. (1990). Persistence of *Borrelia burgdorferi* and histopathological alterations in experimentally infected animals; comparison with histopathological findings in human Lyme disease. Infection *18*, 332–341.

Purser, J.E., Lawrenz, M.B., Caimano, M.J., Howell, J.K., Radolf, J.D., and Norris, S.J. (2003). A plasmid-encoded nicotinamidase (PncA) is essential for infectivity of *Borrelia burgdorferi* in a mammalian host. Mol. Microbiol. *48*, 753–764.

Purser, J.E., and Norris, S.J. (2000). Correlation between plasmid content and infectivity of *Borrelia burgdorferi*. Proc. Natl. Acad. Sci. U.S.A. *97*, 13865–13870.

Qiu, W.G., Schutzer, S.E., Bruno, J.F., Attie, O., Xu, Y., Dunn, J.J., Fraser, C.M., Casjens, S.R., and Luft, B.J. (2004). Genetic exchange and plasmid transfers in *Borrelia burgdorferi* sensu stricto revealed by three-way genome comparisons and multilocus sequence typing. Proc. Natl. Acad. Sci. U.S.A. *101*, 14150–14155.

Raveche, E.S., Schutzer, S.E., Fernandes, H., Bateman, H., McCarthy, B.A., Nickell, S.P., and Cunningham, M.W. (2005). Evidence of *Borrelia* autoimmunity-induced component of Lyme carditis and arthritis. J. Clin. Microbiol. *43*, 850–856.

Rawlings, J.A. (1995). An overview of tick-borne relapsing fever with emphasis on outbreaks in Texas. Tex. Med. *91*, 56–59.

Ray, A., Kumar, D., Shakya, A., Brown, C.R., Cook, J.L., and Ray, B.K. (2004). Serum amyloid A activating factor-1 (SAF-1) transgenic mice are prone to develop a severe form of inflammation-induced arthritis. J. Immunol. *173*, 4684–4691.

Revel, A.T., Blevins, J.S., Almazán, C., Neil, L., Kocan, K.M., de la Fuente, J., Hagman, K.E., and Norgard,

M.V. (2005). *bptA* (*bbe16*) is essential for the persistence of the Lyme disease spirochete, *Borrelia burgdorferi*, in its natural host. Proc. Natl. Acad. Sci. U.S.A. *102*, 6972–6977.

Revel, A.T., Talaat, A.M., and Norgard, M.V. (2002). DNA microarray analysis of differential gene expression in *Borrelia burgdorferi*, the Lyme disease spirochete. Proc. Natl. Acad. Sci. U.S.A. *99*, 1562–1567.

Roberts, E.D., Bohm Jr, R.P., Cogswell, F.B., Lanners, H.N., Lowrie Jr, R.C., Povinelli, L., Piesman, J., and Philipp, M.T. (1995). Chronic Lyme disease in the rhesus monkey. Lab. Invest. *72*, 146–160.

Roberts, E.D., Bohm Jr, R.P., Lowrie Jr, R.C., Habicht, G., Katona, L., Piesman, J., and Philipp, M.T. (1998). Pathogenesis of Lyme neuroborreliosis in the rhesus monkey: the early disseminated and chronic phases of disease in the peripheral nervous system. J. Infect. Dis. *178*, 722–732.

Roehrig, J.T., Piesman, J., Hunt, A.R., Keen, M.G., Happ, C.M., and Johnson, B.J.B. (1992). The hamster immune response to tick transmitted *Borrelia burgdorferi* differs from the response to needle-inoculated, cultured organisms. J. Immunol. *149*, 3648–3653.

Roper, R.J., Weis, J.J., McCracken, B.A., Green, C.B., Ma, Y., Weber, K.S., Fairbairn, D., Butterfield, R.J., Potter, M.R., Zachary, J.F., *et al.* (2001). Genetic control of susceptibility to experimental Lyme arthritis is polygenic and exhibits consistent linkage to multiple loci on chromosome 5 in four independent mouse crosses. Genes Immun. *2*, 388–397.

Rousselle, J.C., Callister, S.M., Schell, R.F., Lovrich, S.D., Jobe, D.A., Marks, J.A., and Wieneke, C.A. (1998). *Borrelia*cidal antibody production against outer surface protein C of *Borrelia burgdorferi*. J. Infect. Dis. *178*, 733–741.

Ruderman, E.M., Kerr, J.S., Telford, S.R.I., Spielman, A., Glimcher, L.H., and Gravallese, E.M. (1995). Early murine Lyme carditis has a macrophage predominance and is independent of major histocompatibility complex class II-CD4+ T-cell interactions. J. Infect. Dis. *171*, 362–370.

Rupprecht, T.A., Pfister, H.W., Angele, B., Kastenbauer, S., Wilske, B., and Koedel, U. (2005). The chemokine CXCL13 (BLC): a putative diagnostic marker for neuroborreliosis.. Neurology *65*, 448–450.

Saba, S., VanderBrink, B.A., Perides, G., Glickstein, L.J., Link, M.S., Homoud, M.K., Bronson, R.T., Estes 3rd, M., and Wang, P.J. (2001). Cardiac conduction abnormalities in a mouse model of Lyme borreliosis. J. Interv. Card. Electrophysiol. *5*, 137–143.

Satoskar, A.R., Elizondo, J., Monteforte, G.M., Stamm, L.M., Bluethmann, H., Katavolos, P., and Telford, S.R. III (2000). Interleukin-4-deficient BALB/c mice develop and enhanced Th1-like response but control cardiac inflammation following *Borrelia burgdorferi* infection. FEMS Microbiol. Lett. *183*, 319–325.

Schaible, U., Kramer, M., Museteanu, C., Zimmer, G., Mossman, H., and Simon, M.M. (1989a). The severe combined immunodeficient mouse: A laboratory model for the analysis of Lyme arthritis and carditis. J. Exp. Med. *170*, 1427–1432.

Schaible, U.E., Gern, L., Wallich, R., Kramer, M.D., Prester, M., and Simon, M.M. (1993). Distinct patterns of protective antibodies are generated against *Borrelia burgdorferi* in mice experimentally inoculated with high and low doses of antigen. Immunol. Lett. *36*, 219–226.

Schaible, U.E., Kramer, M.D., Justus, C.W.E., Musteanu, C., and Simon, M.M. (1989b). Demonstration of antigen-specific T-cells and histopathological alterations in mice experimentally inoculated with *Borrelia burgdorferi*. Infect. Immun. *57*, 41–47.

Schaible, U.E., Kramer, M.D., Museteanu, C., Zimmer, G., Mossmann, H., and Simon, M. (1990). Lyme borreliosis in the severe combined immunodeficiency (*scid*) mouse manifests predominantly in the joints, heart and liver. Am J. Pathol. *137*, 811–820.

Schaible, U.E., Kramer, M.D., Wallich, R., Tran, T., and Simon, M.M. (1991). Experimental *Borrelia burgdorferi* infection in inbred mouse strains: antibody response and association of H-2 genes with resistance and susceptibility to development of arthritis. Eur J. Immunol. *21*, 2397–2405.

Schaible, U.E., Wallich, R., Kramer, M.D., Nerz, G., Stehle, T., Museteanu, C., and Simon, M.M. (1994). Protection against *Borrelia burgdorferi* infection in SCID mice is conferred by presensitized spleen cells and partially by B- but not T-cells alone. Intern. Immun. *6*, 671–681.

Schmitz, J.L., Lovrich, S.D., Callister, S.M., and Schell, R.F. (1991a). Depletion of complement and effects on passive transfer of resistance to infection with *Borrelia burgdorferi*. Infect. Immun. *59*, 1831–3818.

Schmitz, J.L., Schell, R.F., Callister, S.M., Lovich, S.D., Day, S.P., and Coe, J.E. (1992). Immunoglobulin G2 confers protection against *Borrelia burgdorferi* infection in LSH hamsters. Infect. Immun. *60*, 2677–2682.

Schmitz, J.L., Schell, R.F., Hejka, A.G., and England, D.M. (1990). Passive immunization prevents induction of Lyme arthritis in LSH hamsters. Infect. Immun. *58*, 144–148.

Schmitz, J.L., Schell, R.F., Hejke, A., England, D.M., Callister, S.M., and Schell, R.F. (1989). Induction of Lyme arthritis in LSH hamsters. Am J. Pathol. *134*, 1113–1123.

Schmitz, J.L., Schell, R.F., Lovrich, S.D., Callister, S.M., and Coe, J.E. (1991b). Characterization of the protective antibody response to *Borrelia burgdorferi* in experimentally infected LSH hamsters. Infect. Immun. *59*, 1916–1921.

Schuhardt, V., and O'Bryan, B. (1945). Effect of intracranial penicillin therapy on brain involvement in experimental relapsing fever. J. Bacteriol. *49*, 312–313.

Schwan, T.G., Burgdorfer, W., and Garon, C.F. (1988a). Changes in infectivity and plasmid profile of the Lyme disease spirochete, *Borrelia burgdorferi*, as a result of *in vitro* cultivation. Infect. Immun. *56*, 1831–1836.

Schwan, T.G., Burgdorfer, W., Schrumpf, M.E., and Karstens, R.H. (1988b). The urinary bladder: a consistent source of *Borrelia burgdorferi* in experimentally infected white-footed mice (*Peromyscus leucopus*). J. Clin. Microbiol. *26*, 893–895.

Schwan, T.G., Karstens, R.H., Schrumpf, M.E., and Simpson, W.J. (1991). Changes in antigenic reactivity of *Borrelia burgdorferi*, the Lyme disease spirochete, during persistent infection in mice. Can. J. Microbiol. *37*, 450–454.

Seiler, K.P., Ma, Y., Weis, J.H., Frenette, P.S., Hynes, R.O., Wagner, D.D., and Weis, J.J. (1998). E and P selectins are not required for resistance to severe murine Lyme arthritis. Infect. Immun. *66*, 4557–4559.

Seshu, J., Esteve-Gassent, M.D., Labandeira-Rey, M., Kim, J.H., Trzeciakowski, J.P., Hook, M., and Skare, J.T. (2006). Inactivation of the fibronectin-binding adhesin gene *bbk32* significantly attenuates the infectivity potential of *Borrelia burgdorferi*. Mol. Microbiol. *59*, 1591–1601.

Sethi, N., Sondey, M., Bai, Y., Kim, K.S., and Cadavid, D. (2006). Interaction of a neurotropic strain of *Borrelia turicatae* with the cerebral microcirculation system. Infect. Immun. *74*, 6408–6418.

Shadick, N.A., Phillips, C.B., Logigian, E.L., Steere, A.C., Kaplan, R.F., Berardi, V.P., Duray, P.H., Larson, M.G., Wright, E.A., Ginsburg, K.S., *et al.* (1994). The long-term clinical outcomes of Lyme disease. A population-based retrospective cohort study. Ann. Intern. Med. *121*, 560–567.

Shamaei-Tousi, A., Burns, M.J., Benach, J.L., Furie, M.B., Gergel, E.I., and Bergström, S. (2000). The relaping fever spirochete, *Borrelia crocidurae*, activates human endothelial cells and promotes the transendothelial migration of neutrophils. Cell. Microbiol. 2, 591–599.

Shamaei-Tousi, A., Collin, O., Bergh, A., and Bergström, S. (2001). Testicular damage by microcirculatory disruption and colonization of an immuine-privileged site during *Borrelia crocidurae* infection.. J. Exp. Med. *193*, 995–1004.

Shamaei-Tousi, A., Martin, P., Bergh, A., Burman, N., Brännström, T., and Bergström, S. (1999). Erythrocyte-aggregating relapsing fever spirochete *Borrelia crocidurae* induces formation of microemboli. J. Infect. Dis. *180*, 1929–1938.

Shang, E., Wu, X.-Y., Lovett, M., Miller, J.N., and Blanco, D.R. (2001). Homologous and heterologous *Borrelia burgdorferi* challenge of infection-derived immune rabbits using host-adapted organisms. Infect. Immun. *69*, 593–598.

Shang, E.S., Champion, C.I., Wu, X.-Y., Skare, J.T., Blanco, D.R., Miller, J.N., and Lovett, M.A. (2000). Comparison of protection in rabbits against host-adapted and cultivated *Borrelia burgdorferi* following infection-derived immunity or immunization with outer membrane vesicles or outer surface protein A. Infect. Immun. *68*, 4189–4199.

Shi, C., Wolfe, J., Russell, J.Q., Fortner, K., Hardin, N., Anguita, J., and Budd, R.C. (2006). Fas ligand deficiency impairs host inflammatory response against infection with the spirochete *Borrelia burgdorferi*.. Infect. Immun. *74*, 1156–1160.

Shi, Y., Xu, Q., McShan, K., and Liang, F.T. (2008). Both decorin-binding proteins A and B are critical for the overall virulence of *Borrelia burgdorferi*. Infect. Immun. *76*, 1239–1246.

Shih, C.-M., and Liu, L.-P. (1998). Differential efficacy of passive immunisation against infection by Lyme disease spirochaetes transmitted by partially fed vector ticks. J. Med. Microbiol. *47*, 773–779.

Shih, C.-M., Pollack, R.J., Telford, S.R., and Spielman, A. (1992). Delayed dissemination of Lyme disease spirochetes from the site of deposition in the skin of mice. J. Infect. Dis. *166*, 827–831.

Shih, C.M., Spielman, A., and Telford, S.R. (1995). Mode of action of protective immunity to Lyme disease spirochetes. Am. J. Trop. Med. Hyg. *52*, 72–74.

Shih, C.M., Telford, S.R., Pollack, R.J., and Spielman, A. (1993). Rapid dissemination by the agent of Lyme disease in hosts that permit fulminating infection. Infect. Immun. *61*, 2396–2399.

Siebers, A., Zhong, W., Wallich, R., and Simon, M.M. (1999). Loss of pathogenic potential after cloning of the low-passage *Borrelia burgdorferi* ZS7 tick isolate: a cautionary note. Med. Microbiol. Immunol. *188*, 125–130.

Silver, R.M., Yang, L.M., Daynes, R.A., Branch, D.W., Salafia, C.M., and Weis, J.J. (1995). Fetal outcome of murine Lyme disease. Infect. Immun. *63*, 66–72.

Simon, M.M., Wallich, R., and Kramer, M.D. (1996). *Borrelia burgdorferi* infection of inbred strains of mice provides insights into cellular and molecular parameters of pathogenesis and protection of Lyme disease: a viewpoint. J Spirochetal Tick Borne Dis 3, 45–52.

Simpson, W.J., Burgdorfer, W., Schrumpf, M.E., Karstens, R.H., and Schwan, T.G. (1991). Antibody to a 39-kilodalton *Borrelia burgdorferi* antigen (P39) as a marker for infection in experimentally and naturally inoculated animals. J. Clin. Microbiol. *29*, 236–243.

Sinsky, R.J., and Piesman, J. (1989). Ear punch biopsy method for detection and isolation of *Borrelia burgdorferi* from rodents. J. Clin. Microbiol. 27, 1723–1727.

Skotarczak, B., Wodecka, B., Rymaszewska, A., Sawczuk, M., Maciejewska, A., Adamska, M., Hermanowska-Szpakowicz, T., and Swierzbinska, R. (2005). Prevalence of DNA and antibodies to *Borrelia burgdorferi* sensu lato in dogs suspected of borreliosis. Ann Agric Environ Med *12*, 199–205.

Smismans, A., Goossens, V.J., Nulens, E., and Bruggeman, C.A. (2006). Comparison of five different immunoassays for the detection of *Borrelia burgdorferi* IgM and IgG antibodies. Clin Microbiol Infect *12*, 648–655.

Sonnesyn, S.W., Manivel, J.C., Johnson, R.C., and Goodman, J.L. (1993). A guinea pig model for Lyme disease. Infect. Immun. *61*, 4777–4784.

Staprans, S.I., and Feinberg, M.B. (2004). The roles of nonhuman primates in the preclinical evaluation of candidate AIDS vaccines. Expert Rev. Vaccines 3, S5–S32.

Steere, A.C. (1989). Lyme disease. N. Engl. J. Med. *321*, 586–596.

Steere, A.C. (2001). Lyme disease. N Eng J Med 345, 115–125.

Steere, A.C., Coburn, J., and Glickstein, L. (2004). The emergence of Lyme disease. J. Clin. Invest. *113*, 1093–1101.

Steere, A.C., Duray, P.H., and Butcher, E.C. (1988). Spirochetal antigens and lymphoid cell surface markers in Lyme synovitis. Comparison with rheumatoid synovium and tonsillar lymphoid tissue.. Arthritis Rheum. *31*, 487–495.

Steere, A.C., Levin, R.E., Moloy, P.J., Kalish, R.A., Abraham3rd, J.H., Liu, N.Y., and Schmid, C.H. (1994). Treatment of Lyme arthritis. Arthritis Rheum. *37*, 878–888.

Steere, A.C., and Sikand, V.K. (2003). The presenting manifestations of Lyme disease and the outcomes of treatment. N. Engl. J. Med. *348*, 2472–2474.

Stewart, P.E., Wang, X., Bueschel, D.M., Clifton, D.R., Grimm, D., Tilly, K., Carroll, J.A., Weis, J.J., and Rosa, P.A. (2006). Delineating the requirement for the *Borrelia burgdorferi* virulence factor OspC in the mammalian host. Infect. Immun. *74*, 3547–3553.

Stoenner, H.G., Dodd, T., and Larsen, C. (1982). Antigenic variation of *Borrelia hermsii*. J. Exp. Med. *156*, 1297–1311.

Straubinger, R.K. (2000). PCR-based quantification of *Borrelia burgdorferi* organisms in canine tissue over a 500-day postinfection period. J. Clin. Microbiol. *38*, 2191–2199.

Straubinger, R.K., Chang, Y.F., Jacobson, R.H., and Appel, M.J.G. (1995). Sera from OspA-vaccinated dogs, but not those from tick-infected dogs, inhibit *in vitro* growth of *Borrelia burgdorferi*. J. Clin. Microbiol. 33, 2745–2751.

Straubinger, R.K., Rao, T.D., Davidson, E., Summers, B.A., Jacobson, R.H., and Frey, A.B. (2002). Protection against tick-transmitted Lyme disease in dogs vaccinated with a multiantigenic vaccine. Vaccine *20*, 181–193.

Straubinger, R.K., Straubinger, A.F., Harter, L., Jacobson, R.H., Chang, Y.F., Summerss, B.A., Erb, H.N., and Appel, M.J.G. (1997a). *Borrelia burgdorferi* migrates into joint capsules and causes an up-regulation of interleukin-8 in synovial membranes of dogs experimentally infected with ticks. Infect. Immun. *65*, 1273–1285.

Straubinger, R.K., Straubinger, A.F., Summers, B.A., Erb, H.N., Harter, L., and Appel, M.J.G. (1998a). *Borrelia burgdorferi* induces the production and release of proinflammatory cytokines in canine synovial explant cultures. Infect. Immun. *66*, 247–258.

Straubinger, R.K., Straubinger, A.F., Summers, B.A., and Jacobson, R.H. (2000). Status of *Borrelia burgdorferi* infection after antibiotic treatment and the effects of corticosteroids: An experimental study. J. Infect. Dis. *181*, 1069–1081.

Straubinger, R.K., Straubinger, A.F., Summers, B.A., Jacobson, R.H., and Erb, H.N. (1998b). Clinical manifestations, pathogenesis, and effect of antibiotic treatment on Lyme borreliosis in dogs. Wien Klin. Wochednschr. *110*, 874–881.

Straubinger, R.K., Summers, B.A., Chang, Y.F., and Appel, M.J.G. (1997b). Persistence of *Borrelia burgdorferi* in experimentally infected dogs after antibiotic treatment. J. Clin. Microbiol. 35, 111–116.

Strother, K.O., Hodzic, E., Barthold, S.W., and deSilva, A.M. (2007). Infecction of mice with Lyme disease spirochetes constitutively producing outer surface proteins A and B. Infect. Immun. *75*, 2786–2794.

Suk, K., Das, S., Sun, W., Jwang, B., Barthold, S.W., Flavell, R.A., and Fikrig, E. (1995). *Borrelia burgdorferi* genes selectively expressed in the infected host. Proc. Natl. Acad. Sci. U.S.A. *92*, 4269–4273.

Summers, B.A., Straubinger, A.F., Jacobson, R.H., Chang, Y.F., Appel, M.J.G., and Straubinger, R.K. (2005). Histopathological studies of experimental Lyme disease in dogs. J Comp Path *133*, 1–13.

Thomas, N.J., Bunikis, J., Barbour, A.G., and Wolcott, M.J. (2002). Fatal spirochetosis due to a relapsing fever-like *Borrelia* sp. in a northern spotted owl. J. Wildl. Dis. *38*, 187–193.

Thomas, V., Anguita, J., Samanta, S., Rosa, P.A., Stewart, P., Barthold, S.W., and Fikrig, E. (2001). Dissociation of infectivity and pathogenicity in *Borrelia burgdorferi*. Infect. Immun. *69*, 3507–3509.

Tilly, K., Bestor, A., Jewett, M.W., and Rosa, P. (2007). Rapid clearance of Lyme disease spirochetes lacking OspC from skin. Infect. Immun. *75*, 1517–1519.

Tilly, K., Krum, J.G., Bestor, A., Jewett, M.W., Grimm, D., Bueschel, D., Byram, R., Dorward, D., Vanraden, M.J., Stewart, P., *et al.* (2006). *Borrelia burgdorferi* OspC protein required exclusively in a crucial early stage of mammalian infection. Infect. Immun. *74*, 3554–3564.

Tjernberg, I., Kruger, G., and Eliasson, I. (2007). C6 peptide ELISA test in the serodiagnosis of Lyme borreliosis in Sweden. Eur J. Clin. Microbiol. Infect Dis *26*, 37–42.

Topfer, K.H., and Straubinger, R.K. (2007). Characterization of the humoral immune response in dogs after vaccination against the Lyme borreliosis agent. A study with five commercial vaccines using two different vaccination schedules. Vaccine *25*, 314–326.

Trape, J.P., Duplantier, J.M., Bouganali, H., Godeluck, B., Legros, F., Cornet, J.P., and Camicas, J.L. (1991). Tick-borne borreliosis in west Africa. Lancet *337*, 473–475.

Tuomi, J., Rantamaki, L.K., and Tanskanen, R. (1998). Experimental infection of cattle with several *Borrelia burgdorferi* sensu lato strains; immunological heterogeneity of strains as revealed in serological tests. Vet. Microbiol. *60*, 27–43.

Tuomi, J., Rantamaki, L.K., and Tanskanen, R. (2002). Experimental infection of laboratory mice and rabbits with several isolates of *Borrelia burgdorferi* sensu lato; comparison of antigens from different genospecies in serological measurement of immune response. Comp. Immunol. Microbiol. Infect. Dis. *25*, 109–125.

VanHoecke, C., Comberbach, M., DeGrave, D., Desmons, P., Fu, D., Hauser, P., Lebacq, E., Lobet, Y., and Voet, P. (1996). Evaluation of the safety, reactogenicity and immunogenicity of three recombinant outer surface protein (OspA) Lyme vaccines in healthy adults. Vaccine *14*, 1620–1626.

Wallich, R., Brenner, C., Kramer, M.D., and Simon, M.M. (1995). Molecular cloning and immunological characterization of a novel linear-plasmid-encoded gene, *pG*, of *Borrelia burgdorferi* expressed only in vivo. Infect. Immun. *63*, 3327–3335.

Wang, G., Ma, Y., Buyuk, A., McClain, S., Weis, J.J., and Schwartz, I. (2004). Impaired host defense to infection and Toll-like receptor 2-independent killing of

Borrelia burgdorferi clinical isolates in TLR2-deficient C3H/HeJ mice. FEMS Microbiol. Lett. *231*, 219–225.

Wang, G., Ojaimi, C., Iyer, R., Saksenberg, V., McCain, S.A., Wormser, G.P., and Schwartz, I. (2001). Impact of genotypic variation of *Borrelia burgdorferi* sensu stricto on kinetics of dissemination and severity of disease in C3H/HeJ mice. Infect. Immun. *69*, 4303–4312.

Wang, G., Ojaimi, C., Wu, H., Saksenberg, V., Iyer, R., Liveris, D., McClain, S.A., Wormser, G.P., and Schwartz, I. (2002). Disease severity in a murine model of Lyme borreliosis is associated with the genotype of the infecting *Borrelia burgdorferi* sensu stricto strain. J. Infect. Dis. *186*, 782–791.

Weis, J.J., McCracken, B.A., Ma, Y., Fairbairn, D., Roper, R.J., Morrison, T.B., Weis, J.H., Zachary, J.F., Doerge, R.W., and Teuscher, C. (1999). Identification of quantitative trait loci governing arthritis severity and humoral responses in the murine model of Lyme disease. J. Immunol. *162*, 948–956.

Weis, J.J., Yang, L., PetriSeiler, K., and Silver, R.M. (1997). Pathological manifestations in murine Lyme disease: association with tissue invasion and spirochete persistence. Clin. Infect. Dis. *25(suppl)*, S18–S24.

Woodman, M.E., Cooley, A.E., Miller, J.C., Lazarus, J.J., Tucker, K., Bykowski, T., Botto, M., Hellwege, J., Wooten, R.M., and Stevenson, B. (2007). *Borrelia burgdorferi* binding of host complement regulator factor H is not required for efficient mammalian infection. Infect. Immun. *75*, 3131–3139.

Woodrum, J.E., and Oliver, J.H. (1999). Investigation of venereal, transplacental, and contact transmission of the Lyme disease spirochete, *Borrelia burgdorferi*, in Syrian hamsters. J. Parasitol. *85*, 426–430.

Wooten, R.M., Ma, Y., Yoder, R.A., Brown, J.P., Weis, J.H., Zachary, J.F., Kirschning, C.J., and Weis, J.J. (2002). Toll-like receptor 2 is required for innate, but not acquired, host defense to *Borrelia burgdorferi*. J. Immunol. *168*, 348–355.

Wormser, G.P., Dattwyler, R.J., Shapiro, E.D., Halperin, J.J., Steere, A.C., Klempner, M.S., Krause, P.J., Bakken, J.S., Strle, F., Stanek, G., *et al.* (2006). The clinical assessment, treatment, and prevention of Lyme disease, human granulocytic anaplasmosis, and babesiosis: clinical practice guidelines by the Infectious Diseases Society of America. Erratum in: Clin. Infect. Dis. 45:941, 2007. Clin. Infect. Dis. *43*, 1089–1134.

Wright, D.J., and Woodrow, D.F. (1980). Terminal changes in mice experimentally infected with *Borrelia duttoni*. J. Pathol. *130*, 83–90.

Wright, S.D., and Nielsen, S.W. (1990). Experimental infecction of the white-footed mouse with *Borrelia burgdorferi*. Am. J. Vet. *Res. 51*, 1980–1987.

Xu, Q., McShan, K., and Liang, F.T. (2008). Essential protective role attributed to the surface lipoproteins of *Borrelia burgdorferi* against innate defences. Mol. Microbiol. *69*, 15–29.

Xu, Q., seemanapalli, S.V., McShan, K., and Liang, F.T. (2006). Constitutive expression of outer surface protein C diminishes the ability of *Borrelia burgdorferi* to evade specific humoral immunity. Infect. Immun. *74*, 5177–5184.

Xu, Q., Seemanaplli, S.V., McShan, D., and Liang, F.T. (2007). Increasing the interaction of *Borrelia burgdorferi* with decorin significantly reduces the 50 percent infectious dose and severely impairs dissemination. Infect. Immun. *75*, 4272–4281.

Xu, Y.N., Kodner, C., Coleman, L., and Johnson, R.C. (1996). Correlation of plasmids with infectivity of *Borrelia burgdorferi* sensu stricto type strain B31. Infect. Immun. *64*, 3870–3876.

Yang, L.M., Ma, Y., Schoenfeld, R., Griffiths, M., Eichwald, E., Aranco, B., and Weis, J.J. (1992). Evidence for lymphocyte-B mitogen activity in *Borrelia burgdorferi*-infected mice. Infect. Immun. *60*, 3033–3041.

Yang, L.M., Weis, J.H., Eichwald, E., Kolbert, C.P., Persing, D.H., and Weis, J.J. (1994). Heritable susceptibility to severe *Borrelia burgdorferi*-induced arthritis is dominant and is associated with persistence of large numbers of spirochetes in tissues. Infect. Immun. *62*, 492–500.

Yang, X.F., Pal, U., Alani, S.M., Fikrig, E., and Norgard, M.V. (2004). Essential role for OspA/B in the life cycle of the Lyme disease spirochete. J. Exp. Med. *199*, 641–648.

Yrjänäinen, H., Hytönen, J., Söderström, K.-O., Oksi, J., Hartiala, K., and Viljanen, M.K. (2006). Persistent joint swelling and borrelia-specific antibodies in *Borrelia garinii*-infected mice after eradication of vegetative spirochetes with antibiotic treatment. Microbes Infect. *8*, 2044–2051.

Yrjänäinen, H., Hyotenen, J., Song, S.R., Oksi, J., Hartiala, K., and Viljanen, M.K. (2007). Anti-tumor necrosis factor-α treatment activates *Borrelia burgdorferi* spirochetes in 4 weeks after ceftriaxone treatment in C3H/He mice. J. Infect. Dis. *195*, 1489–1496.

Zeidner, N.S., Brandt, K.S., Dadey, E., Dolan, M.C., Happ, C., and Piesman, J. (2004). Sustained-release formulation of doxycycline hyclate for prophylaxis of tick bite infection in a murine model of Lyme borreliosis. Antimicrob. Agents Chemother. *48*, 2697–2699.

Zeidner, N.S., Nuncio, M.S., Schneider, B.S., Gern, L., Piesman, J., Brandao, O., and Filipe, A.R. (2001). A portuguese isolate of *Borrelia lusitaniae* induces disease in C3H/HeN mice. J Med Microbiol *50*, 1055–1060.

Zhang, J.-R., Hardham, J.M., Barbour, A.G., and Norris, S.J. (1997). Antigenic variation in Lyme disease borreliae by promiscuous recombination of VMP-like sequence cassettes. Cell *89*, 1–20.

Zhong, W., Gern, L., Stehle, T., Museteanu, C., Kramer, M., Wallich, R., and Simon, M.M. (1999). Resolution of experimental and tick-borne *Borrelia burgdorferi* infection in mice by passive, but not active immunization using recombinant OspC. Eur J. Immunol. *29*, 946–957.

Zhong, W., Stehle, T., Museteanu, C., Siebers, A., Gern, L., Kramer, M., Wallich, R., and Simon, M.M. (1997). Therapeutic passive vaccination against chronic Lyme disease in mice. Proc. Natl. Acad. Sci. U.S.A. *94*, 12533–12538.

Zimmer, G., Schaible, U.E., Kramer, M.D., Mall, G., Museteanu, C., and Simon, M.M. (1990). Lyme carditis in immunodeficient mice during experimental infection of *Borrelia burgdorferi*. Virchows Arch. A Pathol. Anat. Histopathol. *417*, 129–135.

Host Response

15

Janis J. Weis and Linda K. Bockenstedt

Abstract

The goal of this chapter is to provide an understanding of the dynamics of host responses during infection with *Borrelia burgdorferi*. Pertinent literature will be reviewed with emphasis on studies that influenced our understanding of both the earliest responses of the mammalian host to infection by *B. burgdorferi*, and those that are associated with chronic infection and disease. Discussion of host responses in this chapter has direct implications for the pathogenesis of Lyme disease and will refer to aspects of the Pathobiology chapter (Chapter 12). Animal models, which are more thoroughly discussed in Chapter 14, have been critical to our understanding of the mechanistic aspects of the host response to *B. burgdorferi*. This chapter will focus on the most widely studied animal model of Lyme borreliosis, the mouse model, with comparison made to human disease. The initial response to infection occurs before the spirochaete and mammalian host have made direct contact, that is, during tick feeding, and for this reason, reference also will be made to the Tick Interactions chapter (Chapter 11). The host response is important to two features of Lyme borreliosis: the control of the spirochaete and the development of tissue damage and clinical manifestations. Reference to the chapter on Lyme Disease in Humans (Chapter 18) will direct the reader to assessment of these responses in the context of the patient experience.

15.1 Introductory comments

The pathogenesis of Lyme borreliosis has been the subject of intense study since its recognition in the late 1970s as an emerging infectious disease. The clinical aspects of the disorder arise as a consequence of *B. burgdorferi* infection and persistence in tissues, which can lead to inflammatory pathologies. Although *B. burgdorferi* has rarely been sighted inside cells *ex vivo* or *in vitro*, studies of its genome and mechanisms for evasion of host immunity, which are directed towards subverting phagocyte, antibody, and complement-mediated destruction, point overwhelmingly towards its classification as an extracellular pathogen. When visualized in tissues, the organism has a predilection for extracellular matrix and connective tissue. Histopathological examination of diseased tissues during acute infection reveals an exuberant inflammatory response that appears disproportionate to the number of organisms present. For this reason, much of the pathology associated with Lyme disease is believed to be due to the host immune response to proinflammatory spirochaete components, rather than to direct tissue damage from the organism itself. Recent advances have led to an emerging picture whereby regulation of the host response to limit pathology may come at the expense of pathogen survival, resulting in a state of 'concomitant immunity' that characterizes the persistent phase of untreated *B. burgdorferi* infection.

Lyme disease begins when *B. burgdorferi* is introduced into the skin of the vertebrate host during tick feeding (see Chapters 10 and 11). The clinical manifestations of Lyme disease reflect the immune response to the spirochaete as it multiples in the skin and disseminates through the blood and tissues to establish foci of infection in other organ systems (see Chapter 18). In humans, clinically apparent disease is most often seen in the skin (erythema migrans), the heart (conduction system abnormalities and myocarditis), the joints (arthritis) and the nervous system (especially cranial nerve abnormalities, meningitis/meningoencephalitis, and radiculoneuropathies). These manifestations can remit without specific intervention, only to recur in the same or different site. Older studies of Lyme disease in human subjects, many of whom had the infection for several years prior to diagnosis, have documented the extent of pathology that can occur in various organs during the course of untreated *B. burgdorferi* infection (Duray, 1988). These important clinical/pathological correlates have allowed a comparison to be made between the cellular infiltrates in human tissues and those of *B. burgdorferi*-infected laboratory mice, the most widely studied animal model of Lyme borreliosis. Mice develop two of the major clinical manifestations of human Lyme disease, arthritis and myocarditis, which undergo immune mediated regression that requires T and B lymphocytes. It is important to note that lymphocytes dominate the cellular infiltrates of all human tissues examined by pathology, whereas innate immune cells are more prevalent in inflammatory lesions seen in mice, with T and B lymphocytes infiltrating organs as disease resolves. These differences may reflect, in part, the duration of infection at which samples are evaluated for pathology, as lymphocytes and plasma cells can be seen infiltrating diseased sites during the resolution phase of murine Lyme borreliosis. Mice also develop chronic inflammation in tissues, characterized by segmental perivascular and perineural infiltrates of lymphocytes and plasma cells that are often in association with visible spirochaetes in adjacent tissues.

Many investigators have contributed to our understanding of the responses of the vertebrate host that lead to or modulate the inflammatory pathologies and which can prevent or attenuate infection. Some studies have focused on the interaction of *B. burgdorferi* with cultured cells, while others have taken advantage of mice to assess *in vivo* processes associated with disease. Additional reports have employed human specimens to assess the progression of cell types recruited to sites of infection. Taken together, these studies provide insight into the initial responses of the host, and provide direction for the quest to identify factors required to limit the expansion of *B. burgdorferi* during infection. These studies have also provided insight into the pathways involved in the development of pathological conditions such as arthritis, myocarditis and to a lesser extent, neurological disease. In this chapter the contributions of the innate and acquired host responses to infecting *B. burgdorferi* in controlling bacterial growth *in vivo* and in pathological developments will be discussed. Aspects of the response that are unique to different sites, such as the skin, joint, heart, and CNS, will be reviewed, as well as the involvement of localized responses in regulation of pathology.

15.2 Establishment of infection: the role of the tick vector in Lyme borreliosis

Lyme disease begins with dermal introduction of *B. burgdorferi* into the mammalian host during tick feeding. The skin constitutes the first barrier to tick-borne infection, and breeching of this barrier at the tick feeding site creates a critical portal for passage of the pathogen between the tick and the blood meal host (Frischknecht, 2007). As with other vector-borne pathogens, *B. burgdorferi* benefits from mechanisms that *Ixodes* ticks have evolved to obtain the blood meal and to prevent host immunity to tick refeeding. Blood feeding of *Ixodes* ticks is a highly ordered process that occurs over 3–10 days, with the majority of the blood ingested during the final 24 hours of attachment (Kemp *et al.*, 1982). During this time, tick saliva bathes the feeding pit with pharmacologically active substances that impair haemostasis, host immunity and wound healing, and which can secondarily facilitate pathogen transmission to the blood meal host. Survival of *B. burgdorferi* in the new host depends on its ability to exploit these effects of tick saliva and the unique ecological niche of the tick feeding site (Nuttall and Labuda, 2004).

Several studies have shown that tick saliva can enhance infectivity of *B. burgdorferi*. *I. ricinus* salivary gland extracts impede macrophage production of reactive oxygen species and killing of *B. afzelii* by mouse macrophages *in vitro* (Kuthejlova *et al.*, 2001), suggesting a mechanism whereby tick saliva could enhance *Borrelia* survival after deposition within the host. Co-inoculation of *I. scapularis* salivary gland lysate with *B. burgdorferi* enhances pathogen burden in multiple organs in mice in comparison to syringe inoculation with *B. burgdorferi* alone (Zeidner *et al.*, 2002). The recent mapping of the *I. scapularis* sialotranscriptome and analysis of genes that are differentially expressed during tick feeding have identified salivary components with a broad range of activities, some of which could impact *B. burgdorferi* transmission and survival in mammals (Narasimhan *et al.*, 2002b; Ribeiro *et al.*, 2006). These molecules have been catalogued into functional groups, such as the basic tail family proteins (e.g. salp14) that impair blood clotting (Narasimhan *et al.*, 2002a; Ribeiro *et al.*, 2006), the tissue factor pathway inhibitors ixolaris and penthalaris (Francischetti *et al.*, 2004; Francischetti *et al.*, 2002), vasodilators such as prostacyclin and prostaglandin E_2 (Ribeiro *et al.*, 1988; Sa-Nunes *et al.*, 2007), and inhibitors of angiogenesis (Francischetti *et al.*, 2005). More relevant to the survival of *B. burgdorferi* in the skin are factors that retard local inflammation at the tick bite site, especially those that interfere with the activity of phagocytes, antibody and complement. Tick saliva contains an abundance of protease inhibitors that impair neutrophil functions such as chemotaxis, integrin expression, and reactive oxygen generation by sialostatin L, ILS919, and ISL 1373 (Kotsyfakis *et al.*, 2006; Guo *et al.*, 2009), respectively. Other factors impair macrophage activation or interfere with the alternative pathway of complement activation, including Isac and salp20 (Tyson *et al.*, 2007; Valenzuela *et al.*, 2000). The actions of these molecules at the tick feeding site may help pathogens like *B. burgdorferi* evade the initial skin immune defence and establish infection in the mammal.

Modulation of adaptive immunity by tick feeding may also enhance *B. burgdorferi* survival within the host. Tick saliva contains an IL-2 binding protein that inhibits T-cell proliferation *in vitro* (Gillespie *et al.*, 2001). In addition, PGE_2 in tick saliva increases IL-10 and reduces IL-12 and TNFα production by dendritic cells, and suppresses dendritic cell-induced $CD4^+$ T-cell proliferation and IL-2 production *in vitro* (Sa-Nunes *et al.*, 2007). Repeated tick infestation of C3H/HeN mice leads to reduced IL-2 and IFN-γ production by splenocytes after conconavalin A stimulation, and augments IL-4 and IL-10 production. The increase in anti-inflammatory cytokine production, especially IL-10, may help suppress phagocyte activation and clearance of *B. burgdorferi*. In support of this, treatment of C3H/HeJ mice with TNFα, or with IL-2 or IFN-γ for 10 days after onset of *B. burgdorferi*-infected tick feeding reduced mouse infection rates by 95% and 30–45%, respectively (Zeidner *et al.*, 1996). Langerhans cells are critical for tick-mediated suppression of the conconavalin A-induced Th1 response in lymph nodes draining the tick bite site, but their impact is attenuated when *B. burgdorferi* is introduced by the feeding ticks (Vesely *et al.*, 2009). *Borrelia burgdorferi* also has exploited a tick salivary protein, salp15, to colonize the blood meal host (Ramamoorthi *et al.*, 2005). The expression of this protein is upregulated in infected tick salivary glands and may interact with the outer surface protein (Osp) C on spirochaetes as they enter the host. Recombinant salp15 injected with spirochaetes enhances their survival in mice, and siRNA depletion of the protein in infected ticks reduces *B. burgdorferi* infectivity after vector transmission (Ramamoorthi *et al.*, 2005). Salp15 can shield *B. burgdorferi* from borrelicidal antibodies *in vitro*, suggesting that it could interfere with antibody-mediated killing *in vivo*. Although salp15 binds to the T-cell coreceptor CD4 and inhibits T-cell receptor signal transduction (Garg *et al.*, 2006; Juncadella *et al.*, 2007), it is unclear how this function potentiates the survival of a pathogen for which phagocytes and T-cell independent antibody production are the principal modes of destruction.

15.2.1 Tissue adaptation of *Borrelia burgdorferi* and early dissemination

Several studies have documented a delay in dissemination of *B. burgdorferi* from the site of the tick bite or syringe inoculation. The most detailed kinetic studies were performed in mice and showed that spirochaetes remain locally in the skin for several days after the tick bite, as excision of a 0.6 mm area of skin around the feeding site or topical application of an antibiotic up to 2 days after tick detachment can abort infection (Shih *et al.*, 1992; Shih and Spielman, 1993). This pause in dissemination is not overcome by a large inoculum dose achieved with infection introduced by syringe inoculation, nor is it unique to murine hosts as supported by observations with dogs, rabbits, and humans. The discovery of regulated expression of lipoproteins, now extensively studied *in vitro* and *in vivo*, has led to the concept that the bacteria disseminating from the site of infection may be expressing a distinct set of antigenic proteins when compared with bacteria in culture (Pal *et al.*, 2004) or in the tick (Das *et al.*, 1997; Grimm *et al.*, 2004; Miller *et al.*, 2003; Nowalk *et al.*, 2006; Schwan *et al.*, 1995; Suk *et al.*, 1995). This has led to the hypothesis that part of the delay in dissemination represents a 're-programming' of gene expression by bacteria in the skin (de Silva and Fikrig, 1997; Schwan, 2003). The host response plays a critical role in the profile of outer surface protein expression on spirochaete populations infecting the host. The expression of several outer surface proteins and of OspA and OspC in particular, has been extensively studied in the mouse, the tick, as well as with different culture conditions (de Silva and Fikrig, 1997; Ohnishi *et al.*, 2001; Schwan and Piesman, 2000; Stewart *et al.*, 2006; Tilly *et al.*, 2006). Direct *in vivo* evidence that OspA was downregulated and OspC upregulated on spirochaetes infecting the host was provided over a decade ago (Montgomery *et al.*, 1996), explaining the absence of OspA antibodies in mice infected with low inocula of spirochaetes. In addition, both immune serum and anti-OspA antibodies were found to be effective at preventing infection for only 24–48 hours following inoculation, and host-adapted bacteria in skin transplants were not cleared by anti-OspA antibody, indicating loss of expression this protein (Barthold *et al.*, 1995; de Silva *et al.*, 1998; Hodzic *et al.*, 2003; Hodzic *et al.*, 2005). OspA plays a critical role for spirochaetes within the tick vector, providing protection from antibody acquired during the blood meal of infected hosts (Battisti *et al.*, 2008). Although OspC expression is upregulated on spirochaetes in feeding ticks and as they enter the host, recent studies show that OspC is expressed for only a short time following inoculation into the skin of mice (Liang *et al.*, 2004; Tilly *et al.*, 2007; Tilly *et al.*, 2006). Extended expression of OspC is selected against, presumably by antibody, as shown by the decreased survival of bacteria forced to express OspC in wild type mice relative to *scid* mice (Liang *et al.*, 2002; Tilly *et al.*, 2006; Xu *et al.*, 2006). This raises the question: what cues are provided by the host that leads to the downregulation in expression of some bacterial genes and upregulation of others? Much speculation surrounds this question, and data have been provided to support involvement of the following in host regulation of *Borrelia* gene expression: changes in temperature and pH, presence of oxidative stress, products of the inflammatory response, and steroid hormones (Anguita *et al.*, 2001; Carroll *et al.*, 1999; Ramamoorthy and Scholl-Meeker, 2001; Scheckelhoff *et al.*, 2007; Seshu *et al.*, 2004; Stevenson *et al.*, 1995; Yang *et al.*, 2000). Changes in protein expression could serve several purposes, including facilitating nutrient acquisition by replicating spirochaetes, evasion of innate and adaptive immune-mediated destruction, and upregulation of proteins necessary for the spirochaete to disseminate and persist within the mammal.

15.2.2 The early host defence against *Borrelia burgdorferi*: a critical role for innate immunity

Review of the literature immediately identifies numerous studies pointing to the involvement of the innate host defence in the response to *B. burgdorferi*. As discussed in later sections, both the acquired and innate defences are important in the control of spirochaetes in tissues, with the humoral response particularly important in resolution of disease and protection from infection. Innate immune mechanisms are probably engaged to some extent at all stages of infection and wherever spirochaetes are found in the mammal. Their greatest impact, however, would be during phases where *B. burgdorferi* establishes infection, replicates and disseminates throughout the host. *B. burgdorferi* can grow in the presence of human serum, a feature attributed to its ability to resist complement activation and complement-mediated lysis, (Kochi *et al.*, 1991; Patarakul *et al.*, 1999). This property is common to many human pathogens, and is consistent with the need for this organism to survive extracellularly and disseminate haematogenously. Studies with mice have indicated involvement of complement in response to *B. burgdorferi*. The fifth component of complement, C5, which is essential for formation of the membrane attack complex, is not required for antibody-mediated protection from infecting *B. burgdorferi* (Bockenstedt *et al.*, 1993). In contrast, mice deficient in C3, central to complement pathway mediated opsonization, harboured greater numbers of *B. burgdorferi* in some tissues and required five-fold fewer *B. garinii* spirochaetes to establish infection (Lawrenz *et al.*, 2003; Woodman *et al.*, 2007). Together, these studies in mice suggest a role for complement opsonization for clearance of the organism by phagocytes, rather than in lysis of the bacteria. The survival of *B. burgdorferi* in human serum has been attributed to the presence of outer surface proteins that bind factor H of human complement. Five proteins have been identified as complement regulator-acquiring surface proteins, CRASPs (Alitalo *et al.*, 2002; Brooks *et al.*, 2005; Hartmann *et al.*, 2006; Kraiczy *et al.*, 2004; McDowell *et al.*, 2003) and their expression varies, with some expressed by spirochaetes during tick transmission while others are expressed during infection of the mammalian host (Bykowski *et al.*, 2007; Miller *et al.*, 2003). Biochemical analysis of the interaction of the CRASPS with human factor H indicates high affinity and specificity (Alitalo *et al.*, 2004; Kenedy *et al.*, 2009) and this interaction has been shown to prevent the activation and deposition of complement via the alternative pathway. However, to date the importance of this interaction in an infection model has not been demonstrated, as mice deficient in factor H displayed normal control of *B. burgdorferi* in tissues (Woodman *et al.*, 2007). This may reflect the lower affinity interaction between CRASPs and murine factor H, or point to additional mechanisms by which the spirochaete resists complement. Regardless of the precise role of the CRASPs, it is apparent that the ability of *B. burgdorferi* to avoid complement-mediated clearance in the human and the mouse is essential to the establishment and persistence of infection.

Many studies have pointed to involvement of cellular host defences in control of *B. burgdorferi* in tissues. Phagocytic cells such as monocytes, macrophages, neutrophils, and glial cells, have been found to kill *B. burgdorferi in vitro* (Benach *et al.*, 1984a; Cruz *et al.*, 2008; Kuhlow *et al.*, 2005; Liu *et al.*, 2004; Lusitani *et al.*, 2002; Malawista *et al.*, 1996; Montgomery *et al.*, 1996; Montgomery *et al.*, 1993; Moore *et al.*, 2007; Shin *et al.*, 2008). Early investigations established antibody dependent and independent phagocytosis and killing of *B. burgdorferi* by human monocytes and PMNs (Benach *et al.*, 1984a,b). Other studies revealed that while macrophage-mediated killing did not require IgG or complement opsonization of *B. burgdorferi* for internalization and killing, phagocytosis and killing by human PMNs required IgG opsonization of the bacteria (Lusitani *et al.*, 2002). This requirement could be circumvented *in vitro* by lysis of the PMNs and use of cytosolic components (Lusitani *et al.*, 2002). Phagocytes were found to directly recognize *B. burgdorferi* via the beta integrin CR3, and two types of phagocytosis associated with actin filaments have been reported: coiling phagocytosis and FcγR-dependent conventional phagocytosis (Cinco *et al.*, 1997; Linder *et al.*, 2001; Montgomery *et al.*, 1996; Rittig *et al.*, 1992). Our current understanding is that the phagocyte-mediated internalization and killing that are critical for control of the spirochaete *in vitro* are also effective *in vivo*. Several components of phagocytic cells have been identified with the potential to kill or suppress growth of *B. burgdorferi*. These included the granule components elastase, bactericidal/permeability-increasing protein, human neutrophil peptide-1, and the active fragment of the cytosolic component calprotectin, LL37 (Garcia *et al.*, 1998; Lusitani *et al.*, 2002, 2003). *Borrelia burgdorferi* can also be killed *in vitro* by components of the oxidative pathways classic to anti-bacterial defences. Interestingly, mice deficient in NADPH oxidase or iNOS display unimpaired control of infection (Brown and Reiner, 1999a; Crandall *et al.*, 2005; Lusitani *et al.*, 2002; Seiler *et al.*, 1995). These observations might be explained at least in part, by the fact that *B. burgdorferi* lacks both free iron as well as the iron-containing enzymes that in other bacteria are targets for phagocyte-derived reactive oxygen species (Posey and Gherardini, 2000). Another important outcome of phagocyte interaction with *B. burgdorferi* is the activation of inflammatory responses. Recent *in vitro* studies with human mononuclear cells have found that uptake of living bacteria leads to production of a broader spectrum of cytokines than uptake of non viable organisms, a finding that may directly impact activation of adaptive responses (Cruz *et al.*, 2008; Gebbia *et al.*, 2001; Moore *et al.*, 2007). Studies with murine macrophages deficient in individual elements of TLR signalling pathways revealed involvement of parallel signalling pathways, some of which also required phagocytic uptake of bacteria for optimal cytokine production (Liu *et al.*, 2004; Shin *et al.*, 2008). Although some differences have been found with different phagocytic cell types and *in vitro* model systems, the overall implication is that the interaction of *B. burgdorferi* with phagocytic cells results in uptake and killing, and can direct the activation of adaptive responses.

15.2.3 Innate immune cell recognition of *Borrelia burgdorferi*: PAMPS and PRRs

The delineation of mechanisms whereby innate immune cells recognize and respond to *B. burgdorferi* evolved from early studies that identified a panoply of cells capable of producing a diverse spectrum of inflammatory mediators when exposed to *B. burgdorferi* or its products. Responsive cells included haematopoietic cells (PMNs, monocytes, B lymphocytes, mast cells, dendritic cells), as well as vascular and organ-specific cells such as endothelial cells and glial cells (Habicht *et al.*, 1985; Hertz *et al.*, 2001; Ma and Weis, 1993; Marietta *et al.*, 1997; Morrison *et al.*, 1997; Norgard *et al.*, 1996; Radolf *et al.*, 1995; Sellati *et al.*, 1995; Tatro *et al.*, 1994; Wooten *et al.*, 1996). Macrophages and endothelial cells displayed particularly robust responses in products considered to be downstream of NF-κB translocation to the nucleus, such as cytokines (TNF, IL-6, IL-12, IL-1), adhesion molecules (ICAMs), and chemokines (IL-8, etc.) (Norgard *et al.*, 1995, 1996; Sellati *et al.*, 1995; Thoma-Uszynski *et al.*, 2000; Wooten *et al.*, 1996; Wooten and Weis, 2001). Other types of responses not strictly reliant on NF-κB were also documented, including the rapid generation of reactive oxygen intermediates in PMNs, the generation of nitric oxide by macrophages, the mitogenic induction of B-cell proliferation, and maturation of populations such as dendritic cells (Ma *et al.*, 1994; Ma and Weis, 1993; Tatro *et al.*, 1994; Thoma-Uszynski *et al.*, 2000). The outer surface lipoproteins, containing the tripalmitoyl-*S*-glyceryl-cysteine (Pam_3Cys) modification, were potent stimulants for most of these responses. Although *B. burgdorferi* lack lipopolysaccharide, Pam_3Cys-modified proteins are abundantly expressed during infection of ticks and mammals, and lipoproteins were concluded to comprise the major bacterial inflammatory stimulant in cell culture experiments (Sellati *et al.*, 1999).

In 1999, shortly after the characterization of Toll like receptors (TLRs) as pathogen associated pattern recognition receptors (PRRs) in mammalian cells (Poltorak *et al.*, 1998), Toll like receptor type 2 (TLR2) was recognized as playing an obligatory role in signalling by Pam_3Cys –containing lipoproteins (Aliprantis *et al.*, 1999; Brightbill *et al.*, 1999; Hirschfeld *et al.*, 1999; Kochi *et al.*, 1991; Lien *et al.*, 1999; Patarakul *et al.*, 1999). This finding provided insight into how such a broad variety of cell types could be activated by *B. burgdorferi*: most responses were NF-κB dependent and involved the adapter molecule MyD88 (Akira *et al.*, 2000). CD14 facilitated the interaction of purified lipoproteins with TLR2, while TLR1 was required for optimal TLR2-dependent responses to lipoproteins (Alexopoulou *et al.*, 2002; Giambartolomei *et al.*, 1999; Sellati *et al.*, 1999; Sellati *et al.*, 1998; Wooten *et al.*, 1998). The *in vivo* importance of TLR signalling in humans was demonstrated by assessment of patients with low responsiveness to the OspA vaccine, as macrophages from these individuals produced lower levels of the inflammatory cytokines IL-6 and TNF in response to the lipoproteins (Alexopoulou *et al.*, 2002). This defect was linked to low expression of TLR1, documenting its importance in the innate and adaptive responses to Pam_3Cys-modified lipoproteins. TLR2-deficient cells had limited residual ability to activate NF-κB signalling in response to *B. burgdorferi*, which was further depressed in mice deficient in MyD88 (Bolz *et al.*, 2004). Antibody production in $TLR2^{-/-}$ mice was relatively normal, or even elevated, following *B. burgdorferi* infection (Alexopoulou *et al.*, 2002; Wooten *et al.*, 2002; Yoder *et al.*, 2003). MyD88-deficient mice displayed an enhanced and prolonged IgM response to *B. burgdorferi* infection with some alteration in relative abundance of IgG isotypes (Behera *et al.*, 2006a; Bolz *et al.*, 2004; Liu *et al.*, 2004; Woods *et al.*, 2008). The responses of macrophages from TLR2 and MyD88-deficient mice suggested that other cell surface molecules could contribute to the TLR2-independent, MyD88-dependent activation. The receptors for IL-1β and IL-18 both require MyD88 to transmit intracellular signals and these cytokines are produced during *B. burgdorferi* infection. Absence of IL-18, however, does not alter host susceptibility to *B. burgdorferi* infection and disease (Behera *et al.*, 2006a). A recent study has further demonstrated that the caspase 1-dependent inflammasome, required for *in vitro* production of IL-1β by murine macrophages stimulated with *B. burgdorferi,* is not critical for host defence (Liu *et al.*, 2009). Additional evidence for the existence of pathways involving other pathogen-associated molecular patterns (PAMPs) and pattern recognition receptors (PRR) has come from reports demonstrating that a greater complexity of responses are triggered in human and mouse mononuclear cells cultured with living *B. burgdorferi* than with purified lipoproteins or with sonicated preparations of spirochaetes (Benhnia *et al.*, 2005; Cruz *et al.*, 2008; Lazarus *et al.*, 2008; Moore *et al.*, 2007) (Dennis *et al.*, 2009; Salazar *et al.*, 2009). These ligands may be lost from non-viable bacteria or their configuration after destruction of the spirochaete may not optimally stimulate cells. Alternatively, novel endocytic compartments may be accessible only to living *Borrelia,* but these interesting interactions have not yet been characterized. Interestingly, both virulent and attenuated *B. burgdorferi* displayed similar ability to promote production of proinflammatory cytokines by macrophages (Wang *et al.*, 2008a).

The importance of MyD88-dependent TLR pathways in host defence against *B. burgdorferi* has been underscored by studies in TLR2 and MyD88-deficient mice, which revealed a severe defect in control of bacteria in tissues that was not linked to defective production of immunoglobulin (Behera *et al.*, 2006a; Bolz *et al.*, 2004; Liu *et al.*, 2004; Wang *et al.*, 2002, 2004). Thus, TLR-dependent activation of phagocytic cells appears necessary for effective antibody-mediated destruction of bacteria. *In vitro* studies with macrophages from MyD88-deficient mice revealed slower kinetics for opsonin-independent destruction of *B. burgdorferi* (Liu *et al.*, 2004) and reduced phagocyte-mediated killing (Shin *et al.*, 2008). PMNs and monocytes do gain access to infected tissues of TLR2 and MyD88-deficient mice, thus the extraordinary defect in control of *B. burgdorferi* during infection may depend on failure to effectively phagocytose or activate killing pathways at the site of infection. Numerous anti-bacterial compounds are induced following TLR/MyD88 dependent signalling and could be critical to control in a localized environment (Akira *et al.*, 2000; Akira and Takeda, 2004). Clearly there is a need for quantitative experiments to understand the dynamics of *B. burgdorferi* stasis and death during infection.

An unexpected finding in TLR2 and MyD88-deficient mice infected with *B. burgdorferi* was the appearance of both arthritis and carditis. The identification of TLR2 as the receptor for the major stimulatory component of *B. burgdorferi* led to the prediction that *B. burgdorferi* infection of mice deficient in TLR2 would fail to display arthritis. In fact, these mice developed severe arthritis: TLR2$^{-/-}$ C57BL/6 mice display more severe arthritis than mildly arthritic C57BL/6 mice and increased severity could even be demonstrated on the disease-susceptible C3H background (Alexopoulou *et al.*, 2002; Wang *et al.*, 2004; Wang *et al.*, 2008b). Joint tissues of TLR2$^{-/-}$ mice harbour increased numbers of *B. burgdorferi,* and this was initially proposed to be responsible for the increase in arthritis severity. However, studies with TLR2$^{-/-}$/*scid* double mutant mice demonstrated that the increased arthritis severity in TLR2$^{-/-}$C57BL/6 mice was suppressed by the simultaneous presence of the *scid* mutation: TLR2$^{-/-}$/*scid* mice had mild arthritis while harbouring more *B. burgdorferi* in joint tissue than severely arthritic TLR2$^{-/-}$ mice (Wang *et al.*, 2005). Therefore, the increased arthritis severity in TLR2$^{-/-}$ mice was not solely due to the increase in bacterial number, underscoring the interpretation that TLR2 is not required for Lyme arthritis development as long as lymphocyte-dependent pathways are intact. C3H mice deficient in CD14, a molecule shown to facilitate lipoprotein-mediated activation of cells, also displayed increased severity of arthritis and greater spirochaete burden in tissues (Benhnia *et al.*, 2005). Additionally, CD14$^{-/-}$ mice displayed heightened inflammatory responses *in vivo*, suggesting a previously unappreciated modulatory role for CD14 in lipoprotein-mediated signalling. In *in vitro* studies, live *Borrelia* was required for the enhanced CD14-independent signalling, revealing TLR2-independent signalling pathways in murine macrophages. Infected MyD88-deficient mice on the C57BL/6 background also developed varying severities of arthritis and carditis, although in this case the compromised ability to control bacteria in tissues was extreme (Behera *et al.*, 2006a; Bolz *et al.*, 2004; Liu *et al.*, 2004). These surprising results suggested the existence of additional signalling pathways involved in host resistance and disease that are not as robust *in vitro* as the well characterized TLR-dependent pathways, and which may become more prominent *in vivo* as part of the immune response to host adapted spirochaetes. Additionally, disruption of regulatory pathways normally induced by lipoproteins may contribute to the increased inflammatory responses of CD14$^{-/-}$ mice and increased arthritis severity of TLR2$^{-/-}$ mice.

An important outcome of the interaction with phagocytes is the generation of proinflammatory cytokines with the ability to amplify innate host defences and to activate the adaptive immune response (Cruz *et al.*, 2008; Liu *et al.*, 2004; Moore *et al.*, 2007; Shin *et al.*, 2008). However, anti-inflammatory cytokine responses to *B. burgdorferi* have also been observed, both *in vivo* and *in vitro*. IL-10 is produced by macrophages from mouse, human, and rhesus monkeys incubated with *B. burgdorferi* and its lipoproteins, with optimal signalling dependent on both TLR2 and CD14 (Benhnia *et al.*, 2005; Brown *et al.*, 1999; Dennis *et al.*, 2006; Giambartolomei *et al.*, 1998; Wooten *et al.*, 2002). Interestingly, it has recently been shown that living *B. burgdorferi* induce greater production of IL-10 by murine macrophages than do equivalent numbers of non-viable organisms (Lazarus *et al.*, 2008). Addition of IL-10 to cell cultures has been shown to modulate proinflammatory cytokine production by macrophages and B-cells in response to *B. burgdorferi* and the *B. burgdorferi*-induced production of IL-10 is greater in macrophages from mildly arthritic C57BL/6 mice than severely arthritic C3H mice (Brown *et al.*, 1999; Lazarus *et al.*, 2006). The role of IL-10 in modulating Lyme arthritis severity was shown, as infection of either IL-10$^{-/-}$ C57BL/6 mice or IL-10$^{-/-}$ C3H mice results in more severe disease than was observed in the appropriate wild type parent strain (Brown *et al.*, 1999, 2008). The suppression of arthritis by IL-10 is not due to better control of spirochaetes in tissues as IL-10$^{-/-}$ mice harbour reduced numbers of bacteria in their tissues. The enhanced control of bacteria *in vivo* is dependent on innate host defences, and not due to an observed heightened production of IgG (Lazarus *et al.*, 2006). Thus, although release of the tight control of inflammation by IL-10 deficiency results in more severe arthritis, this is coupled with more effective control of the bacteria. The role of IL-10 in controlling arthritis development is not well understood, as regulation of IL-10 production is complicated, it depends on the cell type, and it is generally considered homeostatic for proinflammatory cytokines (Moore *et al.*, 2001). Mechanistic insight has come from *in vitro* studies in which SOCS1 and SOCS3 have been implicated in the IL-10 mediated suppression of *B. burgdorferi* induced inflammatory responses in macrophages (Dennis *et al.*, 2006), while others have found that IL-10 effects on the endothelium can influence inflammatory cell infiltration (Lisinski and Furie, 2002). Apoptosis was also observed in human monocytes following treatment with lipoproteins or after phagocytosis of *B. burgdorferi*, suggesting another means of limiting inflammatory responses (Aliprantis *et al.*, 1999; Cruz *et al.*, 2008).

15.2.4 MMPs and MAP kinase pathways

Borrelia burgdorferi induction of host matrix metalloproteinases has been characterized at all stages of Lyme disease, including dissemination from the site of the tick bite, invasion of tissues, and in the response of arthritic joints (Behera *et al.*, 2005; Gebbia *et al.*, 2001; Hu *et al.*, 2001). Chondrocytes in the joint can produce a variety of products associated with the tissue destruction and remodelling seen in many forms of arthritis (Hu *et al.*, 2001). The MMP responses to *B. burgdorferi* have been studied in joint tissue from the mouse, in synovial fluid from humans, and in cultures of murine and human chondrocytes (Behera *et al.*, 2005; Gebbia *et al.*, 2001; Hu *et al.*, 2001). MMP9 has been isolated from erythema migrans lesions of the skin whereas MMP2, MMP3 and MMP9 have been identified in synovial fluid of Lyme arthritis patients (Hu *et al.*, 2001; Zhao *et al.*, 2003). Expression analysis of joint tissue from infected C3H mice has indicated induction of transcripts for MMP3 and MMP19 (Behera *et al.*, 2005), while microarray analysis also revealed up-regulation of MMP2, MMP9, MMP13, MMP23, and TIMP1 (Crandall *et al.*, 2006). *In vitro* studies with cultured human chondrocytes revealed *B. burgdorferi* induction MMP1, MMP3, MMP13 and MMP19 proteins, and transcription activation of MMP10 and TIMP1 (Behera *et al.*, 2005). The activation of human chondrocytes by *B. burgdorferi* is mediated by integrin α3β1 and the resulting induction of MMP-1 and MMP-3 synthesis is dependent on activation of JNK and p38 kinases (Behera *et al.*, 2006c; Behera *et al.*, 2004). MAP kinase activation by *B. burgdorferi* has also been characterized in cultured astrocytes and T-cells, indicating this to be a commonly activated pathway in response to *B. burgdorferi* (Hedrick *et al.*, 2006; Perides *et al.*, 1999). Furthermore, mice deficient in the p38 MAP kinase display reduced incidence of *B. burgdorferi*-induced arthritis, (Anguita *et al.*, 2002). In contrast, MMP9 and MMP1 production by human monocytes exposed to *B. burgdorferi* is partially dependent on TLR2 (Gebbia *et al.*, 2004). Together these studies suggest a model by which both MAP kinase-dependent and TLR2-dependent responses of chondrocytes to *B. burgdorferi* participate in the release of MMPs of the joint tissue, and suggest this to be a general feature of the response to *B. burgdorferi* in tissues.

15.3 Responses of acquired host defences

15.3.1 B-cell responses

The importance of B-cell activation and Ig production in the control of *B. burgdorferi* is well accepted, with many studies demonstrating the ability of immune serum or monoclonal antibodies to provide protective immunity, reduce pathogen burden in tissues as well as modulate disease. Mice deficient in antibody production, including SCID, B6*rag1*$^{-/-}$, B6*Igh*$^{-/-}$, and B6μMT mice, exhibit a markedly elevated pathogen burden in blood and tissues that can be reduced by passive transfer of immune mouse serum or B-cells, but not T-cells alone (Barthold *et al.*, 1992b; Bockenstedt *et al.*, 2001; McKisic and Barthold, 2000; Schaible *et al.*, 1989). *In vitro*, the polyclonal activation of B lymphocytes by *B. burgdorferi* and lipoproteins is dependent on TLR2; however, the robust antibody response to infection in TLR2 and MyD88-deficient mice indicates activation can be supported by other signalling pathways *in vivo*. Antibodies that arise in the absence of T-cell help are sufficient to prevent infection and to reduce pathogen burden in infected antibody-deficient mice (Fikrig *et al.*, 1996; McKisic and Barthold, 2000). These antibodies derive from three B-cell populations: B1 B-cells, which are a source of natural antibodies present in pre-immune serum; marginal zone B-cells, a B2 B-cell subset which produce both pathogen-specific antibodies and antibodies to T independent antigens, and follicular B2 B-cells, which produce IgM prior to isotype switching after interaction with activated T-cells. Humoral defence begins prior to deposition of spirochaetes within the host, when ticks ingest interstitial fluid containing IgG or haemorrhaged blood, which contains all Ig subclasses, at the feeding site (Belperron and Bockenstedt, 2001). Infected ticks that feed on B-cell-deficient mice (B6*Igh*$^{-/-}$) harbour more spirochaetes than those that feed on wild type mice, and passive transfer of normal mouse serum into B6*Igh*$^{-/-}$ mice reduces spirochaete burden in feeding ticks. Natural antibodies of the IgM subclass are sufficient to lead to this reduction, whereas IgG antibodies have no effect. IgG can pass into the tick haemolymph where it is bound by immunoglobulin-binding proteins and is excreted in the tick saliva back into the host, a mechanism by which the tick may interrupt the host immune response to tick feeding (Wang and Nuttall, 1999). The tick salivary protein salp15 has been shown *in vitro* to protect *B. burgdorferi* from the borreliacidal effects of immune mouse serum, and it is possible that salp15, by binding to Osp C expressed by spirochaetes in the salivary gland, may shield spirochaetes from natural or pathogen-specific antibodies as they are deposited in the feeding pit (Ramamoorthi *et al.*, 2005; Wang and Nuttall, 1999). This may provide one reason why Osp C is indispensable for initiating vector-borne infection in mice, but the mechanism does not only involve evasion of B-cell responses as mutant spirochaetes lacking Osp C cannot infect SCID mice (Grimm *et al.*, 2004).

Spirochaetes that become blood borne can encounter marginal zone B (MZB) cells, a B2 B-cell subset located near the marginal sinuses of the spleen in mice and also in lymphoid tissue of humans (MacLennan and Liu, 1991), resulting in their expansion (Malkiel *et al.*, 2009). These innate B-cells express Fc and complement receptors and are uniquely positioned to sample opsonized pathogens (Oliver *et al.*, 1997). MZB-cells produce antibodies to T-independent antigens and are an early source of *B. burgdorferi*-specific IgM. Depletion of MZB-cells results in elevated pathogen burden and enhanced arthritis at 2 weeks of infection in disease-susceptible C3H mice (Belperron *et al.*, 2007). MZB-cells are known to assist in the priming of naïve T-cells by transporting IgM-containing immune complexes to the follicular dendritic cells (Ferguson *et al.*, 2004). Not unexpectedly, then, pathogen specific IgG was also reduced in *B. burgdorferi*-infected MZB-cell depleted mice, which correlated with decreased splenic $CD4^{+}$ T-cell proliferation and cytokine production. Passive transfer of immune mouse serum to *B. burgdorferi*-infected MZB-cell-depleted mice was sufficient to reduce pathogen burden, but did not alter T-cell activation.

The early activation of B-cells following infection has also been documented in draining lymph nodes, and engagement of the adaptive immune response results in T-cell activation and follicular B2 B-cell production of isotype switched antibodies to *B. burgdorferi*. Although not required for protective or disease-resolving immunity, *B. burgdorferi*-specific IgG probably contributes to both during the course of natural infection. Polyclonal antisera from experimentally infected mice and from humans with Lyme disease can prevent challenge infection in mice, as can select monoclonal IgG produced from infected mice (Fikrig *et al.*, 1994). Among the antigens targeted by protective antibodies are the surface exposed lipoproteins Osp A, Osp B, Osp C, p35, p37, p39, BmpA, BmpB and decorin binding proteins (Dbp) A and B (Fikrig *et al.*, 1990; Fikrig *et al.*, 1992; Fikrig *et al.*, 1997; Hanson *et al.*, 1998; Pal *et al.*, 2008). The ability of some of these antigens to elicit protective immunity after immunization appears to be *B. burgdorferi* strain dependent, as neither Osp C nor BmpA from *B. burgdorferi* N40 can elicit protective immunity (Barthold *et al.*, 1997; Bockenstedt, 2009). Passive transfer of immune serum from infected mice can also reduce pathogen burden and attenuate arthritis and carditis in SCID, *rag* and B-cell-deficient mice. In contrast, serum from mice immunized with specific antigens can attenuate disease without appreciably changing the pathogen burden. Polyclonal antisera to arthritis-related protein (Arp) or decorin binding protein A (DbpA) passively transferred to *B. burgdorferi*-infected SCID mice can attenuate disease by selectively eliminating spirochaetes in the synovium and myocardium, but not in the adjacent connective tissue of tendons, ligaments and cardiac blood vessel walls (Barthold *et al.*, 2006). This may be due to an inability of IgG to penetrate these relatively avascular structures in sufficient quantities to kill spirochaetes, or to an alteration in the expression of the target antigens by spirochaetes at these sites.

There is a growing body of evidence indicating that *B. burgdorferi*-specific IgG can also contribute to disease pathogenesis. IgG subclasses may influence the severity of inflammation by their differential engagement of activating and inhibitory Fcγ receptors (FcγR) on vasculature and immune cells within the host (Nimmerjahn *et al.*, 2005). In particular, IgG2a engagement of FcγRIV on innate immune cells results in greater inflammatory cytokine production than other isotypes. The relative expression of activating versus inhibitory FcγR depends on the cytokine milieu, with proinflammatory cytokines such as TNFα and IFNγ enhancing activating FcγR expression. In disease susceptible mouse strains that develop pathogen-specific antibodies largely of the IgG2a subclass, opsonization of spirochaetes and FcγR-mediated clearance may induce a greater inflammatory response in comparison to similar infection of mouse strains in which other subclasses are dominant. The human B-cell response to *B. burgdorferi* infection has been less well characterized according to responding B-cell subsets in comparison to the murine response. As expected for exposure to new antigens, IgM responses are the first detected in serum, usually by the 3rd to 6th week of the illness, whereas specific IgG responses increase more slowly (Steere *et al.*, 1983). As infection progresses beyond 6 weeks, IgM responses may remain elevated or may return to normal, but the *B. burgdorferi*-specific IgG titres increase and are highest during the late phase of disease, when arthritis is present (Aguero-Rosenfeld *et al.*, 2005; Craft *et al.*, 1986). IgG titres may remain elevated for years after treatment despite clinical remission. As assessed by immunoblot, antibodies in human serum react to a similar array of *B. burgdorferi* proteins as those in serum from experimentally infected mice, indicating that spirochaetes undergo similar antigenic changes in humans and mice. In one study, the IgG response to select *B. burgdorferi* antigens (OspA and OspB) correlated with more severe and prolonged arthritis, whereas a higher IgG response to p35 correlated with milder episodes and shorter duration of arthritis (Akin *et al.*, 1999). It should be noted that humans with more severe Lyme arthritis typically have Th1 dominant responses to the pathogen, but a relationship between IgG subclasses and arthritis severity has not been investigated.

15.3.2 T-cell response

As with B-cells, a variety of innate and adaptive T-cell subsets participate in the immune response to *B. burgdorferi*. T-cells bearing the γδ T-cell receptor have a limited diversity and are considered part of the innate immune system. Subsets of γδ T-cells, termed dendritic epidermal T-cells, reside in the skin and are present at the earliest stages of *B. burgdorferi* infection. γδ T-cells promote production of the Th1 cell cytokine IFNγ and the Th1 associated IgG2a isotype after syringe-borne infection with *B. burgdorferi*, but vector-transmission overrides these effects (Bockenstedt *et al.*, 2003). In humans, synovial fluid V-delta 1-expressing γδ T-cells have been shown to respond to *B. burgdorferi* lipoproteins and lipopeptides, and to induce maturation of dendritic cells through Fas/Fas ligand interaction, resulting in upregulation of costimulatory molecules and IL-12 secretion (Collins *et al.*, 2005; Roessner *et al.*, 2003; Vincent *et al.*, 1998). Activation of γδ T-cells *in vitro* by *B. burgdorferi* is indirect and requires both TLR and caspase mediated signalling by dendritic cells or monocytes (Collins *et al.*, 2008). Dendritic cells express high levels of the Fas inhibitor c-FLIP and are therefore resistant to Fas signalling induced cell death. Synovial γδ T-cells also exhibit potent cytotoxic activity towards a variety of synovial cell types in the presence of *B. burgdorferi*, which may serve to help attenuate the inflammatory response.

NKT cells represent a second innate T-cell subset involved in host defence against *B. burgdorferi*. These cells express an invariant T-cell receptor and respond to lipid and glycolipid antigens presented by the MHC-like restriction element CD1d. CD1d can bind the *B. burgdorferi* diacylglycerol BbGL-II to activate NKT-cells *in vitro* (Kinjo *et al.*, 2006), and these cells can in turn activate other cells that can contribute to immune defence, including NK cells, dendritic cells, T-cells and B-cells. Absence of CD1d (and therefore NKT cells) results in higher bacterial burden and enhanced arthritis early during the course of *B. burgdorferi* infection in mice (Kumar *et al.*, 2000). The impaired host defence is in part mediated through reduced pathogen-specific antibody production by MZB-cells, which express high levels of CD1d. MZB-cells constitutively express T-cell costimulatory molecules, suggesting that their responses could be modulated by T-cells, including NKT-cells. However, MZB-cells can proliferate and produce IgM in response to viable *B. burgdorferi* in the absence of NK T-cells *in vitro*. NKT-cells have been found to contribute to localized host defence in the heart and joints, by a mechanism at least partially dependent on IFNγ enhanced phagocytosis of the spirochaete (Olson *et al.*, 2009; Tupin *et al.*, 2008). Taken together, these findings suggest that NKT-cells may not be required for the *B. burgdorferi* antibody response, but nevertheless could modulate it *in vivo*.

The observation that SCID and $rag^{-/-}$ mice develop persistent arthritis and myocarditis has underscored an important role for adaptive immunity in disease resolution. Adoptive transfer experiments confirm that B-cells and pathogen-specific antibody are required for pathogen elimination and disease regression, whereas the contribution of αβ T-cells to host defence remains incompletely defined. Early studies in which $CD4^+$ and $CD8^+$ T-cell subsets were depleted in disease-resistant BALB/c mice and disease-susceptible C3H mice suggested that $CD4^+$ T-cells helped reduce pathogen burden and decrease arthritis severity, whereas $CD8^+$ T-cells promoted the disease process. Notably, dissociation was found between pathogen-specific IgG levels and pathogen burden, implicating antibody-independent mechanisms in host defence (Keane-Myers and Nickell, 1995b). Three subsequent studies examining $CD4^+$ T helper (Th)1 and Th2 cytokines correlated IFNγ with elevated pathogen burden and increased arthritis and IL-4 with milder disease (Keane-Myers *et al.*, 1996; Keane-Myers and Nickell, 1995a; Matyniak and Reiner, 1995). Treatment of mice with mAb to IL-12p40, which interferes with Th1 cell priming, reduced arthritis in C3H mice, providing further support for a role for Th1 cells in arthritis severity (Anguita *et al.*, 1996). This view that Th1 responses were deleterious and Th2 responses beneficial was modified by a kinetic study of cytokine induction, which showed that BALB/c mice produce IFNγ earlier in the course of infection than C3H mice, and IL-4 production heralded arthritis resolution. When the Th1/Th2 cytokine balance was altered in favour of IFNγ production by interrupting the B7/CD28 T-cell costimulatory pathway using mAb to the B7 molecules CD80 and CD86 or by treatment with the CD28 ligand CTLA4Ig, no exacerbation of arthritis was seen (Shanafelt *et al.*, 1998). Subsequently, C3H $IFN\gamma^{-/-}$ and DBA $IL\text{-}4^{-/-}$ mice were shown to develop arthritis of comparable severity and duration to their wild-type counterparts, as did normally disease-resistant 129/SvEv mice deficient in the IFNγ receptor, and C57BL/6 and BALB/c mice deficient in IL-4 or the IL-4 receptor alpha chain (Brown and Reiner, 1999b; Glickstein *et al.*, 2001; Potter *et al.*, 2000). Taken together, these studies indicate that susceptibility or resistance to Lyme arthritis in mice is not determined by the balance between Th1 and Th2 cells and their associated cytokines.

A role for two related T-cell subsets – Th17 and $CD4^+CD25^+Foxp3^+$T regulatory (Treg) cells – has been suggested in the development and control of Lyme arthritis. These cells derive from a common precursor cell and represent lineages distinct from Th1 and Th2 cells. Th17 cells produce the cytokine IL-17, which induces stromal cell production of inflammatory cytokines and chemokines that recruit leucocytes, especially neutrophils, to tissue sites (Harrington *et al.*, 2006). Treg cells, in contrast, function primarily to attenuate inflammation through the production of anti-inflammatory cytokines such as IL-10 and TGFβ (Belkaid and Rouse, 2005; Sakaguchi, 2004). Th17 cells have been shown to play a role in host defence against extracellular pathogens and fungi, and also in the pathogenesis of several forms of autoimmune arthritis in mice and humans (Tesmer *et al.*, 2008). *Borrelia burgdorferi* lipopeptides can induce T-cells to produce IL-17 *in vitro* (Infante-Duarte *et al.*, 2000), and *B. burgdorferi* has been shown to induce bone marrow derived dendritic cell production of IL-23, a cytokine involved in potentiating IL-17 release by T-cells (Knauer *et al.*, 2007). The only direct evidence that IL-17 may play a role in Lyme arthritis derives from a model of destructive arthropathy developed in IFNγ-deficient C57BL/6 or wild-type C57BL/6 mice vaccinated with one strain of *B. burgdorferi s.s.* and then infected with *B. bissettii*, a related *B. burgdorferi* sensu lato species. In this model, development of arthritis is dependent on *Borrelia*-specific $CD4^+$ T-cells and can be alleviated by treatments that interfere with IL-17 production (anti-IL-15 mAb) or its effector functions (anti-IL-17 or anti-IL-17 receptor mAb). Reduction in IL-17 was associated with an increase in $CD4^+CD25^+$ T-cells in lymph nodes draining arthritic joints, and depletion of these cells with anti-CD25 mAb exacerbated arthritis. Although the sources of IL-17 and the full characterization of the arthritis-alleviating $CD4^+CD25^+$ T-cell population has not been reported, these observations suggest that in this model system, the balance between Th17 cells and Treg cells could contribute to arthritis severity. Further evidence that Treg cells may modulate the course of Lyme arthritis can be seen with *B. burgdorferi* infection of $CD28^{-/-}$ mice, which are deficient in natural Tregs (Iliopoulou *et al.*, 2007). *Borrelia burgdorferi*-infected $CD28^{-/-}$ mice have an increased incidence, duration and severity of arthritis in comparison to control mice. However, $CD28^{-/-}$ mice have impaired humoral immune responses (Shahinian *et al.*, 1993) and lower pathogen-specific antibody titres after infection, which complicates interpretation of results. *Borrelia burgdorferi*-infected $TLR2^{-/-}$ mice develop arthritis that is associated with an increase in T-cell infiltration into joints in comparison to wild-type mice at similar stages of infection. Although not examined in that study, an increased T-cell response in the joints may be a consequence of deficiency in $CD4^+CD25^+$ Treg cells, which are reduced in number in $TLR2^{-/-}$ mice.

15.4 Organ-specific responses

In humans, *B. burgdorferi* first establishes infection in the skin, and if not dealt with by the local immune response, it may disseminate to other organ sites to cause disease. In some cases, patients may present with clinically apparent disease in multiple sites, such as with multiple EM lesions, carditis and/or neurological disease; in others, especially those with late manifestations, systemic illness is not apparent and disease is confined to a specific site, such as the joint or the skin (acrodermatitis chronicum atrophicans). Insight into the mechanisms of disease has been obtained through murine studies of specific organs and more limited human studies of skin, cerebrospinal and joint fluids, and synovial tissue.

15.4.1 Murine Lyme arthritis

Studies in the mouse have provided an opportunity to study the localized response of the joint, as inflammation follows and is dependent on bacterial invasion of the joint tissue. Barthold and colleagues initially identified arthritic lesions in C3H mice, and identified features common to arthritic joints of Lyme patients (Barthold *et al.*, 1992a). Early inflammatory infiltrates were dominated by PMNs, with oedema, and hyperproliferation of the synovium surrounding the tendon consistently associated with severe disease. This group also demonstrated that different mouse strains developed distinct levels of arthritis severity following infection with the N40 isolate of *B. burgdorferi*, an observation that has formed the foundation of many studies (Barthold *et al.*, 1990). *Borrelia burgdorferi* reach the joint tissue by 1 week of intradermal infection, with inflammatory cell infiltrate evident by 2 weeks of infection (Barthold *et al.*, 1991). The peak of histopathological changes was found at 4 weeks of infection. Interestingly, the occasional recurrent episodes of self-limited arthritis that occurred in mice infected for many months continued to display an infiltrate dominated by PMNs, perhaps reflecting the transient immunological challenge of new antigenic variants of bacteria which fail to persist long enough for the arthritis to progress to a bone-degradative disease. The generation of cytokines has been followed in joint tissue from *B. burgdorferi*-infected mice. Increased production of both the PMN recruiting chemokine KC and the monocyte chemokine MCP1 was found at 2 weeks of infection, and was greater in the severely arthritic C3H mice than in mildly arthritic C57BL/6 mice (Brown *et al.*, 2003). Mice deficient in the major receptor for KC, CXCR2, were protected from arthritis, whereas deficiency of the CCR2 chemokine receptor of monocytes had no effect on arthritis development. These results are consistent with the PMN domination of the inflammatory cell infiltrate in murine Lyme arthritis, and suggest that *B. burgdorferi* induces the localized production of chemokines that orchestrates the type of inflammatory cells recruited to the site.

Although *B. burgdorferi* invasion of the joint tissue is clearly required for the generation of chemokines and the development of arthritis, these findings do not explain the difference between strains of mice displaying severe disease (C3H, AKR) and those developing milder disease (B6, DBA). Quantitative PCR has revealed that the mouse strains with limited, milder arthritis harbour as many spirochaetes in their ankle tissues as the strains displaying severe disease, and therefore similar amounts of proinflammatory lipoproteins (Brown and Reiner, 1998; Ma *et al.*, 1998). Rather, this finding suggests that hosts with distinct genetic backgrounds have qualitatively distinct responses to the invading bacteria. To address this, genetic linkage analysis was performed with intercross populations derived from crosses of severely arthritic C3H mice with mildly arthritic C57BL/6 or BALB/c mice. Quantitative trait loci (QTL) regulating several parameters of *B. burgdorferi*-induced arthritis have been identified on five chromosomes of the mouse, 1, 4, 5, 11, and 12 (Roper *et al.*, 2001; Weis *et al.*, 1999). Identification of these QTL demonstrates multigenic regulation of the severity of Lyme arthritis. Genes within the QTL on chromosomes 5 and 11 were found to reciprocally transfer arthritis phenotype, resulting in mice with significantly altered *B. burgdorferi* induced arthritis severity when compared with parental inbred mice (Crandall *et al.*, 2005; Ma *et al.*, 2009). These results formally link the differences in Lyme arthritis severity in C3H and C58BL/6 mice with genes encoded within the *Bbaa2Bbaa3* and *Bbaa4* loci on chromosomes 5 and 11, respectively.

The identity of pathways responsible for the increased severity of Lyme arthritis in C3H mice was sought in the whole ankle tissue. In an unbiased approach, the induction of genes following infection with B. *burgdorferi* was followed over time by microarray analysis (Crandall *et al.*, 2006). A strong upregulation of transcripts from numerous interferon-inducible genes was reported in the joint tissue of severely arthritic C3H mice at 1 week of infection: a time point associated with bacterial invasion but prior to the appearance of inflammatory cell infiltrate. Although the profile was strongly upregulated at 1 week of infection, it was greatly reduced by two weeks of infection, and remained low at 4 weeks, the peak of arthritis. Transcripts for the interferon genes themselves (type I or II) were not upregulated, making it difficult to assign responsibility to a particular IFN. However, a previous report demonstrating that Type II IFN is not required for severe Lyme arthritis in C3H mice suggested the involvement of type I IFN should be considered (Brown and Reiner, 1999b). This response appears to be localized to the joint tissue as induction of the IFN profile was not found in transcripts from the spleen or ear of infected C3H mice. Arthritis resistant C57BL/6 mice did not share the upregulation of IFN inducible genes at 1 week of infection. Instead, a discrete response to infection of joint tissue was found in C57BL/6 mice characterized by transcripts associated with epidermal differentiation and wound repair. Finally, two different knockout mice on the C57BL6 background previously shown to have increased arthritis severity, IL-10$^{-/-}$ and TLR2$^{-/-}$, also displayed dramatic upregulation of several IFN-inducible transcripts at 2 weeks of infection (Crandall *et al.*, 2006; Wang *et al.*, 2008b). The subsequent finding that mice receiving antibody that blocked the type I IFN receptor prior to infection displayed a significant reduction in arthritis severity provided strong support for involvement of this pathway in Lyme arthritis development (Miller *et al.*, 2008). These findings link the development of severe arthritis in the mouse to the exuberant upregulation of IFN inducible genes in response to invading *B. burgdorferi*. Both murine bone marrow-derived macrophages and human monocytes generate an interferon response when incubated with *B. burgdorferi*, indicating a potential cellular source during infection (Miller *et al.*, 2008; Salazar *et al.*, 2009). Macrophages derived from mice with targeted disruptions indicated that the IFN response to *B. burgdorferi* was independent of MyD88, TLR2, TLR4, and TLR9, but entirely dependent on feedback through the type I IFN receptor (Miller *et al.*, 2008).

The host transcriptional response to *B. burgdorferi* at 2 weeks of infection, when bacterial levels reach their peak in joint tissues and intense inflammatory cell infiltrates are obvious in arthritis susceptible mice, was dominated by markers of host defence, with strong induction of chemokine genes in both C3H and C57BL/6 mice (Crandall *et al.*, 2006). The peak of arthritis in C3H mice, at 4 weeks of infection, revealed numerous transcripts previously identified in the end stage in rodent models of rheumatoid arthritis (Ibrahim *et al.*, 2002). These signature expression profiles included transcripts for numerous MMPs, procollagens, and cathepsins. As mentioned earlier *B. burgdorferi* activates human chondrocytes through the α3β1 receptor, resulting in MMP production via a MAP kinase-dependent, TLR2-independent pathway (Behera *et al.*, 2006c). Microarray assessment provides additional *in vivo* evidence for the contribution of these molecules to arthritis development. In particular, MMP-9 induced cleavage of type I collagen enhances human monocyte chemotaxis *in vitro*, and *B. burgdorferi*-induced arthritis is less severe in MMP-9-deficient mice despite pathogen burdens in joints that are comparable to those in WT mice (Heilpern *et al.*, 2009).

The antibody response localized to the joint tissue has also been studied, with recent publications identifying the evolution of antigenic targets and affinity maturation of immunoglobulin genes in B-cells of synovial tissue. Antibody to cytokeratin has been identified in synovial tissue/fluid from Lyme arthritis patients (Ghosh *et al.*, 2006). Importantly, although acute inflammation in the mouse, and presumably the human, can arise in the absence of B and T-cells, it is likely that in immunocompetent hosts, T and B-cell responses contribute to disease expression. This is particularly relevant to the situation of chronic disease that has been refractory to antibiotic treatment (Drouin *et al.*, 2008; Kannian *et al.*, 2007). In murine studies antibodies to certain antigens (arthritis resolving protein, decorin binding protein, and basic membrane proteins A and B) are effective at resolving arthritis despite persistence of the pathogen in relatively avascular areas, suggesting that reducing spirochaetes accessible to immune cells and their products is central to disease resolution (Barthold *et al.*, 2006; Feng *et al.*, 2000; Pal *et al.*, 2008).

15.4.2 Murine Lyme carditis

Although it has not been possible to study the pathogenesis of Lyme carditis in humans, biopsies taken from early patients with Lyme carditis yielded insights into the changes in cellularity and inflammatory infiltrate associated with symptoms. Myocardial infiltration was noted to consist of lymphocytes, plasma cells, and macrophages. Neutrophils were not major components of the infiltrate, but could be identified in some patients (Duray and Steere, 1988). Carditis also develops in mice, where it is characterized by bacterial invasion of connective tissue at the base of the heart surrounding the aortic valve and at the base of the aorta (Barthold *et al.*, 1991, 1992a). Leucocytic infiltrates consisted predominantly of monocytes, with PMNs and lymphocytes also present, and peaked at 15 days of infection. By 30 days of infection there was evidence of disease resolution and spirochaete clearance. Immunohistochemical analysis revealed the majority of cells to express markers for macrophages and about 5% of cells to be positive for CD4 or CD8 (Ruderman *et al.*, 1995). Carditis developed in *scid* mice, lacking B- and T-lymphocytes, and in mice deficient in class II MHC, indicating acquired host defences are not required for this pathology (Armstrong *et al.*, 1992; Ruderman *et al.*, 1995). Early studies with inbred strains of mice revealed distinct patterns of carditis development: hearts of C3H mice display more severe lesions of carditis than those of C57BL/6 mice (Barthold *et al.*, 1990). The level of spirochaetes is greater in the hearts of C3H mice with severe carditis than in C57BL/6 mice with mild carditis, consistent with the concept that disease is a consequence of the host response to tissue invasion by *B. burgdorferi* (Barthold *et al.*, 1990; Ma *et al.*, 1998).

The predominance of monocytes and T-cells in carditis in C3H mice certainly suggests that an early host response to infection in this tissue is the production of monocyte recruiting chemokines (Ruderman *et al.*, 1995). In fact, both MIP-2 and MCP-1 are produced in response to infection. Mice deficient in CCR2, the major chemokine receptor for MCP-1, had reduced monocyte infiltration but sustained carditis with increased presence of PMNs in heart tissue (Montgomery *et al.*, 2007). These mice displayed heightened production of the chemokine MCP-1 and increased presence of *B. burgdorferi*, and revealed a possible compensatory role for PMNs in the inflammatory infiltrate. In contrast, mice hypomorphic for the β_2 integrin CD18 have more severe myocarditis with increased production of MCP-1 and recruitment of macrophages (Guerau-de-Arellano *et al.*, 2005). Recent studies have identified a role for NKT-cells and IFNγ in control of spirochaetes in the hearts of the disease resistant C57BL/6 mice (Olson *et al.*, 2009). Together, studies in the mouse indicate involvement of monocyte/macrophages in the development of carditis. Antibodies play a critical role in resolution of carditis, however, $CD4^+$ T-cells are required for optimal resolution of inflammation (Bockenstedt *et al.*, 2001; McKisic and Barthold, 2000). This is consistent with the observed prevalence of T-cells in lesions of the heart, but this activity also reflects a delicate balance as administration of antigen-specific $CD4^+$ T-cells in the absence of B-cells results in severe myocarditis, pulmonary disease and death (McKisic *et al.*, 2000). Thus, the early host response of the heart may feature fundamentally distinct pathways from those observed in the joint tissue.

Several features of carditis in mice set it apart from arthritis and suggest different mechanisms may be involved in pathogenesis in the two organs: (1) the peak of carditis is at 2 weeks rather than at 4 weeks as seen for arthritis (Barthold *et al.*, 1991); (2) the inflammatory infiltrate into heart tissue is dominated by macrophages and lymphocytes rather than PMNs which dominate the joint tissue (Barthold *et al.*, 1992a; Ruderman *et al.*, 1995); (3) there is a direct correlation between the severity of carditis and spirochaetes burden in cardiac tissue which is not seen in arthritis (Barthold *et al.*, 1992a; Ma *et al.*, 1998); (4) both Ab and $CD4^+$ T-cells are required for optimal resolution of carditis, whereas Ab is sufficient for arthritis (Bockenstedt *et al.*, 2001; McKisic and Barthold, 2000); and (5) differential antigens may be associated with homing and persistence of spirochaetes in cardiac and joint tissues (Antonara *et al.*, 2007; McKisic and Barthold, 2000; Pal *et al.*, 2008).

15.5 Human disease

While the mouse model of Lyme borreliosis has provided much information about the host immune response to *B. burgdorferi*, many features of human Lyme disease remain to be explained (see Chapter 18). Mice do not develop the hallmark skin lesion erythema migrans (EM) even though the skin is a preferred site for *B. burgdorferi* infection and persistence within the murine host. The reason for this is unclear, given that EM lesions are comprised of lymphocytes, dendritic cells, macrophages and plasma cells, and these cell subsets are part of the host immune response in mice. Characterization of the EM lesions by immunohistochemistry shows that the infiltrates are largely T-cells with $CD68^+$ histiocytes and a few B-cells (Boer *et al.*, 2007). These findings are supported by flow cytometry analysis of dermal leucocytes extracted from EM lesions using a blister technique, which revealed a predominance of T-cells, monocytes/macrophages and dendritic cells, with occasional neutrophils (Salazar *et al.*, 2003). T-cells had an effector phenotype, and dendritic cells had upregulated HLA-DR, CD14, and TLRs 1, 2 and 4 expression, consistent with a mature phenotype capable of presenting antigens to T-cells. The dominant cytokines extracted from the lesions by the blister method were IL-6 and IFN-γ, and the levels were higher in patients with isolated EM in comparison with those with multiple EM lesions. These observations suggest that a strong innate and adaptive immune response to *B. burgdorferi* in the skin may effectively localize the infection to the skin.

Dissemination of *B. burgdorferi* occurs via direct tissue invasion and through a blood-borne route, however relatively little is known about the *in vivo* interaction of spirochaetes with the vasculature. Intravital microscopy in the mouse has revealed multistage interactions with the vasculature that are mediated by bacterial ligands and host receptors (Norman *et al.*, 2008). Activation of vascular endothelium and upregulation of adhesion molecules is required for transmigration of inflammatory cells to tissue sites of infection, and may be necessary also for *B. burgdorferi* to enter or exit blood vessels (Szczepanski *et al.*, 1990). Most studies investigating the response of vascular endothelium to *B. burgdorferi* have employed *in vitro* assays using human umbilical vein endothelial cells (HUVECs). These have shown that exposure of HUVECs to *B. burgdorferi* and its lipoproteins leads to upregulation of adhesion molecules (VLA-4, E-selectin, VCAM-1 and ICAM-1) and chemokines (IL-8) that promote the transmigration of neutrophils, monocytes and T-cells (Burns and Furie, 1998; Burns *et al.*, 1997; Sellati *et al.*, 1995). Th1 cells producing IFN-γ appear to be preferentially transferred over Th2 cells in this process, and the anti-inflammatory cytokine IL-10 can impair the response (Dame *et al.*, 2007; Gergel and Furie, 2004; Lisinski and Furie, 2002). The degree to which host adapted spirochaetes activate vascular endothelium is unknown, but it is likely that much of this response is mediated by TLR recognition of spirochaete components as $MyD88^{-/-}$ mice have comparatively few responding immune cells in tissues despite the markedly increased bacterial load.

Neurological disease is a second example of how mice and humans differ in the pathology that arises after *B. burgdorferi* infection, and is most likely a consequence of blood-borne infection. Although spirochaetes can be seen in the mouse perineurium, they do not incite inflammation and can only be cultured from mouse brains early in infection, during periods when blood cultures are also positive (Barthold *et al.*, 1992a). In contrast, human Lyme neuroborreliosis can manifest as peripheral or central nervous system involvement, with meningoencephalitis, cranial nerve palsies and peripheral radiculoneuropathies most commonly seen (Halperin *et al.*, 1988; Pachner and Steiner, 2007). *Ex vivo* assessment of the response of neuronal slices from a rhesus macaque revealed both proinflammatory and apoptotic responses to *B. burgdorferi* (Ramesh *et al.*, 2008). Analysis of the cerebrospinal fluid (CSF) of patients with acute neuroborreliosis reveals a lymphocytic infiltrate with elevated protein levels and antibodies to *B. burgdorferi*, but the CSF can also be normal. Studies conducted primarily in Europe where Lyme neuroborreliosis is more prevalent have partially characterized the immune response in the CSF. Induction of MMPs may permit *B. burgdorferi* invasion into neural tissue, and production of chemokines, including CXCL11 and CXCL13, contributes to cell recruitment. The CSF lymphocytic pleocytosis is largely due to B-cells, with T-cells and monocytes comprising a smaller fraction of the cellular infiltrate. The predominance of B-cells may be explained by the fact that *B. garinii*, which is most commonly associated with neurological disease, and other strains of *B. burgdorferi* can induce human monocytes to produce the B-cell chemoattractant CXCL13 *in vitro*, and monocytes with increased expression of CXCL13 are found in the CSF (Narayan *et al.*, 2005; Rupprecht *et al.*, 2007). *Borrelia burgdorferi*-specific antibodies to a number of *Borrelia* antigens can be detected and an elevated ratio of pathogen specific IgG in the CSF compared to the peripheral blood is indicative of local antibody production. C1q, C4, C3 and C3a levels are elevated in the CSF, without associated findings in the plasma, suggesting that the inflammatory response is localized to the CSF (Rupprecht *et al.*, 2007). In early neuroborreliosis, $CD8^+$ T-cells dominate the T-cell infiltrates in CSF, but give way to a dominant $CD4^+$ Th1 type response (Jacobsen *et al.*, 2003; Lunemann *et al.*, 2007). Analysis of the Th1 cells isolated from a single US patient with acute neuroborreliosis defined peptides of *B. burgdorferi* heat shock protein 90, 30S ribosomal protein, and proteins involved with porphin and lipoprotein biosynthesis as relevant T-cell antigens (Lunemann *et al.*, 2007). Interestingly, a number of T-cells had specificities for human peptide sequences, including CNPase, a structural protein involved in synaptic transmission, but whether these self-reactive T-cells play a role in the clinical signs and symptoms remains unknown.

In humans, arthritis is considered a late manifestation of Lyme disease and is becoming less common as the infection is recognized and treated at earlier stages. About 60% of untreated patients develop arthritis, usually months after the onset of infection (Steere *et al.*, 1987). Children are more likely to develop arthritis than adults. Arthritis onset is often at a time when other systemic manifestations of Lyme disease have waned, suggesting that the inflammatory response is confined largely to the joint. Lyme arthritis can take two forms, an acute arthritis that is reproduced by the murine model of Lyme borreliosis, and a more protracted form evolving from the acute arthritis that histopathologically resembles rheumatoid arthritis (Steere and Angelis, 2006). In acute arthritis, *B. burgdorferi* DNA can be detected in the synovial fluid and tissue, consistent with infection driving the host response, and these patients generally respond to antibiotic therapy. As with disease in other organ systems, innate immune cell activation via TLRs and induction of MMPs may initiate the inflammatory process. *Borrelia burgdorferi* elicits MMPs and aggrecaneses from human chondrocytes *in vitro* and MMP-2, MMP-3, and MMP-9 are found in synovial fluid (Behera *et al.*, 2006b; Hu *et al.*, 2001). As noted above, MMP-9 expression has recently been linked to the development of Lyme arthritis in a murine model (Heilpern *et al.*, 2009). The synovial fluid contains primarily polymorphonuclear leucocytes, but both αβ- and γδT-cells have been isolated and characterized. αβT-cells include both Th1 and Th2 subsets, and the dominance of Th1 cells correlates with larger effusions. The degree of synovial proliferation is dependent upon the number of episodes of arthritis and the duration of arthritis prior to treatment. Biopsies of the synovium show lymphocytic infiltrates comprised of $CD4^+$ T-cells, often in lymphoid aggregates with B-cells, and scattered macrophages, dendritic cells and plasma cells (Duray and Steere, 1988). In untreated patients, rare spirochaetes and globular antigen deposits can be seen in perivascular areas where lymphocytes are present. Immunohistochemistry reveals upregulation of adhesion molecules (P-selectin, vascular adhesion protein-1, intercellular adhesion molecule-1 and vascular cell adhesion molecule-1) on synovial lining cells, endothelial cells, and cell infiltrates (Akin *et al.*, 2001). The spontaneous remission of an episode of acute arthritis, which can last for only a few days or for several months, is probably immune-mediated but organisms can persist in the absence of antibiotic therapy and give rise to recurrent episodes of acute arthritis. In one study of patients who did not receive antibiotic therapy for EM, arthritis developed in more than half, but the number of patients with recurrent episodes of arthritis declined by 10–20% each year even in the absence of therapy (Steere *et al.*, 1987).

About 10% of patients with Lyme arthritis develop chronic synovitis, usually in a single joint, after episodes of acute arthritis at the same site. In early studies, such patients typically had arthritis present for >12 months, and even though antibiotics were delayed for several years, inflammatory arthritis generally responded to their institution (Steere *et al.*, 1987). With time, a second subgroup of patients has been identified in whom arthritis does not resolve within >3 months of completion of antibiotic therapy, and often patients have received multiple antibiotic courses (Steere *et al.*, 2006). Inflammation persists even though *B. burgdorferi* can no longer be detected in synovial fluid or synovial tissue (Nocton *et al.*, 1994). Significant synovial proliferation in the current era typically occurs only in this 'antibiotic-refractory' form of Lyme arthritis, and probably reflects the intensity and duration of inflammation. A retrospective analysis has compared cytokine and chemokine expression in archived synovial fluid samples from patients who resolved arthritis versus those who had antibiotic-refractory arthritis (Shin *et al.*, 2007). Samples from patients in whom arthritis persisted after antibiotic therapy had higher levels of TNFα, IL-1β and IFNγ in synovial fluid and increased levels of T-cell and monocyte chemoattractants (CXCL9, CXCL10 and CCL4), indicating a more exuberant inflammatory response. Most of the Th1 cytokines and chemokines declined over time in the antibiotic-refractory group once PCR for spirochaete DNA became negative, although antibiotic-refractory arthritis patients continued to have elevated levels in comparison to controls. Therapies typically used in autoimmune forms of arthritis (non-steroidal anti-inflammatory medications, hydroxychloroquine, or methotrexate) can resolve antibiotic-refractory arthritis within 5 years, although some patients undergo synovectomy, which is curative (Steere and Angelis, 2006).

15.5.1 Autoimmunity and Lyme disease

Several theories have been proposed to explain the pathogenesis of antibiotic-refractory Lyme arthritis, including persistence of *B. burgdorferi* or its inflammatory components, defective immune regulation, and infection-induced autoimmunity. The autoimmunity theory has been explored in greatest detail to date, in part because of the similarities between the synovitis seen in antibiotic-refractory Lyme arthritis and rheumatoid arthritis. Patients who develop protracted Lyme arthritis are more likely than those with antibiotic-responsive arthritis to bear HLA-DRB1 molecules, especially DRB1*04 and DRB1*0101 alleles, which are also prevalent among RA patients (Steere *et al.*, 2006). An association has been made between the onset of antibiotic-refractory Lyme arthritis and T and B-cell responses to *B. burgdorferi* Osp A. An immunodominant Osp A T-cell epitope (Osp $A_{163-175}$) was subsequently identified that could elicit T-cell responses in the peripheral blood of patients with Lyme arthritis who possessed DRB1*04 and 0101 alleles, and T-cells specific for this peptide were more prevalent in inflamed joints (Chen *et al.*, 1999; Meyer *et al.*, 2000). Osp $A_{163-175}$ was subsequently found to have partial sequence homology with a peptide of human LFA-1α (LFA-$1_{L332-340}$), an adhesion molecule that is abundantly expressed at sites of tissue inflammation, including Lyme arthritis joints. It was proposed that T-cells reactive with LFA-$1_{L332-340}$ could perpetuate joint inflammation once infection had been cleared. However, this peptide was found to be only a weak agonist for Osp $A_{163-175}$-specific T-cells and did not elicit the Th1 cytokine pattern (Trollmo *et al.*, 2001). Other *B. burgdorferi* peptides could stimulate self-reactive T-cell responses, but whether these are relevant to pathology remains uncertain as some degree of self-reactivity is required for positive selection of the T-cell receptor repertoire during thymic development. A failure to regulate the immune response once the pathogen is cleared could explain the persistence of T-cells with self-reactivity.

15.5.2 Genetics and Lyme disease

As noted above, individuals bearing select DRB1 alleles may be predisposed to prolonged arthritis after *B. burgdorferi* infection. Given the important role of innate immune cells in *B. burgdorferi*-induced inflammation, recent studies have focused on the role of single nucleotide polymorphisms in innate immune response genes in the susceptibility to *B. burgdorferi* disease. In humans, soluble CD14 in the serum is a marker of more severe inflammation with *B. burgdorferi* infection (Lin *et al.*, 2000). A single nucleotide polymorphism (SNP) in the CD14 gene (CD14-550 C/T) is related to the levels of soluble CD14 (Inoue *et al.*, 2007), suggesting that this response is in part genetically determined. In contrast to SNPs that might promote inflammation, a heterozygous SNP in TLR2 (Arg753Gln) has been found to impair innate immune responsiveness to *B. burgdorferi in vitro* and appears to protect against symptomatic late stage Lyme disease, but whether this SNP permits pathogen persistence is not known (Schroder *et al.*, 2005). Clearly, additional studies are needed to further define the relationship between individual variations in the host immune response to *B. burgdorferi* that are genetically determined and the outcome from human Lyme disease.

15.5.3 Host defence and persistent infection

A critical consequence of the host response to *B. burgdorferi* is the clearance or control of infection. The persistent nature of this infection suggests the existence of an intricate dynamics of bacterial infection and tissue invasion coupled with a response designed to limit and eliminate the pathogen. This bacterium employs elaborate mechanisms to avoid immune clearance while the host has an arsenal of tools to control *B. burgdorferi* growth. As discussed in this chapter, there is much evidence for the importance of both the innate and acquired host defences in controlling infection, from *in vitro* studies and from infection models.

Borrelia burgdorferi is unusual in that it is an extracellular bacterial pathogen that causes persistent infection. As such, antibody, complement and phagocytes would be predicted to be critical mechanisms of host defence. It is important to recognize that neither antibodies nor phagocytes alone are capable of reducing pathogen burden in tissues: both are required to reduce pathogen burden in tissues to the level seen in chronic persistent infection in immunocompetent mice (Wang *et al.*, 2005). In some tissue sites, such as the joint, escape from immune mediated destruction may in part be due to the spirochaete predilection for relatively avascular connective tissue, which may not be penetrated well by phagocytes and antibody, or even antibiotics (Hodzic *et al.*, 2008; Mullegger *et al.*, 2007; Salazar *et al.*, 2003). The persistence of spirochaetes in the highly vascular skin of immunocompetent mice during the chronic phase of untreated infection, visualized without an associated immune cell infiltrate, is less easily explained (Barthold *et al.*, 1991; Hodzic *et al.*, 2003). In this stage of 'dormancy', spirochaetes must be viewed as relatively innocuous by the immune system, possibly because they are camouflaged by host proteins or have diminished surface expression of antigenic targets or proinflammatory molecules. An intriguing possibility is that spirochaetes in the persistent phase of infection actively induce an alteration in their environment that suppresses the recruitment of inflammatory cells. Some evidence to support all of these theories has been provided, but additional studies are needed.

References

Aguero-Rosenfeld, M.E., Wang, G., Schwartz, I., and Wormser, G.P. (2005). Diagnosis of Lyme borreliosis. Clin. Microbiol. Rev. *18*, 484–509.

Akin, E., Aversa, J., and Steere, A.C. (2001). Expression of adhesion molecules in synovia of patients with treatment-resistant Lyme arthritis. Infect. Immun. *69*, 1774–1780.

Akin, E., McHugh, G.L., Flavell, R.A., Fikrig, E., and Steere, A.C. (1999). The immunoglobulin (IgG) antibody response to OspA and OspB correlates with severe and prolonged Lyme arthritis and the IgG response to P35 correlates with mild and brief arthritis. Infect. Immun. *67*, 173–181.

Akira, S., Hoshino, K., and Kaisho, T. (2000). The role of Toll-like receptors and MyD88 in innate immune responses. J. Endotoxin Res. *6*, 383–387.

Akira, S., and Takeda, K. (2004). Toll-like receptor signalling. Nat. Rev. Immunol. *4*, 499–511.

Alexopoulou, L., Thomas, V., Schnare, M., Lobet, Y., Anguita, J., Schoen, R.T., Medzhitov, R., Fikrig, E., and Flavell, R.A. (2002). Hyporesponsiveness to vaccination with *Borrelia burgdorferi* OspA in humans and in TLR1- and TLR2-deficient mice. Nat. Med. *8*, 878–884.

Aliprantis, A.O., Yang, R.B., Mark, M.R., Suggett, S., Devaux, B., Radolf, J.D., Klimpel, G.R., Godowski, P., and Zychlinsky, A. (1999). Cell activation and apoptosis by bacterial lipoproteins through Toll-like receptor-2. Science *285*, 736–739.

Alitalo, A., Meri, T., Chen, T., Lankinen, H., Cheng, Z.Z., Jokiranta, T.S., Seppala, I.J., Lahdenne, P., Hefty, P.S., Akins, D.R., *et al.* (2004). Lysine-dependent multipoint binding of the *Borrelia burgdorferi* virulence factor outer surface protein E to the C terminus of factor H. J. Immunol. *172*, 6195–6201.

Alitalo, A., Meri, T., Lankinen, H., Seppala, I., Lahdenne, P., Hefty, P.S., Akins, D., and Meri, S. (2002). Complement inhibitor factor H binding to Lyme disease spirochetes is mediated by inducible expression of multiple plasmid-encoded outer surface protein E paralogs. J. Immunol. *169*, 3847–3853.

Anguita, J., Barthold, S.W., Persinski, R., Hedrick, M.N., Huy, C.A., Davis, R.J., Flavell, R.A., and Fikrig, E. (2002). Murine Lyme arthritis development mediated by p38 mitogen-activated protein kinase activity. J. Immunol. *168*, 6352–6357.

Anguita, J., Persing, D.H., Rincon, M., Barthold, S.W., and Fikrig, E. (1996). Effect of anti-interleukin 12 treatment on murine Lyme borreliosis. J. Clin. Invest. *97*, 1028–1034.

Anguita, J., Thomas, V., Samanta, S., Persinski, R., Hernanz, C., Barthold, S.W., and Fikrig, E. (2001). *Borrelia burgdorferi*-induced inflammation facilitates spirochete adaptation and variable major protein-like sequence locus recombination. J. Immunol. *167*, 3383–3390.

Antonara, S., Chafel, R.M., LaFrance, M., and Coburn, J. (2007). *Borrelia burgdorferi* adhesins identified using *in vivo* phage display. Mol. Microbiol. *66*, 262–276.

Armstrong, A.L., Barthold, S.W., Persing, D.H., and Beck, D.S. (1992). Carditis in Lyme disease susceptible and resistant strains of laboratory mice infected with *Borrelia burgdorferi*. Am. J. Trop. Med. Hyg. *47*, 249–258.

Barthold, S.W., Beck, D.S., Hansen, G.M., Terwilliger, G.A., and Moody, K.D. (1990). Lyme borreliosis in selected strains and ages of laboratory mice. J. Infect. Dis. *162*, 133–138.

Barthold, S.W., de Souza, M., Fikrig, E., and Persing, D.H. (1992a). Lyme borreliosis in the laboratory mouse. In Lyme Disease: Molecular and Immunologic Approaches, S.E. Schutzer, ed. (Plainview (New York), Cold Spring Harbor Laboratory Press), pp. 223–242.

Barthold, S.W., Feng, S., Bockenstedt, L.K., Fikrig, E., and Feen, K. (1997). Protective and arthritis-resolving activity in sera of mice infected with *Borrelia burgdorferi*. Clin. Infect. Dis. *25 Suppl 1*, S9–17.

Barthold, S.W., Fikrig, E., Bockenstedt, L.K., and Persing, D.H. (1995). Circumvention of outer surface protein A immunity by host-adapted *Borrelia burgdorferi*. Infect. Immun. *63*, 2255–2261.

Barthold, S.W., Hodzic, E., Tunev, S., and Feng, S. (2006). Antibody-mediated disease remission in the mouse model of Lyme borreliosis. Infect. Immun. *74*, 4817–4825.

Barthold, S.W., Persing, D.H., Armstrong, A.L., and Peeples, R.A. (1991). Kinetics of *Borrelia burgdorferi* dissemination and evolution of disease after intradermal inoculation of mice. Am. J. Pathol. *139*, 263–273.

Barthold, S.W., Sidman, C.L., and Smith, A.L. (1992b). Lyme borreliosis in genetically resistant and susceptible mice with severe combined immunodeficiency. Am. J. Trop. Med. Hyg. *47*, 605–613.

Battisti, J.M., Bono, J.L., Rosa, P.A., Schrumpf, M.E., Schwan, T.G., and Policastro, P.F. (2008). Outer surface protein A protects Lyme disease spirochetes from acquired host immunity in the tick vector. Infect. Immun. *76*, 5228–5237.

Behera, A.K., Hildebrand, E., Bronson, R.T., Perides, G., Uematsu, S., Akira, S., and Hu, L.T. (2006a). MyD88 deficiency results in tissue-specific changes in cytokine induction and inflammation in interleukin-18-independent mice infected with *Borrelia burgdorferi*. Infect. Immun. *74*, 1462–1470.

Behera, A.K., Hildebrand, E., Scagliotti, J., Steere, A.C., and Hu, L.T. (2005). Induction of host matrix metalloproteinases by *Borrelia burgdorferi* differs in human and murine Lyme arthritis. Infect. Immun. *73*, 126–134.

Behera, A.K., Hildebrand, E., Szafranski, J., Hung, H.H., Grodzinsky, A.J., Lafyatis, R., Koch, A.E., Kalish, R., Perides, G., Steere, A.C., *et al.* (2006b). Role of aggrecanase 1 in Lyme arthritis. Arthritis Rheum. *54*, 3319–3329.

Behera, A.K., Hildebrand, E., Uematsu, S., Akira, S., Coburn, J., and Hu, L.T. (2006c). Identification of a TLR-independent pathway for *Borrelia burgdorferi*-induced expression of matrix metalloproteinases and inflammatory mediators through binding to integrin $\alpha_3\beta_1$. J. Immunol. *177*, 657–664.

Behera, A.K., Thorpe, C.M., Kidder, J.M., Smith, W., Hildebrand, E., and Hu, L.T. (2004). *Borrelia burgdorferi*-induced expression of matrix metalloproteinases from human chondrocytes requires mitogen-activated protein kinase and Janus kinase/signal transducer and activator of transcription signaling pathways. Infect. Immun. *72*, 2864–2871.

Belkaid, Y., and Rouse, B.T. (2005). Natural regulatory T-cells in infectious disease. Nat. Immunol. *6*, 353–360.

Belperron, A.A., and Bockenstedt, L.K. (2001). Natural antibody affects survival of the spirochete *Borrelia burgdorferi* within feeding ticks. Infect. Immun. *69*, 6456–6462.

Belperron, A.A., Dailey, C.M., Booth, C.J., and Bockenstedt, L.K. (2007). Marginal zone B-cell depletion impairs murine host defense against *Borrelia burgdorferi* infection. Infect. Immun. *75*, 3354–3360.

Benach, J.L., Fleit, H.B., Habicht, G.S., Coleman, J.L., Bosler, E.M., and Lane, B.P. (1984a). Interactions of phagocytes with the Lyme disease spirochete: role of the Fc receptor. J. Infect. Dis. *150*, 497–507.

Benach, J.L., Habicht, G.S., Gocinski, B.L., and Coleman, J.L. (1984b). Phagocytic cell responses to *in vivo* and *in vitro* exposure to the Lyme disease spirochete. Yale J. Biol. Med. *57*, 599–605.

Benhnia, M.R., Wroblewski, D., Akhtar, M.N., Patel, R.A., Lavezzi, W., Gangloff, S.C., Goyert, S.M., Caimano, M.J., Radolf, J.D., and Sellati, T.J. (2005). Signaling through CD14 attenuates the inflammatory response to *Borrelia burgdorferi*, the agent of Lyme disease. J. Immunol. *174*, 1539–1548.

Bockenstedt, L.K. (2009). Lyme disease. In Kelley's Textbook of Rheumatology, 8th edition, S. Firestein, R.C. Budd, J. Harris, E.D., I.B. McInnes, S. Ruddy, and J.S. Sergent, eds. (Philadelphia, Saunders Elsevier), pp. 1715–1727.

Bockenstedt, L.K., Barthold, S., Deponte, K., Marcantonio, N., and Kantor, F.S. (1993). *Borrelia burgdorferi* infection and immunity in mice deficient in the fifth component of complement. Infect. Immun. *61*, 2104–2107.

Bockenstedt, L.K., Kang, I., Chang, C., Persing, D., Hayday, A., and Barthold, S.W. (2001). $CD4^+$ T helper 1 cells facilitate regression of murine Lyme carditis. Infect. Immun. *69*, 5264–5269.

Bockenstedt, L.K., Shanafelt, M.C., Belperron, A., Mao, J., and Barthold, S.W. (2003). Humoral immunity reflects altered T helper cell bias in *Borrelia burgdorferi*-infected γδ T-cell-deficient mice. Infect. Immun. *71*, 2938–2940.

Boer, A., Bresch, M., Dayrit, J., and Falk, T.M. (2007). Erythema migrans: a reassessment of diagnostic criteria for early cutaneous manifestations of borreliosis with particular emphasis on clonality investigations. Br. J. Dermatol. *156*, 1263–1271.

Bolz, D.D., Sundsbak, R.S., Ma, Y., Akira, S., Kirschning, C.J., Zachary, J.F., Weis, J.H., and Weis, J.J. (2004). MyD88 plays a unique role in host defense but not arthritis development in Lyme disease. J. Immunol. *173*, 2003–2010.

Brightbill, H.D., Libraty, D.H., Krutzik, S.R., Yang, R.B., Belisle, J.T., Bleharski, J.R., Maitland, M., Norgard, M.V., Plevy, S.E., Smale, S.T., *et al.* (1999). Host defense mechanisms triggered by microbial lipoproteins through toll-like receptors. Science *285*, 732–736.

Brooks, C.S., Vuppala, S.R., Jett, A.M., Alitalo, A., Meri, S., and Akins, D.R. (2005). Complement regulator-acquiring surface protein 1 imparts resistance to human serum in *Borrelia burgdorferi*. J. Immunol. *175*, 3299–3308.

Brown, C.R., Blaho, V.A., and Loiacono, C.M. (2003). Susceptibility to experimental Lyme arthritis correlates with KC and monocyte chemoattractant protein-1 production in joints and requires neutrophil recruitment via CXCR2. J. Immunol. *171*, 893–901.

Brown, C.R., Lai, A.Y., Callen, S.T., Blaho, V.A., Hughes, J.M., and Mitchell, W.J. (2008). Adenoviral delivery of interleukin-10 fails to attenuate experimental Lyme disease. Infect. Immun. *76*, 5500–5507.

Brown, C.R., and Reiner, S.L. (1998). Clearance of *Borrelia burgdorferi* may not be required for resistance to experimental Lyme arthritis. Infect. Immun. *66*, 2065–2071.

Brown, C.R., and Reiner, S.L. (1999a). Development of Lyme arthritis in mice deficient in inducible nitric oxide synthase. J. Infect. Dis. *179*, 1573–1576.

Brown, C.R., and Reiner, S.L. (1999b). Experimental Lyme arthritis in the absence of interleukin-4 or gamma interferon. Infect. Immun. *67*, 3329–3333.

Brown, J.P., Zachary, J.F., Teuscher, C., Weis, J.J., and Wooten, R.M. (1999). Dual role of interleukin-10 in murine Lyme disease: regulation of arthritis severity and host defense. Infect. Immun. *67*, 5142–5150.

Burns, M.J., and Furie, M.B. (1998). *Borrelia burgdorferi* and interleukin-1 promote the transendothelial migration of monocytes in vitro by different mechanisms. Infect. Immun. *66*, 4875–4883.

Burns, M.J., Sellati, T.J., Teng, E.I., and Furie, M.B. (1997). Production of interleukin-8 (IL-8) by cultured endothelial cells in response to *Borrelia burgdorferi* occurs independently of secreted IL-1 and tumor necrosis factor alpha and is required for subsequent transendothelial migration of neutrophils. Infect. Immun. *65*, 1217–1222.

Bykowski, T., Woodman, M.E., Cooley, A.E., Brissette, C.A., Brade, V., Wallich, R., Kraiczy, P., and Stevenson, B. (2007). Coordinated expression of *Borrelia burgdorferi* complement regulator-acquiring surface proteins during the Lyme disease spirochete's mammal-tick infection cycle. Infect. Immun. *75*, 4227–4236.

Carroll, J.A., Garon, C.F., and Schwan, T.G. (1999). Effects of environmental pH on membrane proteins in *Borrelia burgdorferi*. Infect. Immun. *67*, 3181–3187.

Chen, J., Field, J.A., Glickstein, L., Molloy, P.J., Huber, B.T., and Steere, A.C. (1999). Association of antibiotic treatment-resistant Lyme arthritis with T-cell responses to dominant epitopes of outer surface protein A of *Borrelia burgdorferi*. Arthritis Rheum. *42*, 1813–1822.

Cinco, M., Murgia, R., Presani, G., and Perticarari, S. (1997). Integrin CR3 mediates the binding of non-specifically opsonized *Borrelia burgdorferi* to human phagocytes and mammalian cells. Infect. Immun. *65*, 4784–4789.

Collins, C., Shi, C., Russell, J.Q., Fortner, K.A., and Budd, R.C. (2008). Activation of γδ T-cells by *Borrelia burgdorferi* is indirect via a TLR- and caspase-dependent pathway. J. Immunol. *181*, 2392–2398.

Collins, C., Wolfe, J., Roessner, K., Shi, C., Sigal, L.H., and Budd, R.C. (2005). Lyme arthritis synovial gammadelta T-cells instruct dendritic cells via fas ligand. J. Immunol. *175*, 5656–5665.

Craft, J.E., Fischer, D.K., Shimamoto, G.T., and Steere, A.C. (1986). Antigens of *Borrelia burgdorferi* recognized during Lyme disease: appearance of a new immunoglobulin M response and expansion of the immunoglobulin G response late in the illness. J. Clin. Invest. *78*, 934–939.

Crandall, H., Dunn, D.M., Ma, Y., Wooten, R.M., Zachary, J.F., Weis, J.H., Weiss, R.B., and Weis, J.J. (2006). Gene expression profiling reveals unique pathways associated with differential severity of Lyme arthritis. J. Immunol. *177*, 7930–7942.

Crandall, H., Ma, Y., Dunn, D.M., Sundsbak, R.S., Zachary, J.F., Olofsson, P., Holmdahl, R., Weis, J.H., Weiss, R.B., Teuscher, C., *et al.* (2005). *Bb2Bb3* regulation of murine Lyme arthritis is distinct from *Ncf1* and independent of the phagocyte nicotinamide adenine dinucleotide phosphate oxidase. Am. J. Pathol. *167*, 775–785.

Cruz, A.R., Moore, M.W., La Vake, C.J., Eggers, C.H., Salazar, J.C., and Radolf, J.D. (2008). Phagocytosis of *Borrelia burgdorferi*, the Lyme disease spirochete, potentiates innate immune activation and induces apoptosis in human monocytes. Infect. Immun. *76*, 56–70.

Dame, T.M., Orenzoff, B.L., Palmer, L.E., and Furie, M.B. (2007). IFN-γ alters the response of *Borrelia burgdorferi*-activated endothelium to favor chronic inflammation. J. Immunol. *178*, 1172–1179.

Das, S., Barthold, S.W., Giles, S.S., Montgomery, R.R., Telford, S.R., 3rd, and Fikrig, E. (1997). Temporal pattern of *Borrelia burgdorferi p21* expression in ticks and the mammalian host. J. Clin. Invest. *99*, 987–995.

de Silva, A.M., and Fikrig, E. (1997). Arthropod- and host-specific gene expression by *Borrelia burgdorferi*. J. Clin. Invest. *99*, 377–379.

de Silva, A.M., Fikrig, E., Hodzic, E., Kantor, F.S., Telford, S.R., 3rd, and Barthold, S.W. (1998). Immune evasion by tickborne and host-adapted *Borrelia burgdorferi*. J. Infect. Dis. *177*, 395–400.

Dennis, V.A., Dixit, S., O'Brien, S.M., Alvarez, X., Pahar, B., and Philipp, M.T. (2009). Live *Borrelia burgdorferi* spirochetes elicit inflammatory mediators from human monocytes via the Toll-like receptor signaling pathway. Infect. Immun. *77*, 1238–1245.

Dennis, V.A., Jefferson, A., Singh, S.R., Ganapamo, F., and Philipp, M.T. (2006). Interleukin-10 anti-inflammatory response to *Borrelia burgdorferi*, the agent of Lyme disease: a possible role for suppressors of cytokine signaling 1 and 3. Infect. Immun. *74*, 5780–5789.

Drouin, E.E., Glickstein, L., Kwok, W.W., Nepom, G.T., and Steere, A.C. (2008). Searching for borrelial T-cell epitopes associated with antibiotic-refractory Lyme arthritis. Mol. Immunol. *45*, 2323–2332.

Duray, P.H., and Steere, A.C. (1988). Clinical pathologic correlations of Lyme disease by stage. Ann. N.Y. Acad. Sci. *539*, 65–79.

Feng, S., Hodzic, E., and Barthold, S.W. (2000). Lyme arthritis resolution with antiserum to a 37-kilodalton *Borrelia burgdorferi* protein. Infect. Immun. *68*, 4169–4173.

Ferguson, A.R., Youd, M.E., and Corley, R.B. (2004). Marginal zone B-cells transport and deposit IgM-containing immune complexes onto follicular dendritic cells. Int. Immunol. *16*, 1411–1422.

Fikrig, E., Barthold, S.W., Chen, M., Grewal, I.S., Craft, J., and Flavell, R.A. (1996). Protective antibodies in murine Lyme disease arise independently of CD40 ligand. J. Immunol. *157*, 1–3.

Fikrig, E., Barthold, S.W., Kantor, F.S., and Flavell, R.A. (1990). Protection of mice against the Lyme disease agent by immunizing with recombinant OspA. Science *250*, 553–556.

Fikrig, E., Barthold, S.W., Marcantonio, N., Deponte, K., Kantor, F.S., and Flavell, R.A. (1992). Roles of OspA, OspB, and flagellin in protective immunity to Lyme borreliosis in laboratory mice. Infect. Immun. *60*, 657–661.

Fikrig, E., Barthold, S.W., Sun, W., Feng, W., Telford, S.R., 3rd, and Flavell, R.A. (1997). Borrelia burgdorferi P35 and P37 proteins, expressed in vivo, elicit protective immunity. Immunity *6*, 531–539.

Fikrig, E., Bockenstedt, L.K., Barthold, S.W., Chen, M., Tao, H., Ali-Salaam, P., Telford, S.R., and Flavell, R.A. (1994). Sera from patients with chronic Lyme disease protect mice from Lyme borreliosis. J. Infect. Dis. *169*, 568–574.

Francischetti, I.M., Mather, T.N., and Ribeiro, J.M. (2004). Penthalaris, a novel recombinant five-Kunitz tissue factor pathway inhibitor (TFPI) from the salivary gland of the tick vector of Lyme disease, *Ixodes scapularis*. Thromb. Haemost. *91*, 886–898.

Francischetti, I.M., Mather, T.N., and Ribeiro, J.M. (2005). Tick saliva is a potent inhibitor of endothelial cell proliferation and angiogenesis. Thromb. Haemost. *94*, 167–174.

Francischetti, I.M., Valenzuela, J.G., Andersen, J.F., Mather, T.N., and Ribeiro, J.M. (2002). Ixolaris, a novel recombinant tissue factor pathway inhibitor (TFPI) from the salivary gland of the tick, *Ixodes scapularis*: identification of factor X and factor Xa as scaffolds for the inhibition of factor VIIa/tissue factor complex. Blood *99*, 3602–3612.

Frischknecht, F. (2007). The skin as interface in the transmission of arthropod-borne pathogens. Cell. Microbiol. *9*, 1630–1640.

Garcia, R., Gusmani, L., Murgia, R., Guarnaccia, C., Cinco, M., and Rottini, G. (1998). Elastase is the only human neutrophil granule protein that alone is responsible for in vitro killing of *Borrelia burgdorferi*. Infect. Immun. *66*, 1408–1412.

Garg, R., Juncadella, I.J., Ramamoorthi, N., Ashish, Ananthanarayanan, S.K., Thomas, V., Rincon, M., Krueger, J.K., Fikrig, E., Yengo, C.M., *et al.* (2006). Cutting edge: CD4 is the receptor for the tick saliva immunosuppressor, Salp15. J. Immunol. *177*, 6579–6583.

Gebbia, J.A., Coleman, J.L., and Benach, J.L. (2001). *Borrelia* spirochetes up-regulate release and activation of matrix metalloproteinase gelatinase B (MMP-9) and collagenase 1 (MMP-1) in human cells. Infect. Immun. *69*, 456–462.

Gebbia, J.A., Coleman, J.L., and Benach, J.L. (2004). Selective induction of matrix metalloproteinases by *Borrelia burgdorferi* via Toll-like receptor 2 in monocytes. J. Infect. Dis. *189*, 113–119.

Gergel, E.I., and Furie, M.B. (2004). Populations of human T-lymphocytes that traverse the vascular endothelium stimulated by *Borrelia burgdorferi* are enriched with cells that secrete gamma interferon. Infect. Immun. *72*, 1530–1536.

Ghosh, S., Seward, R., Costello, C.E., Stollar, B.D., and Huber, B.T. (2006). Autoantibodies from synovial lesions in chronic, antibiotic treatment-resistant Lyme arthritis bind cytokeratin-10. J. Immunol. *177*, 2486–2494.

Giambartolomei, G.H., Dennis, V.A., Lasater, B.L., and Philipp, M.T. (1999). Induction of pro- and anti-inflammatory cytokines by *Borrelia burgdorferi* lipoproteins in monocytes is mediated by CD14. Infect. Immun. *67*, 140–147.

Giambartolomei, G.H., Dennis, V.A., and Philipp, M.T. (1998). *Borrelia burgdorferi* stimulates the production of interleukin-10 in peripheral blood mononuclear cells from uninfected humans and rhesus monkeys. Infect. Immun. *66*, 2691–2697.

Gillespie, R.D., Dolan, M.C., Piesman, J., and Titus, R.G. (2001). Identification of an IL-2 binding protein in the saliva of the Lyme disease vector tick, *Ixodes scapularis*. J. Immunol. *166*, 4319–4326.

Glickstein, L., Edelstein, M., and Dong, J.Z. (2001). Gamma interferon is not required for arthritis resistance in the murine Lyme disease model. Infect. Immun. *69*, 3737–3743.

Grimm, D., Tilly, K., Byram, R., Stewart, P.E., Krum, J.G., Bueschel, D.M., Schwan, T.G., Policastro, P.F., Elias, A.F., and Rosa, P.A. (2004). Outer-surface protein C of the Lyme disease spirochete: a protein induced in ticks for infection of mammals. Proc. Natl. Acad. Sci. U.S.A. *101*, 3142–3147.

Guerau-de-Arellano, M., Alroy, J., Bullard, D., and Huber, B.T. (2005). Aggravated Lyme carditis in CD11a$^{-/-}$ and CD11c$^{-/-}$ mice. Infect. Immun. *73*, 7637–7643.

Guo, X., Booth, C.J., Paley, M.A., Wang, X., DePonte, K., Fikrig, E., Narasimhan, S., and Montgomery, R.R. (2009). Inhibition of neutrophil function by two tick salivary proteins. Infect. Immun. *77*, 2320–2329.

Habicht, G.S., Beck, G., Benach, J.L., Coleman, J.L., and Leichtling, K.D. (1985). Lyme disease spirochetes induce human and murine interleukin 1 production. J. Immunol. *134*, 3147–3154.

Halperin, J.J., Pass, H.L., Anand, A.K., Luft, B.J., Volkman, D.J., and Dattwyler, R.J. (1988). Nervous system abnormalities in Lyme disease. Ann. N.Y. Acad. Sci. *539*, 24–34.

Hanson, M.S., Cassatt, D.R., Guo, B.P., Patel, N.K., McCarthy, M.P., Dorward, D.W., and Hook, M. (1998). Active and passive immunity against *Borrelia burgdorferi* decorin binding protein A (DbpA) protects against infection. Infect. Immun. *66*, 2143–2153.

Harrington, L.E., Mangan, P.R., and Weaver, C.T. (2006). Expanding the effector CD4 T-cell repertoire: the Th17 lineage. Curr. Opin. Immunol. *18*, 349–356.

Hartmann, K., Corvey, C., Skerka, C., Kirschfink, M., Karas, M., Brade, V., Miller, J.C., Stevenson, B., Wallich, R., Zipfel, P.F., *et al.* (2006). Functional characterization of BbCRASP-2, a distinct outer membrane protein of *Borrelia burgdorferi* that binds host complement regulators factor H and FHL-1. Mol. Microbiol. *61*, 1220–1236.

Hedrick, M.N., Olson, C.M., Jr., Conze, D.B., Bates, T.C., Rincon, M., and Anguita, J. (2006). Control of *Borrelia burgdorferi*-specific CD4$^+$-T-cell effector function by interleukin-12- and T-cell receptor-induced p38 mitogen-activated protein kinase activity. Infect. Immun. *74*, 5713–5717.

Heilpern, A.J., Wertheim, W., He, J., Perides, G., Bronson, R.T., and Hu, L.T. (2009). Matrix metalloproteinase-9 plays a key role in Lyme arthritis but not dissemination of *Borrelia burgdorferi*. Infect. Immun. *77*, 2643–2649.

Hertz, C.J., Kiertscher, S.M., Godowski, P.J., Bouis, D.A., Norgard, M.V., Roth, M.D., and Modlin, R.L. (2001). Microbial lipopeptides stimulate dendritic cell maturation via Toll-like receptor 2. J. Immunol. *166*, 2444–2450.

Hirschfeld, M., Kirschning, C.J., Schwandner, R., Wesche, H., Weis, J.H., Wooten, R.M., and Weis, J.J. (1999). Cutting edge: inflammatory signaling by *Borrelia burgdorferi* lipoproteins is mediated by toll-like receptor 2. J. Immunol. *163*, 2382–2386.

Hodzic, E., Feng, S., Freet, K.J., and Barthold, S.W. (2003). *Borrelia burgdorferi* population dynamics and prototype gene expression during infection of immunocompetent and immunodeficient mice. Infect. Immun. *71*, 5042–5055.

Hodzic, E., Feng, S., Holden, K., Freet, K.J., and Barthold, S.W. (2008). Persistence of *Borrelia burgdorferi* following antibiotic treatment in mice. Antimicrob. Agents Chemother. *52*, 1728–1736.

Hodzic, E., Tunev, S., Feng, S., Freet, K.J., and Barthold, S.W. (2005). Immunoglobulin-regulated expression of *Borrelia burgdorferi* outer surface protein A in vivo. Infect. Immun. *73*, 3313–3321.

Hu, L.T., Eskildsen, M.A., Masgala, C., Steere, A.C., Arner, E.C., Pratta, M.A., Grodzinsky, A.J., Loening, A., and Perides, G. (2001). Host metalloproteinases in Lyme arthritis. Arthritis Rheum. *44*, 1401–1410.

Ibrahim, S.M., Koczan, D., and Thiesen, H.J. (2002). Gene-expression profile of collagen-induced arthritis. J. Autoimmun. *18*, 159–167.

Iliopoulou, B.P., Alroy, J., and Huber, B.T. (2007). CD28 deficiency exacerbates joint inflammation upon *Borrelia burgdorferi* infection, resulting in the development of chronic Lyme arthritis. J. Immunol. *179*, 8076–8082.

Infante-Duarte, C., Horton, H.F., Byrne, M.C., and Kamradt, T. (2000). Microbial lipopeptides induce the production of IL-17 in Th cells. J. Immunol. *165*, 6107–6115.

Inoue, Y., Shimojo, N., Suzuki, Y., Campos Alberto, E.J., Yamaide, A., Suzuki, S., Arima, T., Matsuura, T., Tomiita, M., Aoyagi, M., *et al.* (2007). *CD14* –550 C/T, which is related to the serum level of soluble CD14, is associated with the development of respiratory syncytial virus bronchiolitis in the Japanese population. J. Infect. Dis. *195*, 1618–1624.

Jacobsen, M., Zhou, D., Cepok, S., Nessler, S., Happel, M., Stei, S., Wilske, B., Sommer, N., and Hemmer, B. (2003). Clonal accumulation of activated CD8$^+$ T-cells in the central nervous system during the early phase of neuroborreliosis. J. Infect. Dis. *187*, 963–973.

Juncadella, I.J., Garg, R., Ananthnarayanan, S.K., Yengo, C.M., and Anguita, J. (2007). T-cell signaling pathways inhibited by the tick saliva immunosuppressor, Salp15. FEMS Immunol. Med. Microbiol. *49*, 433–438.

Kannian, P., McHugh, G., Johnson, B.J., Bacon, R.M., Glickstein, L.J., and Steere, A.C. (2007). Antibody responses to *Borrelia burgdorferi* in patients with antibiotic-refractory, antibiotic-responsive, or non-antibiotic-treated Lyme arthritis. Arthritis Rheum. *56*, 4216–4225.

Keane-Myers, A., Maliszewski, C.R., Finkelman, F.D., and Nickell, S.P. (1996). Recombinant IL-4 treatment augments resistance to *Borrelia burgdorferi* infections in both normal susceptible and antibody-deficient susceptible mice. J. Immunol. *156*, 2488–2494.

Keane-Myers, A., and Nickell, S.P. (1995a). Role of IL-4 and IFN-γ in modulation of immunity to *Borrelia burgdorferi* in mice. J. Immunol. *155*, 2020–2028.

Keane-Myers, A., and Nickell, S.P. (1995b). T-cell subset-dependent modulation of immunity to *Borrelia burgdorferi* in mice. J. Immunol. *154*, 1770–1776.

Kemp, D.H., Stone, B.F., and Binnington, K.C. (1982). Tick attachment and feeding: role of the mouthparts, feeding apparatus, salivary gland secretions and the host response. In Physiology of Ticks, F.D. Obenchain, and R. Galun, eds. (Elmsford, NY, Pergamon Press), pp. 119–168.

Kenedy, M.R., Vuppala, S.R., Siegel, C., Kraiczy, P., and Akins, D.R. (2009). CspA-mediated binding of human factor H inhibits complement deposition and confers serum resistance in *Borrelia burgdorferi*. Infect. Immun. *77*, 2773–2782.

Kinjo, Y., Tupin, E., Wu, D., Fujio, M., Garcia-Navarro, R., Benhnia, M.R., Zajonc, D.M., Ben-Menachem, G., Ainge, G.D., Painter, G.F., *et al.* (2006). Natural killer T-cells recognize diacylglycerol antigens from pathogenic bacteria. Nat. Immunol. 7, 978–986.

Knauer, J., Siegemund, S., Muller, U., Al-Robaiy, S., Kastelein, R.A., Alber, G., and Straubinger, R.K. (2007). *Borrelia burgdorferi* potently activates bone marrow-derived conventional dendritic cells for production of IL-23 required for IL-17 release by T-cells. FEMS Immunol. Med. Microbiol. *49*, 353–363.

Kochi, S.K., Johnson, R.C., and Dalmasso, A.P. (1991). Complement-mediated killing of the Lyme disease spirochete *Borrelia burgdorferi*: role of antibody in formation of an effective membrane attack complex. J. Immunol. *146*, 3964–3970.

Kotsyfakis, M., Sa-Nunes, A., Francischetti, I.M., Mather, T.N., Andersen, J.F., and Ribeiro, J.M. (2006). Antiinflammatory and immunosuppressive activity of sialostatin L, a salivary cystatin from the tick *Ixodes scapularis*. J. Biol. Chem. *281*, 26298–26307.

Kraiczy, P., Hellwage, J., Skerka, C., Becker, H., Kirschfink, M., Simon, M.M., Brade, V., Zipfel, P.F., and Wallich, R. (2004). Complement resistance of *Borrelia burgdorferi* correlates with the expression of BbCRASP-1, a novel linear plasmid-encoded surface protein that interacts with human factor H and FHL-1 and is unrelated to Erp proteins. J. Biol. Chem. *279*, 2421–2429.

Kuhlow, C.J., García-Moncó, J.C., Coleman, J.L., and Benach, J.L. (2005). Murine microglia are effective phagocytes for *Borrelia burgdorferi*. J. Neuroimmunol. *168*, 183–187.

Kumar, H., Belperron, A., Barthold, S.W., and Bockenstedt, L.K. (2000). Cutting edge: CD1d deficiency impairs murine host defense against the spirochete, *Borrelia burgdorferi*. J. Immunol. *165*, 4797–4801.

Kuthejlova, M., Kopecky, J., Stepanova, G., and Macela, A. (2001). Tick salivary gland extract inhibits killing of *Borrelia afzelii* spirochetes by mouse macrophages. Infect. Immun. *69*, 575–578.

Lawrenz, M.B., Wooten, R.M., Zachary, J.F., Drouin, S.M., Weis, J.J., Wetsel, R.A., and Norris, S.J. (2003). Effect of complement component C3 deficiency on experimental Lyme borreliosis in mice. Infect. Immun. *71*, 4432–4440.

Lazarus, J.J., Kay, M.A., McCarter, A.L., and Wooten, R.M. (2008). Viable *Borrelia burgdorferi* enhances interleukin-10 production and suppresses activation of murine macrophages. Infect. Immun. *76*, 1153–1162.

Lazarus, J.J., Meadows, M.J., Lintner, R.E., and Wooten, R.M. (2006). IL-10 deficiency promotes increased *Borrelia burgdorferi* clearance predominantly through enhanced innate immune responses. J. Immunol. *177*, 7076–7085.

Liang, F.T., Jacobs, M.B., Bowers, L.C., and Philipp, M.T. (2002). An immune evasion mechanism for spirochetal persistence in Lyme borreliosis. J. Exp. Med. *195*, 415–422.

Liang, F.T., Yan, J., Mbow, M.L., Sviat, S.L., Gilmore, R.D., Mamula, M., and Fikrig, E. (2004). *Borrelia burgdorferi* changes its surface antigenic expression in response to host immune responses. Infect. Immun. *72*, 5759–5767.

Lien, E., Sellati, T.J., Yoshimura, A., Flo, T.H., Rawadi, G., Finberg, R.W., Carroll, J.D., Espevik, T., Ingalls, R.R., Radolf, J.D., *et al.* (1999). Toll-like receptor 2 functions as a pattern recognition receptor for diverse bacterial products. J. Biol. Chem. *274*, 33419–33425.

Lin, B., Noring, R., Steere, A.C., Klempner, M.S., and Hu, L.T. (2000). Soluble CD14 levels in the serum, synovial fluid, and cerebrospinal fluid of patients with various stages of Lyme disease. J. Infect. Dis. *181*, 1185–1188.

Linder, S., Heimerl, C., Fingerle, V., Aepfelbacher, M., and Wilske, B. (2001). Coiling phagocytosis of *Borrelia burgdorferi* by primary human macrophages is controlled by CDC42Hs and Rac1 and involves recruitment of Wiskott-Aldrich syndrome protein and Arp2/3 complex. Infect. Immun. *69*, 1739–1746.

Lisinski, T.J., and Furie, M.B. (2002). Interleukin-10 inhibits proinflammatory activation of endothelium in response to *Borrelia burgdorferi* or lipopolysaccharide but not interleukin-1β or tumor necrosis factor α. J. Leukoc. Biol. *72*, 503–511.

Liu, N., Belperron, A.A., Booth, C.J., and Bockenstedt, L.K. (2009). The caspase 1 inflammasome is not required for control of murine Lyme borreliosis. Infect. Immun. *77*, 3320–3327.

Liu, N., Montgomery, R.R., Barthold, S.W., and Bockenstedt, L.K. (2004). Myeloid differentiation antigen 88 deficiency impairs pathogen clearance but does not alter inflammation in *Borrelia burgdorferi*-infected mice. Infect. Immun. *72*, 3195–3203.

Lunemann, J.D., Gelderblom, H., Sospedra, M., Quandt, J.A., Pinilla, C., Marques, A., and Martin, R. (2007). Cerebrospinal fluid-infiltrating $CD4^+$ T-cells recognize *Borrelia burgdorferi* lysine-enriched protein domains and central nervous system autoantigens in early Lyme encephalitis. Infect. Immun. *75*, 243–251.

Lusitani, D., Malawista, S.E., and Montgomery, R.R. (2002). *Borrelia burgdorferi* are susceptible to killing by a variety of human polymorphonuclear leukocyte components. J. Infect. Dis. *185*, 797–804.

Lusitani, D., Malawista, S.E., and Montgomery, R.R. (2003). Calprotectin, an abundant cytosolic protein from human polymorphonuclear leukocytes, inhibits the growth of *Borrelia burgdorferi*. Infect. Immun. *71*, 4711–4716.

Ma, Y., Miller, J.C., Crandall, H., Larsen, E.T., Dunn, D.M., Weiss, R.B., Subramanian, M., Weis, J.H., Zachary, J.F., Teuscher, C., *et al.* (2009). Interval-specific congenic lines reveal QTL with penetrant Lyme arthritis phenotypes on chromosomes 5, 11, and 12. Infect. Immun. *77*, 3302–3311.

Ma, Y., Seiler, K.P., Eichwald, E.J., Weis, J.H., Teuscher, C., and Weis, J.J. (1998). Distinct characteristics of resistance to *Borrelia burgdorferi*-induced arthritis in C57BL/6N mice. Infect. Immun. *66*, 161–168.

Ma, Y., Seiler, K.P., Tai, K.F., Yang, L., Woods, M., and Weis, J.J. (1994). Outer surface lipoproteins of *Borrelia burgdorferi* stimulate nitric oxide production by the cytokine-inducible pathway. Infect. Immun. *62*, 3663–3671.

Ma, Y., and Weis, J.J. (1993). *Borrelia burgdorferi* outer surface lipoproteins OspA and OspB possess B-cell mitogenic and cytokine-stimulatory properties. Infect. Immun. *61*, 3843–3853.

MacLennan, I.C., and Liu, Y.J. (1991). Marginal zone B-cells respond both to polysaccharide antigens and protein antigens. Res. Immunol. *142*, 346–351.

Malawista, S.E., Montgomery, R.R., and Van Blaricom, G. (1996). Microbial killing by human neutrophil cytokineplasts: similar suppressive effects of reversible and irreversible inhibitors of nitric oxide synthase. J. Leukoc. Biol. *60*, 753–757.

Malkiel, S., Kuhlow, C.J., Mena, P., and Benach, J.L. (2009). The loss and gain of marginal zone and peritoneal B-cells is different in response to relapsing fever and Lyme disease *Borrelia*. J. Immunol. *182*, 498–506.

Marietta, E.V., Weis, J.J., and Weis, J.H. (1997). CD28 expression by mouse mast cells is modulated by lipopolysaccharide and outer surface protein A lipoprotein from *Borrelia burgdorferi*. J. Immunol. *159*, 2840–2848.

Matyniak, J.E., and Reiner, S.L. (1995). T helper phenotype and genetic susceptibility in experimental Lyme disease. J. Exp. Med. *181*, 1251–1254.

McDowell, J.V., Wolfgang, J., Tran, E., Metts, M.S., Hamilton, D., and Marconi, R.T. (2003). Comprehensive analysis of the factor H binding capabilities of *Borrelia* species associated with Lyme disease: delineation of two distinct classes of factor H binding proteins. Infect. Immun. *71*, 3597–3602.

McKisic, M.D., and Barthold, S.W. (2000). T-cell-independent responses to *Borrelia burgdorferi* are critical for protective immunity and resolution of Lyme disease. Infect. Immun. *68*, 5190–5197.

McKisic, M.D., Redmond, W.L., and Barthold, S.W. (2000). Cutting edge: T-cell-mediated pathology in murine Lyme borreliosis. J. Immunol. *164*, 6096–6099.

Meyer, A.L., Trollmo, C., Crawford, F., Marrack, P., Steere, A.C., Huber, B.T., Kappler, J., and Hafler, D.A. (2000). Direct enumeration of *Borrelia*-reactive CD4 T-cells *ex vivo* by using MHC class II tetramers. Proc. Natl. Acad. Sci. U.S.A. *97*, 11433–11438.

Miller, J.C., Ma, Y., Bian, J., Sheehan, K.C., Zachary, J.F., Weis, J.H., Schreiber, R.D., and Weis, J.J. (2008). A critical role for type I IFN in arthritis development following *Borrelia burgdorferi* infection of mice. J. Immunol. *181*, 8492–8503.

Miller, J.C., von Lackum, K., Babb, K., McAlister, J.D., and Stevenson, B. (2003). Temporal analysis of *Borrelia burgdorferi* Erp protein expression throughout the mammal-tick infectious cycle. Infect. Immun. *71*, 6943–6952.

Montgomery, R.R., Booth, C.J., Wang, X., Blaho, V.A., Malawista, S.E., and Brown, C.R. (2007). Recruitment of macrophages and polymorphonuclear leukocytes in Lyme carditis. Infect. Immun. *75*, 613–620.

Montgomery, R.R., Malawista, S.E., Feen, K.J., and Bockenstedt, L.K. (1996). Direct demonstration of antigenic substitution of *Borrelia burgdorferi* ex vivo: exploration of the paradox of the early immune response to outer surface proteins A and C in Lyme disease. J. Exp. Med. *183*, 261–269.

Montgomery, R.R., Nathanson, M.H., and Malawista, S.E. (1993). The fate of *Borrelia burgdorferi*, the agent for Lyme disease, in mouse macrophages: destruction, survival, recovery. J. Immunol. *150*, 909–915.

Moore, K.W., de Waal Malefyt, R., Coffman, R.L., and O'Garra, A. (2001). Interleukin-10 and the interleukin-10 receptor. Annu. Rev. Immunol. *19*, 683–765.

Moore, M.W., Cruz, A.R., LaVake, C.J., Marzo, A.L., Eggers, C.H., Salazar, J.C., and Radolf, J.D. (2007). Phagocytosis of *Borrelia burgdorferi* and *Treponema pallidum* potentiates innate immune activation and induces gamma interferon production. Infect. Immun. *75*, 2046–2062.

Morrison, T.B., Weis, J.H., and Weis, J.J. (1997). *Borrelia burgdorferi* outer surface protein A (OspA) activates and primes human neutrophils. J. Immunol. *158*, 4838–4845.

Mullegger, R.R., Means, T.K., Shin, J.J., Lee, M., Jones, K.L., Glickstein, L.J., Luster, A.D., and Steere, A.C. (2007). Chemokine signatures in the skin disorders of Lyme borreliosis in Europe: predominance of CXCL9 and CXCL10 in erythema migrans and acrodermatitis and CXCL13 in lymphocytoma. Infect. Immun. *75*, 4621–4628.

Narasimhan, S., Koski, R.A., Beaulieu, B., Anderson, J.F., Ramamoorthi, N., Kantor, F., Cappello, M., and Fikrig, E. (2002a). A novel family of anticoagulants from the saliva of *Ixodes scapularis*. Insect Mol. Biol. *11*, 641–650.

Narasimhan, S., Santiago, F., Koski, R.A., Brei, B., Anderson, J.F., Fish, D., and Fikrig, E. (2002b). Examination of the *Borrelia burgdorferi* transcriptome in *Ixodes scapularis* during feeding. J. Bacteriol. *184*, 3122–3125.

Narayan, K., Dail, D., Li, L., Cadavid, D., Amrute, S., Fitzgerald-Bocarsly, P., and Pachner, A.R. (2005). The nervous system as ectopic germinal center: CXCL13 and IgG in Lyme neuroborreliosis. Ann. Neurol. *57*, 813–823.

Nimmerjahn, F., Bruhns, P., Horiuchi, K., and Ravetch, J.V. (2005). FcγRIV: a novel FcR with distinct IgG subclass specificity. Immunity *23*, 41–51.

Nocton, J.J., Dressler, F., Rutledge, B.J., Rys, P.N., Persing, D.H., and Steere, A.C. (1994). Detection of *Borrelia burgdorferi* DNA by polymerase chain reaction in synovial fluid from patients with Lyme arthritis. N. Engl. J. Med. *330*, 229–234.

Norgard, M.V., Arndt, L.L., Akins, D.R., Curetty, L.L., Harrich, D.A., and Radolf, J.D. (1996). Activation of human monocytic cells by *Treponema pallidum* and *Borrelia burgdorferi* lipoproteins and synthetic lipopeptides proceeds via a pathway distinct from that of lipopolysaccharide but involves the transcriptional activator NF-κB. Infect. Immun. *64*, 3845–3852.

Norgard, M.V., Riley, B.S., Richardson, J.A., and Radolf, J.D. (1995). Dermal inflammation elicited by synthetic analogs of *Treponema pallidum* and *Borrelia burgdorferi* lipoproteins. Infect. Immun. *63*, 1507–1515.

Norman, M.U., Moriarty, T.J., Dresser, A.R., Millen, B., Kubes, P., and Chaconas, G. (2008). Molecular mechanisms involved in vascular interactions of the Lyme disease pathogen in a living host. PLoS Pathog. *4*, e1000169.

Nowalk, A.J., Gilmore, R.D., Jr., and Carroll, J.A. (2006). Serologic proteome analysis of *Borrelia burgdorferi* membrane-associated proteins. Infect. Immun. *74*, 3864–3873.

Nuttall, P.A., and Labuda, M. (2004). Tick–host interactions: saliva-activated transmission. Parasitology *129 Suppl.*, S177–189.

Ohnishi, J., Piesman, J., and de Silva, A.M. (2001). Antigenic and genetic heterogeneity of *Borrelia burgdorferi* populations transmitted by ticks. Proc. Natl. Acad. Sci. U.S.A. *98*, 670–675.

Oliver, A.M., Martin, F., Gartland, G.L., Carter, R.H., and Kearney, J.F. (1997). Marginal zone B-cells exhibit unique activation, proliferative and immunoglobulin secretory responses. Eur J. Immunol. *27*, 2366–2374.

Olson, C.M., Jr., Bates, T.C., Izadi, H., Radolf, J.D., Huber, S.A., Boyson, J.E., and Anguita, J. (2009). Local production of IFN-γ by invariant NKT-cells modulates acute Lyme carditis. J. Immunol. *182*, 3728–3734.

Pachner, A.R., and Steiner, I. (2007). Lyme neuroborreliosis: infection, immunity, and inflammation. Lancet Neurol. *6*, 544–552.

Pal, U., Wang, P., Bao, F., Yang, X., Samanta, S., Schoen, R., Wormser, G.P., Schwartz, I., and Fikrig, E. (2008). *Borrelia burgdorferi* basic membrane proteins A and B participate in the genesis of Lyme arthritis. J. Exp. Med. *205*, 133–141.

Pal, U., Yang, X., Chen, M., Bockenstedt, L.K., Anderson, J.F., Flavell, R.A., Norgard, M.V., and Fikrig, E. (2004). OspC facilitates *Borrelia burgdorferi* invasion of *Ixodes scapularis* salivary glands. J. Clin. Invest. *113*, 220–230.

Patarakul, K., Cole, M.F., and Hughes, C.A. (1999). Complement resistance in *Borrelia burgdorferi* strain 297: outer membrane proteins prevent MAC formation at lysis susceptible sites. Microb. Pathol. *27*, 25–41.

Perides, G., Tanner-Brown, L.M., Eskildsen, M.A., and Klempner, M.S. (1999). *Borrelia burgdorferi* induces matrix metalloproteinases by neural cultures. J. Neurosci. Res. *58*, 779–790.

Poltorak, A., He, X., Smirnova, I., Liu, M.Y., Van Huffel, C., Du, X., Birdwell, D., Alejos, E., Silva, M., Galanos, C., *et al.* (1998). Defective LPS signaling in C3H/HeJ and C57BL/10ScCr mice: mutations in *Tlr4* gene. Science *282*, 2085–2088.

Posey, J.E., and Gherardini, F.C. (2000). Lack of a role for iron in the Lyme disease pathogen. Science *288*, 1651–1653.

Potter, M.R., Noben-Trauth, N., Weis, J.H., Teuscher, C., and Weis, J.J. (2000). Interleukin-4 (IL-4) and IL-13 signaling pathways do not regulate *Borrelia burgdorferi*-induced arthritis in mice: IgG1 is not required for host control of tissue spirochetes. Infect. Immun. *68*, 5603–5609.

Radolf, J.D., Arndt, L.L., Akins, D.R., Curetty, L.L., Levi, M.E., Shen, Y., Davis, L.S., and Norgard, M.V. (1995). *Treponema pallidum* and *Borrelia burgdorferi* lipoproteins and synthetic lipopeptides activate monocytes/macrophages. J. Immunol. *154*, 2866–2877.

Ramamoorthi, N., Narasimhan, S., Pal, U., Bao, F., Yang, X.F., Fish, D., Anguita, J., Norgard, M.V., Kantor, F.S., Anderson, J.F., *et al.* (2005). The Lyme disease agent exploits a tick protein to infect the mammalian host. Nature *436*, 573–577.

Ramamoorthy, R., and Scholl-Meeker, D. (2001). *Borrelia burgdorferi* proteins whose expression is similarly affected by culture temperature and pH. Infect. Immun. *69*, 2739–2742.

Ramesh, G., Borda, J.T., Dufour, J., Kaushal, D., Ramamoorthy, R., Lackner, A.A., and Philipp, M.T. (2008). Interaction of the Lyme disease spirochete *Borrelia burgdorferi* with brain parenchyma elicits inflammatory mediators from glial cells as well as glial and neuronal apoptosis. Am. J. Pathol. *173*, 1415–1427.

Ribeiro, J.M., Alarcon-Chaidez, F., Francischetti, I.M., Mans, B.J., Mather, T.N., Valenzuela, J.G., and Wikel, S.K. (2006). An annotated catalog of salivary gland transcripts from *Ixodes scapularis* ticks. Insect Biochem. Mol. Biol. *36*, 111–129.

Ribeiro, J.M., Makoul, G.T., and Robinson, D.R. (1988). *Ixodes dammini*: evidence for salivary prostacyclin secretion. J. Parasitol. *74*, 1068–1069.

Rittig, M.G., Krause, A., Haupl, T., Schaible, U.E., Modolell, M., Kramer, M.D., Lutjen-Drecoll, E., Simon, M.M., and Burmester, G.R. (1992). Coiling phagocytosis is the preferential phagocytic mechanism for *Borrelia burgdorferi*. Infect. Immun. *60*, 4205–4212.

Roessner, K., Wolfe, J., Shi, C., Sigal, L.H., Huber, S., and Budd, R.C. (2003). High expression of Fas ligand by synovial fluid-derived γδ T-cells in Lyme arthritis. J. Immunol. *170*, 2702–2710.

Roper, R.J., Weis, J.J., McCracken, B.A., Green, C.B., Ma, Y., Weber, K.S., Fairbairn, D., Butterfield, R.J., Potter, M.R., Zachary, J.F., *et al.* (2001). Genetic control of susceptibility to experimental Lyme arthritis is polygenic and exhibits consistent linkage to multiple loci on chromosome 5 in four independent mouse crosses. Genes Immun. *2*, 388–397.

Ruderman, E.M., Kerr, J.S., Telford, S.R., 3rd, Spielman, A., Glimcher, L.H., and Gravallese, E.M. (1995). Early murine Lyme carditis has a macrophage predominance and is independent of major histocompatibility complex class II-CD4+ T-cell interactions. J. Infect. Dis. *171*, 362–370.

Rupprecht, T.A., Kirschning, C.J., Popp, B., Kastenbauer, S., Fingerle, V., Pfister, H.W., and Koedel, U. (2007). *Borrelia garinii* induces CXCL13 production in human monocytes through Toll-like receptor 2. Infect. Immun. *75*, 4351–4356.

Sa-Nunes, A., Bafica, A., Lucas, D.A., Conrads, T.P., Veenstra, T.D., Andersen, J.F., Mather, T.N., Ribeiro, J.M., and Francischetti, I.M. (2007). Prostaglandin E2 is a major inhibitor of dendritic cell maturation and function in *Ixodes scapularis* saliva. J. Immunol. *179*, 1497–1505.

Sakaguchi, S. (2004). Naturally arising CD4+ regulatory T-cells for immunologic self-tolerance and negative control of immune responses. Annu. Rev. Immunol. *22*, 531–562.

Salazar, J.C., Duhnam-Ems, S., La Vake, C., Cruz, A.R., Moore, M.W., Caimano, M.J., Velez-Climent, L., Shupe, J., Krueger, W., and Radolf, J.D. (2009). Activation of human monocytes by live *Borrelia burgdorferi* generates TLR2-dependent and -independent responses which include induction of IFN-beta. PLoS Pathog. *5*, e1000444.

Salazar, J.C., Pope, C.D., Sellati, T.J., Feder, H.M., Jr., Kiely, T.G., Dardick, K.R., Buckman, R.L., Moore, M.W., Caimano, M.J., Pope, J.G., *et al.* (2003). Coevolution of markers of innate and adaptive immunity in skin and peripheral blood of patients with erythema migrans. J. Immunol. *171*, 2660–2670.

Schaible, U.E., Kramer, M.D., Museteanu, C., Zimmer, G., Mossmann, H., and Simon, M.M. (1989). The severe combined immunodeficiency (*scid*) mouse: a laboratory model for the analysis of Lyme arthritis and carditis. J. Exp. Med. *170*, 1427–1432.

Scheckelhoff, M.R., Telford, S.R., Wesley, M., and Hu, L.T. (2007). *Borrelia burgdorferi* intercepts host hormonal signals to regulate expression of outer surface protein A. Proc. Natl. Acad. Sci. U.S.A. *104*, 7247–7252.

Schroder, N.W., Diterich, I., Zinke, A., Eckert, J., Draing, C., von Baehr, V., Hassler, D., Priem, S., Hahn, K., Michelsen, K.S., *et al.* (2005). Heterozygous Arg753Gln polymorphism of human TLR-2 impairs immune activation by *Borrelia burgdorferi* and protects from late stage Lyme disease. J. Immunol. *175*, 2534–2540.

Schwan, T.G. (2003). Temporal regulation of outer surface proteins of the Lyme-disease spirochaete *Borrelia burgdorferi*. Biochem. Soc. Trans. *31*, 108–112.

Schwan, T.G., and Piesman, J. (2000). Temporal changes in outer surface proteins A and C of the Lyme disease-associated spirochete, *Borrelia burgdorferi*, during the chain of infection in ticks and mice. J. Clin. Microbiol. *38*, 382–388.

Schwan, T.G., Piesman, J., Golde, W.T., Dolan, M.C., and Rosa, P.A. (1995). Induction of an outer surface protein on *Borrelia burgdorferi* during tick feeding. Proc. Natl. Acad. Sci. U.S.A. *92*, 2909–2913.

Seiler, K.P., Vavrin, Z., Eichwald, E., Hibbs, J.B., Jr., and Weis, J.J. (1995). Nitric oxide production during murine Lyme disease: lack of involvement in host resistance or pathology. Infect. Immun. *63*, 3886–3895.

Sellati, T.J., Bouis, D.A., Caimano, M.J., Feulner, J.A., Ayers, C., Lien, E., and Radolf, J.D. (1999). Activation of human monocytic cells by *Borrelia burgdorferi* and *Treponema pallidum* is facilitated by CD14 and correlates with surface exposure of spirochetal lipoproteins. J. Immunol. *163*, 2049–2056.

Sellati, T.J., Bouis, D.A., Kitchens, R.L., Darveau, R.P., Pugin, J., Ulevitch, R.J., Gangloff, S.C., Goyert, S.M., Norgard, M.V., and Radolf, J.D. (1998). *Treponema pallidum* and *Borrelia burgdorferi* lipoproteins and synthetic lipopeptides activate monocytic cells via a CD14-dependent pathway distinct from that used by lipopolysaccharide. J. Immunol. *160*, 5455–5464.

Sellati, T.J., Burns, M.J., Ficazzola, M.A., and Furie, M.B. (1995). *Borrelia burgdorferi* up-regulates expression of adhesion molecules on endothelial cells and promotes transendothelial migration of neutrophils in vitro. Infect. Immun. *63*, 4439–4447.

Seshu, J., Boylan, J.A., Gherardini, F.C., and Skare, J.T. (2004). Dissolved oxygen levels alter gene expression and antigen profiles in *Borrelia burgdorferi*. Infect. Immun. 72, 1580–1586.

Shahinian, A., Pfeffer, K., Lee, K.P., Kundig, T.M., Kishihara, K., Wakeham, A., Kawai, K., Ohashi, P.S., Thompson, C.B., and Mak, T.W. (1993). Differential T-cell costimulatory requirements in CD28-deficient mice. Science *261*, 609–612.

Shanafelt, M.C., Kang, I., Barthold, S.W., and Bockenstedt, L.K. (1998). Modulation of murine Lyme borreliosis by interruption of the B7/CD28 T-cell costimulatory pathway. Infect. Immun. *66*, 266–271.

Shih, C.M., Pollack, R.J., Telford, S.R., 3rd, and Spielman, A. (1992). Delayed dissemination of Lyme disease spirochetes from the site of deposition in the skin of mice. J. Infect. Dis. *166*, 827–831.

Shih, C.M., and Spielman, A. (1993). Topical prophylaxis for Lyme disease after tick bite in a rodent model. J. Infect. Dis. *168*, 1042–1045.

Shin, J.J., Glickstein, L.J., and Steere, A.C. (2007). High levels of inflammatory chemokines and cytokines in joint fluid and synovial tissue throughout the course of antibiotic-refractory Lyme arthritis. Arthritis Rheum. *56*, 1325–1335.

Shin, O.S., Isberg, R., Akira, S., Uematsu, S., Behera, A.K., and Hu, L.T. (2008). Distinct roles for MyD88, TLR2, TLR5, and TLR9 in phagocytosis of *Borrelia burgdorferi* and inflammatory signaling. Infect. Immun. *76*, 2341–2351.

Steere, A.C., and Angelis, S.M. (2006). Therapy for Lyme arthritis: strategies for the treatment of antibiotic-refractory arthritis. Arthritis Rheum. *54*, 3079–3086.

Steere, A.C., Grodzicki, R.L., Kornblatt, A.N., Craft, J.E., Barbour, A.G., Burgdorfer, W., Schmid, G.P., Johnson, E., and Malawista, S.E. (1983). The spirochetal etiology of Lyme disease. N. Engl. J. Med. *308*, 733–740.

Steere, A.C., Klitz, W., Drouin, E.E., Falk, B.A., Kwok, W.W., Nepom, G.T., and Baxter-Lowe, L.A. (2006). Antibiotic-refractory Lyme arthritis is associated with

HLA-DR molecules that bind a *Borrelia burgdorferi* peptide. J. Exp. Med. *203*, 961–971.

Steere, A.C., Schoen, R.T., and Taylor, E. (1987). The clinical evolution of Lyme arthritis. Ann. Intern. Med. *107*, 725–731.

Stevenson, B., Schwan, T.G., and Rosa, P.A. (1995). Temperature-related differential expression of antigens in the Lyme disease spirochete, *Borrelia burgdorferi*. Infect. Immun. *63*, 4535–4539.

Stewart, P.E., Wang, X., Bueschel, D.M., Clifton, D.R., Grimm, D., Tilly, K., Carroll, J.A., Weis, J.J., and Rosa, P.A. (2006). Delineating the requirement for the *Borrelia burgdorferi* virulence factor OspC in the mammalian host. Infect. Immun. *74*, 3547–3553.

Suk, K., Das, S., Sun, W., Jwang, B., Barthold, S.W., Flavell, R.A., and Fikrig, E. (1995). *Borrelia burgdorferi* genes selectively expressed in the infected host. Proc. Natl. Acad. Sci. U.S.A. *92*, 4269–4273.

Szczepanski, A., Furie, M.B., Benach, J.L., Lane, B.P., and Fleit, H.B. (1990). Interaction between *Borrelia burgdorferi* and endothelium in vitro. J. Clin. Invest. *85*, 1637–1647.

Tatro, J.B., Romero, L.I., Beasley, D., Steere, A.C., and Reichlin, S. (1994). *Borrelia burgdorferi* and *Escherichia coli* lipopolysaccharides induce nitric oxide and interleukin-6 production in cultured rat brain cells. J. Infect. Dis. *169*, 1014–1022.

Tesmer, L.A., Lundy, S.K., Sarkar, S., and Fox, D.A. (2008). Th17 cells in human disease. Immunol. Rev. *22*, 87–113.

Thoma-Uszynski, S., Kiertscher, S.M., Ochoa, M.T., Bouis, D.A., Norgard, M.V., Miyake, K., Godowski, P.J., Roth, M.D., and Modlin, R.L. (2000). Activation of Toll-like receptor 2 on human dendritic cells triggers induction of IL-12, but not IL-10. J. Immunol. *165*, 3804–3810.

Tilly, K., Bestor, A., Jewett, M.W., and Rosa, P. (2007). Rapid clearance of Lyme disease spirochetes lacking OspC from skin. Infect. Immun. *75*, 1517–1519.

Tilly, K., Krum, J.G., Bestor, A., Jewett, M.W., Grimm, D., Bueschel, D., Byram, R., Dorward, D., Vanraden, M.J., Stewart, P., *et al.* (2006). *Borrelia burgdorferi* OspC protein required exclusively in a crucial early stage of mammalian infection. Infect. Immun. *74*, 3554–3564.

Trollmo, C., Meyer, A.L., Steere, A.C., Hafler, D.A., and Huber, B.T. (2001). Molecular mimicry in Lyme arthritis demonstrated at the single cell level: LFA-1α_L is a partial agonist for outer surface protein A-reactive T-cells. J. Immunol. *166*, 5286–5291.

Tupin, E., Benhnia, M.R., Kinjo, Y., Patsey, R., Lena, C.J., Haller, M.C., Caimano, M.J., Imamura, M., Wong, C.H., Crotty, S., *et al.* (2008). NKT-cells prevent chronic joint inflammation after infection with *Borrelia burgdorferi*. Proc. Natl. Acad. Sci. U.S.A. *105*, 19863–19868.

Tyson, K., Elkins, C., Patterson, H., Fikrig, E., and de Silva, A. (2007). Biochemical and functional characterization of Salp20, an *Ixodes scapularis* tick salivary protein that inhibits the complement pathway. Insect Mol. Biol. *16*, 469–479.

Valenzuela, J.G., Charlab, R., Mather, T.N., and Ribeiro, J.M. (2000). Purification, cloning, and expression of a novel salivary anticomplement protein from the tick, *Ixodes scapularis*. J. Biol. Chem. *275*, 18717–18723.

Vesely, D.L., Fish, D., Shlomchik, M.J., Kaplan, D.H., and Bockenstedt, L.K. (2009). Langerhans cell deficiency impairs *Ixodes scapularis* suppression of Th1 responses in mice. Infect. Immun. *77*, 1881–1887.

Vincent, M.S., Roessner, K., Sellati, T., Huston, C.D., Sigal, L.H., Behar, S.M., Radolf, J.D., and Budd, R.C. (1998). Lyme arthritis synovial γδ T-cells respond to *Borrelia burgdorferi* lipoproteins and lipidated hexapeptides. J. Immunol. *161*, 5762–5771.

Wang, G., Ma, Y., Buyuk, A., McClain, S., Weis, J.J., and Schwartz, I. (2004). Impaired host defense to infection and Toll-like receptor 2-independent killing of *Borrelia burgdorferi* clinical isolates in TLR2-deficient C3H/HeJ mice. FEMS Microbiol. Lett. *231*, 219–225.

Wang, G., Petzke, M.M., Iyer, R., Wu, H., and Schwartz, I. (2008a). Pattern of proinflammatory cytokine induction in RAW264.7 mouse macrophages is identical for virulent and attenuated *Borrelia burgdorferi*. J. Immunol. *180*, 8306–8315.

Wang, H., and Nuttall, P.A. (1999). Immunoglobulin-binding proteins in ticks: new target for vaccine development against a blood-feeding parasite. Cell. Mol. Life Sci. *56*, 286–295.

Wang, X., Ma, Y., Weis, J.H., Zachary, J.F., Kirschning, C.J., and Weis, J.J. (2005). Relative contributions of innate and acquired host responses to bacterial control and arthritis development in Lyme disease. Infect. Immun. *73*, 657–660.

Wang, X., Ma, Y., Yoder, A., Crandall, H., Zachary, J.F., Fujinami, R.S., Weis, J.H., and Weis, J.J. (2008b). T-cell infiltration is associated with increased Lyme arthritis in TLR2$^{-/-}$ mice. FEMS Immunol. Med. Microbiol. *52*, 124–133.

Weis, J.J., McCracken, B.A., Ma, Y., Fairbairn, D., Roper, R.J., Morrison, T.B., Weis, J.H., Zachary, J.F., Doerge, R.W., and Teuscher, C. (1999). Identification of quantitative trait loci governing arthritis severity and humoral responses in the murine model of Lyme disease. J. Immunol. *162*, 948–956.

Woodman, M.E., Cooley, A.E., Miller, J.C., Lazarus, J.J., Tucker, K., Bykowski, T., Botto, M., Hellwage, J., Wooten, R.M., and Stevenson, B. (2007). *Borrelia burgdorferi* binding of host complement regulator factor H is not required for efficient mammalian infection. Infect. Immun. *75*, 3131–3139.

Woods, A., Soulas-Sprauel, P., Jaulhac, B., Arditi, B., Knapp, A.M., Pasquali, J.L., Korganow, A.S., and Martin, T. (2008). MyD88 negatively controls hypergammaglobulinemia with autoantibody production during bacterial infection. Infect. Immun. *76*, 1657–1667.

Wooten, R.M., Ma, Y., Yoder, R.A., Brown, J.P., Weis, J.H., Zachary, J.F., Kirschning, C.J., and Weis, J.J. (2002). Toll-like receptor 2 plays a pivotal role in host defense and inflammatory response to *Borrelia burgdorferi*. Vector Borne Zoonotic Dis. 2, 275–278.

Wooten, R.M., Modur, V.R., McIntyre, T.M., and Weis, J.J. (1996). *Borrelia burgdorferi* outer membrane protein A induces nuclear translocation of nuclear factor-κB and inflammatory activation in human endothelial cells. J. Immunol. *157*, 4584–4590.

Wooten, R.M., Morrison, T.B., Weis, J.H., Wright, S.D., Thieringer, R., and Weis, J.J. (1998). The role of CD14 in signaling mediated by outer membrane lipoproteins of *Borrelia burgdorferi*. J. Immunol. *160*, 5485–5492.

Wooten, R.M., and Weis, J.J. (2001). Host–pathogen interactions promoting inflammatory Lyme arthritis: use of mouse models for dissection of disease processes. Curr. Opin. Microbiol. *4*, 274–279.

Xu, Q., Seemanapalli, S.V., McShan, K., and Liang, F.T. (2006). Constitutive expression of outer surface protein C diminishes the ability of *Borrelia burgdorferi* to evade specific humoral immunity. Infect. Immun. *74*, 5177–5184.

Yang, X., Goldberg, M.S., Popova, T.G., Schoeler, G.B., Wikel, S.K., Hagman, K.E., and Norgard, M.V. (2000). Interdependence of environmental factors influencing reciprocal patterns of gene expression in virulent *Borrelia burgdorferi*. Mol. Microbiol. *37*, 1470–1479.

Yoder, A., Wang, X., Ma, Y., Philipp, M.T., Heilbrun, M., Weis, J.H., Kirschning, C.J., Wooten, R.M., and Weis, J.J. (2003). Tripalmitoyl-S-glyceryl-cysteine-dependent OspA vaccination of Toll-like receptor 2-deficient mice results in effective protection from *Borrelia burgdorferi* challenge. Infect. Immun. *71*, 3894–3900.

Zeidner, N., Dreitz, M., Belasco, D., and Fish, D. (1996). Suppression of acute *Ixodes scapularis*-induced *Borrelia burgdorferi* infection using tumor necrosis factor-α, interleukin-2, and interferon-γ. J. Infect. Dis. *173*, 187–195.

Zeidner, N.S., Schneider, B.S., Nuncio, M.S., Gern, L., and Piesman, J. (2002). Coinoculation of *Borrelia* spp. with tick salivary gland lysate enhances spirochete load in mice and is tick species-specific. J. Parasitol. *88*, 1276–1278.

Zhao, Z., Chang, H., Trevino, R.P., Whren, K., Bhawan, J., and Klempner, M.S. (2003). Selective up-regulation of matrix metalloproteinase-9 expression in human erythema migrans skin lesions of acute Lyme disease. J. Infect. Dis. *188*, 1098–1104.

Detection of *Borrelia burgdorferi*

16

Guiqing Wang, Maria E. Aguero-Rosenfeld, Gary P. Wormser and Ira Schwartz

Abstract

Numerous methods have been developed for detection of *Borrelia burgdorferi* in a variety of specimen sources. These can be broadly divided into direct and indirect approaches. *Borrelia burgdorferi* can be visualized by microscopy, cultivated from numerous clinical and environmental specimens and detected by PCR-based modalities. These direct detection techniques are extensively used in experimental and epidemiological studies, but have not been widely employed as diagnostic approaches in the clinical setting. Detection of antibodies to *B. burgdorferi* by ELISA and immunoblot is currently the primary tool for laboratory diagnosis of Lyme disease in humans. Recent advances in serological testing include the development of assays based on recombinant *B. burgdorferi* antigens or derived peptides, rather than sonicated whole cell lysates.

16.1 Introduction

Detection of *Borrelia burgdorferi* [unless otherwise noted, this refers to *B. burgdorferi* sensu lato (s.l.)] is central to experimental and clinical studies of Lyme borreliosis. Methods vary from direct observation of spirochaetes by microscopy to indirect detection by serological approaches. This chapter reviews the current state of detection methodology and describes applications to patient specimens, experimental animal tissues and vector ticks.

16.2 Methods for detection of *Borrelia burgdorferi* infection

16.2.1 Direct detection of *Borrelia burgdorferi* or its components

Several methods have been used for direct detection of *B. burgdorferi* in infected vectors, host reservoirs or clinical samples from patients with Lyme borreliosis. These include microscope-based observation of intact spirochaetes, detection of *B. burgdorferi*-specific antigens, *in vitro* cultivation of *B. burgdorferi*, and polymerase chain reaction (PCR) analysis of *B. burgdorferi* nucleic acids.

Direct microscopic detection

Borrelia burgdorferi can be visualized directly in infected *Ixodes* ticks and animal tissues using dark-field microscopy or after non-specific or immunological staining of specimens. This is rarely possible in tissues or body fluids of patients.

Dark-field microscopy is useful only for specimens in which a large number of spirochaetes are expected, such as suspensions of the midguts of infected ticks or spirochaetes grown in culture media. Individual spirochaetes may be visualized after Giemsa, carbol-fuchsin and silver staining in ticks and in animal or human tissues (Aberer and Duray, 1991; Marcus *et al.*, 1985; de Koning *et al.*, 1989; de Koning and Duray, 1993). However, immunological staining methods are more sensitive for direct detection of *B. burgdorferi* in ticks and animal tissues. Borrelia may be detected by direct immunofluorescence (DFA) using fluorescein isothiocyanate (FITC)-labelled immune sera against *B. burgdorferi* or by indirect immunofluorescence (IFA) using unlabeled *B. burgdorferi*-specific antibodies and secondary FITC-labelled antibodies.

In general, microscope-based assays are employed mainly for detection of *B. burgdorferi* in midgut and salivary glands of the tick vectors in ecological and epidemiological surveys or in experimental studies of tick infection (Piesman *et al.*, 1990; de Silva and Fikrig, 1995; Brunet *et al.*, 1995). The diagnostic value of microscope-based assays in the clinical laboratory is limited, as the spirochaete density in clinical samples is very low. In addition, spirochaete morphology is variable, making these microorganisms difficult to distinguish from host tissue structures (Aberer and Duray, 1991).

Antigen detection

Two types of direct protein antigen detection assays are now available. The antigen-capture enzyme-linked immunosorbent assay (ELISA) and the reverse Western blot or dot blot. The antigen capture ELISA has been used to detect *B. burgdorferi* antigens in cerebrospinal fluid of patients with neuroborreliosis (Coyle *et al.*, 1995) and in urine samples of patients with erythema migrans (EM) (Hyde *et al.*, 1989; Dorward *et al.*, 1991; Klempner *et al.*, 2001b).

These antigen detection assays can be easily performed and, if urine samples are used, they may be readily obtained. The limitations of the antigen detection methods are their relatively lower sensitivity (Burkot *et al.*, 1994) and poor specificity and reproducibility (Klempner *et al.*, 2001b). As a result, these assays are rarely used.

Culture

The liquid media currently used to grow *B. burgdorferi* were derived by various modifications made over time (Preac-Mursic *et al.*, 1986; Stoenner, 1974; Stoenner *et al.*, 1982; Barbour, 1984; Barbour *et al.*, 1983) from the original medium formulated by Kelly for cultivation of relapsing fever borreliae (Kelly, 1971). Current versions of this medium [Barbour–Stoenner–Kelly II medium (BSK II) (Barbour, 1984), BSK-H (Pollack *et al.*, 1993), and a modified Kelly–Pettenkofer medium (MKP) (Preac-Mursic *et al.*, 1986)] yield better recovery of *B. burgdorferi* from specimens containing small numbers of bacteria, and result in more rapid growth and greater bacterial densities (~ 10^8 to 10^9/ml) following inoculation. Key ingredients of BSK II include CMRL-1066, which is a standard medium used for growing various types of mammalian cells; bovine serum albumin fraction V, which serves as a rich source of protein and lipid and also stabilizes pH; N-acetylglucosamine, a precursor for bacterial cell wall biosynthesis; rabbit serum; citrate; and pyruvate. This complex medium is necessary because *B. burgdorferi* is unable to synthesize amino acids, nucleotides, fatty acids, or most other cellular building blocks (Fraser *et al.*, 1997).

The growth-promoting capability of BSK II and related media depends upon the careful selection of certain key components, particularly the specific preparations of BSA (Callister *et al.*, 1990) and rabbit serum (Pollack *et al.*, 1993), that may be highly variable in composition depending on their source (Pollack *et al.*, 1993). For example, a minority of preparations of rabbit serum contain anti-spirochaetal antibodies that will reduce or entirely abrogate growth of *B. burgdorferi* (Pollack *et al.*, 1993). Consequently, as a necessary quality control measure, each new batch of liquid medium must be tested for its ability to adequately support the growth of laboratory-adapted strains of *B. burgdorferi*. Once prepared, BSK-H medium can be preserved for future use at – 20°C for at least 8 months (Pollack *et al.*, 1993). Although one report suggested that the MKP medium was superior to BSK II for primary recovery of *B. burgdorferi* from erythema migrans skin lesions (Picken *et al.*, 1997), this result could not be confirmed in a later study by some of the same investigators (Ruzic-Sabljic *et al.*, 2006).

Borrelia burgdorferi, B. afzelii, and *B. garinii* grow well in both MKP and BSK II media with good preservation of motility and morphology (Ruzic-Sabljic and Strle, 2004). However, the specific formulation of culture medium used to grow *B. burgdorferi* will affect gene expression and may influence both infectivity and pathogenicity of the spirochaete, a particularly important concern in research studies that rely on needle inoculation of the spirochaete (Wang *et al.*, 2004; Yang *et al.*, 2001).

Liquid cultures are usually incubated at 30–34°C under microaerophilic conditions. Incubation at temperatures of 39°C or higher may reduce or prevent growth (Barbour, 1984). Cultures are incubated for up to 12 weeks, which is much longer than is necessary to grow most other human bacterial pathogens. This is due in part to the spirochaete's prolonged generation time (7–20 hours or longer) during log-phase growth (Preac-Mursic *et al.*, 1991; Wang *et al.*, 2001; Barbour, 1984). Detection of growth is accomplished by periodic examination of an aliquot of culture supernatant for the presence of spirochaetes by dark-field microscopy or by fluorescence microscopy after staining with acridine orange or a specific fluorescence-labelled antibody (Wormser *et al.*, 2000a; Reed, 2002). Visualized spirochaete-like structures should be confirmed as *B. burgdorferi* by demonstration of reactivity with specific monoclonal antibodies or by detection of specific DNA sequences using PCR methodology (Wormser *et al.*, 2000a; Stanek *et al.*, 1990; Nadelman *et al.*, 1996; Wang *et al.*, 2001). Lack of experience in the microscopic detection of *B. burgdorferi* can lead to false-positive readings, as other materials, such as cellular debris, may be mistaken for *B. burgdorferi* (Greene *et al.*, 1991). In addition, laboratory contamination of cultures with *B. burgdorferi* is well documented (Wormser *et al.*, 2006; Fallon *et al.*, 2008), necessitating excellent quality control procedures.

Borrelia burgdorferi can also be grown on solid media using agarose to solidify the liquid media described above and incubation under microaerophilic or anaerobic conditions (Preac-Mursic *et al.*, 1991; Barbour, 1984; Kurtti *et al.*, 1987). An advantage of using solid media is that individual colonies can be identified as a means to select particular clonal strains of *B. burgdorferi*. Anaerobic conditions may enhance growth of *B. afzelii* and *B. garinii* (De Martino *et al.*, 2006; Preac-Mursic *et al.*, 1991).

Laboratory-propagated strains of *B. burgdorferi* can be successfully co-cultivated with tick cell lines (Munderloh *et al.*, 1994; Obonyo *et al.*, 1999; Kurtti *et al.*, 1988) and with certain mammalian cell lines (Tylewska-Wierzbanowska and Chmielewski, 1997; Speck *et al.*, 2002; Hechemy *et al.*, 1992). Co-cultivation techniques may prove to be useful for primary isolation of *B. burgdorferi* from clinical specimens as well (Tylewska-Wierzbanowska and Chmielewski, 1997; Speck *et al.*, 2002). Co-cultivation with a tick cell line led to the first successful cultivation of *B. lonestari* (Varela *et al.*, 2004), a spirochaete present in approximately 2% of *Amblyomma americanum* ticks. *B. lonestari* was responsible for a single well-documented case of an erythema migrans-like skin lesion (James *et al.*, 2001).

Molecular methods for detection of Borrelia burgdorferi *DNA*

PCR is now a standard approach for direct detection of *B. burgdorferi* infection in tick vectors, host reservoirs, experimentally infected laboratory animals, and clinical specimens. PCR methods that are available for detection of *B. burgdorferi* DNA in various specimens include standard PCR, nested PCR, competitive PCR and real-time PCR (Schmidt, 1997; Sparagano *et al.*, 1999; Bunikis and Barbour, 2002; Aguero-Rosenfeld *et al.*, 2005).

The efficiency of a PCR assay is determined by several factors. Among these, the selection of an appropriate gene target and primer set for PCR amplification are the most important. In general, a PCR primer set yielding an amplicon of 100–300 bp is recommended, as it has high amplification efficiency under standard PCR conditions and can attentuate the effects of DNA fragmentation during sample processing.

Although PCR assays targeting numerous *B. burgdorferi* genes have been employed extensively in research settings, only a few of these genes have been widely utilized for PCR –based detection of *B. burgdorferi*. These include the chromosomally encoded genes *rrs*, *flaB*, *recA* and *p66*, and the plasmid-encoded gene *ospA* (Schmidt, 1997; Bunikis and Barbour, 2002).

Given that the number of spirochaetes in infected tissues or body fluids of patients is generally very low, appropriate procedures for sample collection, transport and DNA preparation from clinical samples are critical for yielding reliable and consistent PCR results. Sensitivity of PCR assays may be reduced by degradation of the *B. burgdorferi* DNA during sample transport, storage, and processing. If the tissue is kept in BSK medium for over 24 h, some spirochaetes may have migrated from the tissue to the culture medium. In this case, DNA prepared from both the tissue and the medium will increase the positivity rate of the PCR assay (Wang *et al*, unpublished data).

Clinical specimens collected from patients should be subjected to DNA extraction and PCR analysis shortly after collection; if this is not possible, specimens should be kept frozen until analysis. Studies on infected animal tissues suggest that PCR with DNA prepared from fresh frozen tissues has higher yields than that from paraffin-embedded, formalin-fixed tissues (Lebech *et al.*, 1995). As host DNA can interfere with PCR detection of *B. burgdorferi* in clinical and tick samples (Schwartz *et al.*, 1997; Cogswell *et al.*, 1996), an optimized DNA extraction procedure is essential to yield reliable PCR results for certain samples (Bergmann *et al.*, 2002). Also, PCR inhibitors may be present in various biological materials (blood, urine, synovial fluid and cerebrospinal fluid) obtained from patients (Brettschneider *et al.*, 1998).

Real-time PCR has been used for detection, quantification and genotyping of pathogenic *B. burgdorferi* species in ticks and patients (Rauter *et al.*, 2002; Wang, 2002; Exner and Lewinski, 2003; Portnoi *et al.*, 2006; Gaumond *et al.*, 2006; Ivacic *et al.*, 2007). The sensitivity of DNA extraction using manual or automated nucleic acid preparation systems is comparable for a variety of specimen types (Exner and Lewinski, 2003). Only a few genes have been used as targets in real-time PCR (e.g. *ospA, rrs, hbb, recA*) and these assays usually have used smaller fragments (100–200 bp) than those targeted in standard PCR tests. The sensitivity, specificity and reproducibility of real-time PCR and real-time quantitative PCR (qPCR) can be affected by a number of factors, including the properties of selected primers or probe, the quality of template DNA, the detection format (SYBR Green I-based or hybridization probe-based), and the amplification conditions (Ong and Irvine, 2002).

16.2.2 Indirect methods of detection: antibody detection

Serological confirmation of Lyme borreliosis has many challenges mainly due to the antigenic complexity of *B. burgdorferi* and the differential expression of antigens according to the environment in which the organism resides. *Borrelia burgdorferi* antigens are encoded on the chromosome, as well as on many plasmids (Casjens *et al.*, 2000). *Borrelia burgdorferi* expresses certain antigens in infected unfed, *Ixodes* ticks and switches to a different set of antigens during a blood meal (Schwan and Piesman, 2000). Therefore, the antigens expressed by *B. burgdorferi* during *in vivo* stages of infection in mammals may be quite different from those expressed under *in vitro* culture conditions. Many of these differentially expressed antigens are plasmid encoded. Several studies have identified antigens of importance during *in vivo* conditions (Fikrig *et al.*, 2000; Miller *et al.*, 2003) and some of these antigens have been used as either recombinant proteins or peptide antigens in immunoassays.

Borrelia burgdorferi antigens that have been employed in serological assays are listed in Table 16.1. Early assays using *in vitro* cultured *B. burgdorferi* identified the 41-kDa FlaB as an immunodominant antigen eliciting both IgM and IgG antibodies (Craft *et al.*, 1986), but this antigen was soon shown to be highly cross-reactive among different bacteria and with mammalian tissues (Fawcett *et al.*, 1992; Luft *et al.*, 1993). Further studies on immunogenicity led to the cloning of the internal fragment (41-i) that retained its immunogenicity but had less cross-reactivity (Rasiah *et al.*, 1992). Recombinant internal fragment of FlaB has been used by several investigators as a source of antigen in immunoassays (Hauser and Wilske, 1997; Magnarelli *et al.*, 2000).

The outer surface protein C (OspC), a 21–25-kDa plasmid-encoded antigen is highly immunogenic during early infection (Wilske *et al.*, 1986a; Dressler *et al.*, 1993). Patients with erythema migrans develop strong IgM reactivity against this antigen very early after onset of disease (Aguero-Rosenfeld *et al.*, 1993, 1996). Expression of OspC is induced under a number of environmental conditions (e.g. temperature, pH) and is expressed in tissues of infected animals during the earliest stages of infection (Schwan *et al.*, 1995; Crother *et al.*, 2003). As antibodies against this antigen are produced, its expression is downregulated (Liang *et al.*,

Table 16.1 *Borrelia burgdorferi* antigens of diagnostic significance

Antigen	Other designation	Molecular size (kDa)	Gene location
93	P83/100	93	Chromosome
66		66	Chromosome
58		58	Chromosome
BBK32	P47, Fibronectin binding protein	47	Linear plasmid 36
45		45	Chromosome
VlsE		35–43	Linear plasmid 28–1
41	Flagellin, FlaB	41	Chromosome
39	BmpA	39	Chromosome
34	OspB	34	Linear plasmid 54
31	OspA	31	Linear plasmid 54
30		30	
28	OspD	28	Linear plasmid 38
23	OspC	21–25	Circular plasmid 26
DbpA	P17, Decorin binding protein A	17–18	Linear plasmid 54
RevA		17	Circular plasmid 32
ErpA	OspE related protein	17	Circular plasmid 32

2004). Although highly immunogenic, OspC is the most heterogeneous protein in *B. burgdorferi* at the amino acid sequence level. This results in different OspC serotypes (Wilske *et al.*, 1995) and numerous OspC groups have been characterized based on sequence variability (Seinost *et al.*, 1999; Baranton *et al.*, 2001).

Another highly immunogenic antigen is the chromosomally encoded 39-kDa antigen (P39), also known as BmpA. The gene is encoded in a chromosomal cluster that also includes BmpB-D; the exact role and immunogenicity of BmpB-D have not yet been characterized (Fraser *et al.*, 1997). IgM and IgG antibodies to BmpA antigen appear very early after infection and its importance as an antigen in serological testing was described by Engstrom *et al.* using *B. burgdorferi* strain 297 (Engstrom *et al.*, 1995). Expression of this antigen in culture probably varies in different strains of *B. burgdorferi* since its immunogenicity was not evident in serological studies of patients with Lyme borreliosis described by Dressler *et al.* (1993) using *B. burgdorferi* strain G39/40 as the source of antigen in immunoblots.

VlsE, a surface protein encoded on the linear plasmid 28–1 (lp28-1), is also highly immunogenic and expressed during infection. Antigenic recombination occurs between an expression locus of *vlsE* and silent *vls* cassettes leading to antigenic variation. The *vls* locus consists of an expression site (*vlsE*) and 15 silent cassettes (Zhang *et al.*, 1997). Since it is mostly or preferentially expressed *in vivo*, this antigen may not be present in assay formats using *in vitro* cultured *B. burgdorferi* strains, but immunoassays using recombinant VlsE antigen have been used by many investigators.

Many other *B. burgdorferi* antigens have been found to elicit antibodies in patients with Lyme borreliosis. As time elapses after infection, antibodies particularly of the IgG class are developed against other antigens, such as those with a molecular mass of 93-, 66 and 45-kDa (Dressler *et al.*, 1993; Aguero-Rosenfeld *et al.*, 1996). Antigens of particular importance are the decorin binding proteins DbpA and B and RevA. Several of these proteins are induced during mammalian infection (Crother *et al.*, 2003; Carroll *et al.*, 2001).

Outer surface proteins A and B (OspA and OspB) are encoded on linear plasmid 54 (lp54). They are not expressed during early stages of natural infection because expression of *ospA* and *ospB*, which are co-transcribed, is downregulated in the tick midgut during the blood meal. Since OspA and B are produced during cultivation, OspA/B antibodies are elicited in experimental animals if infection is accomplished by inoculation of more than 1000 culture-grown spirochaetes. Antibodies to OspA and B may be observed in sera of patients at later stages of Lyme borreliosis, especially in certain patients with Lyme arthritis (Dressler *et al.*, 1993).

Methods

Several methods have been employed to detect antibodies to *B. burgdorferi*. Earlier methods included IFA, and its modification FIAX. These have been replaced by enzyme immunoassays, ELISA and western blots (immunoblots). In addition, a few groups have used functional antibody assays aimed at detecting borreliacidal antibodies (Sadziene *et al.*, 1994; Callister *et al.*, 1991). Although these functional assays appear to have greater specificity than matrix-based assays, they are cumbersome and may yield false-positive results in patients receiving antibiotic therapy (Jobe *et al.*, 1999).

Although more than 50 enzyme immunoassays have been cleared by the FDA for diagnostic use, many have been discontinued and are no longer marketed. Most of the FDA-cleared commercial enzyme immunoassays are ELISA-based and fewer than five western immunoblots kits are currently in use (http://www.accessdata.fda.gov/scripts/cdrh/cfdocs/cfPMN/pmn.cfm; insert LSR in 'product code' field).

IFA

This method has been largely replaced by enzyme immunoassays since its cumbersome nature is not conducive to routine use in clinical laboratories. However, IFA is commonly used in experimental studies of *B. burgdorferi* infection in ticks (see section 16.3).

ELISA

Most commercially available ELISA assays use whole *B. burgdorferi* lysates as the source of antigen to detect IgG and IgM or polyclonal antibodies (Aguero-Rosenfeld, 2004). Some manufacturers have added purified or recombinant antigens to the antigenic mixture (Aguero-Rosenfeld, 2004). Advantages of ELISA are the objective determination of antibodies using a numeric value (OD, Index) and the capability for automation. Among the disadvantages is lack of specificity because of its use of whole cell lysates, which include a number of antigens that cross-react with antigens in other bacteria. To improve performance of these assays, some formats have employed serum diluents that include Reiter treponeme or other bacteria in order to adsorb non-specific antibodies. Recent assays employ recombinant antigens or peptide antigens instead of whole cell lysates. These are discussed further in section 16.6.4.

Immunoblots

These assays allow the assessment of antibodies to individual antigens of *B. burgdorferi* after separation by SDS-PAGE. These assays have been useful in determining the immunodominant antigens of *B. burgdorferi* at different time points post infection and at various stages of the disease. Wilske *et al* first reported that OspC elicited strong IgM reactivity in patients with early Lyme borreliosis (Wilske *et al.*, 1986a); these findings were later confirmed in patients with Lyme borreliosis in the US (Dressler *et al.*, 1993; Aguero-Rosenfeld *et al.*, 1993). Other important antigens eliciting IgM responses in early disease are flagellin (41-kDa) and P39 (Dressler *et al.*, 1993; Aguero-Rosenfeld *et al.* 1993, 1996; Engstrom *et al.*, 1995). Although IgG antibodies to a few antigens (41, 39-kDa and OspC) may develop during early Lyme borreliosis with erythema migrans, IgG reactivity to most antigens develops later during the course of infection, particularly in patients with disseminated or late stages of Lyme borreliosis (Aguero-Rosenfeld *et al.*, 1993; Aguero-Rosenfeld *et al.*, 1996).

16.3 Detection of *Borrelia burgdorferi* in infected ticks

The incidence of human Lyme disease in endemic foci is correlated with the abundance of *Ixodes* ticks infected with *B. burgdorferi* (Stafford *et al.*, 1998). Numerous studies have been reported on the detection of *B. burgdorferi* in *Ixodes* ticks by immunofluorescent assays, culture and molecular methods (Gern *et al.*, 1998; Burgdorfer *et al.*, 1991; Sparagano *et al.*, 1999). Studies in laboratory animals have suggested that the efficiency of host infection is determined in part by the number of spirochaetes inoculated by needle or deposited by the tick at the time of feeding (Ma *et al.*, 1998; Piesman *et al.*, 1990). Analysis of *B. burgdorferi* density in ticks can, thus, provide more accurate data for assessment of spirochaete transmission between vectors and hosts and the population risk for Lyme borreliosis.

The density and growth dynamics of *B. burgdorferi* in experimentally infected ticks have been measured using either microscope-based assays (de Silva and Fikrig, 1995; Brunet *et al.*, 1995; Burkot *et al.*, 1994; Gern *et al.*, 1996) or real-time PCR (Piesman *et al.*, 2001) in feeding *I. scapularis* or *I. ricinus* ticks. Approximately 500 spirochaetes per tick were present in unfed nymphal *I. scapularis* ticks (Piesman *et al.*, 2001); however, the number of spirochaetes in these ticks increased rapidly upon feeding on hosts and reached a maximum of 166,575 spirochaetes per nymph at 72 h after attachment. This elevated load could be maintained (61,275 spirochaetes per nymph) until day 75 post repletion (Piesman *et al.*, 1990; de Silva and Fikrig, 1995). Similarly, spirochaetes in tick salivary glands increased up to 17-fold, from 1.2 per salivary gland pair before feeding to 20.8 at 72 h post attachment. The period of the most rapid increase in spirochaete number in salivary glands (48–60 h post attachment) coincided with the maximal increase in transmission risk during nymphal tick feeding (Piesman *et al.*, 2001).

Approximately 2000 spirochaetes per tick were detected by microscopy-based assays in the midguts of nymphal ticks collected from three endemic sites in Massachusetts (Brunet *et al.*, 1995), and 3000 spirochaetes per tick were found in adult ticks from Westchester County, New York (Burkot *et al.*, 1994). Using qPCR, the prevalence of ticks infected with *B. burgdorferi* and the spirochaete density in infected ticks collected from four northeastern states (Rhode Island, Connecticut, New York, and New Jersey) were determined (Wang *et al.*, 2003). Ninety-one of 438 (20.7%) nymphal and 15 of 60 (25.0%) adult ticks were positive by qPCR; the sensitivity of qPCR was comparable to that of nested PCR targeting the *B. burgdorferi* 16S–23S ribosomal RNA spacer. The spirochaete numbers in individual ticks varied from 25 to 197,200, with a median of 5351 spirochaetes per adult and 1964 spirochaetes per nymphal tick. These figures are comparable to that reported in *I. ricinus* ticks from southern Germany (~4000 spirochaetes per infected nymph) (Rauter *et al.*, 2002).

Quantitative data on spirochaete loads in host-seeking ticks may have potential applications for studies of Lyme borreliosis ecology. A high degree of *B. burgdorferi* spirochaete aggregation has been observed in questing *I. scapularis* nymphs, with most infected ticks harbouring few spirochaetes (<5000) and a small proportion of ticks with very high spirochaete loads (>20,000). Nymphs with very few spirochaetes may play little or no role in the natural maintenance of *B. burgdorferi*. Levin and Fish (2000)reported that 18.4% of *B. burgdorferi*-infected nymphs did not transmit spirochaetes when fed on white-footed mice (*Peromyscus leucopus*), which are natural reservoirs of infection. The number of organisms found in the 18th percentile of infected ticks was ≤300 spirochaetes (Wang *et al.*, 2003), which may represent a transmission threshold.

Recently, real-time PCR has been employed for genotyping of the three *B. burgdorferi* species in Europe (Pietila *et al.*, 2000; Mommert *et al.*, 2001; Rauter *et al.*, 2002; Portnoi *et al.*, 2006). In one study, melting curve analysis of borrelial *recA* PCR products allowed differentiation of *B. garinii* from *B. burgdorferi* sensu stricto (s.s.) and *B. afzelii* (Pietila *et al.*, 2000). In a second study, differentiation of *B. burgdorferi* s.s. from *B. garinii* and *B. afzelii* was achieved by melting curve analysis of the borrelial *p66* gene amplicons (Mommert *et al.*, 2001). Both studies based their results on melting temperature (T_m) analysis of PCR amplicons using the double-strand DNA binding dye SYBR Green I. The discriminatory power and reliability in differentiating pathogenic species using SYBR Green I may be limited, however, as the T_m variation could be generated by several means including the protocols used to prepare template DNA, the chemical reagents used and interassay fluctuations. A real-time PCR protocol based on a hybridization probe format using a primer internally labelled with a fluorescent dye was reported (Rauter *et al.*, 2002). Melting curve analysis of the amplified *B. burgdorferi ospA* products immediately after thermal cycling allowed clear differentiation in a single PCR run of the three *B. burgdorferi* species (*B. burgdorferi* s.s., *B. garinii*, and *B. afzelii*) that are the most clinically relevant in Europe.

16.4 Detection of *Borrelia burgdorferi* in mammalian host reservoirs and laboratory animals

Culture and PCR of tissue specimens are sensitive and specific methods for detection of *B. burgdorferi* infection in naturally and experimentally infected animals. Laboratory mice are the most widely used animal models for studying various aspects of Lyme borreliosis (Barthold, 1995) (see Chapter 14). Confirmation of *B. burgdorferi* infection in laboratory animals can be achieved by several means, including culture and PCR analysis of *B. burgdorferi* in ear punch biopsy specimens or internal organs and by xenodiagnosis. Cultures of ear punch biopsies from inoculated animals are typically positive in specimens obtained about one week after intradermal inoculation with 10^4 organisms (Barthold *et al.*, 1993). Other tissues with high yield by culture include heart, urinary bladder and, in some species of mice, joints (Barthold *et al.*, 1993; Barthold *et al.*, 1993).

Quantification of the spirochaete loads in various tissues of experimentally infected mice is critical for understanding the pathogenesis of Lyme borreliosis. Spirochaete burdens in mouse tissues have been determined in numerous studies by either competitive PCR (Yang *et al.*, 1994; Brown and Reiner, 1998; Anguita *et al.*, 2000; Germer *et al.*, 1999) or, more recently, real-time qPCR (Morrison *et al.*, 1999; Pahl *et al.*, 1999; Brown *et al.*, 1999; Wooten *et al.*, 2002; Brown *et al.*, 2001; Zeidner *et al.*, 2001; Wang *et al.*, 2002). Direct comparison of the results reported by different laboratories should be made cautiously since the number of spirochaetes in mice could be affected by many variables, such as genetic background of the mice, pathogenic properties of the strain of *B. burgdorferi* inoculated, route of inoculation, size of inoculum, time point of sampling, selection of PCR target, and the protocols used for DNA preparation and PCR amplification. In addition, there is no standardization of external standards among laboratories. Nevertheless, assuming that there are 1×10^4 mouse host gene copies per 200 ng of mouse DNA (Morrison *et al.*, 1999), and that each milligram of mouse tissue could yield ~2 μg of total DNA [both mouse and spirochaetal DNA (G. Wang *et al.*, unpublished data)], studies in most laboratories independently showed that ~10^3–10^4 spirochaetes are present per milligram of heart or ankle joint tissue, which is equivalent to ~10^5–10^6 organisms per heart and/or joint in a C3H mouse infected with *B. burgdorferi* (Yang *et al.*, 1994; Anguita *et al.*, 2000; Morrison *et al.*, 1999; Pahl *et al.*, 1999; Brown *et al.*, 1999; Zeidner *et al.*, 2001; Wang *et al.*, 2002). Spirochaete burdens in tissues from infected dogs (Straubinger, 2000) and rhesus monkeys (England *et al.*, 1997) also have been assessed by qPCR.

16.5 Detection of *Borrelia burgdorferi* in clinical specimens

16.5.1 Culture

Borrelia burgdorferi can be recovered from various tissues and body fluids of patients with Lyme borreliosis, including biopsy specimens of erythema migrans skin lesions, biopsy specimens of acrodermatitis chronicum atrophicans (ACA) skin lesions, biopsy specimens of borrelial lymphocytoma skin lesions, cerebrospinal fluid, and blood (Aguero-Rosenfeld *et al.*, 2005). Anecdotally, recovery of *B. burgdorferi* has also been reported from other tissue or fluid specimens (Maraspin *et al.*, 2001) such as synovial fluid (Schmidli *et al.*, 1988), cardiac tissue (Stanek *et al.*, 1990), and iris (Preac-Mursic *et al.*, 1993). It is interesting to note that although *B. burgdorferi* is a cause of infectious arthritis, and spirochaetes can be readily detected in synovial fluid by PCR (see below), it is nearly impossible to recover *B. burgdorferi* by culturing of synovial fluid.

Erythema migrans

Recovery of *B. burgdorferi* from 2–4 mm skin biopsy samples of an erythema migrans skin lesion can typically be achieved in at least 40% of untreated patients. The highest reported success rates were 86% in a study of 21 US patients from whom 4 mm skin biopsy specimens were cultured (Berger *et al.*, 1992a) and 88% in a study from the Netherlands of 57 patients who underwent 4 mm skin biopsy (van Dam *et al.*, 1993). *Borrelia burgdorferi* can be recovered from both primary (which occurs at the site of a tick bite) and secondary erythema migrans lesions (presumed to arise by haematogenous dissemination to other skin sites) (Mitchell *et al.*, 1993; Munderloh *et al.*, 1994). Despite the clinical tradition of biopsying the advancing border of an erythema migrans lesion, a systematic study of this question conducted in Europe found comparable yields from biopsy specimens taken from the periphery or the centre of erythema migrans lesions (Jurca *et al.*, 1998).

Recovery of *B. burgdorferi* s.s. from solitary erythema migrans lesions is significantly more likely in US patients with skin lesions of shorter duration (Nadelman *et al.*, 1993), suggesting that the patient's immune response is usually effective in clearing the spirochaete from that skin site over a relatively short time interval. The same phenomenon probably explains a much earlier clinical observation that in the absence of antibiotic therapy, erythema migrans lesions will nevertheless resolve in approximately four weeks (Steere *et al.*, 1983a). Clearance of spirochaetes from the skin in the absence of antibiotic treatment has also been documented in a rhesus macaque model of *B. burgdorferi* s.s. infection (Philipp *et al.*, 1993).

Borrelia burgdorferi s.s. usually cannot be recovered from erythema migrans lesions of patients already receiving appropriate antibiotic treatment (Wormser *et al.*, 2000b) and rarely, if ever, from the site of resolved erythema migrans lesions in US patients who have completed a course of appropriate antimicrobial therapy (Berger *et al.*, 1992b; Nadelman *et al.*, 1993). In a European study in which patients who had been treated with antibiotics were recultured systematically to determine persistence of *B. burgdorferi*, the spirochaete was recovered from a skin biopsy sample of normal appearing skin at the site of a resolved erythema migrans lesion in 19 (1.7%) of 1148 patients; all of these patients also had a positive culture of a skin biopsy sample obtained prior to antibiotic therapy (Hunfeld *et al.*, 2005). Of the five cases in which isolates from both the first and second biopsy samples were available for analysis, plasmid and other typing methods suggested that the isolates were not identical for at least four of the pairs (Hunfeld *et al.*, 2005). In one case, the isolates were not even from the same species. The demonstrated strain differences might be interpreted as indicating reinfection, multiplicity of infecting borrelia in the original infection or laboratory contamination (Liveris *et al.*, 1999; Misonne *et al.*, 1998). The authors did not confirm the presence of persistent infection with an independent test method such as PCR.

Whole blood, serum, and plasma

The rate of recovery of *B. burgdorferi* s.s. from blood or blood components of untreated patients with erythema migrans had generally been 5% or less (Benach *et al.*, 1983; Goodman *et al.*, 1995; Steere *et al.*, 1983b; Wallach *et al.*, 1993), and until recently this source of culture material was largely abandoned. In past studies on the sensitivity of blood cultures for recovery of *B. burgdorferi* s.s., only a very small volume of blood or blood components (usually less than 1 ml) was cultured (Nadelman *et al.*, 1990; Benach *et al.*, 1983; Goodman *et al.*, 1995; Steere *et al.*, 1984; Steere *et al.*, 1983b; Wallach *et al.*, 1993; Nadelman *et al.*, 1994). In view of the recommendation to culture quantities of blood as large as 20–30 ml in other bacterial infections, the rationale of culturing much smaller volumes for Lyme borreliosis patients was open to challenge. In a series of experiments in adult patients with erythema migrans from the USA, recovery of *B. burgdorferi* s.s. was better from serum than from an identical volume of whole blood (Wormser *et al.*, 1998), and the yield from plasma was significantly greater than that from serum (Wormser *et al.*, 2000a). Yield directly correlated with the volume of material cultured (Wormser *et al.*, 2001; Wormser *et al.*, 2000a; Coulter *et al.*, 2005). Recovery of *B. burgdorferi* s.s. from high-volume cultures of 9 ml of plasma inoculated into a modified BSK II preparation devoid of antimicrobials and gelatin, with a 20:1 ratio of medium to plasma, was consistently above 40% (Wormser *et al.*, 2001; Wormser *et al.*, 2000a). Unfortunately, high volume plasma cultures are not appropriate for young children since obtaining such a large quantity of blood would be unacceptable. Reported recovery rates from blood of patients with erythema migrans in Europe have been < 10% (Arnez *et al.*, 2001; Maraspin *et al.*, 2001). This may be due to a low frequency of haematogenous dissemination of *B. afzelii*, the principal cause of erythema migrans in Europe (Wang *et al.*, 1999), or to culturing an inadequate volume of blood.

Compared to PCR for detection of spirochaetes in cutaneous specimens, culture has proven to be slightly more sensitive in some studies but not in others (Nowakowski *et al.*, 2001; Moter *et al.*, 1994; Schwartz *et al.*, 1992; Zore *et al.*, 2002; Brettschneider *et al.*, 1998; Steere *et al.*, 1998; Picken *et al.*, 1997; Lebech *et al.*, 2000). Such disparate results are likely to be attributable to differences in the PCR protocols employed as well as differences in culture techniques, including size of the skin biopsy sample cultured and/or choice of culture medium.

Cerebrospinal fluid

Evaluation of the sensitivity of culture of cerebrospinal fluid specimens in patients regarded as having neuroborreliosis by clinical and/or other laboratory test methods (e.g. intrathecal production of antibody to borrelia) has been reported principally from Europe (Ornstein *et al.*, 2001; van Dam *et al.*, 1993; Karlsson *et al.*, 1990; Cimperman *et al.*, 1999). The sensitivity of culture was approximately 10%. The quantity of cerebrospinal fluid that was cultured varied from 0.2 to 5 ml (Ornstein *et al.*, 2001; van Dam *et al.*, 1993; Karlsson *et al.*, 1990). None of the studies systematically compared the yields of culture of different volumes of cerebrospinal fluid. The highest yields for *B. garinii* (the principal cause of neuroborreliosis in Europe) were associated with shorter duration of illness (usually < 14 days) (Ornstein *et al.*, 2001; van Dam *et al.*, 1993; Karlsson *et al.*, 1990).

Practical considerations

Culture has been principally used as a diagnostic modality in research studies and has enabled a better understanding of the clinical, laboratory, and pathogenetic aspects of Lyme borreliosis through identification of a group of culture-authenticated patients (Strle *et al.*, 1996; Nadelman *et al.*, 1996; Strle *et al.*, 1999; Aguero-Rosenfeld *et al.*, 1996; Smith *et al.*, 2002). In addition, culturing of ticks, erythema migrans skin lesions, and blood samples was instrumental to demonstrating that different subtypes of *B. burgdorferi* s.s. vary in their potential to spread to extracutaneous sites (Wormser *et al.*, 1999; Seinost *et al.*, 1999). Patients with culture-confirmed Lyme borreliosis have also been a valuable source of specimens to assess the accuracy of other diagnostic tests (Mitchell *et al.*, 1994; Aguero-Rosenfeld *et al.*, 1996). Such well-defined patient populations have also played an important role in studying therapeutic regimens (Wormser *et al.*, 2003) and vaccines (Steere *et al.*, 1998).

For several reasons, however, culture has not been used for diagnostic purposes in routine clinical practice. There has been the lack of consistent availability of high-quality (i.e. borrelial growth promoting) lots of liquid medium for growing *B. burgdorferi*. Furthermore, by most conventional bacteriological standards, borrelial cultures are more labour intensive, expensive, and much slower, requiring up to 12 weeks of incubation before being considered negative. The rapidity of identifying a positive culture, however, is directly dependent on the frequency with which the culture supernatant is examined microscopically, since macroscopic changes in the appearance of the culture medium tend to occur later, if at all. In our research studies, in which *B. burgdorferi* cultures are usually only first examined at 2 weeks, 70% of positive cultures of skin biopsy samples from erythema migrans lesions and approximately 85% of positive cultures of plasma from erythema migrans patients are already positive at 2 weeks; approximately 95% of samples of both types are positive by 4 weeks (Aguero-Rosenfeld *et al.*, 2005). Time to detection of a positive specimen can be reduced to a clinically relevant time frame if cultures are examined on a daily basis (Mitchell *et al.*, 1994). For example, in one study in which cultures were examined daily, 74 (82%) of 90 positive cultures were identified within the first 7 days of incubation and 32 (35%) within the first 3 days; the longest time to detection of a positive culture was 16 days (Reed, 2002). Another limiting factor is that culture is only useful for untreated patients. Culture positivity is rapidly aborted by even a few doses of appropriate antibiotic treatment (Schwartz *et al.*, 1992; Nadelman *et al.*, 1993).

Perhaps the most fundamental limitation is that culture is far too insensitive in patients with extracutaneous manifestations of Lyme borreliosis, which is the group of patients who pose the greatest diagnostic challenge (Stanek and Strle, 2003; Nadelman and Wormser, 1998). Culture has been most successful for patients with early Lyme borreliosis who have erythema migrans, but visual recognition of the characteristic appearance of the skin lesion is usually sufficient for an accurate diagnosis (Nadelman and Wormser, 1995; Nadelman and Wormser, 1998). Culture would only be helpful in a minority of cases in which the skin lesion is atypical or the patient was not known to have had tick exposure in an area endemic for Lyme borreliosis.

16.5.2 PCR detection

Infection with *B. burgdorferi* can result in dermatological, neurological, cardiac and musculoskeletal symptoms. The diagnostic value of PCR for laboratory diagnosis of Lyme borreliosis varies depending on the clinical manifestations and the PCR method being used. A few reviews on the applications and limitations of PCR in the clinical laboratory have been published (Aguero-Rosenfeld *et al.*, 2005; Schmidt, 1997; Bunikis and Barbour, 2002). In general, PCR-based molecular techniques can be employed for (i) confirmation of the clinical diagnosis of suspected Lyme borreliosis; (ii) molecular species identification and/or typing of the infecting spirochaetes in clinical specimens or on cultured isolates; and (iii) detection of co-infection of *B. burgdorferi* and other tick-borne pathogens.

Skin biopsy specimens

PCR is highly sensitive for detection of *B. burgdorferi* DNA in skin biopsy samples from patients with erythema migrans (Dumler, 2001). For laboratory diagnosis of suspected Lyme borreliosis, detection of *B. burgdorferi* DNA in skin biopsy samples will provide direct evidence supporting the clinical diagnosis, especially in patients with atypical erythematous skin lesions who are seronegative. However, this is not necessary in routine practice, since clinical diagnosis can be easily made in patients with a characteristic skin lesion.

Borrelia burgdorferi DNA was first detected by PCR in skin biopsies from patients with ACA (Melchers *et al.*, 1991) and erythema migrans (Melchers *et al.*, 1991; Schwartz *et al.*, 1992). PCR targets that have been employed for detection of *B. burgdorferi* DNA in skin biopsy specimens include *p66*, 16S rRNA gene, *flaB*, 23S rRNA gene, 5S rRNA–23S rRNA gene spacer, *recA*, and *ospA* (Schmidt, 1997; Aguero-Rosenfeld *et al.*, 2005).

The reported sensitivity of PCR for detection of *B. burgdorferi* DNA in skin lesions is usually high; the median percent positivity is 69% for patients with erythema migrans and 76% for those with ACA (Aguero-Rosenfeld *et al.*, 2005). A prospective study in the USA showed that 85 of 132 (64%) skin biopsies taken from erythema migrans patients during a phase III vaccine trial were positive by PCR (Steere *et al.*, 1998). Sensitivity of PCR for detection of *B. burgdorferi* in skin biopsy samples from ACA patients in Europe appears to be dependent on the target sequences selected; assays targeting *ospA* and *p66* are reported to be more sensitive than tests based on the 5S–23S rDNA intergenic spacer (Rijpkema *et al.*, 1997; Brettschneider *et al.*, 1998).

Blood or plasma

Borrelia burgdorferi DNA has been detected by PCR in blood samples from patients with erythema migrans (Goodman *et al.*, 1995; Oksi *et al.*, 2001) and early disseminated disease such as neuroborreliosis and carditis (Liebling *et al.*, 1993; Demaerschalck *et al.*, 1995). In a prospective study of US patients with erythema migrans, *B. burgdorferi* s.s. DNA was detected by PCR in 14 of 76 (18.4%) plasma samples (Goodman *et al.*, 1995).

In general, the sensitivity of PCR for detection of *B. burgdorferi* DNA in blood, plasma or serum samples from patients with Lyme borreliosis is low with a median percent positivity of 18% for patients in the USA and 10% for European patients (Aguero-Rosenfeld *et al.*, 2005). The low yield could be a reflection of lack of spirochaetaemia or transient spirochaetaemia (Benach *et al.*, 1983), a low level of spirochaetes in blood (Goodman *et al.*, 1995; Wormser *et al.*, 2001), and/or the presence of PCR inhibitors in host blood (Wallach *et al.*, 1993; Cogswell *et al.*, 1996; Abu Al-Soud and Radstrom, 2001). *Borrelia burgdorferi* could not be detected in any of 843 specimens of blood or cerebrospinal fluid that were tested by either culture or PCR from the 129 patients with post-Lyme syndromes (musculoskeletal pains, neurocognitive symptoms, dysaesthesia, fatigue, or malaise, headache, or sleep disturbance) (Klempner, 2002; Klempner *et al.*, 2001a).

Cerebrospinal fluid

Borrelia burgdorferi DNA has been detected by PCR in cerebrospinal fluid specimens from patients with a variety of neurological symptoms in the US and in Europe (Aguero-Rosenfeld *et al.*, 2005). However, the lack of a gold standard method to support the diagnosis of neuroborreliosis makes assessing the performance of PCR on cerebrospinal fluid difficult.

Sensitivity of PCR for detection of *B. burgdorferi* DNA in cerebrospinal fluid specimens may be dependent on the clinical presentation, cerebrospinal fluid cell counts, disease duration, and prior antibiotic treatment. In a study of 60 US patients with neuroborreliosis (16 with early and 44 with late neuroborreliosis), the sensitivity of PCR in cerebrospinal fluid was 38% in early and 25% in late neuroborreliosis, and an inverse correlation was found between duration of antimicrobial treatment and PCR results (Nocton *et al.*, 1996). PCR sensitivity is greater in cerebrospinal fluid specimens from patients with suspected neuroborreliosis of less than 2 weeks duration and with pleocytosis compared to patients without these characteristics (Lebech *et al.*, 2000; Ornstein *et al.*, 2002; Gooskens *et al.*, 2006).

Synovial fluid

PCR analysis of synovial fluid is a much more sensitive approach than culture for detection of *B. burgdorferi* in affected joints of patients with Lyme arthritis. A positive PCR result in a synovial fluid specimen of a patient with exposure in a Lyme borreliosis endemic area, who also has a positive *B. burgdorferi* ELISA and IgG immunoblot, is strongly supportive of a diagnosis of Lyme arthritis. Untreated patients with Lyme arthritis, as well as patients during the first days to weeks of therapy, may exhibit a positive PCR in synovial fluid.

In a study of 88 patients in the USA with Lyme arthritis, *B. burgdorferi* DNA was detected in the synovial fluid of 75 (85%) patients (Nocton *et al.*, 1994). The PCR positivity rate was lower in patients who had received antibiotic therapy compared to untreated patients. Of 73 patients who were untreated or treated with only short courses of oral antibiotics, 70 (96%) had a positive PCR in synovial fluid samples. In contrast, *B. burgdorferi* s.s. DNA was demonstrated in only 7 of 19 (37%) patients who received either parenteral antibiotics or oral antibiotics for more than one month (Nocton *et al.*, 1994). The yield of PCR in synovial samples was also shown to depend on the primer/probe set(s) employed. (Nocton *et al.*, 1994). *Borrelia burgdorferi* DNA has been detected in synovial membrane samples of patients who were PCR negative in synovial fluid specimens after antibiotic treatment (Priem *et al.*, 1998).

Target imbalance has been reported in synovial fluid specimens based on the observation that higher sensitivity of PCR was achieved with primers targeting plasmid-encoded *ospA* than with those targeting the chromosome-encoded 16S ribosomal RNA genes (Persing *et al.*, 1994). *Borrelia burgdorferi* present in the synovium may selectively shed OspA DNA segments into the synovial fluid.

PCR sensitivity

Table 16.2 summarizes the sensitivity and specificity of PCR assays for the detection of *B. burgdorferi* DNA in different clinical samples based on studies published in MEDLINE-indexed periodicals during the years 1991 to 2003. Of the 24 studies in which *B. burgdorferi* DNA in skin biopsies was examined, sensitivity of the PCR assays varied from 36 to 88% for patients with erythema migrans and from 54 to 100% for patients with ACA. The median sensitivities of the reported PCR assays for detection of *B. burgdorferi* DNA in skin biopsies from patients with erythema migrans and ACA were 69% and 76%, respectively. The sensitivity of PCR assays for detection of *B. burgdorferi* in whole blood (plasma or serum) and cerebrospinal fluid specimens is low. By contrast, high PCR sensitivities were reported in both US and European studies with synovial fluid samples from patients with Lyme arthritis.

A published meta-analysis also demonstrated that PCR is a very sensitive approach when it is employed to detect *B. burgdorferi* DNA in skin

Table 16.2 Sensitivities and specificities of PCR assays for detection of *B. burgdorferi* DNA in different clinical specimens from patients with Lyme borreliosis[a]

Clinical specimen	No. of studies	Median sensitivity (%) (range)	Reported specificity (%)
Skin biopsy, EM	17	69 (36–88)	98–100
Skin biopsy, ACA	8	76 (54–100)	100
Blood/plasma/serum	10	21 (0–100)	
Cerebrospinal fluid	18	32 (5–100)	93–100
Synovial fluid	8	78 (42–100)	100

a Updated from Aguero-Rosenfeld *et al.* (2005) with addition of data from Avery *et al.* (2005), Coulter *et al.* (2005), Picha *et al.* (2005), Kondrusik *et al.* (2007), and Santino *et al.* (2008).

biopsy or synovial fluid specimens from patients with Lyme borreliosis; whereas the diagnostic value of PCR assays for detection of *B. burgdorferi* DNA in blood (plasma, serum) and cerebrospinal fluid specimens is low (Dumler, 2001).

Spirochaete load in clinical specimens of patients with Lyme borreliosis

The number of spirochaetes in clinical specimens of patients with Lyme borreliosis has been determined using real-time qPCR. In one study, *B. burgdorferi*-specific *recA* DNA was detected by qPCR in 40 (80%) skin biopsy samples taken from 50 consecutive, untreated adult patients with erythema migrans. The number of spirochaetes in a 2-mm biopsy ranged from 10 to 11,000, with a mean number of 2462 spirochaetes (Liveris *et al.*, 2002). Significantly higher numbers of spirochaetes were detected in culture-positive than in culture-negative specimens (3940 vs. 1642 spirochaetes, $p < 0.01$). In a second study, *B. burgdorferi flaB* DNA was detected using a TaqMan probe in five of 28 (17.9%) synovial fluid specimens and one of five (20%) synovial membrane biopsies obtained from 31 patients with arthropathies in Switzerland (Schwaiger *et al.*, 2001). The numbers of spirochaetes varied from 20 to 41,000/ml of synovial fluid.

16.6 Serological testing

16.6.1 Two-tier testing

In order to improve specificity of serological testing for Lyme borreliosis, the Centers for Disease Control and Prevention (CDC) in conjunction with other agencies, such as the Association of State and Territorial Public Health Directors (ASPHLD), convened the Second Meeting on Standardization of Lyme testing in 1994 and concluded that Lyme serology should be performed using a two-step modality (Centers for Disease Control and Prevention, 1995). Samples are first tested using a sensitive method such as ELISA or IFA that should detect both IgG and IgM antibodies to *B. burgdorferi*. Samples testing negative by the first step should be reported as negative, but samples that test equivocal or positive by these methods should be tested by separate IgG and IgM immunoblots.

16.6.2 Immunoblot interpretation

Criteria for interpretation of immunoblots have been recommended in the US. The IgM criteria of Engstrom (Engstrom *et al.*, 1995) require the presence of two of three significant bands (41, 39 or 23-kDa (OspC)) and the IgG criteria of Dressler (Wilske *et al.*, 1986a) require five of 10 bands (93-, 66-, 58-, 45-, 41-, 39-, 30-, 28-, 23, or 18-kDa). Furthermore, the CDC recommendations indicate that the IgM criteria should only be used in samples of patients presenting with disease of 4 weeks or less, whereas IgG criteria are applicable in samples of patients at any stage of Lyme borreliosis (Centers for Disease Control and Prevention, 1995).

Criteria used for immunoblot interpretation in Europe are more complicated due to the diversity of *B. burgdorferi* genospecies causing

Lyme borreliosis (Hauser *et al.*, 1997). European criteria for a positive IgG immunoblot require two bands of p14, Osp17 (DbpA, 18 kDa), p21, OspC, 30, 39, 41, 59, or 93 kDa. European criteria for positive IgM blots require reactivity to at least OspC (Wilske *et al.*, 2007).

16.6.3 Performance of two-tier testing in the USA

Relativity few studies have been published on the performance of the two-tier testing approach that has been in use now for over 10 years in the US. Bacon *et al.* (2003) published on the use of the two-tier approach as well as on the use of recombinant and peptide antigens to detect antibodies to *B. burgdorferi* in 280 patients with different manifestations of Lyme borreliosis. These authors found that the two-tier approach had a sensitivity of 38% in the acute sera of patients with erythema migrans, which increased to 67% during convalescence. Sensitivity was 87% in patients with early neuroborreliosis and 97% in patients with late manifestations such as Lyme arthritis (Bacon *et al.*, 2003). In another study of patients with erythema migrans, the two-tier protocol had a sensitivity of 40% in the acute phase and 67% during convalescence (Nowakowski *et al.*, 2001). A two-tier approach has also been recommended in Europe (Wilske *et al.*, 2007).

16.6.4 Enzyme immunoassays with recombinant or peptide antigens

Several proteins and immunoreactive epitopes have been synthesized as either recombinant proteins or peptide antigens for use in diagnostic assays. Unfortunately, to date, no single antigen has demonstrated sufficient sensitivity and specificity to warrant replacing two-tier testing. Antigenic variability among *B. burgdorferi* species and the temporal appearance of antibodies to different antigens at various stages of Lyme borreliosis make the choice of a single antigen a difficult task. Recombinant antigens have included those containing the internal portion of flagellin (P41-G or 41-i), FlaA, BBK32, P39, P35, VlsE, DbpA and outer surface proteins A, B, C, E, and F (Hauser and Wilske, 1997; Magnarelli *et al.*, 2000; Fawcett *et al.*, 1993; Fikrig *et al.*, 1997; Fikrig *et al.*, 1992; Panelius *et al.*, 2002; Heikkila *et al.*, 2002; Jauris-Heipke *et al.*, 1999). These antigens have been used in ELISA and immunoblot formats, most commonly in Europe. IgM antibodies were detected in 62% of US patients with erythema migrans using a recombinant OspC derived from a *B. burgdorferi* s.s. strain (Padula *et al.*, 1994). However, some European studies have shown lower (30–44%) IgM reactivity to recombinant OspC in sera of patients with erythema migrans (Panelius *et al.*, 2002; Mathiesen *et al.*, 1998a).

Among those peptide antigens showing promising results as a candidate is the pepC10 peptide, which contains the conserved C terminal 10 amino acids of OspC (Mathiesen *et al.*, 1998b). Although this antigen is less heterogeneous than the parent antigen OspC, it only appears to bind IgM antibodies. Evaluations on the performance of pepC10 in immunoassays have shown that it is highly specific, with sensitivities ranging from 45 to >50% in sera of patients with erythema migrans and neuroborreliosis, respectively (Bacon *et al.*, 2003).

C6 is a peptide derived from the invariable region 6 (IR6) of VlsE, a highly immunogenic and conserved region of VlsE (Liang *et al.*, 1999a). Great enthusiasm was generated with the development of assays using C6 since this antigen was reported to be highly immunogenic and broadly cross-reactive among all *B. burgdorferi* genospecies (Heikkila *et al.*, 2003; Liang *et al.*, 2000). Several reports have indicated that assays using C6 peptide offer greater sensitivity and specificity than other antibody methods when testing US and European patients with Lyme borreliosis (Liang *et al.*, 1999b; Skarpaas *et al.*, 2007). Bacon *et al.* (2003), however, reported that recombinant VlsE had a higher sensitivity than C6 in detecting *B. burgdorferi* antibodies in a small number of patients with acute neuroborreliosis in a study of US patients with Lyme borreliosis.

Comparison of the results of C6 immunoassays has been difficult because of the different origins of such assays. Many investigators have used in-house developed assays and others have employed a commercial assay recently approved by the FDA as a first step assay in the US (Bacon *et al.*, 2003). Furthermore, some European authors have employed a different commercial C6 assay, the QuickC6 assay, not currently approved for use in the US (Marangoni *et al.*, 2005). Owing to its high specificity, the C6 assay has been used in Europe as a second step replacing immunoblot (Nyman *et al.*, 2006). Other studies have found C6 to have a greater sensitivity than two-tier testing and have suggested that C6 assays could replace two-tier testing entirely (Mogilyansky *et al.*, 2004).

Although C6 was originally reported to be highly cross-reactive, recent studies appear to indicate that differences exist between sequences of this peptide in different *B. burgdorferi* genospecies such that assays based on sequences from a single genospecies may be suboptimal for detecting antibodies in European and American patients with Lyme borreliosis (Gomes-Solecki *et al.*, 2007).

16.6.5 Detection of intrathecal production of antibodies (cerebrospinal fluid testing)

Measurement of intrathecal antibodies to *B. burgdorferi* in patients with neuroborreliosis has been a very useful and sensitive method in European patients (Hansen and Lebech, 1991; Blanc *et al.*, 2007). These methods have, however, been of lesser utility in US patients with suspected neuroborreliosis. This difference could be attributed to the different genospecies causing neuroborreliosis on both continents; whereas *B. burgdorferi* s.s. is the only genospecies reported to cause Lyme borreliosis in the US, the most frequent genospecies causing neuroborreliosis in Europe is *B. garinii* (Ruzic-Sabljic *et al.*, 2001).

Methods used to detect intrathecal antibody production include capture immunoassays, antibody index and immunoblot (Blanc *et al.*, 2007; Coyle *et al.*, 1995; Halperin *et al.*, 1991; Wilske *et al.*, 1986b). Determinations of antibody index require the testing of cerebrospinal fluid and serum at equivalent immunoglobulin concentrations using *B. burgdorferi* antibody assays. A positive cerebrospinal fluid antibody index (intrathecal production) is obtained when a cerebrospinal fluid/serum OD ratio is greater than 1.3 (Blanc *et al.*, 2007; Wilske *et al.*, 1986b). Recombinant antigens have also been used in Europe to detect intrathecal *B. burgdorferi* antibody production (Panelius *et al.*, 2003). The performance of these methods to confirm the diagnosis of neuroborreliosis is difficult to evaluate due to the lack of a gold standard for diagnosis of neuroborreliosis. There are currently no commercially available and approved methods to detect antibodies in cerebrospinal fluid in the US.

16.7 Conclusion

Numerous studies on the laboratory detection of *B. burgdorferi* in infected vectors, animal reservoirs and clinical specimens have been performed since the original descriptions of Lyme borreliosis. The diagnostic value of microscope-based methods and antigen detection assays is limited because of low sensitivity, specificity, and reproducibility. When specificity can be assured, cultivation of *B. burgdorferi* from clinical samples is the gold standard to confirm a diagnosis of Lyme borreliosis. Culture is most successful for detection of *B. burgdorferi* when applied to skin biopsy specimens of erythema migrans or ACA lesions and to plasma blood samples of US patients with erythema migrans. Unfortunately, culture is insensitive in extracutaneous manifestations of Lyme borreliosis. At present this diagnostic modality is principally regarded as a research tool.

PCR is a sensitive and specific approach for detection of *B. burgdorferi* in skin biopsy specimens from patients with erythema migrans or ACA. It is also useful for identification of *B. burgdorferi* DNA in synovial fluid from patients with suspected Lyme arthritis. In addition, PCR has the capability for quantification and direct typing of spirochaetes in clinical specimens. Further optimization of various PCR protocols to be used for different sources of clinical samples should significantly improve the laboratory diagnosis of Lyme borreliosis. One of the limitations of nucleic acid amplification methods is the generation of false positive results due to contamination. Adherence to rigorous quality control steps is of utmost importance.

PCR and culture have high sensitivity with skin samples of patients with erythema migrans whose diagnosis is mostly based on clinical recognition of the lesion. PCR on material obtained from extracutaneous sites is, in general, of low sensitivity with the exception of synovial fluid. PCR on synovial fluid has a sensitivity of up to 90% (when using four different primer sets) in patients with untreated or partially treated Lyme arthritis, making it a helpful confirmatory test in these patients. Currently, the best use of PCR is for confirmation of the clinical diagnosis of suspected Lyme arthritis in patients who are IgG immunoblot positive. PCR should not be used as the sole laboratory modality to support a clinical diagnosis of extracutaneous Lyme borreliosis. PCR positivity in seronegative patients suspected of having late manifestations of Lyme borreliosis most likely represents a false positive result.

Borrelia burgdorferi has been frequently recovered from culture of urinary bladder specimens of experimentally infected laboratory animals, suggesting that this spirochaete could be excreted into the urine. However, the sensitivity and specificity of urine PCR is highly variable in patients (Liebling *et al.*, 1993; Karch *et al.*, 1994; Schmidt *et al.*, 1996; Brettschneider *et al.*, 1998). Thus, urine PCR is not appropriate for routine laboratory diagnosis of Lyme borreliosis.

Although serological methods are the main diagnostic tool in the laboratory confirmation of Lyme borreliosis, there are many limitations. Understanding the complexity of *B. burgdorferi* protein expression and the evolution of the antibody response to different antigenic components of *B. burgdorferi* are essential. As in most infectious diseases, antibodies are not fully developed during the acute phase of the illness and confirmation of their presence requires the testing of acute- and convalescent-phase sera. Currently accepted guidelines require the use of two-tier testing, where the first step involves the use of a sensitive EIA or IFA, followed by separate IgG and IgM immunoblot on those specimens testing reactive by the first step. Although this approach may not be optimal, it has provided an improved degree of standardization in serological testing while new approaches are being evaluated. Whether single-antigen antibody tests will be useful for diagnosis of all stages of Lyme borreliosis is currently unknown.

References

Aberer, E., and Duray, P.H. (1991). Morphology of *Borrelia burgdorferi*: structural patterns of cultured borreliae in relation to staining methods. J. Clin. Microbiol. *29*, 764–772.

Abu Al-Soud, W., and Radstrom, P. (2001). Purification and characterization of PCR-inhibitory components in blood cells. J. Clin. Microbiol. *39*, 485–493.

Aguero-Rosenfeld, M.E. (2004). Detection of *Borrelia burgdorferi* antibodies. In H.D. Isenberg (Ed.), Clinical Microbiology Procedures Handbook (pp. 11.6.1–11.6.10). Washington D.C.: ASM Press.

Aguero-Rosenfeld, M.E., Nowakowski, J., Bittker, S., Cooper, D., Nadelman, R.B., and Wormser, G.P. (1996). Evolution of the serologic response to *Borrelia burgdorferi* in treated patients with culture-confirmed erythema migrans. J. Clin. Microbiol. *34*, 1–9.

Aguero-Rosenfeld, M.E., Nowakowski, J., McKenna, D.F., Carbonaro, C.A., and Wormser, G.P. (1993). Serodiagnosis in early Lyme disease. J. Clin. Microbiol. *31*, 3090–3095.

Aguero-Rosenfeld, M.E., Wang, G., Schwartz, I., and Wormser, G.P. (2005). Diagnosis of Lyme borreliosis. Clin. Microbiol. Rev. *18*, 484–509.

Anguita, J., Samanta, S., Revilla, B., Suk, K., Das, S., Barthold, S.W., and Fikrig, E. (2000). *Borrelia burgdorferi* gene expression in vivo and spirochete pathogenicity. Infect. Immun. *68*, 1222–1230.

Arnez, M., Ruzic-Sabljic, E., Ahcan, J., Radsel-Medvescek, A., Pleterski-Rigler, D., and Strle, F. (2001). Isolation of *Borrelia burgdorferi* sensu lato from blood of children with solitary erythema migrans. Pediatr. Infect. Dis. J. *20*, 251–255.

Avery, R.A., Frank, G., and Eppes, S.C. (2005). Diagnostic utility of *Borrelia burgdorferi* cerebrospinal fluid polymerase chain reaction in children with Lyme meningitis. Pediatr. Infect. Dis. J. *24*, 705–708.

Bacon, R.M., Biggerstaff, B.J., Schriefer, M.E., Gilmore, R.D., Jr., Philipp, M.T., Steere, A.C., Wormser, G.P., Marques, A.R., and Johnson, B.J. (2003). Serodiagnosis of Lyme disease by kinetic enzyme-linked immunosorbent assay using recombinant VlsE1 or peptide antigens of *Borrelia burgdorferi* compared with 2-tiered testing using whole-cell lysates. J. Infect. Dis. *187*, 1187–1199.

Baranton, G., Seinost, G., Theodore, G., Postic, D., and Dykhuizen, D. (2001). Distinct levels of genetic diversity of *Borrelia burgdorferi* are associated with different aspects of pathogenicity. Res. Microbiol. *152*, 149–156.

Barbour, A.G. (1984). Isolation and cultivation of Lyme disease spirochetes. Yale. J. Biol. Med. *57*, 521–525.

Barbour, A.G., Burgdorfer, W., Hayes, S.F., Peter, O., and Aeschilimann, A. (1983). Isolation of a cultivable spirochete from *Ixodes ricinus* ticks of Switzerland. Curr. Microbiol. *8*, 123–126.

Barthold, S.W. (1995). Animal models for Lyme disease. Lab. Invest. *72*, 127–130.

Barthold, S.W., de Souza, M.S., Janotka, J.L., Smith, A.L., and Persing, D.H. (1993). Chronic Lyme borreliosis in the laboratory mouse. Am. J. Pathol. *143*, 959–971.

Benach, J.L., Bosler, E.M., Hanrahan, J.P., Coleman, J.L., Habicht, G.S., Bast, T.F., Cameron, D.J., Ziegler, J.L., Barbour, A.G., Burgdorfer, W., Edelman, R., and Kaslow, R.A. (1983). Spirochetes isolated from the blood of two patients with Lyme disease. N. Engl. J. Med. *308*, 740–742.

Berger, B.W., Johnson, R.C., Kodner, C., and Coleman, L. (1992a). Cultivation of *Borrelia burgdorferi* from erythema migrans lesions and perilesional skin. J. Clin. Microbiol. *30*, 359–361.

Berger, B.W., Johnson, R.C., Kodner, C., and Coleman, L. (1992b). Failure of *Borrelia burgdorferi* to survive in the skin of patients with antibiotic-treated Lyme disease. J. Am. Acad. Dermatol. *27*, 34–37.

Bergmann, A.R., Schmidt, B.L., Derler, A.M., and Aberer, E. (2002). Importance of sample preparation for molecular diagnosis of Lyme borreliosis from urine. J. Clin. Microbiol. *40*, 4581–4584.

Blanc, F., Jaulhac, B., Fleury, M., De Seze, J., De Martino, S.J., Remy, V., Blaison, G., Hansmann, Y., Christmann, D., and Tranchant, C. (2007). Relevance of the antibody index to diagnose Lyme neuroborreliosis among seropositive patients. Neurology *69*, 953–958.

Brettschneider, S., Bruckbauer, H., Klugbauer, N., and Hofmann, H. (1998). Diagnostic value of PCR for detection of *Borrelia burgdorferi* in skin biopsy and urine samples from patients with skin borreliosis. J. Clin. Microbiol. *36*, 2658–2665.

Brown, C.R., and Reiner, S.L. (1998). Clearance of *Borrelia burgdorferi* may not be required for resistance to experimental Lyme arthritis. Infect. Immun. *66*, 2065–2071.

Brown, E.L., Wooten, R.M., Johnson, B.J., Iozzo, R.V., Smith, A., Dolan, M.C., Guo, B.P., Weis, J.J., and Hook, M. (2001). Resistance to Lyme disease in decorin-deficient mice. J. Clin. Invest *107*, 845–852.

Brown, J.P., Zachary, J.F., Teuscher, C., Weis, J.J., and Wooten, R.M. (1999). Dual role of interleukin-10 in murine Lyme disease: regulation of arthritis severity and host defense. Infect. Immun. *67*, 5142–5150.

Brunet, L.R., Spielman, A., and Telford, S.R. (1995). Short report: density of Lyme disease spirochetes within deer ticks collected from zoonotic sites. Am. J Trop. Med. Hyg. *53*, 300–302.

Bunikis, J., and Barbour, A.G. (2002). Laboratory testing for suspected Lyme disease. Med. Clin. North Am. *86*, 311–340.

Burgdorfer, W., Anderson, J.F., Gern, L., Lane, R.S., Piesman, J., and Spielman, A. (1991). Relationship of *Borrelia burgdorferi* to its arthropod vectors. Scand. J. Infect. Dis. Suppl. *77*, 35–40.

Burkot, T.R., Piesman, J., and Wirtz, R.A. (1994). Quantitation of the *Borrelia burgdorferi* outer surface protein A in *Ixodes scapularis*: fluctuations during the tick life cycle, doubling times, and loss while feeding. J. Infect. Dis. *170*, 883–889.

Callister, S.M., Case, K.L., Agger, W.A., Schell, R.F., Johnson, R.C., and Ellingson, J.L. (1990). Effects of bovine serum albumin on the ability of Barbour-Stoenner-Kelly medium to detect *Borrelia burgdorferi*. J. Clin. Microbiol. *28*, 363–365.

Callister, S.M., Schell, R.F., and Lovrich, S.D. (1991). Lyme disease assay which detects killed *Borrelia burgdorferi*. J. Clin. Microbiol. *29*, 1773–1776.

Carroll, J.A., El Hage, N., Miller, J.C., Babb, K., and Stevenson, B. (2001). *Borrelia burgdorferi* RevA antigen is a surface-exposed outer membrane protein whose expression is regulated in response to environmental temperature and pH. Infect. Immun. *69*, 5286–5293.

Casjens, S., Palmer, N., van Vugt, R., Mun, H.W., Stevenson, B., Rosa, P., Lathigra, R., Sutton, G., Peterson, J., Dodson, R.J., Haft, D., Hickey, E., Gwinn, M., White, O., and Fraser, M. (2000). A bacterial genome in flux: the twelve linear and nine circular extrachromosomal DNAs in an infectious isolate of the Lyme disease spirochete *Borrelia burgdorferi*. Mol. Microbiol. *35*, 490–516.

Centers for Disease Control and Prevention (1995). Recommendations for test performance and interpretation from the Second National Conference on Serologic Diagnosis of Lyme Disease. MMWR *44*, 590–591.

Cimperman, J., Maraspin, V., Lotric-Furlan, S., Ruzic-Sabljic, E., and Strle, F. (1999). Lyme meningitis: a one-year follow up controlled study. Wien. Klin. Wochenschr. *111*, 961–963.

Cogswell, F.B., Bantar, C.E., Hughes, T.G., Gu, Y., and Philipp, M.T. (1996). Host DNA can interfere with detection of *Borrelia burgdorferi* in skin biopsy specimens by PCR. J. Clin. Microbiol. *34*, 980–982.

Coulter, P., Lema, C., Flayhart, D., Linhardt, A.S., Aucott, J.N., Auwaerter, P.G., and Dumler, J.S. (2005). Two-year evaluation of *Borrelia burgdorferi* culture and supplemental tests for definitive diagnosis of Lyme disease. J. Clin. Microbiol. *43*, 5080–5084.

Coyle, P.K., Schutzer, S.E., Deng, Z., Krupp, L.B., Belman, A.L., Benach, J.L., and Luft, B.J. (1995). Detection of *Borrelia burgdorferi*-specific antigen in antibody- negative cerebrospinal fluid in neurologic Lyme disease. Neurology *45*, 2010–2015.

Craft, J.E., Fischer, D.K., Shimamoto, G.T., and Steere, A.C. (1986). Antigens of *Borrelia burgdorferi* recognized during Lyme disease. Appearance of a new immunoglobulin M response and expansion of the immunoglobulin G response late in the illness. J. Clin. Invest *78*, 934–939.

Crother, T.R., Champion, C.I., Wu, X.Y., Blanco, D.R., Miller, J.N., and Lovett, M.A. (2003). Antigenic composition of *Borrelia burgdorferi* during infection of SCID mice. Infect. Immun. *71*, 3419–3428.

de Koning, J., and Duray, P.H. (1993). Histopathology of human Lyme borreliosis. In K. Weber and W. Burgdorfer (Eds.), Aspects of Lyme borreliosis (pp. 70–92). Berlin: Springer-Verlag.

de Koning, J., Hoogkamp-Korstanje, J.A., van der Linde, M.R., and Crijns, H.J. (1989). Demonstration of spirochetes in cardiac biopsies of patients with Lyme disease. J. Infect. Dis. *160*, 150–153.

De Martino, S.J., Sordet, C., Piemont, Y., Ruzic-Sabljic, E., Thaddee, V.M., Monteil, H., Sibilia, J., and Jaulhac, B. (2006). Enhanced culture of *Borrelia garinii* and *Borrelia afzelii* strains on a solid BSK-based medium in anaerobic conditions. Res. Microbiol. *157*, 726–729.

de Silva, A.M., and Fikrig, E. (1995). Growth and migration of *Borrelia burgdorferi* in *Ixodes* ticks during blood feeding. Am. J Trop. Med. Hyg. *53*, 397–404.

Demaerschalck, I., Ben Messaoud, A., De Kesel, M., Hoyois, B., Lobet, Y., Hoet, P.P., Bigaignon, G., Bollen, A., and Godfroid, E. (1995). Simultaneous presence of different *Borrelia burgdorferi* genospecies in biological fluids of Lyme disease patients. J. Clin. Microbiol. *33*, 602–608.

Dorward, D.W., Schwan, T.G., and Garon, C.F. (1991). Immune capture and detection of *Borrelia burgdorferi* antigens in urine, blood, or tissues from infected ticks, mice, dogs, and humans. J. Clin. Microbiol. *29*, 1162–1170.

Dressler, F., Whalen, J.A., Reinhardt, B.N., and Steere, A.C. (1993). Western blotting in the serodiagnosis of Lyme disease. J. Infect. Dis. *167*, 392–400.

Dumler, J.S. (2001). Molecular diagnosis of Lyme disease: Review and meta-analysis. Mol. Diagn. *6*, 1–11.

England, J.D., Bohm, R.P., Jr., Roberts, E.D., and Philipp, M.T. (1997). Lyme neuroborreliosis in the rhesus monkey. Semin. Neurol. *17*, 53–56.

Engstrom, S.M., Shoop, E., and Johnson, R.C. (1995). Immunoblot interpretation criteria for serodiagnosis of early Lyme disease. J. Clin. Microbiol. 33, 419–427.

Exner, M.M., and Lewinski, M.A. (2003). Isolation and detection of *Borrelia burgdorferi* DNA from cerebral spinal fluid, synovial fluid, blood, urine, and ticks using the Roche MagNA Pure system and real-time PCR. Diagn. Microbiol. Infect. Dis. *46*, 235–240.

Fallon, B.A., Keilp, J.G., Corbera, K.M., Petkova, E., Britton, C.B., Dwyer, E., Slavov, I., Cheng, J., Dobkin, J., Nelson, D.R., and Sackeim, H.A. (2008). A randomized, placebo-controlled trial of repeated IV antibiotic therapy for Lyme encephalopathy. Neurology *70*, 992–1003.

Fawcett, P.T., Gibney, K.M., Rose, C.D., Dubbs, S.B., and Doughty, R.A. (1992). Frequency and specificity of antibodies that crossreact with *Borrelia burgdorferi* antigens. J. Rheumatol. *19*, 582–587.

Fawcett, P.T., Rose, C., Gibney, K.M., Chase, C.A., Kiehl, B., and Doughty, R.A. (1993). Detection of antibodies to the recombinant P39 protein of *Borrelia burgdorferi* using enzyme immunoassay and immunoblotting. J. Rheumatol. *20*, 734–738.

Fikrig, E., Barthold, S.W., Sun, W., Feng, W., Telford, S.R. III, and Flavell, R.A. (1997). Borrelia burgdorferi P35 and P37 proteins, expressed in vivo, elicit protective immunity. Immunity. *6*, 531–539.

Fikrig, E., Feng, W., Barthold, S.W., Telford, S.R. III, and Flavell, R.A. (2000). Arthropod- and host-specific *Borrelia burgdorferi bbk32* expression and the inhibition of spirochete transmission. J. Immunol. *164*, 5344–5351.

Fikrig, E., Huguenel, E.D., Berland, R., Rahn, D.W., Hardin, J.A., and Flavell, R.A. (1992). Serologic diagnosis of Lyme disease using recombinant outer surface proteins A and B and flagellin. J. Infect. Dis. *165*, 1127–1132.

Fraser, C.M., Casjens, S., Huang, W.M., Sutton, G.G., Clayton, R., Lathigra, R., White, O., Ketchum, K.A., Dodson, R., Hickey, E.K., Gwinn, M., Dougherty, B., Tomb, J.F., Fleischmann, R.D., Richardson, D., Peterson, J., Kerlavage, A.R., Quackenbush, J., Salzberg, S., Hanson, M., van Vugt, R., Palmer, N., Adams, M.D., Gocayne, J., and Venter, J.C. (1997). Genomic sequence of a Lyme disease spirochaete, *Borrelia burgdorferi*. Nature *390*, 580–586.

Gaumond, G., Tyropolis, A., Grodzicki, S., and Bushmich, S. (2006). Comparison of direct fluorescent antibody staining and real-time polymerase chain reaction for the detection of *Borrelia burgdorferi* in *Ixodes scapularis* ticks. J. Vet. Diagn. Invest *18*, 583–586.

Germer, J., Ryckmann, B., Moro, M., Hofmeister, E., Barthold, S.W., Bockenstedt, L., and Persing, D.H. (1999). Quantitative detection of *Borrelia burgdorferi* with a microtiter-based competitive polymerase chain reaction assay. Mol. Diagn. *4*, 185–193.

Gern, L., Estrada-Pena, A., Frandsen, F., Gray, J.S., Jaenson, T.G., Jongejan, F., Kahl, O., Korenberg, E., Mehl, R., and Nuttall, P.A. (1998). European reservoir hosts of *Borrelia burgdorferi* sensu lato. Zentralbl. Bakteriol. *287*, 196–204.

Gern, L., Lebet, N., and Moret, J. (1996). Dynamics of *Borrelia burgdorferi* infection in nymphal *Ixodes ricinus* ticks during feeding. Exp. Appl. Acarol. *20*, 649–658.

Gomes-Solecki, M.J., Meirelles, L., Glass, J., and Dattwyler, R.J. (2007). Epitope length, genospecies dependency, and serum panel effect in the IR6 enzyme-linked immunosorbent assay for detection of antibodies to *Borrelia burgdorferi*. Clin. Vaccine Immunol. *14*, 875–879.

Goodman, J.L., Bradley, J.F., Ross, A.E., Goellner, P., Lagus, A., Vitale, B., Berger, B.W., Luger, S., and Johnson, R.C. (1995). Bloodstream invasion in early Lyme disease: results from a prospective, controlled, blinded study using the polymerase chain reaction. Am. J. Med. *99*, 6–12.

Gooskens, J., Templeton, K.E., Claas, E.C., and van Dam, A.P. (2006). Evaluation of an internally controlled real-time PCR targeting the *ospA* gene for detection of *Borrelia burgdorferi sensu lato* DNA in cerebrospinal fluid. Clin. Microbiol. Infect. *12*, 894–900.

Greene, R.T., Walker, R.L., and Greene, C.E. (1991). Pseudospirochetes in animal blood being cultured for *Borrelia burgdorferi*. J. Vet. Diagn. Invest. *3*, 350–352.

Halperin, J.J., Volkman, D.J., and Wu, P. (1991). Central nervous system abnormalities in Lyme neuroborreliosis. Neurology *41*, 1571–1582.

Hansen, K., and Lebech, A.M. (1991). Lyme neuroborreliosis: a new sensitive diagnostic assay for intrathecal synthesis of *Borrelia burgdorferi*-specific immunoglobulin G, A, and M. Ann. Neurol. *30*, 197–205.

Hauser, U., Lehnert, G., Lobentanzer, R., and Wilske, B. (1997). Interpretation criteria for standardized Western blots for three European species of *Borrelia burgdorferi* sensu lato. J. Clin. Microbiol. *35*, 1433–1444.

Hauser, U., and Wilske, B. (1997). Enzyme-linked immunosorbent assays with recombinant internal flagellin fragments derived from different species of *Borrelia burgdorferi* sensu lato for the serodiagnosis of Lyme neuroborreliosis. Med. Microbiol. Immunol. (Berl) *186*, 145–151.

Hechemy, K.E., Samsonoff, W.A., Harris, H.L., and McKee, M. (1992). Adherence and entry of *Borrelia burgdorferi* in Vero cells. J. Med. Microbiol. *36*, 229–238.

Heikkila, T., Huppertz, H.I., Seppala, I., Sillanpaa, H., Saxen, H., and Lahdenne, P. (2003). Recombinant or peptide antigens in the serology of Lyme arthritis in children. J. Infect. Dis. *187*, 1888–1894.

Heikkila, T., Seppala, I., Saxen, H., Panelius, J., Peltomaa, M., Julin, T., Carlsson, S.A., and Lahdenne, P. (2002). Recombinant BBK32 protein in serodiagnosis of early and late Lyme borreliosis. J. Clin. Microbiol. *40*, 1174–1180.

Hunfeld, K.P., Ruzic-Sabljic, E., Norris, D.E., Kraiczy, P., and Strle, F. (2005). In vitro susceptibility testing of *Borrelia burgdorferi* sensu lato isolates cultured from patients with erythema migrans before and after antimicrobial chemotherapy. Antimicrob. Agents Chemother. *49*, 1294–1301.

Hyde, F.W., Johnson, R.C., White, T.J., and Shelburne, C.E. (1989). Detection of antigens in urine of mice and humans infected with *Borrelia burgdorferi*, etiologic agent of Lyme disease. J. Clin. Microbiol. *27*, 58–61.

Ivacic, L., Reed, K.D., Mitchell, P.D., and Ghebranious, N. (2007). A LightCycler TaqMan assay for detection of *Borrelia burgdorferi* sensu lato in clinical samples. Diagn. Microbiol. Infect. Dis. *57*, 137–143.

James, A.M., Liveris, D., Wormser, G.P., Schwartz, I., Montecalvo, M.A., and Johnson, B.J. (2001). *Borrelia lonestari* infection after a bite by an *Amblyomma americanum* tick. J. Infect. Dis. *183*, 1810–1814.

Jauris-Heipke, S., Rößle, B., Wanner, G., Habermann, C., Rössler, D., Fingerle, V., Lehnert, G., Lobentanzer, R., Pradel, I., Hillenbrand, B., Schulte-Spechtel, U., and Wilske, B. (1999). Osp17, a novel immunodominant outer surface protein of *Borrelia afzelii*: recombinant expression in *Escherichia coli* and its use as a diagnostic antigen for serodiagnosis of Lyme borreliosis. Med. Microbiol. Immunol. (Berl.) *187*, 213–219.

Jobe, D.A., Rawal, N., Schell, R.F., and Callister, S.M. Detection of borreliacidal antibodies in Lyme Borreliosis patient sera containing antimicrobial agents. 1999. Clin. Diagn. Lab. Immunol. 6:930–933

Jurca, T., Ruzic-Sabljic, E., Lotric-Furlan, S., Maraspin, V., Cimperman, J., Picken, R.N., and Strle, F. (1998). Comparison of peripheral and central biopsy sites for the isolation of *Borrelia burgdorferi* sensu lato from erythema migrans skin lesions. Clin. Infect. Dis. *27*, 636–638.

Karch, H., Huppertz, H.I., Bohme, M., Schmidt, H., Wiebecke, D., and Schwarzkopf, A. (1994). Demonstration of *Borrelia burgdorferi* DNA in urine samples from healthy humans whose sera contain *B. burgdorferi*-specific antibodies. J. Clin. Microbiol. *32*, 2312–2314.

Karlsson, M., Hovind-Hougen, K., Svenungsson, B., and Stiernstedt, G. (1990). Cultivation and characterization of spirochetes from cerebrospinal fluid of patients with Lyme borreliosis. J. Clin. Microbiol. *28*, 473–479.

Kelly, R. (1971). Cultivation of *Borrelia hermsii*. Science *173*, 443–444.

Klempner, M.S. (2002). Controlled trials of antibiotic treatment in patients with post-treatment chronic Lyme disease. Vector. Borne. Zoonotic. Dis. *2*, 255–263.

Klempner, M.S., Hu, L.T., Evans, J., Schmid, C.H., Johnson, G.M., Trevino, R.P., Norton, D., Levy, L., Wall, D., McCall, J., Kosinski, M., and Weinstein, A. (2001a). Two controlled trials of antibiotic treatment in patients with persistent symptoms and a history of Lyme disease. N. Engl. J. Med. *345*, 85–92.

Klempner, M.S., Schmid, C.H., Hu, L., Steere, A.C., Johnson, G., McCloud, B., Noring, R., and Weinstein, A. (2001b). Intralaboratory reliability of serologic and urine testing for Lyme disease. Am. J. Med. *110*, 217–219.

Kondrusik, M., Grygorczuk, S., Skotarczak, B., Wodecka, B., Rymaszewska, A., Pancewicz, S., Zajkowska, J., Swierzbinska, R., and Hermanowska-Szpakowicz, T. (2007). Molecular and serological diagnosis of *Borrelia burgdorferi* infection among patients with diagnosed *erythema migrans*. Ann Agric. Environ. Med *14*, 209–213.

Kurtti, T.J., Munderloh, U.G., Ahlstrand, G.G., and Johnson, R.C. (1988). *Borrelia burgdorferi* in tick cell culture: growth and cellular adherence. J. Med. Entomol. *25*, 256–261.

Kurtti, T.J., Munderloh, U.G., Johnson, R.C., and Ahlstrand, G.G. (1987). Colony formation and morphology in *Borrelia burgdorferi*. J. Clin. Microbiol. *25*, 2054–2058.

Lebech, A.M., Clemmensen, O., and Hansen, K. (1995). Comparison of *in vitro* culture, immunohistochemical staining, and PCR for detection of *Borrelia burgdorferi* in tissue from experimentally infected animals. J. Clin. Microbiol. *33*, 2328–2333.

Lebech, A.M., Hansen, K., Brandrup, F., Clemmensen, O., and Halkier-Sorensen, L. (2000). Diagnostic value of PCR for detection of *Borrelia burgdorferi* DNA in clinical specimens from patients with erythema migrans and Lyme neuroborreliosis. Mol. Diagn. *5*, 139–150.

Levin, M.L., and Fish, D. (2000). Acquisition of coinfection and simultaneous transmission of *Borrelia burgdorferi* and *Ehrlichia phagocytophila* by *Ixodes scapularis* ticks. Infect. Immun. *68*, 2183–2186.

Liang, F.T., Aberer, E., Cinco, M., Gern, L., Hu, C.M., Lobet, Y.N., Ruscio, M., Voet Jr, P.E., Weynants, V.E., and Philipp, M.T. (2000). Antigenic conservation of an immunodominant invariable region of the VlsE lipoprotein among european pathogenic genospecies of *Borrelia burgdorferi* SL. J. Infect. Dis. *182*, 1455–1462.

Liang, F.T., Alvarez, A.L., Gu, Y., Nowling, J.M., Ramamoorthy, R., and Philipp, M.T. (1999a). An immunodominant conserved region within the variable domain of VlsE, the variable surface antigen of *Borrelia burgdorferi*. J. Immunol. *163*, 5566–5573.

Liang, F.T., Steere, A.C., Marques, A.R., Johnson, B.J., Miller, J.N., and Philipp, M.T. (1999b). Sensitive and specific serodiagnosis of Lyme disease by enzyme-linked immunosorbent assay with a peptide based on an immunodominant conserved region of *Borrelia burgdorferi* VlsE. J. Clin. Microbiol. *37*, 3990–3996.

Liang, F.T., Yan, J., Mbow, M.L., Sviat, S.L., Gilmore, R.D., Mamula, M., and Fikrig, E. (2004). *Borrelia burgdorferi* changes its surface antigenic expression in response to host immune responses. Infect. Immun. *72*, 5759–5767.

Liebling, M.R., Nishio, M.J., Rodriguez, A., Sigal, L.H., Jin, T., and Louie, J.S. (1993). The polymerase chain reaction for the detection of *Borrelia burgdorferi* in human body fluids. Arthritis Rheum. *36*, 665–675.

Liveris, D., Varde, S., Iyer, R., Koenig, S., Bittker, S., Cooper, D., McKenna, D., Nowakowski, J., Nadelman, R.B., Wormser, G.P., and Schwartz, I. (1999). Genetic diversity of *Borrelia burgdorferi* in Lyme disease patients as determined by culture versus direct PCR with clinical specimens. J. Clin. Microbiol. *37*, 565–569.

Liveris, D., Wang, G., Girao, G., Byrne, D.W., Nowakowski, J., McKenna, D., Nadelman, R., Wormser, G.P., and Schwartz, I. (2002). Quantitative detection of *Borrelia burgdorferi* in 2-millimeter skin samples of erythema migrans lesions: correlation of results with clinical and laboratory findings. J. Clin. Microbiol. *40*, 1249–1253.

Luft, B.J., Dunn, J.J., Dattwyler, R.J., Gorgone, G., Gorevic, P.D., and Schubach, W.H. (1993). Cross-reactive antigenic domains of the flagellin protein of *Borrelia burgdorferi*. Res. Microbiol. *144*, 251–257.

Ma, Y., Seiler, K.P., Eichwald, E.J., Weis, J.H., Teuscher, C., and Weis, J.J. (1998). Distinct characteristics of resistance to *Borrelia burgdorferi*-induced arthritis in C57BL/6N mice. Infect. Immun. *66*, 161–168.

Magnarelli, L.A., Ijdo, J.W., Padula, S.J., Flavell, R.A., and Fikrig, E. (2000). Serologic diagnosis of Lyme borreliosis by using enzyme-linked immunosorbent assays with recombinant antigens. J. Clin. Microbiol. *38*, 1735–1739.

Marangoni, A., Sparacino, M., Cavrini, F., Storni, E., Mondardini, V., Sambri, V., and Cevenini, R. (2005). Comparative evaluation of three different ELISA methods for the diagnosis of early culture-confirmed Lyme disease in Italy. J. Med Microbiol. *54*, 361–367.

Maraspin, V., Ruzic-Sabljic, E., Cimperman, J., Lotric-Furlan, S., Jurca, T., Picken, R.N., and Strle, F. (2001). Isolation of *Borrelia burgdorferi* sensu lato from blood of patients with erythema migrans. Infection *29*, 65–70.

Marcus, L.C., Steere, A.C., Duray, P.H., Anderson, A.E., and Mahoney, E.B. (1985). Fatal pancarditis in a patient with coexistent Lyme disease and babesiosis. Demonstration of spirochetes in the myocardium. Ann. Intern. Med. *103*, 374–376.

Mathiesen, M.J., Christiansen, M., Hansen, K., Holm, A., Asbrink, E., and Theisen, M. (1998a). Peptide-based OspC enzyme-linked immunosorbent assay for serodiagnosis of Lyme borreliosis. J. Clin. Microbiol. *36*, 3474–3479.

Mathiesen, M.J., Holm, A., Christiansen, M., Blom, J., Hansen, K., Ostergaard, S., and Theisen, M. (1998b). The dominant epitope of *Borrelia garinii* outer surface protein C recognized by sera from patients with neuroborreliosis has a surface-exposed conserved structural motif. Infect. Immun. *66*, 4073–4079.

Melchers, W., Meis, J., Rosa, P., Claas, E., Nohlmans, L., Koopman, R., Horrevorts, A., and Galama, J. (1991). Amplification of *Borrelia burgdorferi* DNA in skin biopsies from patients with Lyme disease. J. Clin. Microbiol. *29*, 2401–2406.

Miller, J.C., von Lackum, K., Babb, K., McAlister, J.D., and Stevenson, B. (2003). Temporal analysis of *Borrelia burgdorferi* Erp protein expression throughout the mammal-tick infectious cycle. Infect. Immun. *71*, 6943–6952.

Misonne, M.C., van Impe, G., and Hoet, P.P. (1998). Genetic heterogeneity of *Borrelia burgdorferi* sensu lato in *Ixodes ricinus* ticks collected in Belgium. J. Clin. Microbiol. *36*, 3352–3354.

Mitchell, P.D., Reed, K.D., Aspeslet, T.L., Vandermause, M.F., and Melski, J.W. (1994). Comparison of four immunoserologic assays for detection of antibodies to *Borrelia burgdorferi* in patients with culture-positive erythema migrans. J. Clin. Microbiol. *32*, 1958–1962.

Mitchell, P.D., Reed, K.D., Vandermause, M.F., and Melski, J.W. (1993). Isolation of *Borrelia burgdorferi* from skin biopsy specimens of patients with erythema migrans. Am. J. Clin. Pathol. *99*, 104–107.

Mogilyansky, E., Loa, C.C., Adelson, M.E., Mordechai, E., and Tilton, R.C. (2004). Comparison of Western immunoblotting and the C6 Lyme antibody test for laboratory detection of Lyme disease. Clin. Diagn. Lab Immunol. *11*, 924–929.

Mommert, S., Gutzmer, R., Kapp, A., and Werfel, T. (2001). Sensitive detection of *Borrelia burgdorferi* sensu lato DNA and differentiation of *Borrelia* species by LightCycler PCR. J. Clin. Microbiol. *39*, 2663–2667.

Morrison, T.B., Ma, Y., Weis, J.H., and Weis, J.J. (1999). Rapid and sensitive quantification of *Borrelia burgdorferi*-infected mouse tissues by continuous fluorescent monitoring of PCR. J. Clin. Microbiol. *37*, 987–992.

Moter, S.E., Hofmann, H., Wallich, R., Simon, M.M., and Kramer, M.D. (1994). Detection of *Borrelia burgdorferi* sensu lato in lesional skin of patients with erythema migrans and acrodermatitis chronica atrophicans by *ospA*-specific PCR. J. Clin. Microbiol. *32*, 2980–2988.

Munderloh, U.G., Liu, Y., Wang, M., Chen, C., and Kurtti, T.J. (1994). Establishment, maintenance and description of cell lines from the tick *Ixodes scapularis*. J. Parasitol. *80*, 533–543.

Nadelman, R.B., Nowakowski, J., Forseter, G., Bittker, S., Cooper, D., Goldberg, N., McKenna, D., and Wormser, G.P. (1993). Failure to isolate *Borrelia burgdorferi* after antimicrobial therapy in culture-documented Lyme borreliosis associated with erythema migrans: report of a prospective study. Am. J. Med. *94*, 583–588.

Nadelman, R.B., Nowakowski, J., Forseter, G., Goldberg, N.S., Bittker, S., Cooper, Aguero-Rosenfeld, M., and Wormser, G.P. (1996). The clinical spectrum of early Lyme borreliosis in patients with culture-confirmed erythema migrans. Am. J. Med. *100*, 502–508.

Nadelman, R.B., Pavia, C.S., Magnarelli, L.A., and Wormser, G.P. (1990). Isolation of *Borrelia burgdorferi* from the blood of seven patients with Lyme disease. Am. J. Med. *88*, 21–26.

Nadelman, R.B., Schwartz, I., and Wormser, G.P. (1994). Detecting *Borrelia burgdorferi* in blood from patients with Lyme disease. J. Infect. Dis. *169*, 1410–1411.

Nadelman, R.B., and Wormser, G.P. (1995). Erythema migrans and early Lyme disease. Am. J. Med. *98*, 15S-23S.

Nadelman, R.B., and Wormser, G.P. (1998). Lyme borreliosis. Lancet 352, 557–565.

Nocton, J.J., Bloom, B.J., Rutledge, B.J., Persing, D.H., Logigian, E.L., Schmid, G.P., and Steere, A.C. (1996). Detection of *Borrelia burgdorferi* DNA by polymerase chain reaction in cerebrospinal fluid in Lyme neuroborreliosis. J. Infect. Dis. *174*, 623–627.

Nocton, J.J., Dressler, F., Rutledge, B.J., Rys, P.N., Persing, D.H., and Steere, A.C. (1994). Detection of *Borrelia burgdorferi* DNA by polymerase chain reaction in synovial fluid from patients with Lyme arthritis. N. Engl. J. Med. *330*, 229–234.

Nowakowski, J., Schwartz, I., Liveris, D., Wang, G., Aguero-Rosenfeld, M.E., Girao, G., McKenna, D., Nadelman, R.B., Cavaliere, L.F., and Wormser, G.P. (2001). Laboratory diagnostic techniques for patients with early Lyme disease associated with erythema migrans: a comparison of different techniques. Clin. Infect. Dis. 33, 2023–2027.

Nyman, D., Willen, L., Jansson, C., Carlsson, S.A., Granlund, H., and Wahlberg, P. (2006). VlsE C6 peptide and IgG ELISA antibody analysis for clinical diagnosis of Lyme borreliosis in an endemic area. Clin. Microbiol. Infect. *12*, 496–497.

Obonyo, M., Munderloh, U.G., Fingerle, V., Wilske, B., and Kurtti, T.J. (1999). *Borrelia burgdorferi* in tick cell culture modulates expression of outer surface proteins A and C in response to temperature. J. Clin. Microbiol. *37*, 2137–2141.

Oksi, J., Marttila, H., Soini, H., Aho, H., Uksila, J., and Viljanen, M.K. (2001). Early dissemination of *Borrelia burgdorferi* without generalized symptoms in patients with erythema migrans. APMIS *109*, 581–588.

Ong, Y.L., and Irvine, A. (2002). Quantitative real-time PCR: a critique of method and practical considerations. Hematology. 7, 59–67.

Ornstein, K., Berglund, J., Bergström, S., Norrby, R., and Barbour, A.G. (2002). Three major Lyme *Borrelia* genospecies (*Borrelia burgdorferi* sensu stricto, *B. afzelii* and *B. garinii*) identified by PCR in cerebrospinal fluid from patients with neuroborreliosis in Sweden. Scand. J. Infect. Dis. *34*, 341–346.

Ornstein, K., Berglund, J., Nilsson, I., Norrby, R., and Bergström, S. (2001). Characterization of Lyme borreliosis isolates from patients with erythema migrans and neuroborreliosis in southern Sweden. J. Clin. Microbiol. *39*, 1294–1298.

Padula, S.J., Dias, F., Sampieri, A., Craven, R.B., and Ryan, R.W. (1994). Use of recombinant OspC from *Borrelia burgdorferi* for serodiagnosis of early Lyme disease. J. Clin. Microbiol. *32*, 1733–1738.

Pahl, A., Kuhlbrandt, U., Brune, K., Rollinghoff, M., and Gessner, A. (1999). Quantitative detection of *Borrelia burgdorferi* by real-time PCR. J. Clin. Microbiol. *37*, 1958–1963.

Panelius, J., Lahdenne, P., Heikkila, T., Peltomaa, M., Oksi, J., and Seppala, I. (2002). Recombinant OspC from *Borrelia burgdorferi* sensu stricto, *B. afzelii* and *B. garinii* in the serodiagnosis of Lyme borreliosis. J. Med. Microbiol. *51*, 731–739.

Panelius, J., Lahdenne, P., Saxen, H., Carlsson, S.A., Heikkila, T., Peltomaa, M., Lauhio, A., and Seppala, I. (2003). Diagnosis of Lyme neuroborreliosis with antibodies to recombinant proteins DbpA, BBK32, and OspC, and VlsE IR6 peptide. J. Neurol. *250*, 1318–1327.

Persing, D.H., Rutledge, B.J., Rys, P.N., Podzorski, D.S., Mitchell, P.D., Reed, K.D. X.L. B., Fikrig, E., and Malawista, S.E. (1994). Target imbalance: disparity of *Borrelia burgdorferi* genetic material in synovial fluid from Lyme arthritis patients. J. Infect. Dis. *169*, 668–672.

Philipp, M.T., Aydintug, M.K., Bohm, R.P., Jr., Cogswell, F.B., Dennis, V.A., Lanners, H.N., Lowrie, R.C., Jr., Roberts, E.D., Conway, M.D., Karacorlu, M., and *et al.* (1993). Early and early disseminated phases of Lyme disease in the rhesus monkey: a model for infection in humans. Infect. Immun. *61*, 3047–3059.

Picha, D., Moravcova, L., Zdarsky, E., Maresova, V., and Hulinsky, V. (2005). PCR in Lyme neuroborreliosis: a prospective study. Acta Neurol. Scand. *112*, 287–292.

Picken, M.M., Picken, R.N., Han, D., Cheng, Y., Ruzic-Sabljic, E., Cimperman, J., Maraspin, V., Lotric-Furlan, S., and Strle, F. (1997). A two year prospective study to compare culture and polymerase chain reaction amplification for the detection and diagnosis of Lyme borreliosis. Mol. Pathol. *50*, 186–193.

Piesman, J., Oliver, J.R., and Sinsky, R.J. (1990). Growth kinetics of the Lyme disease spirochete (*Borrelia burgdorferi*) in vector ticks (*Ixodes dammini*). Am. J. Trop. Med. Hyg. *42*, 352–357.

Piesman, J., Schneider, B.S., and Zeidner, N.S. (2001). Use of quantitative PCR to measure density of *Borrelia burgdorferi* in the midgut and salivary glands of feeding tick vectors. J. Clin. Microbiol. *39*, 4145–4148.

Pietila, J., He, Q., Oksi, J., and Viljanen, M.K. (2000). Rapid differentiation of *Borrelia garinii* from *Borrelia afzelii* and *Borrelia burgdorferi* sensu stricto by Lightcycler fluorescence melting curve analysis of a PCR product of the *recA* gene. J. Clin. Microbiol. *38*, 2756–2759.

Pollack, R.J., Telford, S.R. III, and Spielman, A. (1993). Standardization of medium for culturing Lyme disease spirochetes. J. Clin. Microbiol. *31*, 1251–1255.

Portnöi, D., Sertour, N., Ferquel, E., Garnier, M., Baranton, G., and Postic, D. (2006). A single-run, real-time PCR for detection and identification of *Borrelia burgdorferi sensu lato* species, based on the *hbb* gene sequence. FEMS Microbiol. Lett. *259*, 35–40.

Preac-Mursic, V., Pfister, H.W., Spiegel, H., Burk, R., Wilske, B., Reinhardt, S., and Bohmer, R. (1993). First isolation of *Borrelia burgdorferi* from an iris biopsy. J. Clin. Neuroophthalmol. *13*, 155–161.

Preac-Mursic, V., Wilske, B., and Reinhardt, S. (1991). Culture of *Borrelia burgdorferi* on six solid media. Eur. J. Clin. Microbiol. Infect. Dis. *10*, 1076–1079.

Preac-Mursic, V., Wilske, B., and Schierz, G. (1986). European *Borrelia burgdorferi* isolated from humans and ticks culture conditions and antibiotic susceptibility. Zentralbl. Bakteriol. Mikrobiol. Hyg. A *263*, 112–118.

Priem, S., Burmester, G.R., Kamradt, T., Wolbart, K., Rittig, M.G., and Krause, A. (1998). Detection of *Borrelia burgdorferi* by polymerase chain reaction in synovial membrane, but not in synovial fluid from patients with persisting Lyme arthritis after antibiotic therapy. Ann. Rheum. Dis *57*, 118–121.

Rasiah, C., Schiltz, E., Reichert, J., and Vogt, A. (1992). Purification and characterization of a tryptic peptide of *Borrelia burgdorferi* flagellin, which reduces cross-reactivity in immunoblots and ELISA. J. Gen. Microbiol. *138*, 147–154.

Rauter, C., Oehme, R., Diterich, I., Engele, M., and Hartung, T. (2002). Distribution of clinically relevant borrelia genospecies in ticks assessed by a novel, single-run, real-time PCR. J. Clin. Microbiol. *40*, 36–43.

Reed, K.D. (2002). Laboratory testing for Lyme disease: possibilities and practicalities. J. Clin. Microbiol. *40*, 319–324.

Rijpkema, S.G., Tazelaar, D.J., Molkenboer, M., Noordhoek, G.T., Plantinga, G., Schouls, L.M., and Schellekens, J.F. (1997). Detection of *Borrelia afzelii, Borrelia burgdorferi* sensu stricto, *Borrelia garinii* and group VS116 by PCR in skin biopsies of patients with erythema migrans and acrodermitis chronica atrophicans. Clin. Microbiol. Infect. *3*, 109–116.

Ruzic-Sabljic, E., Lotric-Furlan, S., Maraspin, V., Cimperman, J., Logar, M., Jurca, T., and Strle, F. (2006). Comparison of isolation rate of *Borrelia burgdorferi* sensu lato in MKP and BSK-II medium. Int. J. Med Microbiol. *296 Suppl. 40*, 267–273.

Ruzic-Sabljic, E., Lotric-Furlan, S., Maraspin, V., Cimperman, J., Pleterski-Rigler, D., and Strle, F. (2001). Analysis of *Borrelia burgdorferi* sensu lato isolated from cerebrospinal fluid. APMIS *109*, 707–713.

Ruzic-Sabljic, E., and Strle, F. (2004). Comparison of growth of *Borrelia afzelii, B. garinii*, and *B. burgdorferi* sensu stricto in MKP and BSK-II medium. Int. J. Med Microbiol. *294*, 407–412.

Sadziene, A., Jonsson, M., Bergström, S., Bright, R.K., Kennedy, R.C., and Barbour, A.G. (1994). A bactericidal antibody to *Borrelia burgdorferi* is directed against a variable region of the OspB protein. Infect. Immun. *62*, 2037–2045.

Santino, I., Berlutti, F., Pantanella, F., Sessa, R., and Del Piano, M. (2008). Detection of *Borrelia burgdorferi* sensu lato DNA by PCR in serum of patients with clinical symptoms of Lyme borreliosis. FEMS Microbiol. Lett. *283*, 30–35.

Schmidli, J., Hunziker, T., Moesli, P., and Schaad, U.B. (1988). Cultivation of *Borrelia burgdorferi* from joint fluid three months after treatment of facial palsy due to Lyme borreliosis [letter]. J. Infect. Dis. *158*, 905–906.

Schmidt, B., Muellegger, R.R., Stockenhuber, C., Soyer, H.P., Hoedl, S., and Luger, A.X. K.H. (1996). Detection of *Borrelia burgdorferi*-specific DNA in urine specimens from patients with erythema migrans before and after antibiotic therapy. J. Clin. Microbiol. *34*, 1359–1363.

Schmidt, B.L. (1997). PCR in laboratory diagnosis of human *Borrelia burgdorferi* infections. Clin. Microbiol. Rev. *10*, 185–201.

Schwaiger, M., Peter, O., and Cassinotti, P. (2001). Routine diagnosis of *Borrelia burgdorferi* (sensu lato) infections using a real-time PCR assay. Clin. Microbiol. Infect. *7*, 461–469.

Schwan, T.G., and Piesman, J. (2000). Temporal changes in outer surface proteins A and C of the Lyme disease-associated spirochete, *Borrelia burgdorferi*, during the chain of infection in ticks and mice. J. Clin. Microbiol. *38*, 382–388.

Schwan, T.G., Piesman, J., Golde, W.T., Dolan, M.C., and Rosa, P.A. (1995). Induction of an outer surface protein on *Borrelia burgdorferi* during tick feeding. Proc. Natl. Acad. Sci. U.S.A. *92*, 2909–2913.

Schwartz, I., Varde, S., Nadelman, R.B., Wormser, G.P., and Fish, D. (1997). Inhibition of efficient polymerase chain reaction amplification of *Borrelia burgdorferi* DNA in blood-fed ticks. Am. J. Trop. Med. Hyg. *56*, 339–342.

Schwartz, I., Wormser, G.P., Schwartz, J.J., Cooper, D., Weissensee, P., Gazumyan, A., Zimmermann, E., Goldberg, N.S., Bittker, S., Campbell, G.L., and *et al.* (1992). Diagnosis of early Lyme disease by polymerase chain reaction amplification and culture of skin biopsies from erythema migrans lesions. J. Clin. Microbiol. *30*, 3082–3088.

Seinost, G., Dykhuizen, D.E., Dattwyler, R.J., Golde, W.T., Dunn, J.J., Wang, I.N., Wormser, G.P., Schriefer, M.E., and Luft, B.J. (1999). Four clones of *Borrelia burgdorferi* sensu stricto cause invasive infection in humans. Infect. Immun. *67*, 3518–3524.

Skarpaas, T., Ljostad, U., Sobye, M., and Mygland, A. (2007). Sensitivity and specificity of a commercial C6 peptide enzyme immuno assay in diagnosis of acute Lyme neuroborreliosis. Eur. J. Clin. Microbiol. Infect. Dis. *26*, 675–677.

Smith, R.P., Schoen, R.T., Rahn, D.W., Sikand, V.K., Nowakowski, J., Parenti, D.L., Holman, M.S., Persing, D.H., and Steere, A.C. (2002). Clinical characteristics and treatment outcome of early Lyme disease in patients with microbiologically confirmed erythema migrans. Ann. Intern. Med. *136*, 421–428.

Sparagano, O.A., Allsopp, M.T., Mank, R.A., Rijpkema, S.G., Figueroa, J.V., and Jongejan, F. (1999). Molecular detection of pathogen DNA in ticks (Acari: Ixodidae): a review. Exp. Appl Acarol. *23*, 929–960.

Speck, S., Failing, K., Reiner, B., and Wittenbrink, M.M. (2002). Evaluation of different media and a BGM cell culture assay for isolation of *Borrelia burgdorferi* sensu lato from ticks and dogs. Vet. Microbiol. *89*, 291–302.

Stafford, K.C., Cartter, M.L., Magnarelli, L.A., Ertel, S.H., and Mshar, P.A. (1998). Temporal correlations between tick abundance and prevalence of ticks infected with *Borrelia burgdorferi* and increasing incidence of Lyme disease. J. Clin. Microbiol. *36*, 1240–1244.

Stanek, G., Klein, J., Bittner, R., and Glogar, D. (1990). Isolation of *Borrelia burgdorferi* from the myocardium of a patient with longstanding cardiomyopathy. N. Engl. J. Med. *322*, 249–252.

Stanek, G., and Strle, F. (2003). Lyme borreliosis. Lancet *362*, 1639–1647.

Steere, A.C., Bartenhagen, N.H., and Craft, J.E. (1983a). The early clinical manifestations of Lyme disease. Ann. Intern. Med. *99*, 76–82.

Steere, A.C., Grodzicki, R.L., Craft, J.E., Shrestha, M., Kornblatt, A.N., and Malawista, S.E. (1984). Recovery of Lyme disease spirochetes from patients. Yale J. Biol. Med. *57*, 557–560.

Steere, A.C., Grodzicki, R.L., Kornblatt, A.N., Craft, J.E., Barbour, A.G., Burgdorfer, W., Schmid, G.P., Johnson, E., and Malawista, S.E. (1983b). The spirochetal etiology of Lyme disease. N. Engl. J. Med. *308*, 733–740.

Steere, A.C., Sikand, V.K., Meurice, F., Parenti, D.L., Fikrig, E., Schoen, R.T., Nowakowski, J., Schmid, C.H.,

Laukamp, S., Buscarino, C., and Krause, D.S. (1998). Vaccination against Lyme disease with recombinant *Borrelia burgdorferi* outer-surface lipoprotein A with adjuvant. N. Engl. J. Med. *339*, 209–215.

Stoenner, H.G. (1974). Biology of *Borrelia hermsii* in Kelly medium. Appl. Microbiol. *28*, 540–543.

Stoenner, H.G., Dodd, T., and Larsen, C. (1982). Antigenic variation of *Borrelia hermsii*. J. Exp. Med. *156*, 1297–1311.

Straubinger, R.K. (2000). PCR-based quantification of *Borrelia burgdorferi* organisms in canine tissues over a 500-day postinfection period. J. Clin. Microbiol. *38*, 2191–2199.

Strle, F., Nadelman, R.B., Cimperman, J., Nowakowski, J., Picken, R.N., Schwartz, I., Maraspin, V., Aguero-Rosenfeld, M.E., Varde, S., Lotric-Furlan, S., and Wormser, G.P. (1999). Comparison of culture-confirmed erythema migrans caused by *Borrelia burgdorferi* sensu stricto in New York State and by *Borrelia afzelii* in Slovenia. Ann. Intern. Med. *130*, 32–36.

Strle, F., Nelson, J.A., Ruzic-Sabljic, E., Cimperman, J., Maraspin, V., Lotric-Furlan, S., Cheng, Y., Picken, M.M., Trenholme, G.M., and Picken, R.N. (1996). European Lyme borreliosis: 231 culture-confirmed cases involving patients with erythema migrans. Clin. Infect. Dis. *23*, 61–65.

Tylewska-Wierzbanowska, S., and Chmielewski, T. (1997). The isolation of *Borrelia burgdorferi* spirochetes from clinical material in cell line cultures. Zentralbl. Bakteriol. *286*, 363–370.

van Dam, A.P., Kuiper, H., Vos, K., Widjojokusumo, A., de Jongh, B.M., Spanjaard, L., Ramselaar, A.C., Kramer, M.D., and Dankert, J. (1993). Different genospecies of *Borrelia burgdorferi* are associated with distinct clinical manifestations of Lyme borreliosis. Clin. Infect. Dis. *17*, 708–717.

Varela, A.S., Luttrell, M.P., Howerth, E.W., Moore, V.A., Davidson, W.R., Stallknecht, D.E., and Little, S.E. (2004). First culture isolation of *Borrelia lonestari*, putative agent of southern tick-associated rash illness. J. Clin. Microbiol. *42*, 1163–1169.

Wallach, F.R., Forni, A.L., Hariprashad, J., Stoeckle, M.Y., Steinberg, C.R., Fisher, L., Malawista, S.E., and Murray, H.W. (1993). Circulating *Borrelia burgdorferi* in patients with acute Lyme disease: results of blood cultures and serum DNA analysis. J. Infect. Dis. *168*, 1541–1543.

Wang, G. (2002). Direct detection methods for the Lyme borrelia, including the use of quantitative assays. Vector. Borne. Zoonotic. Dis. *2*, 227–235.

Wang, G., Iyer, R., Bittker, S., Cooper, D., Small, J., Wormser, G.P., and Schwartz, I. (2004). Variations in Barbour-Stoenner-Kelly culture medium modulate infectivity and pathogenicity of *Borrelia burgdorferi* clinical isolates. Infect. Immun. *72*, 6702–6706.

Wang, G., Liveris, D., Brei, B., Wu, H., Falco, R.C., Fish, D., and Schwartz, I. (2003). Real-time PCR for simultaneous detection and quantification of *Borrelia burgdorferi* in field-collected *Ixodes scapularis* ticks from the north-eastern United States. Appl. Environ. Microbiol. *69*, 4561–4565.

Wang, G., Ojaimi, C., Wu, H., Saksenberg, V., Iyer, R., Liveris, D., McClain, S.A., Wormser, G.P., and Schwartz, I. (2002). Disease severity in a murine model of Lyme borreliosis is associated with the genotype of the infecting *Borrelia burgdorferi* sensu stricto strain. J. Infect. Dis. *186*, 782–791.

Wang, G., van Dam, A.P., Schwartz, I., and Dankert, J. (1999). Molecular typing of *Borrelia burgdorferi* sensu lato: taxonomic, epidemiological, and clinical implications. Clin. Microbiol. Rev. *12*, 633–653.

Wang, G., Wormser, G.P., and Schwartz, I. (2001). *Borrelia burgdorferi*. In Sussman M. (Ed.), Molecular Medical Microbiology (pp. 2059–2092). London: Academic Press.

Wilske, B., Jauris-Heipke, S., Lobentanzer, R., Pradel, I., Preac-Mursic, V., Rössler, D., Soutschek, E., and Johnson, R.C. (1995). Phenotypic analysis of outer surface protein C (OspC) of *Borrelia burgdorferi* sensu lato by monoclonal antibodies: Relationship to genospecies and OspA serotype. J. Clin. Microbiol. *33*, 103–109.

Wilske, B., Johnson, B.J., and Schriefer, M.E. (2007). Borrelia. In P.R. Murray, E.J. Baron, J.H. Jorgensen, M.A. Pfaller, and R.H. Yolken (Eds.), Manual of Clinical Microbiology (pp. 971–986). Washington D.C.: ASM Press.

Wilske, B., Preac-Mursic, V., Schierz, G., and Busch, K.V. (1986a). Immunochemical and immunological analysis of European *Borrelia burgdorferi* strains. Zentralbl. Bakteriol. A. *263*, 92–102.

Wilske, B., Schierz, G., Preac-Mursic, V., Von Busch, K., Kuhbeck, R., Pfister, H.W., and Einhaupl, K. (1986b). Intrathecal production of specific antibodies against *Borrelia burgdorferi* in patients with lymphocytic meningoradiculitis (Bannwarth's syndrome). J. Infect. Dis. *153*, 304–314.

Wooten, R.M., Ma, Y., Yoder, R.A., Brown, J.P., Weis, J.H., Zachary, J.F., Kirschning, C.J., and Weis, J.J. (2002). Toll-like receptor 2 is required for innate, but not acquired, host defense to *Borrelia burgdorferi*. J. Immunol. *168*, 348–355.

Wormser, G.P., Bittker, S., Cooper, D., Nowakowski, J., Nadelman, R.B., and Pavia, C. (2000a). Comparison of the yields of blood cultures using serum or plasma from patients with early Lyme disease. J. Clin. Microbiol. *38*, 1648–1650.

Wormser, G.P., Bittker, S., Cooper, D., Nowakowski, J., Nadelman, R.B., and Pavia, C. (2001). Yield of large-volume blood cultures in patients with early Lyme disease. J. Infect. Dis. *184*, 1070–1072.

Wormser, G.P., Dattwyler, R.J., Shapiro, E.D., Halperin, J.J., Steere, A.C., Klempner, M.S., Krause, P.J., Bakken, J.S., Strle, F., Stanek, G., Bockenstedt, L., Fish, D., Dumler, J.S., and Nadelman, R.B. (2006). The clinical assessment, treatment, and prevention of Lyme disease, human granulocytic anaplasmosis, and babesiosis: clinical practice guidelines by the Infectious Diseases Society of America. Clin. Infect. Dis. *43*, 1089–1134.

Wormser, G.P., Liveris, D., Nowakowski, J., Nadelman, R.B., Cavaliere, L.F., McKenna, D., Holmgren, D., and Schwartz, I. (1999). Association of specific subtypes of *Borrelia burgdorferi* with hematogenous dissemination in early Lyme disease. J Infect. Dis *180*, 720–725.

Wormser, G.P., Nadelman, R.B., Dattwyler, R.J., Dennis, D.T., Shapiro, E.D., Steere, A.C., Rush, T.J., Rahn, D.W., Coyle, P.K., Persing, D.H., Fish, D., and Luft, B.J. (2000b). Practice guidelines for the treatment of Lyme disease. The Infectious Diseases Society of America. Clin. Infect. Dis. *31 Suppl 1*, 1–14.

Wormser, G.P., Nowakowski, J., Nadelman, R.B., Bittker, S., Cooper, D., and Pavia, C. (1998). Improving the yield of blood cultures for patients with early Lyme disease. J. Clin. Microbiol. *36*, 296–298.

Wormser, G.P., Ramanathan, R., Nowakowski, J., McKenna, D., Holmgren, D., Visintainer, P., Dornbush, R., Singh, B., and Nadelman, R.B. (2003). Duration of antibiotic therapy for early Lyme disease. A randomized, double-blind, placebo-controlled trial. Ann. Intern. Med. *138*, 697–704.

Yang, L., Weis, J.H., Eichwald, E., Kolbert, C.P., Persing, D.H., and Weis, J.J. (1994). Heritable susceptibility to severe *Borrelia burgdorferi*-induced arthritis is dominant and is associated with persistence of large numbers of spirochetes in tissues. Infect. Immun. *62*, 492–500.

Yang, X., Popova, T.G., Goldberg, M.S., and Norgard, M.V. (2001). Influence of cultivation media on genetic regulatory patterns in *Borrelia burgdorferi*. Infect. Immun. *69*, 4159–4163.

Zeidner, N.S., Schneider, B.S., Dolan, M.C., and Piesman, J. (2001). An analysis of spirochete load, strain, and pathology in a model of tick-transmitted Lyme borreliosis. Vector Borne Zoonotic Dis. *1*, 35–44.

Zhang, J.R., Hardham, J.M., Barbour, A.G., and Norris, S.J. (1997). Antigenic variation in Lyme disease borreliae by promiscuous recombination of VMP-like sequence cassettes. Cell *89*, 275–285.

Zore, A., Ruzic-Sabljic, E., Maraspin, V., Cimperman, J., Lotric-Furlan, S., Pikelj, A., Jurca, T., Logar, M., and Strle, F. (2002). Sensitivity of culture and polymerase chain reaction for the etiological diagnosis of erythema migrans. Wiener Klinische Wochenschrift *114*, 606–609.

Lyme Disease Vaccines

17

Richard T. Marconi and Christopher G. Earnhart

Abstract

The incidence of Lyme disease is increasing and its endemic regions are expanding. Preventative strategies that have focused primarily on increased public awareness and tick avoidance have proven ineffective. Vaccination remains a viable approach to be pursued in the prevention of Lyme disease. This chapter summarizes past, present and future efforts directed at the development of Lyme disease vaccines.

17.1 Introduction

The quest for a Lyme disease vaccine began shortly after the identification of *Borrelia burgdorferi* as the primary causative agent of this debilitating infection. Due to its universal and high level expression among isolates during *in vitro* cultivation, outer surface protein A (OspA) became the natural focus of early vaccine development efforts. Sixteen years after the demonstration of the spirochaetal aetiology of Lyme disease (Benach *et al.*, 1983; Burgdorfer *et al.*, 1982), and the initial description of OspA (Barbour *et al.*, 1983), an OspA-based vaccine (LYMErix; SmithKline Beecham) was licensed for use in adults. However, the tenure of this vaccine proved relatively short, as it was voluntarily removed from the market by its manufacturer in 2002. Today, considerable research effort continues towards the development of a broadly protective Lyme disease vaccine. Because of the significant health and economic impacts of Lyme disease, reducing the number of annual cases through improved prevention strategies is considered a priority, as detailed in the Healthy People 2010 program of the US Department of Health and Human Services (Poland, 2001). Describing all previous work in the area of Lyme disease vaccine development is beyond the scope of this chapter. Rather, the goals of this chapter are to summarize the history of and need for Lyme vaccines, to describe some special considerations unique to Lyme vaccine development, and to review recent advances in the field.

17.2 Justification for Lyme disease vaccine development

There is a clear and compelling justification for the development of new and improved Lyme disease vaccines. Over the period of 2002 to 2007, an average of 22,600 cases of Lyme disease were reported per year (Bacon *et al.*, 2008). However, this is most certainly a significant underestimate since physician compliance with reporting has been demonstrated to range from 7% to 36% (Campbell *et al.*, 1998; Coyle *et al.*, 1996; Matteson *et al.*, 1992; Meek *et al.*, 1996; Naleway *et al.*, 2002; Orloski *et al.*, 2000; Young, 1998). Demographic profiles of persons with Lyme disease indicate largely a suburban and rural risk. In the USA, the highest per capita incidence of Lyme disease occurs in Delaware, Connecticut, New Jersey, Massachusetts, Pennsylvania, and New York (McNabb *et al.*, 2007). There is also evidence of a significant expansion of the endemic regions of Lyme disease. Over 4500 cases of Lyme disease were reported for the South Atlantic region of the USA in 2007. This is up sharply from previous years. The total number of reported cases in the USA in 2007 (27,699) represents a ~40% increase over 2006.

Lyme disease is also widely endemic in certain regions of Europe. While accurate numbers describing the incidence of infection in Europe are difficult to obtain, at least 85,000 cases are documented each year (Anonymous, 1995). However, as in the USA, this figure is probably a substantial underestimate. Some studies suggest the incidence may be as high as 3% in endemic regions (Lindgren and Jaenson, 2006; Smith and Takkinen, 2006). Hence, as discussed below, vaccination for Lyme disease is clearly justified in certain regions in at-risk populations.

Preventative strategies for Lyme disease have traditionally focused on tick avoidance and early diagnosis and treatment. While 'avoidance' is a standard and important component in the control of vector-borne diseases, in the case of Lyme disease, it has not proven particularly effective at the population level. Estimates of the effectiveness of self-monitoring for exposure, the use of protective clothing, and/or the use of tick repellents varies (Vazquez *et al.*, 2008). Protective clothing measures have been reported to be ~40% effective under the best of circumstances, but often these barrier approaches are impractical. The routine use of tick repellents on skin or clothing has proven to be only ~20% effective (Vazquez *et al.*, 2008). Based on these findings, more aggressive approaches are clearly required to prevent Lyme disease.

The costs associated with the diagnosis of Lyme disease and its management are high. Lengthy antibiotic treatment regimens are commonly employed, although not always in accord with commonly accepted treatment guidelines. In addition, the persistent nature of this infection results in significant loss of productivity and diminished quality of life. Diagnostic and treatment costs are key factors in assessing the cost-effectiveness of potential vaccines. Recently, the direct medical costs of early Lyme disease treatment were calculated to be $464 per patient; for late stage disease, this average cost per patient nearly triples to approximately $1380 (Zhang *et al.*, 2006). When direct medical costs and indirect costs (non-medical costs and productivity losses) were considered, the average Lyme disease case cost approximately $8200 (Zhang *et al.*, 2006). Based on these calculations, the total annual cost for Lyme disease in the USA alone conservatively approaches $230 million. In another study, the total costs were estimated to be approximately $500 million per year (in 1996 dollars, based on an estimated national incidence of 4.73 per 100,000 persons) (Maes *et al.*, 1998). As alluded to above, these costs are significant underestimates, as a large majority of cases are not reported, and there is a long list of hidden costs that are not included in these calculations (Campbell *et al.*, 1998; Coyle *et al.*, 1996; Matteson *et al.*, 1992; Meek *et al.*, 1996; Naleway *et al.*, 2002; Orloski *et al.*, 2000; Young, 1998). Hidden costs are, by definition, more difficult to quantify. As a case in point, the fears and concerns over contracting Lyme disease are commonly known to lead to alterations in recreational and behavioural patterns. Such changes in social behaviour can lead to altered spending patterns that can have significant economic impacts.

Pharmacoeconomic studies indicate that the benefit of vaccination against Lyme disease will be greatest when based on individual risk assessments. Calculations have indicated that vaccine cost-effectiveness occurs when used for persons with an annual risk of contracting Lyme disease that exceeds 1% (Meltzer *et al.*, 1999; Shadick *et al.*, 2001). Previous calculations of the disease incidence at which vaccination becomes cost effective were based on assumptions that may not be valid for future Lyme disease vaccines. As an example, calculation models based on the use of an OspA-based vaccine assume a requirement for three successive vaccinations with yearly boosters. Such calculations may be significantly different for a vaccine based on a different antigen or vaccine formulation that would not require multiple initial boosts and yearly re-administration. In addition, an effective vaccination program would have other, less tangible benefits that are not included in typical cost/benefit analyses, including resurgence in outdoor recreation activities in endemic areas and reduced use of prophylactic antibiotic treatment following an uncomplicated tick bite.

17.3 Unique aspects of Lyme disease vaccine development

17.3.1 Diversity of the spirochaetes that cause Lyme disease

Borrelia burgdorferi was the first *Borrelia* species to be demonstrated as a causative agent of Lyme disease (Benach *et al.*, 1983; Burgdorfer *et al.*, 1982). Subsequent molecular phylogenetic analyses led to the delineation of multiple distinct species that collectively form the *B. burgdorferi* sensu lato complex. This complex consists of the species *B. burgdorferi, B. garinii* (Baranton *et al.*, 1992), *B. afzelii* (Busch *et al.*, 1996), *B. valaisiana* (Wang *et al.*, 1999b), *B. lusitaniae* (Le Fleche *et al.*, 1997), *B. andersonii* (Marconi *et al.*, 1995), *B. bissettii* (Postic *et al.*, 1998), *B. japonica* (Kawabata *et al.*, 1993; Marconi *et al.*, 1995), *B. tanukii* (Fukunaga *et al.*, 1996) and *B. turdae* (Fukunaga *et al.*, 1996), *B. sinica* (Masuzawa *et al.*, 2001). *B. lonestari*, a species which may be more closely related to the relapsing fever borrelia is also thought to be a potential causative agent of Lyme disease (Barbour *et al.*, 1996). A significant challenge in the development of future Lyme disease vaccines will be to provide broad coverage against this diverse group of bacteria.

In addition to the phylogenetic diversity described above, there is also significant intra-species variation among Lyme disease spirochaete isolates. This results in an additional level of complexity that must be considered in vaccine development. The segmented genome of *Borrelia* consists of a collection of 20 to 25 linear and circular genetic elements (Barbour and Garon, 1987; Barbour and Garon, 1988). The B31-MI genome has 1516 open reading frames encoded on a linear chromosome (910 kbp), 12 linear plasmids (lp; 5 to 54 kbp), and nine circular plasmids (cp; 9 to 32 kbp), some of which appear to be prophage (Casjens *et al.*, 2000; Eggers *et al.*, 2000; Eggers and Samuels, 1999; Fraser *et al.*, 1997; Zhang and Marconi, 2005) (see Chapter 2). Plasmid composition can change due to genetic rearrangement and/or loss or acquisition of plasmid/prophage (Eggers *et al.*, 2001). Polymorphisms in overall plasmid profiles have been demonstrated in isolates recovered from both humans and animals. These polymorphisms develop in both the Lyme disease and relapsing fever *Borrelia* during *in vitro* cultivation and *in vivo* passage through mice (Iyer *et al.*, 2003; McDowell *et al.*, 2001a; Wang *et al.*, 2001). Hence, plasmid composition can be markedly different among isolates (Barbour, 1988; Iyer *et al.*, 2003; McDowell *et al.*, 2001a; Persing *et al.*, 1994; Schwan *et al.*, 1988; Terekhova *et al.*, 2006). Recently it has been demonstrated that the majority of antigens that elicit antibody responses during infection are in fact plasmid-encoded lipoproteins (Barbour *et al.*, 2008). Hence, variation in the genomic plasmid composition can significantly influence the host–pathogen interaction. An ideal vaccine candidate would therefore be one encoded by a stable component of the genome. The core Lyme disease spirochaete genome has been defined and consists of the linear chromosome and the cp26 and lp54 plasmids (Terekhova *et al.*, 2006). This portion of the *Borrelia* genome is generally highly stable. While some of the non-core plasmids encode proteins that appear to be necessary for infection and/or maximal virulence (e.g. lp28-1, lp25) (Byram *et al.*, 2004; Embers *et al.*, 2008; Grimm *et al.*, 2004; Labandeira-Rey *et al.*, 2003; Labandeira-Rey and Skare, 2001; Marconi *et al.*, 1993; Purser and Norris, 2000; Tilly *et al.*, 1997), other plasmids encode proteins that appear to be dispensable and are genetically unstable. The inherent variability of the *Borrelia* genome composition has significant implications for Lyme disease vaccine development, and must be considered in selection of potential vaccine antigens.

17.3.2 Immunodominant outer surface proteins and their expression and regulation during the infectious cycle

Differential and environmentally mediated changes in expression of immunodominant antigens of the Lyme disease spirochaetes are well documented (Caimano *et al.*, 2007; Liang *et al.*, 2002a; Narasimhan *et al.*, 2003; Ojaimi *et al.*, 2005; Revel *et al.*, 2002; Schwan *et al.*, 1995). The differential expression of antigens is an important consideration in vaccine development. As *B. burgdorferi* progresses through its enzootic cycle, it must respond rapidly to the radically different environmental conditions that it encounters, and it does so in part by significantly altering its transcriptome. The study of gene expression profiles of the Lyme disease spirochaetes directly within the mammalian environment has been difficult to assess due in part to the low density of spirochaetes in the host. Important tools have been developed to overcome these technical challenges. The rat dialysis membrane chamber model has facilitated the analysis of *in vivo* gene expression (Akins *et al.*, 1998). In this model, a culture of *Borrelia* is placed in a dialysis membrane that is then implanted into peritoneal cavity of host mammals (Akins *et al.*, 1998). During the implant stage, the spirochaetes are exposed to conditions that approximate the natural host environment. After explanting the chambers, the 'host adapted' spirochaetes can be harvested and the transcriptome and proteome assessed. While this model has its limitations, the expression patterns in host-adapted bacteria appear to be similar to those seen in infected mice. For example, OspA, a leading vaccine candidate, is expressed by spirochaetes residing in the tick midgut and is rapidly downregulated during transmission from the tick to the mammalian host (Schwan *et al.*, 1995). Similarly, OspA is downregulated in the host-adapted spirochaete model (Brooks *et al.*, 2006). The expression of OspC in host-adapted bacteria also mimics that seen in infected mammals. OspC expression is stimulated by the initiation of the blood meal and is maintained through the early stages of mammalian infection (Schwan *et al.*, 1995). Consistent with this, OspC expression is significantly upregulated in host-adapted spirochaetes (Brooks *et al.*, 2006). The DMC model represents an important tool that can be used to assess the production patterns of potential vaccine candidates. The identification of surface-exposed proteins that are expressed in the tick is also an important aspect of vaccine development. Antibodies to proteins expressed in the tick environment have the potential to block or significantly decrease the efficiency of transmission of spirochaetes from the tick to mammals. OspA-based vaccines serve as an example. Vaccination with OspA has proven to be highly effective in blocking transmission (de Silva *et al.*, 1996). However, a caveat specific to transmission blocking vaccines is that because transmission occurs within 48 hours, high circulating antibody titres are required. In light of the narrow window of time for the successful blockage of transmission, anamnestic responses occur too slowly to provide protection.

17.4 Selection of potential candidate vaccine targets

17.4.1 Immune response and immunodominant antigens during natural infection

The Lyme disease spirochaete elicits a slowly developing but strong humoral immune response during infection in mammals. The spirochaetes produce a remarkable repertoire of surface antigens and several have been assessed for their utility in the development of a Lyme disease vaccine. In spite of the fact that a number of antigens that are expressed *in vivo* have been demonstrated to elicit bactericidal antibody in mice upon vaccination, natural infection does not culminate in immune clearance or lead to the development of protective immunity. In untreated naturally or experimentally infected mammals, the Lyme spirochaetes can persist long-term, and perhaps indefinitely. This ability is essential for completion of the enzootic cycle and for population maintenance in nature (see Chapter 10). The spirochaetes must survive long enough within their mammalian hosts to allow transmission into feeding ticks. Several divergent and potentially synergistic mechanisms that contribute to immune evasion and persistence have been described. As discussed above and elsewhere in this book, differential expression of the expansive repertoire of surface protein encoding genes over the course of infection may allow the spirochaetes to alter their antigenic profile (Caimano *et al.*, 2007; Ojaimi *et al.*, 2005; Revel *et al.*, 2002) (see Chapter 4). This alteration could prove to be essential for establishing infection in immunologically primed hosts. Antigenic variation also appears to be important in evasion of antibody responses and may serve to ensure persistence within an individual host (Liang and Philipp, 1999; McDowell *et al.*, 2002; Norris, 2006; Zhang and Norris, 1998). Successful long-term colonization of mammals also requires mechanisms that allow evasion of innate immune mechanisms. Serum-resistant strains of *Borrelia* can evade complement through the binding of negative regulators of the complement cascade such as factor H, factor H-like protein 1, and possibly C4b binding protein (Alitalo *et al.*, 2002; Brooks *et al.*, 2005; Hellwage *et al.*, 2001; Kraiczy *et al.*, 2001, 2004a,b; McDowell *et al.*, 2003; Rogers and Marconi, 2007). A strong correlation has been noted between the binding of negative regulators of the complement system and increased serum resistance. *Borrelia* proteins that bind complement regulators have been proposed to be excellent candidates for vaccine development in that they may elicit separate but synergistic correlates of protection (Metts *et al.*, 2003). Antibodies elicited by these proteins may be bactericidal and may also sterically inhibit the binding of negative regulators of complement, making the spirochaetes more susceptible to complement-mediated killing. While the potential protective effects of vaccination with virulence factors in this class has been demonstrated for the *Neisseria* (Beernink *et al.*, 2006; Madico *et al.*, 2007), it has not yet been realized for the Lyme disease spirochaetes.

17.4.2 Proteins that have been explored in Lyme disease vaccine development

The ideal Lyme disease vaccine candidate will be either constitutively expressed or upregulated by tick feeding, and universally expressed by the spirochaete population during infection in mammals. Antibody to vaccine candidates with these properties has the potential to both block transmission of spirochaetes from the tick vector to the mammal and/or lead to antibody-mediated killing of spirochaetes that have transited into vaccinated mammals. In addition, as with any vaccine candidate, the antigen should display conservation within the naturally exposed epitopes that are presented by intact spirochaetes. The corresponding antigen gene should be genetically stable and encoded by an essential plasmid or the chromosome. Of those genes upregulated in the mammal, only a small subset will exhibit all of these properties and thus ultimately be of utility in vaccine development. Some of the proteins that have been assessed as potential vaccine candidates to date include OspC, DbpA, OspE (p21), Elp (paralogous gene family 163), OspF (*bbk2.10*, pG), Mlp proteins, p35 (BbK32), and p37 (BbK50) (Akin *et al.*, 1995, 1998, 1999;; Brooks *et al.*, 2003; Fikrig *et al.*, 1997; Hefty *et al.*, 2001, 2002; Liang *et al.*, 2002a,b; Marconi *et al.*, 1996; McDowell *et al.*, 2001b; Ojaimi *et al.*, 2003; Schwan *et al.*, 1995; Stevenson *et al.*, 1995; Yang *et al.*, 2000, 2003a,b). OspA, OspB, OspC, OspF, DbpA, BBK32, and VraA have been shown to provide some protection against challenge in mice (de Silva *et al.*, 1996; Feng *et al.*, 1998; Fikrig *et al.*, 1997; Hanson *et al.*, 1998; Labandeira-Rey *et al.*, 2001; Liang *et al.*, 2001; Nguyen *et al.*, 1994; Probert and LeFebvre, 1994; Telford *et al.*, 1993; Wallich *et al.*, 1994). Immunogenic, *in vivo*-expressed antigens that did not result in protective immune responses include OspD, OspE, BmpA, and VlsE (Gilmore *et al.*, 1996; Liang *et al.*, 2001; Nguyen *et al.*, 1994; Probert and LeFebvre, 1994). In spite of the large number of potential vaccine candidates that have been investigated, OspA and OspC remain the primary foci of subunit vaccine development. The discussion below will focus on past and present efforts to develop these proteins as vaccines.

17.5 History of the first-generation Lyme disease vaccines

The initial focus of Lyme disease vaccine development centred on outer surface protein A (OspA). As eluded to above, vaccination with OspA induces an antibody response that blocks transmission of spirochaetes from the tick to the mammalian host by killing or immobilizing the spirochaetes within the tick midgut (de Silva *et al.*, 1996; de Silva *et al.*, 1999). OspA is not expressed in the mammal, hence OspA-based vaccines exert their protective effect through a unique mechanism of action. Preclinical trials of OspA vaccination demonstrated protection in animal models of Lyme disease, either as a recombinant protein or lipoprotein, or expressed on the surface of *E. coli*, *Salmonella typhimurium* or *Mycobacterium bovis* (BCG) (Aydintug *et al.*, 1994; Chang *et al.*, 1995; Dunne *et al.*, 1995; Fikrig *et al.*, 1990, 1991; Gomes-Solecki *et al.*, 2006; Langermann *et al.*, 1994; Lovrich *et al.*, 1995; Luke *et al.*, 1997b; Philipp *et al.*, 1997; Probert and LeFebvre, 1994; Schaible *et al.*, 1990; Simon *et al.*, 1991; Stover *et al.*, 1993). Two OspA-based vaccine candidates, ImuLyme (Pasteur Merieux-Connaught) and LYMErix (SmithKlineBeecham), were subsequently developed and tested. Both vaccines consisted of recombinant, lipidated OspA; in the LYMErix formulation, the OspA lipoprotein was adsorbed to alum adjuvant (Erdile *et al.*, 1993; Sigal *et al.*, 1998; Steere *et al.*, 1998). The presence of the lipid motif enhances vaccine immunogenicity by interaction with TLR2 (Yoder *et al.*, 2003). The OspA protein has been demonstrated to display moderate diversity. The sequence chosen for vaccine development represented the most prevalent OspA phyletic type in North American Lyme disease isolates (Wilske *et al.*, 1996). Initial clinical testing demonstrated safety and efficacy in human subjects (Keller *et al.*, 1994; Schoen *et al.*, 1995; Sigal *et al.*, 1998; Steere *et al.*, 1998; Van Hoecke *et al.*, 1996; Wallich *et al.*, 1994). Phase III studies had an enrolment of 11,000 for each of the vaccine trials. Two doses (30 μg) of the vaccine were administered one month apart with a booster vaccination at 12 months. Vaccine efficacy following the first two doses was 68 and 49% for ImuLyme and LYMErix, respectively. Following the 12-month booster, vaccine efficacy rose to 92% for ImuLyme and 76% for LYMErix. Both OspA-based vaccines appeared to be safe and well tolerated as side-effects only consisted of those typically associated with vaccination (e.g. minor soreness, redness, swelling at the injection site, low grade fever, chills). FDA approval was given for the LYMErix vaccine at the end of 1998. An application for ImuLyme approval was submitted but withdrawn, and hence this formulation has not been marketed (Nigrovic and Thompson, 2007). The LYMErix vaccine was subsequently recommended by the CDC Advisory Committee on Immunization Practices for use in populations at high risk for contracting Lyme disease (Anonymous, 1999). It was not approved for use in children.

Shortly after the introduction of LYMErix, concerns regarding a possible connection between administration of LYMErix and the development of arthritis began to emerge. A class-action lawsuit alleging vaccine-related harm was subsequently filed in 1999. There is some evidence that the immune response to OspA may be associated with development of antibiotic-refractory arthritis and that there may be an association between arthritis development and particular MHC II alleles (HLA-DR4) (Chen *et al.*, 1999; Drouin *et al.*, 2008; Gross *et al.*, 1998; Gross and Huber, 2000; Kalish *et al.*, 1993; Kamradt *et al.*, 1996; Meyer *et al.*, 2000; Steere and Glickstein, 2004; Steere *et al.*, 2001; Trollmo *et al.*, 2001) (see Chapter 15). A predicted OspA T-cell epitope that is similar to a sequence found in lymphocyte function antigen 1 (LFA-1) was subsequently identified. This prompted speculation that this epitope could be the basis of an autoimmune arthritis (Gross *et al.*, 1998; Trollmo *et al.*, 2001). However, a definitive correlation between vaccination with OspA and autoimmune arthritis has not been demonstrated (Chen *et al.*, 1999; Drouin *et al.*, 2008; Gross *et al.*, 1998; Hemmer *et al.*, 1999; Meyer *et al.*, 2000; Nocton *et al.*, 1994; Rose *et al.*, 2001; Steere *et al.*, 1990; Steere *et al.*, 2001; Trollmo *et al.*, 2001).

By July, 2000 nearly 1.4 million doses of the vaccine had been administered. There were 905 reported adverse events, of which 102 were coded as arthritis, arthrosis, or rheumatoid arthritis and 12 as facial paralysis (Lathrop *et al.*, 2002). The reported rate of both arthritis and facial paralysis proved to be below background (Lathrop *et al.*, 2002). The CDC and FDA evaluated the data filed with the Vaccine Adverse Event Reporting System (VAERS) and concluded that there was no apparent temporal association between the development of arthritic problems and vaccine administration. Post-approval phase IV studies of LYMErix safety were originally designed to monitor 25,000 vaccine recipients, and 75,000 age- and sex-matched unvaccinated controls; however, only 10% of the planned enrolment was achieved. The results of the phase IV trial to date were presented to an FDA advisory panel in January, 2001. The data did not support an association between vaccination and development of autoimmunity (Anonymous, 2001a; Anonymous, 2001b), and the panel did not suggest removal of the vaccine from the market or modification of its labelling. However, public perceptions of the potential vaccine-associated adverse events, in combination with the above described limitations of the vaccine, led to declining sales and the voluntary removal of the vaccine from the market in 2002.

17.6 Second-generation Lyme disease vaccine candidates

17.6.1 Outer surface protein A (OspA)

OspA structure, regulation and function

OspA was one of the first proteins of the Lyme disease spirochaetes to be studied in detail. OspA is the first gene in an operon that also contains OspB. The operon is present on a linear plasmid of 54 kb (lp54), which is part of the core *B. burgdorferi* genome (Terekhova *et al.*, 2006). The 28 kDa OspA protein is surface-exposed and lipidated (Barbour and Garon, 1988; Howe *et al.*, 1986). Structurally, it consists of two globular domains composed almost entirely of beta-sheets that are connected to each other by a unique single-layer beta-sheet (Li *et al.*, 1997; Makabe *et al.*, 2006). A putative ligand-binding site has been tentatively identified in the C-terminal globular domain of OspA (Li *et al.*, 1997; Makabe *et al.*, 2006). OspA was being intensively investigated as a vaccine candidate even before its functional roles and expression were defined. Interestingly, *ospA* is expressed in the tick but not during infection in mammals (Fingerle *et al.*, 1995; Leuba-Garcia *et al.*, 1998). OspA may play a role in the adherence of spirochaetes to the tick midgut (Pal *et al.*, 2004a). Down-regulation of OspA expression appears to occur during the passage of the spirochaetes from the midgut to the salivary glands (Schwan *et al.*, 1995).

OspA phylogenetics

Seven distinct serotypes of OspA have been identified and designated as serotypes 1 through 7. Serotype 1 is associated with *B. burgdorferi*, serotype 2 with *B. afzelii*, and serotypes 3–7 with *B. garinii* (Wilske *et al.*, 1993). Consistent with the antigenic diversity of OspA, considerable genetic diversity has been described. In one study, 11 OspA variants were characterized in *B. burgdorferi* (Bunikis *et al.*, 2004). In an attempt to broaden potential protection conveyed by an OspA-based vaccine, multivalent-chimeric OspA proteins have been developed that incorporate the protective epitopes from three or more OspA serotypes (Gern *et al.*, 1997). A trivalent vaccine that included the dominant OspA type from each of the major Lyme disease genospecies provided broad protection in mice challenged by tick infection (Gern *et al.*, 1997). However, this trivalent OspA-based vaccine has not been further assessed or tested in humans.

Immune responses to OspA and second generation OspA vaccines

Using monoclonal antibodies, the protective epitopes of OspA have been mapped to the C-terminal domain of the protein. Some of the epitopes appear to be conformationally defined (e.g. the LA-2 epitope) (Ding *et al.*, 2000; Huang *et al.*, 1998; Sears *et al.*, 1991). In an effort to reduce vaccine size and eliminate peptides potentially associated with autoimmune reactions, recombinant OspA variants lacking the N-terminal domain have been generated. These recombinant OspA variants have been demonstrated to elicit protective antibody responses (Bockenstedt *et al.*, 1996). In a separate study, immune responses to recombinant OspA fragments in which specific charged amino acids in the C-terminal domain of the protein were replaced with hydrophobic amino acids were assessed. The replacements increased the stability of the C-terminal domain and the resulting proteins had antigenic properties similar to wild-type OspA (Koide *et al.*, 2005). In another study, a modified recombinant *B. burgdorferi* OspA in which the potential cross-reactive epitope was altered was generated and tested for protective efficacy (Willett *et al.*, 2004). This modified OspA protein elicited protective antibodies in mice at a level similar to that of the wild-type OspA protein, but unlike the wild type OspA, it did not stimulate HLA-DR4-positive T-cells (Willett *et al.*, 2004). These analyses indicate that OspA can be modified to prevent potential autoimmune reactions. In spite of the promise of second-generation OspA vaccines, such vaccines may not be able to overcome the stigma associated with earlier OspA vaccines. In addition, OspA may not be the best vaccine candidate due to issues surrounding its mechanism of action and the inherent need to administer frequents boosts to ensure high circulating antibody titres.

17.6.2 Outer surface protein C (OspC)

OspC structure, regulation and function

OspC has also been at the centre of Lyme disease vaccine development efforts. OspC is a 22 kDa immunodominant lipoprotein (Fuchs *et al.*, 1992) that is encoded by a highly stable 26 kbp circular plasmid (cp26) that is a universal component of the *B. burgdorferi* genome (Marconi *et al.*, 1993; Sadziene *et al.*, 1993; Terekhova *et al.*, 2006). The *ospC* gene, although variable in sequence among strains, is highly stable during infection (Hodzic *et al.*, 2000; Stevenson and Barthold, 1994). Expression of OspC is upregulated at 37°C relative to that observed at 33°C (Schwan *et al.*, 1995). The protein is anchored to the outer membrane by an N-terminal tripalmitoyl-S-glyceryl-cysteine (Brooks *et al.*, 2006). Structural analyses have demonstrated that OspC is largely helical in structure and consists of five alpha helices connected by intervening loops (Eicken *et al.*, 2001; Kumaran *et al.*, 2001). OspC appears to form homodimers and the extensive buried interface of the dimer supports the hypothesis that the dimer is the biologically active form of the protein (Bunikis and Barbour, 1999; Earnhart *et al.*, 2005; Eicken *et al.*, 2001; Huang *et al.*, 1999; Kumaran *et al.*, 2001; Zückert *et al.*, 2001). The function of OspC has been the focus of considerable attention and controversy. There is disagreement over whether the critical role for OspC occurs within the tick or the mammal. In either case, the majority of studies published to date suggest that OspC is a virulence factor that is required for the establishment of mammalian infection (Byram *et al.*, 2004; Grimm *et al.*, 2004; Margolis and Rosa, 1993; Pal *et al.*, 2004b; Ramamoorthi *et al.*, 2005; Rosa, 2005; Stewart *et al.*, 2006; Tilly *et al.*, 1997, 2006, 2007).

OspC phylogenetics

OspC is one of the most highly diverse proteins of the Lyme disease spirochaetes (Theisen *et al.*, 1993; Wilske *et al.*, 1996). In spite of the robust immune response elicited by OspC during early infection, its potential in vaccine development has been hampered by its inherent diversity. However, in recent years much has been learned about the molecular evolution of *ospC* and this information has been exploited in vaccine development. OspC sequences form at least 38 distinct OspC phyletic clusters or OspC types (Earnhart and Marconi, 2007c; Seinost *et al.*, 1999a; Wang *et al.*, 1999c). Within a phyletic cluster (i.e. at the intra-type level), the amino acid sequences are highly conserved (>95% identity) while between types the percent identity values range from 55% to 80% (Attie *et al.*, 2007; Earnhart and Marconi, 2007c; Wang *et al.*, 1999c). In contrast, the most divergent OspA sequences display percent identity values of 65–70% (Wallich *et al.*, 1992; Will *et al.*, 1995; Wilske *et al.*, 1996). The diversity within OspC has been demonstrated to map primarily within distinct surface-exposed domains (Earnhart *et al.*, 2005; Earnhart *et al.*, 2007). While an individual strain carries only a single *ospC* allele, multiple OspC types have been demonstrated in spirochaetes originating from tightly defined geographic areas (Alghaferi *et al.*, 2005; Anderson and Norris, 2006; Earnhart *et al.*, 2005; Lin *et al.*, 2002; Wang *et al.*, 1999c). The maintenance of multiple *ospC* alleles within a local population has been postulated to contribute to the ability of the Lyme spirochaetes to infect immunologically primed reservoir animals (Seinost *et al.*, 1999a; Wang *et al.*, 1999c). In addition, initial analyses suggested that there may also be a strict correlation between specific *ospC* types and invasive disease in humans (Lagal *et al.*, 2003; Seinost *et al.*, 1999a). However, more recent studies have demonstrated that strains producing other OspC types are also associated with invasive infection (Alghaferi *et al.*, 2005; Dykhuizen *et al.*, 2008; Earnhart *et al.*, 2005; Wormser *et al.*, 2008). Of relevance to vaccine development, OspC variation arises primarily through apparently rare genetic exchange and recombination events, and not by hypermutation with concomitant immune selection (Attie *et al.*, 2007; Bunikis *et al.*, 2004; Dykhuizen and Baranton, 2001; Earnhart and Marconi, 2007c; Hanincova *et al.*, 2008; Livey *et al.*, 1995; Marti Ras *et al.*, 1997; Wang *et al.*, 1999a). With the recent significant advances in our understanding of the sequence diversity of OspC, devising strategies to exploit the variation within OspC in vaccine development is now possible.

Immune response to OspC

OspC is immunodominant during early infection and can elicit protective antibody responses; however, the protective range is narrow (Bockenstedt *et al.*, 1997; Brown *et al.*, 2005; Gilmore *et al.*, 1996; Gilmore and Mbow, 1999; Jobe *et al.*, 2003; Mathiesen *et al.*, 1998; Mbow *et al.*, 1999; Probert *et al.*, 1997; Probert and LeFebvre, 1994; Rousselle *et al.*, 1998; Scheiblhofer *et al.*, 2003; Seinost *et al.*, 1999b; Wallich *et al.*, 2001). Consistent with this, the antibody response elicited during early infection is OspC type-specific. As infection progresses, the type-specificity of the antibody response decreases suggesting that a broader array of OspC epitopes is eliciting antibody responses (Earnhart *et al.*, 2005; Earnhart *et al.*, 2007). This phenomenon is most likely due to the antibody-mediated lysis of cells, the release of OspC, and the recognition of epitopes that are normally shielded in viable, intact spirochaetes. Importantly, antibody elicited to the type-specific domains of OspC have type-specific bactericidal activity (Buckles *et al.*, 2006; Earnhart *et al.*, 2007; Earnhart and Marconi, 2007b). The identification of epitopes that are presented by native OspC during the early stages of infection, and that elicit bactericidal antibody, is an important step towards the design of an effective and broadly protective OspC-based vaccine.

Second-generation vaccines based on OspC

Several approaches have been taken as a means to overcome the obstacle to vaccine development created by the genetic variability of OspC. In one approach, Baxter Vaccines developed an OspC recombinant protein cocktail consisting of 14 full-length OspC proteins of different types. However, these formulations proved ineffective as they did not elicit sufficient antibody titres against the type-specific epitopes of each OspC variant included in the vaccine. Furthermore, unacceptable reactigenicity was reported (Hanson and Edelman, 2004). The inclusion of numerous full-length proteins in the vaccine formulation may have resulted in the misdirection of the immune response by presenting immunodominant but non-protective epitopes. In an attempt to address this problem, Earnhart *et al.* have developed polyvalent, recombinant chimeric proteins consisting entirely of protective linear epitopes (Earnhart *et al.*, 2007; Earnhart and Marconi, 2007a; Earnhart and Marconi, 2007b). To accomplish this, a systematic analysis of the protective epitopes of diverse OspC types was undertaken. By screening a series of truncations and subfragments with sera derived from mice experimentally infected with clonal *B. burgdorferi* isolates, two regions of OspC were identified in multiple OspC types that contain type-specific linear epitopes (Earnhart *et al.*, 2005; Earnhart *et al.*, 2007). These were designated as the 'loop 5' and 'helix 5' epitope containing regions. The loop 5 region is composed of the C-terminal portion of alpha helix 3 and loop 5 (Eicken *et al.*, 2001; Kumaran *et al.*, 2001), while the helix 5 region encompasses the fifth alpha helix and a small region C-terminal to this helix (Earnhart *et al.*, 2005; Earnhart *et al.*, 2007; Yang *et al.*, 2006). Subsequent analyses demonstrated that these epitopes are also recognized by antibody that develops during natural infection in humans (Buckles *et al.*, 2006; Earnhart *et al.*, 2005; Earnhart *et al.*, 2007). Antibody targeting the type A OspC loop 5 region is bactericidal and mediates killing through a complement-dependent mechanism (Buckles *et al.*, 2006). Further studies using a tetravalent construct demonstrated that a chimeric protein containing these type-specific epitope regions can elicit bactericidal antibody effective against spirochaetes expressing each of the component types (Earnhart *et al.*, 2007). Analyses to date suggest that a broadly protective OspC vaccine will require the inclusion of epitopes from approximately 28 OspC types (Earnhart and Marconi, 2007c). Such a construct may provide protection against all major Lyme disease spirochaetes species associated with human Lyme disease, and therefore be effective in both Europe and North America.

17.6.3 Combination subunit vaccines

A limited number of studies have investigated the potential of combination subunit Lyme disease vaccines. Synergy in immune responses was reported using a combination of OspA and DbpA (Hanson *et al.*, 2000). Both OspA and DbpA have been demonstrated to elicit protection against needle challenge in the murine infection model (Feng *et al.*, 1998; Hanson *et al.*, 1998). The OspA-DbpA combination protected against a 100-fold greater challenge dose than that observed with the individual vaccine components. In addition, the combination provided greater protection against heterologous strains. However, the true potential of this combination approach has not been determined, as essential tick challenge experiments have not yet been conducted. A separate study has demonstrated that immunization with DbpA alone does not protect against tick challenge (Hagman *et al.*, 2000). In another analysis, a combination of recombinant OspC, DbpA, and fibronectin binding protein (BBK32) were administered as a cocktail (Brown *et al.*, 2005). While the triple combination vaccine was shown to be more effective than single or paired vaccination using the needle challenge model, the efficacy of this vaccine against tick challenge has not been assessed.

17.6.4 DNA-based vaccines

DNA vaccines have been assessed in the effort to improve preventative strategies for Lyme disease. Plasmid, vaccinia virus or pCMV vector constructs expressing OspA, OspC, OspA-C fusion or OspC with a human tissue plasminogen activator (hTPA) leader sequence have been tested in mice and found to exhibit varying degrees of protection (Luke *et al.*, 1997a; Scheckelhoff *et al.*, 2006; Scheiblhofer *et al.*, 2003; Simon *et al.*, 1996; Wallich *et al.*, 2001; Zhong *et al.*, 1996). In the case of the OspC-hTPA construct, the vaccine was administered by either intradermal needle inoculation or gene gun. Intradermal needle inoculation generally induced a Th1 response, while gene gun delivery induced a Th2 response. Both approaches proved to be equally effective in eliciting protective responses, suggesting that a successful Lyme disease vaccine is not restricted to induction of a certain Th type of immune response (Scheiblhofer *et al.*, 2003). However, no studies employing DNA-based Lyme disease vaccines have been published to date that demonstrate broad protective capability.

17.7 Veterinary and wildlife vaccine strategies

Companion animals are susceptible to infection by the Lyme spirochaetes, and if infected may develop serious arthritic sequelae (reviewed in Skotarczak, 2002). Several whole killed bacteria (bacterin) canine Lyme disease vaccines are currently available. Those marketed in the USA contain killed *B. burgdorferi* (e.g. Merilym, Merial, Germany; LymeVax, Fort Dodge, USA; Galaxy Lyme, Schering Plough, USA) (Chu *et al.*, 1992; Levy, 2002; Wiedemann and Milward, 1999). In Europe, to allow for broader coverage, some bacterin-type vaccines contain a mixture of *B. burgdorferi, B. garinii* and *B. afzelii* cells (Biocan B, Bioveta, Czech Republic) (Nepereny *et al.*, 2007; Rapuntean *et al.*, 2003). Subunit vaccines have also been used in the veterinary setting. ProLyme (Intervet, USA) and Recombitek Lyme (Merial, USA) are both OspA subunit vaccines (Conlon *et al.*, 2000; Levy *et al.*, 1993; Wikle *et al.*, 2006). There is little efficacy data available for veterinary Lyme disease vaccines. Evidence suggests that available vaccine formulations do not provide broad and/or long-term protection (Gern *et al.*, 1997; Johnson *et al.*, 1988; Ma *et al.*, 1995; Shang *et al.*, 2001; Topfer and Straubinger, 2007). Bacterin vaccines might be improved by increasing the number of strains included in the vaccine formulation. Alternatively, as discussed above, subunit-chimeric recombinant protein-based vaccines designed for broad protection may also be applied in the veterinary setting.

An interesting conceptual approach in the management of Lyme disease is to reduce the prevalence of the causative agent in its natural wildlife reservoir mammals. A pilot experiment in which approximately 1000 wild *Peromyscus* mice were immunized intraperitoneally with up to three doses of OspA or a control protein showed a reduction in Lyme disease transmission of 48%, 92%, and 99% for one, two, or three doses of OspA, compared with control mice (Tsao *et al.*, 2004). However, infection prevalence in the test study areas decreased by only 16% (Brisson *et al.*, 2008; Tsao *et al.*, 2004). This is most certainly due to the fact that a diverse group of mammals can served as reservoirs for the Lyme disease spirochaetes (Brisson *et al.*, 2008). To facilitate delivery of wildlife vaccines, efforts are under way to develop orally delivered bait vaccines. Such vaccines could allow for the immunization of a broader subset of mammalian reservoirs. To date, OspA has been employed in most bait vaccine development efforts, and several delivery platforms have been investigated including the use of bacterial and viral platforms. The ability of OspA expressed in *E. coli* (Fikrig *et al.*, 1991; Gomes-Solecki *et al.*, 2006), *Salmonella typhimurium* (Dunne *et al.*, 1995), *Lactobacillus plantarum* (del Rio *et al.*, 2008), and recombinant vaccinia virus to convey protection has also been assessed (Scheckelhoff *et al.*, 2006). Orally delivered purified, lipidated OspA has also been tested (Luke *et al.*, 1997b). In each case, the vaccine induced an immune response against OspA, indicating that oral delivery of this antigen is feasible. What remains unclear is the degree of protection that will ultimately be realized in humans and companion animals. The inherent difficulty in delivering bait vaccines in a manner that will allow for immunization of the large and diverse range of animals associated with Lyme disease remains significant. In addition, the requirement to deliver bait vaccines over a large geographic area on a yearly basis may prove to be cost prohibitive.

There is renewed interest in developing vaccines that specifically target the tick vectors of Lyme disease. This approach offers the advantage of protecting against multiple pathogens with one vaccine. Two general classes of candidate tick vaccine antigens have been identified (Nuttall *et al.*, 2006). 'Exposed' antigens are those that are secreted in tick saliva during attachment and feeding. The second group, referred to as 'concealed' antigens, are those that are not delivered into the host and thus are normally hidden from the host immune system. At the present time, a vaccine derived from the midgut membrane-bound Bm86 protein (a concealed antigen) of the cattle tick, *Boophilus microplus*, is the only commercially available ectoparasite vaccine (de la Fuente *et al.*, 2007; Jonsson *et al.*, 2000; Trimnell *et al.*, 2005). Progress towards the development of additional anti-tick vaccines has been hampered by the difficulty associated with the identification of appropriate targets for vaccine development. There are several exposed and concealed candidate antigens that are currently being studied for prevention of *Ixodes* tick infestation. Exposed antigens under investigation include several tick salivary proteins such as Salp14, Salp15, and Iris proteins. Concealed antigens being studied include the tick midgut ligand that is thought to bind to OspA (TROSPA), the 64P protein (Anguita *et al.*, 2002; de la Fuente and Kocan, 2006; Pal *et al.*, 2004a; Pedra *et al.*, 2006; Prevot *et al.*, 2007; Ramamoorthi *et al.*, 2005; Soares *et al.*, 2005; Xu *et al.*, 2005), 4D8 (subolesin) and 4F8 (5′ nucleotidase). Vaccination with the 4D8 and 4F8 proteins have been demonstrated to reduce tick infestation of sheep (Almazan *et al.*, 2005a; Almazan *et al.*, 2003; Almazan *et al.*, 2005b). Collectively, if these vaccines prove to be effective in significantly reducing the egg-bearing adult tick population, then a reduction in tick populations may be realized.

17.8 Prospects for the future

In summary, there has been considerable progress in the development of potentially broadly protective Lyme disease vaccines. However, the experiences with the OspA-based LYMErix vaccine have clearly impacted attitudes in the commercial setting about the financial feasibility of future Lyme disease vaccines that are intended for humans. The recent focus on alternative vaccine candidates that may require fewer boosts and will conceivably provide long term protection is an exciting development. While the successful development of an effective vaccine seems highly likely, the greatest challenges that may be faced may be those associated with bringing such a vaccine to the market.

References

Akin, E., McHugh, G.L., Flavell, R.A., Fikrig, E., and Steere, A.C. (1999). The immunoglobulin (IgG) antibody response to OspA and OspB correlates with severe and prolonged Lyme arthritis and the IgG response to P35 correlates with mild and brief arthritis. Infect. Immun. *67*, 173–181.

Akins, D.R., Bourell, K.W., Caimano, M.J., Norgard, M.V., and Radolf, J.D. (1998). A new animal model for studying Lyme disease spirochetes in a mammalian host-adapted state. J. Clin. Invest. *101*, 2240–2250.

Akins, D.R., Porcella, S.F., Popova, T.G., Shevchenko, D., Baker, S.I., Li, M., Norgard, M.V., and Radolf, J.D. (1995). Evidence for *in vivo* but not *in vitro* expression of a *Borrelia burgdorferi* outer surface protein F (OspF) homologue. Mol. Microbiol. *18*, 507–520.

Alghaferi, M.Y., Anderson, J.M., Park, J., Auwaerter, P.G., Aucott, J.N., Norris, D.E., and Dumler, J.S. (2005). *Borrelia burgdorferi ospC* heterogeneity among human and murine isolates from a defined region of northern Maryland and southern Pennsylvania: lack of correlation with invasive and noninvasive genotypes. J. Clin. Microbiol. *43*, 1879–1884.

Alitalo, A., Meri, T., Lankinen, H., Seppala, I., Lahdenne, P., Hefty, P.S., Akins, D., and Meri, S. (2002). Complement inhibitor factor H binding to Lyme disease spirochetes is mediated by inducible expression of multiple plasmid-encoded outer surface protein E paralogs. J. Immunol. *169*, 3847–3853.

Almazan, C., Blas-Machado, U., Kocan, K.M., Yoshioka, J.H., Blouin, E.F., Mangold, A.J., and de la Fuente, J. (2005a). Characterization of three *Ixodes scapularis* cDNAs protective against tick infestations. Vaccine 23, 4403–4416.

Almazan, C., Kocan, K.M., Bergman, D.K., Garcia-Garcia, J.C., Blouin, E.F., and de la Fuente, J. (2003). Identification of protective antigens for the control of *Ixodes scapularis* infestations using cDNA expression library immunization. Vaccine *21*, 1492–1501.

Almazan, C., Kocan, K.M., Blouin, E.F., and de la Fuente, J. (2005b). Vaccination with recombinant tick antigens for the control of *Ixodes scapularis* adult infestations. Vaccine *23*, 5294–5298.

Anderson, J.M., and Norris, D.E. (2006). Genetic diversity of *Borrelia burgdorferi* sensu stricto in *Peromyscus leucopus*, the primary reservoir of Lyme disease in a region of endemicity in southern Maryland. Appl. Environ. Microbiol. *72*, 5331–5341.

Anguita, J., Ramamoorthi, N., Hovius, J.W., Das, S., Thomas, V., Persinski, R., Conze, D., Askenase, P.W., Rincon, M., Kantor, F.S., and Fikrig, E. (2002). Salp15, an *Ixodes scapularis* salivary protein, inhibits $CD4^+$ T-cell activation. Immunity *16*, 849–859.

Anonymous (1995). WHO workshop on Lyme borreliosis diagnosis and surveillance (Warsaw, Poland, World Health Organization), pp. 1–21.

Anonymous (1999). Recommendations for the use of Lyme disease vaccine. Recommendations of the Advisory Committee on Immunization Practices (ACIP). MMWR Recomm Rep 48, 1–17, 21–15.

Anonymous (2000). Section 14.8 Lyme Disease. In Healthy People 2010: Understanding and improving health and objectives for improving health, 2nd edition (Washington, DC, U.S. Government Printing Office), pp. 14.11–14.49.

Anonymous (2001a). Synopsis of LYMErix phase IV observational study. In United States Food and Drug Administration Center for Biologics Evaluation and Research (Bethesda, MD).

Anonymous (2001b). Vaccines and related biological products advisory committee meeting, January 31, 2001. In United States Food and Drug Administration Center for Biologics Evaluation and Research (Bethesda, MD).

Attie, O., Bruno, J.F., Xu, Y., Qiu, D., Luft, B.J., and Qiu, W.G. (2007). Co-evolution of the outer surface protein C gene (*ospC*) and intraspecific lineages of *Borrelia burgdorferi* sensu stricto in the north-eastern United States. Infect. Genet. Evol. *7*, 1–12.

Aydintug, M.K., Gu, Y., and Philipp, M.T. (1994). *Borrelia burgdorferi* antigens that are targeted by antibody-dependent, complement-mediated killing in the rhesus monkey. Infect. Immun. *62*, 4929–4937.

Bacon, R.M., Kugeler, K.J., and Mead, P.S. (2008). Surveillance for Lyme disease – United States, 1992–2006. MMWR Surveill. Summ. *57*, 1–9.

Baranton, G., Postic, D., Saint Girons, I., Boerlin, P., Piffaretti, J.C., Assous, M., and Grimont, P.A. (1992). Delineation of *Borrelia burgdorferi* sensu stricto, *Borrelia garinii* sp. nov., and group VS461 associated with Lyme borreliosis. Int. J. Syst. Bacteriol. *42*, 378–383.

Barbour, A.G. (1988). Plasmid analysis of *Borrelia burgdorferi*, the Lyme disease agent. J. Clin. Microbiol. *26*, 475–478.

Barbour, A.G., and Garon, C.F. (1987). Linear plasmids of the bacterium *Borrelia burgdorferi* have covalently closed ends. Science *237*, 409–411.

Barbour, A.G., and Garon, C.F. (1988). The genes encoding major surface proteins of *Borrelia burgdorferi* are located on a plasmid. Ann N Y Acad Sci *539*, 144–153.

Barbour, A.G., Jasinskas, A., Kayala, M.A., Davies, D.H., Steere, A.C., Baldi, P., and Felgner, P.L. (2008). A genome-wide proteome array reveals a limited set of immunogens in natural infections of humans and white-footed mice with *Borrelia burgdorferi*. Infect. Immun. *76*, 3374–3389.

Barbour, A.G., Maupin, G.O., Teltow, G.J., Carter, C.J., and Piesman, J. (1996). Identification of an uncultivable *Borrelia* species in the hard tick *Amblyomma americanum*: possible agent of a Lyme disease-like illness. J. Infect. Dis. *173*, 403–409.

Barbour, A.G., Tessier, S.L., and Todd, W.J. (1983). Lyme disease spirochetes and ixodid tick spirochetes share a common surface antigenic determinant defined by a monoclonal antibody. Infect. Immun. *41*, 795–804.

Beernink, P.T., Leipus, A., and Granoff, D.M. (2006). Rapid genetic grouping of factor h-binding protein (genome-derived neisserial antigen 1870), a promising group B meningococcal vaccine candidate. Clin. Vaccine Immuno.l *13*, 758–763.

Benach, J.L., Bosler, E.M., Hanrahan, J.P., Coleman, J.L., Habicht, G.S., Bast, T.F., Cameron, D.J., Ziegler, J.L., Barbour, A.G., Burgdorfer, W., *et al.* (1983). Spirochetes isolated from the blood of two patients with Lyme disease. N. Engl. J. Med. *308*, 740–742.

Bockenstedt, L.K., Fikrig, E., Barthold, S.W., Flavell, R.A., and Kantor, F.S. (1996). Identification of a *Borrelia burgdorferi* OspA T-cell epitope that promotes anti-OspA IgG in mice. J. Immunol. *157*, 5496–5502.

Bockenstedt, L.K., Hodzic, E., Feng, S., Bourrel, K.W., de Silva, A., Montgomery, R.R., Fikrig, E., Radolf, J.D., and Barthold, S.W. (1997). *Borrelia burgdorferi* strain-specific OspC-mediated immunity in mice. Infect. Immun. *65*, 4661–4667.

Brisson, D., Dykhuizen, D.E., and Ostfeld, R.S. (2008). Conspicuous impacts of inconspicuous hosts on the Lyme disease epidemic. Proc Biol Sci *275*, 227–235.

Brooks, C.S., Hefty, P.S., Jolliff, S.E., and Akins, D.R. (2003). Global analysis of *Borrelia burgdorferi* genes regulated by mammalian host-specific signals. Infect. Immun. *71*, 3371–3383.

Brooks, C.S., Vuppala, S.R., Jett, A.M., and Akins, D.R. (2006). Identification of *Borrelia burgdorferi* outer surface proteins. Infect. Immun. *74*, 296–304.

Brooks, C.S., Vuppala, S.R., Jett, A.M., Alitalo, A., Meri, S., and Akins, D.R. (2005). Complement Regulator-Acquiring Surface Protein 1 Imparts Resistance to Human Serum in *Borrelia burgdorferi*. J. Immunol. *175*, 3299–3308.

Brown, E.L., Kim, J.H., Reisenbichler, E.S., and Hook, M. (2005). Multicomponent Lyme vaccine: three is not a crowd. Vaccine *23*, 3687–3696.

Buckles, E.L., Earnhart, C.G., and Marconi, R.T. (2006). Analysis of antibody response in humans to the type A OspC loop 5 domain and assessment of the potential utility of the loop 5 epitope in Lyme disease vaccine development. Clin. Vaccine Immunol. *13*, 1162–1165.

Bunikis, J., and Barbour, A.G. (1999). Access of antibody or trypsin to an integral outer membrane protein (P66) of *Borrelia burgdorferi* is hindered by Osp lipoproteins. Infect. Immun. *67*, 2874–2883.

Bunikis, J., Garpmo, U., Tsao, J., Berglund, J., Fish, D., and Barbour, A.G. (2004). Sequence typing reveals extensive strain diversity of the Lyme borreliosis agents *Borrelia burgdorferi* in North America and *Borrelia afzelii* in Europe. Microbiology *150*, 1741–1755.

Burgdorfer, W., Barbour, A.G., Hayes, S.F., Benach, J.L., Grunwaldt, E., and Davis, J.P. (1982). Lyme disease-a tick-borne spirochetosis? Science *216*, 1317–1319.

Busch, U., Hizo-Teufel, C., Boehmer, R., Fingerle, V., Nitschko, H., Wilske, B., and Preac-Mursic, V. (1996). Three species of *Borrelia burgdorferi* sensu lato (*B. burgdorferi* sensu stricto, *B. afzelii*, and *B. garinii*) identified from cerebrospinal fluid isolates by pulsed-field gel electrophoresis and PCR. J. Clin. Microbiol. *34*, 1072–1078.

Byram, R., Stewart, P.E., and Rosa, P. (2004). The essential nature of the ubiquitous 26-kilobase circular replicon of *Borrelia burgdorferi*. J. Bacteriol. *186*, 3561–3569.

Caimano, M.J., Iyer, R., Eggers, C.H., Gonzalez, C., Morton, E.A., Gilbert, M.A., Schwartz, I., and Radolf, J.D. (2007). Analysis of the RpoS regulon in *Borrelia burgdorferi* in response to mammalian host signals provides insight into RpoS function during the enzootic cycle. Mol. Microbiol. *65*, 1193–1217.

Campbell, G.L., Fritz, C.L., Fish, D., Nowakowski, J., Nadelman, R.B., and Wormser, G.P. (1998). Estimation of the incidence of Lyme disease. Am. J. Epidemiol. *148*, 1018–1026.

Casjens, S., Palmer, N., van Vugt, R., Huang, W.M., Stevenson, B., Rosa, P., Lathigra, R., Sutton, G., Peterson, J., Dodson, R.J., *et al.* (2000). A bacterial genome in flux: the twelve linear and nine circular extrachromosomal DNAs in an infectious isolate of the Lyme disease spirochete *Borrelia burgdorferi*. Mol. Microbiol. *35*, 490–516.

Chang, Y.F., Appel, M.J., Jacobson, R.H., Shin, S.J., Harpending, P., Straubinger, R., Patrican, L.A., Mohammed, H., and Summers, B.A. (1995). Recombinant OspA protects dogs against infection and disease caused by *Borrelia burgdorferi*. Infect. Immun. *63*, 3543–3549.

Chen, J., Field, J.A., Glickstein, L., Molloy, P.J., Huber, B.T., and Steere, A.C. (1999). Association of antibiotic treatment-resistant Lyme arthritis with T-cell responses to dominant epitopes of outer surface protein A of *Borrelia burgdorferi*. Arthritis Rheum. *42*, 1813–1822.

Chu, H.J., Chavez, L.G., Jr., Blumer, B.M., Sebring, R.W., Wasmoen, T.L., and Acree, W.M. (1992). Immunogenicity and efficacy study of a commercial *Borrelia burgdorferi* bacterin. J. Am. Vet. Med. Assoc. *201*, 403–411.

Conlon, J.A., Mather, T.N., Tanner, P., and Gallo, G. (2000). Efficacy of a nonadjuvanted, outer surface protein A, recombinant vaccine in dogs after challenge by ticks naturally infected with *Borrelia burgdorferi*. Vet Ther *1*, 96–107.

Coyle, B.S., Strickland, G.T., Liang, Y.Y., Pena, C., McCarter, R., and Israel, E. (1996). The public health impact of Lyme disease in Maryland. J. Infect. Dis. *173*, 1260–1262.

de la Fuente, J., and Kocan, K.M. (2006). Strategies for development of vaccines for control of Ixodid tick species. Parasite Immunol. *28*, 275–283.

de la Fuente, J., Kocan, K.M., and Blouin, E.F. (2007). Tick vaccines and the transmission of tick-borne pathogens. Vet. Res. Commun. *31 Suppl. 1*, 85–90.

de Silva, A.M., Telford, S.R., 3rd, Brunet, L.R., Barthold, S.W., and Fikrig, E. (1996). *Borrelia burgdorferi* OspA is an arthropod-specific transmission-blocking Lyme disease vaccine. J. Exp. Med. *183*, 271–275.

de Silva, A.M., Zeidner, N.S., Zhang, Y., Dolan, M.C., Piesman, J., and Fikrig, E. (1999). Influence of outer surface protein A antibody on *Borrelia burgdorferi* within feeding ticks. Infect. Immun. *67*, 30–35.

del Rio, B., Dattwyler, R.J., Aroso, M., Neves, V., Meirelles, L., Seegers, J.F., and Gomes-Solecki, M. (2008). Oral immunization with recombinant *Lactobacillus plantarum* induces a protective immune response in mice with Lyme disease. Clin. Vaccine Immunol. *15*, 1429–1435.

Ding, W., Huang, X., Yang, X., Dunn, J.J., Luft, B.J., Koide, S., and Lawson, C.L. (2000). Structural identification of a key protective B-cell epitope in Lyme disease antigen OspA. J. Mol. Biol. *302*, 1153–1164.

Drouin, E.E., Glickstein, L., Kwok, W.W., Nepom, G.T., and Steere, A.C. (2008). Human homologues of a *Borrelia* T-cell epitope associated with antibiotic-refractory Lyme arthritis. Mol Immunol 45, 180–189.

Dunne, M., al-Ramadi, B.K., Barthold, S.W., Flavell, R.A., and Fikrig, E. (1995). Oral vaccination with an attenuated *Salmonella typhimurium* strain expressing *Borrelia burgdorferi* OspA prevents murine Lyme borreliosis. Infect. Immun. 63, 1611–1614.

Dykhuizen, D.E., and Baranton, G. (2001). The implications of a low rate of horizontal transfer in *Borrelia*. Trends Microbiol. 9, 344–350.

Dykhuizen, D.E., Brisson, D., Sandigursky, S., Wormser, G.P., Nowakowski, J., Nadelman, R.B., and Schwartz, I. (2008). The propensity of different *Borrelia burgdorferi* sensu stricto genotypes to cause disseminated infections in humans. Am. J. Trop. Med. Hyg. 78, 806–810.

Earnhart, C.G., Buckles, E.L., Dumler, J.S., and Marconi, R.T. (2005). Demonstration of OspC type diversity in invasive human Lyme disease isolates and identification of previously uncharacterized epitopes that define the specificity of the OspC murine antibody response. Infect. Immun. 73, 7869–7877.

Earnhart, C.G., Buckles, E.L., and Marconi, R.T. (2007). Development of an OspC-based tetravalent, recombinant, chimeric vaccinogen that elicits bactericidal antibody against diverse Lyme disease spirochete strains. Vaccine 25, 466–480.

Earnhart, C.G., and Marconi, R.T. (2007a). Construction and analysis of variants of a polyvalent Lyme disease vaccine: approaches for improving the immune response to chimeric vaccinogens. Vaccine 25, 3419–3427.

Earnhart, C.G., and Marconi, R.T. (2007b). An octavalent Lyme disease vaccine induces antibodies that recognize all incorporated OspC type-specific sequences. Hum. Vaccin. 3, 281–289.

Earnhart, C.G., and Marconi, R.T. (2007c). OspC phylogenetic analyses support the feasibility of a broadly protective polyvalent chimeric Lyme disease vaccine. Clin. Vaccine Immunol. *14*, 628–634.

Eggers, C.H., Casjens, S., Hayes, S.F., Garon, C.F., Damman, C.J., Oliver, D.B., and Samuels, D.S. (2000). Bacteriophages of spirochetes. J. Mol. Microbiol. Biotechnol. 2, 365–373.

Eggers, C.H., Kimmel, B.J., Bono, J.L., Elias, A.F., Rosa, P., and Samuels, D.S. (2001). Transduction by φBB-1, a bacteriophage of *Borrelia burgdorferi*. J. Bacteriol. *183*, 4771–4778.

Eggers, C.H., and Samuels, D.S. (1999). Molecular evidence for a new bacteriophage of *Borrelia burgdorferi*. J. Bacteriol. *181*, 7308–7313.

Eicken, C., Sharma, V., Klabunde, T., Owens, R.T., Pikas, D.S., Höök, M., and Sacchettini, J.C. (2001). Crystal structure of Lyme disease antigen outer surface protein C from *Borrelia burgdorferi*. J. Biol. Chem. *276*, 10010–10015.

Embers, M.E., Alvarez, X., Ooms, T., and Philipp, M.T. (2008). The failure of immune response evasion by linear plasmid 28-1-deficient *Borrelia burgdorferi* is attributable to persistent expression of an outer surface protein. Infect. Immun. *76*, 3984–3991.

Erdile, L.F., Brandt, M.A., Warakomski, D.J., Westrack, G.J., Sadziene, A., Barbour, A.G., and Mays, J.P. (1993). Role of attached lipid in immunogenicity of *Borrelia burgdorferi* OspA. Infect. Immun. *61*, 81–90.

Feng, S., Hodzic, E., Stevenson, B., and Barthold, S.W. (1998). Humoral immunity to *Borrelia burgdorferi* N40 decorin binding proteins during infection of laboratory mice. Infect. Immun. *66*, 2827–2835.

Fikrig, E., Barthold, S.W., Kantor, F.S., and Flavell, R.A. (1990). Protection of mice against the Lyme disease agent by immunizing with recombinant OspA. Science *250*, 553–556.

Fikrig, E., Barthold, S.W., Kantor, F.S., and Flavell, R.A. (1991). Protection of mice from Lyme borreliosis by oral vaccination with *Escherichia coli* expressing OspA. J. Infect. Dis. *164*, 1224–1227.

Fikrig, E., Barthold, S.W., Sun, W., Feng, W., Telford, S.R., 3rd, and Flavell, R.A. (1997). Borrelia burgdorferi P35 and P37 proteins, expressed in vivo, elicit protective immunity. Immunity *6*, 531–539.

Fingerle, V., Hauser, U., Liegl, G., Petko, B., Preac-Mursic, V., and Wilske, B. (1995). Expression of outer surface proteins A and C of *Borrelia burgdorferi* in *Ixodes ricinus*. J. Clin. Microbiol. *33*, 1867–1869.

Fraser, C.M., Casjens, S., Huang, W.M., Sutton, G.G., Clayton, R., Lathigra, R., White, O., Ketchum, K.A., Dodson, R., Hickey, E.K., *et al.* (1997). Genomic sequence of a Lyme disease spirochaete, *Borrelia burgdorferi*. Nature *390*, 580–586.

Fuchs, R., Jauris, S., Lottspeich, F., Preac-Mursic, V., Wilske, B., and Soutschek, E. (1992). Molecular analysis and expression of a *Borrelia burgdorferi* gene encoding a 22 kDa protein (pC) in *Escherichia coli*. Mol. Microbiol. *6*, 503–509.

Fukunaga, M., Hamase, A., Okada, K., and Nakao, M. (1996). *Borrelia tanukii* sp. nov. and *Borrelia turdae* sp. nov. found from Ixodid ticks in Japan: rapid species identification by 16S rRNA gene-targeted PCR analysis. Microbiol. Immunol. *40*, 877–881.

Gern, L., Hu, C.M., Voet, P., Hauser, P., and Lobet, Y. (1997). Immunization with a polyvalent OspA vaccine protects mice against *Ixodes ricinus* tick bites infected by *Borrelia burgdorferi* ss, *Borrelia garinii* and *Borrelia afzelii*. Vaccine *15*, 1551–1557.

Gilmore, R.D., Jr., Kappel, K.J., Dolan, M.C., Burkot, T.R., and Johnson, B.J. (1996). Outer surface protein C (OspC), but not P39, is a protective immunogen against a tick-transmitted *Borrelia burgdorferi* challenge: evidence for a conformational protective epitope in OspC. Infect. Immun. *64*, 2234–2239.

Gilmore, R.D., Jr., and Mbow, M.L. (1999). Conformational nature of the *Borrelia burgdorferi* B31 outer surface protein C protective epitope. Infect. Immun. *67*, 5463–5469.

Gomes-Solecki, M.J., Brisson, D.R., and Dattwyler, R.J. (2006). Oral vaccine that breaks the transmission cycle of the Lyme disease spirochete can be delivered via bait. Vaccine *24*, 4440–4449.

Grimm, D., Tilly, K., Byram, R., Stewart, P.E., Krum, J.G., Bueschel, D.M., Schwan, T.G., Policastro, P.F., Elias, A.F., and Rosa, P.A. (2004). Outer-surface protein C of the Lyme disease spirochete: a protein induced in ticks for infection of mammals. Proc. Natl. Acad. Sci. U.S.A. *101*, 3142–3147.

Gross, D.M., Forsthuber, T., Tary-Lehmann, M., Etling, C., Ito, K., Nagy, Z.A., Field, J.A., Steere, A.C., and Huber, B.T. (1998). Identification of LFA-1 as a candidate autoantigen in treatment-resistant Lyme arthritis. Science *281*, 703–706.

Gross, D.M., and Huber, B.T. (2000). Cellular and molecular aspects of Lyme arthritis. Cell. Mol. Life Sci. *57*, 1562–1569.

Hagman, K.E., Yang, X., Wikel, S.K., Schoeler, G.B., Caimano, M.J., Radolf, J.D., and Norgard, M.V. (2000). Decorin-binding protein A (DbpA) of *Borrelia burgdorferi* is not protective when immunized mice are challenged via tick infestation and correlates with the lack of DbpA expression by *B. burgdorferi* in ticks. Infect. Immun. *68*, 4759–4764.

Hanincova, K., Liveris, D., Sandigursky, S., Wormser, G.P., and Schwartz, I. (2008). *Borrelia burgdorferi* sensu stricto is clonal in patients with early Lyme borreliosis. Appl. Environ. Microbiol. *74*, 5008–5014.

Hanson, M.S., Cassatt, D.R., Guo, B.P., Patel, N.K., McCarthy, M.P., Dorward, D.W., and Hook, M. (1998). Active and passive immunity against *Borrelia burgdorferi* decorin binding protein A (DbpA) protects against infection. Infect. Immun. *66*, 2143–2153.

Hanson, M.S., and Edelman, R. (2004). Vaccines against Lyme disease. In New Generation Vaccines, M. Levine, J.B. Kaper, R. Rappuoli, M.A. Liu, and M.F. Good, eds. (New York, Marcel Dekker AG), pp. 487–498.

Hanson, M.S., Patel, N.K., Cassatt, D.R., and Ulbrandt, N.D. (2000). Evidence for vaccine synergy between *Borrelia burgdorferi* decorin binding protein A and outer surface protein A in the mouse model of Lyme borreliosis. Infect. Immun. *68*, 6457–6460.

Hefty, P.S., Jolliff, S.E., Caimano, M.J., Wikel, S.K., and Akins, D.R. (2002). Changes in temporal and spatial patterns of outer surface lipoprotein expression generate population heterogeneity and antigenic diversity in the Lyme disease spirochete, *Borrelia burgdorferi*. Infect. Immun. *70*, 3468–3478.

Hefty, P.S., Jolliff, S.E., Caimano, M.J., Wikel, S.K., Radolf, J.D., and Akins, D.R. (2001). Regulation of OspE-related, OspF-related, and Elp lipoproteins of *Borrelia burgdorferi* strain 297 by mammalian host-specific signals. Infect. Immun. *69*, 3618–3627.

Hellwage, J., Meri, T., Heikkila, T., Alitalo, A., Panelius, J., Lahdenne, P., Seppala, I.J., and Meri, S. (2001). The complement regulator factor H binds to the surface protein OspE of *Borrelia burgdorferi*. J. Biol. Chem. *276*, 8427–8435.

Hemmer, B., Gran, B., Zhao, Y., Marques, A., Pascal, J., Tzou, A., Kondo, T., Cortese, I., Bielekova, B., Straus, S.E., *et al.* (1999). Identification of candidate T-cell epitopes and molecular mimics in chronic Lyme disease. Nat. Med. *5*, 1375–1382.

Hodzic, E., Feng, S., and Barthold, S.W. (2000). Stability of *Borrelia burgdorferi* outer surface protein C under immune selection pressure. J. Infect. Dis. *181*, 750–753.

Howe, T.R., LaQuier, F.W., and Barbour, A.G. (1986). Organization of genes encoding two outer membrane proteins of the Lyme disease agent *Borrelia burgdorferi* within a single transcriptional unit. Infect. Immun. *54*, 207–212.

Huang, X., Link, K., Koide, A., Dunn, J.J., Luft, B.J., and Koide, S. (1999). ^{1}H, ^{13}C, and ^{15}N NMR backbone assignments of 37 kDa surface antigen OspC from *Borrelia burgdorferi*. J. Biomol. NMR *14*, 283–284.

Huang, X., Yang, X., Luft, B.J., and Koide, S. (1998). NMR identification of epitopes of Lyme disease antigen OspA to monoclonal antibodies. J. Mol. Biol. *281*, 61–67.

Iyer, R., Kalu, O., Purser, J., Norris, S., Stevenson, B., and Schwartz, I. (2003). Linear and circular plasmid content in *Borrelia burgdorferi* clinical isolates. Infect. Immun. *71*, 3699–3706.

Jobe, D.A., Lovrich, S.D., Schell, R.F., and Callister, S.M. (2003). C-terminal region of outer surface protein C binds borreliacidal antibodies in sera from patients with Lyme disease. Clin. Diagn. Lab. Immunol. *10*, 573–578.

Johnson, R.C., Kodner, C., Russell, M., and Duray, P.H. (1988). Experimental infection of the hamster with *Borrelia burgdorferi*. Ann N Y Acad Sci *539*, 258–263.

Jonsson, N.N., Matschoss, A.L., Pepper, P., Green, P.E., Albrecht, M.S., Hungerford, J., and Ansell, J. (2000). Evaluation of tickGARD(PLUS), a novel vaccine against *Boophilus microplus*, in lactating Holstein-Friesian cows. Vet Parasitol *88*, 275–285.

Kalish, R.A., Leong, J.M., and Steere, A.C. (1993). Association of treatment-resistant chronic Lyme arthritis with HLA-DR4 and antibody reactivity to OspA and OspB of *Borrelia burgdorferi*. Infect. Immun. *61*, 2774–2779.

Kamradt, T., Lengl-Janssen, B., Strauss, A.F., Bansal, G., and Steere, A.C. (1996). Dominant recognition of a *Borrelia burgdorferi* outer surface protein A peptide by T helper cells in patients with treatment-resistant Lyme arthritis. Infect. Immun. *64*, 1284–1289.

Kawabata, H., Masuzawa, T., and Yanagihara, Y. (1993). Genomic analysis of *Borrelia japonica* sp. nov. isolated from *Ixodes ovatus* in Japan. Microbiol. Immunol. 37, 843–848.

Keller, D., Koster, F.T., Marks, D.H., Hosbach, P., Erdile, L.F., and Mays, J.P. (1994). Safety and immunogenicity of a recombinant outer surface protein A Lyme vaccine. JAMA *271*, 1764–1768.

Koide, S., Yang, X., Huang, X., Dunn, J.J., and Luft, B.J. (2005). Structure-based design of a second-generation Lyme disease vaccine based on a C-terminal fragment of *Borrelia burgdorferi* OspA. J. Mol. Biol. *350*, 290–299.

Kraiczy, P., Hartmann, K., Hellwage, J., Skerka, C., Kirschfink, M., Brade, V., Zipfel, P.F., Wallich, R., and Stevenson, B. (2004a). Immunological characterization of the complement regulator factor H-binding CRASP and Erp proteins of *Borrelia burgdorferi*. Int. J. Med. Microbiol. *293 Suppl. 37*, 152–157.

Kraiczy, P., Hellwage, J., Skerka, C., Becker, H., Kirschfink, M., Simon, M.M., Brade, V., Zipfel, P.F., and Wallich, R. (2004b). Complement resistance of *Borrelia burgdorferi* correlates with the expression of BbCRASP-1, a novel linear plasmid-encoded surface protein that interacts with human factor H and FHL-1 and is unrelated to Erp proteins. J. Biol. Chem. *279*, 2421–2429.

Kraiczy, P., Skerka, C., and Zipfel, P.F. (2001). Further characterization of complement regulator-acquiring surface proteins of *Borrelia burgdorferi*. Infect. Immun. *69*, 7800–7809.

Kumaran, D., Eswaramoorthy, S., Luft, B.J., Koide, S., Dunn, J.J., Lawson, C.L., and Swaminathan, S. (2001). Crystal structure of outer surface protein C (OspC) from the Lyme disease spirochete, *Borrelia burgdorferi*. EMBO J. *20*, 971–978.

Labandeira-Rey, M., Baker, E.A., and Skare, J.T. (2001). VraA (BBI16) protein of *Borrelia burgdorferi* is a surface-exposed antigen with a repetitive motif that confers partial protection against experimental Lyme borreliosis. Infect. Immun. *69*, 1409–1419.

Labandeira-Rey, M., Seshu, J., and Skare, J.T. (2003). The absence of linear plasmid 25 or 28-1 of *Borrelia burgdorferi* dramatically alters the kinetics of experimental infection via distinct mechanisms. Infect. Immun. *71*, 4608–4613.

Labandeira-Rey, M., and Skare, J.T. (2001). Decreased infectivity in *Borrelia burgdorferi* strain B31 is associated with loss of linear plasmid 25 or 28-1. Infect. Immun. *69*, 446–455.

Lagal, V., Postic, D., Ruzic-Sabljic, E., and Baranton, G. (2003). Genetic diversity among *Borrelia* strains determined by single-strand conformation polymorphism analysis of the *ospC* gene and its association with invasiveness. J. Clin. Microbiol. *41*, 5059–5065.

Langermann, S., Palaszynski, S., Sadziene, A., Stover, C.K., and Koenig, S. (1994). Systemic and mucosal immunity induced by BCG vector expressing outer-surface protein A of *Borrelia burgdorferi*. Nature 372, 552–555.

Lathrop, S.L., Ball, R., Haber, P., Mootrey, G.T., Braun, M.M., Shadomy, S.V., Ellenberg, S.S., Chen, R.T., and Hayes, E.B. (2002). Adverse event reports following vaccination for Lyme disease: December 1998-July 2000. Vaccine *20*, 1603–1608.

Le Fleche, A., Postic, D., Girardet, K., Peter, O., and Baranton, G. (1997). Characterization of *Borrelia lusitaniae* sp. nov. by 16S ribosomal DNA sequence analysis. Int. J. Syst. Bacteriol. *47*, 921–925.

Leuba-Garcia, S., Martinez, R., and Gern, L. (1998). Expression of outer surface proteins A and C of *Borrelia afzelii* in *Ixodes ricinus* ticks and in the skin of mice. Zentralbl. Bakteriol. *287*, 475–484.

Levy, S.A. (2002). Use of a C6 ELISA test to evaluate the efficacy of a whole-cell bacterin for the prevention of naturally transmitted canine *Borrelia burgdorferi* infection. Vet. Ther. 3, 420–424.

Levy, S.A., Lissman, B.A., and Ficke, C.M. (1993). Performance of a *Borrelia burgdorferi* bacterin in borreliosis-endemic areas. J. Am. Vet. Med. Assoc. *202*, 1834–1838.

Li, H., Dunn, J.J., Luft, B.J., and Lawson, C.L. (1997). Crystal structure of Lyme disease antigen outer surface protein A complexed with an Fab. Proc. Natl. Acad. Sci. U.S.A. *94*, 3584–3589.

Liang, F.T., Jacobs, M.B., and Philipp, M.T. (2001). C-terminal invariable domain of VlsE may not serve as target for protective immune response against *Borrelia burgdorferi*. Infect. Immun. *69*, 1337–1343.

Liang, F.T., Nelson, F.K., and Fikrig, E. (2002a). DNA microarray assessment of putative *Borrelia burgdorferi* lipoprotein genes. Infect. Immun. *70*, 3300–3303.

Liang, F.T., Nelson, F.K., and Fikrig, E. (2002b). Molecular adaptation of *Borrelia burgdorferi* in the murine host. J. Exp. Med. *196*, 275–280.

Liang, F.T., and Philipp, M.T. (1999). Analysis of antibody response to invariable regions of VlsE, the variable surface antigen of *Borrelia burgdorferi*. Infect. Immun. *67*, 6702–6706.

Lin, T., Oliver, J.H., Jr., and Gao, L. (2002). Genetic diversity of the outer surface protein C gene of southern *Borrelia* isolates and its possible epidemiological, clinical, and pathogenetic implications. J. Clin. Microbiol. *40*, 2572–2583.

Lindgren, E., and Jaenson, T.G. (2006). Lyme borreliosis in Europe: Influences of climate and climate change, epidemiology, ecology and adaptation measures (World Health Organization), pp. 1–35.

Livey, I., Gibbs, C.P., Schuster, R., and Dorner, F. (1995). Evidence for lateral transfer and recombination in OspC variation in Lyme disease *Borrelia*. Mol. Microbiol. *18*, 257–269.

Lovrich, S.D., Callister, S.M., DuChateau, B.K., Lim, L.C., Winfrey, J., Day, S.P., and Schell, R.F. (1995). Abilities of OspA proteins from different seroprotective groups of *Borrelia burgdorferi* to protect hamsters from infection. Infect. Immun. *63*, 2113–2119.

Luke, C.J., Carner, K., Liang, X., and Barbour, A.G. (1997a). An OspA-based DNA vaccine protects mice against infection with *Borrelia burgdorferi*. J. Infect. Dis. *175*, 91–97.

Luke, C.J., Huebner, R.C., Kasmiersky, V., and Barbour, A.G. (1997b). Oral delivery of purified lipoprotein OspA protects mice from systemic infection with *Borrelia burgdorferi*. Vaccine *15*, 739–746.

Ma, J., Bulger, P.A., Dante, S., Davis, D.R., Perilli-Palmer, B., and Coughlin, R.T. (1995). Characterization of canine humoral immune responses to outer surface protein subunit vaccines and to natural infection by Lyme disease spirochetes. J. Infect. Dis. *171*, 909–915.

Madico, G., Ngampasutadol, J., Gulati, S., Vogel, U., Rice, P.A., and Ram, S. (2007). Factor H binding and function in sialylated pathogenic neisseriae is influenced by gonococcal, but not meningococcal, porin. J. Immunol. *178*, 4489–4497.

Maes, E., Lecomte, P., and Ray, N. (1998). A cost-of-illness study of Lyme disease in the United States. Clin Ther 20, 993–1008; discussion 1992.

Makabe, K., Tereshko, V., Gawlak, G., Yan, S., and Koide, S. (2006). Atomic-resolution crystal structure of *Borrelia burgdorferi* outer surface protein A via surface engineering. Protein Sci. *15*, 1907–1914.

Marconi, R.T., Liveris, D., and Schwartz, I. (1995). Identification of novel insertion elements, restriction fragment length polymorphism patterns, and discontinuous 23S rRNA in Lyme disease spirochetes: phylogenetic analyses of rRNA genes and their intergenic spacers in *Borrelia japonica* sp. nov. and genomic group 21038 (*Borrelia andersonii* sp. nov.) isolates. J. Clin. Microbiol. *33*, 2427–2434.

Marconi, R.T., Samuels, D.S., and Garon, C.F. (1993). Transcriptional analyses and mapping of the *ospC* gene in Lyme disease spirochetes. J. Bacteriol. *175*, 926–932.

Marconi, R.T., Sung, S.Y., Hughes, C.A., and Carlyon, J.A. (1996). Molecular and evolutionary analyses of a variable series of genes in *Borrelia burgdorferi* that are related to *ospE* and *ospF* constitute a gene family, and share a common upstream homology box. J. Bacteriol. *178*, 5615–5626.

Margolis, N., and Rosa, P.A. (1993). Regulation of expression of major outer surface proteins in *Borrelia burgdorferi*. Infect. Immun. *61*, 2207–2210.

Marti Ras, N., Postic, D., Foretz, M., and Baranton, G. (1997). *Borrelia burgdorferi* sensu stricto, a bacterial species 'made in the U.S.A.'? Int. J. Syst. Bacteriol. *47*, 1112–1117.

Masuzawa, T., Takada, N., Kudeken, M., Fukui, T., Yano, Y., Ishiguro, F., Kawamura, Y., Imai, Y., and Ezaki, T. (2001). *Borrelia sinica* sp. nov., a Lyme disease-related *Borrelia* species isolated in China. Int J Syst Evol Microbiol *51*, 1817–1824.

Mathiesen, M.J., Holm, A., Christiansen, M., Blom, J., Hansen, K., Ostergaard, S., and Theisen, M. (1998). The dominant epitope of *Borrelia garinii* outer surface protein C recognized by sera from patients with neuroborreliosis has a surface-exposed conserved structural motif. Infect. Immun. *66*, 4073–4079.

Matteson, E.L., Beckett, V.L., O'Fallon, W.M., Melton, L.J., 3rd, and Duffy, J. (1992). Epidemiology of Lyme disease in Olmsted County, MN, 1975–1990. J. Rheumatol. *19*, 1743–1745.

Mbow, M.L., Gilmore, R.D., Jr., and Titus, R.G. (1999). An OspC-specific monoclonal antibody passively protects mice from tick-transmitted infection by *Borrelia burgdorferi* B31. Infect. Immun. *67*, 5470–5472.

McDowell, J.V., Sung, S.Y., Hu, L.T., and Marconi, R.T. (2002). Evidence that the variable regions of the central domain of VlsE are antigenic during infection with Lyme disease spirochetes. Infect. Immun. *70*, 4196–4203.

McDowell, J.V., Sung, S.Y., Labandeira-Rey, M., Skare, J.T., and Marconi, R.T. (2001a). Analysis of mechanisms associated with loss of infectivity of clonal populations of *Borrelia burgdorferi* B31MI. Infect. Immun. *69*, 3670–3677.

McDowell, J.V., Sung, S.Y., Price, G., and Marconi, R.T. (2001b). Demonstration of the genetic stability and temporal expression of select members of the Lyme disease spirochete OspF protein family during infection in mice. Infect. Immun. *69*, 4831–4838.

McDowell, J.V., Wolfgang, J., Tran, E., Metts, M.S., Hamilton, D., and Marconi, R.T. (2003). Comprehensive analysis of the factor H binding capabilities of *Borrelia* species associated with Lyme disease: delineation of two distinct classes of factor H binding proteins. Infect. Immun. *71*, 3597–3602.

McNabb, S.J., Jajosky, R.A., Hall-Baker, P.A., Adams, D.A., Sharp, P., Anderson, W.J., Javier, A.J., Jones, G.J., Nitschke, D.A., Worshams, C.A., and Richard, R.A., Jr. (2007). Summary of notifiable diseases – United States, 2005. MMWR *54*, 1–92.

Meek, J.I., Roberts, C.L., Smith, E.V., Jr., and Cartter, M.L. (1996). Underreporting of Lyme disease by Connecticut physicians, 1992. J. Publ. Health Manag. Pract. *2*, 61–65.

Meltzer, M.I., Dennis, D.T., and Orloski, K.A. (1999). The cost effectiveness of vaccinating against Lyme disease. Emerg. Infect. Dis. *5*, 321–328.

Metts, M.S., McDowell, J.V., Theisen, M., Hansen, P.R., and Marconi, R.T. (2003). Analysis of the OspE determinants involved in binding of factor H and OspE-targeting antibodies elicited during *Borrelia burgdorferi* infection in mice. Infect. Immun. *71*, 3587–3596.

Meyer, A.L., Trollmo, C., Crawford, F., Marrack, P., Steere, A.C., Huber, B.T., Kappler, J., and Hafler, D.A. (2000). Direct enumeration of *Borrelia*-reactive CD4 T-cells ex vivo by using MHC class II tetramers. Proc. Natl. Acad. Sci. U.S.A. *97*, 11433–11438.

Naleway, A.L., Belongia, E.A., Kazmierczak, J.J., Greenlee, R.T., and Davis, J.P. (2002). Lyme disease incidence in Wisconsin: a comparison of state-reported rates and rates from a population-based cohort. Am. J. Epidemiol. *155*, 1120–1127.

Narasimhan, S., Caimano, M.J., Liang, F.T., Santiago, F., Laskowski, M., Philipp, M.T., Pachner, A.R., Radolf, J.D., and Fikrig, E. (2003). *Borrelia burgdorferi* transcriptome in the central nervous system of non-human primates. Proc. Natl. Acad. Sci. U.S.A. *100*, 15953–15958.

Nepereny, J., Chumela, J., Tuhackova, J., and Vrzal, V. (2007). Verification of vaccine effectivity against Lyme borreliosis in dogs by challenge test. Veterinarstvi *57*, 213–217.

Nguyen, T.P., Lam, T.T., Barthold, S.W., Telford, S.R., 3rd, Flavell, R.A., and Fikrig, E. (1994). Partial destruction of *Borrelia burgdorferi* within ticks that engorged on OspE- or OspF-immunized mice. Infect. Immun. *62*, 2079–2084.

Nigrovic, L.E., and Thompson, K.M. (2007). The Lyme vaccine: a cautionary tale. Epidemiol. Infect. *135*, 1–8.

Nocton, J.J., Dressler, F., Rutledge, B.J., Rys, P.N., Persing, D.H., and Steere, A.C. (1994). Detection of *Borrelia burgdorferi* DNA by polymerase chain reaction in synovial fluid from patients with Lyme arthritis. N. Engl. J. Med. *330*, 229–234.

Norris, S.J. (2006). Antigenic variation with a twist – the *Borrelia* story. Mol. Microbiol. *60*, 1319–1322.

Nuttall, P.A., Trimnell, A.R., Kazimirova, M., and Labuda, M. (2006). Exposed and concealed antigens as vaccine targets for controlling ticks and tick-borne diseases. Parasite Immunol *28*, 155–163.

Ojaimi, C., Brooks, C., Casjens, S., Rosa, P., Elias, A., Barbour, A., Jasinskas, A., Benach, J., Katona, L., Radolf, J., *et al.* (2003). Profiling of temperature-induced changes in *Borrelia burgdorferi* gene expression by using whole genome arrays. Infect. Immun. *71*, 1689–1705.

Ojaimi, C., Mulay, V., Liveris, D., Iyer, R., and Schwartz, I. (2005). Comparative transcriptional profiling of *Borrelia burgdorferi* clinical isolates differing in capacities for hematogenous dissemination. Infect. Immun. *73*, 6791–6802.

Orloski, K.A., Hayes, E.B., Campbell, G.L., and Dennis, D.T. (2000). Surveillance for Lyme disease – United States, 1992–1998. MMWR CDC Surveill. Summ. *49*, 1–11.

Pal, U., Li, X., Wang, T., Montgomery, R.R., Ramamoorthi, N., Desilva, A.M., Bao, F., Yang, X., Pypaert, M., Pradhan, D., *et al.* (2004a). TROSPA, an *Ixodes scapularis* receptor for *Borrelia burgdorferi*. Cell *119*, 457–468.

Pal, U., Yang, X., Chen, M., Bockenstedt, L.K., Anderson, J.F., Flavell, R.A., Norgard, M.V., and Fikrig, E. (2004b). OspC facilitates *Borrelia burgdorferi* invasion of *Ixodes scapularis* salivary glands. J. Clin. Invest. *113*, 220–230.

Pedra, J.H., Narasimhan, S., Deponte, K., Marcantonio, N., Kantor, F.S., and Fikrig, E. (2006). Disruption of the salivary protein 14 in *Ixodes scapularis* nymphs and impact on pathogen acquisition. Am. J. Trop. Med. Hyg. *75*, 677–682.

Persing, D.H., Mathiesen, D., Podzorski, D., and Barthold, S.W. (1994). Genetic stability of *Borrelia burgdorferi* recovered from chronically infected immunocompetent mice. Infect. Immun. *62*, 3521–3527.

Philipp, M.T., Lobet, Y., Bohm, R.P., Jr., Roberts, E.D., Dennis, V.A., Gu, Y., Lowrie, R.C., Jr., Desmons, P., Duray, P.H., England, J.D., *et al.* (1997). The outer surface protein A (OspA) vaccine against Lyme disease: efficacy in the rhesus monkey. Vaccine *15*, 1872–1887.

Poland, G.A. (2001). Prevention of Lyme disease: a review of the evidence. Mayo Clin. Proc. *76*, 713–724.

Postic, D., Ras, N.M., Lane, R.S., Hendson, M., and Baranton, G. (1998). Expanded diversity among Californian *Borrelia* isolates and description of *Borrelia bissettii* sp. nov. (formerly *Borrelia* group DN127). J. Clin. Microbiol. *36*, 3497–3504.

Prevot, P.P., Couvreur, B., Denis, V., Brossard, M., Vanhamme, L., and Godfroid, E. (2007). Protective immunity against *Ixodes ricinus* induced by a salivary serpin. Vaccine *25*, 3284–3292.

Probert, W.S., Crawford, M., Cadiz, R.B., and LeFebvre, R.B. (1997). Immunization with outer surface protein (Osp) A, but not OspC, provides cross-protection of mice challenged with North American isolates of *Borrelia burgdorferi*. J. Infect. Dis. *175*, 400–405.

Probert, W.S., and LeFebvre, R.B. (1994). Protection of C3H/HeN mice from challenge with *Borrelia burgdorferi* through active immunization with OspA, OspB, or OspC, but not with OspD or the 83-kilodalton antigen. Infect. Immun. *62*, 1920–1926.

Purser, J.E., and Norris, S.J. (2000). Correlation between plasmid content and infectivity in *Borrelia burgdorferi*. Proc. Natl. Acad. Sci. U.S.A. *97*, 13865–13870.

Ramamoorthi, N., Narasimhan, S., Pal, U., Bao, F., Yang, X.F., Fish, D., Anguita, J., Norgard, M.V., Kantor, F.S., Anderson, J.F., *et al.* (2005). The Lyme disease agent exploits a tick protein to infect the mammalian host. Nature *436*, 573–577.

Rapuntean, G., Boldizsar, E., Brudasca, E., Fartan, S., and Popescu, R. (2003). Consideration on immunogenic capacity of Biocan B vaccine against borreliosis. Buletinul Universitatii de Stiinte Agricole si Medicina Veterinara Cluj-Napoca Seria Medicina Veterinara *60*, 304–306.

Revel, A.T., Talaat, A.M., and Norgard, M.V. (2002). DNA microarray analysis of differential gene expression in *Borrelia burgdorferi*, the Lyme disease spirochete. Proc. Natl. Acad. Sci. U.S.A. *99*, 1562–1567.

Rogers, E.A., and Marconi, R.T. (2007). Delineation of species-specific binding properties of the CspZ protein (BBH06) of the Lyme disease spirochetes: evidence for new contributions to the pathogenesis of *Borrelia* spp. Infect. Immun. *75*, 5272–5281.

Rosa, P. (2005). Lyme disease agent borrows a practical coat. Nat. Med. *11*, 831–832.

Rose, C.D., Fawcett, P.T., and Gibney, K.M. (2001). Arthritis following recombinant outer surface protein A vaccination for Lyme disease. J Rheumatol *28*, 2555–2557.

Rousselle, J.C., Callister, S.M., Schell, R.F., Lovrich, S.D., Jobe, D.A., Marks, J.A., and Wieneke, C.A. (1998). Borreliacidal antibody production against outer surface protein C of *Borrelia burgdorferi*. J. Infect. Dis. *178*, 733–741.

Sadziene, A., Wilske, B., Ferdows, M.S., and Barbour, A.G. (1993). The cryptic *ospC* gene of *Borrelia burgdorferi* B31 is located on a circular plasmid. Infect. Immun. *61*, 2192–2195.

Schaible, U.E., Kramer, M.D., Eichmann, K., Modolell, M., Museteanu, C., and Simon, M.M. (1990). Monoclonal antibodies specific for the outer surface protein A (OspA) of *Borrelia burgdorferi* prevent Lyme borreliosis in severe combined immunodeficiency (*scid*) mice. Proc. Natl. Acad. Sci. U.S.A. *87*, 3768–3772.

Scheckelhoff, M.R., Telford, S.R., and Hu, L.T. (2006). Protective efficacy of an oral vaccine to reduce carriage of *Borrelia burgdorferi* (strain N40) in mouse and tick reservoirs. Vaccine *24*, 1949–1957.

Scheiblhofer, S., Weiss, R., Durnberger, H., Mostbock, S., Breitenbach, M., Livey, I., and Thalhamer, J. (2003). A DNA vaccine encoding the outer surface protein C from *Borrelia burgdorferi* is able to induce protective immune responses. Microbes Infect. *5*, 939–946.

Schoen, R.T., Meurice, F., Brunet, C.M., Cretella, S., Krause, D.S., Craft, J.E., and Fikrig, E. (1995). Safety and immunogenicity of an outer surface protein A vaccine in subjects with previous Lyme disease. J. Infect. Dis. *172*, 1324–1329.

Schwan, T.G., Burgdorfer, W., and Garon, C.F. (1988). Changes in infectivity and plasmid profile of the Lyme disease spirochete, *Borrelia burgdorferi*, as a result of in vitro cultivation. Infect. Immun. *56*, 1831–1836.

Schwan, T.G., Piesman, J., Golde, W.T., Dolan, M.C., and Rosa, P.A. (1995). Induction of an outer surface protein on *Borrelia burgdorferi* during tick feeding. Proc. Natl. Acad. Sci. U.S.A. *92*, 2909–2913.

Sears, J.E., Fikrig, E., Nakagawa, T.Y., Deponte, K., Marcantonio, N., Kantor, F.S., and Flavell, R.A. (1991). Molecular mapping of Osp-A mediated immunity against *Borrelia burgdorferi*, the agent of Lyme disease. J. Immunol. *147*, 1995–2000.

Seinost, G., Dykhuizen, D.E., Dattwyler, R.J., Golde, W.T., Dunn, J.J., Wang, I.N., Wormser, G.P., Schriefer, M.E., and Luft, B.J. (1999a). Four clones of *Borrelia burgdorferi* sensu stricto cause invasive infection in humans. Infect. Immun. *67*, 3518–3524.

Seinost, G., Golde, W.T., Berger, B.W., Dunn, J.J., Qiu, D., Dunkin, D.S., Dykhuizen, D.E., Luft, B.J., and Dattwyler, R.J. (1999b). Infection with multiple strains of *Borrelia burgdorferi* sensu stricto in patients with Lyme disease. Arch. Dermatol. *135*, 1329–1333.

Shadick, N.A., Liang, M.H., Phillips, C.B., Fossel, K., and Kuntz, K.M. (2001). The cost-effectiveness of vaccination against Lyme disease. Arch Intern Med *161*, 554–561.

Shang, E.S., Wu, X.Y., Lovett, M.A., Miller, J.N., and Blanco, D.R. (2001). Homologous and heterologous *Borrelia burgdorferi* challenge of infection-derived immune rabbits using host-adapted organisms. Infect. Immun. *69*, 593–598.

Sigal, L.H., Zahradnik, J.M., Lavin, P., Patella, S.J., Bryant, G., Haselby, R., Hilton, E., Kunkel, M., Adler-Klein, D., Doherty, T., *et al.* (1998). A vaccine consisting of recombinant *Borrelia burgdorferi* outer-surface protein A to prevent Lyme disease. Recombinant Outer-Surface Protein A Lyme Disease Vaccine Study Consortium. N. Engl. J. Med. *339*, 216–222.

Simon, M.M., Gern, L., Hauser, P., Zhong, W., Nielsen, P.J., Kramer, M.D., Brenner, C., and Wallich, R. (1996). Protective immunization with plasmid DNA containing the outer surface lipoprotein A gene of *Borrelia burgdorferi* is independent of an eukaryotic promoter. Eur J. Immunol. *26*, 2831–2840.

Simon, M.M., Schaible, U.E., Wallich, R., and Kramer, M.D. (1991). A mouse model for *Borrelia burgdorferi* infection: approach to a vaccine against Lyme disease. Immunol Today *12*, 11–16.

Skotarczak, B. (2002). Canine borreliosis – epidemiology and diagnostics. Ann. Agric. Environ Med. *9*, 137–140.

Smith, R., and Takkinen, J. (2006). Lyme borreliosis: Europe-wide coordinated surveillance and action needed? Eur. Surveill. *11*, E060622, 060621.

Soares, C.A., Lima, C.M., Dolan, M.C., Piesman, J., Beard, C.B., and Zeidner, N.S. (2005). Capillary feeding of specific dsRNA induces silencing of the *isac* gene in nymphal *Ixodes scapularis* ticks. Insect Mol. Biol. *14*, 443–452.

Steere, A.C., Dwyer, E., and Winchester, R. (1990). Association of chronic Lyme arthritis with HLA-DR4 and HLA-DR2 alleles. N. Engl. J. Med. *323*, 219–223.

Steere, A.C., and Glickstein, L. (2004). Elucidation of Lyme arthritis. Nat. Rev. Immunol. *4*, 143–152.

Steere, A.C., Gross, D., Meyer, A.L., and Huber, B.T. (2001). Autoimmune mechanisms in antibiotic treatment-resistant Lyme arthritis. J Autoimmun *16*, 263–268.

Steere, A.C., Sikand, V.K., Meurice, F., Parenti, D.L., Fikrig, E., Schoen, R.T., Nowakowski, J., Schmid, C.H., Laukamp, S., Buscarino, C., and Krause, D.S. (1998). Vaccination against Lyme disease with recombinant *Borrelia burgdorferi* outer-surface lipoprotein A with adjuvant. Lyme Disease Vaccine Study Group. N. Engl. J. Med. *339*, 209–215.

Stevenson, B., and Barthold, S.W. (1994). Expression and sequence of outer surface protein C among North American isolates of *Borrelia burgdorferi*. FEMS Microbiol. Lett. *124*, 367–372.

Stevenson, B., Schwan, T.G., and Rosa, P.A. (1995). Temperature-related differential expression of antigens in the Lyme disease spirochete, *Borrelia burgdorferi*. Infect. Immun. *63*, 4535–4539.

Stewart, P.E., Wang, X., Bueschel, D.M., Clifton, D.R., Grimm, D., Tilly, K., Carroll, J.A., Weis, J.J., and Rosa, P.A. (2006). Delineating the requirement for the *Borrelia burgdorferi* virulence factor OspC in the mammalian host. Infect. Immun. *74*, 3547–3553.

Stover, C.K., Bansal, G.P., Hanson, M.S., Burlein, J.E., Palaszynski, S.R., Young, J.F., Koenig, S., Young, D.B., Sadziene, A., and Barbour, A.G. (1993). Protective immunity elicited by recombinant bacille Calmette-Guerin (BCG) expressing outer surface protein A (OspA) lipoprotein: a candidate Lyme disease vaccine. J. Exp. Med. *178*, 197–209.

Telford, S.R., 3rd, Fikrig, E., Barthold, S.W., Brunet, L.R., Spielman, A., and Flavell, R.A. (1993). Protection against antigenically variable *Borrelia burgdorferi* conferred by recombinant vaccines. J. Exp. Med. *178*, 755–758.

Terekhova, D., Iyer, R., Wormser, G.P., and Schwartz, I. (2006). Comparative genome hybridization reveals substantial variation among clinical isolates of *Borrelia burgdorferi* sensu stricto with different pathogenic properties. J. Bacteriol. *188*, 6124–6134.

Theisen, M., Frederiksen, B., Lebech, A.M., Vuust, J., and Hansen, K. (1993). Polymorphism in *ospC* gene of *Borrelia burgdorferi* and immunoreactivity of OspC protein: implications for taxonomy and for use of OspC protein as a diagnostic antigen. J. Clin. Microbiol. *31*, 2570–2576.

Tilly, K., Bestor, A., Jewett, M.W., and Rosa, P. (2007). Rapid clearance of Lyme disease spirochetes lacking OspC from skin. Infect. Immun. *75*, 1517–1519.

Tilly, K., Casjens, S., Stevenson, B., Bono, J.L., Samuels, D.S., Hogan, D., and Rosa, P. (1997). The *Borrelia burgdorferi* circular plasmid cp26: conservation of plasmid structure and targeted inactivation of the *ospC* gene. Mol. Microbiol. *25*, 361–373.

Tilly, K., Krum, J.G., Bestor, A., Jewett, M.W., Grimm, D., Bueschel, D., Byram, R., Dorward, D., Vanraden, M.J., Stewart, P., and Rosa, P. (2006). *Borrelia burgdorferi* OspC protein required exclusively in a crucial early stage of mammalian infection. Infect. Immun. *74*, 3554–3564.

Topfer, K.H., and Straubinger, R.K. (2007). Characterization of the humoral immune response in dogs after vaccination against the Lyme borreliosis agent: a study with five commercial vaccines using two different vaccination schedules. Vaccine 25, 314–326.

Trimnell, A.R., Davies, G.M., Lissina, O., Hails, R.S., and Nuttall, P.A. (2005). A cross-reactive tick cement antigen is a candidate broad-spectrum tick vaccine. Vaccine 23, 4329–4341.

Trollmo, C., Meyer, A.L., Steere, A.C., Hafler, D.A., and Huber, B.T. (2001). Molecular mimicry in Lyme arthritis demonstrated at the single cell level: LFA-1α_L is a partial agonist for outer surface protein A-reactive T-cells. J. Immunol. *166*, 5286–5291.

Tsao, J.I., Wootton, J.T., Bunikis, J., Luna, M.G., Fish, D., and Barbour, A.G. (2004). An ecological approach to preventing human infection: vaccinating wild mouse reservoirs intervenes in the Lyme disease cycle. Proc. Natl. Acad. Sci. U.S.A. *101*, 18159–18164.

Van Hoecke, C., Comberbach, M., De Grave, D., Desmons, P., Fu, D., Hauser, P., Lebacq, E., Lobet, Y., and Voet, P. (1996). Evaluation of the safety, reactogenicity and immunogenicity of three recombinant outer surface protein (OspA) lyme vaccines in healthy adults. Vaccine *14*, 1620–1626.

Vazquez, M., Muehlenbein, C., Cartter, M., Hayes, E.B., Ertel, S., and Shapiro, E.D. (2008). Effectiveness of personal protective measures to prevent Lyme disease. Emerg. Infect. Dis. *14*, 210–216.

Wallich, R., Helmes, C., Schaible, U.E., Lobet, Y., Moter, S.E., Kramer, M.D., and Simon, M.M. (1992). Evaluation of genetic divergence among *Borrelia burgdorferi* isolates by use of OspA, *fla*, HSP60, and HSP70 gene probes. Infect. Immun. *60*, 4856–4866.

Wallich, R., Kramer, M.D., and Simon, M.M. (1994). A recombinant vaccine for Lyme disease. Behring Inst Mitt, 106–108.

Wallich, R., Siebers, A., Jahraus, O., Brenner, C., Stehle, T., and Simon, M.M. (2001). DNA vaccines expressing a fusion product of outer surface proteins A and C from *Borrelia burgdorferi* induce protective antibodies suitable for prophylaxis but not for resolution of Lyme disease. Infect. Immun. *69*, 2130–2136.

Wang, G., Ojaimi, C., Iyer, R., Saksenberg, V., McClain, S.A., Wormser, G.P., and Schwartz, I. (2001). Impact of genotypic variation of *Borrelia burgdorferi* sensu stricto on kinetics of dissemination and severity of disease in C3H/HeJ mice. Infect. Immun. *69*, 4303–4312.

Wang, G., van Dam, A.P., and Dankert, J. (1999a). Evidence for frequent OspC gene transfer between *Borrelia valaisiana* sp. nov. and other Lyme disease spirochetes. FEMS Microbiol. Lett. *177*, 289–296.

Wang, G., van Dam, A.P., Schwartz, I., and Dankert, J. (1999b). Molecular typing of *Borrelia burgdorferi* sensu lato: taxonomic, epidemiological, and clinical implications. Clin. Microbiol. Rev. *12*, 633–653.

Wang, I.N., Dykhuizen, D.E., Qiu, W., Dunn, J.J., Bosler, E.M., and Luft, B.J. (1999c). Genetic diversity of *ospC* in a local population of *Borrelia burgdorferi sensu stricto*. Genetics *151*, 15–30.

Wiedemann, C., and Milward, F. (1999). Efficacy and safety studies with a new vaccine against Lyme borreliosis in dogs (Merilyme(R)). Tierarztl Umsch 54, 242–249.

Wikle, R.E., Fretwell, B., Jarecki, M., and Jarecki-Black, J.C. (2006). Canine Lyme disease: one-year duration of immunity elicited with a canine OspA monovalent Lyme disease vaccine. Int. J. Appl. Res. Vet. Med. *4*, 23–28.

Will, G., Jauris-Heipke, S., Schwab, E., Busch, U., Rossler, D., Soutschek, E., Wilske, B., and Preac-Mursic, V. (1995). Sequence analysis of *ospA* genes shows homogeneity within *Borrelia burgdorferi* sensu stricto and *Borrelia afzelii* strains but reveals major subgroups within the *Borrelia garinii* species. Med. Microbiol. Immunol. *184*, 73–80.

Willett, T.A., Meyer, A.L., Brown, E.L., and Huber, B.T. (2004). An effective second-generation outer surface protein A-derived Lyme vaccine that eliminates a potentially autoreactive T-cell epitope. Proc. Natl. Acad. Sci. U.S.A. *101*, 1303–1308.

Wilske, B., Busch, U., Fingerle, V., Jauris-Heipke, S., Preac Mursic, V., Rossler, D., and Will, G. (1996). Immunological and molecular variability of OspA and OspC. Implications for *Borrelia* vaccine development. Infection *24*, 208–212.

Wilske, B., Preac-Mursic, V., Gobel, U.B., Graf, B., Jauris, S., Soutschek, E., Schwab, E., and Zumstein, G. (1993). An OspA serotyping system for *Borrelia burgdorferi* based on reactivity with monoclonal antibodies and OspA sequence analysis. J. Clin. Microbiol. *31*, 340–350.

Wormser, G.P., Brisson, D., Liveris, D., Hanincova, K., Sandigursky, S., Nowakowski, J., Nadelman, R.B., Ludin, S., and Schwartz, I. (2008). *Borrelia burgdorferi* genotype predicts the capacity for hematogenous dissemination during early Lyme disease. J. Infect. Dis. *198*, 1358–1364.

Xu, Y., Bruno, J.F., and Luft, B.J. (2005). Identification of novel tick salivary gland proteins for vaccine development. Biochem. Biophys. Res. Commun. *326*, 901–904.

Yang, X., Goldberg, M.S., Popova, T.G., Schoeler, G.B., Wikel, S.K., Hagman, K.E., and Norgard, M.V. (2000). Interdependence of environmental factors influencing reciprocal patterns of gene expression in virulent *Borrelia burgdorferi*. Mol. Microbiol. *37*, 1470–1479.

Yang, X., Li, Y., Dunn, J.J., and Luft, B.J. (2006). Characterization of a unique borreliacidal epitope on the outer surface protein C of *Borrelia burgdorferi*. FEMS Immunol. Med. Microbiol. *48*, 64–74.

Yang, X.F., Alani, S.M., and Norgard, M.V. (2003a). The response regulator Rrp2 is essential for the expression of major membrane lipoproteins in *Borrelia burgdorferi*. Proc. Natl. Acad. Sci. U.S.A. *100*, 11001–11006.

Yang, X.F., Hubner, A., Popova, T.G., Hagman, K.E., and Norgard, M.V. (2003b). Regulation of expression of the paralogous Mlp family in *Borrelia burgdorferi*. Infect. Immun. *71*, 5012–5020.

Yoder, A., Wang, X., Ma, Y., Philipp, M.T., Heilbrun, M., Weis, J.H., Kirschning, C.J., Wooten, R.M., and Weis, J.J. (2003). Tripalmitoyl-S-glyceryl-cysteine-dependent OspA vaccination of Toll-like receptor 2-deficient mice results in effective protection from *Borrelia burgdorferi* challenge. Infect. Immun. *71*, 3894–3900.

Young, J.D. (1998). Underreporting of Lyme disease. N. Engl. J. Med. *338*, 1629.

Zhang, H., and Marconi, R.T. (2005). Demonstration of cotranscription and 1-methyl-3-nitroso-nitroguanidine induction of a 30-gene operon of *Borrelia burgdorferi*: evidence that the 32-kilobase circular plasmids are prophages. J. Bacteriol. *187*, 7985–7995.

Zhang, J.R., and Norris, S.J. (1998). Genetic variation of the *Borrelia burgdorferi* gene *vlsE* involves cassette-specific, segmental gene conversion. Infect. Immun. *66*, 3698–3704.

Zhang, X., Meltzer, M.I., Pena, C.A., Hopkins, A.B., Wroth, L., and Fix, A.D. (2006). Economic impact of Lyme disease. Emerg. Infect. Dis. *12*, 653–660.

Zhong, W., Wiesmuller, K.H., Kramer, M.D., Wallich, R., and Simon, M.M. (1996). Plasmid DNA and protein vaccination of mice to the outer surface protein A of *Borrelia burgdorferi* leads to induction of T helper cells with specificity for a major epitope and augmentation of protective IgG antibodies *in vivo*. Eur J. Immunol. *26*, 2749–2757.

Zückert, W.R., Kerentseva, T.A., Lawson, C.L., and Barbour, A.G. (2001). Structural conservation of neurotropism-associated VspA within the variable *Borrelia* Vsp-OspC lipoprotein family. J. Biol. Chem. *276*, 457–463.

Lyme Disease in Humans

18

Justin D. Radolf, Juan C. Salazar and Raymond J. Dattwyler

Abstract

Lyme disease is a rapidly emerging tick-borne, complex, multisystem infectious disorder caused by the spirochaetal bacterium *Borrelia burgdorferi*. The ailment, which affects adults and children alike, is widespread in the northern hemisphere and continues to expand as humans encroach on the sylvatic habitat of the spirochaete's mammalian reservoirs. Since first identified in the 1970s the incidence of Lyme disease has increased more than 30-fold and it is now considered the most prevalent arthropod-transmitted infection in both the USA and Europe. *Borrelia burgdorferi* is transmitted by ticks of the *Ixodes ricinus* complex, including *I. scapularis, I. ricinus,* and *I. persulcatus*. In North America, *B. burgdorferi* sensu stricto is the only species proven to be pathogenic for humans. In Europe, both *B. afzelii* and *B. garinii* are most commonly associated with human disease. The spirochaete's genomic features, as well as its unique molecular architecture, are considered to have a seminal role not only in how it is transmitted from ticks to humans, but also how it triggers immune responses in afflicted individuals. Inflammatory manifestations associated with the disease result from the host's innate and adaptive immune responses to the bacterium, rather than from toxigenic molecules, which borrelia cannot produce. Indeed, the deposition of spirochaetes into human dermal tissue generates a local inflammatory response that becomes manifest as erythema migrans (EM), the hallmark skin lesion of Lyme disease in North America. In Europe, two additional dermatological disorders, borrelial lymphocytoma and acrodermatitis chronicum atrophicans (ACA) are frequently associated with infection. EM is frequently accompanied by 'flu-like' symptoms, including myalgias, arthralgias, and fever, which are generally believed to be cytokine-mediated in response to haematogenous spread of the bacterium. If treated appropriately, the prognosis is excellent; however, if untreated, patients may develop a wide range of inflammatory clinical manifestations, most commonly involving the central nervous system, joints and heart. Within days of treatment, the signs and symptoms associated with the disease typically begin to subside, although in some individuals a complete recovery can take several weeks or even months. A minority of treated patients may go on to develop a poorly defined fibromyalgia-like illness, which is not responsive to prolonged antimicrobial therapy. Below we integrate current knowledge regarding the ecological, epidemiological, microbiological, and immunological facets of Lyme disease into a conceptual framework that sheds light on the disorder that healthcare providers encounter.

18.1 Introduction

Lyme disease is the prototype of an emerging infectious disease. The isolation of its aetiological agent, *Borrelia burgdorferi*, in 1982 (Burgdorfer *et al.*, 1982; Burgdorfer and Hayes, 1986) capped an intensive hunt for a pathogen that just a short time before had been cultured from a new species of deer tick, *Ixodes dammini*, which subsequently was designated *Ixodes scapularis*. Critical in the chain of events that led to the discovery of the bacterium was the observation that many patients involved in the epidemic of oligoarthritis geographically clustered about Lyme, Connecticut, also had a skin rash, now called erythema migrans (Steere *et al.*, 1977a,b), which was closely associated in northern Europe with the bite of the sheep tick *Ixodes ricinus* (Lipschütz, 1923). As is typical of emerging infections, the isolation of the causative agent sparked an explosive increase in our knowledge of the bacterium, the disease it causes, and the enzootic cycle that sustains it. We now know that Lyme disease is widespread in the Northern Hemisphere and that its range is expanding as landscapes denuded during the Industrial Revolution become re-forested and humans encroach on the sylvatic habit of the spirochaete's mammalian reservoirs. Although the protean clinical manifestations of Lyme disease are legendary and continue to be a source of considerable controversy, a relatively small number of syndromes dominate the clinical picture and the vast majority of patients present with easily treatable acute illness. Serological surveys conducted in high prevalence areas indicate that asymptomatic infection also is relatively common; thus, despite the bacterium's notorious reputation, benign outcomes occur frequently (Hanrahan *et al.*, 1984a; Steere *et al.*, 1986; Gustafson *et al.*, 1995).

The genomic sequence of *B. burgdorferi* revealed that the spirochaete lacks genes encoding known toxigenic molecules as well as the secretory apparatus required to deliver them to the extracellular milieu it inhabits within its mammalian host (Fraser *et al.*, 1997; Casjens *et al.*, 2000;). This finding, in concert with an extensive corpus in experimental animal models, primarily the mouse, buttress the prevailing viewpoint that tissue damage and symptomatology in Lyme disease result from the inflammatory response elicited by the bacterium and its constituents and is perpetuated by the organism's propensity for evading those responses (see also Chapters 14 and 15). Beyond this reductionist view, however, we still have only a limited understanding of the microbial factors and pathogenic mechanisms that sway the balance following the adventitious encounter of humans with this peculiar zoonotic microorganism.

Over the years, a number of excellent clinical reviews of Lyme disease have been published in journals and textbooks, so there is no need to reiterate all of this information herein. In addition, the Infectious Diseases Society of America (IDSA) recently issued comprehensive guidelines for the diagnosis and management of this infection (Wormser *et al.*, 2006). Rather, the primary objective of this chapter is to coalesce the present body of knowledge regarding the ecological, epidemiological, microbiological, and immunological facets of Lyme disease into a conceptual framework that illuminates the actual disorder that physicians encounter. As will be seen, many of the factors that determine the level of risk for populations or individuals lie outside the conventional perimeter of the healthcare system. Since our goal is to develop a picture of the human disorder, we intend to rely as much as possible upon data obtained from human studies or human systems. Extrapolation from *in vitro* and animal models, however, is unavoidable given the constraints of human experimentation.

18.2 Historical overview

The history of what we now call Lyme disease dates back to the early part of the twentieth century. However, the discoveries that led to the contemporary view of this infectious disease occurred in 1982, when Willy Burgdorfer and his colleagues identified what turned out to be a new *Borrelia* species from *Ixodes* ticks collected on eastern Long Island (Burgdorfer *et al.*, 1982; Burgdorfer and Hayes, 1986). Within a year, two separate groups, led by Jorge Benach and Allen Steere, isolated the newly described spirochaete from patients, thereby establishing it as the aetiological agent of Lyme disease (Benach *et al.*, 1983; Steere *et al.*, 1983b). The pathogen was named *Borrelia burgdorferi* after Willy Burgdorfer, the renowned medical entomologist who played a critical role in its discovery. The discovery of *B. burgdorferi* as the cause of Lyme disease led directly to the realization that this disorder was not restricted to North America. Indeed, over the ensuing years it became apparent that a range of clinical syndromes described in the European medical literature beginning in the early twentieth century were, in fact, manifestations of a complex disorder caused by members of what eventually came to be known as the *B. burgdorferi* sensu lato complex (Burgdorfer and Hayes, 1986). Erythema migrans (EM), the classic skin lesion associated with this infectious disease, was first publicly described by Afzelius, a Swedish dermatologist, in 1910 (Afzelius, 1910). Afzelius also correctly hypothesized that EM resulted from the tick-borne transmission to humans of a zoonotic pathogen (Afzelius, 1921); in 1921, Lipschütz identified *Ixodes ricinus* as the vector (Lipschütz, 1923). By the beginning of World War II, European investigators had recognized that EM was associated with several neurological and dermatological disorders and with musculoskeletal complaints (Buchwald, 1883; Herxheimer and Hartmann, 1902; Garin and Bujadoux, 1922; Bannwarth, 1941; Bannwarth, 1944). After the war, Lennhoff, a Swedish pathologist, described what he believed were spirochaetes in biopsies of EM lesions (Lennhoff, 1948), while Hollström demonstrated that penicillin was effective for its treatment (Hollström, 1951).

In 1970, a Wisconsin dermatologist, Scrimenti, reported the first well-documented case of EM acquired in the USA (Scrimenti, 1970). Ironically, the patient was a deer hunter. The first reports of EM in the northeastern USA were by Mast and Burrows, who reported a cluster of cases of EM in southeastern Connecticut in 1976 (Mast and Burrows, Jr., 1976). Soon afterwards, Steere and his colleagues at Yale began investigating an unusual cluster of arthritis in a small group of patients, many of whom were children, in and around Old Lyme, Connecticut; they called this mysterious ailment Lyme arthritis (Steere *et al.*, 1977a,b). With the realization that most of the arthritis patients had previously had EM and that heart block, facial nerve palsy or meningitis also was associated with this skin lesion, the name subsequently was changed to Lyme disease (Steere *et al.*, 1977a). Detailed narratives of the history of Lyme disease and the events surrounding the discovery of *Borrelia burgdorferi* can be found in Chapter 1 in this volume and in *Bull's Eye: Unraveling the Medical Mystery of Lyme Disease* (Erdlow, 2003).

18.3 Epidemiology

Lyme disease accounts for more than 90% of all vector-borne disease in the US, making it far and away the most prevalent arthropod-transmitted infection (Dennis and Hayes, 2002). Since the first identification of *B. burgdorferi* in 1982, the annual number of reported cases of Lyme disease has increased more than 30-fold (Dennis and Hayes, 2002). Statistics reported by the Centers for Disease Control and Prevention for the 16-year period from 1992–2007 underscore the continuing surge in new infections (Fig. 18.1) (Bacon *et al.*, 2008). During this interval, a total of 275,773 cases of Lyme disease were reported to the CDC. The number of cases in 2007 (27,444) represented a 176% increase over that reported in 1992 (9908). Cases occur mostly among persons living in coastal northeastern, mid-Atlantic, or north-central regions of the US (Bacon *et al.*, 2008). During the 1992–2006 reporting period, 93% of cases were reported from 10 states: Connecticut, Delaware, Massachusetts, Maryland, Minnesota, New Jersey, New York, Pennsylvania, Rhode Island and Wisconsin. Connecticut has the highest incidence of disease in the country (73.6 cases per 100,000 persons). Physicians, particularly those in non-endemic areas, need to appreciate the focal geographic distribution of the disease when considering the diagnosis. Even within endemic states, the distribution of the disease is extremely non-uniform. For example, in New York, the state with the largest number of cases, more than 80% are reported from just 5 of 62 counties (Bacon *et al.*, 2007). Lyme disease has a bimodal age distribution, with peaks at 5–9 years old and 55–59 years old. It has no significant gender predilection (53% of the cases occur in males) and occurs overwhelmingly (94%) in whites (Bacon *et al.*, 2008). Most cases occur from June through August, when nymphal ticks feed and humans are most likely to engage in outdoor activities (Bacon *et al.*, 2008). As would be expected, risk is proportional to the time spent outdoors, whether recreationally or occupationally, in or near tick-infested woods (Dennis and Hayes, 2002; Hengge *et al.*, 2003). The cumulative risk for acquiring the disease for individuals residing in hyper-endemic foci was demonstrated in a 1989 study that showed that, within an 8-year observation period, 66 of 190 persons living near a coastal Massachusetts nature preserve developed Lyme disease (Lastavica *et al.*, 1989). Serological surveys have revealed rates of asymptomatic or subclinical infections among persons living in endemic areas that equal or

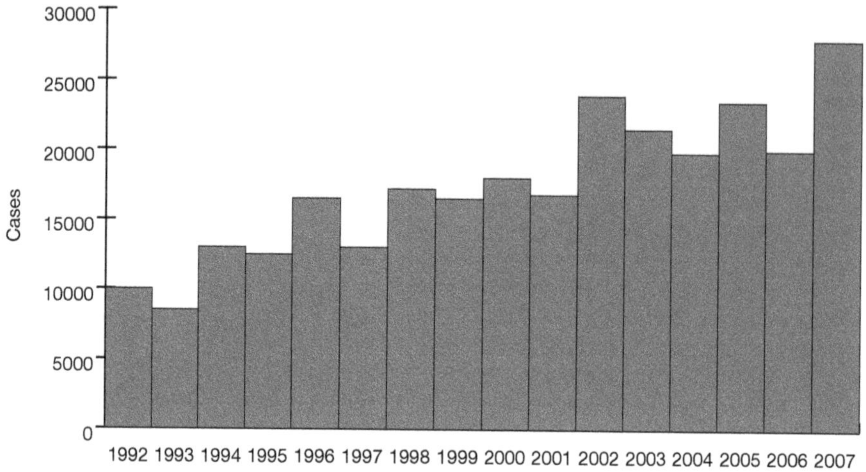

Figure 18.1 Number of newly reported Lyme disease cases by year of diagnosis, USA, 1992–2007 (From http://www.cdc.gov/ncidod/dvbid/lyme/ld_UpClimbLymeDis.htm).

exceed those of persons who become ill (Hanrahan *et al.*, 1984a; Steere *et al.*, 1986; Gustafson *et al.*, 1995).

Lyme disease also is the most prevalent vector-borne illness in Europe, where it is widely distributed, except in the warmer south (e.g. Sicily and southern Spain) and northern extremities of Scandinavia and Russia (Derdakova and Lencakova, 2005). Although methods used to acquire surveillance data vary widely throughout Europe, in 2005 an estimated 19,000 cases were reported from 18 countries, with the highest incidence rates in Slovenia, Austria, and Netherlands (206, 135 and 103 per 100,000, respectively) (Smith and Takkinen, 2006). These estimates, if accurate, are well above the rates seen in many endemic areas in the USA (Bacon *et al.*, 2008).

18.4 Phylogenetic diversity of Lyme disease spirochaetes

Phylogenetically, *B. burgdorferi* is a spirochaete, a member of an ancient and deeply branching bacterial phylum, distantly related to the proteobacteria that includes the true Gram-negative bacteria (Paster *et al.*, 1991). *Borrelia* species fall into two major phyletic groups. One contains the agent of Lyme disease and related spirochaetes. The other contains the relapsing fever spirochaetes that, except for the louse-borne agent of endemic relapsing fever (*B. recurrentis*), are transmitted by soft-bodied, argasid ticks (see Chapters 10 and 13). Barbour and colleagues (Barbour *et al.*, 1985, 1986) were the first to report on the heterogeneity of *B. burgdorferi* strains based on differences in apparent molecular masses of OspA and OspB by SDS-PAGE and differential reactivity with specific monoclonal antibodies. Since then, there has been increasing recognition of the phylogenetic diversity of Lyme disease spirochaetes and its importance for pathogenesis, diagnosis, and vaccine development (van Dam, 2002; van Dam *et al.*, 1993; Nadelman and Wormser, 1998; Wilske *et al.*, 2007; Piesman and Gern, 2004; Stanek and Strle, 2008). A variety of molecular and immunological methods have been developed for differentiating *Borrelia* species and strains (Wang *et al.*, 1999b; Wilske *et al.*, 2007; Rauter *et al.*, 2002; Postic *et al.*, 2007) (see also Chapter 16). Currently, there are at least 13 recognized species in the sensu lato complex: *B. burgdorferi* sensu stricto, *B. afzelii*, *B. garinii*, *B. japonica*, *B. valaisiana*, *B. lusitaniae*, *B. andersonii*, *B. bissettii*, *B. tanukii*, *B. turdi*, *B. sinica*, *B. californiensis*, and *B. spielmanii* along with isolates whose taxonomic classifications are not yet determined (Postic *et al.*, 1998; Wang *et al.*, 1999a; P; Piesman and Gern, 2004; Richter *et al.*, 2004; Postic *et al.*, 2007) (see also Chapter 9).

Borrelia burgdorferi sensu stricto was thought to be the sole genospecies in the USA until 1995, when Marconi and associates isolated *B. andersonii* from cottontail rabbits and *I. dentatus* ticks (Marconi *et al.*, 1995). It soon became evident that considerable diversity exists among

North American spirochaetes and that secondary enzootic cycles involving *B. burgdorferi* sensu lato strains other than *B. burgdorferi* sensu stricto exist in geographically widespread but confined loci (Postic *et al.*, 1998; Oliver, Jr. *et al.*, 2003). The best characterized of these involves *B. bissettii*, which has been recovered in California, Colorado, Illinois and Connecticut (Maupin *et al.*, 1994; Piesman and Schneider, 2002). *B. garinii* and *B. afzelii* are the dominant genospecies in Europe and have overlapping distributions throughout the continent (Hubalek and Halouzka, 1997; Piesman and Gern, 2004; Derdakova and Lencakova, 2005; Kurtenbach *et al.*, 2006). *Borrelia burgdorferi* sensu stricto, the least prevalent of the three major genospecies in Europe, is confined to western regions of the continent (Piesman and Schneider, 2002; Derdakova and Lencakova, 2005). Originally thought to be limited to Switzerland and Ireland, *B. valaisiana* is now known to be widespread in Europe (Derdakova and Lencakova, 2005; Kurtenbach *et al.*, 2006). *B. lusitaniae* appears to be limited to the Iberian Peninsula (Piesman and Gern, 2004), while *B. spielmanii* has been isolated from ticks in the Czech Republic, Germany and France (Derdakova and Lencakova, 2005; Fingerle *et al.*, 2008).

In North America, *B. burgdorferi* sensu stricto is the only species proven to be pathogenic for humans (Piesman and Gern, 2004). The vast majority of European infections are attributable to *B. burgdorferi* sensu stricto, *B. afzelii* and *B. garinii* (Hubalek and Halouzka, 1997; Piesman and Gern, 2004), although *B. bissetti*, *B. lusitaniae*, and *B. valaisiana* have been isolated from a handful of patients (Picken *et al.*, 1996; Rijpkema *et al.*, 1997; Collares-Pereira *et al.*, 2004; Derdakova and Lencakova, 2005). In Europe, most dermatological manifestations are attributed to *B. afzelii*, while neuroborreliosis is mostly caused by *B. garinii* (Hubalek and Halouzka, 1997; Wilske *et al.*, 2007; Strle *et al.*, 2006; Stanek and Strle, 2008). Mixed infections have been reported in ticks, mice, and, sometimes, patients (Guttman *et al.*, 1996; Tazelaar *et al.*, 1997; Seinost *et al.*, 1999b; Jouda *et al.*, 2004; Rauter *et al.*, 2002; Ruzic-Sabljic *et al.* 2005; Anderson and Norris, 2006;). The more widespread distribution of *B. afzelii* and *B. garinii* in Europe, as compared with *B. burgdorferi* sensu stricto, is believed to explain why disease manifestations that are commonly seen in Europeans are rarely, if ever, encountered in North America and why arthritis, which tends to be caused by *B. burgdorferi* sensu stricto, is less frequent in Europe (Steere and Glickstein, 2004; Piesman and Gern, 2004; Stanek and Strle, 2008). While studies in mice and non-human primates do not fully replicate the organotropisms deduced from clinical and epidemiological studies (Anthonissen *et al.*, 1994; van Dam, 2002), they do support the clinical impression that *B. garinii* and *B. afzelii* have less invasive potential than *B. burgdorferi* sensu stricto and that *B. afzelii* has a predilection for skin (Barthold, 1999; Pachner *et al.*, 2004; Cadavid *et al.*, 2004).

18.5 Vector ecology

Borrelia burgdorferi is transmitted by ticks of the *Ixodes ricinus* complex (Burgdorfer *et al.*, 1991; Piesman and Gern, 2004) (see also Chapter 10). *I. scapularis*, *I. ricinus*, and *I. persulcatus*, the principal vectors in the northeastern USA, Europe, and Asia, respectively, feed on many different vertebrate species. This generalist feeding behaviour enhances the opportunities for transmission of spirochaetes amongst infection-competent host species, linkage of ecological niches, and expansion of the geographic range of the disease (Kurtenbach *et al.*, 2006). *Ixodes* ticks have a two-year life cycle with four life-stages: egg, larva, nymph, and adult. They feed only three times, once at each non-egg life stage (Anderson *et al.*, 1989; Bosler, 1993). Ticks are born uninfected, and only become infected with *B. burgdorferi* after feeding upon an infected animal. In competent vectors, borrelial infection persists through moulting to the next stage or instar, a phenomenon called transstadial transmission (Gern and Humair, 2002). Naïve larvae acquire the spirochaete by feeding on an infected reservoir host, typically a rodent, and, after moulting to the nymphal stage, transmit the pathogen when they feed on an uninfected host. The dependence of *B. burgdorferi* on efficient transstadial transmission for long-term survival is a major point of distinction from the relapsing fever spirochaetes, which can be maintained within their argasid tick vectors by vertical or transovarial transmission (Magnarelli and Anderson, 1987; Mather *et al.*, 1991). Nymphal ticks are believed to be responsible for the majority of human infections, although adult ticks are capable of transmitting the pathogen (Falco *et al.*, 1999). In addition to having high infection rates, nymphs quest during the summer months when humans are most likely to be outdoors and, because of their small size (approximately that of a poppy seed), are difficult to detect on body surfaces, clothing, and pets by visual inspection. While adult ticks tend to have the highest prevalence of infection, they quest in late autumn through early spring, when humans are less likely to encounter them and more apt to wear clothing that provides a protective barrier; they also are considerably larger and, therefore, easier to detect (Dennis and Hayes, 2002). Not surprisingly, the risk of infection for humans correlates with infection rates in vectors and reservoir hosts as well as tick density (Mather *et al.*, 1996; Falco *et al.*, 1999; Daniels *et al.*, 1998). It is important to emphasize that humans are incidental and dead-end hosts for the spirochaete, mere targets of opportunity for questing nymphs attracted to body heat and expired carbon dioxide, features we share with the vertebrate species they normally feed upon.

In the northeastern and north central USA, *B. burgdorferi* is maintained in an efficient cycle involving *I. scapularis* (formerly *I. dammini*) and *Peromyscus leucopus*, the white-footed mouse (Spielman *et al.*, 1985; Peisman, 2002). *I. scapularis* infection rates in endemic areas are commonly over 25% (Peisman, 2002). Serosurveys of mice trapped in endemic areas have yielded infection rates as high as 80% during the summer and as many as 10% of animals have active infection based on PCR analysis of blood (Anderson *et al.*, 1987; Bunikis *et al.*, 2004). *P. leukopus* are ubiquitous in forested environments and, once infected with *B. burgdorferi*, can remain infected for life (Levine *et al.*, 1985; Donahue *et al.*, 1987). Interestingly, infection in *Peromyscus* is unaccompanied by end organ histopathological changes (Brown and Lane, 1992; Oliver, Jr. *et al.*, 2003), indicating a high degree of immunological adaptation to the presence of the spirochaete. Adult ticks play no role as vectors because adult males do not feed and adult females usually feed on large animals, typically white-tailed deer that are not reservoir-competent (Kurtenbach *et al.*, 2006). White-tailed deer are, however, the preferred reproductive hosts for adult ticks. The explosive increase in the deer population in restored woodlands, along with the encroachment of humans on areas heavily infested with *B. burgdorferi*-infected ticks, has been a pivotal factor in the dramatic expansion of Lyme disease in the northeastern USA during the latter part of the twentieth century (Maupin *et al.*, 1991; Thompson *et al.*, 2001; Piesman and Schneider, 2002).

Borrelia burgdorferi sensu stricto is capable of infecting phylogenetically distant vertebrate hosts, including mammalian and avian species, and there is strong evidence for multiple reservoirs other than *P. leucopus* in endemic areas (Piesman and Schneider, 2002; Brisson and Dykhuizen, 2004; Tsao *et al.*, 2004; Kurtenbach *et al.*, 2006). Identification of reservoirs other than *P. leucopus* is of enormous importance in formulating strategies for controlling Lyme disease that depend upon vaccination of wildlife hosts (Tsao *et al.*, 2004; Gomes-Solecki *et al.*, 2006; Scheckelhoff *et al.*, 2006). While longitudinal OspA vaccination studies focusing on *P. leucopus* have shown decreases in nymphal tick infection rates, the decrements were small, most likely because the cycle of transmission was not interrupted in other reservoirs (Tsao *et al.*, 2004). Human infections in Pacific coastal sites are attributed to a cycle in which *I. pacificus*, which bites humans, serves as a bridge vector for a cycle involving woodrats, mice, and *I. spinipalis*, which does not attach to humans (Brown and Lane, 1992). *I. pacificus* is a relatively poor vector for transmission of *B. burgdorferi* since only 1–3% of this tick species are infected (Hengge *et al.*, 2003). Complex sylvatic transmission cycles involving 'atypical' (i.e. non-sensu stricto) *B. burgdorferi* species (e.g. *B. andersoni* and *B. bissettii*) are now known to exist throughout the USA, including areas not considered endemic for Lyme disease (Peisman, 2002). These cycles appear to be of lesser importance for humans because they involve reservoir hosts in remote geographic areas, *Ixodes* species that are less vector-competent or tend not to bite humans, and/or *Borrelia* species with limited infectivity for humans (Maupin *et al.*, 1994; Peisman, 2002; Oliver, Jr. *et al.*, 2003).

The vector ecology of Lyme disease in Europe is even more complicated than in North America and can only be briefly mentioned here (see Gern and Humair, 2002 and Piesman and Gern, 2004 for comprehensive reviews). *I. ricinus*, the sheep tick, is the principal vector for all three pathogenic species of *B. burgdorferi* sensu lato commonly isolated from European patients and also has been implicated in transmission of two species, *B. valaisiana* and *B. bissettii*, whose overall contribution to human disease is thought to be limited (Piesman and Gern, 2004) (see below). *I. ricinus* is known to feed on as many as 300 vertebrate species (Gern and Humair, 2002). Small mammals, ground-foraging birds, and reptiles are common hosts for the larval and nymphal stages of *I. ricinus*, while adults (females) feed on medium- to large-sized mammals; both immature and adult stages will attach to humans. As *B. burgdorferi* sensu lato has been found in all *I. ricinus* populations examined (Hubalek *et al.*, 1998), the risk of infection to humans, therefore, is appreciable wherever *I. ricinus* occurs. The widespread distribution of *B. burgdorferi* in Europe was underscored by a 1997 database analysis that found that *B. burgdorferi* sensu lato had been isolated by culture or detected by PCR in 26 European countries spanning the length and breadth of the continent (Hubalek and Halouzka, 1997). Furthermore, the authors conjectured that the distribution is more reflective of the effort expended by a particular country in identifying spirochaetes rather than the bacterium's actual geographic distribution. Because *I. ricinus* can be found in every European country, they speculated that *B. burgdorferi* sensu lato and human cases of Lyme disease were likely to occur in every country in Europe. Two additional tick species, *I. hexagonus* and *I. uriae*, maintain the spirochaete in secondary transmission cycles separate from those in *I. ricinus*. *I. hexagonus* has a more restricted host range than *I. ricinus*, feeding primarily on carnivores, such as foxes and mustelids, occasionally on lagomorphs, and rarely on birds (Gern and Humair, 2002). *I. uriae*, which is involved in the transmission of some *B. garinii* serotypes and *B. valaisiana*, infests more than 50 species of seabirds, as well as passerines resting in marine habitats, and is believed to bear primary responsibility for global dispersal of spirochaetes (Olsen *et al.*, 1993; Olsen *et al.*, 1995).

18.6 The spirochaete

18.6.1 Morphology, ultrastructure and composition

Whereas most non-borrelial spirochaetes are helical or spiral in shape and translate as undulating corkscrews (Holt, 1978), *B. burgdorferi* translates via posteriorly propagated planar waves (Goldstein *et al.*, 1994). A distinctive morphological feature of *B. burgdorferi*, compared to non-borrelial spirochaetes, is that its periplasmic flagella not only transfer motive force to the cell cylinder but also serve a skeletal function in dividing as well as non-dividing cells (Wolgemuth *et al.*, 2006). Lyme disease spirochaetes lacking the major flagellar filament protein, FlaB, form elongated rods and grow as long chains with incomplete division planes (Motaleb *et al.*, 2000). By contrast, flagella-less treponemes retain the helical shape of the wild-type cell, which is imparted by cell wall constituents, most likely peptidoglycan (Wolgemuth *et al.*, 2006).

Because of its double-membrane ultrastructure, *B. burgdorferi* often has been likened to Gram-negative bacteria. This analogy is inaccurate from the standpoints of phylogeny (see Chapter 9), ultrastructure, composition (see Chapter 6), and, most recently, comparative genomics (Daubin *et al.*, 2002; Platanias, 2003). Early reports that *B. burgdorferi* contains lipopolysaccharide (Beck *et al.*, 1985; Habicht *et al.*, 1986) were disproved, initially by careful chemical analysis (Takayama *et al.*, 1987), and subsequently by genomic sequencing, which failed to identify orthologues for enzymes known to be involved in LPS biosynthesis (Fraser *et al.*, 1997). The absence of LPS has important clinical ramifications inasmuch as spirochaetaemic Lyme disease patients rarely, if ever, manifest sepsis-like pathophysiology comparable to that seen in patients with Gram-negative bacteraemia (Steere, 2001; Smith *et al.*, 2002; Wormser *et al.*, 2005). Consequently, the presence of a sepsis syndrome in a Lyme disease patient should prompt a search for other infections, including babesiosis and anaplasmosis, which also are transmitted by ixodid ticks.

The outer membrane of *B. burgdorferi* differs in other important respects from its Gram-negative counterparts:

1 It is much more easily damaged during routine laboratory manipulations (e.g. centrifugation and resuspension), a property that has often led to the incorrect surface localization of cytoplasmic membrane-associated lipoproteins, and it is far more susceptible to detergent solubilization (Brusca *et al.*, 1991; Cox *et al.*, 1996).

2 It contains a much lower density of proteins with membrane-spanning domains, as assessed by freeze-fracture electron microscopic analysis (Radolf *et al.*, 1994; Walker *et al.*, 1991). Although proteins with porin-like properties have been identified in the *B. burgdorferi* outer membrane (Östberg *et al.*, 2002; Pinne *et al.*, 2004; Pinne *et al.*, 2007), the bacterium does not contain orthologues for well-characterized Gram-negative porins (Fraser *et al.*, 1997; Nikaido, 2003). The physical properties of the outer membrane permeability barrier and the mechanisms by which nutrients traverse it either *in vitro* or *in vivo* are poorly understood. The spirochaete does, however, contain orthologues for TolC and the other components of an ATP-dependent, multidrug resistance efflux system that has been shown to contribute to the bacterium's inherent antimicrobial resistance (Bunikis *et al.*, 2008).

3 While lacking true LPS, *B. burgdorferi* outer membranes do contain large amounts of immunogenic but non-inflammatory monogalactosyl diacylglycerolipids (Wheeler *et al.*, 1993; Norgard *et al.*, 1996; Kinjo *et al.*, 2006; Schroder *et al.*, 2008), one of which is a unique cholesterol-containing galactoside (Ben-Menachem *et al.*, 2003).

4 Arguably the most notable difference is the number and variety of lipoproteins that adorn the borrelial surface (Fraser *et al.*, 1997). In Gram-negative bacteria, lipoproteins are typically anchored to the inner leaflet of the outer membrane or the periplasmic leaflet of the cytoplasmic membrane and are not exported to the bacterial surface (Pugsley, 1993; Narita *et al.*, 2004). The mechanism for delivery of lipoproteins to the borrelial surface is believed to be distinct from the secretory pathway by which lipoproteins are anchored to the outer leaflet of the outer membrane in Gram-negative bacteria (Schulze and Zückert, 2006; Bos *et al.*, 2007) (see also Chapter 6). Given the rapid alteration of surface lipoproteins at particular stages of the borrelial life cycle (see below), the secretory pathway also has to be closely integrated with the genetic mechanisms that regulate lipoprotein expression.

Although the tertiary structures of borrelial outer surface lipoproteins differ considerably (Li *et al.*, 1997; Kumaran *et al.*, 2001; Eicken *et al.*, 2001; Eicken *et al.*, 2002; Becker *et al.*, 2005; Radolf and Caimano, 2008), their overall topologies are identical. All consist of soluble polypeptides tethered to the external leaflet of the outer membrane by N-terminal lipids (Jones *et al.*, 1995; Cox *et al.*, 1996). This topological configuration enables borrelia to display on their surface a wide variety of structures capable of interacting with diverse molecules in both the arthropod and vertebrate hosts. As discussed below and throughout this volume, the lipoprotein composition of the *B. burgdorferi* outer membrane is dynamic, undergoing frequent modification in response to environmental signals and immunological pressures. Presumably, lipid-anchored outer surface proteins provide a combination of functional and architectural versatility that cannot be attained with conventional outer membrane spanning proteins.

18.6.2 Genomic complexity

When *B. burgdorferi* was first isolated, the morphological resemblance between it and *Treponema pallidum*, along with the clinical similarities between Lyme disease and syphilis, suggested that the Lyme disease spirochaete might be a new pathogenic treponeme (Steere *et al.*, 1983b; Benach *et al.*, 1983). DNA–DNA hybridization studies, however, disclosed that the new bacterium was much more closely related to the agents of relapsing fever, but sufficiently distinct to merit designation as a separate species (Hyde and Johnson, 1984). The topological complexity of the *B. burgdorferi* genome came to light in a series of classic studies conducted in the late 1980s and early 1990s by Barbour and colleagues (Barbour, 1988; Ferdows and Barbour, 1989; Hinnebusch *et al.*, 1990; Hinnebusch and Barbour, 1991; Kitten and Barbour, 1992; Sadziene *et al.*, 1993b). They reported that both Lyme disease and relapsing fever *Borrelia* contain linear chromosomes of approximately 1 MB along with a complement of low-copy number circular and linear plasmids. The linear plasmids and chromosome were found to have covalently closed ends (hairpin telomeres), which are rare in prokaryotes (Stewart *et al.*, 2005), the end products of an unusual and essential enzyme designated ResT (Huang *et al.*, 2004; Chaconas *et al.*, 2001). Two other seminal pre-genomics observations relating to the spirochaete's plasmids laid the groundwork for pathogenesis-related research themes that are being pursued intensely in the post-genomics era. Loss of plasmids during serial propagation was found to result in diminished infectivity and changes in antigenic profile (Schwan and Burgdorfer, 1987; Schwan *et al.*, 1988; Simpson *et al.*, 1990; Xu *et al.*, 1996; Purser and Norris, 2000; Labandeira-Rey and Skare, 2001). In addition, many groups reported that *B. burgdorferi* harbour a bewildering number of plasmid-encoded repetitive sequences, including multiple paralogous families of differentially expressed lipoproteins (Akins *et al.*, 1995; Porcella *et al.*, 1996; Marconi *et al.*, 1996; Stevenson *et al.*, 1996; Casjens *et al.*, 1997; Guo *et al.*, 1998; Zückert *et al.*, 1999; Yang *et al.*, 1999; Roberts *et al.*, 2000; Sung *et al.*, 2000; Caimano *et al.*, 2000; Porcella *et al.*, 2000). The availability of differentially expressed surface lipoproteins was hypothesized to confer functional diversity for different host and/or tissue milieus, and/or to facilitate the spirochaete's ability to escape antibody-mediated clearance (Akins *et al.*, 1995; Porcella *et al.*, 1996; Casjens *et al.*, 1997; Das *et al.*, 1997). It is noteworthy that this hypothesis has not yet been proven.

The complete genomic sequence of the prototype B31 strain of *B. burgdorferi* (Fraser *et al.*, 1997; Casjens *et al.*, 2000), published in two instalments, in 1997 and 2000, provided the Rosetta stone needed to interpret this bewildering plethora of sequence and mapping data (see Chapter 2 for a detailed presentation). Several general features highlight the uniqueness of the spirochaete's genome. First, it does not contain orthologues for any recognized virulence factors of either Gram-negative or Gram-positive organisms (see also Chapter 12). This has provided strong support for the generally accepted notion that tissue damage and, ultimately, clinical manifestations in patients with Lyme disease result from the host response to the bacterium rather than secretion of toxigenic molecules. Second, an unusually large percentage of the genome (40% of the 1.6 MB total) is composed of plasmids (12 linear and 9 circular). Third, the plasmids contain a remarkable redundancy of genetic information. The linear plasmids, though substantially different from one another, contain numerous paralogous sequences, mostly encoding lipoproteins, as well as proteins required for plasmid replication and segregation (Eggers *et al.*, 2002; Stewart *et al.*, 2003; Grimm *et al.*, 2004a; Beaurepaire and Chaconas, 2005; Deneke and Chaconas, 2008). Many of the paralogous sequences on linear plasmids encode mutationally inactivated or truncated polypeptides (i.e. pseudogenes) that, collectively, are believed to indicate that the genome is in a rapid state of evolutionary flux (Casjens *et al.*, 2000; Kobryn and Chaconas, 2005). The degree of redundancy among circular plasmids is truly unprecedented among prokaryotes. Eight of the nine are derivatives of the same plasmid family and seven of these contain large, virtually identical, stretches of DNA with interspersed lipoprotein-encoding hypervariable islands (Caimano *et al.*, 2000; Stevenson *et al.*, 2000; Eggers *et al.*, 2001). Fourth is the compartmentalization of genetic information between chromosome and plasmids. Close to 60% of chromosomal genes encode database orthologues for proteins with housekeeping or chemotaxis/motility-related functions, as opposed to the mere 10% of plasmid genes encoding non-borrelial orthologues. The relatively small size of the borrelial chromosome explains the spirochaete's limited biosynthetic and intermediary metabolic capacity that, in turn, accounts for its fastidious growth requirements and, ultimately, the difficulty in recovering *B. burgdorferi* from clinical materials. Lyme disease spirochaetes do not contain genes for oxidative phosphorylation, depending instead on the Embden–Meyerhoff glycolytic pathway as their primary energy source (see also Chapter 5). By forsaking oxygen-based energy metabolism, spirochaetes appear to have made an interesting evolutionary trade-off to facilitate their parasitic lifestyle. Lacking cytochromes, they have virtually no iron requirement (Posey and Gherardini, 2000) and, thus, are impervious to host defences based on limiting the ability of pathogens to scavenge iron (Andrews *et al.*, 2003). Interestingly, although their genomes are markedly different in structure, *B. burgdorferi* and *T. pallidum* are quite similar in terms of the minimalist metabolic and biosynthetic functions they encode (Norris, 2006). Lastly, the genome encodes an extensive repertoire (>150) of mostly uncharacterized and non-orthologous lipoproteins. Because of the large number of differentially expressed lipoproteins they encode, the plasmids are believed to be the primary repository of the genetic information for the surface interactions required to sustain the spirochaete throughout the tick–mouse cycle.

The advantages to the spirochaete of maintaining a highly segmented genome are an open and potentially endless source of speculation. The most obvious explanation, which is that individual plasmids functionally segregate to the vector or mammalian phases of the enzootic cycle, is not supported by available evidence. On the other hand, genes vital for survival in one host or the other have been identified on a number of plasmids (Stewart *et al.*, 2005) (see also Chapter 2). lp54 contains the genes for OspA/B, lipoproteins required for attachment of spirochaetes to tick midgut epithelium during larval feeding (Pal *et al.*, 2000, 2004a; Yang *et al.*, 2004;Neelakanta *et al.*, 2007) as well as DbpB/A, decorin-binding lipoproteins that mediate attachment to extracellular matrix following dissemination within the mouse (Guo *et al.*, 1998; Brown *et al.*, 2001; Pikas *et al.*, 2003; Shi *et al.*, 2008). cp26 harbours the gene for OspC, a lipoprotein essential for tick–mammal transmission (Schwan *et al.*, 1995; Grimm *et al.*, 2004b; Pal *et al.*, 2004b; Radolf and Caimano, 2008) (discussed further below). lp25 contains the gene for PncA, an enzyme (and non-lipoprotein) essential for synthesis of nicotinic acid within mammals (Purser *et al.*, 2003). Lastly, lp28-1, one of four unrelated 28-kb linear plasmids, contains the *vls* system for antigenic variation that fends off borreliacidal antibodies during persistent infection (Zhang *et al.*, 1997; Labandeira-Rey *et al.*, 2003; Norris, 2006; Bankhead and Chaconas, 2007).

Comparative genomics studies using either high-throughput sequencing or microarray technology for DNA–DNA hybridization have been performed to identify genes potentially responsible for the divergent clinical manifestations elicited by different borrelial isolates and genospecies. The *B. burgdorferi* chromosome, lp54, cp26, and the cp32 family of plasmids appear to be conserved in content and gene organization, although they do have a few highly variable loci. The remaining linear plasmids, which are either highly divergent or, in the case of lp36 and lp38 for *B. garinii* (Glockner *et al.*, 2004), completely missing, would seem to be the most fertile ground for genomic mining forays to identify genes that contribute to differential strain infectivity or specific tissue tropisms. A detailed discussion of the fledgling field of comparative genomics of Lyme disease spirochaetes can be found in Chapter 2.

18.6.3 Differential gene expression and human disease

A genomic sequence is in reality a bacterial parts list and, as such, is only a starting point on the road to fulfilling the demanding criteria of the molecular version of Koch's postulates (Falkow, 1988; Falkow, 2004). While the task of deciphering how the individual parts interdigitate to bring a prokaryotic blueprint to life is theoretically no different for *B. burgdorferi* than for other bacterial pathogens, in the case of the Lyme disease spirochaete, it is compounded by the organism's dual-host lifestyle. Because animals syringe-inoculated with *in vitro*-cultivated organisms develop manifestations that resemble those in humans (see Chapters 14 and 15), the arthropod phases of the bacterial life cycle could be considered irrelevant to disease pathogenesis. For many years, this was a prevalent viewpoint among investigators, including the authors, who had little or no entomological experience. In fact, syringe inoculation remains a methodological staple for studying *B. burgdorferi* virulence determinants and assessing protective immunity. On the other hand, in the real world, Lyme disease spirochaetes are transmitted by ticks and the tick–mammal interface should be considered the starting point for the infectious process (Pal and Fikrig, 2003) (see also Chapter 11). Indeed, there is now a compelling body of evidence that spirochaetes in feeding ticks undergo complex alterations in their transcriptional and polypeptide profiles that are only partially reproduced by manipulation of *in vitro* culture conditions, including the addition of human blood to the culture medium (Hagman *et al.*, 2000; Carroll *et al.*, 2000; Cugini *et al.*, 2003; Tokarz *et al.*, 2004; Caimano *et al.*, 2007). Collectively, these changes (i) trigger a burst of replication that marks the end of the quiescent state within the flat tick (de Silva and Fikrig, 1995; Piesman *et al.*, 2003); (ii) downregulate genes/proteins, such as OspA, that are required for the tick phase of the enzootic cycle, while up-regulating the expression of others, such as OspC, required to establish a foothold once within the mammalian host (Schwan *et al.*, 1995; Grimm *et al.*, 2004b; Pal *et al.*, 2004b; Yang *et al.*, 2004; Tilly *et al.*, 2006); and (iii) enable the bacterium to bind serum proteins to its surface, including plasminogen (Fuchs *et al.*, 1994; Hu *et al.*, 1995; Coleman *et al.*, 1997; Lagal *et al.*, 2006), which exploits the mammalian host proteolytic systems to facilitate dissemination from midgut to salivary glands, and Factor H and Factor H-like protein-1, which protect the bacterium against complement-mediated lysis (Kraiczy *et al.*, 2001b; Kurtenbach *et al.*, 2002; McDowell *et al.*, 2003; Miller and Stevenson, 2004; Kraiczy *et al.*, 2004; Brooks *et al.*, 2005; Cordes *et al.*, 2006; Hartmann *et al.*, 2006). The obligatory time course for this intricate, programmatic sequence explains why spirochaetes are infrequently transmitted to mice (Piesman *et al.*, 1987; Ohnishi *et al.*, 2001) and humans (Berger *et al.*, 1995; Sood *et al.*, 1997; Nadelman *et al.*, 2001) when ticks are attached for less than 48 h. It is critical for physicians to be aware of this time frame because it establishes the window of opportunity for antimicrobial prophylaxis (Nadelman *et al.*, 2001) and can be used to provide reassurance to individuals who recently removed ticks acquired in endemic areas that their risk of becoming infected is low.

In the 1990s, a series of reports appeared describing *B. burgdorferi* genes and/or proteins that, based upon comparisons with *in vitro*-cultivated organisms, appeared to be expressed exclusively during infection (Champion *et al.*, 1994; Suk *et al.*, 1995; Akins *et al.*, 1995, 1998; Hefty *et al.*, 2001, 2002; Wallich *et al.*, 1995; Miller and Stevenson, 2003; Fikrig *et al.*, 1997). At least two of these, the OspE and OspF variants p21 and Bbk2.10, respectively, subsequently were shown not to be expressed within engorged nymphal ticks (Das *et al.*, 1997; Hefty *et al.*, 2001), indicating that spirochaetes receive and respond to mammalian signals in addition to those present in the blood meal. The *vls* system for antigenic variation is a particularly interesting case in point because it does not become active until several days post-infection (Zhang and Norris, 1998; Bykowski *et al.*, 2006; Norris, 2006) when antibodies against the bacterium begin to be generated. Evidence for tissue-specific expression of *B. burgdorferi* genes, believed to be an important contributing factor to organ system invasion and pathology, also has emerged in recent years (Fikrig *et al.*, 2000; Feng *et al.*, 2000; Narasimhan *et al.*, 2003; Hodzic *et al.*, 2003; Pal *et al.*, 2008). Down-regulation of *B. burgdorferi* lipoprotein genes (e.g. *ospC*) as infection progresses is due, at least in part, to the pressure exerted by specific antibodies and the selection of spirochaetal 'escape' variants (Liang *et al.*, 2002a,2004; Tilly *et al.*, 2006, 2007). Sorting out the continuum of differentially expressed genes as spirochaetes disseminate from the site of inoculation to diverse tissues and delineating the underlying regulatory mechanisms will be a major challenge for years to come (see Chapter 4).

18.7 Infection of humans

18.7.1 The site of inoculation

The seminal microbiological event in the natural history of Lyme disease is, of course, the deposition of spirochaetes into the skin. Surprisingly, the actual inoculum delivered by ticks has not been precisely measured. Available evidence, however, indicates that it is small. Spirochaetes have to transit through salivary glands en route to the skin. A recent qPCR analysis yielded values of approximately 20 spirochaetes per gland at peak infectivity (Piesman *et al.*, 2001); a similar spirochaete burden in salivary glands was determined by immunofluorescence analysis (Ohnishi *et al.*, 2001). While these values would appear to be well below many published ID_{50} values (Sadziene *et al.*, 1993a; Xu *et al.*, 1996), one should bear in mind that spirochaetes delivered by the natural route of infection are able to exploit the pharmacological and immunomodulatory properties of the tick salivary cocktail (Nuttall *et al.*, 2000; Fikrig and Narasimhan, 2006). Even so, the spirochaetes that do complete the journey from midgut to skin appear to have a precarious existence at the outset. In mice, dissemination does not occur if the inoculation site is excised within the first several days of tick detachment (Shih *et al.*, 1992). Moreover, spirochaetes inoculated into skin are antigenically heterogeneous and do not appear to have a uniform ability to establish infection (Ohnishi *et al.*, 2001). Genetic determinants that correlate with the ability of spirochaetes to survive at the inoculation site and disseminate have been identified (see below and Chapter 12).

At some point within the first several days, spirochaetes establish a foothold and begin to replicate, eventually reaching sizable numbers that make them relatively easy to detect by culture and by PCR (Strle *et al.*, 1996; German *et al.*, 1996; Schmidt, 1997; Nowakowski *et al.*, 2001). Spirochaetes often can be visualized by light microscopy in the superficial dermis, in association with collagen fibrils and sporadically in the lumens and walls of blood vessels (de Koning, 1993; Duray, 1989a). The spirochaetes also migrate outwards from the feeding site and downwards towards the dermal microvasculature, en route to disseminating haematogenously (Fig. 18.2) (Duray, 1989b). These movements trigger local danger signals that result in the accumulation of resident innate immune cells from the dermis, primarily macrophages and dendritic cells, as well as the recruitment of circulating immune cells of various types with skin-homing capacity (Salazar *et al.*, 2003). The ensuing local inflammatory response, which is amplified as *B. burgdorferi*-specific T-cells are primed in draining lymph nodes, enter the circulation, and traffic back to the inflamed site, becomes manifest as the hallmark skin lesion, erythema migrans (EM). One of the cardinal differences between Lyme disease in humans and mice is that the latter do not develop EM (Barthold, 1996). Like humans, tick or needle-inoculated non-human primates develop EM (Philipp and Johnson, 1994) as do rabbits (Wheeler *et al.*, 1989; Foley *et al.*, 1995), another non-reservoir species.

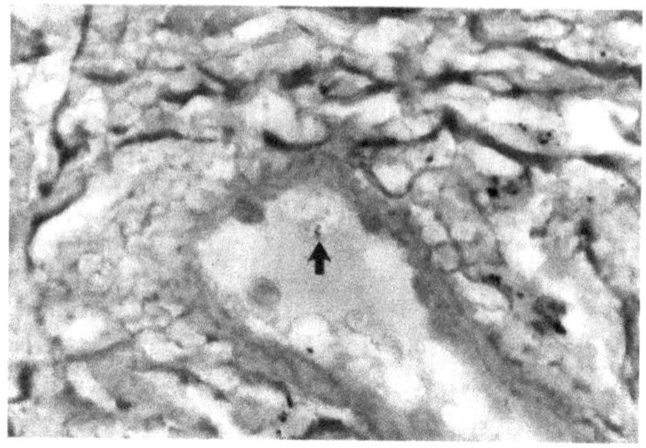

Figure 18.2 Histopathology of erythema migrans. Arrow points to a spirochaete that has entered a dermal venule. (With permission from Duray, 1989b.)

18.7.2 Dissemination

That spirochaetes disseminate haematogenously during early infection has been known since *B. burgdorferi* was first isolated in the early 1980s (Steere *et al.*, 1983b; Benach *et al.*, 1983). For many years, it was believed, based on the low blood culture isolation and/or PCR positivity rates, that only a small percentage of EM patients become spirochaetaemic (Nadelman *et al.*, 1990; Goodman *et al.*, 1995; Schmidt, 1997; Wormser *et al.*, 2006). Wormser and colleagues disproved this notion (Wormser *et al.*, 2001a). They showed that spirochaetaemia occurs in approximately one-half of patients with EM but that the extremely low spirochaete concentrations in blood (estimated to be approx. 0.1 bacterium per ml of blood) necessitate culturing large volumes of plasma (at least 9 ml) (Wormser *et al.*, 2001a) (see also Chapter 16). In a large retrospective review of spirochaetaemic EM patients evaluated using the large-volume culture technique, haematogenous dissemination was not associated with duration or size of EM (Wormser *et al.*, 2005), suggesting that spirochaetaemia is more dependent on borrelial than on host immunological factors (discussed below). *Borrelia burgdorferi* shares a propensity to disseminate haematogenously during early infection with *T. pallidum*, which has long been known to gain access to the bloodstream soon after inoculation, often well before the appearance of the primary lesion (Radolf *et al.*, 2006). Approximately 20% of EM patients have secondary EM-like skin lesions, a clinical indicator of haematogenous dissemination, which are analogous to the rash of secondary syphilis (Wormser *et al.*, 2006). The high incidence of constitutional symptoms in EM patients also is considered to reflect the systemic nature of early disease (Steere, 2001; Smith *et al.*, 2002; Hengge *et al.*, 2003; Wormser *et al.*, 2005).

The wide variety of manifestations attributed to Lyme disease over the years emphasizes the invasive properties of the spirochaete, although scepticism has been voiced about the diagnostic accuracy of many of these reports (Nadelman and Wormser, 1998; Steere *et al.*, 2004). Aside from skin, the nervous system, heart, and joints are the most commonly affected metastatic sites (see below). How frequently do spirochaetes gain access to these tissues? This complex question, which is at the crux of an assessment of the invasive potential of Lyme disease spirochaetes for humans, is difficult to answer since one has to rely on clinical manifestations as the indicator of invasive disease and/or extrapolate from animal models that do not fully replicate the human disease (see Chapters 14 and 15). Furthermore, because EM is now widely recognized in endemic areas and effectively treated (Smith *et al.*, 2002), most forms of disseminated disease that follow EM are now less common.

T. pallidum has long been considered to set the gold standard for neuroinvasiveness of pathogenic spirochaetes (Merritt *et al.*, 1946). In early syphilis, 25% or more of untreated patients without neurological symptoms have microbiological evidence for invasion of the CNS by *T. pallidum* and CNS complications are characteristic of untreated early and late syphilitic infection (Radolf *et al.*, 2006; Merritt *et al.*, 1946; Rolfs *et al.*, 1997). In a report describing 12 US patients with acute disseminated Lyme disease (six with acute cranial neuritis, three with multiple EM and three with both cranial neuritis and multiple EM), eight had spirochaetal DNA in their cerebrospinal fluid (CSF) detectable by PCR and four of these lacked neurological symptoms (Luft *et al.*, 1992). These findings, however, have not been reproduced and, judging by numerous subsequent studies using PCR to detect spirochaetes in patients with documented acute neurological involvement (Schmidt, 1997), more than likely considerably over-estimate the incidence of direct CNS invasion during early Lyme disease. As stated earlier, recovery of *B. garinii* and *B. afzelii* from CSF is well-recognized (Hubalek and Halouzka, 1997; Strle *et al.*, 2006; Wilske *et al.*, 2007; Stanek and Strle, 2008), but these studies do not enable one to gauge the incidence of CNS infection during early infection with these genospecies. Furthermore, in the rhesus macaque model, penetration of the blood–brain barrier is rare in immunocompetent animals (Philipp *et al.*, 1993; Pachner *et al.*, 2001; Cadavid *et al.*, 2000; Pachner *et al.*, 2004). In immunocompromised monkeys, *B. burgdorferi* has been shown to infiltrate leptomeninges, sensory and motor nerve roots, and the dorsal root ganglion, a distribution mimicking that in humans, but not the brain parenchyma (Cadavid *et al.*, 2000). Also worth noting is that CNS infection is not characteristic of the murine model, even in severely immunocompromised animals (Schaible *et al.*, 1990). Neurological complications in Lyme disease often result from inflammation of blood vessels supplying nerve roots and peripheral nerves as opposed to penetration of the blood–brain barrier and/or invasion of brain parenchyma by the bacterium (Halperin, 2008). Combinations of these two pathologies give rise to a wide spectrum of neurological disorders in Lyme disease, involving one or more elements of the classic triad of lymphocytic meningitis, cranial neuritis, and radiculoneuritis (Halperin, 2008). As a whole, these studies suggest that, compared to *T. pallidum*, *B. burgdorferi* possesses a broader tropism for neural tissues but a lesser capacity for direct invasion of the CNS.

Before the widespread usage of antibiotics, carditis was reported to occur in approximately five percent of patients, making it the least common of the three major non-cutaneous forms of disseminated disease (Steere, 2001). Incidence rates of less than 1% in the US and less than 4% in Europe are now reported (Hengge *et al.*, 2003). In contrast to laboratory mice, in which cardiac involvement occurs with diverse strains and in most animals (Barthold *et al.*, 1990) (see also Chapter 14), these low percentages suggest that *B. burgdorferi* does not have a strong tropism for human cardiac tissue. This conclusion is supported by studies with rhesus macaques that found that carditis was absent or mild unless animals were immunosuppressed (Philipp *et al.*, 1993' Cadavid *et al.*, 2004). On the other hand, a classic natural history study provided evidence that, as in mice, Lyme disease spirochaetes (*B. burgdorferi* sensu stricto) do have an extraordinary tropism for joints of humans (Steere *et al.*, 1987). Of 55 untreated patients with EM who were followed over a 6-year period, 28 developed intermittent episodes of monoarticular or oligoarticular arthritis, primarily in large joints, and six developed chronic arthritis in one or more large joints. One of the long-standing paradoxes in Lyme disease is that spirochaetes are notoriously difficult to isolate from synovial fluid (Steere and Glickstein, 2004). PCR detection of *B. burgdorferi* DNA in synovial fluids from the large majority of untreated arthritis patients (Nocton *et al.*, 1994; Aguero-Rosenfeld *et al.*, 2005) provided the key evidence that joint inflammation is usually a direct result of infection. A small percentage of patients continue to have arthritis post-therapy. Synovial fluids in these individuals are usually PCR-negative, prompting the hypothesis that *B. burgdorferi* infection can trigger an autoimmune process in individuals with the appropriate genetic background (Steere and Glickstein, 2004). PCR analysis of synovial fluid for the presence of *B. burgdorferi* DNA is a useful tool for management of patients whose arthritis continues or relapses following therapy; PCR positivity is most likely indicative of persistent infection and thus the need for additional antimicrobial therapy (Puius and Kalish, 2008).

Very little is known about the mechanisms by which circulating spirochaetes recognize and invade target organs. Studies with cultured human umbilical vein endothelial cells have shown that they rapidly attach to vascular endothelium and negotiate their way through intercellular junctions, subsequently attaching to subendothelial matrix components (Szczepanski *et al.*, 1990). Penetration of vascular endothelium by spirochaetes *in vivo* recently has been demonstrated by real-time confocal microscopy to be a multi-step process engaged in by only a small percentage of spirochaetes introduced intravenously into mice (Moriarty *et al.*, 2008; Norman *et al.*, 2008). This result may reflect, in part, the limited degree of mammalian host adaptation that the spirochaetes underwent in the *in vitro* cultivation conditions utilized to generate the inoculum. Several mammalian receptors for Lyme disease spirochaetes, including glycosaminoglycans, decorin, fibronectin, and integrins, have been identified, although much more needs to be learned about the temporal sequence and locales in which these receptors are engaged (Coburn *et al.*, 2005; Antonara *et al.*, 2007; Coburn *et al.*, 2005). Norman and co-workers recently presented *in vivo* evidence, obtained using confocal microscopy, that attachment of intravenously inoculated spirochaetes to microvascular endothelium is initiated by the binding of BBK32 to glycosaminoglycans and fibronectin (Norman *et al.*, 2008). These intriguing results, however, appear to be at variance with mutagenesis studies that show that loss of *bbk32* has only a limited effect on spirochaetal virulence (Seshu *et al.*, 2006). Several spirochaete adhesins for host molecules have been characterized *in vitro* (Coburn *et al.*, 2005); using phage display, additional spirochaetal lipoproteins that mediate attachment to vascular enedothelium *in vivo* also have been identified (Antonara *et al.*, 2007). A comprehensive discussion of borrelial factors implicated in attachment, dissemination, and invasion can be found in Chapter 12.

18.7.3 Determinants of invasive disease in humans

Over the years, a number of methods have been utilized to molecularly type *B. burgdorferi* isolates for taxonomic, epidemiological, and pathogenesis-related purposes (Wang *et al.*, 1999a; see also Chapter 16). One of the most powerful schemes is that developed by Schwartz and co-workers (Liveris *et al.*, 1995; Liveris *et al.*, 1996), which exploits sequence variability in the spacer separating the 16S-23S rRNA genes, to classify *B. burgdorferi* sensu stricto isolates into three PCR-restriction fragment length polymorphism (PCR-RFLP) genotypes, designated RST1, RST2, and RST3. In a study conducted at New York Medical College in Westchester, NY, an area hyperendemic for Lyme disease, all three genotypes were isolated from skin biopsies with roughly equal frequencies; RST1 isolates, however, were nearly 14-fold more likely to be recovered from blood than were RST3 strains (Wormser *et al.*, 1999). This finding subsequently was corroborated by another group in a cohort of Lyme disease patients diagnosed and followed in Connecticut and Rhode Island (Jones *et al.*, 2006). qPCR analysis revealed that the number of spirochaetes in skin biopsies from patients infected with RST1 strains was nearly three-fold greater than those in specimens from patients infected with RST2 or RST3 strains (Liveris *et al.*, 2002). The robustness of this typing scheme as a predictor of invasive capacity has also been corroborated in the murine model

(Wang *et al.*, 2001; Wang *et al.*, 2002; Dolan *et al.*, 2004).

How could a non-coding sequence determine strain infectivity? The likely explanation, of course, is that the rRNA spacer sequence polymorphism is a genetic surrogate for another *B. burgdorferi* locus that encodes a gene product involved in virulence expression. The correlated locus appears to be *ospC*. As noted earlier, spirochaetes in flat tick midguts do not express *ospC*, but the gene is rapidly and intensely upregulated during the blood meal (Schwan *et al.*, 1995; Schwan and Piesman, 2000). Mutant spirochaetes lacking *ospC* are avirulent, although there is a discrepancy as to whether the deficiency affects the ability of spirochaetes to penetrate salivary glands or to evade clearance by the host shortly after inoculation (Pal *et al.*, 2004b; Grimm *et al.*, 2004b; Tilly *et al.*, 2006; Tilly *et al.*, 2007). *ospC* genes and corresponding proteins are highly diverse and it is believed that this high degree of sequence variability reflects selection of allelic variants due to immunological pressure from the mammalian host (Wang *et al.*, 1999b) (see Chapter 9). Ten years ago, Seinost *et al.* (1999a) reported that the frequency distribution of *ospC* types among tick isolates of *B. burgdorferi* sensu stricto was significantly different from that of skin isolates from humans and that only four of the 21 known *ospC* alleles were associated with invasive disease. Earnhart *et al.* (2005) subsequently reported an additional three *ospC* alleles correlated with invasive disease and Dykhuizen *et al.* (2008) added an eighth. The correlation of invasiveness in humans with specific *ospC* alleles subsequently was extended to *B. afzelii* and *B. garinii* (Baranton *et al.*, 2001). The convergence of the RST and *ospC* genotyping schemes occurred with the finding that invasive *ospC* alleles are associated with RST1 and RST2 strains. Comparison of sequences of rRNA spacer regions and *ospC* alleles in a large number of strains revealed that the two loci are in strong linkage disequilibrium due to the clonal nature of *B. burgdorferi* clinical isolates (Hanincova *et al.*, 2008; Wormser *et al.*, 2008). Particularly significant was that only two *ospC* alleles, both of which are strongly associated with dissemination, were found in the RST1 group. The fact that only a minority of *ospC* alleles confers invasiveness for humans is consistent with the notion that *B. burgdorferi* has not evolved to infect humans and should not be construed to mean that strains harbouring 'non-invasive' alleles are less virulent in the natural setting. Also noteworthy is that the frequency distributions of RST types in ticks differed from those in skin and blood, suggesting that other loci are responsible for the transmissibility of a particular genotype from tick to skin (Wormser *et al.*, 2008).

18.7.4 Asymptomatic infection

As noted previously, seroepidemiological studies of populations in Lyme disease endemic areas have shown that substantial proportions of persons with antibodies to *B. burgdorferi* are asymptomatic and that asymptomatic or subclinical infection may be as common as clinically apparent disease (Hanrahan *et al.*, 1984b; Steere *et al.*, 1986; Gustafson *et al.*, 1995). In the OspA vaccine trial, 30 of the 269 patients who met the criteria for Lyme disease were classified as having asymptomatic IgG seroconversion to *B. burgdorferi* (Steere *et al.*, 1998). Upon subsequent investigation, 14 of these individuals were found to have actually had symptoms and/or a rash compatible with EM that was not appreciated during their participation in the study. Eight patients, however, were truly asymptomatic and went untreated. Surprisingly, only one of these eight individuals developed a late complication, arthritis (Steere *et al.*, 2003b). How does one reconcile studies showing what appears to be a relatively benign course for asymptomatic Lyme disease with other studies indicating that the natural history of the disease is to develop late manifestations, usually arthritis, when EM is untreated? An appealing hypothesis proposed by Wormser and co-workers is that some asymptomatic infections may be attributed to non-invasive strains (Wormser *et al.*, 2001b). Unfortunately, methodologies to distinguish patients who have cleared 'benign' asymptomatic infection from those who may be persistently infected and at risk for subsequent late complications do not exist. The management of asymptomatic persons found to be seroreactive for Lyme disease, therefore, is an important unresolved issue for practitioners. Current treatment guidelines do not include recommendations for this common situation (Wormser *et al.*, 2006). Many practitioners base their clinical decision-making on algorithms for treatment of syphilis, which hold that persons with untreated asymptomatic infection are presumed to have latent infection and must be treated because they are at risk for recrudescent disease (Workowski and Berman, 2006).

18.8 The Immune response

The widely held belief from the pre-genomic era that *B. burgdorferi* is unable to elaborate cytotoxins was confirmed when the genome sequence failed to identify toxin orthologues or the genetic components of a secretory apparatus required to deliver them externally (Fraser *et al.*, 1997). Thus, the notion that tissue damage and clinical manifestations in Lyme disease are caused by the host response to spirochaetes and spirochaetal constituents is now generally accepted (Wooten and Weis, 2001). A detailed examination of this complex subject can be found in Chapter 15. Our purpose below is to provide an overview of the innate and adaptive responses that are believed to be relevant to disease in humans.

18.8.1 Pathogen recognition/innate immunity

The discovery that *B. burgdorferi* contains an abundance of lipoproteins (Brandt *et al.*, 1990), eventually confirmed by the genomic sequence (Fraser *et al.*, 1997; Casjens *et al.*, 2000), was followed shortly thereafter by reports that borrelial lipoproteins can stimulate innate immune cells of human and murine origin (Ma *et al.*, 1994; Radolf *et al.*, 1991; Radolf *et al.*, 1995; Morrison *et al.*, 1997; Sellati *et al.*, 1996). As discussed earlier, the significance of these findings was underscored by the demonstration, first biochemically and subsequently from the genomic sequence, that *B. burgdorferi* lacks lipopolysaccharide (Takayama *et al.*, 1987; Fraser *et al.*, 1997), the prototypical pathogen-associated molecular pattern (PAMP). The use of synthetic lipopeptides and recombinant proteins enabled investigators to prove unambiguously that lipoproteins are *bona fide* innate immune agonists, that this property resides in their lipid-modified N-termini, and that it is shared by acylated polypeptides with diverse amino acid sequences (Akins *et al.*, 1993; De-Ogny *et al.*, 1994; Weis *et al.*, 1994; Radolf *et al.*, 1995). By the mid-1990s, it was evident that spirochaetal lipoproteins activate macrophages via a pathway distinct from that used by LPS even though their biological effects appeared to be highly similar (Radolf *et al.*, 1991; Radolf *et al.*, 1995; Norgard *et al.*, 1996). These results were notable, but also puzzling, once it was found that interaction of both LPS and lipoproteins with CD14 is the initial and obligatory step for macrophage activation (Sellati *et al.*, 1998; Wooten *et al.*, 1998; Giambartolomei *et al.*, 1999). Their meaning was clarified, however, with the discovery of Toll-like receptors (Medzhitov *et al.*, 1997) and the demonstration that LPS signals via TLR4 (Poltorak *et al.*, 1998), whereas spirochaetal lipoproteins utilize heterodimers composed of TLR2 and TLR1 (Aliprantis *et al.*, 1999; Brightbill *et al.*, 1999; Hirschfeld *et al.*, 1999; Lien *et al.*, 1999; Ozinsky *et al.*, 2000; Alexopoulou *et al.*, 2002). In addition to marking a major conceptual advance in our understanding of spirochaete recognition in infected tissues, the practical significance of these results was brought home by the observation that immunization against Lyme disease with OspA is much more effective when a lipid-modified polypeptide is used as antigen (Erdile *et al.*, 1993).

Because macrophages from mice deficient in TLR2, CD14, or MyD88 are unresponsive to lipoproteins, one would predict that the knockout mice would display a markedly attenuated inflammatory response to infection with live *B. burgdorferi*. Quite surprisingly, this is not the case. In fact, histopathological changes in these animals equal or surpass those in the wild type; at the same time, the knockouts have higher spirochaete burdens (Wang *et al.*, 2004; Bolz *et al.*, 2004; Liu *et al.*, 2004; Benhnia *et al.*, 2005; Behera *et al.*, 2006a). Taken together these results suggest that (i) TLR signalling has moderating as well as activating effects on the inflammatory response to the spirochaete, (ii) lipoproteins are unlikely to be the sole component of the spirochaete capable of activating innate immune mechanisms, and (iii) there exist TLR-independent pathways for spirochaete recognition that acting alone are less effective at clearing spirochaetes. Interestingly, the greater efficacy of TLR-dependent responses for eradicating spirochaetes does not appear to be due to antibodies. Evidence for these deductions is emerging, although a cohesive mechanistic picture seems far off. Integrin-mediated, MyD88-independent cell activation by spirochaetes has been demonstrated (Behera *et al.*, 2006b; Behera *et al.*, 2008), albeit not for macrophages. TLR2-deficiency diminishes but does not abrogate the response of murine macrophages to live spirochaetes (Shin *et al.*, 2008; Salazar *et al.*, 2009). Live spirochaetes induce responses in human monocytes that are quantitatively and qualitatively different than those elicited by equivalent bacterial lysates, and these enhanced responses are dependent upon phagocytic uptake of live organisms (Moore *et al.*, 2007; Shin *et al.*, 2008; Cruz *et al.*, 2008; Salazar *et al.*, 2009) (Fig. 18.3). The identification of a prominent type I interferon response in human monocytes stimulated with live Bb (Salazar *et al*, 2009) and synovial tissues from *B. burgdorferi*-infected mice (Miller *et al.*, 2008b; Miller *et al.*, 2008a) also points to the existence of borrelial PAMPs other than lipoproteins, since TLR2 signalling does not result in stimulation of type I interferons (Toshchakov *et al.*, 2002). Indeed, Petzke *et al.* (2009) recently demonstrated that live Bb induces human plasmacytoid dendritic cells to secrete IFN-α through endosomal TLR7 and TLR9 activation. Lastly, it is also becoming clear that other arms of the innate immune system are mobilized as part of the early response to the pathogen. It recently was shown that a *B. burgdorferi* diacylglycerol glycolipid triggers a CD1d-dependent natural killer (NK) T-cell response, the first demonstration of an NK T-cell antigen expressed by a human pathogen (Kinjo *et al.*, 2006), while in a human PBMC system, live, but not lysed, spirochaetes induced NK cells to produce IFN-gamma, an effect that required dendritic cell uptake of the bacteria (Moore *et al.*, 2007; Cruz *et al.*, 2008).

18.8.2 Adaptive immunity

Antibodies

The need for accurate serological tests for diagnosis of Lyme disease has been the major driving force behind efforts to elucidate the humoral response to the spirochaete (Aguero-Rosenfeld *et al.*, 2005; Wilske *et al.*, 2007). Consequently, the antibody response to *B. burgdorferi* has been extensively dissected over the years with increasingly sophisticated methodologies, most recently proteomics (Nowalk *et al.*, 2006), in order to identify borrelial polypeptides that are not only highly immunogenic in a large percentage of

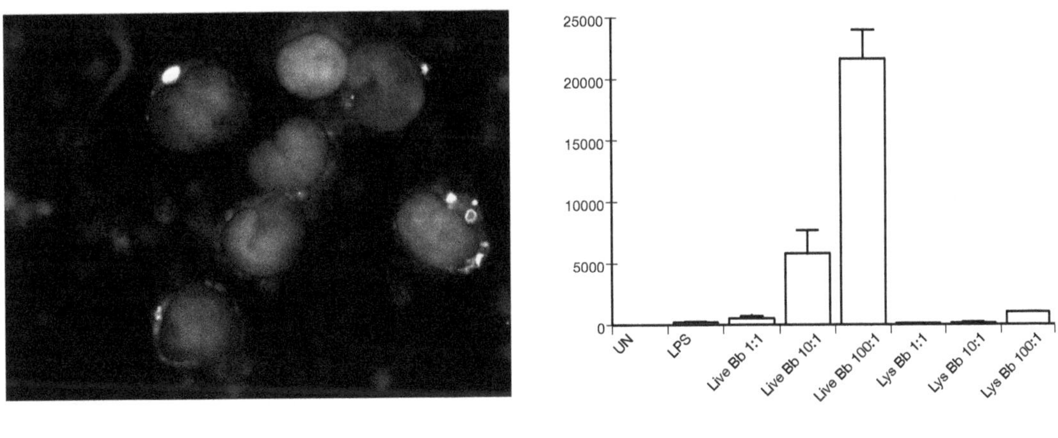

Figure 18.3 Shown on the left are epifluorescence images (100′) acquired from human monocytes incubated for a total of four hours with temperature shifted *B. burgdorferi* expressing green fluorescent protein (Bb-GFP depicted in both white and light grey colour) at MOI 100:1, and labelled with nuclear dye DAPI (shown in dull grey). Phagocytosed (both coiled and degraded) Bb-GFP and fragmented nuclei are observed. Shown on the right are monocytes that were either unstimulated (UN), or incubated for four hours with live or lysed Bb at various MOIs or with lipopolysacharide (LPS). Secretion of cleaved IL-1β was measured in culture supernatants by using a bead array assay. The bars show mean cytokine concentrations (pg/ml) +/– standard error of the mean ($n = 4$). Secreted IL-1β values were significantly greater in response to internalized live Bb, than equivalent amounts of bacterial lysates.

patients, particularly during early infection, but also specific for *B. burgdorferi* (Aguero-Rosenfeld *et al.*, 2005). This combination has not been easy to find. Cross-reactivity between *B. burgdorferi* FlaB, the major flagellar sheath protein, and antigens of other bacteria is the primary reason why this polypeptide, which induces a robust antibody response early in infection (Nowalk *et al.*, 2006), has fallen out of favour as a diagnostic antigen (Aguero-Rosenfeld *et al.*, 2005). OspC also induces a strong antibody response during early infection (Fung *et al.*, 1994), consistent with murine studies demonstrating it is critical for establishing infection following tick inoculation, but immunodominant epitopes of OspC proteins from different isolates are extremely heterogeneous (Theisen *et al.*, 1993; Wilske *et al.*, 1993). A number of other borrelial proteins, many, but not all, lipoproteins, are strong immunogens but are hyper-variable, expressed relatively late in the course of disease, and/or recognized by an insufficient percentage of patients (Aguero-Rosenfeld *et al.*, 2005; Wilske *et al.*, 2007). Ironically, the antigenically variable VlsE protein, which in mice facilitates evasion of antibody responses, has a conserved epitope that is highly immunogenic in early as well as late Lyme disease patients and has great potential for improved serodiagnosis (Aguero-Rosenfeld *et al.*, 2005; Wilske *et al.*, 2007).

The general picture to have emerged over the years is that IgG and IgM antibodies to the spirochaete develop slowly and are directed against an increasingly diverse array of proteins as infection progresses (Craft *et al.*, 1986; Dressler *et al.*, 1993; Nowalk *et al.*, 2006). The earliest responses are to flagellin B (FlaB) and p66, followed by OspC (25 kDa) with responses to a number of additional antigens, such as VlsE, fibronectin-binding protein (BBK32), FlaA (37 kDa), BmpA (39 kDa), and decorin-binding protein A (DbpA) developing as *B. burgdorferi* disseminates (Coleman and Benach, 1987; Engstrom *et al.*, 1995; Bacon *et al.*, 2003; Aguero-Rosenfeld *et al.*, 2005; Wilske *et al.*, 2007). This temporal pattern is consistent with the notion that the bacterium draws upon an expanding repertoire of differentially expressed proteins once within its vertebrate host, including the phased expression of paralogous surface-exposed lipoproteins. Recently, Carroll and co-workers (Nowalk *et al.*, 2006) completed an extensive proteomics-based survey of murine and human antibody responses to *B. burgdorferi* proteins, concluding that the contours of the two antibody profiles share many similarities. Thus, the spirochaete appears to employ similar transcriptional programs in both vertebrate hosts. The clinical ramifications of these observations are significant. Because approximately half of patients with EM do not mount detectable antibody responses to the pathogen, lack of seroreactivity cannot be used to rule out the diagnosis of EM (Steere, 2001; Dandache and Nadelman, 2008). Seroreactivity increases substantially following therapy for EM (Vaz *et al.*, 2001; Dandache and Nadelman, 2008), indicating that killing of the bacterium enhances processing of spirochaetal antigens. By one month, the majority of patients have detectable antibodies (Steere, 2001; Aguero-Rosenfeld *et al.*, 2005). Accordingly, clinical manifestations caused by the dissemination of spirochaetes are, for the most part, accompanied by seroreactivity (Steere, 2001) The flip side of this coin is that IgG antibodies can be present for life and cannot, therefore, be used by themselves as indicators of active infection. While assays using whole-cell lysates are less affected by strain variability and the heterogeneity in patient responses than those using single antigens, they can yield false-positives due to the presence of cross-reactive, background antibodies. This problem has been circumvented by the development of the two-tiered testing algorithm in which immunoblotting is used to confirm serological reactivity by ELISA (Aguero-Rosenfeld *et al.*, 2005). The success of the two-tiered approach rests upon the implementation of strict criteria for the interpretation of immunoblots based on reactivity with a small subset of borrelial polypeptides defined by their SDS-PAGE mobilities (i.e. apparent molecular masses) (Dressler *et al.*, 1993). It is noteworthy, in light of the T-cell reactivity studies discussed below, that reactivity with OspA is not a criterion for immunoblot positivity.

What role, if any, do antibodies play in control of the disease process in humans? This question is far more difficult to answer for humans than for mice because humans cannot be followed serially without treatment. The presence of intense and diverse antibody responses in patients with late manifestations of disease (Steere, 2001) is clear-cut evidence that, as in mice, spirochaetes can persist despite high titres of circulating antibodies (see below). Nevertheless, there also is *in vitro* evidence that the antibodies produced during human infection have bactericidal activity (Pavia *et al.*, 1997; Callister *et al.*, 2002) and they also can passively protect animals against inoculation with *in vitro* cultivated spirochaetes (Fikrig *et al.*, 1994) In the case of syphilis, the delay in the appearance of opsonic antibodies is believed to provide *T. pallidum* with a window of opportunity for local replication and dissemination, while the appearance of opsonic antibodies is thought to be a watershed event in the containment of syphilitic infection (Radolf and Lukehart, 2006). The delayed appearance of antibodies in Lyme disease probably creates a comparable window for *B. burgdorferi*. As stated earlier, infection with less invasive borrelial strains has been proposed as one explanation for the high prevalence of asymptomatic *B. burgdorferi* infection in endemic areas (Wormser *et al.*, 2001b). The capacity of humoral responses to contain and even eliminate spirochaetes provides a second, non-mutually exclusive explanation, which also is in accord with the idea that humans lack reservoir-competence for the bacterium.

T-cells

It was apparent early on from histopathological findings in Lyme disease patients that T-cell responses to the spirochaete are likely to be directly involved in causing tissue damage, leading to organ system dysfunction and clinical manifestations (Duray, 1989b). Studies in the mouse model have shown unambiguously that $CD4^+$ T-cells play a major role in the inflammatory process in both heart and joints (McKisic *et al.*, 1993). One of the historical oddities of Lyme disease research is how efforts to understand the relationship between adaptive cellular responses to the spirochaete and disease pathogenesis became inextricably intertwined with the perceived need for improved diagnostics. In a landmark study published in 1988, Dattwyler and colleagues reported that 17 persistently symptomatic patients who had been treated with oral antibiotics for early Lyme disease, but who were seronegative by conventional ELISA and immunoblotting, manifested a vigorous T-cell proliferative response to borrelial lysates (Dattwyler *et al.*, 1988). They proposed that antibiotics had killed the bulk of the infecting spirochaetes, abrogating an antibody response, while leaving a sufficient number of organisms to prime antigen-specific T-cells; the putative result is the low-level inflammation and symptomatology characteristic of chronic Lyme disease. Implicit in this argument is the inadequacy of serological assays for Lyme disease diagnosis. This line of reasoning has since become entrenched among the proponents of the persistent infection theory of chronic Lyme disease (Erdlow, 2003), even though the laboratory studies upon which it was based have long been called into question (Aguero-Rosenfeld *et al.*, 2005). Multiple groups have found that proliferative responses can be elicited in cells from normal individuals or that such *in vitro* responses correlate poorly with serological and clinical parameters in properly diagnosed Lyme disease patients (Zoschke *et al.*, 1991; Horowitz *et al.*, 1994; Nadelman and Wormser, 1998).

Even so, the literature on systemic T-cell responses of humans to *B. burgdorferi* still remains a difficult one to assess, primarily because of reports purporting to show a dissociation between T and B-cell responses to borrelial antigens, particularly OspA (Lengl-Janssen *et al.*, 1994; Meyer *et al.*, 2000; Vaz *et al.*, 2001). The mechanism proposed to explain how T-cells become sensitized to OspA is essentially no different from that used to support the idea of seronegative Lyme disease, namely, that expression of this tick-phase antigen occurs at a level sufficient to prime T-cells but not B-cells (Vaz *et al.*, 2001). Though difficult to prove in humans, evidence for this mechanism should be sought in the mouse model. It should be noted that it has been shown that spirochaetes inoculated into skin during tick feeding can express OspA (Ohnishi *et al.*, 2001). One also suspects that a number of technical factors may be at work here, including the general difficulty in detecting antigen-specific T-cells, which circulate at very low frequencies in blood, coupled with the mitogenic effects of borrelial lipoproteins (Knigge *et al.*, 1996). Further adding to the complexity of this literature are studies, exclusively from the Steere group (Lengl-Janssen *et al.*, 1994; Kamradt *et al.*, 1996; Meyer *et al.*, 2000), demonstrating that $CD4^+$ T-cell responses to OspA are found predominantly among patients with treatment-resistant Lyme arthritis. As discussed below, the differential responsiveness of persons with certain HLA-types to OspA was long the bedrock for the molecular mimicry theory for the pathogenesis of Lyme arthritis (Gross *et al.*, 1998; Steere and Glickstein, 2004). In any event, while the antigenic specificities of the circulating $CD4^+$ T-cells remains an open question, these uncertainties do not detract from the notion that antigen-specific T-cells recruited to infected sites become activated and contribute to tissue pathology (see below). Finally, limited work in humans (Busch *et al.*, 1996), corroborated by studies in mice (Dong *et al.*, 1997; Beermann *et al.*, 2000), suggests that *B. burgdorferi* infection can also prime $CD8^+$ T-cells. Since *B. burgdorferi*, as an extracellular pathogen, is taken up by phagocytosis (Filgueira *et al.*, 1996), priming of $CD8^+$ T-cells would have to occur by some form of cross-presentation. Presumably peptides presented in the context of class I MHC molecules at the site of inflammation could trigger these cells.

18.8.3 Immune evasion

As a vector-borne disease agent, *B. burgdorferi* differs from most other bacterial pathogens in that it must contend with the defences of its intended vertebrate host even before it has gained access to the host proper. In addition to the innate leucocytic elements in the blood meal that might respond to, and possibly kill, spirochaetes within the midgut of the feeding tick, are potent fluid phase antibacterial systems that do not require prior sensitization to the pathogen in order to become active. Among these innate humoral defences, the alternative pathway for complement activation has been the most intensely studied (Kurtenbach *et al.*, 2002; LaRocca and Benach, 2008). Collectively, these studies have identified five differentially expressed, plasmid-encoded surface lipoproteins that bind factor H and/or factor H-like protein-1 (FHL-1), plasma glycoproteins that compete with factor B for binding to C3B, thereby preventing inappropriate activation of the lytic complement cascade (Kraiczy *et al.*, 2001b; LaRocca and Benach, 2008) (see also Chapter 12). These spirochaetal lipoproteins, collectively designated complement regulator-acquiring surface proteins (CRASPs), have been divided into three groups according to whether they bind both Factor H and FHL-1, FHL-1 alone, or just Factor H (Kraiczy *et al.*, 2001a; Alitalo *et al.*, 2002; McDowell *et al.*, 2003; Kraiczy *et al.*, 2003, 2004; Metts *et al.*, 2003; Brooks *et al.*, 2005; Cordes *et al.*, 2006; Salazar *et al.*, 2009). There is ongoing debate about the contributions of individual proteins to the spirochaete's anti-complement armamentarium as well as their expression patterns (McDowell *et al.*, 2006; Woodman *et al.*, 2007; Coleman *et al.*, 2008). Suffice it to say that CRASPs are poorly expressed during residence in flat ticks and that one or more are expressed during the phases of the enzootic cycle when spirochaetes are exposed to blood (Hefty *et al.*, 2002; Stevenson *et al.*, 2002; Miller *et al.*, 2003). It remains to be determined by mutagenesis of individual CRASPs whether their factor H and FHL-1 binding activities are redundant or, alternatively, whether subtle differences are critical for complement evasion by the bacterium. One of the most important conceptual advances to emanate from the discovery of CRASPs is the hypothesis that the CRASP repertoire is a prime determinant of the host range of genospecies and individual spirochaete strains (Kurtenbach *et al.*, 2002; Stevenson *et al.*, 2002). A correlate of this theory is that CRASPs influence human Lyme disease both directly and indirectly. They do so directly by determining whether a given isolate is resistant or sensitive to human serum. They do so indirectly by determining reservoir competence of mammalian species in an endemic area and, thus, the likelihood that humans will be infected as a result of activities that accidentally place them in harm's way (Oliver, Jr. *et al.*, 2003).

Persistence of spirochaetes in mammalian reservoirs is essential for larval acquisition and, hence, maintenance of the enzootic cycle (Humair and Gern, 2000). Natural hosts for *B. burgdorferi*, as well as laboratory mice, remain infected for life (Oliver, Jr. *et al.*, 2003). Though there is no question that Lyme disease spirochaetes can persist in humans and give rise to disease manifestations well after inoculation, there are no reliable data as to whether persistence is a common or rare occurrence. Nevertheless, in humans as well as rodents, spirochaetes persist, presumably via the same immune evasion strategies. *In vitro* evidence that *B. burgdorferi* can survive in intracellular niches has been reported (Klempner *et al.*, 1993; Montgomery *et al.*, 1993; Girschick *et al.*, 1996). Intracellular spirochaetes, however, have not been observed *in vivo*, while spirochaetes visualized in infected tissues have invariably been extracellular (Duray, 1989b; Barthold *et al.*, 1993). Consequently, it is generally believed that Lyme disease spirochaetes are extracellular throughout the course of infection and, for this reason, investigators have focused on delineating the spirochaete's strategies for evasion of antibody-mediated killing (Embers *et al.*, 2004).

One of the longstanding paradoxes of Lyme disease pathogenesis is that spirochaetes are exquisitely sensitive to antibody-mediated killing (LaRocca and Benach, 2008), yet they are able to persist in animals whose circulating antibodies are capable of conferring protective immunity when administered to naïve animals (Johnson *et al.*, 1986; Barthold *et al.*, 1997). The demonstration that sera from patients is borreliacidal *in vitro* (Pavia *et al.*, 1997; Callister *et al.*, 2002) and also can passively protect animals against inoculation with *in vitro* cultivated spirochaetes (Fikrig *et al.*, 1994) suggests that an identical phenomenon occurs in humans. So, how do extracellular spirochaetes accomplish the remarkable feat of surviving the mammalian host's humoral onslaught? The answer appears to lie in the pathogen's ability to literally outrun the antibody response by constantly generating escape variants that are not susceptible to killing by the antibodies in circulation (Embers *et al.*, 2004). Recent evidence indicates that the bacterium employs two fundamentally different genetic mechanisms to avoid antibody-mediated killing. One is the *vls* system for antigenic variation that is described in detail in Chapter 12. This system, which becomes activated during early infection as the antibody response develops, is capable of generating an enormous number of novel antigenic variants by recombining sequences from a contiguous array of *vls* silent cassettes (Norris, 2006). Spirochaetes lacking lp28-1, the plasmid that harbours the *vls* cassettes, are unable to persist in immunocompetent animals and this defect cannot be restored by a single *vlsE* gene provided *in trans* on a shuttle vector (Purser and Norris, 2000; Ebel *et al.*, 2001; Purser *et al.*, 2003; Grimm *et al.*, 2004a). The other involves antibody-mediated selection of spirochaete subpopulations that have downregulated expression of individual genes whose lipoprotein products are targeted by specific antibodies. OspC is the prototype for this second mechanism. Spirochaetes must express OspC when introduced into the mammalian host in order to evade local clearance, presumably by resident innate surveillance systems in the skin, and disseminate (Grimm *et al.*, 2004b; Tilly *et al.*, 2006; Radolf and Caimano, 2008). Expression of OspC thereafter, however, becomes a liability, and those organisms that continue to do so are killed by the potent antibody response that it evokes (Liang *et al.*, 2002a). While the molecular details of the downregulation of *ospC* expression are not fully elucidated, an operator sequence near the *ospC* promoter required for this to occur has been identified (Xu *et al.*, 2007). It is presumed that an as yet unidentified *trans*-acting factor binds to this sequence, preventing transcription of the gene (see Chapter 4). Admittedly, whether this dual-mechanism scenario for generating escape variants entirely explains how spirochaetes persist is uncertain given the large number of lipoproteins the bacterium expresses at any one time and the continuous expansion of the antibody response that is characteristic of the disease (Liang *et al.*, 2002b; Barbour *et al.*, 2008).

18.9 Clinical manifestations

Much of what is currently viewed as the natural history of Lyme disease is based on observations made in the late 1970s and 1980s that do not accurately reflect the clinical picture of the disease as it exists today. Because of the greater recognition of early infection and the prompt use of the effective antibiotic regimens established in the mid to late 1980s, the clinical manifestations associated with disseminated early and late disease have become much less common (Steere, 2001). Importantly, Lyme arthritis, once emphasized as the key late manifestation of North American Lyme disease (Steere *et al.*, 1987), is now rare. Early studies divided Lyme disease into three distinct stages analogous to those of syphilis: stage 1, erythema migrans; stage 2, neurological or cardiac involvement; and stage 3, arthritis (Steere, 2001). While this staging system has clinical utility as a diagnostic tool, it tends to foster the impression that

Lyme disease patients undergo a stereotypical disease progression. As has long been recognized with syphilis, Lyme disease does not necessarily follow an orderly progression from one stage to the next. Rather it is more logical to view *B. burgdorferi* infection as a progressive, though variable, infectious process with the potential to affect many organ systems, although predominantly the skin, nervous system, heart, and joints.

18.9.1 Dermatological manifestations

Multiple clinical series have established that EM is the most common manifestation of *B. burgdorferi* infection in Europe as well as in the USA (Berglund *et al.*, 1995; Wormser *et al.*, 2006; Steere, 2006; Dandache and Nadelman, 2008). The vast majority of patients presenting with this rash have primary EM, the lesion that develops at the site of tick inoculation, typically within 7 to 14 days after detachment (Steere *et al.*, 2003a; Wormser *et al.*, 2008). Primary EM can be located anywhere but, in adults, it is most commonly found below the waist (Nadelman *et al.*, 1996; Smith *et al.*, 2002). In North American studies, only 14–32% of individuals presenting with EM recall a tick bite (Nadelman *et al.*, 1996; Smith *et al.*, 2002). To distinguish EM from the transient and more localized inflammatory reactions that often develops following an insect bite, the Centers for Disease Control and Prevention designated 5 cm as the minimum lesion diameter required for clinical diagnosis (Fig. 18.4A–D). While this size criterion has improved the specificity of clinical diagnosis, it must be emphasized that bona fide EM lesions can be less than 5 cm, particularly when detected shortly after tick detachment. EM is traditionally described as an expanding, annular, erythematous skin lesion with central clearing, the so-called classic 'bull's eye rash' (Steere, 2001). Central clearing, however, is not as characteristic as once thought: the two largest North American studies of EM found only 37% (Nadelman *et al.*, 1996) and 9% (Smith *et al.*, 2002) of patients had central clearing. Central clearing appears to be more common in Europe, where most cases are due to *B. afzelii* and have a longer duration prior to treatment (Strle *et al.*, 1999; Dandache and Nadelman, 2008). Rapid expansion (approx 20 cm^2 per day) is a key characteristic of EM (Berger, 1993) and is thought to reflect the speed at which spirochaetes migrating away from the bite site are trailed by the local inflammatory response. EM-associated constitutional symptoms can vary from none to mild to moderate flu-like symptoms consisting of arthralgia, malaise, fatigue, headache, and low-grade fever and chills (Nadelman *et al.*, 1996; Smith *et al.*, 2002; Wormser, 2006; Dandache and Nadelman, 2008). Patients can be spirochaetaemic and have low grade fever, arthralgias and malaise without developing EM (Steere *et al.*, 1998; Steere *et al.*, 2003a). Studies of EM patients have reported a difference in the incidence of extracutaneous manifestations between America and Europe. Less than 50% of European patients have extracutaneous manifestations as opposed to more than 75% of patients in the USA (Strle *et al.*, 1999; Dandache and Nadelman, 2008). Regional lymphadenopathy is the most common physical finding associated with EM in both Europe and North America.

EM lesions are analogous to the chancre of primary syphilis, which also tend to have high spirochaete burdens, although the underlying histopathologies differ. Chancres are characterized by an intense inflammatory response in the local microvasculature, endarteritis obliterans, that obstructs blood flow to skin, causing necrosis and ulcer formation (Radolf and Lukehart, 2006). Inflammatory infiltrates in EM also centre about small dermal blood vessels and can be intense, but do not progress to necrosis and ulceration distal from the tick feeding site (Duray, 1989b; de Koning, 1993). Using a novel blister fluid method to elicit skin immune cells for flow cytometric analysis, Salazar and co-workers demonstrated the multicellular nature of the infiltrates in EM and, most notably, that cellular elements derived from both the innate and adaptive arms of the immune system are simultaneously recruited to the infected site where they can theoretically reinforce each other's responses (Salazar *et al.*, 2003). Compared with peripheral blood, lesional infiltrates contained an increased numbers of T-cells, monocytes/macrophages, and dendritic cells, lower proportions of neutrophils, small numbers of plasma cells, and virtually no B-cells. The cutaneous macrophages and DCs showed markedly greater expression of surface activation markers than their peripheral blood counterparts, as well as enhanced expression of TLRs 1, 2, and

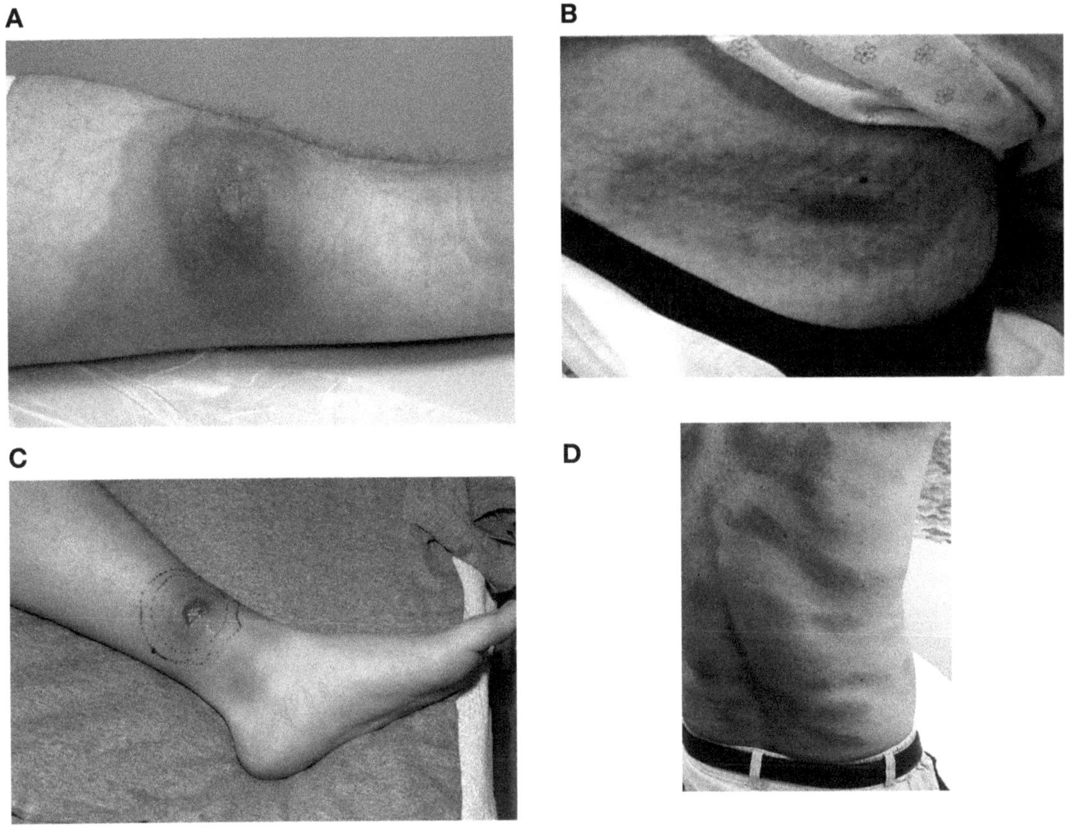

Figure 18.4 Erythema migrans (EM) lesions. (A) Diffuse erythematous EM lesion expanding centripetally over right quadriceps. (B) EM lesion over left iliac crest with surrounding rings consistent with classic target lesion. (C) Ulcerated and necrotic EM lesion. Black lines demarcate area of reseeding erythema following antibiotic therapy. (D) Multiple EM lesions over posterior trunk and lower back.

4; the latter finding suggests that innate immune cells respond to the danger signal conveyed by the spirochaete by lowering their 'set point' for further responses at the site of infection. Although the CD4:CD8 ratios for cutaneous T-cells were similar to those in peripheral blood, lesional T-cells largely consisted of immunophenotypes belonging to memory, memory-effector, and, in the case of $CD8^+$ T-cells, effector subsets as opposed to the predominantly naïve T-cells in circulation. Greater than 80% of lesional T-cells expressed CCR5 and/or CXR3, suggesting a strongly polarized Th1 response, which was confirmed by the high levels of IFN-γ assayed in the cutaneous interstitial fluid. The Th1 nature of the antigen-sensitized T-cells induced during early Lyme disease was supported by Glickstein *et al.* (2003), who found IFN-γ to be the predominant cytokine produced by peripheral blood leucocytes from culture-confirmed EM incubated with borrelial lysates, and by Jones *et al.* (2008), who found high levels of transcripts for IFN-γ and Th1-associated chemokines in RNAs extracted from EM lesions. Although Th1 polarized responses have been associated with chronic inflammation in Lyme disease (Pohl-Koppe *et al.*, 1998; Anguita *et al.*, 1996), these results are in line with evidence from humans (Widhe *et al.*, 2004), and supported by the murine model (Kang *et al.*, 1997; Bockenstedt *et al.*, 2001; Tupin *et al.*, 2008), that early IFN-γ is beneficial to the host. The study by Jones *et al.* (2008) also was noteworthy for its comparison of the chemokine and cytokine mRNA profiles of EM lesions from European and US patients infected with *B. afzelii* and *B. burgdorferi* sensu stricto. Consistent with the generally milder course of EM in European patients, they found that lesions from US patients had significantly higher mRNA levels for chemokines associated with macrophage activation and Th1 cell recruitment.

In Europe, but not in North America, two additional dermatological disorders, borrelial lymphocytoma and acrodermatitis chronicum atrophicans (ACA), are associated with *B. burgdorferi* infection. Borrelial lymphocytoma may occur concurrently with the acute infection or months later (Asbrink and Hovmark, 1988). Borrelial lymphocytomas generally appear as solitary, bluish-red nodules on the ear lobes and in the nipple areas of children (Fig. 18.5) (Gissler and Heininger, 2002) or genital areas of adults (Colli *et al.*, 2004). Histopathologically, it is characterized by a dense lymphocytic infiltrate in the dermis or subcutaneous tissue that can be difficult to differentiate from lymphoma (Colli *et al.*, 2004). Both *B. afzeli* and *B. garinii* have been associated with lymphocytoma, while *B. burgdorferi* sensu stricto has not (Ghislain *et al.*, 2003; Colli *et al.*, 2004). Acrodermatitis chronicum atrophicans (ACA) is a very late manifestation typically developing 10 years after the onset of untreated infection (Asbrink and Hovmark, 1988; Brehmer-Andersson *et al.*, 1998; Duray, 1989b). Occurring throughout northern, central, and eastern Europe, ACA is primarily associated with *B. afzeli* infection (Ohlenbusch *et al.*, 1996; Tazelaar *et al.*, 1997). Initially the skin lesions are distinguished by the presence of inflammation and oedema; however erythema is not always seen and in fact relatively normal-appearing skin may be present, but swelling of soft tissue is characteristic. Early ACA can have an appearance similar to the acute oedematous phase of scleroderma. With time, the lesions eventually become atrophic or sclerotic. ACA typically involves the distal extremities and less commonly the trunk, sparing the face, palms, and soles. Polyneuropathy, small joint arthritis with subluxation, arthritis of the large joints, and periosteal thickening of the bones frequently occur in the same extremity as ACA.

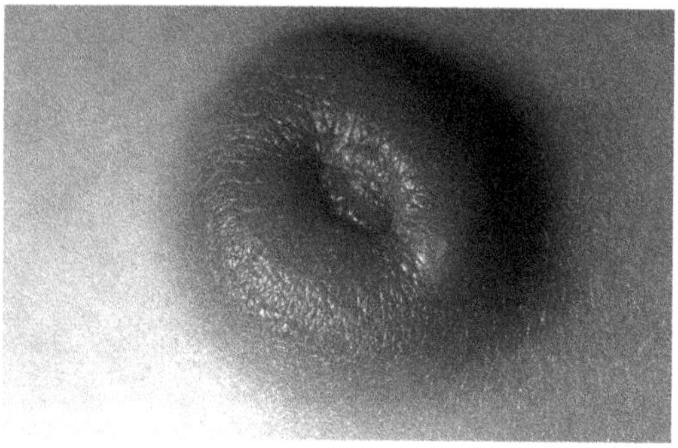

Figure 18.5 Borrelial lymphocytoma ('lymphadenosis benigna cutis'). Micrograph shows a firm, indolent swelling of the left nipple in an eight-year-old boy with prior history of a tick bite and serological evidence of Lyme disease. Symptoms averted following treatment with antibiotics. Photograph courtesy of Professor Ulrich Heininger, University Children's Hospital, Basel, Switzerland.

18.9.2 Nervous system manifestations (neuroborreliosis)

The protean neurological manifestations of Lyme disease, often referred to as neuroborreliosis, reflect the capacity of *B. burgdorferi* to invade diverse targets within the peripheral and central nervous system and to cause neurological complications weeks to years after infection (Halperin, 1995; Halperin, 2008). Early nervous system involvement appears weeks to months after infection and is usually manifested by the involvement of cranial and/or peripheral nerves or nerve roots associated with meningitis. Initial studies of North American patients with untreated EM found that approximately 15% developed meningitis or cranial neuritis within the first three months after presentation (Steere *et al.*, 1983a; Halperin *et al.*, 1989). It is now believed that early neurological involvement is much more common in Europe than in the USA, reflecting the high prevalence of infections by neurotropic genospecies other than *B. burgdorferi* sensu stricto, particularly *B. garinii* (Hengge et al., 2003; Strle *et al.*, 2006; Stanek and Strle, 2008). Although late disseminated *B. burgdorferi* infection has been purportedly associated with a variety of nervous system abnormalities, the neurological entities generally accepted as being associated with late Lyme disease are encephalomyelitis, encephalopathy, and axonal polyneuropathy (Halperin, 1995; Steere, 2001; Hengge *et al.*, 2003; Stanek and Strle, 2003; Halperin, 2008); it must be emphasized that all of these are now considered to be rare. As documented in the recent IDSA Lyme disease treatment guidelines, only one case of encephalomyelitis, nine patients with peripheral neuropathy and seven patients with encephalopathy were seen by the panel members in the five years prior to the publication of the guidelines (Wormser *et al.*, 2006). The entity of Lyme encephalopathy has generated much attention since first described in the late 1980s for a syndrome of cognitive deficits, mood disorders, and somnolence without evidence of focal CNS damage seen almost exclusively in the USA. Originally described in individuals who clearly had active Lyme disease, the entity has been co-opted by those who advocate prolonged courses of antibiotics for persons with chronic fatigue or fibromyalgia-like syndromes and non-reactive serodiagnostic tests (Feder, Jr. *et al.*, 2007; Halperin, 2008). Peripheral nerve involvement in late Lyme disease is an axonopathy that may be distributed symmetrically or asymmetrically and in some cases resemble polyneuritis multiplex (Halperin *et al.*, 1987; Halperin *et al.*, 1990; Logigian and Steere, 1992). Patients with this entity complain of paraesthesia and hyperaesthesia and have abnormal nerve conduction studies. In Europe, peripheral neuropathy can be found in association with ACA but it rarely if ever occurs in patients without ACA (Mygland *et al.*, 2006).

According to Duray (1989b) the basic histopathology of neuroborreliosis, a variable infiltrate of lymphocytes and plasma cells, usually perivascular, is the same regardless of the structure involved. He also maintains that meningovascular syphilis is characterized by more severe infiltrates and vascular changes than neuroborreliosis. Accordingly, vasculopathic involvement of large vessels, the sine qua non of meningovascular syphilis, is rare in Lyme disease; scattered reports notwithstanding, Halperin (2008) has voiced considerable scepticism that Lyme disease is a significant cause of stroke. Not surprisingly, the inflammatory mechanisms underlying CNS neuroborreliosis appear to be complex, involving both innate and adaptive immune systems. Building upon their *in vivo* work with the primate model, Philipp and co-workers (Ramesh *et al.*, 2003, 2008; Bernardino *et al.*, 2008) have demonstrated in an elegant series of studies that spirochaetes induce profound proinflammatory and apoptotic changes in rhesus macaque glial cells, astrocytes, and, most recently, isolated brain slices that involve diverse chemokines, cytokines, and TLR-mediated signals. Analysis of cerebrospinal fluids from neuroborreliosis patients have identified borrelial antigen-specific $CD4^+$ T-cells, $CD8^+$ T-cells, and B-cells (Jacobsen *et al.*, 2003; Cepok *et al.*, 2003; Rupprecht *et al.*, 2005; Lunemann *et al.*, 2007), as well as unusual subsets of dendritic cells (Pashenkov *et al.*, 2001), emphasizing how much *B. burgdorferi* CNS infection differs from that caused by typical pyogenic bacteria.

18.9.3 Cardiac manifestations

As with the other forms of disseminated Lyme disease, the incidence of symptomatic cardiac involvement appears to have decreased dramatically. Early studies reported that 4–10% of untreated EM patients developed Lyme carditis, usually manifested by the acute onset of varying degrees of atrioventricular heart block (Steere *et al.*, 1980; Sigal, 1995). By contrast, in the two prospective OspA vaccine trials, none of the 233 patients who developed Lyme disease showed any evidence of cardiac involvement (Steere *et al.*, 1998; Sigal *et al.*, 1998). How much of this trend is due to prompt recognition and treatment of EM is hard to say. In his classic histopathological study, Duray (1989b) described Lyme carditis as an epi- and transmyocarditis with all regions of the myocardium harbouring interstitial infiltrates of lymphocytes and plasma cells. This stated distribution may have been skewed inadvertently towards the more severe end of the pathological spectrum inasmuch as clinical heart disease usually involves conduction defects, typically heart block proximal to the bundle of His, that resolve with antimicrobial therapy. Indeed, although chronic myocarditis and dilated cardiomyopathy have been reported in the European literature (Klein *et al.*, 1991; Stanek *et al.*, 1990), authorities in North America do not consider *B. burgdorferi* infection to be associated with these forms of cardiac disease (Sonnesyn *et al.*, 1996; Haddad and Nadelman, 2003; Fish *et al.*, 2008). It is interesting to note that the predominance of heart block syndromes correlates with the propensity of spirochaetes to localize to the connective tissue at the base of the heart in the experimental mouse model (Barthold *et al.*, 1991; Armstrong *et al.*, 1992). The nature and timing of cardiovascular involvement are major distinguishing features between syphilis and Lyme disease. Cardiovascular syphilis occurs late in the course of infection and spares the myocardium, instead causing a chronic vasculitis in the proximal aorta leading to aneurysm formation, coronary ostial occlusion, and aortic valvular dysfunction (Jackman and Radolf, 1989).

18.9.4 Lyme arthritis

Lyme arthritis is by far the most studied aspect of late disease and is the most common expression of late untreated infection with *B. burgdorferi* sensu stricto (Steere, 2001; Steere and Glickstein, 2004). Although it was once thought that arthritis was a North American phenomenon and was not associated with *B. burgdorferi* infection in Europe, Lyme arthritis is now recognized in European patients, especially those infected with *B. burgdorferi* sensu stricto (van Dam *et al.*, 1993; Stanek and Strle, 2008). The percentage of individuals who develop arthritis as a complication of Lyme disease is now less than 10% (Steere *et al.*, 1998; Sigal *et al.*, 1998). Physicians must therefore be cautious in making a diagnosis of Lyme arthritis.

Lyme arthritis is a monoarticular or oligoarticular large joint arthritis characterized by episodes of joint inflammation with swelling large effusions and surprisingly little pain; the knee is by far the most commonly affected joint (Steere, 2001; Puius and Kalish, 2008) (Fig. 18.6). Like rheumatoid arthritis, antibiotic-refractory Lyme arthritis can be associated with joint erosions. In the classic study defining the arthritic complication of Lyme disease, 55 patients who presented with EM were observed without antibiotic treatment, 34 of whom developed arthritis (Steere *et al.*, 1987). Twenty-eight of the 34 patients experienced episodic arthritis with the episodes of joint swelling lasting from a few days to several months before spontaneously resolving. Over time, the interval between episodes gradually increased before the arthritis spontaneously remitted entirely. Six patients, however, had a more chronic course consisting of a year or more of unremitting arthritis. What is not generally appreciated is that three of these six remitted after a course of antibiotics and that the other three patients spontaneously remitted after several years. The fact that, even without antibiotic therapy, Lyme arthritis can spontaneously remit is often unrecognized.

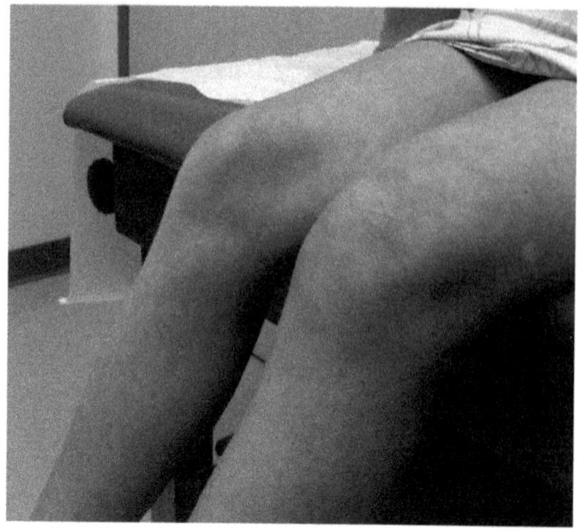

Figure 18.6 Lyme arthritis manifested in a teenager as left knee swelling. Clinically there was no redness, few constitutional symptoms and very little pain. The patient recovered after a 4-week course of oral antibiotics. Photograph courtesy of Dr Jeffrey Thompson, Connecticut Children's Medical Centre.

Given the diminished incidence of Lyme arthritis in recent years (Steere *et al.*, 1998; Sigal *et al.*, 1998), its main importance is as a presumptively autoimmune process in which the aetiological agent is known. Synovial biopsy of untreated Lyme arthritis shows hypertrophy with vascular proliferation, a mixed infiltrate, including T-cells, B-cells, and follicular dendritic cells (Steere *et al.*, 1979; Duray and Steere, 1986), and an abundance of mRNA for IL-1β and TNF-α (Harjacek *et al.*, 2000). Matrix metalloproteinases (MMPs) are believed to contribute to the erosion of cartilage and bone in chronic inflammatory arthritis. In comparing patients with untreated Lyme arthritis to patients with antibiotic-refractory Lyme arthritis, Lin *et al.* (2001) found that MMP-1 and MMP-3 predominated in synovial fluids from patients with untreated Lyme arthritis while synovial fluids from patients with refractory arthritis had higher levels of MMP-8 and MMP-9. Consistent with these results, *B. burgdorferi* was able to induce MMP-1 and MMP-3 production in cultured chrondrocytes but not MMP-8 or MMP-9. A somewhat different result, however, was obtained by Benach's group who found that *B. burgdorferi* was able to induce secretion of MMP-9 as well as MMP-1 in peripheral blood monocytes (Gebbia *et al.*, 2001). Interestingly, only the induction of MMP-9 was TLR2-dependent (Gebbia *et al.*, 2004).

Efforts to elucidate the nature of antibiotic-resistant Lyme arthritis have centred about the hypothesis that it arises from T-cell cross-reactivity between an antigen of *B. burgdorferi* and a host molecule within joints (Gross *et al.*, 2001; Steere and Glickstein, 2004). The groundwork for this hypothesis was laid with genetic population studies showing an association between certain HLA haplotypes and chronic Lyme arthritis (Steere *et al.*, 1990) and the subsequent demonstration that T-cells from patients with antibiotic-resistant arthritis strongly recognize OspA (Lengl-Janssen *et al.*, 1994; Chen *et al.*, 1999). In 1998, a paper published in Science (Gross *et al.*, 1998) reported that the T-cell epitope YVLEGTLTA in OspA from strain B31 (amino acids 165–173) was a molecular mimic for the sequence YVIEGTSKQ in hLFA-1 (amino acids 332–340). The link between these two sequences was supposedly established by the reactivity of patient T-cells with the OspA peptide and the finding that the LFA-1 peptide binds to HLA-DRB1*0401, an HLA molecule strongly associated with arthritis. Initial criticism of this paper centred on the fact that much of the data in it came from a single patient. Objections subsequently were raised on the grounds of T-cell receptor promiscuity; demonstrating that a peptide binds to a particular T-cell receptor is not by itself indicative of its role as a molecular mimic because the average T-cell receptor can bind up to 10^5 to 10^6 different peptides (Benoist and Mathis, 2001). As a demonstration of this principle for *B. burgdorferi*, Maier *et al.* (2000) identified 475 supertope-matching peptides and found that 16 of them stimulated one or more of seven *B. burgdorferi*-specific HLA-DR4-restricted T-cells. They concluded that T-cell cross-reactivity is common. Subsequent publications from Steere's group have backed away from the association of persistent synovitis with T-cell reactivity to either OspA or LFA-1 (Trollmo *et al.*, 2001; Kannian *et al.*, 2007; Drouin *et al.*, 2008).

18.10 Lyme disease in children

The clinical manifestations of Lyme disease in children, as in adults, involve the skin, central nervous system, joints, and heart (Feder, HF Jr., 2008). The primary EM lesion in children is most commonly found on the head and neck (26%), arms or legs (25%), or lower back (24%) (Salazar *et al.*, 1993; Gerber *et al.*, 1996). Less frequently, lesions can be found over the abdomen (9%), the axilla (8%), the groin (5%), or the chest (3%) (Gerber *et al.*, 1996). Most single EM lesions in children are homogenously erythematous; however, they can also have central erythema or clearing, and only in rare instances have painful central ulcerations (Fig. 18.4C). Children also may present with constitutional symptoms, including fever (24%), fatigue (58%), headache (42%), and arthralgias (33%) (Gerber *et al.*, 1996). The presence of multiple EMs has also been documented in up to 20% of children (Gerber *et al.*, 1996) and most of them have associated fever (45%), fatigue (80%), and headache (70%). Similar to adults with Lyme disease, up to 20% of children diagnosed with the disease may present with flu-like symptoms as the only manifestation (Szer *et al.*, 1991; Gerber *et al.*, 1996). Because of this, practitioners in Lyme disease endemic areas need to have a high index of suspicion to avoid missing the diagnosis in such children.

Arthritis may also be the presenting manifestations of Lyme disease in children (Eichenfield *et al.*, 1986; Szer *et al.*, 1991). Unlike reports from the late 1980s and early 1990s, most children who present with Lyme arthritis today do not have a history of erythema migrans. In these children, the rash was either unseen or, perhaps, they never had it at all. The inflammatory changes associated with Lyme arthritis in children usually involve single large joints, most commonly the knee (Gerber *et al.*, 1998). Fever occurs in 25–50% of children with Lyme arthritis, as opposed to adults who are less likely to have fever. Joint fluid obtained from these children may have large numbers of leucocytes (>50,000 cells/mm^3), with a predominance of neutrophils. If not suspected, the disease can be misunderstood for septic arthritis due to pyogenic bacteria (i.e. *Staphylococcus aureus*). Lyme arthritis in children is usually less severe than in adults; if treated appropriately, the prognosis is excellent and, with rare exceptions, a full recovery is usually the final outcome.

Lyme disease in children may also present with neurological (Eppes *et al.*, 1999; Cook *et al.*, 1997) and cardiac manifestations (Silver *et al.*, 2007). Most children who present with evidence of dissemination of the spirochaete to the central nervous system do not have a history of a previously recognized EM lesion. The most common neurological presentations in children include facial nerve palsy, sixth nerve palsy, and lymphocytic meningitis (Eppes *et al.*, 1999; Cook *et al.*, 1997; Feder, Jr., 2008). Children with Lyme meningitis may also have signs of increased intracranial pressure, and some have optic neuritis and papilloedema. More severe neurological manifestations, such as myelitis and severe encephalitis, are exceedingly rare in young children. Lyme carditis may also occur in children, although it is also relatively rare. When it occurs, the most common manifestation in children is transient high-grade atrioventricular block (Silver *et al.*, 2007). Less commonly, children may present with heart failure as a result of myocarditis. When treated appropriately, the prognosis of both neurological and cardiac Lyme disease in children is excellent.

18.11 Reinfection

Recurrent episodes of Lyme disease can be due to either relapse or reinfection (Nadelman and Wormser, 2007). For EM to be recurrent, it has to occur at the same site as the original primary infection and within a relatively short time interval. Reinfections manifest as EM at sites distant from the initial lesion, often have a punctum or eschar indicative of a recent tick bite, and can occur months to years after the initial episode, almost always in people who were previously treated for their initial episode. Two long-term longitudinal studies conducted in highly endemic areas, Westchester, NY and Block Island, RI, found that up to 20% of patients presenting with EM have been reinfected (Nowakowski *et al.*, 2003; Krause *et al.*, 2006). One obvious cause of reinfection is inoculation with a different strain, most likely one with a different *ospC* allele (Probert *et al.*, 1997), that would not be killed by pre-existing antibodies. However, reinfection with the same strain is also possible. Natural infection in the mouse model does induce protective antibody-mediated immunity, but it is of limited duration; Piesman and coworkers found that immunity waned in mice within a year following tick inoculation (Piesman and Happ, 1997). In an elegant study, Barthold and Bockenstedt (1993) showed that antibodies capable of conferring passive protection fall off much more rapidly than antibodies detected by ELISA with whole cell lysates. The implication of this intriguing result is that even strongly reactive serological tests may not be good indicators of immune status against reinfection. Lastly, it must be remembered that a substantial percentage of persons with EM do not have detectable antibodies, even after treatment. One would expect such individuals to be susceptible to reinfection within a relatively short time if they live in an endemic area or frequently engage in high-risk activities. A limitation of the above longitudinal studies (Nowakowski *et al.*, 2003; Krause *et al.*, 2006) is that isolates were not obtained for molecular typing and genetic characterization, leaving unresolved many of the most basic microbiological and immunological questions relating to reinfection.

18.12 Chronic Lyme disease

At one time, the term 'chronic Lyme disease' was used to describe late manifestations of true *B. burgdorferi* infection, particularly neurological disease, such as encephalomyelitis or encephalopathy, and arthritis. During the past two decades, however, the term has come to define a group of individuals with neurocognitive syndromes, mood disorders, fibromyalgia-type symptoms, and chronic fatigue, virtually always in the absence of objective physical findings or laboratory evidence of ongoing inflammation, who are believed to have persistent infection with *B. burgdorferi* (Feder, Jr. *et al.*, 2007; Marques, 2008). Physicians who treat patients with this disorder generally believe that conventional laboratory tests for Lyme disease are insensitive and, therefore, unnecessary for a clinical diagnosis. The therapeutic outcome of this scenario is almost always prolonged antimicrobial therapy, sometimes involving unusual regimens and agents of unproven efficacy for Lyme disease. For unclear reasons, 'chronic Lyme disease' is predominantly a North American phenomenon. Nevertheless, it is noticeably an outgrowth of reports that some patients with true Lyme disease will have persistent or relapsing non-specific symptoms after receiving an adequate course of therapy, the 'post-Lyme disease syndrome' (Dinerman and Steere, 1992; Asch *et al.*, 1994). Reports of Lyme encephalopathy and seronegative Lyme disease (see above) also have contributed to the idea of Lyme disease as a cause of ill-defined syndromes that may not be easily detectable with conventional laboratory tests. University-based studies have shown that physicians who forsake conventional tests often diagnose Lyme disease in persons who have no evidence of *B. burgdorferi* infection or who have illnesses unrelated to it (Steere *et al.*, 1993; Reid *et al.*, 1998; Qureshi *et al.*, 2002). An inappropriate diagnosis of 'chronic Lyme disease' delays or prevents persons from receiving the care they need to address their real problem.

Two highly controversial issues are central to the concept of chronic Lyme disease advanced by its proponents. One, noted above, is that conventional tests are inadequate for diagnosis of Lyme disease. Disproving this negative in humans is virtually impossible, especially to the satisfaction of those practitioners and non-health care workers who are naturally distrusting of the biomedical research community. As noted by Halperin (2008), in order for spirochaetes to persist without provoking discernable inflammation or an immunological response manifested by detectable serum antibodies, *B. burgdorferi* would have to differ from virtually every other known chronic, systemic bacterial pathogen. The other is that currently recommended therapeutic regimens routinely fail to eradicate *B. burgdorferi* infection. In humans, the vast majority of evidence is to the contrary. In one study (Nadelman *et al.*, 1993), 18 patients whose pretreatment biopsies grew *B. burgdorferi* were rebiopsied after treatment; none grew spirochaetes. In another (Berger, 1992), 13 culture-positive patients had a biopsy from the site of the resolved erythema migrans lesion after treatment: all were culture-negative. Moreover, objective evidence of *Borrelia* infection in numerous patients who have post-Lyme disease syndrome has not been obtained using PCR or culture (Klempner *et al.*, 2001; Fallon *et al.*, 2008). Long-term follow-up studies also have shown that patients treated for EM generally become symptom-free and that recurrences are almost always due to reinfection rather than relapse (Nowakowski *et al.*, 2003; Dandache and Nadelman, 2008; Wormser, 2006). Surprisingly, studies in animal models have supported the opposing viewpoint by providing PCR and/or culture evidence that spirochaetes can persist following therapy (Straubinger *et al.*, 1997; Straubinger, 2000; Bockenstedt *et al.*, 2002′ Hodzic *et al.*, 2008). While the clinical relevance of these admittedly perplexing findings is unclear, of greatest importance for the practitioner is that randomized, double-blind treatment trials of 'post Lyme disease' patients have failed to demonstrate lasting benefit for extensive regimens employing intravenous ceftriaxone in combination with oral doxycycline (Fallon *et al.*, 2008; Klempner *et al.*, 2001; Krupp *et al.*, 2003). For a comprehensive discussion of the issues surrounding chronic Lyme disease, the reader is referred to two excellent recent reviews (Feder, Jr. *et al.*, 2007; Marques, 2008).

18.13 Future directions

Lyme disease is now regarded as one of the first in a series of explosive infectious disease phenomena that gave rise to the contemporary concept of emerging infectious diseases, the notion that the infectious agents that impact human welfare are dynamic and strongly influenced by factors both within and outside the normal sphere of human activity. In fact, since the first description of the disease, we have come to appreciate the global dimensions of the threat to human health posed by *B. burgdorferi* and its arthropod vector. Over the years, the notoriety and novelty of the disease have attracted the attention of numerous investigators; as this chapter and the volume that contains it attest, our knowledge of the disorder, its aetiological agent, and the natural world that shapes them has expanded at a meteoric pace. So what remains to be done from the perspective of disease in humans? Vaccine development is an obvious answer, although one presumes that pharmaceutical companies will have little enthusiasm for the development of second-generation vaccines after the commercial failure and precipitous withdrawal of the first-generation vaccine based on OspA. Clearly, alternative prophylactic strategies that focus on preventing enzootic transmission and spread of the pathogen warrant great emphasis. However, the major thrust needs to be in the realms of cellular and molecular pathogenesis. Though cultivatable, *B. burgdorferi* meets anyone's definition of a recalcitrant pathogen. Not only is it hard to introduce DNA into the bacterium, maintaining the stability of its complement of plasmids is a great challenge. Beyond the technical difficulties are the challenges posed by a bacterial genome consisting of literally hundreds of paralogous genes without functionally characterized database orthologues and the need to assign expression and functionality of these genes to two markedly different hosts.

While genetic and bioinformatics breakthroughs undoubtedly will be made, the ultimate question is how to interpret this information in the context of human disease. As any syphilologist will affirm, Lyme disease researchers are blessed not only with a cultivatable organism, but also an excellent and extremely versatile animal model whose full potential has not yet been realized. However, to quote a now clichéd maxim, 'humans are not mice,' and extrapolation is always fraught with danger. Nor are all the major features of human disease recapitulated by the murine model. As examples one need only point to erythema migrans and central nervous system complications, both of which underscore the existence of unique interactions between the spirochaete and human cells. Ironically, while interest in translational Lyme disease research has waned in recent years, perhaps with the exception of AIDS, no infectious disease presents a stronger paradigm for translational research. With the advent of systems biology and related genomics-based technologies, investigators need to build upon findings made with surrogate systems rather than regarding them as ends in themselves. The stakes are high, judging by the intensity of the Lyme disease wars currently being waged on the medical and legal fronts (see Chapter 1). We not only need to understand how *B. burgdorferi* accomplishes its parasitic strategy once introduced into humans, but also how its engagement of the human immune system appears to leave indelible and debilitating footprints long after its progress has been arrested.

Acknowledgements

This work was supported by Public Health Service grants AI-62439 (J.C.S.), AI-38894 (J.D.R.) and AI-29735 (J.D.R.).

References

Afzelius, A. (1910). Verhandlungen der dermatologischen Gesellschaft zu Stockholm. Arch Dermatol. Syph. *101*, 405–406.

Afzelius, A. (1921). Erythema chronicum migrans. *Acta Derm Venereol (Stock)* 2, 120–125.

Aguero-Rosenfeld, M.E., Wang, G., Schwartz, I., and Wormser, G.P. (2005). Diagnosis of Lyme borreliosis. Clin. Microbiol. Rev. *18*, 484–509.

Akins, D.R., Bourell, K.W., Caimano, M.J., Norgard, M.V., and Radolf, J.D. (1998). A new animal model for studying Lyme disease spirochetes in a mammalian host-adapted state. J. Clin. Invest. *101*, 2240–2250.

Akins, D.R., Porcella, S.F., Popova, T.G., Shevchenko, D., Baker, S.I., Li, M., Norgard, M.V., and Radolf, J.D. (1995). Evidence for *in vivo* but not *in vitro* expression of a *Borrelia burgdorferi* outer surface protein F (OspF) homolog. Mol. Microbiol. *18*, 507–520.

Akins, D.R., Purcell, B.K., Mitra, M., Norgard, M.V., and Radolf, J.D. (1993). Lipid modification of the 17-kilodalton membrane immunogen of *Treponema pallidum* determines macrophage activation as well as amphiphilicity. Infect. Immun. *61*, 1202–1210.

Alexopoulou, L., Thomas, V., Schnare, M., Lobet, Y., Anguita, J., Schoen, R.T., Medzhitov, R., Fikrig, E., and Flavell, R.A. (2002). Hyporesponsiveness to vaccination with *Borrelia burgdorferi* OspA in humans and in TLR1- and TLR2-deficient mice. Nat. Med. *8*, 878–884.

Aliprantis, A.O., Yang, R.B., Mark, M.R., Suggett, S., Devaux, B., Radolf, J.D., Klimpel, G.R., Godowski, P., and Zychlinsky, A. (1999). Cell activation and apoptosis by bacterial lipoproteins through Toll-like receptor-2. Science *285*, 736–739.

Alitalo, A., Meri, T., Lankinen, H., Seppala, I., Lahdenne, P., Hefty, P.S., Akins, D., and Meri, S. (2002). Complement inhibitor factor H binding to Lyme disease spirochetes is mediated by inducible expression of multiple plasmid-encoded outer surface protein E paralogs. J. Immunol. *169*, 3847–3853.

Anderson, J.F., Johnson, R.C., and Magnarelli, L.A. (1987). Seasonal prevalence of *Borrelia burgdorferi* in natural populations of white-footed mice, *Peromyscus leucopus*. J. Clin. Microbiol. *25*, 1564–1566.

Anderson, J.F., Magnarelli, L.A., LeFebvre, R.B., Andreadis, T.G., McAninch, J.B., Perng, G.C., and Johnson, R.C. (1989). Antigenically variable *Borrelia burgdorferi* isolated from cottontail rabbits and *Ixodes dentatus* in rural and urban areas. J. Clin. Microbiol. *27*, 13–20.

Anderson, J.M., and Norris, D.E. (2006). Genetic diversity of *Borrelia burgdorferi* sensu stricto in *Peromyscus leucopus*, the primary reservoir of Lyme disease in a region of endemicity in southern Maryland. Appl. Environ. Microbiol. *72*, 5331–5341.

Andrews, S.C., Robinson, A.K., and Rodriguez-Quinones, F. (2003). Bacterial iron homeostasis. FEMS Microbiol. Rev. *27*, 215–237.

Anguita, J., Persing, D.H., Rincon, M., Barthold, S.W., and Fikrig, E. (1996). Effect of anti-interleukin 12 treatment on murine Lyme borreliosis. J. Clin. Invest. *97*, 1028–1034.

Anthonissen, F.M., De Kesel, M., Hoet, P.P., and Bigaignon, G.H. (1994). Evidence for the involvement of different genospecies of *Borrelia* in the clinical outcome of Lyme disease in Belgium. Res. Microbiol. *145*, 327–331.

Antonara, S., Chafel, R.M., LaFrance, M., and Coburn, J. (2007). *Borrelia burgdorferi* adhesins identified using *in vivo* phage display. Mol. Microbiol. *66*, 262–276.

Armstrong, A.L., Barthold, S.W., Persing, D.H., and Beck, D.S. (1992). Carditis in Lyme disease susceptible and resistant strains of laboratory mice infected with *Borrelia burgdorferi*. Am. J. Trop. Med. Hyg. *47*, 249–258.

Asbrink, E., and Hovmark, A. (1988). Early and late cutaneous manifestations in *Ixodes*-borne borreliosis (erythema migrans borreliosis. Lyme borreliosis). Ann. NY Acad. Sci. *539*, 4–15.

Asch, E.S., Bujak, D.I., Weiss, M., Peterson, M.G., and Weinstein, A. (1994). Lyme disease: an infectious and postinfectious syndrome. J. Rheum. *21*, 454–461.

Bacon, R.M., Biggerstaff, B.J., Schriefer, M.E., Gilmore, R.D., Jr., Philipp, M.T., Steere, A.C. *et al.* (2003). Serodiagnosis of Lyme disease by kinetic enzyme-linked immunosorbent assay using recombinant VlsE1 or peptide antigens of *Borrelia burgdorferi* compared with 2-tiered testing using whole-cell lysates. J. Infect. Dis *187*, 1187–1199.

Bacon, R.M., Kugeler, K.J., Griffith, K.S., and Mead, P.S. (2007). Lyme disease--United States, 2003–2005. MMWR Morb. Mortal. Wkly. Rep. *56*, 573–576.

Bacon, R.M., Kugeler, K.J., and Mead, P.S. (2008). Surveillance for Lyme disease--United States, 1992–2006. MMWR Surveill. Summ. *57*, 1–9.

Bankhead, T., and Chaconas, G. (2007). The role of VlsE antigenic variation in the Lyme disease spirochete: persistence through a mechanism that differs from other pathogens. Mol. Microbiol. *65*, 1547–1558.

Bannwarth, A. (1941). Chronische lymphocytare Meningitis, entzundliche Polyneuritis und "Rheumatismus". Ein Beitrag zum Problem "Allergie und Nervensystem". Arch. Psychiatri *113*, 284–376.

Bannwarth, A. (1944). Zur klinik und pathogenese de "chronischen lymphocytaren meningitis". Arch. Psychiatr. Nervenkr. *117*, 161–185.

Baranton, G., Seinost, G., Theodore, G., Postic, D., and Dykhuizen, D. (2001). Distinct levels of genetic diversity of *Borrelia burgdorferi* are associated with different aspects of pathogenicity. Res. Microbiol. *152*, 149–156.

Barbour, A.G. (1988). Laboratory aspects of Lyme borreliosis. Clin. Microbiol. Rev. *1*, 399–414.

Barbour, A.G., Hayes, S.F., Heiland, R.A., Schrumpf, M.E., and Tessier, S.L. (1986). A *Borrelia*-specific monoclonal antibody binds to a flagellar epitope. Infect. Immun. *52*, 549–554.

Barbour, A.G., Heiland, R.A., and Howe, T.R. (1985). Heterogeneity of major proteins in Lyme disease borreliae: a molecular analysis of North American and European isolates. J. Infect. Dis. *152*, 478–484.

Barbour, A.G., Jasinskas, A., Kayala, M.A., Davies, D.H., Steere, A.C., Baldi, P., and Felgner, P.L. (2008). A genome-wide proteome array reveals a limited set of immunogens in natural infections of humans and white-footed mice with *Borrelia burgdorferi*. Infect. Immun. *76*, 3374–3389.

Barthold, S.W. (1996). Lyme borreliosis in the laboratory mouse. J. Spirochet. Tick-Borne Dis. *3*, 22–44.

Barthold, S.W. (1999). Specificity of infection-induced immunity among *Borrelia burgdorferi* sensu lato species. Infect. Immun. *67*, 36–42.

Barthold, S.W., Beck, D.S., Hansen, G.M., Terwilliger, A.A., and Moody, K.D. (1990). Lyme borreliosis in selected strains and ages of laboratory mice. J. Infect. Dis *162*, 133–138.

Barthold, S.W., and Bockenstedt, L.K. (1993). Passive immunizing activity of sera from mice infected with *Borrelia burgdorferi*. Infect. Immun. *61*, 4696–4702.

Barthold, S.W., De Souza, M.S., Janotka, J.L., Smith, A.L., and Persing, D.H. (1993). Chronic Lyme borreliosis in the laboratory mouse. Am J. Pathol. *143*, 959–971.

Barthold, S.W., Feng, S.L., Bockenstedt, L.K., Fikrig, E., and Feen, K. (1997). Protective and arthritis-resolving activity in sera of mice infected with *Borrelia burgdorferi*. Clin. Infect. Dis. *25*, S9–S17.

Barthold, S.W., Persing, D.H., Armstrong, A.L., and Peeples, R.A. (1991). Kinetics of *Borrelia burgdorferi* dissemination and evolution of disease after intradermal inoculation of mice. Am J. Pathol. *139*, 263–273.

Beaurepaire, C. and Chaconas, G. (2005). Mapping of essential replication functions of the linear plasmid lp17 of *B. burgdorferi* by targeted deletion walking. Mol. Microbiol. *57*, 132–142.

Beck, G., Habicht, G.S., Benach, J.L., and Coleman, J.L. (1985). Chemical and biologic characterization of a lipopolysaccharide extracted from the Lyme diseae spirochete (*Borrelia burgdorferi*). J. Infect. Dis *152*, 108–117.

Becker, M., Bunikis, J., Lade, B.D., Dunn, J.J., Barbour, A.G., and Lawson, C.L. (2005). Structural investigation of *B. burgdorferi* OspB, a bactericidal FAB target. J. Biol. Chem. *280*, 17363–17370.

Beermann, C., Wunderli-Allenspach, H., Groscurth, P., and Filgueira, L. (2000). Lipoproteins from *Borrelia burgdorferi* applied in liposomes and presented by dendritic cells induce CD8(+) T-lymphocytes in vitro. Cell. Immunol. *201*, 124–131.

Behera, A.K., Durand, E., Cugini, C., Antonara, S., Bourassa, L., Hildebrand, E., Hu, L.T., and Coburn, J. (2008). *Borrelia burgdorferi* BBB07 interaction with integrin $\alpha_3\beta_1$ stimulates production of pro-inflammatory mediators in primary human chondrocytes. Cell Microbiol. *10*, 320–331.

Behera, A.K., Hildebrand, E., Bronson, R.T., Perides, G., Uematsu, S., Akira, S., and Hu, L.T. (2006a) MyD88 deficiency results in tissue-specific changes in cytokine induction and inflammation in interleukin-18-independent mice infected with *Borrelia burgdorferi*. Infect. Immun. *74*, 1462–1470.

Behera, A.K., Hildebrand, E., Uematsu, S., Akira, S., Coburn, J., and Hu, L.T. (2006b) Identification of a TLR-independent pathway for *Borrelia burgdorferi*-induced expression of matrix metalloproteinases and inflammatory mediators through binding to integrin alpha 3 beta 1. J. Immunol. *177*, 657–664.

Ben-Menachem, G., Kubler-Kielb, J., Coxon, B., Yergey, A., and Schneerson, R. (2003). A newly discovered cholesteryl galactoside from *Borrelia burgdorferi*. Proc. Natl. Acad. Sci. U.S.A. *100*, 7913–7918.

Benach, J.L., Bosler, E.M., Hanrahan, J.P., Coleman, J.L., Habicht, G.S., Bast, T.F., Cameron, D.J., Ziegler, J.L., Barbour, A.G., Burgdorfer, W., Edelman, R., and Kaslow, R.A. (1983). Spirochetes isolated from the blood of two patients with Lyme disease. N. Engl. J. Med. *308*, 740–742.

Benhnia, M.R., Wroblewski, D., Akhtar, M.N., Patel, R.A., Lavezzi, W., Gangloff, S.C. *et al.* (2005). Signaling through CD14 attenuates the inflammatory response to *Borrelia burgdorferi*, the agent of Lyme disease. J. Immunol. *174*, 1539–1548.

Benoist, C., and Mathis, D. (2001). Autoimmunity provoked by infection: how good is the case for T cell epitope mimicry? Nat. Immunol. *2*, 797–801.

Berger, B.W. (1992). Erythema migrans in Lyme disease: a correction. JAMA *268*, 874.

Berger, B.W. (1993). Erythema migrans. Clin. Dermatol. *11*, 359–362.

Berger, B.W., Johnson, R.C., Kodner, C., and Coleman, L. (1995). Cultivation of Borrelia burgdorferi from human tick bite sites: a guide to the risk of infection. J. Am. Acad. Dermatol. *32*, 184–187.

Berglund, J., Eitrem, R., Ornstein, K., Lindberg, A., Ringner, A., Elmrud, H., Carlsson, M., Runehagen, A., Svanborg, C., and Norrby, R. (1995). An epidemiologic study of Lyme disease in southern Sweden. N. Engl. J. Med. *333*, 1319–1324.

Bernardino, A.L., Myers, T.A., Alvarez, X., Hasegawa, A., and Philipp, M.T. (2008). Toll-like receptors: insights into their possible role in the pathogenesis of Lyme neuroborreliosis. Infect. Immun. *76*, 4385–4395.

Bockenstedt, L.K., Kang, I., Chang, C., Persing, D., Hayday, A., and Barthold, S.W. (2001). CD4$^+$ T helper 1 cells facilitate regression of murine Lyme carditis. Infect. Immun. *69*, 5264–5269.

Bockenstedt, L.K., Mao, J., Hodzic, E., Barthold, S.W., and Fish, D. (2002). Detection of attenuated, noninfectious spirochetes in *Borrelia burgdorferi*-infected mice after antibiotic treatment. J. Infect. Dis *186*, 1430–1437.

Bolz, D.D., Sundsbak, R.S., Ma, Y., Akira, S., Kirschning, C.J., Zachary, J.F., Weis, J.H., and Weis, J.J. (2004). MyD88 plays a unique role in host defense but not arthritis development in Lyme disease. J. Immunol. *173*, 2003–2010.

Bos, M.P., Robert, V., and Tommassen, J. (2007). Biogenesis of the gram-negative bacterial outer membrane. Annu. Rev. Microbiol. *61*, 191–214.

Bosler, E. (1993). Tick Vectors and Hosts (St. Louis: Mosby-Year Books, Inc.).

Brandt, M.E., Riley, B.S., Radolf, J.D., and Norgard, M.V. (1990). Immunogenic integral membrane proteins of *Borrelia burgdorferi* are lipoproteins. Infect. Immun. *58*, 983–991.

Brehmer-Andersson, E., Hovmark, A., and Asbrink, E. (1998). Acrodermatitis chronica atrophicans: histopathologic findings and clinical correlations in 111 cases. Acta Dermatol. Venereol. *78*, 207–213.

Brightbill, H.D., Libraty, D.H., Krutzik, S.R., Yang, R.B., Belisle, J.T., Bleharski, J.R. Maitland, M., Norgard, M.V., Plevy, S.E., Smale, S.T., Brennan, P.J., Bloom, B.R., Godowski, P.J., Modlin, R.L. (1999). Host defense mechanisms triggered by microbial lipoproteins through toll-like receptors. Science *285*, 732–736.

Brisson, D., and Dykhuizen, D.E. (2004). OspC diversity in *Borrelia burgdorferi*: different hosts are different niches. Genetics *168*, 713–722.

Brooks, C.S., Vuppala, S.R., Jett, A.M., Alitalo, A., Meri, S., and Akins, D.R. (2005). Complement regulator-acquiring surface protein 1 imparts resistance to human serum in *Borrelia burgdorferi*. J. Immunol. *175*, 3299–3308.

Brown, E.L., Wooten, R.M., Johnson, B.J., Iozzo, R.V., Smith, A., Dolan, M.C., Guo, B.P., Weis, J.J., Hook, M. (2001). Resistance to Lyme disease in decorin-deficient mice. J. Clin. Invest. *107*, 845–852.

Brown, R.N., and Lane, R.S. (1992). Lyme disease in California: a novel enzootic transmission cycle of *Borrelia burgdorferi*. Science *256*, 1439–1442.

Brusca, J.S., McDowall, A.W., Norgard, M.V., and Radolf, J.D. (1991). Localization of outer surface proteins A and B in both the outer membrane and intracellular compartments of *Borrelia burgdorferi*. J. Bacteriol. *173*, 8004–8008.

Buchwald, A. (1883) Ein fall von diffuser idiopathischer haut-atrophie. Arch. Dermatol. Syph. *15*, 553–556.

Bunikis, I., Denker, K., Östberg, Y., Andersen, C., Benz, R., and Bergström, S. (2008). An RND-type efflux system in *Borrelia burgdorferi* is involved in virulence and resistance to antimicrobial compounds. PLoS Pathog. *4*, e1000009.

Bunikis, J., Tsao, J., Luke, C.J., Luna, M.G., Fish, D., and Barbour, A.G. (2004). *Borrelia burgdorferi* infection in a natural population of *Peromyscus leucopus* mice: a longitudinal study in an area where Lyme borreliosis is highly endemic. J. Infect. Dis. *189*, 1515–1523.

Burgdorfer, W., Anderson, J.F., Gern, L., Lane, R.S., Piesman, J., and Spielman, A. (1991). Relationship of *Borrelia burgdorferi* to its arthropod vectors. Scand. J. Infect. Dis. Suppl. *77*, 35–40.

Burgdorfer, W., Barbour, A.G., Hayes, S.F., Benach, J.L., Grunwaldt, E., and Davis, J.P. (1982). Lyme disease-a tick-borne spirochetosis? Science *216*, 1317–1319.

Burgdorfer, W., and Hayes, S.F. (1986). Vector-spirochete relationships in louse-borne ad tick-borne borrelioses with emphasis on Lym.e disease. Adv. Dis. Vec. Res. *6*, 127–150.

Busch, D.H., Jassoy, C., Brinckmann, U., Girschick, H., and Huppertz, H.I. (1996). Detection of *Borrelia burgdorferi*-specific CD8$^+$ cytotoxic T cells in patients with Lyme arthritis. J. Immunol. *157*, 3534–3541.

Bykowski, T., Babb, K., von, L.K., Riley, S.P., Norris, S.J., and Stevenson, B. (2006). Transcriptional regulation of the *Borrelia burgdorferi* antigenically variable VlsE surface protein. J. Bacteriol. *188*, 4879–4889.

Cadavid, D., Bai, Y., Hodzic, E., Narayan, K., Barthold, S.W., and Pachner, A.R. (2004). Cardiac involvement in non-human primates infected with the Lyme disease spirochete *Borrelia burgdorferi*. Lab. Invest. *84*, 1439–1450.

Cadavid, D., O'Neill, T., Schaefer, H., and Pachner, A.R. (2000). Localization of *Borrelia burgdorferi* in the nervous system and other organs in a nonhuman primate model of Lyme disease. Lab. Invest. *80*, 1043–1054.

Caimano, M.J., Iyer, R., Eggers, C.H., Gonzalez, C., Morton, E.A., Gilbert, M.A., Schwartz, I., and Radolf, J.D. (2007). Analysis of the RpoS regulon in *Borrelia burgdorferi* in response to mammalian host signals provides insight into RpoS function during the enzootic cycle. Mol. Microbiol. *65*, 1193–1217.

Caimano, M.J., Yang, X., Popova, T.G., Clawson, M.L., Akins, D.R., Norgard, M.V., and Radolf, J.D. (2000). Molecular and evolutionary characterization of the cp32/18 family of supercoiled plasmids in *Borrelia burgdorferi* 297. Infect. Immun. *68*, 1574–1586.

Callister, S.M., Jobe, D.A., Agger, W.A., Schell, R.F., Kowalski, T.J., Lovrich, S.D., and Marks, J.A. (2002). Ability of the borreliacidal antibody test to confirm lyme disease in clinical practice. Clin. Diagn. Lab. Immunol. *9*, 908–912.

Carroll, J.A., Cordova, R.M., and Garon, C.F. (2000). Identification of 11 pH-regulated genes in *Borrelia burgdorferi* localizing to linear plasmids. Infect. Immun. *68*, 6677–6684.

Casjens, S., Palmer, N., van Vugt, R., Huang, W.M., Stevenson, B., Rosa, P. *et al.* (2000). A bacterial genome in flux: the twelve linear and nine circular extrachromosomal DNAs in an infectious isolate of the Lyme disease spirochete *Borrelia burgdorferi*. Mol. Microbiol. *35*, 490–516.

Casjens, S., van Vugt, R., Tilly, K., Rosa, P.A., and Stevenson, B. (1997). Homology throughout the multiple 32-kilobase circular plasmids present in Lyme disease spirochetes. J. Bacteriol. *179*, 217–227.

Cepok, S., Zhou, D., Vogel, F., Rosche, B., Grummel, V., Sommer, N., and Hemmer, B. (2003). The immune response at onset and during recovery from *Borrelia burgdorferi* meningoradiculitis. Arch. Neurol. *60*, 849–855.

Chaconas, G., Stewart, P.E., Tilly, K., Bono, J.L., and Rosa, P. (2001). Telomere resolution in the Lyme disease spirochete. EMBO J. *20*, 3229–3237.

Champion, C.I., Blanco, D.R., Skare, J.T., Haake, D.A., Giladi, M., Foley, D., Miller, J.N., and Lovett, M.A. (1994). A 9.0 kilobase-pair circular plasmid of *Borrelia burgdorferi* encodes an exported protein: evidence for expression only during infection. Infect. Immun. *62*, 2653–2661.

Chen, J., Field, J.A., Glickstein, L., Molloy, P.J., Huber, B.T., and Steere, A.C. (1999). Association of antibiotic treatment-resistant Lyme arthritis with T cell responses to dominant epitopes of outer surface protein A of *Borrelia burgdorferi*. Arthritis Rheum. *42*, 1813–1822.

Coburn, J., Fischer, J.R., and Leong, J.M. (2005). Solving a sticky problem: new genetic approaches to host cell adhesion by the Lyme disease spirochete. Mol. Microbiol. *57*, 1182–1195.

Coleman, A.S., Yang, X., Kumar, M., Zhang, X., Promnares, K., Shroder, D., Kenedy, M.R., Anderson, J.F., Akins, D.R., and Pal, U. (2008). *Borrelia burgdorferi* complement regulator-acquiring surface protein 2 does not contribute to complement resistance or host infectivity. PLoS ONE 3, 3010e.

Coleman, J.L., and Benach, J.L. (1987). Isolation of antigenic components from the Lyme disease spirochete: their role in early diagnosis. J. Infect. Dis *155*, 756–765.

Coleman, J.L., Gebbia, J.A., Piesman, J., Degen, J.L., Bugge, T.H., and Benach, J.L. (1997). Plasminogen is required for efficient dissemination of *B. burgdorferi* in ticks and for enhancement of spirochetemia in mice. Cell *89*, 1111–1119.

Collares-Pereira, M., Couceiro, S., Franca, I., Kurtenbach, K., Schafer, S.M., Vitorino, L., Goncalves, L., Baptista, S., Vieira, M.L., Cunha, C. (2004). First isolation of *Borrelia lusitaniae* from a human patient. J. Clin. Microbiol. *42*, 1316–1318.

Colli, C., Leinweber, B., Mullegger, R., Chott, A., Kerl, H., and Cerroni, L. (2004). *Borrelia burgdorferi*-associated lymphocytoma cutis: clinicopathologic, immunophenotypic, and molecular study of 106 cases. J. Cutan. Pathol. *31*, 232–240.

Cook, S.P., Macartney, K.K., Rose, C.D., Hunt, P.G., Eppes, S.C., and Reilly, J.S. (1997). Lyme disease and seventh nerve paralysis in children. Am. J. Otolaryngol. *18*, 320–323.

Cordes, F.S., Kraiczy, P., Roversi, P., Simon, M.M., Brade, V., Jahraus, O., Wallis, R., Goodstadt, L., Ponting, C.P., Skerka, C., Zipfel, P.F., Wallich, R., and Lea, S.M. (2006). Structure-function mapping of BbCRASP-1, the key complement factor H and FHL-1 binding protein of *Borrelia burgdorferi*. Int. J. Med. Microbiol. *296 Suppl 40*, 177–184.

Cox, D.L., Akins, D.R., Bourell, K.W., Lahdenne, P., Norgard, M.V., and Radolf, J.D. (1996). Limited surface exposure of *Borrelia burgdorferi* outer surface lipoproteins. Proc. Natl. Acad. Sci. U.S.A. *93*, 7973–7978.

Craft, J.E., Fischer, D.K., Shimamoto, G.T., and Steere, A.C. (1986). Antigens of *Borrelia burgdorferi* recognized during Lyme disease. Appearance of a new immunoglobulin M response and expansion of the immunoglobulin G response late in the illness. J. Clin. Invest. *78*, 934–939.

Cruz, A.R., Moore, M.W., La Vake, C.J., Eggers, C.H., Salazar, J.C., and Radolf, J.D. (2008). Phagocytosis of *Borrelia burgdorferi*, the Lyme disease spirochete, potentiates innate immune activation and induces apoptosis in human monocytes. Infect. Immun. *76*, 56–70.

Cugini, C., Medrano, M., Schwan, T.G., and Coburn, J. (2003). Regulation of expression of the *Borrelia burgdorferi* b(3)-Chain integrin ligand, P66, in ticks and in culture. Infect. Immun. *71*, 1001–1007.

Dandache, P., and Nadelman, R.B. (2008). Erythema migrans. Infect. Dis. Clin. North Am. *22*, 235–60, vi.

Daniels, T.J., Boccia, T.M., Varde, S., Marcus, J., Le, J., Bucher, D.J. *et al.* (1998). Geographic risk for Lyme disease and human granulocytic ehrlichiosis in southern New York state. Appl. Environ. Microbiol. *64*, 4663–4669.

Daniels, T.J., Boccia, T.M., Varde, S., Marcus, J., Le, J., Bucher, D.J., Falco, R.C., and Schwartz, I. (1998). Geographic risk for Lyme disease and human granulocytic ehrlichiosis in southern New York state. Appl. Environ. Microbiol. *64*, 4663–4669.

Das, S., Barthold, S.W., Giles, S.S., Montgomery, R.R., Telford, S.R., III, and Fikrig, E. (1997). Temporal pattern of *Borrelia burgdorferi* p21 expression in ticks and the mammalian host. J. Clin. Invest. *99*, 987–995.

Dattwyler, R.J., Volkman, D.J., Luft, B.J., Halperin, J.J., Thomas, J., and Golightly, M.G. (1988). Seronegative Lyme disease: dissociation of specific T- and B-lymphocyte responses to *Borrelia burgdorferi*. N. Engl. J. Med. *319*, 1441–1446.

Daubin, V., Gouy, M., and Perriere, G. (2002). A phylogenomic approach to bacterial phylogeny: evidence of a core of genes sharing a common history. Genome Res. *12*, 1080–1090.

de Koning, J. (1993). Histopathologic patterns of erythema migrans and borrelial lymphocytoma. Clin. Dermatol. *11*, 377–383.

de Silva, A.M., and Fikrig, E. (1995). Growth and migration of *Borrelia burgdorferi* in *Ixodes* ticks during blood feeding. Am. J. Trop. Med. Hyg. *53*, 397–404.

Deneke, J. and Chaconas, G. (2008). Purification and properties of the plasmid maintenance proteins from the *Borrelia burgdorferi* linear plasmid lp17. J. Bacteriol. *190*, 3992–4000.

Dennis, D.T., and Hayes, E.B. (2002). Epidemiology of Lyme borreliosis. In Lyme Borreliosis. Biology, Epidemiology, and Control. Gray, J.S., Kahl, O., Lane, R.S., and Stanek, G., eds (Wallingford, UK: CABI Publishing), pp. 251–280.

DeOgny, L., Pramanik, B.C., Arndt, L.L., Jones, J.D., Rush, J., Slaughter, C.A., Radolf, J.D., and Norgard, M.V. (1994). Solid-phase synthesis of biologically-active lipopeptides as analogs for spirochetal lipoproteins. Peptide Res. *7*, 91–97.

Derdakova, M., and Lencakova, D. (2005). Association of genetic variability within the *Borrelia burgdorferi* sensu lato with the ecology, epidemiology of Lyme borreliosis in Europe. Ann Agric Environ Med *12*, 165–172.

Dinerman, H., and Steere, A.C. (1992). Lyme disease associated with fibromyalgia. Ann. Intern. Med. *117*, 281–285.

Dolan, M.C., Piesman, J., Schneider, B.S., Schriefer, M., Brandt, K., and Zeidner, N.S. (2004). Comparison of disseminated and nondisseminated strains of *Borrelia burgdorferi* sensu ttricto in mice Naturally Infected by Tick Bite. Infect. Immun. *72*, 5262–5266.

Donahue, J.G., Piesman, J., and Spielman, A. (1987). Reservoir competence of white-footed mice for Lyme disease spirochetes. Am. J. Trop. Med. Hyg. *36*, 92–96.

Dong, Z., Edelstein, M.D., and Glickstein, L.J. (1997). CD8+ T cells are activated during the early Th1 and Th2 immune responses in a murine Lyme disease model. Infect. Immun. *65*, 5334–5337.

Dressler, F., Whalen, J.A., Reinhardt, B.N., and Steere, A.C. (1993). Western blotting in the serodiagnosis of Lyme disease. J. Infect. Dis. *167*, 392–400.

Drouin, E.E., Glickstein, L., Kwok, W.W., Nepom, G.T., and Steere, A.C. (2008). Searching for borrelial T cell epitopes associated with antibiotic-refractory Lyme arthritis. Mol Immunol *45*, 2323–2332.

Duray, P.H. (1989a) Clinical pathologic correlations of Lyme disease. Rev. Infect. Dis. 11 Suppl 6: S1487–S1493.

Duray, P.H. (1989b) Histopathology of clinical phases of human Lyme disease. Rheum. Dis. Clin. North. Am. *15*, 691–710.

Duray, P.H., and Steere, A.C. (1986). The spectrum of organ and systems pathology in human Lyme disease. Zentralbl. Bakteriol. Mikrobiol. Hyg. A *263*, 169–178.

Dykhuizen, D.E., Brisson, D., Sandigursky, S., Wormser, G.P., Nowakowski, J., Nadelman, R.B., and Schwartz, I. (2008). The propensity of different *Borrelia burgdorferi* sensu stricto genotypes to cause disseminated infections in humans. Am. J. Trop. Med. Hyg. *78*, 806–810.

Earnhart, C.G., Buckles, E.L., Dumler, J.S., and Marconi, R.T. (2005). Demonstration of OspC type diversity in invasive human lyme disease isolates and identification of previously uncharacterized epitopes that define the specificity of the OspC murine antibody response. Infect. Immun. *73*, 7869–7877.

Ebel, G.D., Dupuis, A.P., Ngo, K., Nicholas, D., Kauffman, E., Jones, S.A., Young, D., Maffei, J., Shi, P.Y., Bernard, K., and Kramer, L.D. (2001). Partial genetic characterization of west nile virus strains, New York state, 2000. Emerg. Infect. Dis. *7*, 650–653.

Eggers, C.H., Caimano, M.J., Clawson, M.L., Miller, W.G., Samuels, D.S., and Radolf, J.D. (2002). Identification of loci critical for replication and compatibility of a *Borrelia burgdorferi* cp32 plasmid and use of a cp32-based shuttle vector for the expression of fluorescent reporters in the Lyme disease spirochaete. Mol. Microbiol. *43*, 281–295.

Eggers, C.H., Kimmel, B.J., Bono, J.L., Elias, A.F., Rosa, P., and Samuels, D.S. (2001). Transduction by φBB-1, a bacteriophage of *Borrelia burgdorferi*. J. Bacteriol. *183*, 4771–4778.

Eichenfield, A.H., Goldsmith, D.P., Benach, J.L., Ross, A.H., Loeb, F.X., Doughty, R.A., and Athreya, B.H. (1986). Childhood Lyme arthritis: experience in an endemic area. J. Pediatr. *109*, 753–758.

Eicken, C., Sharma, V., Klabunde, T., Lawrenz, M.B., Hardham, J.M., Norris, S.J., and Sacchettini, J.C. (2002). Crystal structure of Lyme disease variable surface antigen VlsE of *Borrelia burgdorferi*. J. Biol. Chem. *277*, 21691–21696.

Eicken, C., Sharma, V., Klabunde, T., Owens, R.T., Pikas, D.S., Hook, M., and Sacchettini, J.C. (2001). Crystal structure of Lyme disease antigen outer surface protein C from *Borrelia burgdorferi*. J. Biol. Chem. *276*, 10010–10015.

Embers, M.E., Ramamoorthy, R., and Philipp, M.T. (2004). Survival strategies of *Borrelia burgdorferi*, the etiologic agent of Lyme disease. Microbes Infect. *6*, 312–318.

Engstrom, S.E., Shoop, E., and Johnson, R.C. (1995). Immunoblot interpretation criteria for serodiagnosis of early Lyme disease. J. Clin. Microbiol. *33*, 419–427.

Eppes, S.C., Nelson, D.K., Lewis, L.L., and Klein, J.D. (1999). Characterization of Lyme meningitis and comparison with viral meningitis in children. Pediatrics *103*, 957–960.

Erdile, L.F., Brandt, M.-A., Warakomski, D.J., Westrack, G.J., Sadziene, A., Barbour, A.G., and Mays, J.P. (1993). Role of attached lipid in immunogenicity of *Borrelia burgdorferi* OspA. Infect. Immun. *61*, 81–90.

Erdlow, J.A. (2003). Bull's Eye. Unraveling the Medical Mystery of Lyme Disease (New Haven: Yale University Press).

Falco, R.C., McKenna, D.F., Daniels, T.J., Nadelman, R.B., Nowakowski, J., Fish, D., and Wormser, G.P. (1999). Temporal relation between *Ixodes scapularis* abundance and risk for Lyme disease associated with erythema migrans. Am. J. Epidemiol. *149*, 771–776.

Falkow, S. (1988). Molecular Koch's postulates applied to microbial pathogenicity. Rev. Infect. Dis. *10 Suppl 2*, S274–S276.

Falkow, S. (2004). Molecular Koch's postulates applied to bacterial pathogenicity--a personal recollection 15 years later. Nat. Rev. Microbiol. *2*, 67–72.

Fallon, B.A., Keilp, J.G., Corbera, K.M., Petkova, E., Britton, C.B., Dwyer, E., Slavov, I., Cheng, J., Dobkin, J., Nelson, D.R., and Sackeim, H.A. (2008). A randomized, placebo-controlled trial of repeated IV antibiotic therapy for Lyme encephalopathy. Neurol. *70*, 992–1003.

Feder, H.M., Jr. (2008). Lyme disease in children. Infect. Dis. Clin. North. Am. *22*, 315–326.

Feder, H.M., Jr., Johnson, B.J., O'Connell, S., Shapiro, E.D., Steere, A.C., Wormser, G.P. *et al.* (2007). A critical appraisal of "chronic Lyme disease". N. Engl. J. Med. *357*, 1422–1430.

Feng, S., Hodzic, E., and Barthold, S.W. (2000). Lyme arthritis resolution with antiserum to a 37-kilodalton *Borrelia burgdorferi* protein. Infect. Immun. *68*, 4169–4173.

Ferdows, M.S., and Barbour, A.G. (1989). Megabase-sized linear DNA in the bacterium *Borrelia burgdorferi*, the Lyme disease agent. Proc. Natl. Acad. Sci. U.S.A. *86*, 5969–5973.

Fikrig, E., Barthold, S.W., Sun, W., Feng, W., Telford, S.R., and Flavell, R.A. (1997). Borrelia burgdorferi p35 and p37 proteins, expressed in vivo, elicit protective immunity. Immunity *6*, 531–539.

Fikrig, E., Bockenstedt, L.K., Barthold, S.W., Chen, M., Tao, H., Ali-Salaam, P. *et al.* (1994). Sera from patients with chronic Lyme disease protect mice from Lyme borreliosis. J. Infect. Dis *169*, 568–574.

Fikrig, E., Feng, W., Barthold, S.W., Telford, S.R., III, and Flavell, R.A. (2000). Arthropod- and host-specific *Borrelia burgdorferi bbk32* expression and the inhibition of spirochete transmission. J. Immunol. *164*, 5344–5351.

Fikrig, E., and Narasimhan, S. (2006). *Borrelia burgdorferi* – traveling incognito? Microbes Infect. *8*, 1390–1399.

Filgueira, L., Groscurth, P., Rittig, M., and Joll, H.I. (1996). Human dendritic cells phagocytose and process *Borrelia burgdorferi*. J. Immunol. *157*, 2998–3005.

Fingerle, V., Schulte-Spechtel, U.C., Ruzic-Sabljic, E., Leonhard, S., Hofmann, H., Weber, K. *et al.* (2008). Epidemiological aspects and molecular characterization of *Borrelia burgdorferi* s.l. from southern Germany with special respect to the new species *Borrelia spielmanii* sp. nov. Int. J. Med. Microbiol. *298*, 279–290.

Fish, A.E., Pride, Y.B., and Pinto, D.S. (2008). Lyme carditis. Infect. Dis. Clin. North. Am. *22*, 275–288.

Foley, D.M., Gayek, R.J., Skare, J.T., Wagar, E.A., Champion, C.I., Blanco, D.R., Lovett, M.A., and Miller, J.N. (1995). Rabbit model of Lyme borreliosis: erythema migrans, infection-derived immunity, and identification of *Borrelia burgdorferi* proteins associated with virulence and protective immunity. J. Clin. Invest *96*, 965–975.

Fraser, C.M., Casjens, S., Huang, W.M., Sutton, G.G., Clayton, R., Lathigra, R., White, O., Ketchum, K.A., Dodson, R., Hickey, E.K., Gwinn, M., *et al.* (1997). Genomic sequence of a Lyme disease spirochaete, *Borrelia burgdorferi*. Nature *390*, 580–586.

Fuchs, H., Wallich, R., Simon, M.M., and Kramer, M.D. (1994). The outer surface protein A of the spirochete *Borrelia burgdorferi* is a plasmin(ogen) receptor. Proc. Natl. Acad. Sci. U.S.A. *91*, 12594–12598.

Fung, B.P., McHugh, G.L., Leong, J.M., and Steere, A.C. (1994). Humoral immune response to outer surface protein C of *Borrelia burgdorferi* in Lyme disease: role of the immunoglobulin M response in the serodiagnosis of early infection. Infect. Immun. *62*, 3213–3221.

Garin, C., and Bujadoux, C. (1922). Paralysie par les tiques. J. Med. Lyon. *3*, 765–767.

Gebbia, J.A., Coleman, J.L., and Benach, J.L. (2001). *Borrelia* spirochetes upregulate release and activation of matrix metalloproteinase gelatinase B (MMP-9) and collagenase 1 (MMP-1) in human cells. Infect. Immun. *69*, 456–462.

Gebbia, J.A., Coleman, J.L., and Benach, J.L. (2004). Selective induction of matrix metalloproteinases by *Borrelia burgdorferi* via Toll-like receptor 2 in monocytes. J. Infect. Dis. *189*, 113–119.

Gerber, M.A., Shapiro, E.D., Burke, G.S., Parcells, V.J., Bell, G.L., and for the Lyme Disease Pediatric Study Group (1996). Lyme disease in children in southeastern Connecticut. N. Engl. J. Med. *335*, 1270–1274.

Gerber, M.A., Zemel, L.S., and Shapiro, E.D. (1998). Lyme arthritis in children: clinical epidemiology and long-term outcomes. Pediatrics *102*, 905–908.

German, R.N., Castellino, F., Han, R., Reis e Sousa, C., Romagnoli, P., Sadegh-Nasseri, S., and Zhong, G.M. (1996). Processing and presentation of endocytically acquired protein antigens by MHC class II and class I molecules. Immunol. Rev. *151*, 5–30.

Gern, L., and Humair, P.-F. (2002). Ecology of *Borrelia burgdorferi* in Europe. In Lyme Borreliosis. Biology, Epidemiology, and Control. J.S. Gray, O. Kahl, R.S. Lane, and G. Stanek, eds (Wallingford, UK: CABI Publishing), pp. 149–174.

Ghislain, P.D., Woestyn, S., Tennstedt, D., Lachapelle, J.M., Bigaignon, G., and Delmee, M. (2003). *Borrelia afzelii* evidenced by polymerase chain reaction in a biopsy of nipple lymphocytoma. Br. J. Dermatol. *148*, 377–379.

Giambartolomei, G.H., Dennis, V.A., Lasater, B.L., and Philipp, M.T. (1999). Induction of pro- and anti-inflammatory cytokines by *Borrelia burgdorferi* lipoproteins in monocytes is mediated by CD14. Infect. Immun. *67*, 140–147.

Girschick, H.J., Huppertz, H.I., Russmann, H., Krenn, V., and Karch, H. (1996). Intracellular persistence of *Borrelia burgdorferi* in human synovial cells. Rheumatol Int *16*, 125–132.

Gissler, S. and Heininger, U. (2002). Borrelia lymphocytoma ("lymphadenosis benigna cutis"). Arch. Dis. Child *87*, 12.

Glickstein, L., Moore, B., Bledsoe, T., Damle, N., Sikand, V., and Steere, A.C. (2003). Inflammatory cytokine production predominates in early Lyme disease in patients with erythema migrans. Infect. Immun. *71*, 6051–6053.

Glockner, G., Lehmann, R., Romualdi, A., Pradella, S., Schulte-Spechtel, U., Schilhabel, M., Wilske, B., Suhnel, J., and Platzer, M. (2004). Comparative analysis of the *Borrelia garinii* genome. Nucleic Acids Res. *32*, 6038–6046.

Goldstein, S.F., Charon, N.W., and Kreiling, J.A. (1994). *Borrelia burgdorferi* swims with a planar waveform similar to that of eukaryotic flagella. Proc. Natl. Acad. Sci. U.S.A. *91*, 3433–3437.

Gomes-Solecki, M.J., Brisson, D.R., and Dattwyler, R.J. (2006). Oral vaccine that breaks the transmission cycle of the Lyme disease spirochete can be delivered via bait. Vaccine *24*, 4440–4449.

Goodman, J.L., Bradley, J.F., Ross, A.E., Goellner, P., Lagus, A., Vitale, B., Berger, B.W., Luger, S., and Johnson, R.C. (1995). Bloodstream invasion in early Lyme disease: results from a prospective, controlled, blinded study using the polymerase chain reaction. Am. J. Med. *99*, 6–12.

Grimm, D., Eggers, C.H., Caimano, M.J., Tilly, K., Stewart, P.E., Elias, A.F., Radolf, J.D., and Rosa, P.A. (2004a). Experimental assessment of the roles of linear plasmids lp25 and lp28-1 of *Borrelia burgdorferi* throughout the infectious cycle. Infect. Immun. *72*, 5938–5946.

Grimm, D., Tilly, K., Byram, R., Stewart, P.E., Krum, J.G., Bueschel, D.M., Schwan, T.G., Policastro, P.F., Elias, A.F., and Rosa, P.A. (2004b). Outer-surface protein C of the Lyme disease spirochete: A protein induced in ticks for infection of mammals. Proc. Natl. Acad. Sci. U.S.A. *101*, 3142–3147.

Gross, D., Huber, B.T., and Steere, A.C. (2001). Molecular mimicry and Lyme arthritis. Curr Dir Autoimmun *3*, 94–111.

Gross, D.M., Forsthuber, T., Tary-Lehmann, M., Etling, C., Ito, K., Nagy, Z.A. *et al.* (1998). Identification of LFA-1 as a candidate autoantigen in treatment- resistant Lyme arthritis. Science *281*, 703–706.

Guo, B.P., Brown, E.L., Dorward, D.W., Rosenberg, L.C., and Höök, M. (1998). Decorin-binding adhesins from *Borrelia burgdorferi*. Mol. Microbiol. *30*, 711–723.

Gustafson, R., Jaenson, T.G., Gardulf, A., Mejlon, H., and Svenungsson, B. (1995). Prevalence of *Borrelia burgdorferi* sensu lato infection in *Ixodes ricinus* in Sweden. Scand J. Infect. Dis *27*, 597–601.

Guttman, D.S., Wang, P.W., Wang, I.N., Bosler, E.M., Luft, B.J., and Dykhuizen, D.E. (1996). Multiple infections of *Ixodes scapularis* ticks by *Borrelia burgdoreri* as revealed by single-strand conformation polymorphism analysis. J. Clin. Microbiol. *34*, 652–656.

Habicht, G.S., Beck, G., Benach, J.L., and Coleman, J.L. (1986). *Borrelia burgdorferi* lipopolysaccharide and its role in the pathogenesis of Lyme disease. *Zentralbl Bakteriol Mikrobiol Hyg [A] 263*, 137–141.

Haddad, F.A., and Nadelman, R.B. (2003). Lyme disease and the heart. Front Biosci *8*, s769–s782.

Hagman, K.E., Yang, X., Wikel, S.K., Schoeler, G.B., Caimano, M.J., Radolf, J.D., and Norgard, M.V. (2000). Decorin-binding protein (Dbp) of *Borrelia burgdorferi* is not protective when immunized mice are challenged via tick infestation and correlates with the lack of DbpA expression by *B. burgdorferi* in ticks. Infect. Immun. *68*, 4759–4764.

Halperin, J., Luft, B.J., Volkman, D.J., and Dattwyler, R.J. (1990). Lyme neuroborreliosis. Peripheral nervous system manifestations. Brain *113*, 1207–1221.

Halperin, J.J. (1995). Neuroborreliosis. Am. J. Med. *98*, 52S–56S.

Halperin, J.J. (2008). Nervous system Lyme disease. Infect. Dis. Clin. North. Am. *22*, 261–74, vi.

Halperin, J.J., Little, B.W., Coyle, P.K., and Dattwyler, R.J. (1987). Lyme disease: cause of a treatable peripheral neuropathy. Neurology *37*, 1700–1706.

Halperin, J.J., Luft, B.J., Anand, A.K., Roque, C.T., Alvarez, O., Volkman, D.J., and Dattwyler, R.J. (1989). Lyme neuroborreliosis: central nervous system manifestations. Neurology*39*, 753–759.

Hanincova, K., Liveris, D., Sandigursky, S., Wormser, G.P., and Schwartz, I. (2008). *Borrelia burgdorferi* sensu stricto is clonal in patients with early Lyme borreliosis. Appl. Environ. Microbiol. *74*, 5008–5014.

Hanrahan, J.P., Benach, J.L., Coleman, J.L., Bosler, E.M., Grabau, J.C., and Morse, D.L. (1984a) Epidemiologic features of Lyme disease in New York. Yale J. Biol. Med. *57*, 643–650.

Hanrahan, J.P., Benach, J.L., Coleman, J.L., Bosler, E.M., Morse, D.L., Cameron, D.J., Edelman, R., and Kaslow, R.A. (1984). Incidence and cumulative frequency of endemic Lyme disease in a community. J. Infect. Dis. *150*, 489–496.

Harjacek, M., Diaz-Cano, S., Alman, B.A., Coburn, J., Ruthazer, R., Wolfe, H., and Steere, A.C. (2000). Prominent expression of mRNA for proinflammatory cytokines in synovium in patients with juvenile rheumatoid arthritis or chronic Lyme arthritis. J. Rheumatol. *27*, 497–503.

Hartmann, K., Corvey, C., Skerka, C., Kirschfink, M., Karas, M., Brade, V., Miller, J.C., Stevenson, B., Wallich, R., Zipfel, P.F., and Kraiczy, P. (2006). Functional characterization of BbCRASP-2, a distinct outer membrane protein of *Borrelia burgdorferi* that binds host complement regulators factor H and FHL-1. Mol. Microbiol. *61*, 1220–1236.

Hefty, P.S., Jolliff, S.E., Caimano, M.J., Wikel, S.K., and Akins, D.R. (2002). Changes in temporal and spatial patterns of outer surface lipoprotein expression generate population heterogeneity and antigenic diversity in the Lyme disease spirochete, *Borrelia burgdorferi*. Infect. Immun. *70*, 3468–3478.

Hefty, P.S., Jolliff, S.E., Caimano, M.J., Wikel, S.K., Radolf, J.D., and Akins, D.R. (2001). Regulation of OspE-related, OspF-related, and Elp lipoproteins of *Borrelia burgdorferi* strain 297 by mammalian host-specific signals. Infect. Immun. *69*, 3618–3627.

Hengge, U.R., Tannapfel, A., Tyring, S.K., Erbel, R., Arendt, G., and Ruzicka, T. (2003). Lyme borreliosis. Lancet Infect. Dis. *3*, 489–500.

Herxheimer, K., and Hartmann, K. (1902). Uber acrodermatitis chronica atrophicans. Arch Dermatol Syph *61*, 57–76.

Hinnebusch, J., and Barbour, A.G. (1991). Linear plasmids of *Borrelia burgdorferi* have a telomeric structure and sequence similar to those of a eukaryotic virus. J. Bacteriol. *173*, 7233–7239.

Hinnebusch, J., Bergström, S., and Barbour, A.G. (1990). Cloning and sequence analysis of linear plasmid telomeres of the bacterium *Borrelia burgdorferi*. Mol. Microbiol. *4*, 811–820.

Hirschfeld, M., Kirschning, C.J., Schwandner, R., Wesche, H., Weis, J.H., Wooten, R.M., and Weis, J.J. (1999). Inflammatory signaling by *Borrelia burgdorferi* lipoproteins is mediated by toll-like receptor 2. J. Immunol. *163*, 2382–2386.

Hodzic, E., Feng, S., Freet, K.J., and Barthold, S.W. (2003). *Borrelia burgdorferi* population dynamics and prototype gene expression during infection of immunocompetent and immunodeficient mice. Infect. Immun. *71*, 5042–5055.

Hodzic, E., Feng, S., Holden, K., Freet, K.J., and Barthold, S.W. (2008). Persistence of *Borrelia burgdorferi* following antibiotic treatment in mice. Antimicrob Agents Chemother *52*, 1728–1736.

Hollström, E. (1951). Successful treatment of erythema migrans Afzelius. Acta Derm. Venereol. *31*, 235–243.

Holt, S.C. (1978). Anatomy and chemistry of spirochetes. Microbiol. Rev. *42*, 114–160.

Horowitz, H.W., Pavia, C.S., Bittker, S., Forseter, G., Cooper, D., Nadelman, R.B., Byrne, D., Johnson, R.C., and Wormser, G.P. (1994). Sustained cellular immune responses to *Borrelia burgdorferi*: lack of correlation with clinical presentation and serology. Clin. Diagn. Lab Immunol. *1*, 373–378.

Hu, L.T., Perides, G., Noring, R., and Klempner, M.S. (1995). Binding of human plasminogen to *Borrelia burgdorferi*. Infect. Immun. *63*, 3491–3496.

Huang, W.M., Joss, L., Hsieh, T., and Casjens, S. (2004). Protelomerase uses a topoisomerase IB/Y-recombinase type mechanism to generate DNA hairpin ends. J. Mol. Biol. *337*, 77–92.

Hubalek, Z., and Halouzka, J. (1997). Distribution of *Borrelia burgdorferi* sensu lato genomic groups in Europe, a review. Eur. J. Epidemiol. *13*, 951–957.

Hubalek, Z., Halouzka, J., and Heroldova, M. (1998). Growth temperature ranges of *Borrelia burgdorferi* sensu lato strains. J. Med. Microbiol. *47*, 929–932.

Humair, P., and Gern, L. (2000). The wild hidden face of Lyme borreliosis in Europe. Microbes Infect. *2*, 915–922.

Hyde, F.W., and Johnson, R.C. (1984). Genetic relationship of Lyme disease spirochetes to *Borrelia*, *Treponema*, and *Leptospira* spp. J. Clin. Microbiol. *20*, 151–154.

Jackman, J.D., and Radolf, J.D. (1989). Cardiovascular syphilis. Am. J. Med. *87*, 425–433.

Jacobsen, M., Zhou, D., Cepok, S., Nessler, S., Happel, M., Stei, S., Wilske, B., Sommer, N., and Hemmer, B. (2003). Clonal accumulation of activated $CD8^+$ T Cells in the central nervous system during the early phase of neuroborreliosis. J. Infect. Dis. *187*, 963–973.

Johnson, R.C., Kodner, C., and Russell, M. (1986). Passive immunization of hamsters against experimental infection with the Lyme disease spirochete. Infect. Immun. *53*, 713–714.

Jones, J.D., Bourell, K.W., Norgard, M.V., and Radolf, J.D. (1995). Membrane topology of *Borrelia burgdorferi* and *Treponema pallidum* lipoproteins. Infect. Immun. *63*, 2424–2434.

Jones, K.L., Glickstein, L.J., Damle, N., Sikand, V.K., McHugh, G., and Steere, A.C. (2006). *Borrelia burgdorferi* genetic markers and disseminated disease in patients with early Lyme disease. J. Clin. Microbiol. *44*, 4407–4413.

Jones, K.L., Muellegger, R.R., Means, T.K., Lee, M., Glickstein, L.J., Damle, N., Sikand, V.K., Luster, A.D., and Steere, A.C. (2008). Higher mRNA levels of chemokines and cytokines associated with macrophage activation in erythema migrans skin lesions in patients from the United States than in patients from Austria with Lyme borreliosis. Clin. Infect. Dis. *46*, 85–92.

Jouda, F., Perret, J.L., and Gern, L. (2004). *Ixodes ricinus* density, and distribution and prevalence of *Borrelia burgdorferi* sensu lato infection along an altitudinal gradient. J. Med. Entomol. *41*, 162–169.

Kamradt, T., Lengl-Janssen, B., Strauss, A.F., Bansal, G., and Steere, A.C. (1996). Dominant recognition of a Borrelia burgdorferi outer surface protein A peptide by T helper cells in patients with treatment-resistant Lyme arthritis. Infect. Immun. *64*, 1284–1289.

Kang, I., Barthold, S.W., Persing, D.H., and Bockenstedt, L.K. (1997). T-helper-cell cytokines in the early evolution of murine Lyme arthritis. Infect. Immun. *65*, 3107–3111.

Kannian, P., Drouin, E.E., Glickstein, L., Kwok, W.W., Nepom, G.T., and Steere, A.C. (2007). Decline in the frequencies of *Borrelia burgdorferi* $OspA_{161,175}$-specific T cells after antibiotic therapy in HLA-DRB1*0401-positive patients with antibiotic-responsive or antibiotic-refractory Lyme arthritis. J. Immunol. *179*, 6336–6342.

Kinjo, Y., Tupin, E., Wu, D., Fujio, M., Garcia-Navarro, R., Benhnia, M.R. *et al.* (2006). Natural killer T cells recognize diacylglycerol antigens from pathogenic bacteria. Nat. Immunol. *7*, 978–986.

Kitten, T., and Barbour, A.G. (1992). The relapsing fever agent *Borrelia hermsii* has multiple copies of its chromosome and linear plasmids. Genetics *132*, 311–324.

Klein, J., Stanek, G., Bittner, R., Horvat, R., Holzinger, C., and Glogar, D. (1991). Lyme borreliosis as a cause of myocarditis and heart muscle disease. *European Heart Journal* 12 Suppl D: 73–75.

Klempner, M.S., Hu, L.T., Evans, J., Schmid, C.H., Johnson, G.M., Trevino, R.P., Norton, D., Levy, L., Wall, D., McCall, J., Kosinski, M., and Weinstein, A. (2001). Two controlled trials of antibiotic treatment in patients with persistent symptoms and a history of Lyme disease. N. Engl. J. Med. *345*, 85–92.

Klempner, M.S., Noring, R., and Rogers, R.A. (1993). Invasion of human skin fibroblasts by the Lyme disease spirochete, *Borrelia burgdorferi*. J. Infect. Dis. *167*, 1074–1081.

Knigge, H., Simon, M.M., Meuer, S.C., Kramer, M.D., and Wallich, R. (1996). The outer surface lipoprotein OspA of *Borrelia burgdorferi* provides co-stimulatory signals to normal human peripheral $CD4^+$ and $CD8^+$ T lymphocytes. Eur J. Immunol. *26*, 2299–2303.

Kobryn, K. and Chaconas, G. (2005). Fusion of hairpin telomeres by the *B. burgdorferi* telomere resolvase ResT implications for shaping a genome in flux. Mol. Cell *17*, 783–791.

Kraiczy, P., Hellwage, J., Skerka, C., Becker, H., Kirschfink, M., Simon, M.M., Brade, V., Zipfel, P.F., and Wallich, R. (2004). Complement resistance of *Borrelia burgdorferi* correlates with the expression of BbCRASP-1, a novel linear plasmid-encoded surface protein that interacts with human factor H and FHL-1 and is unrelated to Erp proteins. J. Biol. Chem. *279*, 2421–2429.

Kraiczy, P., Hellwage, J., Skerka, C., Kirschfink, M., Brade, V., Zipfel, P.F., and Wallich, R. (2003). Immune evasion of *Borrelia burgdorferi*: mapping of a complement-inhibitor factor H-binding site of BbCRASP-3, a novel member of the Erp protein family. Eur J. Immunol. 33, 697–707.

Kraiczy, P., Skerka, C., Kirschfink, M., Brade, V., and Zipfel, P.F. (2001a) Immune evasion of *Borrelia burgdorferi* by acquisition of human complement regulators FHL-1/reconectin and Factor H. Eur J. Immunol. *31*, 1674–1684.

Kraiczy, P., Skerka, C., Kirschfink, M., Zipfel, P.F., and Brade, V. (2001b) Mechanism of complement resistance of pathogenic *Borrelia burgdorferi* isolates. Int Immunopharmacol *1*, 393–401.

Krause, P.J., Foley, D.T., Burke, G.S., Christianson, D., Closter, L., and Spielman, A. (2006). Reinfection and relapse in early Lyme disease. Am. J. Trop. Med. Hyg. *75*, 1090–1094.

Krupp, L.B., Hyman, L.G., Grimson, R., Coyle, P.K., Melville, P., Ahnn, S., Dattwyler, R., and Chandler, B. (2003). Study and treatment of post Lyme disease (STOP-LD): A randomized double masked clinical trial. Neurol. *60*, 1923–1930.

Kumaran, D., Eswaramoorthy, S., Luft, B.J., Koide, S., Dunn, J.J., Lawson, C.L., and Swaminathan, S. (2001). Crystal structure of outer surface protein C (OspC) from the Lyme disease spirochete, *Borrelia burgdorferi*. EMBO J. *20*, 971–978.

Kurtenbach, K., De Michelis, S., Etti, S., Schafer, S.M., Sewell, H.S., Brade, V., and Kraiczy, P. (2002). Host association of *Borrelia burgdorferi* sensu lato--the key role of host complement. Trends Microbiol. *10*, 74–79.

Kurtenbach, K., Hanincova, K., Tsao, J.I., Margos, G., Fish, D., and Ogden, N.H. (2006). Fundamental processes in the evolutionary ecology of Lyme borreliosis. Nat. Rev. Microbiol. *4*, 660–669.

Labandeira-Rey, M., Seshu, J., and Skare, J.T. (2003). The absence of linear plasmid 25 or 28–1 of *Borrelia burgdorferi* dramatically alters the kinetics of experimental infection via distinct mechanisms. Infect. Immun. *71*, 4608–4613.

Labandeira-Rey, M., and Skare, J.T. (2001). Decreased infectivity in *Borrelia burgdorferi* strain B31 is associated with loss of linear plasmid 25 or 28-1. Infect. Immun. *69*, 446–455.

LaRocca, T.J., and Benach, J.L. (2008). The important and diverse roles of antibodies in the host response to *Borrelia* infections. Curr. Top. Microbiol. Immunol. *319*, 63–103.

Lagal, V., Portnoi, D., Faure, G., Postic, D., and Baranton, G. (2006). *Borrelia burgdorferi* sensu stricto invasiveness is correlated with OspC-plasminogen affinity. Microbes. Infect. *8*, 645–652.

Lastavica, C.C., Wilson, M.L., Berardi, V.P., Spielman, A., and Deblinger, R.D. (1989). Rapid emergence of a focal epidemic of Lyme disease in coastal Massacusetts. N. Engl. J. Med. *320*, 133–137.

Lengl-Janssen, B., Strauss, A.F., Steere, A.C., and Kamradt, T. (1994). The T helper cell response in Lyme arthritis: differential recognition of *Borrelia burgdorferi* outer surface protein A in patients with treatment-resistant or treatment-responsive Lyme arthritis. J. Exp. Med. *180*, 2069–2078.

Lennhoff, C. (1948). Spirochaetes in aetiologically obscure diseases. Acta Dermatologica Venereologica *28*, 295–324.

Levine, J.F., Wilson, M.L., and Spielman, A. (1985). Mice as reservoirs of the Lyme disease spirochete. Am. J. Trop. Med. Hyg. *34*, 355–360.

Li, H., Dunn, J.J., Luft, B.J., and Lawson, C.L. (1997). Crystal structure of Lyme disease antigen outer surface protein A complexed with an Fab. Proc. Natl. Acad. Sci. U.S.A. *94*, 3584–3589.

Liang, F.T., Jacobs, M.B., Bowers, L.C., and Philipp, M.T. (2002a) An immune evasion mechanism for spirochetal persistence in Lyme borreliosis. J. Exp. Med. *195*, 415–422.

Liang, F.T., Nelson, F.K., and Fikrig, E. (2002b) Molecular adaptation of *Borrelia burgdorferi* in the murine host. J. Exp. Med. *196*, 275–280.

Liang, F.T., Yan, J., Mbow, M.L., Sviat, S.L., Gilmore, R.D., Mamula, M., and Fikrig, E. (2004). *Borrelia burgdorferi* changes its surface antigenic expression in response to host immune responses. Infect. Immun. *72*, 5759–5767.

Lien, E., Sellati, T.J., Yoshimura, A., Flo, T.H., Rawadi, G., Finberg, R.W., Carroll, J.D., Espevik, T., Ingalls, R.R., Radolf, J.D., and Golenbock, D.T. (1999). Toll-like receptor 2 functions as a pattern recognition receptor for diverse bacterial products. J. Biol. Chem. *274*, 33419–33425.

Lin, B., Kidder, J.M., Noring, R., Steere, A.C., Klempner, M.S., and Hu, L.T. (2001). Differences in synovial fluid levels of matrix metalloproteinases suggest separate mechanisms of pathogenesis in lyme arthritis before and after antibiotic treatment. J. Infect. Dis. *184*, 174–180.

Lipschütz, B. (1923). Weiter Beitgrag zur Kenngtnis des "Erythema chronicum migrans". Arch. Dermatol. Res. *143*, 365–374.

Liu, N., Montgomery, R.R., Barthold, S.W., and Bockenstedt, L.K. (2004). Myeloid differentiation antigen 88 deficiency impairs pathogen clearance but does not alter inflammation in *Borrelia burgdorferi*-infected mice. Infect. Immun. *72*, 3195–3203.

Liveris, D., Gazumyan, A., and Schwartz, I. (1995). Molecular typing of *Borrelia burgdorferi* sensu lato by PCR-restriction fragment length polymorphism analysis. J. Clin. Microbiol. *33*, 589–595.

Liveris, D., Wang, G., Girao, G., Byrne, D.W., Nowakowski, J., McKenna, D., Nadelman, R., Wormser, G.P., and Schwartz, I. (2002). Quantitative detection of *Borrelia burgdorferi* in 2-millimeter skin samples of erythema migrans lesions: correlation of results with clinical and laboratory findings. J. Clin. Microbiol. *40*, 1249–1253.

Liveris, D., Wormser, G.P., Nowakowski, J., Nadelman, R., Bittker, S., Cooper, D., Varde, S., Moy, F.H., Forseter, G., Pavia, C.S., and Schwartz, I. (1996). Molecular typing of *Borrelia burgdorferi* from Lyme disease patients by PCR-restriction fragment length polymorphism analysis. J. Clin. Microbiol. *34*, 1306–1309.

Logigian, E.L., and Steere, A.C. (1992). Clinical and electrophysiologic findings in chronic neuropathy of Lyme disease. Neurol *42*, 303–311.

Luft, B.J., Steinman, C.R., Neimark, H.C., Muraldhar, B., Rush, T., Finkel, M.F., Kunkel, M., and Dattwyler, R.J. (1992). Invasion of the central nervous system by *Borrelia burgdorferi* in acute disseminated infection. JAMA *267*, 1364–1367.

Lunemann, J.D., Gelderblom, H., Sospedra, M., Quandt, J.A., Pinilla, C., Marques, A., and martin, R. (2007). Cerebrospinal fluid-infiltrating CD4+ T cells recognize *Borrelia burgdorferi* lysine-enriched protein domains and central nervous system autoantigens in early lyme encephalitis. Infect. Immun. *75*, 243–251.

Ma, Y., Seiler, K.P., Tai, K.F., Yang, L., Woods, M., and Weis, J.J. (1994). Outer surface lipoproteins of *Borrelia burgdorferi* stimulate nitric oxide production by the cytokine-inducible pathway. Infect. Immun. *62*, 3663–3671.

Magnarelli, L.A., and Anderson, J.F. (1987). Ticks and biting insects infected with the etiologic agent of Lyme disease, *Borrelia burgdorferi*. J. Clin. Microbiol. *25*, 557–558.

Maier, B., Molinger, M., Cope, A.P., Fugger, L., Schneider-Mergener, J., Sonderstrup, G. *et al.* (2000). Multiple cross-reactive self-ligands for *Borrelia burgdorferi*-specific HLA-DR4-restricted T cells. Eur J. Immunol. *30*, 448–457.

Marconi, R.T., Liveris, D., and Schwartz, I. (1995). Identification of novel insertion elements, restriction fragment length polymorphism patterns, and discontinuous 23S rRNA in Lyme disease spirochetes: phylogenetic analyses of rRNA genes and their intergenic spacers in *Borrelia japonica* sp. nov. and genomic group 21038 (*Borrelia andersonii* sp. nov.) isolates. J. Clin. Microbiol. 33, 2427–2434.

Marconi, R.T., Sung, S.Y., Norton Hughes, C.A., and Carlyon, J.A. (1996). Molecular and evolutionary analyses of a variable series of genes in *Borrelia burgdorferi* that are related to *ospE* and *ospF*, constitute a gene family, and share a common upstream homology box. J. Bacteriol. *178*, 5615–5626.

Marques, A. (2008). Chronic Lyme disease: a review. Infect. Dis. Clin. North Am. *22*, 341–360.

Mast, W.E., and Burrows, W.M., Jr. (1976). Erythema chronicum migrans in the United States. JAMA *236*, 859–860.

Mather, T.N., Nicholson, M.C., Donnelly, E.F., and Matyas, B.T. (1996). Entomologic index for human risk of Lyme disease. Am. J. Epidemiol. *144*, 1066–1069.

Mather, T.N., Telford, S.R., and Adler, G.H. (1991). Absence of transplacental transmission of Lyme disease spirochetes from reservoir mice (*Peromyscus leucopus*) to their offspring. J. Infect. Dis *164*, 564–567.

Maupin, G.O., Fish, D., Zultowsky, J., Campos, E.G., and Piesman, J. (1991). Landscape ecology of Lyme disease in a residential area of Westchester County, New York. Am. J. Epidemiol. *133*, 1105–1113.

Maupin, G.O., Gage, K.L., Piesman, J., Montenieri, J., Sviat, S.L., VanderZandenL., Happ, C.M., Dolan, M., and Johnson, B.J. (1994). Discovery of an enzootic cycle of *Borrelia burgdorferi* in *Neotoma mexicana* and *Ixodes spinipalpis* from northern Colorado, an area where Lyme disease is nonendemic. J. Infect. Dis. *170*, 636–643.

McDowell, J.V., Hovis, K.M., Zhang, H., Tran, E., Lankford, J., and Marconi, R.T. (2006). Evidence that the BBA68 protein (BbCRASP-1) of the Lyme disease spirochetes does not contribute to factor H-mediated immune evasion in humans and other animals. Infect. Immun. *74*, 3030–3034.

McDowell, J.V., Wolfgang, J., Tran, E., Metts, M.S., Hamilton, D., and Marconi, R.T. (2003). Comprehensive analysis of the factor H binding capabilities of *Borrelia* species associated with Lyme disease: delineation of two distinct classes of Factor H binding proteins. Infect. Immun. *71*, 3597–3602.

McKisic, M.D., Lancki, D.W., Cronin, D.C., and Fitch, F.W. (1993). Cytolytic activity of murine IL-2-producing $CD4^+$ and $CD8^+$ T cell clones cycles in response to IL-2. J. Immunol. *151*, 4055–4066.

Medzhitov, R., Preston-Hurlburt, P., and Janeway, C.A., Jr. (1997). A human homologue of the *Drosophila* Toll protein signals activation of adaptive immunity. Nature *388*, 394–397.

Merritt, H.H., Adams, R.D., and Solomon, H.C. (1946). Neurosyphilis New York: Oxford University Press.

Metts, M.S., McDowell, J.V., Theisen, M., Hansen, P.R., and Marconi, R.T. (2003). Analysis of the OspE determinants involved in binding of Factor H and OspE-targeting antibodies elicited during *Borrelia burgdorferi* infection in mice. Infect. Immun. *71*, 3587–3596.

Meyer, A.L., Trollmo, C., Crawford, F., Marrack, P., Steere, A.C., Huber, B.T., Kappler, J., and Hafler, D.A. (2000). Direct enumeration of *Borrelia*-reactive CD4 T cells *ex vivo* by using MHC class II tetramers. Proc. Natl. Acad. Sci. U.S.A. *97*, 11433–11438.

Miller, J.C., Ma, Y., Bian, J., Sheehan, K.C., Zachary, J.F., Weis, J.H., Schreiber, R.D., and Weis, J.J. (2008). A critical role for type I IFN in arthritis development following *Borrelia burgdorferi* infection of mice. J. Immunol. *181*, 8492–8503.

Miller, J.C., Ma, Y., Crandall, H., Wang, X., and Weis, J.J. (2008b) Gene expression profiling provides insights into the pathways involved in inflammatory arthritis development: murine model of Lyme disease. Exp. Mol. Pathol. *85*, 20–27.

Miller, J.C., and Stevenson, B. (2003). Immunological and genetic characterization of *Borrelia burgdorferi* BapA and EppA proteins. Microbiology *149*, 1113–1125.

Miller, J.C., and Stevenson, B. (2004). Increased expression of *Borrelia burgdorferi* factor H-binding surface proteins during transmission from ticks to mice. Int. J. Med. Microbiol. *293 Suppl 37*, 120–125.

Miller, J.C., von Lackum, K., Babb, K., McAlister, J.D., and Stevenson, B. (2003). Temporal analysis of *Borrelia burgdorferi* Erp protein expression throughout the mammal-tick infectious cycle. Infect. Immun. *71*, 6943–6952.

Montgomery, R.R., Nathanson, M.H., and Malawista, S.E. (1993). The fate of *Borrelia burgdorferi*, the agent for Lyme disease, in mouse macrophages. Destruction, survival, recovery. J. Immunol. *150*, 909–915.

Moore, M.W., Cruz, A.R., LaVake, C.J., Marzo, A.L., Eggers, C.H., Salazar, J.C., and Radolf, J.D. (2007). Phagocytosis of *Borrelia burgdorferi* and *Treponema pallidum* potentiates innate immune activation and induces gamma interferon production. Infect. Immun. *75*, 2046–2062.

Moriarty, T.J., Norman, M.U., Colarusso, P., Bankhead, T., Kubes, P., and Chaconas, G. (2008). Real-time high resolution 3D imaging of the lyme disease spirochete adhering to and escaping from the vasculature of a living host. PLoS Pathog. *4*, e1000090.

Morrison, T.B., Weis, J.H., and Weis, J.J. (1997). *Borrelia burgdorferi* outer surface protein A (OspA) activates and primes human neutrophils. J. Immunol. *158*, 4838–4845.

Motaleb, M.A., Corum, L., Bono, J.L., Elias, A.F., Rosa, P., Samuels, D.S., and Charon, N.W. (2000). *Borrelia burgdorferi* periplasmic flagella have both skeletal and motility functions. Proc. Natl. Acad. Sci. U.S.A. *97*, 10899–10904.

Mygland, A., Skarpaas, T., and Ljostad, U. (2006). Chronic polyneuropathy and Lyme disease. Eur. J. Neurol. *13*, 1213–1215.

Nadelman, R.B., Nowakowski, J., Fish, D., Falco, R.C., Freeman, K., McKenna, D., Welch, P., Marcus, R., Aguero-Rosenfeld, M.E., Dennis, D.T., and Wormser, G.P. (2001). Prophylaxis with single-dose doxycycline for the prevention of Lyme disease after an *Ixodes scapularis* tick bite. N. Engl. J. Med. *345*, 79–84.

Nadelman, R.B., Nowakowski, J., Forseter, G., Bittker, S., Cooper, D., GoldbergN., McKenna, D., and Wormser, G.P. (1993). Failure to isolate *Borrelia burgdorferi* after antimicrobial therapy in culture-documented Lyme borreliosis associated with erythema migrans: report of a prospective study. Am. J. Med. *94*, 583–588.

Nadelman, R.B., Nowakowski, J., Forseter, G., Goldberg, N.S., Bittker, S., Cooper, D., Aguero-Rosenfeld, M., and Wormser, G.P. (1996). The clinical spectrum of early Lyme borreliosis in patients with culture-confirmed erythema migrans. Am. J. Med. *100*, 502–508.

Nadelman, R.B., Pavia, C.S., Magnarelli, L.A., and Wormser, G.P. (1990). Isolation of *Borrelia burgdorferi* from the blood of seven patients with Lyme disease. Am. J. Med. *88*, 21–26.

Nadelman, R.B., and Wormser, G.P. (1998). Lyme borreliosis. Lancet *352*, 557–565.

Nadelman, R.B., and Wormser, G.P. (2007). Reinfection in patients with Lyme disease. Clin. Infect. Dis. *45*, 1032–1038.

Narasimhan, S., Caimano, M.J., Liang, F.T., Santiago, F., Laskowski, M., Philipp, M.T., Pachner, A.R., Radolf, J.D., and Fikrig, E. (2003). *Borrelia burgdorferi* transcriptome in the central nervous system of non-human primates. Proc. Natl. Acad. Sci. U.S.A. *100*, 15953–15958.

Narita, S., Matsuyama, S., and Tokuda, H. (2004). Lipoprotein trafficking in *Escherichia coli*. Arch. Microbiol. *182*, 1–6.

Neelakanta, G., Li, X., Pal, U., Liu, X., Beck, D.S., Deponte, K., Fish, D., Kantor, F.S., and Fikrig, E. (2007). Outer surface protein B is critical for *Borrelia burgdorferi* adherence and survival within Ixodes ticks. PLoS. Pathog. *3*, e33.

Nikaido, H. (2003). Molecular basis of bacterial outer membrane permeability revisited. Microbiol Mol Biol Rev *67*, 593–656.

Nocton, J.J., Dressler, F., Rutledge, B.J., Rys, P.N., Persing, D.H., and Steere, A.C. (1994). Detection of *Borrelia burgdorferi* DNA by polymerase chain reaction in synovial fluid from patients with Lyme arthritis. N. Engl. J. Med. *330*, 229–234.

Norgard, M.V., Arndt, L.L., Akins, D.R., Curetty, L.L., Harrich, D.A., and Radolf, J.D. (1996). Activation of human monocytic cells by *Treponema pallidum* and *Borrelia burgdorferi* lipoproteins and synthetic lipopeptides proceeds via a pathway distinct from that of lipopolysaccharide but involves the transcriptional activator NF-κB. Infect. Immun. *64*, 3845–3852.

Norman, M.U., Moriarty, T.J., Dresser, A.R., Millen, B., Kubes, P., and Chaconas, G. (2008). Molecular mechanisms involved in vascular interactions of the Lyme disease pathogen in a living host. PLoS Pathog. *4*, e1000169.

Norris, S.J. (2006). Antigenic variation with a twist--the *Borrelia* story. Mol. Microbiol. *60*, 1319–1322.

Norris, S.J., Howell, J.K., Garza, S.A., Ferdows, M.S., and Barbour, A.G. (995) High- and low-infectivity phenotypes of clonal populations of in vitro- cultured *Borrelia burgdorferi*. Infect. Immun. *63*, 2206–2212.

Nowakowski, J., Nadelman, R.B., Sell, R., McKenna, D., Cavaliere, L.F., Holmgren, D., Gaidici, A., and Wormser, G.P. (2003). Long-term follow-up of patients with culture-confirmed Lyme disease. Am. J. Med. *115*, 91–96.

Nowakowski, J., Schwartz, I., Liveris, D., Wang, G., Aguero-Rosenfeld, M.E., Girao, G., McKenna, D., Nadelman, R.B., Cavaliere, L.F., and Wormser, G.P. (2001). Laboratory diagnostic techniques for patients with early Lyme disease associated with erythema migrans: a comparison of different techniques. Clin. Infect. Dis. 33, 2023–2027.

Nowalk, A.J., Gilmore, R.D., Jr., and Carroll, J.A. (2006). Serologic Proteome analysis of *Borrelia burgdorferi* membrane-associated proteins. Infect. Immun. *74*, 3864–3873.

Nuttall, P.A., Paesen, G.C., Lawrie, C.H., and Wang, H. (2000). Vector-host interactions in disease transmission. J. Mol. Microbiol. Biotechnol. *2*, 381–386.

Ohlenbusch, A., Matuschka, F.R., Richter, D., Christen, H.J., Thomssen, R., and Eiffert, H. (1996). Etiology of the acrodermatitis chronica atrophicans lesion in Lyme disease. J. Infect. Dis *174*, 421–423.

Ohnishi, J., Piesman, J., and de Silva, A.M. (2001). Antigenic and genetic heterogeneity of *Borrelia burgdorferi* populations transmitted by ticks. Proc. Natl. Acad. Sci. U.S.A. *98*, 670–675.

Oliver, J.H., Jr., Lin, T., Gao, L., Clark, K.L., Banks, C.W., Durden, L.A. *et al.* (2003). An enzootic transmission cycle of Lyme borreliosis spirochetes in the southeastern United States. Proc. Natl. Acad. Sci. U.S.A. *100*, 11642–11645.

Olsen, B., Duffy, D.C., Jaenson, T.G., Gylfe, A., Bonnedahl, J., and Bergström, S. (1995). Transhemispheric exchange of Lyme disease spirochetes by seabirds. J. Clin. Microbiol. *33*, 3270–3274.

Olsen, B., Jaenson, T.G.T., Noppa, L., Bunikis, J., and Bergström, S. (1993). A Lyme borreliosis cycle in seabirds and *Ixodes uriae* ticks. Nature *362*, 340–342.

Ostberg, Y., Pinne, M., Benz, R., Rosa, P., and Bergström, S. (2002). Elimination of channel-forming activity by insertional inactivation of the p13 gene in *Borrelia burgdorferi*. J. Bacteriol. *184*, 6811–6819.

Ozinsky, A., Underhill, D.M., Fontenot, J.D., Hajjar, A.M., Smith, K.D., Wilson, C.B., Schroeder, L., and Aderem, A. (2000). The repertoire for pattern recognition of pathogens by the innate immune system is defined by cooperation between Toll-like receptors. Proc. Natl. Acad. Sci. U.S.A. *97*, 13766–13771.

Pachner, A.R., Cadavid, D., Shu, G., Dail, D., Pachner, S., Hodzic, E., and Barthold, S.W. (2001). Central and peripheral nervous system infection, immunity, and inflammation in the NHP model of Lyme borreliosis. Ann Neurol *50*, 330–338.

Pachner, A.R., Dail, D., Bai, Y., Sondey, M., Pak, L., Narayan, K., and Cadavid, D. (2004). Genotype determines phenotype in experimental Lyme borreliosis. Ann Neurol *56*, 361–370.

Pal, U., de Silva, A.M., Montgomery, R.R., Fish, D., Anguita, J., Anderson, J.F. *et al.* (2000). Attachment of *Borrelia burgdorferi* within *Ixodes scapularis* mediated by outer surface protein A. J. Clin. Invest. *106*, 561–569.

Pal, U., and Fikrig, E. (2003). Adaptation of *Borrelia burgdorferi* in the vector and vertebrate host. Microbes Infecti. *5*, 659–666.

Pal, U., Li, X., Wang, T., Montgomery, R.R., Ramamoorthi, N., Desilva, A.M., Bao, F., Yang, X., Pypaert, M., Pradhan, D., Kantor, F.S., Telford, S., Anderson, J.F., and Fikrig, E. (2004a). TROSPA, an *Ixodes scapularis* receptor for *Borrelia burgdorferi*. Cell *119*, 457–468.

Pal, U., Wang, P., Bao, F., Yang, X., Samanta, S., Schoen, R., Wormser, G.P., Schwartz, I., and Fikrig, E. (2008). *Borrelia burgdorferi* basic membrane proteins A and B participate in the genesis of Lyme arthritis. J. Exp. Med. *205*, 133–141.

Pal, U., Yang, X., Chen, M., Bockenstedt, L.K., Anderson, J.F., Flavell, R.A., Norgard, M.V., and Fikrig, E. (2004b). OspC facilitates *Borrelia burgdorferi* invasion of *Ixodes scapularis* salivary glands. J. Clin. Invest. *113*, 220–230.

Pashenkov, M., Huang, Y.M., Kostulas, V., Haglund, M., Soderstrom, M., and Link, H. (2001). Two subsets of dendritic cells are present in human cerebrospinal fluid. Brain *124*, 480–492.

Paster, B.J., Dewhirst, F.E., Weisburg, W.G., Tordoff, L.A., Fraser, G.J., Hespell, R.B., Stanton, T.B., Zablen, L., Mandelco, L., and Woese, C.R. (1991). Phylogenetic analysis of the spirochetes. J. Bacteriol. *173*, 6101–6109.

Pavia, C.S., Wormser, G.P., and Norman, G.L. (1997). Activity of sera from patients with Lyme disease against *Borrelia burgdorferi*. Clin. Infect. Dis. 25 Suppl 1: S25–S30.

Peisman, J. (2002). Ecology of *Borrelia burgdorferi* sensu lato in North America. In Lyme Borreliosis. Gray, J.S., Kahl, O., Lane, R.S., and Stanek, G. (eds). New York: CABI Publishing, pp. 223–250.

Petzke, M.M., Brooks, A., Krupna, M.A., Mordue, D., and Schwartz, I. (2009). Recognition of *Borrelia burgdorferi*, the Lyme disease spirochete, by TLR7 and TLR9 induces a type I IFN response by human immune cells. J. Immunol. *183*, 5279–5292.

Philipp, M.T., Aydintug, M.K., Bohm, R.P., Jr., Cogswell, F.B., Dennis, V.A., Lanners, H.N. *et al.* (1993). Early and early disseminated phases of Lyme disease in the rhesus monkey: a model for infection in humans. Infect. Immun. *61*, 3047–3059.

Philipp, M.T., and Johnson, B.J.B. (1994). Animal models of Lyme disease: pathogenesis and immunoprophylaxis. Immunol. Today *2*, 431–437.

Picken, R.N., Cheng, Y., Strle, F., and Picken, M.M. (1996). Patient isolates of *Borrelia burgdorferi* sensu lato with genotypic and phenotypic similarities to strain 25015. J. Infect. Dis. *174*, 1112–1115.

Piesman, J., and Gern, L. (2004). Lyme borreliosis in Europe and North America. Parasitology *129 Suppl* S191–S220.

Piesman, J., and Happ, C.M. (1997). Ability of the Lyme disease spirochete *Borrelia burgdorferi* to infect rodents and three species of human-biting ticks (blacklegged tick, American dog tick, lone star tick) (Acari:Ixodidae). J. Med. Entomol. *34*, 451–456.

Piesman, J., Mather, T.N., Sinsky, R.J., and Spielman, A. (1987). Duration of tick attachment and *Borrelia burgdorferi* transmission. J. Clin. Microbiol. *25*, 557–558.

Piesman, J., and Schneider, B.S. (2002). Dynamic changes in Lyme disease spirochetes during transmission by nymphal ticks. Exp. Appl. Acarol. *28*, 141–145.

Piesman, J., Schneider, B.S., and Zeidner, N.S. (2001). Use of quantitative PCR to measure density of *Borrelia burgdorferi* in the midgut and salivary glands of feeding tick vectors. J. Clin. Microbiol. *39*, 4145–4148.

Piesman, J., Zeidner, N.S., and Schneider, B.S. (2003). Dynamic changes in *Borrelia burgdorferi* populations in *Ixodes scapularis* (Acari: Ixodidae) during transmission: studies at the mRNA level. Vector Borne Zoonotic Dis 3, 125–132.

Pikas, D.S., Brown, E.L., Gurusiddappa, S., Lee, L.Y., Xu, Y., and Hook, M. (2003). Decorin-binding sites in the adhesin DbpA from *Borrelia burgdorferi*: a synthetic peptide approach. J. Biol. Chem. *278*, 30920–30926.

Pinne, M., Östberg, Y., Comstedt, P., and Bergström, S. (2004). Molecular analysis of the channel-forming protein P13 and its paralogue family 48 from different Lyme disease *Borrelia* species. Microbiology *150*, 549–559.

Pinne, M., Thein, M., Denker, K., Benz, R., Coburn, J., and Bergström, S. (2007). Elimination of channel-forming activity by insertional inactivation of the *p66* gene in *Borrelia burgdorferi*. FEMS Microbiol. Lett. *266*, 241–249.

Platanias, L.C. (2003). The p38 mitogen-activated protein kinase pathway and its role in interferon signaling. Pharmacol. Ther. *98*, 129–142.

Pohl-Koppe, A., Balashov, K.E., Steere, A.C., Logigian, E.L., and Hafler, D.A. (1998). Identification of a T cell subset capable of both IFN-γ and IL-10 secretion in patients with chronic *Borrelia burgdorferi* infection. J. Immunol. *160*, 1804–1810.

Poltorak, A., He, X., Smirnova, I., Liu, M.-Y., Van Huffel, C., Du, X. *et al.* (1998). Defective LPS signaling in C3H/HeJ and C57Bl/10ScCr mice: mutations in *Tlr4* gene. Science *282*, 2085–2088.

Porcella, S.F., Fitzpatrick, C.A., and Bono, J.L. (2000). Expression and immunological analysis of the plasmid-borne *mlp* genes of *Borrelia burgdorferi* B31. Infect. Immun. *68*, 4992–5001.

Porcella, S.F., Popova, T.G., Akins, D.R., Li, M., Radolf, J.D., and Norgard, M.V. (1996). *Borrelia burgdorferi* supercoiled plasmids encode multicopy tandem open reading frames and a lipoprotein gene family. J. Bacteriol. *178*, 3293–3307.

Posey, J.E., and Gherardini, F.C. (2000). Lack of a role for iron in the Lyme disease pathogen. Science *288*, 1651–1653.

Postic, D., Garnier, M., and Baranton, G. (2007). Multilocus sequence analysis of atypical *Borrelia burgdorferi* sensu lato isolates--description of *Borrelia californiensis* sp. nov., and genomospecies 1 and 2. Int. J. Med. Microbiol. *297*, 263–271.

Postic, D., Marti Ras, N., Lane, R.S., Hendson, M., and Baranton, G. (1998). Expanded diversity among Californian *Borrelia* isolates and description of *Borrelia bissettii* sp. nov. (formerly *Borrelia* Group DN127). J. Clin. Microbiol. *36*, 3497–3504.

Probert, W.S., Crawford, M., Cadiz, R.B., and LeFebvre, R.B. (1997). Immunization with outer surface protein (Osp) A, but not OspC, provides cross-protection of mice challenged with North American isolates of *Borrelia burgdorferi*. J. Infect. Dis *175*, 400–405.

Pugsley, A.P. (1993). The complete general secretory pathway in gram-negative bacteria. Microbiol. Rev. *57*, 50–108.

Puius, Y.A., and Kalish, R.A. (2008). Lyme arthritis: pathogenesis, clinical presentation, and management. Infect. Dis. Clin. North. Am. *22*, 289–300.

Purser, J.E., Lawrenz, M.B., Caimano, M.J., Howell, J.K., Radolf, J.D., and Norris, S.J. (2003). A plasmid-encoded nicotinamidase (PncA) is essential for infectivity of *Borrelia burgdorferi* in a mammalian host. Mol. Microbiol. *48*, 753–764.

Purser, J.E., and Norris, S.J. (2000). Correlation between plasmid content and infectivity in *Borrelia burgdorferi*. Proc. Natl. Acad. Sci. U.S.A. *97*, 13865–13870.

Qureshi, M.Z., New, D., Zulqarni, N.J., and Nachman, S. (2002). Overdiagnosis and overtreatment of Lyme disease in children. Pediatr. Infect. Dis. J. *21*, 12–14.

Radolf, J.D., Hazlett, K.R.O., and Lukehart, S.A. (2006). Pathogenesis of Syphilis. In Pathogenic Treponemes: Cellular and Molecular Biology. Radolf, J.D., and Lukehart, S.A. (eds). Norfolk, UK: Caister Academic Press, pp. 197–236.

Radolf, J.D., Arndt, L.L., Akins, D.R., Curetty, L.L., Levi, M.E., Shen, Y., Davis, L.S., and Norgard, M.V. (1995). *Treponema pallidum* and *Borrelia burgdorferi* lipoproteins and synthetic lipopeptides activate monocytes/macrophages. J. Immunol. *154*, 2866–2877.

Radolf, J.D., Bourell, K.W., Akins, D.R., Brusca, J.S., and Norgard, M.V. (1994). Analysis of *Borrelia burgdorferi* membrane architecture by freeze-fracture electron microscopy. J. Bacteriol. *176*, 21–31.

Radolf, J.D., and Caimano, M.J. (2008). The long strange trip of *Borrelia burgdorferi* outer-surface protein C. Mol. Microbiol. *69*, 1–4.

Radolf, J.D., and Lukehart, S.A. (2006). Immunology of Syphilis. In Pathogenic Treponemes: Cellular and Molecular Biology. Radolf, J.D., and Lukehart, S.A. (eds). Norfolk, UK: Caister Academic Press, pp. 285–322.

Radolf, J.D., Norgard, M.V., Brandt, M.E., Isaacs, R.D., Thompson, P.A., and Beutler, B. (1991). Lipoproteins of *Borrelia burgdorferi* and *Treponema pallidum* activate cachectin/tumor necrosis factor synthesis: analysis using a CAT reporter construct. J. Immunol. *147*, 1968–1974.

Ramesh, G., Alvarez, A.L., Roberts, E.D., Dennis, V.A., Lasater, B.L., Alvarez, X., and Philipp, M.T. (2003). Pathogenesis of Lyme neuroborreliosis: *Borrelia burgdorferi* lipoproteins induce both proliferation and apoptosis in rhesus monkey astrocytes. Eur J. Immunol. 33, 2539–2550.

Ramesh, G., Borda, J.T., Dufour, J., Kaushal, D., Ramamoorthy, R., Lackner, A.A., and Philipp, M.T. (2008). Interaction of the Lyme disease spirochete *Borrelia burgdorferi* with brain parenchyma elicits inflammatory mediators from glial cells as well as glial and neuronal apoptosis. Am J. Pathol. *173*, 1415–1427.

Rauter, C., Oehme, R., Diterich, I., Engele, M., and Hartung, T. (2002). Distribution of clinically relevant *Borrelia* genospecies in ticks assessed by a novel, single-run, real-time PCR. J. Clin. Microbiol. *40*, 36–43.

Reid, M.C., Schoen, R.T., Evans, J., Rosenberg, J.C., and Horwtiz, R.I. (1998). The consequences of overdiagnosis and overtreatment of Lyme disease: an observational study. Ann. Intern. Med. *128*, 354–362.

Richter, D., Schlee, D.B., Allgower, R., and Matuschka, F.R. (2004). Relationships of a novel Lyme disease spirochete, *Borrelia spielmani* sp. nov., with its hosts in Central Europe. Appl. Environ. Microbiol. *70*, 6414–6419.

Rijpkema, S.G., Tazelaar, D.J., Molkenboer, M.J., Noordhoek, G.T., Plantinga, G., Schouls, L.M., and Schellekens, J.F. (1997). Detection of *Borrelia afzelii*, *Borrelia burgdorferi* sensu stricto, *Borrelia garinii* and group VS116 by PCR in skin biopsies of patients with erythema migrans and acrodermatitis chronica atrophicans. Clin. Microbiol. Infect. 3, 109–116.

Roberts, D.M., Carlyon, J.A., Theisen, M., and Marconi, R.T. (2000). The *bdr* gene families of the Lyme disease and relapsing fever spirochetes: potential influence on biology, pathogenesis, and evolution. *Emer Infect Dis* in press.

Rolfs, R.T., Joesoef, M.R., Hendershot, E.F., Rompalo, A.M., Augenbraun, M.H., Chiu, M., Bolan, G., Johnson, S.C., French, P., Steen, E., *et al.* (1997). A randomized trial of enhanced therapy for early syphilis in patients with and without human immunodeficiency virus infection. The Syphilis and HIV Study Group. N. Engl. J. Med. *337*, 307–314.

Rupprecht, T.A., Pfister, H.W., Angele, B., Kastenbauer, S., Wilske, B., and Koedel, U. (2005). The chemokine CXCL13 (BLC): a putative diagnostic marker for neuroborreliosis. Neurology *65*, 448–450.

Ruzic-Sabljic, E., Arnez, M., Logar, M., Maraspin, V., Lotric-Furlan, S., Cimperman, J., and Strle, F. (2005). Comparison of *Borrelia burgdorferi* sensu lato strains isolated from specimens obtained simultaneously from two different sites of infection in individual patients. J. Clin. Microbiol. *43*, 2194–2200.

Sadziene, A., Barbour, A.G., Rosa, P.A., and Thomas, D.D. (1993a) An OspB mutant of *Borrelia burgdorferi* has reduced invasiveness in vitro and reduced infectivity in vivo. Infect. Immun. *61*, 3590–3596.

Sadziene, A., Wilske, B., Ferdows, M.S., and Barbour, A.G. (1993b) The cryptic *ospC* gene of *Borrelia burgdorferi* B31 is located on a circular plasmid. Infect. Immun. *61*, 2192–2195.

Salazar, J.C., Gerber, M.A., and Goff, C.W. (1993). Long-term outcome of Lyme disease in children given early treatment. *J Pediatr 122*, 591–593.

Salazar, J.C., Pope, C.D., Sellati, T.J., Feder, H.M., Jr., Kiely, T.G., Dardick, K.R., Buckman, R.L., Moore, M.W., Caimano, M.J., Pope, J.G., *et al.* (2003). Coevolution of markers of innate and adaptive immunity in skin and peripheral blood of patients with erythema migrans. J. Immunol. *171*, 2660–2670.

Salazar, J.C., Dunham-Ems, S., LaVake, C.J., Cruz, A.R., Moore, M.W., Caimano, M.J., Velez-Climent, L., Shupe, J., Krueger, W., and Radolf, J.D. (2009). Activation of human monocytes by live *Borrelia burgdorferi* generates TLR2-dependent and -independent responses which include induction of IFN-β. PLoS. Pathog. *5*, e1000444.

Schaible, U.E., Gay, S., Museteanu, C., Kramer, M.D., Zimmer, G., Eichmann, K., Museteanu, U., and Simon, M.M. (1990). Lyme borreliosis in the severe combined immunodeficiency (*scid*) mouse manifests predominantly in the joints, heart, and liver. Am. J. Pathol. *137*, 811–820.

Scheckelhoff, M.R., Telford, S.R., and Hu, L.T. (2006). Protective efficacy of an oral vaccine to reduce carriage of *Borrelia burgdorferi* (strain N40) in mouse and tick reservoirs. Vaccine *24*, 1949–1957.

Schmidt, B.L. (1997). PCR in laboratory diagnosis of human *Borrelia burgdorferi* infections. Clin. Microbiol. Rev. *10*, 185–201.

Schroder, N.W., Eckert, J., Stubs, G., and Schumann, R.R. (2008). Immune responses induced by spirochetal outer membrane lipoproteins and glycolipids. Immunobiology *213*, 329–340.

Schulze, R.J., and Zückert, W.R. (2006). *Borrelia burgdorferi* lipoproteins are secreted to the outer surface by default. Mol. Microbiol. *59*, 1473–1484.

Schwan, T.G., and Burgdorfer, W. (1987). Antigenic changes of *Borrelia burgdorferi* as a result of in vitro cultivation. J. Infect. Dis. *156*, 852–853.

Schwan, T.G., Burgdorfer, W., and Garon, C.F. (1988). Changes in infectivity and plasmid profile of the Lyme disease spirochete, *Borrelia burgdorferi*, as a result of *in vitro* cultivation. Infect. Immun. *56*, 1831–1836.

Schwan, T.G., and Piesman, J. (2000). Temporal changes in outer surface proteins A and C of the Lyme disease-associated spirochete, *Borrelia burgdorferi*, during the chain of infection in ticks and mice. J. Clin. Microbiol. *38*, 382–388.

Schwan, T.G., Piesman, J., Golde, W.T., Dolan, M.C., and Rosa, P.A. (1995). Induction of an outer surface protein on *Borrelia burgdorferi* during tick feeding. Proc. Natl. Acad. Sci. U.S.A. *92*, 2909–2913.

Scrimenti, R.J. (1970). Erythem chronicum migrans. Arch Dermatol *102*, 104–105.

Seinost, G., Dykhuizen, D.E., Dattwyler, R.J., Golde, W.T., Dunn, J.J., Wang, I.-N., Wormser, G.P., Schriefer, M.E., and Luft, B.J. (1999). Four clones of *Borrelia burgdorferi* sensu stricto cause invasive infection in humans. Infect. Immun. *67*, 3518–3524.

Seinost, G., Golde, W.T., Berger, B.W., Dunn, J.J., Qiu, D., Dunkin, D.S., Dykhuizen, D.E., and Luft, B.J. (1999). Infection with multiple strains of *Borrelia burgdorferi* sensu stricto in patients with Lyme disease. Arch. Dermatol. *135*, 1329–1333.

Sellati, T.J., Abrescia, L.D., Radolf, J.D., and Furie, M.B. (1996). Outer surface lipoproteins of *Borrelia burgdorferi* activate vascular endothelium in vitro. Infect. Immun. *64*, 3180–3187.

Sellati, T.J., Bouis, D.A., Kitchens, R.L., Darveau, R.P., Pugin, J., Ulevitch, R.J., Gangloff, S.C., Goyert, S.M., Norgard, M.V., and Radolf, J.D. (1998). *Treponema pallidum* and *Borrelia burgdorferi* lipoproteins and synthetic lipopeptides activate monocytic cells via a CD14-dependent pathway distinct from that used by lipopolysaccharide. J. Immunol. *160*, 5455–5464.

Seshu, J., Esteve-Gassent, M.D., Labandeira-Rey, M., Kim, J.H., Trzeciakowski, J.P., Höök, M., and Skare, J.T. (2006). Inactivation of the fibronectin-binding adhesin gene *bbk32* significantly attenuates the infectivity potential of *Borrelia burgdorferi*. Mol. Microbiol. *59*, 1591–1601.

Shi, Y., Xu, Q., McShan, K., and Liang, F.T. (2008). Both decorin-binding proteins A and B are critical for the overall virulence of *Borrelia burgdorferi*. Infect. Immun. *76*, 1239–1246.

Shih, C.M., Pollack, R.J., Telford, S.R., III, and Spielman, A. (1992). Delayed dissemination of Lyme disease spirochetes from the site of deposition in the skin of mice. J. Infect. Dis *166*, 827–831.

Shin, O.S., Isberg, R., Akira, S., Uematsu, S., Behera, A.K., and Hu, L.T. (2008). Distinct roles for MyD88, TLR2, TLR5, and TLR9 in phagocytosis of *Borrelia burgdorferi* and inflammatory signaling. Infect. Immun. *76*, 2341–2351.

Sigal, L.H. (1995). Early disseminated Lyme disease: cardiac manifestations. Am. J. Med. 98, 25S-28S; discu.

Sigal, L.H., Zahradnik, J.M., Lavin, P., Patella, S.J., Bryant, G., Haselby, R., Hilton, E., Kunkel, M., Adler-Klein, D., Doherty, T., *et al.* (1998). A vaccine consisting of recombinant *Borrelia burgdorferi* outer-surface protein A to prevent Lyme disease. Recombinant Outer-Surface Protein A Lyme Disease Vaccine Study Consortium. N. Engl. J. Med. *339*, 216–222.

Silver, E., Pass, R.H., Kaufman, S., Hordof, A.J., and Liberman, L. (2007). Complete heart block due to Lyme carditis in two pediatric patients and a review of the literature. Congenit. Heart Dis. *2*, 338–341.

Simpson, W.J., Schrumpf, M.E., and Schwan, T.G. (1990). Reactivity of human Lyme borreliosis sera with a 39-kilodalton antigen specific to *Borrelia burgdorferi*. J. Clin. Microbiol. *28*, 1329–1337.

Smith, R., and Takkinen, J. (2006). Lyme borreliosis: Europe-wide coordinated surveillance and action needed? Euro Surveill *11*, E060622.

Smith, R.P., Schoen, R.T., Rahn, D.W., Sikand, V.K., Nowakowski, J., Parenti, D.L., Holman, M.S., Persing, D.H., and Steere, A.C. (2002). Clinical characteristics and treatment outcome of early Lyme disease in patients with microbiologically confirmed erythema migrans. Ann. Intern. Med. *136*, 421–428.

Sonnesyn, S., Diehl, S.C., Johnson, R.C., Kubo, S.H., and Goodman, J.L. (1996). A prospective study of the seroprevalence of *Borrelia burgdorferi* infection in patients with severe heart failure. Am. J. Cardiol. *76*, 97–100.

Sood, S.K., Salzman, M.B., Johnson, B.J., Happ, C.M., Feig, K., Carmody, L., Rubin, L.G., Hilton, E., and Piesman, J. (1997). Duration of tick attachment as a predictor of the risk of Lyme disease in an area in which Lyme disease is endemic. J. Infect. Dis. *175*, 996–999.

Spielman, A., Wilson, M.L., Levine, J.F., and Piesman, J. (1985). Ecology of *Ixodes dammini*-borne human babesiosis and Lyme disease. Ann. Rev. Entomol. *30*, 439–460.

Stanek, G., Klein, J., Bittner, R., and Glogar, D. (1990). Isolation of *Borrelia burgdorferi* from the myocardium of a patient with longstanding cardiomyopathy. N. Engl. J. Med. 322, 249–252.

Stanek, G., and Strle, F. (2003). Lyme borreliosis. Lancet *362*, 1639–1647.

Stanek, G., and Strle, F. (2008). Lyme disease: European perspective. Infect. Dis. Clin. North. Am. 22, 327–39, vii.

Steere, A.C. (2001). Lyme disease. N. Engl. J. Med. *345*, 115–125.

Steere, A.C. (2006). Lyme borreliosis in 2005, 30 years after initial observations in Lyme Connecticut. Wien Klin. Wochenschr. *118*, 625–633.

Steere, A.C., Bartenhagen, N.H., Craft, J.E., Hutchinson, G.J., Newman, J.H., Rahn, D.W., Sigal, L.H., Spieler, P.N., Stenn, K.S., and Malawista, S.E. (1983a). The early clinical manifestations of Lyme disease. Ann. Intern. Med. *99*, 76–82.

Steere, A.C., Batsford, W.P., Weinberg, M., Alexander, J., Berger, H.J., Wolfson, S., and Malawista, S.E. (1980). Lyme carditis: cardiac abnormalities of Lyme disease. Ann. Intern. Med. *93*, 8–16.

Steere, A.C., Coburn, J., and Glickstein, L. (2004). The emergence of Lyme disease. J. Clin. Invest. *113*, 1093–1101.

Steere, A.C., Dhar, A., Hernandez, J., Fischer, P.A., Sikand, V.K., Schoen, R.T., Nowakowski, J., McHugh, G., and Persing, D.H. (2003). Systemic symptoms without erythema migrans as the presenting picture of early Lyme disease. Am. J. Med. *114*, 58–62.

Steere, A.C., Dwyer, E., and Winchester, R. (1990). Association of chronic Lyme arthritis with HLA-DR4 and HLA-DR2 alleles. N. Engl. J. Med. 323, 219–223.

Steere, A.C., Gibofsky, A., Patarroyo, M.E., Winchester, R.J., Hardin, J.A., and Malawista, S.E. (1979). Chronic Lyme arthritis:clinical and immunogenetic differentiation from rheumatoid arthritis. Ann. Intern. Med. *90*, 896–901.

Steere, A.C., and Glickstein, L. (2004). Elucidation of Lyme arthritis. Nat. Rev. Immunol. *4*, 143–152.

Steere, A.C., Grodzicki, R.L., Kornblatt, A.N., Craft, J.E., Barbour, A.G., Burgdorfer, W., Schmid, G.P., Johnson, E., and Malawista, S.E. (1983b). The spirochetal etiology of Lyme disease. N. Engl. J. Med. *308*, 733–740.

Steere, A.C., Malawista, S.E., Hardin, J.A., Ruddy, S., Askenase, W., and Andiman, W.A. (1977a) Erythema chronicum migrans and Lyme arthritis: the enlarging clinical spectrum. Ann. Intern. Med. *86*, 685–698.

Steere, A.C., Malawista, S.E., Syndman, D.R., Shope, R.E., Andiman, W.A., Ross, M.R., and Steele, F.M. (1977b) An epidemic of oliogarticular arthritis in children and adults in three Connecticut communities. Arthritis Rheum. *20*, 7–17.

Steere, A.C., Schoen, R.T., and Taylor, E. (1987). The clinical evolution of Lyme arthritis. Ann. Intern. Med. *107*, 725–731.

Steere, A.C., Sikand, V.K., Meurice, F., Parenti, D.L., Fikrig, E., Schoen, R.T., Nowakowski, J., Schmid, C.H., Laukamp, S., Buscarino, C., and Krause, D.S. (1998). Vaccination against Lyme disease with recombinant *Borrelia burgdorferi* outer-surface lipoprotein A with adjuvant. Lyme Disease Vaccine Study Group. N. Engl. J. Med. *339*, 209–215.

Steere, A.C., Sikand, V.K., Schoen, R.T., and Nowakowski, J. (2003b) Asymptomatic infection with *Borrelia burgdorferi*. Clin. Infect. Dis. 37, 528–532.

Steere, A.C., Taylor, E., McHugh, G.L., and Logigian, E.L. (1993). The overdiagnosis of Lyme disease. JAMA *269*, 1812–1816.

Steere, A.C., Taylor, E., Wilson, M.L., Levine, J.F., and Spielman, A. (1986). Longitudinal assessment of the clinical and epidemiological features of Lyme disease in a defined population. J. Infect. Dis *154*, 295–300.

Stevenson, B., El Hage, N., Hines, M.A., Miller, J.C., and Babb, K. (2002). Differential binding of host complement inhibitor factor H by *Borrelia burgdorferi* Erp surface proteins: a possible mechanism underlying the expansive host range of Lyme Disease spirochetes. Infect. Immun. *70*, 491–497.

Stevenson, B., Tilly, K., and Rosa, P.A. (1996). A family of genes located on four separate 32-kilobase circular plasmids in *Borrelia burgdorferi*. J. Bacteriol. *178*, 3508–3516.

Stevenson, B., Zückert, W.R., and Akins, D.R. (2000). Repetition, conservation, and variation: the multiple cp32 plasmids of *Borrelia* species. J. Mol. Microbiol. Biotechnol 2, 411–422.

Stewart, P.E., Byram, R., Grimm, D., Tilly, K., and Rosa, P.A. (2005). The plasmids of *Borrelia burgdorferi*: essential genetic elements of a pathogen. Plasmid *53*, 1–13.

Stewart, P.E., Chaconas, G., and Rosa, P. (2003). Conservation of plasmid maintenance functions between linear and circular plasmids in *Borrelia burgdorferi*. J. Bacteriol. *185*, 3202–3209.

Straubinger, R.K. (2000). PCR-Based quantification of *Borrelia burgdorferi* organisms in canine tissues over a 500-Day postinfection period. J. Clin. Microbiol. *38*, 2191–2199.

Straubinger, R.K. (2000). PCR-Based quantification of *Borrelia burgdorferi* organisms in canine tissues over a 500-Day postinfection period. J. Clin. Microbiol. *38*, 2191–2199.

Straubinger, R.K., Summers, B.A., Chang, Y.F., and Appel, M.J. (1997). Persistence of *Borrelia burgdorferi* in experimentally infected dogs after antibiotic treatment. J. Clin. Microbiol. *35*, 111–116.

Strle, F., Nadelman, R.B., Cimperman, J., Nowakowski, J., Picken, R.N., Schwartz, I., Maraspin, V., Aguero-Rosenfeld, M.E., Varde, S., Lotric-Furlan, S., and Wormser, G.P. (1999). Comparison of culture-confirmed erythema migrans caused by *Borrelia burgdorferi* sensu stricto in New York State and by *Borrelia afzelii* in Slovenia. Ann. Intern. Med. *130*, 32–36.

Strle, F., Nelson, J.A., Ruzic-Sabljic, E., Cimperman, J., Maraspin, V., Lotric-Furlan, S., Cheng, Y., Picken, M.M., Trenholme, G.M., and Picken, R.N. (1996). European Lyme borreliosis: 231 culture-confirmed cases involving patients with erythema migrans. Clin. Infect. Dis. *23*, 61–65.

Strle, F., Ruzic-Sabljic, E., Cimperman, J., Lotric-Furlan, S., and Maraspin, V. (2006). Comparison of findings for patients with *Borrelia garinii* and *Borrelia afzelii* isolated from cerebrospinal fluid. Clin. Infect. Dis. *43*, 704–710.

Suk, K., Das, S., Sun, W., Jwang, B., Barthold, S.W., Flavell, R.A., and Fikrig, E. (1995). *Borrelia burgdorferi* genes selectively expressed in the infected host. Proc. Natl. Acad. Sci. U.S.A. *92*, 4269–4273.

Sung, S.Y., McDowell, J.V., Carlyon, J.A., and Marconi, R.T. (2000). Mutation and recombination in the upstream homology box-flanked *ospE*-related genes of the Lyme disease spirochetes result in the development of new antigenic variants during infection. Infect. Immun. *68*, 1319–1327.

Szczepanski, A., Furie, M.B., Benach, J.L., Lane, B.P., and Fleit, H.B. (1990). Interaction between *Borrelia burgdorferi* and endothelium in vitro. J. Clin. Invest. *85*, 1637–1647.

Szer, I.S., Taylor, E., and Steere, A.C. (1991). The long-term course of Lyme arthritis in children. N. Engl. J. Med. *325*, 159–163.

Takayama, K., Rothenberg, R.J., and Barbour, A.G. (1987). Absence of lipopolysaccharide in the Lyme disease spirochete, *Borrelia burgdorferi*. Infect. Immun. *55*, 2311–2313.

Tazelaar, D.J., Molkenboer, M.J., Noordhoek, G.T., Plantinga, G., Schouls, L.M., and Schellekens, J.F. (1997). Detection of *Borrelia afzelii*, *Borrelia burgdorferi* sensu stricto, *Borrelia garinii* and group VS116 by PCR in skin biopsies of patients with erythema migrans and acrodermatitis chronica atrophicans. Clin. Microbiol. Infect. 3, 109–116.

Theisen, M., Frederiksen, B., Lebech, A.M., Vuust, J., and Hansen, K. (1993). Polymorphism in *ospC* gene of Borrelia burgdorferi and immunoreactivity of OspC protein: implications for taxonomy and for use of OspC protein as a diagnostic antigen. J. Clin. Microbiol. *31*, 2570–2576.

Thompson, C., Spielman, A., and Krause, P.J. (2001). Coinfecting deer-associated zoonoses: Lyme disease, babesiosis, and ehrlichiosis. Clin. Infect. Dis. *33*, 676–685.

Tilly, K., Bestor, A., Jewett, M.W., and Rosa, P. (2007). Rapid clearance of Lyme disease spirochetes lacking OspC from skin. Infect. Immun. *75*, 1517–1519.

Tilly, K., Krum, J.G., Bestor, A., Jewett, M.W., Grimm, D., Bueschel, D., Byram, R., Dorward, D., Vanraden, M.J., Stewart, P., and Rosa, P. (2006). *Borrelia burgdorferi* OspC protein required exclusively in a crucial early stage of mammalian infection. Infect. Immun. *74*, 3554–3564.

Tokarz, R., Anderton, J.M., Katona, L.I., and Benach, J.L. (2004). Combined effects of blood and temperature shift on *Borrelia burgdorferi* gene expression as determined by whole genome DNA array. Infect. Immun. *72*, 5419–5432.

Toshchakov, V., Jones, B.W., Perera, P.Y., Thomas, K., Cody, M.J., Zhang, S., Williams, B.R., Major, J., Hamilton, T.A., Fenton, M.J., and Vogel, S.N. (2002). TLR4, but not TLR2, mediates IFN-β-induced STAT1αβ-dependent gene expression in macrophages. Nat. Immunol. 3, 392–398.

Trollmo, C., Meyer, A.L., Steere, A.C., Hafler, D.A., and Huber, B.T. (2001). Molecular mimicry in Lyme arthritis demonstrated at the single cell level: LFA-1α_L is a partial agonist for outer surface protein A- reactive T cells. J. Immunol. *166*, 5286–5291.

Tsao, J.I., Wootton, J.T., Bunikis, J., Luna, M.G., Fish, D., and Barbour, A.G. (2004). An ecological approach to preventing human infection: vaccinating wild mouse reservoirs intervenes in the Lyme disease cycle. Proc. Natl. Acad. Sci. U.S.A. *101*, 18159–18164.

Tupin, E., Benhnia, M.R., Kinjo, Y., Patsey, R., Lena, C.J., Haller, M.C., Caimano, M.J., Imamura, M., Wong, C.H., Crotty, S., *et al.* (2008). NKT cells prevent chronic joint inflammation after infection with *Borrelia burgdorferi*. Proc. Natl. Acad. Sci. U.S.A.

van Dam, A.P. (2002). Diversity of *Ixodes*-borne *Borrelia* species – clinical, pathogenetic, and diagnostic implications and impact on vaccine development. Vector Borne Zoonotic Dis 2, 249–254.

van Dam, A.P., Kuiper, H., Vos, K., Widjojokusumo, A., de Jongh, B.M., Spanjaard, L., Ramselaar, A.C.P., Kramer, M.D., and Dankert, J. (1993). Different genospecies of *Borrelia burgdorferi* are associated with distinct clinical manifestations of Lyme borreliosis. Clin. Infect. Dis. *17*, 708–717.

Vaz, A., Glickstein, L., Field, J.A., McHugh, G., Sikand, V.K., Damle, N., and Steere, A.C. (2001). Cellular and humoral immune responses to *Borrelia burgdorferi* antigens in patients with culture-positive early Lyme disease. Infect. Immun. *69*, 7437–7444.

Walker, E.M., Borenstein, L.A., Blanco, D.R., Miller, J.N., and Lovett, M.A. (1991). Analysis of outer membrane ultrastructures of pathogenic *Treponema* and *Borrelia* species by freeze-fracture electron microscopy. J. Bacteriol. *173*, 5585–5588.

Wallich, R., Brenner, C., Kramer, M.D., and Simon, M.M. (1995). Molecular cloning and immunological characterization of a novel linear-plasmid-encoded gene, *pG*, of *Borrelia burgdorferi* expressed only in vivo. Infect. Immun. *63*, 3327–3335.

Wang, G., Ma, Y., Buyuk, A., McClain, S., Weis, J.J., and Schwartz, I. (2004). Impaired host defense to infection and Toll-like receptor 2-independent killing of *Borrelia burgdorferi* clinical isolates in TLR2-deficient C3H/HeJ mice. FEMS Microbiol. Lett. *231*, 219–225.

Wang, G., Ojaimi, C., Iyer, R., Saksenberg, V., McClain, S.A., Wormser, G.P., and Schwartz, I. (2001). Impact of genotypic variation of *Borrelia burgdorferi* sensu stricto on kinetics of dissemination and severity of disease in C3H/HeJ mice. Infect. Immun. *69*, 4303–4312.

Wang, G., Ojaimi, C., Wu, H., Saksenberg, V., Iyer, R., Liveris, D., McClain, S.A., Wormser, G.P., and Schwartz, I. (2002). Disease severity in a murine model of Lyme borreliosis is associated with the genotype of the infecting *Borrelia burgdorferi* sensu stricto strain. J. Infect. Dis. *186*, 782–791.

Wang, G., van Dam, A.P., Schwartz, I., and Dankert, J. (1999a) Molecular typing of *Borrelia burgdorferi* sensu lato: taxonomic, epidemiological, and clinical implications. Clin. Microbiol. Rev. *12*, 633–653.

Wang, I.-N., Dykhuizen, D.E., Qiu, W., Dunn, J.L., Bosler, E.M., and Luft, B.J. (1999b) Genetic diversity of *ospC* in a local population of *Borrelia burgdorferi sensu stricto*. Genetics *151*, 15–30.

Weis, J.J., Ma, Y., and Erdile, L.F. (1994). Biological activities of native and recombinant *Borrelia burgdorferi* outer surface protein A: dependence on lipid modification. Infect. Immun. *62*, 4632–4636.

Wheeler, C.M., Coleman, J.L., Habicht, G.S., and Benach, J.L. (1989). Adult *Ixodes dammini* on rabbits: development of acute inflammation in the skin and immune responses to salivary gland, midgut, and spirochetal components. J. Infect. Dis *159*, 265–273.

Wheeler, C.M., García-Moncó, J.C., Benach, J.L., Golightly, M.G., Habicht, G.S., and Steere, A.C. (1993). Nonprotein antigens of *Borrelia burgdorferi*. J. Infect. Dis. *167*, 665–674.

Widhe, M., Jarefors, S., Ekerfelt, C., Vrethem, M., Bergström, S., Forsberg, P., and Ernerudh, J. (2004). *Borrelia*-specific interferon-γ and interleukin-4 secretion in cerebrospinal fluid and blood during Lyme borreliosis in humans: association with clinical outcome. J. Infect. Dis. *189*, 1881–1891.

Wilske, B., Fingerle, V., and Schulte-Spechtel, U. (2007). Microbiological and serological diagnosis of Lyme borreliosis. FEMS Immunol. Med. Microbiol. *49*, 13–21.

Wilske, B., Preac-Mursic, V., Jauris, S., Hofman, A., Pradel, I., Soutschek, E. *et al.* (1993). Immunological and molecular polymorphisms of OspC, an immunodominant major outer surface protein of *Borrelia burgdorferi*. Infect. Immun. *61*, 2182–2191.

Wolgemuth, C.W., Charon, N.W., Goldstein, S.F., and Goldstein, R.E. (2006). The flagellar cytoskeleton of the spirochetes. J. Mol. Microbiol. Biotechnol *11*, 221–227.

Woodman, M.E., Cooley, A.E., Miller, J.C., Lazarus, J.J., Tucker, K., Bykowski, T., Botto, M., Hellwage, J., Wooten, R.M., and Stevenson, B. (2007). *Borrelia burgdorferi* binding of host complement regulator factor H is not required for efficient mammalian infection. Infect. Immun *75*, 3131–3139.

Wooten, R.M., Morrison, T.B., Weis, J.H., Wright, S.D., Thieringer, R., and Weis, J.J. (1998). The role of CD14 in signaling mediated by outer membrane lipoproteins of *Borrelia burgdorferi*. J. Immunol. *160*, 5485–5492.

Wooten, R.M., and Weis, J.J. (2001). Host-pathogen interactions promoting inflammatory Lyme arthritis: use of mouse models for dissection of disease processes. Curr. Opin. Microbiol. *4*, 274–279.

Workowski, K.A., and Berman, S.M. (2006). Sexually transmitted diseases treatment guidelines, 2006. MMWR Recomm. Rep. *55*, 1–94.

Wormser, G.P. (2006). Clinical practice. Early Lyme disease. N. Engl. J. Med. *354*, 2794–2801.

Wormser, G.P., Bittker, S., Cooper, D., Nowakowski, J., Nadelman, R.B., and Pavia, C. (2001a) Yield of large-volume blood cultures in patients with early Lyme disease. J. Infect. Dis *184*, 1070–1072.

Wormser, G.P., Brisson, D., Liveris, D., Hanincova, K., Sandigursky, S., Nowakowski, J., Nadelman, R.B., Ludin, S., and Schwartz, I. (2008). *Borrelia burgdorferi* genotype predicts the capacity for hematogenous dissemination during early Lyme disease. J. Infect. Dis. *198*, 1358–1364.

Wormser, G.P., Dattwyler, R.J., Shapiro, E.D., Halperin, J.J., Steere, A.C., Klempner, M.S., Krause, P.J., Bakken, J.S., Strle, F., Stanek, G. *et al.* (2006). The clinical assessment, treatment, and prevention of Lyme disease, human granulocytic anaplasmosis, and babesiosis: clinical practice guidelines by the Infectious Diseases Society of America. Clin. Infect. Dis. *43*, 1089–1134.

Wormser, G.P., Liveris, D., Nowakowski, J., Nadelman, R.B., Cavaliere, L.F., McKenna, D., Holmgren, D., and Schwartz, I. (1999). Association of specific subtypes of *Borrelia burgdorferi* with hematogenous dissemination in early Lyme disease. J. Infect. Dis. *180*, 720–725.

Wormser, G.P., McKenna, D., Carlin, J., Nadelman, R.B., Cavaliere, L.F., Holmgren, D., Byrne, D.W., and Nowakowski, J. (2005). Brief communication: hematogenous dissemination in early Lyme disease. Ann. Intern. Med. *142*, 751–755.

Wormser, G.P., Nadelman, R.B., Nowakowski, J., and Schwartz, I. (2001b) Asymptomatic *Borrelia burgdorferi* infection. Med. Hypotheses *57*, 435–438.

Xu, Q., McShan, K., and Liang, F.T. (2007). Identification of an *ospC* operator critical for immune evasion of *Borrelia burgdorferi*. Mol. Microbiol. *64*, 220–231.

Xu, Y., Kodner, C., Coleman, L., and Johnson, R.C. (1996). Correlation of plasmids with infectivity of *Borrelia burgdorferi* sensu stricto type strain B31. Infect. Immun. *64*, 3870–3876.

Yang, X., Popova, T.G., Hagman, K.E., Wikel, S.K., Schoeler, G.G., Caimano, M.J., Radolf, J.D., and Norgard, M.V. (1999). Identification, characterization, and expression of three new members of the *Borrelia burgdorferi* Mlp (2.9) lipoprotein gene family. Infect. Immun. *67*, 6008–6018.

Yang, X.F., Pal, U., Alani, S.M., Fikrig, E., and Norgard, M.V. (2004). Essential role for OspA/B in the life cycle of the Lyme disease spirochete. J. Exp. Med. *199*, 641–648.

Zhang, J.R., Hardham, J.M., Barbour, A.G., and Norris, S.J. (1997). Antigenic variation in Lyme disease borreliae by promiscuous recombination of VMP-like sequence cassettes. Cell *89*, 275–285.

Zoschke, D.C., Skemp, A.A., and Defosse, D.L. (1991). Lymphoproliferative responses to *Borrelia burgdorferi* in Lyme disease. Ann. Intern. Med. *114*, 285–289.

Zückert, W.R., Meyer, J., and Barbour, A.G. (1999). Comparative analysis and immunological characterization of the *Borrelia* Bdr protein family. Infect. Immun. *67*, 3257–3266.

Index

In this index f and t refer to figures and tables, respectively.

C

T

U

Current Books of Interest

Molecular Phylogeny of Microorganisms	2010
Nanotechnology in Water Treatment Applications	2010
Iron Uptake and Homeostasis in Microorganisms	2010
Caliciviruses: Molecular and Cellular Virology	2010
Epstein-Barr Virus: Latency and Transformation	2010
Anaerobic Parasitic Protozoa: Genomics and Molecular Biology	2010
Lentiviruses and Macrophages: Molecular and Cellular Interactions	2010
Microbial Population Genetics	2010
Borrelia: Molecular and Cellular Biology	2010
Influenza: Molecular Virology	2010
RNA Interference and Viruses: Current Innovations and Future Trends	2010
Retroviruses: Molecular Biology, Genomics and Pathogenesis	2010
Metagenomics: Theory, Methods and Applications	2010
Aspergillus: Molecular Biology and Genomics	2010
Environmental Molecular Microbiology	2010
Neisseria: Molecular Mechanisms of Pathogenesis	2010
Frontiers in Dengue Virus Research	2010
ABC Transporters in Microorganisms	2009
Pili and Flagella: Current Research and Future Trends	2009
Lab-on-a-Chip Technology: Biomolecular Separation and Analysis	2009
Lab-on-a-Chip Technology: Fabrication and Microfluidics	2009
Bacterial Polysaccharides: Current Innovations and Future Trends	2009
Microbial Toxins: Current Research and Future Trends	2009
Acanthamoeba	2009
Bacterial Secreted Proteins: Secretory Mechanisms and Role in Pathogenesis	2009
Lactobacillus Molecular Biology: From Genomics to Probiotics	2009
Mycobacterium: Genomics and Molecular Biology	2009
Real-Time PCR: Current Technology and Applications	2009
Clostridia: Molecular Biology in the Post-genomic Era	2009
Plant Pathogenic Bacteria: Genomics and Molecular Biology	2009
Microbial Production of Biopolymers and Polymer Precursors	2009
Plasmids: Current Research and Future Trends	2008
Pasteurellaceae: Biology, Genomics and Molecular Aspects	2008
Vibrio cholerae: Genomics and Molecular Biology	2008
Pathogenic Fungi: Insights in Molecular Biology	2008
Helicobacter pylori: Molecular Genetics and Cellular Biology	2008
Corynebacteria: Genomics and Molecular Biology	2008
Staphylococcus: Molecular Genetics	2008
Leishmania: After The Genome	2008
Archaea: New Models for Prokaryotic Biology	2008
RNA and the Regulation of Gene Expression	2008
Legionella Molecular Microbiology	2008

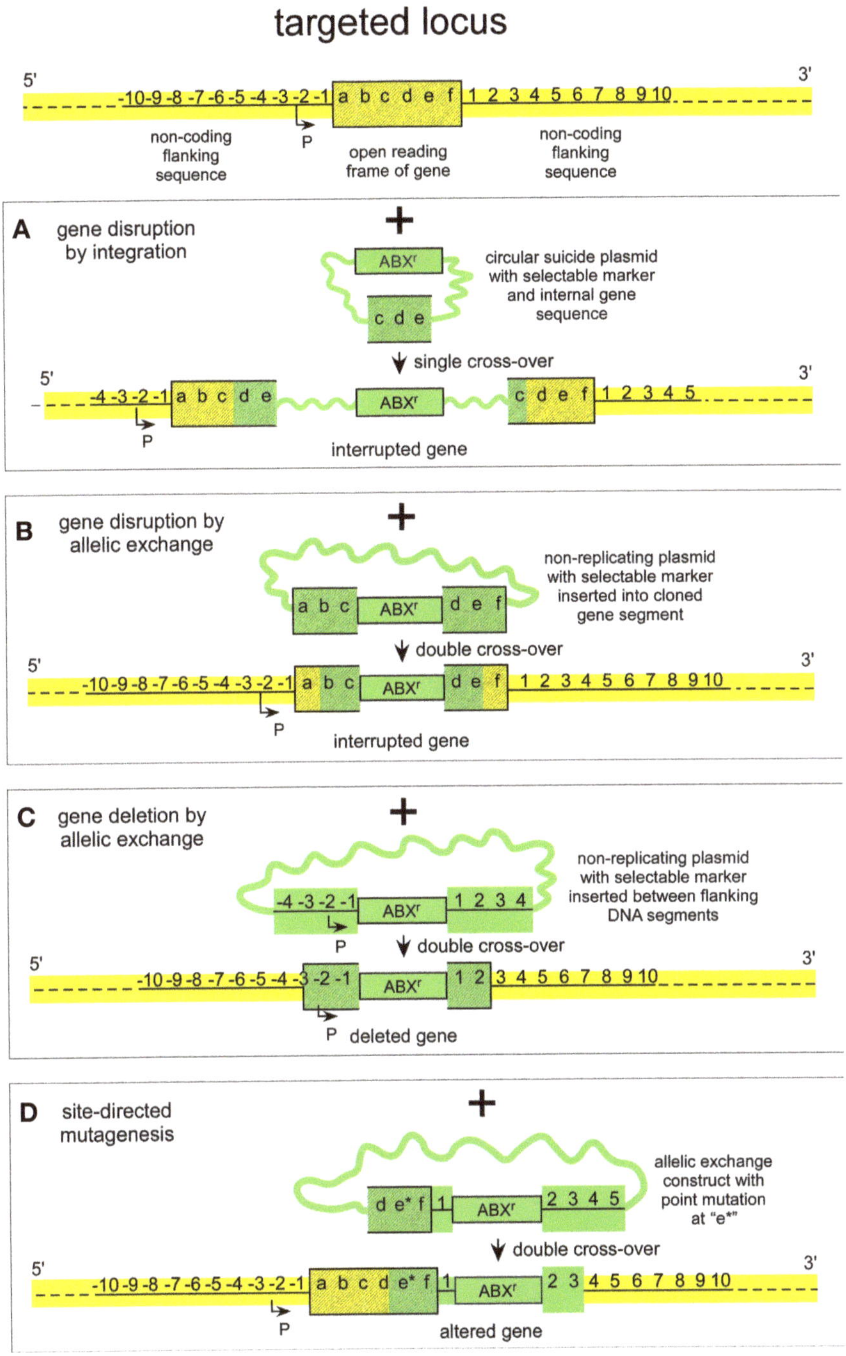

Plate 8.2 Schematic diagram of targeted gene inactivation by homologous recombination. The structure of the cloned inactivation construct determines whether the targeted gene is disrupted (as illustrated in parts a and b), deleted (part c) or altered (part d). In all cases, stable retention of the selectable marker results from homologous recombination between common sequences present at both the targeted locus (shown in yellow) and on the introduced DNA (in green). The precise point of recombination (crossover site) can occur anywhere along these homologous sequences and not only at the positions indicated on these diagrams. If a single rather than double crossover occurs between an allelic exchange construct and the targeted locus, the entire plasmid would integrate and the recovered transformant would carry both wild type and mutant copies of the targeted gene (not illustrated). This outcome can be minimized by linearizing the suicide plasmid carrying the allelic exchange construct prior to electroporation.

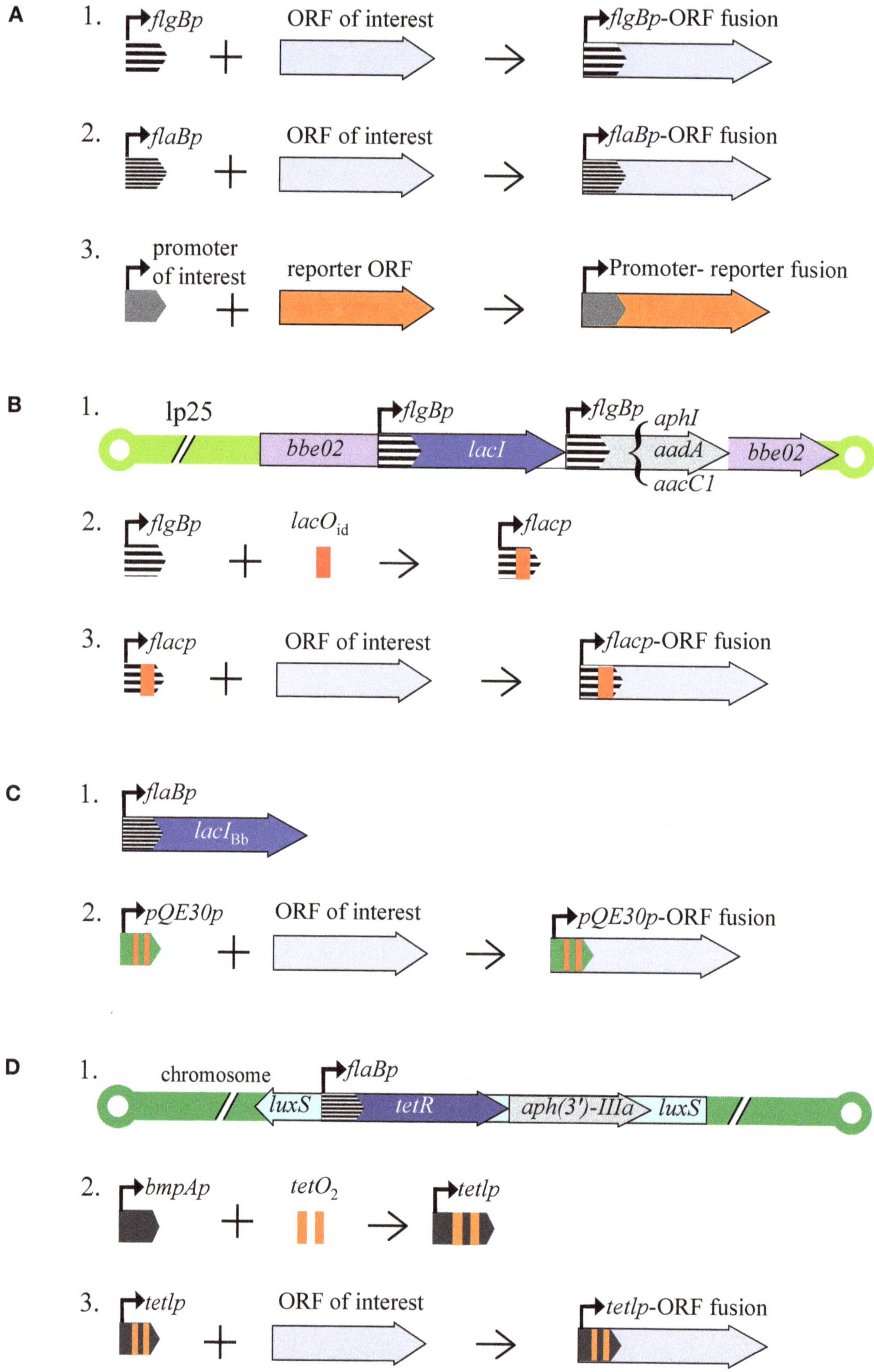
A
1.
flgBp
ORF of interest
flgBp-ORF fusion
2.
flaBp
ORF of interest
flaBp-ORF fusion
3.
promoter of interest
reporter ORF
Promoter- reporter fusion
B
1.
lp25
bbe02
flgBp
lacI
flgBp
aphI
aadA
aacC1
bbe02
2.
flgBp
lacO_id
flacp
3.
flacp
ORF of interest
flacp-ORF fusion
C
1.
flaBp
lacI_Bb
2.
pQE30p
ORF of interest
pQE30p-ORF fusion
D
1.
chromosome
flaBp
luxS
tetR
aph(3')-IIIa
luxS
2.
bmpAp
tetO_2
tetlp
3.
tetlp
ORF of interest
tetlp-ORF fusion

Plate 8.3 (opposite) Schematic diagram of engineered gene expression. (A) Constitutive gene expression (Sohaskey *et al.*, 1997; Bono *et al.*, 2000; Sartakova *et al.*, 2000). The *flgB* promoter (1), *flaB* promoter (2), or other promoters (not shown) can be fused to several open reading frames (or RNA-encoding genes), including antibiotic resistance genes to select for transformants, reporter genes to visualize cells, and endogenous genes to bypass their native regulation or to express them at high levels. Endogenous promoters can be fused to reporter genes to assay transcriptional activity (3). (B) *lac*-based inducible gene expression using the hybrid *flac* promoter (Gilbert *et al.*, 2007). The *flgB* promoter was fused to the *lacI* open reading frame (which encodes the *lac* repressor) and inserted, along with a linked selectable marker, in the *bbe02* gene (a putative barrier to transformation) on lp25 (a plasmid essential for infection); the resulting strains constitutively produce LacI repressor (1). The sequence between the conserved −10 element and the putative ribosome binding site of the *flgB* promoter was replaced with $lacO_{id}$ (a synthetic, 'ideal' lac operator), resulting in the inducible *flac* promoter, which is repressed by LacI repressor (2). The *flac* promoter can be fused to the coding region of any gene, either in *cis* in the genome by replacing the native promoter and linking to a selectable marker or in *trans* on a shuttle vector (3). Dr Mike Gilbert contributed to the design and drawing of this panel. (C) *lac*-based inducible gene expression using the *pQE30* promoter (Blevins *et al.*, 2007). The *flaB* promoter was fused to a *lacI* open reading frame that was codon optimized for *B. burgdorferi* ($lacI_{Bb}$) and cloned onto a shuttle vector (1). The *pQE30* promoter (from plasmid pQE30) is a T5 promoter with two *lacO* sequences that can be fused to the coding region of any gene (2). (D) *tet*-based inducible gene expression (Cabello *et al.*, 2006). The *flaB* promoter was fused to the *tetR* open reading frame (which encodes the tetracycline repressor) and inserted in the *luxS* gene on the chromosome; the resulting strains constitutively produce TetR repressor (1). Two operator (*tetO*) sequences were inserted in the *bmpA* promoter, one upstream of the conserved −35 element and the other between the −35 and the −10 elements, resulting in the inducible *tetI* promoter, which is repressed by TetR repressor (2). The *tetI* promoter can be fused to the coding region of any gene, either in *cis* in the genome by replacing the native promoter and linking to a selectable marker or *in trans* on a shuttle vector (3). Dr Henry Godfrey contributed to the design of this panel.

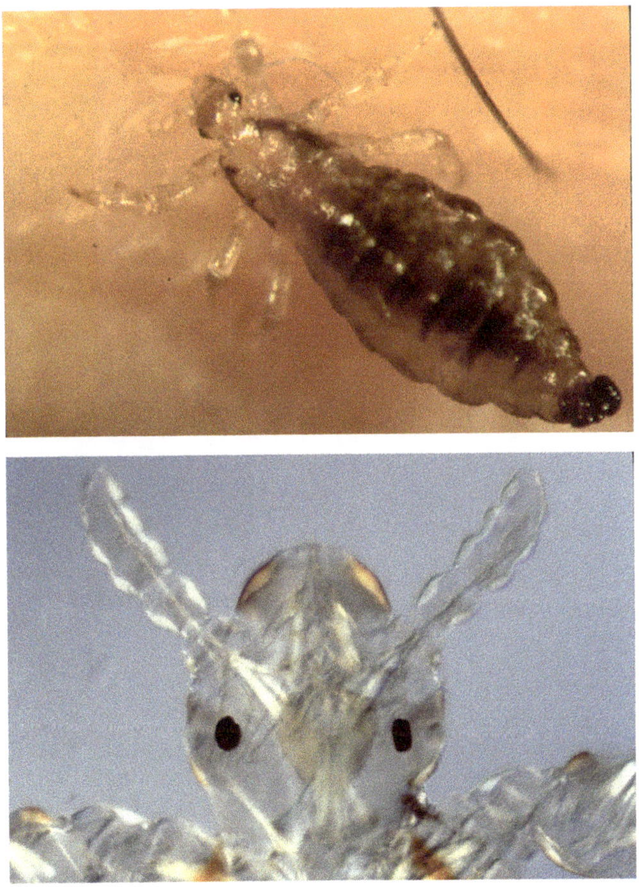

Plate 10.1 Human body louse (*Pediculus humanus*).

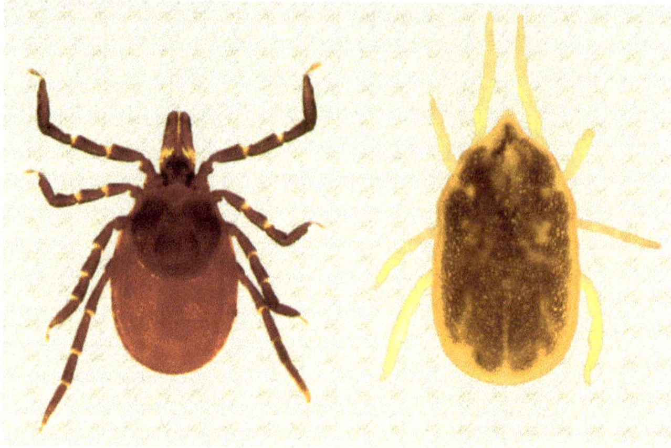

Plate 10.2 *Ixodes scapularis* (left) and *Ornithodoros hermsi* (right).

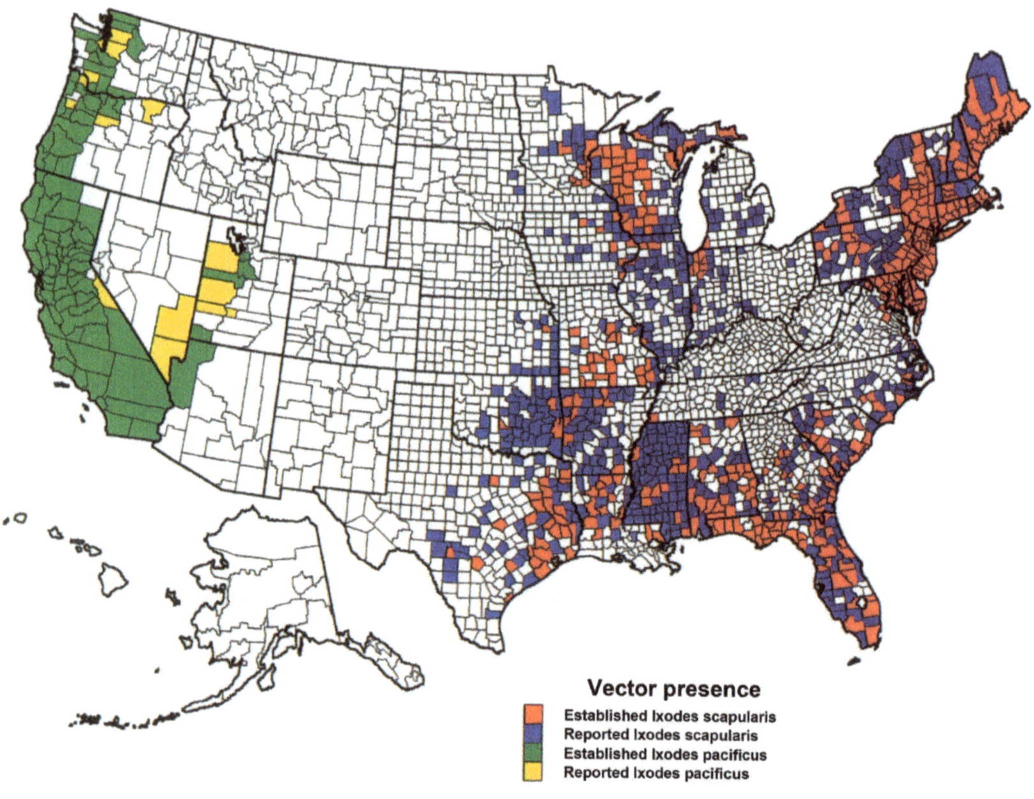

Plate10.5 Distribution of the tick vectors associated with Lyme disease in the USA.

www.ingramcontent.com/pod-product-compliance
Lightning Source LLC
Chambersburg PA
CBHW042013200926
51615CB00029B/5
* 9 7 8 1 9 0 4 4 5 5 5 8 5 *